Monographs in Mathematics
Vol. 105

More information about this series at http://www.springer.com/series/4843

Birkhäuser

Monographs in Mathematics
Vol. 105

More information about this series at http://www.springer.com/series/4843

Jan Prüss · Gieri Simonett

Moving Interfaces and Quasilinear Parabolic Evolution Equations

Birkhäuser

Jan Prüss
Institut für Mathematik
Martin-Luther-Universität Halle-Wittenberg
Halle
Germany

Gieri Simonett
Department of Mathematics
Vanderbilt University
Nashville, TN
USA

ISSN 1017-0480 ISSN 2296-4886 (electronic)
Monographs in Mathematics
ISBN 978-3-319-27697-7 ISBN 978-3-319-27698-4 (eBook)
DOI 10.1007/978-3-319-27698-4

Library of Congress Control Number: 2016940884

Mathematics Subject Classification (2010): 35B35, 35B65, 35J48, 35K41, 35K22, 35K55, 35K59, 35K90, 35K93, 35Q30, 35Q35, 35Q79, 35R35, 42B15, 46E40, 47A60, 47F05, 53C44, 76D03, 76D05, 76D07, 76D45, 76T05, 80A22

Printed on acid-free paper

This book is published under the trade name Birkhäuser
The registered company is Springer International Publishing AG Switzerland
(www.birkhauser-science.com)

Preface

Moving interfaces – and in the stationary case, free boundaries – are ubiquitous in our environment and daily life. They are at the basis of many physical, chemical, and also biological processes.

Typically, a moving boundary problem consists of one or more partial differential equations which have to be solved in a domain that is a priori unknown and that has to be determined as part of the problem. Problems with moving boundaries are in general harder to solve, both analytically and numerically, than the underlying differential equations would be in a prescribed domain. They have an inherent nonlinear structure, as two separate solutions cannot be superposed. Einstein's words

In so far as the theorems of mathematics relate to reality, they are not certain, and in so as far as they are certain, they do not relate to reality

appear to be quite correct in the context of formulating and analyzing problems with moving interfaces or free boundaries. Many simple things, such as a whiskey glass with a melting ice cube or a pot of boiling water with potatoes, are already very difficult to be modeled in a physically accurate way.

But as a matter of fact, mathematicians never give up. If unable to model and analyze a complicated process, we concentrate on simpler ones which already exhibit the important difficulties and characteristics. Mathematicians have followed this route successfully ever since: let us solve model problems in a rigorous way in order to improve our tools and invent new ones, and to sharpen old and design new weapons to tackle real world problems.

The most famous model problem with a moving interface, perhaps, is the *Stefan problem* for the freezing of water, proposed by J. Stefan in the 19th century. This problem has attracted much mathematical research since then, resulting in hundreds of papers; see the biographical remarks at the end of this book. The second historically prominent problem which, likely, has been around for as long as the Stefan-problem, is the *two-phase Navier-Stokes problem*, which describes the motion of droplets of oil in water, for instance. This problem has caused as much mathematical interest as the Stefan problem.

In this book we extend and combine these two historical problems into classes of models for one-component two-phase flows with phase transitions. The proposed models are thermodynamically consistent in the sense that the total energy is preserved, while the total entropy is non-decreasing. The physical derivation of these models and their properties are explained in Chapter 1. A rigorous analysis of the resulting six model problems is presented in Chapters 9, 10, and 11.

Another source of problems with moving interfaces concerns geometric evolution laws which describe the dynamics of hypersurfaces. In these problems, the normal velocity of a surface is given by a law defined by its geometry. Steady states then are special "free boundaries," leading to certain classes of surfaces like *minimal surfaces* or *Willmore surfaces.* Important examples are the *mean curvature flow*, the *surface diffusion flow*, and the *Willmore flow.* On the other hand, some popular quasi-stationary problems like the *Mullins–Sekerka flow*, the *Stokes flow*, or the *Muskat flow*, are determined not only by the geometry of the interface, but also by diffusion in its environment. In this monograph, we extend this list to include what we call the *Stokes flow with phase transition* and the *Muskat flow with phase transition.*

In the last decades, it has become evident that the theory of maximal regularity provides an important tool to tackle problems with moving interfaces. By means of these methods, the quasilinear structure, which is inherent for most problems with moving boundaries that include mean curvature, can be exploited in a mathematically optimal way. Via certain linearizations we may resort to the contraction mapping principle and to the implicit function theorem, as no loss of regularity will occur. This refers to local well-posedness – sometimes called *short-time existence for arbitrary data* – but also to the regularity of solutions and their stability properties near equilibria – sometimes called *long-time existence for small data.* With our methods we can furthermore prove that the interfaces become instantaneously real analytic if the coefficients in the equations are so and the initial values are subject to only mild regularity assumptions. This encodes typical parabolic behaviour.

Techniques relying on variational inequalities and weak solutions have proven successful in analyzing a wide array of problems with moving interfaces that share a particular underlying structure, that is, which can be formulated in a weak sense or in terms of a variational inequality. This enables one to conclude without great efforts that a solution to the free boundary problem exists in some weak sense. One can then proceed to establish the regularity of the solution and then attempt to study the smoothness of the free boundary itself. The advantage of this method is that it provides the existence of a global solution. However, it is often difficult, if not impossible, to derive further information on the location and the qualitative properties of the free boundary. Moreover, problems that include surface tension on the free boundary do not have the luxury of a comparison principle, and this alters the mathematical structure of the equations in a fundamental way. Consequently, methods based on comparison principles, variational inequalities, and viscosity solutions do not seem well-adapted in the presence of surface tension.

The basis of our approach relies on the so-called *direct mapping method* which means that the problem is transformed to a problem with fixed interface. This can be achieved very easily by a Lagrange transform if the interface is advected with the underlying flow, as in the two-phase Navier-Stokes problem. If phase transitions are present, like in the Stefan problem, it is more convenient to employ a Hanzawa transform in which the moving interface is parameterized via a *height function* over a fixed reference hypersurface. This method seems to be better adapted than the Lagrange transform for two-phase flows, as it allows us to prove smoothing of the interface, even if no phase transitions take place but surface tension is present. For this method, some differential geometry of hypersurfaces in Euclidean space is needed as well as advanced knowledge of function spaces.

In this monograph, we employ the theory of maximal L_p-theory throughout. By now, in the L_p-framework, many classical as well as some very recent powerful results for vector-valued harmonic analysis are at our disposal.

To introduce the needed tools, we will explore numerous connections between maximal L_p-regularity, sectorial operators, $\mathcal{H}^\infty$-calculus, Fourier-multipliers, semigroups, and function spaces. Chapters 3 and 4 are devoted to this general theory – a theory that can be used for many other problems besides those with moving boundaries, as is demonstrated in Chapters 5 and 12. Therefore, this book offers many things also to researchers who may not primarily be interested in moving boundaries, but want to learn about parabolic evolution equations.

The monograph is structured as follows. In the introductory Chapter 1, the necessary physical background is introduced and the main problems to be studied are formulated. It is shown that these problems are thermodynamically consistent; their equilibria are identified, and those equilibria which are local maxima of the total entropy are singled out. One major purpose of this book is to show that the latter are precisely the stable ones. We also give an outline of the strategies for their mathematical analysis.

Chapter 2 contains the basic differential geometry of hypersurfaces needed for the direct mapping principle. We investigate the notions of Weingarten tensor, principal curvatures, mean curvature, tubular neighbourhood, surface gradient, surface divergence, and Laplace-Beltrami operator. The main emphasis lies in deriving representations of these quantities for hypersurfaces that are given as parameterized surfaces in normal direction of a fixed reference surface by means of a height function. It is also important to study the mapping properties of these quantities in dependence on the height function, and to derive expressions for their variations. Among other things we study the first and second variations of the area and volume functional. Moreover, we show that C^2-hypersurfaces can be approximated in a suitable topology by smooth (e.g. analytic) hypersurfaces. We then show that the class of compact embedded hypersurfaces in $\mathbb{R}^n$ gives rise to a new manifold whose points are the compact embedded hypersurfaces. Finally, we consider moving hypersurfaces, and we state and prove various transport theorems. While most of the material is well-known, we nevertheless believe that our presentation contains new results and aspects that are also of interest to readers with more advanced knowledge in differential geometry and geometric analysis. This chapter can be read independently from the rest of the book.

In Chapter 3, some elementary results from operator and semigroup theory are recalled. Moreover, some interpolation theory and the concept of maximal L_p-regularity is introduced and discussed. More recent results on vector-valued harmonic analysis, in particular operator-valued Fourier multiplier theorems in L_p-spaces and their implication to maximal L_p-theory, are accounted for in Chapter 4. The results of these chapters form the functional analytic foundation for the maximal regularity results which are at the heart of this book.

To demonstrate the strength and flexibility of the maximal regularity approach, in Chapter 5 an L_p-theory for abstract quasilinear parabolic evolution equations is developed. This includes local well-posedness, regularity, compactness of the induced semiflow, as well as the generalized principle of linearized stability for the analysis of the semiflow near (a manifold of) equilibria. These results can be applied to a wide array of quasilinear parabolic systems, including geometric evolution equations, as shown in Chapter 12. Maximal regularity is also used in an essential way to show the existence of the stable and unstable foliations at normally hyperbolic equilibria. Chapter 5 only relies on the concept of maximal L_p-regularity and, hence, it is also useful for readers who are not primarily interested in problems with moving interfaces, but rather in quasilinear parabolic systems.

Also of independent interest are Chapters 6, 7 and 8 in which maximal L_p-regularity for large classes of linear elliptic and parabolic systems is proved. These classes include transmission problems for two-phase systems, problems with dynamics on the interface, and Stokes problems, needed later on. In these chapters the full strength of the multiplier results from Chapter 4 is used.

Chapters 9, 10, and 11 are devoted to the analysis of the six problems with moving interfaces introduced in Chapter 1. They form the core of this monograph. It turns out that these problems can, unfortunately, not be formulated as abstract evolution equations of the type considered in Chapter 5. This is caused by the presence of nonlinear stationary transmission conditions on the interface. Nevertheless, Chapter 5 is used as a guideline for the analysis of the more involved problems. Chapter 9 deals with their local well-posedness, which can be proved simultaneously for all six problems in question; only the chosen function spaces are specific for each one. The same is also valid for the regularity theory, and for the results on long-time behaviour discussed in Chapter 11. On the other hand, the analysis of the full linearizations at a given equilibrium in Chapter 10 depends on the specific problem.

Miscellaneous applications of the theory developed in Chapter 5 are presented in Chapter 12, which include sections on generalized Newtonian flows, nematic liquid crystal flows, Maxwell–Stefan diffusion, the Stefan problem with variable surface tension, and geometric evolution equations.

The monograph is supplemented with historical and bibliographical remarks, suggestions for extensions of the theory and for further studies and research, a list of symbols and a subject index, and with an extensive bibliography.

This monograph would never have been written and completed without the help of many colleagues, friends, and students. We would like to thank our co-authors Dieter Bothe, Robert Denk, Matthias Hieber, Matthias Köhne, Martin Meyries, Jürgen Saal, Roland Schnaubelt, Yuanzhen Shao, Yoshihiro Shibata, Senjo Shimizu, Mathias Wilke, Rico Zacher, and many other colleagues, for frequent discussions and for their continuous support. We are especially indebted to Martin Herberg, Stefan Meyer, Yuanzhen Shao, Senjo Shimizu, and Mathias Wilke for their efforts in proofreading various sections of the book, and for pointing out many errors and typos in previous versions of the manuscript. Without the skills of Martin Herberg, Stefan Meyer, Yuanzhen Shao, and Mathias Wilke we would not have been able to produce the figures in high quality, and we especially acknowledge their support in this respect. We would also like to thank Jan Bohr for generating the BibTeX-library which was essential for this book, and will prove equally useful for future work.

We thank Herbert Amann for his continuous interest and support.

The first author thanks the *Deutsche Forschungsgemeinschaft* (DFG) for continuous support over the years. The second author thankfully acknowledges support by NSF (DMS-1265579).

The authors cordially thank Bettina, Saskia, Sören, and Helena, Oriana, Seraina for their patience and steady encouragement during the long process of writing.

We are indebted to Birkhäuser, in particular to Dr. Thomas Hempfling and his collaborators, for the smooth process of publishing this book.

Jan Prüss and Gieri Simonett Halle and Nashville, September 2015 of symbols and a subject index, and with an extensive bibliography.

This monograph would never have been written and completed without the help of many colleagues, friends, and students. We would like to thank our coauthors Dieter Bothe, Robert Denk, Matthias Hieber, Matthias Köhne, Martin Meyries, Jürgen Saal, Roland Schnaubelt, Yuanzhen Shao, Yoshihiro Shibata, Senjo Shimizu, Mathias Wilke, Rico Zacher, and many other colleagues, for frequent discussions and for their continuous support. We are especially indebted to Martin Herberg, Stefan Meyer, Yuanzhen Shao, Senjo Shimizu, and Mathias Wilke for their efforts in proofreading various sections of the book, and for pointing out many errors and typos in previous versions of the manuscript. Without the skills of Martin Herberg, Stefan Meyer, Yuanzhen Shao, and Mathias Wilke we would not have been able to produce the figures in high quality, and we especially acknowledge their support in this respect. We would also like to thank Jan Bolz for generating the BibTeX library which was essential for this book, and will prove equally useful for future work.

We thank Herbert Amann for his continuous interest and support.

The first author thanks the Deutsche Forschungsgemeinschaft (DFG) for continuous support over the years. The second author thankfully acknowledges support by NSF (DMS-1265579).

The authors cordially thank Hauke, Marieke, Sebastian, Sonja, and Helene, Oriana, Simone for their patience and steady encouragement during the long process of writing.

We are indebted to Birkhäuser, in particular to Dr. Thomas Hempfling and his collaborators, for the smooth process of publishing this book.

Jan Prüss and Gieri Simonett Halle and Nashville, September 2015

Contents

Basic Notations

Most notations used throughout this book are fairly standard in the modern mathematical literature. So $\mathbb{N}$, $\mathbb{Z}$, $\mathbb{Q}$, $\mathbb{R}$, $\mathbb{C}$ denote the set of natural numbers, integers, rationals, real and complex numbers, respectively, and $\mathbb{N}_0 = \mathbb{N} \cup \{0\}$, $\mathbb{R}_+ = (0, \infty)$, $\dot{\mathbb{R}} = \mathbb{R} \setminus \{0\}$, $\dot{\mathbb{R}}^n = [\dot{\mathbb{R}}]^n$, $\hat{\mathbb{R}}^n = \mathbb{R}^{n-1} \times \dot{\mathbb{R}}$, $\mathbb{R}^n_+ = \mathbb{R}^{n-1} \times \mathbb{R}_+$, $\mathbb{C}_+ = \{z \in \mathbb{C} : \operatorname{Re} z > 0\}$, $a_+ = \max\{a, 0\}$ for $a \in \mathbb{R}$. We also denote by

$$\Sigma_\phi = \{z \in \mathbb{C} : \ z \neq 0, \ |\arg z| < \phi\}$$

the open sector in $\mathbb{C}$ symmetric to $\mathbb{R}_+$ with opening angle $\phi \in (0, \pi]$. We denote the Euclidean norm in $\mathbb{R}^n$ by $|\cdot|$ in case no confusion is likely. Moreover, we use the notation $(x|y)$ or $x \cdot y$ for the inner product of $x, y \in \mathbb{R}^n$.

If (M, d) is a metric space and $N \subset M$, then N^c, $\operatorname{int} N$, $\bar{N}$, ∂N designate the complement, interior, closure, boundary of N, respectively, and $\operatorname{dist}(x, N)$ denotes the distance of x to N, while $B(x_0, r)$ and $\bar{B}(x_0, r)$ are the open resp. closed balls with center x_0 and radius r.

X, Y, Z will always be Banach spaces with norms $|\cdot|_X$, $|\cdot|_Y$, $|\cdot|_Z$; the subscripts will be dropped when there is no danger of confusion. $\mathcal{B}(X, Y)$ denotes the space of all bounded linear operators from X to Y, $\mathcal{B}(X) = \mathcal{B}(X, X)$ for short. The set of all isomorphisms between X and Y is denoted by $\operatorname{Isom}(X, Y)$. The dual space of X is $X^* = \mathcal{B}(X, \mathbb{K})$, where $\mathbb{K} = \mathbb{R}$ or $\mathbb{K} = \mathbb{C}$ is the underlying scalar field; $\langle x|x^*\rangle = \langle x|x^*\rangle_{X,X^*}$ designates the natural pairing between elements $x \in X$ and $x^* \in X^*$. If $(x_n) \subset X$ converges to $x \in X$ we write $x_n \to x$ or $\lim_{n\to\infty} x_n = x$, while $x_n \rightharpoonup x$ or $w - \lim_{n\to\infty} x_n = x$ mean weak convergence; similarly $x_n^* \overset{*}{\rightharpoonup} x^*$ or $w^* - \lim_{n\to\infty} x_n^* = x^*$ stand for weak*-convergence of $(x_n^*) \subset X^*$ to $x^* \in X^*$.

If A is a linear operator in X, $\mathsf{D}(A)$, $\mathsf{R}(A)$, $\mathsf{N}(A)$ denote the domain, range, null space of A, respectively, while $\sigma(A)$ and $\rho(A)$ mean spectrum and resolvent set of A. $\sigma(A)$ is further decomposed into $\sigma_p(A)$, $\sigma_c(A)$, $\sigma_r(A)$, the point spectrum, continuous spectrum, and residual spectrum of A. The operator A^* in X^* denotes the dual of A. If it exists, it is defined by $\langle Ax|x^*\rangle = \langle x|A^*x^*\rangle$. If A is closed, then $\mathsf{D}(A)$ equipped with the graph norm of A, $|x|_A = |x| + |Ax|$, is a Banach space, for which the symbol X_A is employed.

If X is a Hilbert space, $(x|y) = (x \cdot y)_X$ means the inner product in X and $|x| = |x|_X = (x|x)^{1/2}$ the canonical norm. The Hilbert space adjoint of an operator

A will also be denoted by A^*. It is defined by the relation $(Ax|y) = (x|A^*y)$.

Some standard function spaces employed throughout this book are the following.

If (M, d) is a metric space, and X a Banach space, then $C(M; X)$ denotes the space of all continuous functions $f : M \to X$. $C_b(M; X)$ resp. $C_{ub}(M; X)$ designate the spaces of all bounded continuous resp. bounded uniformly continuous functions $f : M \to X$; these spaces become Banach spaces when normed by the sup-norm

$$|f|_0 = \sup_{t \in M} |f(t)|.$$

The space of all functions $f : M \to X$ which are uniformly Lipschitz-continuous is denoted by $\mathrm{Lip}(M; X)$, and

$$|f|_{\mathrm{Lip}} = \sup_{t \neq s} |f(t) - f(s)|/d(t, s).$$

If $(\Omega, \mathcal{A}, \mu)$ is a measure space and X a Banach space, then $L_p((\Omega, \mathcal{A}, \mu); X)$, $1 \le p < \infty$, denotes the space of all Bochner-measurable functions $f : \Omega \to X$ such that $|f(\cdot)|^p$ is integrable. This space is also a well-known Banach space when normed by

$$|f|_p = \Big(\int_\Omega |f(t)|^p d\mu(t) \Big)^{1/p},$$

and functions equal almost everywhere (a.e.) are identified. Similarly, $L_\infty((\Omega, \mathcal{A}, \mu); X)$ denotes the space of (equivalence classes of) Bochner-measurable and essentially bounded functions $f : \Omega \to X$, and the norm is defined according to

$$|f|_\infty = \operatorname*{ess\,sup}_{t \in \Omega} |f(t)|.$$

For $\Omega \subset \mathbb{R}^n$ open, Σ the Lebesgue $\sigma-$algebra, and μ the Lebesgue measure, we abbreviate $L_p((\Omega, \Sigma, \mu); X)$ to $L_p(\Omega; X)$. In this case $H_p^m(\Omega; X)$ is the space of all functions $f : \Omega \to X$ having distributional derivatives $\partial^\alpha f \in L_p(\Omega; X)$ of order $|\alpha| \le m$. The norm in $H_p^m(\Omega; X)$ is given by

$$|f|_{m,p} = \Big(\sum_{|\alpha| \le m} |\partial^\alpha f|_p^p \Big)^{1/p} \quad \text{for } 1 \le p < \infty,$$

and

$$|f|_{m,\infty} = \max_{|\alpha| \le m} |\partial^\alpha f|_\infty \quad \text{for } p = \infty.$$

For $\Omega \subset \mathbb{R}^n$ open, $C^m(\Omega; X)$ denotes the space of all functions $f : \Omega \to X$ which admit continuous partial derivatives $\partial^\alpha f$ in Ω, for each multi-index α with $|\alpha| \le m$. A function f belongs to $C^m(\bar{\Omega}; X)$ if $f \in C^m(\Omega; X)$, and $\partial^\alpha f$ has a continuous extension to $\overline{\Omega}$, for each $|\alpha| \le m$. The norm of its subspaces $C_b^m(\overline{\Omega}; X)$ and $C_{ub}^m(\bar{\Omega}; X)$ is given by

$$|f|_{m,0} = \sup_{|\alpha| \le m} |\partial^\alpha f|_0.$$

By $C_j^{m-}(\overline{\Omega}; X)$ we mean the space of all functions $f \in C_j^{m-1}(\overline{\Omega}; X)$ such that $\partial^\alpha f \in \mathrm{Lip}(\Omega; X)$ for each $|\alpha| = m$. The norm in this space is defined as

$$|f|_{m-,0} = |f|_{m-1,0} + \sup_{|\alpha|=m} |\partial^\alpha f|_{\mathrm{Lip}}.$$

Moreover, if Ω is unbounded we set

$$C_l(\Omega; X) := \{f \in C_b(\Omega) : \lim_{|x|\to\infty} f(x) \text{ exists}\},$$
$$C_0(\Omega; X) := \{f \in C_b(\Omega) : \lim_{|x|\to\infty} f(x) = 0\}.$$

Throughout, we will employ Hörmander's notation for partial derivatives, i.e. we write $D = -i\partial$ and $D^\alpha = (-i)^{|\alpha|}\partial^\alpha$.

For $f \in C(\Omega; X)$ the *support* of f is defined by

$$\operatorname{supp} f = \overline{\{x \in \Omega : f(x) \neq 0\}}.$$

As usual, $\mathcal{D}(\Omega; X)$ means the space of all test functions $\varphi \in \mathcal{D}(\Omega; X) := \bigcap_{m\geq 1} C^m(\Omega; X)$ such that $\operatorname{supp} \varphi \subset \Omega$ is compact. For $X = \mathbb{C}$ we set $\mathcal{D}(\Omega) = \mathcal{D}(\Omega; \mathbb{C})$, the space of scalar test functions.

The subscript 'loc' assigned to any of the above function spaces means membership to the corresponding space when restricted to compact subsets of its domain. Usually, if X is the underlying scalar field $\mathbb{K} = \mathbb{C}$ or $\mathbb{K} = \mathbb{R}$, the image space in the function space notation introduced above will be dropped. For example, $L_{1,loc}(\mathbb{R})$ denotes the space of all measurable scalar-valued functions which are integrable over each compact interval.

A left subscript 0 indicates that the corresponding functions vanish on the boundary of the relevant domain, along with all its derivatives. For instance, we have

$${}_0H_p^m(\Omega; X) = \overline{\mathcal{D}(\Omega; X)}^{H_p^m},$$

and

$${}_0C(J; X) = \{f : J \to X \text{ is continuous, with } \lim_{t\to 0} f(t) = 0\},$$
$${}_0C_0(\mathbb{R}_+; X) = \{f : \mathbb{R}_+ \to X \text{ is continuous, with } \lim_{t\to\{0,\infty\}} f(t) = 0\},$$

where $J = (0, a) \subset \mathbb{R}$.

Other function spaces will be introduced where they are needed for the first time; cf. the list of symbols.

The *Fourier transform* of a function $f \in L_1(\mathbb{R}^n; X)$ is defined by

$$\mathcal{F}f(\xi) := \tilde{f}(\xi) = \int_{\mathbb{R}^n} e^{-i\xi\cdot x} f(x)\, dx, \quad \xi \in \mathbb{R}^n.$$

It is well-known that $\tilde{f} : \mathbb{R}^n \to X$ is uniformly continuous and tends to 0 as $|\xi| \to \infty$, by the Riemann-Lebesgue lemma. If X is a Hilbert space, by Parseval's theorem, the Fourier transform extends to a unitary operator on $L_2(\mathbb{R}^n; X)$. On the *Schwartz space* $\mathcal{S}(\mathbb{R}^n; X)$ of all functions $f \in C^\infty(\mathbb{R}^n; X)$ such that each derivative of f decays faster than any polynomial, the Fourier transform is an isomorphism, and the inversion formula

$$\tilde{\tilde{f}}(x) = (2\pi)^n f(-x), \quad x \in \mathbb{R}^n,$$

holds.

The *Laplace transform* of a function $f \in L_{1,loc}(\mathbb{R}_+; X)$ is denoted by

$$\mathcal{L}f(\lambda) := \widehat{f}(\lambda) := \int_0^\infty e^{-\lambda t} f(t)\, dt, \qquad \operatorname{Re} \lambda > \omega,$$

whenever the integral is absolutely convergent for $\operatorname{Re} \lambda > \omega$. The relation between the Laplace transform of $f \in L_1(\mathbb{R}; X)$, $f(t) \equiv 0$ for $t < 0$, and its Fourier transform is

$$\tilde{f}(\xi) = \hat{f}(i\xi), \quad \xi \in \mathbb{R}.$$

As usual we employ the star $*$ for the *convolution* of scalar functions defined on $\mathbb{R}^n$ or on the half line $\mathbb{R}_+$.

$$(f * g)(x) = \int_{\mathbb{R}^n} f(x-y) g(y)\, dy, \quad x \in \mathbb{R}^n,$$

e.g. for $f, g \in L_1(\mathbb{R}^n)$, and

$$(f * g)(t) = \int_0^t f(t-s) g(s)\, ds, \quad t \in \mathbb{R}_+,$$

e.g. for $f, g \in L_1(\mathbb{R}_+)$. Observe that for $n = 1$, the two representations for $f * g$ above are equivalent for functions that vanish on $t < 0$ and so, there will be no danger of confusion. Recall the convolution theorem for the Fourier transform

$$\mathcal{F}(u * v)(\xi) = \mathcal{F}u(\xi)\mathcal{F}v(\xi), \quad \xi \in \mathbb{R}^n,$$

for $u, v \in L_1(\mathbb{R}^n)$, and that for the Laplace transform

$$\mathcal{L}(u * v)(\lambda) = \mathcal{L}u(\lambda)\mathcal{L}v(\lambda), \quad \operatorname{Re} \lambda > \omega,$$

for functions u, v defined on the half line such that $e^{-\omega t}u, e^{-\omega t}v \in L_1(\mathbb{R}_+)$.

Other symbols and notations are introduced where needed for the first time, and for these the reader should consult the index.

General References

Throughout the book we will use some monographs as standard references without further comments. These are the following.

- **Function Spaces**: Adams [2], Triebel [283, 284];
- **Functional Analysis**: Brezis [50], Rudin [239], Yosida [300];
- **Interpolation Theory**: Bergh-Löfström [38], Lunardi [184], Triebel [282];
- **Nonlinear Analysis**: Deimling [76], Zeidler [305];
- **Operator Theory**: Dunford-Schwartz [91], Kato [157];
- **Semigroup Theory**: Arendt et al. [29], Engel-Nagel [94], Hille-Phillips [147].

Part I

Background

Chapter 1

Problems and Strategies

The purpose of this introductory chapter is to explain the problems to be considered in the main part of this book in some detail. We derive their physical origin from first principles, discuss some of the main structural properties of the models, and describe the strategies of our analytical approach. All the notions and properties relating to differential geometry of hypersurfaces will be introduced and explained in Chapter 2.

1.1 Modeling

Suppose a (fixed) container Ω – a bounded domain in $\mathbb{R}^n$ with smooth boundary – is filled with a material which is present in two phases that occupy the regions $\Omega_1(t)$ and $\Omega_2(t)$. The interface $\Gamma(t)$ separating these two phases will depend on time t, but should not be in contact with the outer boundary $\partial\Omega$ of the container in order to avoid the contact angle problem. Then the so-called *continuous phase* $\Omega_2(t)$ is in contact with the outer boundary, while the *diperse phase* $\Omega_1(t)$ is not, which means that $\partial\Omega_1(t) = \Gamma(t)$ and $\partial\Omega_2(t) = \partial\Omega \cup \Gamma(t)$. The outer unit normal of $\Gamma(t)$ w.r.t. $\Omega_1(t)$ will be denoted by ν_Γ, it depends on $p \in \Gamma(t)$ as well as on t; the outer unit normal of Ω is called ν, it only depends on $p \in \partial\Omega$. The *Weingarten tensor* L_Γ is defined by $L_\Gamma := -\nabla_\Gamma \nu_\Gamma$, where ∇_Γ means the surface gradient, and the $((n-1)$-fold) mean curvature H_Γ of Γ by

$$H_\Gamma = \operatorname{tr} L_\Gamma = -\operatorname{div}_\Gamma \nu_\Gamma,$$

where $\operatorname{div}_\Gamma$ means the surface divergence on Γ. In the sequel, the jump of a physical quantity ϕ across Γ will be denoted by

$$[\![\phi]\!](p) := \lim_{s\to 0+} [\phi(p + s\nu_\Gamma(p)) - \phi(p - s\nu_\Gamma(p))], \quad p \in \Gamma.$$

J. Prüss and G. Simonett, *Moving Interfaces and Quasilinear Parabolic Evolution Equations*, Monographs in Mathematics 105,
DOI 10.1007/978-3-319-27698-4_1

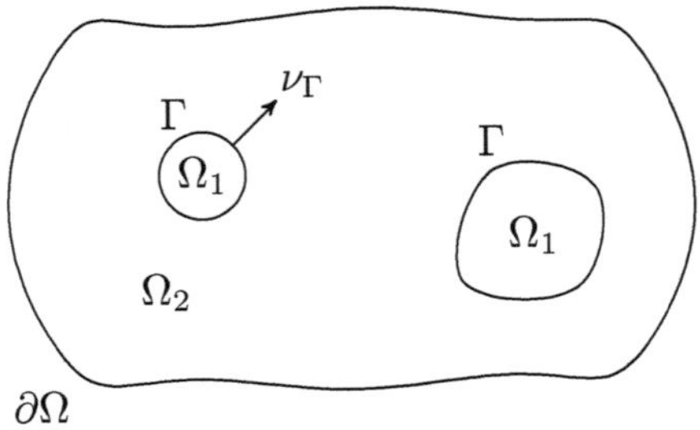

Figure 1.1: A two-phase domain $\Omega = \Omega_1 \cup \Gamma \cup \Omega_2$.

1.1 First Principles in the Bulk

We begin with the basic balance laws in the bulk.

Balance of Mass

Let $\varrho > 0$ denote the density and u the velocity in the bulk phases Ω_j, u_Γ the velocity and $V_\Gamma := u_\Gamma \cdot \nu_\Gamma$ the normal velocity of Γ, respectively. Note that ϱ and u may jump across the interface Γ and that u_Γ is in general not a tangent vector field to Γ. If there are no sources of mass in the bulk, then conservation of mass is given by the *continuity equation*

$$\partial_t \varrho + \operatorname{div}(\varrho u) = 0 \quad \text{in } \Omega \setminus \Gamma(t). \tag{1.1}$$

If there is no surface mass on Γ, we also have the jump condition

$$[\![\varrho(u - u_\Gamma) \cdot \nu_\Gamma]\!] = 0 \quad \text{on } \Gamma(t). \tag{1.2}$$

The interfacial mass flux j_Γ, *phase flux* for short, is defined by means of

$$j_\Gamma := \varrho(u - u_\Gamma) \cdot \nu_\Gamma, \quad \text{i.e.,} \quad [\![\frac{1}{\varrho}]\!] j_\Gamma = [\![u \cdot \nu_\Gamma]\!]. \tag{1.3}$$

Observe that j_Γ is well defined, as (1.2) shows. *Phase Transition* takes place if $j_\Gamma \not\equiv 0$. On the other hand, if $j_\Gamma \equiv 0$, then $u \cdot \nu_\Gamma = u_\Gamma \cdot \nu_\Gamma = V_\Gamma$, and in this case the interface is advected with the velocity field u.

Next we have by the transport theorem for moving domains

$$\begin{aligned}
\frac{d}{dt} \int_{\Omega_1(t)} \varrho \, dx &= \int_{\Gamma(t)} \varrho V_\Gamma \, d\Gamma + \int_{\Omega_1(t)} \partial_t \varrho \, dx \\
&= \int_{\Gamma(t)} \varrho V_\Gamma \, d\Gamma - \int_{\Omega_1(t)} \operatorname{div}(\varrho u) \, dx \\
&= \int_{\Gamma(t)} (\varrho u_\Gamma \cdot \nu_\Gamma - \varrho u \cdot \nu_\Gamma) \, d\Gamma = - \int_{\Gamma(t)} j_\Gamma \, d\Gamma,
\end{aligned}$$

and in case $u \cdot \nu = 0$ on $\partial\Omega$ in the same way

$$\frac{d}{dt}\int_{\Omega_2(t)} \varrho\, dx = \int_{\Gamma(t)} j_\Gamma\, d\Gamma,$$

proving conservation of total mass, i.e.,

$$\frac{d}{dt}\int_{\Omega} \varrho\, dx = 0. \tag{1.4}$$

In this book we mostly consider the *completely incompressible case* , i.e., we assume that the densities are constant in the phases Ω_j. Then conservation of mass reduces to

$$\operatorname{div} u = 0 \quad \text{in} \;\; \Omega \setminus \Gamma(t).$$

If only the latter property holds, we say that the material is *incompressible.* In case both phases are completely incompressible we have

$$\varrho_1|\Omega_1(t)| + \varrho_2|\Omega_2(t)| \equiv \varrho_1|\Omega_1(0)| + \varrho_2|\Omega_2(0)| =: c_0.$$

This implies

$$[\![\varrho]\!]|\Omega_1(t)| = \varrho_2|\Omega| - c_0,$$

hence $|\Omega_1(t)|$ is constant in the case of nonequal densities, i.e., the phase volumes are preserved. On the other hand, there is no preservation of phase volumes in general if one or both phases are compressible, or if the densities are constant and equal.

The Universal Balance Law Let ϕ be any (mass-specific) physical quantity, J its flux, and f its sources. Then the balance law for ϕ in the bulk reads

$$\partial_t(\varrho\phi) + \operatorname{div}(\varrho\phi u + J) = \varrho f \quad \text{in} \;\; \Omega \setminus \Gamma(t), \tag{1.5}$$

and if there is a source f_Γ for ϕ on the interface we have

$$[\![(\varrho\phi(u - u_\Gamma) + J) \cdot \nu_\Gamma]\!] = f_\Gamma \quad \text{on} \;\; \Gamma(t). \tag{1.6}$$

Employing balance of mass and the definition of the phase flux j_Γ this simplifies to

$$\begin{aligned} \varrho(\partial_t\phi + u \cdot \nabla\phi) + \operatorname{div} J &= \varrho f \quad \text{in} \;\; \Omega \setminus \Gamma(t),\\ [\![\phi]\!] j_\Gamma + [\![J \cdot \nu_\Gamma]\!] &= f_\Gamma \quad \text{on} \;\; \Gamma(t). \end{aligned} \tag{1.7}$$

By (2.101), the corresponding *universal transport theorem* becomes

$$\begin{aligned} \frac{d}{dt}\int_\Omega \varrho\phi\, dx &= \int_\Omega \partial_t(\varrho\phi)\, dx - \int_\Gamma [\![\varrho\phi]\!] V_\Gamma\, d\Gamma\\ &= \int_\Omega (\varrho f - \operatorname{div}(\varrho\phi u + J))\, dx - \int_\Gamma [\![\varrho\phi u_\Gamma \cdot \nu_\Gamma]\!]\, d\Gamma\\ &= \int_\Omega \varrho f\, dx + \int_\Gamma [\![(\varrho\phi(u - u_\Gamma) + J) \cdot \nu_\Gamma]\!]\, d\Gamma - \int_{\partial\Omega} (\varrho\phi u + J) \cdot \nu\, d(\partial\Omega)\\ &= \int_\Omega \varrho f\, dx + \int_\Gamma ([\![\phi]\!] j_\Gamma + [\![J \cdot \nu_\Gamma]\!])\, d\Gamma + \int_{\partial\Omega} g\, d(\partial\Omega), \end{aligned}$$

with $g = -(\varrho\phi u + J)\cdot\nu$ on $\partial\Omega$. Therefore, we obtain the conservation law

$$\frac{d}{dt}\int_\Omega \varrho\phi\,dx = \int_\Omega \varrho f\,dx + \int_\Gamma f_\Gamma\,d\Gamma + \int_{\partial\Omega} g\,d(\partial\Omega).$$

In particular, if $(f, f_\Gamma, g) = 0$, then the total amount of ϕ in Ω is conserved.

Balance of Momentum

Let π denote the pressure, T the (symmetric) stress tensor, and let f be a force field, say gravity. Then balance of momentum reads, employing (1.5) with $\phi = u$ and $J = -T$,

$$\partial_t(\varrho u) + \operatorname{div}(\varrho u\otimes u) - \operatorname{div} T = \varrho f \quad \text{in } \Omega\setminus\Gamma(t).$$

Similarly, using (1.6) we get the following jump condition at the interface.

$$[\![(\varrho u\otimes(u-u_\Gamma) - T)\nu_\Gamma]\!] = \operatorname{div}_\Gamma T_\Gamma \quad \text{on } \Gamma(t).$$

Here T_Γ denotes the (symmetric) surface stress, a tensor field on Γ. Using balance of mass and the definition of the phase flux j_Γ we may rewrite these conservation laws as follows.

$$\begin{aligned} \varrho(\partial_t u + u\cdot\nabla u) - \operatorname{div} T &= \varrho f && \text{in } \Omega\setminus\Gamma(t),\\ [\![u]\!]j_\Gamma - [\![T\nu_\Gamma]\!] &= \operatorname{div}_\Gamma T_\Gamma && \text{on } \Gamma(t). \end{aligned} \tag{1.8}$$

By the surface divergence theorem, total conservation of momentum reads as

$$\frac{d}{dt}\int_\Omega \varrho u\,dx = \int_\Omega \varrho f\,dx + \int_{\partial\Omega} g\,d(\partial\Omega),$$

with $g = -(\varrho u u\cdot\nu - T\nu)$ on $\partial\Omega$. Note that total momentum is in general not conserved as the boundary term g on $\partial\Omega$ need not be zero.

Balance of Energy

Let ϵ denote the (mass specific) internal energy density, $\theta > 0$ the absolute temperature, q the heat flux, and r an external (mass specific) heat source. Then with $\phi = |u|^2/2 + \epsilon$, $J = -Tu + q$ we obtain from the universal balance law (1.5) conservation of energy, which in the bulk reads

$$\partial_t\Big(\frac{\varrho}{2}|u|^2 + \varrho\epsilon\Big) + \operatorname{div}\{(\frac{\varrho}{2}|u|^2 + \varrho\epsilon)u\} - \operatorname{div}(Tu - q) = \varrho f\cdot u + \varrho r \quad \text{in } \Omega\setminus\Gamma(t).$$

On the interface we have, in accordance with (1.6),

$$\Big[\!\!\Big[\Big(\frac{\varrho}{2}|u|^2 + \varrho\epsilon\Big)(u - u_\Gamma) - Tu + q\Big]\!\!\Big]\cdot\nu_\Gamma = (\operatorname{div}_\Gamma T_\Gamma)\cdot u_\Gamma + r_\Gamma \quad \text{on } \Gamma(t),$$

where r_Γ denotes a heat source on Γ. Using (1.1), (1.8), and the definition of the phase flux j_Γ we may rewrite this conservation law as follows.

$$\begin{aligned} \varrho(\partial_t\epsilon + u\cdot\nabla\epsilon) + \operatorname{div} q - T:\nabla u = \varrho r \text{ in } \Omega\setminus\Gamma(t),\\ \Big([\![\epsilon]\!] + \Big[\!\!\Big[\frac{1}{2}|u-u_\Gamma|^2\Big]\!\!\Big]\Big)j_\Gamma - [\![T\nu_\Gamma\cdot(u-u_\Gamma)]\!] + [\![q\cdot\nu_\Gamma]\!] = r_\Gamma \text{ on } \Gamma(t). \end{aligned} \tag{1.9}$$

The total bulk energy is given by

$$\mathsf{E}(u,\epsilon,\Gamma) := \frac{1}{2}\int_{\Omega\setminus\Gamma} \varrho|u|^2\,dx + \int_{\Omega\setminus\Gamma} \varrho\epsilon\,dx.$$

For its time derivative we obtain from the universal balance law

$$\partial_t \mathsf{E} = \int_\Omega (\varrho f\cdot u + \varrho r)\,dx + \int_{\partial\Omega} g\,d(\partial\Omega) + \int_\Gamma \{\mathrm{div}_\Gamma T_\Gamma \cdot u_\Gamma + r_\Gamma\}\,d\Gamma,$$

where $g = -\big((\frac{\varrho}{2}|u|^2 + \varrho\epsilon)u\cdot\nu - Tu\cdot\nu + q\cdot\nu\big)$ on $\partial\Omega$. In particular, if $(f,r)=0$ in Ω, $(u\cdot\nu, q\cdot\nu, T\nu\cdot u) = 0$ on $\partial\Omega$ as well as $\mathrm{div}_\Gamma T_\Gamma\cdot u_\Gamma + r_\Gamma = 0$ on Γ, then

$$\frac{d}{dt}\mathsf{E}(u,\epsilon,\Gamma) = 0,$$

which means that the total bulk energy is preserved.

The Entropy
As is common in thermodynamics, we write

$$\epsilon(\varrho,\theta) = \psi(\varrho,\theta) + \theta\eta(\varrho,\theta), \quad \eta(\varrho,\theta) = -\partial_\theta\psi(\varrho,\theta), \tag{1.10}$$

where $\theta > 0$ denotes the absolute temperature, and ψ the *Helmholtz free energy*. In this book it is considered given. η means the (mass specific) *entropy* density. Then the *Clausius–Duhem equation* holds in the bulk, which means

$$\begin{aligned}\partial_t(\varrho\eta) + \mathrm{div}(\varrho\eta u) + \mathrm{div}\,(q/\theta) &= \frac{1}{\theta}S:\nabla u - \frac{1}{\theta^2}q\cdot\nabla\theta \\ &\quad + \frac{\varrho^2\partial_\varrho\psi - \pi}{\theta}\mathrm{div}\,u \quad \text{in } \Omega\setminus\Gamma(t),\end{aligned} \tag{1.11}$$

where $S := T + \pi$ denotes the viscous stress tensor. Therefore, entropy is nondecreasing locally in the bulk provided the right-hand side of (1.11) is nonnegative. This gives the well-known requirements

$$S:\nabla u \ge 0, \quad q\cdot\nabla\theta \le 0, \tag{1.12}$$

and, since in general the last term will not have a sign, either $\mathrm{div}\,u \equiv 0$, which corresponds to the incompressible case, or

$$\pi = p(\varrho,\theta) := \varrho^2\partial_\varrho\psi(\varrho,\theta), \tag{1.13}$$

which is the famous *Maxwell relation* for compressible materials. Note that p should be an increasing function in both variables, ϱ and θ. Hence we require at least

$$\partial_\varrho\partial_\theta\psi \ge 0, \quad 2\partial_\varrho\psi + \varrho\partial_\varrho^2\psi \ge 0, \quad \varrho,\theta > 0,$$

in the compressible case. The total bulk entropy is defined by

$$\mathsf{N}_b(\varrho,\theta,\Gamma) = \int_{\Omega\setminus\Gamma} \varrho\eta(\varrho,\theta)dx.$$

By the universal balance law we then obtain

$$\frac{d}{dt}\mathsf{N}_b(\varrho,\theta,\Gamma) = \int_{\Omega\setminus\Gamma} \Big\{\frac{1}{\theta}S : \nabla u - \frac{1}{\theta^2}q\cdot\nabla\theta\Big\}dx + \int_\Gamma \{[\![\eta]\!]j_\Gamma + [\![q/\theta]\!]\cdot\nu_\Gamma\}d\Gamma,$$

provided $u\cdot\nu = q\cdot\nu = 0$ on $\partial\Omega$. In particular, there is no entropy production on the interface if

$$[\![\eta]\!]j_\Gamma + [\![q/\theta]\!]\cdot\nu_\Gamma = 0 \quad \text{on } \Gamma.$$

1.2 First Principles on the Interface

Throughout we assume that there is no surface mass, and therefore also no surface momentum on Γ. However, due to surface tension we have to take into account surface energy, and then also surface entropy. A basic principle of our approach is conservation of energy and entropy across the interface. We begin with

The Universal Balance Law on the Interface

Suppose ϕ is a scalar physical quantity which also lives on Γ with surface density ϕ_Γ and let J_Γ denote its flux. Thus, J_Γ is a tangent vector field to Γ. The basic balance law for ϕ_Γ reads

$$\frac{D}{Dt}\phi_\Gamma + \phi_\Gamma \mathrm{div}_\Gamma u_\Gamma + \mathrm{div}_\Gamma J_\Gamma = -f_\Gamma. \tag{1.14}$$

Here D/Dt means the Lagrangian derivative with respect to the vector field u_Γ which moves Γ, i.e.,

$$\frac{D}{Dt}\phi_\Gamma(t,\xi) = \frac{d}{ds}\phi_\Gamma(s+t, x(s+t,t,\xi))\Big|_{s=0},$$

with $x(s+t,t,\xi)$ the flow induced by the velocity field u_Γ, i.e.,

$$\frac{d}{ds}x(s+t,t,\xi) = u_\Gamma(s+t, x(s+t,t,\xi)), \quad x(t,t,\xi) = \xi,\ \xi\in\Gamma(t).$$

We emphasize again that the velocity field u_Γ is in general not tangent to Γ. The surface transport theorem then yields

$$\begin{aligned}\frac{d}{dt}\int_\Gamma \phi_\Gamma\, d\Gamma &= \int_\Gamma \Big(\frac{D}{Dt}\phi_\Gamma + \phi_\Gamma \mathrm{div}_\Gamma u_\Gamma\Big)\, d\Gamma \\ &= \int_\Gamma (-\mathrm{div}_\Gamma J_\Gamma - f_\Gamma)\, d\Gamma = -\int_\Gamma f_\Gamma\, d\Gamma,\end{aligned}$$

by the surface divergence theorem. Therefore, we obtain conservation of the total amount of ϕ in Ω, i.e., we have

$$\frac{d}{dt}\Big\{\int_\Omega \varrho\phi\,dx + \int_\Gamma \phi_\Gamma\,d\Gamma\Big\} = 0,$$

provided $(f,g) = 0$. Thus the balance law for ϕ on Γ reads

$$\frac{D}{Dt}\phi_\Gamma + \phi_\Gamma \mathrm{div}_\Gamma u_\Gamma + \mathrm{div}_\Gamma J_\Gamma = -([\![\phi]\!] j_\Gamma + [\![J]\!]\cdot\nu_\Gamma). \tag{1.15}$$

We apply this interface conservation law first to

Conservation of Energy on the Interface
Here we have $\phi_\Gamma = \epsilon_\Gamma$ and $J_\Gamma = -T_\Gamma u_\Gamma + q_\Gamma$. Then balance of surface energy reads

$$\frac{D}{Dt}\epsilon_\Gamma + \epsilon_\Gamma \mathrm{div}_\Gamma\, u_\Gamma + \mathrm{div}_\Gamma\,(q_\Gamma - T_\Gamma u_\Gamma) = -\{(\mathrm{div}_\Gamma\, T_\Gamma)\cdot u_\Gamma + r_\Gamma\}.$$

Hence

$$\frac{D}{Dt}\epsilon_\Gamma + \epsilon_\Gamma \mathrm{div}_\Gamma\, u_\Gamma + \mathrm{div}_\Gamma\, q_\Gamma = T_\Gamma : \nabla_\Gamma u_\Gamma - r_\Gamma.$$

By the conservation laws this implies conservation of total energy

$$\frac{d}{dt}\left\{\int_\Omega \varrho\left(\frac{|u|^2}{2} + \epsilon\right)dx + \int_\Gamma \epsilon_\Gamma\,d\Gamma\right\} = 0,$$

provided $(f,r) = 0$ in Ω, $u\cdot\nu = q\cdot\nu = 0$ and $T\nu\cdot u = 0$ on $\partial\Omega$.

Surface Entropy
As in the bulk we write

$$\epsilon_\Gamma(\theta_\Gamma) = \psi_\Gamma(\theta_\Gamma) + \theta_\Gamma\eta_\Gamma(\theta_\Gamma), \quad \eta_\Gamma(\theta_\Gamma) = -\psi_\Gamma'(\theta_\Gamma),$$

where we consider the free energy ψ_Γ as a given function of surface temperature θ_Γ. Similarly, we decompose

$$T_\Gamma = \sigma(\theta_\Gamma)\mathcal{P}_\Gamma + S_\Gamma,$$

where σ denotes the coefficient of surface tension, $\mathcal{P}_\Gamma = I - \nu_\Gamma\otimes\nu_\Gamma$ the orthogonal projection onto the tangent bundle of Γ, and S_Γ the interface viscous stress. Then surface force becomes

$$\mathrm{div}_\Gamma\, T_\Gamma = \sigma H_\Gamma\nu_\Gamma + \nabla_\Gamma\sigma + \mathrm{div}_\Gamma S_\Gamma.$$

The first term in this decomposition is surface tension which acts in a normal direction. The second is called the *Marangoni force* which acts tangentially, and the last one is the viscous surface force induced by surface viscosity.

The total surface entropy is given by

$$\mathsf{N}_\Gamma = \int_\Gamma \eta_\Gamma d\Gamma.$$

With the surface transport theorem (2.91) we get

$$\begin{aligned}\frac{d}{dt}\mathsf{N}_\Gamma &= \int_\Gamma \Big(\frac{D}{Dt}\eta_\Gamma + \eta_\Gamma \mathrm{div}_\Gamma u_\Gamma\Big)\, d\Gamma = \int_\Gamma \Big(\frac{D}{Dt}\epsilon_\Gamma + \theta_\Gamma\eta_\Gamma \mathrm{div}_\Gamma u_\Gamma\Big)/\theta_\Gamma\, d\Gamma \\ &= \int_\Gamma (-\mathrm{div}_\Gamma q_\Gamma - \psi_\Gamma \mathrm{div}_\Gamma u_\Gamma + T_\Gamma : \nabla_\Gamma u_\Gamma - r_\Gamma)/\theta_\Gamma\, d\Gamma \\ &= \int_\Gamma \big(S_\Gamma : \nabla_\Gamma u_\Gamma/\theta_\Gamma - q_\Gamma \cdot \nabla_\Gamma\theta_\Gamma/\theta_\Gamma^2 + (\sigma - \psi_\Gamma)\mathrm{div}_\Gamma u_\Gamma/\theta_\Gamma - r_\Gamma/\theta_\Gamma\big)\, d\Gamma.\end{aligned}$$

Now we argue as in the bulk case. To ensure entropy production on the interface we should have

$$S_\Gamma : \nabla_\Gamma u_\Gamma \geq 0, \quad q_\Gamma \cdot \nabla_\Gamma \theta_\Gamma \leq 0,$$

as well as

$$\psi_\Gamma = \sigma,$$

which is the analogue of the Maxwell relation on the interface. Thus in the situation considered here, the free energy on the interface is the coefficient of surface tension, which acts as a negative surface pressure.

For the total entropy in Ω we finally obtain

$$\begin{aligned}\frac{d}{dt}\Big(\int_\Omega \varrho\eta\, dx + \int_\Gamma \eta_\Gamma\, d\Gamma\Big) &= \int_\Omega \Big\{\frac{1}{\theta}S : \nabla u - \frac{1}{\theta^2} q \cdot \nabla\theta\Big\}\, dx \\ &\quad + \int_\Gamma \Big\{\frac{1}{\theta_\Gamma} S_\Gamma : \nabla_\Gamma u_\Gamma - \frac{1}{\theta_\Gamma^2} q_\Gamma \cdot \nabla_\Gamma\theta_\Gamma\Big\}\, d\Gamma \qquad (1.16) \\ &\quad + \int_\Gamma \{[\![\eta]\!] j_\Gamma + [\![q/\theta]\!]\cdot \nu_\Gamma - r_\Gamma/\theta_\Gamma\}\, d\Gamma.\end{aligned}$$

Since the integrand in the last integral does not have a sign, we postulate that it vanishes. This means that the only sources for entropy is friction due to viscosity or heat conduction, also on the interface. In case $(S_\Gamma, q_\Gamma) = 0$ it means *conservation of entropy across the interface.*

This assumption implies by (1.9)

$$-\Big(\Big([\![\epsilon]\!] + \Big[\!\!\Big[\frac{1}{2}|u-u_\Gamma|^2\Big]\!\!\Big]\Big) j_\Gamma - [\![T\nu_\Gamma \cdot (u-u_\Gamma)]\!] + [\![q\cdot\nu_\Gamma]\!]\Big)/\theta_\Gamma + [\![\eta]\!] j_\Gamma + [\![q/\theta]\!]\cdot\nu_\Gamma = 0.$$

Assuming $[\![\theta]\!] = 0$, $\theta = \theta_\Gamma$ on Γ, the latter simplifies to

$$\Big([\![\psi]\!] + \Big[\!\!\Big[\frac{1}{2}|u-u_\Gamma|^2\Big]\!\!\Big]\Big) j_\Gamma - [\![T\nu_\Gamma \cdot (u-u_\Gamma)]\!] = 0. \qquad (1.17)$$

This is the *generalized Gibbs–Thomson relation.* Taking it for granted, balance of surface energy becomes

$$\frac{D}{Dt}\epsilon_\Gamma + \epsilon_\Gamma \mathrm{div}_\Gamma u_\Gamma + \mathrm{div}_\Gamma q_\Gamma = S_\Gamma : \nabla_\Gamma u_\Gamma + \sigma \mathrm{div}_\Gamma u_\Gamma - ([\![\theta\eta]\!] j_\Gamma + [\![q \cdot \nu_\Gamma]\!]). \quad (1.18)$$

On the interface the Clausius–Duhem equation reads

$$\begin{aligned}\frac{D}{Dt}\eta_\Gamma + \eta_\Gamma \mathrm{div}_\Gamma u_\Gamma + \mathrm{div}_\Gamma (q_\Gamma/\theta_\Gamma) = \frac{1}{\theta_\Gamma} S_\Gamma : \nabla_\Gamma u_\Gamma - \frac{1}{\theta_\Gamma^2} q_\Gamma \cdot \nabla_\Gamma \theta_\Gamma \quad (1.19)\\ - ([\![\eta]\!] j_\Gamma + [\![q/\theta]\!] \cdot \nu_\Gamma),\end{aligned}$$

showing that surface entropy production is nonnegative, locally on Γ.

Note that in case $(\eta_\Gamma, q_\Gamma, S_\Gamma) = 0$ on Γ this equation implies the famous *Stefan condition*

$$\theta [\![\eta]\!] j_\Gamma + [\![q \cdot \nu_\Gamma]\!] = 0.$$

$\eta_\Gamma \equiv 0$ means $\psi_\Gamma = \sigma \equiv$ *constant* and $\epsilon_\Gamma = \sigma$. In this case total surface energy becomes $\sigma|\Gamma|$, and surface energy balance is trivial.

1.3 Constitutive Laws

In the sequel we assume that there are no external sources for momentum and energy, i.e., $(f, r) = 0$.

Constitutive Laws on the Outer Boundary

$$q \cdot \nu = 0 \quad \text{and} \quad u = 0. \quad (1.20)$$

Actually, we could also consider a condition for u of Navier-type at the outer boundary, which means

$$u \cdot \nu = 0 \quad \text{and} \quad \mathcal{P}_{\partial\Omega} T\nu + ku = 0,$$

where $k \geq 0$, and $\mathcal{P}_{\partial\Omega}$ denotes the projection onto the tangent bundle of the hypersurface $\partial\Omega$. However, here we stay with the simplest case.

Constitutive Laws in the Phases

$$\begin{aligned}&\epsilon(\varrho,\theta) = \psi(\varrho,\theta) + \theta\eta(\varrho,\theta), \quad \eta(\varrho,\theta) = -\partial_\theta \psi(\varrho,\theta),\\ &\quad T = 2\mu(\varrho,\theta) D + \lambda(\varrho,\theta)(\mathrm{div}\, u) I - \pi I, \quad D = \frac{1}{2}(\nabla u + (\nabla u)^{\mathsf{T}}), \quad (1.21)\\ &\quad q = -d(\varrho,\theta)\nabla\theta.\end{aligned}$$

Here μ is called *shear viscosity,* λ *bulk viscosity,* and d is the coefficient of heat conduction or *heat conductivity.* μ, λ, d are functions depending on (ϱ, θ), and on the phase, and hence may jump across the interface $\Gamma(t)$. The second and the third equations are the classical laws of *Newton* and *Fourier.* To meet the requirements (1.12) we assume

$$\mu(\varrho,\theta), d(\varrho,\theta) > 0, \quad \lambda(\varrho,\theta) + 2\mu(\varrho,\theta)/n > 0, \quad \varrho, \theta > 0,$$

and in the compressible case also the Maxwell relation (1.13).

Constitutive Laws on the Interface

$$\begin{aligned}
\epsilon_\Gamma(\theta_\Gamma) &= \sigma(\theta_\Gamma) + \theta_\Gamma \eta_\Gamma(\theta_\Gamma), \quad \eta_\Gamma(\theta_\Gamma) = -\sigma'(\theta_\Gamma),\\
[\![\theta]\!] &= 0, \quad \theta_\Gamma = \theta,\\
\mathcal{P}_\Gamma [\![u]\!] &= 0, \quad \mathcal{P}_\Gamma(u - u_\Gamma) = 0,\\
T_\Gamma &= \sigma(\theta_\Gamma)\mathcal{P}_\Gamma + 2\mu_\Gamma(\theta_\Gamma) D_\Gamma + \lambda_\Gamma(\theta_\Gamma)(\mathrm{div}_\Gamma u_\Gamma)\mathcal{P}_\Gamma,\\
D_\Gamma &= \frac{1}{2}\mathcal{P}_\Gamma(\nabla_\Gamma u_\Gamma + [\nabla_\Gamma u_\Gamma]^{\mathsf{T}})\mathcal{P}_\Gamma, \quad q_\Gamma = -d_\Gamma(\theta_\Gamma)\nabla_\Gamma \theta_\Gamma,\\
0 &= \Big([\![\psi(\theta_\Gamma)]\!] + \Big[\!\!\Big[\frac{1}{2}|u - u_\Gamma|^2\Big]\!\!\Big]\Big) j_\Gamma - [\![T\nu_\Gamma(u - u_\Gamma)]\!].
\end{aligned} \tag{1.22}$$

The coefficient of surface tension σ and the surface viscosities $(\mu_\Gamma, \lambda_\Gamma)$ are functions of θ_Γ, which are subject to

$$\sigma, \mu_\Gamma > 0, \quad \lambda_\Gamma + \frac{2\mu_\Gamma}{n-1} > 0.$$

Recall the relation

$$V_\Gamma := u_\Gamma \cdot \nu_\Gamma = u \cdot \nu_\Gamma - \frac{1}{\varrho} j_\Gamma,$$

for the normal velocity of the interface. In case $[\![\varrho]\!] \neq 0$ this implies

$$[\![u]\!] = [\![1/\varrho]\!] j_\Gamma \nu_\Gamma, \quad j_\Gamma = [\![u \cdot \nu_\Gamma]\!]/[\![1/\varrho]\!], \quad V_\Gamma = [\![\varrho u \cdot \nu_\Gamma]\!]/[\![\varrho]\!], \tag{1.23}$$

and if $[\![\varrho]\!] = 0$ we have $[\![u]\!] = 0$. This shows a fundamental difference between theses cases: if the densities are not equal, then the phase flux enters directly the velocity jump on the interface, inducing what is called *Stefan current*. If the densities are equal, there is no Stefan current and the velocity field is continuous across the interface. On each side of the interface we have the identity

$$u = u_\Gamma + j_\Gamma \nu_\Gamma / \varrho,$$

which, in view of the definition of the phase flux j_Γ, is equivalent to the conditions

$$\mathcal{P}_\Gamma [\![u]\!] = 0, \quad \mathcal{P}_\Gamma(u - u_\Gamma) = 0, \quad [\![\varrho(u - u_\Gamma)\cdot \nu_\Gamma]\!] = 0.$$

Now we may rewrite

$$\begin{aligned}
\Big[\!\!\Big[\frac{1}{2}|u - u_\Gamma|^2\Big]\!\!\Big] &= \Big[\!\!\Big[\frac{1}{2\varrho^2}\Big]\!\!\Big] j_\Gamma^2,\\
-[\![T\nu_\Gamma \cdot (u - u_\Gamma)]\!] &= [\![-T\nu_\Gamma]\!] \cdot \mathcal{P}_\Gamma(u - u_\Gamma) + [\![-T\nu_\Gamma \cdot \nu_\Gamma/\varrho]\!] j_\Gamma\\
&= [\![-T\nu_\Gamma \cdot \nu_\Gamma/\varrho]\!] j_\Gamma,
\end{aligned}$$

hence the generalized Gibbs–Thomson relation becomes

$$\Big([\![\psi]\!] + \Big[\!\!\Big[\frac{1}{2\varrho^2}\Big]\!\!\Big] j_\Gamma^2 - [\![T\nu_\Gamma \cdot \nu_\Gamma/\varrho]\!]\Big) j_\Gamma = 0.$$

It holds trivially if $j_\Gamma \equiv 0$, i.e., if there is *no phase transition*, and otherwise we assume

$$[\![\psi]\!] + \Big[\!\!\Big[\frac{1}{2\varrho^2}\Big]\!\!\Big] j_\Gamma^2 - [\![T\nu_\Gamma \cdot \nu_\Gamma/\varrho]\!] = 0. \tag{1.24}$$

We define the heat capacity κ and the surface heat capacity κ_Γ as usual by

$$\kappa(\varrho,\theta) = \partial_\theta \epsilon(\varrho,\theta) = -\theta\partial_\theta^2\psi(\varrho,\theta), \quad \kappa_\Gamma(\theta_\Gamma) = \epsilon_\Gamma'(\theta_\Gamma) = -\theta_\Gamma\sigma''(\theta_\Gamma)$$

respectively. Moreover, we define the latent heat l and the surface latent heat l_Γ by

$$l(\varrho,\theta) = -[\![\theta\eta(\varrho,\theta)]\!] = [\![\theta\partial_\theta\psi(\varrho,\theta)]\!], \quad l_\Gamma(\theta_\Gamma) = -\theta_\Gamma\eta_\Gamma(\theta_\Gamma) = \theta_\Gamma\sigma'(\theta_\Gamma).$$

The conditions $\partial_\theta^2\psi \leq 0$ as well as $\sigma'' \leq 0$ will be needed for well-posedness.

Remark
(1.24) may be generalized to take into account *kinetic undercooling*. More precisely, we may replace (1.24) by the law

$$[\![\psi]\!] + \Big[\!\!\Big[\frac{1}{2\varrho^2}\Big]\!\!\Big] j_\Gamma^2 - [\![T\nu_\Gamma\cdot\nu_\Gamma/\varrho]\!] = -\gamma j_\Gamma + \mathrm{div}_\Gamma[\alpha\nabla_\Gamma(j_\Gamma/\theta_\Gamma)] + \theta_\Gamma\,\mathrm{div}_\Gamma[\beta\nabla_\Gamma j_\Gamma], \tag{1.25}$$

where $\alpha, \beta, \gamma \geq 0$ may depend on the surface temperature θ_Γ. In this case the entropy production on Γ is increased by

$$\int_\Gamma \{\gamma j_\Gamma^2/\theta_\Gamma + \alpha|\nabla_\Gamma(j_\Gamma/\theta_\Gamma)|_2^2 + \beta|\nabla_\Gamma j_\Gamma|^2\}d\Gamma,$$

and on the right-hand side of the surface energy balance the term

$$j_\Gamma(\gamma j_\Gamma - \mathrm{div}_\Gamma[\alpha\nabla_\Gamma(j_\Gamma/\theta_\Gamma)] - \theta_\Gamma\,\mathrm{div}_\Gamma[\beta\nabla_\Gamma j_\Gamma])$$

has to be added.

1.4 The Resulting Dynamic Problem

Summarizing we obtain the following initial-boundary value problem in the absence of external forces and heat sources.

$$\begin{aligned}
\partial_t\varrho + \mathrm{div}(\varrho u) &= 0 && \text{in } \Omega\setminus\Gamma(t),\\
\varrho(\partial_t u + u\cdot\nabla u) - \mathrm{div}\, S + \nabla\pi &= 0 && \text{in } \Omega\setminus\Gamma(t),\\
u &= 0 && \text{on } \partial\Omega,\\
[\![u]\!] &= [\![1/\varrho]\!] j_\Gamma\nu_\Gamma && \text{on } \Gamma(t),\\
[\![1/\varrho]\!] j_\Gamma^2\nu_\Gamma - [\![T\nu_\Gamma]\!] &= \mathrm{div}_\Gamma T_\Gamma && \text{on } \Gamma(t),\\
\varrho(0) = \varrho_0, \quad u(0) &= u_0 && \text{in } \Omega,
\end{aligned} \tag{1.26}$$

where $S = T + \pi$ and $S_\Gamma = \sigma(\theta_\Gamma)\mathcal{P}_\Gamma - T_\Gamma$ are defined above,

$$\begin{aligned} \varrho\kappa(\partial_t\theta + u\cdot\nabla\theta) - \operatorname{div}(d\nabla\theta) &= S : \nabla u - \theta\partial_\theta p \operatorname{div} u && \text{in } \Omega\setminus\Gamma(t),\\ \partial_\nu\theta &= 0 && \text{on } \partial\Omega,\\ \theta &= \theta_\Gamma && \text{on } \Gamma(t),\\ \theta(0) &= \theta_0 && \text{in } \Omega. \end{aligned} \tag{1.27}$$

On the interface we have

$$\begin{aligned} &\kappa_\Gamma\frac{D}{Dt}\theta_\Gamma - \operatorname{div}_\Gamma(d_\Gamma\nabla_\Gamma\theta_\Gamma)\\ &\qquad = S_\Gamma : \nabla_\Gamma u_\Gamma + \theta_\Gamma\sigma'(\theta_\Gamma)\operatorname{div}_\Gamma u_\Gamma - ([\![\theta\eta]\!]j_\Gamma + [\![q\cdot\nu_\Gamma]\!]) && \text{on } \Gamma(t),\\ &[\![\psi]\!] + [\![1/2\varrho^2]\!]j_\Gamma^2 - [\![T\nu_\Gamma\cdot\nu_\Gamma/\varrho]\!] = 0 && \text{on } \Gamma(t),\\ &V_\Gamma = u_\Gamma\cdot\nu_\Gamma = u\cdot\nu_\Gamma - j_\Gamma/\varrho && \text{on } \Gamma(t)\\ &\Gamma(0) = \Gamma_0. \end{aligned} \tag{1.28}$$

This system has to be supplemented with the constitutive laws for T and T_Γ from the previous subsection. Here the first system should be read as a problem for u and ϱ, resp. π, the second as one for θ, while the last set determines θ_Γ, the free boundary Γ, and the phase flux j_Γ. Note that in the absence of phase transitions, the Gibbs–Thomson relation has to be replaced by $j_\Gamma = 0$.

1.2 Entropy and Equilibria

2.1 The Entropy
We have seen above that the total entropy

$$\mathsf{N} := \int_\Omega \varrho\eta\,dx + \int_\Gamma \eta_\Gamma\,d\Gamma$$

satisfies

$$\begin{aligned} \frac{d}{dt}\Big(\int_\Omega \varrho\eta\,dx + \int_\Gamma \eta_\Gamma\,d\Gamma\Big) &= \int_\Omega \Big\{\frac{1}{\theta}S : \nabla u - \frac{1}{\theta^2}q\cdot\nabla\theta\Big\}\,dx\\ &\quad + \int_\Gamma \Big\{\frac{1}{\theta_\Gamma}S_\Gamma : \nabla_\Gamma u_\Gamma - \frac{1}{\theta_\Gamma^2}q_\Gamma\cdot\nabla_\Gamma\theta_\Gamma\Big\}\,d\Gamma. \end{aligned}$$

Hence the negative total entropy is a Lyapunov functional for the problem. We show now that it is even a strict one. To see this, assume that N is constant on some interval (t_1, t_2). Then $d\mathsf{N}/dt = 0$ in (t_1, t_2), hence $D = 0$ and $\nabla\theta = 0$ in $(t_1, t_2)\times\Omega$. Therefore, θ is constant, which implies $[\![d\partial_\nu\theta]\!] = 0$, and then from the interfacial boundary condition we obtain $j_\Gamma = 0$, provided $[\![\eta]\!] \neq 0$ on Γ; we assume this for the moment. This implies $[\![u]\!] = 0$, hence by Korn's inequality we have $\nabla u = 0$ and then $u = 0$ by the no-slip condition on $\partial\Omega$. Hence $V_\Gamma = 0$,

$u_\Gamma = 0$, and $(\partial_t\theta, \partial_t u, \partial_t \varrho, D\theta_\Gamma/Dt) = 0$, which means that we are at equilibrium. Further, $\nabla\pi = 0$, i.e., the pressure is constant in the components of the phases. If one or both phases are compressible, then assuming p_j to be strictly increasing in ϱ, we conclude that ϱ is constant in the components of $\Omega_j(t)$ as well. Actually ϱ is even constant in each phase. To see this, employing Maxwell's relation we rewrite the Gibbs–Thomson condition $[\![\psi]\!] + [\![\pi/\rho]\!] = 0$ as

$$\partial_\varrho(\varrho\psi_1(\varrho)) = \partial_\varrho(\varrho\psi_2(\varrho)).$$

Suppose ϱ_2 is known; then ϱ_1 is uniquely determined by ϱ_2 (and θ) since $\partial_\varrho(\varrho\psi_j(\varrho))$ is strictly increasing, for, by assumption, p_j has this property. Since θ is continuous across the interface, the last relation shows that π, and therefore ϱ, are constant in all of Ω_1, even if it is not connected. From this we finally deduce by the Young-Laplace law $[\![\pi]\!] = \sigma H_\Gamma$ that Ω_1 is a ball if it is connected, or otherwise a finite union of non-intersecting balls of equal radii, since Ω_1 is bounded by assumption.

If, by chance, $[\![\eta]\!] = 0$ on Γ, or only on part of it, we are not allowed to use Korn's inequality since u may have a jump across the interface. Nevertheless, $u = 0$ holds in this case as well, but the proof is a little more involved. For this we need

Lemma 1.2.1. *Suppose $u \in H^2_2(\Omega\setminus\Gamma)$ satisfies $u = 0$ on $\partial\Omega$ and $\mathcal{P}_\Gamma[\![u]\!] = 0$ on Γ. Then $D = 0$ implies $u = 0$ in Ω.*

Proof. Integrating by parts twice we obtain

$$\begin{aligned} 2|D|^2_{L_2(\Omega)} &= |\nabla u|^2_{L_2(\Omega)} + |\operatorname{div} u|^2_{L_2(\Omega)} + \int_\Gamma [\![u\cdot\nu_\Gamma \operatorname{div} u - \nu_\Gamma\cdot(u\cdot\nabla)u]\!]\, d\Gamma \\ &= |\nabla u|^2_{L_2(\Omega)} + |\operatorname{div} u|^2_{L_2(\Omega)} + \int_\Gamma 2[\![u\cdot\nu_\Gamma]\!]\operatorname{div}_\Gamma \mathcal{P}_\Gamma u - [\![(u\cdot\nu_\Gamma)^2]\!]H_\Gamma\, d\Gamma, \end{aligned}$$

since $u = 0$ on $\partial\Omega$ and $\mathcal{P}_\Gamma[\![u]\!] = 0$ on Γ. Here we employed the identities

$$\begin{aligned} \operatorname{div} u &= \operatorname{div}_\Gamma(\mathcal{P}_\Gamma u) - (u\cdot\nu_\Gamma)H_\Gamma + \nu_\Gamma\cdot\partial_\nu u, \\ \nu_\Gamma\cdot(u\cdot\nabla)u &= (\mathcal{P}_\Gamma u\cdot\nabla_\Gamma)u\cdot\nu_\Gamma + (u\cdot\nu_\Gamma)(\nu_\Gamma\cdot\partial_\nu u) + L_\Gamma\mathcal{P}_\Gamma u\cdot\mathcal{P}_\Gamma u \end{aligned}$$

on Γ as well as the surface divergence theorem. Now, if $D = 0$, then $\nu_\Gamma\cdot\partial_\nu u = 0$, and so the equation for the divergence of u on Γ yields

$$\operatorname{div}_\Gamma\mathcal{P}_\Gamma u = (u\cdot\nu_\Gamma)H_\Gamma,$$

hence $[\![(u\cdot\nu_\Gamma)^2]\!]H_\Gamma = 0$ which implies $\nabla u = 0$ in Ω. Therefore, u is constant in the phases, which yields $u = 0$ in Ω_2 by the no-slip condition on the outer boundary $\partial\Omega$. Further, $[\![u]\!] = \alpha\nu_\Gamma$ is constant on Γ which implies $\alpha = 0$, hence $[\![u]\!] = 0$ and so $u = 0$ in Ω_1, as well. □

Having shown that $u = 0$ we may proceed as before, provided $\varrho_1 \neq \varrho_2$. Actually, there is a problem if $[\![\varrho]\!] = [\![\eta]\!] = 0$; then we cannot conclude $j_\Gamma = 0$

which means that V_Γ may be nontrivial. We exclude this pathology in the sequel. It is absent anyway if kinetic undercooling is included.

If there is *no phase transition*, i.e., $j_\Gamma \equiv 0$, then $[\![u]\!] = 0$, and we obtain directly $u \equiv 0$ by Korn's inequality. In this case we conclude as above that the pressures are constant in the components of the phases, hence the densities are so as well, assuming as before that p_j is increasing. We further conclude from the interface stress condition that H_Γ is constant on each component of the interface, which implies that these components are spheres. But they may have differing sizes, as the Gibbs–Thomson relation is no longer available. If a phase transition is absent, constant temperature does no longer ensure that the spheres have equal size!

2.2 Equilibria as Critical Points of the Entropy

We want to determine the critical points of the total entropy N under the constraints of given total mass M_0 and given total energy E_0. With

$$\mathsf{M} = \int_\Omega \varrho\, dx, \quad \mathsf{E} = \int_\Omega \varrho(|u|^2/2 + \epsilon)\, dx + \int_\Gamma \epsilon_\Gamma\, d\Gamma,$$

the method of Lagrange multipliers then yields

$$\mathsf{N}' + \lambda \mathsf{M}' + \mu \mathsf{E}' = 0.$$

We compute the derivatives of the involved functionals, where $z = (\tau, v, \vartheta, \vartheta_\Gamma, h)$.

$$\begin{aligned}
\langle \mathsf{N}'|z\rangle &= \int_\Omega \{\partial_\varrho(\varrho\eta)\tau + \varrho\partial_\theta\eta\vartheta\}\, dx - \int_\Gamma \{[\![\varrho\eta]\!]h - \eta_\Gamma'\vartheta_\Gamma + \eta_\Gamma H_\Gamma h\}\, d\Gamma,\\
\langle \mathsf{M}'|z\rangle &= \int_\Omega \tau\, dx - \int_\Gamma [\![\varrho]\!]h\, d\Gamma,\\
\langle \mathsf{E}'|z\rangle &= \int_\Omega \{\varrho u\cdot v + \varrho\partial_\theta\epsilon\vartheta + (|u|^2/2 + \epsilon + \varrho\partial_\varrho\epsilon)\tau\}\, dx\\
&\quad - \int_\Gamma \{[\![\varrho|u|^2/2 + \varrho\epsilon]\!]h - \epsilon_\Gamma'\vartheta_\Gamma + \epsilon_\Gamma H_\Gamma h\}\, d\Gamma.
\end{aligned}$$

Varying first ϑ and ϑ_Γ this yields

$$\varrho\partial_\theta\eta + \mu\varrho\partial_\theta\epsilon = 0,$$

and

$$\eta_\Gamma' + \mu\epsilon_\Gamma' = 0,$$

hence $\partial_\theta\epsilon = \theta\partial_\theta\eta = \kappa > 0$ and $\epsilon_\Gamma' = \theta_\Gamma\eta_\Gamma' = \kappa_\Gamma > 0$ imply $\theta_\Gamma = \theta = -1/\mu > 0$ constant. Next we vary v to obtain $u = 0$ since $\mu \neq 0$. Variation of τ (when ϱ is not a priori constant) implies similarly

$$\eta + \varrho\partial_\varrho\eta + \lambda + \mu(\epsilon + \varrho\partial_\varrho\epsilon) = 0,$$

hence $\lambda = (\psi + \varrho\partial_\varrho\psi)/\theta$. As a consequence ϱ is constant, since

$$0 < \partial_\varrho p(\varrho,\theta)/\varrho = 2\partial_\varrho\psi(\varrho,\theta) + \varrho\partial_\varrho^2\psi(\varrho,\theta) = \partial_\varrho(\psi(\varrho,\theta) + \varrho\partial_\varrho\psi(\varrho,\theta))$$

in a phase where ϱ is not a priori constant. In particular, if both phases are compressible this yields $[\![\psi + p/\varrho]\!] = 0$, which is the generalized Gibbs–Thomson relation at equilibrium. Finally, we vary h to obtain

$$-[\![\varrho\eta]\!] - \eta_\Gamma H_\Gamma - \lambda[\![\varrho]\!] + ([\![\varrho\epsilon]\!] + \epsilon_\Gamma H_\Gamma)/\theta = 0,$$

which by the definition of ϵ and $\psi_\Gamma = \sigma$ yields

$$\sigma H_\Gamma + [\![\varrho\psi]\!] = \lambda\theta[\![\varrho]\!]$$

on the interface Γ. This implies that H_Γ is constant, hence Ω_1 consists of a finite number of balls with the same radius. If both phases are compressible we may further conclude $\sigma H_\Gamma = [\![p]\!]$, which is the normal stress condition on the interface.

In this derivation we assumed $\kappa_\Gamma > 0$. If instead $\kappa_\Gamma \equiv 0$, then $\eta_\Gamma' \equiv 0$ as well, hence we do not obtain information on θ_Γ. However, the remaining conclusions are valid as before. In this case $\sigma(\theta_\Gamma)$ is linear, and as there is no surface heat capacity it makes sense then to ignore surface diffusion as well.

In summary, we see that the critical points of the total entropy with the constraints of given mass and prescribed total energy are precisely the equilibria of the system.

2.3 Equilibria which are Maxima of Total Entropy

Suppose we have an equilibrium $e := (\varrho, u, \theta, \theta_\Gamma, \Gamma)$ where the total entropy has a local maximum, w.r.t. the constraints $\mathsf{M} = \mathsf{M}_0$ and $\mathsf{E} = \mathsf{E}_0$ constant. Then $\mathcal{D} := [\mathsf{N} + \lambda\mathsf{M} + \mu\mathsf{E}]''$ is negative semi-definite on the kernel of M' intersected with that of E', where (λ,μ) are the fixed Lagrange multipliers found in the previous subsection. The kernel of $\mathsf{M}'(e)$ is easily found to be characterized by the relation

$$\int_\Omega \tau\,dx = [\![\varrho]\!]\int_\Gamma h\,d\Gamma, \tag{1.29}$$

and that of $\mathsf{E}'(e)$ by

$$\int_\Omega \partial_\varrho(\varrho\eta)\tau\,dx + \int_\Omega (\varrho\kappa/\theta)\vartheta\,dx + (\kappa_\Gamma/\theta)\int_\Gamma \vartheta_\Gamma\,d\Gamma = ([\![\varrho\eta]\!] + \eta_\Gamma H_\Gamma)\int_\Gamma h\,d\Gamma. \tag{1.30}$$

On the other hand, a straightforward but somewhat lengthy calculation yields

$$\begin{aligned}-\theta\langle\mathcal{D}z|z\rangle = &\int_\Omega \varrho|v|^2\,dx + \int_\Omega \partial_\varrho^2(\varrho\psi)\tau^2\,dx + \int_\Omega (\varrho\kappa/\theta)\vartheta^2\,dx \\ &+ (\kappa_\Gamma/\theta)\int_\Gamma \vartheta_\Gamma^2\,d\Gamma - \sigma\int_\Gamma (H_\Gamma' h)h\,d\Gamma.\end{aligned} \tag{1.31}$$

As ϱ, κ, κ_Γ and

$$\partial_\varrho^2(\varrho\psi) = 2\partial_\varrho\psi + \varrho\partial_\varrho^2 = [\partial_\varrho p(\varrho)]/\varrho$$

are nonnegative, we see that the form $\langle\mathcal{D}z|z\rangle$ is negative semi-definite as soon as H'_Γ is negative semi-definite. We will see in the next chapter that

$$H'_\Gamma = (n-1)/R^2 + \Delta_\Gamma,$$

where Δ_Γ denotes the Laplace-Beltrami operator on Γ and R means the radius of the equilibrium spheres.

We want to derive necessary conditions for an equilibrium e to be a local maximum of entropy.

1. Suppose that Γ is not connected, i.e., Γ consists of a finite union of spheres Γ_k. Set $(\tau, v, \vartheta, \vartheta_\Gamma) = 0$, and let $h = h_k$ constant on Γ_k with $\sum_k h_k = 0$. Then the constraints (1.30) and (1.31) hold and

$$\langle\mathcal{D}z|z\rangle = \frac{\sigma(\theta)(n-1)}{\theta R^2}\sum_k \Gamma_k h_k^2 > 0,$$

hence $\mathcal{D}$ is not negative semi-definite in this case. Thus if e is an equilibrium with local maximal total entropy, then Γ must be connected, hence both phases are connected. This is related to the so-called *Ostwald ripening* effect.

2. Assume that Γ is connected and $\varrho_1 \neq \varrho_2$ are a priori constant. Then $\tau = 0$ and the first constraint (1.30) implies $\int_\Gamma h\, d\Gamma = 0$. As $H'_\Gamma(h)$ is negative semi-definite for functions with average zero, we see that in this case $\mathcal{D}$ is negative semi-definite.

3. Assume that Γ is connected and $\varrho_1 = \varrho_2 =: \varrho$ is constant. Then $\tau = 0$, but the first constraint gives no information. Setting $v = 0$, $\vartheta = \vartheta_\Gamma$ constant, as well as h constant, we see that $\mathcal{D}$ negative semi-definite on the kernel of $\mathsf{E}'(e)$ implies the condition

$$\frac{\sigma(\theta)(n-1)}{R^2} \leq \frac{l_0^2|\Gamma|}{\theta((\kappa|\varrho)_\Omega + \kappa_\Gamma|\Gamma|)}, \tag{1.32}$$

where $l_0 = l_0(\theta) = -\theta(\varrho[\![\eta]\!] + \eta_\Gamma H_\Gamma)$.

4. If e is an equilibrium which (locally) maximizes the total entropy, it is generically not isolated. If the sphere Γ does not touch the outer boundary, we may move it inside of Ω without changing the total entropy. This fact is reflected in $\mathcal{D}$ by choosing $\tau = \vartheta = \vartheta_\Gamma = 0$ and $h = Y_j$, the spherical harmonics for Γ, which satisfy $H'_\Gamma Y_j = 0$.

It is one of our purposes in this book to prove in the completely incompressible case that an equilibrium is stable if and only if the total entropy at this equilibrium is maximal. Thus in case $\varrho_1 \neq \varrho_2$ are a priori constant, an equilibrium is stable if and only if the interface is connected, and in case $\varrho_1 = \varrho_2$ if in addition the stability condition (1.32) is satisfied with strict inequality. (Here we exclude the limiting case where in (1.32) equality holds.)

2.4 The Manifold of Equilibria

As we have seen above, the equilibria of the system (1.26), (1.27), (1.28) are zero velocities, constant pressures in the phases, constant temperature, vanishing phase flux, and the dispersed phase Ω_1 consists of finitely many non-intersecting balls with the same radius if phase transition is present. We call an equilibrium *non-degenerate* if the balls do not touch the outer boundary $\partial\Omega$ and also do not touch each other. This set will be denoted by $\mathcal{E}$. We want to show that $\mathcal{E}$ is a manifold; it is not connected but has infinitely many finite dimensional components, the components are given by the number of spheres. The dimension of the component consisting of m spheres is $m(n+1)$, where n comes from the center and 1 from the radius of a particular sphere.

To show that $\mathcal{E}$ is a manifold, we just have to show how a neighbouring sphere is parameterized over a given one. In fact, let us assume that $\Sigma = S_R(0)$ is centered at the origin of $\mathbb{R}^n$. Suppose $\mathcal{S} \subset \Omega$ is a sphere that is sufficiently close to Σ. Denote by $(y_1, \ldots, y_n)$ the coordinates of its center and let y_0 be such that $R + y_0$ corresponds to its radius. Then the sphere $\mathcal{S}$ can be parameterized over Σ by the distance function

$$\delta(y) = \sum_{j=1}^{n} y_j Y_j - R + \sqrt{(\sum_{j=1}^{n} y_j Y_j)^2 + (R + y_0)^2 - \sum_{j=1}^{n} y_j^2},$$

where Y_j are the spherical harmonics of degree one. Obviously, this is a real analytic parametrization.

We summarize our considerations in

Theorem 1.2.2. **(a)** *The total mass* M *and the total energy* E *are preserved for smooth solutions.*

(b) *The negative total entropy* $-\mathsf{N}$ *is a strict Lyapunov functional except on the pathological points* (ϱ, θ) *constant,* $[\![\varrho]\!] = [\![\eta]\!] = 0$.

(c) *The critical points of the entropy functional for prescribed total mass and total energy are precisely the equilibria of the system.*

(d) *The* non-degenerate equilibria *are zero velocities, constant pressures in the components of the phases, and the interface is a union of non-intersecting spheres which do not touch the outer boundary* $\partial\Omega$. *If phase transition is present, then the spheres are of equal size.*

(e) *If the total entropy at an equilibrium is locally maximal, then the phases are connected and, in addition, in the case of equal constant densities the stability condition* (1.32) *holds.*

(f) *The set* $\mathcal{E}$ *of non-degenerate equilibria forms a real analytic manifold.*

This result shows that the models are thermodynamically consistent, hence are physically reasonable.

2.5 Equilibrium Temperatures
To determine θ, R and π at equilibrium, we have to solve the system

$$\begin{aligned} |\Omega_1|\varrho_1\epsilon_1 + |\Omega_2|\varrho_2\epsilon_2 + \epsilon_\Gamma|\Gamma| &= \mathsf{E}_0,\\ [\![\pi]\!] &= \sigma H_\Gamma,\\ [\![\psi]\!] &= -[\![\pi/\varrho]\!]. \end{aligned} \tag{1.33}$$

In addition, there is conservation of mass

$$\varrho_1|\Omega_1| + \varrho_2|\Omega_2| = c_0.$$

If the equilibrium densities are not equal, this equation can be employed to compute the radius of the balls, i.e., with $\omega_n = |\partial B(0,1)|$ we have

$$m(\omega_n/n)R^n = (\varrho_2|\Omega| - c_0)/[\![\varrho]\!]$$

in case there are m balls with common radius R. The energy equation then uniquely determines θ since ϵ_Γ is non-decreasing and $\epsilon_j(\theta)$ are strictly increasing. Finally, the last two conditions in (1.33) determine the pressures in the phases.

If there is no phase transition, then the dimension of the component $\mathcal{E}_m$ of $\mathcal{E}$, with $m \in \mathbb{N}$ the number of spheres, is $\dim \mathcal{E}_m = m(n+1)+1$. Here the variables are the centers of the balls, their radia, and the temperature. Prescribing total energy and individual volumes of the components of the dispersed phase reduces the dimension to mn.

On the other hand, if phase transition takes place and $\varrho_1 \neq \varrho_2$, then $\dim \mathcal{E}_m = mn+2$. The variables are the centers of the balls, the common radius, and the temperature. If we prescribe phase volumes and total energy, then the radius of the balls and the temperature are fixed, resulting into $\dim \mathcal{E}_m = nm$.

But if the equilibrium densities are equal, $\varrho_1 = \varrho_2 =: \varrho$, then conservation of mass determines merely the value of the density ϱ, no information on the phase volumes at equilibrium is available. Hence only θ, R and the pressure jump

$$[\![\pi]\!] = [\![\pi]\!](\theta) = \sigma(\theta)H_\Gamma = -\frac{\sigma(\theta)(n-1)}{R(\theta)}$$

can be obtained from (1.33). This implies that the dimension of $\mathcal{E}_m$ is $mn+1$, and if we prescribe the total energy, then it will be nm.

In this case we get

$$R = R(\theta) = \frac{\sigma(\theta)(n-1)}{\varrho[\![\psi(\theta)]\!]}$$

for the radius $R > 0$, and system (1.33) reduces to a single equation for the temperature θ:

$$\mathsf{E}_e(\theta) := |\Omega|\varrho\epsilon_2(\theta) - m(\omega_n/n)R^n(\theta)\varrho[\![\epsilon(\theta)]\!] + \epsilon_\Gamma m\omega_n R^{n-1}(\theta) = \mathsf{E}_0.$$

We call the function $\mathsf{E}_e(\theta)$ the *equilibrium energy* function.

Note that only the temperature range $[\![\psi(\theta)]\!]/\sigma(\theta) > 0$ is relevant due to the requirement $R > 0$, and with

$$R^*_m = \sup\{R > 0 : \Omega \text{ contains } m \text{ disjoint balls of radius } R\}$$

we must also have $R < R^*_m$, i.e., with $\varphi(\theta) = \varrho[\![\psi(\theta)]\!]$

$$0 < \frac{\sigma(\theta)}{\varphi(\theta)} < \frac{R^*_m}{n-1}.$$

With $\epsilon(\theta) = \psi(\theta) - \theta\psi'(\theta)$ and $\epsilon_\Gamma = \sigma(\theta) - \theta\sigma'(\theta)$, after some calculations $\mathsf{E}_e(\theta)$ may be rewritten as

$$\begin{aligned}\mathsf{E}_e(\theta) &= |\Omega|\varrho\epsilon_2(\theta) + c_n\Big(\frac{\sigma(\theta)^n}{\varphi(\theta)^{n-1}} - \theta\frac{d}{d\theta}\frac{\sigma(\theta)^n}{\varphi(\theta)^{n-1}}\Big)\\ &= |\Omega|\varrho\epsilon_2(\theta) + c_n\Big(\frac{\sigma(\theta)^n}{\varphi(\theta)^{n-1}} + (n-1)\theta\frac{\sigma(\theta)^n\varphi'(\theta)}{\varphi(\theta)^n} - n\theta\frac{\sigma(\theta)^{n-1}\sigma'(\theta)}{\varphi(\theta)^{n-1}}\Big),\end{aligned}$$

where we have set $c_n = m\frac{\omega_n}{n}(n-1)^{n-1}$. Observe that the equilibrium energy function $\mathsf{E}_e(\theta)$ has the form

$$\mathsf{E}_e(\theta) = \Psi(\theta) - \theta\Psi'(\theta),$$

where $\Psi(\theta) = |\Omega|\varrho\psi_2(\theta) + c_n\sigma(\theta)^n/\varphi(\theta)^{n-1}$ plays the role of the equilibrium free energy. We have then

$$\mathsf{E}_e'(\theta) = -\theta\Psi''(\theta),$$

hence with

$$R'(\theta) = \frac{(n-1)\sigma'(\theta)}{\varphi(\theta)} - \frac{\sigma(\theta)(n-1)\varphi'(\theta)}{\varphi^2(\theta)},$$

after some more calculations

$$\mathsf{E}_e'(\theta) = (\kappa(\theta)|\varrho)_\Omega + \kappa_\Gamma(\theta)|\Gamma| - \frac{R(\theta)^2 l_0(\theta)^2|\Gamma|}{\theta\sigma(\theta)(n-1)},$$

with $l_0(\theta)$ defined in the previous subsection. Now recall the stability condition (1.32) to see that $\mathsf{E}_e'(\theta)$ is non-positive if and only if the stability condition holds. Thus, loosely speaking, total entropy is maximal at an equilibrium if and only if $\mathsf{E}_e'(\theta) \leq 0$. We may write $\mathsf{E}_e'(\theta)$ yet in another form, namely

$$\mathsf{E}_e'(\theta) = (\kappa(\theta)|\varrho)_\Omega + \kappa_\Gamma(\theta)|\Gamma| - (n-1)|\Gamma|\sigma\theta(\frac{\varphi'(\theta))}{\varphi(\theta)} - \frac{\sigma'(\theta)}{\sigma(\theta)})^2.$$

In general it is not a simple task to analyze the equation for the temperature

$$\mathsf{E}_e(\theta) = \Psi(\theta) - \theta\Psi'(\theta) = \mathsf{E}_0,$$

unless more properties of the functions $\epsilon_j(\theta)$ and in particular of $\varphi(\theta)$ and $\sigma(\theta)$ are known. A natural assumption is that φ has exactly one positive zero $\theta_m > 0$, the so called *melting temperature*. Therefore we look at two examples.

Example 1. Suppose that ϵ_2 is increasing and convex, $\varrho = 1$, $\eta_\Gamma \equiv 0$, i.e., σ is constant, and that the heat capacities are identical, i.e., $[\![\kappa]\!] \equiv 0$. This implies

$$\theta\varphi''(\theta) = \theta[\![\psi''(\theta)]\!] = -[\![\kappa(\theta)]\!] \equiv 0,$$

which means that $\varphi(\theta) = \varphi_0 + \varphi_1\theta$ is linear. The melting temperature then is $0 < \theta_m = -\varphi_0/\varphi_1$, hence we have two cases.

Case 1. $\varphi_0 < 0$, $\varphi_1 > 0$. This means $l(\theta_m) > 0$.
Then the relevant temperature range is $\theta > \theta_m$ as φ is positive there. As $\theta \to \theta_m+$ we have $\varphi(\theta) \to 0$ hence $\mathsf{E}_e(\theta) \to \infty$, and also $\mathsf{E}_e(\theta) \to \infty$ for $\theta \to \infty$ as $\epsilon_2(\theta)$ is increasing and convex. Further, we have

$$\begin{aligned}\mathsf{E}_e'(\theta) &= |\Omega|\epsilon_2'(\theta) - n(n-1)c_n\sigma(\theta)^n\frac{\varphi_1^2\theta}{(\varphi_0+\varphi_1\theta)^{n+1}},\\ \mathsf{E}_e''(\theta) &= |\Omega|\epsilon_2''(\theta) + n(n-1)c_n\sigma(\theta)^n\varphi_1^2\frac{-\varphi_0+n\varphi_1\theta}{(\varphi_0+\varphi_1\theta)^{n+2}} > 0,\end{aligned}$$

which shows that $\mathsf{E}_e(\theta)$ is strictly convex for $\theta > \theta_m$. Thus $\mathsf{E}_e(\theta)$ has a unique minimum $\theta_0 > \theta_m$, $\mathsf{E}_e(\theta)$ is decreasing for $\theta_m < \theta < \theta_0$ and increasing for $\theta > \theta_0$. Thus there are precisely two equilibrium temperatures $\theta_*^+ \in (\theta_0, \infty)$ and $\theta_*^- \in (\theta_m, \theta_0)$ provided $\mathsf{E}_0 > \phi(\theta_0)$ and none if $\mathsf{E}_0 < \mathsf{E}(\theta_0)$. The smaller temperature leads to stable equilibria while the larger to unstable ones.

Case 2. $\varphi_0 > 0$, $\varphi_1 < 0$. This means $l(\theta_m) < 0$.
Then the relevant temperature range is $0 < \theta < \theta_m$ as h is positive there. As $\theta \to \theta_m-$ we have $\varphi(\theta) \to 0+$ hence $\mathsf{E}_e(\theta) \to -\infty$, and as $\theta \to 0+$ we have $\mathsf{E}_e(\theta) \to \mathsf{E}(0) = |\Omega|\epsilon_2(0) + c_n\sigma^n/\varphi_0^{n-1} > 0$, assuming that $\epsilon_2(0) = \lim_{s\to 0+}\epsilon_2(s)$ exists. Further, for θ close to 0 this implies $\mathsf{E}_e'(\theta) > 0$ and $\mathsf{E}_e'(\theta) \to -\infty$ as $\theta \to \theta_m-$. Therefore $\mathsf{E}_e'(\theta)$ admits at least one zero in $(0, \theta_m)$. But there may be more than one unless $\epsilon_2(\theta)$ is concave, so let us assume this. Let $\theta_0 \in (0, \theta_m)$ denote the absolute maximum of $\mathsf{E}_e(\theta)$ in $(0, \theta_m)$. Then there is exactly one equilibrium temperature $\theta_* \in (\theta_0, \theta_m)$ if $\mathsf{E}_0 < \mathsf{E}_e(0+)$ and it is stable; there are exactly two equilibria $\theta_*^- \in (0, \theta_0)$ and $\theta_*^+ \in (\theta_0, \theta_m)$ if $\mathsf{E}_e(0+) < \mathsf{E}_0 < \mathsf{E}_e(\theta_0)$, the first one is unstable the second stable. If $\mathsf{E}_0 > \mathsf{E}_e(\theta_0)$ there are no equilibria.

Note that in both cases these equilibrium temperatures give rise to equilibria only if the corresponding radius is smaller than R^*.

Example 2. Suppose $\eta_\Gamma \equiv 0$, $\varrho = 1$, and that the internal energies $\epsilon_j(\theta)$ are linear increasing, i.e.,

$$\epsilon_j(\theta) = a_j + \kappa_j\theta, \quad j = 1, 2,$$

where $\kappa_j > 0$, and now $[\![\kappa]\!] \neq 0$. The identity $\epsilon_j = \psi_j - \theta\psi_j'$ then leads to

$$\psi_j(\theta) = a_j + b_j\theta - \kappa_j\theta\log\theta, \quad j = 1, 2,$$

where b_j are arbitrary. This yields, with $\alpha = [\![a]\!]$, $\beta = [\![b]\!]$ and $\gamma = [\![\kappa]\!]$,

$$\varphi(\theta) = \alpha + \beta\theta - \gamma\theta\log\theta.$$

Scaling the temperature by $\theta = \theta_0\vartheta$ with $\beta - \gamma\log\theta_0 = 0$ and scaling φ we may assume $\beta = 0$ and $\gamma = \pm 1$. Then we have to investigate the equation $\mathsf{E}_e(\vartheta) = \mathsf{E}_1$, where

$$\mathsf{E}_e(\vartheta) = \delta\vartheta + \Big\{\frac{1}{\varphi^{n-1}(\vartheta)} + (n-1)\vartheta\frac{\varphi'(\vartheta)}{\varphi^n(\vartheta)}\Big\}, \quad \varphi(\vartheta) = \pm(\alpha + \vartheta\log\vartheta),$$

with $\delta > 0$ and $\alpha, \mathsf{E}_1 \in \mathbb{R}$. The requirement of existence of a melting temperature $\vartheta_m > 0$, i.e., a zero of $\varphi(\vartheta)$, leads to the restriction $\alpha \leq 1/e$. Also here we have to distinguish two cases, namely that of a plus-sign for φ where the relevant temperature range is $\vartheta > \vartheta_m$, and in case of a minus-sign it is $(0, \vartheta_m)$. Note that φ is convex in the first, and concave in the second case. In the case of $\varphi(\vartheta) = (\alpha + \vartheta\log\vartheta)$ we get

$$\mathsf{E}_e'(\vartheta) = \delta + (n-1)\Big\{\frac{\varphi(\vartheta) - n\vartheta\varphi'(\vartheta)^2}{\varphi^{n+1}(\vartheta)}\Big\},$$

$$\mathsf{E}_e''(\vartheta) = n(n-1)\frac{\varphi'(\vartheta)}{\varphi^{n+2}(\vartheta)}\Big\{(n+1)\vartheta\varphi'(\vartheta)^2 - \varphi(\vartheta)(3 + \varphi'(\vartheta))\Big\}.$$

We have $\mathsf{E}_e(\vartheta) \to \infty$ for $\vartheta \to \infty$ and for $\vartheta \to \vartheta_m+$, hence $\mathsf{E}_e(\vartheta)$ has a global minimum θ_0 in (θ_m, ∞). Furthermore, $\mathsf{E}_e''(\vartheta) > 0$ in (θ_m, ∞), hence the minimum is unique and there are precisely two equilibrium temperatures $\vartheta_*^- \in (\vartheta_m, \vartheta_0)$ and $\vartheta_*^+ \in (\vartheta_0, \infty)$, provided $\mathsf{E}_1 > \mathsf{E}_e(\vartheta_0)$, the first one is stable, the second unstable.

To prove convexity of E_e we write

$$(n+1)\vartheta\varphi'(\vartheta)^2 - 3\varphi(\vartheta) - \varphi(\vartheta)\varphi'(\vartheta) = (n-1)\vartheta\varphi'(\vartheta)^2 + f(\vartheta),$$

where

$$f(\vartheta) = 2\vartheta\varphi'(\vartheta)^2 - \varphi(\vartheta)(3 + \varphi'(\vartheta)) = 2\vartheta(1 + \log\vartheta)^2 - (\alpha + \vartheta\log\vartheta)(4 + \log\vartheta).$$

We then have $f(\vartheta_m) = 2\vartheta_m(1 + \log\vartheta_m)^2 > 0$, and

$$f'(\theta) = (1 + \log\vartheta)^2 + 1 - \alpha/\vartheta > 1 - \alpha/\vartheta \geq 0,$$

for $\alpha \leq 1/e < \vartheta_m \leq \vartheta$.

Actually, the requirement that the melting temperature is unique, i.e., that φ has exactly one positive zero, implies $\alpha < 0$. Indeed, for $\alpha \in (0, 1/e)$ there is a second zero $\vartheta_- > 0$ of φ, and φ is positive in $(0, \vartheta_-)$. Equilibrium temperatures in this range would not make sense physically.

Let us illustrate the sign in φ for the water-ice system, ignoring the density jump of water at freezing temperature. So suppose that Ω_2 consists of ice and Ω_1 of water. In this case we have $\kappa_1 > \kappa_2$, and hence $\gamma < 0$, which implies the plus-sign for φ. Here we obtain $\theta_*^\pm > \theta_m$, i.e., the ice is overheated. Equilibria only exist if ψ_0 is large enough, which means that there is enough energy in the system. If the energy in the system is very large, then the stable equilibrium temperature θ_*^- comes close to the melting temperature ϑ_m and then $R(\vartheta)$ will become large, eventually larger than R^*. This excludes equilibria in Ω, the physical interpretation being that everything will eventually melt.

On the other hand, if Ω_1 consists of ice and Ω_2 of water, we have the minus sign, which we want to consider next. Here we expect under-cooling of the water-phase, existence of equilibria only for low values of energy, and if the energy in the system is too small everything will freeze.

So assume that $\varphi(\vartheta) = -(\alpha + \vartheta \log \vartheta)$ and let $\alpha < 0$. Then the relevant temperature range is $(0, \vartheta_m)$. Here we have $\mathsf{E}_e(\vartheta) \to -\infty$ as $\vartheta \to \vartheta_m-$ and $\mathsf{E}_e(\vartheta) \to 1/|\alpha|^{n-1} > 0$. Moreover we have $\mathsf{E}_e'(0) = \delta + (n-1)/|\alpha|^n > 0$, and $\mathsf{E}_e'(\vartheta) \to -\infty$ for $\vartheta \to \vartheta_m-$. Therefore, $\mathsf{E}_e(\theta)$ has an absolute maximum in ϑ_0 in the interval $(0, \vartheta_m)$. If $\mathsf{E}_e(\vartheta)$ would be concave in $(0, \vartheta_m)$, then this maximum would be unique and there would be precisely two equilibrium temperatures $\vartheta_*^- \in (0, \vartheta_0)$ and $\vartheta_*^+ \in (\vartheta_0, \vartheta_m)$, provided $\mathsf{E}_1 \in (-\infty, \mathsf{E}_e(\vartheta_0))$, the first one unstable and the second stable. However, as we will see things are not as simple.

To investigate concavity of E_e in the interval $(0, \vartheta_m)$, we recompute the derivatives of E_e.

$$\begin{aligned}
\mathsf{E}_e'(\vartheta) &= \delta - (n-1)\Big\{\frac{1}{\varphi^n(\vartheta)} + n\frac{\vartheta \varphi'(\vartheta)^2}{\varphi^{n+1}(\vartheta)}\Big\}, \\
\mathsf{E}_e''(\vartheta) &= n(n-1)\frac{\varphi'(\vartheta)}{\varphi^{n+2}(\vartheta)}\Big\{(n+1)\vartheta\varphi'(\vartheta)^2 + \varphi(\vartheta)(3 - \varphi'(\vartheta))\Big\}.
\end{aligned}$$

Setting $\vartheta_+ = 1/e$, for $\vartheta \in (\vartheta_+, \vartheta_m)$ we have $\varphi(\vartheta) > 0$ and $\varphi'(\vartheta) < 0$, and hence $\mathsf{E}_e''(\vartheta) < 0$. On the other hand, for $\vartheta \in (0, \vartheta_+)$, both $\varphi(\vartheta)$ and $\varphi'(\vartheta)$ are positive. Then we rewrite

$$(n+1)\vartheta\varphi'(\vartheta)^2 + 3\varphi(\vartheta) - \varphi(\vartheta)\varphi'(\vartheta) = (n-1)\vartheta(1 + \log\vartheta)^2 + f(\vartheta),$$

where

$$\begin{aligned}
f(\vartheta) &= 2\vartheta\varphi'(\vartheta)^2 + \varphi(\vartheta)(3 - \varphi'(\vartheta)) \\
&= 2\vartheta(1 + \log\vartheta)^2 - (\alpha + \vartheta\log\vartheta)(4 + \log\vartheta) \\
&= \vartheta(2 + \log^2(\theta)) - \alpha(4 + \log\vartheta), \\
f'(\vartheta) &= 2 + \log^2\vartheta + 2\log\vartheta - \alpha/\vartheta = (1 + \log\vartheta)^2 + 1 - \alpha/\vartheta \geq 0,
\end{aligned}$$

provided $\alpha \leq 0$. This shows that f is increasing, $f(\vartheta) \to -\infty$ as $\vartheta \to 0$, and $f(1/e^3) = 11/e^3 - \alpha > 0$. On the other hand, the function $\vartheta(1+\log\vartheta)^2$ is increasing in $(0, 1/e^3)$, hence $\psi''(\vartheta)$ has a unique zero $\vartheta_- \in (0, 1/e^3)$. Therefore, E_e is concave in $(0, \vartheta_-) \cup (\vartheta_+, \vartheta_m)$ and convex in $(\vartheta_-, \vartheta_+)$, and E_e' has a minimum at ϑ_- and a maximum at ϑ_+. Observe that $\mathsf{E}_e'(\vartheta) < \delta$, $\mathsf{E}_e'(\vartheta) \to -\infty$ for $\vartheta \to \vartheta_m-$ and $\mathsf{E}_e'(0+) = \delta - (n-1)/|\alpha|^n < \psi'(\vartheta_+)$. Therefore, E_0' may have no, one, two, or three zeros in $(0, \vartheta_m)$, depending on the value of $\delta > 0$. However, if $\delta > 0$ is large enough, then E_e' has only one zero ϑ_1 which lies in $(\vartheta_+, \vartheta_m)$. In this case E_e is increasing in $(0, \vartheta_1)$ and decreasing in $(\vartheta_1, \vartheta_m)$, hence for $\mathsf{E}_e \in (\psi(0), \psi(\vartheta_1))$ there are precisely two equilibrium temperatures, the smaller leads to unstable, the larger to a stable equilibrium. If $\mathsf{E}_1 < \mathsf{E}_e(0+)$ there is a unique equilibrium which is stable, and in case $\mathsf{E}_1 > \mathsf{E}_e(\vartheta_1)$ there is none. However, in general there may be up to four equilibrium temperatures.

1.3 Goals and Strategies

In this book we will consider only the *completely incompressible* case, i.e., the densities ϱ_1 and ϱ_2 are assumed to be constant. Throughout we neglect viscous surface stress, so we set $S_\Gamma \equiv 0$. Thus the only surface stress acting is the surface tension $T_\Gamma = \sigma \mathcal{P}_\Gamma$. We always assume the constitutive laws

$$T = S - \pi I, \quad S := 2\mu(\theta)D, \quad D = (\nabla u + [\nabla u]^\mathsf{T})/2.$$

In this book we want to consider the following main problems which are ordered by complexity. The main hypotheses for these problems are formulated as well. Throughout, Ω will be a bounded domain with boundary $\partial\Omega$ of class C^3.

3.1 The Main Models

Problem 1. The Stefan Problem with Surface Tension.
Here we assume $\varrho_1 = \varrho_2 =: \varrho > 0$, $\sigma > 0$, and $u \equiv 0$.
Then we have

$$V_\Gamma = -j_\Gamma/\varrho, \quad [\![-T\nu_\Gamma]\!] = \sigma H_\Gamma \nu_\Gamma,$$

hence the Gibbs–Thomson law becomes

$$[\![\psi(\theta)]\!] = \frac{1}{\varrho}[\![T\nu_\Gamma \cdot \nu_\Gamma]\!] = -\frac{\sigma}{\varrho} H_\Gamma,$$

and we have the Stefan law $-\varrho[\![\theta\eta(\theta)]\!]V_\Gamma - [\![d(\theta)\partial_\nu\theta]\!] = 0$ on Γ. Observing that at melting temperature θ_m there holds $[\![\psi(\theta_m)]\!] = 0$, by linearization of ψ one obtains with the relative temperature $\vartheta = (\theta - \theta_m)/\theta_m$

$$\vartheta = -\frac{\sigma}{l_m \varrho} H_\Gamma, \quad l_m = -\theta_m [\![\eta(\theta_m)]\!],$$

which is the standard constitutive relation for the classical Stefan problem with surface tension. Here l_m is the latent heat at melting temperature. Similarly, the linearized Stefan law becomes $\varrho l_m V_\Gamma - [\![d(\theta)\partial_\nu\theta]\!] = 0$, which is the classical one. Note that these relations are only valid near melting temperature, and in particular exclude large curvatures of Γ. In this model, surface entropy is zero and balance of surface energy is trivial. The model equations read

$$\begin{aligned} \varrho\kappa(\theta)\partial_t\theta - \operatorname{div}(d(\theta)\nabla\theta) &= 0 && \text{in } \Omega\setminus\Gamma(t),\\ \partial_\nu\theta &= 0 && \text{on } \partial\Omega,\\ [\![\theta]\!] = 0, \quad \varrho[\![\psi(\theta)]\!] + \sigma H_\Gamma &= 0 && \text{on } \Gamma(t),\\ \theta(0) &= \theta_0 && \text{in } \Omega. \end{aligned} \tag{1.34}$$

$$\begin{aligned} -\varrho[\![\theta\eta(\theta)]\!]V_\Gamma - [\![d(\theta)\partial_\nu\theta]\!] &= 0 \quad \text{on } \Gamma(t),\\ \Gamma(0) &= \Gamma_0. \end{aligned} \tag{1.35}$$

Concerning ψ and d we assume

(H1) $\psi \in C^3(0,\infty)$, $d \in C^2(0,\infty)$, $-\psi''(s), d(s) > 0$ for all $s > 0$.

Remark 1.3.1. If $\kappa \equiv 0$, i.e., if ψ is linear, we obtain the so-called quasi-stationary Stefan problem with surface tension, also called *Mullins–Sekerka problem* or *Mullins–Sekerka flow* in the literature. It has the same equilibria as in the case $\kappa \not\equiv 0$, but their stability properties are different.

Problem 2. The Two-Phase Navier–Stokes Problem with Surface Tension.
Here we assume $j_\Gamma \equiv 0$, $\sigma > 0$ constant.
This is the case without phase transitions. Then

$$[\![u]\!] = 0, \quad V_\Gamma = u\cdot\nu_\Gamma, \quad -[\![T\nu_\Gamma]\!] = \sigma H_\Gamma\nu_\Gamma,$$

which leads to the classical model for *incompressible two-phase flow without phase transitions.*

$$\begin{aligned} \varrho(\partial_t u + u\cdot\nabla u) - \operatorname{div} S + \nabla\pi &= 0 && \text{in } \Omega\setminus\Gamma(t),\\ \operatorname{div} u &= 0 && \text{in } \Omega\setminus\Gamma(t),\\ u &= 0 && \text{on } \partial\Omega,\\ [\![u]\!] = 0, \quad -[\![S\nu_\Gamma]\!] + [\![\pi]\!]\nu_\Gamma &= \sigma H_\Gamma\nu_\Gamma && \text{on } \Gamma(t),\\ u(0) &= u_0 && \text{in } \Omega. \end{aligned} \tag{1.36}$$

$$\begin{aligned} \varrho\kappa(\theta)(\partial_t\theta + u\cdot\nabla\theta) - \operatorname{div}(d(\theta)\nabla\theta) &= 2\mu(\theta)|D|_2^2 && \text{in } \Omega\setminus\Gamma(t),\\ \partial_\nu\theta &= 0 && \text{on } \partial\Omega,\\ [\![\theta]\!] = 0, \quad [\![d(\theta)\partial_\nu\theta]\!] &= 0 && \text{on } \Gamma(t),\\ \theta(0) &= \theta_0 && \text{in } \Omega. \end{aligned} \tag{1.37}$$

$$V_\Gamma = u\cdot\nu_\Gamma \quad \text{on} \quad \Gamma(t), \quad \Gamma(0) = \Gamma_0. \tag{1.38}$$

Here we suppose

(H2) $\psi \in C^3(0,\infty)$, $d, \mu \in C^2(0,\infty)$, $-\psi''(s), d(s), \mu(s) > 0$ for all $s > 0$.

Remark 1.3.2. **(i)** If μ is constant, then the Navier-Stokes problem decouples from the heat problem. More generally, in the *isothermal case*, the temperature is assumed to be constant and the equation for the temperature, i.e., energy balance, is ignored. This means that the friction term $2\mu|D|_2^2$ is neglected. In this case the reduced energy E_0 defined by

$$\mathsf{E}_0(u,\Gamma) := \frac{1}{2}\int_{\Omega\setminus\Gamma} \varrho|u|^2\,dx + \sigma|\Gamma|$$

is a strict Lyapunov functional, as the identity

$$\frac{d}{dt}\mathsf{E}_0(u(t),\Gamma(t)) = -2\int_\Omega \mu|D|_2^2\,dx$$

and Korn's inequality show. Also in this case the equilibria are zero velocity and constant pressures in the components of the phases. The disperse phase Ω_1 is an at most countable union of disjoint balls, and the radia of the balls are related to the pressures according to the *Young-Laplace law*

$$[\![\pi]\!] = \sigma H_\Gamma = -\frac{\sigma(n-1)}{R}.$$

(ii) If θ is constant and ignoring inertia (i.e., the term $\varrho(\partial_t u + u\cdot\nabla u)$) we are left with a quasi-stationary problem, the *two-phase Stokes problem*, which generates the so-called *two-phase Stokes flow*. More precisely, this problem reads

$$\begin{aligned}
-\mathrm{div}\, S + \nabla\pi &= 0 && \text{in } \Omega\setminus\Gamma(t),\\
\mathrm{div}\, u &= 0 && \text{in } \Omega\setminus\Gamma(t),\\
u &= 0 && \text{on } \partial\Omega,\\
[\![u]\!] = 0, \quad -[\![S\nu_\Gamma]\!] + [\![\pi]\!]\nu_\Gamma &= \sigma H_\Gamma\nu_\Gamma && \text{on } \Gamma(t),\\
V_\Gamma &= u\cdot\nu_\Gamma && \text{on } \Gamma(t),\\
\Gamma(0) &= \Gamma_0.
\end{aligned} \tag{1.39}$$

(iii) If $\sigma = 0$, then $u = 0$ is a solution of the Navier–Stokes problem. Then we end up with the standard transmission problem for the heat equation with fixed domain.

(iv) Modeling flows in porous media frequently relies on *Darcy's law*, which reads

$$u = -k\nabla\pi,$$

where $k = k(\pi) > 0$ may depend on π, and depends on the phases. The interface velocity then becomes

$$V_\Gamma = u\cdot\nu_\Gamma = -k(\pi)\partial_\nu\pi.$$

This is meaningful, provided

$$-[\![k(\pi)\partial_\nu\pi]\!] = [\![u\cdot\nu_\Gamma]\!] = 0.$$

Furthermore, the driving force for the evolution of the interface is surface tension, hence we require

$$[\![\pi]\!] = \sigma H_\Gamma,$$

where $\sigma > 0$ is constant. Finally, we have to take into account conservation of mass which results in the *porous medium equation*

$$\partial_t \varrho(\pi) - \operatorname{div}\left(\varrho(\pi)k(\pi)\nabla\pi\right) = 0.$$

Here $\varrho > 0$ is non-decreasing w.r.t. π, and depends on the phases. Summarizing we obtain the problem

$$\begin{aligned} \varrho'(\pi)\partial_t\pi - \operatorname{div}\left(\varrho(\pi)k(\pi)\nabla\pi\right) &= 0 && \text{in } \Omega\setminus\Gamma(t),\\ \partial_\nu\pi &= 0 && \text{on } \partial\Omega,\\ [\![\pi]\!] &= \sigma H_\Gamma && \text{on } \Gamma(t),\\ [\![k(\pi)\partial_\nu\pi]\!] &= 0 && \text{on } \Gamma(t),\\ V_\Gamma + k(\pi)\partial_\nu\pi &= 0 && \text{on } \Gamma(t),\\ \Gamma(0) = \Gamma_0, \quad \pi(0) &= \pi_0. \end{aligned} \tag{1.40}$$

This problem is called the *Verigin problem* in the literature, and its quasi-steady (i.e., incompressible) version, where ϱ is constant in the phases, is known as the *Muskat problem* or the *Muskat flow*, a geometric evolution equation.

(v) A variant of Darcy's law is *Forchheimer's law* which reads

$$g(|u|)u = -\nabla\pi,$$

where the function g is strictly positive and $s \mapsto sg(s)$ is strictly increasing. Solving this equation for u we obtain

$$u = -k(|\nabla\pi|^2)\nabla\pi,$$

where k is strictly positive and satisfies $k(t)+2tk'(t) > 0$ on $\mathbb{R}_+$. These conditions ensure strong ellipticity of the operator $-\operatorname{div}(k(|\nabla\pi|^2)\nabla\pi)$.

Problem 3. Incompressible Two-Phase Fluid Flow with Phase Transition I.
Here we assume $\varrho_1 = \varrho_2 =: \varrho$, $\sigma > 0$ constant.
In this situation the Navier–Stokes problem is only weakly coupled to a Stefan problem. It can be treated by combining the methods developed for Problems 1 and 2. We call this case *temperature dominated.*

$$\begin{aligned} \varrho(\partial_t u + u\cdot\nabla u) - \operatorname{div} S + \nabla\pi &= 0 && \text{in } \Omega\setminus\Gamma(t),\\ \operatorname{div} u &= 0 && \text{in } \Omega\setminus\Gamma(t),\\ u &= 0 && \text{on } \partial\Omega,\\ [\![u]\!] = 0, \quad -[\![S\nu_\Gamma]\!] + [\![\pi]\!]\nu_\Gamma &= \sigma H_\Gamma\nu_\Gamma && \text{on } \Gamma(t),\\ u(0) &= u_0 && \text{in } \Omega. \end{aligned} \tag{1.41}$$

$$
\begin{aligned}
\varrho\kappa(\theta)(\partial_t\theta + u\cdot\nabla\theta) - \operatorname{div}(d(\theta)\nabla\theta) &= 2\mu(\theta)|D|_2^2 && \text{in } \Omega\setminus\Gamma(t),\\
\partial_\nu\theta &= 0 && \text{on } \partial\Omega,\\
[\![\theta]\!] = 0, \quad [\![\theta\eta(\theta)]\!]j_\Gamma - [\![d(\theta)\partial_\nu\theta]\!] &= 0 && \text{on } \Gamma(t),\\
\theta(0) &= \theta_0 && \text{in } \Omega.
\end{aligned}
\tag{1.42}
$$

$$
\begin{aligned}
\varrho[\![\psi(\theta)]\!] + \sigma H_\Gamma &= 0 && \text{on } \Gamma(t),\\
V_\Gamma = u_\Gamma\cdot\nu_\Gamma &= u\cdot\nu_\Gamma - j_\Gamma/\varrho && \text{on } \Gamma(t),\\
\Gamma(0) &= \Gamma_0.
\end{aligned}
\tag{1.43}
$$

We set hypothesis **(H3)** := **(H2)**. Recall that we can eliminate the phase flux j_Γ by

$$j_\Gamma = -[\![d(\theta)\partial_\nu\theta]\!]/l(\theta),$$

provided $l(\theta)\neq 0$. This will be one restriction for well-posedness of this model.

Remark 1.3.3. We will see that the Navier–Stokes problem is only weakly coupled to the Stefan problem with surface tension. Setting $u = 0$ and ignoring the Navier–Stokes problem it reduces to Problem **(P1)**.

Problem 4. Incompressible Two-Phase Fluid Flow with Phase Transition II. Here we assume $\varrho_1\neq\varrho_2$, $\sigma>0$ constant.
This case is more difficult than the previous one. Here the problem for θ is only weakly coupled with that for (u,π,h). We call this case *velocity dominated.*

$$
\begin{aligned}
\varrho(\partial_t u + u\cdot\nabla u) - \operatorname{div} S + \nabla\pi &= 0 && \text{in } \Omega\setminus\Gamma(t),\\
\operatorname{div} u &= 0 && \text{in } \Omega\setminus\Gamma(t),\\
u &= 0 && \text{on } \partial\Omega,\\
[\![u]\!] &= [\![1/\varrho]\!]j_\Gamma\nu_\Gamma && \text{on } \Gamma(t),\\
[\![1/\varrho]\!]j_\Gamma^2\nu_\Gamma - [\![T\nu_\Gamma]\!] &= \sigma H_\Gamma\nu_\Gamma && \text{on } \Gamma(t),\\
u(0) &= u_0 && \text{in } \Omega.
\end{aligned}
\tag{1.44}
$$

$$
\begin{aligned}
\varrho\kappa(\theta)(\partial_t\theta + u\cdot\nabla\theta) - \operatorname{div}(d(\theta)\nabla\theta) &= 2\mu(\theta)|D|_2^2 && \text{in } \Omega\setminus\Gamma(t),\\
\partial_\nu\theta &= 0 && \text{on } \partial\Omega,\\
[\![\theta]\!] &= 0 && \text{on } \Gamma(t),\\
[\![\theta\eta(\theta)]\!]j_\Gamma - [\![d(\theta)\partial_\nu\theta]\!] &= 0 && \text{on } \Gamma(t),\\
\theta(0) &= \theta_0 && \text{in } \Omega.
\end{aligned}
\tag{1.45}
$$

$$
\begin{aligned}
[\![\psi(\theta)]\!] + [\![1/2\varrho^2]\!]j_\Gamma^2 - [\![T\nu_\Gamma\cdot\nu_\Gamma/\varrho]\!] &= 0 && \text{on } \Gamma(t),\\
V_\Gamma = u_\Gamma\cdot\nu_\Gamma &= u\cdot\nu_\Gamma - j_\Gamma/\varrho && \text{on } \Gamma(t),\\
\Gamma(0) &= \Gamma_0.
\end{aligned}
\tag{1.46}
$$

The main hypothesis here is **(H4)** := **(H2)**. Here we can eliminate j_Γ as explained before by means of the identities

$$j_\Gamma = [\![u \cdot \nu_\Gamma]\!]/[\![1/\rho]\!], \quad V_\Gamma = [\![\rho u \cdot \nu_\Gamma]\!]/[\![\rho]\!].$$

Remark 1.3.4. (i) A variant of this problem concerns the situation where heat conduction is taken into account in both phases but only one phase is moving, the model for *melting and solidification*. This problem formally results by letting $\mu_1 \to \infty$. To obtain this model, for finite μ_1, let T_j denote the stress tensor in Ω_j. Set $u \equiv \pi \equiv 0$ in Ω_1, maintain the jump condition for u, drop the stress jump condition on the interface, but replace $T_1\nu_\Gamma \cdot \nu_\Gamma$ in the Gibbs–Thomson law from the normal stress jump, according to

$$T_1\nu_\Gamma \cdot \nu_\Gamma = T_2\nu_\Gamma \cdot \nu_\Gamma + \sigma H_\Gamma - [\![1/\varrho]\!]j_\Gamma^2$$

to the result

$$u_2 = [\![1/\varrho]\!]j_\Gamma\nu_\Gamma, \quad V_\Gamma = -j_\Gamma/\varrho_1,$$

and

$$[\![\psi(\theta)]\!] + (1/2)[\![1/\varrho]\!]^2 j_\Gamma^2 - [\![1/\varrho]\!]T_2\nu_\Gamma \cdot \nu_\Gamma + (\sigma/\varrho_1)H_\Gamma = 0.$$

These conditions on the interface do not contain the viscosity μ_1, hence we may formally pass to the limit $\mu_1 \to \infty$. Therefore, the resulting model reads

$$\begin{aligned}
\varrho(\partial_t u + u \cdot \nabla u) - \operatorname{div} S + \nabla\pi &= 0 && \text{in } \Omega_2(t),\\
\operatorname{div} u &= 0 && \text{in } \Omega_2(t),\\
u &= 0 && \text{on } \partial\Omega,\\
u &= [\![1/\varrho]\!]j_\Gamma\nu_\Gamma && \text{on } \Gamma(t),\\
u(0) &= u_0 && \text{in } \Omega.
\end{aligned}$$

$$\begin{aligned}
\varrho\kappa(\theta)(\partial_t\theta + u \cdot \nabla\theta) - \operatorname{div}(d(\theta)\nabla\theta) &= 2\mu(\theta)|D|_2^2 && \text{in } \Omega \setminus \Gamma(t),\\
\partial_\nu\theta &= 0 && \text{on } \partial\Omega,\\
[\![\theta]\!] &= 0 && \text{on } \Gamma(t),\\
[\![\theta\eta(\theta)]\!]j_\Gamma - [\![d(\theta)\partial_\nu\theta]\!] &= 0 && \text{on } \Gamma(t),\\
\theta(0) &= \theta_0 && \text{in } \Omega_2.
\end{aligned}$$

$$\begin{aligned}
[\![\psi(\theta)]\!] + (1/2)[\![1/\varrho]\!]^2 j_\Gamma^2 - [\![1/\varrho]\!]T_2\nu_\Gamma \cdot \nu_\Gamma + (\sigma/\varrho_1)H_\Gamma &= 0 && \text{on } \Gamma(t),\\
V_\Gamma &= -j_\Gamma/\varrho_1 && \text{on } \Gamma(t),\\
\Gamma(0) &= \Gamma_0.
\end{aligned}$$

This model also has conservation of total energy and production of total entropy is nonnegative, hence it is consistent with thermodynamics. Note, however, that momentum is not conserved across the interface, as at the outer boundary $\partial\Omega$. Furthermore, if the densities are equal, the viscosity is constant, and the initial velocity is zero also in Ω_2, then $u \equiv 0$ and π is constant in Ω_2. In this situation the model reduces to Problem 1.

(ii) In the *isothermal case* the temperature θ is assumed to be constant and the heat problem is ignored. Then we obtain a model for isothermal two-phase flows with surface tension and phase transition, the latter is driven by pressure, only.

(iii) Again in the incompressible, isothermal case, ignoring inertia and j_Γ^2, we obtain the equations for the *Stokes flow with phase transition* which reads

$$\begin{aligned} -\operatorname{div} S + \nabla\pi &= 0 && \text{in } \Omega\setminus\Gamma(t),\\ \operatorname{div} u &= 0 && \text{in } \Omega\setminus\Gamma(t),\\ u &= 0 && \text{on } \partial\Omega,\\ [\![u]\!] &= [\![1/\varrho]\!] j_\Gamma \nu_\Gamma && \text{on } \Gamma(t),\\ -[\![T\nu_\Gamma]\!] &= \sigma H_\Gamma \nu_\Gamma && \text{on } \Gamma(t),\\ -[\![T\nu_\Gamma\cdot\nu_\Gamma/\varrho]\!] &= c && \text{on } \Gamma(t),\\ V_\Gamma = u_\Gamma\cdot\nu_\Gamma &= u\cdot\nu_\Gamma - j_\Gamma/\varrho && \text{on } \Gamma(t),\\ \Gamma(0) &= \Gamma_0. \end{aligned} \tag{1.47}$$

Here $c = -[\![\psi]\!]$ is constant. The phase flux j_Γ can be eliminated from the normal component of the velocity jump, and so we have a transmission problem for the Stokes equation with $(n-1)$ jump conditions for the velocity and $(n+1)$ for the normal stresses. This leads to a geometric evolution equation where the interface is moved by surface tension as well as by stationary phase transitions due to the different densities.

(iv) Employing again Darcy's (or Forchheimer's) law $u = -k(\pi)\nabla\pi$, we obtain the *Verigin problem with phase transition*

$$\begin{aligned} \varrho'(\pi)\partial_t\pi - \operatorname{div}(\varrho(\pi)k(\pi)\nabla\pi) &= 0 && \text{in } \Omega\setminus\Gamma(t),\\ \partial_\nu\pi &= 0 && \text{on } \partial\Omega,\\ [\![\pi]\!] &= \sigma H_\Gamma && \text{on } \Gamma(t),\\ [\![\psi + \pi/\varrho]\!] &= 0 && \text{on } \Gamma(t),\\ [\![\varrho]\!]V_\Gamma + [\![\varrho(\pi)k(\pi)\partial_\nu\pi]\!] &= 0 && \text{on } \Gamma(t),\\ \Gamma(0) = \Gamma_0, \quad \pi(0) &= \pi_0. \end{aligned} \tag{1.48}$$

Note that here the pressure π is the independent variable, and Maxwell's law then reads $\psi'(\pi) = \pi\varrho'(\pi)/\varrho^2(\pi)$. Its quasi-steady version, where ϱ is constant in the phases, is the *Muskat flow with phase transition*, another geometric evolution equation.

Problem 5. Marangoni Forces I. Here we assume $\varrho_1 = \varrho_2 =: \varrho$, σ nonconstant. Experience shows that σ is strictly decreasing and positive at melting temperature

θ_m, and as σ is also concave, it has a unique zero $\theta_c > \theta_m$; we call θ_c the *critical temperature*. As beyond the critical temperature there is no phase separation anymore, we restrict to the temperature range $\theta \in (0, \theta_c)$.

Here the model equations read

$$
\begin{aligned}
\varrho(\partial_t u + u \cdot \nabla u) - \operatorname{div} S + \nabla \pi &= 0 && \text{in } \Omega \setminus \Gamma(t), \\
\operatorname{div} u &= 0 && \text{in } \Omega \setminus \Gamma(t), \\
u &= 0 && \text{on } \partial\Omega, \\
[\![u]\!] = 0, \quad \mathcal{P}_\Gamma u_\Gamma &= \mathcal{P}_\Gamma u && \text{on } \Gamma(t), \\
-[\![T\nu_\Gamma]\!] &= \sigma(\theta_\Gamma) H_\Gamma \nu_\Gamma + \sigma'(\theta_\Gamma)\nabla_\Gamma \theta_\Gamma && \text{on } \Gamma(t), \\
u(0) &= u_0 && \text{in } \Omega.
\end{aligned}
\tag{1.49}
$$

$$
\begin{aligned}
\varrho\kappa(\theta)(\partial_t\theta + u \cdot \nabla\theta) - \operatorname{div}(d(\theta)\nabla\theta) &= 2\mu(\theta)|D|_2^2 && \text{in } \Omega \setminus \Gamma(t), \\
\partial_\nu \theta &= 0 && \text{on } \partial\Omega, \\
\theta &= \theta_\Gamma && \text{on } \Gamma(t), \\
\theta(0) &= \theta_0 && \text{in } \Omega.
\end{aligned}
\tag{1.50}
$$

$$
\begin{aligned}
&\kappa_\Gamma(\theta_\Gamma)\frac{D}{Dt}\theta_\Gamma - \operatorname{div}_\Gamma(d_\Gamma(\theta_\Gamma)\nabla_\Gamma\theta_\Gamma) \\
&\qquad = \theta_\Gamma\sigma'(\theta_\Gamma)\operatorname{div}_\Gamma u_\Gamma - ([\![\theta\eta(\theta)]\!] j_\Gamma - [\![d(\theta)\partial_\nu\theta]\!]) && \text{on } \Gamma(t), \\
&\varrho[\![\psi(\theta)]\!] + \sigma(\theta) H_\Gamma = 0 && \text{on } \Gamma(t), \\
&V_\Gamma = u_\Gamma \cdot \nu_\Gamma = u \cdot \nu_\Gamma - j_\Gamma/\varrho && \text{on } \Gamma(t), \\
&\Gamma(0) = \Gamma_0.
\end{aligned}
\tag{1.51}
$$

We assume

(H5) $\quad \psi, \sigma \in C^3(0, \theta_c), \; d, d_\Gamma, \mu \in C^2(0, \theta_c),$
$\quad -\psi''(s), -\sigma''(s), -\sigma'(s), d(s), d_\Gamma(s), \mu(s) > 0$ for all $s \in (0, \theta_c)$.

In this problem, the Navier-Stokes problem is again only weakly coupled with a Stefan problem, modified by energy conservation on the interface. Note that

$$\operatorname{div}_\Gamma u_\Gamma = \operatorname{div}_\Gamma \mathcal{P}_\Gamma u - H_\Gamma V_\Gamma,$$

which eliminates u_Γ, but here it is not so easy to eliminate j_Γ, as for this problem it really is an implicit variable!

Remark 1.3.5. Setting $u = 0$ and ignoring the Navier-Stokes problem, the latter becomes the Stefan problem with surface tension and surface heat capacity, which reads

$$
\begin{aligned}
\varrho\kappa(\theta)\partial_t\theta - \operatorname{div}(d(\theta)\nabla\theta) &= 0 && \text{in } \Omega \setminus \Gamma(t), \\
\partial_\nu\theta &= 0 && \text{on } \partial\Omega, \\
\theta &= \theta_\Gamma && \text{on } \Gamma(t), \\
\theta(0) &= \theta_0 && \text{in } \Omega.
\end{aligned}
\tag{1.52}
$$

$$
\begin{aligned}
\kappa_\Gamma(\theta_\Gamma)\frac{D}{Dt}\theta_\Gamma - \mathrm{div}_\Gamma(d_\Gamma(\theta_\Gamma)\nabla_\Gamma\theta_\Gamma)& \\
= -\theta_\Gamma\sigma'(\theta_\Gamma)H_\Gamma V_\Gamma + \varrho[\![\theta\eta(\theta)]\!]V_\Gamma + [\![d(\theta)\partial_\nu\theta]\!] &\quad \text{on } \Gamma(t), \\
\varrho[\![\psi(\theta)]\!] + \sigma(\theta)H_\Gamma = 0 &\quad \text{on } \Gamma(t), \\
\Gamma(0) = \Gamma_0. &
\end{aligned} \tag{1.53}
$$

This problem will be studied in Chapter 12.

Problem 6. Marangoni Forces II. Here we assume $\varrho_1 \neq \varrho_2$, σ nonconstant. This is the model of highest complexity considered in this book.

$$
\begin{aligned}
\varrho(\partial_t u + u\cdot\nabla u) - \mathrm{div}\, S + \nabla\pi &= 0 && \text{in } \Omega\setminus\Gamma(t), \\
\mathrm{div}\, u &= 0 && \text{in } \Omega\setminus\Gamma(t), \\
u &= 0 && \text{on } \partial\Omega, \\
\mathcal{P}_\Gamma u_\Gamma = \mathcal{P}_\Gamma u, \quad [\![u]\!] &= [\![1/\varrho]\!] j_\Gamma \nu_\Gamma && \text{on } \Gamma(t), \\
[\![1/\varrho]\!] j_\Gamma^2 \nu_\Gamma - [\![T\nu_\Gamma]\!] &= \sigma(\theta_\Gamma)H_\Gamma + \sigma'(\theta_\Gamma)\nabla_\Gamma\theta_\Gamma && \text{on } \Gamma(t), \\
u(0) &= u_0, && \text{in } \Omega.
\end{aligned} \tag{1.54}
$$

$$
\begin{aligned}
\varrho\kappa(\theta)(\partial_t\theta + u\cdot\nabla\theta) - \mathrm{div}(d(\theta)\nabla\theta) &= 2\mu(\theta)|D|_2^2 && \text{in } \Omega\setminus\Gamma(t), \\
\partial_\nu\theta &= 0 && \text{on } \partial\Omega, \\
\theta &= \theta_\Gamma && \text{on } \Gamma(t), \\
\theta(0) &= \theta_0 && \text{in } \Omega.
\end{aligned} \tag{1.55}
$$

$$
\begin{aligned}
\kappa_\Gamma(\theta_\Gamma)\frac{D}{Dt}\theta_\Gamma - \mathrm{div}_\Gamma(d_\Gamma(\theta_\Gamma)\nabla_\Gamma\theta_\Gamma) =& \\
= \theta_\Gamma\sigma'(\theta_\Gamma)\mathrm{div}_\Gamma u_\Gamma - ([\![\theta\eta(\theta)]\!] j_\Gamma - [\![d(\theta)\partial_\nu\theta]\!]) &\quad \text{on } \Gamma(t), \\
V_\Gamma = u_\Gamma\cdot\nu_\Gamma = u\cdot\nu_\Gamma - j_\Gamma/\varrho &\quad \text{on } \Gamma(t), \\
[\![\psi(\theta)]\!] + [\![1/2\varrho^2]\!] j_\Gamma^2 - [\![T\nu_\Gamma\cdot\nu_\Gamma/\varrho]\!] = 0 &\quad \text{on } \Gamma(t), \\
\Gamma(0) = \Gamma_0. &
\end{aligned} \tag{1.56}
$$

The main assumption on the coefficients is **(H6)** := **(H5)**. Here j_Γ can be eliminated as in Problem **(P4)**, and $\mathrm{div}_\Gamma u_\Gamma$ as in Problem **(P5)**.

3.2 Transformation to a Fixed Domain

A basic idea is to transform Problems **(P1)**–**(P6)** to a domain with a fixed interface Σ, where $\Gamma(t)$ is parameterized over Σ by means of a height function $h(t)$. For this we rely on the so-called *Hanzawa transform* which we will now explain.

(a) The Hanzawa Transform

We assume, as before, that $\Omega \subset \mathbb{R}^n$ is a bounded domain with boundary $\partial\Omega$ of class C^2, and that $\Gamma \subset \Omega$ is a hypersurface of class C^2, i.e., a C^2-manifold which is the boundary of a bounded domain $\Omega_1 \subset \Omega$. As above, we set $\Omega_2 = \Omega \backslash \bar{\Omega}_1$. Note that Ω_2 typically is connected, while Ω_1 may be disconnected. In the later case, Ω_1 consists of finitely many components, since $\partial\Omega_1 = \Gamma \subset \Omega$ by assumption is a manifold, at least of class C^2. As will be shown in Section 2.4, the hypersurface Γ can be approximated by a real analytic hypersurafce Σ, in the sense that the Hausdorff distance of the second-order normal bundles is as small as we please. More precisely, given $\eta > 0$, there exists an analytic hypersurface Σ such that $d_H(\mathcal{N}^2\Sigma, \mathcal{N}^2\Gamma) \leq \eta$. If $\eta > 0$ is small enough, then Σ bounds a domain Ω_1^Σ with $\overline{\Omega_1^\Sigma} \subset \Omega$ and we set $\Omega_2^\Sigma = \Omega \setminus \overline{\Omega_1^\Sigma} \subset \Omega$.

In the sequel we will freely use results that are established in Chapter 2. In particular, it is shown in Section 2.3 that the C^2-hypersurface Σ admits a tubular neighbourhood, which means that there is $a_0 > 0$ such that the map

$$\Lambda : \Sigma \times (-a_0, a_0) \to \mathbb{R}^n,$$
$$\Lambda(p, r) := p + r\nu_\Sigma(p)$$

is a diffeomorphism from $\Sigma \times (-a_0, a_0)$ onto $\mathrm{im}(\Lambda)$, the image of Λ. The inverse

$$\Lambda^{-1} : \mathrm{im}(\Lambda) \to \Sigma \times (-a_0, a_0)$$

of this map is conveniently decomposed as

$$\Lambda^{-1}(x) = (\Pi_\Sigma(x), d_\Sigma(x)), \quad x \in \mathrm{im}(\Lambda).$$

Here $\Pi_\Sigma(x)$ means the metric projection of x onto Σ and $d_\Sigma(x)$ the signed distance from x to Σ; so $|d_\Sigma(x)| = \mathrm{dist}(x, \Sigma)$ and $d_\Sigma(x) < 0$ if and only if $x \in \Omega_1^\Sigma$. In particular we have $\mathrm{im}(\Lambda) = \{x \in \mathbb{R}^n : \mathrm{dist}(x, \Sigma) < a_0\}$. The maximal number a_0 is given by the radius $r_\Sigma > 0$, defined as the largest number r such that the exterior and interior ball conditions for Σ in Ω hold. In the following, we choose

$$a_0 = r_\Sigma/2 \quad \text{and} \quad a = a_0/3.$$

The derivatives of $\Pi_\Sigma(x)$ and $d_\Sigma(x)$ are given by

$$\nabla d_\Sigma(x) = \nu_\Sigma(\Pi_\Sigma(x)), \quad \partial\Pi_\Sigma(x) = M_0(d_\Sigma(x))\mathcal{P}_\Sigma(\Pi_\Sigma(x)),$$

where, as before, $\mathcal{P}_\Sigma(p) = I - \nu_\Sigma(p) \otimes \nu_\Sigma(p)$ denotes the orthogonal projection onto the tangent space $T_p\Sigma$ of Σ at $p \in \Sigma$, and $M_0(r) = (I - rL_\Sigma)^{-1}$, with L_Σ the Weingarten tensor. Then

$$|M_0(r)| \leq 1/(1 - r|L_\Sigma|) \leq 3 \quad \text{for all } |r| \leq 2r_\Sigma/3.$$

If $\mathrm{dist}(\Gamma, \Sigma)$ is small enough, we may use the map Λ to parameterize the unknown free boundary $\Gamma(t)$ over Σ by means of a *height function* $h(t)$ via

$$\Gamma(t) = \{p + h(t, p)\nu_\Sigma(p) : p \in \Sigma\}, \quad t \geq 0,$$

for small $t \geq 0$, at least. Extend this diffeomorphism to all of $\bar{\Omega}$ by means of

$$\Xi_h(t,x) = x + \chi(d_\Sigma(x)/a)h(t,\Pi_\Sigma(x))\nu_\Sigma(\Pi_\Sigma(x)) =: x + \xi_h(t,x).$$

Here χ denotes a suitable cut-off function. More precisely, let $\chi \in \mathcal{D}(\mathbb{R})$, $0 \leq \chi \leq 1$, $\chi(r) = 1$ for $|r| < 1$, and $\chi(r) = 0$ for $|r| > 2$. (We may choose χ in such a way that $1 < |\chi'|_\infty \leq 3$.) Note that $\Xi_h(t,x) = x$ for $|d_\Sigma(x)| > 2a$, and

$$\Pi_\Sigma(\Xi_h(t,x)) = \Pi_\Sigma(x), \quad |d_\Sigma(x)| < a,$$

as well as

$$d_\Sigma(\Xi_h(t,x)) = d_\Sigma(x) + \chi(d_\Sigma(x)/a)h(t,\Pi_\Sigma(x)), \quad |d_\Sigma(x)| < 2a.$$

This yields

$$\Xi_h^{-1}(t,x) = x - h(t,\Pi_\Sigma(x))\nu_\Sigma(\Pi_\Sigma(x)) \quad \text{for} \quad |d_\Sigma(x)| < a,$$

in particular,

$$\Xi_h^{-1}(t,x) = x - h(t,x)\nu_\Sigma(x) \quad \text{for} \quad x \in \Sigma.$$

Furthermore, we obtain

$$\partial\,\Xi_h = I + \partial\xi_h, \qquad (\partial\,\Xi_h)^{-1} = I - [I + \partial\xi_h]^{-1}\partial\xi_h =: I - M_1^{\mathsf{T}}(h),$$

where $\partial := \partial_x$ denotes the derivative with respect to $x \in \mathbb{R}^n$, and

$$\begin{aligned}&\partial\xi_h(t,x)\\ &\quad = \nu_\Sigma(\Pi_\Sigma(x)) \otimes M_0(d_\Sigma(x))\nabla_\Sigma h(t,\Pi_\Sigma(x)) - h(t,\Pi_\Sigma(x))M_0(d_\Sigma(x))L_\Sigma(\Pi_\Sigma(x))\end{aligned}$$

for $|d_\Sigma(x)| < a$, $\xi_h'(t,x) = 0$ for $|d_\Sigma(x)| > 2a$, and in general

$$\begin{aligned}\partial\xi_h(t,x) = {}& \frac{1}{a}\chi'(d_\Sigma(x)/a)h(t,\Pi_\Sigma(x))\nu_\Sigma(\Pi_\Sigma(x)) \otimes \nu_\Sigma(\Pi_\Sigma(x))\\ &+ \chi(d_\Sigma(x)/a)\nu_\Sigma(\Pi_\Sigma(x)) \otimes M_0(d_\Sigma(x))\nabla_\Sigma h(t,\Pi_\Sigma(x))\\ &- \chi(d_\Sigma(x)/a)h(t,\Pi_\Sigma(x))M_0(d_\Sigma(x))L_\Sigma(\Pi_\Sigma(x)).\end{aligned}$$

It is a matter of simple algebra to determine the inverse of $\partial\Xi_h$, to the result

$$\begin{aligned}&(\partial\Xi_h(t,x))^{-1}\\ &= I - \Big(\chi h L_\Sigma - \frac{\chi' h/a}{1+\chi' h/a}\nu_\Sigma \otimes \nu_\Sigma - \frac{\chi}{1+\chi' h/a}\nu_\Sigma \otimes \nabla_\Sigma h\Big)M_0(d_\Sigma + \chi h),\end{aligned}$$

where we dropped the obvious arguments. This implies

$$M_1(h) = \chi M_0(d_\Sigma + \chi h)\big(\frac{\nabla_\Sigma h \otimes \nu_\Sigma}{1+\chi' h/a} - hL_\Sigma\big) + \frac{\chi' h/a}{1+\chi' h/a}\nu_\Sigma \otimes \nu_\Sigma.$$

Note that $M_1(h)$ depends linearly on $\nabla_\Sigma h$. On the interface we then have

$$M_1(h) = M_0(h)\big(\nabla_\Sigma h \otimes \nu_\Sigma - hL_\Sigma\big).$$

In particular, $\partial\Xi_h$ is invertible, provided $M_0(d_\Sigma + \chi h) = (I - (d_\Sigma + \chi h)L_\Sigma)^{-1}$ exists, and $1 + \chi' h/a > 0$. This certainly holds if

$$|d_\Sigma + \chi h||L_\Sigma| \leq 2/3 \quad \text{and} \quad |\chi'|_\infty |h|/a \leq 1/2,$$

which leads to the restriction $|h|_\infty \leq h_\infty := a/2|\chi'|_\infty$; note that $|\chi'|_\infty > 1$. Observe that at this place no restrictions on $\nabla_\Sigma h$ are required.

Next we have

$$\partial_t \Xi_h(t,x) = \chi(d_\Sigma(x)/a)\partial_t h(t, \Pi_\Sigma(x))\nu_\Sigma(\Pi_\Sigma(x)), \quad x \in \bar{\Omega},$$

hence the relation $\Xi_h^{-1}(t, \Xi_h(t,x)) = x$ implies

$$\partial_t \Xi_h^{-1}(t, \Xi_h(t,x)) = -m_0(h)\partial_t h(t, \Pi_\Sigma(x))\nu_\Sigma(\Pi_\Sigma(x)), \quad x \in \bar{\Omega},$$

where

$$m_0(h)(t,x) = \frac{\chi(d_\Sigma(x)/a)}{(1 + h(t, \Pi_\Sigma(x))\chi'(d_\Sigma(x)/a)/a}.$$

With the Weingarten tensor L_Σ and the surface gradient ∇_Σ we further have

$$\begin{aligned} \nu_\Gamma(h) &= \beta(h)(\nu_\Sigma - a(h)), & a(h) &= M_0(h)\nabla_\Sigma h, \\ M_0(h) &= (I - hL_\Sigma)^{-1}, & \beta(h) &= (1 + |a(h)|^2)^{-1/2}, \end{aligned}$$

and

$$V_\Gamma = \partial_t \Xi_h \cdot \nu_\Gamma = (\nu_\Sigma \cdot \nu_\Gamma)\partial_t h = \beta(h)\partial_t h.$$

It will be shown in Section 2.2 that the surface gradient of a function ϕ on Γ is given by

$$\nabla_\Gamma \phi = \mathcal{P}_\Gamma(h)M_0(h)\nabla_\Sigma \bar{\phi} =: \mathcal{G}_\Gamma(h)\bar{\phi},$$

where $\bar{\phi} = \phi \circ \Xi_h$, the surface divergence of a vector field f on Γ becomes

$$\mathrm{div}_\Gamma f = \mathrm{tr}[\mathcal{P}_\Gamma(h)M_0(h)\nabla_\Sigma \bar{f}],$$

and the Laplace–Beltrami operator Δ_Γ reads

$$\Delta_\Gamma \varphi = \mathrm{tr}[\mathcal{P}_\Gamma(h)M_0(h)\nabla_\Sigma \mathcal{P}_\Gamma(h)M_0(h)\nabla_\Sigma \bar{\varphi}].$$

Finally, for the mean curvature $H_\Gamma(h)$ we have

$$H_\Gamma(h) = \beta(h)\{\mathrm{tr}[M_0(h)(L_\Sigma + \nabla_\Sigma a(h))] - \beta^2(h)(M_0(h)a(h)|[\nabla_\Sigma a(h)]a(h))\},$$

a differential expression involving second-order derivatives of h only linearly. We may write

$$H_\Gamma(h) = \mathcal{C}_0(h) : \nabla_\Sigma^2 h + \mathcal{C}_1(h),$$

where $\mathcal{C}_0(h)$ and $\mathcal{C}_1(h)$ depend on h and $\nabla_\Sigma h$, provided $|h| \leq h_\infty$ holds. The linearization of $H_\Gamma(h)$ at $h = 0$ is given by

$$H'_\Gamma(0) = \operatorname{tr} L_\Sigma^2 + \Delta_\Sigma.$$

Here Δ_Σ denotes the Laplace–Beltrami operator on Σ.

(b) The Transformed Problem
Now we define the transformed quantities

$$\begin{aligned} &\bar\varrho(t,x) = \varrho(t,\Xi_h(t,x)), && \bar u(t,x) = u(t,\Xi_h(t,x)) && \text{in } \Omega\backslash\Sigma,\\ &\bar\pi(t,x) = \pi(t,\Xi_h(t,x)), && \bar\theta(t,x) = \theta(t,\Xi_h(t,x)) && \text{in } \Omega\backslash\Sigma,\\ &\bar u_\Gamma(t,p) = u_\Gamma(t,\Xi_h(t,p)), && \bar j_\Gamma(t,p) = j_\Gamma(t,\Xi_h(t,p)) && \text{on } \Sigma, \end{aligned} \tag{1.57}$$

the *pull back* of $(\varrho, u, \pi, \theta, u_\Gamma, j_\Gamma)$. This way we have transformed the time varying regions $\Omega \setminus \Gamma(t)$ to the fixed region $\Omega \setminus \Sigma$. This transforms the general problem (1.26), (1.27), (1.28) to the following problem for $(\bar\varrho, \bar u, \bar\pi, \bar\theta, \bar u_\Gamma, \bar j_\Gamma, h)$.

$$\begin{aligned} \partial_t\bar\varrho + \mathcal{G}(h)\cdot\bar\varrho\bar u &= m_0(h)\partial_t h(\nu_\Sigma\cdot\nabla)\bar\varrho) && \text{in } \Omega\setminus\Sigma,\\ \bar\varrho\partial_t\bar u - \mathcal{G}(h)\cdot\bar S + \mathcal{G}(h)\bar\pi &= \bar\varrho\mathcal{R}_u(\bar u,\bar\theta,h) && \text{in } \Omega\setminus\Sigma,\\ \bar u &= 0 && \text{on } \partial\Omega,\\ [\![1/\bar\varrho]\!]\bar j_\Gamma^2\nu_\Gamma(h) - [\![\bar S\nu_\Gamma(h)]\!] + [\![\bar\pi]\!]\nu_\Gamma(h) &= \mathcal{G}_\Gamma(h)\cdot(\sigma(\bar\theta_\Gamma)\mathcal{P}_\Gamma(h) + \bar S_\Gamma) && \text{on } \Sigma,\\ [\![\bar u]\!] - [\![1/\bar\varrho]\!]\bar j_\Gamma\nu_\Gamma(h) &= 0 && \text{on } \Sigma,\\ \bar\varrho(0) = \bar\varrho_0, \quad \bar u(0) &= \bar u_0, \end{aligned} \tag{1.58}$$

where

$$\begin{aligned} \bar S &= \mu(\bar\theta,\bar\varrho)(\mathcal{G}(h)\bar u + [\mathcal{G}(h)\bar u]^\mathsf{T}) + \lambda(\bar\theta,\bar\varrho)(\mathcal{G}(h)\cdot\bar u)I,\\ \bar S_\Gamma &= \mu_\Gamma(\theta_\Gamma)\mathcal{P}_\Gamma(h)(\mathcal{G}_\Gamma(h)\bar u_\Gamma + [\mathcal{G}_\Gamma(h)\bar u_\Gamma]^\mathsf{T})\mathcal{P}_\Gamma(h) + \lambda(\bar\theta_\Gamma)(\mathcal{G}_\Gamma(h)\cdot\bar u_\Gamma)\mathcal{P}_\Gamma(h), \end{aligned}$$

$$\begin{aligned} \bar\varrho\kappa(\bar\theta,\bar\varrho)\partial_t\bar\theta - \mathcal{G}(h)\cdot d(\bar\theta,\bar\varrho)\mathcal{G}(h)\bar\theta &= \bar\varrho\kappa(\bar\theta,\bar\varrho)\mathcal{R}_\theta(\bar u,\bar\theta,h) && \text{in } \Omega\backslash\Sigma,\\ \partial_\nu\bar\theta &= 0 && \text{on } \partial\Omega,\\ [\![\bar\theta]\!] = 0, \quad \bar\theta &= \bar\theta_\Gamma && \text{on } \Sigma,\\ \bar\theta(0) &= \bar\theta_0 && \text{in } \Omega, \end{aligned} \tag{1.59}$$

$$\begin{aligned} &\kappa_\Gamma(\bar\theta_\Gamma)\partial_t\bar\theta_\Gamma - (\mathcal{G}_\Gamma(h)|d_\Gamma(\bar\theta_\Gamma)\mathcal{G}_\Gamma(h)\bar\theta_\Gamma) - [\![\bar\theta\eta(\bar\theta,\bar\rho)]\!]\bar j_\Gamma\\ &+[\![d(\bar\theta,\bar\rho)\mathcal{G}(h)\bar\theta\cdot\nu_\Gamma(h)]\!] = \bar S_\Gamma : \mathcal{G}_\Gamma(h)\bar u_\Gamma + \sigma(\bar\theta)\mathcal{G}_\Gamma(h)\cdot\bar u_\Gamma + \mathcal{R}_\Gamma(\bar\theta_\Gamma,h) && \text{on } \Sigma\\ &[\![\psi(\bar\theta,\bar\rho)]\!] + [\![1/2\bar\rho^2]\!]\bar j_\Gamma^2 - [\![\bar S\nu_\Gamma\cdot\nu_\Gamma/\bar\rho]\!] + [\![\bar\pi/\bar\rho]\!]\nu_\Gamma(h) = 0 && \text{on } \Sigma,\\ &\beta(h)\partial_t h - (\bar u|\nu_\Gamma) + \bar j_\Gamma/\bar\rho = 0, && \text{on } \Sigma,\\ &\bar\theta_\Gamma(0) = \bar\theta_0, \quad h(0) = h_0. \end{aligned} \tag{1.60}$$

Here $\mathcal{G}(h)$ and $\mathcal{G}_\Gamma(h)$ denote the transformed gradient resp. the transformed surface gradient. More precisely, we have the relations

$$[\nabla\pi]\circ\Xi_h = \mathcal{G}(h)\bar{\pi} = [(\partial\,\Xi_h^{-1})^{\mathsf{T}}\circ\Xi_h]\nabla\bar{\pi} = (I - M_1(h))\nabla\bar{\pi}$$

and

$$[\nabla\theta]\circ\Xi_h = (I - M_1(h))\nabla\bar{\theta},$$

as well as

$$(\nabla\cdot u)\circ\Xi_h = (\mathcal{G}(h)|\bar{u}) = ((I - M_1(h))\nabla|\bar{u}).$$

Furthermore,

$$\frac{D}{Dt}\theta_\Gamma\circ\Xi_h = \partial_t\bar{\theta}_\Gamma + \bar{u}_\Gamma\cdot\nabla_\Sigma\bar{\theta}_\Gamma - \bar{u}_\Gamma\cdot M_1(h)\nabla_\Sigma\bar{\theta}_\Gamma,$$

and

$$[\partial_t u]\circ\Xi_h = \partial_t\bar{u} + \partial\bar{u}[(\partial_t\Xi_h^{-1})\circ\Xi_h] = \partial_t\bar{u} - m_0(h)\partial_t h(\nu_\Sigma\cdot\nabla)\bar{u},$$

hence

$$\mathcal{R}_u(\bar{u},\bar{\theta},h) = -\bar{u}\cdot\mathcal{G}(h)\bar{u} + m_0(h)\partial_t h(\nu_\Sigma\cdot\nabla)\bar{u}.$$

Similarly we have

$$[\partial_t\theta]\circ\Xi_h = \partial_t\bar{\theta} - m_0(h)\partial_t h(\nu_\Sigma\cdot\nabla)\bar{\theta},$$

and so

$$\mathcal{R}_\theta(\bar{u},\bar{\theta},h) = -\bar{u}\cdot\mathcal{G}(h)\bar{\theta} + m_0(h)\partial_t h(\nu_\Sigma\cdot\nabla)\bar{\theta}.$$

In the same way we get

$$R_\Gamma(\bar{\theta}_\Gamma,h) = -\bar{u}_\Gamma\cdot\nabla_\Sigma\bar{\theta}_\Gamma + \bar{u}_\Gamma\cdot M_1(h)\nabla_\Sigma\bar{\theta}_\Gamma + \bar{\theta}_\Gamma\sigma'(\bar{\theta}_\Gamma)\mathcal{G}_\Gamma(h)\cdot\bar{u}_\Gamma.$$

It is convenient to decompose the stress boundary condition into tangential and normal parts; here we set $S_\Gamma = 0$. For this purpose let $\mathcal{P}_\Sigma = I - \nu_\Sigma\otimes\nu_\Sigma$ denote the projection onto the tangent space of Σ. Multiplying the stress interface condition with ν_Σ/β we obtain

$$\begin{aligned}[\![1/\varrho]\!]\bar{j}_\Gamma^2 + [\![\bar{\pi}]\!] - \sigma H_\Gamma(h) = ([\![\bar{S}]\!](\nu_\Sigma - M_0(h)\nabla_\Sigma h)|\nu_\Sigma)\\ + \sigma'\beta(M_0\nabla_\Sigma h|M_0\nabla_\Sigma\bar{\theta}_\Gamma)\end{aligned}\tag{1.61}$$

for the normal part of the stress boundary condition. Substituting this expression into the stress interface condition and then applying the projection $\mathcal{P}_\Sigma$ yields, after some computation,

$$\begin{aligned}\mathcal{P}_\Sigma[\![\bar{S}]\!](\nu_\Sigma - M_0(h)\nabla_\Sigma h) = ([\![\bar{S}]\!](\nu_\Sigma - M_0(h)\nabla_\Sigma h)|\nu_\Sigma)M_0(h)\nabla_\Sigma h\\ + (\sigma'/\beta)M_0(h)\nabla_\Sigma\bar{\theta}_\Gamma\end{aligned}\tag{1.62}$$

for the tangential part. Note that the latter neither contains the phase flux nor the pressure jump nor the curvature!

3.3 Goals and Strategies

The goal of this monograph is the exposition of a general theory for the models introduced above. We present in detail a rigorous analysis of these problems. It will become clear that the scope of our approach is much wider. It can be used for many other problems with moving interfaces, such as phase transitions driven by chemical potentials, two-phase flow problems with surface viscosities, multi-component two phase flows, as well as similar quasi-steady problems or purely geometric ones, to mention a few more applications. The essential restriction is that the problems in question ought to be of *parabolic nature*. In this book we will employ L_p-theory since it avoids higher order compatibility conditions. In addition, deep results of harmonic analysis are at our disposal. Nevertheless, one could also use other frameworks where maximal regularity is available, e.g. C^α-theory.

In particular, we address the following topics.

a) Local well-posedness and local semiflow;

b) Stability analysis of equilibria;

c) Long-time behaviour of solutions.

We now outline our approach, explaining the main ideas and tools to be employed.

a) Local-Well-posedness and Local Semiflow

To obtain local well-posedness we write the transformed problem in the form

$$\mathcal{L}z = (N(z), z_0).$$

Here $\mathcal{L}$ is the principal linear part of the problem in question, and N is the remaining nonlinear part which is small in the sense that N collects all lower order terms and contains only highest order terms which carry a factor $|\nabla_\Sigma h|$ which is small on small time intervals due to the choice of the Hanzawa transform. The variable z with initial value z_0 collects all essential variables of the problem under consideration.

The *first step* is to find function spaces $\mathbb{E}(J)$ and $\mathbb{F}(J)$, $J = (0, a)$ or $J = \mathbb{R}_+$, such that $\mathcal{L} : \mathbb{E}(J) \to \mathbb{F}(J) \times \mathbb{E}_\gamma$ is an isomorphism. Here $\mathbb{E}_\gamma$ denotes the time-trace space of $\mathbb{E}(J)$ which the initial value z_0 should belong to. This is the question of *maximal regularity.* These spaces differ from problem to problem and the question of maximal regularity has to be studied separately for each one. Here we will use the framework of L_p-spaces and rely on deep results from vector-valued harmonic analysis and operator theory which will be introduced and discussed in Chapter 4.

The *second step* then employs the contraction mapping principle to obtain local solutions, and the implicit function theorem to obtain smooth dependence of the solutions on the data. For this, estimates of the nonlinearity N are needed, eventually showing that $N : \mathbb{E}(J) \to \mathbb{F}(J)$ is continuously Fréchet-differentiable, at least. This requires some smoothness of the coefficient functions in the constitutive laws. If these are, say, even real analytic then N will be so as well, and by a scaling argument and the implicit function theorem we will show that the solutions are real analytic jointly in time and space as well. In particular, the interface will become instantaneously real analytic, which shows the strong regularizing effect, characteristic for parabolic problems.

The *third step* consists in setting up the state manifold $\mathcal{SM}$ of the untransformed problem. It will be a truly nonlinear manifold which comes from the generic nonlinear structure, due to geometry and the involved nonlinear compatibility conditions of the problem. Charts for the state manifold are induced by the Hanzawa transform mentioned above. The local existence and regularity results for the transformed problem induce a local semiflow on the proper state manifold $\mathcal{SM}$ for the problem in question.

b) Stability Analysis of Equilibria

For the stability analysis of equilibria it is natural to employ again the Hanzawa transform, where the reference manifold Σ now is the equilibrium interface Γ_*, a union of finitely many disjoint spheres contained in Ω. As the linearized problem enjoys maximal L_p-regularity, an abstract result shows that the operator L associated with the fully linearized problem is the negative generator of a compact analytic C_0-semigroup. Therefore, the spectrum of L consists only of countably many isolated eigenvalues of finite algebraic multiplicity. Thus, it is natural to study these eigenvalues and to apply the *principle of linearized stability* for the nonlinear problem.

However, a major difficulty of this approach lies in the fact that the equilibria are not isolated in the state manifold, but form a finite-dimensional submanifold $\mathcal{E}$ of $\mathcal{SM}$. For the linearization of the transformed problem this implies that the kernel of L is nontrivial, i.e., the imaginary axis is not in the resolvent set of L, and so the standard principle of linearized stability is not applicable. Fortunately, 0 is the only eigenvalue of L on $i\mathbb{R}$ and it is nicely behaved: the kernel $\mathsf{N}(L)$ is isomorphic to the tangent space of $\mathcal{E}$ at this equilibrium, and 0 is semi-simple. This shows that 0 is *normally stable* if the remaining eigenvalues of L have positive real parts, and *normally hyperbolic* if some of them have negative real parts; these are only finitely many. Therefore, we can employ what we call the *generalized principle of linearized stability*, a method which is adapted to such a situation and has been worked out recently for quasilinear parabolic evolution equations by the authors. So our stability analysis of equilibria proceeds in two steps.

In the *first step* we analyze the eigenvalues of L and find conditions, if possible necessary and sufficient, which ensure that all eigenvalues of L except 0 have positive real parts; this is the *normally stable case*. In the *normally hyperbolic case* we determine the dimension of the unstable subspace of L. And of course, we have to show that 0 is semi-simple, to determine the kernel of L, and to prove that $\mathsf{N}(L)$ is isomorphic to the tangent space of $\mathcal{E}$.

In the *second step* we employ the generalized principle of linearized stability to the nonlinear problem. This can be done simultaneously for all six problems in question, as the proof only uses the general structure of the problems under consideration. Here we employ once more the implicit function theorem.

c) Long-Time behaviour of Solutions
In general, solutions in $\mathcal{SM}$ will exist on a maximal time interval $[0, t_+(z_0))$ which typically will be finite, due to several obstructions, such as missing a priori bounds, loss of well-posedness, or topological changes in the moving interface. However, if a solution *does not develop singularities* in a sense to be specified, then we will prove that the solution exists globally, i.e., $t_+(z_0) = \infty$, and it converges in the topology of $\mathcal{SM}$ to an equilibrium. This essentially relies on a method using time weights to improve regularity and on compact Sobolev embeddings. Actually, we are able to *characterize* solutions which exist globally and converge as $t \to \infty$. This result is also proved simultaneously for all problems under consideration, as the proof only relies on general properties of semiflows, relative compactness of bounded orbits, the existence of a strict Lyapunov functional (the negative entropy), and the results on stability of equilibria.

On our way of presenting the tools which are needed to achieve these goals we will frequently discuss other problems to illustrate the main ideas. For example, the Laplacian, the Laplace–Beltrami operator, the heat operator, the Stokes operator, and several Dirichlet-to-Neumann operators will be studied in various frameworks. In Chapter 5 we develop an L_p-theory of abstract quasilinear parabolic evolution equations which serves as a guide for the more complex problems to be studied later on. In Chapter 12 we will present several applications of the main results of Chapter 5 to problems arising from generalized Newtonian flows, nematic liquid crystal flows, Maxwell-Stefan diffusion, and the Stefan problem with surface tension and surface heat capacity, as well as to geometric evolutions equations like the averaged mean curvature flow, the surface diffusion flow, the Mullins–Sekerka flow, the Muskat flow, the Stokes flow, and the Stokes flow with phase transition.

Chapter 2

Tools from Differential Geometry

In this chapter we introduce the necessary background in differential geometry of closed compact hypersurfaces in $\mathbb{R}^n$. We investigate the differential geometric properties of embedded hypersurfaces in n-dimensional Euclidean space, introducing the notions of Weingarten tensor, principal curvatures, mean curvature, tubular neighbourhood, surface gradient, surface divergence, and Laplace-Beltrami operator. The main emphasis lies in deriving representations of these quantities for hypersurfaces $\Gamma = \Gamma_\rho$ that are given as parameterized surfaces in normal direction of a fixed reference surface Σ by means of a height function ρ. We derive all of the aforementioned geometric quantities for Γ_ρ in terms of ρ and Σ. It is also important to study the mapping properties of these quantities in dependence of ρ, and to derive expressions for their variations. For instance, we show that

$$\kappa'(0) = \operatorname{tr} L_\Sigma^2 + \Delta_\Sigma,$$

where $\kappa = \kappa(\rho)$ denotes the mean curvature of Γ_ρ, L_Σ the Weingarten tensor of Σ, and Δ_Σ the Laplace-Beltrami operator on Σ. This is done in Section 2. We also study the first and second variations of the area and volume functional, respectively. In Section 3 we show, among other things, that C^2-hypersurfaces can be approximated in a suitable topology by smooth (i.e., analytic) hypersurfaces. This leads, in particular, to the existence of parameterizations. In Section 4 we show that the class of compact embedded hypersurfaces in $\mathbb{R}^n$ gives rise to a new manifold (whose points are the compact embedded hypersufaces). We also show that the class $\mathcal{M}^2(\Omega, r)$ of all compact embedded hypersurfaces contained in a bounded domain $\Omega \subset \mathbb{R}^n$, and satisfying a uniform ball condition with radius $r > 0$, can be identified with a subspace of $C^2(\bar{\Omega})$. This is important, as it allows us to derive compactness and embedding properties for $\mathcal{M}^2(\Omega, r)$. Finally, in Section 5 we consider moving hypersurfaces and prove various transport theorems.

J. Prüss and G. Simonett, *Moving Interfaces and Quasilinear Parabolic Evolution Equations*, Monographs in Mathematics 105,
DOI 10.1007/978-3-319-27698-4_2

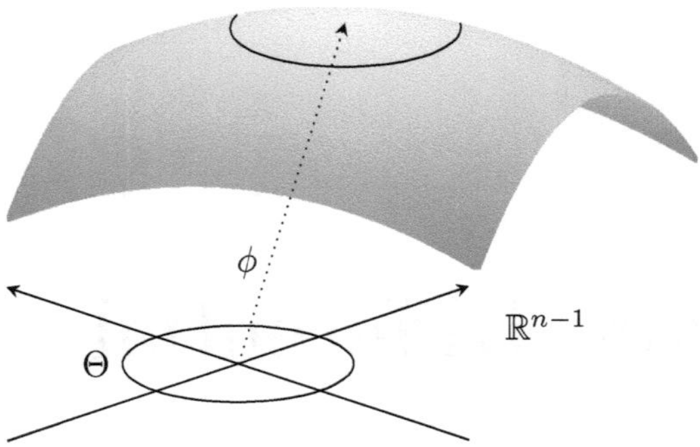

Figure 2.1: A typical chart for Σ.

2.1 Differential Geometry of Hypersurfaces

We consider a closed embedded hypersurface Σ of class C^k, $k \geq 3$, enclosing a bounded domain Ω in $\mathbb{R}^n$. Thus for each point $p \in \Sigma$ there is a ball $B(p,r) \subset \mathbb{R}^n$ and a diffemorphism $\Phi : B(p,r) \to U \subset \mathbb{R}^n$ such that $\Phi(p) = 0 \in U$ and

$$\Phi^{-1}(U \cap (\mathbb{R}^{n-1} \times \{0\})) = B(p,r) \cap \Sigma.$$

We may assume that Σ is connected; otherwise we would concentrate on one of its components. The points of Σ are denoted by p, and $\nu_\Sigma = \nu_\Sigma(p)$ means the outer unit normal of Σ at p. Locally at $p \in \Sigma$ we have the parameterization

$$p = \phi(\theta) := \Phi^{-1}(\theta, 0),$$

where θ runs through an open parameter set $\Theta \subset \mathbb{R}^{n-1}$. We denote the tangent vectors generated by this parameterization by

$$\tau_i = \tau_i(p) = \frac{\partial}{\partial \theta_i}\phi(\theta) = \partial_i \phi, \quad i = 1, \dots, n-1. \tag{2.1}$$

These vectors τ_i form a basis of the *tangent space* $T_p\Sigma$ of Σ at p. Note that $(\tau_i|\nu_\Sigma) = 0$ for all i, where $(\cdot|\cdot) := (\cdot|\cdot)_{\mathbb{R}^n}$ denotes the Euclidean inner product in $\mathbb{R}^n$. Similarly, we set $\tau_{ij} = \partial_i\partial_j\phi$, $\tau_{ijk} = \partial_i\partial_j\partial_k\phi$, and so on. In the sequel we employ Einstein's summation convention, which means that equal lower and upper indices are to be summed, and δ_j^i are the entries of the unit matrix I. For two vectors $a, b \in \mathbb{R}^n$ the tensor product $a \otimes b \in \mathcal{B}(\mathbb{R}^n)$ is defined by $[a \otimes b](x) = (b|x)a$ for $x \in \mathbb{R}^n$. If a belongs to the tangent space $T_p\Sigma$, we may represent a as a linear combination of the basis vectors of $T_p\Sigma$, i.e., $a = a^i\tau_i$. The coefficients a^i are called the *contravariant components* of a. On the other hand, this vector a is also uniquely characterized by its *covariant components*, a_i defined by $a_i = (a|\tau_i)$, which means that the covariant components are the coefficients of the representation of a in the basis $\{\tau^i\}$ dual to the basis $\{\tau_j\}$, defined by the relations $(\tau^i|\tau_j) = \delta^i_j$. Similarly, if $K \in \mathcal{B}(T_p\Sigma)$ is a tensor we have the representations

$$K = k^{ij}\tau_i \otimes \tau_j = k_{ij}\tau^i \otimes \tau^j = k^i_j\tau_i \otimes \tau^j = k^j_i\tau^i \otimes \tau_j,$$

with e.g. $k_{ij} = (\tau_i|K\tau_j)$ and $k^i_j = (\tau^i|K\tau_j)$. Moreover, $\operatorname{tr} K$, the *trace* of K, is given by

$$\operatorname{tr} K = (K\tau_i|\tau^i) = (K\tau^i|\tau_i). \tag{2.2}$$

In particular, $\operatorname{tr}[a \otimes b] = (a|b) = a_i b^i = a^i b_i$.

1.1 The First Fundamental Form

Define

$$g_{ij} = g_{ij}(p) = (\tau_i(p)|\tau_j(p)) = (\tau_i|\tau_j), \quad i, j = 1, \ldots, n-1. \tag{2.3}$$

The matrix $G = [g_{ij}]$ is called *the first fundamental form* of Σ. Note that G is symmetric and also positive definite, since

$$(G\xi|\xi) = g_{ij}\xi^i\xi^j = (\xi^i\tau_i|\xi^j\tau_j) = |\xi^i\tau_i|^2 > 0, \quad \text{for all } \xi \in \mathbb{R}^{n-1},\ \xi \neq 0.$$

We let $G^{-1} = [g^{ij}]$, hence $g_{ik}g^{kj} = \delta^j_i$, and $g^{il}g_{lj} = \delta^i_j$. The determinant $g := \det G$ is positive. Let a be a tangent vector. Then $a = a^i\tau_i$ implies

$$a_k = (a|\tau_k) = a^i(\tau_i|\tau_k) = a^i g_{ik} \quad \text{and} \quad a^i = g^{ik}a_k.$$

Thus the fundamental form G allows for the passage from contra- to covariant components of a tangent vector and vice versa. If a, b are two tangent vectors, then

$$(a|b) = a^i b^j(\tau_i|\tau_j) = g_{ij}a^i b^j = a_j b^j = a^i b_i = g^{ij}a_i b_j =: (a|b)_\Sigma$$

defines an inner product on $T_p\Sigma$ in the canonical way, the *Riemannian metric*. By means of the identity

$$(g^{ik}\tau_k|\tau_j) = g^{ik}g_{kj} = \delta^i_j$$

we further see that

$$\tau^i = g^{ij}\tau_j \quad \text{and} \quad \tau_j = g_{ij}\tau^i.$$

This implies the relations

$$k^i_j = g^{ir}k_{rj} = g_{jr}k^{ri}, \quad k^{ij} = g^{ir}k^j_r, \quad k_{ij} = g_{ir}k^r_j,$$

for any tensor $K \in \mathcal{B}(T_p\Sigma)$. We set for the moment $\mathcal{G} = g^{ij}\tau_i \otimes \tau_j$ and have equivalently

$$\mathcal{G} = g^{ij}\tau_i \otimes \tau_j = g_{ij}\tau^i \otimes \tau^j = \tau_i \otimes \tau^i = \tau^j \otimes \tau_j.$$

Let $u = u^k\tau_k + (u|\nu_\Sigma)\nu_\Sigma$ be an arbitrary vector in $\mathbb{R}^n$. Then

$$\mathcal{G}u = g^{ij}\tau_i(\tau_j|u) = g^{ij}\tau_i u^k g_{jk} = u^k\tau_k,$$

i.e., $\mathcal{G}$ equals the orthogonal projection $\mathcal{P}_\Sigma = I - \nu_\Sigma \otimes \nu_\Sigma$ of $\mathbb{R}^n$ onto the tangent space $T_p\Sigma$ at $p \in \Sigma$. Therefore, we have the relation

$$\mathcal{P}_\Sigma = I - \nu_\Sigma \otimes \nu_\Sigma = \tau_i \otimes \tau^i = \tau^i \otimes \tau_i,$$

where I denotes the identity map on $\mathbb{R}^n$. These properties explain the meaning of the first fundamental form $[g_{ij}]$.

1.2 The Second Fundamental Form
Define

$$l_{ij} = l_{ij}(p) = (\tau_{ij}|\nu_\Sigma), \quad L = [l_{ij}]. \tag{2.4}$$

L is called *the second fundamental form* of Σ. Note that L is symmetric, and differentiating the relations $(\tau_i|\nu_\Sigma) = 0$ we derive

$$l_{ij} = (\tau_{ij}|\nu_\Sigma) = -(\tau_i|\partial_j\nu_\Sigma) = -(\tau_j|\partial_i\nu_\Sigma). \tag{2.5}$$

The matrix K with entries l^i_j, defined by

$$l^i_j = g^{ir} l_{rj}, \quad K = G^{-1}L,$$

is called the *shape matrix* of Σ. The eigenvalues κ_i of K are called the *principal curvatures* of Σ at p, and the corresponding eigenvectors η_i determine the *principal curvature directions.* Observe that $K\eta_i = \kappa_i\eta_i$ is equivalent to $L\eta_i = \kappa_i G\eta_i$, hence the relation

$$(L\eta_i|\eta_i) = \kappa_i(G\eta_i|\eta_i)$$

and symmetry of L and G show that the principal curvatures κ_i are real. Moreover,

$$\kappa_i(G\eta_i|\eta_j) = (L\eta_i|\eta_j) = (\eta_i|L\eta_j) = \kappa_j(\eta_i|G\eta_j) = \kappa_j(G\eta_i|\eta_j)$$

implies that principal directions corresponding to different principal curvatures are orthogonal with respect to the inner product $(G\,\cdot\,|\cdot)_{\mathbb{R}^{n-1}}$. We can always assume that eigenvectors associated to an eigenvalue κ_i are orthogonal w.r.t. $(G\,\cdot\,|\cdot)_{\mathbb{R}^{n-1}}$ in case κ_i has geometric multiplicity greater than one. The eigenvalues κ_i are semi-simple, i.e., $\mathsf{N}((\kappa_i - K)^2) = \mathsf{N}(\kappa_i - K)$. In fact, suppose $x \in \mathsf{N}((\kappa_i - K)^2)$. Then

$$(\kappa_i - K)x = \sum_{r=1}^{m_i} t_r \eta_{i,r},$$

with $t_r \in \mathbb{R}$, where $\{\eta_{i,r} : 1 \le r \le m_i\}$ is an (orthogonal) basis of $\mathsf{N}(\kappa_i - K)$. Therefore,

$$t_k(G\eta_{i,k}|\eta_{i,k}) = \Big(\sum_{r=1}^{m_i} t_r G\eta_{i,r}|\eta_{i,k}\Big) = (G(\kappa_i - K)x|\eta_{i,k}) = (x|(\kappa_i G - L)\eta_{i,k}) = 0$$

for $1 \le k \le m_i$. Since G is positive definite, $t_k = 0$, and hence $x \in \mathsf{N}(\kappa_i - K)$. This shows that K is diagonalizable.

The trace of K, i.e., the first invariant of K, is called the $(n-1)$-fold *mean curvature* κ of Σ at p, i.e., we have

$$\kappa_\Sigma := \operatorname{tr} K = l_i^i = g^{ij} l_{ij} = \sum_{i=1}^{n-1} \kappa_i. \tag{2.6}$$

The *Gaussian curvature* $\mathcal{K}_\Sigma$ is defined as the last invariant of K, i.e.,

$$\mathcal{K}_\Sigma = \det K = g^{-1} \det L = \Pi_{i=1}^{n-1} \kappa_i.$$

We define the *Weingarten tensor* L_Σ by means of

$$L_\Sigma = L_\Sigma(p) = l^{ij} \tau_i \otimes \tau_j = l_j^i \tau_i \otimes \tau^j = l_i^j \tau^i \otimes \tau_j = l_{ij} \tau^i \otimes \tau^j. \tag{2.7}$$

L_Σ is symmetric with respect to the inner product $(\cdot|\cdot)$ in $\mathbb{R}^n$. We note that $L_\Sigma \in \mathcal{B}(\mathbb{R}^n)$ leaves the tangent space $T_p\Sigma$ invariant and, moreover, $L_\Sigma \nu_\Sigma = 0$. This shows that L_Σ enjoys the decomposition

$$L_\Sigma = \begin{bmatrix} L_\Sigma|_{T_p\Sigma} & 0 \\ 0 & 0 \end{bmatrix} : T_p\Sigma \oplus \mathbb{R}\nu_\Sigma \to T_p\Sigma \oplus \mathbb{R}\nu_\Sigma. \tag{2.8}$$

In particular, we note

$$\operatorname{tr} L_\Sigma(p) = \operatorname{tr}[L_\Sigma|_{T_p\Sigma}], \quad \det[I + rL_\Sigma(p)] = \det[(I + rL_\Sigma(p))|_{T_p\Sigma}] \tag{2.9}$$

for $r \in \mathbb{R}$. We will in the following not distinguish between L_Σ and its restriction to $T_p\Sigma$. Observe that

$$\operatorname{tr} L_\Sigma = l_i^i = g^{ij} l_{ij} = \kappa_\Sigma, \tag{2.10}$$

and the eigenvalues of L_Σ in $T_p\Sigma$ are the principal curvatures, since

$$L_\Sigma \eta_k = l_j^i (\tau^j | \eta_k) \tau_i = l_j^i \eta_k^j \tau_i = \kappa_k \eta_k^i \tau_i = \kappa_k \eta_k.$$

The remaining eigenvalue of L_Σ in $\mathbb{R}^n$ is 0 with eigenvector ν_Σ.

1.3 The Third Fundamental Form

To obtain another property of the shape matrix K we differentiate the identity $|\nu_\Sigma|^2 = 1$ to the result $(\partial_i \nu_\Sigma | \nu_\Sigma) = 0$. This means that $\partial_i \nu_\Sigma$ belongs to the tangent space, hence $\partial_i \nu_\Sigma = \gamma_i^k \tau_k$ for some numbers γ_i^k. Taking the inner product with τ_j we get

$$\gamma_i^k g_{kj} = \gamma_i^k (\tau_k | \tau_j) = (\partial_i \nu_\Sigma | \tau_j) = -(\tau_{ij} | \nu_\Sigma) = -l_{ij},$$

hence

$$\gamma_i^r = \gamma_i^k g_{kj} g^{jr} = -l_{ij} g^{jr} = -g^{rj} l_{ji} = -l_i^r,$$

where we used symmetry of L and G. Therefore we have

$$\partial_i \nu_\Sigma = -l_i^r \tau_r = -L_\Sigma \tau_i, \quad i = 1, \dots, n-1, \tag{2.11}$$

the *Weingarten relations*. Furthermore,

$$0 = \partial_i(\nu_\Sigma|\partial_j\nu_\Sigma) = (\partial_i\nu_\Sigma|\partial_j\nu_\Sigma) + (\nu_\Sigma|\partial_i\partial_j\nu_\Sigma)$$

implies

$$-(\partial_i\partial_j\nu_\Sigma|\nu_\Sigma) = (\partial_i\nu_\Sigma|\partial_j\nu_\Sigma) = l_i^r l_j^s(\tau_r|\tau_s) = l_i^r g_{rs} l_j^s = l_{is}g^{sr}l_{rj} = l_i^r l_{rj}, \quad (2.12)$$

which are the entries of the matrix $LG^{-1}L$, i.e., the covariant components of L_Σ^2. This is the so-called *third fundamental form* of Σ. In particular this implies the relation

$$\operatorname{tr} L_\Sigma^2 = (L_\Sigma\tau^i|L_\Sigma\tau_i) = -g^{ij}(\partial_i\partial_j\nu_\Sigma|\nu_\Sigma), \quad (2.13)$$

which will be useful later on. Moreover, we deduce from (2.12)

$$\operatorname{tr} L_\Sigma^2 = (L_\Sigma\tau^i|L_\Sigma\tau_i) = g^{ij}l_i^r l_{rj} = l_i^r l_r^i = \sum_{i=1}^{n-1}\kappa_i^2. \quad (2.14)$$

1.4 The Christoffel Symbols

The *Christoffel symbols* are defined according to

$$\Lambda_{ij|k} = (\tau_{ij}|\tau_k), \quad \Lambda_{ij}^k = g^{kr}\Lambda_{ij|r}. \quad (2.15)$$

Their importance stems from the representation of τ_{ij} in the basis $\{\tau_k, \nu_\Sigma\}$ of $\mathbb{R}^n$ via

$$\tau_{ij} = \Lambda_{ij}^k\tau_k + l_{ij}\nu_\Sigma. \quad (2.16)$$

Indeed, suppose $\tau_{ij} = a_{ij}^k\tau_k + b_{ij}\nu_\Sigma$. Then $l_{ij} = (\tau_{ij}|\nu_\Sigma) = b_{ij}$ and

$$\Lambda_{ij|k} = (\tau_{ij}|\tau_k) = (a_{ij}^r\tau_r|\tau_k) = g_{kr}a_{ij}^r.$$

Therefore, $a_{ij}^s = g^{sk}g_{kr}a_{ij}^r = g^{sk}\Lambda_{ij|k} = \Lambda_{ij}^s$. To express the Christoffel symbols in terms of the fundamental form G we use the identities

$$\begin{aligned}
\partial_k g_{ij} &= \partial_k(\tau_i|\tau_j) = (\tau_{ik}|\tau_j) + (\tau_i|\tau_{jk}),\\
\partial_i g_{kj} &= \partial_i(\tau_k|\tau_j) = (\tau_{ik}|\tau_j) + (\tau_k|\tau_{ij}),\\
\partial_j g_{ik} &= \partial_j(\tau_i|\tau_k) = (\tau_{ij}|\tau_k) + (\tau_i|\tau_{jk}),
\end{aligned}$$

which yield

$$\partial_i g_{jk} + \partial_j g_{ik} - \partial_k g_{ij} = 2(\tau_{ij}|\tau_k),$$

i.e.,

$$\Lambda_{ij|k} = \frac{1}{2}[\partial_i g_{jk} + \partial_j g_{ik} - \partial_k g_{ij}]. \quad (2.17)$$

Another important identity follows by differentiation of the relations $(\tau^j|\tau_k) = \delta_k^j$ and $(\tau^j|\nu_\Sigma) = 0$. We have

$$(\partial_i\tau^j|\tau_k) = -(\tau^j|\tau_{ik}) = -\Lambda_{ik}^r(\tau^j|\tau_r) = -\Lambda_{ik}^j,$$

and

$$(\partial_i \tau^j | \nu_\Sigma) = -(\tau^j | \partial_i \nu_\Sigma) = (\tau^j | L_\Sigma \tau_i) = l_i^j,$$

hence

$$\partial_i \tau^j = -\Lambda_{ik}^j \tau^k + l_i^j \nu_\Sigma. \tag{2.18}$$

This gives another interpretation of the Christoffel symbols and of the second fundamental form.

1.5 The Surface Gradient

Let ρ be a scalar field on Σ. The *surface gradient* $\nabla_\Sigma \rho$ at p is a vector which belongs to the tangent space of Σ at p. Thus it can be characterized by its

- *covariant components* a_i, i.e., $\nabla_\Sigma \rho = a_i \tau^i$, or by its
- *contravariant components* a^i, i.e., $\nabla_\Sigma \rho = a^i \tau_i$.

The chain rule

$$\partial_i(\rho \circ \phi) = (\nabla_\Sigma \rho | \tau_i)$$

yields $a_i = \partial_i(\rho \circ \phi) = \partial_i \rho$. This implies

$$a_i = (\nabla_\Sigma \rho | \tau_i) = a^k (\tau_k | \tau_i) = a^k g_{ki},$$

hence

$$\nabla_\Sigma \rho = \tau^i \partial_i \rho = (g^{ij} \partial_j \rho) \tau_i. \tag{2.19}$$

Suppose $\tilde{\rho}$ is a C^1-extension of ρ in a neighbourhood of Σ. We then have

$$\nabla \tilde{\rho} = (\nabla \tilde{\rho} | \nu_\Sigma) \nu_\Sigma + (\nabla \tilde{\rho} | \tau_i) \tau^i = (\nabla \tilde{\rho} | \nu_\Sigma) \nu_\Sigma + (\nabla_\Sigma \rho | \tau_i) \tau^i,$$

and hence, the surface gradient $\nabla_\Sigma \rho$ is the projection of $\nabla \tilde{\rho}$ onto $T_p \Sigma$, that is,

$$\nabla_\Sigma \rho = \mathcal{P}_\Sigma \nabla \tilde{\rho}. \tag{2.20}$$

For a vector field $f : \Sigma \to \mathbb{R}^m$ of class C^1 we define similarly

$$\nabla_\Sigma f := g^{ij} \tau_i \otimes \partial_j f = \tau^j \otimes \partial_j f. \tag{2.21}$$

In particular, this yields for the identity map id_Σ on Σ

$$\nabla_\Sigma \, \mathrm{id}_\Sigma = g^{ij} \tau_i \otimes \partial_j \phi = g^{ij} \tau_i \otimes \tau_j = \mathcal{P}_\Sigma,$$

and by the Weingarten relations

$$\nabla_\Sigma \nu_\Sigma = g^{ij} \tau_i \otimes \partial_j \nu_\Sigma = -g^{ij} l_j^r \tau_i \otimes \tau_r = -l^{ij} \tau_i \otimes \tau_j = -L_\Sigma.$$

For the surface gradient of tangent vectors we have

$$\begin{aligned}
\nabla_\Sigma \tau_k &= g^{ij} \tau_i \otimes \partial_j \tau_k = g^{ij} \tau_i \otimes \tau_{jk} = g^{ij} \tau_i \otimes (\Lambda_{jk}^r \tau_r + l_{jk} \nu_\Sigma) \\
&= g^{ij} \Lambda_{jk}^r \tau_i \otimes \tau_r + l_k^i \tau_i \otimes \nu_\Sigma = \Lambda_{kj}^r \tau^j \otimes \tau_r + (L_\Sigma \tau_k) \otimes \nu_\Sigma.
\end{aligned}$$

Finally we note that the *surface gradient for tensors* is defined according to

$$\nabla_\Sigma K = \tau^j \otimes \partial_j K. \tag{2.22}$$

1.6 The Surface Divergence

Let f be a tangential vector field on Σ. As before, $f^i = (f|\tau^i)$ denote the contravariant components of f, and $f_i = (f|\tau_i)$ the covariant components, respectively. The *surface divergence* of f is defined by

$$\operatorname{div}_\Sigma f = \frac{1}{\sqrt{g}}\partial_i(\sqrt{g}f^i) = \frac{1}{\sqrt{g}}\partial_i(\sqrt{g}g^{ij}f_j). \tag{2.23}$$

As before, $g := \det G$ denotes the determinant of $G = [g_{ij}]$. This definition ensures that partial integration can be carried out as usual, i.e., that the *surface divergence theorem* holds for tangential C^1-vector fields f:

$$\int_\Sigma (\nabla_\Sigma \rho|f)_\Sigma \, d\Sigma = -\int_\Sigma \rho \operatorname{div}_\Sigma f \, d\Sigma. \tag{2.24}$$

In fact, if e.g. ρ has support in a chart $\phi(\Theta)$ at p, then

$$\begin{aligned}\int_\Sigma (\nabla_\Sigma \rho|f)_\Sigma \, d\Sigma &= \int_\Theta \partial_i(\rho\circ\phi)[(f^i\circ\phi)\sqrt{g})]\, d\theta \\ &= -\int_\Theta (\rho\circ\phi)\frac{1}{\sqrt{g}}\partial_i[\sqrt{g}(f^i\circ\phi)]\sqrt{g}\, d\theta = -\int_\Sigma \rho \operatorname{div}_\Sigma f\, d\Sigma.\end{aligned}$$

Here we used that the surface measure in local coordinates is given by $d\Sigma = \sqrt{g}d\theta$. The general case follows from this argument by using a partition of unity. There is another useful representation of surface divergence, given by

$$\operatorname{div}_\Sigma f = g^{ij}(\tau_j|\partial_i f) = (\tau^i|\partial_i f). \tag{2.25}$$

It comes from

$$\operatorname{div}_\Sigma f = \frac{1}{\sqrt{g}}\partial_i(\sqrt{g}g^{ij}f_j) = \frac{1}{\sqrt{g}}\partial_i[\sqrt{g}g^{ij}(\tau_j|f)],$$

since

$$(\partial_i(\sqrt{g}g^{ij}\tau_j)|\tau_k) = 0, \quad k = 1,\dots,n-1. \tag{2.26}$$

Here (2.26) follows from

$$\begin{aligned}(\partial_i(\sqrt{g}g^{ij}\tau_j)|\tau_k) &= \partial_i(\sqrt{g}g^{ij}(\tau_j|\tau_k)) - \sqrt{g}g^{ij}(\tau_j|\tau_{ki}) = \partial_k\sqrt{g} - \sqrt{g}g^{ij}(\tau_j|\tau_{ki}) \\ &= \partial_k\sqrt{g} - \frac{1}{2}\sqrt{g}g^{ij}\partial_k(\tau_j|\tau_i) = \frac{1}{2\sqrt{g}}\left(\partial_k g - g g^{ij}\partial_k g_{ij}\right)\end{aligned}$$

and the relation

$$\partial_k g = g \,\mathrm{tr}\,[G^{-1}\partial_k G] = g g^{ij}\partial_k g_{ij}. \tag{2.27}$$

The last assertion can be verified as follows:

$$\begin{aligned}\partial_k g = \partial_k \det G &= \sum_{j=1}^{n-1} \det\,[G_1,\cdots,\partial_k G_j,\cdots G_{n-1}]\\ &= (\det G)\sum_{j=1}^{n-1} \det\big(G^{-1}[G_1,\cdots,\partial_k G_j,\cdots,G_{n-1}]\big) = g\,\mathrm{tr}\,[G^{-1}\partial_k G],\end{aligned}$$

where $G = [g_{ij}] = [G_1,\cdots,G_{n-1}]$, with G_j the j-th column of G. From (2.25) follows

$$\mathrm{div}_\Sigma \tau_k = g^{ij}(\tau_j|\tau_{ki}) = g^{ij}\Lambda_{ki|j} = \Lambda^i_{ik}.$$

Equation (2.25) can be used as a definition of surface divergence for general, not necessarily tangential vector fields f, i.e., we have

$$\mathrm{div}_\Sigma f := g^{ij}(\tau_j|\partial_i f) = (\tau^i|\partial_i f), \quad f\in C^1(\Sigma,\mathbb{R}^n). \tag{2.28}$$

For example, consider $f=\nu_\Sigma$. Then $\partial_i\nu_\Sigma = -l_i^k\tau_k$ by the Weingarten relations and we obtain

$$\mathrm{div}_\Sigma\nu_\Sigma = g^{ij}(\tau_j|\partial_i\nu_\Sigma) = -g^{ij}l_{ij} = -\kappa_\Sigma.$$

This way we have derived the important relation

$$\kappa_\Sigma = -\mathrm{div}_\Sigma\nu_\Sigma. \tag{2.29}$$

With this in hand, we can now deduce the relation

$$\mathrm{div}_\Sigma f = \mathrm{div}_\Sigma\mathcal{P}_\Sigma f - (f|\nu_\Sigma)\kappa_\Sigma. \tag{2.30}$$

We remind that the surface divergence theorem (2.24) only holds for tangential vector fields. The *surface divergence theorem* for general vector fields reads as

$$\int_\Sigma (\nabla_\Sigma\rho|f)\,d\Sigma = -\int_\Sigma \rho(\,\mathrm{div}_\Sigma f + (f|\nu_\Sigma)\kappa_\Sigma)\,d\Sigma, \quad f\in C^1(\Sigma,\mathbb{R}^n). \tag{2.31}$$

This follows from (2.24) and (2.30) by means of

$$\int_\Sigma (\nabla_\Sigma\rho|f)_\Sigma\,d\Sigma = \int_\Sigma (\nabla_\Sigma\rho|\mathcal{P}_\Sigma f)_\Sigma\,d\Sigma = -\int_\Sigma \rho\,\mathrm{div}_\Sigma\mathcal{P}_\Sigma f\,d\Sigma.$$

Another representation of the surface divergence of a general vector field f is given by

$$\mathrm{div}_\Sigma f = (\tau^i|\partial_i f) = \mathrm{tr}[\tau^i\otimes\partial_i f] = \mathrm{tr}\,\nabla_\Sigma f. \tag{2.32}$$

Suppose that $f \in C^1(\Sigma, \mathbb{R}^n)$ admits a C^1-extension $\tilde{f}$ in a neighbourhood of Σ. Then

$$\mathrm{div}_\Sigma f = \mathrm{div}_x \tilde{f} - (\nu_\Sigma | [\nabla_x \tilde{f}]^\mathsf{T} \nu_\Sigma) = \mathrm{div}_x \tilde{f} - \Big(\nu_\Sigma \Big| \frac{\partial \tilde{f}}{\partial \nu_\Sigma}\Big),$$

as can be deduced from

$$\begin{aligned}\mathrm{div}_\Sigma f &= (\tau^i | \partial_i f) = (\tau^i | [\nabla_x \tilde{f}]^\mathsf{T} \tau_i) \\ &= (\tau^i | [\nabla_x \tilde{f}]^\mathsf{T} \tau_i) + (\nu_\Sigma | [\nabla_x \tilde{f}]^\mathsf{T} \nu_\Sigma) - (\nu_\Sigma | [\nabla_x \tilde{f}]^\mathsf{T} \nu_\Sigma).\end{aligned}$$

Suppose now that ν_Σ admits a C^1-extension $\tilde{\nu}_\Sigma$ in a neighbourhood of Σ such that $|\tilde{\nu}_\Sigma| = 1$ is this neighbourhood. Then we have

$$2(\nu_\Sigma(p) | [\nabla_x \tilde{\nu}_\Sigma(p)]^\mathsf{T} \nu_\Sigma(p)) = \frac{d}{dt}(\tilde{\nu}_\Sigma(p + t\nu_\Sigma(p) | \tilde{\nu}(p + t\nu_\Sigma(p))\Big|_{t=0} = 0,$$

and we obtain

$$\mathrm{div}_x \tilde{\nu}_\Sigma = \mathrm{div}_\Sigma \nu_\Sigma = -\kappa_\Sigma. \tag{2.33}$$

Consequently, if Σ is given as the zero set of a C^2-level function φ with $\nabla_x \varphi \neq 0$, with $\nabla_x \varphi$ pointing in the direction of ν_Σ, we have the well-known formula

$$\kappa_\Sigma = -\mathrm{div}_x \left(\frac{\nabla_x \varphi}{|\nabla_x \varphi|} \right).$$

Finally, the *surface divergence for tensors* is given by

$$\mathrm{div}_\Sigma K = (\tau^j | \partial_j K) := (\partial_j K)^\mathsf{T} \tau^j. \tag{2.34}$$

This immediately yields the important relation

$$\mathrm{div}_\Sigma \mathcal{P}_\Sigma = \kappa_\Sigma \nu_\Sigma. \tag{2.35}$$

1.7 The Laplace-Beltrami Operator

The *Laplace-Beltrami* operator on Σ is defined for scalar fields by means of

$$\Delta_\Sigma \rho = \mathrm{div}_\Sigma \nabla_\Sigma \rho,$$

which in local coordinates reads

$$\Delta_\Sigma \rho = \frac{1}{\sqrt{g}} \partial_i [\sqrt{g} g^{ij} \partial_j \rho].$$

Another representation of Δ_Σ is given by

$$\Delta_\Sigma \rho = g^{ij} \partial_i \partial_j \rho - g^{ij} \Lambda_{ij}^k \partial_k \rho. \tag{2.36}$$

This follows from (2.19), (2.25) and (2.18). Since at each point $p \in \Sigma$ we may choose a chart such that $g_{ij} = \delta_{ij}$ and $\Lambda_{ij}^k = 0$ at p, we see from this representation that the Laplace-Beltrami operator is equivalent to the Laplacian at the point p; see also Section 2.1.8 below.

To obtain another representation of Δ_Σ, for a scalar C^2-function we compute

$$\nabla_\Sigma^2\rho = \nabla_\Sigma(\tau^j\partial_j\rho) = \tau^i\otimes\partial_i(\tau^j\partial_j\rho).$$

This yields with (2.18)

$$\begin{aligned}\nabla_\Sigma^2\rho =&(\partial_i\partial_j\rho)\tau^i\otimes\tau^j + (\partial_j\rho)\tau^i\otimes\partial_i\tau^j\\ =&(\partial_i\partial_k\rho - \Lambda_{ik}^j\partial_j\rho)\tau^i\otimes\tau^k + (L_\Sigma\nabla_\Sigma\rho)\otimes\nu_\Sigma.\end{aligned}$$

Taking traces gives

$$\Delta_\Sigma\rho = \operatorname{tr}\nabla_\Sigma^2\rho.$$

Similarly, the Laplace-Beltrami operator applies to general vector fields f according to

$$\Delta_\Sigma f = g^{ij}(\partial_i\partial_j f - \Lambda_{ij}^r\partial_r f).$$

For example, this yields, for the identity map id_Σ on Σ,

$$\Delta_\Sigma\,\mathrm{id}_\Sigma = g^{ij}(\partial_i\partial_j\phi - \Lambda_{ij}^r\partial_r\phi) = g^{ij}(\tau_{ij} - \Lambda_{ij}^r\tau_r),$$

and hence by (2.16)

$$\Delta_\Sigma\,\mathrm{id}_\Sigma = g^{ij}l_{ij}\nu_\Sigma = \kappa_\Sigma\nu_\Sigma.$$

Finally, we prove the important formula

$$\Delta_\Sigma\nu_\Sigma = -\nabla_\Sigma\kappa_\Sigma - [\operatorname{tr}L_\Sigma^2]\nu_\Sigma. \tag{2.37}$$

In fact, we have from (2.12)

$$(\Delta_\Sigma\nu_\Sigma|\nu_\Sigma) = g^{ij}(\partial_{ij}\nu_\Sigma - \Lambda_{ij}^r\partial_r\nu_\Sigma|\nu_\Sigma) = g^{ij}(\partial_{ij}\nu_\Sigma|\nu_\Sigma) = -\operatorname{tr}L_\Sigma^2.$$

Next observe that

$$\begin{aligned}&(\partial_k\partial_j\nu_\Sigma|\tau_i) - (\partial_i\partial_j\nu_\Sigma|\tau_k) = \partial_k(\partial_j\nu_\Sigma|\tau_i) - \partial_i(\partial_j\nu_\Sigma|\tau_k)\\ &= -\partial_k(\nu_\Sigma|\tau_{ij}) + \partial_i(\nu_\Sigma|\tau_{kj}) = \partial_k(\partial_i\nu_\Sigma|\tau_j) - \partial_i(\partial_k\nu_\Sigma|\tau_j)\\ &= (\partial_i\nu_\Sigma|\tau_{kj}) - (\partial_k\nu_\Sigma|\tau_{ij}) = \Lambda_{kj}^r(\partial_i\nu_\Sigma|\tau_r) - \Lambda_{ij}^r(\partial_k\nu_\Sigma|\tau_r)\\ &= \Lambda_{kj}^r(\partial_r\nu_\Sigma|\tau_i) - \Lambda_{ij}^r(\partial_r\nu_\Sigma|\tau_k),\end{aligned}$$

hence

$$(\partial_k\partial_j\nu_\Sigma - \Lambda_{kj}^r\partial_r\nu_\Sigma|\tau_i) = (\partial_i\partial_j\nu_\Sigma - \Lambda_{ij}^r\partial_r\nu_\Sigma|\tau_k).$$

This implies

$$(\Delta_\Sigma\nu_\Sigma|\tau_i) = g^{jk}(\partial_k\partial_j\nu_\Sigma - \Lambda_{kj}^r\partial_r\nu_\Sigma|\tau_i) = (\partial_i\partial_j\nu_\Sigma - \Lambda_{ij}^r\partial_r\nu_\Sigma|\tau^j).$$

On the other hand,

$$\begin{aligned}-(\nabla_\Sigma\kappa_\Sigma|\tau_i) &= -\partial_i\kappa_\Sigma = \partial_i(\partial_j\nu_\Sigma|\tau^j)\\ &= (\partial_i\partial_j\nu_\Sigma|\tau^j) + (\partial_r\nu_\Sigma|\partial_i\tau^r)\\ &= (\partial_i\partial_j\nu_\Sigma - \Lambda_{ij}^r\partial_r\nu_\Sigma|\tau^j).\end{aligned}$$

This proves formula (2.37).

1.8 The Case of a Graph over $\mathbb{R}^{n-1}$

Suppose that Σ is a graph over $\mathbb{R}^{n-1}$, i.e., there is a function $h \in C^2(\mathbb{R}^{n-1})$ such that the hypersurface Σ is given by the chart $\phi(x) = [x^\mathsf{T}, h(x)]^\mathsf{T}$, $x \in \mathbb{R}^{n-1}$. Then the tangent vectors are given by $\tau_i = [e_i^\mathsf{T}, \partial_i h]^\mathsf{T}$, where $\{e_i\}$ denotes the standard basis in $\mathbb{R}^{n-1}$. The (upward pointing) normal ν_Σ is given by

$$\nu_\Sigma(x) = \beta(x)[-\nabla_x h(x)^\mathsf{T}, 1]^\mathsf{T}, \quad \beta(x) = 1/\sqrt{1+|\nabla_x h(x)|^2}.$$

The first fundamental form becomes $g_{ij} = \delta_{ij} + \partial_i h\partial_j h$, hence

$$g^{ij} = \delta^{ij} - \beta^2\partial_i h\partial_j h.$$

This yields

$$\tau^i = [[e_i - \beta^2\partial_i h\nabla_x h]^\mathsf{T}, \beta^2\partial_i h]^\mathsf{T},$$

and with $\tau_{ij} = [0, \partial_i\partial_j h]^\mathsf{T}$,

$$l_{ij} = (\tau_{ij}|\nu_\Sigma) = \beta\partial_i\partial_j h,$$

and therefore

$$\kappa_\Sigma = g^{ij}l_{ij} = \beta[\Delta_x h - \beta^2(\nabla_x^2 h\nabla_x h|\nabla_x h)] = \mathrm{div}_x\left(\frac{\nabla_x h}{\sqrt{1+|\nabla_x h|^2}}\right).$$

The Christoffel symbols in this case are given by

$$\Lambda_{ij|k} = \partial_i\partial_j h\partial_k h, \quad \Lambda_{ij}^k = \beta^2\partial_i\partial_j h\partial_k h.$$

Suppose that $\mathbb{R}^{n-1}\times\{0\}$ is the tangent plane at $\phi(0) = 0 \in \Sigma$. Then $h(0) = 0$ and $\nabla_x h(0) = 0$, hence at this point we have $g_{ij} = \delta_{ij}$, $\tau_i = [e_i^\mathsf{T}, 0]^\mathsf{T}$, $\nu_\Sigma = [0,1]^\mathsf{T}$, $\beta = 1$, and $l_{ij} = \partial_i\partial_j h$. Thus the principal curvatures $\kappa_i(0)$ are the eigenvalues of $\nabla_x^2 h(0)$, the mean curvature is $\kappa_\Sigma(0) = \Delta_x h(0)$, and $\Lambda_{ij}^k(0) = 0$.

To obtain a representation of the surface gradient, let $\rho : \Sigma \to \mathbb{R}$. Then

$$\nabla_\Sigma\rho = \tau^j\partial_j\rho = [[\nabla_x\rho - \beta^2(\nabla_x\rho|\nabla_x h)\nabla_x h]^\mathsf{T}, \beta^2(\nabla_x\rho|\nabla_x h)]^\mathsf{T}.$$

Similarly, for $f = (\bar f, f^n) : \Sigma \to \mathbb{R}^{n-1}\times\mathbb{R}$ we obtain

$$\mathrm{div}_\Sigma f = (\tau^i|\partial_i f) = \mathrm{div}_x\bar f + \beta^2(\nabla_x h|\nabla_x f^n - (\nabla_x h\cdot\nabla_x)\bar f),$$

and for the Laplace-Beltrami

$$\Delta_\Sigma\rho = \Delta_x\rho - \beta^2(\nabla_x^2\rho\nabla_x h|\nabla_x h) - \beta^2[\Delta_x h - \beta^2(\nabla_x^2 h\nabla_x h|\nabla_x h)](\nabla_x h|\nabla_x\rho).$$

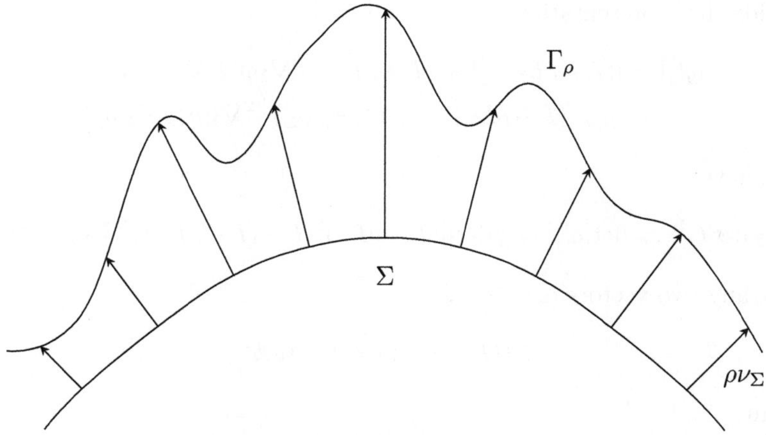

Figure 2.2: Parameterization of Γ over Σ.

2.2 Parameterized Hypersurfaces

We consider now a hypersurface $\Gamma = \Gamma_\rho$ which is parameterized over a fixed hypersurface Σ according to

$$q = \psi_\rho(p) = p + \rho(p)\nu_\Sigma(p), \quad p \in \Sigma, \tag{2.38}$$

where as before $\nu_\Sigma = \nu_\Sigma(p)$ denotes the outer unit normal of Σ at $p \in \Sigma$.

We want to derive the basic geometric quantities of Γ in terms of ρ and those of Σ. In the sequel we assume that ρ is of class C^1 and small enough. A precise bound on ρ will be given below.

2.1 The Fundamental Form
Differentiating (2.38) we obtain with the Weingarten relations (2.11)

$$\tau_i^\Gamma = \partial_i\psi_\rho = \tau_i + \rho\partial_i\nu_\Sigma + (\partial_i\rho)\nu_\Sigma = (I - \rho L_\Sigma)\tau_i + (\partial_i\rho)\nu_\Sigma. \tag{2.39}$$

We may then compute the fundamental form $G^\Gamma = [g_{ij}^\Gamma]$ of Γ to the result

$$\begin{aligned} g_{ij}^\Gamma = (\tau_i^\Gamma|\tau_j^\Gamma) &= ((I - \rho L_\Sigma)\tau_i + \partial_i\rho\nu_\Sigma|(I - \rho L_\Sigma)\tau_j + \partial_j\rho\nu_\Sigma) \\ &= ((I - \rho L_\Sigma)\tau_i|(I - \rho L_\Sigma)\tau_j) + \partial_i\rho\partial_j\rho, \\ &= (\tau_i|(I - \rho L_\Sigma)^2\tau_j) + (\tau_i|[\nabla_\Sigma\rho \otimes \nabla_\Sigma\rho]\tau_j) \end{aligned}$$

where we used that $((I - \rho L_\Sigma)\tau_k|\nu_\Sigma) = 0$. Hence

$$[g_{ij}^\Gamma] = [(I - \rho L_\Sigma)^2 + \nabla_\Sigma\rho \otimes \nabla_\Sigma\rho]_{ij}.$$

This yields the representation

$$\begin{aligned}[g^\Gamma_{ij}] &= [(I-\rho L_\Sigma)^2[I+(I-\rho L_\Sigma)^{-2}\nabla_\Sigma\rho\otimes\nabla_\Sigma\rho]]_{ij}\\ &= [g_{ik}][(I-\rho L_\Sigma)^2[I+(I-\rho L_\Sigma)^{-2}\nabla_\Sigma\rho\otimes\nabla_\Sigma\rho]]^k_j.\end{aligned}\tag{2.40}$$

We then have

$$g^\Gamma := \det G^\Gamma := \det[g^\Gamma_{ij}] = g\det[[(I-\rho L_\Sigma)^2[I+(I-\rho L_\Sigma)^{-2}\nabla_\Sigma\rho\otimes\nabla_\Sigma\rho]].$$

Since for any two vectors $a,b\in\mathbb{R}^n$,

$$\det(I+a\otimes b) = 1+(a|b),$$

we obtain

$$g^\Gamma = g\alpha^2(\rho)\mu^2(\rho),\tag{2.41}$$

where

$$\alpha(\rho) = \det(I-\rho L_\Sigma) = \det(I-\rho K) = \Pi_{i=1}^{n-1}(1-\rho\kappa_i),$$

and

$$\begin{aligned}\mu(\rho) &= (1+((I-\rho L_\Sigma)^{-2}\nabla_\Sigma\rho|\nabla_\Sigma\rho))^{1/2}\\ &= (1+((I-\rho L_\Sigma)^{-1}\nabla_\Sigma\rho|(I-\rho L_\Sigma)^{-1}\nabla_\Sigma\rho))^{1/2}.\end{aligned}$$

This yields for the surface measure $d\Gamma$ on Γ_ρ,

$$d\Gamma = \sqrt{g^\Gamma}d\theta = \alpha(\rho)\mu(\rho)\sqrt{g}\,d\theta = \alpha(\rho)\mu(\rho)\,d\Sigma,\tag{2.42}$$

hence

$$|\Gamma_\rho| = \int_{\Gamma_\rho} d\Gamma = \int_\Sigma \alpha(\rho)\mu(\rho)\,d\Sigma,$$

where $|\Gamma_\rho|$ denotes the surface area of Γ_ρ. Since

$$(I+a\otimes b)^{-1} = I - \frac{a\otimes b}{1+(a|b)},$$

we obtain for $[G^\Gamma]^{-1}$ the identity

$$[G^\Gamma]^{-1} = [g^{ij}_\Gamma] = [[I-\mu^{-2}(\rho)(I-\rho L_\Sigma)^{-2}\nabla_\Sigma\rho\otimes\nabla_\Sigma\rho](I-\rho L_\Sigma)^{-2}]^i_k[g^{kj}_\Sigma].$$

All of this makes sense only for functions ρ such that $I-\rho K$ is invertible, i.e., $\alpha(\rho)$ should not vanish. Thus the precise bound for ρ is determined by the principle curvatures of Σ, and we assume here and in the sequel that

$$|\rho|_\infty < \frac{1}{\max\{|\kappa_i(p)| : i=1,\dots,n-1,\ p\in\Sigma\}} =: \rho_0.\tag{2.43}$$

2.2 The Normal at Γ
We next compute the outer unit normal at Γ. For this purpose we set

$$\nu_\Gamma = \beta(\rho)(\nu_\Sigma - a(\rho)),$$

where $\beta(\rho)$ is a scalar and $a(\rho) \in T_p\Sigma$. Then $\beta(\rho) = (1 + |a(\rho)|^2)^{-1/2}$ and

$$0 = (\nu_\Gamma|\tau_i^\Gamma) = \beta(\rho)(\nu_\Sigma - a(\rho)|(I - \rho L_\Sigma)\tau_i + \nu_\Sigma \partial_i \rho),$$

which yields

$$0 = \partial_i \rho - (a(\rho)|(I - \rho L_\Sigma)\tau_i) = \partial_i \rho - ((I - \rho L_\Sigma)a(\rho)|\tau_i),$$

by symmetry of L_Σ. But this implies $(I - \rho L_\Sigma)a(\rho) = \nabla_\Sigma \rho$, i.e., we have

$$\nu_\Gamma = \beta(\rho)(\nu_\Sigma - a(\rho)) \tag{2.44}$$

with

$$a(\rho) = M_0(\rho)\nabla_\Sigma \rho, \quad M_0(\rho) = (I - \rho L_\Sigma)^{-1}, \quad \beta(\rho) = (1 + |a(\rho)|^2)^{-1/2}. \tag{2.45}$$

Note that $\mu(\rho) = \beta^{-1}(\rho)$, where $\mu(\rho)$ was introduced in the last section. By means of $a(\rho)$, $\beta(\rho)$ and $M_0(\rho)$ this leads to another representation of G^Γ and G_Γ^{-1}, namely

$$[g_{ij}^\Gamma] = [(I - \rho L_\Sigma)[I + a(\rho) \otimes a(\rho)](I - \rho L_\Sigma)]_{ij},$$

and

$$[g_\Gamma^{ij}] = [M_0(\rho)[I - \beta^2(\rho)a(\rho) \otimes a(\rho)]M_0(\rho)]^{ij}.$$

2.3 The Surface Gradient and the Surface Divergence on Γ
It is of importance to have a representation for the surface gradient on Γ in terms of Σ. For this purpose recall that

$$\mathcal{P}_\Gamma = I - \nu_\Gamma \otimes \nu_\Gamma = g_\Gamma^{ij} \tau_i^\Gamma \otimes \tau_j^\Gamma,$$

where $\nu_\Gamma = \beta(\rho)(\nu_\Sigma - M_0(\rho)\nabla_\Sigma \rho)$, and

$$\tau_i^\Gamma = (I - \rho L_\Sigma)\tau_i^\Sigma + \partial_i \rho \nu_\Sigma.$$

By virtue of $L_\Sigma \nu_\Sigma = 0$, the latter implies

$$\tau_i^\Gamma = (I - \rho L_\Sigma)(\tau_i^\Sigma + \partial_i \rho \nu_\Sigma).$$

As remarked before we do not distinguish between $L_\Sigma \in \mathcal{B}(\mathbb{R}^n)$ and its restriction to $T_p\Sigma$. With this identification, and by the fact that $(I - \rho L_\Sigma) = I$ on $\mathbb{R}\nu_\Sigma$, we have

$$(I - \rho L_\Sigma)(p) \in \text{Isom}\,(\mathbb{R}^n, \mathbb{R}^n) \cap \text{Isom}\,(T_p\Sigma, T_p\Sigma),$$

provided ρ satisfies (2.43). As before, ρL_Σ is the short form for $\rho(p)L_\Sigma(p)$. Hence,

$$M_0(\rho)(p) \in \mathrm{Isom}(\mathbb{R}^n, \mathbb{R}^n) \cap \mathrm{Isom}(T_p\Sigma, T_p\Sigma).$$

We conclude that

$$M_0(\rho)\tau_i^\Gamma = \tau_i^\Sigma + (\partial_i \rho)\nu_\Sigma,$$

and therefore

$$\mathcal{P}_\Sigma M_0(\rho)\tau_i^\Gamma = \tau_i^\Sigma. \tag{2.46}$$

On the other hand, we have

$$\mathcal{P}_\Gamma M_0(\rho)\tau_\Sigma^r = g_\Gamma^{ij}\tau_i^\Gamma \otimes \tau_j^\Gamma M_0(\rho)\tau_\Sigma^r = \tau_\Gamma^j(\tau_j^\Gamma | M_0(\rho)\tau_\Sigma^r),$$

hence

$$\mathcal{P}_\Gamma M_0(\rho)\tau_\Sigma^r = \tau_\Gamma^r. \tag{2.47}$$

(2.46) and (2.47) allow for an easy change between the bases of $T_p\Sigma$ and $T_q\Gamma$, where $q = \psi_\rho(p) = p + \rho(p)\nu_\Sigma(p)$. (2.47) implies for a scalar function φ on Γ,

$$\nabla_\Gamma \varphi = \tau_\Gamma^r \partial_r \varphi = \mathcal{P}_\Gamma M_0(\rho)\tau_\Sigma^r \partial_r \varphi_* = \mathcal{P}_\Gamma M_0(\rho)\nabla_\Sigma \varphi_*, \quad \varphi_* = \varphi \circ \psi_\rho$$

which leads to the identity

$$\nabla_\Gamma \varphi = \mathcal{P}_\Gamma M_0(\rho)\nabla_\Sigma \varphi_*.$$

Similarly, if f denotes a vector field on Γ, then

$$\nabla_\Gamma f = \mathcal{P}_\Gamma M_0(\rho)\nabla_\Sigma f_*,$$

and so

$$\mathrm{div}_\Gamma f = (\tau_\Gamma^r | \partial_r f) = (\mathcal{P}_\Gamma M_0(\rho)\tau_\Sigma^r | \partial_r f) = \mathrm{tr}\,[\mathcal{P}_\Gamma M_0(\rho)\nabla_\Sigma f_*].$$

As a consequence, we obtain for the Laplace-Beltrami operator on Γ,

$$\Delta_\Gamma \varphi = \mathrm{tr}\,[\mathcal{P}_\Gamma M_0(\rho)\nabla_\Sigma(\mathcal{P}_\Gamma M_0(\rho)\nabla_\Sigma \varphi_*)],$$

which can be written as

$$\Delta_\Gamma \varphi = M_0(\rho)\mathcal{P}_\Gamma M_0(\rho) : \nabla_\Sigma^2 \varphi_* + (b(\rho, \nabla_\Sigma \rho, \nabla_\Sigma^2 \rho) | \nabla_\Sigma \varphi_*),$$

with $b = \partial_i(M_0(\rho)\mathcal{P}_\Gamma)\mathcal{P}_\Gamma M(\rho)\tau_\Sigma^i$. One should note that the structure of the Laplace-Beltrami operator on Γ in local coordinates is

$$\Delta_\Gamma \varphi = a^{ij}(\rho, \partial\rho)\partial_i\partial_j \varphi_* + b^k(\rho, \partial\rho, \partial^2\rho)\partial_k \varphi_*$$

with

$$a^{ij}(\rho, \partial\rho) = (\mathcal{P}_\Gamma M_0(\rho)\tau_\Sigma^i | \mathcal{P}_\Gamma M_0(\rho)\tau_\Sigma^j) = (\tau_\Gamma^i | \tau_\Gamma^j) = g_\Gamma^{ij}$$

and

$$b^k(\rho,\partial\rho,\partial^2\rho) = (\mathcal{P}_\Gamma M_0(\rho)\tau_\Sigma^i|\partial_i(M_0(\rho)\mathcal{P}_\Gamma\tau^k)) = (\tau_\Gamma^i|\partial_i\tau_\Gamma^k) = -g_\Gamma^{ij}\Lambda_{\Gamma ij}^k.$$

This shows that $-\Delta_\Gamma$ is strongly elliptic on the reference manifold Σ as long as $|\rho|_\infty < \rho_0$.

2.4 Normal Variations

For $\rho, h \in C(\Sigma)$ sufficiently smooth and $F(\rho) : \Sigma \to \mathbb{R}^k$ we define

$$F'(\rho)h := \frac{d}{d\varepsilon}F(\rho+\varepsilon h)\Big|_{\varepsilon=0}.$$

First we have

$$M_0'(\rho) = M_0(\rho)L_\Sigma M_0(\rho), \quad M_0'(0) = L_\Sigma,$$

as $M_0(0) = I$. Next

$$\beta'(\rho)h = -\beta(\rho)^3\big(M_0(\rho)\nabla_\Sigma\rho\big|M_0'(\rho)h\nabla_\Sigma\rho + M_0(\rho)\nabla_\Sigma h\big),$$

which yields $\beta'(0) = 0$, as $\beta(0) = 1$. From this we get for the normal

$$\nu(\rho) = \nu_\Gamma = \beta(\rho)(\nu_\Sigma - M_0(\rho)\nabla_\Sigma\rho)$$

the relation

$$\nu'(\rho)h = \beta'(\rho)h(\nu_\Sigma - M_0(\rho)\nabla_\Sigma\rho) - \beta(\rho)(M_0'(\rho)h\nabla_\Sigma\rho + M_0(\rho)\nabla_\Sigma h),$$

which yields

$$\nu'(0)h = -\nabla_\Sigma h.$$

This in turn implies for the projection $P(\rho) := \mathcal{P}_\Gamma$

$$P'(\rho)h = -\nu'(\rho)h\otimes\nu(\rho) - \nu(\rho)\otimes\nu'(\rho)h,$$

hence

$$P'(0)h = \nabla_\Sigma h\otimes\nu_\Sigma + \nu_\Sigma\otimes\nabla_\Sigma h =: [\nabla_\Sigma\otimes\nu_\Sigma + \nu_\Sigma\otimes\nabla_\Sigma]h.$$

Applying these relations to $\nabla(\rho) := \nabla_\Gamma = P(\rho)M_0(\rho)\nabla_\Sigma$ yields

$$\begin{aligned}(\nabla'(0)h)\varphi &= [P'(0)h + P(0)M'(0)h]\nabla_\Sigma\varphi \\ &= [\nabla_\Sigma h\otimes\nu_\Sigma + \nu_\Sigma\otimes\nabla_\Sigma h + hL_\Sigma]\nabla_\Sigma\varphi = [\nu_\Sigma\otimes\nabla_\Sigma h + hL_\Sigma]\nabla_\Sigma\varphi,\end{aligned}$$

and for a not necessarily tangent vector field f

$$(\nabla'(0)h)f = \nu_\Sigma\otimes(\nabla_\Sigma h|\nabla_\Sigma)f + hL_\Sigma\nabla_\Sigma f$$

where $(\nabla_\Sigma h|\nabla_\Sigma)f := (\nabla_\Sigma h|\tau^j)\partial_j f$. For the divergence of the vector field f this implies

$$[\operatorname{div}'(0)h]f = (\nu_\Sigma|(\nabla_\Sigma h|\nabla_\Sigma)f) + h\,\mathrm{tr}[L_\Sigma\nabla_\Sigma f].$$

Finally, the variation of the Laplace-Beltrami operator $\Delta(\rho) := \Delta_\Gamma$ becomes

$$(\Delta'(0)h)\varphi = h\,\mathrm{tr}[L_\Sigma\nabla_\Sigma^2\varphi + \nabla_\Sigma(L_\Sigma\nabla_\Sigma\varphi)] + 2(L_\Sigma\nabla_\Sigma h|\nabla_\Sigma\varphi) - \kappa(\nabla_\Sigma h|\nabla_\Sigma\varphi).$$

Note that in local coordinates we have

$$\mathrm{tr}[L_\Sigma\nabla_\Sigma^2\varphi] = l_\Sigma^{ij}(\partial_i\partial_j\varphi - \Lambda_{ij}^k\partial_k\varphi),$$

hence with

$$\mathrm{tr}[\nabla_\Sigma(L_\Sigma\nabla_\Sigma\varphi)] = \mathrm{tr}[L_\Sigma\nabla_\Sigma^2\varphi] + (\mathrm{div}_\Sigma L_\Sigma|\nabla_\Sigma\varphi),$$

we may write alternatively

$$(\Delta'(0)h)\varphi = 2h\,\mathrm{tr}[L_\Sigma\nabla_\Sigma^2\varphi] + (h\,\mathrm{div}_\Sigma L_\Sigma + [2L_\Sigma - \kappa_\Sigma]\nabla_\Sigma h|\nabla_\Sigma\varphi).$$

2.5 The Weingarten Tensor and the Mean Curvature of Γ

In invariant formulation we have

$$L(\rho) := L_\Gamma = -\nabla_\Gamma\nu_\Gamma = -P(\rho)M_0(\rho)\nabla_\Sigma\{\beta(\rho)(\nu_\Sigma - M_0(\rho)\nabla_\Sigma\rho)\}.$$

Thus for the variation of L_Γ at $\rho = 0$ we obtain with $P(0) = \mathcal{P}_\Sigma$, $\beta(0) = 1$, $M_0(0) = I$, and $P'(0) = \nabla_\Sigma \otimes \nu_\Sigma + \nu_\Sigma \otimes \nabla_\Sigma$, $\beta'(0) = 0$, $M_0'(0) = L_\Sigma$,

$$L'(0) = \nu_\Sigma \otimes L_\Sigma\nabla_\Sigma + L_\Sigma^2 + \nabla_\Sigma^2.$$

In particular, for $\kappa(\rho) := \kappa_\Gamma$ we have

$$\kappa(\rho) = -\mathrm{tr}[\nabla_\Gamma\nu_\Gamma] = \mathrm{tr}\,L(\rho),$$

hence

$$\kappa'(0) = \mathrm{tr}\,L_\Sigma^2 + \Delta_\Sigma. \tag{2.48}$$

Let us take another look at the mean curvature $\kappa(\rho) := \kappa_\Gamma$. By the relations $\tau_\Gamma^r = \mathcal{P}_\Gamma M_0(\rho)\tau_\Sigma^r$ and $\nu_\Gamma = \beta(\rho)(\nu_\Sigma - a(\rho))$ we obtain

$$\begin{aligned}\kappa(\rho) &= -(\tau_\Gamma^j|\partial_j\nu_\Gamma) = -(\mathcal{P}_\Gamma M_0(\rho)\tau_\Sigma^j|(\partial_j\beta(\rho)/\beta(\rho))\nu_\Gamma + \beta(\rho)(\partial_j\nu_\Sigma - \partial_j a(\rho)))\\ &= \beta(\rho)(\mathcal{P}_\Gamma M_0(\rho)\tau_\Sigma^j|L_\Sigma\tau_j^\Sigma + \partial_j a(\rho))\\ &= \beta(\rho)(M_0(\rho)\tau_\Sigma^j|L_\Sigma\tau_j^\Sigma + \partial_j a(\rho)) - \beta(\rho)(\nu_\Gamma|M_0(\rho)\tau_\Sigma^j)(\nu_\Gamma|L_\Sigma\tau_j^\Sigma + \partial_j a(\rho)).\end{aligned}$$

Since $(M_0(\rho)\tau_\Sigma^j|L_\Sigma\tau_j^\Sigma) = \mathrm{tr}[M_0(\rho)L_\Sigma]$ as well as

$$(M_0(\rho)\tau_\Sigma^j|\partial_j a(\rho)) = \mathrm{tr}[M_0(\rho)\nabla_\Sigma a(\rho)],$$

and $(\nu_\Gamma|M_0(\rho)\tau_\Sigma^j) = -\beta(\rho)[M_0(\rho)a(\rho)]^j$, we obtain

$$\begin{aligned}\kappa(\rho) &= \beta(\rho)\big\{\mathrm{tr}\big[M_0(\rho)(L_\Sigma + \nabla_\Sigma a(\rho))\big]\\ &\qquad + \beta^2(\rho)\big[M_0(\rho)a(\rho)\big]^j\big[(\nu_\Sigma|\partial_j a(\rho)) - (a(\rho)|\partial_j a(\rho)) - (a(\rho)|L_\Sigma\tau_j^\Sigma)\big]\big\}\\ &= \beta(\rho)\big\{\mathrm{tr}\big[M_0(\rho)(L_\Sigma + \nabla_\Sigma a(\rho))\big] - \beta^2(\rho)(M_0(\rho)a(\rho)|\nabla_\Sigma a(\rho)a(\rho))\big\},\end{aligned}$$

as $(\nu_\Sigma|a(\rho)) = 0$ implies

$$(\nu_\Sigma|\partial_j a(\rho)) = -(\partial_j \nu_\Sigma|a(\rho)) = (L_\Sigma \tau_j^\Sigma|a(\rho)).$$

This yields the final form for the mean curvature of Γ.

$$\kappa(\rho) = \beta(\rho)\big\{\mathrm{tr}\big[M_0(\rho)(L_\Sigma + \nabla_\Sigma a(\rho))\big] - \beta^2(\rho)(M_0(\rho)a(\rho)|[\nabla_\Sigma a(\rho)]a(\rho))\big\}. \quad (2.49)$$

Recall that $a(\rho) = M_0(\rho)\nabla_\Sigma \rho$.

We can write the curvature of Γ in local coordinates in the following form.

$$\kappa(\rho) = c^{ij}(\rho, \partial\rho)\partial_i\partial_j\rho + g(\rho, \partial\rho),$$

with

$$c^{ij}(\rho, \partial\rho) = \beta(\rho)[M_0^2(\rho)]^{ij} - \beta^3(\rho)[M_0^2(\rho)\nabla_\Sigma\rho]^i[M_0^2(\rho)\nabla_\Sigma\rho]^j.$$

A simple computation yields for the symbol $c(\rho, \xi) = c^{ij}(\rho, \partial\rho)\xi_i\xi_j$ of the principal part of $-\kappa(\rho)$

$$c(\rho, \xi) = \beta(\rho)\{|M_0(\rho)\xi|^2 - \beta^2(\rho)(a(\rho)|M_0(\rho)\xi)^2\} \geq \beta^3(\rho)|M_0(\rho)\xi|^2 \geq \eta|\xi|^2,$$

for $\xi = \xi_k\tau_\Sigma^k \in T_p\Sigma$, as long as $|\rho|_\infty < \rho_0$. Therefore, $-\kappa(\rho)$ is a quasilinear strongly elliptic differential operator on Σ, acting on the parameterization ρ of Γ over Σ.

2.6 The Area Functional

As shown before, the area functional for the surface $\Gamma_\rho = \{p + \rho(p)\nu_\Sigma(p) : p \in \Sigma\}$ is given by

$$\Phi(\rho) = \int_{\Gamma_\rho} d\Gamma = \int_\Sigma \alpha(\rho)\mu(\rho)\, d\Sigma.$$

Here we use the notation

$$\alpha(\rho) = \det(I - \rho K) = \Pi_{i=1}^{n-1}(1 - \rho\kappa_i), \quad \mu(\rho) = (1 + |a(\rho)|^2)^{1/2},$$

with $a(\rho)$ defined in (2.45).

We compute its first variation to the result

$$\langle\Phi'(\rho)|h\rangle = \int_\Sigma [(\mu(\rho)\alpha'(\rho) + \alpha(\rho)\mu'(\rho)]h\, d\Sigma.$$

For the derivatives of α and μ we get

$$\alpha'(\rho) = \alpha(\rho)\sum_{i=1}^{n-1}\frac{-\kappa_i}{1 - \rho\kappa_i}, \quad \mu'(\rho)h = \mu(\rho)^{-1}(a(\rho)|a'(\rho)h).$$

In particular, at $\rho = 0$ we get with $\alpha(0) = \mu(0) = 1$ and $a(0) = 0$

$$\alpha'(0) = -\kappa_\Sigma, \quad \mu'(0) = 0.$$

This implies for the *first variation* of Φ at $\rho = 0$,

$$\langle \Phi'(0)|h\rangle = -\int_\Sigma \kappa_\Sigma h\, d\Sigma. \tag{2.50}$$

This shows, in particular, that the critical points of the area functional Φ are hypersurfaces with mean curvature $\kappa_\Sigma = 0$. Such surfaces are called *minimal surfaces.*

Similarly, the second variation becomes

$$\begin{aligned}\langle \Phi''(\rho)h|k\rangle = &\int_\Sigma [\mu(\rho)\alpha''(\rho) + \alpha(\rho)\mu''(\rho)]hk\, d\Sigma\\ &+ \int_\Sigma [\alpha'(\rho)h\mu'(\rho)k + \alpha'(\rho)k\mu'(\rho)h]\, d\Sigma.\end{aligned}$$

Since $\alpha(0) = \mu(0) = 1$ and $\mu'(0) = 0$ we get

$$\langle \Phi''(0)h|k\rangle = \int_\Sigma [\alpha''(0) + \mu''(0)]hk\, d\Sigma.$$

We have

$$\alpha''(\rho) = \alpha(\rho)\Big[\Big(\sum_{i=1}^{n-1} \frac{-\kappa_i}{1-\rho\kappa_i}\Big)^2 - \sum_{i=1}^{n-1}\frac{\kappa_i^2}{(1-\rho\kappa_i)^2}\Big],$$

hence

$$\alpha''(0) = \Big(\sum_{i=1}^{n-1}\kappa_i\Big)^2 - \sum_{i=1}^{n-1}\kappa_i^2 = (\operatorname{tr} K)^2 - \operatorname{tr} K^2,$$

which is the second invariant of the shape operator K.

In particular, in case Σ is a sphere of radius R we have $\kappa_i = -1/R$, hence $\alpha''(0) = (n-1)(n-2)/R^2$.

For the second derivative of μ at $\rho = 0$ we obtain

$$\mu''(0)hk = (a'(0)h|a'(0)k) = (\nabla_\Sigma h|\nabla_\Sigma k).$$

This yields the following representation for the *second variation* of Φ at $\rho = 0$,

$$\langle \Phi''(0)h|k\rangle = \int_\Sigma \{[(\operatorname{tr} K)^2 - \operatorname{tr} K^2]hk + (\nabla_\Sigma h|\nabla_\Sigma k)\}d\Sigma. \tag{2.51}$$

By means of the surface divergence theorem (2.24), this representation can be rewritten as

$$\langle \Phi''(0)h|k\rangle = \int_\Sigma \{[(\operatorname{tr} K)^2 - \operatorname{tr} K^2]h - \Delta_\Sigma h\}k\, d\Sigma,$$

and therefore

$$\Phi''(0)h = [(\operatorname{tr} K)^2 - \operatorname{tr} K^2]h - \Delta_\Sigma h, \tag{2.52}$$

i.e., $\Phi''(0)$ is the *Jacobi operator*, (sometimes also called the stability operator). Thus we see that

$$\Phi''(0) = -\kappa'(0) + \kappa_\Sigma^2.$$

In the next section we will come back to this relation.

2.7 The Volume Functional

Let Ω_ρ denote the domain bounded by the surface $\Gamma_\rho = \{p + \rho(p)\nu_\Sigma(p) : p \in \Sigma\}$. We define the volume functional Ψ by means of

$$\Psi(\rho) := |\Omega_\rho|. \tag{2.53}$$

In order to obtain the variation of $\Psi(\rho)$ we rewrite the volume functional by means of the divergence theorem as

$$n\Psi(\rho) = \int_{\Omega_\rho} \operatorname{div} x\, dx = \int_{\Gamma_\rho} (x|\nu_\Gamma)\, d\Gamma = \int_\Sigma (\mathrm{id}_\Sigma + \rho\nu_\Sigma|\nu_\Gamma)\alpha(\rho)\mu(\rho)\, d\Sigma,$$

which yields, with $\nu_\Gamma = \beta(\rho)(\nu_\Sigma - a(\rho))$,

$$n\Psi(\rho) = \int_\Sigma [\rho + (\mathrm{id}_\Sigma|\nu_\Sigma - a(\rho))]\alpha(\rho)\, d\Sigma,$$

where as before $\alpha(\rho) = \det(I - \rho K) = \Pi_{i=1}^{n-1}(1 - \rho\kappa_i)$. The first variation of Ψ then is

$$n\langle\Psi'(\rho)|h\rangle = \int_\Sigma \{[\rho + (\mathrm{id}_\Sigma|\nu_\Sigma - a(\rho))]\alpha'(\rho)h + [h - (\mathrm{id}_\Sigma|a'(\rho)h)]\alpha(\rho)\}\, d\Sigma.$$

From $\alpha(0) = 1$, $\alpha'(0) = -\kappa_\Sigma$ and $a'(0)h = \nabla_\Sigma h$ follows

$$\begin{aligned} n\langle\Psi'(0)|h\rangle &= \int_\Sigma [1 - (\mathrm{id}_\Sigma|\nu_\Sigma)\kappa_\Sigma]h\, d\Sigma - \int_\Sigma (\mathrm{id}_\Sigma|\nabla_\Sigma h)\, d\Sigma \\ &= \int_\Sigma (1 + \operatorname{div}_\Sigma \mathrm{id}_\Sigma)h\, d\Sigma, \end{aligned}$$

where we used the surface divergence theorem (2.31) in the last step. From

$$\operatorname{div}_\Sigma \mathrm{id}_\Sigma = (\tau^i|\partial_i\, \mathrm{id}_\Sigma) = (\tau^i|\tau_i) = (n-1)$$

follows the well-known formula for the *first variation* of the volume functional

$$\langle\Psi'(0)|h\rangle = \int_\Sigma h\, d\Sigma. \tag{2.54}$$

Now we reconsider the area functional Φ. We want to minimize surface area of Σ under the constraint that the volume of the domain bounded by Σ is a given constant Ψ_0. The method of Lagrange multipliers yields a number $\lambda \in \mathbb{R}$ such that $\Phi' - \lambda\Psi' = 0$. According to (2.50) and (2.54), this means

$$0 = \langle \Phi' - \lambda\Psi' | h \rangle = -\int_\Sigma (\kappa_\Sigma + \lambda) h \, d\Sigma = 0,$$

for all functions h. This implies $\kappa_\Sigma \equiv -\lambda$, i.e., Σ must be a sphere since Σ is an embedded closed and compact hypersurface. But then the value Φ is given by the constraint, i.e.,

$$\Phi(S_R(x_0)) = \omega_n R^{n-1}, \quad \kappa_\Sigma = -(n-1)/R, \quad \lambda = (n-1)/R, \quad (\omega_n/n)R^n = \Psi_0.$$

The *second variation* of Ψ can be computed as follows.

$$\begin{aligned} n\langle \Psi''(0)h|k\rangle = &\int_\Sigma (\mathrm{id}_\Sigma|\nu_\Sigma)\alpha''(0)hk \, d\Sigma \\ &+ \int_\Sigma \{[k - (\mathrm{id}_\Sigma|\nabla_\Sigma k)]h + [(h - (\mathrm{id}_\Sigma|\nabla_\Sigma h)]k\}\alpha'(0) \, d\Sigma \\ &- \int_\Sigma (\mathrm{id}_\Sigma|a''(0)hk)\alpha(0) \, d\Sigma. \end{aligned}$$

We observe that

$$(\mathrm{id}_\Sigma|\nabla_\Sigma k)h + (\mathrm{id}_\Sigma|\nabla_\Sigma h)]k = (\mathrm{id}_\Sigma|\nabla_\Sigma(hk))$$

and

$$a''(0)hk = M_0'(0)k\nabla_\Sigma h + M_0'(0)h\nabla_\Sigma k = L_\Sigma[k\nabla_\Sigma h + h\nabla_\Sigma k] = L_\Sigma \nabla_\Sigma(hk).$$

Collecting terms this yields

$$\begin{aligned} \langle \Psi''(0)h|k\rangle = &\frac{1}{n}\int_\Sigma [(\mathrm{id}_\Sigma|\nu_\Sigma)\alpha''(0) + 2\alpha'(0)]hk \, d\Sigma \\ &- \frac{1}{n}\int_\Sigma (\mathrm{id}_\Sigma|[\alpha'(0)I + L_\Sigma]\nabla_\Sigma(hk)) \, d\Sigma. \end{aligned} \tag{2.55}$$

Here we recall that $\alpha'(0) = -\kappa_\Sigma$ and $\alpha''(0) = (\mathrm{tr}\, L_\Sigma)^2 - \mathrm{tr}\, L_\Sigma^2$.

In particular, for a sphere of radius R centered at the origin we get $\mathrm{id}_\Sigma = R\nu_\Sigma$, and hence

$$\langle \Psi''(0)h|k\rangle = \frac{1}{n}\int_\Sigma \left[\frac{R(n-1)(n-2)}{R^2} + \frac{2(n-1)}{R}\right] hk \, d\Sigma = \frac{n-1}{R}\int_\Sigma hk \, d\Sigma.$$

This implies at a stationary point of the surface functional Φ with constraint $\Psi(\rho) = c$ with $\Phi' + \lambda\Psi' = 0$ and $\lambda = \kappa_\Sigma$,

$$\Phi'' + \lambda\Psi'' = -\Delta_\Sigma - (n-1)/R^2 = -\kappa'(0).$$

2.3 Approximation of Hypersurfaces

3.1 The Tubular Neighbourhood of a Hypersurface
Let Σ be a compact connected C^2-hypersurface bounding a domain $\Omega \subset \mathbb{R}^n$, and let ν_Σ be the outer unit normal field on Σ with respect to Ω. Then Σ satisfies the *uniform interior and exterior ball condition* , i.e., there is a number $a > 0$ such that for each point $p \in \Sigma$ there are balls $B(x_1, a) \subset \Omega$ and $B(x_2, a) \subset \bar{\Omega}^c$, such that $\Sigma \cap \bar{B}(x_i, a) = \{p\}$. Choosing the radius a_0 maximal, we set $a = a_0/2$ in the sequel. Consider the mapping

$$\Lambda : \Sigma \times (-a, a) \to \mathbb{R}^n, \quad \Lambda(p, r) := p + r\nu_\Sigma(p). \tag{2.56}$$

We claim that Λ is a C^1-diffeomorphism onto its image

$$U_a := \mathrm{im}(\Lambda) = \{x \in \mathbb{R}^n : \mathrm{dist}(x, \Sigma) < a\}.$$

Note that the centers of the balls $B(x_i, a)$ necessarily are equal to $x_1 = p - a\nu_\Sigma(p)$ and $x_2 = p + a\nu_\Sigma(p)$. To prove injectivity of Λ, suppose

$$p_1 + r_1\nu_\Sigma(p_1) = p_2 + r_2\nu_\Sigma(p_2),$$

where we may assume w.l.o.g. that $r_2 \le r_1 < a$. But then

$$p_2 - (p_1 + r_1\nu_\Sigma(p_1)) = -r_2\nu_\Sigma(p_2),$$

hence $p_2 \in \bar{B}(p_1 + r_1\nu_\Sigma(p_1), r_1) \cap \Sigma = \{p_1\}$, which shows $p_1 = p_2$ and then also $r_1 = r_2$. The set U_a will be called the *tubular neighbourhood* of Σ of of width a. To prove that Λ is a diffeomorphism, fix a point $(p_0, r_0) \in \Sigma \times (-a, a)$ and a chart ϕ for p_0. Then the function $f(\theta, r) = \Lambda(\phi(\theta), r)$ has derivative

$$Df(0, r_0) = [[I - r_0 L_\Sigma(p_0)]\phi'(0), \nu_\Sigma(p_0)].$$

It follows from (2.58) that $[I - r_0 L_\Sigma(p_0)] \in \mathcal{B}(T_{p_0}\Sigma)$ is invertible, and consequently, $Df(0, r_0) \in \mathcal{B}(\mathbb{R}^n)$ is invertible as well. The inverse function theorem implies that Λ is locally invertible with inverse of class C^1.

It will be convenient to decompose the inverse of Λ into $\Lambda^{-1} = (\Pi_\Sigma, d_\Sigma)$ such that

$$\Pi_\Sigma \in C^1(U_a, \Sigma), \quad d_\Sigma \in C^1(U_a, (-a, a)). \tag{2.57}$$

$\Pi_\Sigma(x)$ is the nearest point on Σ to x, $d_\Sigma(x)$ is the signed distance from x to Σ.

From the uniform interior and exterior ball condition follows that the number $1/a_0$ bounds the principal curvatures of Σ, i.e.,

$$\max\{\kappa_i(p) : p \in \Sigma,\ i = 1, \cdots, n-1\} \le 1/a_0. \tag{2.58}$$

A remarkable fact is that the signed distance d_Σ is even of class C^2. To see this, we use the identities

$$x - \Pi_\Sigma(x) = d_\Sigma(x)\nu_\Sigma(\Pi_\Sigma(x)), \quad d_\Sigma(x) = (x - \Pi_\Sigma(x)|\nu_\Sigma(\Pi_\Sigma(x)).$$

Differentiating w.r.t. x_k this yields

$$\begin{aligned}\partial_{x_k} d_\Sigma(x) &= (e_k - \partial_{x_k}\Pi_\Sigma(x)|\nu_\Sigma(\Pi_\Sigma(x))) + (x - \Pi_\Sigma(x)|\partial_{x_k}(\nu_\Sigma \circ \Pi_\Sigma)(x))\\ &= \nu_\Sigma^k(\Pi_\Sigma(x)) + d_\Sigma(x)(\nu_\Sigma(\Pi_\Sigma(x))|\partial_{x_k}(\nu_\Sigma \circ \Pi_\Sigma(x)))\\ &= \nu_\Sigma^k(\Pi_\Sigma(x)),\end{aligned}$$

since $\partial_{x_k}\Pi_\Sigma(x)$ belongs to the tangent space $T_{\Pi_\Sigma(x)}\Sigma$, as does $\partial_{x_k}(\nu_\Sigma \circ \Pi_\Sigma(x))$, since $|\nu_\Sigma \circ \Pi_\Sigma(x)| \equiv 1$. Thus we have the formula

$$\nabla_x d_\Sigma(x) = \nu_\Sigma(\Pi_\Sigma(x)), \quad x \in U_a. \tag{2.59}$$

This shows, in particular, that d_Σ is of class C^2.

It is useful to also have a representation for the derivatve $\partial\Pi_\Sigma(x)$ of $\Pi_\Sigma(x)$. With

$$I - \partial\Pi_\Sigma(x) = \nu_\Sigma(\Pi_\Sigma(x)) \otimes \nabla_x d_\Sigma(x) + d_\Sigma(x)\partial\nu_\Sigma(\Pi_\Sigma(x))\partial\Pi_\Sigma(x),$$

and (2.59), we obtain

$$\partial\Pi_\Sigma(x) = M_0(d_\Sigma(x))(\Pi_\Sigma(x))\mathcal{P}_\Sigma(\Pi_\Sigma(x)), \tag{2.60}$$

where $M_0(r)(p) := (I - rL_\Sigma(p))^{-1}$. This shows that $\partial\Pi_\Sigma(p) = \nabla_x\Pi_\Sigma(p) = \mathcal{P}_\Sigma(p)$, the orthogonal projection onto the tangent space $T_p\Sigma$.

3.2 The Level Function

Let Σ be a compact connected hypersurface of class C^2 bounding the domain Ω in $\mathbb{R}^n$. According to the previous section, Σ admits a tubular neighbourhood U_a of width $a > 0$. We may assume w.l.o.g. $a \le 1$. The signed distance function $d_\Sigma(x)$ in this tubular neighbourhood is of class C^2 as well, and since

$$\nabla_x d_\Sigma(x) = \nu_\Sigma(\Pi_\Sigma(x)), \quad x \in U_a,$$

we can view $\nabla_x d_\Sigma(x)$ as a C^1-extension of the normal field $\nu_\Sigma(x)$ from Σ to the tubular neighbourhood U_a of Σ. Computing the second derivatives $\nabla_x^2 d_\Sigma$ we obtain

$$\begin{aligned}\nabla_x^2 d_\Sigma(x) = \nabla_x \nu_\Sigma(\Pi_\Sigma(x)) &= -L_\Sigma(\Pi_\Sigma(x))\mathcal{P}_\Sigma(\Pi_\Sigma(x))(I - d_\Sigma(x)L_\Sigma(\Pi_\Sigma(x)))^{-1}\\ &= -L_\Sigma(\Pi_\Sigma(x))(I - d_\Sigma(x)L_\Sigma(\Pi_\Sigma(x)))^{-1},\end{aligned}$$

for $x \in U_a$, as $L_\Sigma(p) = L_\Sigma(p)\mathcal{P}_\Sigma(p)$. Taking traces then yields

$$\Delta d_\Sigma(x) = -\sum_{i=1}^{n-1} \frac{\kappa_i(\Pi_\Sigma(x))}{1 - d_\Sigma(x)\kappa_i(\Pi_\Sigma(x))}, \quad x \in U_a. \tag{2.61}$$

In particular, this implies

$$\nabla_x^2 d_\Sigma(p) = -L_\Sigma(p), \quad \Delta_x d_\Sigma(p) = -\kappa_\Sigma(p), \quad p \in \Sigma. \tag{2.62}$$

Therefore the norm of $\nabla_x^2 d_\Sigma$ is equivalent to the maximum of the moduli of the curvatures of Σ at a fixed point. Hence we find a constant c, depending only on n, such that

$$c|\nabla_x^2 d_\Sigma|_\infty \leq \max\{|\kappa_i(p)| : i = 1, \dots, n-1,\, p \in \Sigma\} \leq c^{-1}|\nabla_x^2 d_\Sigma|_\infty.$$

Next we extend d_Σ as a function φ to all of $\mathbb{R}^n$. For this purpose we choose a C^∞-function $\chi(s)$ such that $\chi(s) = 1$ for $|s| \leq 1$, $\chi(s) = 0$ for $|s| \geq 2$, $0 \leq \chi(s) \leq 1$. Then we set

$$\varphi(x) := \begin{cases} d_\Sigma(x)\chi(3d_\Sigma(x)/a) + \operatorname{sign}(d_\Sigma(x))(1 - \chi(3d_\Sigma(x)/a)), & x \in U_a, \\ \chi_{\Omega_{\rm ex}}(x) - \chi_{\Omega_{\rm in}}(x), & x \notin U_a, \end{cases} \tag{2.63}$$

where $\Omega_{\rm ex}$ and $\Omega_{\rm in}$ denote the exterior and interior component of $\mathbb{R}^n \setminus U_a$, respectively. This function φ is then of class C^2, $\varphi(x) = d_\Sigma(x)$ for $x \in U_{a/3}$, and

$$\varphi(x) = 0 \quad \Leftrightarrow \quad x \in \Sigma.$$

Thus Σ is given as zero-level set of φ, i.e., $\Sigma = \varphi^{-1}(0)$. φ is called a *canonical level function* for Σ. It is a special level function for Σ, as

$$\nabla_x \varphi(x) = \nu_\Sigma(\Pi_\Sigma(x)) \quad \text{for } x \in U_{a/3}.$$

3.3 Existence of Parameterizations

Recall the Hausdorff metric on the set $\mathcal{K}$ of compact subsets of $\mathbb{R}^n$ defined by

$$d_H(K_1, K_2) = \max\{\sup_{x \in K_1} \operatorname{dist}(x, K_2), \sup_{y \in K_2} \operatorname{dist}(y, K_1)\}. \tag{2.64}$$

Suppose Σ is a compact connected closed hypersurface of class C^2 bounding a bounded domain in $\mathbb{R}^n$. As before, let U_a be its tubular neighbourhood, $\Pi_\Sigma : U_a \to \Sigma$ the projection and $d_\Sigma : U_a \to \mathbb{R}$ the signed distance. We want to parameterize hypersurfaces Γ which are close to Σ as

$$\Gamma = \{p + \rho(p)\nu_\Sigma(p) : p \in \Sigma\},$$

where $\rho : \Sigma \to \mathbb{R}$ is then called the *normal parameterization* of Γ over Σ. For this to make sense, Γ must belong to the tubular neighbourhood U_a of Σ. Therefore, a natural requirement would be $d_H(\Gamma, \Sigma) < a$. We then say that Γ and Σ are C^0*-close* (of order ε) if $d_H(\Gamma, \Sigma) < \varepsilon$.

However, this condition is not enough to allow for existence of a normal parameterization, since it is not clear that the map Π_Σ is injective on Γ: small Hausdorff distance does not prevent Γ from folding within the tubular neighbourhood. We need a stronger assumption to prevent this. If Γ is a hypersurface of class C^1 we may introduce the *normal bundle* $\mathcal{N}\Gamma$ defined by

$$\mathcal{N}\Gamma := \{(q, \nu_\Gamma(q)) : q \in \Gamma\} \subset \mathbb{R}^{2n}.$$

Suppose Γ is a compact, connected C^1-hypersurface in $\mathbb{R}^n$. We say that Γ and Σ are C^1-*close* (of order ε) if $d_H(\mathcal{N}\Gamma, \mathcal{N}\Sigma) < \varepsilon$. We are going to show that C^1-hypersurfaces Γ which are C^1-close to Σ can in fact be parametrized over Σ.

For this purpose observe that, in case Γ and Σ are C^1-close of order ε, whenever $q \in \Gamma$, there is $p \in \Sigma$ such that $|q-p| + |\nu_\Gamma(q) - \nu_\Sigma(p)| < \varepsilon$. Hence $|q - \Pi_\Sigma q| < \varepsilon$, with $\Pi_\Sigma q := \Pi_\Sigma(q)$, and

$$|\nu_\Gamma(q) - \nu_\Sigma(\Pi_\Sigma q)| \leq |\nu_\Gamma(q) - \nu_\Sigma(p)| + |\nu_\Sigma(\Pi_\Sigma q) - \nu_\Sigma(p)| \leq \varepsilon + L|\Pi_\Sigma q - p|,$$

which yields with $|\Pi_\Sigma q - p| \leq |\Pi_\Sigma q - q| + |p-q| < 2\varepsilon$,

$$|q - \Pi_\Sigma q| + |\nu_\Gamma(q) - \nu_\Sigma(\Pi_\Sigma q)| \leq 2(1+L)\varepsilon,$$

where L denotes the Lipschitz constant of the normal of Σ. In particular, the tangent space $T_q\Gamma$ is transversal to $\nu_\Sigma(\Pi_\Sigma q)$, for each $q \in \Gamma$, that is,

$$(\nu_\Sigma(\Pi_\Sigma q) \mid \nu_\Gamma(q)) \neq 0, \quad q \in \Gamma.$$

Now fix a point $q_0 \in \Gamma$ and set $p_0 = \Pi_\Sigma q_0$. Since the tangent space $T_{q_0}\Gamma$ is transversal to $\nu_\Sigma(p_0)$, we infer that $\Pi'_\Sigma(q_0) : T_{q_0}\Gamma \to T_{p_0}\Sigma$ is invertible. The inverse function theorem yields an open neighbourhood $V(p_0) \subset \Sigma$ and a C^1-map $g : V(p_0) \to \Gamma$ such that $g(p_0) = q_0$, $g(V(p_0)) \subset \Gamma$, and $\Pi_\Sigma g(p) = p$ in $V(p_0)$. Therefore we obtain

$$q = g(p) = \Lambda \circ (\Pi_\Sigma, d_\Sigma) g(p) = \Pi_\Sigma g(p) + d_\Sigma(g(p))\nu_\Sigma(\Pi_\Sigma g(p)) = p + \rho(p)\nu_\Sigma(p),$$

with

$$\rho(p) := d_\Sigma(g(p)).$$

Thus we have a local normal parameterization of Γ over Σ. We may extend g to a maximal domain $V \subset \Sigma$, e.g. by means of Zorn's lemma. Clearly V is open in Σ and we claim that $V = \Sigma$. If not, then the boundary of V in Σ is nonempty and hence we find a sequence $p_n \in V$ such that $p_n \to p_\infty \in \partial V$. Since $\rho_n = \rho(p_n)$ is bounded, we may assume w.l.o.g. that $\rho_n \to \rho_\infty$. But then $q_\infty = p_\infty + \rho_\infty \nu_\Sigma(p_\infty)$ belongs to Γ as Γ is closed. Now we may apply the inverse function theorem again to see that V cannot be maximal. Since the map $\Phi(p) = p + \rho(p)\nu_\Sigma(p)$ is a local C^1-diffeomorphism, it is also open. Hence $\Phi(\Sigma) \subset \Gamma$ is open and compact, i.e., $\Phi(\Sigma) = \Gamma$ by connectedness of Γ. The map Φ is therefore a C^1-diffeomorphism from Σ to Γ. In case Σ is of class C^{k+1} and Γ is of class C^k for $k \geq 1$ the proof above immediately implies that $\Phi \in \mathrm{Diff}^k(\Sigma, \Gamma)$.

Observe that because of $x = \Pi_\Sigma x + d_\Sigma(x)\nu_\Sigma(\Pi_\Sigma x)$ in U_a we have $x \in \Gamma$ if and only if $d_\Sigma(x) = \rho(\Pi_\Sigma x)$. This property can be used to construct a C^1-function ψ on $\mathbb{R}^n$ such that $\Gamma = \psi^{-1}(0)$, i.e., a level function for Γ. For example we may take

$$\psi(x) = \varphi(x) - \rho(\Pi_\Sigma x)\chi(3d_\Sigma(x)/a), \quad x \in \mathbb{R}^n,$$

provided $\varepsilon < a/3$, where φ and χ are defined in (2.63).

3.4 Approximation of Hypersurfaces
Suppose as before that Σ is a compact connected hypersurface of class C^2 bounding a bounded domain Ω in $\mathbb{R}^n$. We may use the level function $\varphi : \mathbb{R}^n \to \mathbb{R}$ introduced in (2.63) to construct a real analytic hypersurface Σ_ε such that Σ appears as a C^2-graph over Σ_ε. In fact, we show that there is $\varepsilon_0 \in (0, a/3)$ such that for every $\varepsilon \in (0, \varepsilon_0)$ there is an analytic manifold Σ_ε and a function $\rho_\varepsilon \in C^2(\Sigma_\varepsilon)$ with the property that

$$\Sigma = \{p + \rho_\varepsilon(p)\nu_{\Sigma_\varepsilon}(p) : p \in \Sigma_\varepsilon\}$$

and

$$|\rho_\varepsilon|_\infty + |\nabla_{\Sigma_\varepsilon}\rho_\varepsilon|_\infty + |\nabla^2_{\Sigma_\varepsilon}\rho_\varepsilon|_\infty \le \varepsilon.$$

For this purpose, choose $R > 0$ such that $\varphi(x) = 1$ for $|x| > R/2$. Then define

$$\psi_k(x) = c_k\Big(1 - \frac{|x|^2}{R^2}\Big)^k_+, \quad x \in \mathbb{R}^n,$$

where $c_k > 0$ is chosen such that $\int_{\mathbb{R}^n} \psi_k(x)dx = 1$. Then $c_k \sim \alpha k^{n/2}$ as $k \to \infty$, with some number $\alpha = \alpha(n, R)$. Indeed, we have

$$\int_{B(0,R)} \Big(1 - \frac{|x|^2}{R^2}\Big)^k dx = \omega_n R^n \int_0^1 (1 - r^2)^k r^{n-1}\, dr = \frac{\omega_n R^n}{2} \int_0^1 (1 - t)^k t^{n/2-1}\, dt,$$

where $\omega_n = |\partial B(0, 1)|$. Using the well-known relations

$$\int_0^1 (1 - t)^k t^{n/2-1}\, dt = \mathsf{B}(\frac{n}{2}, k + 1) = \frac{\Gamma(\frac{n}{2})\Gamma(k + 1)}{\Gamma(k + 1 + \frac{n}{2})} \sim \Gamma(n/2)k^{-n/2}$$

with B the Beta function and Γ the Gamma function, the claim follows, with $\alpha = ((\omega_n R^n/2)\Gamma(n/2))^{-1} = (\pi R^2)^{-n/2}$.

Then as $k \to \infty$, we have $\psi_k(x) \to 0$, uniformly for $|x| \ge \eta > 0$, since $k^{n/2}q^k \to 0$ for any fixed $q \in (0, 1)$. Consequently, $\psi_k * f \to f$ in $C^m_{ub}(\mathbb{R}^n)$, whenever $f \in C^m_{ub}(\mathbb{R}^n)$. Let $\varphi_k = 1 + \psi_k * (\varphi - 1)$. Then

$$\varphi_k \to \varphi \quad \text{in} \quad C^2_{ub}(\mathbb{R}^n). \tag{2.65}$$

Moreover,

$$\psi_k * (\varphi - 1)(x) = \int_{\mathbb{R}^n} (\varphi(y) - 1)\psi_k(x - y)dy = \int_{|y|\le R/2} (\varphi(y) - 1)\psi_k(x - y)dy.$$

For $|x|, |y| < R/2$ follows $|x-y| < R$, and hence $\psi_k(x-y) = c_k(1-|x-y|^2/R^2)^k$ is polynomial in x, y. But then $\varphi_k(x)$ is a polynomial for such values of x; in particular, φ_k is real analytic in U_a. Choosing k large enough, we have $|\varphi - \varphi_k|_{C^2_b(\mathbb{R}^n)} < \varepsilon$.

Now suppose $\varphi_k(x) = 0$. Then $|\varphi(x)| < \varepsilon$, hence $x \in U_a$ and then $|d_\Sigma(x)| < \varepsilon$. This shows that the set $\Sigma_k := \varphi_k^{-1}(0)$ is in the ε-tubular neighbourhood around

Σ. Moreover, $|\nabla\varphi_k - \nabla\varphi|_\infty < \varepsilon$ yields $\nabla\varphi_k(x) \neq 0$ in U_a, and therefore Σ_k is a manifold, which is real analytic.

Next we show that Σ and Σ_k are C^1-diffeomorphic. For this purpose, fix a point $q_0 \in \Sigma_k$. Then $q_0 = p_0 + r_0\nu_\Sigma(p_0)$, where $p_0 = \Pi_\Sigma q_0 \in \Sigma$ and $r_0 = d_\Sigma(q_0)$. Consider the equation $g_k(p,r) := \varphi_k(p + r\nu_\Sigma(p)) = 0$ near (p_0, r_0). Since

$$\partial_r g_k(p,r) = (\nabla_x\varphi_k(p + r\nu_\Sigma(p))|\nu_\Sigma(p))$$

we have

$$\begin{aligned}\partial_r g_k(p_0,r_0) &= (\nabla_x\varphi_k(q_0)|\nabla_x\varphi(p_0))\\ &\geq 1 - |\nabla_x\varphi_k(q_0) - \nabla_x\varphi(q_0)| - |\nabla_x\varphi(q_0) - \nabla_x\varphi(p_0)|\\ &\geq 1 - |\varphi_k - \varphi|_{C_b^1(\mathbb{R}^n)} - \varepsilon|\nabla_x^2\varphi|_{C_b(\mathbb{R}^n)} > 0.\end{aligned}$$

Therefore, we may apply the implicit function theorem to obtain an open neighbourhood $V(p_0) \subset \Sigma$ and a C^1-function $r_k : V(p_0) \to \mathbb{R}$ such that $r_k(p_0) = r_0$ and $p + r_k(p)\nu_\Sigma(p) \in \Sigma_k$ for all $p \in V(p_0)$. We can now proceed as in the previous subsection to extend $r_k(\cdot)$ to a maximal domain $V \subset \Sigma$, which coincides with Σ by compactness and connectedness of Σ.

Thus we have a well-defined C^1-map $f_k : \Sigma \to \Sigma_k$, $f_k(p) = p + r_k(p)\nu_\Sigma(p)$, which is injective and a diffeomorphism from Σ to its range. We claim that f_k is also surjective. If not, there is some point $q \in \Sigma_k$, $q \notin f_k(\Sigma)$. Set $p = \Pi_\Sigma q$. Then $q = p + d_\Sigma(p)\nu_\Sigma(p)$ with $d_\Sigma(p) \neq r_k(p)$. Thus, there are at least two numbers $\beta_1, \beta_2 \in (-a,a)$ with $p + \beta_i\nu_\Sigma(p) \in \Sigma_k$. This implies with $\nu_\Sigma = \nu_\Sigma(p)$

$$0 = \varphi_k(p+\beta_2\nu_\Sigma) - \varphi_k(p+\beta_1\nu_\Sigma) = (\beta_2-\beta_1)\int_0^1 (\nabla_x\varphi_k(p+(\beta_1+t(\beta_2-\beta_1))\nu_\Sigma)|\nu_\Sigma)\,dt,$$

which yields $\beta_2 - \beta_1 = 0$ since

$$\int_0^1 (\nabla_x\varphi_k(p + (\beta_1 + t(\beta_2 - \beta_1))\nu_\Sigma)|\nu_\Sigma)\,dt \geq 1 - \varepsilon - \varepsilon|\nabla_x^2\varphi|_{C_b(\mathbb{R}^n)} > 0,$$

as above. Therefore, the map f_k is also surjective, and hence $f_k \in \mathrm{Diff}^1(\Sigma, \Sigma_k)$. This implies, in particular, that $\Sigma_k = f_k(\Sigma)$ is connected. For later use we note that

$$|r_k|_\infty + |\nabla_\Sigma r_k|_\infty \to 0 \ \text{ as } \ k \to \infty,$$

as can be inferred from $\partial_i r_k(p) = (\tau_i^{\Sigma_k}(p+r_k(p)\nu_\Sigma(p))|\nu_\Sigma(p))$ for $p \in \Sigma$, see (2.39).

Next we show that the mapping

$$\Lambda_k : \Sigma_k \times (-a/2, a/2) \to U(\Sigma_k, a/2), \quad \Lambda_k(q,s) := q + s\nu_k(q)$$

is a C^1-diffeomorphism for $k \geq k_0$, with $k_0 \in \mathbb{N}$ sufficiently large. In order to see this, we use the diffeomorphism f_k constructed above to rewrite Λ_k as

$$\begin{aligned}\Lambda_k(q,s) &= \Lambda_k(f_k(p), s)\\ &= p + s\,\nu_\Sigma(p) + r_k(p)\nu_\Sigma(p) + s[\nu_k(p + r_k(p)\nu_\Sigma(p)) - \nu_\Sigma(p)]\\ &=: \Lambda(p,s) + G_k(p,s) =: H_k(p,s).\end{aligned}$$

Clearly $H_k \in C^1(\Sigma \times (-a/2, a/2), \mathbb{R}^n)$ and $\Lambda \in \mathrm{Diff}^1(\Sigma \times (-a,a), U(\Sigma, a))$. It is not difficult to see that

$$|G_k(p,s)| + |DG_k(p,s)| \to 0 \text{ as } k \to \infty, \text{ uniformly in } (p,s) \in \Sigma \times [-a/2, a/2].$$

Consequently, $DH_k(p,s) : T_p\Sigma \times (-a/2, a/2) \to \mathbb{R}^n$ is invertible for $k \geq k_0$, and by the inverse function theorem, H_k is a local C^1-diffeomorphism. We claim that H_k is injective for all k sufficiently large. For this purpose, note that due to compactness of $\Sigma \times [-a/2, a/2]$ and injectivity of Λ there exists a constant $c > 0$ such that

$$|\Lambda(p,s) - \Lambda(\bar{p}, \bar{s})| \geq c\big(|p - \bar{p}| + |s - \bar{s}|\big), \quad (p,s),\ (\bar{p}, \bar{s}) \in \Sigma \times [-a/2, a/2].$$

The properties of G_k and compactness of $\Sigma \times [-a/2, a/2]$ imply, in turn, that the estimate above remains true for Λ replaced by H_k, and c replaced by $c/2$, provided $k \geq k_0$ with k_0 sufficiently large. Hence H_k is a C^1-diffeomorphism onto its image for k sufficiently large, as claimed. This shows that Σ_k has a uniform tubular neighbourhood of width $a/2$ for any $k \geq k_0$, and it follows that $\Sigma \subset U(\Gamma_k, a/2)$. Σ and Σ_k are compact connected closed C^1 hypersurfaces, and we may now apply the results of the previous subsection, showing that Σ can be parameterized over Σ_k by means of

$$p \mapsto p + \rho_k(p)\nu_k(p) \text{ with } \rho_k \in C^2(\Sigma_k, \mathbb{R}),$$

with $\nu_k := \nu_{\Sigma_k}$.

Finally we show that $|\rho_k|_\infty + |\nabla_{\Sigma_k}\rho_k|_\infty + |\nabla^2_{\Sigma_k}\rho_k|_\infty \leq \varepsilon$ for k sufficiently large. We already know from the construction that $|\rho_k|_\infty \to 0$ as $k \to \infty$. However, we need the following estimate on the rate of convergence: there exists $k_0 \in \mathbb{N}$ and a constant $C = C(k_0)$ such that

$$|\rho_k|_\infty \leq Ck^{-1/2}, \quad k \geq k_0. \tag{2.66}$$

In order to see this, we first observe that, for $|x| \leq R/2$,

$$\begin{aligned}|\varphi(x) - \varphi_k(x)| &= \Big|\int_{\mathbb{R}^n} [\varphi(x) - \varphi(x-y)]\psi_k(y)\,dy\Big| \leq |\nabla\varphi|_\infty \int_{|y|\leq R} |y|\psi_k(y)dy \\ &= |\nabla\varphi|_\infty C(n,R) c_k \mathsf{B}\big(\frac{n+1}{2}, k+1\big).\end{aligned}$$

Using similar arguments as above for the asymptotics of c_k and $B((n+1)/2, k+1)$ this yields constants $k_0 \in \mathbb{N}$ and $C = C(k_0)$ such that $|\varphi(x) - \varphi_k(x)| \leq Ck^{-1/2}$, whenever $|x| \leq R/2$ and $k \geq k_0$. Let $p \in \Sigma_k$ be given, and let $q = p + \rho_k(p)\nu_k(p)$. Then $|\varphi_k(q)| = |\varphi_k(q) - \varphi(q)| \leq Ck^{-1/2}$ for $k \geq k_0$. On the other hand,

$$|\varphi_k(q)| = |\varphi_k(q) - \varphi_k(p)| = \rho_k(p)\Big|\int_0^1 (\nabla\varphi_k(p + t\rho_k(p)\nu_k(p)|\nu_k(p))dt\Big| \geq \frac{1}{2}\rho_k(p),$$

provided k is sufficiently large, and this implies (2.66).

Next we show that there exists $k_0 \in \mathbb{N}$ and $C = C(k_0)$ such that $|\partial^\alpha \varphi_k(x)| \le Ck^{1/2}$ whenever $k \ge k_0$, $|x| \le R/2$, and $|\alpha| = 3$. Indeed this follows from

$$\begin{aligned}|\partial_\ell \partial_i \partial_j \varphi_k(x)| &= (2/R^2)kc_k \Big| \int_{|y|\le R} \partial_i \partial_j \varphi_k(x-y) y_\ell (1-|y|^2/R^2)^{k-1} dy \Big| \\ &\le Ckc_k \mathsf{B}\Big(\frac{n+1}{2}, k\Big) \sim ck^{1/2},\end{aligned}$$

where c is an appropriate constant. Combining with (2.66) we have shown that there are constants $k_0 \in \mathbb{N}$ and $C = C(k_0)$ such that

$$\rho_k(p)|\partial^\alpha \varphi_k(x)| \le C, \tag{2.67}$$

for $k \ge k_0$, $|\alpha| = 3$, $p \in \Sigma_k$, and $|x| \le R/2$.

In order to show smallness of $|\nabla_{\Sigma_k} \rho_k|_\infty + |\nabla^2_{\Sigma_k} \rho_k|_\infty$, we consider the relation

$$\varphi\big(\psi_k(\theta) + (\rho_k \nu_k)(\psi_k(\theta))\big) = 0, \quad \theta \in \Theta_k, \tag{2.68}$$

where $\psi_k : \Theta_k \to \Sigma_k$ is a C^2-parameterization of Σ_k around a point $p_k = f_k(q)$ for some $q \in \Sigma$. Since $\Sigma_k = \varphi_k^{-1}(0)$ and $\varphi_k \to \varphi$ in $C^2_{ub}(\mathbb{R}^n)$ one shows that $|\partial_j \psi_k(0)|$ is uniformly bounded in k for k sufficiently large.

Let $\tilde{\nu}_k(x) := \nabla_x \varphi_k(x)/|\nabla_x \varphi_k(x)|$ for $x \in \mathbb{R}^n$. Clearly, $\nu_k(\psi_k(\theta)) = \tilde{\nu}_k(\psi_k(\theta))$. Taking partial derivatives in (2.68) and using the orthogonality relation $(\nabla_x \varphi_k(\psi_k(\theta)) \,|\, \partial_j \psi_k(\theta)) = 0$ yields

$$\begin{aligned}&\partial_j(\rho_k \circ \psi_k)(\theta)\big(\nabla_x \varphi(q_k(\theta)) \,|\, (\nu_k \circ \psi_k)(\theta)\big) \\ &\qquad = \big(\nabla_x \varphi(q_k(\theta)) - \nabla_x \varphi_k(\psi_k(\theta)) \,|\, \partial_j \psi_k(\theta)\big) \\ &\qquad\quad - (\rho_k \circ \psi_k)(\theta)\big(\nabla_x \varphi(q_k(\theta)) \,|\, \partial_j(\tilde{\nu}_k \circ \psi_k)(\theta)\big)\end{aligned} \tag{2.69}$$

where, for brevity, we set $q_k(\theta) = \psi_k(\theta) + (\rho_k \nu_k)(\psi_k(\theta))$. It follows from (2.65) and uniform continuity that

$$(\nabla_x \varphi(q_k(\theta))|(\nu_k \circ \psi_k)(\theta)) \ge 1/2, \tag{2.70}$$

provided $k \ge k_0$ with k_0 sufficiently large. The fact that $\partial_j \psi_k(\theta)$ is uniformly bounded for $k \ge k_0$ and (2.65) implies that the right-hand side in (2.69) converges to zero as $k \to \infty$. We have shown that $|\partial_j \rho_k(p_k)| \le \varepsilon$, provided that $k \ge k_0$ with k_0 sufficiently large.

Next, we take an additional derivative $\partial_i = \partial_{\theta_i}$ in (2.69). This will produce the terms

$$\begin{aligned}&\partial_i \partial_j(\rho_k \circ \psi_k)(\theta)\big(\nabla_x \varphi(q_k(\theta)) \,|\, (\nu_k \circ \psi_k)(\theta)\big) \\ &+ \partial_j(\rho_k \circ \psi_k)(\theta)\partial_i\big(\nabla_x \varphi(q_k(\theta)) \,|\, (\nu_k \circ \psi_k)(\theta)\big)\end{aligned}$$

on the left-hand side. From the previous step for $\partial_j(\rho_k \circ \psi_k)$ we conclude that the second term converges to 0 as $k \to \infty$. Thus it follows from (2.70) that

$\partial_i\partial_j(\rho_k \circ \psi_k)$ converges to 0 as $k \to \infty$, provided we can show that the derivatives of the right-hand side in (2.69) converge to zero as $k \to \infty$. A moment of reflection shows that this is indeed the case, with the possible exception of the term $\rho_k(\psi_k(\theta))\big(\nabla_x\varphi(q_k(\theta)) \,|\, \partial_i\partial_j(\tilde\nu_k \circ \psi_k)(\theta)\big)$ which is problematic as $\partial_i\partial_j(\tilde\nu_k \circ \psi_k)$ involves third-order derivatives of φ_k. Since $((\tilde\nu_k \circ \psi_k)(\theta) \,|\, (\tilde\nu_k \circ \psi_k)(\theta)) = 1$ we get $((\tilde\nu_k \circ \psi_k)(\theta) \,|\, \partial_j(\tilde\nu_k \circ \psi_k)(\theta)) = 0$, and hence

$$((\tilde\nu_k \circ \psi_k)(\theta) \,|\, \partial_i\partial_j(\tilde\nu_k \circ \psi_k)(\theta)) = -(\partial_i(\tilde\nu_k \circ \psi_k)(\theta) \,|\, \partial_j(\tilde\nu_k \circ \psi_k)(\theta)).$$

With $\nabla_x\varphi(q_k(\theta)) = \nu_\Sigma(q_k(\theta))$ this yields

$$\begin{aligned}&\rho_k(\psi_k(\theta))(\nabla_x\varphi(q_k(\theta)) \,|\, \partial_i\partial_j(\tilde\nu \circ \psi_k))(\theta)))\\ &= \big(\nu_\Sigma(q_k(\theta)) - \tilde\nu_k(\psi_k(\theta)) \,|\, \rho_k(\psi_k(\theta))\partial_i\partial_j(\tilde\nu_k \circ \psi_k)(\theta)\big)\\ &\quad + \rho_k(\psi_k(\theta))\big(\partial_i(\tilde\nu_k \circ \psi_k)(\theta) \,|\, \partial_j(\tilde\nu_k \circ \psi_k)(\theta)\big).\end{aligned}$$

Convergence to 0 of the first term on the right-hand side follows from (2.67) and (2.65), while the second term converges to 0 since ρ_k has this property.

Since $f_k : \Sigma \to \Sigma_k$ is a bijection, the assertion holds true for any point $p_k \in \Sigma_k$, $k \geq k_0$, and hence the claim follows.

2.4 The Manifold of Hypersurfaces in $\mathbb{R}^n$

4.1 Compact Connected Hypersurfaces of Class C^2

Consider the set $\mathcal{MH}^2$ of all compact connected C^2-hypersurfaces Σ in $\mathbb{R}^n$. Let $\mathcal{N}\Sigma$ denote their associated normal bundles. The second normal bundle of Σ is defined by

$$\mathcal{N}^2\Sigma = \{(p, \nu_\Sigma(p), \nabla_\Sigma\nu_\Sigma(p)) : p \in \Sigma\}.$$

We introduce a metric $d_{\mathcal{MH}^2}$ on $\mathcal{MH}^2$ by means of $d_{\mathcal{MH}^2}(\Sigma_1, \Sigma_2) = d_H(\mathcal{N}^2\Sigma_1, \mathcal{N}^2\Sigma_2)$. This way $\mathcal{MH}^2$ becomes a metric space. We want to show that $\mathcal{MH}^2$ is a Banach manifold.

Fix a hypersurface $\Sigma \in \mathcal{MH}^2$ of class C^3. Then we define a chart over the Banach space $X_\Sigma := C^2(\Sigma, \mathbb{R})$ as follows. Σ has a tubular neighbourhood U_a of width a. For a given function $\rho \in B_{X_\Sigma}(0, a/3)$ we obtain a hypersurface Γ_ρ^Σ by means of the map

$$\Phi_\Sigma(\rho)(p) := p + \rho(p)\nu_\Sigma(p), \quad p \in \Sigma.$$

According to Section 2.3, this yields a hypersurface Γ_ρ^Σ of class C^2, diffeomorphic to Σ. Moreover, with some constant C_a^Σ, we have

$$d_{\mathcal{MH}^2}(\Gamma_\rho^\Sigma, \Sigma) \leq C_a^\Sigma |\rho|_{C_b^2(\Sigma)},$$

which shows that the map $\Phi_\Sigma : B_{X_\Sigma}(0, a/3) \to \mathcal{MH}^2$ is continuous. Conversely, given $\Gamma \in \mathcal{MH}^2$ which is C^2-close to Σ, the results in Section 2.3.3 show that Γ can be parameterized by a function $\rho \in C^2(\Sigma, \mathbb{R})$, such that $|\rho|_{C_b^2(\Sigma)} < a/3$.

We now determine the tangent space $T_\Sigma \mathcal{MH}^2$ at some fixed C^3-hypersurface $\Sigma \in \mathcal{MH}^2$. For this purpose we take a differentiable curve $\Gamma : (-\delta_0, \delta_0) \to \mathcal{MH}^2$ with $\Gamma(0) = \Sigma$. According to Section 2.3.3, there is $\delta \in (0, \delta_0)$ such that for each $t \in (-\delta, \delta)$ we find a normal parameterization $\rho(t) \in C^2(\Sigma, \mathbb{R})$ of $\Gamma(t)$. Then in these coordinates we have

$$v := \frac{d}{dt}\Gamma(0) = \frac{d}{dt}\rho(0)\nu_\Sigma \in C^2(\Sigma, T_\Sigma^\perp \mathcal{MH}^2).$$

On the other hand, if $v = \rho\nu_\Sigma$ is a normal field on Σ with $\rho \in X_\Sigma$ we obtain a curve $\Gamma : (-\delta, \delta) \to \mathcal{MH}^2$ by means of $\Gamma(t)(p) = p + t\rho(p)\nu_\Sigma(p)$. Clearly, $\Gamma(0) = \Sigma$ and $\rho\nu_\Sigma = \frac{d}{dt}\Gamma(0) \in T_\Sigma \mathcal{MH}^2$. In other words, the tangent space $T_\Sigma \mathcal{MH}^2$ consists of all normal fields v on Σ which are of class C^2.

There is one shortcoming with this approach, namely the need to require that $\Sigma \in C^3$. This is due to the fact that we are losing one derivative when forming the normal ν_Σ. However, since we may approximate a given hypersurface of class C^2 by a real analytic one in the second normal bundle, this defect can be avoided by only parameterizing over real analytic hypersurfaces, which will be sufficient below.

4.2 Compact Hypersurfaces with Uniform Ball Condition

Let $\Omega \subset \mathbb{R}^n$ be a bounded domain, and consider a closed compact connected C^2-hypersurface $\Gamma \subset \Omega$. This hypersurface separates Ω into two disjoint open connected sets Ω_1 and Ω_2, the interior and the exterior of Γ w.r.t. Ω. By means of the level function φ_Γ of Γ we have $\Omega_1 = \varphi_\Gamma^{-1}(-\infty, 0)$ and $\Omega_2 = \Omega \setminus \bar{\Omega}_1$. Then $\partial\Omega_1 = \Gamma$ and $\partial\Omega_2 = \partial\Omega \cup \Gamma$.

The hypersurface Γ satisfies the *ball condition*, i.e., there is a radius $r > 0$ such that for each point $p \in \Gamma$ there are balls $B(x_i, r) \subset \Omega_i$ such that $\Gamma \cap \bar{B}(x_i, r) = \{p\}$. The set of hypersurfaces of class C^2 contained in Ω satisfying the ball condition with radius $r > 0$ will be denoted by $\mathcal{MH}^2(\Omega, r)$. Note that hypersurfaces in this class have uniformly bounded principal curvatures.

The elements of $\mathcal{MH}^2(\Omega, r)$ have a tubular neighbourhood of width a larger than r. Therefore the construction of the level function φ_Γ of Γ from Section 2.3.2 can be carried out with the same a and the same cut-off function χ for each $\Gamma \in \mathcal{MH}^2(\Omega, r)$. More precisely, we have

$$\varphi_\Gamma(x) = g(d_\Gamma(x)), \quad x \in \Omega,$$

with

$$g(s) = s\chi(3s/a) + \operatorname{sgn}(s)(1 - \chi(3s/a)), \quad s \in \mathbb{R};$$

note that g is strictly increasing and equals ± 1 for $\pm s > 2a/3$. This induces an injective map

$$\Phi : \mathcal{MH}^2(\Omega, r) \to C^2(\bar{\Omega}), \quad \Phi(\Gamma) := \varphi_\Gamma. \tag{2.71}$$

Φ is in fact a homeomorphism of $\mathcal{MH}^2(\Omega, r)$ onto $\Phi(\mathcal{MH}^2(\Omega, r)) \subset C^2(\bar{\Omega})$.

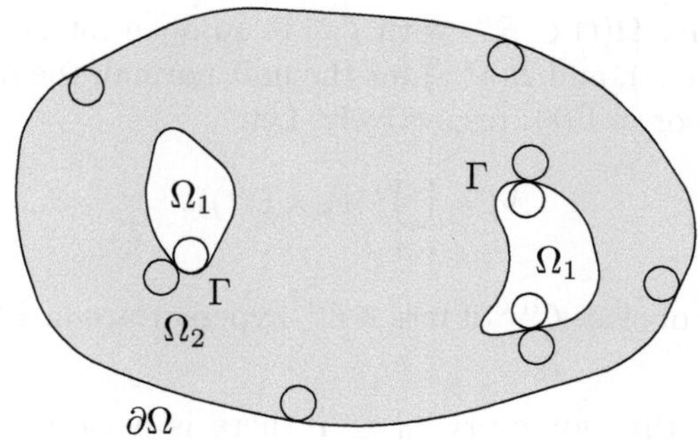

Figure 2.3: Illustration of the ball condition.

This can be seen as follows. Let $\varepsilon > 0$ be small enough. If $|\varphi_{\Gamma_1} - \varphi_{\Gamma_2}|_{2,\infty} \le \varepsilon$, then $d_{\Gamma_1}(x) \le \varepsilon$ on Γ_2 and $d_{\Gamma_2}(x) \le \varepsilon$ on Γ_1, which implies $d_H(\Gamma_1, \Gamma_2) \le \varepsilon$. Moreover, we also have $|\nabla_x \varphi_{\Gamma_1}(x) - \nu_{\Gamma_2}(x)| \le \varepsilon$ on Γ_2 and $|\nabla_x \varphi_{\Gamma_2}(x) - \nu_{\Gamma_1}(x)| \le \varepsilon$ on Γ_1 which yields $d_H(\mathcal{N}\Gamma_1, \mathcal{N}\Gamma_2) \le C_0\varepsilon$. Then the hypersurfaces Γ_j can both be parameterized over a C^3-hypersurface Σ, and therefore $d_H(\mathcal{N}^2\Gamma_1, \mathcal{N}^2\Gamma_2) \le \varepsilon$ if and only if

$$|\rho_1 - \rho_2|_\infty + |\nabla_\Sigma(\rho_1 - \rho_2)|_\infty + |\nabla_\Sigma^2(\rho_1 - \rho_2)|_\infty \le C_1\varepsilon.$$

This in turn is equivalent to $|\varphi_{\Gamma_1} - \varphi_{\Gamma_2}|_{2,\infty} \le C_2\varepsilon$.

Let $s - (n-1)/p > 2$. For $\Gamma \in \mathcal{MH}^2(\Omega, r)$ we then define

$$\Gamma \in W^s_p(\Omega, r) \quad \text{if} \quad \varphi_\Gamma \in W^s_p(\Omega), \tag{2.72}$$

and

$$\mathrm{dist}_{W^s_p(\Omega,r)}(\Gamma_1, \Gamma_2) := |\varphi_{\Gamma_1} - \varphi_{\Gamma_2}|_{W^s_p(\Omega)}. \tag{2.73}$$

In this case the local charts for Γ can be chosen of class W^s_p as well. A subset $A \subset W^s_p(\Omega, r)$ is said to be (relatively) compact, if $\Phi(A) \subset W^s_p(\Omega)$ is (relatively) compact. In particular, it follows from Rellich's theorem that $W^s_p(\Omega, r)$ is a compact subset of $W^\sigma_q(\Omega, r)$, whenever $s - n/p > \sigma - n/q$, and $s > \sigma$.

2.5 Moving Hypersurfaces and Domains

In this section we consider the situation of moving hypersurfaces, that is, hypersurfaces that are time dependent. We first introduce the notion of normal velocity, and we then prove a transport theorem for moving surfaces. A special case is the well-known formula for the change of surface area. In addition, we prove a transport theorem for moving domains, and derive the change of volume formula.

5.1 Moving Hypersurfaces

Let $\{\Gamma(t) : t \in I\}$ be a family of compact connected closed C^2-hypersurfaces in

$\mathbb{R}^n$ bounding domains $\Omega(t) \subset \mathbb{R}^n$, with $I \subset \mathbb{R}$ an open interval. In the following, we write $\nu_\Gamma(t,\cdot)$, $\kappa_\Gamma(t,\cdot)$, and $L_\Gamma(t,\cdot)$ for the unit normal, the mean curvature and the Weingarten tensor of $\Gamma(t)$, respectively. Let

$$\mathcal{M} = \bigcup_{t\in I} (\{t\} \times \Gamma(t)). \tag{2.74}$$

By definition, $\mathcal{M}$ is of class $C^{1,2}$ if it is a C^1-hypersurface in $\mathbb{R}^{n+1}$ and, moreover, $\nu_\Gamma \in C^1(\mathcal{M}, \mathbb{R}^n)$.

We now show that for every $t_0 \in I$ there is a closed, compact, analytic hypersurface Σ, an interval $I_0 := (t_0 - \delta, t_0 + \delta) \subset I$ and a function $\rho : I_0 \times \Sigma \to \mathbb{R}$ with

$$\rho \in C^1(I_0 \times \Sigma), \quad \nabla_\Sigma \rho \in C^1(I_0 \times \Sigma, \mathbb{R}^n) \tag{2.75}$$

such that

$$\Gamma(t) = \{\xi + \rho(t,\xi)\nu_\Sigma(\xi) : t \in I_0,\ \xi \in \Sigma\}. \tag{2.76}$$

This is obtained as follows. Let $t_0 \in I$ be fixed. The assumption that $\mathcal{M}$ is a hypersurface in $\mathbb{R}^{n+1}$ implies that for every $\varepsilon > 0$ there exits $\delta > 0$ such that $d_H(\Gamma(t_0), \Gamma(t)) \le \varepsilon$ whenever $|t - t_0| \le \delta$. In order to prove the assertion, it suffices to show that for every $\varepsilon > 0$ there exists $\delta > 0$ such that

$$\mathrm{dist}(p, \Gamma(t_0)) \le \varepsilon \quad \text{for all } p \in \Gamma(t) \text{ and all } |t - t_0| \le \delta.$$

Suppose the latter assertion is not true. Then there exists $a > 0$, a sequence $(p_n)_{n\in\mathbb{N}}$ in $\Gamma(t)$, and a sequence $(t_n)_{n\in\mathbb{N}}$ with $t_n \to t_0$ such that $\mathrm{dist}(p_n, \Gamma(t_n)) \ge 2a$ for all $n \in \mathbb{N}$. As $\Gamma(t_0)$ is compact, we find $p \in \Gamma(t_0)$ and a subsequence of $(p_n)_{n\in\mathbb{N}}$, again denoted by $(p_n)_{n\in\mathbb{N}}$, such that $p_n \to p$. Therefore, $\mathrm{dist}(p, \Gamma(t_n)) \ge a$ for $n \ge N$, with N sufficiently large. This shows that $\big(\{t_n\} \times \Gamma(t_n)\big) \cap \big(\mathbb{R} \times B_{\mathbb{R}^n}(p,a)\big) = \emptyset$ for $n \ge N$, contradicting the assumption that $\mathcal{M}$ is a manifold. As ν_Γ is continuous on $\mathcal{M}$ we conclude that for every $\varepsilon > 0$ there exists $\delta > 0$ such that

$$d_H(\mathcal{N}\Gamma(t), \mathcal{N}\Gamma(t_0)) \le \varepsilon, \quad \text{whenever } |t - t_0| \le \delta.$$

According to the approximation result in Section 2.3.4 we can find an analytic hypersurface Σ which approximates $\Gamma(t_0)$. We can assume that $\Gamma(t) \subset U_{\varepsilon/2}(\Sigma)$ for $t \in I_0$, that is, $\Gamma(t)$ is contained in the tubular neighbourhood $U_{\varepsilon/2}(\Sigma)$ of Σ of width $\varepsilon/2$. By Section 2.3.3, for every $t \in I_0$ there exists a function $\rho(t,\cdot) \in C^2(\Sigma)$ such that (2.76) holds. It remains to show that ρ satisfies the regularity assumptions claimed in (2.75). In order to see this, let us consider the mapping

$$\hat{\Pi}_\Sigma : \mathcal{M}(I_0) \to I_0 \times \Sigma, \quad \hat{\Pi}_\Sigma(t,p) = (t, \Pi_\Sigma(p)), \text{ where } \mathcal{M}(I_0) := \bigcup_{t\in I_0} (\{t\} \times \Gamma(t)\}.$$

We note that $\hat{\Pi}_\Sigma$ is well-defined, as $\Gamma(t) \subset U_{\varepsilon/2}(\Sigma)$ for each $t \in I_0$. Moreover, we have

$$\hat{\Pi}_\Sigma \in C^1(\mathcal{M}(I_0), I_0 \times \Sigma), \quad \hat{\Pi}_\Sigma(t,\cdot) = \Pi_\Sigma|_{\Gamma(t)}.$$

An inspection of the proof in Section 2.3.3 shows that

$$\hat{\Pi}_\Sigma \in \mathrm{Diff}^1(\mathcal{M}(I_0), I_0 \times \Sigma), \quad (\hat{\Pi}_\Sigma)^{-1}(t,\xi) = \Phi(t,\xi) := (t, \xi + \rho(t,\xi)\nu_\Sigma(\xi)).$$

This yields, in particular, $\rho \in C^1(I_0 \times \Sigma)$ and it remains to show the additional regularity claimed in (2.75). We recall from (2.44) that

$$\nu_\Gamma(\Phi(t,\xi)) = \big(\beta(\rho)(\nu_\Sigma - M_0(\rho)\nabla_\Sigma \rho)\big)(t,\xi), \quad (t,\xi) \in I_0 \times \Sigma. \tag{2.77}$$

This representation, in conjuction with the regularity $\rho \in C^1(I_0 \times \Sigma)$ already established, implies that

$$\nu_\Gamma \in C^1(\mathcal{M}(I_0), \mathbb{R}^n) \iff \nabla_\Sigma \rho \in C^1(I_0 \times \Sigma, \mathbb{R}^n),$$

as we will see next. Clearly, $\nu_\Gamma \in C^1(\mathcal{M}(I_0), \mathbb{R}^n)$ iff $\nu_\Gamma \circ \Phi \in C^1(I_0 \times \Sigma, \mathbb{R}^n)$.

Suppose that $\nu_\Gamma \in C^1(\mathcal{M}(I_0), \mathbb{R}^n)$. Thanks to $\beta(\rho)(t,\xi) = (\nu_\Gamma(\Phi(t,\xi)) \mid \nu_\Sigma(\xi))$ we have $\beta(\rho) \in C^1(I_0 \times \Sigma)$ and this, in turn, implies

$$\nabla_\Sigma \rho = (I - \rho L_\Sigma)\big(\nu_\Sigma - (1/\beta(\rho))(\nu_\Gamma \circ \Phi)\big) \in C^1(I_0 \times \Sigma).$$

On the other hand, if ρ satisfies the regularity assumptions in (2.75) and the family $\{\Gamma(t) : t \in I_0\}$ is given by (2.76), then it is not difficult to verify that $\mathcal{M}(I_0)$ is a hypersurface of class $C^{1,2}$.

We now state a useful variant of (2.76). The result reads as follows: for every fixed $t \in I$ there exists a number $\delta > 0$ and a function $\rho \in C^1((-\delta,\delta) \times \Sigma)$, where $\Sigma = \Gamma(t)$, such that

$$\Gamma(t+s) = \{p + \rho(s,p)\nu_\Sigma(p) : s \in (-\delta,\delta),\ p \in \Sigma\}, \quad \Sigma := \Gamma(t). \tag{2.78}$$

This follows by an obvious modification of the arguments given above. In fact, the proof is less involved, as there is no need to generate a smooth approximation for $\Gamma(t)$.

5.2 The Normal Velocity

Let $\mathcal{M}$ be as above. Suppose I_0 is a subinterval of I and $\gamma : I_0 \to \mathbb{R}^n$ is a C^1-curve. Then γ is called a *C^1-curve on $\mathcal{M}$* if $\gamma(t) \in \Gamma(t)$ for each $t \in I_0$. Hence, γ is a C^1-curve on $\mathcal{M}$ iff $(t,\gamma(t)) \in \mathcal{M}$ for $t \in I_0$. If γ is C^1-curve on $\mathcal{M}$, then

$$V_\Gamma(t,p) := (\gamma'(t) | \nu_\Gamma(t,p)), \quad p = \gamma(t), \tag{2.79}$$

is called the *normal velocity* of $\{\Gamma(t) : t \in I\}$ at the point (t,p). The normal velocity V_Γ is well-defined, that is, $V_\Gamma(t,p)$ does not depend on the choice of a C^1-curve on $\mathcal{M}$ through $p \in \Gamma(t)$. Indeed, let $\gamma : I_0 \to \mathbb{R}^n$ be an arbitrary C^1-curve on $\mathcal{M}$ and let $p = \gamma(t)$. We can assume, by possibly shrinking I_0, that the representation (2.76) holds. Therefore, the curve γ can be expressed by

$$\gamma(t) = \xi(t) + \rho(t,\xi(t))\nu_\Sigma(\xi(t)), \quad t \in I_0, \quad \xi(t) \in \Sigma,$$

and hence,

$$\gamma'(t) = (I - \rho(t,\xi(t))L_\Sigma(\xi(t)))\xi'(t) + (\partial_t\rho(t,\xi(t)) + (\nabla_\Sigma\rho(t,\xi(t))|\xi'(t))\nu_\Sigma(\xi(t)).$$

Using (2.77), and suppressing the variables, we obtain

$$V_\Gamma = (\gamma'|\nu_\Gamma) = \beta(\rho)\{\partial_t\rho + (\nabla_\Sigma\rho|\xi') - ((I-\rho L_\Sigma)\xi'|M_0(\rho)\nabla_\Sigma\rho)\} = \beta(\rho)\partial_t\rho,$$

or in more precise notation, $V_\Gamma(t,p) = (V_\Gamma \circ \Phi)(t,\xi) = \beta(\rho(t))(\xi)\partial_t\rho(t,\xi)$. This expression does not refer to the curve γ, and this shows that the definition (2.79) is independent of a particular curve. Moreover, this also shows that we can, alternatively, define the normal velocity by

$$V_\Gamma = \beta(\rho)\partial_t\rho, \tag{2.80}$$

provided $\{\Gamma(t) : t \in I_0\}$ is represented by (2.76), which can always be assumed.

For later use we note that

$$[1, V_\Gamma\nu_\Gamma]^{\mathsf{T}} \in T_{(t,p)}\mathcal{M}, \tag{2.81}$$

i.e., $[1,(V_\Gamma\nu_\Gamma)(t,p)]^{\mathsf{T}}$ is a tangent vector for $\mathcal{M}$ at the point (t,p). This can be seen as follows. Suppose $\gamma : I_0 \to \mathbb{R}^n$ is a C^1-curve on $\mathcal{M}$. Then $(t,\gamma(t)) \in \mathcal{M}$ for $t \in I_0$ and consequently, $[1,\gamma'(t)]^{\mathsf{T}} \in T_{(t,p)}\mathcal{M}$ with $p = \gamma(t)$. Hence, by (2.79),

$$\begin{aligned} [1,(V_\Gamma\nu_\Gamma)(t,p)]^{\mathsf{T}} &= [1,(\gamma'(t)|\nu_\Gamma(t,p))\nu_\Gamma(t,p)]^{\mathsf{T}} \\ &= [1,\gamma'(t)]^{\mathsf{T}} - [0,\mathcal{P}_{\Gamma(t)}(p)\gamma'(t)]^{\mathsf{T}} \in T_{(t,p)}\mathcal{M}, \end{aligned}$$

as $[0,v]^{\mathsf{T}} \in T_{(t,p)}\mathcal{M}$ for any vector $v \in T_p\Gamma(t)$.

5.3 The Lagrange Derivative for Moving Surfaces

Suppose that $u_\Gamma(t,\cdot) := u_{\Gamma(t)}(\cdot) : \Gamma(t) \to \mathbb{R}^n$ is a vector field for each $t \in I$. Hence u_Γ is defined on $\mathcal{M}$ and we assume that $u_\Gamma \in C^1(\mathcal{M},\mathbb{R}^n)$. Then u_Γ is called a *C^1-velocity field* for the family $\{\Gamma(t) : t \in I\}$ if

$$V_\Gamma = (u_\Gamma|\nu_\Gamma), \tag{2.82}$$

or more precisely, if $V_\Gamma(t,p) = (u_\Gamma(t,p)|\nu_\Gamma(t,p))$ for $(t,p) \in \mathcal{M}$.

A velocity field u_Γ is called a *normal velocity field* for $\{\Gamma(t) : t \in I\}$ if $u_\Gamma(t,\cdot) \in T^\perp\Gamma(t)$, i.e., $u_\Gamma(t,\cdot)$ lies in the normal bundle of $\Gamma(t)$ for each $t \in I$. Hence,

$$u_\Gamma \text{ is a normal velocity field} \iff u_\Gamma = V_\Gamma\nu_\Gamma. \tag{2.83}$$

Although only normal velocity fields matter from a geometric point of view, we nevertheless need to consider general velocity fields in order to treat the motion of fluid particles in fluid flows subject to phase transitions.

We note that if u_Γ is a velocity field for $\{\Gamma(t) : t \in I\}$ then

$$[1, u_\Gamma]^\mathsf{T} \in T\mathcal{M}. \tag{2.84}$$

This can be deduced from (2.81), (2.82), and the decomposition

$$[1, u_\Gamma] = [1, (u_\Gamma|\nu_\Gamma)\nu_\Gamma] + [0, \mathcal{P}_\Gamma u_\Gamma] = [1, V_\Gamma \nu_\Gamma] + [0, \mathcal{P}_\Gamma u_\Gamma],$$

where, as before, we use the fact that $[0, v]^\mathsf{T} \in T\mathcal{M}$ for any vector $v \in T\Gamma(t)$.

Next we show that for every C^1-velocity field u_Γ and every $p \in \Gamma(t)$, with t fixed, there exists $\delta > 0$ and a unique C^1-curve $[s \mapsto x(t+s)] : (-\delta, \delta) \to \mathbb{R}^n$ such that

$$\begin{aligned} &\frac{d}{ds}x(t+s) = u_\Gamma(t+s, x(t+s)), \quad x(t+s) \in \Gamma(t+s), \quad s \in (-\delta, \delta), \\ &x(t) = p. \end{aligned} \tag{2.85}$$

The solution to (2.85), in the sequel denoted by $x(t+s, t, p)$, is then called a *trajectory* or a *flow line* on $\mathcal{M}$ through $p \in \Gamma(t)$, generated by the velocity field u_Γ. The existence of such a trajectory can be seen by the following argument. Setting

$$z(s) := [t+s, x(t+s)]^\mathsf{T}$$

we see that $x(t+s) \in \Gamma(t+s)$ is equivalent to $z(s) \in \mathcal{M}$ for $s \in (-\delta, \delta)$. Therefore, (2.85) has a (unique) solution if and only if the differential equation

$$\dot{z}(s) = [1, u_\Gamma(z(s))]^\mathsf{T}, \quad s \in (-\delta, \delta), \quad z(0) = (t, p), \tag{2.86}$$

has a (unique) solution. Existence and uniqueness of a solution $z(s) = z(s, (t, p))$ to (2.86) follows from the fact that the vector field $[1, u_\Gamma]^\mathsf{T}$ is tangential to $\mathcal{M}$, see (2.84), and well-known results from the theory of differential equations. Moreover, we conclude that

$$[(s, (t, p)) \mapsto z(s, (t, p))] \in C^1((-\delta, \delta) \times \mathcal{M}, \mathcal{M}),$$

and this implies

$$[(s, p) \mapsto x(t+s, t, p)] \in C^1((-\delta, \delta) \times \Gamma(t), \Gamma(t)).$$

We note that

$$u_\Gamma \text{ is a } C^1\text{-velocity field} :\Longleftrightarrow V_\Gamma = (u_\Gamma|\nu_\Gamma) \Longleftrightarrow [1, u_\Gamma]^\mathsf{T} \in T\mathcal{M}. \tag{2.87}$$

The first equivalence follows by definition, while the second implication " $\Rightarrow$ " has been shown above. Suppose that $[1, u_\Gamma]^\mathsf{T} \in T\mathcal{M}$. Then (2.85) admits a C^1-solution $[s \mapsto x(t+s, t, p)]$, and the definition of V_Γ in (2.79) implies

$$V_\Gamma(t, p) = \Big(\frac{d}{ds}x(t+s, t, p)\big|_{s=0} \;\big|\; \nu_\Gamma(t, p)\Big) = (u_\Gamma(t, p)|\nu_\Gamma(t, p)).$$

It is illustrative to point out an alternative way to establish existence of solutions to (2.85). By (2.76) we can assume that $\{\Gamma(t+s) : s \in (-\delta, \delta)\}$ is given by

$$\Gamma(t+s) = \{\xi + \rho(s,\xi)\nu_\Sigma(\xi) : s \in (-\delta,\delta),\ \xi \in \Sigma\},$$

where Σ is a smooth hypersurface. Then the curve $x(s) = \xi(s) + \rho(s,\xi(s))\nu_\Sigma(\xi(s))$, with $\xi(s) \in \Sigma$, satisfies (2.76) if and only if

$$\begin{aligned} \xi'(s) &= (I - \rho L_\Sigma)^{-1}\mathcal{P}_\Sigma(\xi(s))u_\Gamma(s, \xi(s) + \rho(s,\xi(s))\nu_\Sigma(\xi(s))) \\ \xi(t) &= \xi_0, \end{aligned} \tag{2.88}$$

where $(I - \rho L_\Sigma)$ is the short form for $(I - \rho(s,\xi(s))L_\Sigma(\xi(s)))$. Indeed, applying the projection $\mathcal{P}_\Sigma$ to the equation

$$(I - \rho L_\Sigma)\xi'(s) + [\partial_s\rho(s,\xi(s)) + (\nabla_\Sigma\rho(s,\xi(s))|\xi'(s))]\nu_\Sigma(\xi(s)) = u_\Gamma(s,x(s))$$

yields (2.88), while the projection onto $T^\perp\Sigma$ trivializes, i.e., we automatically have

$$\partial_s\rho(s,\xi(s)) + (\nabla_\Sigma\rho(s,\xi(s))|\xi'(s)) = (u_\Gamma(s,x(s))|\nu_\Sigma(\xi(s))).$$

The last assertion follows from

$$\begin{aligned} &\beta(\rho(s))(u_\Gamma(s,x(s))|\nu_\Sigma(\xi(s))) \\ &\quad= (u_\Gamma(s,x(s))|\nu_\Gamma(s,x(s))) + \beta(\rho(s))(u_\Gamma(s,x(s))|M_0(\rho(s))\nabla_\Sigma\rho(s,\xi(s))) \\ &\quad= \beta(\rho(s))[\partial_s\rho(s,\xi(s)) + (M_0(\rho(s))\mathcal{P}_\Sigma(\xi(s))u_\Gamma(s,x(s))|\nabla_\Sigma\rho(s,\xi(s)))) \\ &\quad= \beta(\rho(s))[\partial_s\rho(s,\xi(s)) + (\xi'(s)|\nabla_\Sigma\rho(s,\xi(s))), \end{aligned}$$

where we employed (2.77), (2.80) and (2.88). It remains to observe that the ordinary differential equation (2.88), defined on Σ, admits a unique solution as $(I - \rho L_\Sigma)^{-1}\mathcal{P}_\Sigma u_\Gamma \in T\Sigma$.

Suppose that u_Γ is a C^1-velocity field for $\{\Gamma(t) : t \in I\}$ and $f_\Gamma \in C^1(\mathcal{M}, \mathbb{R})$. Then we define the *Lagrange derivative* of f_Γ (sometimes also called the *material derivative* of f_Γ) with respect to the velocity field u_Γ at the point $(t,p) \in \mathcal{M}$ by

$$\frac{D}{Dt} f_\Gamma(t,p) := \frac{D_{u_\Gamma}}{Dt} f_\Gamma(t,p) := \frac{d}{ds}\, f_\Gamma(t+s, x(t+s,t,p))\Big|_{s=0}$$

where $[s \mapsto x(s+t,t,p)]$ denotes the solution of (2.85). In case u_Γ is a normal C^1-velocity field, in which case $u_\Gamma = V_\Gamma\nu_\Gamma$, the Lagrange derivative is called the *normal derivative*, and we set

$$\frac{D_n}{Dt} := \frac{D_{V_\Gamma\nu_\Gamma}}{Dt}.$$

Then the following relation holds.

$$\frac{D_{u_\Gamma}}{Dt} f_\Gamma(t,p) = \frac{D_n}{Dt} f_\Gamma(t,p) + (u_\Gamma(t,p)|\nabla_{\Gamma(t)})f_\Gamma(t,p). \tag{2.89}$$

In order to see this, let us consider an extension $\tilde{f}_\Gamma$ of f_Γ in an open neighbourhood of $\mathcal{M}$ in $\mathbb{R}^{n+1}$. Such an extension can, for instance, be obtained on the neighbourhood

$$\mathcal{U}_a(\mathcal{M}) := \bigcup_{t\in I} (\{t\} \times U_a(\Gamma(t))),$$

where $U_a(\Gamma(t))$ is a tubular neighbourhood of $\Gamma(t)$ of with a, by setting

$$\tilde{f}_\Gamma(t,x) := f_\Gamma(t,p), \quad (t,x) \in \mathcal{U}_a(\mathcal{M}), \quad p = \Pi_{\Gamma(t)}(x).$$

Then one obtains

$$\frac{D_{u_\Gamma}}{Dt} f_\Gamma(t,p) = \partial_t \tilde{f}_\Gamma(t,p) + (u_\Gamma(t,p)|\nabla_x)\tilde{f}_\Gamma(t,p). \tag{2.90}$$

By the same argument one has

$$\begin{aligned}\frac{D_n}{Dt} f_\Gamma(t,p) &= \frac{d}{ds} f_\Gamma(t+s, y(t+s,t,p))\Big|_{s=0} = \frac{d}{ds}\tilde{f}_\Gamma(t+s, y(t+s,t,p))\Big|_{s=0} \\ &= \partial_t \tilde{f}_\Gamma(t,p) + V_\Gamma(t,p)(\nu_\Gamma(t,p)|\nabla_x)\tilde{f}_\Gamma(t,p),\end{aligned}$$

where $y(\cdot)$ is the solution of (2.85) with respect to the normal velocity field $V_\Gamma\nu_\Gamma$. Using the relation

$$\nabla_x \tilde{f}_\Gamma = (\nabla_x \tilde{f}_\Gamma|\nu_\Gamma)\nu_\Gamma + \mathcal{P}_\Gamma \nabla_x \tilde{f}_\Gamma = (\nabla_x \tilde{f}_\Gamma|\nu_\Gamma)\nu_\Gamma + \nabla_\Gamma f_\Gamma,$$

see (2.20), we conclude with (2.82)

$$\begin{aligned}\frac{D_{u_\Gamma}}{Dt} f_\Gamma(t,p) &= \partial_t \tilde{f}_\Gamma(t,p) + V_\Gamma(t,p)(\nu_\Gamma(t,p)|\nabla_x)\tilde{f}_\Gamma(t,p) + (u_\Gamma(t,p)|\nabla_{\Gamma(t)})f_\Gamma(t,p) \\ &= \frac{D_n}{Dt} f_\Gamma(t,p) + (u_\Gamma(t,p)|\nabla_{\Gamma(t)})f_\Gamma(t,p).\end{aligned}$$

5.4 The Transport Theorem for Moving Hypersurfaces
Suppose u_Γ is a C^1-velocity field for $\{\Gamma(t) : t \in I\}$ and $f_\Gamma \in C^1(\mathcal{M}, \mathbb{R}^n)$. The *transport theorem for moving surfaces* states that

$$\begin{aligned}\frac{d}{dt}\int_{\Gamma(t)} f_\Gamma(t,x)\,d\Gamma &= \int_{\Gamma(t)} \Big[\frac{D_{u_\Gamma}}{Dt} f_\Gamma(t,x) + f_\Gamma(t,x)\,\mathrm{div}_\Gamma u_\Gamma(t,x)\Big]\,d\Gamma \\ &= \int_{\Gamma(t)} \Big[\frac{D_n}{Dt} f_\Gamma(t,x) - f_\Gamma(t,x)\kappa_\Gamma(t,x)V_\Gamma(t,x)\Big]\,d\Gamma.\end{aligned} \tag{2.91}$$

Proof. Let $(t,p) \in \mathcal{M}$ be fixed let $\phi(t,\cdot) : \Theta \subset \mathbb{R}^{n-1} \to \Gamma(t)$ be a sufficiently smooth parameterization of an open neighbourhood of p in $\Gamma(t)$. Then

$$\phi(t+s,\cdot) := x(t+s,t,\phi(t,\cdot)) : \Theta \to \Gamma(t+s), \quad s \in (-\delta,\delta),$$

defines a C^1-parameterization of a neighbourhood of $x(t+s,t,p)$ in $\Gamma(t+s)$. We first suppose that $\operatorname{supp} f_\Gamma \subset\subset U := \{\phi(t+s,\theta) : (s,\theta) \in (-\delta,\delta)\times\Theta\}$. Let

$$g_{ij}(t+s,\theta) := (\partial_i\phi(t+s,\theta) \,|\, \partial_j\phi(t+s,\theta)), \quad G(t+s,\theta) := [g_{ij}(t+s,\theta)].$$

Hence, $G(t+s,\theta)$ is the fundamental matrix of $\Gamma(t+s)$ with respect to the parameterization $\phi(t+s,\cdot)$. With $g(t+s,\cdot) := \det G(t+s,\cdot)$ we obtain

$$\int_{\Gamma(t+s)} f_\Gamma(t+s,y)\,d\Gamma = \int_\Theta f_\Gamma(t+s,\phi(t+s,\theta))\sqrt{g(t+s,\theta)}\,d\theta,$$

and hence

$$\begin{aligned}&\frac{d}{ds}\int_{\Gamma(t+s)} f_\Gamma(t+s,y)\,d\Gamma\,\Big|_{s=0}\\ &= \int_\Theta \Big(\frac{D}{Dt}f_\Gamma(t,\phi(t,\theta))\Big)\sqrt{g(t,\theta)} + f_\Gamma(t,\phi(t,\theta))\frac{\partial}{\partial s}\sqrt{g(t+s,\theta)}\,\Big|_{s=0}\,d\theta.\end{aligned}$$

As in (2.27) we obtain

$$\begin{aligned}\frac{\partial}{\partial s}\sqrt{g(t+s,\theta)} &= \frac{1}{2\sqrt{g(t+s,\theta)}}\,\frac{\partial}{\partial s}g(t+s,\theta)\\ &= \frac{1}{2}\sqrt{g(t+s,\theta)}g^{ij}(t+s,\theta)\frac{\partial}{\partial s}g_{ij}(t+s,\theta).\end{aligned}$$

From

$$\partial_s\partial_i x(t+s,t,\phi(t,\theta)) = \partial_i\partial_s x(t+s,t,\phi(t,\theta)) = \partial_i u_\Gamma(t+s,x(t+s,t,\phi(t,\theta)))$$

follows

$$\begin{aligned}&\frac{1}{2}g^{ij}(t+s,\theta)\frac{\partial}{\partial s}g_{ij}(t+s,\theta)\Big|_{s=0}\\ &= \frac{1}{2}g^{ij}(t,\theta)\left[(\partial_i u_\Gamma(t,\phi(t,\theta)) \,|\, \partial_j\phi(t,\theta)) + (\partial_i\phi(t,\theta) \,|\, \partial_j u_\Gamma(t,\phi(t,\theta)))\right]\\ &= \frac{1}{2}g^{ij}(t,\theta)\Big[\big(\partial_i u_\Gamma(t,\phi(t,\theta)) \,|\, \tau_j^{\Gamma(t)}(\phi(t,\theta))\big) + \big(\tau_i^{\Gamma(t)}(\phi(t,\theta)) \,|\, \partial_j u_\Gamma(t,\phi(t,\theta))\big)\Big]\\ &= \frac{1}{2}\Big[\big(\partial_i u_\Gamma(t,\phi(t,\theta)) \,|\, \tau^i_{\Gamma(t)}(\phi(t,\theta))\big) + \big(\tau^j_{\Gamma(t)}(\phi(t,\theta)) \,|\, \partial_j u_\Gamma(t,\phi(t,\theta))\big)\Big]\\ &= \operatorname{div}_{\Gamma(t)} u_\Gamma(t,\phi(t,\theta)).\end{aligned}$$

Combining all steps yields

$$\begin{aligned}&\frac{d}{ds}\int_{\Gamma(t+s)} f_\Gamma(t+s,y)\,d\Gamma\Big|_{s=0}\\ &= \int_\Theta \left[\frac{D}{Dt}f_\Gamma(t,\phi(t,\theta)) + f_\Gamma(t,\phi(t,\theta))\operatorname{div}_{\Gamma(t)}u_\Gamma(t,\phi(t,\theta))\right]\sqrt{g(t,\theta)}\,d\theta\\ &= \int_{\Gamma(t)} \left[\frac{D}{Dt}f_\Gamma(t,y) + f_\Gamma(t,y)\operatorname{div}_{\Gamma(t)}u_\Gamma(t,y)\right]d\Gamma.\end{aligned}$$

For a more general function f_Γ we can apply the result above in conjunction with a partition of unity for $\mathcal{M}$. Hence, we have shown that

$$\frac{d}{dt}\int_{\Gamma(t)} f_\Gamma(t,x)\,d\Gamma = \int_{\Gamma(t)} \Big[\frac{D_{u_\Gamma}}{Dt} f_\Gamma(t,x) + f_\Gamma(t,x)\,\mathrm{div}_\Gamma u_\Gamma(t,x)\Big]\,d\Gamma.$$

The second assertion in (2.91) follows from the surface divergence theorem (2.31) and (2.89). □

We note that (2.91) implies, in particular, the well-known *change of area formula*

$$\frac{d}{dt}|\Gamma(t)| = -\int_{\Gamma(t)} \kappa_\Gamma V_\Gamma\,d\Gamma. \tag{2.92}$$

It is worthwhile to point out that (2.92) can also be derived from (2.50). This can be obtained as follows. Using the representation (2.78) we have $|\Gamma(t+s)| = \Phi(\rho(s))$ for t fixed, and the change of area formula (2.50) in conjunction with the relation $\rho(0) = 0$ immediately yields

$$\frac{d}{ds}|\Gamma(t+s)|\Big|_{s=0} = \langle \Phi'(0), \partial_s\rho(0)\rangle = -\int_\Sigma \kappa_\Sigma \partial_s\rho(0)\,d\Sigma = -\int_{\Gamma(t)} \kappa_\Gamma V_\Gamma\,d\Gamma.$$

5.5 The Transport Theorem for Moving Domains

Suppose $\{\Gamma(t) : t\in I\}$ is a family of compact connected closed C^2-hypersurfaces in $\mathbb{R}^n$, bounding domains $\Omega(t)\subset\mathbb{R}^n$. We assume again that

$$\mathcal{M} = \bigcup_{t\in I}(\{t\}\times\Gamma(t))$$

is a $C^{1,2}$-hypersurface in $\mathbb{R}^{n+1}$, and we set

$$\mathcal{Q} = \bigcup_{t\in I}(\{t\}\times\Omega(t)).$$

Let $f\in C^1(\overline{\mathcal{Q}})$. Then we have the *transport theorem for moving domains*:

$$\frac{d}{dt}\int_{\Omega(t)} f(t,x)\,dx = \int_{\Omega(t)} \partial_t f(t,x)\,dx + \int_{\Gamma(t)} f(t,x)V_\Gamma(t,x)\,d\Gamma. \tag{2.93}$$

Proof. We first show that for each fixed $t\in I$ there exists a family of mappings

$$\Phi(t+s,\cdot) : \overline{\Omega}(t)\to\overline{\Omega}(t+s), \quad s\in(-\delta,\delta),$$

such that

$$\Phi(t+s,\cdot)\in \mathrm{Diff}^1(\Omega(t),\Omega(t+s))\cap \mathrm{Diff}^1(\Gamma(t),\Gamma(t+s)), \quad s\in(-\delta,\delta), \tag{2.94}$$

where $\mathrm{Diff}^1(U,V)$ denotes the set of all C^1-diffeomorphisms from U into V. The mappings $\Phi(t+s,\cdot)$ can, for instance, be constructed as follows. According to (2.78) we know that

$$\phi(t+s,p) := p + \rho(s,p)\nu_\Sigma(p), \quad p \in \Sigma := \Gamma(t), \ \ s \in (-\delta,\delta), \tag{2.95}$$

satisfies $\phi(t+s,\cdot) \in \mathrm{Diff}^1(\Gamma(t),\Gamma(t+s))$. By means of a Hanzawa transform we can extend $\phi(t+s,\cdot)$ to $\overline{\Omega}(t)$ such that (2.94) holds. In more detail, let

$$\Phi(t+s,x) = x + \chi(d_\Sigma(x)/a)\rho(s,\Pi_\Sigma(x))\nu_\Sigma(\Pi_\Sigma(x)), \quad x \in \overline{\Omega}(t).$$

Here d_Σ and Π_Σ have the same meaning as in (2.57), and χ is a suitable cut-off function, say $\chi \in \mathcal{D}(\mathbb{R})$, $0 \le \chi \le 1$, $\chi(r) = 1$ for $|r| < 1/3$, and $\chi(r) = 0$ for $|r| > 2/3$.

Clearly, $\Phi(t+s,p) = \phi(t+s,p)$ for $p \in \Gamma(t)$. Since $\Phi(t,\cdot) = \mathrm{id}_{\overline{\Omega}(t)}$ we can assume that $\det \partial_x\Phi(t+s,x) > 0$ for $(s,x) \in (-\delta,\delta) \times \Omega(t)$ by choosing δ small enough. Next we observe that by (2.27)

$$\frac{d}{ds}\det \partial_x\Phi(t+s,x) = \det \partial_x\Phi(t+s,x)\mathrm{tr}([\partial_y\Phi(t+s,x)]^{-1}[\partial_x\partial_s\Phi(t+s,x)]),$$

and therefore,

$$\frac{d}{ds}\det \partial_x\Phi(t+s,x)\Big|_{s=0} = \mathrm{tr}[\partial_x\partial_s\Phi(t,x)] = \mathrm{div}_x\, \partial_s\Phi(t,x).$$

Employing the transformation rule for integrals yields

$$\int_{\Omega(t+s)} f(t+s,y)\,dy = \int_{\Omega(t)} f(t+s,\Phi(t+s,x))\det \partial_x\Phi(t+s,x)\,dx,$$

and hence,

$$\begin{aligned}
&\frac{d}{ds}\int_{\Omega(t+s)} f(t+s,y)\,dy\,\Big|_{s=0}\\
&= \int_{\Omega(t)} \Big[\partial_t f(t,x) + (\nabla_x f(t,x)|\partial_s\Phi(t,x)) + f(t,x)\mathrm{div}_x\,\partial_s\Phi(t,x)\Big]\,dx\\
&= \int_{\Omega(t)} \Big[\partial_t f(t,x) + \mathrm{div}_x\,(f(t,x)\partial_s\Phi(t,x))\Big]\,dx\\
&= \int_{\Omega(t)} \partial_t f(t,x)\,dx + \int_{\Gamma(t)} f(t,x)(\partial_s\Phi(t,x)|\nu_\Gamma(t,x))d\Gamma\\
&= \int_{\Omega(t)} \partial_t f(t,x)\,dx + \int_{\Gamma(t)} f(t,x)V_\Gamma(t,x)d\Gamma,
\end{aligned} \tag{2.96}$$

where we used (2.79) in the last step. This completes the proof. □

The relation (2.93) immediately yields the well-known *change of volume formula*

$$\frac{d}{dt}|\Omega(t)| = \int_{\Gamma(t)} V_\Gamma \, d\Gamma. \tag{2.97}$$

We point out that (2.97) can also be derived from (2.54). Indeed, using once more the representation

$$\Gamma(t+s) = \{p + \rho(s,p)\nu_\Sigma(p) : s \in (-\delta,\delta),\ p \in \Sigma\}, \quad \Sigma := \Gamma(t),$$

we have $|\Omega(t+s)| = \Psi(\rho(s))$, with Ψ the volume functional introduced in Section 2.2.7. Then the first variation formula (2.54) and the relation $\rho(0) = 0$ imply

$$\frac{d}{ds}|\Omega(t+s)|\Big|_{s=0} = \langle \Psi'(0), \partial_s\rho(0)\rangle = \int_\Sigma \partial_s\rho(0)\, d\Sigma = \int_{\Gamma(t)} V_\Gamma\, d\Gamma.$$

We now consider the more special case where the moving domain $\Omega(t)$ is transported by a velocity field u. Suppose then that $J \subset \mathbb{R}$ is an open interval, $G \subset \mathbb{R}^n$ is an open set, and $u \in C^1(J \times \Omega, \mathbb{R}^n)$. We assume that solutions to the ordinary differential equation

$$y'(t) = u(t, y(t)), \quad y(\tau) = \xi,$$

exist on I for all $(\tau, \xi) \in J \times G$, and we denote the unique solution with initial value $y(\tau) = \xi$ by $y(t,\tau,\xi)$. Let $\Omega_0 \subset\subset G$ be a C^2-domain, $t_0 \in I$ a fixed number, and suppose that the family $\{\Omega(t) : t \in I\}$ of moving domains is given by

$$\Omega(t) = y(t, t_0, \cdot)|_{\Omega_0} = \{y(t, t_0, x_0) : x_0 \in \Omega_0\}, \quad t \in I.$$

Suppose that $f \in C^1(J \times G)$. Then the *Reynolds transport theorem* states that

$$\frac{d}{dt}\int_{\Omega(t)} f(t,x)\, dx = \int_{\Omega(t)} \left[\partial_t f(t,x) + \operatorname{div}_x\left(f(t,x)u(t,x)\right)\right] dx. \tag{2.98}$$

Proof. Let $t \in I$ be fixed and let $\Phi(s,x) := y(s,t,x)$ for $(s,x) \in J \times G$. From the theory of ordinary differential equations follows that

$$\Phi(t+s,\cdot) \in \operatorname{Diff}^1\big(\Omega(t), \Omega(t+s)\big), \quad s \in (-\delta,\delta), \tag{2.99}$$

with $\phi^{-1}(t+s,\cdot) = y(t,t+s,\cdot)$. We can now follow the computations in (2.93) to the result

$$\begin{aligned}\frac{d}{dt}\int_{\Omega(t)} f(x,t)\, dx &= \int_{\Omega(t)} \Big[\partial_t f(t,x) + \operatorname{div}_x\left(f(t,x)\partial_s\Phi(t,x)\right)\Big] dx \\ &= \int_{\Omega(t)} \Big[\partial_t f(t,x) + \operatorname{div}_x\left(f(t,x)u(t,x)\right)\Big] dx\end{aligned}$$

and this completes the proof. □

5.6 The Transport Theorem for Two-Phase Moving Domains
Let $\Omega \subset \mathbb{R}^n$ be a bounded open domain in $\mathbb{R}^n$ with C^2-boundary $\partial\Omega$. Suppose that $\{\Gamma(t) : t \in I\}$ is a family of closed compact C^2-hypersurfaces with $\Gamma(t) \subset \Omega$, such that $\Gamma(t)$ encloses a region $\Omega_1(t) \subset \Omega$, and such that $\partial\Omega_1(t) = \Gamma(t)$ for each $t \in I$. Let $\Omega_2(t) := \Omega \setminus \bar{\Omega}_1(t)$. Then

$$\overline{\Omega} = \overline{\Omega}_1(t) \cup \overline{\Omega}_2(t), \quad \overline{\Omega}_1(t) \cap \overline{\Omega}_2(t) = \Gamma(t), \quad \partial\Omega_2(t) = \Gamma(t) \cup \partial\Omega, \quad t \in I.$$

Hence, $\Gamma(t)$ separates Ω into an 'inner' region $\Omega_1(t)$ and an 'outer' region $\Omega_2(t)$, with $\Omega_2(t)$ being in contact with the boundary $\partial\Omega$. Then $\nu_{\Gamma(t)}$ denotes the outward pointing unit normal field for $\Omega_1(t)$ on $\Gamma(t)$. Let

$$\mathcal{Q}_j = \bigcup_{t\in I} \big(\{t\} \times \Omega_j(t)\big), \quad j = 1, 2.$$

As above, we assume that $\mathcal{M}$ is a $C^{1,2}$-hypersurface. Let $f_j : \mathcal{Q}_j \to \mathbb{R}$ be given. Then we set

$$f(t,x) := \begin{cases} f_1(t,x), & x \in \Omega_1(t), \\ f_2(t,x), & x \in \Omega_2(t), \end{cases}$$

so that $f : \mathcal{Q}_1 \cup \mathcal{Q}_2 \to \mathbb{R}$. In case f_j admits a continuous extension $\bar{f}_j \in C(\overline{\mathcal{Q}}_j)$ we define the jump of f across $\Gamma(t)$ by means of

$$[\![f(t,p)]\!] := \bar{f}_2(t,p) - \bar{f}_1(t,p), \quad p \in \Gamma(t). \tag{2.100}$$

Suppose that the functions f_j admit extensions $\bar{f}_j \in C^1(\overline{\mathcal{Q}}_j)$, $j = 1, 2$. Then the *transport theorem for two-phase moving domains* states that

$$\frac{d}{dt}\int_{\Omega\setminus\Gamma(t)} f(t,x)\,dx = \int_{\Omega\setminus\Gamma(t)} \partial_t f(t,x)\,dx - \int_{\Gamma(t)} [\![f(t,x)]\!] V_\Gamma(t,x)\,d\Gamma. \tag{2.101}$$

Proof. Let $t \in I$ be fixed. As in the proof of (2.93) we extend the family of diffeormorphisms $\phi(t+s,\cdot) : \Gamma(t) \to \Gamma(t+s)$ given in (2.95) by means of

$$\Phi(t+s,x) = x + \chi(d_\Sigma(x)/a)\rho(s,\Pi_\Sigma(x))\nu_\Sigma(\Pi_\Sigma(x)), \quad x \in \overline{\Omega},$$

to a family of diffeomorphisms $\Phi(t+s,\cdot) : \Omega \to \Omega$ such that

$$\Phi_j(t+s,\cdot) := \Phi(t+s,\cdot)|_{\Omega_j(t+s)} \in \mathrm{Diff}^1\big(\Omega_j(t), \Omega_j(t+s)\big), \quad j = 1, 2, \ \ s \in (-\delta,\delta).$$

By choosing a small enough we can assume that a tubular neighbourhood of $\Gamma(t)$ of width a is contained in Ω, and hence that $\Phi(t+s,\cdot) = \mathrm{id}_{\mathbb{R}^n}$ in a neighbourhood of $\partial\Omega$. We can now proceed as in the proof of (2.93) to obtain

$$\frac{d}{dt}\int_{\Omega_j(t)} f(t,x)\,dx = \int_{\Omega_j(t)} \partial_t f(t,x)\,dx - \int_{\Gamma(t)} (-1)^j \bar{f}_j(t,x) V_\Gamma(t,x)\,d\Gamma,$$

and (2.101) then follows from (2.100). $\square$

Part II

Abstract Theory

Chapter 3

Operator Theory and Semigroups

In this chapter we introduce some basic tools from operator and semigroup theory. The class of sectorial operators is studied in detail, its functional calculus is introduced, leading to analytic semigroups and complex powers. The classes $\mathcal{BIP}(X)$ and $\mathcal{H}^\infty(X)$ are defined and elementary properties are shown. Via trace theory for abstract Cauchy problems the connections to real interpolation are derived, and the relation of complex interpolation to powers of operators is shown. The chapter concludes with a first study of maximal L_p-regularity.

3.1 Sectorial Operators

The concept of sectorial operators introduced in Definition 3.1.1 below is basic in this book. Most closed linear operators appearing in applications have this property, at least after translation and rotation. We will meet many examples of such operators in later sections.

1.1 Sectorial Operators
We begin with the definition of sectorial operators.

Definition 3.1.1. *Let X be a complex Banach space, and A a closed linear operator in X. A is called* **sectorial** *if the following two conditions are satisfied.*

(S1) $\overline{\mathsf{D}(A)} = X$, $\overline{\mathsf{R}(A)} = X$, $(-\infty, 0) \subset \rho(A)$;

(S2) $|t(t+A)^{-1}| \leq M$ *for all* $t > 0$, *and some* $M < \infty$.

The class of sectorial operators in X will be denoted by $\mathcal{S}(X)$. If $(-\infty,0) \subset \rho(A)$ and only **(S2)** *holds then A is said to be* **pseudo-sectorial**. *The class of pseudo-sectorial operators will be denoted by $\mathcal{PS}(X)$.*

J. Prüss and G. Simonett, *Moving Interfaces and Quasilinear Parabolic Evolution Equations*, Monographs in Mathematics 105,
DOI 10.1007/978-3-319-27698-4_3

Suppose that A is a linear operator in X which is pseudo-sectorial. Then the operator family $\{A(t+A)^{-1}\}_{t>0} \in \mathcal{B}(X)$ is uniformly bounded as well. For $x \in \mathsf{D}(A)$ we have

$$t(t+A)^{-1}x - x = -A(t+A)^{-1}x = -(t+A)^{-1}Ax \to_{t\to\infty} 0,$$

hence $\lim_{t\to\infty} t(t+A)^{-1}x = x$ for all $x \in \overline{\mathsf{D}(A)}$, by **(S2)**. In particular, if $\mathsf{D}(A)$ is dense in X then

$$\lim_{t\to\infty} t(t+A)^{-1}x = x \quad \text{for all } x \in X.$$

Similarly, for $y = Ax \in \mathsf{R}(A)$ we have

$$A(t+A)^{-1}Ax - Ax = -t(t+A)^{-1}Ax = -tA(t+A)^{-1}x \to_{t\to 0} 0,$$

hence $\lim_{t\to 0} A(t+A)^{-1}y = y$ for all $y \in \overline{\mathsf{R}(A)}$, employing once more **(S2)**. In particular, if $\mathsf{R}(A)$ is dense in X then

$$\lim_{t\to 0} A(t+A)^{-1}x = x \quad \text{for all } x \in X.$$

On the other hand, if $x \in \mathsf{N}(A)$ then $A(t+A)^{-1}x = 0$, and this shows that we always have $\mathsf{N}(A) \cap \overline{\mathsf{R}(A)} = \{0\}$.

If $\mathsf{D}(A)$ is dense in X, then its dual A^* is well-defined. The relation

$$\mathsf{N}(A^*) = \mathsf{R}(A)^\perp$$

then shows that $A \in \mathcal{S}(X)$ iff $A \in \mathcal{PS}(X)$ and $\mathsf{N}(A^*) = 0$.

Next, let X be reflexive and A be pseudo-sectorial. Then any sequence $(\lambda_n) \subset \rho(A)$, $\lambda_n \to \infty$ contains a subsequence, which may depend on x, such that $\lambda_n(\lambda_n + A)^{-1}x \rightharpoonup y \in X$. This implies $\lambda_n(\lambda_n + A)^{-1}(\lambda + A)^{-1}x \rightharpoonup (\lambda + A)^{-1}y$, for any $\lambda > 0$. But by means of the resolvent equation

$$\lambda_n(\lambda_n + A)^{-1}(\lambda + A)^{-1}x = \frac{\lambda_n}{\lambda_n - \lambda}[(\lambda + A)^{-1} - (\lambda_n + A)^{-1}]x \rightharpoonup (\lambda + A)^{-1}x,$$

hence $(\lambda + A)^{-1}x = (\lambda + A)^{-1}y$, by uniqueness of weak limits. This implies $x = y$, hence $\lambda(\lambda + A)^{-1}x \rightharpoonup x$ as $\lambda \to \infty$. As a consequence of this we see that $\mathsf{D}(A)$ is weakly dense in X, hence also strongly dense, and then by what has been proved before $\lambda(\lambda + A)^{-1}x \to x$ as $\lambda \to \infty$, for each $x \in X$.

At $\lambda = 0$ we proceed similarly. Fix $x \in X$ and choose a sequence $(\lambda_n) \subset \rho(A)$, $\lambda_n \to 0$ such that $A(\lambda_n + A)^{-1}x \rightharpoonup y \in X$. Then $\lambda A(\lambda_n + A)^{-1}(\lambda + A)^{-1}x \rightharpoonup \lambda(\lambda + A)^{-1}y \in X$, hence the resolvent equation yields

$$y - \lambda(\lambda + A)^{-1}y = A(\lambda + A)^{-1}x = x - \lambda(\lambda + A)^{-1}x,$$

for any $\lambda > 0$. This identity shows $x - y \in \mathsf{N}(A)$, in particular $A(\lambda + A)^{-1}x = A(\lambda + A)^{-1}y$, hence $A(\lambda_n + A)^{-1}y \rightharpoonup y$ as well. Writing

$$x = (x - y) + A(\lambda_n + A)^{-1}x + \lambda_n(\lambda_n + A)^{-1}y$$

and observing $\lambda_n(\lambda_n + A)^{-1}y \rightharpoonup 0$ the latter implies that $\mathsf{N}(A) + \mathsf{R}(A)$ is weakly dense in X, hence also strongly dense. But from what we already proved above this implies $A(\lambda + A)^{-1}x \to Px \in X$ as $\lambda \to 0$, for each $x \in X$. Here $P \in \mathcal{B}(X)$, by the Banach-Steinhaus theorem, and $\mathsf{R}(P) \subset \overline{\mathsf{R}(A)}$, as well as $\mathsf{R}(I - P) \subset \mathsf{N}(A)$. Finally, $A(\lambda + A)^{-1}x = A(\lambda + A)^{-1}Px$ for all $x \in X$ implies $P^2 = P$, i.e., P is the projection onto $\overline{\mathsf{R}(A)}$ along $\mathsf{N}(A)$. We have proved in particular the direct sum decomposition $X = \mathsf{N}(A) \oplus \overline{\mathsf{R}(A)}$. Thus in a reflexive space, $\mathsf{R}(A)$ is dense in X if and only if $\mathsf{N}(A) = \{0\}$.

Let us summarize what we have shown above in

Theorem 3.1.2. *Let X be a Banach space and A a pseudo-sectorial operator in X. Then*

(i) $\mathsf{N}(A) \cap \overline{\mathsf{R}(A)} = \{0\}$, *and*

$$\lim_{t\to\infty} t(t + A)^{-1}x = x \quad \textit{for each } x \in \overline{\mathsf{D}(A)},$$
$$\lim_{t\to 0+} A(t + A)^{-1}x = x \quad \textit{for each } x \in \overline{\mathsf{R}(A)}. \tag{3.1}$$

(ii) *If $\mathsf{D}(A)$ is dense in X, then $A \in \mathcal{S}(X)$ if and only if $\mathsf{N}(A^*) = 0$.*

(iii) *If X is reflexive then $\lim_{t\to\infty} t(t + A)^{-1}x = x$ and $\lim_{t\to 0+} A(t + A)^{-1}x = Px$ for each $x \in X$, where P is the projection onto $\overline{\mathsf{R(A)}}$ along $\mathsf{N(A)}$, and $X = \mathsf{N}(A) \oplus \overline{\mathsf{R}(A)}$. Thus, if X is reflexive then any pseudo-sectorial operator A with $\mathsf{N}(A) = \{0\}$ is sectorial.*

(iv) *If X is a general Banach space and A is sectorial, then $\mathsf{D}(A^k) \cap \mathsf{R}(A^k)$ is dense in X, for each $k \in \mathbb{N}$.*

Concerning the last assertion of Theorem 3.1.2, note that $(1 + n^{-1}A)^{-k}A^k(n^{-1} + A)^{-k}$ converges strongly to I as $n \to \infty$ and has range in $\mathsf{D}(A^k) \cap \mathsf{R}(A^k)$.

Let $\Sigma_\theta \subset \mathbb{C}$ denote the open sector with vertex 0, opening angle 2θ, which is symmetric w.r.t. the positive half-axis $\mathbb{R}_+$, i.e.,

$$\Sigma_\theta = \{\lambda \in \mathbb{C} \setminus \{0\} : |\arg \lambda| < \theta\}.$$

If $A \in \mathcal{PS}(X)$ then $\rho(-A) \supset \Sigma_\theta$, for some $\theta > 0$, and

$$\sup\{|\lambda(\lambda + A)^{-1}| : |\arg \lambda| < \theta\} < \infty.$$

In fact, with $(d/dt)^n(t + A)^{-1} = (-1)^n n!(t + A)^{-(n+1)}$, for $t > 0$ the Taylor expansion

$$(\lambda + A)^{-1} = \sum_{n=0}^{\infty}(-1)^n(\lambda - t)^n(t + A)^{-(n+1)}$$

and (**S2**) yield the estimate

$$|(\lambda + A)^{-1}| \le \sum_{n=0}^{\infty} |\lambda - t|^n |(t+A)^{-(n+1)}| \le (M/t) \sum_{n=0}^{\infty} (M|\lambda - t|/t)^n.$$

This bound is finite provided $|\lambda/t - 1| < 1/M$, which by minimization over $t > 0$ yields $|\sin\phi| < 1/M$, where $\lambda = re^{i\phi}$.

Therefore it makes sense to define the *spectral angle* ϕ_A of $A \in \mathcal{PS}(X)$ by

$$\phi_A = \inf\{\phi : \rho(-A) \supset \Sigma_{\pi-\phi},\ \sup_{\lambda\in\Sigma_{\pi-\phi}} |\lambda(\lambda + A)^{-1}| < \infty\}. \tag{3.2}$$

Evidently, we have $\phi_A \in [0, \pi)$ and

$$\phi_A \ge \sup\{|\arg\lambda| : \lambda \in \sigma(A)\}. \tag{3.3}$$

If $A \in \mathcal{PS}(X)$ is bounded and $0 \in \rho(A)$ then there is equality in (3.3). In fact, by holomorphy of $(\lambda - A)^{-1}$ on $\rho(A)$, $\lambda(\lambda - A)^{-1}$ is bounded in $\mathcal{B}(X)$ on each compact subset of $\rho(A)$, and for all $|\lambda| > |A|$ we have

$$|\lambda(\lambda - A)^{-1}| \le \frac{|\lambda|}{|\lambda| - |A|},$$

which is uniformly bounded, say for $|\lambda| \ge 2|A|$. But this implies uniform boundedness of $\lambda(\lambda + A)^{-1}$ on each sector $\Sigma_{\pi-\phi}$ with $\phi > \sup\{|\arg(\lambda)| :\ \lambda \in \sigma(A)\}$.

For $\phi \in (\phi_A, \pi)$ we frequently employ the notations

$$M_{\pi-\phi}(A) = \sup_{\lambda\in\Sigma_{\pi-\phi}} |\lambda(\lambda + A)^{-1}|, \quad C_{\pi-\phi}(A) = \sup_{\lambda\in\Sigma_{\pi-\phi}} |A(\lambda + A)^{-1}|. \tag{3.4}$$

It is not difficult to see that $C_{\pi-\phi}(A) \ge 1$ as well as $M_{\pi-\phi}(A) \ge 1$, for all $\phi \in (\phi_A, \pi]$. Observe the limiting case $\phi = \pi$:

$$M_0(A) = \sup_{r>0} |r(r + A)^{-1}|, \quad C_0(A) = \sup_{r>0} |A(r + A)^{-1}|. \tag{3.5}$$

1.2 Permanence Properties

The class of sectorial operators has a number of nice permanence properties which are summarized in the following

Proposition 3.1.3. *Let X be a complex Banach space. The class $\mathcal{S}(X)$ of sectorial operators has the following permanence properties.*

(i) *$A \in \mathcal{S}(X)$ iff $\mathsf{N}(A) = \{0\}$ and $A^{-1} \in \mathcal{S}(X)$; then $\phi_{A^{-1}} = \phi_A$;*

(ii) *$A \in \mathcal{S}(X)$ implies $rA \in \mathcal{S}(X)$ and $\phi_{rA} = \phi_A$ for all $r > 0$;*

(iii) *$A \in \mathcal{S}(X)$ implies $e^{\pm i\psi}A \in \mathcal{S}(X)$ for all $\psi \in [0, \pi - \phi_A)$, and $\phi_{e^{\pm i\psi}A} = \phi_A + \psi$;*

(iv) *$A \in \mathcal{S}(X)$ implies $(\mu + A) \in \mathcal{S}(X)$ for all $\mu \in \Sigma_{\pi-\phi_A}$, and $\phi_{\mu+A} \le \max\{\phi_A, |\arg\mu|\}$;*

(v) *if* $\mathsf{D}(A)$ *is dense in* X *and* $\mathsf{D}(A^*)$ *dense in* X^*, *then* $A \in \mathcal{S}(X)$ *iff* $A^* \in \mathcal{S}(X^*)$, *and* $\phi_A = \phi_{A^*}$;

(vi) *if* Y *denotes another Banach space and* $T \in \mathcal{B}(X, Y)$ *is bijective, then* $A \in \mathcal{S}(X)$ *iff* $A_1 = TAT^{-1} \in \mathcal{S}(Y)$, *and* $\phi_A = \phi_{A_1}$.

Proof. Assertion (i) follows from the identity

$$\lambda(\lambda + A^{-1})^{-1} = \lambda A(1 + \lambda A)^{-1} = A(\lambda^{-1} + A)^{-1}.$$

Similarly, (ii) is a consequence of

$$\lambda(\lambda + rA)^{-1} = (\lambda/r)((\lambda/r) + A)^{-1}, \quad r > 0,$$

and (iii) follows from $|(\lambda + e^{i\phi}A)^{-1}| = |(\lambda e^{-i\phi} + A)^{-1}|$. If $\mu \in \Sigma_{\pi-\phi_A}$, $|\arg(\mu)| = \psi$, and $\lambda \in \Sigma_{\pi-\phi}$, then for $(\pi - \phi) + \psi < \pi$ we have

$$|\arg(\lambda + \mu)| \le \max\{|\arg(\lambda)|, |\arg(\mu)|\},$$

as well as

$$|\lambda + \mu| \ge c(|\lambda| + |\mu|), \quad \text{where } c = \cos((\pi - \phi + \psi)/2).$$

Therefore, $\phi > \max\{\phi_A, \psi\}$ implies

$$|(\lambda + \mu + A)^{-1}| \le \frac{M_{\pi-\phi}(A)}{|\lambda + \mu|} \le \frac{M_{\pi-\phi}(A)}{c(|\lambda| + |\mu|)}, \quad \text{for all } \lambda \in \Sigma_{\pi-\phi},$$

and this yields (iv). To prove (v) it is enough to observe that an operator $T \in \mathcal{B}(X)$ is invertible if and only if $T^* \in \mathcal{B}(X^*)$ is invertible, and $|T| = |T^*|$. Finally, to prove (vi) we verify that the relation

$$(\lambda + A_1)^{-1} = T(\lambda + A)^{-1}T^{-1}$$

is satisfied. □

Next we introduce approximations of a sectorial operator which are again sectorial, but in addition bounded and invertible. This will be achieved as follows. For a given pseudo-sectorial operator A and $\varepsilon > 0$ set

$$A_\varepsilon = (\varepsilon + A)(1 + \varepsilon A)^{-1}. \tag{3.6}$$

Then A_ε is bounded, invertible with inverse

$$A_\varepsilon^{-1} = (1 + \varepsilon A)(\varepsilon + A)^{-1} = ((1/\varepsilon) + A)(1 + (1/\varepsilon)A)^{-1} = A_{1/\varepsilon},$$

and, more generally,

$$\begin{aligned}(t + A_\varepsilon)^{-1} &= (t + (\varepsilon + A)(1 + \varepsilon A)^{-1})^{-1} \\ &= (1 + \varepsilon A)(t + \varepsilon + (1 + \varepsilon t)A)^{-1} \\ &= \frac{1}{1 + \varepsilon t}(1 + \varepsilon A)(\frac{t + \varepsilon}{1 + \varepsilon t} + A)^{-1}, \quad t, \varepsilon > 0.\end{aligned}$$

This implies $\rho(A_\varepsilon) \supset (-\infty, 0]$, and as $\varepsilon \to 0$, $(t + A_\varepsilon)^{-1} \to (t + A)^{-1}$ in $\mathcal{B}(X)$ for each $t > 0$, $A_\varepsilon x \to Ax$ for each $x \in \mathsf{D}(A)$, $A_\varepsilon^{-1}x \to A^{-1}x$ for each $x \in \mathsf{R}(A)$. Since

$$|t(t + A_\varepsilon)^{-1}| \le \frac{tM_0(A)}{t+\varepsilon} + \frac{\varepsilon t C_0(A)}{1+\varepsilon t} \le M_0(A) + C_0(A), \quad t, \varepsilon > 0,$$

we have $A_\varepsilon \in \mathcal{S}(X)$ for each $\varepsilon > 0$, and there is a constant M for **(S2)** which is independent of ε. Replacing $t > 0$ by $\lambda \in \Sigma_{\pi-\phi}$ and observing that the functions $\varphi_\varepsilon(\lambda) = (\varepsilon+\lambda)/(1+\varepsilon\lambda)$ are leaving all sectors Σ_ϕ invariant, we obtain the following result.

Proposition 3.1.4. *Suppose $A \in \mathcal{PS}(X)$, and let A_ε be defined according to* (3.6). *Then A_ε is bounded, sectorial, and invertible, for each $\varepsilon > 0$. The spectral angle of A_ε satisfies $\phi_{A_\varepsilon} \le \phi_A$, and the bounds $C_{\pi-\phi}(A_\varepsilon)$ and $M_{\pi-\phi}(A_\varepsilon)$ are uniformly bounded w.r.t. $\varepsilon > 0$, for each fixed $\phi > \phi_A$. Moreover,*

$$\lim_{\varepsilon\to 0}(\lambda + A_\varepsilon)^{-1} = (\lambda + A)^{-1} \quad \textit{in } \mathcal{B}(X) \textit{ for each } \lambda \in \Sigma_{\pi-\phi_A}, \tag{3.7}$$

and in case A is sectorial,

$$\begin{aligned} \lim_{\varepsilon\to 0} A_\varepsilon x &= Ax \quad \textit{for each } x \in \mathsf{D}(A), \\ \lim_{\varepsilon\to 0} A_\varepsilon^{-1} x &= A^{-1}x \quad \textit{for each} x \in \mathsf{R}(A). \end{aligned} \tag{3.8}$$

In later sections we shall frequently make use of the approximations A_ε.

1.3 Perturbation Theory

In this section we consider the behaviour of the class $\mathcal{S}(X)$ w.r.t. perturbations. For this purpose, suppose $A \in \mathcal{S}(X)$, and let B be a closed linear operator in X which is subordinate to A in the sense that $\mathsf{D}(A) \subset \mathsf{D}(B)$ and

$$|Bx| \le b|Ax|, \quad \text{for all } x \in \mathsf{D}(A), \tag{3.9}$$

with some constant $b \ge 0$. If $b < 1$ then $A + B$ defined by

$$(A + B)x = Ax + Bx, \quad x \in \mathsf{D}(A + B) = \mathsf{D}(A), \tag{3.10}$$

is also closed, densely defined, and $\mathsf{N}(A + B) = \{0\}$. In fact, if $(A + B)x = 0$ then $|Ax| = |Bx| \le b|Ax|$, hence $Ax = 0$, which by injectivity of A in turn implies $x = 0$. The operator $K := BA^{-1}$ with domain $\mathsf{D}(K) = \mathsf{R}(A)$ is densely defined and bounded by $b < 1$, hence by density of $\mathsf{R}(A)$ in X admits a unique bounded extension to all of X which we again denote by K. Then $A+B$ can be factored as $A+B = (I+K)A$, and $I+K$ is invertible, by $b < 1$. Therefore, if $x^* \perp \mathsf{R}(A+B)$ then $(I+K^*)x^* \perp \mathsf{R}(A)$, hence $(I+K^*)x^* = 0$ by density of $\mathsf{R}(A)$ in X, and then $x^* = 0$, by invertibility of $I + K^*$. This shows that $\mathsf{R}(A + B)$ is also dense in X. Moreover, for $r > 0$ we have

$$r + A + B = (1 + B(r + A)^{-1})(r + A),$$

hence $r+A+B$ is invertible, provided $|B(r+A)^{-1}|<1$, and then

$$(r+A+B)^{-1}=(r+A)^{-1}(1+B(r+A)^{-1})^{-1}. \tag{3.11}$$

This implies that $A+B$ is also sectorial, whenever $bC_0(A)<1$, where $C_0(A)$ is defined by (3.5), and then

$$|r(r+A+B)^{-1}|\le \frac{M_0(A)}{1-bC_0(A)}, \quad \text{for all } r>0, \tag{3.12}$$

with $M_0(A)$ also given by (3.5). Replacing $r>0$ by $\lambda\in\Sigma_{\pi-\phi}$ in the above argument we also obtain an estimate for the spectral angle of $A+B$, namely

$$\phi_{A+B}\le \inf\{\phi>\phi_A:\ bC_{\pi-\phi}(A)<1\}. \tag{3.13}$$

Thus the class of operators B satisfying (3.9) with $bC_0(A)<1$ forms an admissible class of perturbations for $A\in\mathcal{S}(X)$.

Theorem 3.1.5. *Suppose $A\in\mathcal{S}(X)$, B linear with $\mathsf{D}(A)\subset\mathsf{D}(B)$ such that* (3.9) *holds, and let $A+B$ be defined by* (3.10).

Then $bC_0(A)<1$ implies $A+B\in\mathcal{S}(X)$, and the spectral angle ϕ_{A+B} of $A+B$ satisfies (3.13).

Let us next consider perturbations B which instead of (3.9) are subject to

$$|Bx|\le b|Ax|+a|x|, \quad \text{for all } x\in\mathsf{D}(A), \tag{3.14}$$

where $a,b\ge 0$. Then even for small b one cannot expect that $A\in\mathcal{S}(X)$ implies $A+B\in\mathcal{S}(X)$, in general. For example $Bx=-ax$ satisfies (3.14) with $b=0$, but $A+B\notin\mathcal{S}(X)$ unless $\sigma(A)\cap[0,a)=\emptyset$. However, $\mathcal{S}(X)$ is invariant w.r.t. right shifts, and therefore it is reasonable to ask whether $\mu+A+B$ is sectorial, for some $\mu\ge 0$. Now (3.14) implies

$$\begin{aligned}|B(\mu+A)^{-1}| &\le a|(\mu+A)^{-1}|+b|A(\mu+A)^{-1}|\\ &\le \frac{aM_0(A)}{\mu}+bC_0(A),\end{aligned} \tag{3.15}$$

hence $\mu+A+B$ is invertible provided $aM_0(A)/\mu+bC_0(A)<1$, i.e., if $bC_0(A)<1$ and $\mu>\mu_0:=aM_0(A)/(1-bC_0(A))$, and then

$$|(\mu+A+B)^{-1}|\le \frac{M_0(A)}{1-bC_0(A)}\frac{1}{\mu-\mu_0}, \quad \text{for all } \mu>\mu_0. \tag{3.16}$$

This shows that $\mu+A+B\in\mathcal{S}(X)$ if $bC_0(A)<1$ and $\mu\ge\mu_0$.

Similarly, applying Theorem 3.1.5 to the pair $(\mu+A,B)$ instead of (A,B) we obtain the following result.

Corollary 3.1.6. *Suppose $A \in \mathcal{PS}(X)$, B linear with $\mathsf{D}(A) \subset \mathsf{D}(B)$ such that* (3.14) *holds, and let $A + B$ be defined by* (3.10).

Then there are numbers $b_0 > 0$ and $\mu_0 \geq 0$ such that $\mu + A + B \in \mathcal{S}(X)$, whenever $b < b_0$ and $\mu \geq \mu_0$.

It should be mentioned that the condition of *lower order type*

$$|Bx| \leq a|x| + b|A^\alpha x|, \quad \text{for all } x \in \mathsf{D}(A), \tag{3.17}$$

where $a, b \geq 0$ and $\alpha \in [0, 1)$, implies (3.14) via the moment inequality, see (3.55),

$$|A^\alpha x| \leq k|Ax|^\alpha |x|^{1-\alpha}, \quad x \in \mathsf{D}(A), \tag{3.18}$$

for any $b > 0$. For the definition of A^α as well as for (3.18) we refer to the next subsections. In fact, (3.17) and (3.18) yield

$$|Bx| \leq a|x| + b|A^\alpha x| \leq a|x| + bk|Ax|^\alpha |x|^{1-\alpha},$$

hence by means of Young's inequality

$$|Bx| \leq (a + bk(1-\alpha)\varepsilon^{-\alpha/(1-\alpha)})|x| + \alpha bk\varepsilon|Ax|, \quad x \in \mathsf{D}(A).$$

Since ε can be chosen arbitrarily small, Corollary 3.1.6 applies in particular to perturbations satisfying (3.17) without restrictions on a and b, provided $\alpha \in [0, 1)$.

Next we consider A-compact perturbations, i.e., operators B in X such that $B : X_A \to X$ is compact. For such perturbations we have

Lemma 3.1.7. *Let $A \in \mathcal{PS}(X)$, B a linear operator in X such that $B : X_A \to X$ is compact. Furthermore, assume either of the following two conditions*

(i) *B is closable in X,*

(ii) *X is reflexive.*

Then for each $b > 0$ there is $a > 0$ such that (3.14) *is valid.*

Proof. We may assume that A is invertible; replace A by $A+1$ otherwise. Suppose the assertion does not hold. Then there is a constant $b_0 > 0$ and a sequence $(x_n) \subset \mathsf{D}(A)$ with $|Ax_n| = 1$ such that

$$|Bx_n| \geq b_0|Ax_n| + n|x_n| = b_0 + n|x_n|, \quad n \in \mathbb{N}.$$

As B is A-compact, there is a convergent subsequence $Bx_{n_k} \to y$ in X, hence $x_{n_k} \to 0$ in X, and $|y| \geq b_0 > 0$.

If (i) holds, then $y = 0$ as B is closable in X, which yields a contradiction to $y \neq 0$.

If (ii) holds, then there is a weakly-convergent subsequence Ax_{n_k}, its limit is 0 as $x_{n_k} \to 0$ in X. Therefore (x_{n_k}) converges to 0 weakly in X_A, hence $Bx_{n_k} \to y = 0$ strongly in X by compactness, and so we again obtain a contradiction to $y \neq 0$. □

As another consequence of Theorem 3.1.5, let us consider multiplicative perturbations. So let $A \in \mathcal{S}(X)$ and suppose $C \in \mathcal{B}(X)$; then the operator CA with domain $\mathsf{D}(CA) = \mathsf{D}(A)$ is well- and densely defined, and it is closed if in addition C is invertible. Moreover, the latter property of C shows also that $\mathsf{R}(CA)$ is dense in X. It is more difficult to obtain $\rho(CA) \supset (-\infty, 0)$ and (**S2**) for CA. A very simple case arises if we require C to be such that $|C - I| < 1/C_0(A)$. In fact, then we may write $CA = A + (C - I)A$, and $B = (C - I)A$ is subject to the assumption of Theorem 3.1.5. Note that this condition on C necessarily implies that C is bounded but also invertible since $C_0(A) \geq 1$. Observing that $\mathcal{S}(X)$ is invariant under dilations, as a second corollary to Theorem 3.1.5 we obtain

Corollary 3.1.8. *Suppose $A \in \mathcal{S}(X)$ and that $C \in \mathcal{B}(X)$ satisfies the condition*

$$|C - r_C| < r_C/C_0(A), \quad \textit{for some } r_C > 0. \tag{3.19}$$

Then CA and AC with natural domains $\mathsf{D}(CA) = \mathsf{D}(A)$ and $\mathsf{D}(AC) = C^{-1}\mathsf{D}(A)$ belong to $\mathcal{S}(X)$.

The assertion for AC follows by the similarity transform $AC = C^{-1}(CA)C$ of CA.

1.4 The Dunford Functional Calculus

In this subsection we want to develop the functional calculus for pseudo-sectorial operators. For this purpose we first introduce the following function algebras. Let $\phi \in (0, \pi]$ and define the algebra of holomorphic functions on Σ_ϕ

$$H(\Sigma_\phi) = \{f : \Sigma_\phi \to \mathbb{C} \text{ is holomorphic}\}, \tag{3.20}$$

and

$$H^\infty(\Sigma_\phi) = \{f : \Sigma_\phi \to \mathbb{C} : f \text{ is holomorphic and bounded}\}. \tag{3.21}$$

$H^\infty(\Sigma_\phi)$ with norm

$$|f|_{H^\infty(\Sigma_\phi)} = \sup\{|f(\lambda)| : |\arg \lambda| < \phi\} \tag{3.22}$$

is a Banach algebra. First we assume $B \in \mathcal{S}(X)$ to be bounded and invertible, and fix $\phi > \phi_B$. Then the well-known Dunford calculus for bounded linear operators applies. In fact, in this situation the spectrum $\sigma(B)$ is a compact subset of Σ_ϕ, hence choosing a simple closed path Γ_B in Σ_ϕ surrounding $\sigma(B)$ counterclockwise we define

$$f(B) = \frac{1}{2\pi i} \int_{\Gamma_B} f(\lambda)(\lambda - B)^{-1}\, d\lambda, \quad f \in H(\Sigma_\phi). \tag{3.23}$$

Since Γ_B is compact there are no convergence problems with the integral in this formula, and it defines an algebra homomorphism from $H(\Sigma_\phi)$ to $\mathcal{B}(X)$.

(3.23) can be used as a starting point to define the functional calculus for arbitrary pseudo-sectorial operators A in X. To achieve this, a main idea is to take $B = A_\varepsilon$, the approximations of A introduced in (3.6), and to pass to the limit

$\varepsilon \to 0+$. But then we first have to make the integration path Γ_B independent of B. This can be done in several ways at the expense that we have to restrict the function algebra $H(\Sigma_\phi)$.

(i) A natural approach is to deform the integration path Γ_B into Γ defined by $\Gamma = (\infty, 0]e^{i\psi} \cup [0, \infty)e^{-i\psi}$, where $\phi_A < \psi < \phi$. We will do this in two steps. First we deform Γ_B into the path $\Gamma_{r,R}$ defined by

$$\Gamma_{r,R} = e^{-i\psi}[r, R] \cup Re^{i[-\psi,\psi]} \cup e^{i\psi}[R, r] \cup re^{i[\psi,-\psi]}. \tag{3.24}$$

Here the numbers $0 < r < R$ should be chosen such that $R > |B|$ and $r < |B^{-1}|^{-1}$. By means of Cauchy's theorem we then obtain

$$f(B) = \frac{1}{2\pi i} \int_{\Gamma_{r,R}} f(\lambda)(\lambda - B)^{-1}\, d\lambda, \quad f \in H(\Sigma_\phi), \tag{3.25}$$

since $\Gamma_{r,R}$ is also a simple compact Lipschitz curve surrounding $\sigma(B)$ counterclockwise. But we still have the dependence of the integration path in (3.25) on the norms of B and B^{-1}.

Next we let $r \to 0+$ and $R \to \infty$. This cannot be done for arbitrary $f \in H(\Sigma_\phi)$, but by means of Lebesgue's convergence theorem it works for the subspace $H_0(\Sigma_\phi)$ defined according to

$$H_0(\Sigma_\phi) = \bigcup_{\alpha,\beta<0} H_{\alpha,\beta}(\Sigma_\phi), \quad \text{where} \tag{3.26}$$

$$H_{\alpha,\beta}(\Sigma_\phi) = \{f \in H(\Sigma_\phi) : |f|^\phi_{\alpha,\beta} < \infty\}, \quad \text{and} \tag{3.27}$$

$$|f|^\phi_{\alpha,\beta} = \sup_{|\lambda|\le 1} |\lambda^\alpha f(\lambda)| + \sup_{|\lambda|\ge 1} |\lambda^{-\beta} f(\lambda)|. \tag{3.28}$$

With $\Gamma = (\infty, 0]e^{i\psi} \cup [0, \infty)e^{-i\psi}$ this yields (3.23) with Γ_B replaced by the contour Γ which is independent of r, R.

Now let $A \in \mathcal{PS}(X)$ be arbitrary. Employing the approximations A_ε introduced before, setting $B = A_\varepsilon$ and using Proposition 3.1.4, we may pass to the limit $\varepsilon \to 0+$, to obtain the following result.

Proposition 3.1.9. *Let $A \in \mathcal{PS}(X)$, fix any $\phi \in (\phi_A, \pi]$, and let $H_0(\Sigma_\phi)$ be defined as above. Then, with $\Gamma = (\infty, 0]e^{i\psi} \cup [0, \infty)e^{-i\psi}$, the Dunford integral*

$$f(A) = \frac{1}{2\pi i} \int_\Gamma f(\lambda)(\lambda - A)^{-1}\, d\lambda, \quad f \in H_0(\Sigma_\phi), \tag{3.29}$$

defines via $\Phi_A(f) = f(A)$ a functional calculus $\Phi_A : H_0(\Sigma_\phi) \to \mathcal{B}(X)$ which is a bounded algebra homomorphism. Moreover, we have

$$\lim_{\varepsilon \to 0+} f(A_\varepsilon) = f(A) \quad \text{in } \mathcal{B}(X), \tag{3.30}$$

and $\{f(A_\varepsilon)\}_{\varepsilon>0} \subset \mathcal{B}(X)$ is uniformly bounded, for each $f \in H_0(\Sigma_\phi)$.

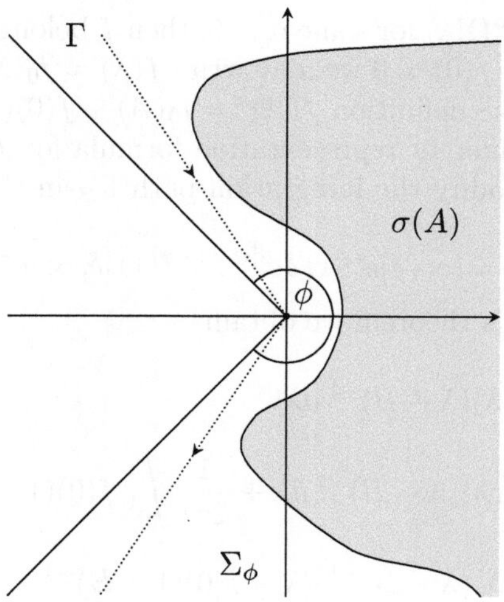

Figure 3.1: Integration path for the Dunford integral.

Observe that boundedness of Φ_A is understood in the sense of inductive limits. This means that we have estimates of the form

$$|f(A)| \leq C|f|^{\phi}_{\alpha,\beta}, \quad \text{for } f \in H_{\alpha,\beta}(\Sigma_\phi),$$

where C depends only on A, ϕ, α, and β. This follows directly from (3.29). In virtue of Proposition 3.1.4, a similar estimate holds also for A_ε, uniformly in $\varepsilon > 0$.

Remark 3.1.10. Consider the map $\varphi(\lambda) = 1/\lambda$ which maps Σ_ϕ onto itself. Then we have the identity

$$(f \circ \varphi)(A) = f(A^{-1}), \quad \text{for each } f \in H_0(\Sigma_\phi), \tag{3.31}$$

in case $\mathsf{N}(A) = 0$. In fact, the change of variable $\lambda \mapsto 1/\lambda$ yields

$$\begin{aligned}(f \circ \varphi)(A) &= \frac{1}{2\pi i}\int_\Gamma f(1/\lambda)(\lambda - A)^{-1}\,d\lambda \\ &= -\frac{1}{2\pi i}\int_\Gamma f(\lambda)(1/\lambda - A)^{-1}\,d\lambda/\lambda^2 \\ &= -\frac{1}{2\pi i}\int_\Gamma f(\lambda)(A^{-1} - \lambda)^{-1}A^{-1}\,d\lambda/\lambda \\ &= \frac{1}{2\pi i}\int_\Gamma f(\lambda)[-1/\lambda + (\lambda - A^{-1})^{-1}]\,d\lambda = f(A^{-1}),\end{aligned}$$

where the last equality follows from Cauchy's theorem.

There is a simple but useful extensions of the Dunford calculus in Proposition 3.1.9. Namely, in case $f \in H(\Sigma_\phi)$ is holomorphic in a neighbourhood of zero and such that $\lambda^\alpha f(\lambda) \in H^\infty(\Sigma_\phi)$ for some $\alpha > 0$, then f belongs to $H_0(\Sigma_\phi)$ if and only if $f(0) = 0$. But in case $f(0) \neq 0$ we may write $f(\lambda) = f_0(\lambda) + f(0)/(1+\lambda)$, where $f_0 \in H_0^\infty(\Sigma_\phi)$, hence the definition $f(A) := f_0(A) + f(0)(1+A)^{-1}$ is reasonable. We want to derive a different representation formula for $f(A)$ in such situations. For this purpose we modify the integration path Γ_B in the representation (3.23) of $f(B)$ into

$$\Gamma_\delta = (\infty, \delta]e^{i\psi} \cup \delta e^{i[\psi, 2\pi-\psi]} \cup [\delta, \infty)e^{-i\psi},$$

and employing Cauchy's theorem we obtain

$$\begin{aligned} f(B) &= \frac{1}{2\pi i}\int_{\Gamma_\delta} f(\lambda)(\lambda - B)^{-1}\, d\lambda \\ &= \frac{1}{2\pi i}\int_{\Gamma_\delta} f_0(\lambda)(\lambda - B)^{-1}\, d\lambda + \frac{1}{2\pi i}\int_{\Gamma_\delta} f(0)(1+\lambda)^{-1}(\lambda - B)^{-1}\, d\lambda \\ &= \frac{1}{2\pi i}\int_{\Gamma} f_0(\lambda)(\lambda - B)^{-1}\, d\lambda + f(0)(1+B)^{-1} \\ &= f_0(B) + f(0)(1+B)^{-1}. \end{aligned}$$

Setting again $B = A_\varepsilon$ and passing to the limit $\varepsilon \to 0+$, we get

$$f(A) = \frac{1}{2\pi i}\int_{\Gamma_\delta} f(\lambda)(\lambda - A)^{-1}\, d\lambda, \tag{3.32}$$

where δ is small enough but arbitrary otherwise. Define the corresponding space by

$$H_a(\Sigma_\phi) = \{f \in \bigcup_{\beta<0} H_{0,\beta}(\Sigma_\phi) : f \text{ is holomorphic in a neighbourhood of } 0\}.$$

Then we have the following result.

Corollary 3.1.11. *Let $A \in \mathcal{PS}(X)$ with spectral angle ϕ_A, fix any $\phi > \phi_A$, and let $H_a(\Sigma_\phi)$ be defined as above.*

Then the Dunford map $\Phi : H_a(\Sigma_\phi) \to \mathcal{B}(X)$ defined via $\Phi(f) = f(A)$, where $f(A)$ is given by the Dunford integral (3.32)*, is well-defined and an algebra homomorphism. It coincides with the Dunford map of Proposition* 3.1.9*, and we have the relation*

$$f(A) = f_0(A) + f(0)(1+A)^{-1},$$

where $f_0(\lambda) = f(\lambda) - f(0)/(1+\lambda)$ belongs to $H_0(\Sigma_\phi)$. In particular, for the functions $g_\mu(\lambda) = 1/(\lambda - \mu)$ with $\mu \notin \overline{\Sigma_\phi}$ we have $g_\mu(A) = (A-\mu)^{-1}$. The convergence assertion (3.30) *of Proposition* 3.1.9 *is also valid for $H_a(\Sigma_\phi)$.*

Remark 3.1.12. (a) A similar result can be obtained for functions $f \in H(\Sigma_\phi)$ which are holomorphic at infinity and decay polynomially at zero. With $f_\infty(\lambda) = f(\lambda) - f(\infty)\lambda/(1+\lambda)$ we then have the relation

$$f(A) = f_\infty(A) + f(\infty)A(I+A)^{-1},$$

and there is an integral representation corresponding to (3.32) which we do not explicitly state here.

(b) If $f \in H(\Sigma_\phi)$ is holomorphic at infinity and at zero we have correspondingly

$$f(A) = f_{0,\infty}(A) + f(0)(I+A)^{-1} + f(\infty)A(I+A)^{-1}.$$

With $\delta > 0$ small and $\rho > \delta$ large one obtains alternatively

$$f(A) = \frac{1}{2\pi i}\int_{\Gamma_\delta^\rho} f(\lambda)(\lambda - A)^{-1}\, d\lambda,$$

where

$$\Gamma_\delta^\rho = [\rho,\delta]e^{i\psi} \cup \delta e^{i[\psi,2\pi-\psi]} \cup [\delta,\rho]e^{-i\psi} \cup \rho e^{i[2\pi-\psi,\psi]}.$$

The proof of these facts is left to the reader.

(c) The functions $\varphi_\varepsilon(\lambda) = (\varepsilon+\lambda)/(1+\varepsilon\lambda)$ map Σ_ϕ into itself, and $\varphi(0) = \varepsilon$, $\varphi(\infty) = 1/\varepsilon$. This means that $f_\varepsilon = f \circ \varphi_\varepsilon$ belongs to $H(\Sigma_\phi)$ and is holomorphic at infinity and at zero, for any $f \in H(\Sigma_\phi)$. Therefore, (b) of this Remark applies and we obtain

$$(f \circ \varphi_\varepsilon)(A) = f(A_\varepsilon).$$

In fact, the identity

$$\begin{aligned}
(\lambda - A_\varepsilon)^{-1} &= (1+\varepsilon A)(\lambda - \varepsilon - (1-\lambda\varepsilon)A)^{-1}\\
&= (1+\varepsilon A)(1-\varepsilon\lambda)^{-1}(\frac{\lambda-\varepsilon}{1-\varepsilon\lambda} - A)^{-1}\\
&= \frac{1-\varepsilon^2}{(1-\varepsilon\lambda)^2}(\frac{\lambda-\varepsilon}{1-\varepsilon\lambda} - A)^{-1} - \frac{\varepsilon}{1-\varepsilon\lambda}
\end{aligned}$$

and the variable transformation $z = (\lambda-\varepsilon)/(1-\varepsilon\lambda)$, i.e., $\lambda = \varphi_\varepsilon(z)$ yield

$$\begin{aligned}
f(A_\varepsilon) &= \frac{1}{2\pi i}\int_{\Gamma_r^R} f(\lambda)(\lambda - A_\varepsilon)^{-1}\, d\lambda\\
&= \frac{1}{2\pi i}\int_{\Gamma_r^R} f(\lambda)(\frac{\lambda-\varepsilon}{1-\varepsilon\lambda} - A)^{-1}\frac{1-\varepsilon^2}{(1-\varepsilon\lambda)^2}\, d\lambda\\
&= \frac{1}{2\pi i}\int_{\varphi_\varepsilon(\Gamma_r^R)} f(\varphi_\varepsilon(z)))(z-A)^{-1}\, dz = (f\circ\varphi_\varepsilon)(A),
\end{aligned}$$

employing once more Cauchy's theorem.

3.2 The Derivation Operator

This section is devoted to the most elementary operator in analysis, the derivation operator d/dt. We will consider this operator on intervals $J = \mathbb{R}$, $J = \mathbb{R}_+$, and on $J = (0, a)$, in various spaces.

2.1. The Whole Line Case

Let $J = \mathbb{R}$. In the sequel we will use the notation $Y_p(\mathbb{R}) = L_p(\mathbb{R}; Y)$, where Y denotes a Banach space and $p \in [1, \infty]$, $Y_b(\mathbb{R}) = C_b(\mathbb{R}; Y)$, $Y_{ub}(\mathbb{R}) = C_{ub}(\mathbb{R}; Y)$, and $Y_0(\mathbb{R}) = C_0(\mathbb{R}; Y)$. Define B_p in $Y_p(\mathbb{R})$ by means of

$$(B_p u)(t) = \dot{u}(t), \quad t \in \mathbb{R}, \quad u \in \mathsf{D}(B_p) = H^1_p(\mathbb{R}; Y), \tag{3.33}$$

for $p \in [1, \infty]$ and $\mathsf{D}(B_p) = C^1_p(\mathbb{R}; Y)$ for $p \in \{0, b, ub\}$. It is easy to see that B_p is closed, and B_p is densely defined except for $p \in \{\infty, b\}$. Since $\dot{u}(t) = 0$ for all $t \in \mathbb{R}$ implies that u is constant, we have $\mathsf{N}(B_p) = \{0\}$ for all $p \in [1, \infty) \cup \{0\}$, while $\mathsf{N}(B_b) = \mathsf{N}(B_{ub}) = \mathsf{N}(B_\infty) \equiv Y$.

Next consider the range of B_p for $p \in (1, \infty) \cup \{0\}$. If $f \in C(\mathbb{R}; Y)$ has compact support and mean value $Mf = \int_{-\infty}^{\infty} f(s)\, ds = 0$, then the solution u of $\dot{u} = f$ on $\mathbb{R}$ belongs to $C^1(\mathbb{R}; Y)$ and has compact support as well. Since the set of such functions f is dense in $Y_p(\mathbb{R})$ for $1 < p < \infty$ and for $p = 0$, by the following lemma, we see that $\mathsf{R}(B_p)$ is dense in $Y_p(\mathbb{R})$, $1 < p < \infty$ and $p = 0$.

Lemma 3.2.1. *Let Y be a Banach space, $\varphi \in L_1(\mathbb{R}) \cap C_0(\mathbb{R})$ such that $\varphi \geq 0$, $\int_{\mathbb{R}} \varphi(t)\, dt = 1$, and define $\varphi_\varepsilon(t) = \varepsilon\varphi(\varepsilon t)$, $t \in \mathbb{R}$, $\varepsilon > 0$.*

Then for $f \in Y_1(\mathbb{R}) + Y_\infty(\mathbb{R})$ the approximations f_ε of f defined by $f_\varepsilon = \varphi_\varepsilon * f$ have the following properties.

(i) *$f_\varepsilon \to_{\varepsilon\to\infty} f$ in $Y_p(\mathbb{R})$, for each $f \in Y_p(\mathbb{R})$, $p \in [1, \infty) \cup \{0, ub\}$;*

(ii) *$f_\varepsilon \to_{\varepsilon\to 0+} 0$ in $Y_p(\mathbb{R})$, for each $f \in Y_p(\mathbb{R})$, $p \in (1, \infty) \cup \{0\}$.*

Proof. (i) Let $T(t)$ denote the translation group defined by

$$[T(t)f](s) = f(t + s), \quad t, s \in \mathbb{R}.$$

Then for $p \in [1, \infty) \cup \{0, ub\}$ we have $T(t)f \to f$ in $Y_p(\mathbb{R})$ as $t \to 0$, for each $f \in Y_p(\mathbb{R})$. Therefore with $\int_{\mathbb{R}} \varphi(t)\, dt = 1$ we obtain

$$\begin{aligned}
|f_\varepsilon - f|_p &= |\int_{\mathbb{R}} \varphi_\varepsilon(s)([T(-s)f] - f)\, ds|_p \\
&\leq \int_{|s|\leq R} \varphi_\varepsilon(s)|T(-s)f - f|_p\, ds + \int_{|s|\geq R} \varphi_\varepsilon(s)(|T(-s)f|_p + |f|_p)\, ds \\
&\leq \sup_{|s|\leq R} |T(-s)f - f|_p + 2|f|_p \int_{|s|\geq R\varepsilon} |\varphi(s)|\, ds.
\end{aligned}$$

Now, given an arbitrary number $\eta > 0$, choose first $R > 0$ such that $|T(s)f - f|_p \leq \eta/2$ for all $|s| \leq R$, and then for this fixed R a number $\varepsilon_\eta > 0$ such that $2|f|_p \int_{|s|\geq R\varepsilon_\eta} |\varphi(s)|\, ds < \eta/2|f|_p$. Then $|f_\varepsilon - f|_p \leq \eta$ for all $\varepsilon \geq \varepsilon_\eta$, which implies assertion (i).

(ii) To prove the second assertion, note that by Young's inequality $|f_\varepsilon|_p \leq |f|_p$, for each $f \in Y_p(\mathbb{R})$. On the other hand, $|f_\varepsilon|_\infty \leq \varepsilon|\varphi|_\infty|f|_1$. This implies $|f_\varepsilon|_\infty \to 0$

as $\varepsilon \to 0+$, for each $f \in Y_1(\mathbb{R})$, hence also

$$|f_\varepsilon|_p \leq |f_\varepsilon|_\infty^{1-1/p}|f_\varepsilon|_1^{1/p} \leq [|\varphi|_\infty \varepsilon]^{1-1/p}|f|_1 \to 0+$$

as $\varepsilon \to 0+$, for each $f \in Y_1(\mathbb{R}) \cap Y_0(\mathbb{R})$. By (i) and a cut off procedure such functions are dense in $Y_p(\mathbb{R})$, $p \in (1,\infty) \cup \{0\}$, and so assertion (ii) follows. □

For $p = 1$, $Mf = 0$ is a necessary condition for $f \in \mathsf{R}(B_1)$, hence $\mathsf{R}(B_1) \subset \mathsf{N}(M)$ and because M is bounded, $\mathsf{N}(M) \neq Y_1(\mathbb{R})$ is closed and so $\mathsf{R}(B_1)$ is not dense in $Y_1(\mathbb{R})$.

The kernel $\mathsf{N}(B_p)$ consists of the constant functions for $p \in \{b, ub, \infty\}$, hence $\dim \mathsf{N}(B_p) = 1$, and B_p is pseudo-sectorial as we shall see below, so $\mathsf{R}(B_p)$ cannot be dense for these p, by Theorem 3.1.2.

To compute the spectrum of B_p, we consider the equation

$$\lambda u(t) + \dot{u}(t) = f(t), \quad t \in \mathbb{R}. \tag{3.34}$$

For $\mathrm{Re}\,\lambda > 0$ a solution is given by

$$u_\lambda(t) = \int_0^\infty e^{-\lambda s} f(t-s)\,ds = \int_{-\infty}^t e^{-\lambda(t-s)} f(s)\,ds, \quad t \in \mathbb{R},$$

and we have the estimate

$$|u_\lambda|_p \leq |f|_p / \mathrm{Re}\,\lambda, \quad \mathrm{Re}\,\lambda > 0.$$

On the other hand, for $\mathrm{Re}\,\lambda < 0$ a solution is

$$u_\lambda(t) = -\int_{-\infty}^0 e^{-\lambda s} f(t-s)\,ds = -\int_t^\infty e^{-\lambda(t-s)} f(s)\,ds, \quad t \in \mathbb{R},$$

and

$$|u_\lambda|_p \leq |f|_p / |\mathrm{Re}\,\lambda|, \quad \mathrm{Re}\,\lambda < 0.$$

Since the general solution of (3.34) is given by $u(t) = u_\lambda(t) + ce^{-\lambda t}$, and for $\mathrm{Re}\,\lambda \neq 0$ the function $e^{-\lambda t}$ is not in $Y_p(\mathbb{R})$, we have $\mathsf{N}(\lambda + B_p) = 0$ for all $\mathrm{Re}\,\lambda \neq 0$. Summarizing we have

Proposition 3.2.2. *Let $J = \mathbb{R}$. Then the operators B_p and $-B_p$ defined above are pseudo-sectorial in $Y_p(\mathbb{R})$ with spectral angles $\phi_{B_p} = \phi_{-B_p} = \pi/2$, for all $p \in [1,\infty] \cup \{0, b, ub\}$. The domains of B_p are dense for all $p \in [1,\infty) \cup \{0, ub\}$, their kernels are trivial for all $p \in [1,\infty) \cup \{0\}$, and $\mathsf{R}(B_p)$ is dense for all $p \in (1,\infty) \cup \{0\}$. Consequently, B_p and $-B_p$ are sectorial iff $p \in (1,\infty) \cup \{0\}$.*

2.2 The Half-Line Case

Next we consider the operator B_p on $J = \mathbb{R}_+$. This time we let $Y_p(\mathbb{R}_+) = L_p(\mathbb{R}_+; Y)$ for $p \in [1,\infty]$, $Y_p(\mathbb{R}_+) = {}_0C_p(\bar{\mathbb{R}}_+; Y)$ for $p \in \{0, b, ub\}$, where the subscript 0 indicates zero trace at $t = 0$. Define

$$(B_p u)(t) = \dot{u}(t),\ t \in J,\ u \in \mathsf{D}(B_p) = {}_0H^1_p(\mathbb{R}_+; Y), \tag{3.35}$$

for $p \in [1,\infty]$ and $\mathsf{D}(B_p) = {}_0C_p(\mathbb{R}_+; Y) \cap C^1_p(\mathbb{R}_+; Y)$ for $p \in \{0, b, ub\}$. As in the case of $J = \mathbb{R}$, it is easy to see that B_p is closed, and that B_p is densely defined except for $p \in \{\infty, b\}$. Since $\dot{u}(t) = 0$ for all $t \in \mathbb{R}_+$ implies that u is constant hence $u(t) \equiv u(0) = 0$, we have $\mathsf{N}(B_p) = 0$ for all $p \in [1,\infty] \cup \{0, b, ub\}$.

To compute the spectrum of B_p for $J = \mathbb{R}_+$, consider the problem

$$\lambda u(t) + \dot{u}(t) = f(t),\ t > 0, \quad u(0) = 0.$$

For all $\lambda \in \mathbb{C}$ its solution is given by

$$u_\lambda(t) = \int_0^t e^{-\lambda s} f(t-s)\, ds, \quad t \in \mathbb{R}_+,$$

and we have the estimate

$$|u_\lambda|_p \le |f|_p / \mathrm{Re}\,\lambda, \quad \mathrm{Re}\,\lambda > 0.$$

Concerning the range of B_p, note that necessarily $(B_p^{-1} f)(t) = \int_0^t f(s)\, ds$ whenever $f \in \mathsf{R}(B_p)$. Since the set of continuous functions f with compact support in $(0,\infty)$ and mean value $Mf = \int_0^\infty f(s)\, ds = 0$ is dense in $Y_p(\mathbb{R}_+)$ for each $p \in (1,\infty) \cup \{0\}$, we see that the range of B_p for such p is dense. On the other hand, as in the case of $J = \mathbb{R}$ we see that $\mathsf{R}(B_1)$ is not dense, and this is also the case for $p \in \{\infty, b\}$. In fact, consider a Hahn-Banach extension of the limit functional $\langle l | f \rangle := \lim_{t\to\infty} f(t)$ from the closed subspace $C_l(\bar{\mathbb{R}}_+; Y)$ of $Y_{ub}(\mathbb{R}_+)$ to $Y_b(\mathbb{R}_+)$. Then for $f \in \mathsf{R}(B_p)$, $p \in \{b, ub\}$, $f \in C_l(\bar{\mathbb{R}}_+; Y)$ we must necessarily have $\langle l | f \rangle = 0$, which means $\mathsf{R}(B_p) \subset \mathsf{N}(l)$. From these considerations we obtain

Proposition 3.2.3. *Let $J = \mathbb{R}_+$. Then the operator B_p defined by (3.35) is injective and pseudo-sectorial in $Y_p(\mathbb{R}_+)$ with spectral angle $\phi_{B_p} = \pi/2$, for all $p \in [1,\infty] \cup \{0, b, ub\}$. The domain of B_p is dense for all $p \in [1,\infty) \cup \{0, ub\}$, and $\mathsf{R}(B_p)$ is dense for all $p \in (1,\infty) \cup \{0\}$. Consequently, B_p is sectorial iff $p \in (1,\infty) \cup \{0\}$.*

2.3 Finite Interval

Here we consider the operator B_p on the finite interval $J = (0,a)$. This time we let $Y_p(J) = L_p(J; Y)$ for $p \in [1,\infty]$, $Y_p(J) = {}_0C_p(\bar{J}; Y)$ for $p \in \{0, b, ub\}$, where as before the subscript 0 indicates trace zero at $t = 0$. Define

$$(B_p u)(t) = \dot{u}(t),\ t \in J,\ u \in \mathsf{D}(B_p) = {}_0H^1_p(J; Y), \tag{3.36}$$

for $p \in [1,\infty]$ and $\mathsf{D}(B_p) = {}_0C_p(J; Y) \cap C^1_p(J; Y)$ for $p \in \{0, b, ub\}$. As in the case of $J = \mathbb{R}_+$, it is easy to see that B_p is closed, injective, and that B_p is densely defined except for $p = \infty$.

This time the spectrum of B_p is empty for each p, in fact we have the relation

$$(\lambda + B_p)^{-1} f(t) = u_\lambda(t) = \int_0^t e^{-\lambda s} f(t-s)\, ds, \quad t \in J,\ \lambda \in \mathbb{C},$$

$$|u_\lambda|_p \le |f|_p (1 - e^{-\operatorname{Re}\lambda a})/\operatorname{Re}\lambda, \quad \operatorname{Re}\lambda \ne 0,$$

and

$$|u_\lambda|_p \le |f|_p a, \quad \operatorname{Re}\lambda = 0.$$

Therefore, although $\sigma(B_p) = \emptyset$, B_p still has spectral angle $\pi/2$. More precisely we have

Proposition 3.2.4. *Let $J = (0, a)$. Then the operator B_p defined by (3.36) is invertible and pseudo-sectorial in $Y_p(J)$ with spectral angle $\phi_{B_p} = \pi/2$, for all $p \in [1, \infty] \cup \{0, b, ub\}$. The domain of B_p is dense for all $p \in [1, \infty) \cup \{0, b, ub\}$, hence, B_p is sectorial iff $p \ne \infty$.*

It is instructive to have a look at the functional calculus for B_p. Since the resolvent of B_p admits the kernel representation

$$(\lambda - B_p)^{-1} w(t) = -\int_J e_\lambda(t-s) w(s)\, ds, \quad t \in J,$$

where $e_\lambda(t) = e^{\lambda t}$ for $t > 0$, $e_\lambda(t) = 0$ for $t \le 0$, for a function $f \in H_0(\Sigma_\phi)$, $\phi > \pi/2$, the operators $f(B_p)$ admit a kernel representation as well, namely

$$[f(B_p)w](t) = \int_J k_f(t-s) w(s)\, ds, \quad t \in J.$$

The kernel $k_f(t)$ is obtained as the contour integral

$$k_f(t) = -\frac{1}{2\pi i} \int_\Gamma f(\lambda) e_\lambda(t)\, d\lambda,$$

in particular $k_f(t) = 0$ for $t \le 0$. The contour Γ is chosen as in Section 3.1.4. This is precisely the inversion formula for the Laplace transform, i.e., f and k_f are related by $\widehat{k}_f(\lambda) = f(\lambda)$, for $\lambda > 0$, say.

The approximations $(B_p)_\varepsilon$ of B_p introduced in Section 3.1.2 also admit a kernel representation. In fact, the functions $f_\varepsilon(\lambda) = (\varepsilon + \lambda)/(1 + \varepsilon\lambda)$ are the Laplace transforms of $k_\varepsilon(t) = \delta_0(t)/\varepsilon + (1 - 1/\varepsilon^2) e^{-t/\varepsilon} \eta_0(t)$, where η_0 denotes the Heaviside function, and δ_0 its derivative, the Dirac measure. This implies

$$[(B_p)_\varepsilon w](t) = \varepsilon^{-1} w(t) + (1 - \varepsilon^{-2}) \int_0^t w(t-s) e^{-s/\varepsilon}\, ds, \quad t \in J,\ \varepsilon > 0,$$

the kernel representation of $(B_p)_\varepsilon$.

2.4 Weighted L_p-Spaces
Let Y be a Banach space and assume that $p \in (1, \infty)$ and $1/p < \mu \le 1$. We set

$$L_{p,\mu}(\mathbb{R}_+; Y) := \{f : \mathbb{R}_+ \to Y : \, t^{1-\mu} f \in L_p(\mathbb{R}_+; Y)\}$$

and equip it with the norm $|f|_{L_{p,\mu}(\mathbb{R}_+;Y)} := (\int_0^\infty |t^{1-\mu} f(t)|^p \, dt)^{1/p}$. We also define

$$H^1_{p,\mu}(\mathbb{R}_+; Y) := \{u \in L_{p,\mu}(\mathbb{R}_+; Y) \cap H^1_{1,\mathrm{loc}}(\mathbb{R}_+; Y) : \, \dot{u} \in L_{p,\mu}(\mathbb{R}_+; Y)\}.$$

$H^1_{p,\mu}(\mathbb{R}_+; Y)$ will always be given the norm

$$|u|_{H^1_{p,\mu}} = |u|^p_{L_{p,\mu}(\mathbb{R}_+;Y)} + |\dot{u}|^p_{L_{p,\mu}(\mathbb{R}_+;Y)})^{1/p},$$

which turns it into a Banach space.

Lemma 3.2.5. *Suppose $p \in (1, \infty)$ and $1/p < \mu \le 1$. Then*

(a) $L_{p,\mu}(\mathbb{R}_+; Y) \hookrightarrow L_{1,\mathrm{loc}}(\bar{\mathbb{R}}_+; Y)$*;*

(b) $H^1_{p,\mu}(\mathbb{R}_+; Y) \hookrightarrow W^1_{1,\mathrm{loc}}(\bar{\mathbb{R}}_+; Y)$*;*

(c) *Every function $u \in H^1_{p,\mu}(\mathbb{R}_+; Y)$ has a well-defined trace, that is, $u(0)$ is well-defined in Y.*

Proof. (a) The first assertion follows from

$$\int_0^T |f(t)| \, dt \le (\int_0^T t^{-p'(1-\mu)} \, dt)^{1/p'} (\int_0^T |t^{1-\mu} f(t)|^p \, dt)^{1/p} \le c|f|_{L_{p,\mu}(\mathbb{R}_+;Y)}$$

which is valid provided that $\mu > 1/p$.
(b) This follows from the definition of $H^1_{p,\mu}(\mathbb{R}_+; Y)$ and from (a).
(c) We conclude from (b) that every function $u \in H^1_{p,\mu}(\mathbb{R}_+; Y)$ is locally absolutely continuous, and this yields the assertion in (c). □

In the following we set

$${}_0H^1_{p,\mu}(\mathbb{R}_+; Y) := \{u \in H^1_{p,\mu}(\mathbb{R}_+; Y) : u(0) = 0\}.$$

Then the derivation operator

$$B_{p,\mu} u(t) := \dot{u}(t) := \frac{d}{dt} u(t), \quad t > 0, \quad \mathsf{D}(B_{p,\mu}) := {}_0H^1_{p,\mu}(\mathbb{R}_+; Y) \tag{3.37}$$

is well-defined on $L_{p,\mu}(\mathbb{R}_+; Y)$. It is natural to introduce the mapping

$$\Phi_\mu : L_{p,\mu}(\mathbb{R}_+; Y) \to L_p(\mathbb{R}_+; Y), \quad (\Phi_\mu u)(t) := t^{1-\mu} u(t), \quad t > 0.$$

Next we show that the operator Φ_μ also maps ${}_0H^1_{p,\mu}(\mathbb{R}_+; Y)$ into ${}_0H^1_p(\mathbb{R}_+; Y)$, provided $\mu > 1/p$.

Proposition 3.2.6. *Let $p \in (1,\infty)$ and let $1/p < \mu \le 1$. Then*

(a) $\Phi_\mu : L_{p,\mu}(\mathbb{R}_+;Y) \to L_p(\mathbb{R}_+;Y)$ *is an isometric isomorphism.*

(b) $\Phi_\mu : {}_0H^1_{p,\mu}(\mathbb{R}_+;Y) \to {}_0H^1_p(\mathbb{R}_+;Y)$ *is a (topological) isomorphism.*

Proof. (a) The assertion in (a) is clear.

(b) (i) We will first show that Φ_μ^{-1} maps ${}_0H^1_p(\mathbb{R}_+;Y)$ into ${}_0H^1_{p,\mu}(\mathbb{R}_+;Y)$. In order to see this, let $v \in {}_0H^1_p(\mathbb{R}_+;Y)$ be given. An easy computation shows that the function $t^{\mu-1}v$ is in $H^1_{p,\mathrm{loc}}(\mathbb{R}_+;Y)$ and that

$$t^{1-\mu}\frac{d}{dt}[t^{\mu-1}v](t) = \dot v(t) - (1-\mu)\frac{v(t)}{t}, \quad t>0. \tag{3.38}$$

By means of Hardy's inequality (see Proposition 3.4.5 below) we can verify that the function v/t belongs to $L_p(\mathbb{R}_+;Y)$. Indeed, we infer from $v(t) = \int_0^t \dot v(s)\,ds$ that

$$\Big(\int_0^\infty |t^{-1}v(t)|^p\,dt\Big)^{1/p} = \Big(\int_0^\infty |t^{-1}\int_0^t \dot v(s)ds|^p\,dt\Big)^{1/p} \le p'\Big(\int_0^\infty |\dot v(s)|^p ds\Big)^{1/p}. \tag{3.39}$$

We conclude from (3.38)–(3.39) that $\Phi_\mu^{-1}v$ belongs to $H^1_{p,\mu}(\mathbb{R}_+;Y)$, and also that the mapping Φ_μ^{-1} is linear and bounded between the indicated spaces.

(ii) Next we show that $u = \Phi_\mu^{-1}v$ has trace zero. Observing that

$$u(t) = t^{\mu-1}v(t) = t^{\mu-1}\int_0^t \dot v(s)\,ds$$

we obtain by Hölder's inequality that $|u(t)| \le t^{\mu-1/p}(\int_0^t |\dot v(s)|^p\,ds)^{1/p}$. This shows that $u(t)\to 0$ as $t\to 0+$.

(iii) Similar arguments show that Φ_μ maps ${}_0H^1_{p,\mu}(\mathbb{R}_+;Y)$ into ${}_0H^1_p(\mathbb{R}_+;Y)$, and that the mapping is bounded and linear. □

We will now consider the derivation operator $B_{p,\mu}$ defined in (3.37). Thanks to Proposition 3.2.6 the operator

$$\bar B_{p,\mu} := \Phi_\mu B_{p,\mu}\Phi_\mu^{-1}, \quad \mathsf{D}(\bar B_{p,\mu}) := {}_0H^1_p(\mathbb{R}_+;Y), \tag{3.40}$$

which acts on the function space $L_p(\mathbb{R}_+;Y)$, is well-defined. It follows from (3.38) that

$$\bar B_{p,\mu} = B_{p,1} + B_0, \quad \text{where} \quad (B_0v)(t) := -(1-\mu)v(t)/t. \tag{3.41}$$

Observe that $\bar B_{p,\mu}$ and $B_{p,\mu}$ coincide if $\mu = 1$. Moreover, note that $B_{\mu,p}$ in $L_{p,\mu}(\mathbb{R}_+;Y)$ is similar to $B_{p,1}+B_0$ in $L_p(\mathbb{R}_+;Y)$. It follows from equation (3.39) that B_0 is relatively bounded with respect to $B_{p,1}$, with bound smaller than 1, provided $(1-\mu)p' < 1$, i.e., for $1 \ge \mu > 1/p$. It is now easy to see that the operators $B_{p,\mu}$ and $\bar B_{p,\mu}$ share the following properties.

Proposition 3.2.7. *Suppose* $1 < p < \infty$ *and* $1/p < \mu \le 1$. *Then*

(i) $\bar{B}_{p,\mu}$ *is closed and densely defined in* $L_p(\mathbb{R}_+;Y)$. *Moreover,* $\mathsf{N}(\bar{B}_{p,\mu}) = 0$, *and* $\mathsf{R}(\bar{B}_{p,\mu})$ *is dense in* $L_p(\mathbb{R}_+;Y)$.

(ii) $B_{p,\mu}$ *is closed and densely defined in* $L_{p,\mu}(\mathbb{R}_+;Y)$. *Moreover,* $\mathsf{N}(B_{p,\mu}) = 0$, *and* $\mathsf{R}(B_{p,\mu})$ *is dense in* $L_{p,\mu}(\mathbb{R}_+;Y)$.

Proof. (i) It has been proved above that $B_{p,1}$ has all the properties listed in the proposition. Since B_0 is relatively bounded with respect to $B_{p,1}$ with relative bound strictly smaller than 1, we obtain from (3.41) that $\bar{B}_{p,\mu}$ enjoys the same properties, see Section 3.1.3.

(ii) The assertions in (ii) follow from (i) by employing the isomorphism Φ_μ. □

In the sequel we take the liberty to work with $B_{p,\mu}$ and $\bar{B}_{p,\mu}$ interchangeably, that is, we will use the representation that is the most convenient one.

Lemma 3.2.8. *Let* $1/p < \mu \le 1$ *and suppose that* $k \in L_1(\mathbb{R}_+;\mathcal{B}(X,Y))$ *satisfies* $|k(t)| \le \kappa(t)$, *where* $\kappa \in L_1(\mathbb{R}_+)$ *is nonnegative and nonincreasing, and where* X, Y *are Banach spaces. Then we have*

(i) $\Big|\int_0^t k(t-s)(t/s)^{1-\mu}v(s)\,ds\Big|_p \le c_{p,\mu}|\kappa|_1|v|_p$ *for* $v \in L_p(\mathbb{R}_+;X)$, *where* $c_{p,\mu} = 2^{1-\mu}[1 + (1-p'(1-\mu))^{-p/p'}]^{1/p}$.

(ii) *The convolution operator* $K := k*$ *belongs to* $\mathcal{B}(L_{p,\mu}(\mathbb{R}_+;X), L_{p,\mu}(\mathbb{R}_+;Y))$ *and* $|K| \le c_{p,\mu}|\kappa|_1$.

Proof. (i) Let $v \in L_p(\mathbb{R}_+;X)$ be given. Then Hölder's inequality implies

$$\begin{aligned}
\Big|\int_0^t k(t-s)(t/s)^{1-\mu}v(s)\,ds\Big|_p^p &\le \int_0^\infty \Big[\int_0^t \kappa(t-s)(t/s)^{1-\mu}|v(s)|ds\Big]^p dt\\
&\le \int_0^\infty \Big[\int_0^t \kappa(t-r)r^{-p'(1-\mu)}dr\Big]^{p/p'} t^{p(1-\mu)}\int_0^t \kappa(t-s)|v(s)|^p\,dsdt\\
&= \int_0^\infty |v(s)|^p\Big\{\int_s^\infty t^{p(1-\mu)}\kappa(t-s)\Big[\int_0^t \kappa(t-r)r^{-p'(1-\mu)}dr\Big]^{p/p'} dt\Big\}\,ds\\
&\le c_{p,\mu}^p|\kappa|_1^p|v|_p^p,
\end{aligned}$$

as the following estimates show. On the one hand, we have

$$\begin{aligned}
&\int_s^\infty t^{p(1-\mu)}\kappa(t-s)\Big[\int_{t/2}^t \kappa(t-r)r^{-p'(1-\mu)}dr\Big]^{p/p'} dt\\
&\le 2^{p(1-\mu)}\int_s^\infty \kappa(t-s)\Big[\int_{t/2}^t \kappa(t-r)dr\Big]^{p/p'} dt\\
&\le 2^{p(1-\mu)}|\kappa|_1^{1+p/p'} = 2^{p(1-\mu)}|\kappa|_1^p.
\end{aligned}$$

Since $\kappa(t)$ is nonincreasing and $(1-\mu)p' < 1$ we have, on the other hand,

$$\begin{aligned}
&\int_s^\infty t^{p(1-\mu)}\kappa(t-s)\Big[\int_0^{t/2}\kappa(t-r)r^{-p'(1-\mu)}dr\Big]^{p/p'}dt\\
&\le \int_s^\infty t^{p(1-\mu)}\kappa(t-s)\Big[\kappa(t/2)\int_0^{t/2}r^{-p'(1-\mu)}dr\Big]^{p/p'}dt\\
&= (1-p'(1-\mu))^{-p/p'}2^{p(1-\mu)}\int_s^\infty \kappa(t-s)[\kappa(t/2)(t/2)]^{p/p'}dt\\
&\le (1-p'(1-\mu))^{-p/p'}2^{p(1-\mu)}|\kappa|_1^p.
\end{aligned}$$

Note that the last inequality follows from

$$\kappa(t/2)(t/2) = \int_0^{t/2}\kappa(t/2)\,d\tau \le \int_0^{t/2}\kappa(\tau)\,d\tau \le |\kappa|_1,$$

where we have once more used that κ is nonincreasing.
(ii) We conclude from (i) that

$$\begin{aligned}
|Kv|_{L_{p,\mu}} &= \Big(\int_0^\infty t^{(1-\mu)p}|Kv(t)|^p\,dt\Big)^{1/p}\\
&= \Big(\int_0^\infty\Big|\int_0^t k(t-s)(t/s)^{1-\mu}s^{1-\mu}v(s)\,ds\Big|^p dt\Big)^{1/p}\\
&\le c_{p,\mu}|\kappa|_1|s^{1-\mu}v|_p = c_{p,\mu}|\kappa|_1|v|_{L_{p,\mu}},
\end{aligned}$$

and the proof of Lemma 3.2.8 is complete. □

We already know that the operator $-B_{p,1}$ generates a positive C_0-semigroup $\{T(t) : t\in\mathbb{R}_+\}$ of contractions on $L_p(\mathbb{R}_+;Y)$ which is given by

$$[T(t)u](s) := \begin{cases} u(s-t) & \text{if } s>t,\\ 0 & \text{if } s<t.\end{cases} \tag{3.42}$$

This implies the resolvent estimate

$$|(\lambda+B_{p,1})^{-1}|_{\mathcal{B}(L_p(\mathbb{R}_+;Y))} \le \frac{1}{\operatorname{Re}\lambda},\quad \operatorname{Re}\lambda>0.$$

However, note that this semigroup is not of class C_0 in $L_{p,\mu}(\mathbb{R}_+;Y)$ for $\mu<1$, as $T(t)$ does not map $L_{p,\mu}(\mathbb{R}_+;Y)$ into $L_{p,\mu}(\mathbb{R}_+;Y)$ for $t>0$. Nevertheless, we now prove a resolvent estimate for $B_{p,\mu}$, which is best possible.

Proposition 3.2.9. *Let $1/p<\mu\le 1$. Then the resolvent set $\rho(B_{p,\mu})$ contains the open negative half-plane $\mathbb{C}_- = -\Sigma_{\pi/2}$, and there is a constant $c_{p,\mu}>1$ such that*

$$|(\lambda+B_{p,\mu})^{-1}|_{\mathcal{B}(L_{p,\mu}(\mathbb{R}_+;Y))} \le \frac{c_{p,\mu}}{\operatorname{Re}\lambda},\quad \operatorname{Re}\lambda>0, \tag{3.43}$$

holds. In particular, $B_{p,\mu}$ is sectorial in $L_{p,\mu}(\mathbb{R}_+;Y)$ with $\phi_{B_{p,\mu}}=\pi/2$.

Proof. (i) Let $\lambda \in \mathbb{C}$ with $\operatorname{Re}\lambda > 0$ be fixed and set

$$(K_\lambda f)(t) := \int_0^t e^{-\lambda(t-s)} f(s)\, ds, \quad f \in L_{p,\mu}(\mathbb{R}_+; Y).$$

Moreover, let $\kappa(t) := e^{-t\operatorname{Re}\lambda}$. Then K_λ satisfies the assertions of Lemma 3.2.8, with $|\kappa|_1 = 1/\operatorname{Re}\lambda$. Consequently, Lemma 3.2.8 shows that K_λ is a bounded linear operator in $L_{p,\mu}(\mathbb{R}_+; Y)$, and that

$$|K_\lambda|_{\mathcal{B}(L_{p,\mu}(\mathbb{R}_+;Y))} \le \frac{c_{p,\mu}}{\operatorname{Re}\lambda}. \tag{3.44}$$

(ii) We verify that $(\lambda + B_{p,\mu}) : \mathsf{D}(B_{p,\mu}) \to L_{p,\mu}(\mathbb{R}_+; Y)$ is invertible for $\operatorname{Re}\lambda > 0$, with

$$[(\lambda + B_{p,\mu})^{-1} f](t) = \int_0^t e^{-\lambda(t-s)} f(s)\, ds, \quad f \in L_{p,\mu}(\mathbb{R}_+; Y). \tag{3.45}$$

Indeed, let $f \in L_{p,\mu}(\mathbb{R}_+; Y)$ be given and recall that $L_{p,\mu}(\mathbb{R}_+; Y)$ is embedded into $L_{1,\mathrm{loc}}(\mathbb{R}_+; Y)$. It is then not difficult to see that the differential equation

$$(\lambda + \frac{d}{dt})u = f, \quad u(0) = 0,$$

has a unique solution $u = u_\lambda$ in $H^1_{1,\mathrm{loc}}(\bar{\mathbb{R}}_+; Y)$. It is given by the right-hand side of equation (3.45). It remains to show $u_\lambda \in \mathsf{D}(B_{p,\mu})$. For this we note that $u_\lambda = K_\lambda f$ and $\dot{u}_\lambda = f - \lambda K_\lambda u_\lambda$. Hence we obtain from (i) that u_λ as well as $\dot{u}_\lambda$ belong to the space $L_{p,\mu}(\mathbb{R}_+; Y)$. Since $u_\lambda(0) = 0$ we conclude $u_\lambda \in \mathsf{D}(B_{p,\mu})$, and this establishes equation (3.45). We have shown that $\rho(B_{p,\mu})$ contains $\mathbb{C}_-$, and the resolvent estimate (3.43) is now a direct consequence of (3.44)–(3.45).

(iii) It follows from (3.43) that $\phi_{B_{p,\mu}} \le \pi/2$. On the other hand, $\phi_{B_{p,\mu}}$ cannot be strictly smaller than $\pi/2$, as this would imply that $B_{p,\mu}$ generates a (strongly continuous analytic) semigroup on $L_{p,\mu}(\mathbb{R}_+; Y)$, which is not possible. The assertion follows now from Proposition 3.2.7. □

3.3 Analytic Semigroups and Fractional Powers

3.1 Holomorphic Semigroups

Typical examples of functions in $H_a(\Sigma_\phi)$ with $\phi < \pi/2$ are the functions $e_t(z) = e^{-zt}$ for each $t > 0$. Provided $\phi_A < \pi/2$, the Dunford calculus from Section 3.1.4 gives rise to the family of operators $e_t(A) =: e^{-tA}$, $t > 0$, which because of the multiplicativity of the the calculus yields the semigroup property

$$e^{-A(t+s)} = e^{-At} e^{-As}, \quad t, s > 0.$$

Therefore it is called a *semigroup of operators.*

Definition 3.3.1. *A family of operators* $\{T(t)\}_{t\geq 0} \subset \mathcal{B}(X)$ *in a Banach space* X *is called a* **semigroup**, *if*

$$T(t+s) = T(t)T(s), \quad t, s > 0, \quad T(0) = I,$$

is satisfied. The semigroup is called of **class** C_0, *if in addition*

$$\lim_{t\to 0+} T(t)x = x, \quad x \in X,$$

holds.

We prove the following result which is basic in semigroup theory and for parabolic partial differential equations.

Theorem 3.3.2. *Let* A *be a closed densely defined operator in a Banach space* X. *Then the following assertions are equivalent.*

(a) A *is pseudo-sectorial with spectral angle less than* $\pi/2$;

(b) $-A$ *generates a* C_0*-semigroup* $T(t)$ *which admits a bounded and holomorphic extension to a sector* Σ_ψ;

(c) $-A$ *generates a* C_0*-semigroup* $T(t)$ *such that* $\mathsf{R}(T(t)) \subset \mathsf{D}(A)$, *and there is a constant* $M_0 > 0$ *such that* $|T(t)| + |tAT(t)| \leq M_0$, *for each* $t > 0$.

Proof. (c) $\Rightarrow$ (b). Suppose $-A$ generates a C_0-semigroup such that the conditions of (c) are satisfied. Define $T(z)$ by means of the power series

$$T(t+z) = \sum_{n=0}^{\infty} \frac{z^n}{n!} T^{(n)}(t).$$

Because of $T^{(n)}(t) = A^n T(t) = [AT(t/n)]^n$ we obtain $|T^{(n)}(t)| \leq [M_0 n/t]^n$, for all $t > 0$ and $n \in \mathbb{N}_0$. These estimates imply

$$|T(t+z)| \leq \sum_{n=0}^{\infty} \frac{[n|z|M_0]^n}{t^n n!} < \infty,$$

provided

$$\overline{\lim}_{n\to\infty} [(n|z|M_0)^n / t^n n!]^{1/n} = M_0|z|e/t < 1,$$

which means $|z| < t/M_0 e$ or $|\arg z| < \psi_T := \arcsin(1/M_0 e)$. On each smaller sector Σ_ψ, $\psi < \psi_T$, $T(z)$ is then holomorphic, bounded, and has the semigroup property $T(z_1)T(z_2) = T(z_1 + z_2)$, and $|T(z)| \leq M_\psi$.

(b) $\Rightarrow$ (a). Now let $T(z)$ be holomorphic on Σ_{ψ_T} and bounded on each smaller sector Σ_ψ. Then for each $\lambda > 0$, Cauchy's theorem applied to the closed contour $\Gamma_R = [0, R] \cup Re^{i[0,\psi]} \cup e^{i\psi}[R, 0]$ implies with $R \to \infty$

$$(\lambda + A)^{-1} = \int_0^\infty e^{-\lambda t} T(t)\, dt = \int_0^\infty e^{-\lambda t e^{i\psi}} T(te^{i\psi})\, dt, \tag{3.46}$$

for each $|\psi| < \psi_T$, by virtue of

$$\Big|\int_0^{\psi} T(Re^{i\varphi})e^{-\lambda Re^{i\varphi}} iRe^{i\varphi}\, d\varphi\Big| \le M_\psi R\int_0^{\psi} e^{-R\lambda\cos\varphi}\, d\varphi \to 0$$

as $R\to\infty$. Because of the estimate

$$\begin{aligned}\Big|\int_0^{\infty} e^{-\lambda te^{i\psi}} T(te^{i\psi})\, dt\Big| &\le M_\psi \int_0^{\infty} e^{-t\mathrm{Re}(\lambda e^{i\psi})}\, dt \\ &\le \frac{M_\psi}{|\lambda|\cos(\psi+\arg\lambda)},\end{aligned} \tag{3.47}$$

formula (3.46) allows for holomorphic extension of the resolvent of A to the sector $-\Sigma_{\pi/2+\psi_T}$, and implies $\sigma(A)\subset\overline{\Sigma}_{\pi/2-\psi_T}$, and (3.46) holds for all $\lambda\in\Sigma_{\pi/2+\psi_T}$. Moreover, estimate (3.47) yields $\sup_{\lambda\in\Sigma_{\pi-\phi}}|\lambda(\lambda+A)^{-1}|<\infty$ for all $\phi>\pi/2-\psi_T$, and therefore $A\in\mathcal{PS}(X)$ and $\phi_A\le\pi/2-\psi_T$.

(a) $\Rightarrow$ (c). Suppose $A\in\mathcal{PS}(X)$ satisfies $\phi_A<\frac{\pi}{2}$, and let $\phi_A<\phi<\frac{\pi}{2}$. Then for $z\in\Sigma_\psi$, the functions $e_z(\lambda)=e^{-z\lambda}$ are holomorphic in $\mathbb{C}$ and belong to $H_a(\Sigma_\phi)$, as long as $\psi<\pi/2-\phi$. Therefore, the functional calculus for pseudo-sectorial operators yields bounded linear operators $T(z)=e_z(A)=e^{-zA}$, which satisfy the semigroup property

$$T(z_1+z_2)=T(z_1)T(z_2),\quad z_1,z_2\in\Sigma_{\frac{\pi}{2}-\phi}.$$

Since the map $z\mapsto f_z$ is holomorphic on $\Sigma_{\frac{\pi}{2}-\phi}$ with derivative $\partial_z e_z(\lambda)=-\lambda e_z(\lambda)$ which even belongs to $H_0(\Sigma_\phi)$, we may conclude that the family $\{T(z)\}_{z\in\Sigma_{\frac{\pi}{2}-\phi}}\subset\mathcal{B}(X)$ is holomorphic and $\frac{d}{dz}T(z)=-AT(z)$. In particular, $-A$ is the generator of $T(z)$ and the operators $T(z)$ have ranges contained in $\mathsf{D}(A)$, for each $z\in\Sigma_{\frac{\pi}{2}-\phi}$. Let us next derive bounds for $|T(z)|$. For this purpose we take the representation of $e_z(A)$ from (3.32).

$$T(z)=\frac{1}{2\pi i}\int_{\Gamma_\delta} e^{-z\lambda}(\lambda-A)^{-1}\, d\lambda.$$

With $|\arg z|\le\psi<\pi/2-\phi$ a straightforward estimate yields

$$\begin{aligned}|T(z)| &\le \frac{M_{\pi-\phi}(A)}{2\pi}\int_{\Gamma_\delta} e^{-\mathrm{Re}(z\lambda)}\frac{|d\lambda|}{|\lambda|}\\ &\le \frac{M_{\pi-\phi}(A)}{\pi}\Big[\int_\delta^\infty e^{-|z|r\cos(\phi+\psi)}\frac{dr}{r}+\int_\psi^\pi e^{|z|\delta}\, d\varphi\Big]\le K_\psi^0(A),\end{aligned}$$

by the choice $\delta=1/|z|$. This shows that the semigroup $T(z)$ is uniformly bounded on Σ_ψ. Similarly, choosing $\delta=0$ we obtain

$$|AT(z)|\le M_{\pi-\phi}(A)\int_0^\infty e^{-|z|r\cos(\phi+\psi)}dr=\frac{K_\psi^1(A)}{|z|},\quad z\in\Sigma_{\frac{\pi}{2}-\phi}.$$

To see that $T(z) \to I$ strongly as $z \to 0$, let $x \in \mathsf{D}(A)$ and fix $\delta > 0$. Then the identity $(\lambda - A)^{-1}x = x/\lambda + (\lambda - A)^{-1}Ax/\lambda$ yields

$$T(z) = \frac{1}{2\pi i}\int_{\Gamma_\delta} e^{-z\lambda}[x + (\lambda - A)^{-1}Ax]\frac{d\lambda}{\lambda}.$$

By means of residue calculus the first part of this integral can be evaluated to the result

$$T(z)x = x + \frac{1}{2\pi i}\int_{\Gamma_\delta} e^{-z\lambda}(\lambda - A)^{-1}Ax\frac{d\lambda}{\lambda},$$

and passing to the limit $z \to 0$, contracting the contour in $-\Sigma_{\pi-\phi}$ we conclude

$$T(z)x \to x + \frac{1}{2\pi i}\int_{\Gamma_\delta} (\lambda - A)^{-1}Ax\frac{d\lambda}{\lambda} = x,$$

by Cauchy's theorem. Since $\mathsf{D}(A)$ is dense in X and $T(z)$ is uniformly bounded we obtain $T(z) \to I$ strongly as $z \to 0$. The theorem is proved. □

3.2 Extended Functional Calculus

We consider now a method to define $f(A)$ for all $A \in \mathcal{PS}(X)$ and all functions $f \in H(\Sigma_\phi)$ which grow at most polynomially at infinity and zero. More precisely, suppose $f \in H_{\alpha,\alpha}(\Sigma_\phi)$ for some $\alpha \in \mathbb{R}_+$. Define $\psi(\lambda) = \lambda/(1+\lambda)^2$; this function is rational and belongs to $H_0(\Sigma_\phi)$. Contracting the contour Γ, by residue calculus we obtain $\psi(A) = A(I + A)^{-2}$. This operator is bounded and injective, its range equals $\mathsf{D}(A) \cap \mathsf{R}(A)$ and its inverse is given by $\psi(A)^{-1} = 2 + A + A^{-1}$. If $k \in \mathbb{N}$ is such that $k > \alpha$ then $\psi^k f \in H_0(\Sigma_\phi)$ and so the Dunford calculus of Proposition 3.1.9 applies and yields a bounded operator $(\psi^k f)(A)$. We then set

$$\begin{aligned} f(A) &= \psi(A)^{-k}(\psi^k f)(A), \quad \text{and} \\ \mathsf{D}(f(A)) &= \{x \in X :\ (\psi^k f)(A)x \in \mathsf{D}(A^k) \cap \mathsf{R}(A^k)\}. \end{aligned} \tag{3.48}$$

This definition of $f(A)$ is independent of $k > \alpha$; in fact, if $l > k > \alpha$ then $\psi^l f = \psi^{l-k}\psi^k f$, hence $(\psi^l f)(A) = \psi^{l-k}(A)(\psi^k f)(A)$ since ψ^{l-k} and also $\psi^k f$ belong to $H_0(\Sigma_\phi)$. Therefore we may always choose $k = [\alpha]+1$, the smallest integer larger than α. $f(A)$ defined this way is closed and densely defined. Moreover, we have

Theorem 3.3.3. *Let X be a complex Banach space and $A \in \mathcal{PS}(X)$. Then the functional calculus Φ_A defined by $\Phi_A(f) = f(A)$ with $f(A)$ given by* (3.48) *is well-defined for all functions in $\bigcup_{\alpha\in\mathbb{R}} H_{\alpha,\alpha}(\Sigma_\phi)$. For $\alpha \geq 0$ and $f \in H_{\alpha,\alpha}(\Sigma_\phi)$, $f(A)$ is a closed linear operator in X with domain*

$$\mathsf{D}(f(A)) = \{x \in X :\ (f\psi^k)(A)x \in \mathsf{D}(A^k) \cap \mathsf{R}(A^k)\},$$

where $k > \alpha$. The inclusion $\mathsf{D}(f(A)) \supset \mathsf{D}(A^k) \cap \mathsf{R}(A^k)$ is valid, and

$$f(A)x = (f\psi^k)(A)\psi^{-k}(A)x, \quad x \in \mathsf{D}(A^k) \cap \mathsf{R}(A^k).$$

In particular, $f(A)$ is densely defined if A is sectorial. Φ_A is an algebra homomorphism in the sense that

$$(af+bg)(A)x = af(A)x+bg(A)x, \quad \text{for all } f,g \in H_{\alpha,\alpha}(\Sigma_\phi),\ x \in \mathsf{D}(A^k)\cap\mathsf{R}(A^k),$$

and all $a,b \in \mathbb{C}$, with $k > \alpha$, and

$$(fg)(A)x = f(A)g(A)x, \quad f \in \mathcal{H}_{\alpha,\alpha}(\Sigma_\phi),\ g \in H_{\beta,\beta}(\Sigma_\phi),\ x \in \mathsf{D}(A^k)\cap\mathsf{R}(A^k),$$

for $k > \alpha+\beta$. The approximations A_ε of A satisfy

$$\lim_{\varepsilon\to 0+} f(A_\varepsilon)x = f(A)x, \quad \text{for all } f \in H_{\alpha,\alpha}(\Sigma_\phi),\ x \in \mathsf{D}(A^k)\cap\mathcal{R}(A^k),\ k>\alpha.$$

It is useful to have a representation of $f(A)x$ as a contour integral, for $f \in H_{\alpha,\beta}(\Sigma_\phi)$ and $x \in \mathsf{D}(A^k)\cap\mathsf{R}(A^l)$, with $k > \alpha$ and $l > \beta$. To this aim we use again (3.25) for a bounded and invertible $B \in \mathcal{S}(X)$. Split the contour as $\Gamma_{r,R} = \Gamma_1^R \cup \Gamma_2^r$, where

$$\Gamma_1^R = e^{-i\psi}[1,R]\cup Re^{i[-\psi,\psi]}\cup e^{i\psi}[R,1], \quad \Gamma_2^r = [1,r]e^{i\psi}\cup re^{i[\psi,-\psi]}\cup[r,1]e^{-i\psi}. \tag{3.49}$$

Fix any $l \in \mathbb{N}_0$. On Γ_1^R we write

$$(\lambda-B)^{-1} = \sum_{j=1}^{l} \lambda^{-j}B^{j-1} + \lambda^{-l}(\lambda-B)^{-1}B^l,$$

and then we have

$$\int_{\Gamma_1^R} f(\lambda)(\lambda-B)^{-1}\,d\lambda = \int_{\Gamma_1^R} \lambda^{-l}f(\lambda)(\lambda-B)^{-1}B^l\,d\lambda + \sum_{j=1}^{l}\int_{\Gamma_1^R} f(\lambda)\lambda^{-j}B^{j-1}\,d\lambda.$$

Deforming the contour Γ_1^R into $\Gamma_0 = e^{i[-\psi,\psi]}$ in Σ_ϕ, we may employ Cauchy's theorem to see that the contributions from the terms $\lambda^{l-j}B^{j-1}$ are independent of R.

The integral over Γ_2^r can be treated similarly. On this path we replace the resolvent $(\lambda-B)^{-1}$ according to the identity

$$(\lambda-B)^{-1} = \lambda^k(\lambda-B)^{-1}B^{-k} - \sum_{j=1}^{k}\lambda^{j-1}B^{-j},$$

to the result

$$\int_{\Gamma_2^r} f(\lambda)(\lambda-B)^{-1}\,d\lambda = \int_{\Gamma_2^r} \lambda^k f(\lambda)(\lambda-B)^{-1}B^{-k}\,d\lambda - \sum_{j=1}^{k}\int_{\Gamma_2^r} f(\lambda)\lambda^{j-1}B^{-j}\,d\lambda.$$

Again by Cauchy's theorem we may deform the contributions from the terms $\lambda^{j-1}B^{-j}$ into an integral over Γ_0 which is independent of $r > 0$.

This way, we obtain the following representation formula for $f(B)$.

$$\begin{aligned} f(B) &= \frac{1}{2\pi i}\int_{\Gamma_1^R} \lambda^{-l} f(\lambda)(\lambda - B)^{-1} B^l \, d\lambda \\ &+ \frac{1}{2\pi i}\int_{\Gamma_2^r} \lambda^{k} f(\lambda)(\lambda - B)^{-1} B^{-k} \, d\lambda \\ &+ \frac{1}{2\pi i}\int_{\Gamma_0} f(\lambda)[\sum_{j=1}^{k} \lambda^{j-1}B^{-j} + \sum_{j=1}^{l} \lambda^{-j}B^{j-1}] \, d\lambda, \end{aligned} \tag{3.50}$$

where the contours Γ_1^R, Γ_2^r are defined by (3.49), and $\Gamma_0 = e^{i[-\psi,\psi]}$. Observe that the last integral is of the form

$$\sum_{j=-k}^{l-1} c_j(f)B^j, \quad \text{with} \tag{3.51}$$

$$c_{-j}(f) = \frac{1}{2\pi i}\int_{\Gamma_0} \lambda^{-(j+1)} f(\lambda)\, d\lambda, \quad c_j(f) = \frac{1}{2\pi i}\int_{\Gamma_0} \lambda^{-(j+1)} f(\lambda)\, d\lambda.$$

This shows that the coefficients $c_j(f)$ depend on f linearly and boundedly, in fact we have

$$|c_j(f)| \le 2\phi \sup\{|f(e^{it})| : |t| \le \phi\}, \quad \text{for all } j \in \mathbb{Z}.$$

For functions $f \in H(\Sigma_\phi)$ which grow at most polynomially at infinity and at zero we may now pass to the limits $R \to \infty$ and $r \to 0+$.

$$\begin{aligned} f(B) &= \frac{1}{2\pi i}\int_{\Gamma_1} \lambda^{-l} f(\lambda)(\lambda - B)^{-1} B^l \, d\lambda \\ &+ \frac{1}{2\pi i}\int_{\Gamma_2} \lambda^{k} f(\lambda)(\lambda - B)^{-1} B^{-k} \, d\lambda + \sum_{j=-k}^{l-1} c_j(f)B^j \end{aligned} \tag{3.52}$$

where $k, l \in \mathbb{N}_0$ denote any numbers such that $\alpha < k$ and $\beta < l$.

Now consider an arbitrary operator $A \in \mathcal{S}(X)$ such that $\phi > \phi_A$. Then for any $\varepsilon > 0$ we let A_ε denote the approximations of A introduced in Section 3.1.2, and we may set $B = A_\varepsilon$ in formula (3.52). With Proposition 3.1.4 we have $(\lambda - A_\varepsilon)^{-1} \to (\lambda - A)^{-1}$ as $\varepsilon \to 0+$ in $\mathcal{B}(X)$, as well as $A_\varepsilon^j x \to A^j x$ for all $x \in \mathsf{D}(A^l)$, $0 \le j \le l$, and $A_\varepsilon^{-j} x \to A^{-j} x$ for all $x \in \mathsf{R}(A^k)$, $0 \le j \le k$. Since the function $|\lambda^{-(l+1)} f(\lambda)|$ is integrable over Γ_1, $|\lambda^{k-1} f(\lambda)|$ has this property on Γ_2, we may pass to the limit $\varepsilon \to 0+$ to the result

$$\begin{aligned} f(A)x &= \frac{1}{2\pi i}\int_{\Gamma_1} \lambda^{-l} f(\lambda)(\lambda - A)^{-1} A^l x \, d\lambda \\ &+ \frac{1}{2\pi i}\int_{\Gamma_2} \lambda^{k} f(\lambda)(\lambda - A)^{-1} A^{-k} x \, d\lambda + \sum_{j=-k}^{l-1} c_j(f)A^j x, \end{aligned} \tag{3.53}$$

for any $x \in \mathsf{D}(A^l) \cap \mathsf{R}(A^k)$. This is the representation formula of $f(A)x$ we have been looking for.

3.3 Complex Powers of Sectorial Operators

For $z \in \mathbb{C}$ the functions $h_z(\lambda) = \lambda^z$ are holomorphic on Σ_π, the sliced complex plane and the estimate

$$|h_z(\lambda)| = |e^{z \log \lambda}| = e^{\mathrm{Re}\, z \log|\lambda| - \mathrm{Im} z \arg \lambda} \leq |\lambda|^{\mathrm{Re}\, z} e^{\phi|\mathrm{Im} z|}, \quad \lambda \in \Sigma_\phi,$$

shows that h_z belongs to $H_{\alpha,\alpha}(\Sigma_\phi)$ for $\alpha = \mathrm{Re}\, z$. Therefore, we may apply the extended functional calculus for sectorial operators to obtain the following result.

Proposition 3.3.4. *Suppose $A \in \mathcal{S}(X)$, let A^z be defined by $A^z = h_z(A)$, and $|\mathrm{Re} z| < k$, $k \in \mathbb{N}$. Then*

(i) *$A^z x$ is holomorphic on the strip $|\mathrm{Re}\, z| < k$, for each $x \in \mathsf{D}(A^k) \cap \mathsf{R}(A^k)$;*

(ii) *A^z is closed for each $z \in \mathbb{C}$;*

(iii) *$A^{z+w}x = A^z A^w x$ for all $z, w \in \mathbb{C}$, $x \in \mathsf{D}(A^k) \cap \mathsf{R}(A^k)$, where $k > |\mathrm{Re}\, z|, |\mathrm{Re}\, w|, |\mathrm{Re}\,(z+w)|$;*

(iv) *$A^z x = \lim_{\varepsilon \to 0} A_\varepsilon^z x$, $x \in \mathsf{D}(A^k) \cap \mathsf{R}(A^k)$, $|\mathrm{Re}\, z| < k$.*

Because of Proposition 3.3.4, the operators A^z are linear, closed, densely defined and, because of $A^z A^{-z} x = x = A^{-z} A^z x$ for x in a dense subset of X, have also dense ranges and trivial kernels. If $A \in \mathcal{S}(X)$ is invertible then $\{A^{-z}$, Re $z > 0\}$ forms a bounded holomorphic C_0-semigroup on $\Sigma_{\pi/2}$. This can be seen from formula (3.53) with $l = 0$ and $k = 1$ which in this case makes sense for all $x \in X$.

It turns out that for real α with $|\alpha| < \pi/\phi_A$ the powers A^α are sectorial as well, and the power law $(A^\alpha)^z x = A^{\alpha z} x$ is valid.

Theorem 3.3.5. *Let $A \in \mathcal{S}(X)$ and $\alpha \in \mathbb{R}$ be such that $|\alpha| < \pi/\phi_A$. Then A^α is also sectorial and $\phi_{A^\alpha} \leq |\alpha|\phi_A$. If $z \in \mathbb{C}$ and $k > |\mathrm{Re}\, z||\alpha|$, then*

$$(A^\alpha)^z x = A^{\alpha z} x, \quad \textit{for all } x \in \mathsf{D}(A^k) \cap \mathsf{R}(A^k). \tag{3.54}$$

For any real numbers $\alpha < \beta < \gamma$ with $\gamma - \alpha < \pi/\phi_A$, the **moment inequality**

$$|A^\beta x| \leq k|A^\alpha x|^{\frac{\gamma-\beta}{\gamma-\alpha}} |A^\gamma x|^{\frac{\beta-\alpha}{\gamma-\alpha}}, \quad x \in \mathsf{D}(A^\alpha) \cap \mathcal{D}(A^\gamma), \tag{3.55}$$

is valid, where k denotes a constant depending only on α, β, γ and A.

Proof. Since $A^{-\alpha} = (A^{-1})^\alpha$, it is enough to consider positive α. So let $\alpha \in (0, \pi/\phi_A)$ be fixed. We want to show that the operators $\mu + A^\alpha$ are invertible for $\mu \in \Sigma_{\pi - \alpha\phi_A}$, and that the resolvent estimate

$$\sup_{\mu \in \Sigma_{\phi_\alpha}} |\mu(\mu + A^\alpha)^{-1}| \leq M_{\phi_\alpha} < \infty$$

is valid for each $\phi_\alpha < \pi - \alpha\phi_A$. For this purpose we consider the functions $g_\mu(\lambda) = \mu/(\mu + \lambda^\alpha)$, which are holomorphic and bounded on Σ_ϕ, uniformly w.r.t. μ, as long as $\mu \in \Sigma_{\phi_\alpha}$, and $\phi_\alpha + \alpha\phi < \pi$. By means of the extended functional calculus we have $g_\mu(A) = \mu(\mu + A^\alpha)^{-1}$, the problem is to show that these operators are bounded with a bound which is uniform in $\mu \in \Sigma_{\phi_\alpha}$. Observe that although the functions $g_\mu(\lambda)$ are uniformly bounded, they are neither holomorphic at zero nor at infinity, due to the presence of the power λ^α.

As a starting point we use formula (3.29) for the approximations A_ε of A which are bounded and invertible. Contract the contour Γ by means of Cauchy's theorem and by residue calculus to the halfray $\Gamma_\alpha = [0, \infty)e^{i\theta}$, with $\pi \geq \theta \geq \phi > \phi_A$, where the branch cut of λ^α is put on this ray. This is possible if the function $\mu + \lambda^\alpha$ has no zeros on this ray, which means that with $\varphi = \arg \mu$ we have $\varphi - \alpha\theta \neq (2k+1)\pi$ and $\varphi + 2\alpha\pi - \alpha\theta \neq (2k+1)\pi$, for all $k \in \mathbb{Z}$. Let λ_j, $j = 1, \ldots, n$ denote the zeros of $\mu + \lambda^\alpha$; note that there are only finitely many of them, and $n = 0$ means that there are none. n is bounded from above in terms of α and ϕ_A. Then we obtain

$$
\begin{aligned}
g_\mu(A_\varepsilon) &= \mu \frac{1}{2\pi i} \int_0^\infty \Big[\frac{e^{i(\theta - 2\pi)}}{\mu + r^\alpha e^{i\alpha(\theta - 2\pi)}} - \frac{e^{i\theta}}{\mu + r^\alpha e^{i\alpha\theta}}\Big] (re^{i\theta} - A_\varepsilon)^{-1}\, dr \\
&\quad + \mu \sum_{j=1}^n \lambda_j^{1-\alpha} (\lambda_j - A_\varepsilon)^{-1}/\alpha \\
&= \frac{\mu e^{i\theta}}{2\pi i} \int_0^\infty \Big[\frac{e^{i(\theta\alpha)} - e^{i(\theta - 2\pi)\alpha}}{(\mu + r^\alpha e^{i\alpha(\theta - 2\pi)})(\mu + r^\alpha e^{i\alpha\theta})}\Big] r^\alpha (re^{i\theta} - A_\varepsilon)^{-1}\, dr \\
&\quad + \mu \sum_{j=1}^n \lambda_j^{1-\alpha} (\lambda_j - A_\varepsilon)^{-1}/\alpha.
\end{aligned}
$$

Estimating this expression we get

$$
\begin{aligned}
|g_\mu(A_\varepsilon)| &\leq C|\mu| \int_0^\infty \frac{r^{\alpha-1}\, dr}{|\mu e^{-i\alpha\theta} + r^\alpha||\mu e^{i\alpha(2\pi - \theta)} + r^\alpha|} + C \\
&\leq C\Big\{1 + \int_0^\infty \frac{dr}{|e^{i(\varphi - \alpha\theta)} + r||e^{i(\varphi - \alpha\theta + 2\alpha\pi)} + r|}\Big\} \leq C.
\end{aligned}
$$

Therefore we have uniform bounds on $g_\mu(A_\varepsilon)$, hence with $\varepsilon \to 0+$ also on $g_\mu(A)$, in virtue of $g_\mu(A_\varepsilon)x \to g_\mu(A)x$ as $\varepsilon \to 0+$ on a dense subset of X, and of the Banach-Steinhaus theorem. This proves that A^α is sectorial and $\phi_{A^\alpha} \leq \alpha\phi_A$ if $\alpha < \pi/\phi_A$.

The identity $(A_\varepsilon^\alpha)^z = A_\varepsilon^{\alpha z}$ is obviously valid, hence passing to the limit we obtain (3.54).

To prove the moment inequality, let us observe that it is enough to consider the case $\alpha = 0$ and $\gamma = 1$; in fact, replace x by $A^\alpha x$, β by $(\beta - \alpha)/(\gamma - \alpha)$, A by $A^{\gamma - \alpha}$, to see this; observe that by the restriction $\gamma - \alpha < \pi/\phi_A$, the operator

$A^{\gamma-\alpha}$ is again sectorial, by the first part of this proof. Contracting the contour Γ in the representation of $A_\varepsilon^{\beta-1}$ to the negative half-axis we obtain

$$A_\varepsilon^{\beta-1} = \frac{\sin(\beta\pi)}{\pi}\int_0^\infty r^{\beta-1}(r+A_\varepsilon)^{-1}\,dr.$$

Application of this formula to Ax for $x \in \mathsf{D}(A)$ and passing to the limit $\varepsilon \to 0+$ leads to

$$A^\beta x = A^{\beta-1}Ax = \frac{\sin(\beta\pi)}{\pi}\int_0^\infty r^{\beta-1}(r+A)^{-1}Ax\,dr;$$

observe that this integral is absolutely convergent. We split the range of integration at $\delta > 0$ and estimate as follows.

$$\begin{aligned}|A^\beta x| &\le C\int_0^\delta r^{\beta-1}\,dr|x| + C\int_\delta^\infty r^{\beta-2}\,dr|Ax|\\ &= C|x|\delta^\beta/\beta + C|Ax|\delta^{\beta-1}/(1-\beta) = C|x|^{1-\beta}|Ax|^\beta,\end{aligned}$$

by the choice $\delta = |Ax|/|x|$. This completes the proof of Theorem 3.3.5. □

3.4 Operators with Bounded Imaginary Powers

Proposition 3.3.4 shows that the following definition makes sense.

Definition 3.3.6. *Suppose $A \in \mathcal{S}(X)$. Then A is said to admit* **bounded imaginary powers** *if $A^{is} \in \mathcal{B}(X)$ for each $s \in \mathbb{R}$, and there is a constant $C > 0$ such that $|A^{is}| \le C$ for $|s| \le 1$. The class of such operators will be denoted by $\mathcal{BIP}(X)$.*

Since by Proposition 3.3.4, A^{is} has the group property, it is clear that A admits bounded imaginary powers if and only if $\{A^{is} : s \in \mathbb{R}\}$ forms a strongly continuous group of bounded linear operators in X. The growth bound θ_A of this group, i.e.,

$$\theta_A = \overline{\lim}_{|s|\to\infty}\frac{1}{|s|}\log|A^{is}| \tag{3.56}$$

will be called the **power angle** of A. Then for each $\omega > \theta_A$ there is a constant $M \ge 1$ such that

$$|A^{it}|_{\mathcal{B}(X)} \le Me^{\omega|t|}, \quad t \in \mathbb{R}.$$

It is in general not easy to verify that a given $A \in \mathcal{S}(X)$ belongs to $\mathcal{BIP}(X)$, although quite a few classes of operators are known for which the answer is positive; cf. the next subsections.

For a first application of the class $\mathcal{BIP}(X)$, consider the fractional power spaces

$$X_\alpha = X_{A^\alpha} = (\mathsf{D}(A^\alpha), |\cdot|_\alpha), \quad |x|_\alpha = |x| + |A^\alpha x|, \quad 0 < \alpha < 1,$$

where $A \in \mathcal{S}(X)$; the embeddings

$$X_A \hookrightarrow X_\beta \hookrightarrow X_\alpha \hookrightarrow X, \quad 1 > \beta > \alpha > 0,$$

are well-known. If A belongs to $\mathcal{BIP}(X)$, a characterization of X_α in terms of *complex interpolation spaces* can be derived.

Theorem 3.3.7. *Suppose $A \in \mathcal{BIP}(X)$. Then*

$$X_\theta \cong (X, X_A)_\theta, \quad \theta \in (0,1), \tag{3.57}$$

where $(X, X_A)_\theta$ denotes the complex interpolation space between X and $X_A \hookrightarrow X$ of order θ.

We recall the definition of the complex interpolation space $(X, X_A)_\theta$, $\theta \in (0,1)$. Consider the strip $S \subset \mathbb{C}$ given by $S := \{z \in \mathbb{C} : 0 < \operatorname{Re} z < 1\}$. Then $x \in (X, X_A)_\theta$ iff there is an $f \in H^\infty(S; X) \cap C(\bar{S}; X)$ with $\sup_{t\in\mathbb{R}} |f(1+it)|_{X_A} < \infty$, such that $f(\theta) = x$. The norm in $(X, X_A)_\theta$ is defined in the canonical way. More precisely,

$$|x|_{(X,X_A)_\theta} := \inf\{|h(i\cdot)|_{L_\infty(\mathbb{R};X)} + |h(1+i\cdot)|_{L_\infty(\mathbb{R};X_A)} : h \in H^\infty(S;X),\ h(\theta) = x\}.$$

The spaces $(X, X_A)_\theta$ are well-known to be Banach spaces such that $X_A \hookrightarrow (X, X_A)_\theta \hookrightarrow X$, with both embeddings dense if $\mathsf{D}(A)$ is dense in X.

Proof. We may assume w.l.o.g. that $A \in \mathcal{BIP}(X)$ is invertible. In fact, the functions $h_1(z) = (1+z)^\alpha(1+z^\alpha)^{-1} - 1$ and $h_2(z) = (1+z^\alpha)/(1+z)^\alpha - 1$ both belong to $H_0(\Sigma_\phi)$, for any $\phi < \pi$. This implies that $(1+A)^\alpha(1+A^\alpha)^{-1}$ and $(1+A^\alpha)(1+A)^{-\alpha}$ are bounded, and so $\mathsf{D}(A^\alpha) = \mathsf{D}((A+1)^\alpha)$.

Let $x \in \mathsf{D}(A)$ and let

$$f(z) = e^{z^2-\theta^2} A^{-z+\theta} x, \quad z \in S.$$

Then f is continuous on $\bar{S}$, holomorphic in S and bounded in X, since

$$|f(\sigma + it)| \le M e^{1-\theta^2} e^{\omega|t|-t^2} |A^{-\sigma+\theta} x| \le C|Ax|,$$

with some constant $C > 0$, as by assumption $A \in \mathcal{BIP}(X)$ is invertible, and employing the moment inequality. Moreover, for $\sigma = 0, 1$ we have

$$|f(it)|_X \le C|A^\theta x|, \quad |Af(1+it)| \le C|A^\theta x|,$$

hence

$$|x|_{(X,X_A)_\theta} \le C|A^\theta x|,$$

by definition of the complex interpolation spaces. As $\mathsf{D}(A)$ is dense in $\mathsf{D}(A^\theta)$ as well as in $(X, X_A)_\theta$, this yields the embedding $\mathsf{D}(A^\theta) \hookrightarrow (X, X_A)_\theta$.

To obtain the converse inclusion, fix $x \in \mathsf{D}(A)$, and let $f : \bar{S} \to X$ be bounded, continuous, and holomorphic in S, $f(\theta) = x$, and such that

$$|f(i\cdot)|_\infty, |Af(1+i\cdot)|_\infty \le 2|x|_{(X,X_A)_\theta}.$$

Set $g_\varepsilon(z) = e^{z^2-\theta^2} A^z (1+\varepsilon A)^{-1} f(z)$, $z \in S$. Then

$$g_\varepsilon(\theta) = A^\theta(1+\varepsilon A)^{-1} f(\theta) = A^\theta (1+\varepsilon A)^{-1} x \to A^\theta x \text{ as } \varepsilon \to 0,$$

as A^θ is closed and commutes with the resolvent of A. Obviously, g_ε is continuous and bounded on $\bar{S}$, holomorphic in S and

$$|g_\varepsilon(it)| \le Me^{\omega|t|-t^2}|(1+\varepsilon A)^{-1}||f(it)| \le C|x|_{(X,X_A)_\theta},$$

as well as

$$|g_\varepsilon(1+it)| \le Me\, e^{\omega|t|-t^2}|(1+\varepsilon A)^{-1}||Af(1+it)| \le C|x|_{(X,X_A)_\theta}.$$

Hadamard's three lines theorem then implies

$$|A^\theta(1+\varepsilon A)^{-1}x| = |g_\varepsilon(\theta)| \le |g_\varepsilon(i\cdot)|_\infty^{1-\theta}|g_\varepsilon(1+i\cdot)|_\infty^\theta \le C|x|_{(X,X_A)_\theta}.$$

Passing to the limit $\varepsilon \to 0$, this yields the inclusion $(X, X_A)_\theta \hookrightarrow \mathsf{D}(A^\theta)$, using once more density of $\mathsf{D}(A)$ in $\mathsf{D}(A^\theta)$ and in $(X, X_A)_\theta$. □

The importance of Theorem 3.3.7 is twofold. It shows on one hand that X_α is largely independent of A; for instance if $A, B \in \mathcal{BIP}(X)$ are such that $\mathsf{D}(A) = \mathsf{D}(B)$ then $\mathsf{D}(A^\alpha) = \mathsf{D}(B^\alpha)$ for all $\alpha \in (0,1)$. On the other hand, (3.57) makes the tools of complex interpolation theory available for fractional power spaces and it becomes possible to characterize X_α in many cases. For example, the reiteration theorem yields the relation

$$(X_\alpha, X_\beta)_\theta = X_{\alpha(1-\theta)+\theta\beta}, \qquad \text{for all } 0 \le \alpha < \beta \le 1,\ \theta \in (0,1),$$

for complex interpolation of fractional power spaces of operators $A \in \mathcal{BIP}(X)$.

Some permanence properties for the class $\mathcal{BIP}(X)$ are collected in the next proposition.

Proposition 3.3.8. *Let X be a complex Banach space. The class $\mathcal{BIP}(X)$ has the following permanence properties.*

(i) *$A \in \mathcal{BIP}(X)$ iff $A^{-1} \in \mathcal{BIP}(X)$; then $\theta_{A^{-1}} = \theta_A$;*

(ii) *$A \in \mathcal{BIP}(X)$ implies $rA \in \mathcal{BIP}(X)$ and $\theta_{rA} = \theta_A$ for all $r > 0$;*

(iii) *$A \in \mathcal{BIP}(X)$ implies $e^{\pm i\psi}A \in \mathcal{BIP}(X)$ for all $\psi \in [0, \pi - \theta_A)$, and $\theta_{e^{\pm i\psi}A} \le \theta_A + \psi$;*

(iv) *$A \in \mathcal{BIP}(X)$ implies $(\mu + A) \in \mathcal{BIP}(X)$ for all $\mu \in \Sigma_{\pi-\phi_A}$, and $\theta_{\mu+A} \le \max\{\theta_A, |\arg\mu|\}$;*

(v) *if $\mathsf{D}(A^*)$ is dense in X^*, then $A \in \mathcal{BIP}(X)$ iff $A^* \in \mathcal{BIP}(X^*)$, and $\theta_A = \theta_{A^*}$;*

(vi) *if Y denotes another Banach space and $T \in \mathcal{B}(X, Y)$ is bijective, then $A \in \mathcal{BIP}(X)$ iff $A_1 = TAT^{-1} \in \mathcal{BIP}(Y)$, and $\theta_A = \theta_{A_1}$.*

Proof. Using the extended functional calculus and suitable variable transformations these permanence properties are abtained as in the proof of Proposition 3.1.3, except for (iv) which is a little more tricky. In fact, (iv) is very much related to the perturbation theory for the class $\mathcal{BIP}(X)$, it follows from our next proposition with $B = \mu$ and $h(z) = z^{is}$. □

Proposition 3.3.9. *Suppose $A \in \mathcal{S}(X)$, B is a linear operator in X with $\mathsf{D}(B) \supset \mathsf{D}(A^\alpha)$, and*

$$|Bx| \leq a|x| + b|A^\alpha x|, \quad x \in \mathsf{D}(A^\alpha),$$

holds with constants $a, b > 0$ and $\alpha \in [0, 1)$. Assume that $A + B$ is sectorial and invertible.

Then $h(A) \in \mathcal{B}(X)$ implies $h(A + B) \in \mathcal{B}(X)$, for any $h \in H^\infty(\Sigma_\phi)$, where $\phi > \phi_A, \phi_{A+B}$. In particular, if $A \in \mathcal{BIP}(X)$ then $A + B \in \mathcal{BIP}(X)$, and

$$\theta_{A+B} \leq \max\{\theta_A, \phi_{A+B}\}.$$

Proof. Fix h according to the assumptions of this proposition and let $f = \psi h$ with ψ as in Section 3.2.2. Then

$$h(A + B) = \psi^{-1}(A + B)f(A + B) = (2 + (A + B)^{-1} + A + B)f(A + B),$$

and with $B = B(1 + A)^{-1}(1 + A)$ this gives

$$h(A + B) = (2 + (A + B)^{-1} + B(1 + A)^{-1} + (1 + B(1 + A)^{-1})A)f(A + B).$$

Now, $(A + B)^{-1}$ and $B(1 + A)^{-1}$ are bounded by assumption and $f(A + B)$ is bounded since $f \in H_0(\Sigma_\phi)$, hence we only need to show that $Af(A+B)$ is bounded. Choosing a standard contour Γ, the resolvent equation implies

$$Af(A + B) = Af(A) + \frac{1}{2\pi i}\int_\Gamma f(\lambda)A(\lambda - A)^{-1}B(\lambda - (A + B))^{-1}\, d\lambda.$$

Since by assumption $h(A)$ is bounded, $Af(A) = A\psi(A)h(A)$ is bounded as well, and the integral is absolutely convergent since B is of lower order. □

In connection with operators with bounded imaginary powers another functional calculus is very useful and will be crucial. For this purpose recall the *Mellin transform* defined by

$$F(z) = \int_0^\infty f(t)t^{z-1}\, dt.$$

Mellin's inversion formula reads

$$f(t) = \frac{1}{2\pi i}\int_{c-i\infty}^{c+i\infty} F(z)t^{-z}\, dz.$$

The inverse Mellin transform can be used to define a functional calculus for $A \in \mathcal{BIP}(X)$ as follows. Set

$$M_\theta(\mathbb{R}) = \{\mu \in M_0(\mathbb{R}) : |\mu|_\theta := \frac{1}{2\pi}\int_\mathbb{R} e^{\theta|s|}|\, d\mu(s)| < \infty\},$$

where $M_0(\mathbb{R})$ denotes the space of all finite complex Borel measures on $\mathbb{R}$. $M_\theta(\mathbb{R})$ becomes a Banach algebra with unit, the convolution of measures, scaled by the factor $1/2\pi$ as multiplication. Evidently the Dirac masses δ_s with unit mass in $s \in \mathbb{R}$ belong to $M_\theta(\mathbb{R})$, and $2\pi\delta_0$ is the unit. For measures $\mu \in M_\theta(\mathbb{R})$ we define

$$f(z) = \frac{1}{2\pi}\int_{\mathbb{R}} z^{-is}\, d\mu(s), \quad z \in \Sigma_\theta.$$

This yields an algebra homomorphism from $M_\theta(\mathbb{R})$ into the Banach algebra $H^\infty(\Sigma_\theta)$, and it gives rise to the algebra homomorphism from $M_\theta(\mathbb{R})$ to $\mathcal{B}(X)$ defined by the formula

$$f(A) = \frac{1}{2\pi}\int_{\mathbb{R}} A^{-is}\, d\mu(s),$$

for any operator $A \in \mathcal{BIP}(X)$ with $\theta_A < \theta$. In fact, this formula is precisely the Phillips calculus for the C_0-group A^{-is}. We summarize these observations as

Theorem 3.3.10. *Let $A \in \mathcal{BIP}(X)$ and $\theta > \theta_A$. Then the formula*

$$f(A) = \frac{1}{2\pi}\int_{\mathbb{R}} A^{-is}\, d\mu(s)$$

defines an algebra homomorphism from $M_\theta(\mathbb{R})$ to $\mathcal{B}(X)$, where f and μ are related by

$$f(z) = \frac{1}{2\pi}\int_{\mathbb{R}} z^{-is}\, d\mu(s).$$

In particular, $f(z) = z^{-is}$ is mapped to A^{-is}, for each $s \in \mathbb{R}$. Moreover, there is a constant $K > 0$ such that

$$|f(A)|_{\mathcal{B}(X)} \le K|\mu|_\theta, \quad \textit{for all } \mu \in \mathcal{M}_\theta(\mathbb{R}),$$

where $K = \sup_{s\in\mathbb{R}} e^{-\theta|s|}|A^{is}|_{\mathcal{B}(X)}$.

Proof. The only thing left to prove is the multiplication property. Here we need to recall the convolution theorem for the Mellin transform, i.e., if $f_j(t) = \frac{1}{2\pi}\int_{-\infty}^{\infty} t^{-is}\, d\mu_j(s)$, then

$$f_1(t)f_2(t) = \frac{1}{2\pi}\int_{-\infty}^{\infty} d(\mu_1 * \mu_2)(s), \quad t > 0.$$

This identity implies

$$\begin{aligned}
(f_1f_2)(A) &= \frac{1}{2\pi}\int_{\mathbb{R}} A^{-is}\, d(\mu_1 * \mu_2)(s)\\
&= \frac{1}{(2\pi)^2}\int_{R} A^{-is}\int_{\mathbb{R}} d\mu_1(s-\tau)\, d\mu_2(\tau)\\
&= \frac{1}{(2\pi)^2}\int_{\mathbb{R}} A^{-is}\, d\mu_1(s)\int_{\mathbb{R}} A^{-i\tau}\, d\mu_2(\tau)\\
&= f_1(A)f_2(A).
\end{aligned}$$

□

It is not obvious how to get the resolvent of an operator A from its imaginary powers. This is due to the fact that the Mellin transform of the function $1/(1+t)$ has poles at 0 and 1. However, since such representations are useful and in particular show that the functional calculus from Theorem 3.3.10 is consistent with the Dunford calculus, we comment on this.

For this purpose observe that

$$(1+t)^{-1} = \frac{1}{2i}\int_{c-i\infty}^{c+i\infty} t^{-z}\frac{dz}{\sin(\pi z)}, \quad t>0,$$

where $0<c<1$ is arbitrary. Therefore,

$$Tx = \frac{1}{2i}\int_{c-i\infty}^{c+i\infty} A^{-z}x\frac{dz}{\sin(\pi z)}$$

is well-defined since the integral is absolutely convergent for $x \in \mathsf{D}(A)\cap\mathsf{R}(A)$. By Cauchy's theorem, the integral is independent of c. Using again Cauchy's theorem, we obtain by an easy computation $T=(1+A)^{-1}$. In fact, apply $1+A$ to Tx to the result

$$(1+A)Tx = \frac{1}{2\pi i}\int_{c-i\infty}^{c+i\infty} A^{-z}x\frac{\pi dz}{\sin(\pi z)} + \frac{1}{2\pi i}\int_{c-i\infty}^{c+i\infty} A^{1-z}x\frac{\pi dz}{\sin(\pi z)}.$$

Deforming the contour in the first integral to

$$\Gamma_0 = (-i\infty,-i\varepsilon]\cup\varepsilon e^{i[-\pi/2,\pi/2]}\cup[i\varepsilon,i\infty)$$

and the second one to

$$\Gamma_1 = (1-i\infty,1-i\varepsilon]\cup(1-\varepsilon e^{i[-\pi/2,\pi/2]})\cup[1+i\varepsilon,1+i\infty),$$

observing that the contributions on the straight lines cancel, and passing to the limit $\varepsilon\to 0+$ there follows $(1+A)Tx = x$ for each $x\in\mathsf{D}(A)\cap\mathsf{R}(A)$. Since by assumption A is sectorial this implies $Tx=(1+A)^{-1}x$ for each $x\in\mathsf{D}(A)\cap\mathsf{R}(A)$.

Replacing A by sA, $s>0$, and shifting the contour to the imaginary axis we get the formula

$$(1+sA)^{-1}x = \frac{1}{2}x + \frac{1}{2i}\mathrm{PV}\int_{-\infty}^{\infty}(sA)^{-i\rho}\frac{d\rho}{\sinh(\pi\rho)}, \quad s>0, \tag{3.58}$$

where PV means the principal value.

To deduce the second formula, recall the identity

$$\frac{1}{1+\lambda t} = \frac{1}{1+rt} + \frac{1}{2i}\int_{-\infty}^{\infty}(rt)^{-i\rho}\frac{(e^{\phi\rho}-1)}{\sinh(\pi\rho)}\,d\rho,$$

where $\lambda = re^{i\phi}$, $|\phi| < \pi$. Since the measure with density $(e^{i\phi\rho} - 1)r^{-i\rho}/\sinh(\pi\rho)$ belongs to $M_\theta(\mathbb{R})$, provided $|\phi| < \pi - \theta$, we get by Theorem 3.3.10 the identity

$$(1 + \lambda A)^{-1} = (1 + |\lambda| A)^{-1} + \frac{1}{2i}\int_{-\infty}^{\infty} (|\lambda| A)^{-i\rho} \frac{(e^{\phi\rho} - 1)}{\sinh(\pi\rho)}\, d\rho, \tag{3.59}$$

whenever $\phi = \arg(\lambda) \in (-\pi + \theta, \pi - \theta)$. As a consequence we have

Corollary 3.3.11. *Suppose $A \in \mathcal{BIP}(X)$, $\theta_A < \pi$. Then $\phi_A \leq \theta_A$.*

3.5 Operators with Bounded $\mathcal{H}^\infty$-Calculus

There is another important concept related to the Dunford calculus for a sectorial operator.

Definition 3.3.12. *A sectorial operator A is said to admit a* **bounded $\mathcal{H}^\infty$-calculus** *if there are $\phi > \phi_A$ and a constant $K_\phi < \infty$ such that*

$$|f(A)| \leq K_\phi |f|_{H^\infty(\Sigma_\phi)}, \quad \textit{for all } f \in H_0(\Sigma_\phi). \tag{3.60}$$

The class of sectorial operators A which admit an $\mathcal{H}^\infty$-calculus will be denoted by $\mathcal{H}^\infty(X)$. The **$\mathcal{H}^\infty$-angle** *of A is defined by*

$$\phi_A^\infty = \inf\{\phi > \phi_A : \ (3.60) \textit{ is valid}\}. \tag{3.61}$$

If this is the case, then the functional calculus for A on $H_0(\Sigma_\phi)$ extends uniquely to $H^\infty(\Sigma_\phi)$. This can be seen by formula (3.53) with $k = l = 1$, which is valid for $x \in \mathsf{D}(A) \cap \mathsf{R}(A)$. If $f \in H^\infty(\Sigma_\phi)$ and $(f_n) \subset H_0(\Sigma_\phi)$ is uniformly bounded and converges to f, uniformly on compact subsets of Σ_ϕ, then (3.53) for f_n and Lebesgue's dominated convergence theorem show $f_n(A)x \to f(A)x$ as $n \to \infty$, for each $x \in \mathsf{D}(A) \cap \mathsf{R}(A)$. Since $\mathsf{D}(A) \cap \mathsf{R}(A)$ is dense in X, (3.53) and the Banach-Steinhaus theorem then yield $f_n(A) \to f(A)$ in the strong operator topology. This is a special case of the so-called convergence lemma.

Lemma 3.3.13. *Let $A \in \mathcal{S}(X)$ and $\phi > \phi_A$. Suppose $(f_n)_{n\geq 0} \subset H^\infty(\Sigma_\phi)$ is such that $f_n \to f_0$ uniformly on compact subsets of Σ_ϕ.*

Then $\sup_{n\geq 1} |f_n(A)|_{\mathcal{B}(X)} < \infty$ implies $f_n(A) \to f_0(A)$ strongly. In particular, this assertion holds if $|f_n|_{H^\infty(\Sigma_\phi)} \leq M < \infty$ and A admits a bounded $\mathcal{H}^\infty$-calculus on Σ_ϕ.

Well-known examples for general classes of sectorial operators with bounded $\mathcal{H}^\infty$-calculus are

(a) normal sectorial operators in Hilbert spaces;

(b) m-accretive operators in Hilbert spaces;

(c) generators of bounded C_0-groups on L_p-spaces;

(d) negative generators of positive contraction semigroups in L_p-spaces.

Here (a) follows from the functional calculus for normal operators in Hilbert spaces, see e.g. Dunford-Schwartz [91], while by the Cayley transform, (b) is a consequence of the Foias-Nagy calculus for contractions in Hilbert spaces; see Foias-Nagy [273]. (c) and (d) and some vector-valued extensions are implied by the theory of Coifman and Weiss [69].

Since the functions $f_s(z) = z^{is}$ belong to $H^\infty(\Sigma_\phi)$, for any $s \in \mathbb{R}$ and $\phi \in (0, \pi)$, we obviously have the inclusions

$$\mathcal{H}^\infty(X) \subset \mathcal{BIP}(X) \subset \mathcal{S}(X), \tag{3.62}$$

and the inequalities

$$\phi_A^\infty \geq \theta_A \geq \phi_A \geq \sup\{|\arg\lambda| : \lambda \in \sigma(A)\}. \tag{3.63}$$

The permanence properties of the class $\mathcal{H}^\infty(X)$ are like those for general sectorial operators.

Proposition 3.3.14. *Let X be a complex Banach space. The class $\mathcal{H}^\infty(X)$ has the following permanence properties.*

(i) $A \in \mathcal{H}^\infty(X)$ *iff* $A^{-1} \in \mathcal{H}^\infty(X)$*; then* $\phi_{A^{-1}}^\infty = \phi_A^\infty$*;*

(ii) $A \in \mathcal{H}^\infty(X)$ *implies* $rA \in \mathcal{H}^\infty(X)$ *and* $\phi_{rA}^\infty = \phi_A^\infty$ *for all* $r > 0$*;*

(iii) $A \in \mathcal{H}^\infty(X)$ *implies* $e^{\pm i\psi}A \in \mathcal{H}^\infty(X)$ *for all* $\psi \in [0, \pi - \phi_A^\infty)$*, and* $\phi_{e^{\pm i\psi}A}^\infty = \phi_A^\infty + \psi$*;*

(iv) $A \in \mathcal{H}^\infty(X)$ *implies* $(\mu + A) \in \mathcal{H}^\infty(X)$ *for all* $\mu \in \Sigma_{\pi-\phi_A}$*, and* $\phi_{\mu+A}^\infty \leq \max\{\phi_A^\infty, |\arg\mu|\}$*;*

(v) *if* $\mathsf{D}(A^*)$ *is dense in* X^**, then* $A \in \mathcal{H}^\infty(X)$ *iff* $A^* \in \mathcal{H}^\infty(X^*)$*, and* $\phi_A^\infty = \phi_{A^*}^\infty$*;*

(vi) *if Y denotes another Banach space and* $T \in \mathcal{B}(X, Y)$ *is bijective, then* $A \in \mathcal{H}^\infty(X)$ *iff* $A_1 = TAT^{-1} \in \mathcal{H}^\infty(Y)$*, and* $\phi_A^\infty = \phi_{A_1}^\infty$*.*

Following the lines of the proof of Proposition 3.1.3, the proof of this result is evident. Concerning perturbations, we have the following result which is a direct consequence of Proposition 3.3.9.

Corollary 3.3.15. *Suppose* $A \in \mathcal{H}^\infty(X)$*, B is a linear operator in X with* $\mathsf{D}(B) \supset \mathsf{D}(A^\alpha)$*, and*

$$|Bx| \leq a|x| + b|A^\alpha x|, \quad x \in \mathsf{D}(A^\alpha),$$

holds with constants $a, b > 0$ and $\alpha \in [0, 1)$. Assume that $A + B$ is sectorial and invertible.

Then $A + B \in \mathcal{H}^\infty(X)$*, and* $\phi_{A+B}^\infty \leq \max\{\phi_A^\infty, \phi_{A+B}\}$*.*

3.4 Trace Spaces: Real Interpolation

4.1 Trace Spaces of L_p-Type
Consider the homogeneous Cauchy problem

$$\dot{u} + Au = 0, \quad t > 0, \quad u(0) = x, \tag{3.64}$$

in a Banach space X, where A is a densely defined pseudo-sectorial operator with spectral angle $\phi_A < \pi/2$. Then $-A$ generates a bounded holomorphic C_0-semigroup in X and the solution $u(t)$ of (3.64) is given by $u(t) = T(t)x$, for all $t \geq 0$, where $T(t) = e^{-At}$ denotes the semigroup generated by $-A$. In this subsection, we study again regularity properties of $u(t)$. More specifically, we ask for which initial values x the solution $u(t)$ is such that $u(t) \in \mathsf{D}(A)$ for a.a. $t > 0$ and $Au \in L_{p,\mu}(\mathbb{R}_+; X)$, $\mu \in (1/p, 1]$. In virtue of (3.64) this is equivalent to $u \in W^1_{p,loc}(\mathbb{R}_+; X)$ and $\dot{u} \in L_{p,\mu}(\mathbb{R}_+; X)$.

Suppose that u has this property. Then the initial value $x \in X$ satisfies $\int_0^\infty |AT(t)x|^p t^{p(1-\mu)}\, dt < \infty$. Let us introduce the following trace spaces.

Definition 3.4.1. *Let A be a densely defined pseudo-sectorial operator in X with spectral angle $\phi_A < \pi/2$, let $\alpha \in (0,1)$ and $p \in [1,\infty)$. The spaces $D_A(\alpha,p)$ are defined by means of*

$$D_A(\alpha,p) = \Big\{x \in X : \, [x]_{\alpha,p} := \Big(\int_0^\infty |t^{1-\alpha} AT(t)x|^p \, dt/t\Big)^{1/p} < \infty\Big\}.$$

When equipped with the norm

$$|x|_{\alpha,p} := |x| + [x]_{\alpha,p}, \quad x \in D_A(\alpha,p),$$

$D_A(\alpha,p)$ becomes a Banach space. For $k \in \mathbb{N}$ the spaces $D_A(k+\alpha,p)$ are defined by

$$D_A(k+\alpha,p) := \{x \in \mathsf{D}(A^k) : \, A^k x \in D_A(\alpha,p)\}.$$

We can now give a complete answer to the question raised at the beginning of this subsection.

Proposition 3.4.2. *Suppose A is a densely defined invertible sectorial operator in X with spectral angle $\phi_A < \pi/2$, $p \in (1,\infty)$ and $\mu \in (1/p, 1]$.*

Then for the solution u of (3.64) the following assertions are equivalent.

(a) *$u(t) \in \mathsf{D}(A)$ for a.a. $t > 0$, and $u \in L_{p,\mu}(\mathbb{R}_+; X_A)$;*
(b) *$u \in H^1_{p,\mu}(\mathbb{R}_+; X)$;*
(c) *$x \in D_A(\mu - 1/p, p)$.*

In this case there is a constant $C_{p,\mu} > 0$ depending only on A, p and μ, such that

$$|\dot{u}|_{L_{p,\mu}(\mathbb{R}_+;X)} + |Au|_{L_{p,\mu}(\mathbb{R}_+;X)} \leq C_{p,\mu} |x|_{\mu-1/p,p},$$

for all $x \in D_A(\mu - 1/p, p)$.

Proof. By assumption, $-A$ generates the holomorphic semigroup $T(t) = e^{-At}$ which is bounded on $\mathbb{R}_+$, satisfies $T(t)X \subset \mathsf{D}(A)$ and, with some $\omega > 0$,

$$|T(t)| + t|AT(t)| \le Me^{-\omega t}, \quad t > 0.$$

Let $x \in X$ and $u(t) = T(t)x$. Then $u(t) \in \mathsf{D}(A)$ for $t > 0$. By definition, $x \in D_A(\mu - 1/p, p)$ implies $Au \in L_{p,\mu}(\mathbb{R}_+; X)$, hence (c) implies (a). Since $T(t)$ is holomorphic and $\dot{T}(t) = AT(t)$ for $t > 0$, (a) implies (b). On the other hand, (b) yields $Au = -\dot{u} \in L_{p,\mu}(\mathbb{R}_+; X)$, hence

$$[x]^p_{\mu-1/p,p} = |Au|^p_{L_{p,\mu}(\mathbb{R}_+;X)}$$

shows that (b) implies (c). □

We will also use frequently the following result which extends the previous proposition to fractional orders.

Proposition 3.4.3. *Suppose A is a densely defined invertible sectorial operator in X with spectral angle $\phi_A < \pi/2$, $p \in (1, \infty)$, $\mu \in (1/p, 1]$, and $\alpha - 1 + \mu - 1/p > 0$.*

Then for the solution u of (3.64) the following assertions are equivalent.

(a) $u \in L_{p,\mu}(\mathbb{R}_+; D_A(\alpha, p))$;

(b) $x \in D_A(\alpha - 1 + \mu - 1/p, p)$.

In this case, we have in addition

(c) $u \in W^\alpha_{p,\mu}(\mathbb{R}_+; X) \cap H^\alpha_{p,\mu}(\mathbb{R}_+; X) \cap L_{p,\mu}(\mathbb{R}_+; \mathsf{D}(A^\alpha))$, *and there is a constant $C_{p,\mu} > 0$ depending only on A, p and μ, such that*

$$\begin{aligned} &|u|_{W^\alpha_{p,\mu}(\mathbb{R}_+;X)} + |u|_{H^\alpha_{p,\mu}(\mathbb{R}_+;X)} + |u|_{L_{p,\mu}(\mathbb{R}_+;D_A(\alpha,p))} + |u|_{L_{p,\mu}(\mathbb{R}_+;\mathsf{D}(A^\alpha))} \\ &\qquad \le C_{p,\mu}|x|_{\alpha-1+\mu-1/p,p}, \quad \text{for all } x \in D_A(\alpha - 1 + \mu - 1/p, p). \end{aligned}$$

Note that for $\alpha - 1 + \mu - 1/p < 0$ assertions (a) and (c) hold for all $x \in X$. The spaces W^α and H^α are defined via interpolation; see Section 3.4.5 below.

Proof. Observe that (a) holds if and only if $I := \int_0^\infty |u(t)|^p_{D_A(\alpha,p)} t^{p(1-\mu)}\, dt < \infty$. We have by Fubini's theorem

$$\begin{aligned} I &= \int_0^\infty \int_0^\infty |\tau^{1-\alpha} Ae^{-A\tau} u(t)|^p \frac{d\tau}{\tau} t^{p(1-\mu)}\, dt \\ &= \int_0^\infty \int_0^\infty |Ae^{-A(\tau+t)}x|^p t^{p(1-\mu)}\, dt \tau^{p(1-\alpha)-1} d\tau \\ &= \int_0^\infty \int_\tau^\infty |Ae^{-As}x|^p (s-\tau)^{p(1-\mu)}\, ds \tau^{p(1-\alpha)-1} d\tau, \end{aligned}$$

therefore applying Fubini another time

$$\begin{aligned} I &= \int_0^\infty |Ae^{-As}x|^p \int_0^s (s-\tau)^{p(1-\mu)}\tau^{p(1-\alpha)-1} d\tau ds \\ &= C_0(\alpha,\mu,p)\int_0^\infty |Ae^{-As}x|^p s^{p(1-\alpha+1-\mu)}\, ds \\ &\le C_0(\alpha,\mu,p)|x|^p_{D_A(\alpha-1+\mu-1/p,p)}, \end{aligned}$$

with $C_0(\alpha,\mu,p) = \mathsf{B}(p(1-\alpha), p(1-\mu)+1)$, where B denotes the Beta function.

The assertions in **(c)** will be proved in Section 3.4.6. □

4.2 Trace Spaces and Real Interpolation

We present now some other characterizations of the trace spaces $D_A(\alpha,p)$.

For this, we first recall the definition of the *real interpolation spaces* $(X,X_A)_{\alpha,p}$ of order $\alpha\in(0,1)$ and exponent $p\in[1,\infty)$. $x\in(X,X_A)_{\alpha,p}$ iff there exist a function $w\in C([0,1];X)\cap C((0,1];X_A)\cap C^1((0,1];X)$ with $w(0)=x$, such that

$$[[w]]_{\alpha,p} := \Big[\int_0^1 |t^{1-\alpha}\dot w(t)|^p\, dt/t\Big]^{1/p} + \Big[\int_0^1 |t^{1-\alpha}Aw(t)|^p\, dt/t\Big]^{1/p} < \infty. \tag{3.65}$$

The norm in $(X,X_A)_{\alpha,p}$ is then defined as $|x|_{(X,X_A)_{\alpha,p}} := |x| + \inf[[w]]_{\alpha,p}$, where the infimum is taken over all functions w with the described properties.

Proposition 3.4.4. *Let A be a densely defined pseudo-sectorial operator in a Banach space X with spectral angle $\phi_A<\pi/2$, let $\alpha\in(0,1)$, and $p\in[1,\infty)$. Then for $x\in X$ the following assertions are equivalent.*

(a) $x\in D_A(\alpha,p)$;

(b) $[x]'_{\alpha,p} := [\int_0^\infty |t^{-\alpha}(T(t)x-x)|^p\, dt/t]^{1/p} < \infty$;

(c) $[x]''_{\alpha,p} := [\int_0^\infty |\lambda^\alpha A(\lambda+A)^{-1}x|^p\, d\lambda/\lambda]^{1/p} < \infty$;

(d) $x\in(X,X_A)_{\alpha,p}$.

The norms

$$|\cdot|_{\alpha,p},\ |\cdot|'_{\alpha,p} = |\cdot| + [\cdot]'_{\alpha,p},\ |\cdot|''_{\alpha,p} = |\cdot| + [\cdot]''_{\alpha,p},\ |\cdot|_{(X,X_A)_{\alpha,p}}$$

are equivalent.

To prove this result we need some preparation. Firstly, Note that (d) in the proposition makes sense for all closed linear operators in X, while (c) is well-defined if A is pseudo-sectorial, in contrast to (a) which requires $\phi_A<\pi/2$, and (b) where $-A$ must be the generator of a bounded C_0-semigroup.

Secondly, recall Jensen's inequality

$$\phi\Big(\int_\Omega g(\omega)\, d\mu(\omega)\Big) \le \int_\Omega \phi(g(\omega))\, d\mu(\omega), \tag{3.66}$$

which is valid for each probability measure μ on Ω, for each integrable function g on Ω, and $\phi : \mathbb{R} \to \mathbb{R}$ convex.

Thirdly, we shall need *Hardy's inequality.*

Lemma 3.4.5 (Hardy's inequality). *Let $p \in [1, \infty)$, $0 < T \leq \infty$, and $f : \mathbb{R}_+ \to X$ be measurable and such that $\int_0^T |t^\beta f(t)|^p\, dt < \infty$, for some $\beta < 1/p' = 1 - 1/p$. Then*

$$\int_0^T \left| t^{\beta-1} \int_0^t f(s)ds \right|^p dt \leq c(\beta, p)^p \int_0^T |t^\beta f(t)|^p\, dt < \infty,$$

where $c(\beta, p) = (1/p' - \beta)^{-1}$.

Proof. The change of variables $t = e^\tau$, $s = e^\sigma$ yields

$$\begin{aligned}\int_0^T \left| t^{\beta-1} \int_0^t f(s)ds \right|^p dt &= \int_{-\infty}^{\log(T)} \left| e^{(\beta-1)\tau} \int_{-\infty}^\tau f(e^\sigma)e^\sigma d\sigma \right|^p e^\tau d\tau \\ &\leq \int_{-\infty}^{\log(T)} \Big[\int_{-\infty}^\tau |f(e^\sigma)| e^{(\beta+1/p)\sigma} \cdot e^{(\beta-1+1/p)(\tau-\sigma)} d\sigma \Big]^p d\tau,\end{aligned}$$

hence by Young's inequality for convolutions

$$\begin{aligned}\int_0^T \left| t^{\beta-1} \int_0^t f(s)ds \right|^p dt &\leq \Big[\int_0^\infty e^{(\beta-1/p')\sigma} d\sigma \Big]^p \cdot \Big[\int_{-\infty}^{\log(T)} |f(e^\tau)e^{(\beta+1/p)\tau}|^p d\tau \Big] \\ &= (1/p' - \beta)^{-p} \Big[\int_0^T |t^\beta f(t)|^p\, dt \Big],\end{aligned}$$

which proves the lemma. □

Proof of Proposition 3.4.4.
(a) ⇒ (b). Let $x \in D_A(\alpha, p)$; then the identity

$$T(t)x - x = -\int_0^t AT(s)x\, ds$$

and Lemma 3.4.5 with $\beta = 1 - \alpha - 1/p$ yield

$$\begin{aligned}\int_0^\infty |t^{-\alpha}(T(t)x - x)|^p\, dt/t &= \int_0^\infty t^{(\beta-1)p} \left| \int_0^t AT(s)x\, ds \right|^p dt \\ &\leq \alpha^{-p} \int_0^\infty s^{\beta p} |AT(s)x|^p\, ds \\ &= \alpha^{-p} \int_0^\infty |t^{1-\alpha} AT(t)x|^p\, dt/t \\ &= \alpha^{-p} [x]_{\alpha,p}^p.\end{aligned}$$

This implies $[x]'_{\alpha,p} \leq \alpha^{-1}[x]_{\alpha,p}$.

(b) $\Rightarrow$ (c). To prove this implication we employ the identity

$$A(\lambda+A)^{-1}x = x - \lambda(\lambda+A)^{-1}x = \int_0^\infty \lambda e^{-\lambda t}[x - T(t)x]\,dt, \quad \lambda > 0,$$

which yields by Jensen's inequality (3.66) and Fubini's theorem

$$\begin{aligned}\int_0^\infty |\lambda^\alpha A(\lambda+A)^{-1}x|^p\,d\lambda/\lambda &= \int_0^\infty \lambda^{\alpha p}\Big|\int_0^\infty (T(t)x - x)\lambda e^{-\lambda t}\,dt\Big|^p\,d\lambda/\lambda \\ &\le \int_0^\infty \lambda^{\alpha p}\Big[\int_0^\infty |T(t)x - x|^p \lambda e^{-\lambda t}\,dt\Big]\,d\lambda/\lambda \\ &= \int_0^\infty |T(t)x - x|^p\Big[\int_0^\infty \lambda^{\alpha p} e^{-\lambda t}\,d\lambda\Big]\,dt \\ &= \int_0^\infty |T(t)x - x|^p \Gamma(\alpha p + 1) t^{-\alpha p - 1}\,dt\end{aligned}$$

where $\Gamma(z)$ denotes the Gamma function. This yields $[x]''_{\alpha,p} \le (\Gamma(\alpha p + 1))^p [x]'_{\alpha,p}$.

(c) $\Rightarrow$ (d). Suppose $[x]''_{\alpha,p} < \infty$. Define $u(t) = (1 + tA)^{-1}x$ for $t \in [0,1]$; then $u \in C([0,1];X) \cap C((0,1];X_A) \cap C^1((0,1];X)$, $u(0) = x$, and $\dot u(t) = -A(1+tA)^{-2}x$ for $t \in (0,1]$. The variable transformation $t = 1/\lambda$ gives

$$\begin{aligned}[[u]]_{\alpha,p} &= \Big[\int_0^1 |t^{1-\alpha}A(1+tA)^{-2}x|^p\,dt/t\Big]^{1/p} + \Big[\int_0^1 |t^{1-\alpha}A(1+tA)^{-1}x|^p\,dt/t\Big]^{1/p} \\ &\le C\Big[\int_0^1 |t^{1-\alpha}A(1+tA)^{-1}x|^p\,dt/t\Big]^{1/p} \\ &= C\Big[\int_1^\infty |\lambda^\alpha A(\lambda+A)^{-1}x|^p\,d\lambda/\lambda\Big]^{1/p} \\ &\le C[x]''_{\alpha,p}.\end{aligned}$$

This proves $x \in (X, X_A)_{\alpha,p}$ and $|x|_{(X,X_A)_{\alpha,p}} \le C|x|''_{\alpha,p}$.

(d) $\Rightarrow$ (a). Let $x \in (X, X_A)_{\alpha,p}$ and $w \in C([0,1];X) \cap C((0,1];X_A) \cap C^1((0,1];X)$ with $w(0) = x$, be such that

$$[[w]]_{\alpha,p} = \Big[\int_0^1 |t^{1-\alpha}\dot w(t)|^p\,dt/t\Big]^{1/p} + \Big[\int_0^1 |t^{1-\alpha}Aw(t)|^p\,dt/t\Big]^{1/p} < \infty.$$

Then the identity

$$x = w(0) = w(t) - \int_0^t \dot w(s)\,ds$$

implies by Lemma 3.4.5 with $\beta = 1/p' - \alpha$

$$\begin{aligned}
&\Big[\int_0^1 |t^{1-\alpha}AT(t)x|^p\,dt/t\Big]^{1/p}\\
&\le \Big[\int_0^1 |t^{1-\alpha}T(t)Aw(t)|^p\,dt/t\Big]^{1/p} + \Big[\int_0^1 \Big|t^{1-\alpha}AT(t)\int_0^t \dot w(s)ds\Big|^p\,dt/t\Big]^{1/p}\\
&\le C\Big[\int_0^1 |t^{1-\alpha}Aw(t)|^p\,dt/t\Big]^{1/p} + C\Big[\int_0^1 \Big|t^{-\alpha}\int_0^t \dot w(s)ds\Big|^p\,dt/t\Big]^{1/p}\\
&\le C\Big[\int_0^1 |t^{1-\alpha}Aw(t)|^p\,dt/t\Big]^{1/p} + C\alpha^{-p}\Big[\int_0^1 |t^{1-\alpha-1/p}\dot w(t)|^p\,dt\Big]^{1/p}\\
&\le C\Big[\int_0^1 |t^{1-\alpha}Aw(t)|^p\,dt/t\Big]^{1/p} + C\Big[\int_0^1 |t^{1-\alpha}\dot w(t)|^p\,dt/t\Big]^{1/p}.
\end{aligned}$$

Because of boundedness of $tAT(t)$ on $\mathbb{R}_+$ we also have

$$\int_1^\infty |t^{1-\alpha}AT(t)x|^p\,dt/t \le C|x|^p\int_1^\infty t^{-\alpha p-1}\,dt = C|x|^p/\alpha p,$$

hence we obtain $[x]_{\alpha,p} \le C(|x| + [[w]]_{\alpha,p})$, and since w has been arbitrary it is also clear that $[x]_{\alpha,p} \le C|x|_{(X,X_A)_{\alpha,p}}$ holds, for some constant C independent of x. The proof is complete. □

4.3 Embeddings

We continue the study of the trace spaces $D_A(\alpha,p)$ with some essential embedding results. For this purpose we extend the definition of $D_A(\alpha,p)$ to the cases $p = \infty, 0$.

$$D_A(\alpha,\infty) := \{x \in X : [x]_{D_A(\alpha,\infty)} := \sup_{\lambda>0} \lambda^\alpha|A(\lambda + A)^{-1}x| < \infty\},$$

and

$$D_A(\alpha,0) := \{x \in D_A(\alpha,\infty) : \lim_{\lambda\to\infty} \lambda^\alpha A(\lambda + A)^{-1}x = 0\}.$$

These definitions make sense for any pseudo-sectorial operator A in X. The norm in these spaces are

$$|x|_{D_A(\alpha,\infty)} = |x| + [x]_{D_A(\alpha,\infty)}.$$

Obviously the *continuous interpolation space* $D_A(\alpha,0)$ is a closed subspace of $D_A(\alpha,\infty)$.

Proposition 3.4.6. *Let A be a pseudo-sectorial operator in X with dense domain. Then for all $0 < \alpha < \beta < 1$, $1 \le p < q < \infty$, $r \in [1,\infty] \cup \{0\}$, we have*

(i) $\mathsf{D}(A) \hookrightarrow D_A(\beta,r) \hookrightarrow D_A(\alpha,r) \hookrightarrow X$;

(ii) $D_A(\beta,\infty) \hookrightarrow D_A(\alpha,1)$;

(iii) $D_A(\alpha,1) \hookrightarrow D_A(\alpha,p) \hookrightarrow D_A(\alpha,q) \hookrightarrow D_A(\alpha,0) \hookrightarrow D_A(\alpha,\infty)$;

(iv) $D_A(\alpha,1) \hookrightarrow \mathsf{D}(A^\alpha) \hookrightarrow D_A(\alpha,0)$;

(v) $\mathsf{D}(A) \subset D_A(\alpha,r)$ *is dense for each* $r \neq \infty$;

(vi) *if* $-A$ *generates a bounded* C_0*-semigroup in* X*, then its restriction to* $D_A(\alpha,r)$ *is also a bounded* C_0*-semigroup, for each* $r \neq \infty$.

Proof. (i) Since for $x \in \mathsf{D}(A)$, $t > 0$, we have

$$t^\alpha |A(t+A)^{-1}x| \leq Ct^{\alpha-1}|Ax|,$$

so the first inclusion is obvious. The second one follows from assertion (ii) and (iii), while the third one is trivial by definition of $D_A(\alpha,p)$.

(ii) Let $x \in D_A(\beta,\infty)$, $\beta > \alpha$; then

$$\int_1^\infty t^\alpha |A(t+A)^{-1}x| \frac{dt}{t} \leq |x|_{\beta,\infty} \int_1^\infty t^{\alpha-\beta-1}\,dt = \frac{|x|_{\beta,\infty}}{\beta-\alpha},$$

which implies assertion (ii).

(iii) Let $p \in [1,\infty)$, $x \in D_A(\alpha,p)$; then choosing a standard contour we obtain

$$t^\alpha A(t+A)^{-1}x = \frac{1}{2\pi i}\int_\Gamma \frac{t^\alpha \lambda^{1-\alpha}}{t+\lambda} \cdot \lambda^\alpha A(\lambda-A)^{-1}x \frac{d\lambda}{\lambda}.$$

For $p > 1$, by means of Hölder's inequality this gives

$$t^\alpha |A(t+A)^{-1}x| \leq \frac{1}{2\pi}\left[\int_\Gamma \left|\frac{t^\alpha \lambda^{1-\alpha}}{t+\lambda}\right|^{p'} \left|\frac{d\lambda}{\lambda}\right|\right]^{1/p'} \left[\int_\Gamma |\lambda^\alpha A(\lambda-A)^{-1}x|^p \left|\frac{d\lambda}{\lambda}\right|\right]^{1/p}.$$

Next observe that from the resolvent equation

$$(\lambda-A)^{-1} = (|\lambda|+A)^{-1}[-1+(\lambda+|\lambda|)(\lambda-A)^{-1}]$$

we obtain

$$|A(\lambda-A)^{-1}x| \leq (1+2|\lambda(\lambda-A)^{-1}|)|A(|\lambda|+A)^{-1}x| \leq C|A(|\lambda|+A)^{-1}x|.$$

Since by the variable transformation $\lambda = tz$

$$\int_\Gamma \left|\frac{t^\alpha\lambda^{1-\alpha}}{t+\lambda}\right|^{p'} \left|\frac{d\lambda}{\lambda}\right| = \int_\Gamma \left|\frac{z^{1-\alpha}}{1+z}\right|^{p'} \left|\frac{dz}{z}\right| < \infty,$$

we conclude

$$|t^\alpha A(t+A)^{-1}x| \leq C|x|_{\alpha,p},$$

which yields the embedding $D_A(\alpha,p) \hookrightarrow D_A(\alpha,\infty)$ in case $p > 1$. For $p = 1$ we use boundedness of $t^\alpha|\lambda|^{1-\alpha}/|t+\lambda|$ instead.

For $q > p$ we have from this

$$([x]''_{\alpha,q})^q = \int_0^\infty |t^\alpha A(t+A)^{-1}x|^q \frac{dt}{t}$$
$$\le \sup_{t>0} |t^\alpha A(t+A)^{-1}x|^{q-p} \int_0^\infty |t^\alpha A(t+A)^{-1}x|^p \frac{dt}{t}$$
$$\le [x]^{q-p}_{D_A(\alpha,\infty)}([x]''_{\alpha,p})^p \le C[x]^q_{\alpha,p},$$

which yields $D_A(\alpha,p) \hookrightarrow D_A(\alpha,q)$.

Finally, since $D_A(\alpha,0) \subset D_A(\alpha,\infty)$ is closed, the embedding $D_A(\alpha,p) \subset D_A(\alpha,0)$ follows from (v).

(iv) Let $x \in \mathsf{D}(A)$; then we know from Section 3.3.3

$$A^\alpha x = \frac{\sin(\alpha\pi)}{\pi} \int_0^\infty r^\alpha A(r+A)^{-1}x \frac{dr}{r}.$$

This easily implies the first inclusion in (iv), as $\mathsf{D}(A)$ is dense in $\mathsf{D}(A^\alpha)$.

On the other hand, for $x \in \mathsf{D}(A^\alpha)$ and $r > 0$ we have by the moment inequality

$$r^\alpha|A(r+A)^{-1}x| = r^\alpha|A^{1-\alpha}(r+A)^{-1}A^\alpha x| \le r^\alpha C r^{-\alpha}|A^\alpha x|.$$

This proves the second embedding in (iv), by density of $\mathsf{D}(A)$ in $D_A(\alpha,0)$.

(v) Since $\mathsf{D}(A) \subset X$ is dense by assumption, we have $x_\varepsilon := (1+\varepsilon A)^{-1}x \to x$ as $\varepsilon \to 0$, for each $x \in X$. Therefore $t^\alpha A(t+A)^{-1}(x - x_\varepsilon) \to 0$ for each $t > 0$. Since

$$|t^\alpha A(t+A)^{-1}(x-x_\varepsilon)| \le C|t^\alpha A(t+A)^{-1}x|,$$

for $x \in D_A(\alpha,p)$, Lebesgue's theorem implies $x_\varepsilon \to x$ also in $D_A(\alpha,p)$, i.e., $\mathsf{D}(A)$ is dense in $D_A(\alpha,p)$. To prove density of $\mathsf{D}(A)$ in $D_A(\alpha,0)$, observe that the set $\{t^\alpha A(t+A)^{-1}x :\ t > 0\}$ is relatively compact in X, in case $x \in D_A(\alpha,0)$. But this implies

$$t^\alpha A(t+A)^{-1}x_\varepsilon = (1+\varepsilon A)^{-1}t^\alpha A(t+A)^{-1}x \to t^\alpha A(t+A)^{-1}x$$

uniformly in $t > 0$, which shows $x_\varepsilon \to x$ also in $D_A(\alpha,0)$.

(vi) If $-A$ generates a bounded C_0-semigroup in X, it follows from the definition of the spaces $D_A(\alpha,r)$ that $T(t)$ is also bounded in $D_A(\alpha,p)$. Since $T(\cdot)x$ is continuous in $\mathsf{D}(A)$ for each $x \in \mathsf{D}(A)$, the density of the embedding $\mathsf{D}(A) \hookrightarrow D_A(\alpha,r)$ for $r \neq \infty$ implies that $T(t)$ is strongly continuous also in $D_A(\alpha,r)$, $r \neq \infty$. □

4.4 Interpolation of Intersections

The following result on real interpolation of intersections is very useful.

Theorem 3.4.7. *Let $A, B \in \mathcal{PS}(X)$ be densely defined and resolvent-commuting, $\alpha \in (0,1)$, $1 \le p < \infty$.*

Then $(X, \mathsf{D}(A) \cap \mathsf{D}(B))_{\alpha,p} \cong (X, \mathsf{D}(A))_{\alpha,p} \cap (X, \mathsf{D}(B))_{\alpha,p}$.

In particular, if $A+B$ with natural domain $\mathsf{D}(A+B) = \mathsf{D}(A) \cap \mathsf{D}(B)$ is pseudo-sectorial then

$$D_{A+B}(\alpha, p) \cong D_A(\alpha, p) \cap D_B(\alpha, p).$$

Proof. We may assume that A, B are sectorial and invertible. The inclusion " $\subset$ " is trivial. To prove the converse inclusion, let $x \in (X, \mathsf{D}(A))_{\alpha,p} \cap (X, \mathsf{D}(B))_{\alpha,p}$ be given. Define $u(t) = (I + tA)^{-1}(I + tB)^{-1}x$. As the resolvents of A and B commute, it is clear that $u \in C([0,1]; X) \cap C((0,1]; \mathsf{D}(A) \cap \mathsf{D}(B))$, and

$$|t^{1-\alpha-1/p} Au(t)|_p = |t^{1-\alpha-1/p}(I + tB)^{-1}A(I + tA)^{-1}x|_p \le M_B |x|_{D_A(\alpha,p)},$$

as well as

$$|t^{1-\alpha-1/p} Bu(t)|_p = |t^{1-\alpha-1/p}(I + tA)^{-1}B(I + tB)^{-1}x|_p \le M_A |x|_{D_B(\alpha,p)}.$$

Next we have $\dot{u}(t) = -(I+tB)^{-1}(I+tA)^{-1}(A(I+tA)^{-1}x + B(I+tB)^{-1}x)$, hence in the same way as above we obtain

$$|t^{1-\alpha-1/p}\dot{u}(t)|_p \le M_A M_B (|x|_{D_A(\alpha,p)} + |x|_{D_B(\alpha,p)}).$$

This shows the converse inclusion. □

4.5 Vector-Valued Fractional Sobolev, Besov and Bessel-Potential Spaces

(i) Let Y be a Banach space and $1 < p < \infty$, $\omega > 0$. Then B_p is sectorial in $X_0 := L_p(\mathbb{R}_+; Y)$ with domain $X_1 = {}_0H^1_p(\mathbb{R}_+; Y)$, and spectral angle $\pi/2$, according to Section 3.2.3. Then we define the *vector-valued Besov spaces* by

$$ {}_0B^\alpha_{pq}(\mathbb{R}_+; Y) := D_{B_p}(\alpha, q) = (X_0, X_1)_{\alpha,q}, \quad \alpha \in (0,1),\ q \in [1,\infty] \cup \{0\}, \quad (3.67)$$

and the *vector-valued fractional Sobolev spaces* by

$$ {}_0W^\alpha_p(\mathbb{R}_+; Y) := {}_0B^\alpha_{pp}(\mathbb{R}_+; Y) = D_{B_p}(\alpha, p) = (X_0, X_1)_{\alpha,p}, \quad \alpha \in (0,1). \quad (3.68)$$

(ii) This definition extends to the weighted spaces $X_{0,\mu} = L_{p,\mu}(\mathbb{R}_+; Y)$ for $1/p < \mu \le 1$, as $B_{p,\mu}$ is also sectorial in this space, with domain $X_{1,\mu} = {}_0H^1_{p,\mu}(\mathbb{R}_+; Y)$, by Proposition 3.2.9. So we set

$$ {}_0B^\alpha_{pq,\mu}(\mathbb{R}_+; Y) := D_{B_{p,\mu}}(\alpha, q) = (X_{0,\mu}, X_{1,\mu})_{\alpha,q}, \quad (3.69)$$

for $\alpha \in (0,1)$, $q \in [1,\infty] \cup \{0\}$, and

$$ {}_0W^\alpha_{p,\mu}(\mathbb{R}_+; Y) := {}_0B^\alpha_{pp,\mu}(\mathbb{R}_+; Y) = D_{B_{p,\mu}}(\alpha, p) = (X_{0,\mu}, X_{1,\mu})_{\alpha,p} \quad (3.70)$$

for $\alpha \in (0,1)$. We recall the isomorphism Φ_μ from Section 3.2.4 defined by $\Phi_\mu(u)(t) = t^{1-\mu}u(t)$ which maps $X_{j,\mu}$ onto X_j for $j = 0,1$, by Proposition 3.2.6. Interpolating these isomorphisms by the real method implies that

$$\Phi_\mu : {}_0B^\alpha_{pq,\mu}(\mathbb{R}_+;Y) \to {}_0B^\alpha_{pq}(\mathbb{R}_+;Y)$$

is an isomorphism as well, hence we have the characterizations

$$u \in {}_0B^\alpha_{pq,\mu}(\mathbb{R}_+;Y) \quad \Leftrightarrow \quad t^{1-\mu}u \in {}_0B^\alpha_{pq}(\mathbb{R}_+;Y),$$

and

$$u \in {}_0W^\alpha_{p,\mu}(\mathbb{R}_+;Y) \quad \Leftrightarrow \quad t^{1-\mu}u \in {}_0W^\alpha_{p}(\mathbb{R}_+;Y),$$

for all $\alpha \in (0,1)$, $q \in [1,\infty] \cup \{0\}$.

(iii) Similarly, as B_p is also sectorial in $L_p(\mathbb{R};Y)$, we define

$$B^\alpha_{pq}(\mathbb{R};Y) := (L_p(\mathbb{R};Y), H^1_p(\mathbb{R};Y))_{\alpha,q}, \quad W^\alpha_p(\mathbb{R};Y) := B^\alpha_{pp}(\mathbb{R};Y),$$

for $p \in (1,\infty)$, $\alpha \in (0,1)$, and $q \in [1,\infty] \cup \{0\}$. Next we let $B^\alpha_{pq,\mu}(\mathbb{R}_+;Y)$ be defined by

$$B^\alpha_{pq,\mu}(\mathbb{R}_+;Y) = (L_{p,\mu}(\mathbb{R}_+;Y), H^1_{p,\mu}(\mathbb{R}_+;Y))_{\alpha,q}.$$

(iv) The *vector-valued Bessel-potential spaces* $H^\alpha_p(\mathbb{R};Y)$, $H^\alpha_p(\mathbb{R}_+;Y)$, as well as ${}_0H^\alpha_p(\mathbb{R}_+;Y)$ and ${}_0H^\alpha_{p,\mu}(\mathbb{R}_+;Y)$ are defined in an analogous way, employing the complex interpolation method. From the isomorphism Φ_μ we deduce

$$u \in {}_0H^\alpha_{p,\mu}(\mathbb{R}_+;Y) \quad \Leftrightarrow \quad t^{1-\mu}u \in {}_0H^\alpha_{p}(\mathbb{R}_+;Y),$$

for all $p \in (1,\infty)$ and $\alpha \in (0,1)$.

(v) *Sobolev Embeddings.* Consider the operator $B = -d/dt$ in $X_0 = L_{p,\mu}(\mathbb{R}_+;Y)$ with maximal domain

$$X_1 = \mathsf{D}(B) = H^1_{p,\mu}(\mathbb{R}_+;Y).$$

Here we take $p \in (1,\infty)$, $\mu \in (1/p,1]$, $\alpha \in (0,1]$ and set $\beta := \alpha - 1 + \mu - 1/p$. Then for $\beta > 0$ the *Sobolev embedding* $\mathsf{D}(B^\alpha) \hookrightarrow C_0(\bar{\mathbb{R}}_+;Y)$ is valid. More precisely, there a is constant $C > 0$ such that

$$|u(t)|_Y \le C|u|_{\mathsf{D}(B^\alpha)}, \quad t \ge 0, \quad u \in \mathsf{D}(B^\alpha).$$

By Section 3.4.3 and general interpolation theory, this shows that $K^\alpha_{p,\mu}(\mathbb{R}_+;Y) \hookrightarrow C_0(\bar{\mathbb{R}}_+;Y)$ for $K \in \{W,H\}$, as long as $\beta > 0$.

In fact, it is easy to verify the identity

$$u(t) = \int_t^\infty e^{-(s-t)} \frac{(s-t)^{\alpha-1}}{\Gamma(\alpha)} (B+1)^\alpha u(s)\, ds, \quad s > 0,$$

for, say, $u \in \mathsf{D}(B)$. Applying Hölder's inequality, this relation implies

$$|u(t)|_Y \le \varphi_0(t)|(B+1)^\alpha u|_{X_0} \le C\varphi_0(t)|u|_{\mathsf{D}(B^\alpha)},$$

where

$$\varphi_0(t) = [\Gamma(\alpha)^{-1} \int_t^\infty e^{-p'(s-t)}(s-t)^{p'(\alpha-1)} s^{p'(\mu-1)}\, ds]^{1/p'}.$$

In case $\beta > 0$, an easy estimate yields

$$\sup_{t\ge 0}(1+t)^{(1-\mu)}\varphi_0(t) < \infty,$$

which proves the assertion, by density of $\mathsf{D}(B)$ in $\mathsf{D}(B^\alpha)$, and the embedding $H^1_p(\mathbb{R}_+;Y) \hookrightarrow C_0(\bar{\mathbb{R}}_+;Y)$.

We note that in case $\mu < 1$, $u(t)$ has even uniform polynomial decay as $t \to \infty$.

(vi) *Hölder Embeddings.* For $\beta > 0$ the *Hölder embedding* $\mathsf{D}(B^\alpha) \hookrightarrow \dot{C}^\beta_b(\bar{\mathbb{R}}_+;Y)$ is valid. More precisely, there is a constant $C > 0$ such that

$$|u(t+h) - u(t)|_Y \le Ch^\beta |B^\alpha u|_{X_0}, \quad t \ge 0, \quad u \in \mathsf{D}(B^\alpha).$$

By Section 3.4.3 and general interpolation theory, this shows $K^\alpha_{p,\mu}(\mathbb{R}_+;Y) \hookrightarrow C^{\beta-\varepsilon}_b(\bar{\mathbb{R}}_+;Y)$ for $K \in \{W, H\}$, as long as $\beta > \varepsilon > 0$. We observe that in case Y belongs to the class $\mathcal{HT}$, we may set $\varepsilon = 0$. In fact, in this case $\mathsf{D}(B^\alpha) = (X_0, X_1)_\alpha$ by Theorems 3.3.7 and by the analogue of Theorem 4.3.14 for B.

To prove the claim, as in **(v)** we use the identity

$$u(t) = \int_t^\infty \frac{(s-t)^{\alpha-1}}{\Gamma(\alpha)} B^\alpha u(s)\, ds, \quad s > 0,$$

where $u \in \mathsf{D}(B)$. Then for $t, h \ge 0$,

$$\begin{aligned} u(t+h) - u(t) &= \Gamma(\alpha)^{-1} \int_{t+h}^\infty [(s-(t+h))^{\alpha-1} - (s-t)^{\alpha-1}] B^\alpha u(s)\, ds \\ &\quad - \Gamma(\alpha)^{-1} \int_t^{t+h} (s-t)^{\alpha-1} B^\alpha u(s) ds =: I_1 + I_2. \end{aligned}$$

We estimate separately by Hölder's inequality.

$$\begin{aligned} |I_1| &\le [\Gamma(\alpha)^{-1} \int_{t+h}^\infty |(s-(t+h))^{\alpha-1} - (s-t)^{\alpha-1}|^{p'} s^{p'(\mu-1)}\, ds]^{1/p'} |B^\alpha u|_{X_0} \\ &=: \varphi_1(h)|B^\alpha u|_{X_0}, \end{aligned}$$

and

$$|I_2| \le [\Gamma(\alpha)^{-1} \int_t^{t+h} (s-t)^{p'(\alpha-1)} s^{p'(\mu-1)}\, ds]^{1/p'} |B^\alpha u|_{X_0} =: \varphi_2(h)|B^\alpha u|_{X_0}.$$

Next, we have

$$\varphi_1(h) \le c[\int_0^\infty (\tau^{\alpha-1} - (\tau+h)^{\alpha-1})^{p'}(\tau+h)^{p'(\mu-1)}\, d\tau]^{1/p'}$$
$$= ch^\beta[\int_0^\infty (r^{\alpha-1} - (r+1)^{\alpha-1})^{p'}(r+1)^{p'(\mu-1)}\, dr]^{1/p'},$$

and

$$\varphi_2(h) \le [\int_0^h \tau^{p'(\alpha+\mu-2)}\, d\tau]^{1/p'} = ch^\beta.$$

Both integrals are absolutely convergent as $p'(\alpha+\mu-2) = p'(\beta - 1/p') > -1$, provided $\beta > 0$. This proves the assertion.

4.6 A General Trace Theorem
We consider functions in the class $K^\alpha_{p,\mu}(\mathbb{R}_+;Y) \cap L_{p,\mu}(\mathbb{R}_+;D_A(\alpha,p))$, where $K \in \{W,H\}$, $1 \ge \mu > 1/p$, and $\alpha \in (0,1]$ (recall that $W^1_p = H^1_p$ for $p \in (1,\infty)$). For $\beta := \alpha-1+\mu-1/p > 0$ we have $K^\alpha_{p,\mu}(\mathbb{R}_+;Y) \hookrightarrow C(\bar{\mathbb{R}}_+;Y)$, so the question is what regularity the initial value $u_0 := u(0)$ of the function u enjoys. We want to prove the following result, which is employed at many places in subsequent sections.

Theorem 3.4.8. *Suppose A is a densely defined invertible sectorial operator in Y with spectral angle $\phi_A < \pi/2$, $p \in (1,\infty)$, $\mu \in (1/p,1]$, and $\beta := \alpha-1+\mu-1/p > 0$. Let $K \in \{H,W\}$, and set $Y_\alpha = D_A(\alpha,p)$ or $Y_\alpha = \mathsf{D}(A^\alpha)$.*

Then the trace map

$$\mathsf{tr} : K^\alpha_{p,\mu}(\mathbb{R}_+;Y) \cap L_{p,\mu}(\mathbb{R}_+;Y_\alpha) \to D_A(\beta,p), \quad \mathsf{tr} : u \mapsto u(0),$$

is linear and bounded. In particular, if $u \in K^\alpha_{p,\mu}(\mathbb{R}_+;Y)$ then the function $v = u - e^{-At}u_0$ belongs to ${}_0K^\alpha_{p,\mu}(\mathbb{R}_+;Y)$, and the trace map tr is surjective.

Note that the second assertion follows from Proposition 3.4.3.

Proof. (i) Observe that Hardy's inequality implies

$${}_0H^1_{p,\mu}(\mathbb{R}_+;Y) \hookrightarrow L_{p,\mu+1}(\mathbb{R}_+;Y),$$

hence interpolating with the trivial embedding

$$L_{p,\mu}(\mathbb{R}_+;Y) \hookrightarrow L_{p,\mu}(\mathbb{R}_+;Y)$$

we obtain by the complex method

$${}_0H^\alpha_{p,\mu}(\mathbb{R}_+;Y) \hookrightarrow L_{p,\mu+\alpha}(\mathbb{R}_+;Y),$$

and by the real method

$${}_0W^\alpha_{p,\mu}(\mathbb{R}_+;Y) \hookrightarrow L_{p,\mu+\alpha}(\mathbb{R}_+;Y),$$

for all $\alpha \in (0,1)$ and $1 \geq \mu > 1/p$.

(ii) We can now prove assertion **(c)** of Proposition 3.4.3. For this purpose, let $x \in D_A(\alpha - 1 + \mu - 1/p, p)$; then $u(t) = e^{-At}x - e^{-t}x \in {}_0H^1_{p,\mu+\alpha-1}(\mathbb{R}_+; X)$. Step (i) implies $u \in L_{p,\mu+\alpha}(\mathbb{R}_+; X)$, hence by complex interpolation $u \in {}_0H^\alpha_{p,\mu}(\mathbb{R}_+; X)$, hence $e^{-At}x \in H^\alpha_{p,\mu}(\mathbb{R}_+; X)$. On the other hand, using real interpolation of type (α, p) we obtain $u \in {}_0W^\alpha_{p,\mu}(\mathbb{R}_+; X)$, hence $e^{-At}x \in W^\alpha_{p,\mu}(\mathbb{R}_+; X)$. For the last assertion, observe that $v(t) = e^{-At}x - e^{-t}A^{-1}x$ as before belongs to $L_{p,\mu+\alpha}(\mathbb{R}_+; X)$, but it is also in $L_{p,\mu+\alpha-1}(\mathbb{R}_+; X_A)$ by Proposition 3.4.2. Hence complex interpolation yields $u \in L_{p,\mu}(\mathbb{R}_+; \mathsf{D}(A^\alpha))$, which proves the last statement in **(c)** of Proposition 3.4.3.

(iii) Let $u \in K^\alpha_{p,\mu}(\mathbb{R}_+; Y) \cap L_{p,\mu}(\mathbb{R}_+; D_A(\alpha, p))$ be given and set $u_0 := u(0)$. We decompose u_0 as

$$u_0 = \frac{1}{t}\int_0^t u(s)\,ds + \frac{1}{t}\int_0^t (u_0 - u(s))\,ds = u_1 + u_2.$$

This decomposition leads to

$$|u_0|_{D_A(\beta,p)} \leq |u_1|_{D_A(\beta,p)} + |u_2|_{D_A(\beta,p)} = I_1^{1/p} + I_2^{1/p}.$$

We first estimate I_1.

$$\begin{aligned}
I_1 &\leq \int_0^1 t^{-1-\beta p}\Big[\int_0^t |Ae^{-At}u(s)|ds\Big]^p\,dt \\
&\leq \int_0^1 t^{-1-\beta p}\Big[\int_0^t s^{p'(\mu-1)}\,ds\Big]^{p/p'}\int_0^t s^{p(1-\mu)}|Ae^{-At}u(s)|^p\,ds]dt \\
&= c_{p,\mu}\int_0^1 t^{-1-\beta p+p/p'+p\mu-p}\int_0^t s^{p(1-\mu)}|Ae^{-At}u(s)|^p\,ds]dt \\
&= c_{p,\mu}\int_0^1 s^{p(1-\mu)}\Big[\int_s^1 (t^{1-\alpha}|Ae^{-At}u(s)|)^p\,dt/t\Big]\,ds \leq c_{p,\mu}|u|^p_{L_{p,\mu}(\mathbb{R}_+;D_A(\alpha,p))},
\end{aligned}$$

where $c_{p,\mu} = (1 + p'(\mu-1))^{-p/p'}$.

In case $Y_\alpha = \mathsf{D}(A^\alpha)$, we use the moment inequality to obtain the estimate $|t^{1-\alpha}A^{1-\alpha}e^{-At}| \leq C$, and employ once more Hardy's inequality, to the result

$$\begin{aligned}
I_1 &\leq C\int_0^1 t^{-\mu p}\Big[\int_0^t |A^\alpha u(s)|\,ds\Big]^p\,dt \\
&\leq C\int_0^1 |A^\alpha u(s)|^p s^{p(1-\mu)}\,ds = C|u|_{L_{p,\mu}(\mathbb{R}_+;\mathsf{D}(A^\alpha))}.
\end{aligned}$$

Next we estimate I_2 by the bound C for tAe^{-At} and Hardy's inequality

$$I_2 = \int_0^1 t^{p(1-\beta)} \Big| Ae^{-At} t^{-1} \int_0^t (u(s) - u_0)\, ds \Big|^p \, dt/t$$
$$\leq C \int_0^1 t^{-1-\beta p - p} \Big[\int_0^t |u(s) - u_0|\, ds \Big]^p dt \leq C \int_0^1 |u(s) - u_0|^p \frac{ds}{s^{1+\beta p}}.$$

By the embeddings in part (i), the last term is bounded by $|u - u_0|^p_{K^\alpha_{p,\mu}((0,1);Y)}$. This completes the proof. □

Example 3.4.9. In this example Σ will always denote a compact sufficiently smooth hypersurface.

(i) Consider as a base space Y the space $Y = L_p(\Sigma)$. Let $A = 1 - \Delta_\Sigma$, $\mu \in (1/p, 1]$. Then for all $\alpha \in (0,1]$ we have

$$\mathsf{tr}[W^\alpha_{p,\mu}(\mathbb{R}_+; L_p(\Sigma)) \cap L_{p,\mu}(\mathbb{R}_+; W^{2\alpha}_p(\Sigma))] = W_p^{2\alpha-2+2\mu-2/p}(\Sigma).$$

This will later on be used for $\alpha = 1$, $\alpha = 1 - 1/2p$, and $\alpha = 1/2 - 1/2p$.

(ii) Consider as a base space Y again the space $Y = L_p(\Sigma)$. Let $A = (1 - \Delta_\Sigma)^2$, $\mu \in (1/p, 1]$. Then we have

$$\mathsf{tr}[W^{1/2-1/2p}_{p,\mu}(\mathbb{R}_+; L_p(\Sigma)) \cap L_{p,\mu}(\mathbb{R}_+; W^{2-2/p}_p(\Sigma))] = W_p^{4\mu-2-6/p}(\Sigma).$$

This result will be used in Section 6.6.

(iii) Consider as a base space Y the space $Y = H^2_p(\Sigma)$. Let $A = 1 - \Delta_\Sigma$, $\mu \in (1/p, 1]$. Then we have

$$\mathsf{tr}[W^{1-1/2p}_{p,\mu}(\mathbb{R}_+; H^2_p(\Sigma)) \cap L_{p,\mu}(\mathbb{R}_+; W^{4-1/p}_p(\Sigma))] = W_p^{2+2\mu-3/p}(\Sigma).$$

This result will be also used in Section 6.6.

(iv) Consider as a base space Y the space $Y = W^{2-1/p}_p(\Sigma)$. Let $A = (1 - \Delta_\Sigma)^{1/2}$, $\mu \in (1/p, 1]$. Then we have

$$\mathsf{tr}[H^1_{p,\mu}(\mathbb{R}_+; W^{2-1/p}_p(\Sigma)) \cap L_{p,\mu}(\mathbb{R}_+; W^{3-1/p}_p(\Sigma))] = W_p^{2+\mu-2/p}(\Sigma).$$

This result will be used in Chapter 8.

3.5 Maximal L_p-Regularity

5.1 Maximal L_p-Regularity

Let $J = \mathbb{R}_+$ or $(0, a)$ for some $a > 0$ and let $f : J \to X$. We consider the inhomogeneous initial value problem

$$\dot{u}(t) + Au(t) = f(t), \quad t \in J, \quad u(0) = u_0, \tag{3.71}$$

in $L_p(J; X)$ for $p \in (1, \infty)$.

The definition of *maximal L_p-regularity* for (3.71) is as follows.

Definition 3.5.1. *Suppose $A : \mathsf{D}(A) \subset X \to X$ is closed and densely defined. Then A is said to belong to the class $\mathcal{MR}_p(J;X)$ – and we say that there is* **maximal L_p-regularity** *for* (3.71) *– if for each $f \in L_p(J;X)$ there exists a unique $u \in H^1_p(J;X) \cap L_p(J;X_A)$ satisfying* (3.71) *a.e. in J, with $u_0 = 0$.*

The closed graph theorem implies then that there exists a constant $C > 0$ such that

$$|u|_{L_p(J;X)} + |\dot{u}|_{L_p(J;X)} + |Au|_{L_p(J;X)} \le C|f|_{L_p(J;X)}. \tag{3.72}$$

Combining L_p-maximal regularity with Section 3.4.1 we then obtain for the solution of (3.71) the estimate

$$|u|_{L_p(J;X)} + |\dot{u}|_{L_p(J;X)} + |Au|_{L_p(J;X)} \le C\big(|u_0|_{D_A(1-1/p,p)} + |f|_{L_p(J;X)}\big). \tag{3.73}$$

We denote the solution operator $f \mapsto u$ by $\mathcal{R}$. It is well known that there is maximal L_p regularity for (3.71) only if $-A$ generates an analytic semigroup. If $J = \mathbb{R}_+$, then the semigroup is even of negative exponential type. We state this as

Proposition 3.5.2. *Let $A \in \mathcal{MR}_p(J;X)$ for some $p \in (1,\infty)$.*

Then the following assertions are valid.

(i) *If $J = (0,a)$ then there are constants $\omega \ge 0$ and $M \ge 1$ such that*

$$\{z \in \mathbb{C} : \operatorname{Re} z \le -\omega\} \subset \rho(A) \quad \text{and} \quad |z(z+A)^{-1}|_{\mathcal{B}(X)} \le M, \quad \operatorname{Re} z \ge \omega,$$

is valid. In particular, $\omega + A$ is sectorial with spectral angle $< \pi/2$.

(ii) *If $J = \mathbb{R}_+$ then $\mathbb{C}_- := \{z \in \mathbb{C} : \operatorname{Re} z < 0\} \subset \rho(A)$ and there is a constant $M \ge 1$ such that*

$$|(z+A)^{-1}|_{\mathcal{B}(X)} \le \frac{M}{1+|z|}, \quad \operatorname{Re} z > 0,$$

is valid. In particular, A is sectorial with spectral angle $< \pi/2$ and $0 \in \rho(A)$.

Proof. Consider first the case $J = (0,a)$. We show that there are constants $\omega_1 \ge 0$ and $M \ge 1$ such that

$$|\mu||x|_X + |x|_{X_A} \le M|(\mu + A)x|_X, \quad x \in \mathsf{D}(A), \ \operatorname{Re}\mu > \omega_1. \tag{3.74}$$

In particular, $\mu + A$ is injective for each $\operatorname{Re}\mu > \omega_1$. Indeed, choose $\mu \in \mathbb{C}_+$, $x \in \mathsf{D}(A)$, and and let $v_\mu(t) := e^{\mu t}x$. Then v_μ satisfies $\dot{v}_\mu + Av_\mu = g_\mu(t)$ and $v_\mu(0) = x$, where $g_\mu(t) = e^{\mu t}(\mu + A)x \in L_p(J;X)$. The maximal regularity estimate (3.73) implies

$$|e^{t\operatorname{Re}\mu}|_{L_p(J;X)}\big(\mu|x|_X + |x|_{X_A}\big) \le C\big(|e^{t\operatorname{Re}\mu}|_{L_p(J;X)}|(\mu+A)x|_X + |x|_{X_A}\big).$$

Choosing ω_1 large enough such that $2C \le |e^{t\operatorname{Re}\mu}|_{L_p(J;X)}$ yields (3.74).

In a next step, which is more involved, we show that there is a constant $\omega_2 \ge 0$ such that $\mu + A$ is surjective for $\operatorname{Re}\mu > \omega_2$. Choose $\mu \in \mathbb{C}_+$, $x \in X$, and define $f_\mu \in L_p(\mathbb{R}_+;X)$ by $f_\mu(t) = e^{-\mu t}x$. Let $u_\mu(t;x) = \mathcal{R}(f_\mu)(t)$, where $\mathcal{R}$ denotes the solution operator for (3.71) with $u_0 = 0$. Set

$$U_\mu x := 2\mathrm{Re}\,\mu \int_0^a e^{-\bar{\mu}t} u_\mu(t;x)\,dt = \frac{2\mathrm{Re}\,\mu}{\bar{\mu}}\Big[\int_0^a e^{-\bar{\mu}t}\dot{u}_\mu(t;x)\,dt - e^{-\bar{\mu}a}u_\mu(a;x)\Big].$$

The maximal regularity property for (3.71) implies that there exists a constant $C > 0$ such that

$$|U_\mu|_{\mathcal{B}(X)} \le C(1+|\mu|)^{-1}, \quad \mathrm{Re}\,\mu > 0,$$

where ω is sufficiently large. In fact, we have with Hölder's inequality and the maximal regularity estimate (3.72)

$$|U_\mu x| \le 2(p'\mathrm{Re}\,\mu)^{1-1/p'}|u_\mu|_{L_p(J;X)} \le C(\mathrm{Re}\,\mu)^{1/p}|f_\mu|_{L_p(J;X)} \le C|x|,$$

as well as

$$\begin{aligned} |U_\mu x| &\le 2\mathrm{Re}\,\mu|\mu|^{-1}\big[(p'\mathrm{Re}\,\mu)^{-1/p'} + e^{-a\mathrm{Re}\,\mu}a^{1/p'}\big]|\dot{u}_\mu|_{L_p(J;X)} \\ &\le C|\mu|^{-1}(\mathrm{Re}\,\mu)^{1/p}|f_\mu|_{L_p(J;X)} \le |\mu|^{-1}C|x|. \end{aligned}$$

Next we multiply (3.71) with $f = f_\mu$ by $e^{-\bar{\mu}t}$ and integrate over J. This yields by closedness of A and an integration by parts

$$\begin{aligned} (1-e^{-2a\mathrm{Re}\,\mu})x &= 2\mathrm{Re}\,\mu\int_0^a e^{-\bar{\mu}t}f_\mu(t)\,dt = 2\mathrm{Re}\,\mu\int_0^a e^{-\bar{\mu}t}[\dot{u}_\mu(t;x) + Au_\mu(t;x)]\,dt \\ &= (\bar{\mu}+A)U_\mu x + 2(\mathrm{Re}\,\mu)e^{-\bar{\mu}a}u_\mu(a;x), \end{aligned}$$

which after rearrangement becomes

$$(\bar{\mu}+A)U_\mu x = x - V_\mu x, \quad V_\mu x := e^{-2a\mathrm{Re}\,\mu}x + 2(\mathrm{Re}\,\mu)e^{-\bar{\mu}a}u_\mu(a;x).$$

Estimating as before we obtain

$$|V_\mu x| \le \big[e^{-2a\mathrm{Re}\,\mu} + Ce^{-a\mathrm{Re}\,\mu}(a\mathrm{Re}\,\mu)^{1/p'}\big]|x|,$$

from which we see that there is $\omega_2 > 0$ such that $|V_\mu|_{\mathcal{B}(X)} \le 1/2$, for each $\mathrm{Re}\,\mu \ge \omega_2$. This then shows that $\bar{\mu}+A$ is surjective for all such μ. Setting $\omega = \max\{\omega_1,\omega_2\}$ we conclude that $\mu + A : D(A) \to X$ is invertible, and

$$(\bar{\mu}+A)^{-1} = U_\mu(1-V_\mu)^{-1}, \quad \mathrm{Re}\,\mu > \omega.$$

The estimate on U_μ (or the a priori estimate in (3.74)) then shows that $\omega + A$ is sectorial with spectral angle $< \pi/2$.

For the case $J = \mathbb{R}_+$ the proof is simpler; one deduces in the same way the relation $(\bar{\mu}+A)^{-1} = U_\mu$ with $\omega = 0$. □

There is variant of maximal L_p-regularity if one requires for the solution of (3.71) only $u \in C(\bar{\mathbb{R}}_+; X)$ and $\dot{u}, Au \in L_p(\mathbb{R}_+; X)$. We call the class of operators with this weaker property ${}_0\mathcal{MR}_p(\mathbb{R}_+; X)$. The proof of Proposition 3.5.2 shows that then in (ii) the condition $0 \in \rho(A)$ is dropped. More precisely we have

Corollary 3.5.3. *Suppose $A \in {}_0\mathcal{MR}_p(\mathbb{R}_+; X)$.*

Then A is pseudo-sectorial in X with spectral angle $< \pi/2$.
Moreover, $A \in \mathcal{MR}_p(\mathbb{R}_+; X)$ if and only if $A \in {}_0\mathcal{MR}_p(\mathbb{R}_+; X)$ and $0 \in \rho(A)$.

Proposition 3.5.2 shows that for a finite interval $J = (0, a)$ its length $a > 0$ plays no role for maximal L_p-regularity, and up to a shift of A, without loss of generality, we may consider $J = \mathbb{R}_+$ and may assume that $-A$ is the generator of an analytic semigroup of negative exponential type. Therefore, in the sequel we mostly consider $J = \mathbb{R}_+$ and abbreviate $\mathcal{MR}_p(X) = \mathcal{MR}_p(\mathbb{R}_+; X)$ as well as ${}_0\mathcal{MR}_p(X) = {}_0\mathcal{MR}_p(\mathbb{R}_+; X)$.

Unfortunately, the converse of Proposition 3.5.2 is false. Actually, it is a formidable task to prove that a given operator A belongs to $\mathcal{MR}_p(X)$. We want to explain the difficulty in more detail. Obviously, the variation of parameters formula

$$u(t) = e^{-At}u_0 + \int_0^t e^{-A(t-s)} f(s)\, ds, \quad t \geq 0,$$

implies that there is maximal L_p-regularity for (3.71) if and only if the operator $\mathcal{R}$ defined by

$$\mathcal{R}f := A \int_0^t e^{-A(t-s)} f(s)\, ds$$

acts as a bounded operator on $L_p(\mathbb{R}_+; X)$. It is nontrivial to show this since the kernel of this convolution operator on the half-line is Ae^{-At} which has a non-integrable singularity near $t = 0$, behaving like $1/t$, as follows from the well-known, best possible estimate

$$|Ae^{-At}|_{\mathcal{B}(X)} \leq \frac{Me^{-\eta t}}{t}, \quad t > 0,$$

valid for exponentially stable analytic semigroups. Therefore, $\mathcal{R}$ is a *singular* integral operator on $L_p(\mathbb{R}_+; X)$ with operator-valued kernel. This calls for vector-valued harmonic analysis and we take up this topic in the next chapter.

5.2 Maximal Regularity in Weighted L_p-Spaces

We next study maximal regularity in spaces $L_{p,\mu}$. The main result of this section reads as follows.

Theorem 3.5.4. *Let X be a Banach space, $p \in (1, \infty)$, and $1/p < \mu \leq 1$. Then*

$$A \in \mathcal{MR}_p(X) \text{ if and only if } A \in \mathcal{MR}_{p,\mu}(X).$$

Proof. In the following we shall use the notation $X_0 := X$ and $X_1 := X_A$. It follows that X_1 is a Banach space which is densely embedded in X_0.
(i) Suppose that $A \in \mathcal{MR}_p(X)$. Then we know by Proposition (3.5.2) that $-A$ generates an exponentially stable analytic semigroup $\{e^{-tA} : t \geq 0\}$ on X_0. Let $f \in L_{p,\mu}(\mathbb{R}_+; X_0)$ be given. Let us consider the function u defined by the variation of constants formula

$$u(t) := \int_0^t e^{-(t-s)A} f(s)\, ds, \quad t > 0. \tag{3.75}$$

It follows from Lemma 3.2.5(a) that this integral exists in X_0. We will now rewrite equation (3.75) in the following way

$$\begin{aligned} u(t) &= t^{\mu-1} \int_0^t e^{-(t-s)A} s^{1-\mu} f(s)\, ds + t^{\mu-1} \int_0^t e^{-(t-s)A} [(t/s)^{1-\mu} - 1] s^{1-\mu} f(s)\, ds \\ &= \Phi_\mu^{-1}[(B_p + A)^{-1}\Phi_\mu f + T_A \Phi_\mu f] = \Phi_\mu^{-1}[v_1 + v_2]. \end{aligned}$$

Here we use the same notation for A and its canonical extension on $L_p(\mathbb{R}_+; X_0)$, given by $(Au)(t) := Au(t)$ for $t > 0$. By definition, T_A is the integral operator

$$(T_A g)(t) := \int_0^t e^{-(t-s)A} [(t/s)^{1-\mu} - 1] g(s)\, ds, \quad g \in L_p(\mathbb{R}_+; X_0).$$

Observe that the kernel $K_A(t) := Ae^{-tA}$ satisfies the assumptions of Proposition 4.3.13 below with $Y = X_1$. We conclude that

$$T_A \in \mathcal{B}(L_p(\mathbb{R}_+; X_0), L_p(\mathbb{R}_+; X_1)). \tag{3.76}$$

It is a consequence of (3.76) that v_2 has a derivative almost everywhere, given by

$$\dot{v}_2 = -AT_A\Phi_\mu f + (1-\mu)t^{-\mu} \int_0^t e^{-(t-s)A} f(s)\, ds.$$

It follows from Hardy's inequality, Lemma 3.4.5, that

$$\int_0^\infty \Big| t^{-\mu} \int_0^t e^{-(t-s)A} f(s)\, ds \Big|^p dt \leq M \int_0^\infty \Big(t^{-\mu} \int_0^t |f(s)|\, ds \Big)^p dt \leq cM |f|^p_{L_{p,\mu}}$$

and we infer that

$$v_2 \in {}_0H^1_p(\mathbb{R}_+; X_0) \cap L_p(\mathbb{R}_+; X_1). \tag{3.77}$$

It follows from our assumption that v_1 enjoys the same regularity properties as v_2 and consequently, v satisfies (3.77) as well. Proposition 3.2.6 then shows that

$$u \in {}_0H^1_{p,\mu}(\mathbb{R}_+; X_0) \cap L_{p,\mu}(\mathbb{R}_+; X_1). \tag{3.78}$$

It is now easy to verify that u is in fact a solution of the Cauchy problem (3.71) with initial value 0. We have thus shown that $A \in \mathcal{MR}_{p,\mu}(X)$.

(b) Suppose now that $A \in \mathcal{MR}_{p,\mu}(X_0)$. As in the case $\mu = 1$ one shows that A generates a bounded analytic C_0-semigroup $\{e^{-tA}; t \geq 0\}$ on X_0. Let $f \in L_p(\mathbb{R}_+; X_0)$ be given. Here we use the representation

$$\begin{aligned} u(t) &= t^{1-\mu} \int_0^t e^{-(t-s)A} s^{\mu-1} f(s)\, ds - \int_0^t e^{-(t-s)A}[(t/s)^{1-\mu} - 1] f(s)\, ds \\ &= \Phi_\mu (B_{p,\mu} + A)^{-1} \Phi_\mu^{-1} f - T_A f, \end{aligned}$$

with T_A as above. The assertion follows now by similar arguments as in (a). □

We will now consider the Cauchy problem (3.71) in $L_{p,\mu}(\mathbb{R}_+; X)$. Define the function spaces

$$\begin{aligned} \mathbb{E}_{0,\mu} &:= \mathbb{E}_{0,\mu}(\mathbb{R}_+) : = L_{p,\mu}(\mathbb{R}_+; X_0), \\ \mathbb{E}_{1,\mu} &:= \mathbb{E}_{1,\mu}(\mathbb{R}_+) : = H^1_{p,\mu}(\mathbb{R}_+; X_0) \cap L_{p,\mu}(\mathbb{R}_+; X_1), \end{aligned}$$

where $X_0 := X$ and $X_1 := X_A$. It is not difficult to verify that the norm

$$|u|_{\mathbb{E}_{1,\mu}} := \left(|u|^p_{L_{p,\mu}(\mathbb{R}_+;X_1)} + |\dot{u}|^p_{L_{p,\mu}(\mathbb{R}_+;X_0)}\right)^{1/p} \tag{3.79}$$

turns $\mathbb{E}_{1,\mu}(\mathbb{R}_+)$ into a Banach space. The result reads as follows

Theorem 3.5.5. *Let $p \in (1, \infty)$ and $1/p < \mu \leq 1$. Suppose that $A \in \mathcal{MR}_p(X)$. Then*

$$\left(\frac{d}{dt} + A, \mathsf{tr}\right) \in \mathrm{Isom}(\mathbb{E}_{1,\mu}(\mathbb{R}_+), \mathbb{E}_{0,\mu}(\mathbb{R}_+) \times X_{\gamma,\mu}),$$

where $\mathsf{tr}(u) := u(0)$ denotes the trace operator, and $X_{\gamma,\mu} = D_A(\mu - 1/p, p)$.

Proof. We observe that $(\frac{d}{dt} + A) \in \mathcal{B}(\mathbb{E}_{1,\mu}, \mathbb{E}_{0,\mu})$ and $\mathsf{tr} \in \mathcal{B}(\mathbb{E}_{1,\mu}, X_{\gamma,\mu})$ yield boundedness of $(\frac{d}{dt} + A, \mathsf{tr})$. Theorem 3.5.4 shows that the operator $(B_{p,\mu} + A)$ with domain

$$\mathsf{D}(B_{p,\mu} + A) = \mathsf{D}(B_{p,\mu}) \cap \mathsf{D}(A) = \{u \in \mathbb{E}_{1,\mu}(\mathbb{R}_+) : u(0) = 0\}$$

is invertible. Let $(f, u_0) \in \mathbb{E}_{0,\mu} \times X_{\gamma,\mu}$ be given and let

$$u := (B_{p,\mu} + A)^{-1} f + e^{-tA} u_0. \tag{3.80}$$

Clearly, u solves the Cauchy problem (3.71). Therefore, $(\frac{d}{dt} + A, \mathsf{tr})$ is surjective. The assertion follows now from the open mapping theorem. □

If $1 < p < \infty$ and $\mu = 1$ the semigroup of translations $T(\tau)u(t) = u(t + \tau)$ is strongly continuous in $\mathbb{E}_{1,1}$, which implies that the time-trace tr maps $\mathbb{E}_{1,1}$ into $C(\bar{\mathbb{R}}_+; X_{\gamma,1})$, with bound

$$\sup_{t \geq \tau} |u(t)|_{X_{\gamma,1}} \leq C |T(\tau)u|_{\mathbb{E}_{1,1}} \to 0 \quad \text{as } \tau \to \infty.$$

Therefore, we have the embedding

$$\mathbb{E}_{1,1}(\mathbb{R}_+) \hookrightarrow C_0(\bar{\mathbb{R}}_+; X_{\gamma,1}). \tag{3.81}$$

On the other hand, as the time weights $t^{1-\mu}$ act only near $t = 0$ we obtain

$$\mathbb{E}_{1,\mu}(\mathbb{R}_+) \hookrightarrow \mathbb{E}_{1,1}(\delta, \infty), \quad \text{for each } \delta > 0.$$

This implies

$$\mathbb{E}_{1,\mu}(\mathbb{R}_+) \hookrightarrow C(\bar{\mathbb{R}}_+; X_{\gamma,\mu}) \cap C_0(\mathbb{R}_+; X_{\gamma,1}), \tag{3.82}$$

which shows parabolic regularization. This will be very useful in later chapters.

It is sometimes important to also have solvability results for the non-autonomous problem

$$\dot{u} + A(t)u = f(t), \quad t > 0, \quad u(0) = u_0.$$

This is the content of the next proposition.

Proposition 3.5.6. *Suppose $A \in C(J, \mathcal{B}(X_1, X_0))$ and $A(t) \in \mathcal{M}_p(J, X_0)$ for each $t \in J = [0, a]$. Then*

$$\Big(\frac{d}{dt} + A(\cdot), \mathsf{tr}\Big) \in \mathrm{Isom}(\mathbb{E}_{1,\mu}(J), \mathbb{E}_{0,\mu}(J) \times X_{\gamma,\mu}).$$

In particular, the non-autonomous problem

$$\dot{u} + A(t)u = f(t), \quad t \in \dot{J}, \quad u(0) = u_0,$$

admits for each $(f, u_0) \in \mathbb{E}_{0,\mu}(J) \times X_{\gamma,\mu}$ a unique solution $u \in \mathbb{E}_{1,\mu}(J)$.

Proof. (ii) As $(\frac{d}{dt} + A(\cdot), \mathsf{tr}) \in \mathcal{B}(\mathbb{E}_{1,\mu}(J), \mathbb{E}_{0,\mu}(J) \times X_{\gamma,\mu})$ it suffices to show that $(\frac{d}{dt} + A(\cdot), \mathsf{tr})$ is bijective, thanks to the open mapping theorem. By a perturbation and compactness argument one shows that there is a constant M such that

$$\Big|\Big(\frac{d}{dt} + A(s), \mathsf{tr}\Big)^{-1}\Big|_{\mathcal{B}(\mathbb{E}_{1,\mu}(J), \mathbb{E}_{0,\mu}(J) \times X_{\gamma,\mu})} \leq M, \quad s \in J.$$

By compactness of J we can choose points $0 = s_0 < s_1 \cdots < s_{m+2} = a$ such that

$$\max_{s_j \leq t \leq s_{j+2}} |A(t) - A(s_j)|_{\mathcal{B}(X_1, X_0)} \leq 1/2M, \quad j = 0, \ldots, m.$$

A Neumann series argument then yields with $J_j = (s_j, s_{j+1})$

$$\Big(\frac{d}{dt} + A(\cdot), \mathsf{tr}\Big) \in \mathrm{Isom}(\mathbb{E}_{1,\mu}(J_j), \mathbb{E}_{0,\mu}(J_j) \times X_{\gamma,\mu}), \quad j = 0, \ldots, m. \tag{3.83}$$

Let $(f, x) \in \mathbb{E}_{0,\mu}(J) \times X_{\gamma,\mu}$ be given. Then we solve the problem with maximal $L_{p,\mu}$-regularity on the first interval J_0. The final value $u(s_1)$ then belongs to X_γ, hence we solve the problem on J_1 with this initial value and maximal L_p-regularity, and then by induction on all of the remaining intervals. □

5.3 Maximal $L_{2,\mu}$-Regularity in Hilbert Spaces

Let X be a Hilbert space and let A be pseudo-sectorial with $\phi_A < \frac{\pi}{2}$. Then $-A$ is the generator of a bounded holomorphic C_0-semigroup, in particular the domain of A is also dense in X. In this subsection we want to consider the L_2-theory of the abstract Cauchy problem

$$\dot{u}(t) + Au(t) = f(t), \quad t > 0, \quad u(0) = u_0, \tag{3.84}$$

where $f \in L_{2,\mu}(\mathbb{R}_+; X)$. It is the purpose of this subsection to give a simple proof of maximal-L_2-regularity in this case.

Theorem 3.5.7. *Let X be a Hilbert space and $A \in \mathcal{PS}(X)$ and such that $\phi_A < \frac{\pi}{2}$. Then $A \in {}_0\mathcal{MR}_2(X)$.*

Proof. The proof of the result follows by the vector-valued Paley-Wiener theorem on the halfline which is valid in a Hilbert space setting. This result states that in case X is a Hilbert space, the Laplace transform is an isometric isomorphism from $L_2(\mathbb{R}_+; X)$ onto the vector-valued Hardy space $H_2(\mathbb{C}_+; X)$ equipped with the norm

$$|u|^2_{H_2(\mathbb{C}_+;X)} = \frac{1}{2\pi}\int_{\mathbb{R}} |u(i\rho)|^2 d\rho.$$

Let $f \in \mathcal{D}(\mathbb{R}_+; X)$ first. Then (3.84) admits a unique strong solution u. Laplace transform yields

$$\widehat{u}(\lambda) = (\lambda + A)^{-1}\widehat{f}(\lambda), \quad \operatorname{Re}\lambda > 0.$$

Uniform boundedness of $\lambda(\lambda + A)^{-1}$ on $\mathbb{C}_+$ then implies

$$|\lambda\widehat{u}(\lambda)| + |A\widehat{u}(\lambda)| \leq C|\widehat{f}(\lambda)|, \quad \operatorname{Re}\lambda > 0,$$

with a constant $C > 0$ depending only on A, hence by the Paley-Wiener theorem

$$|\dot{u}|_{L_2(\mathbb{R}_+;X)} + |Au|_{L_2(\mathbb{R}_+;X)} \leq C|f|_{L_2(\mathbb{R}_+;X)}. \tag{3.85}$$

Now $\mathcal{D}(\mathbb{R}_+; X)$ is dense in $L_2(\mathbb{R}_+; X)$, hence a standard approximation argument applies to obtain this estimate also for arbitrary $f \in L_2(\mathbb{R}_+; X)$. □

5.4 Maximal L_p-Regularity in Real Interpolation Spaces

It is a remarkable fact that maximal L_p-regularity holds in the real interpolation spaces $D_A(\alpha, p)$ if $-A$ generates an analytic C_0-semigroup in X. This is the content of the following result.

Theorem 3.5.8. *Let X be a Banach space, $A \in \mathcal{S}(X)$ invertible with $\phi_A < \pi/2$, let $\alpha \in (0, 1)$, and $p \in [1, \infty)$.*

Then $A \in \mathcal{MR}_p(D_A(\alpha, p))$.

Proof. Let $f \in L_p(\mathbb{R}_+; D_A(\alpha, p))$ be given and set $u = e^{-At} * f$; we have to prove

$$|Au|_{L_p(\mathbb{R}_+;D_A(\alpha,p))} \leq C|f|_{L_p(\mathbb{R}_+;D_A(\alpha,p))},$$

for some constant $C > 0$ independent of f. For this purpose, note that

$$|Ae^{-A\tau}Au(t)| \leq \int_0^t |A^2 e^{-A(\tau+s)} f(t-s)|\, ds \leq M \int_0^t |Ae^{-A(\tau+s)} f(t-s)|(\tau+s)^{-1}\, ds,$$

hence by Hölder's inequality

$$|Ae^{-A\tau}Au(t)|^p \leq M\Big[\int_0^t (\tau+s)^{-ap'}\, ds\Big]^{p/p'} \int_0^t |Ae^{-A(\tau+s)} f(t-s)|^p (\tau+s)^{-bp}\, ds,$$

where $a + b = 1$ and $a > 1/p'$ to ensure

$$\Big[\int_0^t (\tau+s)^{-ap'} ds\Big]^{p/p'} \leq \Big[\int_0^\infty (\tau+s)^{-ap'} ds\Big]^{p/p'} = c_1 \tau^{p(1/p'-a)} < \infty.$$

Integrating over $t > 0$ and using Fubini's theorem, this yields

$$\begin{aligned}
|Ae^{-\tau A}Au|^p_{L_p(\mathbb{R}_+;X)} &\leq c_1 M \tau^{p(1/p'-a)} \int_0^\infty \int_s^\infty |Ae^{-A(\tau+s)} f(t-s)|^p (\tau+s)^{-bp}\, dtds\\
&= c_1 M \tau^{p(1/p'-a)} \int_0^\infty \int_0^\infty |Ae^{-A(\tau+s)} f(t)|^p (\tau+s)^{-bp}\, dtds.
\end{aligned}$$

From this estimate we obtain integrating over $\tau > 0$ with weight $\tau^{p(1-\alpha)-1}$, using again Fubini's theorem

$$\begin{aligned}
|Au|^p_{L_p(\mathbb{R}_+;D_A(\alpha,p))} &\leq c_1 M \int_0^\infty \int_0^\infty \int_0^\infty \tau^{\beta-1} |Ae^{-A(\tau+s)} f(t)|^p (\tau+s)^{-bp} dsd\tau dt\\
&= c_1 M \int_0^\infty \int_0^\infty \int_\tau^\infty \tau^{\beta-1} |Ae^{-As} f(t)|^p s^{-bp} dsd\tau dt\\
&= c_1 M \int_0^\infty \int_0^\infty |Ae^{-As} f(t)|^p \int_0^s \tau^{\beta-1} d\tau s^{-bp} dsdt\\
&= c_1 M \beta^{-1} \int_0^\infty \int_0^\infty |Ae^{-As} f(t)|^p s^{\beta-bp} dsdt\\
&= c_1 M \beta^{-1} |f|_{L_p(\mathbb{R}_+;D_A(\alpha,p))},
\end{aligned}$$

with $\beta = (1-\alpha)p + p/p' - ap > 0$ provided $a < 1 - \alpha + 1/p'$, and then $\beta - bp = (1-\alpha)p - 1$. The argument for $p = 1$ is similar and even simpler. □

Chapter 4

Vector-Valued Harmonic Analysis

In this chapter, operator-valued Fourier multiplier results for vector-valued L_p-spaces are derived and discussed. These form the basic tools for the proof of various results on maximal L_p-regularity which are needed for the nonlinear problems.

4.1 $\mathcal{R}$-Boundedness

A central concept in modern analysis is $\mathcal{R}$-boundedness of families of operators. By means of this notion stochastic analysis is introduced into operator theory.

1.1 $\mathcal{R}$-Bounded Families of Operators
We begin with the definition of *$\mathcal{R}$-boundedness.*

Definition 4.1.1. *Let X and Y be Banach spaces. A family of operators $\mathcal{T} \subset \mathcal{B}(X,Y)$ is called* $\mathcal{R}$**-bounded**, *if there is a constant $C > 0$ and $p \in [1,\infty)$ such that for each $N \in \mathbb{N}$, $T_j \in \mathcal{T}$, $x_j \in X$ and for all independent, symmetric, $\{-1,1\}$-valued random variables ε_j on a probability space $(\Omega, \mathcal{A}, \mu)$ the inequality*

$$\Big|\sum_{j=1}^{N} \varepsilon_j T_j x_j\Big|_{L_p(\Omega;Y)} \leq C\Big|\sum_{j=1}^{N} \varepsilon_j x_j\Big|_{L_p(\Omega;X)} \tag{4.1}$$

is valid. The smallest such C is called $\mathcal{R}$-bound *of $\mathcal{T}$, we denote it by $\mathcal{R}(\mathcal{T})$.*

Example 4.1.2. As a prototype for the random variables ε_k, consider $\Omega = [0,1]$, $\mathcal{A}$ the Borel sets in $[0,1]$ and μ the Lebesgue measure. The *Rademacher functions* $r_k(t) = \mathrm{sgn}(\sin(2^k\pi t))$, $k \geq 1$, are independent, symmetric, $\{1,-1\}$-valued random variables on $[0,1]$.

J. Prüss and G. Simonett, *Moving Interfaces and Quasilinear Parabolic Evolution Equations*, Monographs in Mathematics 105,
DOI 10.1007/978-3-319-27698-4_4

Remark 4.1.3. **(a)** If $\mathcal{T} \subset \mathcal{B}(X,Y)$ is $\mathcal{R}$-bounded, then it is uniformly bounded, with

$$\sup\{|T| : T \in \mathcal{T}\} \leq \mathcal{R}(\mathcal{T}).$$

This follows from the definition of $\mathcal{R}$-bounded with $N = 1$, since $|\varepsilon_1|_{L_p(\Omega)} = 1$.

(b) The definition of $\mathcal{R}$-boundedeness is independent of $p \in [1, \infty)$. This follows from *Kahane's inequality.* For any Banach space X and $1 \leq p, q < \infty$ there is a constant $C(p,q,X)$ such that

$$\Big|\sum_{j=1}^{N} \varepsilon_j x_j\Big|_{L_p(\Omega;X)} \leq C(p,q,X)\Big|\sum_{j=1}^{N} \varepsilon_j x_j\Big|_{L_q(\Omega;X)}, \tag{4.2}$$

for each $N \in \mathbb{N}$, $x_j \in X$, and for all independent, symmetric, $\{-1,1\}$-valued random variables ε_j on a probability space $(\Omega, \mathcal{A}, \mu)$. However, one should keep in mind that the $\mathcal{R}$-bound does depend on p. For convenience, we drop this p-dependence of $\mathcal{R}$.

(c) In case X and Y are Hilbert spaces, then $\mathcal{T} \subset \mathcal{B}(X,Y)$ is $\mathcal{R}$-bounded if and only if $\mathcal{T}$ is uniformly bounded. In fact, let $\mathcal{T}$ be uniformly bounded by $C > 0$. Then choosing $p = 2$ we obtain

$$\begin{aligned}
\Big|\sum_{j=1}^{N} \varepsilon_j T_j x_j\Big|^2_{L_2(\Omega;Y)} &= \sum_{j,k=1}^{N} \Big[\int_\Omega \varepsilon_j(\omega)\varepsilon_k(\omega)\, d\mu\Big](T_j x_j|T_k x_k) \\
&= \sum_{j=1}^{N} \Big[\int_\Omega \varepsilon_j^2(\omega)\, d\mu\Big]|T_j x_j|^2_Y \leq C^2 \sum_{j=1}^{N} \Big[\int_\Omega \varepsilon_j^2(\omega)\, d\mu\Big]|x_j|^2_X \\
&= C^2 \sum_{j,k=1}^{N} \Big[\int_\Omega \varepsilon_j(\omega)\varepsilon_k(\omega)\, d\mu\Big](x_j|x_k) = C^2\Big|\sum_{j=1}^{N} \varepsilon_j x_j\Big|^2_{L_2(\Omega;X)},
\end{aligned}$$

since the ε_j are independent, hence orthogonal in $L_2(\Omega)$.

(d) Let $X = Y = L_p(G)$ for some open $G \subset \mathbb{R}^n$. Then $\mathcal{T} \subset \mathcal{B}(X,Y)$ is $\mathcal{R}$-bounded if and only if there is a constant $M > 0$ such that the following *square function estimate* holds:

$$\Big|\Big(\sum_{j=1}^{N} |T_j f_j|^2\Big)^{1/2}\Big|_{L_p(G)} \leq M\Big|\Big(\sum_{j=1}^{N} |f_j|^2\Big)^{1/2}\Big|_{L_p(G)}, \tag{4.3}$$

for all $N \in \mathbb{N}$, $f_j \in L_p(G)$, and $T_j \in \mathcal{T}$.
This is a consequence of the *Khintchine inequality.* For each $p \in [1, \infty)$ there is a constant $K_p > 0$ such that

$$K_p^{-1}\Big|\sum_{j=1}^{N} \varepsilon_j a_j\Big|_{L_p(\Omega)} \leq \Big(\sum_{j=1}^{N} |a_j|^2\Big)^{1/2} \leq K_p\Big|\sum_{j=1}^{N} \varepsilon_j a_j\Big|_{L_p(\Omega)}, \tag{4.4}$$

for all $N \in \mathbb{N}$, $a_j \in \mathbb{C}$, and for all independent, symmetric, $\{-1,1\}$-valued random variables ε_j on a probability space $(\Omega, \mathcal{A}, \mu)$. Note that in case X is a Hilbert space,

$$\Big|\sum_{j=1}^N \varepsilon_j x_j\Big|^2_{L_2(\Omega;X)} = \sum_{j=0}^N |x_j|^2_X,$$

by orthogonality of ε_j in L_2. So Khintchine's inequality is the scalar version of Kahane's inequality; it extends to Hilbert spaces.

To prove the assertion, if (4.3) holds, we have by (4.4)

$$\begin{aligned}
\Big|\sum_{j=1}^N \varepsilon_j T_j f_j\Big|_{L_p(\Omega,L_p(G))} &= \Big|\sum_{j=1}^N \varepsilon_j T_j f_j\Big|_{L_p(G,L_p(\Omega))}\\
\le K_p\Big|\Big(\sum_{j=1}^N |T_j f_j|^2\Big)^{1/2}\Big|_{L_p(G)} &\le K_p M\Big|\Big(\sum_{j=1}^N |f_j|^2\Big)^{1/2}\Big|_{L_p(G)}\\
\le K_p^2 M\Big|\sum_{j=1}^N \varepsilon_j f_j\Big|_{L_p(G,L_p(\Omega))} &= K_p^2 M\Big|\sum_{j=1}^N \varepsilon_j f_j\Big|_{L_p(\Omega,L_p(G))}.
\end{aligned}$$

The proof of the converse is similar.

Part (d) of the above remark gives a very useful sufficient condition for $\mathcal{R}$-boundedness of kernel operators in $L_p(G)$, which we state as

Proposition 4.1.4. *Let $G \subset \mathbb{R}^n$ be open and consider a family $\mathcal{T} = \{T_\lambda : \lambda \in \Lambda\} \subset \mathcal{B}(L_p(G;\mathbb{R}^m))$ of kernel operators*

$$[T_\lambda f](x) = \int_G k_\lambda(x,y) f(y)dy, \quad x \in G,\ f \in L_p(G;\mathbb{R}^m),$$

with kernels dominated by a kernel k_0, i.e.,

$$|k_\lambda(x,y)| \le k_0(x,y), \quad \text{for a.a. } x,y \in G, \text{ and all } \lambda \in \Lambda.$$

Then $\mathcal{T} \subset \mathcal{B}(L_p(G;\mathbb{R}^m))$ is $\mathcal{R}$-bounded, provided T_0 is bounded in $L_p(G)$.

Proof. By Remark 4.1.3**(d)** we only have to verify the square function estimate (4.3). But this is easy; due to L_p-boundedness of the dominating operator T_0, we have

$$\begin{aligned}
\Big|\Big(\sum_{j=1}^N |T_j f_j|^2\Big)^{1/2}\Big|_{L_p(G)} &\le \Big|\Big(\sum_{j=1}^N (T_0|f_j|)^2\Big)^{1/2}\Big|_{L_p(G)}\\
\le \Big|T_0\Big(\sum_{j=1}^N |f_j|^2\Big)^{1/2}\Big|_{L_p(G)} &\le |T_0|_{\mathcal{B}(L_p(G))}\Big|\Big(\sum_{j=1}^N |f_j|^2\Big)^{1/2}\Big|_{L_p(G)}.
\end{aligned}$$

□

A considerable extension of this result reads as follows.

Proposition 4.1.5. *Let X and Y be Banach spaces, $G \subset \mathbb{R}^n$ open, and $1 < p < \infty$. Suppose $\mathcal{K} \subset \mathcal{B}(L_p(G;X), L_p(G;Y))$ is a family of kernel operators in the sense that*

$$Kf(x) = \int_G k(x,x')f(x')\,dx', \quad x \in G,\ f \in L_p(G;X),$$

for each $K \in \mathcal{K}$, where the kernels $k : G \times G \to \mathcal{B}(X,Y)$ are measurable, with

$$\mathcal{R}\{k(x,x') : k \in \mathcal{K}\} \le k_0(x,x'), \quad x, x' \in G,$$

and the operator K_0 with scalar kernel k_0 is bounded in $L_p(G)$.

Then $\mathcal{K} \subset \mathcal{B}(L_p(G;X), L_p(G;Y))$ is $\mathcal{R}$-bounded and $\mathcal{R}(\mathcal{K}) \le |K_0|_{L_p(G)}$.

Proof. This follows easily from the estimate

$$\begin{aligned}
\Big|\sum_{j=1}^N \varepsilon_j K_j f_j\Big|_{L_p(\Omega;L_p(G;Y))} &\le \Big|\int_G \Big|\sum_{j=1}^N \varepsilon_j k_j(\cdot,x')f_j(x')\Big|_{L_p(\Omega;X)} dx'\Big|_{L_p(G)} \\
&\le \Big|\int_G k_0(\cdot,x')\Big|\sum_{j=1}^N \varepsilon_j f_j(x')\Big|_{L_p(\Omega;X)} dx'\Big|_{L_p(G)} \\
&\le |K_0|_{L_p(G)}\Big|\sum_{j=1}^N \varepsilon_j f_j\Big|_{L_p(\Omega;L_p(G;X))}.
\end{aligned}$$

□

The next proposition shows that $\mathcal{R}$-bounds behave like norms.

Proposition 4.1.6. **(a)** *Let X, Y be Banach spaces, and $\mathcal{T}, \mathcal{S} \subset \mathcal{B}(X,Y)$ be $\mathcal{R}$-bounded. Then*

$$\mathcal{T} + \mathcal{S} = \{T + S : T \in \mathcal{T},\ S \in \mathcal{S}\}$$

is $\mathcal{R}$-bounded as well, and $\mathcal{R}(\mathcal{T} + \mathcal{S}) \le \mathcal{R}(\mathcal{T}) + \mathcal{R}(\mathcal{S})$.

(b) *Let X, Y, Z be Banach spaces, and $\mathcal{T} \in \mathcal{B}(X,Y)$ and $\mathcal{S} \subset \mathcal{B}(Y,Z)$ be $\mathcal{R}$-bounded. Then*

$$\mathcal{S}\mathcal{T} = \{ST : T \in \mathcal{T},\ S \in \mathcal{S}\}$$

is $\mathcal{R}$-bounded, and $\mathcal{R}(\mathcal{S}\mathcal{T}) \le \mathcal{R}(\mathcal{S})\mathcal{R}(\mathcal{T})$.

Proof. The first assertion is a consequence of the triangle inequality.

$$\begin{aligned}
\Big|\sum_{j=1}^N \varepsilon_j (T_j + S_j)x_j\Big|_{L_p(\Omega;Y)} &\le \Big|\sum_{j=1}^N \varepsilon_j T_j x_j\Big|_{L_p(\Omega;Y)} + \Big|\sum_{j=1}^N \varepsilon_j S_j x_j\Big|_{L_p(\Omega;Y)} \\
&\le \mathcal{R}(\mathcal{T})\Big|\sum_{j=1}^N \varepsilon_j x_j\Big|_{L_p(\Omega;X)} + \mathcal{R}(\mathcal{S})\Big|\sum_{j=1}^N \varepsilon_j x_j\Big|_{L_p(\Omega;X)}.
\end{aligned}$$

The second assertion follows from

$$\Big|\sum_{j=1}^N \varepsilon_j(S_jT_j)x_j\Big|_{L_p(\Omega;Z)} \le \mathcal{R}(\mathcal{S})\Big|\sum_{j=1}^N \varepsilon_j T_j x_j\Big|_{L_p(\Omega;Y)}$$
$$\le \mathcal{R}(\mathcal{S})\mathcal{R}(\mathcal{T})\Big|\sum_{j=1}^N \varepsilon_j x_j\Big|_{L_p(\Omega;X)}. \qquad \square$$

1.2 The Contraction Principle

A very useful device in connection with $\mathcal{R}$-boundedness is the *contraction principle of Kahane*, which we state as a lemma. For the sake of completeness, a proof is given here, too.

Lemma 4.1.7. *Let X be a Banach space, $N \in \mathbb{N}$, $x_j \in X$, ε_j independent, symmetric, $\{-1,1\}$-valued random variables on a probability space $(\Omega, \mathcal{A}, \mu)$, and $\alpha_j, \beta_j \in \mathbb{C}$ such that $|\alpha_j| \le |\beta_j|$, for each $j = 1, \ldots, N$. Then*

$$\Big|\sum_{j=1}^N \alpha_j\varepsilon_j x_j\Big|_{L_p(\Omega;X)} \le 2\Big|\sum_{j=1}^N \beta_j\varepsilon_j x_j\Big|_{L_p(\Omega,X)}.$$

The constant 2 can be omitted in case α_j and β_j are real.

Proof. Replacing x_j by $\beta_j x_j$ if necessary, we may assume w.l.o.g. $\beta_j = 1$ for each j. Decomposing α_j into real and imaginary parts the triangle inequality yields the assertion once the claim holds for real α_j with constant 1. Let e_k, $k = 1, \ldots, 2^N$, be any enumeration of the extreme points of the cube $[-1,1]^N$. Then because $\alpha = (\alpha_1, \ldots, \alpha_N) \in [-1,1]^N$ we find numbers $\lambda_k \in [0,1]$ with $\sum_{k=1}^{2^N} \lambda_k = 1$ such that $\alpha = \sum_{k=1}^{2^N} \lambda_k e_k$. This implies

$$\Big|\sum_{j=1}^N \alpha_j\varepsilon_j x_j\Big|_{L_p(\Omega,X)} \le \sum_{k=1}^{2^N} \lambda_k \Big|\sum_{j=1}^N e_{kj}\varepsilon_j x_j\Big|_{L_p(\Omega,X)} = \Big|\sum_{j=1}^N \varepsilon_j x_j\Big|_{L_p(\Omega;X)},$$

since the random variables $\{\varepsilon_j : j = 1, \ldots, N\}$ and $\{e_{kj}\varepsilon_j : j = 1, \ldots, N\}$ have the same joint distribution. $\square$

It is an easy consequence of Lemma 4.1.7 that any *finite* family $\mathcal{T} \subset \mathcal{B}(X,Y)$ is $\mathcal{R}$-bounded. Another simple application of the contraction principle deals with pointwise scalar multipliers in L_p-spaces. We have

Proposition 4.1.8. *Suppose X is a Banach space, $G \subset \mathbb{R}^n$ open, and $1 \le p < \infty$. Then the set*

$$\{m_\phi \in \mathcal{B}(L_p(G;X)) : \phi \in L_\infty(G),\ |\phi|_\infty \le r\}$$

of pointwise multipliers defined by

$$[m_\phi f](x) = \phi(x)f(x), \quad x \in G,\ f \in L_p(G;X), \tag{4.5}$$

is $\mathcal{R}$-bounded for each $r > 0$ with $\mathcal{R}$-bound $\le 2r$.

Proof. This follows by the contraction principle

$$\Big|\sum_{j=1}^N \varepsilon_j m_{\phi_j} f_j\Big|_{L_p(\Omega;L_p(G;X))} = \Big|\sum_{j=1}^N \varepsilon_j \phi_j f_j\Big|_{L_p(G;L_p(\Omega;X))}$$
$$\le 2r\Big|\sum_{j=1}^N \varepsilon_j f_j\Big|_{L_p(G;L_p(\Omega;X))} = 2r\Big|\sum_{j=1}^N \varepsilon_j f_j\Big|_{L_p(\Omega;L_p(G;X))}. \qquad \square$$

We quote a simple corollary which will be employed below frequently. It follows directly from Propositions 4.1.6 and 4.1.8.

Corollary 4.1.9. *Let $1 \le p < \infty$, X, Y be Banach spaces, $G \subset \mathbb{R}^n$ open, and let $\mathcal{T} \subset \mathcal{B}(L_p(G;X), L_p(G;Y))$ be $\mathcal{R}$-bounded with $\mathcal{R}$-bound τ. Then*

$$\{m_\phi T m_\psi : T \in \mathcal{T}, \phi, \psi \in L_\infty(G), |\phi|_\infty \le r, |\psi|_\infty \le s\} \subset \mathcal{B}(L_p(G;X), L_p(G;Y))$$

is $\mathcal{R}$-bounded with $\mathcal{R}$-bound $\le 4rs\tau$.

Another important consequence of the contraction principle is the following very useful property, known as *convexity of $\mathcal{R}$-bounds.*

Proposition 4.1.10. *Let X, Y be a Banach spaces and $\mathcal{T} \subset \mathcal{B}(X,Y)$ be $\mathcal{R}$-bounded. Then the closure in the strong operator topology of the absolute convex hull of $\mathcal{T}$ is also $\mathcal{R}$-bounded, and the inequality*

$$\mathcal{R}(\overline{\mathrm{aco}}^s(\mathcal{T})) \le 2\mathcal{R}(\mathcal{T})$$

is valid.

Proof. (a) We first show that the convex hull $\mathrm{co}(\mathcal{T})$ of $\mathcal{T}$ is $\mathcal{R}$-bounded. For this purpose choose $N \in \mathbb{N}$, $x_j \in X$, $T_j \in \mathrm{co}(\mathcal{T})$, and N independent, symmetric $\{-1,1\}$-valued random variables ε_j on a probability space $(\Omega, \mathcal{A}, \mu)$. Then there are numbers $\lambda_{kj} \in [0,1]$, with $\sum_{j=1}^{m_k} \lambda_{kj} = 1$, and $T_{kj} \in \mathcal{T}$ such that

$$T_k = \sum_{j=1}^{m_k} \lambda_{kj} T_{kj}, \quad k = 1, \dots, N.$$

Set $\lambda_{kj} = 0$ for $j > m_k$, $l = (l_1, \dots, l_N)$, $T_{kl} = T_{kl_k}$, and $\lambda_l = \Pi_{i=1}^N \lambda_{il_i}$. Then we have

$$T_k = \sum_{l \in \mathbb{N}^N} \lambda_l T_{kl}, \quad k = 1, \dots, N,$$

and

$$\lambda_l \in [0,1], \quad l \in \mathbb{N}^N, \quad \sum_{l \in \mathbb{N}^N} \lambda_l = 1.$$

Note that the sums are finite, i.e., only finitely many terms are nonzero. The triangle inequality now yields

$$\Big|\sum_{k=1}^{N} \varepsilon_k T_k x_k\Big|_{L_p(\Omega;Y)} \leq \sum_{l\in\mathbb{N}^N} \lambda_l \Big|\sum_{k=1}^{N} \varepsilon_k T_{kl} x_k\Big|_{L_p(\Omega;Y)}$$

$$\leq \mathcal{R}(\mathcal{T}) \sum_{l\in\mathbb{N}^N} \lambda_l \Big|\sum_{k=1}^{N} \varepsilon_k x_k\Big|_{L_p(\Omega;X)} = \mathcal{R}(\mathcal{T}) \Big|\sum_{k=1}^{N} \varepsilon_k x_k\Big|_{L_p(\Omega;X)}.$$

This proves

$$\mathcal{R}(\mathrm{co}\mathcal{T}) \leq \mathcal{R}(\mathcal{T})$$

for the convex hull of $\mathcal{T}$.

(b) The contraction principle shows that with $\mathcal{T}$ the set

$$\mathcal{T}_0 = \{\lambda T : T \in \mathcal{T},\ \lambda \in \mathbb{C},\ |\lambda| \leq 1\}$$

is also $\mathcal{R}$-bounded and $\mathcal{R}(\mathcal{T}_0) \leq 2\mathcal{R}(\mathcal{T})$. The absolute convex hull of $\mathcal{T}$ can be written as

$$\mathrm{aco}(\mathcal{T}) = \Big\{\sum_{j=1}^{m} \lambda_j T_j : T_j \in \mathcal{T},\ m \in \mathbb{N},\ \sum_{j=1}^{m} |\lambda_j| \leq 1\Big\} = \mathrm{co}(\mathcal{T}_0),$$

hence the assertion for the absolute convex hull follows.

(c) Finally, it is obvious that $\mathcal{R}$-boundedness is preserved by convergence in the strong operator topology. □

One simple, but nevertheless useful, consequence of Proposition 4.1.10 is

Corollary 4.1.11. *Suppose $(\Omega, \mathcal{A}, \mu)$ is a finite measure space, and let $T : \Omega \times \Lambda \to \mathcal{B}(X,Y)$ be such that $T(\cdot,\lambda)$ is μ-integrable in $\mathcal{B}(X,Y)$, for each $\lambda \in \Lambda$, and assume that $T(\Omega \times \Lambda)$ is $\mathcal{R}$-bounded.*

Then $\{\int_\Omega T(\omega,\lambda)\, d\mu(\omega) : \lambda \in \Lambda\}$ is $\mathcal{R}$-bounded.

Proof. Without loss of generality we may assume $\mu(\Omega) = 1$. In virtue of Proposition 4.1.10 we only need to show that $\int_\Omega T(\omega,\lambda)\, d\mu(\omega)$ belongs to $\overline{\mathrm{co}}^s T(\Omega,\lambda)$, for each fixed $\lambda \in \Lambda$. But this is an easy consequence of the fact that $T(\cdot,\lambda)$ can be uniformly approximated by countably-valued functions $T_{\lambda,n} = \sum_k \chi_{A_{\lambda,k}} T_{\lambda,k}$ with $T_{\lambda,k} \in \mathcal{T}(\Omega,\lambda)$. □

Another very useful result about $\mathcal{R}$-boundedness is contained in

Proposition 4.1.12. *Let $G \subset \mathbb{C}^n$ be open, $K \subset G$ compact, and suppose $F : G \to \mathcal{B}(X,Y)$ is holomorphic.*

Then $F(K) \subset \mathcal{B}(X,Y)$ is $\mathcal{R}$-bounded.

Proof. Let $z_0 \in K$ be fixed. Since F is holomorphic in G there is a ball $B(z_0, r) \subset G$ such that the power series representation

$$F(z) = \sum_{\alpha \in \mathbb{N}_0^n} \frac{\partial^\alpha F(z_0)}{\alpha!}(z - z_0)^\alpha, \quad |z - z_0| \leq r,$$

is absolutely convergent, we have

$$\rho_0 := \sum_{\alpha \in \mathbb{N}_0^n} \frac{|\partial^\alpha F(z_0)|_{\mathcal{B}(X,Y)}}{\alpha!} r^{|\alpha|} < \infty.$$

Proposition 4.1.6 and Lemma 4.1.7 imply $\mathcal{R}(F(B(z_0, r)) \leq 2\rho_0$. Covering K by a finite set of such balls we obtain the assertion. □

4.2 Unconditionallity

Unconditional convergence is one important ingredient for the Fourier multiplier theorems we intend to prove. This section serves as an introduction to this topic.

2.1 Unconditional Convergence
We begin with

Definition 4.2.1. *Let X be a Banach space and $(x_n) \subset X$. The series $\sum_{n=1}^\infty x_n$ is called* **unconditionally convergent** *if $\sum_{n=1}^\infty x_{\sigma(n)}$ is convergent in norm, for every permutation $\sigma : \mathbb{N} \to \mathbb{N}$.*

Recall that in the finite-dimensional case, a series is unconditionally convergent if and only if it is absolutely convergent. A famous theorem of Dvoretzky and Rogers states that this equivalence is even characteristic for finite-dimensional spaces. The standard example of unconditionally convergent series is

$$x = \sum_{n=1}^\infty (x, e_n)e_n,$$

where $\{e_n\}_{n\in\mathbb{N}}$ is an orthonormal basis in an infinite-dimensional Hilbert space. Note that such a series is absolutely convergent if and only if $\sum_n |(x, e_n)| < \infty$.

Let us first observe that in case $\sum_{n=1}^\infty x_n$ is unconditionally convergent, then the sums $\sum_{n=1}^\infty x_{\sigma(n)}$ are independent of the permutation. In fact, if $\sum_{n=1}^\infty x_n$ is unconditionally convergent, then so are the scalar series $\sum_{n=1}^\infty < x^*, x_n >$, for each $x^* \in X^*$. Since X^* separates points in X and the assertion holds in the scalar case, it is also true in X.

The following proposition contains some characterizations of unconditional convergence.

Proposition 4.2.2. *Let X be a Banach space and $(x_n) \subset X$. Then the following are equivalent.*

(a) *The series* $\sum_{n=1}^{\infty} x_n$ *is unconditionally convergent;*

(b) *there is an* $x \in X$ *such that for each* $\varepsilon > 0$ *there is a finite set* $A_\varepsilon \subset \mathbb{N}$ *with*

$$\Big|x - \sum_{n \in B} x_n\Big| \le \varepsilon, \quad \textit{for all finite sets } B \supset A_\varepsilon;$$

(c) *for each* $\varepsilon > 0$ *there is a finite subset* $A_\varepsilon \subset \mathbb{N}$ *such that*

$$\Big|\sum_{n \in B} x_n\Big| \le \varepsilon, \quad \textit{for each finite set } B \subset \mathbb{N},\ B \cap A_\varepsilon = \emptyset;$$

(d) $\sum_{n=0}^{\infty} \varepsilon_n x_n$ *is convergent for all choices of* $\varepsilon_n \in \{-1, 1\}$*;*

(e) $\sum_{n=0}^{\infty} \delta_n x_n$ *is convergent for all choices of* $\delta_n \in \{0, 1\}$*;*

(f) *the series* $\sum_{n=1}^{\infty} \lambda_n x_n$ *is convergent, for every bounded sequence* $(\lambda_n) \subset \mathbb{C}$*.*

Proof. (a) $\Rightarrow$ (*b*). Suppose (a) holds but (b) is violated; then for each $x \in X$ there is $\varepsilon_0 > 0$ such that for each finite $A \subset \mathbb{N}$ there is a finite $B \supset A$ with $|x - \sum_{n \in B} x_n| > \varepsilon_0$. Choosing $x = \sum_{n=1}^{\infty} x_n$ we now construct a permutation σ such that $\sum_{n=1}^{\infty} x_{\sigma(n)} \neq x$, which contradicts uniqueness of the sum. In fact, let $m_1 = n_1 = 1$ and $A_1 = \{1\}$; then there is a finite $A_2 \supset A_1$ such that $|x - \sum_{n \in A_2} x_n| > \varepsilon_0$. Set $n_2 = \#A_2$, $m_2 = \max A_2$, and define σ on the set $\{m_1, \ldots, m_2\}$ by first enumerating A_2 and then enumerating $\{m_1, \ldots, m_2\} \setminus A_2$. Suppose we have constructed $m_{k-1} \le n_k \le m_k$ and a permutation

$$\sigma : \{1, \ldots m_k\} \to \{1, \ldots, m_k\}$$

with the property $|x - \sum_{n=1}^{n_k} x_n| > \varepsilon_0$. Then there is a finite set $A_{k+1} \supset \{1, \ldots, m_k\}$ such that $|x - \sum_{n \in A_{k+1}} x_n| > \varepsilon_0$. Set then $n_{k+1} = \#A_{k+1}$, $m_{k+1} = \max A_{k+1}$ and extend σ to the set $\{1, \ldots, m_{k+1}\}$ by first enumerating $A_{k+1} \setminus \{1, \ldots, m_k\}$ and then the remaining elements of $\{1, \ldots, m_{k+1}\}$. This way we obtain a permutation σ of $\mathbb{N}$ and an increasing sequence n_k such that $|x - \sum_{l=1}^{n_k} x_{\sigma(l)}| > \varepsilon_0$. But this contradicts $\sum_{j=1}^{\infty} x_{\sigma(j)} = x$.

(b) $\Rightarrow$ (c). Suppose (b) holds; then for each $\varepsilon > 0$ there is $A_\varepsilon \subset \mathbb{N}$ such that $|x - \sum_{n \in B} x_n| \le \varepsilon/2$, for each finite $B \supset A_\varepsilon$. If D is finite and such that $A_\varepsilon \cap D = \emptyset$, then with $B = A_\varepsilon \cup D$ we obtain

$$\Big|\sum_{n \in D} x_n\Big| = \Big|\Big[\sum_{n \in B} x_n - x\Big] + \Big[x - \sum_{n \in A_\varepsilon} x_n\Big]\Big| \le \varepsilon,$$

i.e., (c) holds.

(c) $\Rightarrow$ (a). Let $\varepsilon > 0$ and a permutation σ of $\mathbb{N}$ be given. Choose A_ε according to (c), and let $N \in \mathbb{N}$ be such that $A_\varepsilon \subset \sigma\{1, \ldots, N\}$. Then for $k \ge l > N$ we have $A_\varepsilon \cap \sigma\{l, \ldots, k\} = \emptyset$, hence from (c) we get

$$\Big|\sum_{n=l}^{k} x_{\sigma(n)}\Big| \le \varepsilon, \quad \text{for all } k \ge l > N.$$

But this shows convergence of the series $\sum_{n=1}^{\infty} x_{\sigma(n)}$, by completeness of X.

(c) $\Rightarrow$ (d). Let $\varepsilon_n \in \{-1,1\}$ for $n \in \mathbb{N}$ and $\varepsilon > 0$ be given and choose A_ε according to (c). For $k \geq l \geq \max A_\varepsilon$, we then set

$$B_+ = \{n \in \mathbb{N} : l \leq n \leq k, \varepsilon_n = 1\}, \quad B_- = \{n \in \mathbb{N} : l \leq n \leq k, \varepsilon_n = -1\}.$$

Then we obtain from (c)

$$\Big|\sum_{n=k}^{l} \varepsilon_n x_n\Big| \leq \Big|\sum_{n\in B_+} x_n\Big| + \Big|\sum_{n\in B_-} x_n\Big| \leq 2\varepsilon,$$

i.e., the series $\sum_{j=1}^{\infty} \varepsilon_n x_n$ is convergent since X is complete.

(d) $\Rightarrow$ (e). Let $\delta_n \in \{0,1\}$ be given and set $\varepsilon_n = 2\delta_n - 1$. Then

$$\sum_{n=1}^{\infty} \delta_n x_n = \frac{1}{2}\Big[\sum_{n=1}^{\infty} x_n + \sum_{n=1}^{\infty} \varepsilon_n x_n\Big]$$

converges since the series on the right-hand side converge.

(e) $\Rightarrow$ (c). Suppose that (e) holds but (c) is not valid. Then there is $\varepsilon_0 > 0$ such that for each finite set $A \subset \mathbb{N}$ there is a finite set $B := B(A)$ with $A \cap B = \emptyset$ and $|\sum_{n\in B} x_n| \geq \varepsilon_0$. Choose $A_1 = \{1\}$, set $n_1 = 1$, let $A_2 = B(A_1)$, and set $n_2 = \max A_2$. We may proceed inductively to obtain a sequence of sets A_k, an increasing sequence n_k such that $\{1, \dots n_{k-1}\} \cap A_k = \emptyset$, $n_k = \max A_k$, and $|\sum_{n\in A_k} x_n| \geq \varepsilon_0$. Define $\delta_n = 1$ if $n \in A_k$, $\delta_n = 0$ if $n \notin A_k$ for $n_{k-1} < n \leq n_k$, $k \in \mathbb{N}$. But then the series $\sum_{n=1}^{\infty} \delta_n x_n$ cannot be convergent since

$$\sum_{n_{k-1}+1}^{n_k} \delta_n x_n = \sum_{n\in A_k} x_n$$

does not tend to zero as $k \to \infty$, a contradiction.

(c) $\Rightarrow$ (f). Let $\varepsilon > 0$ be given and choose a set A_ε according to (c). For an arbitrary bounded sequence $\lambda_n \geq 0$ we then obtain with $k \geq l > N_\varepsilon := \max A_\varepsilon$

$$\sum_{n=l}^{k} \lambda_n x_n = \mu_1 \sum_{n=l}^{k} x_n + \sum_{n=l}^{k} \lambda_{1,n} x_n,$$

where $\mu_1 = \min\{\lambda_l, \dots, \lambda_k\}$ and $\lambda_{1,n} = \lambda_n - \mu_1$ for all $l \leq n \leq k$. Proceeding inductively we then obtain finitely many $\mu_j \geq 0$, with $\sum_j \mu_k \leq |\lambda|_\infty$, and sets $A_j \subset \{l, \dots, k\}$ such that

$$\sum_{n=l}^{k} \lambda_n x_n = \sum_j \mu_j \sum_{n\in A_j} x_n.$$

Estimating we get by (c)

$$\Big|\sum_{n=l}^{k}\lambda_n x_n\Big| \le \sum_j \mu_j\Big|\sum_{n\in A_j} x_n\Big| \le \varepsilon\sum_j \mu_j \le \varepsilon|\lambda|_\infty.$$

This shows that the series $\sum_{n=1}^{\infty}\lambda_n x_n$ are convergent, even uniformly in λ, provided $\lambda_n \ge 0$ are bounded. For the general case we decompose λ_j into real and imaginary parts and these into positive and negative parts, to reduce to the case $\lambda_j \ge 0$.

(f) $\Rightarrow$ (d). This is trivial. □

Note that according to the proof of Proposition 4.2.2, the operator

$$L : l_\infty \to X, \quad L(\lambda_n) = \sum_{n=1}^{\infty}\lambda_n x_n$$

is compact and uniformly approximated by the finite rank operators

$$L_N : l_\infty \to X, \quad L_N(\lambda_n) = \sum_{n=1}^{N}\lambda_n x_n.$$

2.2 Schauder Decompositions

Next we introduce Schauder decompositions.

Definition 4.2.3. *A sequence of projections $(\Delta_n)_{n\in\mathbb{N}} \subset \mathcal{B}(X)$ is called a* **Schauder decomposition** *of X if*

$$\Delta_n\Delta_m = 0 \text{ for all } m \ne n, \text{ and } \sum_{n=1}^{\infty}\Delta_n x = x \text{ for each } x \in X.$$

If a Schauder decomposition satisfies in addition $\dim\mathcal{R}(\Delta_n) = 1$ for each $n \in \mathbb{N}$ we may choose $0 \ne e_n \in \mathcal{R}(\Delta_n)$; then $(e_n)_{n\in\mathbb{N}}$ is called a **Schauder basis** *of X. A Schauder decomposition (or Schauder basis) is called* **unconditional** *if the series $\sum_{n=1}^{\infty}\Delta_n x$ converges unconditionally, for each $x \in X$.*

The next result contains some characterizations of unconditional Schauder decompositions.

Proposition 4.2.4. *Let $(\Delta_n)_{n\in\mathbb{N}} \subset \mathcal{B}(X)$ be a Schauder decomposition of the Banach space X. Then the following are equivalent.*

(i) *The decomposition $(\Delta_n)_{n\in\mathbb{N}}$ is unconditional;*

(ii) *there is a constant $C > 0$ such that*

$$\Big|\sum_{n=1}^{N}\varepsilon_n\Delta_n x\Big| \le C\Big|\sum_{n=1}^{N}\Delta_n x\Big|, \quad \text{for all } N\in\mathbb{N},\ x\in X,\ \varepsilon_n\in\{-1,1\}; \quad (4.6)$$

(iii) *there is a constant $C > 0$ such that*

$$\Big|\sum_{n=1}^{N}\varepsilon_n x_n\Big| \le C\Big|\sum_{n=1}^{N} x_n\Big|, \quad \textit{for all } N\in\mathbb{N},\ x_n\in\mathcal{R}(\Delta_n),\ \varepsilon_n\in\{-1,1\}.$$

Proof. (i) $\Rightarrow$ (ii). Since (Δ_n) is an unconditional Schauder decomposition, for every $x \in X$ and $\bar\varepsilon = (\varepsilon_n) \subset \{-1,1\}$ the series

$$M_{\bar\varepsilon}x := \sum_{n=1}^{\infty}\varepsilon_n\Delta_n x$$

is convergent. This defines linear operators $M_{\bar\varepsilon} : X \to X$, which are also closed, hence bounded by the closed graph theorem.

Next, by Proposition 4.2.2 the family $\{M_{\bar\varepsilon} : \bar\varepsilon \in \{-1,1\}^{\mathbb{N}}\} \subset \mathcal{B}(X)$ is pointwise bounded, hence uniformly bounded by the uniform boundedness principle. This implies (ii).

(ii) $\Rightarrow$ (iii). Apply (ii) to $x = \sum_{n=1}^{N} x_n$.

(iii) $\Rightarrow$ (i). Since (Δ_n) is a Schauder decomposition, we have $x = \sum_{n=1}^{\infty}\Delta_n x$. For arbitrary $k \ge l$ set $x_n = \Delta_n x$ if $l \le n \le k$, $x_n = 0$ elsewhere. Then (iii) yields

$$\Big|\sum_{n=l}^{k}\varepsilon_n\Delta_n x\Big| \le C\Big|\sum_{n=l}^{k}\Delta_n x\Big|,$$

for arbitrary signs $\varepsilon_n \in \{-1,1\}$. Since the right-hand side of this inequality tends to zero as $k,l \to \infty$, we see that the series $\sum_{n=1}^{\infty}\varepsilon_n\Delta_n x$ converge for all choices of signs, hence $\sum_{n=1}^{\infty}\Delta_n x$ converges unconditionally by Proposition 4.2.2 for each $x \in X$. □

The smallest constant C such that (4.6) holds is called the *unconditional constant* of the Schauder decomposition (Δ_n), we denote it by C_Δ.

The estimates in Proposition 4.2.4 are actually two-sided. In fact, applying e.g. (iii) to $\varepsilon_n x_n$ instead of x_n yields

$$\Big|\sum_{n=1}^{N} x_n\Big| \le C\Big|\sum_{n=1}^{N}\varepsilon_n x_n\Big|, \quad \text{for all } N\in\mathbb{N},\ x_n\in\mathcal{R}(\Delta_n),\ \varepsilon_n\in\{-1,1\}.$$

This is the first part of following corollary.

Corollary 4.2.5. *Suppose $(\Delta_n)_{n\in\mathbb{N}} \subset \mathcal{B}(X)$ is an unconditional Schauder decomposition of the Banach space X. Then*

(a) *for all $N \in \mathbb{N}$, $x_n \in \mathcal{R}(\Delta_n)$ we have*

$$C_\Delta^{-1}\Big|\sum_{n=1}^{N} x_n\Big| \le \Big|\sum_{n=1}^{N}\varepsilon_n x_n\Big| \le C_\Delta\Big|\sum_{n=1}^{N} x_n\Big|; \tag{4.7}$$

(b) *for each sequence $(\lambda_n) \in l_\infty$, the linear operator M_λ defined by*

$$M_\lambda x = \sum_{n=1}^{\infty} \lambda_n \Delta_n x, \quad x \in X,$$

is bounded in X with $|M_\lambda|_{\mathcal{B}(X)} \leq 2C_\Delta |\lambda|_\infty$. The factor 2 can be omitted if all λ_n are real.

Proof. We know from Proposition 4.2.2 that $M_\lambda x$ exists for each $x \in X$ and the truncations $M_\lambda^N := \sum_{n=1}^N \lambda_n \Delta_n$ converge strongly to M_λ, even uniformly for $|\lambda| \leq 1$. Fix N and write $\lambda^N = (\lambda_1, \ldots, \lambda_N)$ as a convex combination of the extreme points $\bar{\varepsilon}_k$ of $[-1,1]^N$, in case all $\lambda_n \in [-1,1]$. Then (4.7) gives

$$|M_\lambda^N x| = \Big|\sum_k \alpha_k \sum_{n=1}^N \varepsilon_{kn} \Delta_n x\Big| \leq C_\Delta \Big|\sum_{n=1}^N \Delta_n x\Big|,$$

hence with $N \to \infty$ the assertion follows for real λ. For the general case decompose λ into real and imaginary parts. □

Suppose that $(\Delta_n)_{n\in\mathbb{N}} \subset \mathcal{B}(X)$ is an unconditional Schauder decomposition. Let $\varepsilon_n : \Omega \to \{-1,1\}$ be a sequence of random variables on a probability space $(\Omega, \mathcal{A}, \mu)$ which are symmetric, i.e., their means are zero, and let them be independent. Then from (4.7) we get for each $\omega \in \Omega$ and all $x_n \in \mathcal{R}(\Delta_n)$

$$C_\Delta^{-1}\Big|\sum_{n=1}^N x_n\Big| \leq \Big|\sum_{n=1}^N \varepsilon_n(\omega) x_n\Big| \leq C_\Delta \Big|\sum_{n=1}^N x_n\Big|,$$

hence we also have

$$C_\Delta^{-1}\Big|\sum_{n=1}^N x_n\Big| \leq \Big|\sum_{n=1}^N \varepsilon_n x_n\Big|_{L_p(\Omega;X)} \leq C_\Delta \Big|\sum_{n=1}^N x_n\Big|, \quad x_n \in \mathcal{R}(\Delta_n),\ N \in \mathbb{N}, \tag{4.8}$$

for each $p \in [1, \infty)$. In connection with estimate (4.8) one speaks about *randomizing the norm* in X. An important consequence of this randomization is the following

Theorem 4.2.6. *Suppose $(\Delta_n)_{n\in\mathbb{N}} \subset \mathcal{B}(X)$ is an unconditional Schauder decomposition of the Banach space X, and let $\{M_n\}_{n\in\mathbb{N}} \subset \mathcal{B}(X)$ be $\mathcal{R}$-bounded, and such that $M_n\Delta_n = \Delta_n M_n \Delta_n$ for all $n \in \mathbb{N}$. Then the operator T_M defined by*

$$T_M x := \sum_{n=1}^{\infty} M_n \Delta_n x, \quad x \in X,$$

is well-defined and bounded. We have

$$|T_M|_{\mathcal{B}(X)} \leq C_\Delta^2 \mathcal{R}\{M_n : n \in \mathbb{N}\},$$

where C_Δ means the unconditional constant of the decomposition (Δ_n).

Proof. Fix an $x \in X$; since $M_n\Delta_n x = \Delta_n M_n \Delta_n x \in \mathcal{R}(\Delta_n)$ we have by (4.8) for $k \geq l$,

$$\Big|\sum_{n=l}^{k} M_n\Delta_n x\Big|_X \leq C_\Delta \Big|\sum_{n=l}^{k} \varepsilon_n M_n \Delta_n x\Big|_{L_p(\Omega;X)}$$
$$\leq C_\Delta \mathcal{R}\{M_n : n \in \mathbb{N}\}\Big|\sum_{n=l}^{k} \varepsilon_n \Delta_n x\Big|_{L_p(\Omega;X)} \leq C_\Delta^2 \mathcal{R}\{M_n : n \in \mathbb{N}\}\Big|\sum_{n=l}^{k} \Delta_n x\Big|_X.$$

Therefore the series defining $T_M x$ is convergent in X, for each $x \in X$, hence T_M is well-defined and passing to the limit as $N \to \infty$, the asserted bound for $|T_M|_{\mathcal{B}(X)}$ follows. □

2.3 Property (α)
A very useful property of a Banach space is defined as follows.

Definition 4.2.7. *A Banach space X is said to have* **property** (α) *if there exists a constant $\alpha > 0$ such that*

$$\Big|\sum_{i,j=1}^{N} \alpha_{ij}\varepsilon_i\varepsilon_j' x_{ij}\Big|_{L_2(\Omega\times\Omega';X)} \leq \alpha \Big|\sum_{i,j=1}^{N} \varepsilon_i\varepsilon_j' x_{ij}\Big|_{L_2(\Omega\times\Omega';X)},$$

for all $\alpha_{ij} \in \{-1,1\}$, $x_{ij} \in X$, $N \in \mathbb{N}$, and all symmetric independent $\{-1,1\}$-valued random variables ε_i resp. ε_j' on a probability space $(\Omega, \mathcal{A}, \mu)$ resp. $(\Omega', \mathcal{A}', \mu')$.

We note that every Hilbert space has property (α), and if a Banach space E has property (α), then $L_p(S;E)$ has as well, for every sigma-finite measure space (S, Σ, σ) and $1 \leq p < \infty$.

The importance of property (α) in connection with $\mathcal{R}$-boundedness lies in the following fact.

Proposition 4.2.8. *Let X be a Banach space with property (α), and $\mathcal{T} \subset \mathcal{B}(X)$ be $\mathcal{R}$-bounded.*

Then there is a constant $K > 0$ such that

$$\Big|\sum_{i,j=1}^{N} \varepsilon_i\varepsilon_j' T_{ij} x_{ij}\Big|_{L_2(\Omega\times\Omega';X)} \leq K \Big|\sum_{i,j=1}^{N} \varepsilon_i\varepsilon_j' x_{ij}\Big|_{L_2(\Omega\times\Omega';X)}$$

for all $x_{ij} \in X$, $T_{ij} \in \mathcal{T}$, $N \in \mathbb{N}$, and all symmetric independent $\{-1,1\}$-valued random variables ε_i resp. ε_j' on probability spaces $(\Omega, \mathcal{A}, \mu)$ resp. $(\Omega', \mathcal{A}', \mu')$.

Proof. Let α_{ij} be independent, symmetric, $\{-1,1\}$-valued random variables on a probability space $(\Omega'', \mathcal{A}'', \mu'')$. Since X has by assumption property (α), we have, replacing x_{ij} by $\alpha_{ij}(\omega'')T_{ij}x_{ij}$ in the definition of (α),

$$\Big|\sum_{i,j=1}^{N} \varepsilon_i\varepsilon_j' T_{ij} x_{ij}\Big|_{L_2(\Omega\times\Omega';X)} \leq \alpha \Big|\sum_{i,j=1}^{N} \alpha_{ij}(\omega'')\varepsilon_i\varepsilon_j' T_{ij} x_{ij}\Big|_{L_2(\Omega\times\Omega';X)},$$

for each $\omega'' \in \Omega''$. Squaring and integrating over Ω'' this yields

$$\Big| \sum_{i,j=1}^{N} \varepsilon_i \varepsilon_j' T_{ij} x_{ij} \Big|_{L_2(\Omega\times\Omega';X)} \le \alpha \Big| \sum_{i,j=1}^{N} \alpha_{ij} \varepsilon_i \varepsilon_j' T_{ij} x_{ij} \Big|_{L_2(\Omega'';L_2(\Omega\times\Omega';X))}.$$

For fixed $\omega \in \Omega$, $\omega' \in \Omega'$, the set $\{\alpha_{ij}\varepsilon_i(\omega)\varepsilon_j'(\omega')\}$ consists of independent, symmetric, $\{-1,1\}$-valued random variables on the probability space $(\Omega'', \mathcal{A}'', \mu'')$, hence interchanging integration by Fubini's theorem, we may use $\mathcal{R}$-boundedness of $\mathcal{T}$ to estimate further

$$\Big| \sum_{i,j=1}^{N} \varepsilon_i \varepsilon_j' T_{ij} x_{ij} \Big|_{L_2(\Omega\times\Omega';X)} \le \alpha \mathcal{R}(\mathcal{T}) \Big| \sum_{i,j=1}^{N} \alpha_{ij} \varepsilon_i \varepsilon_j' x_{ij} \Big|_{L_2(\Omega'';L_2(\Omega\times\Omega';X))}.$$

Employing property (α) another time we arrive at

$$\Big| \sum_{i,j=1}^{N} \varepsilon_i \varepsilon_j' T_{ij} x_{ij} \Big|_{L_2(\Omega\times\Omega';X)} \le \alpha^2 \mathcal{R}(\mathcal{T}) \Big| \sum_{i,j=1}^{N} \varepsilon_i \varepsilon_j' x_{ij} \Big|_{L_2(\Omega\times\Omega';X)},$$

which is the assertion. □

This proposition has a remarkable consequence.

Theorem 4.2.9. *Let X be a Banach space with property (α), let $\Delta = \{\Delta_k\}_{k=0}^{\infty}$ be an unconditional Schauder decomposition, and let $\mathcal{T} \subset \mathcal{B}(X)$. Suppose that $\mathcal{T}_\Delta = \{T\Delta_j : T \in \mathcal{T},\ j \in \mathbb{N}_0,\ T\Delta_j = \Delta_j T \Delta_j\}$ is $\mathcal{R}$-bounded.*

Then the set $\mathcal{S} := \{\sum_{k=0}^{\infty} T_k \Delta_k : T_k \in \mathcal{T}_\Delta\}$ is $\mathcal{R}$-bounded.

Proof. We may assume $0 \in \mathcal{T}$. Let $S_1, \dots S_n$ be of the form $S_j = \sum_{k=0}^{N} T_{jk}\Delta_k$, with $T_{jk} \in \mathcal{T}_\Delta$, and fix $x_1, \dots, x_n \in X$. Then by Corollary 4.2.5 and the above proposition

$$\begin{aligned}
\Big| \sum_{j=1}^{n} \varepsilon_j S_j x_j \Big|_{L_2(\Omega;X)} &= \Big| \sum_{k=0}^{N} \Big(\sum_{j=1}^{n} \varepsilon_j T_{jk} \Delta_k x_j \Big) \Big|_{L_2(\Omega;X)} \\
&\le C_\Delta \Big| \sum_{k=0}^{N} \varepsilon_k' \Big(\sum_{j=1}^{n} \varepsilon_j T_{jk} \Delta_k x_j \Big) \Big|_{L_2(\Omega\times\Omega';X)} \\
&\le K C_\Delta \Big| \sum_{k=0}^{N} \varepsilon_k' \Big(\sum_{j=1}^{n} \varepsilon_j \Delta_k x_j \Big) \Big|_{L_2(\Omega\times\Omega';X)} \\
&= K C_\Delta \Big| \sum_{k=0}^{N} \varepsilon_k' \Delta_k \Big(\sum_{j=1}^{n} \varepsilon_j x_j \Big) \Big|_{L_2(\Omega';L_2(\Omega;X))} \\
&\le K C_\Delta^2 \Big| \sum_{j=1}^{n} \varepsilon_j x_j \Big|_{L_2(\Omega;X)},
\end{aligned}$$

where in the last step we used once more Corollary 4.2.5. With $N \to \infty$ the assertion follows from Proposition 4.1.10. □

4.3 Operator-Valued Fourier Multipliers

We are in position to consider operator-valued Fourier multipliers in vector-valued L_p-spaces.

3.1 Banach Spaces of Class $\mathcal{HT}$
We consider once more the derivation operator $B_p = d/dt$ from Section 3.2 in $Y_p := L_p(\mathbb{R}; Y)$, where Y is a Banach space. The basic question we want to address here is whether B_p admits bounded imaginary powers or even a bounded $\mathcal{H}^\infty$-calculus.

For this purpose let $h \in H_0(\Sigma_\phi)$ for some $\phi > \pi/2$. Then $h(B_p)$ is given by the Dunford integral

$$h(B_p) = \frac{1}{2\pi i} \int_\Gamma h(z)(z - B_p)^{-1}\, dz,$$

where $\Gamma = (\infty, 0]e^{i\psi} \cup [0, \infty)e^{-i\psi}$ with $\pi/2 < \psi < \phi$. With the representation of the resolvent of B_p from Section 3.2 this gives for a test function u on $\mathbb{R}$

$$h(B_p)u(t) = \frac{1}{2\pi i} \int_\Gamma h(z)(z - B_p)^{-1} u(t)\, dz = \frac{1}{2\pi i} \int_\Gamma h(z) \int_0^\infty e^{zs} u(t-s)\, ds dz,$$

i.e., Fubini's theorem yields

$$h(B_p)u(t) = \int_0^\infty k_h(s) u(t-s)\, ds, \quad t \in \mathbb{R},$$

where the kernel $k_h(s)$ is defined by

$$k_h(s) = \frac{1}{2\pi i} \int_\Gamma h(z) e^{zs}\, dz = \mathcal{L}^{-1} h(s), \quad s > 0,$$

i.e., it is the inverse Laplace transform of h. This kernel evidently satisfies an estimate of the form $s|k_h(s)| \le c|h|_{H^\infty(\Sigma_\phi)}$, but nothing more, in general, which means that we end up with singular integrals.

A different viewpoint uses Fourier transforms. The Fourier transform of $B_p u$ is given by $\mathcal{F} B_p u(\xi) = i\xi \mathcal{F} u(\xi)$, hence $\mathcal{F}(z - B_p)^{-1} u(\xi) = (z - i\xi)^{-1} \mathcal{F} u(\xi)$. This implies by Cauchy's theorem

$$\mathcal{F}[h(B_p)u](\xi) = \frac{1}{2\pi i} \int_\Gamma h(z)(z - i\xi)^{-1} \mathcal{F} u(\xi)\, dz = h(i\xi) \mathcal{F} u(\xi), \tag{4.9}$$

i.e., $h(B_p)u = \mathcal{F}^{-1}[h(i\cdot)\mathcal{F} u]$. Thus the question is whether such functions $h(i\cdot)$ are *Fourier multipliers* for $L_p(\mathbb{R}; Y)$.

At this point we recall the classical *Mikhlin Fourier multiplier theorem* in the scalar one-dimensional case, which states that for a function $m \in \dot{W}^1_{1,loc}(\mathbb{R})$, the condition

$$|m(\xi)| + |\xi m'(\xi)| \leq M, \quad \text{for almost all } \xi \in \mathbb{R}, \tag{4.10}$$

is sufficient for the operator $T_m = \mathcal{F}^{-1}m(\cdot)\mathcal{F}$ to be $L_p(\mathbb{R})$-bounded, for each $p \in (1, \infty)$.

If $h \in H^\infty(\Sigma_\phi)$ for some $\phi > \pi/2$, then by the Cauchy estimate for holomorphic functions, we also have $zh(z)$ uniformly bounded on each smaller sector Σ_ψ. In particular, the function $m(\xi) = h(i\xi)$ satisfies the Mikhlin condition (4.10). We may therefore conclude that B_p admits an $\mathcal{H}^\infty$-caluculus in $L_p(\mathbb{R})$, for each $p \in (1, \infty)$, with angle $\phi^\infty_{B_p} = \pi/2$.

Now we turn to the vector-valued case. Obviously, if the Mikhlin multiplier theorem holds in $Y_p := L_p(\mathbb{R}; Y)$, then B_p admits a bounded $\mathcal{H}^\infty$-calculus in Y_p as well and its $\mathcal{H}^\infty$-angle equals $\pi/2$ as in the scalar case. So let us look for necessary conditions. For this purpose assume that $B_p \in \mathcal{BIP}(Y_p)$. Then $-B_p \in \mathcal{BIP}(Y_p)$ as well since $-B_p = RB_pR$ is similar to B_p via the reflection $Ru(t) = u(-t)$ which is bounded in Y_p and invertible with inverse $R^{-1} = R$. Define next the operator

$$K := \frac{1}{i\sinh(\pi)}[B_p^{-i}(-B_p)^i - \cosh(\pi)).$$

It is not difficult to compute the symbol of K which is given by

$$k(\xi) = \frac{1}{i\sinh(\pi)}[(i\xi)^{-i}(-i\xi)^i - \cosh(\pi)] = -i\,\mathrm{sgn}(\xi),$$

which is precisely the symbol of the Hilbert transform H, defined by

$$Hu(t) = \lim_{R\to\infty}\int_{R^{-1}\leq|s|\leq R} f(t-s)\frac{ds}{\pi s}, \quad t \in \mathbb{R}. \tag{4.11}$$

Note that Hu is well-defined pointwise for all test functions u such that $0 \notin \operatorname{supp}\mathcal{F}(u)$, a dense subset of Y_p in case $1 < p < \infty$.

Thus, as a necessary condition for $B_p \in \mathcal{BIP}(Y_p)$ we find that the Hilbert transform H must be bounded on $L_p(\mathbb{R}; Y)$. This gives rise to the following definition.

Definition 4.3.1. *A Banach space Y is said to belong to the* **class** $\mathcal{HT}$ *if the Hilbert transform defined by* (4.11) *is bounded on $L_2(\mathbb{R}; Y)$. The* **class** $\mathcal{HT}(\alpha)$ *denotes the set of all Banach spaces which belong to $\mathcal{HT}$ and have property* (α).

We note that each Hilbert space E is of class $\mathcal{HT}$, hence also in $\mathcal{HT}(\alpha)$, and if (S, Σ, σ) is a sigma-finite measure space, then $L_p(S; E)$ is of class $\mathcal{HT}$, hence in $\mathcal{HT}(\alpha)$, as well, for each $p \in (1, \infty)$.

We shall see below that – surprisingly – the Mikhlin multiplier theorem remains valid in $L_p(\mathbb{R}; Y)$ in case Y is a Banach space of class $\mathcal{HT}$.

3.2 Necessary Conditions

Let X be a Banach space and consider the spaces $L_p(\mathbb{R}^n; X)$ for $1 < p < \infty$. Given a Fourier multiplier $M \in C(\dot{\mathbb{R}}^n; \mathcal{B}(X,Y))$, where Y denotes another Banach space, we may define an operator $T_M : \mathcal{F}^{-1}\mathcal{D}(\mathbb{R}^n; X) \to \mathcal{S}'(\mathbb{R}^n; Y)$ by means of

$$T_M\phi := \mathcal{F}^{-1}M\mathcal{F}\phi, \quad \text{for all } \mathcal{F}\phi \in \mathcal{D}(\mathbb{R}^n; X), \tag{4.12}$$

where $\mathcal{F}$ denotes the Fourier transform. Since $\mathcal{F}^{-1}\mathcal{D}(\mathbb{R}^n; X)$ is dense in $L_p(\mathbb{R}^n; X)$, we see that T_M is well-defined and linear on a dense subset of $L_p(\mathbb{R}^n; X)$.

The main question about such Fourier multipliers is their boundedness in L_p-norm. This is a classical subject treated in many books; here we want to study the vector-valued case. We show first that $\mathcal{R}$-boundedness of the operator family $\mathcal{T} := \{M(\xi) : \xi \in \dot{\mathbb{R}}^n\} \subset \mathcal{B}(X,Y)$ is necessary for L_p-boundedness of T_M.

For this purpose we have to show that there is a constant C such that for given N points $\xi_j \in \dot{\mathbb{R}}^n$, N vectors $x_j \in X$, N independent $\{-1,1\}$-valued random variables ε_j on a probability space $(\Omega, \mathcal{A}, \mu)$, the inequality

$$\int_\Omega \Big|\sum_j \varepsilon_j M(\xi_j)x_j\Big|_Y^p d\mu \le C^p \int_\Omega \Big|\sum_j \varepsilon_j x_j\Big|_X^p d\mu$$

is valid. Choose a function $\psi \in \mathcal{D}(\mathbb{R}^n)$, nonnegative, radially symmetric, $0 \le \psi \le 1$, such that $\int_{\mathbb{R}^n} \psi(\xi)^2\, d\xi = 1$, and let $\psi_k(\xi) = \psi(\xi k)$, and $\phi_k(x) = \phi(x/k)$; set $\phi = \mathcal{F}\psi$. Then $\mathcal{F}\phi_k(\xi) = k^n\psi(\xi k)$ by symmetry of ψ, and $\mathcal{F}\phi_k \cdot \psi_k \in \mathcal{D}(\mathbb{R}^n)$, as well as

$$\int_{\mathbb{R}^n} \mathcal{F}\phi_k(\xi)\psi_k(\xi)\, d\xi = 1.$$

Therefore we have

$$\lim_{k\to\infty} \int_{\mathbb{R}^n} M(\xi)[\mathcal{F}\phi_k](\xi - \xi_0)\psi_k(\xi - \xi_0)\, d\xi = M(\xi_0)$$

in $\mathcal{B}(X,Y)$, for each $\xi_0 \in \dot{\mathbb{R}}^n$. We estimate as follows.

$$\begin{aligned}
&\int_\Omega \Big|\sum_j \varepsilon_j \int_{\mathbb{R}^n} M(\xi)\mathcal{F}[\phi_k](\xi - \xi_j)x_j\psi_k(\xi - \xi_j)\, d\xi\Big|_Y^p d\mu \\
&\le \int_\Omega \Big|\sum_j \varepsilon_j \int_{\mathbb{R}^n} \mathcal{F}T_M[e^{i\xi_j\cdot}\phi_k(\cdot)x_j](\xi)\cdot\psi_k(\xi - \xi_j)\, d\xi\Big|_Y^p d\mu \\
&= \int_\Omega \Big|\sum_j \varepsilon_j \int_{\mathbb{R}^n} T_M[e^{i\xi_j\cdot}\phi_k(\cdot)x_j](x)\mathcal{F}[\psi_k(\cdot - \xi_j)](x)\, dx\Big|_Y^p d\mu \\
&= \int_\Omega \Big|\sum_j \varepsilon_j \int_{\mathbb{R}^n} e^{-i\xi_j\cdot x}T_M[e^{i\xi_j\cdot}\phi_k(\cdot)x_j](x)[\mathcal{F}\psi_k](x)\, dx\Big|_Y^p d\mu \\
&\le \int_\Omega \Big[\int_{\mathbb{R}^n} \Big|\sum_j \varepsilon_j e^{-i\xi_j\cdot x}T_M[e^{i\xi_j\cdot}\phi_k(\cdot)x_j](x)\Big|_Y |\mathcal{F}\psi_k(x)|\, dx\Big]^p d\mu
\end{aligned}$$

$$\leq \Big[\int_\Omega\int_{\mathbb{R}^n}\Big|\sum_j \varepsilon_j e^{-i\xi_j\cdot x}T_M[e^{i\xi_j\cdot}\phi_k(\cdot)x_j](x)\Big|_Y^p\,dx\Big]\cdot\Big[\int_{\mathbb{R}^n}|\mathcal{F}\psi_k(t)|^{p'}\,dt\Big]^{p/p'}\,d\mu;$$

here $1/p+1/p'=1$. Since T_M is bounded from $L_p(\mathbb{R}^n;X)$ into $L_p(\mathbb{R}^n;Y)$, the set $\{e^{i\sigma\cdot}T_M e^{i\tau\cdot}:\ \sigma,\tau\in\mathbb{R}^n\}$ is $\mathcal{R}$-bounded in $\mathcal{B}(L_p(\mathbb{R}^n;X),L_p(\mathbb{R}^n;Y))$ by Corollary 4.1.9, hence we may continue

$$\begin{aligned}
&\leq C\Big[\int_{\mathbb{R}^n}|\mathcal{F}\psi_k(x)|^{p'}\,dx\Big]^{p/p'}\int_\Omega\int_{\mathbb{R}^n}\Big|\sum_j\varepsilon_j\phi_k(x)x_j\Big|_X^p\,dx\,d\mu\\
&= C\Big[\int_{\mathbb{R}^n}|\mathcal{F}\psi_k(x)|^{p'}\,dx\Big]^{p/p'}\cdot\int_\Omega\Big|\sum_j\varepsilon_j x_j\Big|_X^p\,d\mu\cdot\int_{\mathbb{R}^n}|\phi_k(x)|^p\,dx\\
&= C\Big[\int_{\mathbb{R}^n}|\mathcal{F}\psi_k(x)|^{p'}\,dx\Big]^{p/p'}\int_{\mathbb{R}^n}|\phi_k(x)|^p\,dx]\cdot\int_\Omega\Big|\sum_j\varepsilon_j x_j\Big|_X^p\,d\mu.
\end{aligned}$$

By Fatou's lemma this yields

$$\int_\Omega\Big|\sum_j\varepsilon_j M(\xi_j)x_j\Big|_Y^p\,d\mu\leq C\Big[\int_{\mathbb{R}^n}|\phi(x)|^p\,dx\Big]\Big[\int_{\mathbb{R}^n}|\phi(x)|^{p'}\,dx\Big]^{p/p'}\int_\Omega\Big|\sum_j\varepsilon_j x_j\Big|_X^p\,d\mu.$$

This shows that $\mathcal{T}$ is necessarily $\mathcal{R}$-bounded, in particular M is bounded a.e.

Proposition 4.3.2. *Suppose X, Y are Banach spaces, $1<p<\infty$, and let T_M defined by (4.12) be bounded from $L_p(\mathbb{R}^n;X)$ into $L_p(\mathbb{R}^n;Y)$, for some given $M\in C(\dot{\mathbb{R}}^n;\mathcal{B}(X,Y))$.*

Then $\{M(\xi):\xi\in\dot{\mathbb{R}}^n\}$ is $\mathcal{R}$-bounded in $\mathcal{B}(X,Y)$, in particular we necessarily have $M\in L_\infty(\mathbb{R}^n;\mathcal{B}(X,Y))$, and thus $T_M\mathcal{F}^{-1}\mathcal{D}(\mathbb{R}^n;X)\subset L_\infty(\mathbb{R}^n;Y)$.

3.3. The One-Dimensional Case

The following theorem is the *operator-valued Mikhlin Fourier multiplier theorem* in one variable.

Theorem 4.3.3. *Suppose X, Y are spaces of class $\mathcal{HT}$, let $1<p<\infty$, and $M\in C^1(\dot{\mathbb{R}};\mathcal{B}(X,Y))$ be such that the following conditions are satisfied.*

(i) *$\mathcal{R}(\{M(\xi):\xi\in\dot{\mathbb{R}}\})=:\kappa_0<\infty$;*

(ii) *$\mathcal{R}(\{\xi M'(\xi):\xi\in\dot{\mathbb{R}})=:\kappa_1<\infty$.*

Then the operator T defined by (4.12) is bounded from $L_p(\mathbb{R};X)$ into $L_p(\mathbb{R};Y)$ with norm $|T|_{\mathcal{B}(L_p(\mathbb{R};X),L_p(\mathbb{R};Y))}\leq C(\kappa_0+\kappa_1)$, where the constant $C>0$ depends only on p, X, and Y.

Proof. Before we proof this result, let us note that w.l.o.g. we may assume $X=Y$. Otherwise, we consider the multiplier

$$\tilde{M}(\xi)(x,y)=(0,M(\xi)x),\quad \xi\in\dot{\mathbb{R}},\ (x,y)\in X\times Y,$$

on the product space $X \times Y$.

W.l.o.g. we may restrict attention to multipliers which vanish on $(-\infty, 0)$. In fact, if $M(\xi)$ is given, set $M_+(\xi) = M(\xi)\chi_{\mathbb{R}_+}(\xi)$, $M_-(\xi) = M(\xi)\chi_{\mathbb{R}_-}(\xi)$, where χ_A denotes the characteristic function of the set A. Once we know that the operator corresponding to the multiplier M_+ is bounded, by reflection we also get boundedness of that one with symbol M_-, hence also the operator with symbol M is bounded.

Let R denote the Riesz projection, i.e., the operator in L_p with symbol $\chi_{[0,\infty)}$. Since by assumption X and Y belong to the class $\mathcal{HT}$, R is bounded in $L_p(\mathbb{R}; X)$. Define $R_s = e_s R e_{-s}$ where e_s denotes multiplication with e^{ist}. Then we know by Corollary 4.1.9 that the set $\{R_s : s \in \mathbb{R}\}$ is $\mathcal{R}$-bounded. Therefore, also the set of projections $\mathcal{P} = \{R_\sigma - R_\tau : \sigma, \tau \in \mathbb{R}, \tau > \sigma\}$ is $\mathcal{R}$-bounded. Note that the symbol of R_σ is $\chi_{[\sigma,\infty)}$, hence for $\sigma < \tau$ the symbol of $R_\sigma - R_\tau$ equals $\chi_{[\sigma,\tau)}$.

Next we define the dyadic Schauder decomposition $\Delta = \{\Delta_j : j \in \mathbb{Z}\}$ by means of $\Delta_j = R_{2^j} - R_{2^{j+1}}$. A deep result of Bourgain shows that it is unconditional in $L_p(\mathbb{R}; X)$, provided the Banach space X belongs to the class $\mathcal{HT}$.

Relying on Theorem 4.2.6, we have to show that the set $\{T_M \Delta_m : m \in \mathbb{Z}\}$ is $\mathcal{R}$-bounded. To see this, for $\xi \in [2^j, 2^{j+1})$ write

$$\begin{aligned} M(\xi) &= M(2^j) + \int_{2^j}^{\xi} M'(s)\, ds \\ &= M(2^j) + \int_{2^j}^{2^{j+1}} M'(s)\chi_{[2^j,\xi)}(s)\, ds \\ &= M(2^j) + \int_0^1 2^j M'(2^j(1+r))\chi_{[2^j,\xi)}(2^j(1+r))\, dr. \end{aligned}$$

Taking inverse Fourier transforms, this means

$$T_M \Delta_j = M(2^j)\Delta_j + \int_0^1 2^j M'(2^j(1+r))[R_{2^j(1+r)} - R_{2^{j+1}}]\, dr \Delta_j. \tag{4.13}$$

By means of this representation we estimate as follows.

$$\begin{aligned} &\mathcal{R}\{T_M \Delta_j\} \\ &\leq \mathcal{R}\{M(2^j)\Delta_j\} + \mathcal{R}\Big\{ \int_0^1 2^j M'(2^j(1+r))[R_{2^j(1+r)} - R_{2^{j+1}}]\Delta_j\, dr \Big\} \\ &\leq \kappa_0 \mathcal{R}(\mathcal{P}) + \mathcal{R}\{2^j M'(2^j(1+r))[R_{2^j(1+r)} - R_{2^{j+1}}]\Delta_j :\ j \in \mathbb{Z}, r \in [0,1]\} \\ &\leq \mathcal{R}(\mathcal{P})[\kappa_0 + \mathcal{R}\{2^j M'(2^j(1+r)) :\ j \in \mathbb{Z}, r \in [0,1]\}] \\ &\leq \mathcal{R}(\mathcal{P})[\kappa_0 + \mathcal{R}\{s M'(s) :\ s \in \mathbb{R}, s \neq 0\}] \\ &\leq \mathcal{R}(\mathcal{P})[\kappa_0 + \kappa_1], \end{aligned}$$

where we used convexity of $\mathcal{R}$-bounds and the contraction principle. □

We may strengthen the result in case X and Y also have property (α), employing Theorem 4.2.9.

Theorem 4.3.4. *Suppose X, Y are spaces of class $\mathcal{HT}(\alpha)$, let $1 < p < \infty$, and $\mathcal{M} \subset C^1(\dot{\mathbb{R}}; \mathcal{B}(X,Y))$ be such that the following conditions are satisfied.*

(i) $\mathcal{R}(\{M(\xi) : \xi \in \dot{\mathbb{R}},\ M \in \mathcal{M}\}) =: \kappa_0 < \infty$;

(ii) $\mathcal{R}(\{\xi M'(\xi) : \xi \in \dot{\mathbb{R}},\ M \in \mathcal{M}\}) =: \kappa_1 < \infty$.

Then the family $\mathcal{T} = \{T_M : M \in \mathcal{M}\} \subset \mathcal{B}(L_p(\mathbb{R}; X), L_p(\mathbb{R}; Y))$, with T_M defined by (4.12), is $\mathcal{R}$-bounded with $\mathcal{R}$-bound $\mathcal{R}(\mathcal{T}) \le C(\kappa_0 + \kappa_1)$, where the constant $C > 0$ depends only on p, X, and Y.

Proof. Define the operator family $\mathcal{T}$ as $\mathcal{T} = \{T_M : M \in \mathcal{M}\}$. By Theorem 4.3.3 we know $\mathcal{T} \subset \mathcal{B}(L_p(\mathbb{R}; X); L_p(\mathbb{R}; Y))$, therefore by Theorem 4.2.9 we only need to show that

$$\mathcal{R}(\mathcal{T}_\Delta) < \infty, \quad \text{with } \mathcal{T}_\Delta = \{T_M \Delta_j : j \in \mathbb{Z}, M \in \mathcal{M}\}.$$

This can be proved in exactly the same way as in the last step of the proof of Theorem 4.3.3. □

As in Hilbert spaces $\mathcal{R}$-boundedness is equivalent to uniform boundedness, if X and Y are Hilbert spaces, Theorem 4.3.4 reduces to

Corollary 4.3.5. *Suppose X, Y are Hilbert spaces, let $1 < p < \infty$, and $\mathcal{M} \subset C^1(\dot{\mathbb{R}}; \mathcal{B}(X,Y))$ be such that the following conditions are satisfied.*

(i) $\sup\{|M(\xi)|_{\mathcal{B}(X,Y)} : \xi \in \dot{\mathbb{R}},\ M \in \mathcal{M}\} =: \kappa_0 < \infty$;

(ii) $\sup\{|\xi M'(\xi)|_{\mathcal{B}(X,Y)} : \xi \in \dot{\mathbb{R}},\ M \in \mathcal{M}\} =: \kappa_1 < \infty$.

Then the operator family $\mathcal{T} := \{T_M : M \in \mathcal{M}\} \subset \mathcal{B}(L_p(\mathbb{R}; X), L_p(\mathbb{R}; Y))$ is $\mathcal{R}$-bounded with $\mathcal{R}$-bound $\mathcal{R}(\mathcal{T}) \le C(\kappa_0 + \kappa_1)$.

In particular, this corollary covers and strengthens the finite-dimensional case, i.e., the classical Mikhlin theorem. For the case of scalar multipliers the assumptions of Theorem 4.3.3 can be relaxed.

Corollary 4.3.6. *Suppose X belongs to the class $\mathcal{HT}$ and let $p \in (1, \infty)$. Let $m \in L_\infty(\mathbb{R}) \cap BV_{loc}(\dot{\mathbb{R}})$ be such that $\sup_{R>0} \frac{1}{R} \int_{-R}^{R} |\xi||dm(\xi)| < \infty$. Then for $M = mI$, the operator T_M is bounded in $L_p(\mathbb{R}; X)$, and*

$$|T_M|_{\mathcal{B}(L_p(\mathbb{R};X))} \le C[|m|_\infty + \sup_{R>0} \frac{1}{R} \int_{-R}^{R} |\xi|\, |dm(\xi)|],$$

where C only depends on X and p.

Proof. In the proof of Theorem 4.3.3 we write instead

$$T_M = \sum_j \Big\{m(2^j) - \int_{2^j}^{2^{j+1}} [R_{2^{j+1}} - R_s]\, dm(s)\Big\}\Delta_j.$$

Since $\sup_j \operatorname{Var} m|_{2^j}^{2^{j+1}} < \infty$ if and only if $\sup_{R>0} \frac{1}{R}\int_{-R}^{R} |\xi||\, dm(\xi)| < \infty$, the result follows from unconditionality of the Schauder decomposition $\Delta = \{\Delta_j : j \in \mathbb{Z}\}$ and $\mathcal{R}$-boundedness of the family of projections $\mathcal{P}$; cf. Proposition 4.2.6. □

In case X has also property (α) we obtain the following improvement.

Corollary 4.3.7. *Suppose X belongs to the class $\mathcal{HT}(\alpha)$ and let $p \in (1, \infty)$. Let $\mathcal{M} \subset L_\infty(\mathbb{R}) \cap BV_{loc}(\dot{\mathbb{R}})$ be such that*

$$\sup\{|m|_\infty : m \in \mathcal{M}\} < \infty,$$

and

$$\sup\Big\{\frac{1}{R}\int_{-R}^{R} |\xi||dm(\xi)| : R > 0, m \in \mathcal{M}\Big\} < \infty.$$

Then the operator family $\mathcal{T} = \{T_{mI} : m \in \mathcal{M}\} \subset \mathcal{B}(L_p(\mathbb{R}; X))$ is $\mathcal{R}$-bounded.

The proof follows the above line of arguments and is therefore omitted.

3.4. The Multi-Dimensional Case

If $X, Y \in \mathcal{HT}(\alpha)$, then an operator-valued Fourier multiplier theorem of Lizorkin-type can be deduced form the one-dimensional case Theorem 4.3.4 by induction.

Lemma 4.3.8. *Let $1 < p_1, p_2 < \infty$, $X, Y \in \mathcal{HT}(\alpha)$, and suppose that the family of multipliers $\mathcal{M} \subset C^2(\dot{\mathbb{R}}^2; \mathcal{B}(X, Y))$ satisfies*

$$\mathcal{R}(\{\xi^\alpha D_\xi^\alpha M(\xi) : \xi \in \dot{\mathbb{R}}^2, \alpha \in \{0,1\}^2, |\alpha| \le 2, M \in \mathcal{M}\} =: \kappa < \infty.$$

Then the family of bounded linear operators $\mathcal{T} := \{T_M : M \in \mathcal{M}\}$ is $\mathcal{R}$-bounded in $\mathcal{B}(L_{p_1}(\mathbb{R}; L_{p_2}(\mathbb{R}; X)), L_{p_1}(\mathbb{R}; L_{p_2}(\mathbb{R}; Y)))$, with $\mathcal{R}$-bound less than $C\kappa$, where $C > 0$ only depends on p_1, p_2, X, Y.

Proof. As before, let $X = Y$ w.l.o.g. We first consider the family $\{M(\xi_1, \xi_2)\}$ in dependence on ξ_2, with parameters $\xi_1 \in \dot{\mathbb{R}}$ and $M \in \mathcal{M}$, and apply Theorem 4.3.4 to obtain a family of multipliers

$$\mathcal{K}_0 := \{K_M(\xi_1) = T_{M(\xi_1,\cdot)} : \xi_1 \in \dot{\mathbb{R}}, M \in \mathcal{M}\}$$

which is $\mathcal{R}$-bounded in $\mathcal{B}(L_{p_2}(\mathbb{R}; X))$. Here we use $\mathcal{R}$-bounedness of M and $\xi_2\partial_2 M$. As $L_{p_2}(\mathbb{R}; X)$ belongs to $\mathcal{HT}(\alpha)$, we may apply Theorem 4.3.4 another time. For this we have to verify that the set

$$\mathcal{K}_1 := \{\xi_1\partial_1 K_M(\xi_1) : \xi_1 \in \dot{\mathbb{R}}, M \in \mathcal{M}\}$$

is also $\mathcal{R}$-bounded. But as $\xi_1\partial_1 K_M(\xi_1) = T_{\xi_1\partial_1 M(\xi_1,\cdot)}$, this follows from $\mathcal{R}$-boundedness of $\xi_1\partial_1 M$ and $\xi_1\xi_2\partial_1\partial_2 M$. □

By means of this lemma, by induction there follows the *operator-valued Lizorkin Fourier multiplier theorem* in n dimensions.

Theorem 4.3.9. *Let $1 < p < \infty$, $X, Y \in \mathcal{HT}(\alpha)$, and suppose that the family of multipliers $\mathcal{M} \subset C^n(\dot{\mathbb{R}}^n; \mathcal{B}(X,Y))$ satisfies*

$$\mathcal{R}(\{\xi^\alpha D_\xi^\alpha M(\xi) : \xi \in \dot{\mathbb{R}}^n,\ \alpha \in \{0,1\}^n,\ M \in \mathcal{M}\} =: \kappa < \infty. \tag{4.14}$$

Then the family of operators $\mathcal{T} := \{T_M : M \in \mathcal{M}\} \subset \mathcal{B}(L_p(\mathbb{R}^n; X), L_p(\mathbb{R}^n; Y))$ is $\mathcal{R}$-bounded with $\mathcal{R}$-bound less than $C\kappa$, where $C > 0$ only depends on p, X, Y.

To verify the Lizorkin condition, the following observation is very useful.

Proposition 4.3.10. *Let X, Y be Banach spaces and suppose that, for some $\varphi > 0$, the family of multipliers $\mathcal{M} \subset H^\infty((\Sigma_\varphi \cup -\Sigma_\varphi)^n; \mathcal{B}(X,Y))$ satisfies*

$$\mathcal{R}\{M(z) : z \in (\Sigma_\varphi \cup -\Sigma_\varphi)^n, M \in \mathcal{M}\} =: \kappa < \infty.$$

Then

$$\mathcal{R}(\{\xi^\alpha D_\xi^\alpha M(\xi) : \xi \in \dot{\mathbb{R}}^n, |\alpha| = k, M \in \mathcal{M}\} \leq \kappa/(\sin\varphi)^k,$$

for each $k \in \mathbb{N}_0$.

Proof. Fix $\xi \in \dot{\mathbb{R}}^n$, w.l.o.g. $\xi_j > 0$ for all j, and choose contours $\Gamma_j = \partial B_{r_j}(\xi_j)$ with $r_j = \xi_j \sin\varphi$. Then we have

$$M(\xi) = \frac{1}{(2\pi i)^n} \int_{\Gamma_1} \cdots \int_{\Gamma_n} M(z) \Pi_{j=1}^n (z_j - \xi_j)^{-1}\, dz_n \ldots dz_1,$$

hence taking α derivatives w.r.t. ξ and parameterizing the contours Γ_j by means of $z_j = \xi_j + \xi_j e^{i\theta_j} \sin\varphi$ we obtain

$$\xi^\alpha D_\xi^\alpha M(\xi) = \frac{1}{(2\pi)^n (\sin\varphi)^{|\alpha|}} \int_{(-\pi,\pi)^n} M(z) e^{-i\sum_j \theta_j \alpha_j}\, d\theta.$$

Employing the result on convexity of $\mathcal{R}$-bounds and the contraction principle this yields the assertion. □

We conclude with the n-dimensional analogue of Theorem 4.3.3.

Theorem 4.3.11. *Suppose X, Y are spaces of class $\mathcal{HT}$, $1 < p < \infty$, let $M \in C^n(\mathbb{R}^n \setminus \{0\}; \mathcal{B}(X,Y))$ be such that the following $\mathcal{R}$-boundedness condition is satisfied.*

$$\mathcal{R}(|\xi|^{|\alpha|} |D_\xi^\alpha M(\xi)| : \xi \in \mathbb{R}^n \setminus \{0\}, \alpha \in \{0,1\}^n) =: \kappa < \infty. \tag{4.15}$$

Then the operator T_M defined by (4.12) is bounded from $L_p(\mathbb{R}^n; X)$ into $L_p(\mathbb{R}^n; Y)$ with norm $|T_M|_{\mathcal{B}(L_p(\mathbb{R};X), L_p(\mathbb{R};Y))} \leq C\kappa$, where the constant $C > 0$ depends only on p, X, and Y.

Observe that the Condition (4.15) is stronger than (4.3.9), as the scalar $2D$-example $m(\xi) = \xi_1/(i\xi_1 + \xi_2^2)$ shows. However, in Theorem 4.3.11 we do not need property (α). Since we rarely use this result in the sequel, we omit the proof here; cf. the Bibliographical Comments.

3.5 The Derivation Operator

We consider once more the derivation operator B_p in $L_p(\mathbb{R}; Y)$. As we have seen in (4.9), the Dunford functional calculus for B_p in terms of Fourier-transforms is given by

$$\mathcal{F}\{h(B_p)u\}(\xi) = h(i\xi)\mathcal{F}u(\xi), \quad \xi \in \mathbb{R},$$

where $h \in H_0^\infty(\Sigma_\phi)$ for some $\phi > \pi/2$. As such functions $h(i\cdot)$ satisfy the Mikhlin condition, the operator-valued Fourier multiplier theorem 4.3.3 implies that there is a constant $c = c(p, Y, \phi) > 0$ such that

$$|h(B_p)|_{L_p(\mathbb{R};Y)} \leq c|h|_{H^\infty(\Sigma_\phi)}, \quad h \in H^\infty(\Sigma_\phi),$$

provided Y belongs to the class $\mathcal{HT}$. The same assertion remains valid by causality on intervals $J = \mathbb{R}_+$ or $J = (0, a)$. This yields

Corollary 4.3.12. *Let $p \in (1, \infty)$ and $Y \in \mathcal{HT}$.*

Then $B_p \in \mathcal{H}^\infty(L_p(J; Y))$, for each interval $J = \mathbb{R}, \mathbb{R}_+, (0, a)$. In particular we have $\mathsf{D}(B_p^\alpha) = {}_0H_p^\alpha(\mathbb{R}_+; Y)$ for $J = \mathbb{R}_+$.

We want to extend this result to the case of weighted L_p-spaces. For this purpose we need the following result.

Proposition 4.3.13. *Let $p \in (1, \infty)$ and let $1/p < \mu \leq 1$. Let X, Y be Banach spaces and suppose that $\mathcal{K} \subset C(\mathbb{R}_+; \mathcal{B}(X, Y))$ satisfies $\mathcal{R}\{K(t) : K \in \mathcal{K}\} \leq M/t$ for $t > 0$, where M is a positive constant. Let*

$$(T_K f)(t) := \int_0^t K(t-s)[(t/s)^{1-\mu} - 1]f(s)\,ds, \quad f \in L_p(\mathbb{R}_+; X),\ K \in \mathcal{K}. \tag{4.16}$$

Then $\{T_K : K \in \mathcal{K}\} \subset \mathcal{B}(L_p(\mathbb{R}_+; X), L_p(\mathbb{R}_+; Y))$ is $\mathcal{R}$-bounded, with $\mathcal{R}$-bound $\mathcal{R}\{T_K : K \in \mathcal{K}\} \leq cM$, where $c = c(p, \mu)$.

Proof. Let $f \in L_p(\mathbb{R}_+; X)$ be given. To shorten notation we set

$$\varphi(r) := (1 + r)^{1-\mu} - 1.$$

It is not difficult to establish the elementary estimate

$$\varphi(r) \leq \min\{r^{1-\mu}, (1-\mu)r\}, \quad r > 0. \tag{4.17}$$

Observe that by assumption

$$\mathcal{R}\{K(t-s)[(t/s)^\mu - 1] : K \in \mathcal{K}\} \leq M(t-s)^{-1}[(t/s)^\mu - 1] = \frac{M}{(t-s)}\varphi(\frac{t-s}{s}).$$

Therefore, the $\mathcal{R}$-bound of the kernels of T_K is bounded pointwise by the kernel of the scalar integral operator S given by

$$(Su)(t) := M \int_0^t \frac{1}{(t-s)} \varphi(\frac{t-s}{s}) u(s)\, ds, \quad u \in L_p(\mathbb{R}_+).$$

To apply Proposition 4.1.4, we have to show that this operator S is L_p-bounded.

For this purpose, we use Hölder's inequality to obtain

$$\begin{aligned}|(Su)(t)| &\leq M|u|_p \Big(\int_0^t \Big[\varphi\Big(\frac{t-s}{s}\Big)\frac{1}{(t-s)}\Big]^{p'} ds\Big)^{1/p'} \\ &= M|u_p\Big(\int_0^1 \Big[\varphi\Big(\frac{1-\sigma}{\sigma}\Big)\frac{1}{(1-\sigma)}\Big]^{p'} d\sigma\Big)^{1/p'} \cdot t^{-1/p}\end{aligned}$$

for any $u \in L_p(\mathbb{R}^+)$. Here we have to observe that the integral $\int_0^1 [\varphi(\frac{1-\sigma}{\sigma})\frac{1}{(1-\sigma)})]^{p'} d\sigma$ is finite. In fact, this follows from (4.17) due to

$$\int_0^{1/2} \Big[\frac{(1-\sigma)^{1-\mu}}{\sigma^{1-\mu}} \cdot \frac{1}{1-\sigma}\Big]^{p'} d\sigma + (1-\mu) \int_{1/2}^1 \Big[\frac{1-\sigma}{\sigma} \cdot \frac{1}{1-\sigma}\Big]^{p'} d\sigma \leq c(p,\mu).$$

We conclude that $S : L_p(\mathbb{R}_+) \to L_{p,weak}(\mathbb{R}_+)$ is bounded for each $p > 1/\mu$. By the Marcinkiewicz interpolation theorem, S is bounded in $L_p(\mathbb{R}_+)$ for each $p > 1/\mu$, with bound dominated by $c(p,\mu)M$, where $c(p,\mu)$ depends only on p and μ. □

We are now in position to prove the main result of this subsection.

Theorem 4.3.14. *Let $p \in (1,\infty)$ and $1/p < \mu \leq 1$. Suppose that Y is of class $\mathcal{HT}$. Then $B_{p,\mu}$ admits an $\mathcal{H}^\infty$-calculus in $L_{p,\mu}(\mathbb{R}_+;Y)$ with $\mathcal{H}^\infty$-angle $\phi^\infty_{B_{p,\mu}} = \pi/2$. In particular, we have $\mathsf{D}(B^\alpha_{p,\mu}) = {}_0H^\alpha_{p,\mu}(\mathbb{R}_+;Y)$.*

Proof. Let $\phi > \pi/2$ be fixed and let $h \in H_0(\Sigma_\phi)$ be given. As we have seen in Section 3.2, $h(B_{p,\mu})$ is also represented by the convolution

$$[h(B_{p,\mu})v](t) = \int_0^t k_h(t-s)v(s)\, ds, \quad t > 0, \tag{4.18}$$

where the kernel k_h belongs to $C(\mathbb{R}_+) \cap L_1(\mathbb{R}_+)$ and is given by the inverse Laplace transform of h,

$$k_h(t) = \frac{1}{2\pi i}\int_\Gamma h(\lambda)e^{\lambda t}\, d\lambda, \quad t > 0.$$

To prove the assertion we have to estimate this convolution in $L_{p,\mu}(\mathbb{R}_+;Y)$, i.e., we have to prove an inequality of the form

$$\Big| \int_0^t k_h(t-s)(t/s)^{1-\mu} v(s)\, ds \Big|_p \leq C_\phi |h|_\infty |v|_p \tag{4.19}$$

for $v \in L_p(\mathbb{R}_+;Y)$ and $h \in H_0(\Sigma_\phi)$, with a constant C_ϕ independent of h.

This will be achieved by comparing $h(B_{p,\mu})$ with the functional calculus of B_p in $L_p(\mathbb{R}_+;Y)$ which, by Corollary 4.3.12, is bounded as Y is of class $\mathcal{HT}$. So we know that there is a constant M_ϕ independent of h such that

$$|h(B_p)v|_p = \left| \int_0^t k_h(t-s)v(s)\,ds \right|_p \le M_\phi |h|_{H^\infty(\Sigma_\phi)} |v|_p \tag{4.20}$$

for any $v \in L_p(\mathbb{R}_+;Y)$ and $h \in H_0(\Sigma_\phi)$. One easily verifies that

$$\begin{aligned}\Phi_\mu h(B_p)\Phi_\mu^{-1} &= \frac{1}{2\pi i}\int_\Gamma h(\lambda)\Phi_\mu(\lambda - B_p)^{-1}\Phi_\mu^{-1}\,d\lambda \\ &= \frac{1}{2\pi i}\int_\Gamma h(\lambda)(\lambda-(B_p+B_0))^{-1}\,d\lambda = h(B_p+B_0).\end{aligned}$$

Consequently,

$$(T_h v)(t) := [h(B_p+B_0) - h(B_p)]v(t) = \int_0^t k_h(t-s)[(t/s)^{1-\mu}-1]v(s)\,ds, \tag{4.21}$$

where $v \in L_p(\mathbb{R}_+;Y)$. Observe that

$$|k_h(t)| \le \frac{|h|_\infty}{\pi}\int_0^\infty e^{tr\cos\psi}\,dr \le \frac{C_\phi |h|_{H^\infty(\Sigma_\phi)}}{t}, \quad h \in H_0(\Sigma_\phi).$$

Therefore, the kernel k_h satisfies the assumptions of Proposition 4.3.13 and we conclude that $T_h \in \mathcal{B}(L_p(\mathbb{R}_+;Y))$ with

$$|T_h|_{\mathcal{B}(L_p(\mathbb{R}_+;Y))} \le c(p,\mu,\phi)|h|_{H^\infty(\Sigma_\phi)}, \quad h \in H_0(\Sigma_\phi),$$

where the constant $c(p,\mu,\phi)$ does not depend on h. We can now conclude that $B_{p,\mu}$ has an $\mathcal{H}^\infty$-calculus and that the $\mathcal{H}^\infty$-angle equals $\pi/2$. □

4.4 $\mathcal{R}$-Sectoriality

It is a natural idea to replace the uniform norm-bound on the resolvent of a sectorial operator by the $\mathcal{R}$-bound. Surprisingly, this leads to characterizations of maximal L_p-regularity.

4.1 $\mathcal{R}$-sectorial Operators

The concept of $\mathcal{R}$-bounded families of operators leads immediately to the notion of $\mathcal{R}$-sectorial operators.

Definition 4.4.1. *A sectorial operator is called* **$\mathcal{R}$-sectorial** *if*

$$\mathcal{R}_A(0) := \mathcal{R}\{t(t+A)^{-1} : t > 0\} < \infty.$$

The $\mathcal{R}$-angle ϕ_A^R of A is defined by means of

$$\phi_A^R := \inf\{\theta \in (0,\pi) : \mathcal{R}_A(\pi-\theta) < \infty\},$$

where

$$\mathcal{R}_A(\theta) := \mathcal{R}\{\lambda(\lambda+A)^{-1} : |\arg\lambda| \leq \theta\}.$$

The class of $\mathcal{R}$-sectorial operators will be denoted by $\mathcal{RS}(X)$.

This definition makes sense for the following reason. Suppose $\mathcal{R}_A(0) < \infty$. Then the Taylor series

$$(\lambda+A)^{-1} = \sum_{n=0}^{\infty}(t-\lambda)^n(t+A)^{-n-1}$$

is convergent for $|\lambda - t| < t/M$. With $\theta = \arg(\lambda)$, $t = |\lambda|\cos(\theta)$, and $M := \sup\{t|(t+A)^{-1}| : t > 0\}$, we obtain by the convexity of $\mathcal{R}$-bounds

$$\begin{aligned}\mathcal{R}\{\lambda(\lambda+A)^{-1}:\ |\arg\lambda|\leq\theta\} &\leq |\cos(\theta)|^{-1}\sum_{n=0}^{\infty}|\tan\theta|^n\mathcal{R}_A(0)^{n+1}\\ &= |\cos(\theta)|^{-1}\mathcal{R}_A(0)/(1-|\tan\theta|\mathcal{R}_A(0)).\end{aligned}$$

Thus whenever $|\tan\theta|\mathcal{R}_A(0) < 1$, then $\mathcal{R}_A(\theta) < \infty$. This shows that the $\mathcal{R}$-angle of an $\mathcal{R}$-sectorial operator A is well-defined and it is always not smaller than the spectral angle of A.

The argument we just presented shows also that $\mathcal{R}$-sectorial operators are well-behaved under perturbations, like sectorial operators. The classes of operators with bounded imaginary powers or $\mathcal{H}^\infty$-calculus do not have this property, as we have seen in the previous section. This makes the concept of $\mathcal{R}$-sectorial operators particularly useful.

Proposition 4.4.2. *Suppose A is sectorial in a Banach space X, and let B be closed linear, such that $\mathsf{D}(A) \subset \mathsf{D}(B)$ and $|Bx| \leq b|Ax|$, $x \in \mathsf{D}(A)$, for some $b > 0$. Assume*

$$\mathcal{R}\{A(\lambda+A)^{-1}:\ \lambda\in\Sigma_\theta\} =: c < \infty.$$

Then

$$\mathcal{R}\{\lambda(\lambda+A+B)^{-1}:\ \lambda\in\Sigma_\theta\} < \infty,$$

whenever $b < 1/c$.

Proof. We have

$$(\lambda+A+B)^{-1} = (\lambda+A)^{-1}(1+B(\lambda+A)^{-1})^{-1} \tag{4.22}$$

$$= (\lambda+A)^{-1}\sum_{n=0}^{\infty}[-B(\lambda+A)^{-1}]^n, \tag{4.23}$$

and by induction

$$\begin{aligned}\mathcal{R}\{\lambda(\lambda+A)^{-1}[B(\lambda+A)^{-1}]^n\} &\leq \mathcal{R}\{\lambda(\lambda+A)^{-1}\}[\mathcal{R}\{BA^{-1}A(\lambda+A)^{-1}\}]^n\\ &\leq \mathcal{R}\{\lambda(\lambda+A)^{-1}\}|BA^{-1}|^n\mathcal{R}\{A(\lambda+A)^{-1}\}^n\\ &= \gamma^n\mathcal{R}\{\lambda(\lambda+A)^{-1}\},\end{aligned}$$

where $\gamma = |BA^{-1}|\mathcal{R}\{A(\lambda+A)^{-1}\}$. Now, if $\gamma \leq b\mathcal{R}\{A(\lambda+A)^{-1}\} < 1$, then

$$\mathcal{R}\{(\lambda+A+B)^{-1} : \lambda \in \Sigma_\theta\} \leq \mathcal{R}\{\lambda(\lambda+A)^{-1} : \lambda \in \Sigma_\theta\}/(1-\gamma),$$

which implies the result. □

In the case of relatively bounded perturbations

$$\mathsf{D}(A) \subset \mathsf{D}(B),\ |Bx| \leq b|Ax| + a|x|,\ x \in \mathsf{D}(A), \tag{4.24}$$

with small relative bound b, as usual in perturbation theory for sectorial operators, we have to shift $A+B$.

Proposition 4.4.3. *Suppose A is sectorial in a Banach space X, and let B be closed linear, such that* (4.24) *holds with some constants a and b. Assume*

$$\mathcal{R}\{\lambda(\lambda+A)^{-1} :\ \lambda \in \Sigma_\theta\} =: c < \infty.$$

Then

$$\mathcal{R}\{\lambda(\lambda+\mu+A+B)^{-1} :\ \lambda \in \Sigma_\theta\} < \infty,$$

whenever $b < 1/(1+c)C_0(A)$ and $\mu > aM_0(A)(1+c)/(1-bC_0(A)(1+c))$, where

$$C_0(A) = \sup_{\mu>0}|A(\mu+A)^{-1}| \quad \textit{and} \quad M_0(A) = \sup_{\mu>0}|\mu(\mu+A)^{-1}|.$$

Proof. As in the proof of Proposition 4.4.2 we have

$$(\lambda+\mu+A+B)^{-1} = (\lambda+\mu+A)^{-1}(1+B(\lambda+\mu+A)^{-1})^{-1} \tag{4.25}$$

$$= (\lambda+\mu+A)^{-1}\sum_{n=0}^{\infty}[-B(\lambda+\mu+A)^{-1}]^n, \tag{4.26}$$

and by induction

$$\begin{aligned}&\mathcal{R}\{\lambda(\lambda+\mu+A)^{-1}[B(\lambda+\mu+A)^{-1}]^n\}\\ &\leq \mathcal{R}\{\lambda(\lambda+\mu+A)^{-1}\}[\mathcal{R}\{B(\mu+A)^{-1}(\mu+A)(\lambda+\mu+A)^{-1}\}]^n\\ &\leq \mathcal{R}\{\lambda(\lambda+\mu+A)^{-1}\}|B(\mu+A)^{-1}|^n\mathcal{R}\{(\mu+A)(\lambda+\mu+A)^{-1}\}^n\end{aligned}$$

(4.24) implies

$$|B(\mu+A)^{-1}| \leq bC_0(A) + aM_0(A)/\mu.$$

On the other hand, by convexity of $\mathcal{R}$-bounds,

$$\mathcal{R}\{\lambda(\lambda+\mu+A)^{-1} : \lambda \in \Sigma_\theta\} \le \mathcal{R}\{\lambda(\lambda+A)^{-1} : \lambda \in \Sigma_\theta\} = c,$$

hence

$$\mathcal{R}\{(\mu+A)(\lambda+\mu+A)^{-1} : \lambda \in \Sigma_\theta\} \le 1 + \mathcal{R}\{\lambda(\lambda+\mu+A)^{-1} : \lambda \in \Sigma_\theta\} \le 1 + c.$$

Thus we obtain

$$\mathcal{R}\{\lambda(\lambda+\mu+A+B)^{-1} : \lambda \in \Sigma_\theta\} < \infty,$$

if $(bC_0(A) + aM_0(A)/\mu)(1+c) < 1$. □

4.2 Maximal L_p-regularity

Consider now the Cauchy problem

$$\dot{u}(t) + Au(t) = f(t), \quad t > 0, \quad u(0) = 0, \tag{4.27}$$

where A denotes a sectorial operator in a Banach space X with spectral angle $\phi_A < \pi/2$. Then for a given function $f \in L_p(\mathbb{R}_+; X)$ the solution is represented by the variation of parameters formula

$$u(t) = \int_0^t e^{-As} f(t-s)\, ds, \quad t \ge 0.$$

Maximal regularity of type L_p is then equivalent to $Au \in L_p(\mathbb{R}_+; X)$, for each $f \in L_p(\mathbb{R}_+; X)$. Looking at the problem (4.27) on the whole line instead of the half-line, the question then becomes whether the convolution operator with kernel

$$K(t) = Ae^{-At}\chi_{(0,\infty)}(t), \quad t \in \mathbb{R},$$

is L_p-bounded. The symbol of this convolution operator is given by

$$M(\xi) = A(i\xi + A)^{-1}, \quad \xi \in \mathbb{R},$$

and so by Proposition 4.3.2, $\mathcal{R}$-boundedness of the family $\{A(i\xi + A)^{-1} : \xi \in \mathbb{R}\}$ is necessary for maximal regularity of (4.27) of type L_p, even in a general Banach space. However, in spaces X of class $\mathcal{HT}$ the converse also holds. This explains the importance of $\mathcal{R}$-sectorial operators.

Theorem 4.4.4. *Let X be a Banach space of class $\mathcal{HT}$, $1 < p < \infty$, and let A be a sectorial operator in X with spectral angle $\phi_A < \pi/2$.*

Then (4.27) *has maximal regularity of type L_p if and only if A is $\mathcal{R}$-sectorial with $\phi_A^R < \pi/2$. More precisely, the following statements are equivalent.*

(i) *The Cauchy problem* (4.27) *has maximal regularity of type L_p;*

(ii) *the set $\{A(i\xi + A)^{-1} : \xi \in \mathbb{R}\}$ is $\mathcal{R}$-bounded;*

(iii) *the set $\{A(\lambda + A)^{-1} : \lambda \in \Sigma_\theta\}$ is $\mathcal{R}$-bounded, for some $\theta > \pi/2$;*

(iv) *the set* $\{e^{-Az} : z \in \Sigma_\vartheta\}$ *is* $\mathcal{R}$*-bounded for some* $\vartheta > 0$;

(v) *the sets* $\{e^{-At} : t > 0\}$ *and* $\{tAe^{-At} : t > 0\}$ *are* $\mathcal{R}$*-bounded.*

Proof. We have already seen that (i) implies (ii). For the converse we employ Theorem 4.3.3 with $M(\xi) = A(i\xi + A)^{-1}$. To see that $\{\xi M'(\xi) : \xi \in \mathbb{R},\ \xi \neq 0\}$ is $\mathcal{R}$-bounded, we only have to observe

$$\xi M'(\xi) = -i\xi A(i\xi + A)^{-2} = -A(i\xi + A)^{-1} + [A(i\xi + A)^{-1}]^2,$$

and to apply Proposition 4.1.6.

(ii)⇒(iii) Since $H(\lambda) = A(\lambda + A)^{-1}$ is holomorphic and bounded in a sector Σ_θ for some $\theta > \pi/2$, we may employ the Poisson formula to write for $\lambda = \sigma + i\tau$

$$H(\lambda) = \int_{\mathbb{R}} p_\sigma(\tau - \rho) H(i\rho)\, d\rho,$$

where

$$p_\sigma(\tau) = \frac{1}{\pi} \frac{\sigma}{\sigma^2 + \tau^2}$$

denotes the Poisson kernel in one space dimension. By means of a scaling we may alternatively write

$$H(\lambda) = \int_{\mathbb{R}} p_1(\rho) H(i(\tau - \rho\sigma))\, d\rho.$$

But since $\int_{\mathbb{R}} p_1(\rho)\, d\rho = 1$, and $p_1(\rho) \geq 0$ we obtain

$$\{H(\lambda) : \lambda \in \Sigma_{\pi/2}\} \subset \overline{\text{co}}\{H(i\rho) : \rho \in \mathbb{R}\}.$$

Thus the $\mathcal{R}$-angle must be smaller than $\pi/2$ by the Neumann series argument from the previous subsection.

(iii)⇒(v) Fix any $\varphi \in (\pi/2, \theta)$ and denote by Γ_r the contour

$$\Gamma_r = \{z \in \mathbb{C} : |z| \geq r,\ |\arg(z)| = \varphi\} \cup \{z \in \mathbb{C} : |z| = r,\ |\arg(z)| \leq \varphi\},$$

oriented properly, where $r \geq 0$. Then we have as in Section 3.1.4 the representation formulae

$$e^{-At} = \frac{1}{2\pi i} \int_{\Gamma_1} e^z \frac{z}{t} \left(\frac{z}{t} + A\right)^{-1} \frac{dz}{z}, \quad t > 0,$$

and

$$tAe^{-At} = \frac{1}{2\pi i} \int_{\Gamma_0} e^z A \left(\frac{z}{t} + A\right)^{-1} dz, \quad t > 0.$$

Thus we obtain

$$\{e^{-At} : t > 0\} \subset c_0 \cdot \overline{aco}(\{\lambda(\lambda + A)^{-1} : \lambda \in \Sigma_\theta\}),$$

where

$$c_0 = (2\pi)^{-1} \int_{\Gamma_1} |e^z| \frac{|dz|}{|z|} < \infty,$$

and similarly

$$\{tAe^{-At} : t > 0\} \subset c_1 \cdot \overline{aco}(\{A(\lambda + A)^{-1} : \lambda \in \Sigma_\theta\}),$$

with

$$c_1 = (2\pi)^{-1} \int_{\Gamma_0} |e^z||dz| < \infty.$$

Applying Proposition 4.1.10 on the absolute convex hull the implication in question follows.

(v)⇒(iv) We use the power series expansion for e^{-Az} according to

$$e^{-Az} = \sum_{k=0}^{\infty} \frac{1}{k!} [rA]^k e^{-Ar} \left(1 - \frac{z}{r}\right)^k.$$

We may choose e.g. $r = |z|$, for $|\arg(z)| \le \vartheta$. The series then converges absolutely provided ϑ is small enough. Then we obtain from Proposition 4.1.6 and Lemma 4.1.7

$$\begin{aligned}\mathcal{R}(\{e^{-Az} : |\arg(z)| \le \vartheta\}) &\le \mathcal{R}(\{e^{-Ar} : r > 0\}) \\ &\quad + \sum_{k=1}^{\infty} \frac{k^k}{k!} (2\sin(\vartheta/2))^k \mathcal{R}(\{rAe^{-Ar} : r > 0\})^k < \infty,\end{aligned}$$

in case $2e\sin(\vartheta/2)\mathcal{R}(\{rAe^{-Ar} : r > 0\}) < 1$.

(iv)⇒(ii) Here we employ the Laplace transform to obtain with $\arg(\lambda) = \psi$

$$\begin{aligned}\lambda(\lambda + A)^{-1} &= \lambda \int_0^\infty e^{-\lambda t} e^{-At}\, dt \\ &= e^{-i\varphi} \int_0^\infty \lambda e^{-\lambda e^{-i\varphi} t} e^{-Ae^{-i\varphi} t}\, dt \\ &= e^{i(\psi-\varphi)} \int_0^\infty e^{-e^{i(\psi-\varphi)} s} e^{-Ae^{-i\varphi} s/|\lambda|}\, ds,\end{aligned}$$

where we used Cauchy's theorem and the scaling $s = |\lambda| t$. Now $\psi = \pm\pi/2$, so fix an angle φ such that $|\varphi| < \vartheta$ and $\cos(\psi - \varphi) > 0$. Then the integral is absolutely convergent and

$$\{i\xi(i\xi + A)^{-1} : \xi \in \mathbb{R}\} \subset c_2 \cdot \overline{aco}(\{e^{-Az} : |\arg(z)| < \vartheta\}),$$

where $c_2 = 1/\cos(\psi - \varphi)$. Therefore, Proposition 4.1.10 yields the claim. □

4.3 A Sufficient Condition for R-Sectoriality

The class of operators with bounded imaginary powers is contained in the class of $\mathcal{R}$-sectorial operators, at least in case the underlying Banach space X belongs to the class $\mathcal{HT}$.

Theorem 4.4.5. *Suppose X is a space of class $\mathcal{HT}$ and let $A \in \mathcal{BIP}(X)$ with power angle θ_A. Then A is $\mathcal{R}$-sectorial and $\phi_A^R \leq \theta_A$.*

Proof. The proof is based on the representation formulae (3.58) and (3.59) which have been obtained in Section 3.3. Suppose $A \in \mathcal{BIP}(X)$ with power angle $\theta_A := \overline{\lim}_{|s|\to\infty}|s|^{-1}\log|A^{is}|_{\mathcal{B}(X)}$. Then the first identity reads

$$(1+rA)^{-1}x = \frac{1}{2\pi i}\Big[PV\int_{\mathbb{R}}(rA)^{-is}x\frac{\pi ds}{\sinh(\pi s)}\Big] + \frac{1}{2}x, \quad x \in X,\ r > 0, \tag{4.28}$$

where PV means principal value. The second one is

$$(1+re^{i\phi}A)^{-1} = (1+rA)^{-1}+\frac{1}{2\pi i}\int_{\mathbb{R}}(rA)^{-is}(e^{\phi s}-1)\frac{\pi ds}{\sinh(\pi s)}, \quad |\phi| < \pi-\theta_A,\ r > 0. \tag{4.29}$$

Suppose $n \in \mathbb{N}$, $x_j \in X$, $\lambda_j = r_j e^{i\phi_j}$, $r_j > 0$, $|\phi_j| \leq \pi - \theta$ are given, where $\theta > \theta_A$ is fixed. We have to prove that there is a constant $C > 0$, depending only on θ, such that

$$\Big|\sum_{j=1}^{n}\varepsilon_j(1+r_je^{i\phi_j}A)^{-1}x_j\Big|_{L_p(\Omega;X)} \leq C\Big|\sum_{j=1}^{n}\varepsilon_j x_j\Big|_{L_p(\Omega;X)}$$

is valid, where the functions ε_j are independent symmetric random variables on some probability space $(\Omega, \mathcal{A}, \mu)$ with values in $\{-1, 1\}$. For this purpose we decompose

$$(1+r_je^{i\phi_j}A)^{-1} = \frac{1}{2} + T_j + S_j + R_j,$$

where

$$T_j = \int_{\mathbb{R}}(r_jA)^{-is}\psi_j^T(s)\,ds, \quad \psi_j^T(s) = (e^{\phi_j s}-1)\frac{1}{2\pi i\sinh \pi s},$$

$$S_j = \int_{\mathbb{R}}(r_jA)^{-is}\psi_j^S(s)\,ds, \quad \psi_j^S(s) = \frac{1}{2\pi i}\Big(\frac{\pi}{\sinh(\pi s)} - \frac{\chi(s)}{s}\Big),$$

$$R_jx = \frac{1}{2\pi i}PV\int_{-1}^{1}(r_jA)^{-is}x\,\frac{ds}{s}, \quad x \in X.$$

Here χ means the characteristic function of the the interval $[-1, 1]$. By the triangle inequality we estimate separately.

Using Kahane's contraction principle, in virtue of

$$\big|\psi_j^T(s)\big| \leq ce^{(|\phi_j|-\pi)|s|} \leq ce^{-\theta|s|},$$

we get for the first term

$$\begin{aligned}\Big|\sum_j \varepsilon_j T_j x_j\Big|_{L_p(\Omega;X)} &\leq \int_{\mathbb{R}} \Big|A^{-is}\sum_j r_j^{-is}\varepsilon_j x_j \psi_j^T(s)\Big|_{L_p(\Omega;X)} ds\\ &\leq M_\eta \int_{\mathbb{R}} e^{(\theta_A+\eta)|s|}\Big|\sum_j r_j^{-is}\psi_j^T(s)\varepsilon_j x_j\Big|_{L_p(\Omega;X)} ds\\ &\leq M_\eta c\int_{\mathbb{R}} e^{(\theta_A+\eta-\theta)|s|}\Big|\sum_j \varepsilon_j x_j\Big|_{L_p(\Omega;X)} ds = M\Big|\sum_j \varepsilon_j x_j\Big|_{L_p(\Omega;X)},\end{aligned}$$

because of $\theta > \theta_A$, and for $\eta < \theta - \theta_A$.

The same type of estimate applies to the term involving the S_j. This time we have

$$|\psi_j^S(s)| \leq ce^{-\pi|s|}, \quad s \in \mathbb{R},$$

and so

$$\begin{aligned}\Big|\sum_j \varepsilon_j S_j x_j\Big|_{L_p(\Omega;X)} &\leq \int_{\mathbb{R}} \Big|A^{-is}\sum_j \varepsilon_j x_j r_j^{-is}\psi_j^S(s)\Big|_{L_p(\Omega;X)} ds\\ &\leq M_\eta \int_{\mathbb{R}} e^{(\theta_A+\eta)|s|}\Big|\sum_j r_j^{-is}\psi_j^S(s)\varepsilon_j x_j\Big|_{L_p(\Omega;X)} ds\\ &\leq M_\eta c\int_{\mathbb{R}} e^{(\theta_A+\eta-\pi)|s|}\Big|\sum_j \varepsilon_j x_j\Big|_{L_p(\Omega;X)} ds = M\Big|\sum_j \varepsilon_j x_j\Big|_{L_p(\Omega;X)},\end{aligned}$$

when ever $\eta < \pi - \theta_A$.

The third term is more sophisticated, it is only here where we use the $\mathcal{HT}$-property of the underlying space X.

We begin with Kahane's contraction principle and also apply the boundedness of A^{is} for $|s| \leq 1$. For $t \in [-1,1]$ we have

$$\Big|\sum_{j=1}^N \varepsilon_j R_j x_j\Big|_{L_p(\Omega;X)} \leq C\Big|A^{it}\sum_{j=1}^N \varepsilon_j r_j^{it} R_j x_j\Big|_{L_p(\Omega;X)},$$

hence integrating over $t \in [-1,1]$

$$\begin{aligned}2\Big|\sum_{j=1}^N \varepsilon_j R_j x_j\Big|^p_{L_p(\Omega;X)} &\leq C\int_{-1}^1 \Big|A^{it}\sum_{j=1}^N \varepsilon_j r_j^{it} R_j x_j\Big|^p_{L_p(\Omega;X)} dt\\ &= C\int_\Omega\int_{-1}^1 \Big|A^{it}\sum_{j=1}^N \varepsilon_j(\omega) r_j^{it} R_j x_j\Big|^p_X dtd\omega\\ &= C\int_\Omega\int_{-1}^1 \Big|\int_{-1}^1 \Big[A^{i(t-s)}\sum_j \varepsilon_j(\omega) r_j^{i(t-s)} x_j\Big]\frac{ds}{s}\Big|^p_X dtd\omega\end{aligned}$$

$$\le C \int_\Omega \int_{-1}^1 \Big| \sum_j A^{it} \varepsilon_j(\omega) r_j^{it} x_j \Big|_X^p \, dt d\omega$$

$$\le C \int_\Omega \int_{-1}^1 \Big| \sum_{j=1}^N r_j^{it} \varepsilon_j x_j \Big|_X^p \, dt d\omega \le C \Big| \sum_{j=1}^N \varepsilon_j x_j \Big|_{L_p(\Omega;X)}^p,$$

where in the next to last step we used the boundedness of the Hilbert transform, and then once more Kahane's contraction principle. These estimates prove the theorem. □

4.5 Operators with $\mathcal{R}$-Bounded Functional Calculus

The previous subsections have shown that the concept of $\mathcal{R}$-boundedness is important. We now want to connect this idea to the $\mathcal{H}^\infty$-calculus of operators.

5.1 The Class $\mathcal{RH}^\infty(X)$
This class is given by

Definition 4.5.1. *Let X be a Banach space and suppose $A \in \mathcal{H}^\infty(X)$. The operator A is said to admit an $\mathcal{R}$-**bounded H^∞-calculus** if the set*

$$\{h(A) : h \in H^\infty(\Sigma_\theta),\ |h|_{H^\infty(\Sigma_\theta)} \le 1\}$$

*is $\mathcal{R}$-bounded, for some $\theta > 0$. We denote the class of such operators by $\mathcal{RH}^\infty(X)$, and define the $\mathcal{RH}^\infty$-**angle** $\phi_A^{R\infty}$ of A as the infimum of such angles $\theta > \phi_A$.*

Note that in Hilbert spaces we have the relations

$$\mathcal{RH}^\infty(X) = \mathcal{H}^\infty(X) = \mathcal{BIP}(X) \subset \mathcal{RS}(X) = \mathcal{S}(X),$$

and counterexamples show that the inclusion is strict, in general. For general Banach spaces the equalities will become strict inclusions, too. The same can be said about the corresponding angles, we have

$$\phi_A^{R\infty} \ge \phi_A^\infty \ge \theta_A \ge \phi_A^R \ge \phi_A,$$

but in general, the inequalities may be strict.

In the next subsection we will see that $\mathcal{RH}^\infty(X) = \mathcal{H}^\infty(X)$ and $\phi_A^{R\infty} = \phi_A^\infty$ for each $A \in \mathcal{H}^\infty(X)$, provided the underlying Banach space has property (α).

The importance of this class of operators lies in the following fact.

Proposition 4.5.2. *Let X be a Banach space, $A \in \mathcal{RH}^\infty(X)$ and suppose that $\{h_\lambda\}_{\lambda\in\Lambda} \subset H^\infty(\Sigma_\theta)$ is uniformly bounded, for some $\theta > \phi_A^{R\infty}$, where Λ is an arbitrary index set.*

Then $\{h_\lambda(A) : \lambda \in \Lambda\}$ is $\mathcal{R}$-bounded.

The proof of this result follows directly from the definition of operators with $\mathcal{R}$-bounded functional calculus. Nevertheless, this result is useful since it allows us to verify $\mathcal{R}$-boundedness conditions like that in Theorem 4.3.3. In particular, we may this way obtain quite directly a joint $\mathcal{H}^\infty$-calculus for two commuting sectorial operators; see the next subsection.

Next we consider the derivation operator $B_{p,\mu} = d/dt$ in $L_{p,\mu}(\mathbb{R}_+;Y)$, $1 < p < \infty$, $1/p < \mu \leq 1$. We have seen in Section 4.3.5 that $B_{p,\mu}$ has bounded $\mathcal{H}^\infty$-calculus, provided $Y \in \mathcal{HT}$. If, moreover, Y has property (α), then the $\mathcal{H}^\infty$-calulus of $B_{p,\mu}$ is also $\mathcal{R}$-bounded.

Theorem 4.5.3. *Let $1 < p < \infty$, $1/p < \mu \leq 1$, and let $Y \in \mathcal{HT}(\alpha)$. Then $B_{p,\mu} \in \mathcal{RH}^\infty(L_{p,\mu}(\mathbb{R}_+;Y))$ with $\phi^{\mathcal{R}\infty}_{B_{p,\mu}} = \pi/2$.*

Proof. We first apply Theorem 4.3.4 to see that $B_p := B_{p,1} \in \mathcal{RH}^\infty(L_p(\mathbb{R}_+;Y))$ with $\mathcal{RH}^\infty$-angle $\pi/2$. Then we proceed as in the proof of Theorem 4.3.14, comparing $h(B_p + B_0)$ with $h(B_p)$ in L_p. Here we use the full strength of Proposition 4.3.13 to obtain $\mathcal{R}$-boundedness of the set

$$\{h(B_p + B_0) - h(B_p) : |h|_\infty \leq 1\} = \{T_h : |h|_\infty \leq 1\}$$

in $\mathcal{B}(L_p(\mathbb{R}_+;Y))$. □

This result will also be a consequence of Theorem 4.5.6. The second goal here is the following surprising result.

Theorem 4.5.4. *Let X be a Banach space, $A \in \mathcal{S}(X)$ be invertible, $\alpha \in (0,1)$, and $1 \leq q < \infty$.*

Then $A \in \mathcal{RH}^\infty(D_A(\alpha,q))$ with $\mathcal{RH}^\infty$-angle equal to ϕ_A.

Proof. Let ε_j be independent, symmetric $\{-1,1\}$-valued random variables on a probability space $(\Omega, \mathcal{A}, \mu)$, $x_j \in X$, and fix $h_j \in H_0(\Sigma_\phi)$, where $j = 1, \ldots, N$, and $\phi > \phi_A$. We have to prove that there is a constant $C > 0$ such that

$$\Big|\sum_{j=1}^N \varepsilon_j h_j(A) x_j\Big|_{L_q(\Omega;D_A(\alpha,q))} \leq C \Big|\sum_{j=1}^N \varepsilon_j x_j\Big|_{L_q(\Omega;D_A(\alpha,q))}.$$

For this purpose we choose a standard contour Γ with angle $\theta \in (\theta_A, \phi)$, and note that, by Cauchy's theorem,

$$\begin{aligned} A(t+A)^{-1}h_j(A) &= \frac{1}{2\pi i}\int_\Gamma h_j(z)A(t+A)^{-1}(z-A)^{-1}\,dz \\ &= \frac{1}{2\pi i}\int_\Gamma h_j(z)A(z-A)^{-1}\,dz/(t+z). \end{aligned}$$

First we take $1 < q < \infty$. We estimate using Hölder's inequality, Fubini's theorem several times, and the contraction principle.

$$\begin{aligned}
&\Big|\sum_{j=1}^N \varepsilon_j h_j(A)x_j\Big|^q_{L_q(\Omega;D_A(\alpha,q))}\\
&\le \int_\Omega\int_0^\infty \Big|t^\alpha \sum_j \varepsilon_j \frac{1}{2\pi i}\int_\Gamma h_j(z)A(z-A)^{-1}x_j\frac{dz}{t+z}\Big|^q dt/t\,d\mu\\
&\le C\int_0^\infty t^{\alpha q-1}\Big[\int_\Gamma |z|^{-\beta q'}\frac{|dz|}{t+|z|}\Big]^{q/q'}\cdot\\
&\quad\cdot\int_\Gamma\int_\Omega\Big|\sum_j \varepsilon_j h_j(z) z^\beta A(z-A)^{-1})x_j\Big|^q d\mu\frac{|dz|}{t+|z|}\,dt\\
&\le C\sup_j |h_j|^q_\infty \int_0^\infty t^{\alpha q-1-\beta q}\int_\Gamma\int_\Omega\Big|\sum_j \varepsilon_j z^\beta A(z-A)^{-1}x_j\Big|^q d\mu\frac{|dz|}{t+|z|}\,dt\\
&= C\sup_j |h_j|^q_\infty \int_\Omega\int_\Gamma\Big[\int_0^\infty t^{\alpha q-1-\beta q}\frac{dt}{t+|z|}\Big]\Big|z^\beta A(z-A)^{-1}\sum_j\varepsilon_j x_j\Big|^q|dz|\,d\mu\\
&\le C\sup_j |h_j|^q_\infty \int_\Omega\int_\Gamma\Big|z^\alpha A(z-A)^{-1}\sum_j\varepsilon_j x_j\Big|^q|dz|/|z|\,d\mu\\
&= C\sup_j |h_j|^q_\infty \Big|\sum_j \varepsilon_j x_j\Big|^q_{L_q(\Omega;D_A(\alpha,q))}.
\end{aligned}$$

Here we chose $\beta > 0$ such that $\alpha - 1/q < \beta < \alpha$. This proves the theorem for $1 < q < \infty$. For $q = 1$ the argument is similar and simpler. □

Specializing to the case $X = L_p(\mathbb{R}_+; Y)$ we obtain

Corollary 4.5.5. *Let Y be a Banach space, $\alpha \in (0,1)$, $p, q \in [1,\infty)$.*
Then $B_{p,\mu}+\omega \in \mathcal{RH}^\infty({}_0B^\alpha_{pq,\mu}(\mathbb{R}_+;Y))$ with $\mathcal{RH}^\infty$-angle $\pi/2$, for each $\omega > 0$.

One should compare this result with Theorem 4.5.3.

5.2 The Operator-Valued Functional Calculus

In this section we prove the following result which extends the scalar $\mathcal{H}^\infty$-calculus of a sectorial operator to the $\mathcal{R}$-bounded operator-valued case.

Theorem 4.5.6 (Kalton-Weis theorem). *Let X be a Banach space, $A \in \mathcal{H}^\infty(X)$, $\phi > \phi_A$, and let $\mathcal{F}$ be an operator-valued family $\mathcal{F} \subset H^\infty(\Sigma_\phi; \mathcal{B}(X))$ such that*

$$F(\lambda)(\mu - A)^{-1} = (\mu - A)^{-1}F(\lambda), \quad \mu \in \rho(A),\ \lambda \in \Sigma_\phi,\ F \in \mathcal{F}.$$

Then there is a constant $C_A > 0$ depending only on A and X such that
(i) *If $\sup_{F\in\mathcal{F}} \mathcal{R}(F(\Sigma_\phi)) < \infty$, then $\mathcal{F}(A) := \{F(A) : F \in \mathcal{F}\} \subset \mathcal{B}(X)$ and*

$$|F(A)|_{\mathcal{B}(X)} \le C_A\mathcal{R}(F(\Sigma_\phi)), \quad F \in \mathcal{F}.$$

(ii) *If X has property (α) and $\mathcal{R}\{F(z) : z \in \Sigma_\phi, F \in \mathcal{F}\} < \infty$, then the operator family $\mathcal{F}(A)$ is $\mathcal{R}$-bounded, and*

$$\mathcal{R}(\mathcal{F}(A)) \leq C_A \mathcal{R}\{F(z) : z \in \Sigma_\phi, F \in \mathcal{F}\}.$$

(iii) *In particular, if X has property (α), then $A \in \mathcal{RH}^\infty(X)$ with $\phi_A^{R\infty} = \phi_A^\infty$.*

Before we are going into its proof let us discuss some of its consequences. The first two corollaries concern the so-called *joint functional calculus* of sectorial operators.

Corollary 4.5.7. *Suppose $A \in \mathcal{H}^\infty(X)$ and $B \in \mathcal{S}(X)$ are commuting, $\mathcal{F} \subset H^\infty(\Sigma_\phi \times \Sigma_\psi)$ with $\phi > \phi_A^\infty$, $\psi > \phi_B$, and set $\mathcal{F}(A, B) = \{f(A, B) : f \in \mathcal{F}\}$. Then*
(i) *If $\sup_{f \in \mathcal{F}} \mathcal{R}(f(\Sigma_\phi, B)) < \infty$, then $\mathcal{F}(A, B) \subset \mathcal{B}(X)$ and*

$$|f(A, B)|_{\mathcal{B}(X)} \leq C_A \mathcal{R}(f(\Sigma_\phi, B)), \quad f \in \mathcal{F}.$$

(ii) *If X has property (α), and $\mathcal{R}\{f(z, B) : z \in \Sigma_\phi, f \in \mathcal{F}\} < \infty$, then the operator family $\mathcal{F}(A, B) \subset \mathcal{B}(X)$ is $\mathcal{R}$-bounded and*

$$\mathcal{R}(\mathcal{F}(A, B)) \leq C_A \mathcal{R}\{f(z, B) : z \in \Sigma_\phi, f \in \mathcal{F}\} < \infty.$$

Corollary 4.5.7 follows from Theorem 4.5.6 by setting $F(\lambda) = f(\lambda, B)$. This corollary takes an especially nice form in case $B \in \mathcal{RH}^\infty(X)$.

Corollary 4.5.8 (Joint functional calculus). *Suppose $A \in \mathcal{H}^\infty(X)$ and $B \in \mathcal{RH}^\infty(X)$ are commuting in the resolvent sense, $\mathcal{F} \subset H^\infty(\Sigma_\phi \times \Sigma_\psi)$ with $\phi > \phi_A^\infty$, $\psi > \phi_B$, and let $\mathcal{F}(A, B) = \{f(A, B) : f \in \mathcal{F}\}$. Then*
(i) *$\mathcal{F}(A, B) \subset \mathcal{B}(X)$ and*

$$\sup_{f \in \mathcal{F}} |f(A, B)|_{\mathcal{B}(X)} \leq C_{A,B} \sup_{f \in \mathcal{F}} |f|_{H^\infty(\Sigma_\phi)}.$$

(ii) *If X has property (α), then $\mathcal{F}(A, B) \subset \mathcal{B}(X)$ is $\mathcal{R}$ bounded, and*

$$\mathcal{R}(\mathcal{F}(A, B)) \leq C_{A,B} \sup_{f \in \mathcal{F}} |f|_{H^\infty(\Sigma_\phi)}.$$

This result follows from Corollary 4.5.7 since $f(\Sigma_\phi, B)$ is $\mathcal{R}$-bounded because of $B \in \mathcal{RH}^\infty(X)$.

Corollary 4.5.9 (Dore-Venni theorem). *Suppose $A \in \mathcal{H}^\infty(X)$ and $B \in \mathcal{RS}(X)$ are commuting, and such that $\phi_A^\infty + \phi_B^R < \pi$.*

Then $A + B$ with domain $\mathsf{D}(A + B) = \mathsf{D}(A) \cap \mathsf{D}(B)$ is closed, $A + B \in \mathcal{S}(X)$ with $\phi_{A+B} \leq \max\{\phi_A^\infty, \phi_B^R\}$, and

$$|Ax| + |Bx| \leq C|(A + B)x|, \quad x \in \mathsf{D}(A) \cap \mathsf{D}(B),$$

for some constant $C > 0$. $(A + B)$ is invertible if A or B are so. If in addition X has property (α), then $A + B \in \mathcal{RS}(X)$, and $\phi_{A+B}^R \leq \max\{\phi_A^\infty, \phi_B^R\}$.

This result follows from Corollary 4.5.7 by setting $f(\lambda,\mu)=\mu/(\lambda+\mu)$ resp. $f_z(\lambda,\mu)=z/(z+\lambda+\mu)$.

Corollary 4.5.10 (Mixed derivative theorem). *Suppose $A\in\mathcal{H}^\infty(X)$ and $B\in\mathcal{RS}(X)$ are commuting, and such that $\phi_A^\infty+\phi_B^R<\pi$.*

Then $A^\alpha B^{1-\alpha}(A+B)^{-1}$ is bounded, for each $\alpha\in(0,1)$. In particular,

$$\mathsf{D}(A)\cap\mathsf{D}(B)=\mathsf{D}(A+B)\hookrightarrow\mathsf{D}(A^\alpha B^{1-\alpha}),$$

for each $\alpha\in(0,1)$.

Here we choose $F(\lambda)=\lambda^\alpha B^{1-\alpha}(\lambda+B)^{-1}$ and employ Theorem 4.5.6. In fact, with an appropriate contour Γ, the representation

$$F(\lambda)=\frac{1}{2\pi i}\int_\Gamma\frac{z^{-\alpha}}{1+z}\lambda z(\lambda z-B)^{-1}\,dz$$

shows that $F(\Sigma_\phi)$ is $\mathcal{R}$-bounded as B is $\mathcal{R}$-sectorial, by convexity of $\mathcal{R}$-bounds.

Corollary 4.5.11. *Suppose $A\in\mathcal{H}^\infty(X)$ and $B\in\mathcal{RH}^\infty(X)$ are commuting, and such that $\phi_A^\infty+\phi_B^{R\infty}<\pi$, $\alpha\in(0,1)$.*

Then $A+B\in\mathcal{H}^\infty(X)$ and $\phi_{A+B}^\infty\le\max\{\phi_A^\infty,\phi_B^{R\infty}\}$.
Moreover, $\mathsf{D}((A+B)^\alpha)=\mathsf{D}(A^\alpha)\cap\mathsf{D}(B^\alpha)$.

To see this, choose $f(\lambda,\mu)=h(\lambda+\mu)$ resp. $f(\lambda,\mu)=(\lambda+\mu)^\alpha/(\lambda^\alpha+\mu^\alpha)$ and apply Corollary 4.5.8.

Corollary 4.5.12. *Suppose $A\in\mathcal{H}^\infty(X)$ and $B\in\mathcal{RS}(X)$ are commuting, $0\in\rho(A)$, and such that $\phi_A^\infty+\phi_B^R<\pi$. Then*
(i) *AB with domain $\mathsf{D}(AB)=\{x\in\mathsf{D}(B):Bx\in\mathsf{D}(A)\}$ is closed, $AB\in\mathcal{S}(X)$ with $\phi_{AB}\le\phi_A^\infty+\phi_B^R$.*
(ii) *In case X has property (α), then $AB\in\mathcal{RS}(X)$ and $\phi_{AB}^R\le\phi_A^\infty+\phi_B^R$.*
(iii) *If $B\in\mathcal{RH}^\infty$ and $\phi_A^\infty+\phi_B^{R\infty}<\pi$, then $AB\in\mathcal{H}^\infty(X)$ and $\phi_{AB}^\infty\le\phi_A^\infty+\phi_B^{R\infty}$.*

This result follows from Corollary 4.5.7 by setting $F(\lambda)=z(z+\lambda B)^{-1}$ and $F(\lambda)=h(\lambda B)$, respectively.

5.3 Proof of Theorem 4.5.6.
For the proof of Theorem 4.5.6 we shall use the following lemma on unconditionallity which is interesting in itself.

Lemma 4.5.13. *Suppose $A\in\mathcal{H}^\infty(X)$, $h\in H_0(\Sigma_\phi)$, $\phi>\phi_A^\infty$.*

Then there is a constant $C>0$ such that

$$\Big|\sum_{k\in\mathbb{Z}}\alpha_k h(2^k tA)\Big|_{\mathcal{B}(X)}\le C\sup_{k\in\mathbb{Z}}|\alpha_k|$$

for all $\alpha_k\in\mathbb{C}$ and $t>0$.

Proof. $h \in H_0(\Sigma_\phi)$ implies

$$|h(z)| \leq c\frac{|z|^\beta}{1+|z|^{2\beta}}, \quad z \in \Sigma_\phi,$$

for some $\beta > 0$. Set $f(z) = \sum_{k\in\mathbb{Z}} \alpha_k h(2^k tz)$; this series is absolutely convergent as can be seen from the estimate

$$|f(z)| \leq |\alpha|_\infty \sum_k |h(2^k tz)| \leq C|\alpha|_\infty,$$

since

$$\sum_k |h(2^k tz)| \leq c \sum_k \frac{(r2^k)^\beta}{1+(r2^k)^{2\beta}} \leq \frac{2c}{1-2^{-\beta}}, \quad r = t|z|.$$

Therefore $f \in H^\infty(\Sigma_\phi)$ and so by $A \in \mathcal{H}^\infty(X)$, $\phi > \phi_A^\infty$ we obtain

$$\Big|\sum_{k\in\mathbb{Z}} \alpha_k h(2^k tA)\Big|_{\mathcal{B}(X)} = |f(A)|_{\mathcal{B}(X)} \leq C_A|f|_{\mathcal{H}^\infty} \leq C|\alpha|_\infty. \qquad \square$$

Proof of Theorem 4.5.6. (a) Suppose first $F \in H_0(\Sigma_\phi; \mathcal{B}(X))$. Then we have

$$F(A) = \frac{1}{2\pi i}\int_\Gamma F(\lambda)(\lambda - A)^{-1}\,d\lambda = \frac{1}{2\pi i}\int_\Gamma F(\lambda)\lambda^{-1/2}A^{1/2}(\lambda - A)^{-1}\,d\lambda,$$

where Γ denotes the contour $\Gamma = \{re^{\pm i\theta} : r \in \mathbb{R}_+\}$, properly oriented, with $\phi_A^\infty < \theta < \phi$. Since the integral defining $F(A)$ is absolutely convergent, we also have

$$F(A) = \lim_{N\to\infty} \frac{1}{2\pi i}\int_{\Gamma_N} F(\lambda)\lambda^{-1/2}A^{1/2}(\lambda - A)^{-1}\,d\lambda = \lim_{N\to\infty} F^N,$$

where $\Gamma_N = \{\lambda \in \Gamma : 2^{-N} \leq |\lambda| \leq 2^N\}$. We write $F^N = F^{+N} + F^{-N}$, with

$$\begin{aligned}
F^{\pm N} &= \frac{e^{\pm i\theta/2}}{2\pi i}\int_{2^{-N}}^{2^N} F(re^{\pm i\theta})A^{1/2}(re^{\pm i\theta} - A)^{-1}\,dr/\sqrt{r} \\
&= \frac{e^{\pm i\theta/2}}{2\pi i}\sum_{k=-N}^{N-1}\int_{2^k}^{2^{k+1}} F(re^{\pm i\theta})(A/r)^{1/2}(e^{\pm i\theta} - A/r)^{-1}\,dr/r \\
&= \frac{e^{\pm i\theta/2}}{2\pi i}\sum_{k=-N}^{N-1}\int_1^2 F(2^k te^{\pm i\theta})(A/2^k t)^{1/2}(e^{\pm i\theta} - A/2^k t)^{-1}\,dt/t \\
&= \frac{e^{\pm i\theta/2}}{2\pi i}\int_1^2 \sum_{k=-N}^{N-1} F(2^k te^{\pm i\theta})h_\pm(A/2^k t)\,dt/t \\
&= \frac{e^{\pm i\theta/2}}{2\pi i}\int_1^2 T_F^{\pm N}(t)\,dt/t,
\end{aligned}$$

where $h_\pm(z) = \sqrt{z}/(e^{\pm i\theta} - z)$ belongs to $H_0(\Sigma_\pi)$. So we have to estimate $T_F^{\pm N}(t)$.

(b) Next we randomize and estimate as follows

$$
\begin{aligned}
|\langle T_F^{\pm N}(t)x|x^*\rangle| &= \Big|\sum_{k=-N}^{N-1} \langle F(2^k te^{\pm i\theta})h_\pm(A/t2^k)x|x^*\rangle\Big| \\
&= \Big|\int_\Omega \sum_{k=-N}^{N-1} \varepsilon_k^2 \langle F(2^k te^{\pm i\theta})h_\pm(A/t2^k)x|x^*\rangle\, d\mu\Big| \\
&= \Big|\int_\Omega < \sum_{k=-N}^{N-1} \varepsilon_k F(2^k te^{\pm i\theta})h_\pm^{1/2}(A/t2^k)x, \sum_{k=-N}^{N-1} \varepsilon_k h_\pm^{1/2}(A^*/t2^k)x^* > d\mu\Big| \\
&\le \Big|\sum_{k=-N}^{N-1} \varepsilon_k F(2^k te^{\pm i\theta})h_\pm^{1/2}(A/t2^k)x\Big|_{L_2(\Omega;X)} \Big|\Big(\sum_{k=-N}^{N-1} \varepsilon_k h_\pm^{1/2}(A/t2^k)\Big)^* x^*|_{L_2(\Omega;X^*)} \\
&\le \mathcal{R}(F(\Sigma_\phi))\Big|\sum_{k=-N}^{N-1} \varepsilon_k h_\pm^{1/2}(A/t2^k)x\Big|_{L_2(\Omega;X)} \Big|\Big(\sum_{k=-N}^{N-1} \varepsilon_k h_\pm^{1/2}(A/t2^k)\Big)^* x^*\Big|_{L_2(\Omega;X^*)} \\
&\le C^2\mathcal{R}(F(\Sigma_\phi))|x||x^*|,
\end{aligned}
$$

by Lemma 4.5.13. This shows that $T_F^{\pm N}(t)$ is uniformly bounded in $t \in [1, 2]$ and in $N \in \mathbb{N}$, hence so is $F(A)$, with

$$|F(A)|_{\mathcal{B}(X)} \le C^2\mathcal{R}(F(\Sigma_\phi)).$$

For the general case, replace F by $F_\varepsilon(z) = F(z)z^\varepsilon/(1 + z^{2\varepsilon})$ and let $\varepsilon \to 0$.

(c) We can improve this estimate if X has property (α). Fix independent symmetric $\{-1, 1\}$-valued random variables ε_k reps. ε'_j on probability spaces $(\Omega, \mathcal{A}, \mu)$ resp. $(\Omega', \mathcal{A}', \mu')$, $F_k \in \mathcal{F}$, and $x_k \in X$. Then as in **(b)** we have with $T_k = T_{F_k}^{\pm N}$

$$
\begin{aligned}
\Big|\langle\sum_k \varepsilon_k T_k x_k|x^*\rangle\Big| &= \Big|\sum_j\sum_k \langle \varepsilon_k F_k(2^j te^{\pm i\theta})h_\pm(A/t2^j)x_k|x^*\rangle\Big| \\
&= \Big|\int_{\Omega'} \sum_j (\varepsilon'_j)^2 \sum_k \langle \varepsilon_k F_k(2^j te^{\pm i\theta})h_\pm(A/t2^j)x_k|x^*\rangle\, d\mu'\Big| \\
&\le \Big|\sum_{k,j} \varepsilon'_j\varepsilon_k F_k(2^j te^{\pm i\theta})h_\pm^{1/2}(A/t2^j)x_k\Big|_{L_2(\Omega';X)} \Big|\Big(\sum_j \varepsilon'_j h_\pm^{1/2}(A/t2^j)\Big)^* x^*\Big|_{L_2(\Omega';X^*)} \\
&\le C|x^*|\Big|\sum_{k,j} \varepsilon'_j\varepsilon_k F_k(2^j te^{\pm\theta})h_\pm^{1/2}(A/t2^j)x_k\Big|_{L_2(\Omega')},
\end{aligned}
$$

where we employed Lemma 4.5.13 in the last step. Taking the sup over $|x^*| \le 1$, squaring, and integrating over Ω this implies by Proposition 4.2.8

$$\begin{aligned}\Big|\sum_k \varepsilon_k T_k x_k\Big|_{L_p(\Omega;X)} &\le C\Big|\sum_{k,j} \varepsilon_j'\varepsilon_k F_k(2^j t e^{\pm\theta}) h_\pm^{1/2}(A/t2^j)x_k\Big|_{L_2(\Omega\times\Omega';X)}\\ &\le C\Big|\sum_{k,j} \varepsilon_j'\varepsilon_k h_\pm^{1/2}(A/t2^j)x_k\Big|_{L_2(\Omega\times\Omega';X)}\\ &= C\Big|\sum_j \varepsilon_j' h_\pm^{1/2}(A/t2^j)\sum_k \varepsilon_k x_k\Big|_{L_2(\Omega\times\Omega';X)}\\ &\le \Big|\sum_k \varepsilon_k x_k\Big|_{L_2(\Omega;X)},\end{aligned}$$

where we used Lemma 4.5.13 in the last step. Together with Proposition 4.1.9, the second assertion of the theorem follows. The last assertion follows by specializing to the scalar case. □

5.4 Fractional Evolution Equations.
For an illustration of the strength of the results proved above, we consider *fractional evolution equations.* In the sequel, we assume that we are given a Banach space X_0 of class $\mathcal{HT}$, $\alpha \in (0,2)$ and $A \in \mathcal{RS}(X_0)$ is invertible and has $\mathcal{RS}$-angle $\phi_A^R < \pi(1-\alpha/2)$. We set as usual $X_1 = \mathsf{D}(A)$ equipped with the graph norm of A. Consider the *fractional evolution equation*

$$\partial_t^\alpha u + Au = f, \quad t>0, \quad u(0)=0 \tag{4.30}$$

in the space $\mathbb{E}_{0,\mu} := L_{p,\mu}(\mathbb{R}_+;X_0)$. For this purpose, extend A in the canonical way to $\mathbb{E}_{0,\mu}$ with natural domain $L_{p,\mu}(\mathbb{R}_+;X_1)$, and define $B_{p,\mu} = d/dt$ with domain ${}_0H^1_{p,\mu}(\mathbb{R}_+;X_0)$, as in Section 3.2.4. Then $B_{p,\mu}$ belongs to $\mathcal{H}^\infty(\mathbb{E}_{0,\mu})$ and commutes with A. Setting $B = B^\alpha_{p,\mu}$, the Kalton-Weis theorem implies that $A+B$ is closed and invertible on its natural domain

$$\mathsf{D}(A)\cap\mathsf{D}(B) = {}_0H^\alpha_{p,\mu}(\mathbb{R}_+;X_0)\cap L_{p,\mu}(\mathbb{R}_+;X_1).$$

Moreover, $A+B$ is sectorial with angle $\alpha\pi/2$.

Therefore, (4.30) admits a unique solution $u \in {}_0H^\alpha_{p,\mu}(\mathbb{R}_+;X_0)\cap L_{p,\mu}(\mathbb{R}_+;X_1)$ whenever $f \in L_{p,\mu}(\mathbb{R}_+;X_0)$, and the solution map is bounded between the corresponding spaces.

Note that the same result is true in the case of the line, where we consider $L_p(\mathbb{R};X_0)$ as a base space. If $f \in L_p(\mathbb{R};X_0)$, then the unique solution of $\partial_t^\alpha u + Au = f$ will belong to $H^\alpha_p(\mathbb{R};X_0)\cap L_p(\mathbb{R};X_1)$.

In the half-line situation, the initial value of u is vanishing if u has a trace, provided $\alpha > 1-\mu+1/p$. Moreover, $\partial_t u$ also has a trace if $\alpha > 2-\mu+1/p$. What about the case when these traces are nontrivial at zero? To understand this situation we rewrite (4.30) as the *evolutionary integral equation*

$$u(t) + k * Au(t) = k * f(t), \quad t>0,$$

where $k(t) = t^{\alpha-1}/\Gamma(\alpha)$. The simplest way to see the equivalence of this equation with (4.30) is to use Laplace transforms. To admit nontrivial initial data in case $\alpha > 1-\mu+1/p$ and a nontrivial velocity trace if $\alpha > 2-\mu+1/p$, we add u_0+tu_1 to the right-hand side of this convolution equation, which yields the problem

$$\partial_t^\alpha(u - u_0 - tu_1) + Au = f, \quad t > 0, \quad u(0) = u_0,\ \partial_t u(0) = u_1. \tag{4.31}$$

Here u_0 is redundant if $\alpha < 1-\mu+1/p$, and u_1 is so in case $\alpha < 2-\mu+1/p$; we exclude the exceptional cases below. By linearity, $u = v + w$, where v solves (4.30) and w solves (4.31) with $f = 0$. Then $w(t) = S(t)u_0 + 1 * S(t)u_1$, where $S(t)$ denotes the *resolvent family* of the problem. Using Laplace transforms, $S(t)$ is seen to be defined by the relation

$$H(\lambda) := \mathcal{L}S(\lambda) = \frac{1}{\lambda}(1 + \lambda^{-\alpha}A)^{-1}, \quad \lambda > 0.$$

Note that by the assumption $\phi_A^R + \alpha\pi/2 < \pi$, the holomorphic family $H(\lambda)$ is well-defined on a sector Σ_ϕ with angle $\phi > \pi/2$. Therefore, using the standard contour $\Gamma = (-\infty, 0]e^{-i\theta} \cup [0,\infty)e^{i\theta}$ with $\pi/2 < \theta < \phi$ we obtain by the inverse Laplace transform the representation

$$S(t) = \frac{1}{2\pi i}\int_\Gamma H(\lambda)e^{\lambda t}\,d\lambda.$$

The function $S : \mathbb{R}_+ \to \mathcal{B}(X_0)$ is easily seen to be bounded and holomorphic in a sector Σ_φ, strongly continuous on its closure and $S(0) = I$. In the particular case $\alpha = 1$ we have $S(t) = e^{-At}$, i.e., this is the semigroup case.

The question in case $\alpha > 1-\mu+1/p$ now is for which initial values u_0 does $w = S(t)u_0$ belong to the space $H^\alpha_{p,\mu}(\mathbb{R}_+; X_0) \cap L_{p,\mu}(\mathbb{R}_+; X_1)$? This is a again a trace space problem and the answer is given in the next proposition.

Proposition 4.5.14. *Let X_0 be a Banach space of class $\mathcal{HT}$, $1 < p < \infty$, $1/p < \mu \le 1$, $\alpha \in (0,2)$, and suppose $A \in \mathcal{RS}(X_0)$ is invertible and $\phi_A^R + \alpha\pi/2 < \pi$. Let $S(t)$ denote the resolvent of* (4.31) *as defined above, and let $\alpha \neq 1-\mu+1/p, 2-\mu-1/p$. Then*

(i) *if $\alpha < 1-\mu+1/p$, then $S(\cdot)x \in H^\alpha_{p,\mu}(\mathbb{R}_+; X_0) \cap L_{p,\mu}(\mathbb{R}_+; X_1)$; for $\alpha > 1-\mu+1/p$ we have $S(\cdot)x \in H^\alpha_{p,\mu}(\mathbb{R}_+; X_0) \cap L_{p,\mu}(\mathbb{R}_+; X_1)$ if and only if $x \in D_A(1-(1-\mu+1/p)/\alpha, p)$;*

(ii) *if $\alpha < 2-\mu+1/p$, then $(1 * S)(\cdot)x \in H^\alpha_{p,\mu}(\mathbb{R}_+; X_0) \cap L_{p,\mu}(\mathbb{R}_+; X_1)$; for $\alpha > 2-\mu+1/p$, we have $1 * S(t)x \in H^\alpha_{p,\mu}(\mathbb{R}_+; X_0) \cap L_{p,\mu}(\mathbb{R}_+; X_1)$ if and only if $x \in D_A(1-(2-\mu+1/p)/\alpha, p))$.*

The corresponding maps $x \mapsto S(\cdot)x$ resp. $x \mapsto (1 * S)(\cdot)x$ are continuous between the relevant spaces

Proof. We only consider the first statement, as the second one is proved in a similar way. By Laplace transform we have the relation

$$r^{\alpha-1}(r^\alpha + A)^{-1}x = \int_0^\infty e^{-rt} S(t)x\, dt, \quad r > 0.$$

Setting $w(t) = AS(t)x$ this yields by Hölder's inequality and Fubini's theorem for $\beta = (1-\mu+1/p)/\alpha < 1$

$$\begin{aligned}
&\int_0^\infty r^{p\alpha(1-\beta)}|A(r^\alpha + A)^{-1}x|^p\, dr/r = \int_0^\infty r^{p\mu-2}|r^{\alpha-1}A(r^\alpha+A)^{-1}x|^p\, dr \\
&\le C\int_0^\infty r^{p\mu-2}\Big[\int_0^\infty |w(t)|e^{-rt}\, dt\Big]^p dr \\
&\le C\int_0^\infty t^{\gamma p} r^{p\mu-2}\Big[\int_0^\infty e^{-rt}|w(t)|^p\, dt\Big]\Big[\int_0^\infty t^{-\gamma p'}e^{-rt}\, dt\Big]^{p/p'} dr \\
&= C\int_0^\infty r^{p(\mu-1+\gamma-1/p)}\Big[\int_0^\infty e^{-rt}|w(t)|^p\, dt\Big] dr = C\int_0^\infty |w(t)|^p t^{p(1-\mu)}\, dt,
\end{aligned}$$

where $\gamma \in (1-\mu, 1-1/p)$. This shows the first implication.

To obtain the converse implication, we observe

$$S(t) = \frac{1}{2\pi i}\int_\Gamma e^{zt}z^{\alpha-1}(z^\alpha - A)^{-1}x\, dz, \quad t > 0,$$

where Γ denotes a standard contour. Observe also that

$$|z(z-A)^{-1}x| \le C|z||(|z|+A)^{-1}x|$$

on Γ, as in Section 3.4.3. This yields with some constants $c, C > 0$, using again Hölder and Fubini

$$\begin{aligned}
&\int_0^\infty |AS(t)x|^p t^{p(1-\mu)}\, dt \le C\int_0^\infty t^{p(1-\mu)}\Big[\int_\Gamma e^{\mathrm{Re}\, zt}|z|^{\alpha-1}|A(z^\alpha - A)^{-1}x||dz|\Big]^p dt \\
&\le C\int_0^\infty t^{p(1-\mu)}\Big[\int_0^\infty e^{-crt}r^{\alpha-1}|A(r^\alpha+A)^{-1}x|\, dr\Big]^p dt \\
&\le C\int_0^\infty t^{p(1-\mu)}\int_0^\infty e^{-crt}r^{p(\alpha-1+\gamma)}|A(r^\alpha+A)^{-1}x|^p\, dr\Big[\int_0^\infty r^{-\gamma p'}e^{-crt}\, dr\Big]^{p/p'} dt \\
&= C\int_0^\infty r^{p\alpha(1-\beta)}|A(r^\alpha+A)^{-1}x|^p\, dr/r = C\int_0^\infty r^{p(1-\beta)}|A(r+A)^{-1}x|^p\, dr/r,
\end{aligned}$$

where $\gamma \in (\mu - 2/p, 1-1/p)$. The case $\beta > 1$ is simpler and can be proved by a direct estimate, similar to the last step. □

Having this proposition at our disposal we can now state the final result on the fractional evolution equation (4.31).

Theorem 4.5.15. *Let $1 < p < \infty$, $1/p < \mu \leq 1$, $0 < \alpha < 2$, X_0 a Banach space of class $\mathcal{HT}$. Suppose that $A \in \mathcal{RS}(X_0)$ is invertible, $X_1 = X_A$, $\phi_A^R + \alpha\pi/2 < \pi$.*

Then (4.31) admits a unigue solution $u \in H^\alpha_{p,\mu}(\mathbb{R}_+, X_0) \cap L_{p,\mu}(\mathbb{R}_+; X_1)$ if and only if $f \in L_{p,\mu}(\mathbb{R}_+; X_0)$ and $u_0 \in D_A(1-(1-\mu+1/p)/\alpha)$ if $\alpha > 1-\mu+1/p$, and in addition $u_1 \in D_A(1-(2-\mu+1/p)/\alpha)$ in case $\alpha > 2-\mu+1/p$.

5.5 Time-Space Embeddings.
In this subsection, we want to exploit the strength of the Mixed derivative theorem, Corollary 4.5.10, to derive several time-space embedding results. For this, we assume that we are given a Banach space X_0 of class $\mathcal{HT}(\alpha)$, and that $A \in \mathcal{H}^\infty(X_0)$ is invertible and has $\mathcal{H}^\infty$-angle zero; set as usual $X_1 = \mathsf{D}(A)$ equipped with the graph norm of A. Let $B = B^\alpha_{p,\mu}$ be as in the previous section. By Corollary 4.5.11, $A+B$ belongs to $\mathcal{H}^\infty(X_0)$ with the same angle $\alpha\pi/2$, and

$$\mathsf{D}((A+B)^\beta) = \mathsf{D}(B^\beta) \cap \mathsf{D}(A^\beta) = {}_0H^{\alpha\beta}_{p,\mu}(\mathbb{R}_+; X_0) \cap L_{p,\mu}(\mathbb{R}_+; \mathsf{D}(A^\beta)), \quad \beta \in [0,1],$$

by the reiteration theorem. The same result is also valid for the base space $\mathbb{E}_{s,\mu} := \mathsf{D}(B^s) = {}_0H^s_{p,\mu}(\mathbb{R}_+; X_0)$, for any $s \geq 0$, with domain

$$\mathsf{D}((A+B)^\beta) = \mathsf{D}(B^\beta) \cap \mathsf{D}(A^\beta) = {}_0H^{\alpha\beta+s}_{p,\mu}(\mathbb{R}_+; X_0) \cap {}_0H^s_{p,\mu}(\mathbb{R}_+; \mathsf{D}(A^\beta)), \ \ \beta \in [0,1].$$

Next we apply the Mixed derivative theorem 4.5.10 to obtain the embedding

$${}_0H^{s+\alpha}_{p,\mu}(\mathbb{R}_+; X_0) \cap {}_0H^s_{p,\mu}(\mathbb{R}_+; X_1) \hookrightarrow {}_0H^r_{p,\mu}(\mathbb{R}_+; \mathsf{D}(A^{1-\frac{r-s}{\alpha}})), \tag{4.32}$$

which is valid for all all $\alpha \in (0,2)$, $0 \leq s \leq r \leq s+\alpha$.

This is the basic embedding. We may extend it using real interpolation in the following way. In case $s > 0$ we apply real interpolation of type $(1/2, p)$ to (4.32) with s replaced by $s+\varepsilon$ and $s-\varepsilon$, and r by $r+\varepsilon$ and $r-\varepsilon$, respectively, to the result

$${}_0W^{s+\alpha}_{p,\mu}(\mathbb{R}_+; X_0) \cap {}_0W^s_{p,\mu}(\mathbb{R}_+; X_1) \hookrightarrow {}_0W^r_{p,\mu}(\mathbb{R}_+; \mathsf{D}(A^{1-\frac{r-s}{\alpha}})), \tag{4.33}$$

which is valid for all $\alpha \in (0,2)$, $0 < s < r < s+\alpha$.

More surprising is the following embedding which is also obtained by real interpolation of type $(1/2, p)$ to (4.32) with s replaced by $s+\varepsilon$ and $s-\varepsilon$, but keeping r fixed, to the result

$${}_0W^{s+\alpha}_{p,\mu}(\mathbb{R}_+; X_0) \cap {}_0W^s_{p,\mu}(\mathbb{R}_+; X_1) \hookrightarrow {}_0H^r_{p,\mu}\Big(\mathbb{R}_+; D_A\Big(1 - \frac{r-s}{\alpha}, p\Big)\Big), \tag{4.34}$$

which is also valid for all all $\alpha \in (0,2)$, $0 < s < r < s+\alpha$.

In case $s > 0$ we may, once more, apply real interpolation of type $(1/2, p)$ to (4.34) with s replaced by $s+\varepsilon$ and $s-\varepsilon$, and r by $r+\varepsilon$ and $r-\varepsilon$, respectively, to the result

$${}_0W^{s+\alpha}_{p,\mu}(\mathbb{R}_+; X_0) \cap {}_0W^s_{p,\mu}(\mathbb{R}_+; X_1) \hookrightarrow {}_0W^r_{p,\mu}\Big(\mathbb{R}_+; D_A\Big(1 - \frac{r-s}{\alpha}, p\Big)\Big), \tag{4.35}$$

which is valid for all $\alpha \in (0,2)$, $0 < s < r < s+\alpha$.

Another interesting embedding which will be used below comes from

$$ {}_0H^{\alpha}_{p,\mu}(\mathbb{R}_+;X_0) \cap L_{p,\mu}(\mathbb{R}_+;X_1) \hookrightarrow {}_0H^{r}_{p,\mu}(\mathbb{R}_+;\mathsf{D}(A^{1-\frac{r}{\alpha}})), $$

interpolated with the real p-method with the trivial embedding

$$ {}_0H^{\alpha-r}_{p,\mu}(\mathbb{R}_+;X_0) \cap L_{p,\mu}(\mathbb{R}_+;\mathsf{D}(A^{1-\frac{r}{\alpha}})) \hookrightarrow L_{p,\mu}(\mathbb{R}_+;\mathsf{D}(A^{1-\frac{r}{\alpha}})), $$

to the result

$$ {}_0W^{s}_{p,\mu}(\mathbb{R}_+;X_0) \cap L_{p,\mu}\Big(\mathbb{R}_+;D_A\Big(\frac{s}{\alpha},p\Big)\Big) \hookrightarrow {}_0W^{r+s-\alpha}_{p,\mu}(\mathbb{R}_+;\mathsf{D}(A^{(1-\frac{r}{\alpha})})), \tag{4.36} $$

which is valid for all $\alpha \in (0,2)$, $0 < s, r < \alpha < r+s$.

In a similar way, interpolating the embedding

$$ {}_0H^{\alpha}_{p,\mu}(\mathbb{R}_+;X_0) \cap L_{p,\mu}(\mathbb{R}_+;X_1) \hookrightarrow {}_0H^{r}_{p,\mu}(\mathbb{R}_+;\mathsf{D}(A^{1-\frac{r}{\alpha}})), $$

by the real p-method with the trivial embedding

$$ {}_0H^{r}_{p,\mu}(\mathbb{R}_+;X_0) \cap L_{p,\mu}(\mathbb{R}_+;\mathsf{D}(A^{r/\alpha})) \hookrightarrow {}_0H^{r}_{p,\mu}(\mathbb{R}_+;X_0), $$

we obtain

$$ {}_0W^{s}_{p,\mu}(\mathbb{R}_+;X_0) \cap L_{p,\mu}\Big(\mathbb{R}_+;D_A\Big(\frac{s}{\alpha},p\Big)\Big) \hookrightarrow {}_0H^{r}_{p,\mu}\Big(\mathbb{R}_+;D_A\Big(\frac{s-r}{\alpha},p\Big)\Big), \tag{4.37} $$

which is valid for all $\alpha \in (0,2)$, $0 < r < s < \alpha$.

Similar results are obtained when we consider the whole line case $L_p(\mathbb{R};X_0)$. This corresponds to the result for the fractional evolution equation on the line

$$ \partial_t^{\alpha} u + Au = f, \quad t \in \mathbb{R}, $$

which has been discussed in the previous subsection as well. For this problem, similar results as for the half-line case are valid. In the statements derived above, the symbols $L_{p,\mu}(\mathbb{R}_+;\cdot)$ and ${}_0K^{s}_{p,\mu}(\mathbb{R}_+;\cdot)$ for $K \in \{H,W\}$ only have to be replaced by $L_p(\mathbb{R};\cdot)$ resp. $K^s_p(\mathbb{R};\cdot)$.

Example 4.5.16. **(i)** Consider the base space $X_0 = K^{\sigma}_p(\mathbb{R}^n)$ for $K \in \{W,H\}$, $p \in (1,\infty)$, $\sigma \in \mathbb{R}$, and let $A = (I-\Delta)^{m/2}$, $m \geq 0$, with $\mathsf{D}(A) = K^{s+m}_p(\mathbb{R}^n)$. Then $A \in \mathcal{H}^{\infty}(X_0)$ is invertible and has $\mathcal{H}^{\infty}$-angle zero. By the embeddings (4.32)-(4.35) we obtain

$$ {}_0W^{s+\alpha}_{p,\mu}(\mathbb{R}_+;K^{\sigma}_p(\mathbb{R}^n)) \cap {}_0W^{s}_{p,\mu}(\mathbb{R}_+;K^{\sigma+m}_p(\mathbb{R}^n)) \hookrightarrow {}_0W^{r}_{p,\mu}(\mathbb{R}_+;K^{\sigma+m(1-\frac{r-s}{\alpha})}_p(\mathbb{R}^n)), $$

as well as

$$ {}_0H^{s+\alpha}_{p,\mu}(\mathbb{R}_+;K^{\sigma}_p(\mathbb{R}^n)) \cap {}_0H^{s}_{p,\mu}(\mathbb{R}_+;K^{\sigma+m}_p(\mathbb{R}^n)) \hookrightarrow {}_0H^{r}_{p,\mu}(\mathbb{R}_+;K^{\sigma+m(1-\frac{r-s}{\alpha})}_p(\mathbb{R}^n)), $$

valid for all $\alpha \in (0,2)$, $0 < s < r < s+\alpha$.

(ii) Here we chose $X_0 = L_p(\mathbb{R}^n)$, $A = I - \Delta$, $s = 1 - 1/2p$, $\alpha = 1/2$, and $r = 1$. Then the embedding (4.34) (or the second embedding in Example (i) with $\sigma = 0$ and $K = H$) yields

$$ {}_0W^{3/2-1/2p}_{p,\mu}(\mathbb{R}_+; L_p(\mathbb{R}^n)) \cap W^{1-1/2p}_{p,\mu}(\mathbb{R}_+; H^2_p(\mathbb{R}^n)) \hookrightarrow {}_0H^1_{p,\mu}(\mathbb{R}_+; W^{2-2/p}_p(\mathbb{R}^n)). $$

This result will be used for the Stefan problem in Section 6.6.

(iii) Choosing $X_0 = H^{-1}_p(\mathbb{R}^n)$, $A = I - \Delta$, $\alpha = 1$, and $s = 0$, we obtain from (4.32)

$$ {}_0H^1_{p,\mu}(\mathbb{R}_+; H^{-1}_p(\mathbb{R}^n)) \cap L_{p,\mu}(\mathbb{R}_+; H^1_p(\mathbb{R}^n)) \hookrightarrow {}_0H^{1/2}_{p,\mu}(\mathbb{R}_+; L_p(\mathbb{R}^n)). $$

Interpolating this embedding with the trivial embedding

$$ {}_0H^1_{p,\mu}(\mathbb{R}_+; L_p(\mathbb{R}^n)) \cap L_{p,\mu}(\mathbb{R}_+; H^2_p(\mathbb{R}^n)) \hookrightarrow {}_0H^1_{p,\mu}(\mathbb{R}_+; L_p(\mathbb{R}^n)) $$

by the real method of type $(1/p, p)$, this yields

$$ {}_0H^1_{p,\mu}(\mathbb{R}_+; W^{-1/p}_p(\mathbb{R}^n)) \cap L_{p,\mu}(\mathbb{R}_+; W^{2-1/p}_p(\mathbb{R}^n)) \hookrightarrow {}_0W^{1-1/2p}_{p,\mu}(\mathbb{R}_+; L_p(\mathbb{R}^n)). $$

This result will be used in Sections 8.3 and 8.6.

As a summary, starting from the basic embedding (4.32) via interpolation theory one can create a variety of time-space embeddings, which exchange time and space regularity.

The method of time-space embeddings explained above is a general powerful tool in modern analysis, which will be used frequently in subsequent chapters.

Chapter 5

Quasilinear Parabolic Evolution Equations

In this chapter we consider abstract quasilinear parabolic problems of the form

$$\dot{u} + A(u)u = F(u), \quad t > 0, \quad u(0) = u_0, \tag{5.1}$$

where $(A, F) : V_\mu \to \mathcal{B}(X_1, X_0) \times X_0$ and $u_0 \in V_\mu$. The spaces X_1, X_0 are Banach spaces such that $X_1 \hookrightarrow X_0$ with dense embedding, and V_μ is an open subset of the real interpolation space

$$X_{\gamma,\mu} := (X_0, X_1)_{\mu-1/p,p}, \quad \mu \in (1/p, 1].$$

Our goal is to develop a solution theory for (5.1) which parallels that for ODE's.

We are mainly interested in solutions $u(t)$ of (5.1) having *maximal L_p-regularity*, i.e.,

$$u \in H^1_p(J; X_0) \cap L_p(J; X_1).$$

The trace space of this class of functions is given by $X_\gamma := X_{\gamma,1}$. However, to see and exploit the effect of parabolic regularization in the L_p-framework it is also useful to consider solutions in the class of weighted spaces

$$u \in H^1_{p,\mu}(J; X_0) \cap L_{p,\mu}(J; X_1).$$

The trace space for this class of weighted spaces is given by $X_{\gamma,\mu}$. In our approach it is crucial to know that the operators $A(u)$ have the property of maximal L_p-regularity. Observe that we may add a term ωu to both sides of (5.1), so we may always consider the class $\mathcal{MR}_p(X_0)$ of maximal L_p-regularity.

5.1 Local Well-Posedness

We suppose that the nonlinear mappings A and F satisfy

$$(A, F) \in C^{1-}(V_\mu; \mathcal{B}(X_1, X_0) \times X_0). \tag{5.2}$$

J. Prüss and G. Simonett, *Moving Interfaces and Quasilinear Parabolic Evolution Equations*, Monographs in Mathematics 105,
DOI 10.1007/978-3-319-27698-4_5

The main result of this section reads as follows.

Theorem 5.1.1. *Let $p \in (1,\infty)$, $u_0 \in V_\mu$ be given and suppose that (A,F) satisfies (5.2) for some $\mu \in (1/p,1]$. Assume in addition that $A(u_0) \in \mathcal{MR}_p(X_0)$.*

Then there exist $T = T(u_0) > 0$ and $\varepsilon = \varepsilon(u_0) > 0$ with $\bar{B}_{X_{\gamma,\mu}}(u_0,\varepsilon) \subset V_\mu$ such that (5.1) has a unique solution

$$u(\cdot,u_1) \in \mathbb{E}_{1,\mu}(0,T) := H^1_{p,\mu}((0,T);X_0) \cap L_{p,\mu}((0,T);X_1) \cap C([0,T];V_\mu),$$

on $[0,T]$, for any initial value $u_1 \in \bar{B}_{X_{\gamma,\mu}}(u_0,\varepsilon)$. There exists a constant $c = c(u_0) > 0$ such that for all $u_1,u_2 \in \bar{B}_{X_{\gamma,\mu}}(u_0,\varepsilon)$ the estimate

$$|u(\cdot,u_1) - u(\cdot,u_2)|_{\mathbb{E}_{1,\mu}(0,T)} \le c|u_1 - u_2|_{X_{\gamma,\mu}}$$

is valid. Moreover, for each $\delta \in (0,T)$ we have in addition

$$u \in \mathbb{E}_1(\delta,T) := \mathbb{E}_{1,1}(\delta,T) \hookrightarrow C([\delta,T];X_\gamma),$$

i.e., the solution regularizes instantaneously.

Proof. Since $u_0 \in V_\mu$ and by (5.2), there exists $\varepsilon_0 > 0$ and a constant $L > 0$ such that $\bar{B}_{X_{\gamma,\mu}}(u_0,\varepsilon_0) \subset V_\mu$ and

$$|A(w_1)v - A(w_2)v|_{X_0} \le L|w_1 - w_2|_{X_{\gamma,\mu}}|v|_{X_1}, \tag{5.3}$$

as well as

$$|F(w_1) - F(w_2)|_{X_0} \le L|w_1 - w_2|_{X_{\gamma,\mu}} \tag{5.4}$$

hold for all $w_1,w_2 \in \bar{B}_{X_{\gamma,\mu}}(u_0,\varepsilon_0)$, $v \in X_1$. W.l.o.g. we may assume that $e^{-A(u_0)t}$ is exponentially stable. Introduce a reference function $u_0^* \in \mathbb{E}_{1,\mu}(0,T)$ as the solution of the linear problem

$$\dot{w} + A(u_0)w = 0, \quad w(0) = u_0.$$

Define the set $\mathbb{B}_{r,T,u_1} \subset \mathbb{E}_{1,\mu}(0,T)$ by

$$\mathbb{B}_{r,T,u_1} := \{v \in \mathbb{E}_{1,\mu}(0,T) : v(0) = u_1 \text{ and } |v - u_0^*|_{\mathbb{E}_{1,\mu}(0,T)} \le r\}, \quad 0 < r \le 1.$$

Let $u_1 \in \bar{B}_{X_{\gamma,\mu}}(u_0,\varepsilon)$ with $\varepsilon \in (0,\varepsilon_0]$ be given. We will show that $v(t) \in \bar{B}_{X_{\gamma,\mu}}(u_0,\varepsilon_0)$ for all $v \in \mathbb{B}_{r,T,u_1}$ and all $t \in [0,T]$, provided that $r,T,\varepsilon > 0$ are sufficiently small. For this purpose we define $u_1^* \in \mathbb{E}_{1,\mu}(0,T)$ as the unique solution of

$$\dot{w} + A(u_0)w = 0, \quad w(0) = u_1.$$

Note that u_0^* and u_1^* are given by $e^{-A(u_0)t}u_0$ and $e^{-A(u_0)t}u_1$, respectively. Given $v \in \mathbb{B}_{r,T,u_1}$ we estimate as follows.

$$|v - u_0|_{\infty,X_{\gamma,\mu}} \le |v - u_1^*|_{\infty,X_{\gamma,\mu}} + |u_1^* - u_0^*|_{\infty,X_{\gamma,\mu}} + |u_0^* - u_0|_{\infty,X_{\gamma,\mu}}. \tag{5.5}$$

Since u_0 is fixed, there exists a number $T_0 = T_0(u_0) > 0$ such that

$$\sup_{t\in[0,T_0]} |u_0^*(t) - u_0|_{X_{\gamma,\mu}} \le \varepsilon_0/3.$$

Observe that $v(0) - u_1^*(0)) = 0$, hence

$$|v - u_1^*|_{\infty,X_{\gamma,\mu}} \le C_1|v - u_1^*|_{\mathbb{E}_{1,\mu}(0,T)}$$

and the constant $C_1 > 0$ does not depend on T. Therefore,

$$\begin{aligned}|v - u_1^*|_{\infty,X_{\gamma,\mu}} &\le C_1|v - u_1^*|_{\mathbb{E}_{1,\mu}(0,T)} \le C_1\big(|v - u_0^*|_{\mathbb{E}_{1,\mu}(0,T)} + |u_0^* - u_1^*|_{\mathbb{E}_{1,\mu}(0,T)}\big)\\ &\le C_1\big(r + |u_0^* - u_1^*|_{\mathbb{E}_{1,\mu}(0,T)}\big).\end{aligned}$$

Since by assumption the semigroup $e^{-A(u_0)t}$ is exponentially stable it follows that

$$|u_0^* - u_1^*|_{\infty,X_{\gamma,\mu}} + C_1|u_0^* - u_1^*|_{\mathbb{E}_{1,\mu}(0,T)} \le C_\gamma|u_0 - u_1|_{X_{\gamma,\mu}}, \tag{5.6}$$

with a constant $C_\gamma > 0$ which does not depend on T. Choosing $\varepsilon \le \varepsilon_0/(3C_\gamma)$ and $r \le \varepsilon_0/(3C_1)$, we obtain

$$|v - u_0|_{\infty,X_{\gamma,\mu}} \le C_1 r + C_\gamma\varepsilon + |u_0^* - u_0|_{\infty,X_{\gamma,\mu}} \le \varepsilon_0. \tag{5.7}$$

Throughout the remainder of this proof we will assume that $u_1 \in B_{X_{\gamma,\mu}}(u_0,\varepsilon)$, $\varepsilon \le \varepsilon_0/(3C_\gamma)$, $T \in [0,T_0]$, and $r \le \varepsilon_0/(3C_1)$. Under these assumptions, we may define a mapping

$$\mathcal{T}_{u_1} : \mathbb{B}_{r,T,u_1} \to \mathbb{E}_{1,\mu}(0,T), \quad \mathcal{T}_{u_1}v := u,$$

where u is the unique solution of the linear problem

$$\dot{u} + A(u_0)u = F(v) + (A(u_0) - A(v))v, \quad t > 0, \quad u(0) = u_1.$$

In order to apply the contraction mapping principle, we show that $\mathcal{T}_{u_1}(\mathbb{B}_{r,T,u_1}) \subset \mathbb{B}_{r,T,u_1}$, and that $\mathcal{T}_{u_1}$ defines a strict contraction on $\mathbb{B}_{r,T,u_1}$, i.e., there exists $\kappa \in (0,1)$ such that

$$|\mathcal{T}_{u_1}v - \mathcal{T}_{u_1}\bar{v}|_{\mathbb{E}_{1,\mu}(0,T)} \le \kappa|v - \bar{v}|_{\mathbb{E}_{1,\mu}(0,T)}$$

is valid for all $v,\bar{v} \in \mathbb{B}_{r,T,u_1}$. We will first take care of the self-mapping property. Note that for $v \in \mathbb{B}_{r,T,u_1}$ we have

$$(\mathcal{T}_{u_1}v)(t) - u_0^*(t) = u_1^*(t) - u_0^*(t) + \Big(e^{-A(u_0)\cdot} * (F(v) + (A(u_0) - A(v))v)\Big)(t).$$

The assumption $A(u_0) \in \mathcal{MR}_p(X_0)$ then implies

$$\begin{aligned}&|e^{-A(u_0)\cdot} * (F(v) + (A(u_0) - A(v))v)|_{\mathbb{E}_{1,\mu}(0,T)}\\ &\qquad\qquad \le C_0|F(v) + (A(u_0) - A(v))v|_{\mathbb{E}_{0,\mu}(0,T)},\end{aligned}$$

and $C_0 > 0$ does not depend on T. Let us first estimate $(A(u_0) - A(v))v$ in $\mathbb{E}_{0,\mu}(0,T)$. By (5.3) and (5.7) we obtain

$$\begin{aligned} |(A(u_0) - A(v))v|_{\mathbb{E}_{0,\mu}(0,T)} &\le L|v - u_0|_{\infty,X_{\gamma,\mu}}|v|_{\mathbb{E}_{1,\mu}(0,T)} \\ &\le L|v - u_0|_{\infty,X_{\gamma,\mu}}(r + |u_0^*|_{\mathbb{E}_{1,\mu}(0,T)}). \end{aligned}$$

Furthermore, by (5.4)

$$\begin{aligned} |F(v)|_{\mathbb{E}_{0,\mu}(0,T)} &\le |F(v) - F(u_0)|_{\mathbb{E}_{0,\mu}(0,T)} + |F(u_0)|_{\mathbb{E}_{0,\mu}(0,T)} \\ &\le \sigma(T)\big(L|v - u_0|_{\infty,X_{\gamma,\mu}} + |F(u_0)|_{X_0}\big) \end{aligned}$$

with $\sigma(T) := \frac{1}{(1+(1-\mu)p)^{1/p}}T^{1/p+1-\mu}$. Since

$$|u_0^* - u_0|_{\infty,X_{\gamma,\mu}},\ |u_0^*|_{\mathbb{E}_{1,\mu}(0,T)},\ \sigma(T) \to 0 \text{ as } T \to 0,$$

this yields with (5.7)

$$|\mathcal{T}_{u_1}v - u_0^*|_{\mathbb{E}_{1,\mu}(0,T)} \le |u_1^* - u_0^*|_{\mathbb{E}_{1,\mu}(0,T)} + r/2,$$

provided $r > 0, T > 0, \varepsilon > 0$ are chosen sufficiently small. By (5.6) we obtain in addition

$$|\mathcal{T}_{u_1}v - u_0^*|_{\mathbb{E}_{1,\mu}(0,T)} \le (C_\gamma/C_1)|u_1 - u_0|_{X_{\gamma,\mu}} + r/2 \le r/2 + r/2 = r,$$

with a possibly smaller $\varepsilon > 0$. This proves the self-mapping property of $\mathcal{T}_{u_1}$.

Let $u_1, u_2 \in \bar{B}_{X_{\gamma,\mu}}(u_0,\varepsilon)$ be given and let $v_1 \in \mathbb{B}_{r,T,u_1}$, $v_2 \in \mathbb{B}_{r,T,u_2}$. Then, since $A(u_0) \in \mathcal{MR}_p(X_1, X_0)$, we have

$$\begin{aligned} |\mathcal{T}_{u_1}v_1 - \mathcal{T}_{u_2}v_2|_{\mathbb{E}_{1,\mu}(0,T)} &\le |e^{-A(u_0)\cdot}(u_1 - u_2)|_{\mathbb{E}_{1,\mu}(0,T)} + C_0|F(v_1) - F(v_2)|_{\mathbb{E}_{0,\mu}(0,T)} \\ &+ C_0|(A(v_1) - A(u_0))(v_1 - v_2)|_{\mathbb{E}_{0,\mu}(0,T)} + C_0|(A(v_1) - A(v_2))v_2|_{\mathbb{E}_{0,\mu}(0,T)}. \end{aligned} \tag{5.8}$$

For the first term on the right-hand side we can make use of (5.6), where u_0 and u_0^* have to be replaced by u_2 and $e^{-A(u_0)t}u_2$, respectively. The second term can be treated as follows. By (5.4), we obtain

$$|F(v_1) - F(v_2)|_{\mathbb{E}_{0,\mu}(0,T)} \le \sigma(T)L|v_1 - v_2|_{\infty,X_{\gamma,\mu}}.$$

Moreover, by (5.6) and the trace theorem we have

$$\begin{aligned} |v_1 - v_2|_{\infty,X_{\gamma,\mu}} &\le |v_1 - v_2 - e^{-A(u_0)\cdot}(u_1 - u_2)|_{\infty,X_{\gamma,\mu}} \\ &\quad + |e^{-A(u_0)\cdot}(u_1 - u_2)|_{\infty,X_{\gamma,\mu}} \\ &\le C_1|v_1 - v_2 - e^{-A(u_0)\cdot}(u_1 - u_2)|_{\mathbb{E}_{1,\mu}(0,T)} + C_\gamma|u_1 - u_2|_{X_{\gamma,\mu}} \\ &\le C_1|v_1 - v_2|_{\mathbb{E}_{1,\mu}(0,T)} + 2C_\gamma|u_1 - u_2|_{X_{\gamma,\mu}}. \end{aligned} \tag{5.9}$$

This yields

$$|F(v_1) - F(v_2)|_{\mathbb{E}_{0,\mu}(0,T)} \le \sigma(T)L\left(C_1|v_1 - v_2|_{\mathbb{E}_{1,\mu}(0,T)} + 2C_\gamma|u_1 - u_2|_{X_{\gamma,\mu}}\right).$$

For the remaining terms in (5.8) we make use of (5.3) which results in

$$\begin{aligned}|(A(v_1) - A(u_0))(v_1 - v_2)|_{\mathbb{E}_{0,\mu}(0,T)} + |(A(v_1) - A(v_2))v_2|_{\mathbb{E}_{0,\mu}(0,T)}\\ \le L(|v_1 - u_0|_{\infty,X_{\gamma,\mu}}|v_1 - v_2|_{\mathbb{E}_{1,\mu}(0,T)} + |v_1 - v_2|_{\infty,X_{\gamma,\mu}}|v_2|_{\mathbb{E}_{1,\mu}(0,T)}.\end{aligned}$$

By (5.7), the term $|v_1 - u_0|_{\infty,X_{\gamma,\mu}}$ can be made as small as we wish by decreasing $r > 0, T > 0$ and $\varepsilon > 0$. Furthermore, we have

$$|v_2|_{\mathbb{E}_{1,\mu}(0,T)} \le |v_2 - u_0^*|_{\mathbb{E}_{1,\mu}(0,T)} + |u_0^*|_{\mathbb{E}_{1,\mu}(0,T)} \le r + |u_0^*|_{\mathbb{E}_{1,\mu}(0,T)},$$

hence $|v_2|_{\mathbb{E}_{1,\mu}(0,T)}$ is small, provided $r > 0$ and $T > 0$ are small enough. Lastly, the term $|v_1 - v_2|_{\infty,X_{\gamma,\mu}}$ can be estimated by (5.9). In summary, if we choose $r > 0, T > 0$ and $\varepsilon > 0$ sufficiently small, we obtain a constant $c = c(u_0) > 0$ such that the estimate

$$|\mathcal{T}_{u_1}v_1 - \mathcal{T}_{u_2}v_2|_{\mathbb{E}_{1,\mu}(0,T)} \le \frac{1}{2}|v_1 - v_2|_{\mathbb{E}_{1,\mu}(0,T)} + c|u_1 - u_2|_{X_{\gamma,\mu}} \tag{5.10}$$

is valid for all $u_1, u_2 \in \bar{B}_{X_{\gamma,\mu}}(u_0, \varepsilon)$ and $v_1 \in \mathbb{B}_{r,T,u_1}$, $v_2 \in \mathbb{B}_{r,T,u_2}$. In the very special case $u_1 = u_2$, (5.10) yields the contraction mapping property of $\mathcal{T}_{u_1}$ on $\mathbb{B}_{r,T,u_1}$. Now we are in a position to apply Banach's fixed point theorem to obtain a unique fixed point $\tilde{u} \in \mathbb{B}_{r,T,u_1}$ of $\mathcal{T}_{u_1}$, i.e., $\mathcal{T}_{u_1}\tilde{u} = \tilde{u}$. Therefore $\tilde{u} \in \mathbb{B}_{r,T,u_1}$ is the unique local solution to (5.1). Furthermore, if $u(t, u_1)$ and $u(t, u_2)$ denote the solutions of (5.1) with initial values $u_1, u_2 \in \bar{B}_{X_{\gamma,\mu}}(u_0, \varepsilon)$, respectively, the last assertion of the theorem follows from (5.10). This completes the proof. □

The next result provides information about the continuation of local solutions.

Corollary 5.1.2. *Let the assumptions of Theorem* 5.1.1 *be satisfied and assume that* $A(v) \in \mathcal{MR}_p(X_0)$ *for all* $v \in V_\mu$. *Then the solution* $u(t)$ *of* (5.1) *has a maximal interval of existence* $J(u_0) = [0, t_+(u_0))$, *which is characterized by*

(i) *Global existence:* $t_+(u_0) = \infty$;

(ii) $\liminf_{t\to t_+(u_0)} \operatorname{dist}_{X_{\gamma,\mu}}(u(t), \partial V_\mu) = 0$;

(iii) $\lim_{t\to t_+(u_0)} u(t)$ *does not exist in* $X_{\gamma,\mu}$.

Proof. Fix $u_0 \in X_{\gamma,\mu}$, and define

$$t_+(u_0) = \sup\{a > 0 \,:\, (5.1) \text{ has a solution on } [0, a]\}.$$

Suppose that $t_+(u_0)$ is finite, $\operatorname{dist}_{X_{\gamma,\mu}}(u(t), \partial V_\mu) \ge \eta$, for some $\eta > 0$, and assume that the solution $u(t)$ converges to some $u_1 := u(t_+(u_0)) \in V_\mu$ as $t \to t_+(u_0)$.

Then the set $u([0,t_+(u_0)]) \subset V_\mu$ is compact in $X_{\gamma,\mu}$. Hence by Theorem 5.1.1 we find a uniform $\delta > 0$ such that the problem

$$\dot{v} + A(v)v = F(v), \quad v(0) = u(s), \quad s \in [0, t_+(u_0)],$$

has a unique solution in $\mathbb{E}_{1,\mu}(0,\delta)$. Fixing $s_0 \in (t_+(u_0) - \delta, t_+(u_0))$ the corresponding solution $v(\tau)$ coincides with $u(s_0+\tau)$ and extends the solution u beyond $t_+(u_0)$, a contradiction. This proves the result. □

5.2 Regularity

In this section we want to show that additional regularity of A and F induces corresponding regularity of the solutions of (5.1). For this purpose we assume $(A,F) \in C^k(V_\mu; \mathcal{B}(X_1,X_0) \times X_0)$ and $u_0 \in V_\mu$, where $k \in \mathbb{N} \cup \{\infty, \omega\}$; here ω means real analytic.

Suppose that we are given a solution $u_* = u(\cdot, u_0) \in \mathbb{E}_{1,\mu}(J)$, and assume that $A(u_*(t)) \in \mathcal{MR}_p(X_0)$ for each $t \in J = [0,T]$. Then by continuous dependence, there is a ball $B_{X_{\gamma,\mu}}(u_0,r_0) \subset V_\mu$ such that for any $v \in B_{X_{\gamma,\mu}}(u_0,r_0)$ the solutions $u(\cdot, v)$ exist on the same interval J. Introduce new functions

$$u_\lambda(t,v) := u(\lambda t, v), \quad \lambda \in (1-\varepsilon, 1+\varepsilon),\ v \in B_{X_{\gamma,\mu}}(u_0,r_0),\ t \in J_\varepsilon,$$

where $J_\varepsilon = [0, T/(1+\varepsilon)]$, $\varepsilon > 0$ fixed but as small as we please. This new function satisfies $\dot{u}_\lambda(t,v) = \lambda \dot{u}(\lambda t, v)$, and hence

$$\dot{u}_\lambda(t,v) + \lambda A(u_\lambda(t,v))u_\lambda(t,v) = \lambda F(u_\lambda(t,v)), \quad t \in J_\varepsilon,\ u_\lambda(0,v) = v.$$

Now we consider the map

$$H : (1-\varepsilon, 1+\varepsilon) \times B_{X_{\gamma,\mu}}(u_0,r_0) \times \mathbb{E}_{1,\mu}(J_\varepsilon) \to \mathbb{E}_{0,\mu}(J_\varepsilon) \times X_{\gamma,\mu}$$

defined by

$$H(\lambda, v, w)(t) = (\dot{w}(t) + \lambda A(w(t))w(t) - \lambda F(w(t)), w(0) - v), \quad t \in J_\varepsilon,$$

with $\mathbb{E}_{0,\mu}(J_\varepsilon) := L_{p,\mu}(J_\varepsilon; X_0)$. Since A and F are of class C^k and H is polynomial in λ and linear in v, there follows also $H \in C^k$. Furthermore, we know $H(1, u_*, u_0) = 0$ and

$$D_\lambda H(\lambda, v, w) = (A(w)w - F(w), 0),$$

$$D_v H(\lambda, v, w) = (0, -I),$$

and

$$D_w H(\lambda, v, w)h = (\dot{h} + \lambda A(w)h + \lambda (A'(w)h)w - \lambda F'(w)h, h(0)).$$

In particular, $D_wH(1, u_*, u_0) : \mathbb{E}_{1,\mu}(J_\varepsilon) \to \mathbb{E}_{0,\mu} \times X_{\gamma,\mu}$ is given by

$$D_wH((1, u_0), u_*)h = (\dot{h} + A(u_*)h + (A'(u_*)h)u_* - F'(u_*)h, h(0)).$$

Since $A(u_*(t))$ has maximal L_p-regularity for each $t \in J$ and $F'(u_*(t))$ is of lower order, we obtain with Proposition 4.4.3 and Theorem 4.4.4 that

$$A(t) = A(u_*(t)) - F'(u_*(t)), \quad t \in J,$$

satisfies the assumptions of Proposition 3.5.6 as $u_* \in C(J; X_{\gamma,\mu})$. Setting

$$R(t)z := (A'(u_*(t)z))u_*(t), \quad z \in X_{\gamma,\mu}, \ t \in J,$$

we see that $R \in L_{p,\mu}(J; \mathcal{B}(X_{\gamma,\mu}, X_0))$, with norm in $L_{p,\mu}(I; \mathcal{B}(X_{\gamma,\mu}, X_0))$ dominated by $C|u_*|_{L_{p,\mu}(I,X_1)}$ for each subinterval $I = [t_0, t_1] \subset J$, where C is a universal constant. This shows that $R(t)|_I$ is a small perturbation of $A(t)|_I$ in $L_{p,\mu}(I, \mathcal{B}(X_1, X_0)$, provided the length of I is small enough. A similar argument as in the proof of Proposition 3.5.6 shows that $D_wH(1, u_0, u_*)$ is an isomorphism. Then we may apply the implicit function theorem to obtain a C^k-map

$$\Phi : (1 - \delta, 1 + \delta) \times B_{X_{\gamma,\mu}}(u_0, r) \to \mathbb{E}_{1,\mu}(J_\varepsilon)$$

such that $H(\lambda, v, \Phi(\lambda, v)) = 0$ for each $\lambda \in (1 - \delta, 1 + \delta)$, $v \in B_{X_{\gamma,\mu}}(u_0, r)$ and $\Phi(1, u_0) = u_*$. By the definition of H and by uniqueness we then see that $\Phi(\lambda, v) = u_\lambda(\cdot, v)$, and

$$[(\lambda, v) \mapsto u_\lambda(\cdot, v)] \in C^k((1 - \delta, 1 + \delta) \times B_{X_{\gamma,\mu}}(u_0, r), \mathbb{E}_{1,\mu}(J_\varepsilon)).$$

The embedding $\mathbb{E}_{1,\mu}(J_\varepsilon) \hookrightarrow C(J_\varepsilon; X_{\gamma,\mu})$ then shows that the map

$$(\lambda, v) \mapsto u_\lambda(t, v) = u(\lambda t, v)$$

is of class C^k for each fixed t. But this implies $u \in C^k((0, T) \times B_{X_{\gamma,\mu}}(u_0, r); X_{\gamma,\mu})$.

To extract more regularity from the map Φ, note that

$$\frac{\partial}{\partial \lambda} u_\lambda(t, v)|_{\lambda=1} = t\dot{u}(t, v), \quad t \in J_\varepsilon,$$

hence $t\dot{u}(\cdot, v) \in \mathbb{E}_{1,\mu}(J_\varepsilon)$, and by induction $t^k \partial_t^k u(\cdot, v) \in \mathbb{E}_{1,\mu}(J_\varepsilon)$, which means

$$u(\cdot, v) \in H_p^{k+1}((\varepsilon, T); X_0) \cap H_p^k((\varepsilon, T); X_1),$$

for each $\varepsilon \in (0, T)$. By embedding we obtain from this

$$u(\cdot, v) \in C^{k+1-1/p}((0, T); X_0) \cup C^{k-1/p}((0, T); X_1).$$

In particular, if $k = \infty$, then $u(\cdot, v) \in C^\infty((0, T); X_1)$ and in case $k = \omega$ we get real analyticity of $u(t, v) \in X_1$ in $(0, T) \times B_{X_{\gamma,\mu}}(u_0, r)$. Note that in these assertions the parameter λ completely disappeared.

Let us summarize the result in

Theorem 5.2.1. *Let $1 < p < \infty$, $k \in \mathbb{N} \cup \{\infty, \omega\}$, $J = [0,T]$, and assume $(A,F) \in C^k(V_\mu; \mathcal{B}(X_1, X_0) \times X_0)$. Let $u_* = u(\cdot, u_0) \in H^1_{p,\mu}(J; X_0) \cap L_{p,\mu}(J; X_1)$ be the solution of (5.1) with initial value $u_0 \in X_{\gamma,\mu}$. Suppose $A(u_*(t)) \in \mathcal{MR}_p(X_0)$ for each $t \in J$.*

Then there is $r > 0$ such that the maps

$$\phi_j : B_{X_{\gamma,\mu}}(u_0, r) \to H^1_p(J; X_0) \cap L_p(J; X_1), \quad \phi_j(v) = t^{j+1-\mu} \partial_t^j u,$$

are of class C^{k-j}, for each $j \le k$, where $u = u(\cdot, v)$ is the unique solution of (5.1) with initial value v. In particular,

$$\psi : B_{X_{\gamma,\mu}}(u_0, r) \mapsto H^{j+1}_p((\varepsilon, T); X_0) \cap H^j_p((\varepsilon, T); X_1), \quad \psi(v) = u$$

is of class C^{k-j} for each $\varepsilon \in (0,T)$, and

$$\psi : B_{X_{\gamma,\mu}}(u_0, r) \to C^{j+1-1/p}((0,T); X_0) \cap C^{j-1/p}((0,T); X_1)$$

is of class C^{k-j} as well. If $k = \infty$, then $\psi : B_{X_{\gamma,\mu}}(u_0, r) \to C^\infty((0,T); X_1)$ is of class C^∞ and if $k = \omega$, then ψ is also real analytic.

Note that, in particular, the flow map $\varphi : (t, v) \mapsto u(t, v)$ is of class C^k from $(0,a) \times B_{X_{\gamma,\mu}}(u_0, r)$ to $X_{\gamma,\mu}$. Furthermore, we obtain from the derivative of Φ with $w = D_2\Phi(1, u_0)v$ the relation

$$\dot{w}(t) + B(t)w(t) = 0, \quad t \in (0,T), \quad w(0) = v,$$

where $B(t) = A(u_*(t)) + (A'(u_*(t))\cdot)u_*(t) - F'(u_*(t))$. In particular, if u_* is an equilibrium, then $w(t) = e^{-Bt}v$.

One should also observe that once we have regularity of $\partial_t u$, we may write (5.1) as

$$A_0(t)u(t) = f(t) := F(u(t)) - \partial_t u(t), \quad t \in J,$$

with $A_0(t) = A(u(t))$. This is typically an elliptic equation for $u(t)$, where $t \in J$ now serves as a parameter. This reduces the study of regularity "in space" to a linear elliptic problem.

5.3 Normally Stable Equilibria

Here we assume that there is an open set $V \subset X_\gamma$ such that

$$(A, F) \in C^1(V, \mathcal{B}(X_1, X_0) \times X_0). \tag{5.11}$$

Let $\mathcal{E} \subset V \cap X_1$ denote the set of equilibrium solutions of (5.1), which means that

$$u \in \mathcal{E} \quad \text{if and only if} \quad u \in V \cap X_1, \ A(u)u = F(u).$$

Given an element $u_* \in \mathcal{E}$, we assume that u_* is contained in an m-dimensional *manifold of equilibria*. This means that there is an open subset $U \subset \mathbb{R}^m$, $0 \in U$, and a C^1-function $\Psi : U \to X_1$, such that

$$\begin{aligned} &\bullet\ \Psi(U) \subset \mathcal{E} \text{ and } \Psi(0) = u_*,\\ &\bullet\ \text{the rank of } \Psi'(0) \text{ equals } m, \text{ and}\\ &\bullet\ A(\Psi(\zeta))\Psi(\zeta) = F(\Psi(\zeta)), \quad \zeta \in U. \end{aligned} \tag{5.12}$$

For the moment, we assume further that near u_* there are no other equilibria than those given by $\Psi(U)$, i.e., $\mathcal{E} \cap B_{X_1}(u_*, r_1) = \Psi(U)$, for some $r_1 > 0$. Below we show that this assumption is redundant.

We suppose that the operator $A(u_*)$ has the property of maximal L_p-regularity. Introducing the deviation $v = u - u_*$ from the equilibrium u_*, the equation for v then reads as

$$\dot{v}(t) + A_0 v(t) = G(v(t)), \quad t > 0, \quad v(0) = v_0, \tag{5.13}$$

where $v_0 = u_0 - u_*$ and

$$A_0 v = A(u_*)v + (A'(u_*)v)u_* - F'(u_*)v \quad \text{for } v \in X_1. \tag{5.14}$$

The function G can be written as $G(v) = G_1(v) + G_2(v, v)$, where

$$\begin{aligned} G_1(v) &= (F(u_* + v) - F(u_*) - F'(u_*)v) - (A(u_* + v) - A(u_*) - A'(u_*)v)u_*,\\ G_2(v, w) &= -(A(u_* + v) - A(u_*))w, \quad w \in X_1,\ v \in V_*, \end{aligned}$$

with $V_* := V - u_*$. It follows from (5.11) that $G_1 \in C^1(V_*, X_0)$ and also that $G_2 \in C^1(V_* \times X_1, X_0)$. Moreover, we have

$$(G_1(0), G_2(0,0)) = 0, \quad (G_1'(0), G_2'(0,0)) = 0, \tag{5.15}$$

where G_1' and G_2' denote the Fréchet derivatives of G_1 and G_2, respectively.

Setting $\psi(\zeta) = \Psi(\zeta) - u_*$ results in the following equilibrium equation for problem (5.13)

$$A_0 \psi(\zeta) = G(\psi(\zeta)), \quad \text{for all } \zeta \in U. \tag{5.16}$$

Taking the derivative with respect to ζ and using the fact that $G'(0) = 0$ we conclude that $A_0 \psi'(0) = 0$ and this implies that

$$T_{u_*}\mathcal{E} \subset \mathsf{N}(A_0), \tag{5.17}$$

where $T_{u_*}\mathcal{E}$ denotes the tangent space of $\mathcal{E}$ at u_*.

After these preparations we can state the following result on convergence of solutions starting near u_* which will be termed the *generalized principle of linearized stability.*

Theorem 5.3.1. *Let $1 < p < \infty$. Suppose $u_* \in V \cap X_1$ is an equilibrium of* (5.1), *and suppose that the functions (A, F) satisfy* (5.11). *Suppose further that $A(u_*)$ has the property of maximal L_p-regularity. Let A_0, defined in* (5.14), *denote the linearization of* (5.1) *at u_*. Suppose that u_* is* **normally stable**, *which means that*

(i) *near u_* the set of equilibria $\mathcal{E}$ is a C^1-manifold in X_1 of dimension $m \in \mathbb{N}$,*

(ii) *the tangent space for $\mathcal{E}$ at u_* is isomorphic to $\mathsf{N}(A_0)$,*

(iii) 0 *is a* semi-simple eigenvalue *of A_0, i.e.,* $\mathsf{N}(A_0) \oplus \mathsf{R}(A_0) = X_0$,

(iv) $\sigma(A_0) \setminus \{0\} \subset \mathbb{C}_+ = \{z \in \mathbb{C} : \operatorname{Re} z > 0\}$.

Then u_* is stable in X_γ, and there exists $\delta > 0$ such that the unique solution $u(t)$ of (5.1) *with initial value $u_0 \in X_\gamma$ satisfying $|u_0 - u_*|_\gamma < \delta$ exists on $\mathbb{R}_+$ and converges at an exponential rate in X_γ to some $u_\infty \in \mathcal{E}$ as $t \to \infty$.*

Proof. (a) Note first that assumption (iii) implies that 0 is an isolated point of $\sigma(A_0)$, the spectrum of A_0. According to assumption (iv), $\sigma(A_0)$ admits a decomposition into two disjoint nontrivial parts with

$$\sigma(A_0) = \{0\} \cup \sigma_s, \quad \sigma_s \subset \mathbb{C}_+ = \{z \in \mathbb{C} : \operatorname{Re} z > 0\}.$$

The spectral set $\sigma_c := \{0\}$ corresponds to the center part, and σ_s to the stable part of the analytic C_0-semigroup $e^{-A_0 t}$, or equivalently of the Cauchy problem $\dot{w} + A_0 w = f$.

In the following, we let P^l, $l \in \{c, s\}$, denote the spectral projections according to the spectral sets $\sigma_c = \{0\}$ and σ_s, and we set $X_j^l := P^l X_j$ for $l \in \{c, s\}$ and $j \in \{0, 1, \gamma\}$. The spaces X_j^l are equipped with the norms $|\cdot|_j$ for $j \in \{0, 1, \gamma\}$. We have the topological direct decomposition

$$X_1 = X_1^c \oplus X_1^s, \quad X_0 = X_0^c \oplus X_0^s,$$

and this decomposition reduces A_0 into $A_0 = A_c \oplus A_s$, where A_l is the part of A_0 in X_0^l for $l \in \{c, s\}$. Since $\sigma_c = \{0\}$ is compact it follows that $X_0^c \subset X_1$. Therefore, X_0^c and X_1^c coincide as vector spaces. In the following, we will just write $X^c = (X^c, |\cdot|_j)$ for either of the spaces X_0^c and X_1^c. The operator A_s inherits the property of L_p-maximal regularity from A_0. Since $\sigma(A_s) = \sigma_s \subset \mathbb{C}_+$ we obtain that the Cauchy problem

$$\dot{w} + A_s w = f, \quad w(0) = 0, \tag{5.18}$$

also enjoys the property of maximal regularity, even on the interval $J = (0, \infty)$. In fact the following estimates are true. For any $a \in (0, \infty]$ let

$$\mathbb{E}_0(a) = L_p((0, a); X_0), \quad \mathbb{E}_1(a) = H_p^1((0, a); X_0) \cap L_p((0, a); X_1). \tag{5.19}$$

The natural norms in $\mathbb{E}_j(a)$ will be denoted by $|\cdot|_{\mathbb{E}_j(a)}$ for $j = 0, 1$. Then the Cauchy problem (5.18) has for each $f \in L_p((0, a); X_0^s)$ a unique solution

$$w \in H_p^1((0, a); X_0^s) \cap L_p((0, a); X_1^s),$$

and there exists a constant M_0 such that $|w|_{\mathbb{E}_1(a)} \le M_0|f|_{\mathbb{E}_0(a)}$ for every $a > 0$, and every function $f \in L_p((0,a); X_0^s)$. In fact, since $\sigma(A_s - \omega)$ is still contained in $\mathbb{C}_+$ for ω small enough, we see that the operator $A_s - \omega$ enjoys the same properties as A_s. Therefore, every solution of the Cauchy problem (5.18) satisfies the estimate

$$|e^{\sigma t} w|_{\mathbb{E}_1(a)} \le M_0 |e^{\sigma t} f|_{\mathbb{E}_0(a)}, \quad \sigma \in [0,\omega], \quad a > 0, \tag{5.20}$$

for $f \in L_p((0,a); X_0^s)$, where $M_0 = M_0(\omega)$ for $\omega > 0$ fixed. Furthermore, there exists a constant $M_1 > 0$ such that

$$|e^{\omega t} e^{-A_s t} P^s u|_{\mathbb{E}_1(a)} + \sup_{t\in[0,a)} |e^{\omega t} e^{-A_s t} P^s u|_\gamma \le M_1 |P^s u|_\gamma \tag{5.21}$$

for every $u \in X_\gamma$ and $a \in (0,\infty]$. For future use we note that

$$\sup_{t\in[0,a)} |w(t)|_\gamma \le c_0 |w|_{\mathbb{E}_1(a)} \quad \text{for all } w \in \mathbb{E}_1(a) \text{ with } w(0) = 0 \tag{5.22}$$

with a constant c_0 that is independent of $a \in (0,\infty]$.

(b) It follows from the considerations above and assumptions (i)-(iii) that in fact

$$\mathsf{N}(A_0) = X^c \quad \text{and} \quad \dim(X^c) = m.$$

As X^c has finite dimension, the norms $|\cdot|_j$ for $j \in \{0,1,\gamma\}$ are equivalent, and we equip X^c with one of these equivalent norms, say with $|\cdot|_0$. Let us now consider the mapping

$$g : U \subset \mathbb{R}^m \to X^c, \quad g(\zeta) := P^c \psi(\zeta), \quad \zeta \in U.$$

It follows from our assumptions that $g'(0) = P^c\psi'(0) : \mathbb{R}^m \to X^c$ is an isomorphism (between the finite dimensional spaces $\mathbb{R}^m$ and X^c). By the inverse function theorem, g is a C^1-diffeomorphism of a neighbourhood of 0 in $\mathbb{R}^m$ into a neighbourhood, say $B_{X^c}(0,\rho_0)$, of 0 in X^c. Let $g^{-1} : B_{X^c}(0,\rho_0) \to U$ be its inverse mapping. Then g^{-1} is C^1 and $g^{-1}(0) = 0$. Next we set $\Phi(x) := \psi(g^{-1}(x))$ for $x \in B_{X^c}(0,\rho_0)$ and we note that

$$\Phi \in C^1(B_{X^c}(0,\rho_0), X_1), \quad \Phi(0) = 0, \quad \{\Phi(x) + u_* : x \in B_{X^c}(0,\rho_0)\} = \mathcal{E} \cap W,$$

where W is an appropriate neighbourhood of u_* in X_1. One readily verifies that

$$P^c\Phi(x) = ((P^c \circ \psi) \circ g^{-1})(x) = (g \circ g^{-1})(x) = x, \quad x \in B_{X^c}(0,\rho_0),$$

and this yields $\Phi(x) = P^c\Phi(x) + P^s\Phi(x) = x + P^s\Phi(x)$ for $x \in B_{X^c}(0,\rho_0)$. Setting $\phi(x) := P^s\Phi(x)$ we conclude that

$$\phi \in C^1(B_{X^c}(0,\rho_0), X_1^s), \quad \phi(0) = \phi'(0) = 0, \tag{5.23}$$

and that

$$\{x + \phi(x) + u_* : x \in B_{X^c}(0,\rho_0)\} = \mathcal{E} \cap W,$$

where W is a neighbourhood of u_* in X_1. This shows that the manifold $\mathcal{E}$ can be represented as the (translated) graph of the function ϕ in a neighbourhood of u_*. Moreover, the tangent space of $\mathcal{E}$ at u_* coincides with $\mathsf{N}(A_0) = X^c$. By applying the projections P^l, $l \in \{c, s\}$, to equation (5.16) and using that $x + \phi(x) = \psi(g^{-1}(x))$ for $x \in B_{X^c}(0, \rho_0)$, and that $A_c \equiv 0$, we obtain the following equivalent system of equations for the equilibria of (5.13)

$$P^c G(x + \phi(x)) = 0, \quad P^s G(x + \phi(x)) = A_s \phi(x), \quad x \in B_{X_c}(0, \rho_0). \tag{5.24}$$

Finally, let us also agree that ρ_0 has already been chosen small enough so that

$$|\phi'(x)|_{\mathcal{B}(X^c, X_1^s)} \leq 1, \quad |\phi(x)|_1 \leq |x|, \quad x \in B_{X^c}(0, \rho_0). \tag{5.25}$$

This can always be achieved, thanks to (5.23).

(c) Introducing the new variables

$$\begin{aligned} x &= P^c v = P^c(u - u_*), \\ y &= P^s v - \phi(P^c v) = P^s(u - u_*) - \phi(P^c(u - u_*)) \end{aligned}$$

we then obtain the following system of evolution equations in $X^c \times X_0^s$

$$\left\{ \begin{aligned} \dot{x} &= T(x, y), \quad x(0) = x_0, \\ \dot{y} + A_s y &= R(x, y), \quad y(0) = y_0, \end{aligned} \right. \tag{5.26}$$

with $x_0 = P^c v_0$ and $y_0 = P^s v_0 - \phi(P^c v_0)$, where the functions T and R are given by

$$\begin{aligned} T(x, y) &= P^c G(x + \phi(x) + y), \\ R(x, y) &= P^s G(x + \phi(x) + y) - A_s \phi(x) - \phi'(x) T(x, y). \end{aligned}$$

Using the equilibrium equations (5.24), the expressions for R and T can be rewritten as

$$\begin{aligned} T(x, y) &= P^c\big(G(x + \phi(x) + y) - G(x + \phi(x))\big), \\ R(x, y) &= P^s\big(G(x + \phi(x) + y) - G(x + \phi(x))\big) - \phi'(x) T(x, y). \end{aligned} \tag{5.27}$$

Although the term $P^c G(x + \phi(x))$ in T is zero, see (5.24), we include it here for reasons of symmetry, and for justifying the estimates for T below. Equation (5.27) immediately yields

$$(T(x, 0), R(x, 0)) = 0 \quad \text{for all} \quad x \in B_{X^c}(0, \rho_0),$$

showing that the equilibrium set $\mathcal{E}$ of (5.1) near u_* has been reduced to the set $B_{X^c}(0, \rho_0) \times \{0\} \subset X^c \times X_1^s$.

Observe also that there is a unique correspondence between the solutions of (5.1) close to u_* in X_γ and those of (5.26) close to 0. We call system (5.26) the *normal form* of (5.1) near its normally stable equilibrium u_*.

(d) From the representation of G and (5.15) we obtain the following estimates for G_1 and G_2: for given $\eta > 0$ we may choose $r = r(\eta) > 0$ small enough such that

$$|G_1(v_1) - G_1(v_2)|_0 \le \eta |v_1 - v_2|_\gamma, \quad v_1, v_2 \in B_{X_\gamma}(0,r).$$

Moreover, there is a constant $L > 0$ such that

$$\begin{aligned} |G_2(v_1,w) - G_2(v_2,w)|_0 &\le L|w|_1\,|v_1 - v_2|_\gamma, \quad w \in X_1, \quad && v_1, v_2 \in B_{X_\gamma}(0,r),\\ |G_2(v,w_1) - G_2(v,w_2)|_0 &\le L\,r\,|w_1 - w_2|_1, \quad w_1, w_2 \in X_1, \quad && v \in B_{X_\gamma}(0,r). \end{aligned}$$

We remark that L does not depend on $r \in (0, r_0]$ with r_0 appropriately chosen. Combining these estimates we have

$$\begin{aligned} |G(v_1) - G(v_2)|_0 &\le \big(\eta + L|v_2|_1\big)|v_1 - v_2|_\gamma + Lr|v_1 - v_2|_1 \\ &\le C_0\big(\eta + r + |v_2|_1\big)|v_1 - v_2|_1 \end{aligned} \tag{5.28}$$

for all $v_1, v_2 \in B_{X_\gamma}(0,r) \cap X_1$, where C_0 is independent of $r \in (0, r_0]$.

In the following, we will always assume that $r \in (0, r_0]$ and $r_0 \le 3\rho_0$. Taking $v_1 = x + \phi(x) + y$ and $v_2 = x + \phi(x)$ in (5.28) we infer from (5.25) and (5.27) that

$$|T(x,y)|,\ |R(x,y)|_0 \le C_1\big(\eta + r + |x + \phi(x))|_1\big)|y|_1 \le \beta |y|_1, \tag{5.29}$$

for all $x \in \bar{B}_{X^c}(0,\rho)$, $y \in \bar{B}_{X^s_\gamma}(0,\rho) \cap X_1$ and all $\rho \in (0, r/3)$, where $\beta = C_2(\eta + r)$, and where C_1 and C_2 are uniform constants. Suppose that η and, accordingly, r were already chosen small enough so that

$$M_0\beta = M_0 C_2(\eta + r) \le 1/2. \tag{5.30}$$

(e) By Theorem 5.1.1 with $\mu = 1$, Problem (5.13) admits for each $v_0 \in B_{X_\gamma}(0,r)$ a unique local strong solution $v \in \mathbb{E}_1(a) \cap C([0,a]; X_\gamma)$ for some number $a > 0$. This solution can be extended to a maximal interval of existence $[0, t_+)$. If t_+ is finite, then either $v(t)$ leaves the ball $B_{X_\gamma}(0,r)$ at time t_+, or the limit $\lim_{t\to t_+} v(t)$ does not exist in X_γ. We show that this cannot happen for initial values $v_0 \in B_{X_\gamma}(0,\delta)$, with $\delta \le r$ to be chosen later.

Suppose that $x_0 \in B_{X^c}(0, N\delta)$ and $y_0 \in B_{X^s_\gamma}(0, N\delta)$ are given, where the number δ will be determined later and $N := |P^c|_{\mathcal{B}(X_0)} + |P^s|_{\mathcal{B}(X_\gamma)}$. Let t_+ denote the existence time for the solution $(x(t), y(t))$ of System (5.26) with initial values (x_0, y_0), or equivalently, for the solution $v(t)$ of (5.13) with initial value $v_0 = x_0 + \phi(x_0) + y_0$. Let ρ be fixed so that the estimates in (5.29) hold. Set

$$t_1 := t_1(x_0, y_0) := \sup\{t \in (0, t_+) : |x(\tau)|,\ |y(\tau)|_\gamma \le \rho,\ \tau \in [0,t]\}$$

and suppose that $t_1 < t_+$. Due to (5.20)–(5.21) and (5.29) we obtain

$$\begin{aligned} |e^{\omega t} y|_{\mathbb{E}_1(t_1)} &\le M_1 |y_0|_\gamma + M_0 |e^{\omega t} R(x,y)|_{\mathbb{E}_0(t_1)} \\ &\le M_1 |y_0|_\gamma + M_0 \beta |e^{\omega t} y|_{\mathbb{E}_1(t_1)}. \end{aligned}$$

This yields with (5.30)

$$|e^{\sigma t}y|_{\mathbb{E}_1(t_1)} \le 2M_1|y_0|_\gamma\,, \quad \sigma \in [0,\omega]. \tag{5.31}$$

Using this estimate as well as (5.21)–(5.22) we further have for $t \in [0,t_1)$

$$\begin{aligned} |e^{\sigma t}y(t)|_\gamma &\le |e^{\sigma t}y(t) - e^{\sigma t}e^{-A_s t}y_0|_\gamma + |e^{\sigma t}e^{-A_s t}y_0|_\gamma \\ &\le c_0|e^{\sigma t}y(t) - e^{\sigma t}e^{-A_s t}y_0|_{\mathbb{E}_1(t_1)} + M_1|y_0|_\gamma \\ &\le (3c_0M_1 + M_1)|y_0|_\gamma, \end{aligned}$$

which yields with $M_2 = (3c_0 + 1)M_1$,

$$|y(t)|_\gamma \le M_2e^{-\sigma t}|y_0|_\gamma, \quad t \in [0,t_1),\ \sigma \in [0,\omega]. \tag{5.32}$$

We deduce from the equation for x, the estimate for T in (5.29), and Hölder's inequality that

$$\begin{aligned} |x(t)| &\le |x_0| + \int_0^t |T(x(s),y(s))|\,ds \\ &\le |x_0| + \beta\int_0^t |y(s)|_1\,ds \\ &= |x_0| + \beta c_1|e^{\omega t}y|_{\mathbb{E}_1(t_1)} \\ &\le |x_0| + M_3|y_0|_\gamma, \quad t \in [0,t_1), \end{aligned}$$

where $M_3 = 2M_1c_1\beta$ and $c_1 = (1/[\omega p'])^{1/p'}$. Summarizing, we have shown that $|x(t)| + |y(t)|_\gamma \le |x_0| + (M_2 + M_3)|y_0|_\gamma$ for all $t \in [0,t_1)$. By continuity and the assumption $t_1 < t_+$ this inequality also holds for $t = t_1$. Hence

$$|x(t_1)| + |y(t_1)|_\gamma \le |x_0| + (M_2 + M_3)|y_0|_\gamma \le (1 + M_2 + M_3)N\delta < \rho/2,$$

provided $\delta \le \rho/[2N(1 + M_2 + M_3)]$. This contradicts the definition of t_1 and we conclude that $t_1 = t_+$.

In the following, we assume that $\delta \le \rho/[2N(1+M_2+M_3)]$. Then the estimates derived above and (5.25) yield the uniform bounds

$$|v|_{\mathbb{E}_1(a)} + \sup_{t\in[0,a)} |v(t)|_\gamma \le M, \tag{5.33}$$

for every initial value $v_0 \in B_{X_\gamma}(0,\delta)$ and every $a < t_+$. It follows from Corollary 5.1.2 that the solution $v(t)$ of (5.13) exists on $\mathbb{R}_+$.

(f) By repeating the above estimates on the interval $(0,\infty)$ we obtain

$$|x(t)| \le |x_0| + M_3|y_0|_\gamma, \quad |y(t)|_\gamma \le M_2e^{-\omega t}|y_0|_\gamma, \quad t \in [0,\infty), \tag{5.34}$$

for all $x_0 \in B_{X^c}(0,N\delta)$ and $y_0 \in B_{X^s_\gamma}(0,N\delta)$. Moreover,

$$\lim_{t\to\infty} x(t) = x_0 + \int_0^\infty T(x(s),y(s))\,ds =: x_\infty$$

exists since the integral is absolutely convergent. Next observe that we in fact obtain exponential convergence of $x(t)$ towards x_∞, as

$$\begin{aligned}|x(t)-x_\infty| &= \left|\int_t^\infty T(x(s),y(s))\,ds\right|\\ &\le \beta\int_t^\infty |y(s)|_1\,ds\\ &\le \beta\left(\int_t^\infty e^{-\omega s p'}\,ds\right)^{1/p'} |e^{\omega s}y|_{\mathbb{E}_1(\infty)}\\ &\le M_4 e^{-\omega t}|y_0|_\gamma, \quad t\ge 0.\end{aligned}$$

This yields existence of

$$v_\infty := \lim_{t\to\infty} v(t) = \lim_{t\to\infty} x(t)+\phi(x(t))+y(t) = x_\infty+\phi(x_\infty).$$

Clearly, v_∞ is an equilibrium for equation (5.13), and $v_\infty+u_*\in\mathcal{E}$ is an equilibrium for (5.1). Due to (5.25), (5.34) and the exponential estimate for $|x(t)-x_\infty|$ we get

$$\begin{aligned}|v(t)-v_\infty|_\gamma &= |x(t)+\phi(x(t))+y(t)-v_\infty|_\gamma\\ &\le |x(t)-x_\infty|_\gamma + |\phi(x(t))-\phi(x_\infty)|_\gamma + |y(t)|_\gamma\\ &\le (CM_4+M_2)e^{-\omega t}|y_0|_\gamma\\ &\le Me^{-\omega t}|P^s v_0-\phi(P^c v_0)|_\gamma,\end{aligned}\tag{5.35}$$

thereby completing the proof of the second part of Theorem 5.3.1. Concerning stability, note that given $r>0$ small enough we may choose $0<\delta\le r$ such that the solution starting in $B_{X_\gamma}(u_*,\delta)$ exists on $\mathbb{R}_+$ and stays within $B_{X_\gamma}(u_*,r)$. □

Remarks 5.3.2. **(a)** If $m=0$ the equilibrium u_* is isolated and $0\notin\sigma(A_0)$. In this case all solutions starting in a neighbourhood of u_* converge to u_* in X_γ. This is the classical *principle of linearized stability.*

(b) Theorem 5.3.1 shows, given that situation, that near u_* the set of equilibria constitutes the (unique) *center manifold* for (5.1).

(c) It is worthwhile to point out a slightly different way to obtain the function ϕ used in the proof of Theorem 5.3.1. Applying the projections P^s and P^c to the equilibrium equation (5.16) yields the following equivalent system of equations near $v=0$

$$A_s z = P^s G(x+z), \quad A_c x = P^c G(x+z), \tag{5.36}$$

with $z=P^s\psi(\zeta)$ and $x=P^c\psi(\zeta)$. Since $(G(0),G'(0))=0$ and A_s is invertible, by the implicit function theorem we may solve the first equation for z in terms of x, i.e., there is a C^1-function $\phi: B_{X^c}(0,\rho_0)\to X_1^s$ such that

$$\phi(0)=0 \quad\text{and}\quad A_s\phi(x)=P^sG(x+\phi(x)), \quad x\in B_{X^c}(0,\rho_0).$$

As $x+\phi(x)$ is the unique solution of the first equation in (5.36) we additionally have $A_c x = P^c G(x+\phi(x))$, as well as $P^s\psi(\zeta) = \phi(P^c\psi(\zeta))$ for all $\zeta \in U$. Since $G'(0)=0$ we obtain $A_s\phi'(0) = P^s G'(0) = 0$ and this implies $\phi'(0)=0$. This shows that $\mathcal{E} \subset \mathcal{M}$ with $\mathcal{M} = \{x+\phi(x)+u_* : x \in B_{X^c}(0,\rho_0)\}$ in a neighbourhood of u_* in X_1.

$\mathcal{M}$ is a C^1-manifold of dimension $\ell := \dim(X^c)$ with tangent space $T_{u_*}\mathcal{M} = X^c$ and $\mathcal{E}$ is a submanifold in $\mathcal{M}$. In general, $\mathcal{E}$ has lower dimension than $\mathcal{M}$. Our assumptions in Theorem 5.3.1 do in fact exactly amount to asserting that $\mathcal{E}$ and $\mathcal{M}$ are of equal dimension. Since $\mathcal{E} \subset \mathcal{M}$ we can then conclude that they coincide in a neighbourhood of u_*.

(d) An inspection of the argument given above shows that in fact all equilibria of equation (5.1) that are close to the equilibrium u_* are contained in the manifold $\mathcal{M} = \{x+\phi(x)+u_* : x \in B_{X^c}(0,\rho_0)\}$ such that $\phi(0) = \phi'(0) = 0$, with no additional assumptions on the structure of the equilibria. To see this, let us once more consider the equation

$$A_s z = P^s G(x+z), \quad x \in X^c, \quad z \in X_1^s. \tag{5.37}$$

Clearly, $(x,z) = 0$ is a solution. Exactly as in the remark above, we can solve (5.37) by the implicit function theorem for z in terms of x, obtaining a C^1-function $\phi : B_{X^c}(0,\rho_0) \to X_1^s$ with $(\phi(0),\phi'(0)) = 0$. If $v \in X_1$ is an equilibrium for the evolution equation (5.13) close to 0, then the pair $x = P^c v$, $z = P^s v$ necessarily satisfies equation (5.36), and therefore lies on the graph of ϕ.

(e) It is well-known even in the 2D-ODE-case that convergence to equilibria fails if one of the conditions (i)-(iii) in Theorem 5.3.1 does not hold.

5.4 Instability of Equilibria

We consider again the behaviour of (5.1) near an equilibrium $u_* \in \mathcal{E}$. Here we are interested in instability, and we want to prove the second half of the *principle of linearized stability* for (5.1). The result reads as follows.

Theorem 5.4.1. *Let $(A,F) \in C^1(V;\mathcal{B}(X_1,X_0)\times X_0)$, $V \subset X_\gamma$ open, and suppose $u_* \in \mathcal{E}\cap V$ is such that $A(u_*)$ has the property of maximal L_p-regularity, and assume that the linearization of* (5.1) *at $u_* \in \mathcal{E}$*

$$A_0 = A(u_*) + [A'(u_*)\cdot]u_* - F'(u_*)$$

satisfies

$$\sigma(-A_0)\cap[\kappa + i\mathbb{R}] = \emptyset, \quad \sigma(-A_0)\cap\{z\in\mathbb{C} : \operatorname{Re} z > \kappa\} \neq \emptyset,$$

for some $\kappa \geq 0$. Then the equilibrium $u_* \in \mathcal{E}$ is unstable in X_γ.

Proof. (a) As in the previous section, we transform the equilibrium u_* to 0 by setting $v = u - u_*$. Then as there, (5.1) can be rewritten as the problem

$$\dot{v} + A_0 v = G(v), \quad v(0) = v_0, \tag{5.38}$$

where $G: X_1 \to X_0$ satisfies the following property. For each $\eta > 0$ there is $r > 0$ such that

$$|G(v)|_0 \le \eta |v|_1, \quad \text{for all } |v|_\gamma \le r, \quad v \in V \cap X_1.$$

Next we find $\mu > 0$ such that the strip $[\kappa - 2\mu, \kappa + 2\mu] + i\mathbb{R}$ does not intersect the spectrum of $-A_0$, this is a spectral gap for $-A_0$. Let P_+ denote the spectral projection of A_0 corresponding to $\{z \in \mathbb{C} : \operatorname{Re} z > \kappa + 2\mu\} \cap \sigma(-A_0)$ and let $P_- = I - P_+$ be the complementary projection. We set $A_\pm = P_\pm A_0$ and observe that A_+ is nontrivial but bounded. Further, there is a constant $M_0 > 0$ such that

$$|P_- e^{-A_- t}|_0 \le M_0 e^{(\kappa - \mu)t}, \quad |P_+ e^{A_+ t}|_0 \le M_0 e^{-(\kappa + \mu)t}, \quad t > 0.$$

As the operator $\kappa + A_-$ belongs to $\mathcal{MR}_p(\mathsf{R}(P_-))$ we find a constant $M_1 > 0$ such that the solution of

$$\dot{w} + A_- w = P_- f, \quad w(0) = P_- w_0,$$

satisfies the estimate

$$|e^{-\kappa t} w|_{\mathbb{E}_1(a)} \le M_1(|P_- w_0|_\gamma + |e^{-\kappa t} P_- f|_{\mathbb{E}_0(a)}),$$

and $M_1 \ge 1$ is independent of $a > 0$. We may assume w.l.o.g. $\kappa > 0$. Further, we may assume $|v|_0 = |P_- v|_0 + |P_+ v|_0$, as well as $|v|_0 \le |v|_\gamma \le |v|_1$, by proper definition of the norms. Observe further that there is a constant $C_1 \ge 1$ such that

$$|P_+ v|_0 \le |P_+ v|_1 \le C_1 |P_+ v|_0,$$

as A_+ is bounded.

(b) Suppose that u_* is stable in X_γ for (5.1). Then for each $\varepsilon > 0$ there is $\delta > 0$ such that

$$|v_0|_\gamma \le \delta \quad \Rightarrow |v(t)|_\gamma \le \varepsilon, \quad t \ge 0.$$

Fix any $\eta > 0$ with $\eta M_1 \le 1/2$, and let $\varepsilon = \min\{r, \eta(p\kappa)^{1/p}\}$. Applying P_- to (5.38) and using maximal L_p-regularity of A_- we obtain

$$\begin{aligned} |e^{-\kappa t} v|_{\mathbb{E}_1(a)} &\le |e^{-\kappa t} P_- v|_{\mathbb{E}_1(a)} + |e^{-\kappa t} P_+ v|_{\mathbb{E}_1(a)} \\ &\le M_1(|P_- v_0|_\gamma + \eta |e^{-\kappa t} v|_{\mathbb{E}_1(a)}) + |e^{-\kappa t} P_+ v|_{\mathbb{E}_1(a)}, \quad a > 0, \end{aligned}$$

and hence

$$|e^{-\kappa t} v|_{\mathbb{E}_1(a)} \le 2M_1(|P_- v_0|_\gamma + |e^{-\kappa t} P_+ v|_{\mathbb{E}_1(a)}), \quad a > 0. \tag{5.39}$$

For the part $e^{-\kappa t}P_+z$ we use relation

$$\partial_t(e^{-\kappa t}P_+z(t)) = -(\kappa + A_+)e^{-\kappa t}v(t) + e^{-\kappa t}P_+G(v(t))$$

and the fact that A_+ is bounded to the result

$$|e^{-\kappa t}P_+v|_{\mathbb{E}_1(a)} \le C_2(|e^{-\kappa t}P_+v|_{\mathbb{E}_0(a)} + \eta|e^{-\kappa t}v|_{\mathbb{E}_1(a)}). \tag{5.40}$$

Next we have

$$|e^{-\kappa t}P_+v|_{\mathbb{E}_0(a)} \le \varepsilon\Big(\int_0^\infty e^{-p\kappa t}\,dt\Big)^{1/p} = \varepsilon(p\kappa)^{-1/p} \le \eta.$$

This implies with (5.39) and (5.40)

$$|e^{-\kappa t}v|_{\mathbb{E}_1(a)} \le C_3(|P_-v_0|_\gamma + \eta + \eta|e^{-\kappa t}v|_{\mathbb{E}_1(a)}), \quad a > 0$$

and hence, assuming that $C_3\eta \le 1/2$,

$$|e^{-\kappa t}v|_{\mathbb{E}_1(a)} \le 2C_3(\eta + |P_-v_0|_\gamma), \quad a > 0.$$

From this and Hölder's inequality we deduce

$$\begin{aligned}\int_t^\infty |e^{-A_+(t-s)}P_+G(v(s)|_0\,ds &\le M_0\eta\, e^{\kappa t}\int_t^\infty e^{\mu(t-s)}|e^{-\kappa s}v(s)|_1\,ds\\ &\le C_4\eta\, e^{\kappa t}|e^{-\kappa t}v|_{\mathbb{E}_1(\infty)}\end{aligned}$$

where $C_4 = M_0/(\mu p')^{1/p'}$. This shows that the integral $\int_t^\infty e^{-A_+(t-s)}P_+G(v(s)\,ds$ exists in $X_+ = \mathsf{R}(P_+)$ for any $t \ge 0$. Moreover, its norm in X_+ grows no faster than an exponential function $Ce^{\kappa t}$. Therefore, by means of the variation of constants formula we may write

$$\begin{aligned}P_+v(t) &= e^{-A_+t}P_+v_0 + \int_0^t e^{-A_+(t-s)}P_+G(v(s))\,ds\\ &= e^{-A_+t}w_0 - \int_t^\infty e^{-A_+(t-s)}P_+G(v(s))\,ds,\end{aligned}$$

with

$$w_0 = P_+v_0 + \int_0^\infty e^{A_+s}P_+G(v(s))\,ds.$$

The estimate

$$|e^{A_+t}(P_+v(t) + \int_t^\infty e^{-A_+(t-s)}P_+G(v(s))\,ds)|_{X_+} \le C\eta e^{-(\kappa+\mu)t}(1 + e^{\kappa t}|e^{-\kappa t}v|_{\mathbb{E}_1(\infty)})$$

shows that $w_0 = 0$, i.e we have

$$P_+v(t) = -\int_t^\infty e^{-A_+(t-s)}P_+G(v(s))\,ds, \quad t \ge 0, \tag{5.41}$$

hence $|e^{-\kappa t}P_+v(t)|_0 \le \eta M_0 \int_t^\infty e^{\mu(t-s)}e^{-\kappa s}|Pv(s)|_1\, ds$, and so by Young's inequality

$$|e^{-\kappa t}P_+v|_{\mathbb{E}_0(\infty)} \le \frac{\eta M_0}{\mu}|e^{-\kappa t}v|_{\mathbb{E}_1(\infty)}.$$

We may now conclude with (5.39) and (5.40) by similar arguments as above that

$$|e^{-\kappa t}v|_{\mathbb{E}_1(\infty)} \le C_6|P_-v_0|_\gamma.$$

Finally, with (5.41) for $t = 0$ we obtain the inequality

$$\begin{aligned}|P_+v_0|_\gamma &\le C_1|P_+v_0|_0 \le C_1\Big|\int_0^\infty e^{A_+s}P_+G(v(s))\Big|_0 \le C_1C_4\eta|e^{-\kappa t}v|_{\mathbb{E}_1(\infty)}\\ &\le C_7|P_-v_0|_\gamma.\end{aligned}$$

This proves instability of u_* for (5.1). □

Remark 5.4.2. (a) Refining the argument in the above proof, one can even show the following stronger instability result: there exists $\varepsilon_0 > 0$ and a sequence of initial values $u_{0k} \to u_*$ in X_1 and a sequence of times t_k such that $|u(t_k, u_{0k})|_0 \ge \varepsilon_0$. This means that u_* is unstable in a much stronger sense than stated in Theorem 5.4.1.

(b) The proof of Theorem 5.4.1 relies heavily on the existence of a spectral gap in the right half-plane. It is an open question whether this spectral gap is essential: does $\sigma(-A_0) \cap \mathbb{C}_+ \ne \emptyset$ already imply instability? The answer is yes, provided A and F have slightly more regularity: $(A, F) \in C^{1+\alpha}$, for some $\alpha > 0$. This can been shown by a result due to D. Henry [140, Theorem 5.1.5].

5.5 Normally Hyperbolic Equilibria

We return to the setting of Section 5.3 for the case that $\sigma(A_0)$ also contains an unstable part, i.e., we now assume that

$$\sigma(A_0) = \{0\} \cup \sigma_s \cup \sigma_u, \quad \text{with } \sigma_s \subset \mathbb{C}_+,\ \sigma_u \subset \mathbb{C}_-, \tag{5.42}$$

such that $\sigma_u \ne \emptyset$. In this situation we can prove the following result.

Theorem 5.5.1. *Let $1 < p < \infty$. Suppose $u^* \in V \cap X_1$ is an equilibrium of* (5.1), *and suppose that the functions (A, F) satisfy* (5.11). *Suppose further that $A(u^*)$ has the property of maximal L_p-regularity. Let A_0 be the linearization of* (5.1) *at u_*. Suppose that u_* is* **normally hyperbolic**, *which means that*

(i) *near u_* the set of equilibria $\mathcal{E}$ is a C^1-manifold in X_1 of dimension $m \in \mathbb{N}_0$,*

(ii) *the tangent space for $\mathcal{E}$ at u_* is isomorphic to $\mathsf{N}(A_0)$,*

(iii) *0 is a semi-simple eigenvalue of A_0, i.e., $\mathsf{N}(A_0) \oplus \mathsf{R}(A_0) = X_0$,*

(iv) *$\sigma(A_0) \cap i\mathbb{R} = \{0\}$, $\sigma_u := \sigma(A_0) \cap \mathbb{C}_- \ne \emptyset$.*

Then u_* is unstable in X_γ. For each sufficiently small $\rho > 0$ there exists $0 < \delta \le \rho$ such that the unique solution $u(t)$ of (5.1) with initial value $u_0 \in B_{X_\gamma}(u_*, \delta)$ either satisfies

- $\mathrm{dist}_{X_\gamma}(u(t_0), \mathcal{E}) > \rho$ *for some finite time $t_0 > 0$, or*
- *$u(t)$ exists on $\mathbb{R}_+$ and converges at an exponential rate to some $u_\infty \in \mathcal{E}$ in X_γ as $t \to \infty$.*

Proof. The first assertion follows from Theorem 5.4.1, so we only need to prove the second claim.
(a) Let P^l denote the spectral projections corresponding to the spectral sets σ_l, where $\sigma_c = \{0\}$ and σ_s, σ_u are as in (5.42). Let $X_j^l = P^l(X_j)$, $l \in \{c, s, u\}$, where these spaces are equipped with the norms of X_j for $j \in \{0, 1, \gamma\}$. We may assume that X_1 is equipped with the graph norm of A_0, i.e., $|v|_1 := |v|_0 + |A_0 v|_0$ for $v \in X_1$. Since the operator $-A_0$ generates an analytic C_0-semigroup on X_0, σ_u is a compact spectral set for A_0. This implies that $P^u(X_0) \subset X_1$. Consequently, X_0^u and X_1^u coincide as vector spaces. In addition, since A_u, the part of A_0 in X_0^u, is invertible, we conclude that the spaces X_j^u carry equivalent norms. We set $X^u := X_0^u = X_1^u$ and equip X^u with the norm of X_0, that is, $X^u = (X^u, |\cdot|_0)$. As in the proof of Theorem 5.3.1 we obtain the decomposition

$$X_1 = X^c \oplus X_1^s \oplus X^u, \quad X_0 = X^c \oplus X_0^s \oplus X^u,$$

and this decomposition reduces A_0 into $A_0 = A_c \oplus A_s \oplus A_u$, where A_l is the part of A_0 in X_0^l for $l \in \{c, s, u\}$. It follows that $\sigma(A_l) = \sigma_l$ for $l \in \{c, s, u\}$. Moreover, due to assumption (iii), $A_c \equiv 0$. In the sequel, as a norm in X_j we take

$$|v|_j = |P^c v| + |P^s v|_j + |P^u v| \quad \text{for} \quad j \in \{0, \gamma, 1\}. \tag{5.43}$$

We remind that the spaces X_j^l have been given the norm of X_0^l for $l \in \{c, u\}$. We also fix constants $\omega \in (0, \inf \mathrm{Re}\, \sigma(-A_u))$ and $M_5 > 0$ such that $|e^{A_u t}| \le M_5 e^{-\omega t}$ for all $t > 0$. Wlog we may take $\omega \le 1$.

(b) Let Φ be the mapping obtained in step (b) of the proof of Theorem 5.3.1, and set $\phi_l(x) := P^l \Phi(x)$ for $l \in \{s, u\}$ and for $x \in B_{X^c}(0, \rho_0)$. Then

$$\phi_l \in C^1(B_{X^c}(0, \rho_0), X_1^l), \quad (\phi_l(0), \phi_l'(0)) = 0 \quad \text{for } l \in \{s, u\}. \tag{5.44}$$

These mappings parameterize the manifold $\mathcal{E}$ of equilibria near u_* via

$$x \mapsto (x + \phi_s(x) + \phi_u(x) + u_*), \quad x \in B_{X^c}(0, \rho_0).$$

We may assume that ρ_0 has been chosen small enough so that

$$|\phi_l'(x)|_{\mathcal{B}(X^c, X_1^l)} \le 1, \quad x \in B_{X^c}(0, \rho_0), \quad l \in \{s, u\}. \tag{5.45}$$

(c) The equilibrium equation (5.16) now corresponds to the system

$$\begin{aligned} &P^c G(x + \phi_s(x) + \phi_u(x)) = 0, \\ &P^l G(x + \phi_s(x) + \phi_u(x)) = A_l \phi_l(x), \quad x \in B_{X^c}(0, \rho_0), \quad l \in \{s, u\}. \end{aligned} \tag{5.46}$$

The canonical variables are

$$x = P^c v, \quad y = P^s v - \phi_s(x), \quad z = P^u v - \phi_u(x)$$

and the canonical form of the system is given by

$$\begin{cases} \dot{x} = T(x, y, z), & x(0) = x_0, \\ \dot{y} + A_s y = R_s(x, y, z), & y(0) = y_0, \\ \dot{z} + A_u z = R_u(x, y, z), & z(0) = z_0. \end{cases} \tag{5.47}$$

Here the functions T, R_s, and R_u are given by

$$\begin{aligned} T(x, y, z) &= P^c\big(G(x + y + z + \phi_s(x) + \phi_u(x)) - G(x + \phi_s(x) + \phi_u(x))\big), \\ R_l(x, y, z) &= P^l\big(G(x + y + z + \phi_s(x) + \phi_u(x)) - G(x + \phi_s(x) + \phi_u(x))\big) \\ &\quad - \phi_l'(x)T(x, y, z), \quad l \in \{s, u\}, \end{aligned} \tag{5.48}$$

where we have used the equilibrium equations (5.46). Clearly,

$$(R_l(x, 0, 0), T(x, 0, 0)) = 0, \quad x \in B_{X^c}(0, \rho_0), \quad l \in \{s, u\},$$

showing that the equilibrium set $\mathcal{E}$ of (5.1) near u_* has been reduced to the set $B_{X^c}(0, \rho_0) \times \{0\} \times \{0\} \subset X^c \times X^s \times X^u$.

There is a unique correspondence between the solutions of (5.1) close to u_* in X_γ and those of (5.47) close to 0. We again call (5.47) the *normal form* of (5.1) near its *normally hyperbolic* equilibrium u_*.

(d) The estimates for R_l and T are similar to those derived in Section 5.3, and we have

$$|T(x, y, z)|, \ |R_l(x, y, z)|_0 \le \beta(|y|_1 + |z|), \tag{5.49}$$

for all $x, z \in \bar{B}_{X^{\tilde{l}}}(0, \rho)$, $\tilde{l} \in \{c, u\}$, and $y \in \bar{B}_{X^s_\gamma}(0, \rho) \cap X_1$, where $\rho \le \rho_0$, $r = 5\rho$, and $\beta = C_2(\eta + Lr)$.

(e) Let us assume for the moment that ρ is chosen so that $4\rho \le \rho_0$. Let $u(t) = u_* + \Phi(x(t)) + y(t) + z(t)$ be a solution of (5.47) on some maximal time interval $[0, t_+)$ which satisfies $\mathrm{dist}_{X_\gamma}(u(t), \mathcal{E}) \le \rho$. Set

$$t_1 := t_1(x_0, y_0, z_0) := \sup\{t \in (0, t_+) : |u(\tau) - u_*|_\gamma \le 3\rho, \ \tau \in [0, t]\}$$

and suppose that $t_1 < t_+$. Assuming w.l.o.g. that the embedding constant of $X_1 \hookrightarrow X_\gamma$ is less or equal to one it follows from (5.43), (5.45) and the definition of t_1 that

$$|x(t)|, |y(t)|_\gamma, |z(t)| \le 3\rho, \quad t \in [0, t_1], \tag{5.50}$$

so that the estimate (5.49) holds for $(x(t), y(t), z(t))$, $t \in [0, t_1]$.

Since $\mathcal{E}$ is a finite-dimensional manifold, for each $u \in B_{X_\gamma}(u_*, 3\rho)$ there is $\bar{u} \in \mathcal{E}$ such that $\mathrm{dist}_{X_\gamma}(u, \mathcal{E}) = |u - \bar{u}|_\gamma$, and by the triangle inequality $\bar{u} \in B_{X_\gamma}(u_*, 4\rho)$. Thus we may write $u = u_* + \Phi(x) + y + z$ and $\bar{u} = u_* + \Phi(\bar{x})$, and therefore

$$\begin{aligned}\rho \geq \mathrm{dist}_{X_\gamma}(u, \mathcal{E}) = |u - \bar{u}|_\gamma & \\ &= |x - \bar{x}| + |y + \phi_s(x) - \phi_s(\bar{x})|_\gamma + |z + \phi_u(x) - \phi_u(\bar{x})| \\ &\geq |x - \bar{x}| + |z| - |\phi_u(x) - \phi_u(\bar{x})| \geq |z|,\end{aligned}$$

since $x, \bar{x} \in B_{X^c}(0, \rho_0)$ and ϕ_s is non-expansive, see (5.45). Therefore we obtain the improved estimate $|z(t)| \leq \rho$ for all $t \in [0, t_1]$.

We begin the estimates with that for the unstable component $z(t)$. Integrating the equation for z backwards yields

$$z(t) = e^{A_u(t_1 - t)} z(t_1) - \int_t^{t_1} e^{A_u(s-t)} R_u(x(s), y(s), z(s))\, ds. \tag{5.51}$$

With (5.49) and $|z(t_1)| \leq \rho$ we get

$$|z(t)| \leq M_5 e^{-\omega(t_1 - t)} \rho + \beta M_5 \int_t^{t_1} e^{-\omega(s-t)} (|y(s)|_1 + |z(s)|)\, ds$$

for $t \in [0, t_1]$. Gronwall's inequality yields

$$|z(t)| \leq M_5 e^{-\omega_1(t_1 - t)} \rho + \beta M_5 \int_t^{t_1} e^{-\omega_1(s-t)} |y(s)|_1\, ds$$

for $t \in [0, t_1]$, where $\omega_1 = \omega - \beta M_5 > 0$ provided β, i.e., η, r are small enough. In particular, with $M_6 = M_5/\omega_1$, this inequality implies

$$|z|_{L_q(J_1;X_0)} \leq M_6 \rho + \beta M_6 |y|_{L_q(J_1;X_1)}, \tag{5.52}$$

where we have set $J_1 = (0, t_1)$; here $q \in [1, \infty]$ is arbitrary at the moment. A similar estimate holds for the time-derivative of z, namely

$$|\dot{z}|_{L_q(J_1;X_0)} \leq (|A_u| + \beta) |z|_{L_q(J_1;X_0)} + \beta |y|_{L_q(J_1;X_1)}. \tag{5.53}$$

Note that

$$\begin{aligned} |z(t+h) - z(t)| &\leq h^{1/p'} |\dot{z}|_{L_p(J_1;X_0)}, \\ \int_0^{t_1 - h} |z(t+h) - z(t)|\, dt &\leq h |\dot{z}|_{L_1(J_1;X_0)}. \end{aligned} \tag{5.54}$$

Next we consider the equation for x. We have

$$\begin{aligned}|x(t)| &\leq |x_0| + \int_0^t |\dot{x}(s)|\, ds = |x_0| + \int_0^t |T(x(s), y(s), z(s))|\, ds \\ &\leq |x_0| + \beta(|y|_{L_1(J_1;X_1)} + |z|_{L_1(J_1;X_0)}).\end{aligned}$$

Combining this estimate with that for z we obtain

$$\sup_{t\in J_1} |x(t)| \le |x_0| + |\dot{x}|_{L_1(J_1;X_0)},$$
$$|\dot{x}|_{L_q(J_1;X_0)} \le \beta(M_6\rho + (1+\beta M_6)|y|_{L_q(J_1;X_1)}).$$

This estimate is best possible and shows that in order to control $|x(t)|$ we must be able to control $|y|_{L_1(J_1;X_1)}$. Note that

$$\begin{aligned} |x(t+h) - x(t)| &\le h^{1/p'}|\dot{x}|_{L_p(J_1;X_0)}, \\ \int_0^{t_1-h} |x(t+h) - x(t)|\,dt &\le h|\dot{x}|_{L_1(J_1;X_0)}. \end{aligned} \tag{5.55}$$

Now we turn to the equation for y, the stable but infinite dimensional part of the problem. As in the proof of Theorem 5.3.1, part (e), we obtain from (5.49)

$$|y|_{\mathbb{E}_1(t_1)} \le M_1|y_0|_\gamma + \beta M_0(|y|_{\mathbb{E}_1(t_1)} + |z|_{\mathbb{E}_0(t_1)}).$$

Employing (5.52) with $q = p$ we get

$$|y|_{\mathbb{E}_1(t_1)} \le M_1|y_0|_\gamma + \beta M_0 M_6\rho + \beta M_0(1+\beta M_6)|y|_{\mathbb{E}_1(t_1)}.$$

Assuming $\beta M_0(1+\beta M_6)) < 1/2$, this yields

$$|y|_{\mathbb{E}_1(t_1)} \le 2M_1|y_0|_\gamma + 2\beta M_0 M_6\rho. \tag{5.56}$$

Repeating the estimates leading up to (5.32) with $\sigma = 0$ we now get

$$|y(t)|_\gamma \le C_5(|y_0|_\gamma + \beta\rho), \quad t \in [0, t_1], \tag{5.57}$$

where C_5 is a constant independent of ρ, y_0 and t_1. In particular, we see that $|y(t)|_\gamma \le \rho$ for all $t \in J_1$, provided $|y_0|_\gamma$ and β, i.e., η and r are sufficiently small.

For later purposes we need an estimate for $|y(t+h) - y(t)|_\gamma$. We have

$$\begin{aligned} |y(t+h) - y(t)|_\gamma &\le C|y(t+h) - y(t)|_0^{1-\gamma}|y(t+h) - y(t)|_1^\gamma \\ &\le Ch^{(1-\gamma)/p'}|\dot{y}|_{L_p(J_1;X_0)}^{1-\gamma}(|y(t+h)|_1^\gamma + y(t)|_1^\gamma) \end{aligned} \tag{5.58}$$

for all $t \in [0, t_1]$, $t+h \in [0, t_1]$ with $y(t+h), y(t) \in X_1$. We remind that $\gamma = 1 - 1/p$.

Unfortunately, this is not enough to keep $|x(t)|$ small on J_1, for this we need to control $|y|_{L_1(J_1;X_1)}$, and we cannot expect maximal regularity in L_1.

To handle $|y|_{L_1(J_1;X_1)}$, we are forced to use another type of maximal regularity, namely that for the vector-valued Besov spaces $B^\alpha_{11}(J_1; X)$, where $\alpha \in (0, 1)$. Before stating the result we remind that

$$|g|_{B^\alpha_{11}(J_1;X)} := |g|_{L_1(J_1;X)} + [g]_{J_1;\alpha,X},$$
$$[g]_{J_1;\alpha,X} := \int_0^{t_1} h^{-\alpha}\int_0^{a-h} |g(t+h) - g(t)|_X\,dtdh/h$$

defines a norm for $g \in B^\alpha_{11}(J_1;X)$, where $J_1 = (0,t_1)$. The maximal regularity result, which is valid for all exponentially stable analytic C_0-semigroups, reads as follows: there is a constant M_7 depending only on A_s and on $\alpha \in (0,1)$ such that the solution y of

$$\dot{y} + A_s y = f, \quad t \in J, \quad y(0) = y_0, \tag{5.59}$$

satisfies the estimate

$$|y|_{B^\alpha_{11}(J;X^s_1)} \le M_7\big(|y_0|_{D_{A_s}(\alpha,1)} + |f|_{B^\alpha_{11}(J;X^s_0)}\big).$$

In our situation, this estimate is a consequence of Corollary 4.5.9 as $B = d/dt + \omega \in \mathcal{RH}^\infty(B^\alpha_{11}(J;X^s_0))$ for any Banach space by Theorem 4.5.4, and $A_s - \omega \in \mathcal{MR}_p(X^s)$ for $\omega > 0$ sufficiently small implies $A_s - \omega \in \mathcal{RS}(X^s)$ by Theorem 4.4.4. Note that this estimate is in particular independent of $J_1 = (0,a)$, by exponential stability of $e^{-A_s t}$. Furthermore we have $y_0 \in X_\gamma \cap X^s = D_{A_s}(1-1/p,p) \hookrightarrow D_{A_s}(\alpha,1)$, provided $\alpha < 1-1/p$. Another parabolic estimate valid for (5.59) that we shall make use of reads

$$|y|_{B^\alpha_{11}(J;X^s_1)} \le M_8\big(|y_0|_{D_{A_s}(\alpha,1)} + |f|_{L_1(J_1;X^s_1)}\big),$$

provided $\alpha < 1$. Here the constant M_8 is also independent of $J_1 = (0,t_1)$. Actually this result is elementary, it only uses analyticity of the semigroup $e^{-A_s t}$.

We set $R_1(t) = -\phi_s'(x(t))T(x(t),y(t),z(t))$ and recall that $|\phi_s'(x(t))|_{\mathcal{B}(X^c,X_1)} \le 1$ for $t \in [0,t_1]$. Employing the L_1-estimate for z, see (5.52), yields

$$\begin{aligned} |R_1|_{L_1(J_1;X_1)} &\le \int_0^{t_1} |T(x(s),y(s),z(s))|\,ds \le \beta(|y|_{L_1(J_1;X_1)} + |z|_{L_1(J_1;X_0)}) \\ &\le \beta M_6\rho + \beta(1+\beta M_6)|y|_{L_1(J_1;X_1)}. \end{aligned}$$

Therefore, for the solution y_1 of (5.59) with $f = R_1$ we obtain

$$|y_1|_{B^\alpha_{1\infty}(J_1;X_1)} \le M_8\big(|y_0|_\gamma + \beta M_6\rho + \beta(1+M_6\beta)|y|_{L_1(J_1;X_1)}\big).$$

Next let $R_2(t) = P^s(G(\Phi(x)+y+z) - G(\Phi(x)))$. Then by estimate (5.28)

$$\begin{aligned} |R_2|_{L_1(J_1;X_0)} &\le \beta(|y|_{L_1(J_1;X_1)} + |z|_{L_1(J_1;X_0)}) \\ &\le \beta M_6\rho + \beta(1+M_6\beta)|y|_{L_1(J_1;X_1)}, \end{aligned}$$

and with some constant C_6

$$\begin{aligned} |R_2(t) - R_2(\bar{t})|_0 &\le C_6\beta\big(|y(t)-y(\bar{t})|_1 + |z(t)-z(\bar{t})| + |x(t)-x(\bar{t})|\big) \\ &\quad + C_6|y(t)|_1\big(|y(t)-y(\bar{t})|_\gamma + |x(t)-x(\bar{t})| + |z(t)-z(\bar{t})|\big). \end{aligned}$$

Hence we obtain the following estimate

$$\begin{aligned} [R_2]_{\alpha,0} \le C_6\beta\{[y]_{\alpha,1} + [z]_{\alpha,0} + [x]_{\alpha,0}\} + C_6 \int_0^{t_1} h^{-\alpha-1} \int_0^{t_1-h} |y(t)|_1 \\ \cdot \big\{|y(t+h)-y(t)|_\gamma + |x(t+h)-x(t)| + |z(t+h)-z(t)|\big\}\,dtdh, \end{aligned}$$

where we set $[\cdot]_{\alpha,j} := [\cdot]_{J_1;\alpha,X_j}$ for $j = 0, 1$. (5.52)–(5.54) yields for each $\alpha \in (0, 1)$

$$[z]_{\alpha,0} \leq |\dot{z}|_{L_1(J_1;X_0)} \leq C_7(\rho + \beta|y|_{L_1(J_1;X_1)}),$$

with some uniform constant C_7. In the same way we may estimate $[x]_{\alpha,0}$. Next we have again by (5.52)–(5.54)

$$\begin{aligned} h^{-\alpha-1} \int_0^{t_1-h} |y(t)|_1 |z(t+h) - z(t)|\, dt &\leq h^{1/p'-\alpha-1} |\dot{z}|_{L_p(J_1;X_0)} |y|_{L_1(J_1;X_1)} \\ &\leq h^{1/p'-\alpha-1} C_8(|y_0|_\gamma + \rho)|y|_{L_1(J_1;X_1)}, \end{aligned}$$

hence its integral over h is finite, provided $\alpha < 1 - 1/p$, and similarly for the corresponding integral containing the x-difference. Last but not least, for $\alpha < (1-\gamma)(1-1/p)$ we have by (5.58)

$$\begin{aligned} &h^{-\alpha-1} \int_0^{t_1-h} |y(t)|_1 |y(t+h) - y(t)|_\gamma\, dt \\ &\leq 2Ch^{(1-\gamma)/p'-\alpha-1} |\dot{y}|^{1-\gamma}_{L_p(J_1;X_0)} |y|_{L_p(J_1;X_1)} |y|^{\gamma}_{L_1(J_1;X_1)} \\ &\leq h^{(1-\gamma)/p'-\alpha-1} C_9(|y_0|_\gamma + \beta\rho)^{2-\gamma} |y|^{\gamma}_{L_1(J_1;X_1)} \\ &\leq h^{(1-\gamma)/p'-\alpha-1} C_{10}\big((|y_0|_\gamma + \beta\rho)^2 + (|y_0|_\gamma + \beta\rho)|y|_{L_1(J_1;X_1)}\big), \end{aligned}$$

where we used Young's inequality in the last line.

Collecting now all terms and choosing $\alpha = (1-\gamma)/2p' = 1/2pp'$, we find a uniform constant C_{11} such that for $|y_0|_\gamma \leq \delta$

$$|y|_{B^\alpha_{11}(J_1;X^s_1)} \leq C_{11}\big(|y_0|_\gamma + \beta\rho + (\beta + \rho + \delta)|y|_{B^\alpha_{11}(J_1;X^s_1)}\big),$$

hence

$$|y|_{L_1(J_1;X^s_1)} \leq |y|_{B^\alpha_{11}(J_1;X^s_1)} \leq 2C_{11}(|y_0|_\gamma + \beta\rho), \tag{5.60}$$

provided $C_{11}(\beta + \rho + \delta) < 1/2$. Choosing now first β, i.e., η and r small enough, and then ρ and $\delta > 0$, we see that $|u(t_1) - u_*|_\gamma < 3\rho$, a contradiction to $t_1 < t_+$. As in (e) of the proof of Theorem 5.3.1 we may then conclude that $t_+ = \infty$, which means that the solution exists globally and stays in the ball $\bar{B}_{X_\gamma}(u_*, 3\rho)$.

(f) To prove convergence, let $(x(t), y(t), z(t))$ be a global solution of (5.47) that satisfies

$$|x(t)|, |y(t)|_\gamma, |z(t)| \leq 3\rho, \quad \text{for all } t \geq 0,$$

see (5.50). Similarly to the proof of Theorem 5.3.1, part (e), we obtain from (5.49)

$$|e^{\omega t} y|_{\mathbb{E}_1(\infty)} \leq 2M_1|y_0|_\gamma + 2\beta M_0 |e^{\omega t} z|_{\mathbb{E}_0(\infty)}, \tag{5.61}$$

where $\omega \in (0, \inf\{\mathrm{Re}\,\lambda : \lambda \in \sigma(A_s)\})$ is a fixed number and β is given in (5.30). Repeating the estimates leading up to (5.32) we get

$$|e^{\omega t} y(t)|_\gamma \leq M_2|y_0|_\gamma + 2\beta c_0 M_0 |e^{\omega t} z|_{\mathbb{E}_0(\infty)}, \quad t \geq 0. \tag{5.62}$$

From equation (5.51) we infer that

$$z(t) = -\int_t^\infty e^{-A_u(t-s)} R_u(x(s), y(s), z(s))\,ds, \quad t \geq 0, \tag{5.63}$$

since $|z(t_1)| \leq \rho$ for each $t_1 > 0$ and $e^{A_u(t_1-t)}$ is exponentially decaying for $t_1 \to \infty$. Using (5.63) and the estimate for R_u from (5.49) and proceeding as in the proof of Young's inequality for convolution integrals one shows that

$$|e^{\omega t} z|_{\mathbb{E}_0(\infty)} \leq C_{12}\beta\big(|e^{\omega t} y|_{\mathbb{E}_1(\infty)} + |e^{\omega t} z|_{\mathbb{E}_0(\infty)}\big). \tag{5.64}$$

Making β sufficiently small (by decreasing η and, accordingly, r) it follows from (5.61) and (5.64) that

$$|e^{\omega t} y|_{\mathbb{E}_1(\infty)} + |e^{\omega t} z|_{\mathbb{E}_0(\infty)} \leq C_{13}|y_0|_\gamma.$$

This estimate in turn, together with (5.62), implies $|y(t)|_\gamma \to 0$ and $|z(t)| \to 0$ exponentially fast as $t \to \infty$. As in the proof of Theorem 5.3.1 part (f) we get

$$x(t) \to x_\infty := x_0 + \int_0^\infty T(x(s), y(s), z(s))\,ds.$$

This yields existence of the limit

$$u_\infty = u_* + v_\infty := u_* + \lim_{t\to\infty} v(t) = u_* + x_\infty + \phi_s(x_\infty) + \phi_u(x_\infty) \in \mathcal{E}.$$

Similar arguments as in (f) of the proof of Theorem 5.3.1 yield exponential convergence of $u(t)$ to u_∞. □

5.6 The Stable and Unstable Foliations

Our intention in this section is to study the behaviour of the semiflow near $\mathcal{E}$ in more detail. If u_* is normally hyperbolic, then any $w \in \mathcal{E}$ close to u_* in X_γ will be normally hyperbolic as well. Therefore, intuitively, at each point $w \in \mathcal{E}$ near u_* there should be a stable manifold $\mathcal{M}^s_w$ and an unstable manifold $\mathcal{M}^u_w$ such that $\mathcal{M}^s_w \cap \mathcal{M}^u_w \cap B(u_*, r) = \{w\}$, and these manifolds should depend continuously on $w \in B(u_*, r) \cap \mathcal{E}$. The tangent spaces of these manifolds are expected to be the projections with respect to the stable resp. unstable part of the spectrum of the linearization of (5.1) at w.

We prove these assertions below, and call it the **stable** resp. **unstable foliation** of (5.1) near $u_* \in \mathcal{E}$. The stable resp. unstable manifolds $\mathcal{M}^s_w$ and $\mathcal{M}^u_w$ are termed the **leaves** or **fibers** of these foliations. They turn out to be positively and also negatively invariant under the semiflow. As a consequence, the convergent solutions are precisely those which start at initial values sitting on one of the stable fibers.

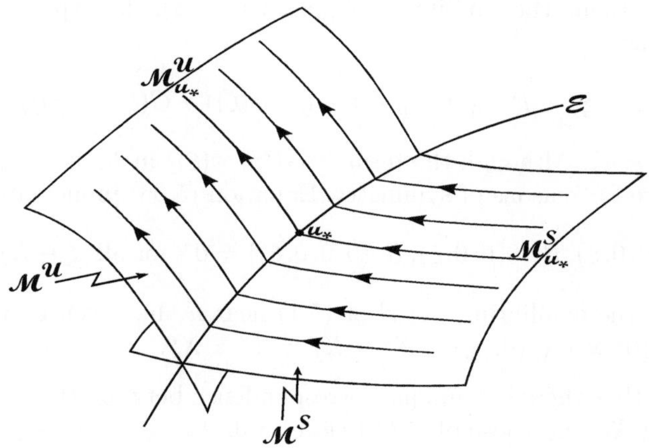

Figure 5.1: Foliation and fibers near $\mathcal{E}$.

In the normally stable case where $\sigma(A_0) \cap \mathbb{C}_- = \emptyset$, this implies that the stable foliation covers a neighbourhood of u_* in V.

The fibers of these foliations will be manifolds of class C^1 as well, however, their dependence on $w \in \mathcal{E}$ is only continuous: here we lose one degree of regularity. This is quite natural and should be compared to the loss of regularity for the normal field ν_Γ of a hypersurface $\Gamma \subset \mathbb{R}^n$. We also show that in case $(A, F) \in C^k$ the dependence on w is of class C^{k-1}, for each $k \in \mathbb{N} \cup \{\infty, \omega\}$, where ω means real analytic.

Following the notation of the previous section, we introduce the new variables

$$\begin{aligned} x &= P^c v - \xi = P^c(u - u_*) - \xi, \\ y &= P^s v - \phi_s(\xi) = P^s(u - u_*) - \phi_s(\xi), \\ z &= P^u v - \phi_u(\xi) = P^u(u - u_*) - \phi_u(\xi), \end{aligned}$$

where $\xi \in B_{X^c}(0, \rho_0)$, to obtain the following system of evolution equations in $X^c \times X_0^s \times X^u$:

$$\begin{cases} \dot{x} = R_c(x, y, z, \xi), & x(0) = x_0 - \xi, \\ \dot{y} + A_s y = R_s(x, y, z, \xi), & y(0) = y_0 - \phi_s(\xi), \\ \dot{z} + A_u z = R_u(x, y, z, \xi), & z(0) = z_0 - \phi_u(\xi), \end{cases} \tag{5.65}$$

with $x_0 = P^c v_0$, $y_0 = P^s v_0$, $z_0 = P^u v_0$, and

$$\begin{aligned} R_c(x, y, z, \xi) &= P^c G(x + y + z + \xi + \phi(\xi)), \\ R_l(x, y, z, \xi) &= P^l G(x + y + z + \xi + \phi(\xi)) - A_l \phi_l(\xi) \end{aligned}$$

for $l \in \{s, u\}$. Using the equilibrium equations (5.24), the expressions for R_l can be rewritten as

$$R_l(x, y, z, \xi) = P^l\big(G(x + y + z + \xi + \phi(\xi)) - G(\xi + \phi(\xi))\big), \tag{5.66}$$

where $l \in \{c, s, u\}$. Although the term $P^c G(\xi + \phi(\xi))$ in R_c is zero, see (5.24), we include it here for reasons of symmetry. Equation (5.66) immediately yields

$$\big(R_c(0,0,0,\xi), R_s(0,0,0,\xi), R_u(0,0,0,\xi)\big) = 0 \quad \text{for all} \quad \xi \in B_{X^c}(0, \rho_0),$$

showing that the equilibrium set $\mathcal{E}$ of (5.1) near u_* has been reduced to the set $\{0\} \times \{0\} \times \{0\} \times B_{X^c}(0, \rho_0) \subset X^c \times X_1^s \times X^u \times X^c$.

Observe also that there is a unique correspondence between the solutions of (5.1) close to u_* in X_γ and those of (5.65) close to 0. As $u_\infty := u_* + \xi + \phi(\xi) \in \mathcal{E}$ will be the limit of $u(t)$ in X_γ as $t \to \infty$, we call system (5.65) the *asymptotic normal form* of (5.1) near its normally hyperbolic equilibrium u_*.

6.1 The Stable Foliation

To motivate our approach for the construction of the stable foliation we formally define a map H_s according to

$$H_s((x, y, z), (y_0, \xi))(t) = \begin{bmatrix} x(t) + \int_t^\infty R_c(x(\tau), y(\tau), z(\tau), \xi)\, d\tau \\ y(t) - L_s(R_s(x, y, z, \xi), y_0 - \phi_s(\xi)) \\ z(t) + \int_t^\infty e^{-A_u(t-\tau)} R_u(x(\tau), y(\tau), z(\tau), \xi)\, d\tau \end{bmatrix}, \tag{5.67}$$

where $t > 0$. Here $w = L_s(f, y_0)$ denotes the unique solution of the problem

$$\dot{w}(t) + A_s w(t) = f(t), \quad t > 0, \quad w(0) = y_0.$$

Obviously, we have $H_s(0, 0) = 0$. Moreover, H_s will be of class C^1 w.r.t. to the variables (x, y, z, y_0), but in general only continuous in ξ. The derivative of H_s w.r.t. (x, y, z) at (0,0) is given by $D_{(x,y,z)} H_s(0, 0) = I$, and hence the implicit function theorem formally applies and yields a map

$$\Lambda^s : (y_0, \xi) \mapsto (x, y, z) \tag{5.68}$$

which is well-defined near 0, such that $H_s(\Lambda^s(y_0, \xi), (y_0, \xi)) = 0$. Λ^s will be continuous in (y_0, ξ), and of class C^1 w.r.t. y_0. Given $(x, y, z) = \Lambda^s(y_0, \xi)$, we set

$$\begin{aligned} x_0 &:= -\int_0^\infty R_c(x(\tau), y(\tau), z(\tau), \xi)\, d\tau + \xi, \\ z_0 &:= -\int_0^\infty e^{A_u \tau} R_u(x(\tau), y(\tau), z(\tau), \xi)\, d\tau + \phi_u(\xi). \end{aligned} \tag{5.69}$$

For $(x, y, z) = \Lambda^s(y_0, \xi)$ given, the first component of H_s yields

$$\dot{x}(t) = R_c(x(t), y(t), z(t), \xi), \quad t > 0,$$

the second component implies

$$\dot{y}(t) + A_s y(t) = R_s(x(t), y(t), z(t), \xi), \quad t > 0, \quad y(0) = y_0 - \phi_s(\xi),$$

and the third one leads to

$$\dot{z}(t) + A_u z(t) = R_u(x(t), y(t), z(t), \xi), \quad t > 0.$$

Moreover, due to (5.69), the initial values of x and z are given by

$$x(0) = x_0 - \xi, \quad z(0) = z_0 - \phi_u(\xi).$$

Assuming, in addition, that $(x(t), y(t), z(t))$ converges to $(0, 0, 0)$ as $t \to \infty$, we conclude that

$$u(t) = u_* + x(t) + y(t) + z(t) + \xi + \phi(\xi), \quad t > 0,$$

is a solution of (5.1) with $\lim_{t\to\infty} u(t) = u_\infty := u_* + \xi + \phi(\xi) \in \mathcal{E}$. The map

$$\lambda^s : (y_0, \xi) \mapsto u(0) = u_* + x(0) + y(0) + z(0) + \xi + \phi(\xi) \tag{5.70}$$

yields a foliation of the stable manifold $\mathcal{M}^s$ near u_*, and

$$\mathcal{M}^s_\xi := \{\lambda^s(y_0, \xi) : y_0 \in B_{X^s}(0, r)\} \tag{5.71}$$

are the fibers over $B_{X^c}(0, r)$, or equivalently over $\mathcal{E}$ near u_*. We note that the fibers are C^1-manifolds, but they will depend only continuously on ξ, or equivalently on $\mathcal{E}$.

The strategy of our approach can be summarized as follows: given a base point $u_*+\xi+\phi(\xi)$ on the manifold $\mathcal{E}$, and an initial value $y_0 \in X^s_\gamma$, we determine with the help of the implicit function theorem an initial value u_0 and a solution $u(t)$ such that $u(t)$ converges to the base point $u_* + \xi + \phi(\xi)$ exponentially fast. Exponential convergence will be obtained by setting up the implicit function theorem in a space of exponentially decaying functions.

After these heuristic considerations we can state our first main result, employing the notation introduced above.

Theorem 5.6.1. *Consider* (5.1) *under the assumption* (5.11), *and let* $u_* \in \mathcal{E}$ *be a normally hyperbolic equilibrium. Suppose further that* $A(u_*)$ *has the property of maximal* L_p*-regularity. Then there is a number* $r > 0$ *and a continuous map*

$$\lambda^s : B_{X^s_\gamma}(0, r) \times B_{X^c}(0, r) \to X_\gamma \quad \textit{with} \quad \lambda^s(0, 0) = u_*,$$

the **stable foliation**, *such that the solution $u(t)$ of (5.1) with initial value $\lambda^s(y_0,\xi)$ exists on $\mathbb{R}_+$ and converges to $u_\infty := u_* + \xi + \phi(\xi)$ in X_γ exponentially fast as $t \to \infty$. The image of λ^s defines the stable manifold $\mathcal{M}^s$ of (5.1) near u_*.*

Furthermore, for fixed $\xi \in B_{X^c}(0,r)$, the function

$$\lambda^s_\xi : B_{X^s_\gamma}(0,r) \to X_\gamma, \quad \textit{given by} \quad \lambda^s_\xi(y_0) = \lambda^s(y_0,\xi),$$

defines the fibers $\mathcal{M}^s_\xi := \lambda^s_\xi(B_{X^s_\gamma}(0,r))$ of the foliation. Moreover, we have

(i) *an initial value $u_0 \in X_\gamma$ near u_* belongs to $\mathcal{M}^s$ if and only if the solution $u(t)$ of (5.1) exists globally on $\mathbb{R}_+$ and converges to some $u_\infty \in \mathcal{E}$ exponentially fast as $t \to \infty$;*

(ii) *λ^s_ξ is of class C^1, and the derivative $D_{y_0}\lambda^s$ is continuous, jointly in (y_0,ξ);*

(iii) *the fibers are C^1-manifolds which are invariant under the semiflow generated by (5.1);*

(iv) *the tangent space of $\mathcal{M}^s_{u_\infty}$ at $u_\infty \in \mathcal{E}$ is precisely the projection of the stable part of the linearization of (5.1) at u_∞;*

(v) *if u_* is normally stable, then $\mathcal{M}^s$ forms a neighbourhood of u_* in X_γ.*

Proof. For fixed $\sigma \in (0,\omega]$ we define the function spaces

$$\begin{aligned}
\mathbb{F}^l_0(\sigma) : &= \{f : e^{\sigma t} f \in L_p(\mathbb{R}_+; X^l_0)\},\\
\mathbb{F}^l_1(\sigma) : &= \{w : e^{\sigma t} w \in H^1_p(\mathbb{R}_+; X^l_0) \cap L_p(\mathbb{R}_+; X^l_1)\}, \quad l \in \{c,s,u\},
\end{aligned}$$

equipped with their natural norms. Then we set

$$\mathbb{Y} := \mathbb{Y}(\sigma) := \mathbb{F}^c_1(\sigma) \times \mathbb{F}^s_1(\sigma) \times \mathbb{F}^u_1(\sigma), \quad Z := X^s_\gamma \times X^c,$$

and we define $H_s : B_{\mathbb{Y}}(0,\rho) \times B_Z(0,\rho) \to \mathbb{Y}$ by (5.67). Observe that H_s is the composition of the substitution operator

$$R(x,y,z,\xi) = [(R_c,R_s,R_u)(x+y+z+\xi+\phi(\xi)), \phi_s(\xi)]^{\mathsf{T}},$$

which maps $B_{\mathbb{Y}}(0,\rho) \times B_Z(0,\rho)$ into $\mathbb{F}^c_0(\sigma) \times \mathbb{F}^s_0(\sigma) \times \mathbb{F}^u_0(\sigma) \times X^s_\gamma$, and the bounded linear operator

$$\mathbb{L}(x,y,z,y_0,\xi,R_1,R_2,R_3,R_4) = \begin{bmatrix} x(t) + \int_t^\infty R_1(\tau), d\tau \\ y(t) - L_s(R_2, y_0 - R_4) \\ z(t) + \int_t^\infty e^{-A_u(t-\tau)} R_3(\tau)\, d\tau \end{bmatrix},$$

which maps

$$\mathbb{Y} \times Z \times \mathbb{F}^c_0(\sigma) \times \mathbb{F}^s_0(\sigma) \times \mathbb{F}^u_0(\sigma) \times X^s_\gamma \quad \text{into} \quad \mathbb{Y}.$$

In order to see that the integral operator $[R_1 \mapsto \int_t^\infty R_1(\tau)\,d\tau]$ maps $\mathbb{F}_0^c(\sigma)$ into $\mathbb{F}_1^c(\sigma)$, we set

$$(KR_1)(t) := \int_t^\infty R_1(\tau)\,d\tau, \quad R_1 \in \mathbb{F}_0^c(\sigma).$$

Clearly, $e^{\delta t}(Kg)(t) = \int_t^\infty e^{\delta(t-\tau)} e^{\delta\tau} g(\tau)\,d\tau$, and Young's inequality for convolution integrals readily yields

$$K \in \mathcal{B}(\mathbb{F}_0^c(\sigma), \mathbb{F}_1^c(\sigma)).$$

Similar arguments also apply for the term $\int_t^\infty e^{-A_u(t-\tau)} R_3(\tau)\,d\tau$.

As G is of class C^1, it is not difficult to see that R is C^1 with respect to the variables (x, y, z, y_0), but in general only continuous with respect to ξ. This, in turn, implies that H_s is of class C^1 w.r.t. (x, y, z, y_0), and continuous in ξ. In addition, as $(G(0), G'(0)) = 0$, we obtain

$$H_s(0,0) = 0, \quad D_{(x,y,z)} H_s(0,0) = I_{\mathbb{Y}}.$$

Therefore, by the implicit function theorem, there is a radius $r > 0$ and a continuous map $\Lambda^s : B_Z(0,r) \to \mathbb{Y}$ such that

$$H_s(\Lambda^s(y_0,\xi), (y_0,\xi)) = 0, \quad \text{for all } (y_0,\xi) \in B_Z(0,r),$$

and there is no other solution of $H_s((x,y,z),(y_0,\xi)) = 0$ in the ball $B_{\mathbb{Y}}(0,r) \times B_Z(0,r)$. Moreover, Λ^s is also C^1 w.r.t. y_0, and one shows that

$$D_{y_0}\Lambda^s(0,0)w_0 = [0, e^{-tA_s}w_0, 0]^{\mathsf{T}}. \tag{5.72}$$

We may now continue as indicated in the heuristic considerations preceding Theorem 5.6.1 to define the stable manifold $\mathcal{M}^s$ and their fibers $\mathcal{M}^s_\xi$ near u_*, see (5.70)–(5.71). Clearly, the fibers are positively and negatively invariant under the semiflow, as (5.1) is invariant concerning time-translations. As λ^s_ξ is of class C^1, it is clear that the fibers are C^1-manifolds, parameterized over X^s_γ. In fact, the fibers are diffeomorphic, which can be seen by interchanging the roles of u_* and $u_\infty = u_* + \xi + \phi(\xi)$.

It follows from (5.72) that $D_{y_0}\lambda^s(0,0) = I_{X^s_\gamma}$. Interchanging the roles of u_* and u_∞, it becomes clear that the tangent space of the fiber $\mathcal{M}^s_\xi$ at 0 is precisely the projection of the stable part of the linearization $A_\infty = A(u_\infty) + [A'(u_\infty)\cdot]u_\infty - F'(u_\infty)$ of (5.1) at $u_\infty \in \mathcal{E}$, yielding assertion (iv).

To obtain the characterization of $\mathcal{M}^s$, observe that we proved that there are balls $B_{X_\gamma}(u_*, r_0)$ and $B_{X_\gamma}(u_*, r_1)$ such that any solution of (5.1) starting in $B_{X_\gamma}(u_*, r_0)$ and staying near $\mathcal{E}$ stays in $B_{X_\gamma}(u_*, r_1)$ and converges to an equilibrium $u_\infty \in \mathcal{E}$ exponentially fast. This implies that its initial value must belong to $\mathcal{M}^s$ by uniqueness of Λ^s.

If $u_* \in \mathcal{E}$ is normally stable, then there are balls $B_{X_\gamma}(u_*, r_0)$ and $B_{X_\gamma}(u_*, r_1)$ such that any solution of (5.1) starting in $B_{X_\gamma}(u_*, r_0)$ stays in $B_{X_\gamma}(u_*, r_1)$ and converges to an equilibrium $u_\infty \in \mathcal{E}$ exponentially fast. This implies that $\mathcal{M}^s$ forms a neighbourhood of u_* in X_γ, thus establishing assertion (v) of the theorem. □

6.2 The Unstable Foliation

Our second main result concerning the unstable foliation of $\mathcal{E}$ reads as follows.

Theorem 5.6.2. *Consider* (5.1) *under the assumption* (5.11), *and let* $u_* \in \mathcal{E}$ *be a normally hyperbolic equilibrium. Suppose further that* $A(u_*)$ *has the property of maximal* L_p*-regularity. Then there is a number* $r > 0$ *and a continuous map*

$$\lambda^u : B_{X^u}(0,r) \times B_{X^c}(0,r) \to X_\gamma \quad \textit{with} \quad \lambda^u(0,0) = u_*,$$

the **unstable foliation**, *such that the solution* $u(t)$ *of* (5.1) *with initial value* $\lambda^u(y_0,\xi)$ *exists on* $\mathbb{R}_-$ *and converges to* $u_\infty := u_* + \xi + \phi(\xi)$ *in* X_γ *exponentially fast as* $t \to -\infty$. *The image of* λ^u *defines the unstable manifold* $\mathcal{M}^u$ *of* (5.1) *near* u_*.

Furthermore, for fixed $\xi \in B_{X^c}(0,r)$, *the function*

$$\lambda^u_\xi : B_{X^u}(0,r) \to X_\gamma, \quad \textit{given by} \quad \lambda^u_\xi(y_0) = \lambda^u(y_0,\xi),$$

defines the fibers $\mathcal{M}^u_\xi := \lambda^u_\xi(B_{X^u}(0,r))$ *of the foliation. Moreover, we have*

(i) *an initial value* $u_0 \in X_\gamma$ *near* u_* *belongs to* $\mathcal{M}^u$ *if and only if the solution* $u(t)$ *of* (5.1) *exists globally on* $\mathbb{R}_-$ *and converges to some* $u_\infty \in \mathcal{E}$ *exponentially fast as* $t \to -\infty$;

(ii) λ^u_ξ *is of class* C^1, *and the derivative* $D_{y_0}\lambda^u$ *is continuous, jointly in* (y_0,ξ);

(iii) *the fibers are* C^1*-manifolds which are invariant under the semiflow generated by* (5.1);

(iv) *the tangent space of* $\mathcal{M}^u_{u_\infty}$ *at* $u_\infty \in \mathcal{E}$ *is precisely the projection of the unstable part of the linearization of* (5.1) *at* u_∞.

Proof. For fixed $\sigma \in (0,\omega]$ we define the function spaces

$$\begin{aligned}
\mathbb{F}^l_0(\sigma) &:= \{f : e^{-\sigma t} f \in L_p(\mathbb{R}_-; X^l_0)\},\\
\mathbb{F}^l_1(\sigma) &:= \{w : e^{-\sigma t} w \in H^1_p(\mathbb{R}_-; X^l_0) \cap L_p(\mathbb{R}_-; X^l_1)\}, \quad l \in \{c,s,u\},
\end{aligned}$$

equipped with their natural norms. We set

$$\mathbb{Y} := \mathbb{Y}(\sigma) := \mathbb{F}^c_1(\sigma) \times \mathbb{F}^s_1(\sigma) \times \mathbb{F}^u_1(\sigma), \quad Z := X^u_\gamma \times X^c,$$

and we define $H_u : B_{\mathbb{Y}}(0,\rho) \times B_Z(0,\rho) \to \mathbb{Y}$ by

$$H_u((x,y,z),(z_0,\xi))(t) = \begin{bmatrix} x(t) - \int_{-\infty}^t R_c(x(\tau),y(\tau),z(\tau),\xi)\,d\tau \\ y(t) - \int_{-\infty}^t e^{-A_s(t-\tau)} R_s(x(\tau),y(\tau),z(\tau),\xi)\,d\tau \\ z(t) - L_u(R_u(x,y,z,\xi), z_0 - \phi_u(\xi)) \end{bmatrix},$$

where $t < 0$ and $w = L_u(f,z_0)$ denotes the unique solution of the backward problem

$$\dot w(t) + A_u w(t) = f(t),\ t \le 0, \quad w(0) = z_0.$$

Again, we have $H_u(0,0)=0$, H_u is of class C^1 w.r.t. (x,y,z,z_0), but only continuous in ξ, and $D_{(x,y,z)}H_u(0,0)=I$. As in the previous proof, the implicit function theorem yields a map $\Lambda^u:(z_0,\xi)\mapsto(x,y,z)$ such that $H_u(\Lambda^u(z_0,\xi),(z_0,\xi))=0$. Then

$$\lambda^u:(z_0,\xi)\mapsto u(0)=u_*+v(0)$$

yields the foliation of the unstable manifold $\mathcal{M}^u$ near u_*, and

$$\mathcal{M}^u_\xi:=\{\lambda^u(z_0,\xi):z_0\in B_{X^u}(0,r)\}$$

are the fibers over $B_{X^c}(0,r)$, or equivalently over $\mathcal{E}$ near u_*. Note that the fibers $\mathcal{M}^u_\xi$ are C^1-manifolds, but they depend only continuously on ξ or equivalently on $\mathcal{E}$. The remainder of the proof then follows along similar lines as that of Theorem 5.6.1 and is therefore left to the interested reader. □

Note that the fibers can be extended to global fibers following the backward or forward flow as long as it exists. Laterally, i.e., along the direction of $\mathcal{E}$, the foliation can be extended up to equilibria $u_\# \in \mathcal{E}$ which are no longer normally hyperbolic.

Concerning regularity of the foliation we note that the implicit function theorem yields the following result.

Corollary 5.6.3. *Under the assumptions of Theorem* 5.6.1 *and Theorem* 5.6.2 *the following regularity result is valid: if*

$$(A,F)\in C^k(V,\mathcal{B}(X_1,X_0)\times X_0),$$

then the foliations satisfy $\lambda^l\in C^{k-1}$ *and the fibers* λ^l_ξ *are of class* C^k, *for all* $k\in\mathbb{N}\cup\{\infty,\omega\}$, *where* $l\in\{s,u\}$ *and* C^ω *means real analytic.*

Proof. It follows from the regularity assumptions that $G\in C^k(V,X_0)$. We can then conclude from the second line in (5.46) that $\phi\in C^k(B_{X^c}(0,\rho_0),X_1^s\oplus X_1^u)$. Indeed, for this it suffices to observe that ϕ is implicitly defined by the second line in (5.46), as A_s and A_u are invertible. The assertions follow now from the proof of Theorems 5.6.1 and 5.6.2. □

5.7 Compactness and Long-Time Behaviour

7.1 Relative Compactness of Orbits

Let $u_0\in V_\mu$ be given. Suppose that (A,F) satisfies (5.2) and $A(v)$ has maximal L_p-regularity for all $v\in V_\mu$ and for some $\mu\in(1/p,1)$. In the sequel we assume that the unique solution of (5.1) satisfies $u\in C_b([\tau,t_+(u_0));V_\mu\cap X_{\gamma,\bar\mu})$ for some $\tau\in(0,t_+(u_0))$, $1\ge\bar\mu>\mu$ and

$$\operatorname{dist}_{X_{\gamma,\mu}}(u(t),\partial V_\mu)\ge\eta>0 \tag{5.73}$$

for all $t \in J(u_0) = [0, t_+(u_0))$. Suppose furthermore that the embedding

$$X_{\gamma,\bar{\mu}} \hookrightarrow X_{\gamma,\mu} \quad \text{is compact.} \tag{5.74}$$

It follows from the boundedness of $u(t)$ in $X_{\gamma,\bar{\mu}}$ that the set $\{u(t)\}_{t\in J(u_0)} \subset V_\mu$ is relatively compact in $X_{\gamma,\mu}$. (5.73) implies that $\mathcal{V} := \overline{\{u(t)\}}_{t\in J(u_0)}$ is a compact subset of V_μ. Applying Theorem 5.1.1 we find for each $v \in \mathcal{V}$ numbers $\varepsilon(v) > 0$ and $\delta(v) > 0$ such that $B_{X_{\gamma,\mu}}(v, \varepsilon(v)) \subset V_\mu$ and all solutions of (5.1) which start in $B_{X_{\gamma,\mu}}(v, \varepsilon(v))$ have the common interval of existence $[0, \delta(v)]$. Therefore, the set

$$\bigcup_{v\in\mathcal{V}} B_{X_{\gamma,\mu}}(v, \varepsilon(v))$$

is an open covering of $\mathcal{V}$ and by compactness of $\mathcal{V}$ there exist $N \in \mathbb{N}$ and $v_k \in \mathcal{V}$, $k = 1, \ldots, N$, such that

$$\mathcal{U} := \bigcup_{k=1}^{N} B_{X_{\gamma,\mu}}(v_k, \varepsilon_k) \supset \mathcal{V} = \overline{\{u(t)\}}_{t\in J(u_0)} \supset \{u(t)\}_{t\in J(u_0)},$$

where $\varepsilon_k := \varepsilon(v_k)$, $k = 1, \ldots, N$. To each of these balls corresponds an interval of existence $[0, \delta_k]$, $\delta_k := \delta(v_k) > 0$, $k = 1, \ldots, N$. Consider the problem

$$\dot{v} + A(v)v = F(v), \; s > 0, \quad v(0) = u(t), \tag{5.75}$$

where $t \in J(u_0)$ is fixed and let $\delta := \min\{\delta_k, \; k = 1, \ldots, N\}$. Since $u(t) \subset \mathcal{U}$, $t \in J(u_0)$, the solution of (5.75) exists at least on the interval $[0, \delta]$. By uniqueness it holds that $v(s) = u(t + s)$ if $t + s \in J(u_0)$, $t \in J(u_0)$, $s \in [0, \delta]$, hence $\sup J(u_0) = +\infty$, i.e., the solution exists globally.

By continuous dependence on the initial data, the solution operator $G_1 : \mathcal{U} \to \mathbb{E}_{1,\mu}(0, \delta)$, which assigns to each initial value $u_1 \in \mathcal{U}$ the unique solution $v(\cdot, u_1) \in \mathbb{E}_{1,\mu}(0, \delta)$, is continuous. Furthermore,

$$(\delta/2)^{1-\mu}|v|_{\mathbb{E}_1(\delta/2,\delta)} \le |v|_{\mathbb{E}_{1,\mu}(\delta/2,\delta)} \le |v|_{\mathbb{E}_{1,\mu}(0,\delta)}, \; \mu \in (1/p, 1),$$

wherefore the mapping $G_2 : \mathbb{E}_{1,\mu}(0, \delta) \to \mathbb{E}_1(\delta/2, \delta)$ with $v \mapsto v$ is continuous. Finally

$$|v(\delta)|_{X_\gamma} \le |v|_{C_b((\delta/2,\delta);X_\gamma)} \le C(\delta)|v|_{\mathbb{E}_1(\delta/2,\delta)},$$

hence the mapping $G_3 : \mathbb{E}_1(\delta/2, \delta) \to X_\gamma$ with $v \mapsto v(\delta)$ is continuous. This yields the continuity of the composition $G = G_3 \circ G_2 \circ G_1 : \mathcal{U} \to X_\gamma$, whence $G(\{u(t)\}_{t\ge 0}) = \{u(t + \delta)\}_{t\ge 0}$ is relatively compact in X_γ, since the continuous image of a relatively compact set is relatively compact. Since the solution has relatively compact range in X_γ, it is an easy consequence that the ω-limit set

$$\omega(u_0) := \{v \in V_\mu \cap X_\gamma : \; \exists\, t_n \nearrow \infty \text{ s.t. } u(t_n; u_0) \to v \text{ in } X_\gamma\} \subset X_\gamma$$

is nonempty, connected and compact. We summarize the preceding considerations in the following

Theorem 5.7.1. *Suppose that $A(v)$ has maximal L_p-regularity for all $v \in V_\mu$ and let* (5.2) *as well as* (5.74) *hold for some fixed $\mu \in (1/p, 1)$ and some fixed $\bar{\mu} \in (\mu, 1]$. Assume furthermore that the solution $u(t)$ of* (5.1) *satisfies*

$$u \in C_b([\tau, t_+(u_0)); V_\mu \cap X_{\gamma,\bar{\mu}})$$

for some $\tau \in (0, t_+(u_0))$, and

$$\operatorname{dist}_{X_{\gamma,\mu}}(u(t), \partial V_\mu) \geq \eta > 0$$

for all $t \in J(u_0)$. Then the solution exists globally, and for each $\delta > 0$ the orbit $\{u(t)\}_{t \geq \delta}$ is relatively compact in X_γ. If, in addition, $u_0 \in V_\mu \cap X_\gamma$, then $\{u(t)\}_{t \geq 0}$ is relatively compact in X_γ.

7.2 Long-Time Behaviour of Solutions

We want to extend the local result on the qualitative behaviour near a normally stable or normally hyperbolic equilibrium to a global one, under the slightly stronger assumption $(A, F) \in C^1(V_\mu; \mathcal{B}(X_1, X_0) \times X_0)$ for some $\mu \in (1/p, 1)$ and provided that (5.74) holds. Assume that $u \in C_b([\tau, \infty); X_{\gamma,\bar{\mu}})$ is a global solution to (5.1), satisfying

$$\operatorname{dist}_{X_{\gamma,\mu}}(u(t), \partial V_\mu) \geq \eta > 0$$

for all $t \geq 0$. The mapping $(t, u_1) \mapsto S(t)u_1$, defined by $S(t)u_1 = u(t, u_1)$, $t \geq 0$, $u_1 \in V_{\mu,\gamma}$ defines a local semiflow in $V_\mu \cap X_\gamma$. Let $\Phi \in C(V_\mu \cap X_\gamma; \mathbb{R})$ be a strict *Lyapunov function* for $\{S(t)\}_{t \geq 0}$, which means that the function $t \mapsto \Phi(S(t)u_0)$ is strictly decreasing along non-constant solutions. Theorem 5.7.1 yields that the orbit $\{u(t)\}_{t \geq 0}$ is relatively compact in X_γ. Hence the ω-limit set

$$\omega(u_0) = \{v \in V_{\mu,\gamma} : \exists\, t_n \nearrow +\infty \text{ s.t. } S(t_n)u_0 \to v \text{ in } X_\gamma, \text{ as } n \to \infty\} \tag{5.76}$$

is nonempty, compact, and connected. Moreover, as Φ is a strict Lyapunov functional, $\operatorname{dist}(S(t)u_0, \omega(u_0)) \to 0$ in X_γ as $t \to \infty$ and $\omega(u_0) \subset \mathcal{E} \subset V_\mu \cap X_1$, hence the set of equilibria is nonempty. Let $u_* \in \omega(u_0)$, then there exists a sequence $t_n \nearrow +\infty$ such that $S(t_n)u_0 \to u_*$ in X_γ as $n \to \infty$. Assuming that u_* is normally hyperbolic and t_n is large enough, Theorems 5.3.1 and 5.5.1 yield convergence of $S(t)u_0$ to some equilibrium $u_\infty \in V_{\mu,\gamma}$ as $t \to \infty$. Uniqueness of the limit finally implies $u_\infty = u_*$. We obtain the following result.

Theorem 5.7.2. *Let $p \in (1, \infty)$, $\mu \in (1/p, 1)$, $\bar{\mu} \in (\mu, 1]$, $V_\mu \subset X_{\gamma,\mu}$ be open. Assume that $(A, F) \in C^1(V_\mu; \mathcal{B}(X_1, X_0) \times X_0)$, and that the embedding $X_{\gamma,\bar{\mu}} \hookrightarrow X_{\gamma,\mu}$ is compact. Suppose furthermore that $u \in C_b([\tau, \infty); V_\mu \cap X_{\gamma,\bar{\mu}})$ is a global solution to* (5.1), *satisfying*

$$\operatorname{dist}_{X_{\gamma,\mu}}(u(t), \partial V_\mu) \geq \eta > 0, \quad \textit{for all } t \geq 0, \tag{5.77}$$

and let $\Phi \in C(V_\mu \cap X_\gamma; \mathbb{R})$ be a strict Lyapunov function for (5.1).

Then the ω-limit set defined by (5.76) *is nonempty, compact, connected, and $\omega(u_0) \subset \mathcal{E}$. If, in addition, there exists $u_* \in \omega(u_0)$ which is normally stable or normally hyperbolic, then $\lim_{t \to \infty} u(t) = u_*$ in X_γ.*

This is the optimal result on the asymptotic behaviour in the presence of a Lyapunov functional. In practice, it remains to verify (5.77), which is trivial in case $V_\mu = X_{\gamma,\mu}$, and to prove boundedness of u in $X_{\gamma,\bar{\mu}}$. This depends on the problem in question.

Part III

Linear Theory

Chapter 6

Elliptic and Parabolic Problems

In this chapter we prove maximal L_p-regularity for various linear parabolic and elliptic problems. These results will be crucial for our study of quasilinear parabolic problems, including those introduced in Chapter 1. The proofs are based on the vector-valued Fourier multiplier theorems and $\mathcal{H}^\infty$-calculi developed in Chapter 4, as well as on arguments involving perturbations, domain transformations, and localizations.

6.1 Elliptic and Parabolic Problems on $\mathbb{R}^n$

We begin with the constant coefficient case.

1.1 Kernel Estimates

Let $\mathcal{A}(\xi)$ denote a $\mathcal{B}(E)$-valued polynomial on $\mathbb{R}^n$ which is homogeneous of degree $m \in \mathbb{N}$, i.e.,

$$\mathcal{A}(\xi) = \sum_{|\alpha|=m} a_\alpha \xi^\alpha, \quad \xi \in \mathbb{R}^n,$$

where we use multi-index notation, and $a_\alpha \in \mathcal{B}(E)$, E a Banach space. We want to consider the vector-valued partial differential equation

$$\lambda u(x) + \mathcal{A}(D)u(x) = f(x), \quad x \in \mathbb{R}^n, \tag{6.1}$$

where the function f is given, $\lambda \in \mathbb{C}$, and $D = -i(\partial_1, \ldots, \partial_n)$. The purpose of this subsection is the derivation of a kernel representation for the solution $u(x)$ of the form

$$u(x) = \int_{\mathbb{R}^n} \gamma_\lambda(x - x') f(x')\, dx', \quad x \in \mathbb{R}^n, \tag{6.2}$$

as well as estimates for the kernel γ_λ.

Homogeneity of $\mathcal{A}$ of degree m implies that γ_λ must be of the form

$$\gamma_\lambda(x) = |\lambda|^{\frac{n}{m}-1} \gamma_\theta(|\lambda|^{1/m} x), \quad x \in \mathbb{R}^n,\ \arg(\lambda) = \theta,\ \lambda \neq 0. \tag{6.3}$$

J. Prüss and G. Simonett, *Moving Interfaces and Quasilinear Parabolic Evolution Equations*, Monographs in Mathematics 105,
DOI 10.1007/978-3-319-27698-4_6

Here γ_θ denotes the fundamental solution of (6.1), i.e., it satisfies the equation

$$e^{i\theta}\gamma_\theta + \mathcal{A}(D)\gamma_\theta = \delta_0$$

in the sense of distributions.

In fact, a formal argument, which will become precise later, is as follows. Taking Fourier transforms we obtain for the solution of (6.1) the expression

$$\mathcal{F}u(\xi) = (\lambda + \mathcal{A}(\xi))^{-1}\mathcal{F}f(\xi), \quad \xi \in \mathbb{R}^n.$$

Taking inverse transforms this yields

$$u(x) = (2\pi)^{-n}\int_{\mathbb{R}^n} (\lambda + \mathcal{A}(\xi))^{-1}\mathcal{F}f(\xi)e^{ix\cdot\xi}\, d\xi.$$

By the convolution theorem we get

$$\gamma_\lambda(x) = (2\pi)^{-n}\int_{\mathbb{R}^n} (\lambda + \mathcal{A}(\xi))^{-1}e^{ix\cdot\xi}\, d\xi,$$

which after the scaling $\xi = |\lambda|^{1/m}\xi'$ leads to the representation (6.3) with

$$\gamma_\theta(x) = (2\pi)^{-n}\int_{\mathbb{R}^n} (e^{i\theta} + \mathcal{A}(\xi'))^{-1}e^{ix\cdot\xi'}\, d\xi', \tag{6.4}$$

where $\theta = \arg(\lambda)$.

For all this to make sense we surely must know that $\lambda + \mathcal{A}(\xi)$ is invertible for all $\xi \in \mathbb{R}^n$ and for all λ in question. This naturally leads to the basic assumption we make here, namely that of parameter-ellipticity.

Definition 6.1.1. *The $\mathcal{B}(E)$-valued polynomial $\mathcal{A}(\xi)$ is called* **parameter-elliptic** *if there is an angle $\phi \in [0, \pi)$ such that the spectrum $\sigma(\mathcal{A}(\xi))$ of $\mathcal{A}(\xi)$ satisfies*

$$\sigma(\mathcal{A}(\xi)) \subset \Sigma_\phi \quad \textit{for all } \xi \in \mathbb{R}^n,\ |\xi| = 1. \tag{6.5}$$

We call

$$\phi_{\mathcal{A}} := \inf\{\phi :\ (6.5)\ \textit{holds}\} = \sup_{|\xi|=1} |\arg \sigma(\mathcal{A}(\xi))|$$

angle of ellipticity *of $\mathcal{A}$. $\mathcal{A}(\xi)$ is called* **normally elliptic** *if it is parameter-elliptic with angle $\phi_{\mathcal{A}} < \pi/2$. We then call the differential operator $\mathcal{A}(D)$ parameter-elliptic resp. normally elliptic as well.*

Some remarks are in order.

Remark 6.1.2. **(i)** It is easy to see that parameter-ellipticity as well as $\phi_{\mathcal{A}}$ are invariant under orthogonal transformations, but even more is true. Consider a coordinate transformation of the form $Tu(x) = u(Qx)$ where $Q \in \mathbb{R}^{n\times n}$ is invertible. Then the transformed differential operator will be

$$\mathcal{A}_Q(D) := T^{-1}\mathcal{A}(D)T = \mathcal{A}(Q^{\mathsf{T}}D).$$

Hence with $\mathcal{A}(\xi)$ also $\mathcal{A}_Q(\xi) = \mathcal{A}(Q^{\mathsf{T}}\xi)$ is parameter-elliptic, and $\phi_{\mathcal{A}_Q} = \phi_{\mathcal{A}}$.

(ii) Note that m is necessarily even in case $\phi_{\mathcal{A}} < \pi/2$. Indeed,

$$\mathcal{A}(-\xi) = -\mathcal{A}(\xi), \quad \xi \in \mathbb{R}^n,$$

in case m is odd, and hence

$$\sigma(\mathcal{A}(\xi)) \subset \Sigma_\phi \cap -\Sigma_\phi = \emptyset, \quad |\xi| = 1,$$

which is impossible.

(iii) On the other hand, there are parameter-elliptic operators of odd order, e.g. for $n = 1$, $m = 1$, $\mathcal{A}(D) = iD$ is parameter-elliptic with $\phi_{\mathcal{A}} = \pi/2$.

(iv) Recall that the symbol $\mathcal{A}(\xi) = \sum_{|\alpha|=m} a_\alpha \xi^\alpha$ is called **elliptic** if $0 \notin \sigma(\mathcal{A}(\xi))$ for all $\xi \in \mathbb{R}^n$, $\xi \neq 0$. Obviously, each parameter-elliptic symbol is also elliptic, but not conversely. A famous counterexample is the Cauchy-Riemann operator $\mathcal{A}(\xi) = \xi_1 + i\xi_2$ with $n = 2$, $E = \mathbb{C}$; for this operator we have $\cup_{|\xi|=1}\sigma(\mathcal{A}(\xi)) = \mathbb{S}^1$, the unit sphere in $\mathbb{C}$.

If E is a Hilbert space, there is another notion of ellipticity.

Definition 6.1.3. *The $\mathcal{B}(E)$-valued polynomial $\mathcal{A}(\xi)$ is called* **strongly elliptic** *if there is a constant $c > 0$ such that*

$$\mathrm{Re}(\mathcal{A}(\xi)v|v)_E \geq c|\xi|^m |v|_E^2, \quad \xi \in \mathbb{R}^n,\ v \in E.$$

The largest such c will be called the **ellipticity constant** *$c_{\mathcal{A}}$ of $\mathcal{A}(D)$. The differential operator $\mathcal{A}(D)$ is then also called strongly elliptic.*

Also for this notion of ellipticity some remarks are in order.

Remark 6.1.4. **(i)** Observe that also strong ellipticity as well as $c_{\mathcal{A}}$ are invariant under orthogonal transformations. More generally, strong ellipticity is invariant also under general coordinate transformations, but the constant $c_{\mathcal{A}}$ does not have this property.

(ii) To understand the condition of strong ellipticity, recall that the **numerical range** $\mathsf{n}(B)$ of an operator $B \in \mathcal{B}(E)$ is defined by

$$\mathsf{n}(B) := \overline{\{z \in \mathbb{C} : z = (Bv|v)_E \text{ for some } v \in E,\ |v|_E = 1\}}.$$

It is easy to see that $\sigma(B) \subset \mathsf{n}(B)$, and that $\mathsf{n}(B) \subset \bar{B}_{\mathbb{C}}(0, |B|)$ holds. Therefore, $\mathcal{A}$ is strongly elliptic if the numerical range of $\mathcal{A}(\xi)$ is contained in the half-space $\mathrm{Re}\, z \geq c > 0$ for each $\xi \in \mathbb{R}^n$, $|\xi| = 1$. Consequently, if $\mathcal{A}$ is strongly elliptic then

$$\sigma(\mathcal{A}(\xi)) \subset \mathsf{n}(\mathcal{A}(\xi)) \subset \Sigma_\phi, \quad \xi \in \mathbb{R}^n,\ |\xi| = 1.$$

In particular, every strongly elliptic polynomial $\mathcal{A}$ is parameter-elliptic with

$$\phi_{\mathcal{A}} \leq \sup\{|\arg(\mathcal{A}(\xi)v|v)_E| : v \in E, |v|_E = 1, \xi \in \mathbb{R}^n, |\xi| = 1\} < \pi/2,$$

hence even normally elliptic.

(iii) The class of strongly elliptic differential operators contains some of the most common elliptic operators arising in applications.

Now assume that $\mathcal{A}$ is parameter-elliptic with angle of ellipticity $\phi_{\mathcal{A}}$ and let $\phi > \phi_{\mathcal{A}}$. We are going to justify the formal procedure from above for $|\theta| \leq \pi - \phi$. For this purpose we consider the Fourier integral

$$\gamma_\theta^\varepsilon(x) = (2\pi)^{-n} \int_{\mathbb{R}^n} (e^{i\theta} + \mathcal{A}(\xi'))^{-1} e^{ix\cdot\xi'} e^{-\varepsilon|\xi'|}\, d\xi', \tag{6.6}$$

with $\varepsilon > 0$ fixed. Note that this integral is absolutely convergent due to the additional exponential factor, in contrast to (6.4). For the moment we restrict attention to the case $n \geq 3$. We will comment at the end of this section on $n = 1, 2$. Fix $x \in \mathbb{R}^n$, $x \neq 0$, and choose a rotation Q such that $Qx = re_1$, where $r = |x|$ and e_1 means the first unit vector in $\mathbb{R}^n$. By means of the variable transformation

$$Q\xi' = (\eta, s\zeta), \quad \eta \in \mathbb{R},\ s > 0,\ \zeta \in \mathbb{S}^{n-2},$$

where $\mathbb{S}^k$ denotes the k-dimensional unit sphere, we obtain the following representation of $\gamma_\theta^\varepsilon$.

$$\gamma_\theta^\varepsilon(x) = \frac{1}{(2\pi)^n} \int_0^\infty s^{n-2} \int_{\mathbb{S}^{n-2}} \int_{\mathbb{R}} (e^{i\theta} + \mathcal{A}(Q^\mathsf{T}(\eta, s\zeta)))^{-1} e^{i|x|\eta} e^{-\varepsilon(\eta^2+s^2)^{1/2}}\, d\eta d\zeta ds.$$

Next we employ the scaling $\eta = (1+s)z$ for η and observe that by homogeneity of $\mathcal{A}$ we have

$$\begin{aligned}
\mathcal{A}(Q^\mathsf{T}(\eta, s\zeta)) &= \sum_{k=0}^m \eta^{m-k} s^k \sum_{|\beta|=k} b_\beta \zeta^\beta \\
&= (1+s)^m \sum_{k=0}^m z^{m-k} \Big(1 - \frac{1}{1+s}\Big)^k b_k(\zeta) \\
&= (1+s)^m P(z, \zeta, 1/(1+s)),
\end{aligned}$$

for some $b_\beta \in \mathcal{B}(E)$, $b_k(\zeta) = \sum_{|\beta|=k} b_\beta \zeta^\beta$. Then we set

$$H(z, \zeta, \sigma, \theta) = (2\pi)^{-n}(e^{i\theta}\sigma^m + P(z, \zeta, \sigma)),$$

and finally obtain the representation

$$\gamma_\theta^\varepsilon(x) = \int_0^\infty \frac{s^{n-2}}{(1+s)^{m-1}} \Big[\int_{\mathbb{S}^{n-2}} h_\varepsilon(s, \zeta, \theta, r) d\zeta\Big]\, ds, \tag{6.7}$$

with

$$h_\varepsilon(s,\zeta,\theta,r) = \int_{\mathbb{R}} H(z,\zeta,1/(1+s),\theta)^{-1} e^{ir(1+s)z} e^{-\varepsilon(1+s)[z^2+(s/(1+s))^2]^{1/2}}\, dz.$$

The function $H(z,\zeta,\sigma,\theta)$ is a $\mathcal{B}(E)$-valued polynomial in z, with coefficients depending continuously on $p = (\zeta,\sigma,\theta) \in P := \mathbb{S}^{n-2} \times [0,1] \times [-\pi+\phi, \pi-\phi]$, a compact set.

By parameter-ellipticity, the set of $z \in \mathbb{C}$ such that $H(z,p)$ is not invertible is compact and does not contain real values. This set is upper-semicontinuous in p, hence the set of singularities of $H(\cdot,p)^{-1}$ is a compact set not intersecting the real line, uniformly for $p \in P$. Since H^{-1} is holomorphic in z we may therefore deform the path of integration to a contour Γ of the form

$$\Gamma := \{z = t + i\kappa(1+|t|) : t \in \mathbb{R}\},$$

where $\kappa > 0$ is small and independent of $p \in P$. Then we obtain by Cauchy's theorem

$$h_\varepsilon(s,\zeta,\theta,r) = \int_{\Gamma} H(z,\zeta,1/(1+s),\theta)^{-1} e^{ir(1+s)z} e^{-\varepsilon(1+s)[z^2+(s/(1+s))^2]^{1/2}}\, dz.$$

Since H^{-1} is bounded on Γ, and

$$|e^{ir(1+s)z}| = e^{-\kappa r(1+s)(1+|t|)},$$

the integral defining h_ε is absolutely convergent and

$$|h_\varepsilon(s,\zeta,\theta,r)| \le Ce^{-\kappa r(1+s)}/[r(1+s)],$$

independently of $\varepsilon > 0$. Hence we may pass to the limit $\varepsilon \to 0$ to the result

$$\gamma_\theta(x) = \int_0^\infty \frac{s^{n-2}}{(1+s)^{m-1}} \Big[\int_{\mathbb{S}^{n-2}} h(s,\zeta,\theta,r)d\zeta\Big]\, ds \tag{6.8}$$

with

$$h(s,\zeta,\theta,r) = \int_{\Gamma} H(z,\zeta,1/(1+s),\theta)^{-1} e^{ir(1+s)z}\, dz.$$

Contracting the contour Γ in the set $\{\mathrm{Im}\, z > \kappa\} \subset \mathbb{C}$ into a smooth Jordan curve Γ_0 surrounding the singularities of H^{-1} in the upper half-plane, we finally get the following representation for h.

$$h(s,\zeta,\theta,r) = \int_{\Gamma_0} H(z,\zeta,1/(1+s),\theta)^{-1} e^{ir(1+s)z}\, dz. \tag{6.9}$$

This implies the estimate

$$|h(s,\zeta,\theta,r)| \le Ce^{-\kappa(1+s)r}, \quad s > 0,\ \zeta \in \mathbb{S}^{n-2},\ |\theta| \le \pi - \phi \tag{6.10}$$

for h. We summarize these considerations in

Theorem 6.1.5. *Let $n, m \in \mathbb{N}$, E a Banach space, $a_\alpha \in \mathcal{B}(E)$, and suppose*

$$\mathcal{A}(\xi) = \sum_{|\alpha|=m} a_\alpha \xi^\alpha, \quad \xi \in \mathbb{R}^n,$$

is parameter-elliptic with angle of ellipticity $\phi_{\mathcal{A}} < \pi$. Then for each $\phi > \phi_{\mathcal{A}}$ there is a constant C_ϕ such that the solution $\gamma_\theta(x)$ of

$$e^{i\theta} u + \mathcal{A}(D)u = \delta_0$$

satisfies the estimate

$$|\gamma_\theta(x)| \leq C_\phi p_0(|x|), \quad x \in \mathbb{R}^n,\ x \neq 0,\ |\theta| \leq \pi - \phi, \tag{6.11}$$

where p_0 is given by

$$p_0(r) = \int_0^\infty \frac{s^{n-2}}{(1+s)^{m-1}} e^{-\kappa r(1+s)}\, ds,$$

for some $\kappa > 0$. The function $p_0 : (0,\infty) \to (0,\infty)$ is completely monotone, and satisfies

$$\int_0^\infty r^{n+\rho-1} p_0(r)\, dr < \infty \quad \textit{if and only if} \quad \rho > -m.$$

Note that we can estimate p_0 further by

$$p_0(r) \leq \begin{cases} ce^{-\kappa r} & \text{if} \quad n < m; \\ ce^{-\kappa r} \log(2 + 1/r) & \text{if} \quad n = m; \\ \frac{c}{r^{n-m}} e^{-\kappa r} & \text{if} \quad n > m. \end{cases}$$

Together with (6.8) and (6.3), Theorem 6.1.5 leads to a Poisson estimate for the kernel γ_λ from (6.2), i.e., for each $\phi > \phi_{\mathcal{A}}$ there is a constant $C_\phi > 0$ such that

$$|\gamma_\lambda(x)| \leq C_\phi |\lambda|^{\frac{n}{m}-1} p_0(|\lambda|^{1/m}|x|), \quad x \in \mathbb{R}^n,\ |\arg \lambda| \leq \pi - \phi. \tag{6.12}$$

However, even more is true. Applying the differential operator D^β to (6.6) and employing the same arguments as above we obtain

Corollary 6.1.6. *In the situation of Theorem* 6.1.5 *for each $k \in \mathbb{N}_0$, we have in addition*

$$|D^\beta \gamma_\theta(x)| \leq C_{\phi,k} p_k(|x|), \quad x \in \mathbb{R}^n,\ x \neq 0,\ |\theta| \leq \pi - \phi,\ |\beta| = k,$$

where p_k is given by

$$p_k(r) = \int_0^\infty \frac{s^{n-2}}{(1+s)^{m-k-1}} e^{-\kappa r(1+s)}\, ds,$$

for some $\kappa > 0$.

Observe that this corollary implies the estimate

$$|D^\beta \gamma_\lambda(x)| \le C_{\phi,k}|\lambda|^{\frac{n+k}{m}-1} p_k(|\lambda|^{1/m}|x|), \quad x \in \mathbb{R}^n,\ |\arg\lambda| \le \pi - \phi, \tag{6.13}$$

for the derivatives of the fundamental solution γ_λ, with $k = |\beta|$. This yields $D^\beta\gamma_\lambda \in L_1(\mathbb{R}^n; \mathcal{B}(E))$ if $|\beta| < m$.

Concluding, some remarks concerning the cases $n = 1, 2$ have to be made. For $n = 1$, instead of the rotation Q we may use reflection; all above arguments remain valid for this case, simply dropping the integrals over s and ζ. In that case the functions p_k should be replaced by

$$p_k(r) = \int_0^\infty \frac{1}{(1+s)^{m-k}} e^{-\kappa r(1+s)}\, ds.$$

For $n = 2$ the arguments are also valid if we interpret $\mathbb{S}^0$ as the set consisting of the two points ± 1. Therefore the above results are valid for all dimensions $n \in \mathbb{N}$.

1.2 L_q-Realizations of Elliptic Differential Operators

Next we consider the L_q-realizations of the differential operator $\mathcal{A}(D)$.

Theorem 6.1.7. *Let $n, m \in \mathbb{N}$, E a Banach space, $a_\alpha \in \mathcal{B}(E)$, $1 < q < \infty$, and suppose $\mathcal{A}(D) = \sum_{|\alpha|=m} a_\alpha D^\alpha$ is parameter-elliptic with angle of ellipticity $\phi_{\mathcal{A}} < \pi$. Define the L_q-realization A of $\mathcal{A}$ in $X_0 = L_q(\mathbb{R}^n; E)$ by means of $A = \overline{A_0}$, where*

$$[A_0 u](x) = \mathcal{A}(D)u(x), \quad x \in \mathbb{R}^n, \quad u \in \mathsf{D}(A_0) := H_q^m(\mathbb{R}^n; E).$$

Then A is sectorial with spectral angle $\phi_A \le \phi_{\mathcal{A}}$, and

$$H_q^m(\mathbb{R}^n; E) \subset \mathsf{D}(A) \subset H_q^{m-1}(\mathbb{R}^n; E).$$

Proof. Obviously, A has dense domain. If $f \in L_q(\mathbb{R}^n; E)$, choose a sequence $f_k \in \mathcal{D}(\mathbb{R}^n; E)$ such that $f_k \to f$ in $L_q(\mathbb{R}^n; E)$. For $\lambda \in \Sigma_{\pi-\phi}$, $\phi > \phi_{\mathcal{A}}$, we obtain $u_k = \gamma_\lambda * f_k \in H_q^m(\mathbb{R}^n; E)$ as well as $\lambda u_k + \mathcal{A}(D)u_k = f_k$, by uniqueness of the Fourier transform. Since $u_k \to u = \gamma_\lambda * f$ in $L_q(\mathbb{R}^n; E)$ as $k \to \infty$, we see that $u \in \mathsf{D}(A)$ and $\lambda u + Au = f$. This shows that $\lambda + A$ is invertible for each $\lambda \in \Sigma_\phi$ and $(\lambda + A)^{-1} f = \gamma_\lambda * f$. Thus by Corollary 6.1.6 we obtain the inclusions

$$H_q^m(\mathbb{R}^n; E) = \mathsf{D}(A_0) \subset \mathsf{D}(A) \subset H_q^{m-1}(\mathbb{R}^n; E),$$

and Theorem 6.1.5 yields $-\Sigma_{\pi-\phi} \subset \rho(A)$, as well as

$$|\lambda(\lambda + A)^{-1}|_{\mathcal{B}(L_q(\mathbb{R}^n;E))} \le M_{\pi-\phi}, \tag{6.14}$$

for each $\phi > \phi_{\mathcal{A}}$.

For $f \in \mathcal{D}(\mathbb{R}^n; E)$, supp $f \subset B(0, R)$, we have by Theorem 6.1.5

$$|\lambda\gamma_\lambda * f(x)| \le \int_{\mathbb{R}^n} p_0(|y|)|f(x - y/|\lambda|^{1/m})|\, dy \to 0$$

as $\lambda \to 0$, uniformly for bounded x. On the other hand, for $|x| \geq 2R$ we have $|x-y| \geq |x| - |y| \geq |y|$. Since p_0 is non-increasing this yields

$$\begin{aligned}|\lambda\gamma_\lambda * f(x)| &\leq |f|_\infty \int_{B_R(0)} |\lambda|^{n/m} p(|\lambda|^{1/m}|x-y|)\,dy \\ &\leq |f|_\infty \int_{B_R(0)} |\lambda|^{n/m} p(|\lambda|^{1/m}|y|)\,dy \\ &= |f|_\infty \int_0^{|\lambda|^{1/m}R} p(r)\,dr \to 0\end{aligned}$$

as $\lambda \to 0$. This implies $|\lambda(\lambda + A)^{-1} f|_\infty \to 0$ for $\lambda \to 0$, for each $f \in \mathcal{D}(\mathbb{R}^n; E)$, but then by interpolation

$$|\lambda\gamma_\lambda * f|_q \leq |\lambda\gamma_\lambda * f|_1^{1/q} |\lambda\gamma_\lambda * f|_\infty^{1/q'} \to 0.$$

Therefore, $A(\lambda + A)^{-1} \to I$ strongly as $\lambda \to 0$, i.e., $\mathsf{R}(A)$ is dense in $L_q(\mathbb{R}^n; E)$ and $\mathsf{N}(A) = 0$, for each $1 < q < \infty$. Thus A is sectorial and $\phi_A \leq \phi_{\mathcal{A}}$. □

One can show that we even have

$$H_q^m(\mathbb{R}^n; E) \hookrightarrow \mathsf{D}(A) \hookrightarrow H_q^s(\mathbb{R}^n; E), \quad \text{for each } s < m.$$

Nevertheless, we cannot prove the elliptic maximal L_q-regularity

$$\mathsf{D}(A) = H_q^m(\mathbb{R}^n; E)$$

unless more is known on the geometry of E. Here harmonic analysis comes into play.

1.3 $\mathcal{H}^\infty$-Calculus for Elliptic Operators

If E is a Banach space of class $\mathcal{HT}$, for differential operators with parameter-elliptic symbols the following result is valid.

Theorem 6.1.8. *Let E be a Banach space of class $\mathcal{HT}$, $n, m \in \mathbb{N}$, and $1 < q < \infty$. Suppose $\mathcal{A}(D) = \sum_{|\alpha|=m} a_\alpha D^\alpha$ with $a_\alpha \in \mathcal{B}(E)$ is a homogeneous differential operator of order m which is parameter-elliptic with angle of ellipticity $\phi_{\mathcal{A}}$. Let A denote its realization in $X_0 = L_q(\mathbb{R}^n; E)$ with domain $\mathsf{D}(A) = H_q^m(\mathbb{R}^n; E)$.*

Then $A \in \mathcal{H}^\infty(X_0)$ with $\mathcal{H}^\infty$-angle $\phi_A^\infty \leq \phi_{\mathcal{A}}$. In particular, A is $\mathcal{R}$-sectorial with $\phi_A^R \leq \phi_{\mathcal{A}}$.

Proof. (i) Observe first that the symbol $\mathcal{A}(\xi)$ is homogeneous of degree m, i.e., $\mathcal{A}(\xi) = \rho^m \mathcal{A}(\zeta)$, $\rho = |\xi|$. Parameter-ellipticity implies that $\mathcal{A}(\zeta)$ is invertible for each $|\zeta| = 1$ and $|\mathcal{A}(\zeta)^{-1}| \leq M_0$, where M_0 is independent of ζ, by compactness of the set $|\zeta| = 1$; this implies in particular $|\mathcal{A}(\xi)^{-1}| \leq M_0 \rho^{-m}$. Hence $\xi^\alpha \mathcal{A}(\xi)^{-1} = \zeta^\alpha \mathcal{A}(\zeta)^{-1}$ is bounded for each $|\alpha| = m$. But since $\mathcal{A}(\zeta)$ is holomorphic, $\zeta^\alpha \mathcal{A}(\zeta)^{-1}$ is so as well, and since $\mathbb{S}^{n-1}$ is compact, $\{\xi^\alpha \mathcal{A}(\xi)^{-1} : \xi \in \mathbb{R}^n \setminus \{0\}\}$ is $\mathcal{R}$-bounded by Proposition 4.1.12. The same holds true for $\{|\xi|^k D^\beta[\xi^\alpha \mathcal{A}(\xi)^{-1}] : \xi \in \mathbb{R}^n \setminus \{0\}\}$, $|\beta| = k \in \mathbb{N}$, as a simple calculation shows. The vector-valued Mikhlin theorem, Theorem 4.3.11, then implies that there is a constant $C > 0$ such that

$$C^{-1}|D^\alpha u|_{X_0} \le |\mathcal{A}(D)u|_{X_0}, \quad \text{for all } u \in H^m_q(\mathbb{R}^n; E),\ |\alpha| = m,$$

holds. In particular, we have $\mathsf{D}(A) = H^m_q(\mathbb{R}^n; E)$, and by (6.14) A is sectorial with spectral angle $\phi_A \le \phi_{\mathcal{A}}$.

(ii) To show that A admits an $\mathcal{H}^\infty$-calculus such that the $\mathcal{H}^\infty$-angle satisfies $\phi^\infty_A \le \phi_{\mathcal{A}}$, let $\phi > \phi_{\mathcal{A}}$ be fixed and choose a function $h \in H_0(\Sigma_\phi)$. Let Γ denote the contour $\Gamma = (\infty, 0]e^{i\theta} \cup (0, \infty)e^{-i\theta}$, where $\phi_{\mathcal{A}} < \theta < \phi$. Then $h(A)$ is well defined as the Dunford integral

$$h(A) = \frac{1}{2\pi i}\int_\Gamma h(\lambda)(\lambda - A)^{-1}\, d\lambda.$$

For $u \in \mathcal{D}(\mathbb{R}^n; E)$, we may take Fourier transforms, to the result

$$\begin{aligned}\mathcal{F}[h(A)u](\xi) &= \frac{1}{2\pi i}\int_\Gamma h(\lambda)(\lambda - \mathcal{A}(\xi))^{-1}\mathcal{F}u(\xi)\, d\lambda \\ &= h(\mathcal{A}(\xi))\mathcal{F}u(\xi),\end{aligned}$$

hence the symbol of $h(A)$ is given by $h(\mathcal{A}(\xi))$. Therefore, it is enough to show that this symbol is a Fourier multiplier for $L_q(\mathbb{R}^n; E)$, with norm $\le C|h|_{H^\infty(\Sigma_\phi)}$. This will be done employing the vector-valued Mikhlin theorem another time.

By means of the rescalings $\xi = \rho\zeta$ and $\mu = \lambda\rho^{-m}$ we obtain the representation

$$h(\mathcal{A}(\xi)) = \frac{1}{2\pi i}\int_\Gamma h(\rho^m \mu)(\mu - \mathcal{A}(\zeta))^{-1}\, d\mu.$$

Since $\sigma_0 = \cup_{|\zeta|=1}\sigma(\mathcal{A}(\zeta))$ is compact and contained in Σ_{ϕ_0}, we may deform the contour Γ within Σ_θ into a compact simple smooth closed path Γ_0 surrounding σ_0 counter-clockwise, and by Cauchy's theorem

$$h(\mathcal{A}(\xi)) = \frac{1}{2\pi i}\int_{\Gamma_0} h(\rho^m \mu)(\mu - \mathcal{A}(\zeta))^{-1}\, d\mu.$$

By compactness of Γ_0 and of $\mathbb{S}^{n-1}$, in virtue of Proposition 4.1.12, $(\mu - \mathcal{A}(\zeta))^{-1}$ is $\mathcal{R}$-bounded on $\Gamma_0 \times \mathbb{S}^{n-1}$, hence this representation of $h(\mathcal{A}(\xi))$ yields

$$\mathcal{R}\{h(\mathcal{A}(\xi)) : \xi \in \mathbb{R}^n\} \le (2\pi)^{-1}|h|_{H^\infty(\Sigma_\phi)} l(\Gamma_0)\mathcal{R}\{(\mu - \mathcal{A}(\zeta))^{-1} : \mu \in \Gamma_0, \zeta \in \mathbb{S}^{n-1}\}$$

where $l(\Gamma_0)$ denotes the length of Γ_0. Thus the symbol of $h(A)$ is $\mathcal{R}$-bounded.

To obtain appropriate bounds for the derivatives of $h(\mathcal{A}(\xi))$, observe the relation

$$D_\xi = -i\zeta\frac{\partial}{\partial\rho} + \frac{1}{\rho}(I - \zeta \otimes \zeta)D_\zeta.$$

With $G_0(\mu,\zeta) = (2\pi i)^{-1}(\mu - \mathcal{A}(\zeta))^{-1}$ we have

$$h(\mathcal{A}(\xi)) = \int_{\Gamma_0} h(\rho^m \mu) G_0(\mu,\zeta)\, d\mu,$$

hence differentiating this expression inductively we get

$$\rho^{|\alpha|} D_\xi^\alpha h(\mathcal{A}(\xi)) = \sum_{k=0}^{|\alpha|} \int_{\Gamma_0} (\rho^m\mu)^k h^{(k)}(\rho^m \mu) G_{\alpha,k}(\mu,\zeta)\, d\mu,$$

where the functions $G_{\alpha,k}(\mu,\zeta)$ are analytic in a neighbourhood of $\Gamma_0 \times \mathbb{S}^{n-1}$. Therefore we obtain

$$\mathcal{R}\{|\xi|^{|\alpha|} D_\xi^\alpha h(\mathcal{A}(\xi)) : \xi \in \mathbb{R}^n\} \le \sum_{k=0}^{|\alpha|} C_{\alpha,k} \sup_{z\in\Sigma_\theta} |z^k h^{(k)}(z)|.$$

Finally, by the Cauchy estimates we have $\sup_{z\in\Sigma_\theta} |z^k h^{(k)}(z)| \le c_k |h|_{H^\infty(\Sigma_\phi)}$, and so for each $\alpha \in \mathbb{N}_0^n$ there is a constant C_α such that

$$\mathcal{R}\{|\xi|^{|\alpha|} D_\xi^\alpha h(\mathcal{A}(\xi)) :\ \xi \in \mathbb{R}^n, \xi \neq 0\} \le C_\alpha |h|_{H^\infty(\Sigma_\phi)}$$

is satisfied. C_α is independent of h, it depends only on $\mathcal{A}(\xi)$, on the contour Γ_0, and on ϕ. By Theorem 4.3.11 we therefore obtain $|h(A)|_{\mathcal{B}(L_q(\mathbb{R}^n;E))} \le M_\phi |h|_{H^\infty(\Sigma_\phi)}$, which implies the assertion. □

In the situation of the last theorem, since $A \in \mathcal{H}^\infty(X_0)$ we have, by Theorem 3.3.7,

$$\mathsf{D}(A^\theta) = (L_q(\mathbb{R}^n;E), \mathsf{D}(A))_\theta = H_q^{m\theta}(\mathbb{R}^n;E)$$

for each $\theta \in [0,1]$, hence $D^\beta A^{-k/m}$ is bounded for each $|\beta| = k \le m$. On the other hand, for each $\nu \in (0,1)$ we have the representation

$$\lambda^{1-\nu} A^\nu (\lambda + A)^{-1} = \int_{-\infty}^{\infty} \frac{\lambda^{is}}{2\sin\pi(\nu + is)} A^{-is}\, ds, \quad \lambda \in \Sigma_{\pi-\phi}, \quad \phi > \phi_A. \tag{6.15}$$

Convexity of $\mathcal{R}$-bounds and the contraction principle then show that the sets

$$\{\lambda^{1-\nu} A^\nu (\lambda + A)^{-1} :\ \lambda \in \Sigma_{\pi-\phi}\}$$

are $\mathcal{R}$-bounded. As a consequence we obtain

Corollary 6.1.9. *Let the assumptions of Theorem* 6.1.7 *be satisfied, and let* $\alpha \in (0,1)$, $q \in (1,\infty)$, $r \in [1,\infty]$. *Then*

(i) *The set*

$$\{\lambda^{1-k/m} D^\beta (\lambda + A)^{-1} :\ \lambda \in \Sigma_{\pi-\phi},\ 0 \le |\beta| = k \le m\}$$

is $\mathcal{R}$*-bounded in* $X_0 = L_q(\mathbb{R}^n;E)$;

(ii) $\mathsf{D}(A^\alpha) = (X_0, \mathsf{D}(A))_\alpha = H_q^{\alpha m}(\mathbb{R}^n; E)$;

(iii) $D_A(\alpha, r) = (X_0, \mathsf{D}(A))_{\alpha,r} = B_{qr}^{\alpha m}(\mathbb{R}^n; E)$.

1.4 Elliptic Operators with Variable Coefficients

Let E be a Banach space of class $\mathcal{HT}$, and consider the differential operator with variable $\mathcal{B}(E)$-valued coefficients

$$[Au](x) = \mathcal{A}(x, D)u(x), \quad x \in \mathbb{R}^n,\ u \in \mathsf{D}(A) = H_p^m(\mathbb{R}^n; E), \tag{6.16}$$

where

$$\mathcal{A}(x, D) = \sum_{\alpha|\leq m} a_\alpha(x) D^\alpha. \tag{6.17}$$

By means of the results on homogeneous parameter-elliptic operators with constant coefficients from the previous sections, perturbation and localization, we will prove the following result.

Theorem 6.1.10. *Let E be a Banach space of class $\mathcal{HT}$, $n, m \in \mathbb{N}$, and $1 < q < \infty$. Suppose $\mathcal{A}(x, D) = \sum_{|\alpha|\leq m} a_\alpha(x) D^\alpha$ with $a_\alpha(x) \in \mathcal{B}(E)$ is a differential operator of order m with variable coefficients. Assume the following* **Condition (ra)***:*

(ra1) *$a_\alpha \in C_l(\mathbb{R}^n; \mathcal{B}(E))$ for each $|\alpha| = m$;*

(ra2) *$\mathcal{A}_\#(x, \xi) = \sum_{|\alpha|=m} a_\alpha(x)\xi^\alpha$ is parameter-elliptic with angle of ellipticity $\leq \phi_{\mathcal{A}}$, for each $x \in \mathbb{R}^n \cup \{\infty\}$;*

(ra3) *$a_\alpha \in [L_{r_k} + L_\infty](\mathbb{R}^n; \mathcal{B}(E))$ for each $|\alpha| = k < m$, with $r_k \geq q$ and $m - k > n/r_k$.*

Let A denote the realization of $\mathcal{A}(x, D)$ in the base space $X_0 = L_q(\mathbb{R}^n; E)$ with domain $\mathsf{D}(A) = H_q^m(\mathbb{R}^n; E)$.

Then for each $\phi > \phi_{\mathcal{A}}$ there is $\mu_\phi \geq 0$ such that $\mu_\phi + A$ is $\mathcal{R}$-sectorial with $\phi^R_{\mu_\phi + A} \leq \phi$.

Proof. (a) Freeze the coefficients a_α, $|\alpha| = m$, at an arbitrary $x_0 \in \mathbb{R}^n \cup \{\infty\}$ and consider the homogeneous differential operator with constant coefficients $\mathcal{A}_\#(x_0, D)$; let A_0 denote its L_q-realization. Then we know from Theorem 6.1.8 that $\mathsf{D}(A_0) = H_q^m(\mathbb{R}^n; E)$ and that A_0 is $\mathcal{R}$-sectorial with $\mathcal{R}$-angle $\phi^R_{A_0} \leq \phi_{\mathcal{A}}$. By assumption **(ra1)** the coefficients a_α belong to a compact subset of $\mathcal{B}(E)$. By Corollary 6.1.9(i) and the perturbation results from Section 4.4, we see that the $\mathcal{R}$-bounds of $\lambda^{1-|\beta|/m} D^\beta (\lambda + A_0)^{-1}$ are upper semi-continuous in the coefficients, where $\lambda \in \Sigma_{\pi-\phi}$ for $\phi > \phi_{\mathcal{A}}$ fixed. Therefore, they are uniform in $x_0 \in \mathbb{R}^n \cup \{\infty\}$.

Applying the perturbation argument from Section 4.4 another time, we see that there is a number $\eta > 0$ independent of x_0 such that the L_q-realization $A_0 + A_1$ of $\mathcal{A}_0(D) + \mathcal{A}_1(x, D)$ is again $\mathcal{R}$-sectorial, whenever

$$\mathcal{A}_1(x, D) = \sum_{|\alpha|=m} a^1_\alpha(x) D^\alpha$$

has L_∞-coefficients subject to $|a^1_\alpha(x)|_{\mathcal{B}(E)} \leq \eta$, uniformly in x, for each $|\alpha| = m$. The corresponding $\mathcal{R}$-bounds are also uniform in x_0, and the domain of $A_0 + A_1$ equals $H^m_q(\mathbb{R}^n; E)$.

(b) Here we assume $a_\alpha \in L_\infty$ for $|\alpha| < m$ and Condition **(ra1)**. Choose a large ball $B(0, r_0)$ such that

$$|a_\alpha(x) - a_\alpha(\infty)|_{\mathcal{B}(E)} \leq \eta, \quad \text{for all } |x| \geq r_0,\ |\alpha| = m,$$

and set $U_0 = \mathbb{R}^n \setminus \bar{B}(0, r_0)$. Cover $\bar{B}(0, r_0)$ by finitely many balls $U_j = B(x_j, r_j)$ such that

$$|a_\alpha(x) - a_\alpha(x_j)|_{\mathcal{B}(E)} \leq \eta, \quad \text{for all } |x - x_j| \leq r_j,\ |\alpha| = m,\ j = 1, \ldots, N.$$

Define coefficients of local operators A_j e.g. by reflection, i.e.,

$$a^0_\alpha(x) = \begin{cases} a_\alpha(x), & x \notin \bar{B}(0, r_0) \\ a_\alpha\big(r_0^2 \frac{x}{|x|^2}\big), & x \in \bar{B}(0, r_0) \end{cases}$$

and

$$a^j_\alpha(x) = \begin{cases} a_\alpha(x) & x \in \bar{B}(x_j, r_j) \\ a_\alpha\big(x_j + r_j^2 \frac{x - x_j}{|x - x_j|^2}\big), & x \notin \bar{B}(x_j, r_j) \end{cases}$$

for each $j = 1, \ldots, N$. Then $|a^j_\alpha(x) - a_\alpha(x_j)|_{\mathcal{B}(E)} \leq \eta$, for each $x \in \mathbb{R}^n$ and $j = 0, \ldots, N$, hence by step (a) above the L_q-realizations A_j of

$$\mathcal{A}_j(x, D) = \sum_{|\alpha| = m} a^j_\alpha(x) D^\alpha$$

are $\mathcal{R}$-sectorial and the $\mathcal{R}$-bounds of the sets

$$\{\lambda^{1-k/m} D^\beta (\lambda + A_j)^{-1} :\ \lambda \in \Sigma_{\pi - \phi},\ |\beta| = k \leq m\}$$

are finite. Next we choose a partition of unity $\varphi_j \in \mathcal{D}(\mathbb{R}^n)$ such that $0 \leq \varphi_j(x) \leq 1$ and $\operatorname{supp} \varphi_j \subset U_j$. We may also choose $\psi_j \in \mathcal{D}(\mathbb{R}^n)$ such that $\operatorname{supp} \psi_j \subset U_j$ and $\psi_j = 1$ on $\operatorname{supp} \varphi_j$. Set $\mathcal{B}(x, D) = \sum_{|\alpha| < m} a_\alpha(x) D^\alpha$. We then obtain a representation of $(\lambda + A)^{-1}$ as follows.

$$\lambda u + Au = f \quad \text{iff} \quad \lambda u + \mathcal{A}_\#(x, D)u = f - \mathcal{B}(x, D)u.$$

Multiply by φ_j to obtain

$$\lambda(\varphi_j u) + \mathcal{A}_\#(x, D)(\varphi_j u) = \varphi_j f + [\mathcal{A}_\#(x, D), \varphi_j]u - \varphi_j \mathcal{B}(x, D)u.$$

Noting that $\mathcal{A}_\#(x, D)(\varphi_j u) = A_j(\varphi u)$, we employ the resolvent of A_j to the result

$$\varphi_j u = (\lambda + A_j)^{-1}(\varphi_j f) + (\lambda + A_j)^{-1}\{[\mathcal{A}_\#(x, D), \varphi_j]u - \varphi_j \mathcal{B}(x, D)u\}.$$

Observing $\psi_j = 1$ on $\operatorname{supp}\varphi_j$, multiplying with ψ_j and summing over j we finally get

$$u = \sum_j \psi_j(\lambda + A_j)^{-1}\varphi_j f + \sum_j \psi_j(\lambda + A_j)^{-1}\mathcal{C}_j(x, D)u, \tag{6.18}$$

where the differential operators

$$\mathcal{C}_j(x, D) := [\mathcal{A}_\#(x, D), \varphi_j] - \varphi_j\mathcal{B}(x, D) = \sum_{|\beta|<m} c^j_\beta(x)D^\beta$$

are in fact operators of order $\leq m - 1$. Hence for each $\varepsilon > 0$ there is $C_\varepsilon > 0$ such that

$$|\mathcal{C}_j(x, D)v|_q \leq \varepsilon|D^m v|_q + C_\varepsilon|v|_q, \quad \text{for all } v \in \mathsf{D}(A),\ j = 0, \ldots, N.$$

By a Neumann series argument, (6.18) implies existence of a left inverse S_λ, which is given by

$$S_\lambda f = (I - \sum_j \psi_j(\lambda + A_j)^{-1}\mathcal{C}_j(x, D))^{-1}\sum_j \psi_j(\lambda + A_j)^{-1}\varphi_j f,$$

for $\lambda \in \Sigma_{\pi-\phi}$, $|\lambda| \geq \lambda_0$ for some sufficiently large λ_0, as well as

$$|\lambda S_\lambda f|_q + |D^m S_\lambda f|_q \leq C|f|_q, \quad \lambda \in \Sigma_{\pi-\phi}, \quad |\lambda| \geq \lambda_0.$$

This shows that $\mu + A$ is sectorial for $\mu \geq \lambda_0$, and $\phi_{\mu+A} \leq \phi$, provided $\lambda + A$ is surjective, i.e., there is also a right inverse.

To show the latter we apply $\lambda + \mathcal{A}_\#(x, D)$ to $u = S_\lambda f$ which yields

$$\begin{aligned}(\lambda + \mathcal{A}_\#(D))S_\lambda &= \sum_j (\lambda + \mathcal{A}_\#(D))\psi_j(\lambda + A_j)^{-1}(\varphi_j + \mathcal{C}_j(x, D)S_\lambda)\\ &= \sum_j \psi_j\{\varphi_j + \mathcal{C}_j(x, D)S_\lambda\} + \sum_j [\mathcal{A}_\#(x, D), \psi_j](\lambda + A_j)^{-1}\{\varphi_j + \mathcal{C}_j(x, D)S_\lambda\}.\end{aligned}$$

Since $\psi_j = 1$ on $\operatorname{supp}\varphi_j$ and $\sum_j \varphi_j = 1$, as well as $\sum_j[\mathcal{A}_\#(x, D), \varphi_j] = 0$, we obtain

$$\sum_j \psi_j\{\varphi_j + \mathcal{C}_j(x, D)S_\lambda\} = \sum_j \{\varphi_j + \mathcal{C}_j(x, D)S_\lambda\} = I - \mathcal{B}(x, D)S_\lambda.$$

This yields the following identity

$$(\lambda + \mathcal{A}(x, D))S_\lambda = I + \sum_j [\mathcal{A}_\#(x, D), \psi_j](\lambda + A_j)^{-1}\{\varphi_j + \mathcal{C}_j(x, D)S_\lambda\}. \tag{6.19}$$

The commutators $[\mathcal{A}(x, D), \psi_j]$ are differential operators of order $m-1$, hence the second term on the right-hand side of (6.19) will be small for large $|\lambda|$ which as above shows that (6.19) gives rise to a right inverse of $\lambda + A$; in particular $\lambda + A$ is surjective for large $|\lambda|$.

Next, with

$$R_0(\lambda) = \sum_{j=0}^{N} \psi_j(\lambda + A_j)^{-1}\varphi_j, \quad R_1(\lambda) = \sum_{j=0}^{N} \psi_j(\lambda + A_j)^{-1}\mathcal{C}_j(x, D),$$

the resolvent of A may be written as the Neumann series

$$(\lambda + A)^{-1} = \sum_{k=0}^{\infty} R_1(\lambda)^k R_0(\lambda), \quad \lambda \in \Sigma_{\pi-\phi}, \ |\lambda| \geq \lambda_0.$$

For $j, k = 0, \ldots, N$ we obtain by the contraction principle

$$\begin{aligned}
&\mathcal{R}\{\mathcal{C}_j(x, D)(\lambda + A_k)^{-1} : \ \lambda \in \Sigma_{\pi-\phi}, \ |\lambda| \geq \lambda_0\} \\
&\quad \leq \sum_{|\beta|<m} |c^j_\beta|_{L_\infty(\mathbb{R}^n;E)} \mathcal{R}\{D^\beta(\lambda + A_k)^{-1}\} \\
&\quad \leq \sum_{|\beta|<m} |c^j_\beta|_{L_\infty(\mathbb{R}^n;E)} \lambda_0^{-1+|\beta|/m} \mathcal{R}\{\lambda^{1-|\beta|/m} D^\beta(\lambda + A_k)^{-1}\} \leq C\varepsilon,
\end{aligned} \tag{6.20}$$

provided λ_0 is sufficiently large. This then implies

$$\begin{aligned}
&\mathcal{R}\{\lambda^{1-|\alpha|/m} D^\alpha(\lambda + A)^{-1} : \ \lambda \in \Sigma_{\pi-\phi}, \ |\lambda| \geq \lambda_0, \ |\alpha| \leq m\} \\
&\quad \leq (N+1)C \sum_{k=0}^{\infty} ((N+1)C\varepsilon)^k = (N+1)C/(1 - (N+1)C\varepsilon) < \infty,
\end{aligned} \tag{6.21}$$

in particular, $\mu + A$ is $\mathcal{R}$-sectorial for all $\mu \geq \lambda_0$.

(c) Let us consider now the case where $a_\beta \in L_{r_k}(\mathbb{R}^n; \mathcal{B}(E))$, with $|\beta| = k < m$ and $r_k \geq q$, $m - k > n/r_k$. Then we estimate the terms $a_\beta(x)D^\beta(\lambda + A_l)^{-1}$ as follows. With $qr = r_k$, $1/r + 1/r' = 1$, the *Gagliardo-Nirenberg inequality* yields

$$\begin{aligned}
&\Big|\sum_j \varepsilon_j a_\beta D^\beta(\lambda_j + A_l)^{-1} f_j\Big|_{L_q(\mathbb{R}^n;E)} \\
&\quad \leq |a_\beta|_{L_{qr}(\mathbb{R}^n;\mathcal{B}(E))} \Big|\sum_j \varepsilon_j D^\beta(\lambda_j + A_l)^{-1} f_j\Big|_{L_{qr'}(\mathbb{R}^n;E))} \\
&\quad \leq C|a_\beta|_{L_{qr}(\mathbb{R}^n;\mathcal{B}(E))} \Big[\sum_{|\alpha|=m} \Big|\sum_j \varepsilon_j D^\alpha(\lambda_j + A_l)^{-1} f_j\Big|_{L_q(\mathbb{R}^n;E))}\Big]^a \cdot \\
&\qquad \cdot \Big|\sum_j \varepsilon_j(\lambda_j + A_l)^{-1} f_j\Big|_{L_q(\mathbb{R}^n;E))}\Big]^{1-a}
\end{aligned}$$

$$\le C|a_\beta|_{L_{qr}(\mathbb{R}^n;\mathcal{B}(E))}\Big[\sum_{|\alpha|=m}\Big|\sum_j \varepsilon_j D^\alpha(\lambda_j+A_l)^{-1}f_j\Big|_{L_q(\mathbb{R}^n;E))}\Big]^a\cdot$$

$$\cdot\,\lambda_0^{-(1-|\beta|/m)(1-a)}\Big[\Big|\sum_j \varepsilon_j\lambda_j^{1-|\beta|/m}(\lambda_j+A_l)^{-1}f_j\Big|_{L_q(\mathbb{R}^n;E))}\Big]^{1-a},$$

where $am-k = n/qr = n/r_k$, in particular $a<1$ by assumption **(ra 3)**. Integrating over Ω this yields

$$\begin{aligned}&\Big|\sum_j \varepsilon_j a_\beta D^\beta(\lambda_j+A_l)^{-1}f_j\Big|_{L_q(\Omega\times\mathbb{R}^n;E)}\\ &\quad\le C\lambda_0^{-(1-|\beta|/m)(1-a)}|a_\beta|_{L_{qr}(\mathbb{R}^n;\mathcal{B}(E))}\Big|\sum_j\varepsilon_j f_j\Big|_{L_q(\mathbb{R}^n;E)}\\ &\quad\le C\varepsilon\Big|\sum_j\varepsilon_j f_j\Big|_{L_q(\mathbb{R}^n;E)},\end{aligned}$$

whenever λ_0 is sufficiently large, and consequently we have

$$\mathcal{R}\{a_\beta(x)D^\beta(\lambda+A_k)^{-1}:\ \lambda\in\Sigma_{\pi-\phi},\ |\lambda|\ge\lambda_0\}\le C\varepsilon.$$

We now may proceed as in Step (b) to obtain the result in the general case. □

As a consequence of the results on maximal regularity from Section 4.5 we obtain for the time-dependent parabolic equation

$$\partial_t u+\omega u+Au=f,\quad t>0,\quad u(0)=u_0, \tag{6.22}$$

the following result.

Theorem 6.1.11. *Let Condition* **(ra)** *hold,* $1<p,q<\infty$, $\mu\in(1/p,1]$, *let* $\mathcal{A}(x,D)$ *be uniformly normally elliptic and,* $\omega\ge\omega_0>\mathsf{s}(-A)=\sup\operatorname{Re}\sigma(-A)$, *the spectral bound of* $-A$.

Then (6.22) *has maximal regularity of type* $L_{p,\mu}-L_q$ *on* $\mathbb{R}_+$. *More precisely,* (6.22) *admits a solution* u *in the class*

$$u\in H^1_{p,\mu}(\mathbb{R}_+;L_q(\mathbb{R}^n;E))\cap L_{p,\mu}(\mathbb{R}_+;H^2_q(\mathbb{R}^n;E))=:\mathbb{E}_{1\mu}$$

if and only if

$$f\in L_{p,\mu}(\mathbb{R}_+;L_q(\mathbb{R}^n;E))=:\mathbb{E}_{0\mu}\quad\text{and}\quad u_0\in B^{m(\mu-1/p)}_{qp}(\mathbb{R}^n;E)=:X_{\gamma,\mu}.$$

Moreover, there is a constant $C>0$ *such that*

$$|u|_{\mathbb{E}_{1\mu}}+\omega|u|_{\mathbb{E}_{0\mu}}\le C(|u_0|_{X_{\gamma,\mu}}+|f|_{\mathbb{E}_{0\mu}}),$$

for all $(f,u_0)\in\mathbb{E}_{0\mu}\times X_{\gamma,\mu}$, *and all* $\omega\ge\omega_0$.

We observe that via the exponential shifts $u_\omega = e^{\omega t}u$ and $f_\omega = e^{\omega t}f$, u is a solution of (6.22) if and only if u_ω solves

$$\partial_t u_\omega + Au_\omega = f_\omega, \quad t > 0, \quad u_\omega(0) = u_0. \tag{6.23}$$

This way the following result is obtained.

Corollary 6.1.12. *Let Condition* **(ra)** *hold,* $1 < p, q < \infty$, $\mu \in (1/p, 1]$, *and let* $\mathcal{A}(x, D)$ *be uniformly normally elliptic and,* $\omega > \mathsf{s}(-A)$.

Then (6.23) *admits a unique solution* u *in the class*

$$e^{-\omega t}u \in H^1_{p,\mu}(\mathbb{R}_+; L_q(\mathbb{R}^n; E)) \cap L_{p,\mu}(\mathbb{R}_+; H^2_q(\mathbb{R}^n; E))$$

if and only if

$$e^{-\omega t}f \in L_{p,\mu}(\mathbb{R}_+; L_q(\mathbb{R}^n; E)) \quad \textit{and} \quad u_0 \in B^{m(\mu-1/p)}_{qp}(\mathbb{R}^n; E).$$

Consequently, on finite intervals (6.22) *has maximal* $L_{p,\mu} - L_q$*-regularity, for each* $\omega \in \mathbb{R}$.

1.5 Different Spatial Orders

Many times one is in need of maximal regularity results with different spatial regularity. In this subsection we briefly discuss this topic. We assume below that $\mathcal{A}(x, D)$ satisfies properties **(ra1), (ra2), (ra3)**.

(i) Higher Order Regularity

Here we want to replace the base space $L_q(\mathbb{R}^n; E)$ by $K^s_q(\mathbb{R}^n; E)$ for $s > 0$ and $K \in \{H, W\}$, where $s \notin \mathbb{N}$ in case $K = W$. For this purpose we fix any $k \in \mathbb{N}$ and consider the operator $\mathcal{A}(x, D)$ in $H^k_q(\mathbb{R}^n; E)$. Differentiating the equations

$$(\lambda + \omega + \mathcal{A}(x, D))u = f \text{ in } \mathbb{R}^n,$$

or

$$(\partial_t + \omega + \mathcal{A}(x; D))u = f, \ t > 0, \ u(0) = 0, \text{ in } \mathbb{R}^n$$

k times in space leads to the problems

$$(\lambda + \omega + \mathcal{A}(x, D))D^\beta u - [\mathcal{A}(x, D), D^\beta]u = D^\beta f \text{ in } \mathbb{R}^n,$$

or

$$(\partial_t + \omega + \mathcal{A}(x; D))D^\beta u - [\mathcal{A}(x, D), D^\beta]u = D^\beta f, \ t > 0, \ D^\beta u(0) = 0, \text{ in } \mathbb{R}^n.$$

As the commutator $[\mathcal{A}(x, D), D^\beta]$ is of lower order, this yields with Proposition 4.4.3 the analogues of Theorems 6.1.10 and 6.1.11 with base space $L_q(\mathbb{R}^n; E)$ replaced by $H^k_q(\mathbb{R}^n; E)$, provided the coefficients of $\mathcal{A}(x, D)$ have enough regularity. Computing the relevant commutator shows that Condition **(ra3)** must be replaced by

(ra3$_\mathbf{k}$) $a_\alpha \in H^k_{r_l}(\mathbb{R}^n; \mathcal{B}(E)) + W^k_\infty(\mathbb{R}^n; \mathcal{B}(E))$, $|\alpha| = l \leq m$, $r_l \geq q$, $m + k - l > n/r_l$.

Then employing real or complex interpolation, we see that Theorems 6.1.10 and 6.1.11 are also valid for the base spaces $K^s_q(\mathbb{R}^n;E)$, for all $0 \le s \le k$, $s \notin \mathbb{N}_0$ in case $K = W$. Note that for the parabolic problem we first choose $p = q$, $\mu = 1$ to obtain $\mathcal{R}$-sectoriality, and then use Theorems 4.4.4 and 3.5.4 for the general case.

(ii) Lower Order Regularity
Here we want to replace the base space $L_q(\mathbb{R}^n;E)$ by $K^{-s}_q(\mathbb{R}^n;E)$ where $s > 0$ and $K \in \{H, W\}$, $s \notin \mathbb{N}$ in case $K = W$. Consider first the space $H^{-2}_q(\mathbb{R}^n;E)$. As $I - \Delta : L_q(\mathbb{R}^n;E) \to H^{-2}_q(\mathbb{R}^n;E)$ is an isomorphism, it is reasonable to apply $(I-\Delta)^{-1}$ to the equations under consideration to obtain problems in L_q. This yields equations for $v = (I-\Delta)^{-1}u$ in $L_q(\mathbb{R}^n;E)$,

$$(\lambda + \omega + \mathcal{A}(x,D))v - [\mathcal{A}(x,D),(I-\Delta)^{-1}]u = (I-\Delta)^{-1}f \ \text{ in } \mathbb{R}^n,$$

or

$$(\partial_t + \omega + \mathcal{A}(x;D))v - [\mathcal{A}(x,D),(I-\Delta)^{-1}]u = (I-\Delta)^{-1}f,\ t>0,\ u(0)=0, \ \text{ in } \mathbb{R}^n.$$

Looking at the commutator we find

$$[\mathcal{A}(x,D),(I-\Delta)^{-1}]u = (I-\Delta)^{-1}[\Delta,\mathcal{A}(x,D)](I-\Delta)^{-1}u = (I-\Delta)^{-1}[\Delta,\mathcal{A}(x,D)]v.$$

Now we have

$$[\Delta, a_\alpha D^\alpha] = \sum_{j=1}^n (\partial_j^2 a_\alpha)D^\alpha + 2(\partial_j a_\alpha)\partial_j D^\alpha,$$

which implies that the commutator is of lower order in $L_q(\mathbb{R}^n;E)$, provided the coefficients a_α are subject to $(\mathbf{ra3_2})$. Therefore, in this case Theorems 6.1.10 and 6.1.11 are also valid for the base space $H^{-2}_q(\mathbb{R}^n;E)$. Induction yields the same result for $H^{-2k}_q(\mathbb{R}^n;E)$ provided the coefficients satisfy $(\mathbf{ra3_{2k}})$, for all $k \in \mathbb{N}$. Interpolation finally shows that Theorems 6.1.10 and 6.1.11 hold for the base space $K^{-s}_q(\mathbb{R}^n;E)$, for all $s \in [0,2k]$, provided $(\mathbf{ra3_{2k}})$ holds; here $s \in \mathbb{N}_0$ is excluded in case $K = W$.

Remark 6.1.13. A more refined analysis shows that Theorems 6.1.10 and 6.1.11 are valid in $K^{\pm s}_q(\mathbb{R}^n;E)$, $s > 0$, if the coefficients merely satisfy

$(\mathbf{a3_s})$ $a_\alpha \in H^s_{r_l}(\mathbb{R}^n;\mathcal{B}(E)) + W^s_\infty(\mathbb{R}^n;\mathcal{B}(E))$, $|\alpha| = l \le m$, $r_l \ge q$, $m+s-l > n/r_l$.

However, this assertion is more elaborate, and so we refrain here from a proof.

6.2 Elliptic and Parabolic Systems on $\mathbb{R}^n_+$

Let E be a Banach space of class $\mathcal{HT}$, and consider the parabolic problem

$$\begin{aligned} \partial_t u + \omega u + \mathcal{A}(x,D)u &= f && \text{in } \mathbb{R}^n_+,\\ \mathcal{B}_j(x,D)u &= g_j && \text{on } \partial\mathbb{R}^n_+,\ j=1,\dots,m,\\ u(0) &= u_0 && \text{in } \mathbb{R}^n_+. \end{aligned} \tag{6.24}$$

Here $\mathcal{A}(x,D) = \sum_{|\alpha|\le 2m} a_\alpha D^\alpha$ is a differential operator of degree $2m$, $\mathcal{B}_j(x,D) = \sum_{|\beta|\le m_j} b_{j\beta} D^\beta$ are differential operators of degree $m_j < 2m$, and the data (f, g_j) and u_0 are given. This problem may be reduced to a homogeneous problem with inhomogeneous boundary conditions as follows. Extend the function $f \in L_{p,\mu}(\mathbb{R}_+; L_q(\mathbb{R}^n_+; E))$ trivially to a function $\bar{f} \in L_{p,\mu}(\mathbb{R}_+; L_q(\mathbb{R}^n; E))$, the coefficients of $\mathcal{A}(x,D)$ by symmetry to all of $\mathbb{R}^n$, and extend the initial value $u_0 \in B^{2m(\mu-1/p)}_{qp}(\mathbb{R}^n_+; E)$ to some $\bar{u}_0 \in B^{2m(\mu-1/p)}_{qp}(\mathbb{R}^n; E)$. Then we may apply the results from the previous section, in particular Theorem 6.1.11, to obtain the solution

$$\bar{u} \in H^1_{p,\mu}(\mathbb{R}_+; L_q(\mathbb{R}^n; E)) \cap L_{p,\mu}(\mathbb{R}_+; H^{2m}_q(\mathbb{R}^n; E)))$$

of the full space problem

$$\begin{aligned} \partial_t \bar{u} + \omega \bar{u} + \mathcal{A}(x,D)\bar{u} &= \bar{f} \quad \text{in } \mathbb{R}^n, \\ \bar{u}(0) &= \bar{u}_0 \quad \text{in } \mathbb{R}^n. \end{aligned} \tag{6.25}$$

Then the function $\tilde{u} = u - \bar{u}$ satisfies (6.24) with $(f, u_0) = 0$ and g_j replaced by $\tilde{g}_j = g_j - \mathcal{B}_j(x,D)\bar{u}$. This way we have reduced the problem to a homogeneous parabolic equation with trivial initial data, but inhomogeneous boundary data. Note that the natural compatibility conditions

$$\mathcal{B}_j(x,D)u_0 = g_j(0), \quad j = 1, \dots m,$$

become $\tilde{g}_j(0) = 0$. Below we will therefore always consider the case $(f, u_0) = 0$.

Similarly for the elliptic problem

$$\begin{aligned} \lambda u + \omega u + \mathcal{A}(x,D)u &= f \quad \text{in } \mathbb{R}^n_+, \\ \mathcal{B}_j(x,D)u &= g_j \quad \text{on } \partial\mathbb{R}^n_+,\ j = 1, \dots, m. \end{aligned} \tag{6.26}$$

We may assume $f = 0$, by Theorem 6.1.10 of the previous section.

2.1 The Boundary Symbol

We begin with the constant coefficient case, i.e., we consider

$$\mathcal{A}(D) = \sum_{|\alpha|=2m} a_\alpha D^\alpha, \quad \mathcal{B}_j(D) = \sum_{|\beta|=m_j} b_{j\beta} D^\beta$$

with coefficients $a_\alpha, b_{j\beta} \in \mathcal{B}(E)$. It is convenient to replace x by (x, y), where $x \in \mathbb{R}^{n-1}$ are tangential variables and $y > 0$ is the normal variable. Taking the Laplace transform in time with covariable λ and Fourier transform in the tangential direction with covariable $\xi \in \mathbb{R}^{n-1}$, with $\nu = e_n$ we obtain the transformed problem

$$\begin{aligned} (\lambda + \omega)v_1(y) + \mathcal{A}(\xi + \nu D_y)v_1(y) &= 0, \quad y > 0, \\ \mathcal{B}_j(\xi + \nu D_y)v_1(0) &= h_j, \quad j = 1, \dots, m. \end{aligned} \tag{6.27}$$

This is a boundary value problem for an ordinary differential equation on $\mathbb{R}_+$, where the covariables λ and ξ are parameters. We may rewrite the differential operators in the following form.

$$\mathcal{A}(\xi + \nu D_y) = \sum_{k=0}^{2m} a_k(\xi) D_y^{2m-k}, \quad \mathcal{B}_j(\xi + \nu D_y) = \sum_{k=0}^{m_j} b_{jk}(\xi) D_y^{m_j-k}.$$

Observe that $a_k(\xi)$ as well as $b_{jk}(\xi)$ are homogeneous polynomials of degree k.

We shall assume from now on that $\mathcal{A}(D)$ is parameter-elliptic with angle $\phi_{\mathcal{A}}$. Then $a_0 = \mathcal{A}(0, \dots, 0, 1)$ is invertible. For $\lambda \in \Sigma_{\pi-\phi}$, $\phi > \phi_{\mathcal{A}}$, we introduce the new variables $v = [v_j]$, and the scaling parameter $\rho = (w + \lambda + |\xi|^{2m})^{1/2m}$

$$v_j(y) = \rho^{-j+1} D_y^{j-1} v_1(y), \quad j = 1, \dots, 2m,$$

we may rewrite the differential equation in (6.27) as

$$\partial_y v(y) = i\rho A_0(b, \sigma) v(y), \quad y > 0,$$

with $\sigma = (\omega + \lambda)/\rho^{2m}$, $b = \xi/\rho$ and

$$A_0(b, \sigma) = \begin{pmatrix} 0 & I & 0 & \dots & 0 \\ 0 & 0 & I & \dots & 0 \\ \vdots & \vdots & \vdots & \vdots & \vdots \\ 0 & 0 & \dots & 0 & I \\ c_{2m}(b, \sigma) & c_{2m-1}(b) & \dots & c_2(b) & c_1(b) \end{pmatrix},$$

where $c_j(b) = -a_0^{-1} a_j(b)$, $j = 1, \dots, 2m - 1$ and $c_{2m}(b, \sigma) = -a_0^{-1}(\sigma + a_{2m}(b))$. Similarly, for homogeneity reasons the boundary conditions become

$$B_j^0(b) v(0) = \rho^{-m_j} h_j =: \tilde{h}_j, \quad j = 1, \dots, m,$$

with $B_j^0(b) : E^{2m} \to E$ defined by

$$B_j^0(b) = (b_{jm_j}(b), \dots, b_{j0}(b), 0, \dots, 0), \quad j = 1, \dots, m.$$

This way the boundary value problem (6.27) is transformed to the first-order system

$$\begin{aligned} \partial_y v(y) &= i\rho A_0(b, \sigma) v(y), & y > 0, \\ B_j^0(b) v(0) &= \tilde{h}_j, & j = 1, \dots, m. \end{aligned} \tag{6.28}$$

To solve this boundary value problem we need some preparation.

Lemma 6.2.1. *Let b, σ and $A_0(b, \sigma)$ be defined as above. Then*

$$\sigma(A_0(b, \sigma)) \cap \mathbb{R} = \emptyset.$$

Proof. We first prove that $\sigma_p(A_0(b,\sigma)) \cap \mathbb{R}$ is empty, where $\sigma_p(A_0(b,\sigma))$ denotes the point spectrum of $A_0(b,\sigma)$. To this end, suppose that $\eta \in \mathbb{R}$ is an eigenvalue of $A_0(b,\sigma)$ with eigenvector $x = [x_0, \dots, x_{2m-1}]^\mathsf{T} \neq 0$. Then

$$\begin{aligned} &\eta x_0 = x_1, \quad \dots \quad \eta x_{2m-2} = x_{2m-1}, \\ &\eta x_{2m-1} = -a_0^{-1}((\sigma + a_{2m}(b))x_0 + a_{2m-1}(b)x_1 + \dots + a_1(b)x_{2m-1}). \end{aligned} \tag{6.29}$$

This implies $(\sigma + \sum_{k=0}^{2m} a_k(b)\eta^{2m-k})x_0 = 0$. It follows from the first line of (6.29) that $x_0 \neq 0$. Therefore, $-\sigma$ is an eigenvalue for $\mathcal{A}(b,\eta)$ with eigenvector x_0. But as $\mathcal{A}$ is parameter-elliptic this implies $-\sigma \in \Sigma_\phi$, which contradicts the assumption $\lambda \in \Sigma_{\pi-\phi}$.

Next, assume that $\eta \in \mathbb{R}$ belongs to the residual spectrum $\sigma_r(A_0(b,\sigma))$. Then $\eta \in \sigma_p(A_0^*(b,\sigma))$, hence there is $x^* = (x_0^*, \dots, x_{2m-1}^*)^\mathsf{T} \neq 0$ such that $A_0^*(b,\sigma)x^* = \eta x^*$. This implies as before $x_{2m-1}^* \neq 0$ and

$$(\sigma + \sum_{k=0}^{2m} a_k(b)^* \eta^{2m-k})x_{2m-1}^* = 0.$$

This shows that $-\sigma$ is an eigenvalue of $\mathcal{A}^*(b,\eta)$, hence belongs to $\sigma_r(\mathcal{A}(b,\eta))$, which is not possible.

Finally, assume that $\eta \in \mathbb{R}$ is in the continuous spectrum $\sigma_c(A_0(b,\sigma))$. Then we find $x_n = (x_{n,0}, \dots, x_{n,2m-1})^\mathsf{T}$ with $|x_n|_{E^{2m}} = 1$ such that $A_0(b,\sigma)x_n = \eta x_n + y_n$, with $y_n \to 0$ as $n \to \infty$. As above this yields

$$(\sigma + \sum_{k=0}^{2m} a_k(b)\eta^{2m-k})x_{n,0} \to 0,$$

hence $-\sigma$ belongs to $\sigma_c(\mathcal{A}(b,\eta))$ which yields a contradiction as before. □

This lemma shows that the spectrum of $iA_0(b,\sigma) \in \mathcal{B}(E^{2m})$ splits into two parts, $s_-(b,\sigma)$ contained in the open left half-plane, and $s_+(b,\sigma)$ contained in the open right half-plane. By compactness, there are constants $c_\pm > 0$ such that

$$\sup \operatorname{Re} s_-(b,\sigma) \leq -c_- < 0 < c_+ \leq \inf \operatorname{Re} s_+(b,\sigma),$$

for all relevant b,σ. Let $P_\pm(b,\sigma) \in \mathcal{B}(E^{2m})$ denote the associated spectral projections of $iA_0(b,\sigma)$; these are holomorphic and bounded, uniformly in (b,σ). The boundary value problem (6.28) admits precisely one solution $v \in C_0(\mathbb{R}_+; E^{2m})$ if and only if the system

$$\begin{aligned} B_j^0(b)w &= \tilde{h}_j, \quad j = 1, \dots, m, \\ P_+(b,\sigma)w &= 0 \end{aligned} \tag{6.30}$$

admits a unique solution $w \in E^{2m}$. The solution v of (6.28) is then given by

$$v(y) = e^{iy\rho A_0(b,\sigma)}w, \quad y \geq 0.$$

To ensure this solvability property we assume the equivalent

Lopatinskii-Shapiro Condition (LS)
For each $\xi, \nu \in \mathbb{R}^n$, $\lambda \in \Sigma_{\pi-\phi}$ for some $\phi > \phi_{\mathcal{A}}$, where $(\lambda, \xi) \neq (0,0)$, $|\nu| = 1$, $(\xi|\nu) = 0$, the problem

$$\begin{aligned} \lambda u(y) + \mathcal{A}(\xi + \nu D_y)u(y) &= 0, \quad y > 0, \\ \mathcal{B}_j(\xi + \nu D_y)u(0) &= g_j, \quad j = 1, \ldots, m, \end{aligned}$$

has exactly one solution $u \in C_0(\mathbb{R}_+; E)$, for any given vectors $g_j \in E$, $j = 1, \ldots, m$.

Remark 6.2.2. **(i)** It is obvious that also the Lopatinskii-Shapiro condition is invariant under orthogonal transformations. But even more, it is invariant w.r.t. general coordinate transformations as well. In fact, under the coordinate transformation $Tu(x) = u(Qx)$ with invertible $Q \in \mathbb{R}^{n\times n}$, the normal ν transforms to $\nu_Q = Q^{-\mathsf{T}}\nu$. Therefore,

$$\mathcal{A}_Q(\xi' + \nu_Q D_y) = \mathcal{A}(Q^{\mathsf{T}}\xi' + \nu D_y) = \mathcal{A}(\xi + \alpha\nu + \nu D_y),$$

where $(\xi|\nu) = 0$ and $\alpha = (\xi'|Q\nu)$. The same applies to the boundary operators $\mathcal{B}_j$. The exponential shift $v(y) = e^{i\alpha y}w(y)$ then shows that we may assume $\alpha = 0$. This reduces **(LS)** for the transformed problem to **(LS)** for the original one.

(ii) The shift argument also shows that the condition $(\xi|\nu) = 0$ in **(LS)** is redundant, only $|\nu| = 1$ is essential.

(iii) There are versions of the Lopatinskii-Shapiro condition for more refined boundary value problems which also appear in applications. Each of the m boundary operators may be split into finitely many ones of different order. More precisely, for fixed $j \in \{1, \ldots, m\}$, we let $E = \oplus_{k=0}^{n_j} E_{jk}$, and replace the condition $\mathcal{B}_j(D)u = g_j$ by

$$\mathcal{B}_{jk}(D)u = g_{jk}, \quad k = 0, \ldots, n_j,$$

where the coefficients of $\mathcal{B}_{jk}(D)$ satisfy $b_{jk\beta} \in \mathcal{B}(E, E_{jk})$, and their orders are $m_{jk} \in \{0, \ldots, 2m-1\}$. Condition **(LS)** extends literally to such cases, and the analysis presented here carries over.

(iv) If $E \simeq \mathbb{C}^N$ is finite-dimensional, then the kernel of P_+ has dimension mN, hence if we prescribe mN scalar boundary conditions, it is enough to have uniqueness in **(LS)**, by a dimensional argument.

The Lopatinskii-Shapiro condition implies the following result.

Proposition 6.2.3. *Suppose that $\mathcal{A}(D)$ is parameter-elliptic with angle $\phi_{\mathcal{A}}$, and assume the Lopatinskii-Shapiro Condition for some $\phi > \phi_{\mathcal{A}}$. Then for each $\tilde{h} = [\tilde{h}_j] \in E^m$, $j = 1, \ldots, m$, problem* (6.30) *admits a unique solution $w \in E^{2m}$. This solution is represented as $w = M_0(b,\sigma)\tilde{h}$, where the map $M_0 : U \to \mathcal{B}(E^m, E^{2m})$ is holomorphic on a neighbourhood $U \subset \mathbb{C}^{n+1}$ of $\{(b,\sigma) : (\lambda, \xi) \in \Sigma_{\pi-\phi} \times \mathbb{R}^{n-1}\}$.*

Proof. Existence, uniqueness and linearity are clear, so we need to show holomorphy of M_0. For this purpose set $z = (b, \sigma) \in U$ and $B(z) = (B_1^0(z), \ldots, B_m^0(z))$. Then $u(z) = M_0(z)g$ defines the unique solution of the system

$$P_+(z)u = 0, \quad B(z)u = g.$$

Let D denote a compact subset of U. By means of the closed graph theorem, we obtain uniform boundedness of the maps $M_0(z) \in \mathcal{B}(E^m, E^{2m})$. In fact, the map $g \mapsto u(z)$ is a closed linear map from E^m into $B(D; E^{2m})$, the space of bounded functions from D to E^{2m}, hence bounded, i.e., $\sup_{z\in D} |M_0(z)| =: C_D < \infty$. By compactness and continuity this also holds on an open neighbourhood – which we again call U – of D.

Next we use the fact that $P_+(z)$ as well as $B(z)$ are holomorphic on U. Fix any $z \in U$, $h \in \mathbb{C}^n$ and let $0 \neq t \in \mathbb{C}$ be small. Then for fixed $g \in E^m$ we have

$$P_+(z + th)w(z + th) = 0 = P_+(z)w(z),$$

and

$$B(z + th)w(z + th) = g = B(z)w(z),$$

hence

$$\begin{aligned} P_+(z + th)[w(z + th) - w(z)] &= -[P_+(z + th) - P_+(z)]w(z) \\ B(z + th)[w(z + th) - w(z)] &= -[B(z + th) - B(z)]w(z). \end{aligned}$$

Now, $P_+(z)^2 = P_+(z)$ implies

$$\begin{aligned} &P_+(z + th) - P_+(z) = P_+(z + th)^2 - P_+(z)^2 \\ &= P_+(z + th)[P_+(z + th) - P_+(z)] + [P_+(z + th) - P_+(z)]P_+(z), \end{aligned}$$

which by $P_+(z)w(z) = 0$ yields

$$[P_+(z + th) - P_+(z)]w(z) = P_+(z + th)[P_+(z + th) - P_+(z)]w(z).$$

From this identity we obtain

$$P_+(z + th)[w(z + th) - w(z) + (P_+(z + th) - P_+(z))w(z) = 0,$$

and

$$\begin{aligned} &B(z + th)[w(z + th) - w(z) + (P_+(z + th) - P_+(z))w(z)] \\ &= B(z + th)[P_+(z + th) - P_+(z)]w(z) - [B(z + th) - B(z)]w(z), \end{aligned}$$

which implies

$$\begin{aligned} &w(z + th) - w(z) + [P_+(z + th) - P(z)]w(z) \\ &= M_0(z + th)[B(z + th)(P_+(z + th) - P_+(z))w(z) - (B(z + th) - B(z))w(z)]. \end{aligned} \tag{6.31}$$

By continuity of P_+ and B as well as boundedness of M_0, this shows continuity of w on complex lines. Thus $M_0(z)$ has this property as well. Dividing (6.31) by t we get

$$\begin{aligned}\frac{w(z+th)-w(z)}{t} &= -\frac{P_+(z+th)-P(z)}{t}w(z)\\ &+ M_0(z+th)B(z+th)\frac{P_+(z+th)-P_+(z)}{t}w(z)\\ &- M_0(z+th)\frac{B(z+th)-B(z)}{t}w(z),\end{aligned}$$

which shows that $w(z)$ is complex differentiable on U, thanks to holomorphy of P_+ and B. Therefore, M_0 is also holomorphic on U. □

2.2 Harmonic Analysis

The last subsection shows that the unique solution v of (6.28) is given by

$$v = e^{iy\rho A_0(b,\sigma)}M_0(b,\sigma)\tilde{h}.$$

To invert the Laplace and Fourier transforms in the right regularity class, we rewrite this equation as

$$\rho^{2m}v = M(y,\rho,b,\sigma)\rho e^{-\eta y\rho}\rho^{2m-1}\tilde{h} = M(y,\rho,b,\sigma)\tilde{g}, \tag{6.32}$$

where $\eta > 0$ is small,

$$M(y,\rho,\sigma,b) = e^{iy\rho A_0(b,\sigma)+\eta y\rho}M_0(b,\sigma)$$

and

$$\tilde{g} = \rho e^{-\eta y\rho}\rho^{2m-1}\tilde{h}.$$

Here we need a result on analytic C_0-semigroups and the vector-valued Triebel-Lizorkin spaces $F^{\alpha}_{pq,\mu}$, which we state now. Define $L_0 = (\omega + \partial_t + (-\Delta_x)^m)$ in the space $X_0 = L_{p,\mu}(\mathbb{R}_+; L_q(\mathbb{R}^{n-1}; E))$ with domain

$$\mathsf{D}(L_0) = {}_0H^1_{p,\mu}(\mathbb{R}_+; L_q(\mathbb{R}^{n-1}; E)) \cap L_{p,\mu}(\mathbb{R}_+; H^{2m}_q(\mathbb{R}^{n-1}; E)).$$

This operator, by the Dore-Venni theorem, belongs to the class $\mathcal{S}(X_0)$ with angle $\pi/2$. Therefore, its root $L_0^{1/2m}$ is also in this class, with angle $\pi/4m < \pi/2$. This implies that $L_0^{1/2m}$ is the negative generator of an analytic C_0-semigroup $e^{-yL_0^{1/2m}}$. In the sequel, we denote by L the canonical extension of L_0 to the space $\mathbb{E}_{0\mu} = L_{p,\mu}(\mathbb{R}_+; L_q(\mathbb{R}^n_+; E))$. We are here interested in the question for which boundary values $g \in X_0$ the extension $u(y) = e^{-yL_0^{1/2m}}g$ satisfies $L^{1/2m}u \in \mathbb{E}_{0\mu}$. The result is surprising; it is the content of the following proposition.

Proposition 6.2.4. *Let $1 < p,q < \infty$, $\mu \in (1/p,1]$, and E be a Banach space with property $\mathcal{HT}(\alpha)$. Moreover, let L_0 and L be defined as above, and let $u(y) = e^{-yL_0^{1/2m}}g$, $g \in X_0$, $y > 0$.*

Then the following assertions are equivalent.

(i) $u \in {}_0H^{1/2m}_{p,\mu}(\mathbb{R}_+; L_q(\mathbb{R}_+ \times \mathbb{R}^{n-1}; E)) \cap L_{p,\mu}(\mathbb{R}_+; H^1_q(\mathbb{R}_+ \times \mathbb{R}^{n-1}; E))$;

(ii) $L^{1/2m}u \in L_{p,\mu}(\mathbb{R}_+; L_q(\mathbb{R}_+ \times \mathbb{R}^{n-1}; E)) = \mathbb{E}_{0\mu}$;

(iii) $g \in {}_0F^{1/2m-1/2mq}_{pq,\mu}(\mathbb{R}_+; L_q(\mathbb{R}^{n-1}; E)) \cap L_{p,\mu}(\mathbb{R}_+; B^{1-1/q}_{qq}(\mathbb{R}^{n-1}; E)) =: {}_0\mathbb{F}_{0\mu}$.

Similar statements are valid on $\mathbb{R}$; replace the symbols $L_{p,\mu}(\mathbb{R}_+;\cdot)$ and ${}_0K_{p,\mu}(\mathbb{R}_+;\cdot)$ by $L_p(\mathbb{R};\cdot)$ and $K_p(\mathbb{R};\cdot)$, respectively.

Proof. (i)⇒(iii). As the trace operator $(\mathrm{tr}\, u)(t,x) := u(t,0,x)$ maps the space $H^1_q(\mathbb{R}_+ \times \mathbb{R}^{n-1}; E)$ boundedly into $B^{1-1/q}_{qq}(\mathbb{R}^{n-1}; E)$ we see that $g \in L_{p,\mu}(\mathbb{R}_+; B^{1-1/q}_{qq}(\mathbb{R}^{n-1}; E))$. To obtain the time regularity of g we may concentrate on the variables (t,y), and hide x in $\tilde{E} = L_q(\mathbb{R}^{n-1}; E)$ which belongs to the class $\mathcal{HT}$ as $E \in \mathcal{HT}$. Then with $\alpha = 1/2m$, we have

$$u \in \mathbb{E}_{\alpha\mu} := {}_0H^{\alpha}_{p,\mu}(\mathbb{R}_+; L_q(\mathbb{R}_+; \tilde{E})) \cap L_{p,\mu}(\mathbb{R}_+; H^1_q(\mathbb{R}_+; \tilde{E})).$$

Define an operator A in $\mathbb{E}_{0,\mu} = L_{p,\mu}(\mathbb{R}_+; L_q(\mathbb{R}_+; \tilde{E}))$ by means of $Au = \partial_y u$ with domain $\mathsf{D}(A) = L_{p,\mu}(\mathbb{R}_+; {}_0H^1_q(\mathbb{R}_+; \tilde{E}))$ and B by means of $Bu = (\omega + \partial_t)^{\alpha} u$ with domain $\mathsf{D}(B) = {}_0H^{\alpha}_{p,\mu}(\mathbb{R}_+; L_q(\mathbb{R}_+; \tilde{E}))$. Both operators are in $\mathcal{H}^\infty$ with $\mathcal{H}^\infty$-angles $\pi/2$, $\alpha\pi/2$, respectively, and B is invertible. They commute in the resolvent sense and $\phi^\infty_A + \phi^\infty_B = (1+\alpha)\pi/2 < \pi$. Therefore, by Corollary 4.5.11, $A+B$ with domain $\mathsf{D}(A+B) = \mathsf{D}(A) \cap \mathsf{D}(B) = \mathbb{E}_{\alpha\mu}$ belongs to the class $\mathcal{H}^\infty$, as well. Next we solve the problem $Av + Bv = \partial_y u + Bu \in \mathbb{E}_{0\mu}$ with maximal regularity to obtain a unique solution $v \in \mathsf{D}(A+B) = \mathbb{E}_{\alpha\mu}$. Then $w = u - v$ satisfies $\partial_y w = -Bw$ hence $w = e^{-By}g \in \mathbb{E}_{\alpha\mu} \subset \mathsf{D}(B)$. Therefore, Lemma 6.7.5 in the Appendix to this section yields $g \in {}_0F^{\alpha}_{pq,\mu}(\mathbb{R}_+; \tilde{E})$, which proves (iii).

(i)⇔(ii). We know that $L = \omega + \partial_t + (-\Delta_x)^m$ belongs to $\mathcal{H}^\infty$ with $\mathcal{H}^\infty$-angle $\pi/2$. Its domain is given by

$$\begin{aligned}\mathsf{D}(L) &= {}_0H^1_{p,\mu}(\mathbb{R}_+; L_q(\mathbb{R}_+; L_q(\mathbb{R}^{n-1}; E))) \cap L_{p,\mu}(\mathbb{R}_+; L_q(\mathbb{R}_+; H^{2m}_q(\mathbb{R}^{n-1}; E)))\\ &= \mathsf{D}(B^{2m}) \cap \mathsf{D}((-\Delta_x)^m).\end{aligned}$$

Then by complex interpolation we have

$$\begin{aligned}\mathsf{D}(L^{1/2m}) &= \mathsf{D}(B) \cap \mathsf{D}((-\Delta_x)^{1/2})\\ &= {}_0H^{\alpha}_{p,\mu}(\mathbb{R}_+; L_q(\mathbb{R}_+; L_q(\mathbb{R}^{n-1}; E))) \cap L_{p,\mu}(\mathbb{R}_+; L_q(\mathbb{R}_+; H^1_q(\mathbb{R}^{n-1}; E))),\end{aligned}$$

hence $L^{1/2m}u \in \mathbb{E}_{0\mu}$ if and only if $u \in \mathsf{D}(L^{1/2m})$. Furthermore, the representation $u = e^{-L_0^{1/2m}y}g$ implies also $\partial_y u \in \mathbb{E}_{0\mu}$. This proves the equivalence in question.

(iii)⇒(ii). Suppose

$$g \in {}_0F^{1/2m-1/2mq}_{pq,\mu}(\mathbb{R}_+; L_q(\mathbb{R}^{n-1}; E)) \cap L_{p,\mu}(\mathbb{R}_+; B^{1-1/q}_{qq}(\mathbb{R}^{n-1}; E)) =: {}_0\mathbb{F}_{0\mu}.$$

Set $A_0 = (-\Delta_x)^{1/2}$ with $\mathsf{D}(A_0) = L_{p,\mu}(\mathbb{R}_+; H^1_q(\mathbb{R}^{n-1}; E))$ and $B_0 = (\omega + \partial_t)^\alpha$ with domain $\mathsf{D}(B_0) = {}_0H^\alpha_{p,\mu}(\mathbb{R}_+; L_q(\mathbb{R}^{n-1}; E))$. These operators are of class $\mathcal{H}^\infty$ in the base space $X_0 = L_{p,\mu}(\mathbb{R}_+; L_q(\mathbb{R}^{n-1}; E))$, with $\mathcal{H}^\infty$ angles 0 and $\alpha\pi/2$, respectively, and they commute in the resolvent sense. Then by Lemma 6.7.5 we see that $e^{-B_0 y} g \in \mathsf{D}(B) = {}_0H^\alpha_{p,\mu}(\mathbb{R}_+; L_q(\mathbb{R}^n_+; E))$. On the other hand, $e^{-A_0 y} g \in L_{p,\mu}(\mathbb{R}_+; H^1_q(\mathbb{R}^n_+; E))$. Define $v = e^{-\eta(A_0+B_0)y} g$; then $(A_0 + B_0)v \in \mathbb{E}_{0,\mu}$, as $e^{-A_0 y}$ and $e^{-B_0 y}$ act boundedly in $\mathbb{E}_{0,\mu}$.
$A_0 + B_0$ is equivalent to $L_0^{1/2m}$ as $\mathsf{D}(L_0^{1/2m}) = \mathsf{D}(A_0) \cap \mathsf{D}(B_0)$. Moreover, by perturbation, $L_0^{1/2m} - \eta(A_0 + B_0)$ is $\mathcal{R}$-sectorial with $\mathcal{R}$-angle $\alpha\pi/2$, provided $\eta > 0$ is sufficiently small. By means of Fourier multipliers it is not difficult to see that $e^{-(L_0^{1/2m} - \eta(A_0+B_0))y}$ acts boundedly on $\mathbb{E}_{0\mu}$.

In fact, we show that the symbol

$$m(\lambda, \xi, y) = e^{-y(\lambda+\omega+|\xi|^{2m})^{1/2m} - \eta((\omega+\lambda)^{1/2m} + |\xi|)}$$

is a Fourier multiplier for $\mathbb{E}_{0\mu}$. To prove this, we first observe that m is uniformly bounded and holomorphic in $(\lambda, \xi) \in \Sigma_{\pi/2+\varepsilon} \times (\Sigma_\varepsilon \cup -\Sigma_\varepsilon)^n$, provided $\eta, \varepsilon > 0$ are small. This implies the Mikhlin-condition w.r.t. ξ, uniformly in (λ, y), hence we first invert the Fourier transform, to obtain an $\mathcal{R}$-bounded family of operators $\mathcal{T}(\lambda, y)$ on $L_q(\mathbb{R}^{n-1}; E)$, provided E is of class $\mathcal{HT}$ and has property (α). Uniformity then shows that the family $T_m(\lambda) = \mathcal{T}_m(\lambda, \cdot)$ is also $\mathcal{R}$-bounded in $L_q(\mathbb{R}^n_+; E)$ and then trivially also in $\mathbb{E}_{0\mu}$. Finally, by the Kalton-Weis theorem, $T(\partial_t + \omega)$ is bounded in $\mathbb{E}_{0\mu}$.

Therefore

$$L^{1/2m} e^{-L_0^{1/2m} y} g = L^{1/2m}(A+B)^{-1} e^{-(L_0^{1/2m} - \eta(A_0+B_0))y} \cdot (A_0 + B_0)v \in \mathbb{E}_{0,\mu},$$

which proves the implication (iii) ⇒ (ii). □

Now we may continue the argumentation preceding Proposition 6.2.4. As h_j is the transform of a function in

$${}_0\mathbb{F}_{j\mu} = {}_0F^{1-m_j/2m-1/2mq}_{pq,\mu}(\mathbb{R}_+; L_q(\mathbb{R}^{n-1}; E)) \cap L_{p,\mu}(\mathbb{R}_+; B^{2m-m_j-1/q}_{qq}(\mathbb{R}^{n-1}; E))$$

we see that $\rho^{2m-1}\tilde{h}_j = \rho^{2m-m_j-1} h_j$ is the transform of a function in ${}_0\mathbb{F}_{0\mu}$, for each $j = 1, \ldots, m$. Proposition 6.2.4 then implies that $\rho e^{-\eta y \rho} \rho^{2m-1} \tilde{h}_j$ is the transform a function $g_j \in \mathbb{E}_{0\mu} := L_{p,\mu}(\mathbb{R}_+; L_q(\mathbb{R}^n_+; E))$.

Therefore, we need to know that $M(y, \rho, b, \sigma)$ is a Fourier multiplier for X_0. To prove this, we first observe that M is uniformly bounded and holomorphic in (λ, ξ). This implies the Mikhlin-condition w.r.t. ξ, hence we first invert the Fourier transform, to obtain an $\mathcal{R}$-bounded family of operators $\mathcal{T}(\lambda, y)$ on $L_q(\mathbb{R}^{n-1}; E)$, provided E is of class $\mathcal{HT}$ and has property (α). Uniformity then shows that the family $T(\lambda) = \mathcal{T}(\lambda, \cdot)$ is also $\mathcal{R}$-bounded in $L_q(\mathbb{R}^n_+; E)$ and then trivially also in X_0. Finally, by the Kalton-Weis theorem, $T(\partial_t + \omega)$ is bounded in X_0.

Summarizing we have proved the sufficiency part of the following result for the original parabolic half-space problem (6.24).

Theorem 6.2.5. *Let $1 < p, q < \infty$, $\omega > 0$, $\mu \in (1/p, 1]$, and E be a Banach space of class $\mathcal{HT}(\alpha)$. Assume that $\mathcal{A}(D)$ is a normally elliptic differential operator of order $2m$, let $\mathcal{B}_j(D)$, $j = 1, \dots, m$, denote differential operators of order $m_j < 2m$, and suppose the Lopatinskii-Shapiro condition* **(LS)** *is satisfied, for some angle $\phi < \pi/2$.*

Then (6.24) admits a unique solution u in the class

$$u \in \mathbb{E}_{1\mu} := H^1_{p,\mu}(\mathbb{R}_+; L_q(\mathbb{R}^n_+; E)) \cap L_{p,\mu}(\mathbb{R}_+; H^{2m}_q(\mathbb{R}^n_+; E)),$$

if and only if the data are subject to the following conditions.

(a) $f \in \mathbb{E}_{0\mu} = L_{p,\mu}(\mathbb{R}_+; L_q(\mathbb{R}^n_+; E))$, $u_0 \in X_{\gamma,\mu} = B^{2m(\mu-1/p)}_{qp}(\mathbb{R}^n_+; E)$;

(b) $g_j \in \mathbb{F}_{j\mu} = F^{\kappa_j}_{pq,\mu}(\mathbb{R}_+; L_q(\mathbb{R}^{n-1}; E)) \cap L_{p,\mu}(\mathbb{R}_+; B^{2m\kappa_j}_{qq}(\mathbb{R}^{n-1}; E))$;

(c) $\mathcal{B}_j(D)u_0 = g_j(0)$ *if* $\kappa_j > 1/p + 1 - \mu$, $j = 1, \dots, m$.

Here $\kappa_j = 1 - m_j/2m - 1/2mq$. The solution depends continuously on the data in the corresponding spaces.

Remark 6.2.6. **(i)** Note that $\kappa_j > 1/p + 1 - \mu$ if and only $m_j < 2m(\mu - 1/p) - 1/q$.
(ii) In the case $p = q$ we have $F^{\kappa_j}_{pp,\mu} = B^{\kappa_j}_{pp,\mu} = W^{\kappa_j}_{p,\mu}$ as well as $B^{2m\kappa_j}_{pp} = W^{2m\kappa_j}_p$.

Proof. Necessity. We still need to prove the necessity part of Theorem 6.2.5. Suppose $u \in H^1_{p,\mu}(\mathbb{R}_+; L_q(\mathbb{R}^n_+; E)) \cap L_{p,\mu}(\mathbb{R}_+; H^{2m}_q(\mathbb{R}^n_+; E))$ is a solution of (6.24). Then inserting u into (6.24) we clearly have $f \in L_{p,\mu}(\mathbb{R}_+; L_q(\mathbb{R}^n_+; E))$. To obtain the regularity of the time trace u_0 of u at time $t = 0$, we extend u in space by means of a usual extension operator to obtain a function $\bar{u} \in H^1_{p,\mu}(\mathbb{R}_+; L_q(\mathbb{R}^n; E)) \cap L_{p,\mu}(\mathbb{R}_+; H^{2m}_q(\mathbb{R}^n; E))$. Applying the trace theorem for the semigroup $e^{-(-\Delta)^m t}$ with base space $L_q(\mathbb{R}^n; E)$ this yields

$$\bar{u}_{|t=0} \in (L_q(\mathbb{R}^n; E), H^{2m}_q(\mathbb{R}^n; E))_{\mu-1/p,p} = B^{2m(\mu-1/p)}_{qp}(\mathbb{R}^n; E),$$

which implies by restriction $u_0 \in B^{2m(\mu-1/p)}_{qp}(\mathbb{R}^n_+; E)$. Next we consider the lateral traces at $y = 0$. For this purpose we first replace u by $v = t^{1-\mu}u$ and extend v in time by symmetry to $\mathbb{R}$. Then $v \in H^1_p(\mathbb{R}; L_q(\mathbb{R}^n_+; E)) \cap L_p(\mathbb{R}; H^{2m}_q(\mathbb{R}^n_+; E))$, hence $w = (\omega + \partial_t)^\alpha \partial^k_y D^\beta_x u$ belongs to $H^{1/2m}_p(\mathbb{R}; L_q(\mathbb{R}^n_+; E)) \cap L_p(\mathbb{R}; H^1_q(\mathbb{R}^n_+; E))$ if $2m\alpha + k + |\beta| = 2m - 1$. Next we solve the problem

$$\partial_y \bar{w} + L_0^{1/2m} \bar{w} = \partial_y w + L_0^{1/2m} w, \; y > 0, \; \bar{w}(0) = 0,$$

with maximal regularity, which shows that $\bar{w}$ has the same regularity as w, hence $w - \bar{w} = e^{-yL_0^{1/2m}} w|_{y=0}$ has as well. Then Proposition 6.2.4 implies that the trace of w at $y = 0$ belongs to $F_{pq}^{1/2m-1/2mq}(\mathbb{R}; L_q(\mathbb{R}^{n-1}; E)) \cap L_p(\mathbb{R}; B_{qq}^{1-1/q}(\mathbb{R}^{n-1}; E))$. By the definition of w and proper choices of β and k, this yields

$$t^{1-\mu} g_j = \mathcal{B}_j(D) t^{1-\mu} v \in {}_0F_{pq}^{\kappa_j}(\mathbb{R}_+; L_q(\mathbb{R}^{n-1}; E)) \cap L_p(\mathbb{R}_+; B_{qq}^{2m\kappa_j}(\mathbb{R}^{n-1}; E)),$$

by restriction to $t > 0$; therefore we finally obtain $g_j \in F_{pq,\mu}^{\kappa_j}(\mathbb{R}_+; L_q(\mathbb{R}^{n-1}; E)) \cap L_{p,\mu}(\mathbb{R}_+; B_{qq}^{2m\kappa_j}(\mathbb{R}^{n-1}; E))$. This proves the necessity of the conditions in Theorem 6.2.5. □

It is of importance to have estimates on the solution which are also uniform in ω. This is the content of

Corollary 6.2.7. *Let the assumptions of Theorem 6.2.5 be satisfied, and fix any $\omega_0 > 0$. Then there is a constant $C > 0$ such that the solution of (6.24) satisfies the estimate*

$$|u|_{\mathbb{E}_{1\mu}} + \omega |u|_{\mathbb{E}_{0\mu}} \le C\Big(|u_0|_{X_{\gamma,\mu}} + |f|_{\mathbb{E}_{0\mu}} + \sum_{j=1}^m (|g_j|_{\mathbb{F}_{j\mu}} + \omega^{1-m_j/2m} |e^{-yL_\omega} g_j|_{\mathbb{E}_{0\mu}}) \Big), \tag{6.33}$$

for all $\omega \ge \omega_0$, $(f, g_j, u_0) \in \mathbb{E}_{0\mu} \times \mathbb{F}_{j\mu} \times X_{\gamma,\mu}$, $j = 1, \ldots, m$. Here L_ω is defined by $L_\omega = (\partial_t + \omega + (-\Delta)^m)^{1/2m}$.

Proof. To derive the inequality (6.33) we proceed in a similar way as in the proof of Theorem 6.2.5. We again work in frequency domain. Recall that the symbol of L_ω is $\rho = (\lambda + \omega + |\xi|^{2m})^{1/2m}$, and set $\rho_0 = (\lambda + \omega_0 + |\xi|^{2m})^{1/2m}$. Here we decompose as

$$\rho^{2m} v = M_1 \cdot M_2 \cdot \rho_0 e^{-\eta y \rho_0} \rho_0^{2m-m_j-1} h_j + M \cdot M_2 \omega^{1-m_j/2m} e^{-\eta y \rho} h_j,$$

with

$$M_1 = e^{i\rho A_0(b,\sigma) + \eta y \rho_0} M_0(b,\sigma), \quad M_2 = \frac{\rho^{2m-m_j}}{\rho_0^{2m-m_j} + \omega^{1-m_j/2m}}.$$

By the arguments at the end of the proof of Theorem 6.2.5, M as well as M_1 and M_2 are bounded Fourier multipliers for $\mathbb{E}_{0\mu}$, uniformly for $\omega \ge \omega_0 > 0$, hence the result follows by the same arguments. □

Estimate (6.33) is sharp for the half-space case. However, the last term involves a norm which is specific for a half-space. Observing that with some $\delta > 0$,

$$|e^{-yL_\omega} g_j|_{\mathbb{E}_{0\mu}} \le C |e^{-\delta \omega^{1/2m} y} g_j|_{\mathbb{E}_{0\mu}} \le C \omega^{-1/2mq} |g_j|_{L_{p,\mu}(L_q)},$$

we obtain the slightly weaker estimate

$$|u|_{\mathbb{E}_{1\mu}} + \omega|u|_{\mathbb{E}_{0\mu}} \le C\big(|u_0|_{X_{\gamma,\mu}} + |f|_{\mathbb{E}_{0\mu}} + \sum_{j=1}^{m}(|g_j|_{\mathbb{F}_{j\mu}} + \omega^{\kappa_j}|g_j|_{L_{p,\mu}(L_q)})\big). \quad (6.34)$$

The advantage of (6.34) lies in the fact that it only involves the norms of the boundary data. It is not good enough to cover boundary perturbations of highest order, but it is well suited to handle such of lower order, and is in particular useful for the localization process in domains.

2.3 Perturbed Coefficients

To consider the case of variable coefficients, on the boundary we have to work in Besov spaces. Here a result on pointwise multipliers is essential. Therefore we begin with this topic.

Lemma 6.2.8. *Let $1 \le p, q \le \infty$, $s > 0$, E a Banach space, and assume*

$$a \in B^s_{rq}(\mathbb{R}^n; \mathcal{B}(E)) + B^s_{\infty q}(\mathbb{R}^n; \mathcal{B}(E)), \quad (6.35)$$

with $r \ge p$ and $s > n/r$.

Then the multiplication operator $v \mapsto av$ is bounded in $B^s_{pq}(\mathbb{R}^n; E)$. Moreover, there are constants $\alpha \in [0, 1)$ and $C > 0$ such that

$$|av|_{B^s_{pq}} \le |a|_{L_\infty}|v|_{B^s_{pq}} + C|v|^{\alpha}_{B^s_{pq}}|v|^{1-\alpha}_{L_p}, \quad (6.36)$$

for all $v \in B^s_{pq}(\mathbb{R}^n; E)$. The constant C depends linearly on the norm of the space of multipliers defined by (6.35).

Proof. We concentrate on the case $s \in (0, 1]$, as the general case can be reduced to this one by differentiation.

We will use the following norm on $B^s_{pq}(\mathbb{R}^n; E)$:

$$|v|_{B^s_{pq}} = |v|_{L_p} + [v]_{s,p,q},$$

where

$$[v]_{s,p,q} = \Big(\int_{|h|\le 1} (|h|^{-s}|\tau_h v - v|_{L_p})^q \, dh/|h|^n\Big)^{1/q}, \quad 1 \le q < \infty,$$

and

$$[v]_{s,p,\infty} = \sup_{|h|\le 1} |h|^{-s}|\tau_h v - v|_{L_p}.$$

Here $\{\tau_h\}_{h\in\mathbb{R}^n}$ denotes the group of translations defined by

$$(\tau_h v)(x) = v(x + h), \quad x, h \in \mathbb{R}^n.$$

Obviously we have $|av|_{L_p} \le |a|_{L_\infty}|v|_{L_p}$, so we concentrate on the estimation of $[av]_{s,p,q}$. The identity

$$\tau_h(av) - av = \tau_h a(\tau_h v - v) + (\tau_h a - a)v$$

yields with Hölder's inequality and Remark 6.2.9(ii)

$$[av]_{s,p,q} \leq |a|_{L_\infty}[v]_{s,p,q} + [a]_{s,r,q}|v|_{L_{p\rho'}},$$

where $r = p\rho$, $1/\rho + 1/\rho' = 1$, and $s - n/p > -n/p\rho'$. The Gagliardo-Nirenberg inequality implies

$$|v|_{L_{p\rho'}} \leq C|v|^{\alpha}_{B^s_{pq}}|v|^{1-\alpha}_{L_p},$$

with some constants $C > 0$ and $\alpha \in [0,1)$. Alternatively, we may estimate like

$$[av]_{s,p,q} \leq |a|_{L_\infty}[v]_{s,p,q} + [a]_{s,\infty,q}|v|_{L_p}.$$

In both cases (6.36) follows. □

Remark 6.2.9. (i) This lemma shows that $B^s_{pq}(\mathbb{R}^n)$ is a Banach algebra w.r.t. pointwise multiplication, provided $s > n/p$, i.e., provided it embeds into L_∞.

(ii) Observe that the multiplier space defined in (6.35) embeds into the uniform Hölder spaces $C_b^{s-n/r}(\mathbb{R}^n; \mathcal{B}(E))$.

We now consider problem (6.24) with variable coefficients, applying perturbation arguments. Thus we look at the case

$$\mathcal{A}(x,D) = \mathcal{A}^0(D) + \mathcal{A}^1(x,D), \quad \mathcal{B}_j(x,D) = \mathcal{B}_j^0(D) + \mathcal{B}_j^1(x,D),$$

where the system $(\mathcal{A}^0(D), \mathcal{B}_1^0(D), \ldots, \mathcal{B}_m^0(D))$ is normally elliptic and subject to the Lopatinskii-Shapiro condition.

For perturbations of $\mathcal{A}^0(D)$ the arguments of Section 6.1.4 apply again, so we require

$$a^1_\alpha \in L_{r_k}(\mathbb{R}^n_+; \mathcal{B}(E)) + L_\infty(\mathbb{R}^n_+; \mathcal{B}(E)), \quad |\alpha| = k < 2m, \ r_k \geq q, \ 2m - k > n/r_k,$$

and in addition the smallness condition

$$|a^1_\alpha|_{L_\infty} \leq \eta, \quad |\alpha| = 2m.$$

The essential perturbations to be considered here are the boundary perturbations. In the sequel we assume

$$b^1_{j\beta} \in B^{2m\kappa_j}_{r_{jk}q}(\mathbb{R}^{n-1}; \mathcal{B}(E)) + B^{2m\kappa_j}_{\infty q}(\mathbb{R}^{n-1}; \mathcal{B}(E)),$$
$$|\beta| = k \leq m_j, \ r_{jk} \geq q, \ 2m\kappa_j > (n-1)/r_{jk},$$

and the smallness condition

$$|b^1_{j\beta}|_{L_\infty} \leq \eta, \quad |\beta| = m_j, \quad j = 1, \ldots, m.$$

Recall the definition $\kappa_j = 1 - m_j/2m - 1/2mq$, and observe that

$$b^1_{j\beta} \in C_b^{2m\kappa_j - (n-1)/r_{jm_j}}(\mathbb{R}^{n-1}; \mathcal{B}(E)), \quad |\beta| = m_j.$$

We estimate the boundary perturbations as follows, employing Lemma 6.2.8. For the highest order terms we get

$$\begin{aligned}|b^1_{j\beta}D^\beta u|_{B^{2m\kappa_j}_{qq}} &\le |b^1_{j\beta}|_{L_\infty}|D^\beta u|_{B^{2m\kappa_j}_{qq}} + C|D^\beta u|^\alpha_{B^{2m\kappa_j}_{qq}}|D^\beta u|^{1-\alpha}_{L_p}\\ &\le 2\eta|u|_{H^{2m}_q} + C_\eta|u|_{L_q}.\end{aligned}$$

This implies

$$|\mathcal{B}^1_{j\#}(x,D)u|_{L_{p,\mu}(\mathbb{R}_+;B^{2m\kappa_j}_{qq})} \le 2\eta|u|_{\mathbb{E}_{1\mu}} + C_\eta|u|_{\mathbb{E}_{0\mu}}.$$

In a similar way we can dominate the lower order terms, without any smallness condition.

Next we need to estimate the terms $|e^{-L_\omega y}b^1_{j\beta}D^\beta v|_{L_q(\mathbb{R}^{n-1})}$, where $v = u_{|y=0}$ denotes the trace of u on the boundary. For this purpose we write

$$\begin{aligned}e^{-\omega^{1/2m}y}b^1_{j\beta}D^\beta v &= -\int_0^\infty \partial_s(e^{-\omega^{1/2m}(y+s)}b^1_{j\beta}D^\beta u(s))\,ds\\ &= \int_0^\infty \omega^{1/2m}e^{-\omega^{1/2m}(y+s)}b^1_{j\beta}D^\beta u(s)\,ds\\ &\quad - \omega^{-1/2m}\int_0^\infty \omega^{1/2m}e^{-\omega^{1/2m}(y+s)}b^1_{j\beta}\partial_s D^\beta u(s)\,ds.\end{aligned}$$

This implies

$$|e^{-\omega^{1/2m}y}b^1_{j\beta}D^\beta v|_{L_q} \le C|b^1_{j\beta}|_{L_\infty}\int_0^\infty (|D^\beta u(s)|_{L_q} + \omega^{-1/2m}|\partial_s D^\beta u(s)|_{L_q})\frac{ds}{y+s},$$

and as the scalar Hilbert transform is bounded in $L_q(\mathbb{R}_+)$,

$$|e^{-\omega^{1/2m}y}b^1_{j\beta}D^\beta v|_{L_q(\mathbb{R}^n_+)} \le C|b^1_{j\beta}|_{L_\infty}(|D^\beta u|_{L_q(\mathbb{R}^n_+)} + \omega^{-1/2m}|\partial_s D^\beta u|_{L_q(\mathbb{R}^n_+)}),$$

which yields by the Gagliardo-Nirenberg inequality

$$\omega^{1-m_j/2m}|e^{-\omega^{1/2m}y}b^1_{j\beta}D^\beta u|_{L_q(\mathbb{R}^n_+)} \le C|b^1_{j\beta}|_{L_\infty}\sum_{i=1,2}|u|^{\gamma_i}_{H^{2m}_q(\mathbb{R}^n_+)}(\omega|u|_{L_q(\mathbb{R}^n_+)})^{1-\gamma_i},$$

with some constants $C>0$ and $\gamma_i \in [0,1]$. As the coefficients $b^1_{j\beta}$ do not depend on time, this estimate implies

$$\omega^{1-m_j/2m}|e^{-\omega^{1/2m}y}\mathcal{B}^1_{j\#}v|_{\mathbb{E}_{0\mu}} \le C\eta[|u|_{\mathbb{E}_{1\mu}} + \omega|u|_{\mathbb{E}_{0\mu}}].$$

Finally, as $e^{-(L_\omega-\delta\omega^{1/2m})y}$ is bounded in $\mathbb{E}_{0\mu}$, for some $\delta>0$, this implies

$$\omega^{1-m_j/2m}|e^{-L_\omega y}\mathcal{B}^1_{j\#}v|_{\mathbb{E}_{0\mu}} \le C\eta[|u|_{\mathbb{E}_{1\mu}} + \omega|u|_{\mathbb{E}_{0\mu}}].$$

We now turn to the perturbed initial-boundary value problem. Without loss of generality, we may assume $u_0 = 0$, solving a whole-space problem. We write the half-space problem in abstract form as

$$L_0 u + L_1 u = F,$$

where

$$L_0 u = (\partial_t u + \omega u + \mathcal{A}^0(D)u, \mathcal{B}_1^0(D)u, \ldots, \mathcal{B}_m^1(D)u)$$

defines an isomorphism between the spaces ${}_0\mathbb{E}_{1,\mu}$ and $\mathbb{E}_{0,\mu} \times \Pi_{j=1}^m {}_0\mathbb{F}_{j\mu}$,

$$L_1 u = (\mathcal{A}^1(x,D)u, \mathcal{B}^1(x,D)u, \ldots, \mathcal{B}_m^1(x,D)u),$$

and $F = (f, g_1, \ldots, g_m) \in \mathbb{E}_{0,\mu} \times \Pi_{j=1}^m {}_0\mathbb{F}_{j\mu}$. If $\eta > 0$ is small enough, choosing $\omega > 0$ large enough, we see by the above estimates that $L_0 + L_1$ is also an isomorphism. This way we obtain the following result on (6.24).

Theorem 6.2.10. *Let E be a Banach space of class $\mathcal{HT}(\alpha)$. Assume that $\mathcal{A}^0(D)$ is a normally elliptic differential operator of order $2m$, let $\mathcal{B}_j^0(D)$, $j = 1, \ldots, m$, denote differential operators of order $m_j < 2m$, and suppose the Lopatinskii-Shapiro condition for $(\mathcal{A}^0(D), \mathcal{B}_j^0(D))$ is satisfied, with some angle $\phi < \pi/2$. Let*

$$\mathcal{A}(x,D) = \mathcal{A}^0(D) + \mathcal{A}^1(x,D), \quad \mathcal{B}_j(x,D) = \mathcal{B}_j^0 + \mathcal{B}_j^1(x,D),$$

where the coefficients $a_\alpha^1(x)$, $b_{j\beta}^1(x)$ satisfy the following conditions.

$$\begin{aligned}
&|a_\alpha^1|_{L_\infty}, |b_{j\beta}^1|_{L_\infty} \le \eta, \quad |\alpha| = 2m, \ |\beta| = m_j, \ j = 1, \ldots, m;\\
&a_\alpha^1 \in L_{r_k}(\mathbb{R}^n_+; \mathcal{B}(E)) + L_\infty(\mathbb{R}^n_+; \mathcal{B}(E)), \quad |\alpha| = k < 2m, \ r_k \ge q, \ 2m - k > n/r_k;\\
&b_{j\beta}^1 \in B^{2m\kappa_j}_{r_{jk}q}(\mathbb{R}^{n-1}; \mathcal{B}(E)) + B^{2m\kappa_j}_{\infty q}(\mathbb{R}^{n-1}; \mathcal{B}(E)),\\
&|\beta| = k, \ r_{jk} \ge q, \ 2m\kappa_j > (n-1)/r_{jk}.
\end{aligned}$$

Then there is $\eta_0 > 0$ such that the assertions of Theorem 6.2.5 *and estimate* (6.33) *remain valid for the perturbed problem, provided $\eta \le \eta_0$.*

2.4 Localization

Here we assume that the top order coefficients a_α with $|\alpha| = 2m$, and $b_{j\beta}$ with $|\beta| = m_j$ are continuous, with limits at infinity. This replaces the smallness condition of the previous subsection. Choose a large ball $B(0,R) \subset \mathbb{R}^n$ such that

$$\begin{aligned}
|a_\alpha(x) - a_\alpha(\infty)| &\le \eta, \quad x \in \bar{\mathbb{R}}^n_+ \setminus B(0,R), \ |\alpha| = 2m,\\
|b_{j\beta}(x) - b_{j\beta}(\infty)| &\le \eta, \quad x \in \mathbb{R}^{n-1}, \ |x| \ge R, \ |\beta| = m_j, \ j = 1, \ldots, m.
\end{aligned}$$

Observe that $R > 0$ exists, as the top order coefficients are continuous and have limits at infinity. Next we cover the boundary $\bar{B}(0,R) \cap \mathbb{R}^{n-1}$ by N_1 balls $B(x_k, r/2) \subset \mathbb{R}^n$ such that

$$\begin{aligned}
|a_\alpha(x) - a_\alpha(x_k)| &\le \eta, \quad x \in B(x_k, 2r), \ |\alpha| = 2m,\\
|b_{j\beta}(x) - b_{j\beta}(x_k)| &\le \eta, \quad x \in B(x_k, 2r) \cap \mathbb{R}^{n-1}, \ |\beta| = m_j, \ j = 1, \ldots, m.
\end{aligned}$$

Finally, we cover the compact set $\bar{B}(0,R)\setminus\big(\cup_{k=1}^{N_1}B(x_k,r/2)\big)$ by balls $B(x_k,r/2)$, $k=N_1+1,\ldots,N_2$. We then set $U_0=\mathbb{R}^n\setminus\bar{B}(0,R)$, and $U_k=B(x_k,r)$, for $k=1,\ldots,N_2$. Then $\{U_k\}_{k=0}^{N_2}$, forms an open covering of $\bar{\mathbb{R}}^n_+$. Fix a partition of unity $\{\varphi_k\}_{k=0}^{N}$ of class C^∞ subordinate to this open covering, and let $\psi_k\in C^\infty(\mathbb{R}^n)$ be such that $\psi_k=1$ on $\operatorname{supp}\varphi_k$ and $\operatorname{supp}\psi_k\subset U_k$.

We assume in the sequel that the operator $\mathcal{A}_\#(x_0,D)$ is normally elliptic, for each $x_0\in\bar{\mathbb{R}}^n_+\cup\{\infty\}$, and that the system $(\mathcal{A}_\#(x_0,D),\mathcal{B}_{j\#}(x_0,D))$ satisfies the Lopatinskii-Shapiro Condition **(LS)**, for each $x_0\in\mathbb{R}^{n-1}\cup\{\infty\}$, with angle $\phi(x_0)<\pi/2$. Then the maximal regularity constants for the problems with frozen coefficients will be uniform in $x_0\in\bar{\mathbb{R}}^n_+\cup\{\infty\}$, by continuity and compactness, hence η_0 in Theorem 6.2.10 will be uniform in x_0 as well. Now we fix any $\eta\in(0,\eta_0]$.

Next we define for each k local operators $\mathcal{A}_k(x,D)$ on the half-space $\mathbb{R}^n_+$ and $\mathcal{B}_{jk}(x,D)$ on the boundary $\mathbb{R}^{n-1}$ in the following way. Choose a function $\chi\in\mathcal{D}(\mathbb{R})$ such that $\chi(s)=1$ for all $|s|\le 1$, $0\le\chi(s)\le 1$ and $\chi(s)=0$ for $|s|\ge 2$. Then we set

$$a^k_\alpha(x)=a_\alpha(x_k+\chi(|x-x_k|^2/r^2)(x-x_k)),\quad x\in\bar{\mathbb{R}}^n_+,\ |\alpha|=2m,\ k=1,\ldots,N_2,$$
$$b^k_{j\beta}(x)=b_{j\beta}(x_k+\chi(|x-x_k|^2/r^2)(x-x_k)),\quad x\in\mathbb{R}^{n-1},\ |\beta|=m_j,$$

$j=1,\ldots,m$, and

$$a^0_\alpha(x)=a_\alpha(\infty)+\chi(R^2/|x|^2)(a_\alpha(x)-a_\alpha(\infty)),\quad x\in\bar{\mathbb{R}}^n_+,\ |\alpha|=2m,$$
$$b^0_{j\beta}(x)=b_{j\beta}(\infty)+\chi(R^2/|x|^2(b_{j\beta}(x)-b_{j\beta}(\infty)),\quad x\in\mathbb{R}^{n-1},\ |\beta|=m_j.$$

Here we set $a^0_\alpha(0)=a_\alpha(\infty)$ and $b^0_{j\beta}(0)=b^0_{j\beta}(\infty)$. Then we define the local operators by means of

$$\mathcal{A}^k(x,D)=\sum_{|\alpha|=2m}a^k_\alpha(x)D^\alpha,\quad \mathcal{B}^k_j(x,D)=\sum_{|\beta|=m_j}b^k_{j\beta}(x)D^\beta.$$

By solving a full space problem, by Theorem 6.1.11, extending all coefficients of $\mathcal{A}(x,D)$ by symmetry to all of $\mathbb{R}^n$, we may assume $u_0=0$. Now let the data g_j be given and let $u\in{}_0\mathbb{E}_{1\mu}$ be a solution of (6.24) in $\mathbb{R}^n_+$. We set $u^k=\varphi_k u$, $f^k=\varphi_k f$, and $g^k_j=\varphi_k g_j$. Then we obtain the following localized problems. For the interior charts $k=N_1+1,\ldots,N_2$, the functions u^k satisfy

$$\begin{aligned}\partial_t u^k+\omega u^k+\mathcal{A}^k(x,D)u^k&=f^k+[\mathcal{A}_\#(x,D),\varphi_k]u-\varphi_k\mathcal{A}_1(x,D)u\ \text{ in }\ \mathbb{R}^n,\\ u^k(0)&=0,\end{aligned}$$

where $\mathcal{A}_1(x,D)=\mathcal{A}(x,D)-\mathcal{A}_\#(x,D)$ denotes the lower order part of $\mathcal{A}(x,D)$. Note that $\mathcal{A}^k(x,D)\varphi_k=\mathcal{A}_\#(x,D)\varphi_k$ by construction, and observe that the commutators $[\mathcal{A}_\#(x,D),\varphi_k]$ are of lower order as well. The boundary charts $k=0,\ldots,N_1$ lead to the following half-space problems.

$$\begin{aligned}\partial_t u^k+\omega u^k+\mathcal{A}^k(x,D)u^k&=f^k+[\mathcal{A}_\#(x,D),\varphi_k]u-\varphi_k\mathcal{A}_1(x,D)u\quad\text{in}\quad\mathbb{R}^n_+,\\ \mathcal{B}^k_j(x,D)u^k&=g^k_j+[\mathcal{B}_{j\#}(x,D),\varphi_k]u-\varphi_k\mathcal{B}_{j1}(x,D)u\quad\text{on}\quad\mathbb{R}^{n-1},\\ u^k(0)&=0,\end{aligned}$$

where $\mathcal{B}_{j1}(x,D) = \mathcal{B}_j(x,D) - \mathcal{B}_{j\#}(x,D)$ as well as the commutator $[\mathcal{B}_{j\#}(x,D), \varphi_k]$ are of order $m_j - 1$, these are trivial in case $m_j = 0$. We write these problems abstractly as

$$L_k u^k = G_k u + F_k, \quad k = 0, \ldots, N_2,$$

where the operators L_k are defined by the left-hand sides of the localized equations, $G_k u$ are the lower order perturbations on the right-hand side, and F_k collects the data coming from the inhomogeneities (f, g_j). More precisely,

$$G_k u = ([\mathcal{A}_\#(x,D), \varphi_k]u - \varphi_k \mathcal{A}_1(x,D)u, [\mathcal{B}_{j\#}(x,D), \varphi_k]u - \varphi_k \mathcal{B}_{j1}(x,D)u)$$

and $F_k = \varphi_k F = \varphi_k(f, g_j)$. By Theorem 6.2.10, the operators L_k are invertible for ω large, hence we obtain

$$u^k = L_k^{-1} F_k + L_k^{-1} G_k u, \quad k = 0, \ldots, N_2, \tag{6.37}$$

and so the following representation of the solution u. We first write

$$u = \sum_{k=0}^{N_2} \varphi_k u = \sum_{k=0}^{N_2} \psi_k \varphi_k u = \sum_{k=0}^{N_2} \psi_k u^k,$$

and then

$$u = \sum_{k=0}^{N_2} \psi_k L_k^{-1} F_k + \Big(\sum_{k=0}^{N_2} \psi_k L_k^{-1} G_k\Big) u.$$

We estimate in the following way, employing Theorem 6.1.11 for the interior charts and (6.34) for the boundary charts.

$$\begin{aligned}&|\psi_k L_k^{-1} G_k u|_{\mathbb{E}_{1\mu}} + \omega |\psi_k L_k^{-1} G_k u|_{\mathbb{E}_{0\mu}} \\ &\le C\Big(|G_k^i u|_{\mathbb{E}_{0\mu}} + \sum_{j=1}^m (|G_{kj}^b u|_{\mathbb{F}_{j\mu}} + \omega^{\kappa_j} |G_{kj}^b u|_{L_{p,\mu}(L_q)}\Big).\end{aligned}$$

Here the boundary terms are absent for the interior charts $k = N_1 + 1, \ldots, N_2$. For the interior operators G_k^i defined by

$$G_k^i u = [\mathcal{A}_\#(x,D), \varphi_k]u - \varphi_k \mathcal{A}_1(x,D)u,$$

we obtain by the Gagliardo-Nirenberg inequality

$$|G_k^i u|_{\mathbb{E}_{0\mu}} \le C |u|_{\mathbb{E}_{1\mu}}^{\gamma} |u|_{\mathbb{E}_{0,\mu}}^{1-\gamma},$$

with some constants $C > 0$ and $\gamma \in (0,1)$, hence

$$|G_k^i u|_{\mathbb{E}_{0\mu}} \le \frac{C}{\omega^{1-\gamma}} \big(|u|_{\mathbb{E}_{1\mu}} + \omega |u|_{\mathbb{E}_{0\mu}}\big).$$

The boundary terms are of the form

$$G_{kj}^b u = [\mathcal{B}_{j\#}(x,D), \varphi_k]u - \varphi_k \mathcal{B}_{j1}(x,D)u.$$

Therefore, as in the previous subsection

$$|G_{kj}^b u|_{\mathbb{F}_{j\mu}} \le C_j |u|_{\mathbb{E}_{1\mu}}^{\gamma_j} |u|_{\mathbb{E}_{0\mu}}^{1-\gamma_j} \le \frac{C_j}{\omega^{1-\gamma_j}} \big(|u|_{\mathbb{E}_{1\mu}} + \omega |u|_{\mathbb{E}_{0\mu}}\big),$$

with constants $C_j > 0$ and $\gamma_j \in (0,1)$. Finally, applying once more arguments of the previous subsection, we also obtain

$$\omega^{\kappa_j} |G_{kj}^b u|_{L_{p,\mu}(L_q)} \le \frac{C_j}{\omega^{1-\gamma_j}} \big(|u|_{\mathbb{E}_{1\mu}} + \omega |u|_{\mathbb{E}_{0\mu}}\big),$$

with possibly different constants $C_j > 0$ and $\gamma_j \in [0,1)$.

Summarizing, we see that for ω sufficiently large, the operator $G^L := \sum_{k=0}^{N_2} \psi_k L_k^{-1} G_k$ on ${}_0\mathbb{E}_{1\mu}$ satisfies the estimate

$$|G^L u|_{\mathbb{E}_{1\mu}} + \omega |G^L u|_{\mathbb{E}_{0\mu}} \le \frac{C}{\omega^{1-\gamma}} \big(|u|_{\mathbb{E}_{1\mu}} + \omega |u_{\mathbb{E}_{0\mu}}\big)$$

with appropriate constants C and γ that do not depend on ω. Equipping ${}_0\mathbb{E}_{1\mu}$ with the parameter-dependent norm $|u|_{\mathbb{E}_{1\mu}}^{\omega} := |u|_{\mathbb{E}_{1\mu}} + \omega |u|_{\mathbb{E}_{0\mu}}$ we conclude that the operator $I - G^L$ is invertible in $({}_0\mathbb{E}_{1\mu}, |\cdot|_{\mathbb{E}_{1\mu}}^{\omega})$, provided ω is sufficiently large. This yields a left inverse S of (6.24), which is given by

$$S(f, g_j) = (I - G^L)^{-1} \sum_{k=0}^{N_2} \psi_k L_k^{-1} \varphi_k (f, g_j).$$

In particular, the operator L defined by the left-hand side of (6.24) is injective and has closed range. So it remains to prove that L is also surjective. To show this we construct a right inverse which then by algebra equals its left inverse.

For this purpose we apply $L_\# := (\partial_t + \omega + \mathcal{A}_\#(x,D), \mathcal{B}_{j\#}(x,D))$ to $u = SF$, observing $L_\# = L_k$ in U_k. This yields with (6.37)

$$L_\# u = L_\# \sum_{k=0}^{N_2} \psi_k u^k = \sum_{k=0}^{N_2} [L_\#, \psi_k] L_k^{-1} (F_k + G_k u) + \sum_{k=0}^{N_2} \psi_k (F_k + G_k u).$$

Next, as $\psi_k = 1$ on the support of φ_k, we may drop ψ_k in the second term, which implies in the interior

$$\sum_{k=0}^{N_2} \psi_k (F_k + G_k u)^i = \sum_k (f_k + [\mathcal{A}_\#(x,D), \varphi_k]u - \varphi_k \mathcal{A}_1(x,D)u) = f - \mathcal{A}_1(x,D)u,$$

and on the boundary

$$\sum_{k=0}^{N_2} \psi_k(F_k+G_k u)^b = \sum_k g_{jk} + [\mathcal{B}_{j\#}(x,D),\varphi_k]u - \varphi_k \mathcal{B}_{j1}(x,D)u = g_j - \mathcal{B}_{j1}(x,D)u.$$

Replacing $u = SF$, this yields

$$LS = I + \Big(\sum_{k=0}^{N_2}[L_\#,\psi_k]L_k^{-1}\varphi_k\Big) + \Big(\sum_{k=0}^{N_2}[L_\#,\psi_k]L_k^{-1}G_k\Big)S =: I + G^R.$$

As the commutator $[L_\#,\psi_k] = ([\mathcal{A}_\#(x,D),\psi_k],[\mathcal{B}_{j\#}(x,D),\psi_k])$ is lower order, we see as above that the norm of G^R in $\mathbb{E}_{0\mu}$ is smaller than 1, provided ω is chosen large. Therefore $I+G^R$ is invertible, and so $R := S(I+G^R)^{-1}$ is a right inverse of L. This implies the following result for the half-space.

Theorem 6.2.11. *Let $1 < p,q < \infty$, $\mu \in (1/p,1]$ and E be a Banach space of class $\mathcal{HT}(\alpha)$. Assume that $\mathcal{A}(x,D)$ is a differential operator of order $2m$, let $\mathcal{B}_j^0(D)$, $j = 1,\ldots,m$, denote differential operators of order $m_j < 2m$. Suppose that the coefficients $a_\alpha(x)$, $b_{j\beta}(x)$ satisfy the following conditions.*

$$\begin{aligned}
&a_\alpha \in C_l(\bar{\mathbb{R}}^n_+;\mathcal{B}(E)), \quad b_{j\beta} \in C_l(\mathbb{R}^{n-1};\mathcal{B}(E)) \quad |\alpha| = 2m,\ |\beta| = m_j,\ j = 1,\ldots,m;\\
&a_\alpha \in L_{r_k}(\mathbb{R}^n_+;\mathcal{B}(E)) + L_\infty(\mathbb{R}^n_+;\mathcal{B}(E)), \quad |\alpha| = k < 2m,\ r_k \ge q,\ 2m-k > n/r_k;\\
&b_{j\beta} \in B^{2m\kappa_j}_{r_{jk}q}(\mathbb{R}^{n-1};\mathcal{B}(E)) + B^{2m\kappa_j}_{\infty q}(\mathbb{R}^{n-1};\mathcal{B}(E)),\\
&|\beta| = k \le m_j,\ r_{jk} \ge q,\ 2m\kappa_j > (n-1)/r_{jk}.
\end{aligned}$$

Assume that $\mathcal{A}_\#(x,D)$ is normally elliptic for each $x \in \bar{\mathbb{R}}^n_+ \cup \{\infty\}$, and that $(\mathcal{A}_\#(x,D),\mathcal{B}_{j\#}(x,D))$ satisfies the Lopatinskii-Shapiro Condition **(LS)** *with some angle $\phi(x) < \pi/2$, for each $x \in \mathbb{R}^{n-1} \cup \{\infty\}$.*

Then the assertions of Theorem 6.2.5 *and Corollary* 6.2.7 *remain valid for the half-space problem with variable coefficients.*

2.5 Normal Strong Ellipticity

We now consider the special case of strongly elliptic second-order operators in a Hilbert space E with so-called *natural boundary conditions*. This means, we consider $\mathcal{A}(D) = a^{ij}D_iD_j$, where $a^{ij} = a^{ji}$, with boundary operator either of Dirichlet type, i.e., $\mathcal{B}(D) = I$, or of co-normal (Neumann) type $\mathcal{B}(D) = \nu_i a^{ij}D_j$; here we employ the Einstein summation convention. Assuming that $\mathcal{A}(D)$ is strongly elliptic, what more conditions are needed for the Lopatinskii-Shapiro condition to be valid for these natural boundary operators?

To answer this question, let $u \in L_2(\mathbb{R}_+;E)$ be a solution of the ODE-boundary value problem

$$\begin{aligned}
\lambda u(y) + \mathcal{A}(\xi+\nu D_y)u(y) &= 0, \quad y > 0,\\
\mathcal{B}(\xi+\nu D_y)u(0) &= 0.
\end{aligned} \tag{6.38}$$

Here $\operatorname{Re}\lambda \geq 0$, $\xi, \nu \in \mathbb{R}^n$ are fixed, with $(\lambda, \xi) \neq (0,0)$, $|\nu| = 1$, $(\xi|\nu) = 0$. Take the inner product with u in E, integrate over $\mathbb{R}_+$, and take real parts. By means of the natural boundary conditions this yields the identity

$$\operatorname{Re}\lambda |u|_2^2 + \int_0^\infty \operatorname{Re}(a^{ij}(\xi_j + \nu_j D_y)u|(\xi_i + \nu_i D_y)u)\, dy = 0. \tag{6.39}$$

To be able to conclude from this identity that $u = 0$, the following condition is natural.

Definition 6.2.12. *A differential operator $\mathcal{A}(D) = a^{ij} D_i D_j$, with $a^{ij} = a^{ji} \in \mathcal{B}(E)$, is called* **normally strongly elliptic**, *if its is strongly elliptic and there is a constant $c > 0$ such that*

$$\operatorname{Re}(a^{ij}(\xi_j u + \nu_j v)|\xi_i u + \nu_i v) \geq c|\operatorname{Im}(u|v)|, \quad u, v \in E,$$

for all $\xi, \nu \in \mathbb{R}^n$, $|\xi| = |\nu| = 1$, $(\xi|\nu) = 0$.

From this condition we may then conclude $\operatorname{Im}(u(y)|D_y u(y)) = 0$ for all $y > 0$, which implies

$$\frac{d}{dy}|u(y)|^2 = 2\operatorname{Re}(u(y)|\partial_y u(y)) = 2\operatorname{Im}(u(y)|D_y u(y)) = 0,$$

hence $|u|$ is constant on $\mathbb{R}_+$, and so must be 0 as $u \in L_2(\mathbb{R}_+; E)$.

In case E is finite-dimensional, we are finished, as by strong ellipticity the dimension of the space of solutions of the homogeneous differential equation (6.38) has dimension $\dim E$. The map $T : u \mapsto \mathcal{B}(\xi+\nu D_y)u(0)$ is injective, hence also surjective, and so the Lopatinskii-Shapiro condition holds. If E is infinite-dimensional we have to work a little harder to obtain this result.

For this purpose observe first that the operator T defined above is injective, but also has dense range, as with $\mathcal{A}(D)$ also $\mathcal{A}^*(D)$ is normally strongly elliptic. Therefore we need to show that the range of T is closed. So let $u \in L_2(\mathbb{R}_+; E)$ be a solution of the ODE-problem

$$\begin{aligned} \lambda u(y) + \mathcal{A}(\xi + \nu D_y)u(y) &= 0, \quad y > 0, \\ \mathcal{B}(\xi + \nu D_y)u(0) &= g \in E. \end{aligned} \tag{6.40}$$

(i) We first consider the Neumann case. Multiplying the equation for u in (6.40) with $u(y)$, integrating over $\mathbb{R}_+$ and integrating by parts, we get by normal strong ellipticity

$$c|u_0|^2 \leq c\int_0^\infty |\partial_y |u(y)|^2|\, dy \leq 2|g||u_0|,$$

where $u_0 = u(0)$. This implies $|u_0| \leq C|g|$. Hence we may restrict our attention to the Dirichlet case, and the goal is to prove that there is a constant $C > 0$ such that $|u|_2 \leq C|u_0|$, for each L_2-solution u of the homogeneous problem

$$\lambda u(y) + \mathcal{A}(\xi + \nu D_y)u(y) = 0, \quad y > 0.$$

(ii) We begin estimating the L_2-norm of $u'(y) := \partial_y u(y)$ as follows, employing an integration by parts.

$$\begin{aligned}|u'|_2^2 = -(u_1|u_0) - (u|u'')_2 &\le |u_1||u_0| + |u|_2|u''|_2\\ &\le |u_1||u_0| + C|u|_2(|u|_2 + |u'|_2).\end{aligned}$$

Here $u_1 = u'(0)$ and we used the equation for u, as well as the fact that the operator $a^{ij}\nu_i\nu_j$ is invertible in E, by strong ellipticity. This implies by Young's inequality

$$|u'|_2^2 \le 2|u_1||u_0| + C_1|u|_2^2. \tag{6.41}$$

(iii) Next we write

$$|u_1|^2 = -2Re\int_0^\infty (u''(y)|u'(y))\,dy,$$

to obtain

$$|u_1|^2 \le 2|u'|_2|u''|_2 \le C|u'|_2(|u|_2 + |u'|_2),$$

hence by Young's inequality and (6.41)

$$|u_1|^2 \le C_2(|u|_2^2 + |u_0|^2). \tag{6.42}$$

(iv) Now we employ once more normal strong ellipticity, to obtain as in (i) the estimate

$$|u(y)|^2 \le C|u_0|(|u_0| + |u_1|) \le (C_\varepsilon|u_0| + \varepsilon|u_1|)^2, \tag{6.43}$$

again using Young's inequality.

The final estimate comes from strong ellipticity. Taking the Laplace transform of $\lambda u(y) + \mathcal{A}(\xi + \nu D_y)u(y) = 0$ w.r.t. the variable y we obtain

$$\mathcal{L}u(z) = -(\lambda + \mathcal{A}(\xi - iz\nu))^{-1}[(a^{kl}\nu_k\nu_l(zu_0 + u_1) + 2ia^{kl}\xi_k\nu_l u_0].$$

As $u \in L_2(\mathbb{R}_+; E)$, by strong ellipticity, the function $\mathcal{L}u(z)$ has only singularities in a compact subset of the negative half-plane, which only depends on (λ, ξ, ν). So choosing a contour Γ_- surrounding these singularities and lying entirely in the left half-plane, we obtain the representation

$$u(y) = \frac{1}{2\pi i}\int_{\Gamma_-} e^{zy}\mathcal{L}u(z)\,dz, \quad y > 0.$$

This implies

$$e^{\omega y}|u(y)| \le C_3(|u_0| + |u_1|), \tag{6.44}$$

with some fixed constants $\omega > 0$ and $C_3 > 0$ independent of u. Interpolating (6.43) and (6.44) and integrating over $y > 0$, this implies

$$\begin{aligned}|u|_2^2 &\le \frac{C_3}{\omega}(|u_0| + |u_1|)(C_\varepsilon|u_0| + \varepsilon|u_1|)\\ &\le \frac{C_3}{\omega}(C_\varepsilon'|u_0|^2 + 2\varepsilon|u_1|^2),\end{aligned}$$

applying once more Young's inequality. Finally, choosing $\varepsilon > 0$ small enough, combining the last estimate with (6.42) yields $|u|_2^2 \leq C|u_0|^2$, which is what we wanted to prove.

(v) Finally we consider *mixed boundary conditions* which are also important in applications. For this purpose let $P \in \mathcal{B}(E)$ be an orthogonal projection, i.e., $P = P^* = P^2$, and consider the boundary conditions

$$Pu(0) = g_0, \quad (I - P)\mathcal{B}(D)u(0) = g_1.$$

Then the energy argument yields an estimate of the form

$$c|u_0|^2 \leq C|g_0|(|u_0| + |u_1|) + |g_1||u_0|,$$

which implies

$$|u_0|^2 \leq C(|g_0|^2 + |g_1|^2) + \varepsilon|u_1|^2,$$

and so by (6.42)

$$|u_0|^2 \leq C(|g_0|^2 + |g_1|^2) + \varepsilon|u|_2^2,$$

and finally

$$|u|_2^2 \leq C(|g_0|^2 + |g_1|^2).$$

This shows that also the case of mixed boundary conditions is covered.

We summarize the result obtained above.

Proposition 6.2.13. *Let E be a Hilbert space and suppose that $\mathcal{A}(D)$ is a second-order, normally strongly elliptic differential operator in E.*

Then the Lopatinskii-Shapiro condition is satisfied for the natural boundary conditions, i.e., for Dirichlet, Neumann, or mixed conditions.

The following proposition deals with a very special case which, however, is frequently met in applications.

Proposition 6.2.14. *Let $a^{ij} = \alpha^{ij}b$, where the matrix $[\alpha^{ij}]$ is real, symmetric, and positive definite, and $b \in \mathcal{B}(E)$ is strongly accretive in the Hilbert space E, i.e.,*

$$\mathrm{Re}(bu|u) \geq c|u|^2, \quad u \in E,$$

for some positive constant $c > 0$.

Then $\mathcal{A}(D)$ is normally strongly elliptic.

We leave the proof of this proposition to the interested reader, as it only involves the Cauchy-Schwarz inequality.

Remark. **(i)** For $E = \mathbb{C}^n$ there is another stronger concept of ellipticity. We say that $a \in \mathcal{B}(E)^{n\times n}$ satisfies the *strong Legendre condition*, if there is a constant $C > 0$ such that

$$\mathrm{Re}\, a_{kl}^{ij} d_j^l \bar{d}_i^k \geq C|d|_2^2, \quad \text{for all } d \in \mathcal{B}(\mathbb{C}^n).$$

This condition means that a is strongly accretive on $\mathcal{B}(\mathbb{C}^n)$.

Obviously, the strong Legendre condition implies normal strong ellipticity, as for $d = \xi \otimes u + \nu \otimes v$ with $\xi \cdot \nu = 0$ we have

$$|d|_2^2 = |\xi|^2|u|^2 + |\nu|^2|v|^2 \geq 2|\xi||\nu||(u|v)|.$$

(ii) For many applications, however, the strong Legendre condition is too strong. This comes from the fact that the tensor a usually has symmetries like

$$a_{kl}^{ij} = a_{ij}^{kl} = a_{kj}^{il} = a_{il}^{kj}.$$

These symmetries are called *hyperelastic* and mean that a only acts on the symmetric part of a matrix and yields again a symmetric matrix. This is quite common in elasticity theory and also in compressible fluids, as there a represents *stress-strain* relations like $S = aD$, where D means the symmetric part of a deformation gradient, or of a velocity gradient. Then the stress S will also be symmetric. In this case the operator a maps the space of symmetric matrices $\mathrm{Sym}(\mathbb{C}^n)$ into itself. For this situation, the appropriate condition – which we call the *Legendre condition* – reads

$$\mathrm{Re}\, a_{kl}^{ij} e_j^l \bar{e}_i^k \geq C|e|_2^2, \quad \text{for all } e \in \mathrm{Sym}(\mathbb{C}^n).$$

This means that a is strongly accretive on $\mathrm{Sym}(\mathbb{C}^n)$, and it will be even selfadjoint in case $a_{kl}^{ij} = \bar{a}_{lk}^{ji}$.

Obviously, the Legendre condition implies strong ellipticity, but also normal strong ellipticity. In fact, for $d = \xi \otimes u + \nu \otimes v$ and $e = (d + d^{\mathsf{T}})/2$ we have with $|\xi| = |\nu| = 1$, $\xi \cdot \nu = 0$, and

$$u = (u|\xi)\xi + (u|\nu)\nu + u_\perp, \quad v = (v|\xi)\xi + (v|\nu)\nu + v_\perp, \quad u_\perp, v_\perp \perp \xi, \nu,$$

the identity

$$|e|_2^2 = \frac{1}{2}\{|u_\perp|^2 + |v_\perp|^2 + 2|(u|\xi)|^2 + 2|(v|\nu)|^2 + |(u|\nu) + (v|\xi)|^2\}.$$

This shows $e = 0$ if and only if $u_\perp = v_\perp = 0$, $(u|\xi) = (v|\nu) = 0$, $(u|\nu) = -(v|\xi)$, which implies $u = (u|\nu)\nu$, $v = (v|\xi)\xi$, in particular $(u|v) = 0$. In other words, if $|\xi| = |\nu| = 1$, $\xi \cdot \nu = 0$, and $\mathrm{Im}(u|v) \neq 0$, then $e \neq 0$. Therefore, the Legendre condition implies normal strong ellipticity.

(iii) In summary, we have the following implications for a second-order differential operator $\mathcal{A}(D) = a^{ij}D_iD_j$, with $a^{ij} = a^{ji} \in \mathcal{B}(\mathbb{C}^n)$:

$\mathcal{A}(D)$ satisfies the strong Legendre condition

$\Rightarrow \mathcal{A}(D)$ satisfies the Legendre condition

$\Rightarrow \mathcal{A}(D)$ is normally strongly elliptic

$\Rightarrow \mathcal{A}(D)$ is strongly elliptic

$\Rightarrow \mathcal{A}(D)$ is normally elliptic.

(iv) As an example we consider the well-known *Lamé operator* L, which is defined by

$$\begin{aligned}\mathsf{L}u :=& -\operatorname{div}[\mu_s(\nabla u + \nabla u^{\mathsf{T}}) + \mu_b(\operatorname{div} u)I]\\ =& -\mu_s\Delta u - (\mu_s + \mu_b)\nabla\operatorname{div} u,\end{aligned}$$

which yields

$$[\mathsf{L}u]_k = -a_{kl}^{ij}\partial_i\partial_j u_l, \quad \text{with } a_{kl}^{ij} = \mu_s(\delta_{ij}\delta_{kl} + \delta_{il}\delta_{jk}) + \mu_b\delta_{ik}\delta_{jl}.$$

The tensor a is easily checked to be hyperelastic and selfadjoint, and the Legendre condition is equivalent to

$$\mu_s > 0, \quad 2\mu_s + n\mu_b > 0.$$

On the other hand, a is strongly elliptic if and only if

$$\mu_s > 0, \quad 2\mu_s + \mu_b > 0,$$

and a is normally strongly elliptic if and only if

$$\mu_s > 0, \quad \mu_s + \mu_b > 0.$$

This can be shown by elementary linear algebra.

6.3 General Domains

Let $\Omega \subset \mathbb{R}^n$ be a domain with compact boundary $\partial\Omega$ of class C^{2m}. So Ω may be an interior or an exterior domain. In this section we consider the following general parabolic initial-boundary problem which is completely inhomogeneous. Let E be a Banach space of class $\mathcal{HT}$, and consider the parabolic problem

$$\begin{aligned}\partial_t u + \omega u + \mathcal{A}(x,D)u &= f && \text{in } \Omega,\\ \mathcal{B}_j(x,D)u &= g_j && \text{on } \partial\Omega,\ j = 1,\dots,m,\\ u(0) &= u_0 && \text{in } \Omega.\end{aligned} \tag{6.45}$$

Here $\mathcal{A}(x,D) = \sum_{|\alpha|\le 2m} a_\alpha(x)D^\alpha$ is a differential operator of order $2m$, $\mathcal{B}_j(x,D) = \sum_{|\beta|\le m_j} b_{j\beta}(x)D^\beta$ are differential operators of order $m_j < 2m$, $\omega \in \mathbb{R}$, and the data (f, g_j, u_0) are given. We are interested in maximal $L_{p,\mu}-L_q$-regularity of (6.45).

3.1 The Main Result

We formulate the assumptions of the main theorem in the following way. The most essential is the ellipticity assumption.

Definition 6.3.1. *We call the system* $(\mathcal{A}(x,D),\mathcal{B}_1(x,D),\ldots,\mathcal{B}_m(x,D))$ **uniformly normally elliptic** *if*

(i) $\mathcal{A}(x,D)$ *is normally elliptic, for each* $x\in\bar{\Omega}\cup\{\infty\}$*;*

(ii) *The Lopatinskii-Shapiro condition* **(LS)** *holds, for each* $x\in\partial\Omega$.

This assumption is crucial, and even necessary, for the main result stated below; see the Bibliographical Comments.

Next we state the regularity assumptions on the coefficients.

Condition (rA)

(rA1) $a_\alpha\in C_l(\bar{\Omega};\mathcal{B}(E))$ for each $|\alpha|=2m$;

(rA2) $a_\alpha\in L_{r_k}(\Omega;\mathcal{B}(E))+L_\infty(\Omega;\mathcal{B}(E))$ for each $|\alpha|=k<2m$, with $r_k\geq q$ and $2m-k>n/r_k$.

For the regularity of the coefficients on the boundary we recall the definition $\kappa_j=1-m_j/2m-1/2mq$.

Condition (rB)

(rB) $b_{j\beta}\in B^{2m\kappa_j}_{r_{jk}q}(\partial\Omega;\mathcal{B}(E))$ for each $|\beta|=k\leq m_j$, with $r_{jk}\geq q$, and $2m\kappa_j>(n-1)/r_{jk}$.

With these assumptions we can state the main theorem of this section.

Theorem 6.3.2. *Let* $\Omega\subset\mathbb{R}^n$ *be open with compact boundary* $\partial\Omega$ *of class* C^{2m}, $1<p,q<\infty$, $\mu\in(1/p,1]$, *and let* E *be a Banach space of class* $\mathcal{HT}(\alpha)$. *Assume that* $(\mathcal{A}(x,D),\mathcal{B}_1(x,D),\ldots\mathcal{B}_m(x,D))$ *is uniformly normally elliptic, and satisfies the regularity conditions* **(rA)** *and* **(rB)**. *Let* $\kappa_j\neq 1/p+1-\mu$ *for all* j.

Then there is $\omega_0\in\mathbb{R}$ *such that for each* $\omega>\omega_0$, *equation* (6.45) *admits a unique solution* u *in the class*

$$u\in\mathbb{E}_{1\mu}:=H^1_{p,\mu}(\mathbb{R}_+;L_q(\Omega;E))\cap L_{p,\mu}(\mathbb{R}_+;H^{2m}_q(\Omega;E)),$$

if and only if the data are subject to the following conditions.

(a) $f\in\mathbb{E}_{0\mu}=L_{p,\mu}(\mathbb{R}_+;L_q(\Omega;E))$, $u_0\in X_{\gamma,\mu}=B^{2m(\mu-1/p)}_{qp}(\Omega;E)$;

(b) $g_j\in\mathbb{F}_{j\mu}=F^{\kappa_j}_{pq,\mu}(\mathbb{R}_+;L_q(\partial\Omega;E))\cap L_{p,\mu}(\mathbb{R}_+;B^{2m\kappa_j}_{qq}(\partial\Omega;E)),\quad j=1,\ldots,m.$

(c) $\mathcal{B}_j(D)u_0=g_j(0)$ *if* $\kappa_j>1/p+1-\mu$, $j=1,\ldots,m$.

The solution depends continuously on the data in the corresponding spaces.

The proof of this result is given in the next subsections.

3.2 Coordinate Transformations

(a) Let $\Phi\in C^{2m}_b(\mathbb{R}^n;\mathbb{R}^n)$ be such that

$$c\leq|\det\partial\Phi(x)|\leq c^{-1},\quad x\in\mathbb{R}^n,$$

for some constant $c > 0$, and $\partial\Phi(x) \to I$ as $|x| \to \infty$. Define the coordinate transform T by means of

$$(Tv)(x) = v(\Phi(x)), \quad x \in \mathbb{R}^n.$$

Then $T : H^k_p(\mathbb{R}^n; E) \to H^k_p(\mathbb{R}^n; E)$ is an isomorphism for each $0 \le k \le 2m$. For the derivative $D = (D_1, \cdots, D_n)$ we obtain the transformation law

$$DTv(x) = \partial\Phi^{\mathsf{T}}(x)(Dv)(\Phi(x)),$$

hence the differential operator $\mathcal{A}(x, D)$ tranforms to $\mathcal{A}^\Phi(y, D)$, given by

$$\mathcal{A}^\Phi(y, D) = T^{-1}\mathcal{A}(x, D)T = \sum_{|\alpha|\le 2m} a^\Phi_\alpha(y)D = \sum_{|\alpha|\le 2m} a_\alpha(\Phi^{-1}(y))(\partial\Phi^{\mathsf{T}}(\Phi^{-1}(y))D)^\alpha.$$

Therefore, the coefficients a^Φ_α enjoy the same regularity conditions as a_α, and the principal symbol of $\mathcal{A}^\Phi$ is given by

$$\mathcal{A}^\Phi_\#(y, \xi) = \mathcal{A}_\#(\Phi^{-1}(y), \partial\Phi^{\mathsf{T}}(\Phi^{-1}(y))\xi), \quad y, \xi \in \mathbb{R}^n.$$

This shows that parameter-ellipticity of $\mathcal{A}^\Phi$ is equivalent to that of $\mathcal{A}$, with the same angle of ellipticity.

(b) We consider now the situation of a *bent half-space*. Replacing the variable $x \in \mathbb{R}^n_+$ by $(x, y) \in \mathbb{R}^{n-1} \times \mathbb{R}_+$, a bent half-space is defined by a coordinate transformation of the form $\Phi(x, y) = (x, y + h(x))$, with

$$h \in C^{2m}_b(\mathbb{R}^{n-1}; \mathbb{R}), \qquad \lim_{|x|\to\infty} \partial h(x) = 0. \tag{6.46}$$

Note that the boundary of the transformed domain is the graph $(x, h(x))$. Clearly, $\Phi \in C^{2m}_b(\mathbb{R}^n; \mathbb{R}^n)$, and with

$$\partial\Phi(x, y) = \begin{bmatrix} I & 0 \\ \partial h(x) & 1 \end{bmatrix}, \quad \partial\Phi(x, y)^{-1} = \begin{bmatrix} I & 0 \\ -\partial h(x) & 1 \end{bmatrix}$$

satisfies $\lim_{|x|+|y|\to\infty} \partial\Phi(x, y) = I$. Moreover, $\det \partial\Phi(x, y) = 1$. Hence we see that (a) applies. In a similar way, the boundary operators $\mathcal{B}_j(x, D)$ are transformed to $\mathcal{B}^\Phi(\cdot, D) = T^{-1}\mathcal{B}_j(\cdot, D)T$, hence their principal parts become

$$\mathcal{B}^\Phi_{j\#}(y, \xi) = \mathcal{B}_{j\#}(\Phi^{-1}(y), \partial\Phi^{\mathsf{T}}(\Phi^{-1}(y))\xi), \quad y, \xi \in \mathbb{R}^n.$$

Note that the normal of $\mathbb{R}^n_+$ at (x, y) transforms to

$$\nu = \frac{-\partial\Phi^{-\mathsf{T}} e_n}{|\partial\Phi^{-\mathsf{T}} e_n|} = \frac{1}{\sqrt{1 + |\nabla_x h|^2}}[\nabla_x h(x), -1]^{\mathsf{T}}.$$

This shows, by the remarks following the definition of the Lopatinskii-Shapiro Condition **(LS)**, that **(LS)** holds for the transformed problem $(\mathcal{A}^\Phi(x,D),\mathcal{B}_1^\Phi(x,D),\ldots,\mathcal{B}_m^\Phi(x,D))$ if and only it holds for the original problem.

(c) As the boundary spaces for the half-space are transformed to the corresponding boundary spaces on the bent half-space, these considerations show that the main result for the half-space, Theorem 6.2.11 as well as the estimate (6.34) remain valid for bent half-spaces.

3.3 Localization
If $\Omega\subset\mathbb{R}^n$ is unbounded, i.e., an exterior domain, we choose a large ball $B(0,R)\supset\Omega^c$ and define $U_0=\mathbb{R}^n\setminus\bar{B}(0,R)$. If Ω is bounded then $U_0=\emptyset$. We cover the compact set $\partial\Omega\subset\mathbb{R}^n$ by balls $B(x_k,r/2)$ with $x_k\in\partial\Omega$, $k=1,\ldots,N_1$, such that each part $\partial\Omega\cap B(x_k,2r)$ of the boundary $\partial\Omega$ can be parameterized by a function $h_k\in C^{2m}$ as a C^{2m}-graph over the tangent space $T_{x_k}\partial\Omega$. We extend this function h_k to a global function on $T_{x_k}\partial\Omega$ by a cut-off procedure, and denote the resulting bent half-space by H_k. This is possible by the regularity assumption $\partial\Omega\in C^{2m}$ as well as by compactness of $\partial\Omega$. We set $U_k=B(x_k,r)\cap\Omega$, $k=1,\ldots,N_1$. We cover the compact set $\bar{\Omega}\setminus\cup_{k=0}^{N_1}U_k$ by finitely many balls $B(x_k,r/2)$, $k=N_1+1,\ldots,N_2$, and set $U_k=B(x_k,r)$. Then $\{U_k\}_{k=0}^{N_2}$ is a finite open covering of $\bar{\Omega}$. Fix a C^∞-partition of unity $\{\varphi_k\}_{k=1}^{N_2}$ subordinate to the open covering $\{U_k\}_{k=0}^{N_2}$ of $\bar{\Omega}$, and let ψ_k denote C^∞-functions with $\psi_k=1$ on $\operatorname{supp}\varphi_k$, $\operatorname{supp}\psi_k\subset U_k$.

To define local operators $\mathcal{A}^k(x,D)$ and $\mathcal{B}_j^k(x,D)$ we proceed as follows. For the interior charts $k=0$, $k=N_1+1,\ldots,N_2$, we define the coefficients of $\mathcal{A}^k(x,D)$ by reflection of the top order coefficients at the boundary of U_k. This is the same trick as in Section 6.1.4. For the boundary charts $k=1,\ldots,N_1$ we first transform the top order coefficients of $\mathcal{A}(x,D)$ and $\mathcal{B}_j(x,D)$ in U_k to a half-space, extend them as in the Section 6.2.4, and then transform them back to the bent half space H_k.

Having defined the local differential operators, we may proceed as in Section 6.2.4, introducing local problems for the functions $u^k=\varphi_k u$, which for the interior charts $k=0$, and $k=N_1+1,\ldots,N_2$ are problems on $\mathbb{R}^n$, and for the boundary charts $k=1,\ldots,N_1$ are problems on the bent half-spaces H_k. For the latter, instead of using Theorem 6.2.10 we employ the extension of Theorem 6.2.11 to bent half-spaces. This completes the proof of Theorem 6.3.2.

3.4 The Semigroup
To define the semigroup associated with (6.45), we introduce the base space $X_0:=L_q(\Omega;E)$, as well as the operator A by means of

$$\begin{aligned}&(Au)(x):=\mathcal{A}(x,D)u(x),\quad x\in\Omega,\\ &u\in\mathsf{D}(A):=\{u\in H_q^{2m}(\Omega;E);\ \mathcal{B}_j(x,D)u=0 \text{ on } \partial\Omega,\ j=1,\ldots,m\},\end{aligned}$$

and we set $X_1=\mathsf{D}(A)$ equipped with the graph norm. Then the problem

$$\dot{u}+Au=f,\quad t>0,\quad u(0)=u_0,$$

has maximal L_p-regularity, by Theorem 6.3.2, hence $\omega_0+A\in\mathcal{MR}_p(X_0)$, for some $\omega_0>0$, and so $-A$ generates an analytic C_0-semigroup in X_0, by Proposition 3.5.2. This implies that $\omega+A$ is $\mathcal{R}$-sectorial for all $\omega>\mathsf{s}(-A)$, the spectral bound of $-A$. We note that the time-trace space $X_{\gamma,\mu}$ is given by

$$X_{\gamma,\mu}=\{u\in B_{qp}^{2m(\mu-1/p)}(\Omega;E);\ \mathcal{B}_j(x,D)u=0,\ \text{if}\ \kappa_j>1/p+1-\mu,\ j=1,\ldots,m\},$$

where we exclude the degenerate cases $\kappa_j=1/p+1-\mu$.

To determine the smallest value ω_0 in Theorem 6.3.2, we fix some large number ω_1 and solve (6.45) with ω replaced by ω_1 which results in some function $\bar{u}\in\mathbb{E}_{1\mu}$. Setting $\tilde{u}=u-\bar{u}$, the new function $\tilde{u}$ must solve the problem

$$\begin{aligned}\partial_t\tilde{u}+\omega\tilde{u}+\mathcal{A}(x,D)\tilde{u}&=(\omega_1-\omega)\bar{u} && \text{in }\Omega,\\ \mathcal{B}_j(x,D)\tilde{u}&=0 && \text{on }\partial\Omega,\ j=1,\ldots,m,\\ \tilde{u}(0)&=0 && \text{in }\Omega,\end{aligned}$$

for $t>0$. But this means

$$\dot{\tilde{u}}+\omega\tilde{u}+A\tilde{u}=(\omega_1-\omega)\bar{u},\quad t>0,\quad \tilde{u}(0)=0,$$

and so we see that $\omega>\mathsf{s}(-A)$ is sufficient, i.e., $\omega_0=\mathsf{s}(-A)$.

3.5 Higher Order Space Regularity
In many problems maximal L_p-regularity in $H_q^s(\Omega;E)$ is required, where $s>0$. In this subsection we consider the case $s=1$, and comment later on other values of s. By localization, coordinate transformation and perturbation, it is again enough to restrict to the half-space case with constant coefficients. We have to distinguish two cases, namely (i) $m_j\geq 1$ for all j, and (ii) $m_j=0$ for at least one j. We begin with the first case.

(i) $m_j\geq 1$ for all $j=1,\ldots,m$.
This case is the easy one. So suppose that we have a solution of (6.24) in the class

$$u\in H_{p,\mu}^1(\mathbb{R}_+;H_q^1(\mathbb{R}_+^n;E))\cap L_{p,\mu}(\mathbb{R}_+;H_q^{2m+1}(\mathbb{R}_+^n;E)). \tag{6.47}$$

Then necessarily

$$f\in L_{p,\mu}(\mathbb{R}_+;H_q^1(\mathbb{R}_+^n;E)),\quad u_0\in B_{qp}^{2m(\mu-1/p)+1}(\mathbb{R}_+^n;E),$$

and

$$D^\beta u\in H_{p,\mu}^{1-k/2m+1/2m}(\mathbb{R}_+;L_q(\mathbb{R}_+^n;E))\cap L_{p,\mu}(\mathbb{R}_+;H_q^{2m+1-k}(\mathbb{R}_+^n;E)),$$

for $|\beta|=k$; hence

$$g_j\in F_{pq,\mu}^{\kappa_j+1/2m}(\mathbb{R}_+;L_q(\mathbb{R}^{n-1};E))\cap L_{p,\mu}(\mathbb{R}_+;B_{qq}^{2m\kappa_j+1}(\mathbb{R}^{n-1};E)),$$

and the compatibility conditions

$$\mathcal{B}_j(D)u_0 = g_j(0), \quad \kappa_j > 1/p + 1 - \mu - 1/2m, \ j = 1, \ldots, m,$$

are satisfied.

Conversely, let data (f, g_j, u_0) with these properties be given, and let $\mathcal{A}(D)$ be normally elliptic and assume that $(\mathcal{A}(D), \mathcal{B}_1(D), \ldots, \mathcal{B}(D))$ satisfies the Lopatinskii-Shapiro condition. Then we can show that (6.24) admits a unique solution in the class (6.47). In fact, extending f and u_0 to all of $\mathbb{R}^n$, we obtain a solution of the full-space problem in the right class. Thus we may restrict attention to the case $(f, u_0) = 0$. Looking at the crucial equation for the half-space (6.32), we see that the solution in this case has regularity (6.47), as we may multiply $\tilde{g}$ in (6.32) by ρ.

Obviously, for variable coefficients and general domains with compact boundary we need to require additional smoothness of the coefficients and Ω. These turn out to be

(rA1+) $a_\alpha \in C_l(\bar{\Omega}; \mathcal{B}(E))$ for each $|\alpha| = 2m$;

(rA2+) $a_\alpha \in H^1_{r_k}(\Omega; \mathcal{B}(E)) + W^1_\infty(\Omega; \mathcal{B}(E))$ for each $|\alpha| = k \leq 2m$, with $r_k \geq q$ and $2m + 1 - k > n/r_k$;

(rB+) $b_{j\beta} \in B^{2m\kappa_j+1}_{r_{jk}q}(\partial\Omega; \mathcal{B}(E))$ for each $|\beta| = k \leq m_j$, with $r_{jk} \geq q$, and $2m\kappa_j + 1 > (n-1)/r_{jk}$.

With these assumptions, we have the following result which parallels Theorem 6.3.2.

Theorem 6.3.3. *Let $\Omega \subset \mathbb{R}^n$ be open with compact boundary $\partial\Omega$ of class C^{2m+1}, $1 < p, q < \infty$, $\mu \in (1/p, 1]$, and let E be a Banach space of class $\mathcal{HT}(\alpha)$. Assume that $(\mathcal{A}(x, D), \mathcal{B}_1(x, D), \ldots \mathcal{B}_m(x, D))$ is uniformly normally elliptic, and satisfies* **(rA1+)**, **(rA2+)** **(rB+)**. *Let $\kappa_j \neq 1/p + 1 - \mu - 1/2m$ for all j, and $m_j \geq 1$.*

Then there is $\omega_0 \in \mathbb{R}$ such that for each $\omega > \omega_0$, equation (6.45) *admits a unique solution u in the class*

$$u \in \mathbb{E}_{1\mu} := H^1_{p,\mu}(\mathbb{R}_+; H^1_q(\Omega; E)) \cap L_{p,\mu}(\mathbb{R}_+; H^{2m+1}_q(\Omega; E)),$$

if and only if the data are subject to the following conditions.

(a) $f \in L_{p,\mu}(\mathbb{R}_+; H^1_q(\Omega; E))$, $u_0 \in B^{2m(\mu-1/p)+1}_{qp}(\Omega; E)$;

(b) $g_j \in F^{\kappa_j+1/2m}_{pq,\mu}(\mathbb{R}_+; L_q(\partial\Omega; E)) \cap L_{p,\mu}(\mathbb{R}_+; B^{2m\kappa_j+1}_{qq}(\partial\Omega; E))$;

(c) $\mathcal{B}_j(D)u_0 = g_j(0)$ *if* $\kappa_j > 1/p + 1 - \mu - 1/2m$, $j = 1, \ldots, m$.

The solution depends continuously on the data in the corresponding spaces.

(ii) $m_j = 0$, for some j.
So let for simplicity $\mathcal{B}_1(D) = I$, a Dirichlet condition, and $m_j \geq 1$ for $j = 2, \ldots, m$. This case is more involved than **(i)**, as an additional compatibility condition shows

up. In fact, we have $\kappa_1 + 1/2m = 1 + (1-1/q)/2m > 1$, hence $\partial_t u$ has a time trace on the boundary, which by taking the time derivative of the first boundary condition yields

$$\partial_t g_1 = \partial_t u = f_{|\partial\Omega} - [\mathcal{A}(D)u]_{|\partial\Omega}.$$

This suggests

$$g_1 \in H^1_{p,\mu}(\mathbb{R}_+; B^{1-1/q}_{qq}(\mathbb{R}^{n-1}; E)) \cap L_{p,\mu}(\mathbb{R}_+; B^{2m+1-1/q}_{qq}(\mathbb{R}^{n-1}; E)).$$

On the other hand, we have

$$\mathcal{A}(D)u \in H^{1/2m}_{p,\mu}(\mathbb{R}_+; L_q(\mathbb{R}^n_+; E)) \cap L_{p,\mu}(\mathbb{R}_+; H^1_q(\mathbb{R}^n_+; E)),$$

which yields for its trace on $\partial\Omega$

$$[\mathcal{A}(D)u]_{|\partial\Omega} \in F^{(1-1/q)/2m}_{pq,\mu}(\mathbb{R}_+; L_q(\mathbb{R}^{n-1}; E)) \cap L_{p,\mu}(\mathbb{R}_+; B^{1-1/q}_{qq}(\mathbb{R}^{n-1}; E)).$$

This implies the additional regularity

$$\partial_t g_1 - f_{|\partial\Omega} \in F^{(1-1/q)/2m}_{pq,\mu}(\mathbb{R}_+; L_q(\mathbb{R}^{n-1}; E)) \cap L_{p,\mu}(\mathbb{R}_+; B^{2m+1-1/q}_{qq}(\mathbb{R}^{n-1}; E)),$$

and the additional compatibility condition

$$\partial_t g_1(0) + [\mathcal{A}(D)u_0]_{|\partial\Omega} = f(0)_{|\partial\Omega}, \quad \text{if } (1-1/q)/2m > 1/p + 1 - \mu.$$

The regularity and compatibility of g_j for $j \geq 2$ is the same as in **(i)**, and $g_1(0) = u_0$ on $\partial\Omega$ must be satisfied, as well.

Having worked out these *higher order compatibilities*, we now may proceed as in **(i)** to see that these conditions yield also sufficiency for solutions of (6.24) in the class (6.47).

Theorem 6.3.4. *Let $\Omega \subset \mathbb{R}^n$ be open with compact boundary $\partial\Omega$ of class C^{2m+1}, $1 < p, q < \infty$, $\mu \in (1/p, 1]$, and let E be a Banach space of class $\mathcal{HT}(\alpha)$. Assume that $(\mathcal{A}(x,D), \mathcal{B}_1(x,D), \ldots \mathcal{B}_m(x,D))$ is uniformly normally elliptic, and satisfies* **(rA1+)**, **(rA2+)**, **(rB1+)**, *for $j = 2, \ldots, m$. Let $\kappa_j \neq 1/p + 1 - \mu - 1/2m$ for all $j \geq 1$. Further assume that $\mathcal{B}_1(x,D)u = u$, i.e., $\mathcal{B}_1$ is a Dirichlet boundary condition.*

Then there is $\omega_0 \in \mathbb{R}$ such that for each $\omega > \omega_0$, equation (6.45) *admits a unique solution u in the class*

$$u \in \mathbb{E}_{1\mu} := H^1_{p,\mu}(\mathbb{R}_+; H^1_q(\Omega; E)) \cap L_{p,\mu}(\mathbb{R}_+; H^{2m+1}_q(\Omega; E)),$$

if and only if the data are subject to the following conditions.

(a) $f \in L_{p,\mu}(\mathbb{R}_+; H^1_q(\Omega; E))$, $u_0 \in B^{2m(\mu-1/p)+1}_{qp}(\Omega; E)$;

(b) $g_j \in F^{\kappa_j+1/2m}_{pq,\mu}(\mathbb{R}_+; L_q(\partial\Omega; E)) \cap L_{p,\mu}(\mathbb{R}_+; B^{2m\kappa_j+1}_{qq}(\partial\Omega; E))$;

(c) $\mathcal{B}_j(D)u_0 = g_j(0)$ *if* $\kappa_j > 1/p + 1 - \mu - 1/2m$, $j = 1, \ldots, m$;

(d) $\partial_t g_1 - f_{|\partial\Omega} \in F^{(1-1/q)/2m}_{pq,\mu}(\mathbb{R}_+; L_q(\mathbb{R}^{n-1}; E)) \cap L_{p,\mu}(\mathbb{R}_+; B^{1-1/q}_{qq}(\mathbb{R}^{n-1}; E))$;

(f) $\partial_t g_1(0) + [\mathcal{A}(D)u_0]_{|\partial\Omega} = f(0)_{|\partial\Omega}$, *if* $(1-1/q)/2m > 1/p + 1 - \mu$.

The solution depends continuously on the data in the corresponding spaces.

(iii) General $s > 0$.
Extending the observations in **(i)** and **(ii)**, we are able to study solutions in the class

$$u \in H^1_{p,\mu}(\mathbb{R}_+; H^s_q(\mathbb{R}^n_+; E)) \cap L_{p,\mu}(\mathbb{R}_+; H^{2m+s}_q(\mathbb{R}^n_+; E)), \tag{6.48}$$

for any $s > 0$ excluding the special values $s_i = m_i + 1/q$, and imposing the natural additional regularities $\partial\Omega \in C^{2m+s}$, as well as

$$a_\alpha \in H^s_{r_k}(\Omega) + H^s_\infty(\Omega), \quad r_k \geq q,\ 2m + s - k > n/r_k,\ 0 \leq |\alpha| = k \leq 2m,$$

and

$$b_{j\beta} \in B^{2m\kappa_j + s}_{r_{jk}q}(\partial\Omega), r_{jk} \geq q,\ 2m\kappa_j + s > (n-1)/r_{jk},\ 0 \leq |\beta| = k \leq m_j,$$

and imposing the higher order compatibilities as explained above. More precisely, let $m^0_1 < m^0_2 < \ldots < m^0_{i_{max}}$ be defined by the different orders m_j. Then for $0 \leq s < m^0_1 + 1/q$ we have no higher order compatibilities, for $m^0_1 + 1/q < s < m^0_2 + 1/q$ we have first (time-) order compatibilities, and with increasing s the number and the order of these higher compatibilities increases, whenever s crosses one the exceptional numbers s_i. So if s is large, this leads to a very complicated set of higher order compatibilities, which one clearly would like to avoid.

As a summary, in parabolic problems, such higher order compatibilities do not occur if $s < \min\{m_j\} + 1/q$, i.e., if the time derivatives of the boundary conditions do not have a space trace. For second-order problems this means in the Dirichlet case if $s < 1/q$, and in the Neumann case if $s < 1 + 1/q$.

(iv) The elliptic case.
Finally, we note that for *elliptic problems* this phenomenon does not occur. If $f \in H^s_q(\Omega)$ and $g_j \in B^{2m\kappa_j+s}_{qq}(\partial\Omega)$, then the solution of the elliptic problem

$$(\omega + \mathcal{A}(x, D)u = f \text{ in } \Omega, \quad \mathcal{B}_j(x, D)u = g_j \text{ on } \partial\Omega,\ j = 1, \ldots, m$$

has a unique solution in $H^{s+2m}_q(\Omega)$, provided $\mathcal{A}(x, D)$ is normally elliptic, the Lopatinskii-Shapiro condition holds, $\omega > \mathsf{s}(-A)$, $\partial\Omega \in C^{2m+s}$, and the coefficients satisfy the regularity conditions in **(iii)**.

3.6 Lower Order Space Regularity
In many problems, maximal L_p-regularity in $H^s_p(\Omega; E)$ is required, where $s < 0$. In this subsection we consider the case $s = -1$, i.e., we want to consider *weak solutions*. By localization, coordinate transformation and perturbation, it is again

enough to prove the results for the half-space case with constant coefficients. For all of this, we make the structural assumption

$$\mathcal{A}(x,D) = -i\sum_{\ell=1}^{n} \partial_\ell \mathcal{A}_\ell(x,D) = -i\,\mathrm{div}\,\mathsf{A}(x,D),$$

where $\mathcal{A}_\ell(x,D) = \sum_{|\alpha|\le 2m-1} a_{\ell\alpha}(x)D^\alpha$ are differential operators of order $2m-1$. We have to distinguish two cases:

(i) $m_j \le 2m-2$ for all $j = 1,\dots,m$.

(ii) $m_j \le 2m-2$ for all $j = 1,\dots,m-1$, but $m_m = 2m-1$; in this case we require

$$\mathcal{B}_m(x,D) = i\nu\cdot \mathsf{A}(x,D).$$

We begin with the first case.

(i) $m_j \le 2m-2$ for all $j = 1,\dots,m$.
We assume that $\mathcal{A}$ is normally elliptic, and that the system $(\mathcal{A},\mathcal{B}_1,\dots,\mathcal{B}_m)$ satisfies the Lopatinskii-Shapiro condition. The operator

$$\mathrm{Grad}_0 : {}_0H^1_{q'}(\Omega) \to L_{q'}(\Omega;\mathbb{C}^n), \quad \mathrm{Grad}_0\,\phi := \nabla\phi,$$

is well-defined, linear, bounded, and injective. Therefore, its dual

$$\mathrm{Div}_0 = -\mathrm{Grad}_0^* : L_q(\Omega;\mathbb{C}^n) \to {}_0H^1_{q'}(\Omega)^* =: H^{-1}_q(\Omega)$$

is also well-defined, bounded and has dense range. Note that in case Ω is bounded, by the Poincaré inequality $\mathsf{R}(\mathrm{Grad}_0)$ is closed, and hence Div_0 is surjective. Problem (6.45) can now be rewritten as

$$\begin{aligned} \partial_t(u|\phi)_\Omega + \omega(u|\phi)_\Omega + i(\mathsf{A}(x,D)u|\nabla\phi)_\Omega &= (f|\phi)_\Omega, \ \ \phi\in {}_0H^1_{q'}(\Omega)\\ \mathcal{B}_j(x,D)u &= g_j \qquad \text{on } \partial\Omega,\ j=1,\dots,m,\\ u(0) &= u_0 \qquad \text{in } \Omega. \end{aligned} \tag{6.49}$$

Abstractly, the first equation in (6.49) can be written as

$$\partial_t u + \omega u - i\,\mathsf{Div}_0(\mathsf{A}(x,D)u) = f \quad \text{in } H^{-1}_q(\Omega;E).$$

So we are looking for solutions in the class

$$u \in H^1_{p,\mu}(\mathbb{R}_+; H^{-1}_q(\Omega;E)) \cap L_{p,\mu}(\mathbb{R}_+; H^{2m-1}_q(\Omega;E)). \tag{6.50}$$

This implies the following necessary regularity conditions for the data.

(a) $f \in L_{p,\mu}(\mathbb{R}_+; H^{-1}_q(\Omega;E))$, $\quad u_0 \in B^{2m(\mu-1/p)-1}_{qp}(\Omega;E)$;

(b) $g_j \in F^{\kappa_j-1/2m}_{pq,\mu}(\mathbb{R}_+; L_q(\partial\Omega;E)) \cap L_{p,\mu}(\mathbb{R}_+; B^{2m\kappa_j-1}_{qq}(\Omega;E))$, for all $j = 1, \ldots, m$. Here we require $1 \geq \mu > 1/p + 1/2m$. The compatibility conditions now read

$$\mathcal{B}_j(x, D)u_0 = g_j(0), \quad \kappa_j > 1/p + 1 - \mu + 1/2m, \ j = 1, \ldots, m.$$

The assumptions on the coefficients are changed slightly, they read

(rA1-) $a_{\ell\alpha} \in C_l(\bar{\Omega}; \mathcal{B}(E))$, $\ell = 1, \ldots, n$, $|\alpha| = 2m - 1$;

(rA2-) $a_{\ell\alpha} \in [L_{r_k} + L_\infty](\Omega; \mathcal{B}(E))$, $\ell = 1, \ldots, n$, $k = |\alpha| < 2m - 1$, with $r_k \geq q$, $2m - k > n/r_k$;

(rB-) $b_{j\beta} \in B^{2m\kappa_j - 1}_{r_{jk}q}(\partial\Omega; \mathcal{B}(E))$, $|\beta| = k \leq m_j$, with $r_{jk} \geq q$, and $2m\kappa_j - 1 > (n-1)/r_{jk}$.

Finally, in this situation we only need to require $\partial\Omega \in C^{2m-1}$ (in case $m > 1$ it is even enough to require $\partial\Omega \in C^{(2m-1)-}$).

Theorem 6.3.5. *Let $\Omega \subset \mathbb{R}^n$ be open with compact boundary $\partial\Omega$ of class C^{2m-1}, $1 < p, q < \infty$, $\mu \in (1/p, 1]$, and let E be a Banach space of class $\mathcal{HT}(\alpha)$. Assume that $(\mathcal{A}(x, D), \mathcal{B}_1(x, D), \ldots \mathcal{B}_m(x, D))$, with $\mathcal{A}(x, D) = -i\sum_{\ell=1}^n \partial_\ell \mathcal{A}_\ell(x, D)$, is uniformly normally elliptic, and* **(rA1-)**, **(rA2-)** *and* **(rB-)**. *Let $m_j \leq 2m-2$ and $\kappa_j \neq 1/p + 1 - \mu + 1/2m$ for all j.*

Then there is $\omega_0 \in \mathbb{R}$ such that for each $\omega > \omega_0$, equation (6.45) *admits a unique solution u in the class*

$$u \in \mathbb{E}_{1\mu} := H^1_{p,\mu}(\mathbb{R}_+; H^{-1}_q(\Omega; E)) \cap L_{p,\mu}(\mathbb{R}_+; H^{2m-1}_q(\Omega; E)),$$

if and only if the data are subject to the following conditions.

(a) $f \in \mathbb{E}_{0\mu} = L_{p,\mu}(\mathbb{R}_+; H^{-1}_q(\Omega; E))$, $u_0 \in B^{2m(\mu-1/p)-1}_{qp}(\Omega; E)$;

(b) $g_j \in \mathbb{F}_{j\mu} = F^{\kappa_j - 1/2m}_{pq,\mu}(\mathbb{R}_+; L_q(\partial\Omega; E)) \cap L_{p,\mu}(\mathbb{R}_+; B^{2m\kappa_j - 1}_{qq}(\partial\Omega; E))$;

(c) $\mathcal{B}_j(D)u_0 = g_j(0)$ *if* $\kappa_j > 1/p + 1 - \mu - 1/2m$, $j = 1, \ldots, m$.

The solution depends continuously on the data in the corresponding spaces.

(ii) $m_j \leq 2m - 2$ for all $j = 1, \ldots, m-1$, $m_m = 2m - 1$.
In this case, as has been said before, we only consider $\mathcal{B}_m = i\nu \cdot \mathsf{A}$. Here we set

$$\mathsf{Grad} : \tilde{H}^1_{q'}(\Omega) \to L_{q'}(\Omega; \mathbb{C}^n), \quad \mathsf{Grad}\,\phi := \nabla\phi,$$

where $\tilde{H}$ means factorization over the constants, and we define

$$-\mathsf{Div} := \mathsf{Grad}^* : L_q(\Omega; \mathbb{C}^n) \to {}_0H^{-1}_q(\Omega) := \tilde{H}^1_{q'}(\Omega)^*.$$

As Grad is bounded, linear, injective, its dual Div is bounded, linear, and has dense range. Note that in case Ω is bounded, by the Poincaré-Wirtinger inequality

$\mathsf{R}(\mathsf{Grad})$ is closed, and hence Div is surjective. Problem (6.45) with f replaced by f_0 can now be rewritten as

$$\begin{aligned}\partial_t(u|\phi)_\Omega+\omega(u|\phi)_\Omega+i(\mathsf{A}(x,D)u|\nabla\phi)_\Omega&=\langle f|\phi\rangle,\quad \phi\in\tilde{H}^1_{q'}(\Omega),\\ \mathcal{B}_j(x,D)u&=g_j\ \text{ on }\partial\Omega,\ j=1,\ldots,m-1,\\ u(0)&=u_0\ \text{ in }\Omega,\end{aligned}\tag{6.51}$$

with the function $f\in L_{p,\mu}(\mathbb{R}_+;{}_0H^{-1}_q(\Omega;E))$ defined by

$$\langle f|\phi\rangle:=(f_0|\phi)_\Omega+(g_m|\phi)_{\partial\Omega}.$$

Abstractly, the first equation in (6.49) can be written as

$$\partial_t u+\omega u-i\,\mathsf{Div}(\mathsf{A}(x,D)u)=f\quad\text{in }{}_0H^{-1}_q(\Omega).$$

So we are looking for solutions in the class

$$u\in H^1_{p,\mu}(\mathbb{R}_+;{}_0H^{-1}_q(\Omega;E))\cap L_{p,\mu}(\mathbb{R}_+;H^{2m-1}_q(\Omega;E)).\tag{6.52}$$

The necessary regularity conditions on the data (g_j,u_0) as well as the compatibility and regularity conditions on the coefficients are the same as in **(i)**, where here $j=1,\ldots,m-1$. The condition for f changes in an obvious way.

Theorem 6.3.6. *Let $\Omega\subset\mathbb{R}^n$ be open with compact boundary $\partial\Omega$ of class C^{2m-1}, $1<p,q<\infty$, $\mu\in(1/p,1]$, and let E a Banach space of class $\mathcal{HT}(\alpha)$. Assume that $(\mathcal{A}(x,D),\mathcal{B}_1(x,D),\ldots\mathcal{B}_m(x,D))$, with $\mathcal{A}(x,D)=-i\sum_{\ell=1}^n\partial_\ell\mathcal{A}_\ell(x,D)$ and $\mathcal{B}_m(x,D)=i\nu\cdot\mathsf{A}(x,D)$, is uniformly normally elliptic, and* **(rA1-)**, **(rA2-)**, **(rB-)**, *$m_j\le 2m-2$, $\kappa_j\neq 1/p+1-\mu+1/2m$ for $j=1,\ldots,m-1$.*

Then there is $\omega_0\in\mathbb{R}$ such that for each $\omega>\omega_0$, equation (6.45) admits a unique solution u in the class

$$u\in\mathbb{E}_{1\mu}:=H^1_{p,\mu}(\mathbb{R}_+;{}_0H^{-1}_q(\Omega;E))\cap L_{p,\mu}(\mathbb{R}_+;H^{2m-1}_q(\Omega;E)),$$

if and only if the data are subject to the following conditions.

(a) $f\in L_{p,\mu}(\mathbb{R}_+;{}_0H^{-1}_q(\Omega;E))$, $u_0\in B^{2m(\mu-1/p)-1}_{qp}(\Omega;E)$;

(b) $g_j\in F^{\kappa_j-1/2m}_{pq,\mu}(\mathbb{R}_+;L_q(\partial\Omega;E))\cap L_{p,\mu}(\mathbb{R}_+;B^{2m\kappa_j-1}_{qq}(\partial\Omega;E))$, $j=1,\ldots,m-1$;

(c) $\mathcal{B}_j(D)u_0=g_j(0)$ *if* $\kappa_j>1/p-1-\mu+1/2m$, $j=1,\ldots,m-1$.

The solution depends continuously on the data in the corresponding spaces.

(iii) Sufficiency of the conditions in Theorems 6.3.5 and 6.3.6 for the half-space case with constant coefficients.

We first reduce to the case $(f,u_0)=0$ in the usual way: extend $u_0\in B^{2m(\mu-1/p)-1}_{qp}(\Omega)$ to all of $\mathbb{R}^n$ and f trivially by zero in case **(i)** and symmetrically in case **(ii)**. Solve the resulting problem in $\mathbb{R}^n$ in the proper class, and subtract this function from u. Then we consider the central identity (6.32) in the form

$$\rho^{2m-1}v = M(y,\rho,b,\sigma)\tilde{g}/\rho,$$

to see that the solution has regularity (6.50) in case **(i)** and (6.52) for **(ii)**. As a result, $\mathsf{A}(D)u \in L_{p,\mu}(\mathbb{R}_+; L_q(\mathbb{R}^n_+; E^n))$, hence by construction

$$\partial_t u = i \operatorname{div} \mathsf{A}(D)u = i \operatorname{\mathsf{Div}}_0 \mathsf{A}(D)u \in L_{p,\mu}(\mathbb{R}_+; H^{-1}_q(\mathbb{R}^n_+; E)),$$

in case **(i)**, and similarly in case **(ii)** we have

$$\partial_t u = i \operatorname{div} \mathsf{A}(D)u = i \operatorname{\mathsf{Div}} \mathsf{A}(D)u \in L_{p,\mu}(\mathbb{R}_+; {}_0H^{-1}_q(\mathbb{R}^n_+; E)).$$

(iv) The corresponding analytic C_0-semigroups.
Having maximal L_p-regularity of the problems (6.49) and (6.51) at our disposal, we may now argue as in Section 6.3.4 to derive the corresponding analytic C_0-semigroups in $H^{-1}_q(\Omega; E)$ resp. in ${}_0H^{-1}_q(\Omega; E)$. We omit the details here, however, note that these semigroups yield also corresponding semigroups in $L_q(\Omega; E)$, defining A_0 as the part of A in $L_q(\Omega; E)$. Note that $\mathsf{D}(A) \subset H^{2m-1}_q(\Omega; E)$, but $\mathsf{D}(A_0)$ is not explicitly known. Therefore it is an interesting question how the spectra of these extensions change, in particular the spectral bound. Then as $L_q(\Omega; E) \subset H^{-1}_q(\Omega; E)$, it is easy to see that $\rho(A) \subset \rho(A_0)$. But the converse is also true. In fact, suppose $f \in H^{-1}_q(\Omega; E)$ is given and $\lambda \in \rho(A_0)$. Set $J_\varepsilon = (I + \varepsilon A)^{-1}$; then $f_\varepsilon = J_\varepsilon f \in H^{2m-1}_q(\Omega; E)$ and $f_\varepsilon \to f$ in $H^{-1}_q(\Omega; E)$ as $\varepsilon \to 0$. Let $u_\varepsilon = (\lambda - A_0)^{-1} f_\varepsilon$, and choose ω large. Then we have

$$\begin{aligned} u_\varepsilon &= (\omega + A_0)^{-1}[-f_\varepsilon + (\omega + \lambda)u_\varepsilon] \\ &= -(\omega + A_0)^{-1} f_\varepsilon + (\omega + \lambda)(\lambda - A_0)^{-1}(\omega + A_0)^{-1} f_\varepsilon \\ &= -(\omega + A)^{-1} f_\varepsilon + (\omega + \lambda)(\lambda - A_0)^{-1}(\omega + A)^{-1} f_\varepsilon, \end{aligned}$$

as $(\omega + A)^{-1} f_\varepsilon = (\omega + A_0)^{-1} f_\varepsilon$. But this implies

$$u_\varepsilon \to u := (-I + (\omega + \lambda)(\lambda - A_0)^{-1})(\omega + A)^{-1} f.$$

Since $\mathsf{D}(A_0) \subset \mathsf{D}(A)$, we obtain $u \in \mathsf{D}(A)$ and then $(\lambda - A)u = f$. Hence $\lambda \in \rho(A)$. Therefore $\rho(A) = \rho(A_0)$ in case **(i)**, and by the same argument also in case **(ii)**.

6.4 Elliptic and Parabolic Problems on Hypersurfaces

Suppose that Σ is a compact hypersurface without boundary in $\mathbb{R}^n$ of class C^l. It is the purpose of this section to derive solvability results for elliptic and parabolic problems on Σ.

Let $\mathcal{A} : C^m(\Sigma; E) \to C(\Sigma; E)$ be a linear operator, where E denotes a Banach space of class $\mathcal{HT}$. Then $\mathcal{A}$ is a differential operator of order m on Σ if all representations of $\mathcal{A}$ in local coordinates (U, φ) are given by

$$\varphi_* \mathcal{A} u = \mathcal{A}_{(U,\varphi)}(x, D)\varphi_* u := \sum_{|\alpha| \le m} a^\alpha_{(U,\varphi)}(x) D^\alpha \varphi_* u, \tag{6.53}$$

where the coefficients $a^\alpha_{(U,\varphi)}$ are defined on the open set $\varphi(U)$ in $\mathbb{R}^{n-1}$, and $\varphi_* v = v \circ \varphi^{-1}$. $\mathcal{A}$ is said to be of class C^k if all coefficients are in this class. We may assume that the charts are normalized in such a way that $\varphi(U) = B_{\mathbb{R}^{n-1}}(0, 1)$.

The typical examples we have in mind, and which are used below, are the negative Laplace-Beltrami operator $-\Delta_\Sigma$ and Δ^2_Σ; see Section 2.1. A more involved operator is

$$\mathcal{A} = -\mathrm{div}_\Sigma(a(x)\nabla_\Sigma), \quad a \in C^1(\Sigma; \mathcal{B}(T\Sigma \otimes E)).$$

By using the language of covariant derivatives one can show that a differential operator defined on Σ is completely determined by the local representations (6.53).

Definition 6.4.1. *A differential operator $\mathcal{A}$ of order m on Σ is called* parameter-elliptic *if all local representations $\mathcal{A}_{(U,\varphi)}$ have this property. This means that for any local representation $\mathcal{A}_{(U,\varphi)}$ there is $\phi < \pi$ such that*

$$\sigma(\mathcal{A}^\#_{(U,\varphi)}(x, \xi)) \subset \Sigma_\phi, \quad (x, \xi) \in B_{\mathbb{R}^{n-1}}(0, 1) \times \mathbb{S}^{n-1}, \tag{6.54}$$

where

$$\mathcal{A}^\#_{(U,\varphi)}(x, \xi) := \sum_{|\alpha|=m} a^\alpha_{(U,\varphi)}(x)\xi^\alpha, \quad (x, \xi) \in B_{\mathbb{R}^{n-1}}(0, 1) \times \mathbb{S}^{n-1}.$$

By compactness, we then obtain

$$\phi_{\mathcal{A}} = \sup_{(U,\varphi)} \inf\{\phi \in (0, \pi) : (6.54) \text{ holds}\} < \pi.$$

$\phi_{\mathcal{A}}$ is called the angle of ellipticity *of $\mathcal{A}$. Finally, $\mathcal{A}$ is called normally elliptic if it is parameter-elliptic with angle $\phi_{\mathcal{A}} < \pi/2$.*

It is not difficult to show that the definition of the angle of ellipticity $\phi_{\mathcal{A}}$ is independent of the local representations. Moreover, $\mathcal{A}_{(U,\varphi)}(x, \xi)$ is continuous and invertible, hence by compactness of Σ, $\mathcal{A}_{(U,\varphi)}(x, \xi)$ as well as $\mathcal{A}_{(U,\varphi)}(x, \xi)^{-1}$ are uniformly bounded on $B_{\mathbb{R}^{n-1}}(0, 1) \times \mathbb{S}^{n-1}$.

By compactness of Σ we find a family of charts $\{(U_j, \varphi_j) : 1 \le j \le N\}$ such that $\{U_j\}_{j=1}^N$ covers Σ. Let $\{\pi_j : 1 \le j \le N\} \subset C^l(\Sigma)$ be a family of functions on Σ such that $\{(U_j, \pi_j^2) : 1 \le j \le N\}$ is a partition of unity subordinate to the open cover $\{U_j : 1 \le j \le N\}$, i.e.,

$$\mathrm{supp}(\pi_j) \subset U_j, \quad \sum_{j=1}^N \pi_j^2 = 1 \quad \text{on } \Sigma. \tag{6.55}$$

Then we call $\{(U_j, \varphi_j, \pi_j) : 1 \le j \le N\}$ a *localization system* for Σ.

Definition 6.4.2. *Given a localization system $\{(U_j, \varphi_j, \pi_j) : 1 \le j \le N\}$ for Σ, let*

$$\begin{aligned} R^c &: L_1(\Sigma; E) \to L_1(\mathbb{R}^{n-1}; E)^N, \quad R^c u := (\psi_j^*(\pi_j u)), \\ R &: L_1(\mathbb{R}^{n-1}; E)^N \to L_1(\Sigma; E), \quad R((u_j)) := \sum_{j=1}^{N} \pi_j \varphi_j^* u_j, \end{aligned} \tag{6.56}$$

where $\varphi_j^* v := v \circ \varphi$ and $\psi_j := \varphi_j^{-1}$. Moreover, we set

$$\mathcal{A}_j := \mathcal{A}_{(U_j, \varphi_j)}(x, D), \quad 1 \le j \le N. \tag{6.57}$$

We extend the coefficients in the usual way (e.g. as in Section 6.2) to obtain an extension of $\mathcal{A}_j$ to all of $\mathbb{R}^{n-1}$ with coefficients which have a limit at infinity, so that we may apply the results of Section 6.1.

It follows that R is a *retraction* with R^c a *co-retraction*, i.e., we have

$$RR^c u = u, \quad u \in L_1(\Sigma; E). \tag{6.58}$$

In the sequel, we set $\mathsf{u} = R^c u$, so $R\mathsf{u} = u$. Moreover,

$$\psi_j^* \mathcal{A} u = \mathcal{A}_j \psi_j^* u, \quad 1 \le j \le N,$$

and

$$\psi_j^* \pi_j \mathcal{A} u = \mathcal{A}_j \psi_j^* \pi_j u + \psi_j^* [\pi_j, \mathcal{A}] u =: \mathcal{A}_j \psi_j^* \pi_j u + B_j u.$$

Set $\mathsf{A} = \operatorname{diag}[\mathcal{A}_j]$ and $\mathsf{B} = [B_j R]$; then we obtain with (6.58)

$$R^c(\lambda + \omega + \mathcal{A})u = (\lambda + \omega + \mathsf{A} + \mathsf{B})\mathsf{u}. \tag{6.59}$$

By Theorem 6.1.10, $\omega + \mathcal{A}_j$ is $\mathcal{R}$-sectorial in $L_q(\mathbb{R}^{n-1}; E)$ for ω sufficiently large, $j = 1, \ldots, N$, and $\omega + \mathsf{A}$ is $\mathcal{R}$-sectorial for such ω as well. As B_j are of lower order, it follows by perturbation arguments (choosing ω even larger) that

$$\lambda + \omega + \mathsf{A} + \mathsf{B} : H_q^m(\mathbb{R}^{n-1}; E)^N \to L_q(\mathbb{R}^{n-1}; E)^N, \quad \lambda \in \Sigma_\phi,$$

is invertible, and $\lambda(\lambda + \omega + \mathsf{A} + \mathsf{B})^{-1}$ is $\mathcal{R}$-bounded in $L_q(\mathbb{R}^{n-1}; E)^N$, where $\phi > \phi_{\mathcal{A}}$ is fixed. Therefore, the operators

$$L_{\lambda,\omega} := R(\lambda + \omega + \mathsf{A} + \mathsf{B})^{-1} R^c : L_q(\Sigma; E) \to H_q^m(\Sigma; E), \quad \lambda \in \Sigma_\phi, \tag{6.60}$$

are well-defined, and with (6.58) and (6.59) we obtain

$$L_{\lambda,\omega}(\lambda + \omega + \mathcal{A})u = RR^c u = u, \quad u \in H_q^m(\Sigma; E),$$

i.e., $L_{\omega,\lambda}$ is a left-inverse for $(\lambda + \omega + \mathcal{A})$ and in addition, the family $\{L_{\lambda,\omega}\}_{\lambda \in \Sigma_\phi}$ is $\mathcal{R}$-bounded in $L_q(\Sigma)$.

On the other hand, we also have

$$\mathcal{A}(\pi_j\varphi_j^* u_j) = \pi_j\varphi_j^*\mathcal{A}_j u_j + \varphi_j^*[\mathcal{A}_j, \psi_j^*\pi_j]u_j =: \pi_j\varphi_j^*\mathcal{A}_j u_j + C_j u_j$$

and this yields

$$(\lambda+\omega+\mathcal{A})R\mathsf{u} = R(\lambda+\omega+\mathsf{A}+\mathsf{C})\mathsf{u}, \quad \mathsf{C}\mathsf{u} := R^c\sum_{j=1}^N C_j u_j. \tag{6.61}$$

For ω sufficiently large, we can again conclude that

$$\lambda+\omega+\mathsf{A}+\mathsf{C} : H_q^m(\mathbb{R}^{n-1};E))^N \to L_q(\mathbb{R}^{n-1};E)^N, \quad \lambda\in\Sigma_\phi,$$

is invertible, and hence

$$R_{\lambda,\omega} := R(\lambda+\omega+\mathsf{A}+\mathsf{C})^{-1}R^c$$

is well-defined. It follows from (6.58) and (6.61) that

$$(\lambda+\omega+\mathcal{A})R_{\lambda,\omega}u = RR^c u = u, \quad u\in H_q^m(\Sigma;E),$$

and this shows that $R_{\lambda,\omega}$ is a right-inverse for $\lambda+\omega+\mathcal{A}$. This implies

$$R_{\lambda,\omega} = L_{\lambda,\omega} = (\lambda+\omega+\mathcal{A})^{-1},$$

and $\{\lambda(\lambda+\omega+\mathcal{A})^{-1} : \lambda\in\Sigma_\phi\}\subset\mathcal{B}(L_q(\Sigma))$ is $\mathcal{R}$-bounded. Therefore $\omega+\mathcal{A}$ is $\mathcal{R}$-sectorial, which in case $\phi_{\mathcal{A}}<\pi/2$ implies, by Theorems 4.4.4 and 3.5.4, $\mathcal{A}\in\mathcal{MR}_{p,\mu}(L_q(\Sigma))$ for all $p,q\in(1,\infty)$, $1/p<\mu\le 1$.

Replacing in the above arguments the base space $L_q(\Sigma;E)$ by $K_q^s(\Sigma;E)$ and the regularity space $H_q^m(\Sigma;E)$ by $K_q^{s+m}(\Sigma;E)$, where $K=H$ or $K=W$, we obtain the same result, provided we have the corresponding result in $\mathbb{R}^{n-1}$. Employing Section 6.1.5, this yields the following maximal regularity result.

Theorem 6.4.3. *Let Σ be a compact hypersurface of class C^l without boundary in $\mathbb{R}^n$, $3\le l\le\infty$, $E\in\mathcal{HT}$, and let $p,q\in(1,\infty)$, $\mu\in(1/p,1]$. Suppose that $\mathcal{A}$ is a differential operator on Σ of order $m\in\mathbb{N}$ with coefficients in C^{2k}, where $k\in\mathbb{N}$, $2k+m\le l$. Define the realization A of $\mathcal{A}$ in $K_q^s(\Sigma;E)$ by means of*

$$Au := \mathcal{A}u \text{ on } \Sigma, \quad \mathrm{u}\in\mathsf{D}(\mathrm{A}) := \mathrm{K}_q^{s+m}(\Sigma;\mathrm{E}),$$

where $K\in\{H,W\}$, $|s|\le 2k$, $s\notin\mathbb{N}_0$ for $K=W$. Then we have

(i) *Suppose that $\mathcal{A}$ is parameter-elliptic. Then there is $\omega_0\ge 0$ such that the equation*

$$(\lambda+\omega+A)u = f \quad \text{in } K_q^s(\Sigma;E)$$

admits a unique solution $u\in K_q^{s+m}(\Sigma;E)$ for each $\omega\ge\omega_0$ and each $f\in K_q^s(\Sigma;E)$. For any $\phi>\phi_{\mathcal{A}}$ there is a constant M_ϕ such that the resolvent estimate

$$|\lambda(\lambda+\omega+A)^{-1}|_{\mathcal{B}(K_q^s(\Sigma;E))}\le M_\phi, \quad \lambda\in\Sigma_\phi,\ \omega\ge\omega_0,\ |s|\le 2k,$$

is valid. In addition, we have $\omega_0 + A \in \mathcal{RS}(K^s_q(\Sigma;E))$ *with* $\phi^R_A \le \phi_A$.

(ii) *Suppose that* $\mathcal{A}$ *is normally elliptic. Then there is* $\omega_0 \ge 0$ *such that the equation*

$$(\partial_t + \omega + A)u = f, \quad t>0, \quad u(0) = 0,$$

admits a unique solution $u \in H^1_{p,\mu}(\mathbb{R}_+ K^s_q(\Sigma;E)) \cap L_{p,\mu}(\mathbb{R}_+; K^{s+m}_q(\Sigma;E))$ *for each* $\omega \ge \omega_0$ *and each* $f \in L_{p,\mu}(\mathbb{R}_+; K^s_q(\Sigma;E))$. *Moreover, there is a constant* $C>0$ *independent of* ω *and* s *such that*

$$\omega |u|_{L_{p,\mu}(K^s_q)} + |\partial_t u|_{L_{p,\mu}(K^s_q)} + |u|_{L_{p,\mu}(K^{s+m}_q)} \le C|u|_{L_{p,\mu}(K^s_q)},$$

for all $f \in L_{p,\mu}(K^s_q(\Sigma;E))$. *In particular,* $\omega_0 + A \in \mathcal{MR}_p(K^s_q(\Sigma;E))$.

This result will be used frequently below, to understand moving boundaries analytically via the Hanzawa transform, and to handle dynamics on moving interfaces.

6.5 Transmission Problems

Elliptic and parabolic transmission conditions are present everywhere in mathematical physics, but one hardly finds citable references on this topic in the literature. For this reason, and also since we need results on transmission problems below, we consider such problems here, restricting to the second-order but vector-valued case.

Suppose that $\Omega \subset \mathbb{R}^n$ is a bounded domain with C^2-boundary, consisting of two parts Ω_1 and Ω_2 which are also open and such that that Ω_1 is separated from the boundary of Ω. Then we call Ω_2 the *continuous phase* and Ω_1 the *disperse phase.* Let $\Sigma = \partial\Omega_1$ be the *interface* separating Ω_1 and Ω_2 such that $\Omega = \Omega_1 \cup \Sigma \cup \Omega_2$. This is the typical two-phase situation. We consider in this section the following *transmission problem.*

$$\begin{aligned} (\partial_t + \omega + \mathcal{A}(x,\nabla_x))u &= f && \text{in } \Omega\setminus\Sigma,\\ \mathcal{B}(x,\nabla_x)u &= 0 && \text{on } \partial\Omega,\\ [\![u]\!] = g_\Sigma, \quad [\![\mathcal{B}(x,\nabla_x)u]\!] &= g && \text{on } \Sigma,\\ u(0) &= u_0 && \text{on } \Omega \end{aligned} \tag{6.62}$$

for $t>0$. Here u lives in a finite-dimensional Hilbert space E and

$$\mathcal{A}(x,\nabla_x) = -\mathrm{div}(a(x)\nabla_x), \quad \mathcal{B}(x,\nabla_x) = -(\nu(x)|a(x)\nabla_x),$$

where $\nu(x)$ denotes the outer unit normal at $x\in\Sigma$ directed into the interior of Ω_2 (resp. the outer unit normal of Ω at $x \in \partial\Omega$) and $a \in C^1_{ub}(\Omega\setminus\Sigma; \mathcal{B}(E))^{n\times n}$. The data (f, g_Σ, g, u_0) are given.

The purpose of this section is to prove the following result.

Theorem 6.5.1. *Let $1 < p, q < \infty$ and $1 \geq \mu > 1/p$, let E be a finite-dimensional Hilbert space, and assume that $a \in C^1_{ub}(\Omega \setminus \Sigma; \mathcal{B}(E))^{n\times n}$ is uniformly normally strongly elliptic.*

Then there is $\omega_0 \in \mathbb{R}$ such that for each $\omega > \omega_0$, problem (6.62) admits exactly one solution u in the class

$$u \in H^1_{p,\mu}(\mathbb{R}_+; L_q(\Omega; E)) \cap L_{p,\mu}(\mathbb{R}_+; H^2_q(\Omega \setminus \Sigma; E)),$$

if and only if

(a) $f \in L_{p,\mu}(\mathbb{R}_+; L_q(\Omega; E))$;

(b) $g_\Sigma \in F^{1-1/2q}_{pq,\mu}(\mathbb{R}_+; L_q(\Sigma; E)) \cap L_{p,\mu}(\mathbb{R}_+; W^{2-1/p}_q(\Sigma; E))$;

(c) $g \in F^{1/2-1/2q}_{pq,\mu}(\mathbb{R}_+; L_q(\Sigma; E)) \cap L_{p,\mu}(\mathbb{R}_+; W^{1-1/p}_q(\Sigma; E))$;

(d) $u_0 \in B^{2\mu-2/p}_{qp}(\Omega \setminus \Sigma; E)$;

(e) $[\![u_0]\!] = g_\Sigma(0)$ *for* $\mu > 3/2p$, *and* $[\![\mathcal{B}(x,\nabla)u_0]\!] = g(0)$ *for* $\mu > 1/2 + 3/2p$.

The solution map is continuous between the corresponding spaces.

The next subsections deal with the proof of this result.

5.1 The Model Problem

We consider the constant coefficient case with flat interface $\Sigma = \mathbb{R}^{n-1} \times \{0\} = \mathbb{R}^{n-1}$, and $\Omega = \mathbb{R}^n \setminus \Sigma$. As before, it is convenient to replace the variable $x \in \mathbb{R}^n$ by $(x, y) \in \mathbb{R}^{n-1} \times \mathbb{R}$. Then the problem reads

$$\begin{aligned} (\partial_t + \omega + \mathcal{A}(\nabla_x + \nu\partial_y))u &= f, \quad && y \neq 0, \\ [\![u]\!] = g_\Sigma, \quad [\![\mathcal{B}(\nabla_x + \nu\partial_y)u]\!] &= g, && y = 0, \\ u(0) &= u_0, && y \neq 0, \end{aligned} \tag{6.63}$$

for $t > 0$, with $\nu = e_n$ the outer unit normal of $\Omega_1 = \mathbb{R}^n_-$. We first verify the Lopatinskii-Shapiro condition for this case. For this purpose let $u \in L_2(\mathbb{R}; E)$ be a solution of the ode-problem

$$\lambda u(y) + \mathcal{A}(i\xi + \nu\partial_y)u(y) = 0, \quad y \neq 0,$$

such that

$$[\![u]\!] = 0, \quad [\![\mathcal{B}(i\xi + \nu\partial_y)u]\!] = 0 \quad \text{for } y = 0.$$

Here $\operatorname{Re}\lambda \geq 0$, $\xi \in \mathbb{R}^n$ and $(\xi|\nu) = 0$. Taking the inner product with $u(y)$, integrating over $\mathbb{R}$, and employing an integration by parts we obtain

$$0 = \lambda|u|_2^2 + \int_{\mathbb{R}} \sum_{k,l=1}^n (a^{kl}(\xi_l u(y) - i\nu_l\partial_y u(y)|(\xi_k u(y) - i\nu_k\partial_y u(y))_E\, dy,$$

as the boundary terms disappear by the jump conditions. Taking real parts, by normal strong ellipticity this yields

$$\operatorname{Re}(a^{kl}(\xi_l u(y) - i\nu_l\partial_y u(y)|(\xi_k u(y) - i\nu_k\partial_y u(y))_E = 0, \quad y \neq 0.$$

Using normal strong ellipticity once more we obtain

$$\partial_y |u(y)|_E^2 = 2\mathrm{Re}(\partial_y u(y)|u(y))_E = 0, \quad y \neq 0,$$

hence u is constant on $(0,\infty)$ and also on $(-\infty,0)$ which implies $u=0$ as $u \in L_2(\mathbb{R};E)$ by assumption. Thus the Lopatinskii-Shapiro condition for the two-phase problem is valid.

To obtain solvability of the problem in the right regularity class, perform a transformation to the half-space case as follows. Set

$$\begin{aligned} \tilde{u}(t,x,y) &= [u(t,x,y), u(t,x,-y)]^{\mathsf{T}}, & \tilde{u}_0(x,y) &= [u_0(x,y), u_0(x,-y)]^{\mathsf{T}}, \\ \tilde{f}(t,x,y) &= [f(t,x,y), f(t,x,-y)]^{\mathsf{T}}, & &\text{for } t \in (0,\infty),\ x \in \mathbb{R}^{n-1},\ y \in (0,\infty), \end{aligned}$$

and consider the problem

$$\begin{aligned} (\partial_t + \omega + \tilde{\mathcal{A}}(\nabla_x + \nu\partial_y))\tilde{u} &= \tilde{f} \quad \text{in } \mathbb{R}^n_+, \\ \tilde{u}(0) &= \tilde{u}_0 \quad \text{on } \mathbb{R}^{n-1}, \end{aligned} \tag{6.64}$$

with $t>0$, where $\tilde{\mathcal{A}}(\nabla_x + \nu\partial_y) = \mathrm{diag}[\mathcal{A}_2(\nabla_x + \nu\partial_y), \mathcal{A}_1(\nabla_x - \nu\partial_y)]$, with subscripts $2,1$ referring to the coefficients in the upper resp. lower half-plane. The boundary conditions now become

$$\begin{aligned} \tilde{u}_2(t,x,0) - \tilde{u}_1(t,x,0) &= g_\Sigma(t,x), \\ \mathcal{B}_2(\nabla_x + \nu\partial_y)\tilde{u}_2(t,x,0) + \mathcal{B}_1(\nabla_x + \nu\partial_y)\tilde{u}_1(t,x,0) &= g(t,x). \end{aligned}$$

Then with these boundary conditions, (6.64) is normally strongly elliptic and satisfies the Lopatinskii-Shapiro condition for the half-space. By the results of the previous section this problem is uniquely solvable in the right class, hence the transmission problem (6.63) has this property as well. This proves Theorem 6.5.1 for the constant coefficient case with flat interface.

5.2 Proof of Theorem 6.5.1

To complete the proof of Theorem 6.5.1, we may now proceed as in the one-phase case.

1. By perturbation, the result for the flat interface with constant coefficients remains valid for variable coefficients with small deviation from constant ones.

2. By another perturbation argument, a proper coordinate transformation transfers the result to the case of a bent interface.

3. The localization technique finally yields the result for the case of general domains and general coefficients.

One may then employ perturbation arguments another time to include lower order terms, at the expense of possibly enlarging ω_0.

5.3 The Steady Case

A result like Theorem 6.5.1 also holds for the steady case, i.e., for elliptic transmission problems. We consider here the corresponding result for the problem

$$\begin{aligned} (\omega + \mathcal{A}(x,\nabla_x))u &= f \quad \text{in } \Omega\setminus\Sigma,\\ \mathcal{B}(x,\nabla)u &= 0 \quad \text{on } \partial\Omega,\\ [\![u]\!] = g_\Sigma, \quad [\![\mathcal{B}(x,\nabla)u]\!] &= g \quad \text{on } \Sigma. \end{aligned} \tag{6.65}$$

Here the data (f, g_Σ, g) are given. For this problem we have

Theorem 6.5.2. *Let $1 < p < \infty$, let E be a finite-dimensional Hilbert space, and assume that $a \in C^1_{ub}(\Omega\setminus\Sigma;\mathcal{B}(E))^{n\times n}$ is uniformly normally strongly elliptic.*

Then there is $\omega_0 \in \mathbb{R}$ such that for each $\omega > \omega_0$, problem (6.65) admits exactly one solution u in the class

$$u \in H^2_p(\Omega\setminus\Sigma;E),$$

if and only if $(f, g_\Sigma, g) \in L_p(\Omega;E)\times W^{2-1/p}_p(\Sigma;E)\times W^{1-1/p}_p(\Sigma;E)$. The solution map is continuous between the corresponding spaces.

Remark 6.5.3. Higher regularity can be obtained for transmission problems in the same way as in Section 6.3.5 for the one-phase case, whereas lower regularity is obtained in the same way as in Section 6.3.6.

A natural question which arises is to determine the minimal value of ω_0. For this purpose, we first solve (6.65) for a large value $\omega = \bar\omega$, to obtain a function $\bar u$. Then we set $\tilde u = u - \bar u$; $\tilde u$ then must satisfy the problem

$$\begin{aligned} (\omega + \mathcal{A}(x,\nabla_x))\tilde u &= (\bar\omega - \omega)\bar u \quad \text{in } \Omega\setminus\Sigma,\\ \mathcal{B}(x,\nabla)\tilde u &= 0 \quad \text{on } \partial\Omega\setminus\Sigma,\\ [\![\tilde u]\!] = 0, \quad [\![\mathcal{B}(x,\nabla)\tilde u]\!] &= 0 \quad \text{on } \Sigma. \end{aligned} \tag{6.66}$$

This means that $-\omega$ should belong to the resolvent set of the operator A in $L_p(\Omega;E)$ defined by

$$\begin{aligned} &Au(x) = \mathcal{A}(x,\nabla_x)u(x), \quad x \in \Omega\setminus\Sigma, \qquad (6.67)\\ &\mathsf{D}(A) = \{u \in H^2_p(\Omega\setminus\Sigma;E) : [\![u]\!] = [\![\mathcal{B}(x,\nabla_x)u]\!] = 0 \text{ on } \Sigma,\ \mathcal{B}(x,\nabla_x)u = 0 \text{ on } \partial\Omega\}. \end{aligned}$$

In virtue of Theorem 6.5.1, this operator has maximal L_p-regularity, hence $-A$ generates an analytic C_0-semigroup. Therefore, ω_0 is the spectral bound $\mathsf{s}(-A)$ of $-A$. By a similar argument, the same is valid for the number ω_0 in Theorem 6.5.1.

5.4 Dirichlet-to-Neumann Operators

Dirichlet-to-Neumann operators appear frequently in mathematical physics and also at several places in this book. Such operators map Dirichlet boundary data to Neumann boundary data in several possible ways, and the goal is to obtain properties of such maps. In this subsection we assume throughout that $\mathcal{A}(x,\nabla_x)$ is uniformly normally strongly elliptic and that $\mathcal{B}$ is the corresponding co-normal derivative, as in the previous subsections.

(i) We begin with the elliptic case. Here there are two types of Dirichlet-to-Neumann operators, namely one- and two-phase operators. In the following, we always consider the elliptic problem

$$\begin{aligned}(\omega+\mathcal{A}(x,\nabla_x))u &= 0 \quad \text{in } \Omega\setminus\Sigma,\\ \mathcal{B}(x,\nabla_x)u &= 0 \quad \text{on } \partial\Omega,\end{aligned} \tag{6.68}$$

where at first $\omega \geq 0$ is sufficiently large. We may now assign Dirichlet data on the interface.

$$[\![u]\!] = 0, \qquad u = g \quad \text{on } \Sigma, \tag{6.69}$$

to obtain a unique solution $u \in H^2_p(\Omega\setminus\Sigma;E)$ provided $g \in W^{2-1/p}_p(\Sigma;E)$. These are actually two one-phase problems, one in Ω_1 and one in Ω_2. We then may compute the Neumann-boundary values $\mathcal{B}(x,\nabla_x)u$ on either side of Σ. We set $u_k = u|_{\Omega_k}$ for $k=1,2$ in the following definition.

Definition 6.5.4. *We call the maps $S_k : W^{2-1/p}_p(\Sigma;E) \to W^{1-1/p}_p(\Sigma;E)$ defined by the one-sided traces of the conormal derivative at Σ*

$$S_1 g := -\mathcal{B}(x,\nabla_x)u_1|_\Sigma, \quad S_2 g := \mathcal{B}(x,\nabla_x)u_2|_\Sigma,$$

the ***one-phase Dirichlet-to-Neumann operators*** *of* (6.68)–(6.69).

The operators S_k for $k=1,2$ are well-defined whenever the corresponding boundary value problem (6.68) with Dirichlet condition on Σ is well-posed. Clearly, S_k only depends on Ω_k, so that these operators are really one-phase. Considering (6.68) in Ω_k with Neumann condition $\mathcal{B}(x,\nabla_x)u = h$ on Σ, it becomes apparent that each S_k, $k=1,2$, is invertible if the corresponding boundary value problem with Neumann condition on Σ is well-posed. So in this situation S_1 and S_2 are isomorphisms.

On the other hand, there are two typical two-phase Dirichlet-to-Neumann operators for (6.68). The first one, called S_{d}, is obtained by solving the transmission problem

$$\begin{aligned}(\omega+\mathcal{A}(x,\nabla_x))u &= 0 \quad \text{in } \Omega\setminus\Sigma,\\ \mathcal{B}(x,\nabla_x)u &= 0 \quad \text{on } \partial\Omega,\\ [\![u]\!] = 0, \quad u &= g \quad \text{on } \Sigma,\end{aligned} \tag{6.70}$$

and setting $S_{\mathsf{d}} g := [\![\mathcal{B}(x,\nabla_x)u]\!]$. Actually we have $S_{\mathsf{d}} = S_1 + S_2$, as the normals of Ω_k on Σ have opposite directions. To obtain the inverse of S_{d}, one has to solve problem (6.68) with transmission conditions

$$[\![u]\!] = 0, \quad [\![\mathcal{B}(x,\nabla_x)u]\!] = h \quad \text{on } \Sigma,$$

yielding $g = u_{|\Sigma} = S_{\mathsf{d}}^{-1}h$. Hence S_{d} is an isomorphism as well.

To define the second two-phase Dirichlet-to-Neumann operator S_{n} we solve the transmission problem

$$\begin{aligned} (\omega + \mathcal{A}(x, \nabla_x))u &= 0 \quad \text{in } \Omega \setminus \Sigma, \\ \mathcal{B}(x, \nabla_x)u &= 0 \quad \text{on } \partial\Omega, \\ [\![u]\!] = g, \quad [\![\mathcal{B}(x, \nabla_x)u]\!] &= 0 \quad \text{on } \Sigma, \end{aligned} \tag{6.71}$$

and set $S_{\mathsf{n}}g := \mathcal{B}(x, \nabla_x)u$. To obtain the inverse of S_{n} we have to solve (6.68) with boundary condition

$$[\![\mathcal{B}(x, \nabla_x)u]\!] = 0, \quad \mathcal{B}(x, \nabla_x)u = h \quad \text{on } \Sigma,$$

yielding $g = [\![u]\!] = S_{\mathsf{n}}^{-1}h$. An easy computation shows the relation

$$S_{\mathsf{n}} = S_1 S_{\mathsf{d}}^{-1} S_2 = S_2 S_{\mathsf{d}}^{-1} S_1.$$

The two-phase Dirichlet-to-Neumann operators

$$S_{\mathsf{d}}, S_{\mathsf{n}} : W_p^{2-1/p}(\Sigma; E) \to W_p^{1-1/p}(\Sigma; E)$$

are well-defined and at the same time isomorphisms if ω is large enough. Observe that S_k, $k \in \{1, 2, \mathsf{d}, \mathsf{n}\}$, are pseudo-differential operators of order 1, while S_k^{-1} typically are integral operators on Σ with weakly singular kernels.

(ii) In the parabolic case one proceeds similarly. We begin with the problem in the bulk

$$\begin{aligned} (\partial_t + \omega + \mathcal{A}(x, \nabla_x))u &= 0 \quad \text{in } \Omega \setminus \Sigma, \\ \mathcal{B}(x, \nabla_x)u &= 0 \quad \text{on } \partial\Omega, \\ u(0) &= 0 \quad \text{in } \Omega, \end{aligned} \tag{6.72}$$

with $t > 0$. Here we have to distinguish the case of a finite interval $J = [0, a]$, from that of the half-line $J = \mathbb{R}_+$. We concentrate on the case of the half-line and assume $\omega \geq 0$ to be sufficiently large. For a finite interval $J = [0, a]$, no restrictions on $\omega \in \mathbb{R}$ are necessary. To avoid compatibility conditions here, we assume initial value $u(0) = 0$.

Imposing conditions on Σ as for the elliptic case in (i), we obtain the corresponding parabolic Dirichlet-to-Neumann operators, which we call again S_k, for $k \in \{1, 2, \mathsf{d}, \mathsf{n}\}$. The same assertions as in (i) are valid, but now the spaces are of course also time-dependent. We have isomorphisms

$$\begin{aligned} S_k : {}_0W_{p,\mu}^{1-1/2p}(\mathbb{R}_+; L_p(\Sigma; E)) &\cap L_{p,\mu}(\mathbb{R}_+; W_p^{2-1/p}(\Sigma; E)) \\ &\to {}_0W_{p,\mu}^{1/2-1/2p}(\mathbb{R}_+; L_p(\Sigma; E)) \cap L_{p,\mu}(\mathbb{R}_+; W_p^{1-1/p}(\Sigma; E)) \end{aligned}$$

for $k \in \{1, 2, \mathsf{d}, \mathsf{n}\}$, provided ω is sufficiently large. Note that in this case S_k are pseudo-differential operators jointly in time and space, of order $1/2$ in time and order 1 in space. These assertions remain valid if $\mathcal{A}$ and $\mathcal{B}$ are perturbed by lower order operators, at the expense that one possibly has to enlarge ω.

(iii) We now look closer at the possible values of ω. If $\mathcal{A}(x,\nabla_x) = -\partial_i a^{ij}(x)\partial_j$ and $\mathcal{B}(x,\nabla_x) = -\nu_i(x)a^{ij}(x)\partial_j$ such that $\mathcal{A}(x,\nabla_x)$ is normally strongly elliptic, uniformly in $x \in \Omega$ and $a^{ij} \in C^1_{ub}(\Omega \setminus \Sigma; \mathcal{B}(E))$, then $\omega > 0$ is sufficient. This follows from the fact that, as E is finite-dimensional, $\mathcal{A}(x,\nabla_x)$ with Neumann condition on $\partial\Omega$ and with each of the interface conditions (6.69), (6.70), (6.71) has compact resolvent, hence its spectrum consists only of discrete eigenvalues of finite multiplicity, and is independent of $p \in (1,\infty)$. By the standard energy argument it follows that the corresponding spectral bounds are in each case 0. The case $\omega = 0$ is more involved, as 0 is an eigenvalue. We postpone this case to Chapter 10, where $\omega = 0$ is essential.

6.6 Linearized Stefan Problems

The following linear problem is essential for the understanding of Problems **(P1)**, **(P3)**, **(P5)** and many other problems with moving interface. For its formulation, let $\Omega \subset \mathbb{R}^n$ be a bounded domain with boundary $\partial\Omega$ of class C^2. As before, we assume that Ω consists of two parts, Ω_1 and Ω_2 such that $\Sigma = \partial\Omega_1$ does not touch $\partial\Omega$. We assume that the hypersurface Σ is a C^3-manifold in $\mathbb{R}^n$. Note that in this section $E = \mathbb{C}$. Consider

$$\begin{aligned} (\partial_t + \omega + \mathcal{A}(x,\nabla_x))u &= f_u && \text{in } \Omega\setminus\Sigma,\\ \mathcal{B}(x,\nabla_x)u &= 0 && \text{on } \partial\Omega,\\ [\![u]\!] = 0, \quad u - \mathcal{C}(x,\nabla_\Sigma)h &= g && \text{on } \Sigma,\\ (\partial_t+\omega)h + [\![\mathcal{B}(x,\nabla_x)u]\!] &= f_h && \text{on } \Sigma,\\ u(0) = u_0 \text{ in } \Omega, \quad h(0) &= h_0 && \text{on } \Sigma. \end{aligned} \tag{6.73}$$

for $t > 0$. Here $\omega \geq 0$,

$$\mathcal{A}(x,\nabla_x) = -\mathrm{div}(a(x)\nabla),\ \mathcal{B}(x,\nabla_x) = -\nu(x)\cdot a(x)\nabla_x,\ \mathcal{C}(x,\nabla_\Sigma) = -\mathrm{div}_\Sigma(c(x)\nabla_\Sigma).$$

We assume that the coefficients $a \in C^1_{ub}(\Omega\setminus\Sigma; \mathcal{B}(\mathbb{R}^n))$ and $c \in C^3(\Sigma; \mathcal{B}(T\Sigma))$ are symmetric and uniformly positive definite. Note that the coefficients of $\mathcal{A}$ are allowed to jump across the interface Σ. The unit normal $\nu(x)$ at $x \in \Sigma$ is pointing from Ω_1 into Ω_2.

For Problems **(P1)**, **(P3)**, and **(P5)**, the prototype operators will be $\mathcal{A} = -\Delta$, $\mathcal{B} = -\partial_\nu$ and $\mathcal{C} = -\Delta_\Sigma$. The main result for this problem in the L_p-setting, $3 < p < \infty$, is the following.

Theorem 6.6.1. *Let $p > 3$ and $1 \geq \mu > 1/2 + 3/2p$. There exists $\omega_0 \in \mathbb{R}$ such that for each $\omega > \omega_0$, Problem (6.73) admits exactly one solution (u,h) in the class*

$$u \in H^1_{p,\mu}(\mathbb{R}_+; L_p(\Omega)) \cap L_{p,\mu}(\mathbb{R}_+; H^2_p(\Omega\setminus\Sigma)) =: \mathbb{E}_u,$$

$$h \in W^{3/2-1/2p}_{p,\mu}(\mathbb{R}_+; L_p(\Sigma)) \cap W^{1-1/2p}_{p,\mu}(\mathbb{R}_+; H^2_p(\Sigma)) \cap L_{p,\mu}(\mathbb{R}_+; W^{4-1/p}_p(\Sigma)) =: \mathbb{E}_h,$$

if and only if the data (f_u, g, f_h, u_0, h_0) are subject to the following conditions:

(a) $f_u \in L_{p,\mu}(\mathbb{R}_+; L_p(\Omega)) =: \mathbb{F}_u$;

(b) $g \in W^{1-1/2p}_{p,\mu}(\mathbb{R}_+; L_p(\Sigma)) \cap L_{p,\mu}(\mathbb{R}_+; W^{2-1/p}_p(\Sigma)) =: \mathbb{F}$;

(c) $f_h \in W^{1/2-1/2p}_{p,\mu}(\mathbb{R}_+; L_p(\Sigma)) \cap L_{p,\mu}(\mathbb{R}_+; W^{1-1/p}_p(\Sigma)) =: \mathbb{F}_h$;

(d) $u_0 \in W^{2\mu-2/p}_p(\Omega \setminus \Sigma)$, $h_0 \in W^{2+2\mu-3/p}_p(\Sigma)$;

(e) $u_0 - \mathcal{C}(x, \nabla_\Sigma)h_0 = g(0)$, $[\![\mathcal{B}(x, \nabla_x)u_0]\!] - f_h(0) \in W^{4\mu-2-6/p}_p(\Sigma)$, $\mathcal{B}(x, \nabla_x)u_0 = 0$ *on* $\partial\Omega$.

The solution map is continuous between the corresponding spaces.

6.1 Solution Spaces

To show necessity of the conditions in Theorem 6.6.1 and to explain the choice of the space for h which is illustrated in Figure 6.1, we begin with the regularity of u, which is the desired regularity in the bulk phases $\Omega \setminus \Sigma$. So let $(u, h) \in \mathbb{E}_u \times \mathbb{E}_h$ be a solution of (6.73). Then $f_u \in \mathbb{F}_u$ and the trace theory for second-order parabolic problems yields $u_0 \in W^{2\mu-2/p}_p(\Omega \setminus \Sigma)$, and

$$\begin{aligned} u_{|\Sigma} &\in W^{1-1/2p}_{p,\mu}(\mathbb{R}_+; L_p(\Sigma)) \cap L_{p,\mu}(\mathbb{R}_+; W^{2-1/p}_p(\Sigma)) = \mathbb{F}, \\ \nabla u_{|\Sigma} &\in W^{1/2-1/2p}_{p,\mu}(\mathbb{R}_+; L_p(\Sigma))^n \cap L_{p,\mu}(\mathbb{R}_+; W^{1-1/p}_p(\Sigma))^n = \mathbb{F}^n_h. \end{aligned}$$

This implies (a), and it is natural to assume $\mathcal{C}(\nabla_\Sigma)h \in \mathbb{F}$ as well, which then implies (b) and suggests

$$h \in W^{1-1/2p}_{p,\mu}(\mathbb{R}_+; H^2_p(\Sigma)) \cap L_{p,\mu}(\mathbb{R}_+; W^{4-1/p}_p(\Sigma)).$$

Example 3.4.9(iii) then yields $h_0 \in W^{2+2\mu-3/p}_p(\Sigma)$. Looking at the equation for h this implies $f_h \in \mathbb{F}_h$, hence (c), and suggests

$$h \in W^{3/2-1/2p}_{p,\mu}(\mathbb{R}_+; L_p(\Sigma)) \cap H^1_{p,\mu}(\mathbb{R}_+; W^{1-1/p}_p(\Sigma)).$$

By Example 4.5.16(ii) we have

$$W^{3/2-1/2p}_{p,\mu}(\mathbb{R}_+; L_p(\Sigma)) \cap W^{1-1/2p}_{p,\mu}(\mathbb{R}_+; H^2_p(\Sigma)) \hookrightarrow H^1_{p,\mu}(\mathbb{R}_+; W^{2-2/p}_p(\Sigma)), \quad (6.74)$$

and we arrive at the natural space $\mathbb{E}_h$ for h.

The first compatibility condition in (e) is obviously necessary if the corresponding traces exist, i.e., if $2\mu > 3/p$. The second compatibility condition is somewhat hidden, coming from the trace of $\partial_t h$. In fact we have by (6.74) and Example 3.4.9(ii)

$$W^{3/2-1/2p}_{p,\mu}(\mathbb{R}_+; L_p(\Sigma)) \cap W^{1-1/2p}_{p,\mu}(\mathbb{R}_+; H^2_p(\Sigma)) \hookrightarrow C^1_{ub}(\mathbb{R}_+; W^{4\mu-2-6/p}_p(\Sigma)),$$

hence the trace of $\partial_t h$ at $t = 0$ exists if $\mu > 1/2 + 3/2p$. This yields the second compatibility condition in (e). Note that the time trace of the class $\mathbb{F}_h$ merely

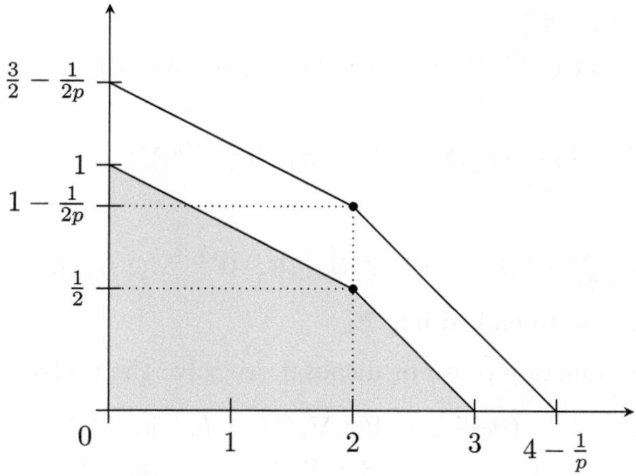

Figure 6.1: Regularity diagram for the Stefan problem.

belongs to $W_p^{2\mu-1-3/p}(\Sigma)$, as follows from Example 3.4.9(i). We remark that later on for the nonlinear problems we even have to require $\mu > 1/2+(n+2)/2p$, hence we cannot avoid this compatibility condition. The next subsections deal with the proof of sufficiency in Theorem 6.6.1.

6.2 Reductions

It is convenient to reduce problem (6.73) to the homogeneous conditions $(u_0, h_0, f_u, g) = 0$ and $f_h \in {}_0\mathbb{F}_h$, to simplify the problem and in particular to trivialize the compatibility conditions. For this purpose we define the operators $A = 1+\omega-\Delta_\Sigma$ and $B = 1+\omega+\Delta_\Sigma^2$; these are negative generators of exponentially stable analytic C_0-semigroups with maximal L_p-regularity on $L_p(\Sigma)$, hence also on $H_p^s(\Sigma)$ and on $W_p^s(\Sigma)$. We then define

$$\bar{h}(t) = (2e^{-At} - e^{-2At})h_0 + (e^{-Bt} - e^{-2Bt})B^{-1}h_1,$$

where $h_0 \in W_p^{2+2\mu-3/p}(\Sigma)$ and $h_1 = f_h(0) - [\![\mathcal{B}(x,\nabla_x)u_0]\!] - \omega h_0 \in W_p^{4\mu-2-6/p}(\Sigma)$. Obviously we have

$$\bar{h}(0) = h_0, \quad (\partial_t+\omega)\bar{h}(0) = h_1 + \omega h_0,$$

hence $\tilde{h} = h - \bar{h}$ has vanishing traces at $t = 0$.

We have to show that $\bar{h}$ belongs to $\mathbb{E}_h$. For this purpose we only need to consider the functions $e^{-At}h_0$ and $e^{-Bt}h_1$.

(i) Choosing as a base space $X_0 = H_p^2(\Sigma)$, Proposition 3.4.3 yields

$$e^{-At}h_0 \in W_{p,\mu}^{1-1/2p}(\mathbb{R}_+; H_p^2(\Sigma)) \cap L_{p,\mu}(\mathbb{R}_+; W_p^{4-1/p}(\Sigma)) \Leftrightarrow h_0 \in W_p^{2+2\mu-3/p}(\Sigma).$$

This then implies

$$\partial_t e^{-At}h_0 = -Ae^{-At}h_0 \in W_{p,\mu}^{1-1/2p}(\mathbb{R}_+; L_p(\Sigma)),$$

which yields $e^{-At}h_0 \in \mathbb{E}_h$.
(ii) Next we look at $e^{-Bt}B^{-1}h_1$ in the base space $X_0 = L_p(\Sigma)$. Proposition 3.4.3 yields

$$e^{-Bt}h_1 \in W^{1/2-1/2p}_{p,\mu}(\mathbb{R}_+; L_p(\Sigma)) \cap L_{p,\mu}(\mathbb{R}_+; W^{2-2/p}_p(\Sigma)) \Leftrightarrow h_1 \in W^{4\mu-2-6/p}_p(\Sigma).$$

This implies

$$e^{-Bt}B^{-1}h_1 \in W^{3/2-1/2p}_{p,\mu}(\mathbb{R}_+; L_p(\Sigma)) \cap H^1_{p,\mu}(\mathbb{R}_+; W^{2-2/p}_p(\Sigma)) \cap L_{p,\mu}(\mathbb{R}_+; W^{6-2/p}_p(\Sigma)),$$

which is easily seen to embed into $\mathbb{E}_h$.

Having the function $\bar{h}$ at our disposal, we solve the problem

$$\begin{aligned} (\partial_t + \omega + \mathcal{A}(x,\nabla_x))\bar{u} &= f_u && \text{in } \Omega\setminus\Sigma,\\ \mathcal{B}(x,\nabla_x)\bar{u} &= 0 && \text{on } \partial\Omega,\\ [\![\bar{u}]\!] = 0, \quad \bar{u} - \mathcal{C}(x,\nabla_\Sigma)\bar{h} &= g && \text{on } \Sigma,\\ \bar{u}(0) &= u_0 && \text{in } \Omega, \end{aligned}$$

in the class $\mathbb{E}_u$. Then the pair $(\tilde{u},\tilde{h}) = (u-\bar{u}, h-\bar{h})$ must satisfy (6.73) with data $(f_u, g, u_0, h_0) = 0$ and f_h replaced by $\tilde{f}_h$, defined by

$$\tilde{f}_h = f_h - [\![\mathcal{B}(x,\nabla_x))\bar{u}]\!] - (\partial_t+\omega)\bar{h} \in {}_0\mathbb{F}_h.$$

6.3 The Boundary Symbol

In this subsection we consider the constant coefficient case in $\Omega = \mathbb{R}^n$ with flat interface $\Sigma = \mathbb{R}^{n-1}\times\{0\} = \mathbb{R}^{n-1}$. This means that we consider the problem

$$\begin{aligned} (\partial_t + \omega + \mathcal{A}(\nabla_x))u &= f_u && \text{in } \dot{\mathbb{R}}^n,\\ [\![u]\!] = 0, \quad u - \mathcal{C}(\nabla_\Sigma)h &= g && \text{on } \mathbb{R}^{n-1},\\ (\partial_t+\omega)h + [\![\mathcal{B}(\nabla_x)u]\!] &= f_h && \text{on } \mathbb{R}^{n-1},\\ u(0) = u_0 \text{ in } \dot{\mathbb{R}}^n,\ h(0) &= h_0 && \text{on } \mathbb{R}^{n-1}. \end{aligned} \tag{6.75}$$

Here once more we use the notation $\dot{\mathbb{R}}^n = \mathbb{R}^{n-1}\times\dot{\mathbb{R}}$. As explained in the previous subsection, we may assume $(f_u, g, u_0, h_0) = 0$. We want to show that this problem admits a unique solution $h \in \mathbb{E}_h$ once we have $f_h \in {}_0\mathbb{F}_h$; then u is determined by its boundary value $u_\Sigma = \mathcal{C}(\nabla_x)h$ as explained in the previous subsection. It is also convenient to replace the variable $x \in \mathbb{R}^n$ by $(x,y) \in \mathbb{R}^{n-1}\times\mathbb{R}$, which means that we split into the tangential variable x and the normal variable y.

Taking Laplace transforms in time and Fourier transforms in the tangential variables we obtain the problem

$$\begin{aligned} (\lambda + a(\xi,\xi))\tilde{u} - 2ia(\xi,\nu)\partial_y\tilde{u} - a(\nu,\nu)\partial_y^2\tilde{u} &= 0, && y>0,\\ [\![\tilde{u}]\!] = 0, \quad \tilde{u} - \mathcal{C}(\xi)\tilde{h} &= 0, && y=0,\\ \lambda\tilde{h} - [\![a(\nu,\nu)\partial_y\tilde{u} + ia(\xi,\nu)\tilde{u}]\!] &= \tilde{f}_h, && y=0, \end{aligned} \tag{6.76}$$

where the tilde indicates Laplace transform in t with λ the co-variable of $\partial_t + \omega$ and Fourier transform in the tangential variable x with co-variable ξ. Here we employed the notation $\nu = e_n$ for the normal at the interface; observe that $\xi \perp \nu$. Note that the coefficients of $\mathcal{A}(\nabla_x)$ may jump across the interface. As the forms a_k, $k = 1, 2$ defining $\mathcal{A}(\nabla_x)$ are real symmetric and positive-definite, given $u_\Sigma = \mathcal{C}(\xi)\tilde{h}$, we may solve the equations in the region $y \neq 0$ to the result

$$\tilde{u}(y) = e^{-yr_2(\lambda,\xi)} u_\Sigma, \quad y > 0,$$

and

$$\tilde{u}(y) = e^{yr_1(\lambda,\xi)} u_\Sigma, \quad y < 0.$$

The symbols $r_k$are defined by $r_k(\lambda, \xi) = a_k(\nu|\nu)^{-1}[n_k(\lambda, \xi) + (-1)^k i a_k(\xi, \nu)]$, with

$$n_k(\lambda, \xi) = \sqrt{(\lambda + a_k(\xi, \xi)) a_k(\nu, \nu) - a_k(\xi, \nu)^2}, \quad k = 1, 2.$$

This implies

$$-[\![a(\nu, \nu)\partial_y \tilde{u} + i a(\xi, \nu)\tilde{u}]\!] = \big(n_1(\lambda, \xi) + n_2(\lambda, \xi)\big) u_\Sigma.$$

For the equation on the boundary this yields

$$s(\lambda, \xi)\tilde{h} = \tilde{f}_h, \quad \text{with} \quad s(\lambda, \xi) = \lambda + \mathcal{C}(\xi)\big(n_1(\lambda, \xi) + n_2(\lambda, \xi)\big). \tag{6.77}$$

So the main task is to show that this boundary symbol is invertible, and to obtain lower bounds of the form

$$|s(\lambda, \xi)| \geq c(|\lambda| + |\xi|^2 \sqrt{\lambda + |\xi|^2}), \quad \lambda \in \Sigma_{\pi/2}, \quad \xi \in \mathbb{R}^{n-1}.$$

Observe that a multiple of the lower bound in the line above yields trivially also an upper bound for $s(\lambda, \xi)$. Actually, as $|a_k(\xi, \nu)|^2 \leq a_k(\xi, \xi) a_k(\nu, \nu)$, with equality only if ξ and ν are linearly dependent - which is not possible as $\xi \perp \nu$ - this is very easy since the second and third terms in the definition of $s(\lambda, \xi)$ lie in the sector $\Sigma_{\pi/4}$ if $\lambda \in \Sigma_{\pi/2}$, and $\mathcal{C}(\xi)$ is positive and scales like $|\xi|^2$. As a consequence, the symbol

$$m(\lambda, \xi) := \frac{\lambda + |\xi|^2 \sqrt{\lambda + |\xi|^2}}{s(\lambda, \xi)}$$

is bounded from above and below even on a larger set

$$\lambda \in \Sigma_{\pi/2+\varepsilon}, \quad \xi \in \Sigma_\varepsilon^{n-1} \cup -\Sigma_\varepsilon^{n-1},$$

and it is a holomorphic function in λ and ξ. Therefore, m satisfies the scalar Mikhlin-condition w.r.t. ξ, uniformly w.r.t. $\lambda \in \Sigma_{\pi/2+\varepsilon}$. Inverting the Fourier transform, we obtain a holomorphic family of operators $M(\lambda)$ on $L_p(\mathbb{R}^{n-1})$, hence also on $W_p^s(\mathbb{R}^{n-1})$ for any real number s. The Kalton-Weis Theorem implies that $M(\partial_t + \omega)$ is bounded in each space ${}_0H^m_{p,\mu}(\mathbb{R}_+; W^s_p(\mathbb{R}^{n-1}))$, $m \geq 0$, hence by real interpolation also on ${}_0W^r_{p,\mu}(\mathbb{R}_+; W^s_p(\mathbb{R}^{n-1}))$, $r > 0$, and so Theorem 6.6.1 is valid for this model problem.

Remark 6.6.2. The argument given above shows that the boundary symbol $s(\lambda,\xi)$ is equivalent to the *essential symbol* of the problem which is given by

$$s_{ess}(\lambda,\xi) = \lambda + |\xi|^2\sqrt{\lambda+|\xi|^2}, \quad \operatorname{Re}\lambda > 0,\ \xi \in \mathbb{R}^{n-1}.$$

The essential symbol is responsible for the 'strange' solution space of h. The symbol does not come from an evolution equation, but from an evolutionary integral equation. In fact, $s_{ess}(\lambda,\xi)$ is the symbol of the pseudo-differential operator

$$L_{ess} = \partial_t + (-\Delta_x)\sqrt{\partial_t - \Delta_x},$$

which in different form may be written as

$$L_{ess} = \partial_t + (-\Delta_x)(\partial_t - \Delta_x)k_t\star,$$

where k_t denotes the heat kernel and $\star$ convolution in space and time.

6.4 General Coefficients and Domains

To complete the proof of Theorem 6.6.1, we may now proceed as before.

1. By perturbation, the result for the flat interface with constant coefficients remains valid for variable coefficients with the required regularity and small deviation from constant ones.
2. By another perturbation argument, the usual coordinate transformation transfers the result to the case of a bent interface.
3. The localization technique yields the case of general domains and general coefficients.
4. Employing perturbation arguments another time, we may include lower order terms, at the expense of possibly enlarging ω_0.

We refrain here from working out details, this is left to the interested reader.

6.5 The Stefan Semigroup

As problem (6.73) is a linear well-posed system of differential equations, there should be an underlying semigroup. However, it is not straightforward to formulate this, and to show that its negative generator has maximal regularity. To extract the semigroup, we indeed need another type of maximal regularity. For this purpose observe that by (6.74)

$$W^{3/2-1/2p}_{p,\mu}(\mathbb{R}_+;L_p(\Sigma)) \cap W^{1-1/2p}_{p,\mu}(\mathbb{R}_+;H^2_p(\Sigma)) \hookrightarrow H^1_{p,\mu}(\mathbb{R}_+;W^{2-2/p}_p(\Sigma)).$$

Therefore it makes sense to consider as the base space

$$(u,h) \in X_0 := L_{p,\mu}(\mathbb{R}_+;L_p(\Omega)) \times L_{p,\mu}(\mathbb{R}_+;W^{2-2/p}_p(\Sigma)),$$

and to ask for solutions

$$(u,h) \in \mathbb{E}_u \times \mathbb{E}_h^{sg}, \quad \text{with } \mathbb{E}_h^{sg} = H^1_{p,\mu}(\mathbb{R}_+; W_p^{2-2/p}(\Sigma)) \cap L_{p,\mu}(\mathbb{R}_+; W_p^{4-1/p}(\Sigma)).$$

This means that, given $(f_u, g, u_0, h_0) = 0$, but now with $f_h \in L_{p,\mu}(\mathbb{R}_+; W_p^{2-2/p}(\Sigma))$ instead of $f_h \in \mathbb{F}_h$, we want to find a unique solution $(u,h) \in \mathbb{E}_u \times \mathbb{E}_h^{sg}$ satisfying (6.73). Clearly, if such a solution exists then the extra condition

$$[\![\mathcal{B}(x,\nabla_x)u]\!] \in L_{p,\mu}(\mathbb{R}_+; W_p^{2-2/p}(\Sigma)) \tag{6.78}$$

must be satisfied. As we also have $[\![\mathcal{B}(x,\nabla_x)u]\!] \in W^{1/2-1/2p}_{p,\mu}(\mathbb{R}_+; L_p(\Sigma))$, by Example 3.4.9(ii) we obtain the compatibility condition $[\![\mathcal{B}(x,\nabla_x)u_0]\!] \in W_p^{4\mu-2-6/p}(\Sigma)$. This property allows again reduction to the case $(f_u, g, u_0, h_0) = 0$, by first solving (6.73) by means of Theorem 6.6.1 with $f_h = 0$ and (f_u, g, u_0, h_0) satisfying the assumptions of the theorem, to obtain functions $(\bar{u}, \bar{h}) \in \mathbb{E}_u \times \mathbb{E}_h$. The residual functions $(\tilde{u}, \tilde{h}) = (u - \bar{u}, h - \bar{h})$ must then satisfy (6.73) with $(f_u, g, u_0, h_0) = 0$, as contemplated. Note that $\bar{u}$ has the property (6.78), hence $\tilde{u}$ will also have this property if $\tilde{h} \in \mathbb{E}_h^{sg}$ and $f_h \in L_{p,\mu}(\mathbb{R}_+; W_p^{2-2/p}(\Sigma))$. Thus we need to show that for such f_h, problem (6.73) admits a unique solution in $\mathbb{E}_u \times \mathbb{E}_h^{sg}$. Actually, this follows immediately from the mapping properties of the symbol $s(\lambda, \xi)$ for the constant coefficient case with flat interface, and by perturbation and localization in general, as in the previous subsections. As a result we obtain

Theorem 6.6.3. *Let $p > 3$ and $1 \geq \mu > 1/2 + 3/2p$. There exists $\omega_0 \in \mathbb{R}$ such that for each $\omega > \omega_0$, Problem* (6.73) *admits exactly one solution (u,h) in the class*

$$\begin{aligned}
&u \in H^1_{p,\mu}(\mathbb{R}_+; L_p(\Omega)) \cap L_{p,\mu}(\mathbb{R}_+; H_p^2(\Omega \setminus \Sigma)) =: \mathbb{E}_u,\\
&[\![\mathcal{B}(x,\nabla_x)u]\!] \in L_{p,\mu}(\mathbb{R}_+; W_p^{2-2/p}(\Sigma)),\\
&h \in H^1_{p,\mu}(\mathbb{R}_+; W_p^{2-2/p}(\Sigma)) \cap W^{1-1/2p}_{p,\mu}(\mathbb{R}_+; H^2_p(\Sigma)) \cap L_{p,\mu}(\mathbb{R}_+; W_p^{4-1/p}(\Sigma)),
\end{aligned}$$

if and only if the data (f_u, g, f_h, u_0, h_0) are subject to the following conditions:

(a) $f_u \in L_{p,\mu}(\mathbb{R}_+; L_p(\Omega)) =: \mathbb{F}_u$;

(b) $g \in W^{1-1/2p}_{p,\mu}(\mathbb{R}_+; L_p(\Sigma)) \cap L_{p,\mu}(\mathbb{R}_+; W_p^{2-1/p}(\Sigma)) =: \mathbb{F}$;

(c) $f_h \in L_{p,\mu}(\mathbb{R}_+; W_p^{2-2/p}(\Sigma)) =: \mathbb{F}_h^{sg}$;

(d) $u_0 \in W_p^{2\mu-2/p}(\Omega \setminus \Sigma)$, $h_0 \in W_p^{2+2\mu-3/p}(\Sigma)$;

(e) $u_0 - \mathcal{C}(x,\nabla_\Sigma)h_0 = g(0)$, $[\![\mathcal{B}(x,\nabla_x)u_0]\!] \in W_p^{4\mu-2-6/p}(\Sigma)$,
$\mathcal{B}(x,\nabla_x)u_0 = 0$ *on* $\partial\Omega$.

The solution map is continuous between the corresponding spaces.

By means of Theorem 6.6.3, we may define the *Stefan semigroup* in X_0 in the following way. We set $z = [u,h]^{\mathsf{T}}$, $X_1 = H_p^2(\Omega \setminus \Sigma) \times W_p^{4-1/p}(\Sigma)$, and define an operator A in $X_0 = L_p(\Omega) \times W_p^{2-2/p}(\Sigma)$ by means of

$$\begin{aligned} A &= \begin{bmatrix} \mathcal{A}(x,\nabla_x) & 0 \\ [\![\mathcal{B}(x,\nabla_x)]\!] & 0 \end{bmatrix}, \\ \mathsf{D}(A) &= \{z \in X_1 : \mathcal{B}(x,\nabla_x)u = 0 \text{ on } \partial\Omega,\ u - \mathcal{C}(x,\nabla_\Sigma)h = 0 \text{ on } \Sigma, \\ &\qquad [\![\mathcal{B}(x,\nabla_x)u]\!] \in W_p^{2-2/p}(\Sigma)\}. \end{aligned} \tag{6.79}$$

Problem (6.73) for $g = 0$ is equivalent to the abstract evolution equation

$$\dot{z} + Az = f, \quad t > 0, \quad z(0) = z_0, \tag{6.80}$$

where we employed the abbreviations $z_0 = [u_0, h_0]^\mathsf{T}$ and $f = [f_u, f_h]^\mathsf{T}$. Then maximal L_p-regularity of (6.80) is equivalent to maximal L_p-regularity of (6.73) for $g = 0$ in the modified setting. Theorem 6.6.3 and Proposition 3.5.2 imply that $-A$ is the generator of an analytic C_0-semigroup with maximal L_p-regularity. This completes the construction of the semigroup.

Again we are interested in the smallest possible value of ω in Theorem 6.6.3. For this purpose we first solve the problem for a large value of ω, say $\bar{\omega}$, to obtain a solution $(\bar{u}, \bar{h}) \in \mathbb{E}_u \times \mathbb{E}_h^{sg}$, and we set $\tilde{u} = u - \bar{u}$, $\tilde{h} = h - \bar{h}$. Then we obtain the reduced system for these new functions

$$\begin{aligned} (\partial_t + \omega + \mathcal{A}(x,\nabla_x))\tilde{u} &= (\bar{\omega} - \omega)\bar{u} && \text{in } \Omega \setminus \Sigma, \\ \mathcal{B}(x,\nabla_x)\tilde{u} &= 0 && \text{on } \partial\Omega, \\ [\![\tilde{u}]\!] = 0, \quad u - \mathcal{C}(x,\nabla_\Sigma)\tilde{h} &= 0 && \text{on } \Sigma, \\ (\partial_t + \omega)\tilde{h} + [\![\mathcal{B}(x,\nabla_x)\tilde{u}]\!] &= (\bar{\omega} - \omega)\bar{h} && \text{on } \Sigma, \\ \tilde{u}(0) = 0 \ \text{ in } \Omega, \quad \tilde{h}(0) &= 0 && \text{on } \Sigma. \end{aligned} \tag{6.81}$$

Employing the semigroup this yields

$$\dot{\tilde{z}} + \omega\tilde{z} + A\tilde{z} = \tilde{f},\ t > 0, \quad \tilde{z}(0) = 0,$$

with $\tilde{z} = [\tilde{u}, \tilde{h}]^\mathsf{T}$ and $\tilde{f} = (\bar{\omega} - \omega)[\bar{u}, \bar{h}]^\mathsf{T}$. Therefore, the lower bound of ω is the spectral bound $\omega_0 = \mathsf{s}(-A)$. We are going to discuss this number in more detail in Chapter 10.

6.6 The Linearized Mullins-Sekerka Problem

In this subsection we consider the quasi-steady problem

$$\begin{aligned} (\eta + \mathcal{A}(x,\nabla_x))u &= f_u && \text{in } \Omega \setminus \Sigma, \\ \mathcal{B}(x,\nabla_x)u &= 0 && \text{on } \partial\Omega, \\ [\![u]\!] = 0, \quad u - \mathcal{C}(x,\nabla_\Sigma)h &= g && \text{on } \Sigma, \\ (\partial_t + \omega)h + [\![\mathcal{B}(x,\nabla_x)u]\!] &= f_h && \text{on } \Sigma, \\ h(0) &= h_0 && \text{on } \Sigma. \end{aligned} \tag{6.82}$$

Here $\omega, \eta \geq 0$, $\mathcal{A}(x, \nabla_x) = -\mathrm{div}(a(x)\nabla_x)$, $\mathcal{B}(x, \nabla_x) = -(\nu(x)|a(x)\nabla_x)$ and $\mathcal{C}(x, \nabla_\Sigma) = -\mathrm{div}_\Sigma(c(x)\nabla_\Sigma)$ are differential operators with $a \in C^1_{ub}(\Omega \setminus \Sigma; \mathcal{B}(\mathbb{R}^n))$, $c \in C^3(\Sigma; \mathcal{B}(T\Sigma))$, with both a and c symmetric and uniformly positive definite. Note that the coefficients of $\mathcal{A}$ are allowed to jump across the interface Σ. Here the unit normal $\nu(x)$ at $x \in \Sigma$ is pointing from Ω_1 into Ω_2.

The main result for this problem in the L_p-setting, $1 < p < \infty$, is the following.

Theorem 6.6.4. *Let $p \in (1, \infty)$ and $1 \geq \mu > 1/p$. There exists $\omega_0, \eta_0 \in \mathbb{R}$ such that for each $\omega > \omega_0$, $\eta > \eta_0$, problem (6.82) admits exactly one solution (u, h) in the class*

$$u \in L_{p,\mu}(\mathbb{R}_+; H^2_p(\Omega \setminus \Sigma)) =: \mathbb{E}_u,$$
$$h \in H^1_{p,\mu}(\mathbb{R}_+; W^{1-1/p}_p(\Sigma)) \cap L_{p,\mu}(\mathbb{R}_+; W^{4-1/p}_p(\Sigma)) =: \mathbb{E}_h,$$

if and only if the data (f_u, g, f_h, h_0) are subject to the following conditions:

(a) $f_u \in L_{p,\mu}(\mathbb{R}_+; L_p(\Omega)) =: \mathbb{F}_u$;

(b) $g \in L_{p,\mu}(\mathbb{R}_+; W^{2-1/p}_p(\Sigma)) =: \mathbb{F}$;

(c) $f_h \in L_{p,\mu}(\mathbb{R}_+; W^{1-1/p}_p(\Sigma)) =: \mathbb{F}_h$;

(d) $h_0 \in W^{1+3\mu-4/p}_p(\Sigma)$.

The solution map is continuous between the corresponding spaces.

This result is proved in the same way as Theorem 6.6.1. As the bulk problem is stationary, the proof is even simpler, so we skip the details here.

We are interested in the parameters η and ω. For this purpose we define an operator A in $X = L_p(\Omega)$ by means of

$$Au(x) = \mathcal{A}(x, \nabla_x)u(x), \quad x \in \Omega \setminus \Sigma, \tag{6.83}$$
$$\mathsf{D}(A) = \{u \in H^2_p(\Omega \setminus \Sigma) : \ u = 0 \text{ on } \Sigma, \ \mathcal{B}(x, \nabla_x)u = 0 \text{ on } \partial\Omega\}.$$

As $\mathcal{A}$ is uniformly strongly elliptic by assumption, Theorem 6.5.1 shows that $-A$ is the generator of an analytic C_0-semigroup with maximal L_p-regularity. Moreover, as Ω is bounded and Σ and $\partial\Omega$ are of class C^2 and do not intersect, the semigroup as well as the resolvent of A are compact. Therefore, the spectrum of A consist only of eigenvalues of finite algebraic multiplicity, and is independent of p. So we only need to consider $p = 2$. If z is an eigenvalue of A with eigenfunction $u \neq 0$, the usual energy argument yields

$$z|u|^2_{L_2} = \int_\Omega a^{ij} \partial_j u \, \overline{\partial_i u} \, dx,$$

we see that z must be real, and employing uniform strong ellipticity,

$$z|u|^2_{L_2} \geq c|\nabla u|^2_{L_2},$$

hence $z \geq 0$. If $z = 0$ then $\nabla u = 0$ in Ω hence u is constant, as Ω is connected, and u has no jump across Σ, and so $u = 0$. This shows that $0 \in \rho(A)$.

We now may proceed as follows. Solve the problem

$$\begin{aligned} (\eta + \mathcal{A}(x, \nabla_x))u &= 0 \quad \text{in } \Omega \setminus \Sigma, \\ \mathcal{B}(x, \nabla_x)u &= 0 \quad \text{on } \partial\Omega, \\ [\![u]\!] = 0, \; u &= g \quad \text{on } \Sigma, \end{aligned}$$

and denote the solution by $u_\eta = T_\eta g$. The Dirichlet-to-Neumann operator for this problem is given by $S_{\mathsf{d},\eta} g = [\![\mathcal{B}(x, \nabla_x) T_\eta g]\!]$. Then we define A_η in $X_0 := W_p^{1-1/p}(\Sigma)$ by means of

$$A_\eta h = S_{\mathsf{d},\eta} \mathcal{C}(x, \nabla_\Sigma) h, \quad X_1 := \mathsf{D}(A_\eta) = W_p^{4-1/p}(\Sigma). \tag{6.84}$$

It is clear that (6.82) with $\eta = 0$, and $(f_u, g) = 0$ is equivalent to the evolution equation

$$\partial_t h + \omega h + A_0 h = f_h, \quad t > 0, \quad h(0) = h_0.$$

We can easily show that $-A_0$ generates an analytic C_0-semigroup with maximal L_p-regularity, the *Mullins-Sekerka semigroup.* In fact, for this purpose note that by Theorem 6.6.4, A_η has maximal L_p-regularity for η large. Now we have the identity

$$T_0 g = T_\eta g + \eta(\eta + A)^{-1} T_0 g,$$

which follows from

$$\begin{aligned} \eta(\eta + A)^{-1} T_0 g &= (\eta + A)(\eta + A)^{-1} T_0 g - A(\eta + A)^{-1} T_0 g \\ &= T_0 g - (\eta + A)^{-1} A (T_0 g - T_\eta g) - A(\eta + A)^{-1} T_\eta g \\ &= T_0 g + (\eta + A)^{-1} \mathcal{A}(x, \nabla_x) T_\eta g - A(\eta + A)^{-1} T_\eta g \\ &= T_0 g - (\eta + A)(\eta + A)^{-1} T_\eta g. \end{aligned}$$

Hence,

$$A_0 = S_{\mathsf{d},0} \mathcal{C}(x, \nabla_\Sigma) = A_\eta + \eta [\![\mathcal{B}(x, \nabla_x)]\!] (\eta + A)^{-1} T_0 \mathcal{C}(x, \nabla_x).$$

As the second term is a compact perturbation of the first one, the claim follows. We summarize these considerations.

Corollary 6.6.5. *The* Mullins-Sekerka operator A_0 *defined above is the negative generator of an analytic C_0-semigroup $e^{-A_0 t}$, the* Mullins-Sekerka semigroup, *with maximal L_p-regularity in the base space $X_0 = W_p^{1-1/p}(\Sigma)$ and domain $X_1 = \mathsf{D}(A_0) = W_p^{4-1/p}(\Sigma)$.*

We note that A_0 is a pseudo-differential operator of order three. The spectrum of this operator will be considered in Chapter 12.

6.7 The Linearized Verigin Problem

The following linear problem arises as the linearization of the *Verigin problem.* It can be treated analytically in the same way as the linearized Stefan problem with surface tension. Therefore we will keep this section quite short. For the formulation, as in the previous section, let $\Omega \subset \mathbb{R}^n$ be a bounded domain with boundary $\partial\Omega$ of class C^2. Ω consists of two parts, Ω_1 and Ω_2 such that $\Sigma = \partial\Omega_1$ does not touch $\partial\Omega$. We assume that the hypersurface Σ is a C^3-manifold in $\mathbb{R}^n$. Consider

$$\begin{aligned}
(\partial_t + \omega + \mathcal{A}(x,\nabla_x))u &= f_u && \text{in } \Omega\setminus\Sigma,\\
\mathcal{B}(x,\nabla_x)u &= 0 && \text{on } \partial\Omega,\\
[\![u]\!] + \mathcal{C}(x,\nabla_\Sigma)h &= g && \text{on } \Sigma,\\
[\![\mathcal{B}(x,\nabla_x)u]\!] &= 0 && \text{on } \Sigma,\\
(\partial_t+\omega)h - \mathcal{B}(x,\nabla_x)u &= f_h && \text{on } \Sigma,\\
u(0) = u_0 \ \text{in } \Omega, \quad h(0) &= h_0 && \text{on } \Sigma.
\end{aligned} \tag{6.85}$$

Here $\omega \geq 0$, $\mathcal{A}(x,\nabla_x) = -\mathrm{div}(a(x)\nabla)$, $\mathcal{B}(x,\nabla_x) = -(\nu(x)|a(x)\nabla_x)$ and $\mathcal{C}(x,\nabla_\Sigma) = -\mathrm{div}_\Sigma(c(x)\nabla_\Sigma)$ are differential operators with $a \in C^1_{ub}(\Omega\setminus\Sigma;\mathcal{B}(\mathbb{R}^n))$, $c \in C^3(\Sigma;\mathcal{B}(T\Sigma))$, where a and c are both symmetric and uniformly positive definite. The coefficients of $\mathcal{A}$ are allowed to jump across the interface Σ. The unit normal $\nu(x)$ at $x \in \Sigma$ is pointing from Ω_1 into Ω_2.

The main result for this problem in the L_p-setting, $3 < p < \infty$, is the following.

Theorem 6.7.1. *Let $p > 3$ and $1 \geq \mu > 1/2 + 3/2p$. There exists $\omega_0 \in \mathbb{R}$ such that for each $\omega \geq \omega_0$, problem* (6.85) *admits exactly one solution (u,h) in the class*

$$u \in H^1_{p,\mu}(\mathbb{R}_+;L_p(\Omega)) \cap L_{p,\mu}(\mathbb{R}_+;H^2_p(\Omega\setminus\Sigma)) =: \mathbb{E}_u,$$
$$h \in W^{3/2-1/2p}_{p,\mu}(\mathbb{R}_+;L_p(\Sigma)) \cap W^{1-1/2p}_{p,\mu}(\mathbb{R}_+;H^2_p(\Sigma)) \cap L_{p,\mu}(\mathbb{R}_+;W^{4-1/p}_p(\Sigma)) =: \mathbb{E}_h,$$

if and only if the data (f_u, g, f_h, u_0, h_0) are subject to the following conditions:

(a) $f_u \in L_{p,\mu}(\mathbb{R}_+;L_p(\Omega)) =: \mathbb{F}_u$;

(b) $g \in W^{1-1/2p}_{p,\mu}(\mathbb{R}_+;L_p(\Sigma)) \cap L_{p,\mu}(\mathbb{R}_+;W^{2-1/p}_p(\Sigma)) =: \mathbb{F}$;

(c) $f_h \in W^{1/2-1/2p}_{p,\mu}(\mathbb{R}_+;L_p(\Sigma)) \cap L_{p,\mu}(\mathbb{R}_+;W^{1-1/p}_p(\Sigma)) =: \mathbb{F}_h$;

(d) $u_0 \in W^{2\mu-2/p}_p(\Omega\setminus\Sigma)$, $h_0 \in W^{2+2\mu-3/p}_p(\Sigma)$;

(e) $[\![u_0]\!] + \mathcal{C}(x,\nabla_\Sigma)h_0 = g(0)$, $\mathcal{B}(x,\nabla_x)u_0 + f_h(0) \in W^{4\mu-2-6/p}_p(\Sigma)$,

$[\![\mathcal{B}(x,\nabla_x)u]\!] = 0$, $\mathcal{B}(x,\nabla_x)u_0 = 0$ *on* $\partial\Omega$.

The solution map is continuous between the corresponding spaces.

There is no need to discuss the solution spaces, as they are the same as in the previous section, similar reductions are available, and the process of localization will also be the same. Therefore we will concentrate on the model problem.

7.1 The Boundary Symbol

In this subsection we consider the constant coefficient case in $\Omega = \mathbb{R}^n$ with flat interface $\Sigma = \mathbb{R}^{n-1} \times \{0\} = \mathbb{R}^{n-1}$, for short. This means that we consider the problem which is already in reduced form

$$\begin{aligned} (\partial_t + \omega + \mathcal{A}(\nabla_x))u &= 0 && \text{in } \hat{\mathbb{R}}^n, \\ [\![u]\!] + \mathcal{C}(\nabla_\Sigma)h &= 0 && \text{on } \mathbb{R}^{n-1}, \\ [\![\mathcal{B}(\nabla_x)u]\!] &= 0 && \text{on } \mathbb{R}^{n-1}, \\ (\partial_t + \omega)h - \mathcal{B}(\nabla_x)u &= f_h && \text{on } \mathbb{R}^{n-1}, \\ u(0) = 0 \ \text{ in } \hat{\mathbb{R}}^n, \quad h(0) &= 0 && \text{on } \mathbb{R}^{n-1}. \end{aligned} \tag{6.86}$$

As in the previous section, it is convenient to replace the variable $x \in \mathbb{R}^n$ by $(x, y) \in \hat{\mathbb{R}} := \mathbb{R}^{n-1} \times \dot{\mathbb{R}}$, which means that we split into the tangential variables x and the normal variable y.

Taking Laplace transform in time and Fourier transform in the tangential variables we obtain the problem

$$\begin{aligned} (\lambda + a(\xi,\xi))\tilde{u} - 2ia(\xi,\nu)\partial_y\tilde{u} - a(\nu,\nu)\partial_y^2\tilde{u} &= 0, && y > 0, \\ [\![\tilde{u}]\!] + \mathcal{C}(\xi)\tilde{h} &= 0, && y = 0, \\ [\![a(\nu,\nu)\partial_y\tilde{u} + ia(\xi,\nu)\tilde{u}]\!] &= 0, && y = 0, \\ \lambda\tilde{h} + (a(\nu,\nu)\partial_y\tilde{u} + ia(\xi,\nu)\tilde{u}) &= \tilde{f}_h, && y = 0, \end{aligned} \tag{6.87}$$

where, as before, the tilde indicates Laplace transform in t with co-variable τ, $\lambda = \tau + \omega$, and Fourier transform in the tangential variable x with co-variable ξ, and $\nu = e_n$ is the normal at the interface. Note that the coefficients of $\mathcal{A}(\nabla_x)$ may jump across the interface. As the forms a_k, $k = 1, 2$, defining $\mathcal{A}(\nabla_x)$ are real symmetric and positive definite, and given $u_\Sigma = \mathcal{C}(\xi)\tilde{h}$, we may solve the equations in the region $y \neq 0$ to the result

$$\tilde{u}(y) = e^{-yr_2(\lambda,\xi)}u_\Sigma^2, \ \ y > 0, \quad \text{and} \quad \tilde{u}(y) = e^{yr_1(\lambda,\xi)}u_\Sigma^1, \ \ y < 0,$$

where u_Σ^k denote the unknown boundary values of u in Ω_k. The symbols r_k, $k = 1, 2$, are defined as in Section 6.6.3. The interface conditions imply

$$u_\Sigma^2 - u_\Sigma^1 = -\mathcal{C}(\xi)\tilde{h},$$

and with the notation

$$n_k(\lambda,\xi) = \sqrt{(\lambda + a_k(\xi,\xi))a_k(\nu,\nu) - a_k(\xi,\nu)^2},$$

the second interface condition reads

$$n_1(\lambda,\xi)u_\Sigma^1 + n_2(\lambda,\xi)u_\Sigma^2 = 0.$$

For the equation on the boundary this yields

$$s(\lambda,\xi)\tilde{h} = \tilde{f}_h, \quad \text{with} \quad s(\lambda,\xi) = \lambda + \mathcal{C}(\xi)\frac{n_1(\lambda,\xi)n_2(\lambda,\xi)}{n_1(\lambda,\xi)+n_2(\lambda,\xi)}. \tag{6.88}$$

As the harmonic mean $n_1n_2/(n_1+n_2) = 1/(1/n_1+1/n_2)$ is leaving each sector Σ_θ, $\theta \le \pi/2$, invariant we may conclude as in Section 6.6.3 that the symbol

$$m(\lambda,\xi) := \frac{\lambda + |\xi|^2\sqrt{\lambda+|\xi|^2}}{s(\lambda,\xi)}$$

is bounded from above and below even on a larger set $\lambda \in \Sigma_{\pi/2+\varepsilon}$, $\xi \in \Sigma_\varepsilon^{n-1} \cup -\Sigma_\varepsilon^{n-1}$, and as in Section 6.6.3 this proves the assertion for the case of constant coefficients and flat interface. Note that the essential symbol of the Verigin problem is the same as that for the Stefan problem considered in the previous section.

7.2 The Verigin Semigroup

As problem (6.85) is a linear well-posed system of differential equations there should be an underlying semigroup. This semigroup can be constructed in a similar way as the Stefan semigroup in the previous section.

Theorem 6.7.2. *Let $p > 3$ and $1 \ge \mu > 1/2 + 3/2p$. There exists $\omega_0 \in \mathbb{R}$ such that for each $\omega \ge \omega_0$, Problem (6.85) admits exactly one solution (u,h) in the class*

$$\begin{aligned}
&u \in H^1_{p,\mu}(\mathbb{R}_+; L_p(\Omega)) \cap L_{p,\mu}(\mathbb{R}_+; H^2_p(\Omega\setminus\Sigma)) =: \mathbb{E}_u,\\
&\mathcal{B}(x,\nabla_x)u \in L_{p,\mu}(\mathbb{R}_+; W^{2-2/p}_p(\Sigma)),\\
&h \in H^1_{p,\mu}(\mathbb{R}_+; W^{2-2/p}_p(\Sigma)) \cap W^{1-1/2p}_{p,\mu}(\mathbb{R}_+; H^2_p(\Sigma)) \cap L_{p,\mu}(\mathbb{R}_+; W^{4-1/p}_p(\Sigma)),
\end{aligned}$$

if and only if the data (f_u, g, f_h, u_0, h_0) are subject to the following conditions:

(a) $f_u \in L_{p,\mu}(\mathbb{R}_+; L_p(\Omega)) =: \mathbb{F}_u$;

(b) $g \in W^{1-1/2p}_{p,\mu}(\mathbb{R}_+; L_p(\Sigma)) \cap L_{p,\mu}(\mathbb{R}_+; W^{2-1/p}_p(\Sigma)) =: \mathbb{F}$;

(c) $f_h \in L_{p,\mu}(\mathbb{R}_+; W^{2-2/p}_p(\Sigma)) =: \mathbb{F}_h$;

(d) $u_0 \in W^{2\mu-2/p}_p(\Omega\setminus\Sigma)$, $h_0 \in W^{2+2\mu-3/p}_p(\Sigma)$;

(e) $[\![u_0]\!] + \mathcal{C}(x,\nabla_\Sigma)h_0 = g(0)$.

The solution map is continuous between the corresponding spaces.

Proof. The proof of this result involves similar ideas as the proof of Theorem 6.6.1 and we will hence skip the details. □

By means of Theorem 6.7.2, we may define the *Verigin semigroup* in X_0 in the following way. We set $z = [u,h]^{\mathsf{T}}$, $X_1 = H^2_p(\Omega\setminus\Sigma) \times W^{4-1/p}_p(\Sigma)$, and define an operator A in $X_0 = L_p(\Omega) \times W_p^{2-2/p}(\Sigma)$ by means of

$$A = \begin{bmatrix} \mathcal{A}(x,\nabla_x) & 0 \\ \mathcal{B}(x,\nabla_x) & 0 \end{bmatrix}, \tag{6.89}$$

$$\mathsf{D}(A) = \{z \in X_1 : \mathcal{B}(x,\nabla_x)u = 0 \text{ on } \partial\Omega,\ [\![u]\!] + \mathcal{C}(x,\nabla_\Sigma)h = 0 \text{ on } \Sigma,$$
$$\mathcal{B}(x,\nabla_x)u \in W_p^{2-2/p}(\Sigma)\}.$$

Then (6.85) for $g = 0$ is equivalent to the abstract evolution equation

$$\dot{z} + Az = f, \quad t > 0, \quad z(0) = z_0, \tag{6.90}$$

where we employed the abbreviations $z_0 = [u_0, h_0]^\mathsf{T}$ and $f = [f_u, f_h]^\mathsf{T}$. Maximal L_p-regularity of (6.90) is equivalent to maximal L_p-regularity of (6.85) for $g = 0$ in the modified setting. Theorem 6.7.2 and Proposition 3.5.2 then imply that $-A$ is the generator of an analytic C_0-semigroup with maximal L_p-regularity. This completes the construction of the Verigin semigroup.

In the same way as in the previous section, employing the semigroup this yields that the lower bound of ω is the spectral bound $\omega_0 = \mathsf{s}(-A)$.

7.3 The Linearized Muskat Problem

In this subsection we consider the quasi-steady problem

$$\begin{aligned}
(\eta + \mathcal{A}(x,\nabla_x))u &= f_u && \text{in } \Omega\setminus\Sigma,\\
\mathcal{B}(x,\nabla_x)u &= 0 && \text{on } \partial\Omega,\\
[\![u]\!] + \mathcal{C}(x,\nabla_\Sigma)h &= g && \text{on } \Sigma,\\
[\![\mathcal{B}(x,\nabla_x)u]\!] &= 0 && \text{on } \Sigma,\\
(\partial_t + \omega)h - \mathcal{B}(x,\nabla_x)u &= f_h && \text{on } \Sigma,\\
h(0) &= h_0 && \text{on } \Sigma.
\end{aligned} \tag{6.91}$$

The main result for this problem in the L_p-setting, $3 < p < \infty$, is the following.

Theorem 6.7.3. *Let $p \in (1,\infty)$ and $1 \geq \mu > 1/p$. There exists $\omega_0, \eta_0 \in \mathbb{R}$ such that for each $\omega > \omega_0$, $\eta > \eta_0$, Problem (6.91) admits exactly one solution (u,h) in the class*

$$u \in L_{p,\mu}(\mathbb{R}_+; H_p^2(\Omega\setminus\Sigma)) =: \mathbb{E}_u,$$
$$h \in H^1_{p,\mu}(\mathbb{R}_+; W_p^{1-1/p}(\Sigma)) \cap L_{p,\mu}(\mathbb{R}_+; W_p^{4-1/p}(\Sigma)) =: \mathbb{E}_h,$$

if and only if the data (f_u, g, f_h, h_0) are subject to the following conditions:

(a) $f_u \in L_{p,\mu}(\mathbb{R}_+; L_p(\Omega)) =: \mathbb{F}_u$;

(b) $g \in L_{p,\mu}(\mathbb{R}_+; W_p^{2-1/p}(\Sigma)) =: \mathbb{F}$;

(c) $f_h \in L_{p,\mu}(\mathbb{R}_+; W_p^{1-1/p}(\Sigma)) =: \mathbb{F}_h$.

(d) $h_0 \in W_p^{1+3\mu-4/p}(\Sigma)$;

The solution map is continuous between the corresponding spaces.

Proof. This result is proved in the same way as Theorem 6.7.1. □

We are interested in the parameters η and ω. For this purpose we define the operator A in $X = L_p(\Omega)$ by means of

$$Au(x) = \mathcal{A}(x, \nabla_x)u(x), \quad x \in \Omega \setminus \Sigma, \tag{6.92}$$
$$\mathsf{D}(A) = \{u \in H_p^2(\Omega \setminus \Sigma) : \mathcal{B}(x, \nabla_x)u = 0 \text{ on } \partial\Omega, \; [\![\mathcal{B}(x, \nabla_x)u]\!] = [\![u]\!] = 0 \text{ on } \Sigma\}.$$

As $\mathcal{A}$ is uniformly strongly elliptic by assumption, Theorem 6.5.1 shows that $-A$ is the generator of an analytic C_0-semigroup e^{-At} with maximal L_p-regularity. The semigroup as well as the resolvent of A are compact. Therefore the spectrum of A consists only of eigenvalues of finite algebraic multiplicity, which do not depend on p. By the energy argument, we obtain $\sigma(A) \subset \mathbb{R}_+$. However, in contrast to the case of the linearized Mullins-Sekerka problem, here 0 is an eigenvalue of A, it is algebraically simple and spanned by the function e which is constant 1, $\mathsf{e} \perp \mathsf{R}(A)$ as the divergence theorem shows. To circumvent this difficulty in the construction of the Muskat semigroup, we observe that in Theorem 6.7.3 the solution u has mean value 0 if f_u has this property. So instead of $X = L_p(\Omega)$ we employ

$$X = L_{p,0}(\Omega) = \{u \in L_p(\Omega) : (u|\mathsf{e})_\Omega = 0\}.$$

This removes 0 from the spectrum of A. Then we proceed as in Section 6.6.6 to construct the *Muskat operator* as follows.

Define the *Muskat operator* A_0 in $X_0 := W_p^{1-1/p}(\Sigma)$ with help of the Dirichlet-to-Neumann operator S_n by means of

$$A_0 h = S_\mathsf{n} \mathcal{C}(x, \nabla_\Sigma)h, \quad X_1 := \mathsf{D}(A_0) = W_p^{4-1/p}(\Sigma). \tag{6.93}$$

Then it is obvious that (6.91) with $\eta = 0$, and $(f_u, g) = 0$ is equivalent to the evolution equation

$$\partial_t h + \omega h + A_0 h = f_h, \quad t > 0, \quad h(0) = h_0.$$

As for the Mullins-Sekerka case, we can show that $-A_0$ generates an analytic C_0-semigroup with maximal L_p-regularity.

Corollary 6.7.4. *The* Muskat operator A_0 *defined above is the negative generator of an analytic C_0-semigroup $e^{-A_0 t}$, the* Muskat semigroup, *with maximal L_p-regularity in $X_0 = W_p^{1-1/p}(\Sigma)$ and domain $X_1 = \mathsf{D}(A_0) = W_p^{4-1/p}(\Sigma)$.*

The spectrum of this operator will be considered in Chapter 12.

Appendix

The Triebel-Lizorkin spaces $F_{pq}^\alpha(\mathbb{R}; E)$ and ${}_0F_{pq,\mu}^\alpha(\mathbb{R}_+; E)$ for $\alpha \in (0,1)$, $p, q \in (1, \infty)$, and $1/p < \mu \leq 1$ can be characterized as follows.

Lemma 6.7.5. *Let $1 < p, q < \infty$, $1/p < \mu \leq 1$, $\alpha \in (0,1)$, and suppose E is a Banach space of class $\mathcal{HT}$. Define $B = (\partial_t)^\alpha$ in $L_{p,\mu}(\mathbb{R}_+; L_q((0,1);E))$ with domain $\mathsf{D}(B) = {}_0H^\alpha_{p,\mu}(\mathbb{R}_+; L_q((0,1);E))$.*

Then, for any $g \in L_{p,\mu}(\mathbb{R}_+; E)$,

$$w := e^{-By} g \in {}_0H^\alpha_{p,\mu}(\mathbb{R}_+; L_q((0,1);E))$$

if and only if $g \in {}_0F^{\alpha(1-1/q)}_{pq,\mu}(\mathbb{R}_+; E)$.

The same result is valid for the whole line case, i.e.,

$$w \in H^\alpha_p(\mathbb{R}; L_q((0,1);E)) \quad \Leftrightarrow \quad g \in F^{\alpha(1-1/q)}_{pq}(\mathbb{R}; E).$$

These results hold for $\mathbb{R}_+$ instead of $(0,1)$ if we replace ∂_t^α by $(\omega + \partial_t)^\alpha$, for some $\omega > 0$.

Actually, we might have taken the assertion of this lemma for the whole line case as a definition for the vector-valued spaces $F^\alpha_{pq}(\mathbb{R}; E)$. However, to draw the connection with the definition of F^α_{pq} given in Triebel [284], we add a proof. Observe that

$$u \in {}_0F^\alpha_{pq,\mu}(\mathbb{R}_+; E) \quad \Leftrightarrow \quad t_+^{1-\mu} u \in F^\alpha_{pq}(\mathbb{R}; E),$$

where $t_+^{1-\mu} = \max\{t^{1-\mu}, 0\}$. Therefore we may concentrate on the whole line case, and we restrict to the case $\omega = 0$.

Proof. For $E = \mathbb{C}$, Theorem 2.4.1 of [284] proves Lemma 6.7.5 with the choices $\phi(x) = (ix)^\alpha e^{-(ix)^\alpha}$ and $\phi_0(x) = 1$, $s_0 = 0$, $s_1 = \alpha$. The proof given there carries over to the vector-valued case since E is assumed to be of class $\mathcal{HT}$, provided $\alpha > a > 1/\min\{p,q\}$. For general $p, q \in (1,\infty)$ Theorem 2.4.1 of [284] does not apply since the moment condition (8) in that reference does not hold.

To see sufficiency of the condition in the general case, assume that $w_0 := Be^{-By} g \in L_p(\mathbb{R}; L_q((0,1);E))$. Using maximal regularity we solve successively the problems

$$\partial_y w_k + Bw_k = Bw_{k-1}, \quad w_k|_{y=0} = 0,$$

to obtain

$$Bw_k = y^k B^{k+1} e^{-yB} g \in L_p(\mathbb{R}; L_q((0,1);E)), \quad k \in \mathbb{N}_0.$$

Now we have with the variable transformation $y = \tau^\alpha$

$$\begin{aligned}\int_0^1 |y^k B^{k+1} e^{-yB} g|_E^q \, dy &= \alpha \int_0^1 \tau^{-q\alpha(1-1/q)} |(\tau^\alpha B)^{k+1} e^{-(\tau^\alpha B)} g|_E^q \, \frac{d\tau}{\tau} \\ &= \alpha \int_0^1 \tau^{-q\alpha(1-1/q)} |\phi(\tau D) g|_E^q \, \frac{d\tau}{\tau},\end{aligned}$$

where we used the notation in [284], Section 2.4.1, with $\phi(\xi) = (i\xi)^{\alpha(k+1)} e^{-(i\xi)^\alpha}$. It is not difficult to check that the relevant conditions (7) and (9) are valid for all $k \in \mathbb{N}_0$ with $s_0 = 0$. On the other hand, (8) holds in case $\alpha k \geq 1$. In fact, the inverse Fourier transform $p^{k+1}(t)$ of $\phi(i\xi)$, with contour $\Gamma = e^{-i\theta}(\infty, 0] \cup e^{\theta}[0,\infty)$, $\theta \in (\pi/2, \pi)$, $\alpha\theta < \pi/2$, becomes

$$p^{k+1}(t) = \frac{1}{2\pi i} \int_\Gamma z^{\alpha k+1} e^{-z^\alpha} e^{zt} \, dz, \quad t \geq 0.$$

Note that the support of p^{k+1} is contained in $\mathbb{R}_+$, thanks to holomorphy. This formula is valid for all $\alpha(k+1) > -1$, and it implies that $p^{k+1}(t)$ is bounded and behaves asymptotically like $t^{-(1+\alpha(k+1))}$ as $t \to \infty$. Therefore $(1+t^a)p^{k+1} \in L_1(\mathbb{R}_+)$ if and only if $a < \alpha(k+1)$. Choosing $s_1 = \alpha$ and $1/\min\{p,q\} < a < 1$, and $k \geq 1/\alpha$, the vector-valued version of Theorem 2.4.1 of [284] implies $g \in F_{pq}^{\alpha(1-1/q)}(\mathbb{R};E)$.

For the converse statement we need to choose $k = 0$. Since the critical condition (8) does not hold, we have to modify Steps 1 and 4 of the proof of Theorem 2.4.1 of [284], the only places where (8) is used. We concentrate on the modification of Step 1, and employ the notation used there. Let $s = \alpha(1-1/q)$ and fix a resolution of unity $\{\rho_j\}_{j\in\mathbb{N}_0}$ in the sense of [284] Section 2.3.1. Then by definition, $g \in {}_0F^s_{pq}(\mathbb{R};E)$ if and only if

$$(2^{sj}\rho_j(D)g)_{j\in\mathbb{N}_0} \in L_p(\mathbb{R}; l_q(\mathbb{N}_0;E)).$$

Now we have as in [284], proof of Theorem 2.4.1, Step 1

$$2^{js}\mathcal{F}^{-1}\mathcal{L}p^1(2^{-j}i\xi)\mathcal{F} = \sum_{l=-\infty}^{\infty} 2^{js}\mathcal{F}^{-1}\mathcal{L}p^1(2^{-j}i\xi)\rho_{l+j}(\xi)\mathcal{F}g.$$

Here $\mathcal{L}$ denotes the Laplace transform. Splitting the sum into two parts, we have to estimate in Step 1 the part running from $l=-\infty$ to $l=k$. We write

$$\begin{aligned}&2^{js}\mathcal{F}^{-1}\mathcal{L}p^1(2^{-j}i\xi)\rho_{l+j}(\xi)\mathcal{F}g\\ &= 2^{\alpha l/q}\mathcal{F}^{-1}\mathcal{L}p^0(2^{-j}i\xi)\cdot(2^{-(j+l)}i\xi)^\alpha\chi(2^{-(j+l)}\xi)\cdot 2^{s(j+l)}\rho_{j+l}\mathcal{F}g,\end{aligned}$$

where $\chi(r)$ denotes a cut off function which is 1 on $|r|\leq 2$. Since $\sum_{l=-\infty}^k 2^{\alpha l/q} < \infty$, it suffices to estimate

$$\mathcal{F}^{-1}\mathcal{L}p^0(2^{-j}i\xi)\cdot(2^{-(j+l)}i\xi)^\alpha\chi(2^{-(j+l)}\xi)\cdot 2^{s(j+l)}\rho_{j+l}\mathcal{F}g$$

in $L_p(\mathbb{R}; l_q(\mathbb{N}_0;E))$, uniformly w.r.t. l. By assumption we have

$$|(2^{s(j+l)}\mathcal{F}^{-1}\rho_{j+l}\mathcal{F}g)_{j\geq 0}|_{L_p(\mathbb{R};l_q(\mathbb{N}_0;E))} \leq |g|_{F^s_{pq}(\mathbb{R}_+;E)},$$

hence its is enough to show that the sequences $(\mathcal{L}p^0(2^{-j}i\xi))_{j\in\mathbb{N}_0}$ and $((2^{-(j+l)}i\xi)^\alpha\chi(2^{-(j+l)}\xi))_{j\in\mathbb{N}_0}$ define Fourier multipliers for $L_p(\mathbb{R}; l_q(\mathbb{N}_0;E))$ with bounds independent of l.

For the first sequence, observe that $\mathcal{L}p^0(\lambda) = e^{-\lambda^\alpha}$ is completely monotonic, hence $p^0(t)$ is nonnegative and integrable with integral equal to 1, i.e., p^0 is a probability density. Therefore, the operator defined by the first sequence is given by

$$(T_1f)_j(t) = 2^j p^0(2^j\cdot) * f_j(t), \quad t>0,\ j\in\mathbb{N}_0.$$

Thus we obtain

$$|(T_1f)_j(t)|_E \leq M|f_j|_E(t), \quad t>0,\ j\in\mathbb{N}_0,$$

where M denotes the usual maximal operator. Since M is bounded in $L_p(\mathbb{R}; l_q(\mathbb{N}_0))$, the assertion follows for the first sequence, i.e., T_1 is bounded in $L_p(\mathbb{R}; l_q(\mathbb{N}_0;E))$.

The second sequence is treated in a similar way. We write

$$(i\xi)^\alpha\chi(\xi) = \frac{(i\xi)^\alpha}{(1+i\xi)^2} + \frac{(i\xi)^\alpha}{(1+i\xi)^2}(\chi(\xi)-1) + \frac{(i\xi)^{1+\alpha}(2+i\xi)}{(1+i\xi)^2}\chi(\xi).$$

The first term belongs to the Hardy space $\mathcal{H}^\infty(\mathbb{C}_+)$ and its derivative belongs to $\mathcal{H}^1(\mathbb{C}_+)$, therefore by Hardy's inequality it is the Laplace transform of a function $k_1 \in L_1(\mathbb{R}_+)$. The second and the third terms belong to $L_2(\mathbb{R})$ as well as their derivatives, hence by means of Bernstein's theorem they are Fourier transforms of functions $k_j \in L_1(\mathbb{R})$, $j = 2, 3$. This shows that $(i\xi)^\alpha \chi(\xi) = \mathcal{F}k(\xi)$, for some $k \in L_1(\mathbb{R})$. Now we may argue as before to see that also the second sequence defines a bounded operator T_2 in $L_p(\mathbb{R}; l_q(\mathbb{N}_0; E))$, with bound independent of l. This completes the proof of Lemma 6.7.5. $\square$

Chapter 7

Generalized Stokes Problems

This chapter is devoted to maximal L_p-regularity of one-phase linear generalized Stokes problems on domains $\Omega \subset \mathbb{R}^n$ which are either $\mathbb{R}^n$, $\mathbb{R}^n_+$, or domains with compact boundary $\partial\Omega$ of class C^3, i.e., interior or exterior domains. Here we only consider the physically natural boundary conditions no-slip, pure slip, outflow, and free. As in Chapter 6, our approach is based on vector-valued Fourier multiplier theory, perturbation, and localization. It turns out that due to the divergence condition (and the pressure), the analysis for the half-space as well as the localization procedure are much more involved than in the previous chapter. Nevertheless, besides some extra compatibility condition which comes from the divergence condition, the main results will parallel those in Chapter 6.

7.1 The Generalized Stokes Problem on $\mathbb{R}^n$

1.1 Constant Coefficients
We consider the problem

$$\begin{aligned} \partial_t u(t,x) + \mathcal{A}(D)u(t,x) + \nabla\pi(t,x) &= f(t,x) \quad \text{in } \mathbb{R}^n, \\ \operatorname{div} u(t,x) &= g(t,x) \quad \text{in } \mathbb{R}^n, \\ u(0,x) &= u_0(x) \quad \text{in } \mathbb{R}^n, \end{aligned} \tag{7.1}$$

Here $\mathcal{A}(D) = \sum_{k,l=1}^n a^{kl} D_k D_l$ denotes a differential operator with constant coefficient matrices a^{kl} acting on $\mathbb{C}^n$-valued functions. We assume that $\mathcal{A}(D)$ is *strongly elliptic*. As we have seen in the previous chapter, this implies that the problem

$$\begin{aligned} \partial_t u(t,x) + \mathcal{A}(D)u(t,x) &= f(t,x) \quad \text{in } \mathbb{R}^n, \\ u(0,x) &= u_0(x) \quad \text{in } \mathbb{R}^n. \end{aligned} \tag{7.2}$$

has maximal $L_{p,\mu} - L_q$-regularity, $1 < p,q < \infty$, $\mu \in (1/p, 1]$. We want to show that the same assertion is valid for the generalized Stokes problem (7.1). More precisely, we have the following result.

J. Prüss and G. Simonett, *Moving Interfaces and Quasilinear Parabolic Evolution Equations*, Monographs in Mathematics 105,
DOI 10.1007/978-3-319-27698-4_7

Theorem 7.1.1. *Let $1 < p, q < \infty$, $\mu \in (1/p, 1]$, and assume that $\mathcal{A}(D)$ is strongly elliptic.*

Then (7.1) has maximal $L_{p,\mu}-L_q$-regularity in the following sense. There is a unique solution (u, π) of (7.1) with $u \in L_{1,loc}(\mathbb{R}_+; H^2_q(\mathbb{R}^n; \mathbb{C}^n))$ such that

$$\partial_t u_k, \partial_i\partial_j u_k \in L_{p,\mu}(\mathbb{R}_+; L_q(\mathbb{R}^n)), \quad \pi \in L_{p,\mu}(\mathbb{R}_+; \dot{H}^1_q(\mathbb{R}^n)),$$

if and only if the data (f, g, u_0) satisfy the subsequent conditions.

(a) $f \in L_{p,\mu}(\mathbb{R}_+; L_q(\mathbb{R}^n; \mathbb{C}^n))$;

(b) $\partial_t g \in L_{p,\mu}(\mathbb{R}_+; \dot{H}^{-1}_q(\mathbb{R}^n))$ *and* $\nabla g \in L_{p,\mu}(\mathbb{R}_+; L_q(\mathbb{R}^n; \mathbb{C}^n))$;

(c) $u_0 \in B^{2(\mu-1/p)}_{qp}(\mathbb{R}^n; \mathbb{C}^n)$ *and* $\operatorname{div} u_0 = g(0)$ *in* $\mathcal{D}'(\mathbb{R}^n)$.

The solution (u, π) depends continuously on the data in the corresponding spaces.

Proof. Necessity follows easily by trace theory. To prove sufficiency of the conditions, note that by the open mapping theorem, the continuity assertion follows as soon as the solvability assertion is proved. So let data (f, g, u_0) be given which are subject to conditions (a), (b), and (c). We first solve the parabolic problem

$$\partial_t v + \mathcal{A}(D)v = f, \quad v(0) = u_0,$$

with maximal $L_{p,\mu}-L_q$-regularity, applying Theorem 6.1.8 and Theorem 4.4.4. Then $w = u - v$ must be a solution of the system

$$\partial_t w + \mathcal{A}(D)w + \nabla\pi = 0, \quad \operatorname{div} w = g_0, \quad w(0) = 0,$$

where $g_0 = g - \operatorname{div} v$ has the same regularity as g and trace 0 at time $t = 0$.

Suppose the pressure π is already known. Taking Fourier transform in the space variables and Laplace transform in the time variable we obtain the system

$$\begin{aligned} \lambda\hat{w} + \mathcal{A}(\xi)\hat{w} &= -i\xi\hat{\pi}, \\ i(\hat{w}|\xi) &= \hat{g}_0. \end{aligned} \tag{7.3}$$

Solving for $\hat{w}$ this yields

$$\hat{w} = -i(\lambda + \mathcal{A}(\xi))^{-1}\xi\hat{\pi},$$

and inserting this relation into the second equation of (7.3) we obtain

$$\hat{g}_0 = ((\lambda + \mathcal{A}(\xi))^{-1}\xi|\xi)\hat{\pi}.$$

Set $\eta = (\lambda + \mathcal{A}(\xi))^{-1}\xi$. Then $\eta \neq 0$ unless $\xi = 0$, and

$$\alpha(\lambda, \xi) := ((\lambda + \mathcal{A}(\xi))^{-1}\xi|\xi) = \overline{\lambda}|\eta|^2 + (\eta|\mathcal{A}(\xi)\eta).$$

Therefore, strong ellipticity of $\mathcal{A}(D)$ implies $\alpha(\lambda, \xi) \neq 0$ for all $\xi \in \mathbb{R}^n$, $\operatorname{Re}\lambda \geq 0$, with $|\xi| + |\lambda| \neq 0$. We may now solve for $\hat{\pi}$ to the result

$$\hat{\pi}(\lambda, \xi) = \hat{g}_0(\lambda, \xi)/\alpha(\lambda, \xi),$$

and for $\hat{w}$ we get

$$\hat{w}(\lambda,\xi) = -i\,\frac{(\lambda+\mathcal{A}(\xi))^{-1}\xi}{\alpha(\lambda,\xi)}\hat{g}_0(\lambda,\xi).$$

Choose $v_0 \in L_{p,loc}(\mathbb{R}_+; H^2_q(\mathbb{R}^n;\mathbb{C}^n))$ such that

$$\partial_t v_{0k}, \partial_i\partial_j v_{0k} \in L_{p,\mu}(\mathbb{R}_+; L_q(\mathbb{R}^n)), \quad \operatorname{div} v_0 = g_0.$$

This is possible by assumption (b) on the function g. In fact, setting

$$g_1 = (-\Delta)^{-1/2}\partial_t g_0 + (-\Delta)^{1/2} g_0 \in L_{p,\mu}(\mathbb{R}_+; L_q(\mathbb{R}^n))$$

we obtain $g_0 = -\operatorname{div} R(\partial_t-\Delta)^{-1}g_1$, where R denotes the Riesz transform defined by the symbol $i\xi/|\xi|$, i.e., we may choose $v_0 = -R(\partial_t-\Delta)^{-1}g_1$. Therefore,

$$(\partial_t-\Delta)w = T_1(\partial_t-\Delta)v_0, \quad \nabla\pi = T_2(\partial_t-\Delta)v_0,$$

where T_j are defined by means of their Fourier-Laplace symbols

$$\hat{T}_1(\lambda,\xi) = \frac{(\lambda+\mathcal{A}(\xi))^{-1}\xi\otimes\xi}{\alpha(\lambda,\xi)}, \quad \hat{T}_2(\lambda,\xi) = -\frac{\xi\otimes\xi}{(\lambda+|\xi|^2)\alpha(\lambda,\xi)}.$$

Thus, to prove the theorem, it is enough to show that the operators T_j with symbols $\hat{T}_j(\lambda,\xi)$ are bounded in $L_{p,\mu}(\mathbb{R}_+; L_q(\mathbb{R}^n;\mathbb{C}^n))$.

This in turn will follow by an application of the Kalton-Weis theorem and $\mathcal{R}$-boundedness of families of Fourier multipliers. By the scaling $\mu = \lambda/|\xi|^2$, $\zeta = \xi/|\xi|$, we may rewrite the symbols as

$$\hat{T}_1(\lambda,\xi) = \frac{(\mu+\mathcal{A}(\zeta))^{-1}\zeta\otimes\zeta}{\alpha(\mu,\zeta)}, \quad \hat{T}_2(\lambda,\xi) = -\frac{\zeta\otimes\zeta}{(1+\mu)\alpha(\mu,\zeta)}.$$

By strong ellipticity, we already know $\alpha(\mu,\zeta)\neq 0$ for all $\zeta\in\mathbb{R}^n$, $|\zeta|=1$, and $\operatorname{Re}\mu\geq 0$. As $|\mu|\to\infty$ we have $\mu\alpha(\mu,\zeta)\to 1$, while $\alpha(\mu,\zeta)\to\alpha(0,\zeta) = (\mathcal{A}(\zeta)^{-1}\zeta|\zeta)\neq 0$ as $\mu\to 0$. This shows that we may extend the range of $\mu\in\mathbb{C}$ to some sector Σ_ϕ, with $\phi>\pi/2$. Furthermore, by compactness, $|(1+\mu)\alpha(\mu,\zeta)|\geq \alpha_0>0$ for all such ζ and μ, where α_0 denotes a constant. This implies boundedness of the symbols $\hat{T}_j(\mu|\xi|^2,\xi)$, uniformly in ξ and μ. Furthermore, $\hat{T}_j(\mu|\xi|^2,\xi)$ are homogeneous in ξ of degree 0, and so $|\xi|^{|\beta|}D^\beta_\xi\hat{T}_j(\mu|\xi|^2,\xi)$ are also uniformly bounded in ξ and μ, for each multi-index $\beta\in\mathbb{N}_0^n$. The Lizorkin multiplier theorem, Theorem 4.3.9, then implies that these symbols are Fourier multipliers in $L_q(\mathbb{R}^n;E_j)$ w.r.t. ξ, which yields a holomorphic $\mathcal{R}$-bounded family of operators $\{T_j(\mu)\}_{\mu\in\Sigma_\phi}\subset\mathcal{B}(L_q(\mathbb{R}^n;E_j))$ for $j=1,2$, where $E_1=\mathbb{C}^n$, $E_2=\mathbb{C}$. By canonical extension, it is also $\mathcal{R}$-bounded in $L_{p,\mu}(\mathbb{R}_+;L_q(\mathbb{R}^n;E_j))$. Since the operator $L:=\partial_t(-\Delta)^{-1}$ admits an $\mathcal{H}^\infty$-calculus in $L_{p,\mu}(\mathbb{R}_+;L_q(\mathbb{R}^n;E_j))$ of angle $\pi/2$, the Kalton-Weis theorem, Theorem 4.5.6, implies boundedness of $T_j(L)$ in $L_{p,\mu}(\mathbb{R}_+;L_q(\mathbb{R}^n;E_j))$. This completes the proof of Theorem 7.1.1. □

1.2 The Generalized Stokes Operator

Let $\mathcal{A}(D)$ be strongly elliptic as in the previous section and consider (7.1) with $(\operatorname{div} f, g, u_0) = 0$. Then, according to Theorem 7.1.1, Problem (7.1) admits a unique solution (u, π) with maximal $L_{p,\mu}-L_q$-regularity, which means

$$u \in L_{1,loc}(\mathbb{R}_+; H_q^2(\mathbb{R}^n;\mathbb{C}^n)), \quad \pi \in L_{p,\mu}(\mathbb{R}_+; \dot{H}_q^1(\mathbb{R}^n)),$$
$$\partial_t u_k, \partial_i\partial_j u_k \in L_{p,\mu}(\mathbb{R}_+; L_q(\mathbb{R}^n)),$$

whenever $f \in L_{p,\mu}(\mathbb{R}_+; L_q(\mathbb{R}^n;\mathbb{C}^n))$.

Define the base space X_0 by means of

$$X_0 = L_{q,\sigma}(\mathbb{R}^n) := \{u \in L_q(\mathbb{R}^n;\mathbb{C}^n) : \operatorname{div} u = 0 \text{ in } \mathcal{D}'(\mathbb{R}^n)\},$$

and let $P_H := I - R\otimes R$ denote the *Helmholtz projection* from $L_q(\mathbb{R}^n;\mathbb{C}^n)$ onto X_0, where R means the Riesz operator defined via its symbol $\tilde{R} = i\xi/|\xi|$, as before. The **generalized Stokes operator** A associated to $\mathcal{A}(D)$ is defined according to

$$(Au)(x) := [P_H\mathcal{A}(D)u](x), \quad x \in \mathbb{R}^n, \tag{7.4}$$

with domain

$$\mathsf{D}(A) := H_q^2(\mathbb{R}^n;\mathbb{C}^n) \cap L_{q,\sigma}(\mathbb{R}^n).$$

Then $u \in L_{1,loc}(\mathbb{R}_+; X_0)$ is the unique solution of the evolution equation

$$\dot{u} + Au = f, \quad t > 0, \quad u(0) = u_0, \tag{7.5}$$

in the base space X_0. It belongs to the maximal regularity class $\partial_t u, Au \in L_{p,\mu}(\mathbb{R}_+; X_0)$, i.e., (7.5) has maximal $L_{p,\mu}$-regularity. Then Theorem 4.4.4 and Proposition 3.5.2 imply that A is $\mathcal{R}$-sectorial with angle $\phi_A < \pi/2$. But even more is true.

Theorem 7.1.2. *Let $1 < p, q < \infty$, $\mu \in (1/p, 1]$, and assume that $\mathcal{A}(D)$ is strongly elliptic. Let A be defined by (7.4) in $X_0 = L_{q,\sigma}(\mathbb{R}^n)$.*

Then $A \in \mathcal{H}^\infty(X_0)$ with $\mathcal{H}^\infty$-angle $\phi_A^\infty \le \phi_{\mathcal{A}}$, where

$$\phi_{\mathcal{A}} \le \max\{|\arg(\mathcal{A}(\xi)v|v)| : \xi \in \mathbb{R}^n,\ v \in \mathbb{C}^n\} < \pi/2.$$

In particular, $A \in \mathcal{RS}(X_0)$ with $\mathcal{R}$-angle $\phi_A^R \le \phi_{\mathcal{A}}$, and (7.5) has maximal $L_{p,\mu}$–L_q-regularity.

Proof. From the previous subsection we have for the resolvent $(\lambda + A)^{-1}$ of A the symbolic representation

$$\mathcal{F}(\lambda + A)^{-1}(\xi) = [I - (\lambda + \mathcal{A}(\xi))^{-1}\xi\otimes\xi/\alpha(\lambda,\xi)](\lambda + \mathcal{A}(\xi))^{-1}, \quad \xi \in \mathbb{R}^n,$$

where $\alpha(\lambda,\xi) = ((\lambda + \mathcal{A}(\xi))^{-1}\xi|\xi)$. We proceed as in the proof of Theorem 6.1.8. So let $h \in H_0(\Sigma_\phi)$ with $\phi > \phi_{\mathcal{A}}$ be given. Then the symbol of $h(A)$ reads

$$\mathcal{F}h(A)(\xi) = \frac{1}{2\pi i}\int_\Gamma h(\lambda)\mathcal{F}(\lambda - A)^{-1}(\xi)\, d\lambda, \quad \xi \in \mathbb{R}^n,$$

where Γ denotes the contour $\Gamma = (\infty, 0]e^{i\theta} \cup [0,\infty)e^{-i\theta}$ with $\theta \in (\phi_{\mathcal{A}}, \phi)$. Employing the scaling $\xi = \rho\zeta$, $\rho = |\xi|$, and $\lambda = \mu\rho^2$, we obtain

$$\mathcal{F}h(A)(\xi) = \frac{1}{2\pi i}\int_\Gamma h(\rho^2\mu)\big(I - (\mu - \mathcal{A}(\zeta))^{-1}\zeta \otimes \zeta/\alpha(\mu,\zeta)\big)(\mu - \mathcal{A}(\zeta))^{-1}\,d\mu.$$

As $\mathsf{n}_0 = \cup_{|\zeta|=1}\mathsf{n}(\mathcal{A}(\zeta))$, where n denotes the numerical range, is compact and contained in $\Sigma_{\phi_{\mathcal{A}}}$, according to Cauchy's theorem, we may deform the contour within Σ_θ into a closed compact contour Γ_0 surrounding n_0 counter-clockwise to obtain the representation

$$\mathcal{F}h(A)(\xi) = \frac{1}{2\pi i}\int_{\Gamma_0} h(\rho^2\mu)\big(I - (\mu - \mathcal{A}(\zeta))^{-1}\zeta \otimes \zeta/\alpha(\mu,\zeta)\big)(\mu - \mathcal{A}(\zeta))^{-1}\,d\mu.$$

By compactness of Γ_0 and $\mathbb{S}^{n-1}$ this implies boundedness of the symbol $\mathcal{F}h(A)(\xi)$ in terms of $|h|_{H^\infty(\Sigma_\phi)}$. As in the proof of Theorem 6.1.8 we also obtain bounds for the derivatives $|\xi|^{|\alpha|}|D_\xi^\alpha \mathcal{F}h(A)(\xi)|$, hence by the classical Mikhlin multiplier theorem we obtain

$$|h(A)|_{\mathcal{B}(L_q)} \le C|h|_{H^\infty(\Sigma_\phi)}, \quad h \in H_0(\Sigma_\phi).$$

Therefore, the generalized Stokes operator A admits a bounded $\mathcal{H}^\infty$-calculus with $\mathcal{H}^\infty$-angle $\phi_A^\infty \le \phi_{\mathcal{A}}$. □

We observe that for the trace spaces $X_{\gamma,\mu}$ of A we obtain

$$\begin{aligned} X_{\gamma,\mu} &:= (X_0, \mathsf{D}(A))_{\mu-1/p,p} = (L_q(\mathbb{R}^n;\mathbb{C}^n) \cap X_0, H_q^2(\mathbb{R}^n;\mathbb{C}^n) \cap X_0)_{\mu-1/p,p} \\ &= (L_q(\mathbb{R}^n;\mathbb{C}^n), H_q^2(\mathbb{R}^n;\mathbb{C}^n))_{\mu-1/p,p} \cap X_0 = B_{qp}^{2(\mu-1/p)}(\mathbb{R}^n;\mathbb{C}^n) \cap X_0, \end{aligned}$$

for $1 < p, q < \infty$ and $\mu \in (1/p, 1]$. For the fractional power spaces we have

$$\begin{aligned} \mathsf{D}(A^\alpha) &= (X_0, \mathsf{D}(A))_\alpha = (L_q(\mathbb{R}^n;\mathbb{C}^n) \cap X_0, H_q^2(\mathbb{R}^n;\mathbb{C}^n) \cap X_0)_\alpha \\ &= (L_q(\mathbb{R}^n;\mathbb{C}^n), H_q^2(\mathbb{R}^n;\mathbb{C}^n))_\alpha \cap X_0 = H_q^{2\alpha}(\mathbb{R}^n;\mathbb{C}^n) \cap X_0, \end{aligned}$$

for each $\alpha \in (0,1)$, as A admits an $\mathcal{H}^\infty$-calculus.

1.3 Variable Coefficients

(i) We can easily extend Theorem 7.1.1 to the case of variable coefficients with small deviation from constant ones. To see this, let $\mathcal{A}(x,D) = \mathcal{A}_0(D) + \mathcal{A}_1(x,D)$, where $\mathcal{A}_1(x,D) = \sum_{k,l} a_1^{kl}(x)D_kD_l$ with

$$\sup\{|a_1^{kl}(x)| : k,l = 1,\dots n,\ x \in \mathbb{R}^n\} \le \eta.$$

Let S denote the solution operator of the generalized Stokes problem (7.1) from Theorem 7.1.1 for $\mathcal{A}_0(D)$, and T that of the perturbed problem. Then we obtain the identity

$$T = S - SBT, \quad \text{where} \quad B = \begin{bmatrix} \mathcal{A}_1(x,D) & 0 \\ 0 & 0 \end{bmatrix}.$$

The norm of B as an operator from $H^2_q(\mathbb{R}^n;\mathbb{C}^n)$ into $L_q(\mathbb{R}^n;\mathbb{C}^n)$ is bounded by $C\eta$, where $C>0$ denotes a constant independent of η. Let $|S|$ stand for the norm of the solution operator from the data space to the maximal regularity space. If $|S|C\eta<1$, then a Neumann series argument shows that $T=(I+SB)^{-1}S$ in fact exists and is bounded as a map from the data space to the maximal regularity space as well. Let us state this as

Corollary 7.1.3. *The assertions of Theorem* 7.1.1 *remain valid in the case of variable coefficients* $\mathcal{A}(x,D)=\mathcal{A}_0(D)+\mathcal{A}_1(x,D)$, *provided the coefficients* $a_1^{kl}(x)$ *of* $\mathcal{A}_1(D)$ *are subject to*

$$\sup\{|a_1^{kl}(x)| : k,l=1,\ldots n,\ x\in\mathbb{R}^n\}\le\eta,$$

for some sufficiently small $\eta>0$, *which only depends on* p,q,μ, $\max_{k,l}|a_0^{kl}|$, *and the ellipticity constant of* $\mathcal{A}_0(D)$.

(ii) Below we will need a certain decomposition of the solution operator. For this purpose observe that from the proof of Theorem 7.1.1 we have the representations

$$\hat{u}=[I-(\lambda+\mathcal{A}(\xi))^{-1}\xi\otimes\xi/\alpha(\lambda,\xi)](\lambda+\mathcal{A}(\xi))^{-1}(\hat{f}+\tilde{u}_0)-i\alpha^{-1}(\lambda+\mathcal{A}(\xi))^{-1}\xi\hat{g},$$

and

$$\hat{\pi}=-i\alpha^{-1}((\lambda+\mathcal{A}(\xi))^{-1}(\hat{f}+\tilde{u}_0)|\xi)+\hat{g}/\alpha.$$

Let us have a closer look at the term $1/\alpha(\lambda,\xi)$. We may write

$$\begin{aligned}\frac{1}{\alpha(\lambda,\xi)}&=(\mu+1)\frac{1}{(\mu+1)((\mu+\mathcal{A}(\zeta))^{-1}\zeta|\zeta)}\\&=\mu+1+(\mu+1)[\frac{1}{(\mu+1)((\mu+\mathcal{A}(\zeta))^{-1}\zeta|\zeta)}-1]\\&=\mu+1+\frac{(\mu+1)[((\mu+\mathcal{A}(\zeta))-(\mu+1)](\mu+\mathcal{A}(\zeta))^{-1}\zeta|\zeta)}{(\mu+1)((\mu+\mathcal{A}(\zeta))^{-1}\zeta|\zeta)}\\&=\mu+1+\frac{([\mathcal{A}(\zeta)-1](\mu+1)(\mu+\mathcal{A}(\zeta))^{-1}\zeta|\zeta)}{(\mu+1)((\mu+\mathcal{A}(\zeta))^{-1}\zeta|\zeta)}\\&=\lambda/|\xi|^2+1+M_{22}(\lambda,\xi),\end{aligned}$$

where we used again the notation $\mu=\lambda/|\xi|^2$ and $\zeta=\xi/|\xi|$. As in the proof of Theorem 7.1.1, $\xi\mapsto M_{22}(\mu|\xi|^2,\xi)$ is homogeneous of degree 0 and bounded, uniformly in $\xi\in\mathbb{R}^n$ and $\lambda\in\Sigma_\phi$. The arguments given there apply again to the result that there is an $L_{p,\mu}(\mathbb{R}_+;L_q(\mathbb{R}^n;\mathbb{C}^n)))$-bounded operator S_{22} with symbol $\hat{S}_{22}=M_{22}$. In a similar way we decompose

$$-i\alpha^{-1}(\lambda+\mathcal{A}(\xi))^{-1}\xi=-i\xi/|\xi|^2+(\lambda+|\xi|^2)^{-1}|\xi|M_{21}(\lambda,\xi),$$

where M_{21} is the symbol of an $L_{p,\mu}-L_q$-bounded operator S_{21}, as well as

$$-i((\lambda+\mathcal{A}(\xi))^{-1}\cdot|\xi)/\alpha(\lambda,\xi)=-i(\xi/|\xi|^2|\cdot)+(\lambda+|\xi|^2)^{-1}|\xi|M_{12}(\lambda,\xi),$$

and M_{12} is the symbol of an $L_{p,\mu}-L_q$-bounded operator S_{12}. Last but not least, in the same way we obtain the decomposition

$$[I-(\lambda+\mathcal{A}(\xi))^{-1}\xi\otimes\xi/\alpha(\lambda,\xi)](\lambda+\mathcal{A}(\xi))^{-1} = (\lambda+|\xi|^2)^{-1}(I-\xi\otimes\xi/(\lambda+|\xi|^2)) + (\lambda+|\xi|^2)^{-2}|\xi|^2M_{11}(\lambda,\xi),$$

with M_{11} the symbol of an $L_{p,\mu}-L_q$-bounded operator S_{11}. Thus the solution operator S of the generalized Stokes problem splits as $S=S_0+S_1$, where the symbols of S_j are given by

$$\hat{S}_0 = \begin{bmatrix} (\lambda+|\xi|^2)^{-1}(I-\xi\otimes\xi/(\lambda+|\xi|^2)) & -i\xi/|\xi|^2 \\ -i\xi^{\mathsf{T}}/|\xi|^2 & (\lambda+|\xi|^2)/|\xi|^2 \end{bmatrix}, \tag{7.6}$$

and

$$\hat{S}_1 = \begin{bmatrix} (\lambda+|\xi|^2)^{-2}|\xi|^2M_{11}(\lambda,\xi) & (\lambda+|\xi|^2)^{-1}|\xi|M_{12}(\lambda,\xi) \\ (\lambda+|\xi|^2)^{-1}|\xi|M_{21}(\lambda,\xi) & M_{22}(\lambda,\xi) \end{bmatrix}. \tag{7.7}$$

It is interesting to note that S_0 is independent of the coefficients of $\mathcal{A}(D)$, in fact, it is the solution of the classical Stokes problem where $\mathcal{A}(D)=-\Delta$. The operator S_1 factors as

$$\hat{S}_1 = \begin{bmatrix} \frac{1}{\lambda+|\xi|^2} & 0 \\ 0 & \frac{1}{|\xi|} \end{bmatrix} \begin{bmatrix} M_{11} & M_{12} \\ M_{21} & M_{22} \end{bmatrix} \begin{bmatrix} \frac{|\xi|^2}{\lambda+|\xi|^2} & 0 \\ 0 & |\xi| \end{bmatrix}.$$

Here $M=[M_{ij}]$ is the symbol of an $L_{p,\mu}-L_q$-bounded operator matrix.

It is a remarkable fact that such a decomposition remains valid in the variable coefficient case of Corollary 7.1.3. This can be seen as follows. We have the Neumann series for T which reads

$$T = S + \sum_{n\geq 1}(SB)^nS = S_0+S_1+\sum_{n\geq 1}(SB)^nS.$$

By induction we obtain

$$(SB)^n = \begin{bmatrix} (S_{11}\mathcal{A}_1)^n & 0 \\ S_{21}\mathcal{A}_1(S_{11}\mathcal{A}_1)^{n-1} & 0 \end{bmatrix},$$

and

$$(SB)^nS = \begin{bmatrix} (S_{11}\mathcal{A}_1)^nS_{11} & (S_{11}\mathcal{A}_1)^nS_{12} \\ S_{21}\mathcal{A}_1(S_{11}\mathcal{A}_1)^{n-1}S_{11} & S_{21}\mathcal{A}_1(S_{11}\mathcal{A}_1)^{n-1}S_{12} \end{bmatrix}.$$

In symbolic notation, using the factorization of S this yields for the first entry

$$(S_{11}\mathcal{A}_1)^nS_{11} = \frac{1}{\lambda+|\xi|^2}(1+\frac{|\xi|^2}{\lambda+|\xi|^2}M_{11})(\mathcal{A}_1(D)S_{11})^{n-1}\mathcal{A}_1(\zeta)(1+\frac{|\xi|^2}{\lambda+|\xi|^2}M_{11})\frac{|\xi|^2}{\lambda+|\xi|^2}.$$

Similarly, for the second entry we get

$$\begin{aligned}&(S_{11}\mathcal{A}_1)^n S_{12}\\ &= \frac{1}{\lambda+|\xi|^2}(1+\frac{|\xi|^2}{\lambda+|\xi|^2}M_{11})(\mathcal{A}_1(D)S_{11})^{n-1}\mathcal{A}_1(\zeta)(\frac{-i\xi}{|\xi|}+\frac{|\xi|^2}{\lambda+|\xi|^2}M_{12})|\xi|.\end{aligned}$$

In the same way the third entry becomes

$$\begin{aligned}&S_{21}(\mathcal{A}_1 S_{11})^n\\ &= \frac{1}{|\xi|}(\frac{-i\xi}{|\xi|}+\frac{|\xi|^2}{\lambda+|\xi|^2}M_{21})(\mathcal{A}_1(D)S_{11})^{n-1}\mathcal{A}_1(\zeta)(1+\frac{|\xi|^2}{\lambda+|\xi|^2}M_{11})\frac{|\xi|^2}{\lambda+|\xi|^2},\end{aligned}$$

and finally the last entry is

$$\begin{aligned}&S_{21}\mathcal{A}_1(S_{11}\mathcal{A}_1)^{n-1}S_{12}\\ &= \frac{1}{|\xi|}(\frac{-i\xi}{|\xi|}+\frac{|\xi|^2}{\lambda+|\xi|^2}M_{12})(\mathcal{A}_1(D)S_{11})^{n-1}\mathcal{A}_1(\zeta)(\frac{-i\xi}{|\xi|}+\frac{|\xi|^2}{\lambda+|\xi|^2}M_{12})|\xi|.\end{aligned}$$

This proves the assertion.

(iii) It is very useful to consider also the shifted Stokes problem

$$\begin{aligned}\partial_t u(t,x)+\omega u(t,x)+\mathcal{A}(D)u(t,x)+\nabla\pi(t,x) &= f(t,x) \quad \text{in } \mathbb{R}^n,\\ \operatorname{div} u(t,x) &= g(t,x) \quad \text{in } \mathbb{R}^n,\\ u(0,x) &= u_0(x) \quad \text{in } \mathbb{R}^n,\end{aligned} \tag{7.8}$$

for $t>0$, where $\omega>0$ is fixed. One should note that the substitutions $u_\omega = e^{-\omega t}u$, $f_\omega = e^{-\omega t}f$, and $g_\omega = e^{-\omega t}g$ transform the system (7.1) into (7.8). The advantage lies in the fact that we also obtain estimates for the $L_{p,\mu}-L_q$-norm. In fact, we get the following estimates for the solution u of (7.8). Setting

$$\mathbb{E}_{0\mu} = L_{p,\mu}(\mathbb{R}_+;L_q(\mathbb{R}^n;\mathbb{C}^n)),\quad \mathbb{G}_{1\mu} = H^1_{p,\mu}(\mathbb{R}_+;\dot{H}^{-1}_q(\mathbb{R}^n))\cap L_{p,\mu}(\mathbb{R}_+;H^1_q(\mathbb{R}^n)),$$

and $X_{\gamma,\mu} = B^{2(\mu-1/p)}_{qp}(\mathbb{R}^n;\mathbb{C}^n)$, there is a constant $C>0$ such that

$$\begin{aligned}&\omega|u|_{\mathbb{E}_{0\mu}} + |\partial_t u|_{\mathbb{E}_{0\mu}} + |\nabla^2 u|_{\mathbb{E}_{0\mu}} + |\nabla\pi|_{\mathbb{E}_{0\mu}}\\ &\qquad \le C\big(|u_0|_{X_{\gamma,\mu}} + |f|_{\mathbb{E}_{0,\mu}} + |g|_{\mathbb{G}_{1\mu}} + \omega|g|_{L_{p,\mu}(\dot{H}^{-1}_q)}\big),\end{aligned} \tag{7.9}$$

for all $(f,g,u_0)\in\mathbb{E}_{0,\mu}\times\mathbb{G}_{1\mu}\times X_{\gamma,\mu}$ such that $\operatorname{div} u_0 = g(0)$ in $\mathcal{D}'(\mathbb{R}^n)$. Here the constant C depends only on p,q,μ and on the symbol $\mathcal{A}(\zeta)$. This result follows directly from the representation of the symbol of S, one only has to observe that the exponential shift replaces λ by $\lambda+\omega$.

(iv) At several places it will be convenient to reduce the full Stokes problem to a problem for the Stokes operator. This can be achieved as follows. We first solve the problem

$$\begin{aligned} \partial_t v + \omega v + \mathcal{A}(D)v + \nabla\pi &= f &&\text{in } \mathbb{R}^n,\\ \operatorname{div} v &= g &&\text{in } \mathbb{R}^n,\\ v(0) &= v_0 &&\text{in } \mathbb{R}^n, \end{aligned} \tag{7.10}$$

for $t > 0$, with ω sufficiently large. Then $w = u - v$ must satisfy

$$\dot{w} + Aw = \omega v, \quad t > 0, \quad w(0) = 0.$$

This reduction will be useful in several situations.

1.4 Localization

Now we are in position for the general case, i.e., we consider the problem

$$\begin{aligned} \partial_t v + \omega v + \mathcal{A}(x, D)v + \nabla q &= f &&\text{in } \mathbb{R}^n,\\ \operatorname{div} v &= g &&\text{in } \mathbb{R}^n,\\ v(0) &= v_0 &&\text{in } \mathbb{R}^n. \end{aligned} \tag{7.11}$$

As before the data (f, g, v_0) are given, and we assume that the differential operator $\mathcal{A}(x, D) = \sum_{k,l} a^{kl}(x) D_k D_l$ has coefficients $a^{kl} \in C_l(\mathbb{R}^n; \mathcal{B}(\mathbb{C}^n))$ and that $\mathcal{A}(x, D)$ is uniformly strongly elliptic, i.e.,

$$\operatorname{Re}(\mathcal{A}(x, \xi)v|v) \geq c_0 |\xi|^2 |v|^2, \quad \xi \in \mathbb{R}^n,\ v \in \mathbb{C}^n,\ x \in \mathbb{R}^n,$$

with some constant $c_0 > 0$. The parameter $\omega \geq 0$ will be chosen later. Observe that maximal regularity on finite intervals does not depend on ω.

First, we reduce the problem as above to the case $(f, u_0) = 0$, employing the results of Chapter 6. To solve the remaining problem we employ the method of localization. Choose a large ball $B(0, R)$ such that

$$\sup\{|a(x) - a(\infty)| : |x| \geq R\} \leq \eta.$$

Cover the ball $\bar{B}(0, R)$ by finitely many balls $B(x_k, r)$, $k = 1, \ldots, N$, such that

$$\sup\{|a(x) - a(x_k)| : x \in B(x_k, r)\} \leq \eta.$$

Fix a C^∞-partition of unity ϕ_k which is subordinate to the covering $\bar{B}(0, R)^c \cup \bigcup_{k=1}^N B(x_k, r)$ of $\mathbb{R}^n$. The index $k = 0$ corresponds to the chart at infinity. Define local operators $\mathcal{A}_k(D) = \mathcal{A}(x, D)$ for each chart $B(x_k, r)$, $k = 1, \ldots, N$, and $\mathcal{A}_0(D) = \mathcal{A}(x, D)$, extend these coefficients to all of $\mathbb{R}^n$, say by reflection at the boundary of the corresponding ball. Corollary 7.1.3 shows that each of these operators has maximal regularity, provided $\eta > 0$ is sufficiently small, but independent of k.

Suppose (v, q) is a solution of (7.11) (with $(f, v_0) = 0$). In the sequel we normalize the pressure by $\int_{B(0,2R)} q(t, x)\, dx = 0$. Setting $v_k = \phi_k v$, $q_k = \phi_k q$,

$g_k = \phi_k g$ we obtain the following problem for the functions v_k and q_k.

$$\begin{aligned} \partial_t v_k + \omega v_k + \mathcal{A}_k(D)v_k + \nabla q_k &= (\nabla\phi_k)q + [\mathcal{A}, \phi_k]v && \text{in } \mathbb{R}^n, \\ \operatorname{div} v_k &= g_k + (\nabla\phi_k|v) && \text{in } \mathbb{R}^n, \\ v_k(0) &= 0 && \text{in } \mathbb{R}^n, \end{aligned} \tag{7.12}$$

where $[\mathcal{A}(x,D), \phi_k]v = \mathcal{A}(x,D)(\phi_k v) - \phi_k\mathcal{A}(x,D)v$ means the commutator of $\mathcal{A}(x,D)$ and ϕ_k. Denote the solution operator of the generalized Stokes problem for $\omega + \mathcal{A}_k$ by S^k. Then we have the representation

$$\begin{bmatrix} v_k \\ q_k \end{bmatrix} = S^k \begin{bmatrix} (\nabla\phi_k)q + [\mathcal{A}, \phi_k]v \\ g_k + (\nabla\phi_k|v) \end{bmatrix}.$$

Summing over all charts k we deduce

$$\begin{bmatrix} v \\ q \end{bmatrix} = \sum_{k=0}^{N} \begin{bmatrix} v_k \\ q_k \end{bmatrix} = \sum_{k=0}^{N} S^k \begin{bmatrix} (\nabla\phi_k)q + [\mathcal{A}, \phi_k]v \\ g_k + (\nabla\phi_k|v) \end{bmatrix}.$$

We decompose this representation of the solution as

$$\begin{bmatrix} v \\ q \end{bmatrix} = \sum_{k=0}^{N} S^k \begin{bmatrix} 0 \\ g_k \end{bmatrix} + T \begin{bmatrix} q \\ v \end{bmatrix} + Rv,$$

where

$$T = \sum_{k=0}^{N} S^k \nabla\phi_k \quad \text{and} \quad R = \sum_{k=0}^{N} S^k \begin{bmatrix} [\mathcal{A}, \phi_k] \\ 0 \end{bmatrix}.$$

We estimate T and R separately. For this purpose, we define the maximal regularity space

$$\mathbb{E}_{1\mu} := [H^1_{p,\mu}(\mathbb{R}_+; L_q(\mathbb{R}^n;\mathbb{C}^n)) \cap L_{p,\mu}(\mathbb{R}_+; H^2_q(\mathbb{R}^n;\mathbb{C}^n))] \times L_{p,\mu}(\mathbb{R}_+; \dot{H}^1_q(\mathbb{R}^n)),$$

and recall the definition of the spaces $\mathbb{E}_{0\mu}$ and $\mathbb{G}_{1\mu}$ from above. To begin with T, recall that each S^k splits into $S^k = S_0 + S^k_1$, with the same S_0 for each k, since the latter does not depend on the coefficients of $\mathcal{A}_k$. Hence

$$T = \sum_{k=0}^{N} S^k \nabla\phi_k = \sum_{k=0}^{N} S^k_1 \nabla\phi_k + S_0 \nabla \sum_{k=0}^{N} \phi_k = \sum_{k=0}^{N} S^k_1 \nabla\phi_k,$$

since ϕ_k forms a partition of unity. Let us decompose T into its components, employing the factorization of S_1 obtained in Section 7.1.3. We have

$$T_{11}q = (\partial_t + \omega - \Delta)^{-1} \sum_k S_{11}^k (-\Delta)(\partial_t + \omega - \Delta)^{-1}(q\nabla\phi_k),$$
$$T_{21}q = (-\Delta)^{-1/2} \sum_k S_{21}^k (-\Delta)(\partial_t + \omega - \Delta)^{-1}(q\nabla\phi_k),$$
$$T_{12}v = (\partial_t + \omega - \Delta)^{-1} \sum_k S_{12}^k (-\Delta)^{1/2}(\nabla\phi_k|v),$$
$$T_{22}v = (-\Delta)^{-1/2} \sum_k S_{22}^k (-\Delta)^{1/2}(\nabla\phi_k|v).$$

Since $\nabla\phi_k$ has compact support also for $k = 0$, we see that $(\nabla\phi_k)q$ belongs to $L_{p,\mu}(\mathbb{R}_+; H^1_q(\mathbb{R}^n))$, and

$$|(-\Delta)^{1/2}(q\nabla\phi_k)|_{\mathbb{E}_{0,\mu}} \leq C|\nabla q|_{\mathbb{E}_{0,\mu}}$$

holds with some constant $C > 0$; recall the normalization of the pressure $\int_{B(0,2R)} q(t,x)\,dx = 0$, hence Poincaré's inequality is valid. Therefore,

$$|(-\Delta)(\partial_t + \omega - \Delta)^{-1}(q\nabla\phi_k)|_{\mathbb{E}_{0,\mu}} \leq \frac{C}{\sqrt{\omega}}|\nabla q|_{\mathbb{E}_{0,\mu}}.$$

Similarly, there is a constant $C > 0$ such that

$$|(-\Delta)^{1/2}(\nabla\phi_k|v)|_{\mathbb{E}_{0\mu}} \leq \frac{C}{\sqrt{\omega}}|(\partial_t + \omega - \Delta)v|_{\mathbb{E}_{0,\mu}}.$$

As a consequence, the operator T satisfies

$$\omega\left|T\begin{bmatrix} q \\ v \end{bmatrix}\right|_{\mathbb{E}_{0\mu}} + \left|T\begin{bmatrix} q \\ v \end{bmatrix}\right|_{\mathbb{E}_{1\mu}} \leq \frac{C}{\sqrt{\omega}}\Big(\left|\begin{bmatrix} v \\ q \end{bmatrix}\right|_{\mathbb{E}_{1\mu}} + \omega\left|\begin{bmatrix} v \\ q \end{bmatrix}\right|_{\mathbb{E}_{0\mu}}\Big).$$

Next, R is given by

$$R\begin{bmatrix} q \\ v \end{bmatrix} = \sum_k S^k \begin{bmatrix} [\mathcal{A}, \phi_k]v \\ 0 \end{bmatrix}.$$

The commutator $[\mathcal{A}(x,D), \phi_k]$ is a differential operator of first order, hence

$$\omega\left|R\begin{bmatrix} q \\ v \end{bmatrix}\right|_{\mathbb{E}_{0\mu}} + \left|R\begin{bmatrix} q \\ v \end{bmatrix}\right|_{\mathbb{E}_{1\mu}} \leq \frac{C}{\sqrt{\omega}}\Big(\left|\begin{bmatrix} v \\ q \end{bmatrix}\right|_{\mathbb{E}_{1\mu}} + \omega\left|\begin{bmatrix} v \\ q \end{bmatrix}\right|_{\mathbb{E}_{0\mu}}\Big).$$

The above arguments show that, choosing first $\eta > 0$ small and then $\omega > 0$ large enough, there is a constant $C > 0$ such that the estimate

$$\omega|v|_{\mathbb{E}_{0\mu}} + |v|_{\mathbb{E}_{1\mu}} + |\nabla\pi|_{\mathbb{E}_{0\mu}} \leq C\big(|g|_{\mathbb{G}_{1\mu}} + \omega|g|_{L_{p,\mu}(\dot{H}^{-1}_q)}\big) \tag{7.13}$$

holds. Therefore, the operator L defined by the first two lines of the left-hand side of (7.11) is injective and has closed range, hence it is semi-Fredholm, for each set of coefficients which are continuous on $\mathbb{R}^n$, admit uniform limits as $|x| \to \infty$, and are uniformly strongly elliptic. Define the family $\mathcal{A}_\tau = \tau\mathcal{A} + (1-\tau)(-\Delta)$. By strong ellipticity, we then may conclude that for each $\tau \in [0,1]$, the corresponding operator L_τ is injective and has closed range. By the continuity of the Fredholm index, it must be constant, i.e., the index is zero for all $\tau \in [0,1]$ since L_0 is bijective by Theorem 7.1.1. This shows that $L = L_1$ is also surjective.

Summarizing, for the problem with variable coefficients

$$\begin{aligned} \partial_t v + \omega v + \mathcal{A}(x,D)v + \nabla\pi &= f \quad \text{in } \mathbb{R}^n,\\ \operatorname{div} v &= g \quad \text{in } \mathbb{R}^n,\\ v(0) &= v_0 \quad \text{in } \mathbb{R}^n, \end{aligned} \tag{7.14}$$

we have proved the following result.

Theorem 7.1.4. *Let $1 < p,q < \infty$, $\mu \in (1/p, 1]$, and assume that $\mathcal{A}(x,D)$ is a second-order differential operator with coefficients $a^{kl} \in C_l(\mathbb{R}^n; \mathcal{B}(\mathbb{C}^n))$ which is uniformly strongly elliptic.*

Then there is $\omega_0 \in \mathbb{R}$ such that for all $\omega > \omega_0$, (7.14) has maximal $L_{p,\mu}$–L_q-regularity in the following sense. There is a unique solution (u,π) of (7.14) with

$$u \in H^1_{p,\mu}(\mathbb{R}_+; L_q(\mathbb{R}^n;\mathbb{C}^n)) \cap L_{p,\mu}(\mathbb{R}_+; H^2_q(\mathbb{R}^n;\mathbb{C}^n)), \quad \pi \in L_{p,\mu}(\mathbb{R}_+; \dot H^1_q(\mathbb{R}^n)),$$

if and only if the data f, g, u_0 satisfy the subsequent conditions.

(a) $f \in L_{p,\mu}(\mathbb{R}_+; L_q(\mathbb{R}^n;\mathbb{C}^n))$;

(b) $g \in H^1_{p,\mu}(\mathbb{R}_+; \dot H^{-1}_q(\mathbb{R}^n)) \cap L_{p,\mu}(\mathbb{R}_+; H^1_q(\mathbb{R}^n))$;

(c) $u_0 \in B^{2(\mu-1/p)}_{qp}(\mathbb{R}^n;\mathbb{C}^n)$ *and* $\operatorname{div} u_0 = g(0)$ *in* $\mathcal{D}'(\mathbb{R}^n)$.

The solution (u,π) depends continuously on the data in the corresponding spaces. Moreover, the estimate (7.9) is valid.

We may now define *the generalized Stokes operator A* in the case of variable coefficients as in Section 1.2, to the result that $\omega + A \in \mathcal{MR}_p(X_0)$ for $\omega > \omega_0$. The lower bound for ω_0 is easily seen to be $s(-A)$, the spectral bound of $-A$.

7.2 Generalized Stokes Problems in a Half-Space

In this section we consider the generalized Stokes problem in $\mathbb{R}^n_+ = \mathbb{R}^{n-1} \times \mathbb{R}_+$ with either one of the four boundary conditions explained below. Thus we consider the problem

$$\begin{aligned} (\partial_t + \omega)u + \mathcal{A}(D)u + \nabla\pi &= f(t,x) \quad \text{in } \mathbb{R}^n_+,\\ \operatorname{div} u &= g(t,x) \quad \text{in } \mathbb{R}^n_+,\\ u(0,x) &= u_0(x) \quad \text{in } \mathbb{R}^n_+, \end{aligned} \tag{7.15}$$

with $t > 0$. Here, as in Section 7.1, $\mathcal{A}(D) = \sum_{k,l=1}^{n} a^{kl} D_k D_l$ denotes a strongly elliptic differential operator with constant coefficients acting on $\mathbb{C}^n$-valued functions, $J = \mathbb{R}_+$, and $\omega \geq 0$.

In the sequel, $\mathcal{P}_\Sigma$ denotes the projection onto the tangent bundle of Σ; more precisely, $\mathcal{P}_\Sigma(p)$ means the orthogonal projection onto the tangent space $T_p\Sigma$. With $\nu = -e_n$, the n-th unit vector in $\mathbb{R}^n$, the boundary conditions are either

(i) no slip

$$u = h_0 \quad \text{on } \partial\mathbb{R}^n_+, \tag{7.16}$$

(ii) pure slip

$$(u|\nu) = h_{0\nu},\ \mathcal{P}_\Sigma \nu_k a^{kl} D_l u = h_\Sigma \quad \text{on } \partial\mathbb{R}^n_+, \tag{7.17}$$

(iii) outflow

$$\mathcal{P}_\Sigma u = h_{0\Sigma},\ (\nu_k a^{kl} D_l u|\nu) + i\pi = h_\nu \quad \text{on } \partial\mathbb{R}^n_+, \tag{7.18}$$

(iv) free

$$\nu_k a^{kl} D_l u \nu + i\pi\nu = h \quad \text{on } \partial\mathbb{R}^n_+. \tag{7.19}$$

Of course, appropriate compatibility conditions have to be satisfied. Assuming normal strong ellipticity, as in Section 6.2.5, it is easily verified that the parabolic problem without pressure and divergence condition satisfies the Lopatinskii-Shapiro condition for these boundary conditions, hence is well-posed and has maximal L_p-regularity for $1 < p < \infty$. The main result of this section states that these properties carry over to the generalized Stokes problem.

For this we need some notation. If $\Omega \subset \mathbb{R}^n$ is a C^1 domain, $\Sigma \subset \partial\Omega$ open, $1 < q < \infty$, we define

$$\dot{H}^1_q(\Omega) = \{w \in L_{1,loc}(\Omega) : \nabla w \in L_q(\Omega)\}.$$

By means of standard arguments in the theory of function spaces, $\dot{H}^1_q(\Omega)$ embeds into $H^1_q(\Omega \cap \mathbb{B}(0,R))$, for each $R > 0$. This shows that traces of functions in $\dot{H}^1_q(\Omega)$ are well defined, and that in this space localization is possible. In fact, if χ is $\mathcal{D}(\mathbb{R}^n)$, then by the Poincaré-Wirtinger inequality, $\chi u \in H^1_q(\Omega)$ for each $u \in \dot{H}^1_q(\Omega)$. In the case of $\Omega = \mathbb{R}^n$ it is true that

$$\dot{H}^1_q(\mathbb{R}^n) = \{u \in \mathcal{S}'(\mathbb{R}^n) :\ \nabla u \in L_q(\mathbb{R}^n)\}.$$

We next define

$$\dot{H}^1_{q,\Sigma}(\Omega) = \{w \in L_{1,loc}(\Omega) : \nabla w \in L_q(\Omega),\ w = 0 \text{ on } \Sigma\};$$

in particular, $\dot{H}^1_{q,\emptyset}(\Omega) = \dot{H}^1_q(\Omega)$. Then $\dot{H}^{-1}_{q,\Sigma}(\Omega)$ is defined as

$$\dot{H}^{-1}_{q,\Sigma}(\Omega) := [\dot{H}^1_{q',\partial\Omega\backslash\Sigma}(\Omega)]^*.$$

Especially,

$$\dot{H}_q^{-1}(\Omega) = \dot{H}_{q,\emptyset}^{-1}(\Omega), \quad {}_0\dot{H}_q^{-1}(\Omega) = \dot{H}_{q,\partial\Omega}^{-1}(\Omega).$$

Observe that $\dot{H}_q^{-1}(\Omega)$ consists solely of distributions in Ω, but ${}_0\dot{H}_q^{-1}(\Omega)$ does not have this property.

Assume that (7.15) admits a solution (u, π) in the regularity class

$$u \in H^1_{p,\mu}(J; L_q(\Omega))^n \cap L_{p,\mu}(J; H^2_q(\Omega))^n, \quad \pi \in L_{p,\mu}(J; \dot{H}^1_q(\Omega)).$$

By trace theory, the conditions for the right-hand side f and for the initial value u_0 are the same as in the previous section. They are collected in *condition* **(D)**

(a) $f \in L_{p,\mu}(\mathbb{R}_+; L_q(\mathbb{R}^n_+; \mathbb{C}^n))$, $u_0 \in B^{2\mu-2/p}_{qp}(\mathbb{R}^n_+; \mathbb{C}^n)$.

For g, trace theory yields

(b) $g \in H^1_{p,\mu}(\mathbb{R}_+; \dot{H}^{-1}_q(\mathbb{R}^n_+)) \cap L_{p,\mu}(\mathbb{R}_+; H^1_q(\mathbb{R}^n_+))$, $\operatorname{div} u_0 = g(0)$.

The boundary data must satisfy

(d0) for no-slip (Dirichlet) boundary conditions:
$h_0 \in F^{1-1/2q}_{pq,\mu}(\mathbb{R}_+; L_q(\mathbb{R}^{n-1}; \mathbb{C}^n)) \cap L_{p,\mu}(\mathbb{R}_+; B^{2-1/q}_{qq}(\mathbb{R}^{n-1}; \mathbb{C}^n))$ and for $\mu > 3/2p$ in addition $h(0) = u_0$.

Similarly, we have

(ds) for pure slip boundary conditions:
$h_{0\nu} \in F^{1-1/2q}_{pq,\mu}(\mathbb{R}_+; L_q(\mathbb{R}^{n-1})) \cap L_{p,\mu}(\mathbb{R}_+; B^{2-1/q}_{qq}(\mathbb{R}^{n-1}))$;
$h_\Sigma \in F^{1/2-1/2q}_{pq,\mu}(\mathbb{R}_+; L_q(\mathbb{R}^{n-1}; \mathbb{C}^{n-1})) \cap L_{p,\mu}(\mathbb{R}_+; B^{1-1/q}_{qq}(\mathbb{R}^{n-1}; \mathbb{C}^{n-1}))$ and $\mathcal{P}_\Sigma \nu_k a^{kl} D_l u_0 = h_\Sigma(0)$ for $\mu > 3/p$;

(do) for outflow boundary conditions:
$h_{0\Sigma} \in F^{1-1/2q}_{pq,\mu}(\mathbb{R}_+; L_q(\mathbb{R}^{n-1}; \mathbb{C}^{n-1})) \cap L_{p,\mu}(\mathbb{R}_+; B^{2-1/q}_{qq}(\mathbb{R}^{n-1}; \mathbb{C}^{n-1}))$;
$h_\nu \in F^{1/2-1/2q}_{pq,\mu}(\mathbb{R}_+; L_q(\mathbb{R}^{n-1})) \cap L_{p,\mu}(\mathbb{R}_+; B^{1-1/q}_{qq}(\mathbb{R}^{n-1}))$ and $\mathcal{P}_\Sigma u_0 = h_{0\Sigma}(0)$ for $\mu > 3/2p$;

(dn) for free (Neumann) boundary conditions:
$h \in F^{1/2-1/2q}_{pq,\mu}(\mathbb{R}_+; L_q(\mathbb{R}^{n-1}; \mathbb{C}^n)) \cap L_{p,\mu}(\mathbb{R}_+; B^{1-1/q}_{qq}(\mathbb{R}^{n-1}; \mathbb{C}^n))$ and $\mathcal{P}_\Sigma \nu_k a^{kl} D_l u_0 = P_\Sigma h(0)$ for $\mu > 3/p$.

In case of outflow or Neumann conditions these are all requirements needed. In case of slip or Dirichlet conditions we have the additional property

(e) $(g, h_{0\nu}) \in H^1_{p,\mu}(\mathbb{R}_+; {}_0\dot{H}_q^{-1}(\mathbb{R}^n_+))$ and $h_{0\nu}(0) = (\nu|u_0)$.

Observe that the last condition is a compatibility condition which comes from the divergence equation, as the identity

$$-\int_{\mathbb{R}^n_+} u\cdot\nabla\phi\, d(x,y) = \int_{\mathbb{R}^n_+} \operatorname{div} u\phi\, d(x,y) - \int_{\mathbb{R}^{n-1}} u\cdot\nu\,\phi\, dx$$
$$= \int_{\mathbb{R}^n_+} g\phi\, d(x,y) - \int_{\mathbb{R}^{n-1}} h_{0\nu}\phi\, dx =: \langle (g,h_{0\nu})|\phi\rangle$$

shows. Here $\phi\in \dot{H}^1_{q'}(\mathbb{R}^n_+)$.

After these preliminaries we can state the main result of this section.

Theorem 7.2.1. *Let* $1<p,q<\infty$, $1\geq\mu>1/p$, $\mu\neq 3/2p, 3/p$, *and assume that* $\mathcal{A}(D)=\sum_{k,l=1}^n a^{kl}D_kD_l$ *is normally strongly elliptic. Then for each* $\omega>0$, (7.15) *with boundary conditions* (7.16) *or* (7.17) *or* (7.18) *or* (7.19) *has maximal* $L_{p,\mu}-L_q$*-regularity in the following sense. There is a unique solution* (u,π) *of* (7.15) *in the class*

$$u\in H^1_{p,\mu}(J;L_q(\mathbb{R}^n_+;\mathbb{C}^n))\cap L_{p,\mu}(J;H^2_q(\mathbb{R}^n_+;\mathbb{C}^n)),\quad \pi\in L_{p,\mu}(J;\dot{H}^1_q(\mathbb{R}^n_+)),$$

satisfying the corresponding boundary condition, and in addition

$$\pi\in F^{1/2-1/2q}_{pq,\mu}(J;L_q(\partial\mathbb{R}^n_+))$$

in case of outflow or Neumann boundary condition, if and only if the data (f,g,h,u_0) *satisfy the conditions* **(D)**. *The solution* u *depends continuously on the data in the corresponding spaces.*

The next subsections are devoted to the proof of this result.

2.1 Reductions

According to the discussion above, we only need to show the sufficiency part. Let data (f,g,u_0) and boundary data h with the corresponding regularity be given. Without loss of generality we may assume $(f,g,u_0)=0$ and trace 0 of h at $t=0$ in case it exists. This can be seen as follows. Firstly, extend the initial value to all of $\mathbb{R}^n$ in the class $B^{2\mu-2/p}_{qp}(\mathbb{R}^n)^n$, and extend f trivially to $f\in L_{p,\mu}(J;L_q(\mathbb{R}^n))^n$. Solving the parabolic initial-boundary value problem without pressure and divergence condition on all of $\mathbb{R}^n$ yields a function u_1 in the right regularity class. Then $u_2:=u-u_1$ and $\pi_2:=\pi$ should solve the problem with $(f,u_0)=0$ and g replaced by $g_1:=g-\operatorname{div} u_1$, which belongs to the same regularity class but has trace 0 at $t=0$. Extend g_1 evenly in x_n to all of $J\times\mathbb{R}^n$, and solve the full-space generalized Stokes problem (7.1) with $(f,u_0)=0$ to obtain a pair (u_3,π_3) in the right regularity class. Then the pair (u_4,π_4) defined by $u_4:=u_2-u_3$, $\pi_4:=\pi_2-\pi_3$ should solve (7.15) with the boundary condition in question, where $(f,g,u_0)=0$ and $h_4=h-\mathcal{B}(D)(u_1+u_3,\pi_3)$; here $\mathcal{B}(D)$ denotes the boundary operator under consideration. Note that the new boundary datum h belongs to the right regularity class and has trace 0 at $t = 0$ whenever it exists. The compatibility condition (e) becomes now

$$h_{0,\nu} \in {}_0H^1_{p,\mu}(J; \dot{W}_q^{-1/q}(\mathbb{R}^{n-1})).$$

So we have to solve the homogeneous problem (7.15) with one of the inhomogeneous boundary conditions. It is convenient to replace the spatial variables by (x, y), where $x \in \mathbb{R}^{n-1}$ and $y > 0$; recall that $\nu = -e_n$. Similarly we decompose $u = (v, w)$, with $v \in \mathbb{R}^{n-1}$ the tangential and $w \in \mathbb{R}$ the normal velocity.

2.2 Fourier-Laplace Transform

Taking Fourier transform in the tangential space directions, Laplace transform in t we obtain the parameter dependent ODE-problem

$$\begin{aligned}
(\lambda + \mathcal{A}_{11}(\xi + e_n D_y))\hat{v} + \mathcal{A}_{12}(\xi + e_n D_y)\hat{w} + i\xi\hat{\pi} &= 0, \qquad y > 0,\\
\mathcal{A}_{21}(\xi + e_n D_y)\hat{v} + (\lambda + \mathcal{A}_{22}(\xi + e_n D_y))\hat{w} + \partial_y\hat{\pi} &= 0, \qquad y > 0,\\
i\xi^{\mathsf{T}}\hat{v} + iD_y\hat{w} &= 0, \qquad y > 0, \qquad (7.20)\\
\mathcal{B}_{11}(\xi + e_n D_y)\hat{v}(0) + \mathcal{B}_{12}(\xi + e_n D_y)\hat{w}(0) &= \hat{h}_v,\\
\mathcal{B}_{21}(\xi + e_n D_y)\hat{v}(0) + \mathcal{B}_{22}(\xi + e_n D_y)\hat{w}(0) + \mathcal{B}_{23}\hat{\pi}(0) &= \hat{h}_w,
\end{aligned}$$

where $\mathcal{B}$ is defined by one of the boundary conditions (7.16), (7.17), (7.18) or (7.19). The parameters ξ and λ satisfy $(\xi, \lambda) \in \mathbb{R}^n \times \Sigma_\phi$, for some $\phi > \pi/2$ and $\xi_n = 0$. Here and below we identify $\xi \in \mathbb{R}^{n-1}$ with $(\xi, 0) \in \mathbb{R}^n$. Introducing the vector

$$\mathsf{x} = [\hat{v}, \hat{w}, \partial_y\hat{v}, \partial_y\hat{w}, \hat{\pi}]^{\mathsf{T}},$$

we rewrite this problem as the first-order system

$$E\partial_y\mathsf{x} + A\mathsf{x} = 0, \quad y > 0, \quad B\mathsf{x}(0) = \hat{h}, \qquad (7.21)$$

where the dependence on (λ, ξ) has been dropped. Here the $(2n+1)$-dimensional square matrix E is defined as

$$E = \begin{bmatrix} 1 & 0 & 0 & 0 & 0 \\ 0 & 1 & 0 & 0 & 0 \\ 0 & 0 & A^0_{11} & A^0_{12} & 0 \\ 0 & 0 & A^0_{21} & A^0_{22} & -1 \\ 0 & 1 & 0 & 0 & 0 \end{bmatrix}$$

and A by

$$A = \begin{bmatrix} 0 & 0 & -1 & 0 & 0 \\ 0 & 0 & 0 & -1 & 0 \\ -(\lambda + A^2_{11}) & -A^2_{12} & A^1_{11} & A^1_{12} & -i\xi \\ -A^2_{21} & -(\lambda + A^2_{22}) & A^1_{21} & A^1_{22} & 0 \\ i\xi^{\mathsf{T}} & 0 & 0 & 0 & 0 \end{bmatrix}.$$

We used the abbreviations

$$A^2 = (a^{kl}\xi_k\xi_l), \quad A^1 = i(a^{kl}\nu_k\xi_l + a^{kl}\nu_l\xi_k), \quad A^0 = (a^{kl}\nu_k\nu_l)$$

recalling the summation convention. Observe that A^k are homogeneous in ξ of order k; in particular A^0 is constant and invertible by ellipticity. Also note that E does neither depend on λ nor on ξ. The boundary matrices B are

$$B = \begin{bmatrix} 1 & 0 & 0 & 0 & 0 \\ 0 & 1 & 0 & 0 & 0 \end{bmatrix}$$

in case of Dirchlet conditions,

$$B = \begin{bmatrix} B^1_{11} & B^1_{12} & B^0_{11} & B^0_{12} & 0 \\ 0 & 1 & 0 & 0 & 0 \end{bmatrix}$$

for slip conditions,

$$B = \begin{bmatrix} 1 & 0 & 0 & 0 & 0 \\ B^1_{21} & B^1_{22} & B^0_{21} & B^0_{22} & -1 \end{bmatrix}$$

for outflow conditions, and

$$B = \begin{bmatrix} B^1_{11} & B^1_{12} & B^0_{11} & B^0_{12} & 0 \\ B^1_{21} & B^1_{22} & B^0_{21} & B^0_{22} & -1 \end{bmatrix}$$

in the case of Neumann conditions. Here B^k_{ij} are homogeneous of order k in ξ, and $B^0 = A^0$. Recall that the Lopatinskii-Shapiro condition means that system (7.21) admits at most one solution $\mathsf{x} \in C_0(\mathbb{R}_+;\mathbb{C}^{2n+1})$, for each $\hat{h} \in \mathbb{C}^n$ and $\xi \in \mathbb{R}^{n-1}$, $\operatorname{Re}\lambda \geq 0$, $\xi \neq 0$. This follows from normal strong ellipticity as in Section 6.2.5, as the crucial identity (6.39) holds also in the Stokes cases for the four boundary conditions under consideration

2.3 The DAE-System

It is our purpose to derive a representation formula of the function x in terms of the given data $\hat{h}$, which is accessible to inversion of the Fourier and Laplace transform.

So, assume that $\mathsf{x} \in C_0(\mathbb{R}_+;\mathbb{C}^{2n+1})$ is a solution of (7.21). Taking Laplace transform $\mathcal{L}$ in y, this yields

$$(zE + A)\mathcal{L}\mathsf{x}(z) = E\mathsf{x}^0, \quad \operatorname{Re} z > 0, \quad B\mathsf{x}^0 = \hat{h},$$

where $\mathsf{x}^0 = \mathsf{x}(0)$ denotes the initial value of x. To obtain a representation of x we have to study the operator pencil $zE + A$. To this end note that E is not invertible but its kernel $\mathsf{N}(E)$ is one-dimensional, and $\mathsf{N}(E^2) = \mathsf{N}(E)$, hence $\mathsf{N}(E) \oplus \mathsf{R}(E) = \mathbb{C}^{2n+1}$. Therefore, (7.21) is a differential-algebraic system of index ≥ 1. This implies that the characteristic polynomial $p(z) = \det(zE + A)$ has at most order $2n$. Let us show that it is precisely of order $2n$, i.e., that the index is 1. This can be seen as follows. Expand $\det(zE+A)$ first w.r.t. the last column and the last row and then w.r.t. the second row. This yields up to a sign

$$p(z) = z^2 \det \begin{bmatrix} z & -1 \\ -(\lambda + A^2_{11}) & zA^0_{11} + A^1_{11} \end{bmatrix} + q(z),$$

where $q(z)$ is of order less than $2n$. Asymptotically this yields for large z

$$p(z) \sim z^2 \det \begin{bmatrix} z & 0 \\ 0 & zA^0_{11} \end{bmatrix} = z^{2n} \det A^0_{11},$$

and $\det A^0_{11} \neq 0$ by strong ellipticity. Therefore, $p(z)$ is of order $2n$. Ellipticity shows also that $p(z)$ has no zeros on the imaginary axis, for $\xi \neq 0$. Now consider the case $\xi = 0$. Then we see by the same procedure that $p(z)$ is in fact a function of z^2, i.e., if z_0 is a zero of p then $-z_0$ is one as well. Unfortunately, $z = 0$ is a solution in case $\xi = 0$, here the degeneracy of the Stokes problem shows up. We have to look at this zero more closely.

The eigenvalue problem for these small zeros $z(\xi)$ for small ξ (or large λ) becomes

$$(A(z,\xi) - \lambda) \begin{bmatrix} \mathsf{x}_1 \\ \mathsf{x}_2 \end{bmatrix} = \begin{bmatrix} i\xi \\ z \end{bmatrix}, \quad (i\xi|\mathsf{x}_1) + z\mathsf{x}_2 = 0,$$

where

$$A(z,\xi) = z^2 A^0 + zA^1(\xi) - A^2(\xi).$$

Since by $\lambda \neq 0$ we have invertibility of $A(z,\xi) - \lambda$, this implies the condition

$$\left(\begin{bmatrix} i\xi \\ z \end{bmatrix} \Big| (A(z,\xi) - \lambda)^{-1} \begin{bmatrix} i\xi \\ z \end{bmatrix} \right) = 0$$

for the small eigenvalues. Writing $(A(z,\xi) - \lambda)^{-1}$ as a Neumann series, this condition becomes

$$z^2 - |\xi|^2 + O((|\xi| + |z|)^4) = 0,$$

which shows that $z = \pm|\xi| + O(|\xi|^2)$ near $\xi = 0$. Therefore, the double zero $z(0) = 0$ for $\xi = 0$ splits into two simple real zeros which behave like $z_1^{\pm}(\xi) \sim \pm|\xi|$ near $\xi = 0$.

Varying now ξ we may conclude that $p(z)$ has exactly n roots with positive real parts, counting with multiplicity, for each $\xi \in \mathbb{R}^{n-1}$, $\operatorname{Re}\lambda > 0$, $\xi \neq 0$, since none of them can cross the imaginary axis by ellipticity.

We may now write

$$\mathcal{L}\mathsf{x}(z) = (zE + A)^{-1} E\mathsf{x}^0, \quad B\mathsf{x}^0 = \hat{h},$$

for the Laplace transform of x. The initial value x^0 thus must be chosen in such a way that $\mathcal{L}\mathsf{x}(z)$ has no poles in the right half-plane, and $B\mathsf{x}^0 = \hat{h}$ holds.

Define the projection P^+ by means of

$$P^+ = \frac{1}{2\pi i}\int_{\Gamma_+} (zE+A)^{-1}E\,dz,$$

where Γ_+ denotes a closed simple contour in the right half-plane surrounding the poles of $(zE+A)^{-1}$, i.e., the zeros of $p(z)$ in the right half-plane. Let z_k, $k=1,\ldots,m^+$, denote the zeros of $p(z)$ in the right and for $k=-m^-,\ldots,-1$ in the left half-plane. Set

$$P_k = \frac{1}{2\pi i}\int_{|z-z_k|=r} (zE+A)^{-1}E\,dz.$$

These operators are mutually disjoint projections and by Cauchy's theorem we have

$$P^+ = \sum_{k=1}^{m^+} P_k.$$

It can be seen e.g. by Cramer's rule that $(zE+A)^{-1}$ is a rational function which is bounded at ∞, hence admits a limit as $|z|\to\infty$. Therefore

$$z(zE+A)^{-1}E = I-(zE+A)^{-1}A$$

is bounded at ∞ as well and admits the limit

$$Q_0 = \lim_{z\to\infty} z(zE+A)^{-1}E,$$

which is a projection, too. We set $P_0 = I - Q_0$. Obviously, $Q_0\mathsf{x}=0$ for each $\mathsf{x}\in \mathsf{N}(E)$, and on the other hand, we have

$$EQ_0 = \lim_{z\to\infty} zE(zE+A)^{-1}E = \lim_{z\to\infty}(E-A(zE+A)^{-1}E) = E.$$

This implies that P_0 projects onto the kernel of E. Moreover,

$$\sum_k P_k = P_0 + \lim_{R\to\infty}\frac{1}{2\pi i}\int_{|z|=R}(zE+A)^{-1}E\,dz = P_0+Q_0 = I,$$

which also shows that $P_0P_k = P_kP_0 = 0$ for all $k\neq 0$. Linear algebra implies further that the dimension of the range of P_k is m_k, hence P^+ has dimension n. Since

$$\mathsf{x}^0 = \mathsf{x}(0) = \lim_{t\to 0+}\mathsf{x}(t) = \lim_{\mathbb{R}\ni z\to\infty} z\mathcal{L}\mathsf{x}(z) = \lim_{z\to\infty} z(zE+A)^{-1}E\mathsf{x}^0 = Q_0\mathsf{x}^0,$$

we must have $P_0\mathsf{x}^0=0$. It is not difficult to compute the projection P_0, it is given by

$$P_0\mathsf{x} = \frac{\mathsf{x}_4 + (i\xi|\mathsf{x}_1)}{\alpha_0}\begin{bmatrix} 0 \\ A^{0^{-1}}\begin{bmatrix} 0 \\ -1\end{bmatrix} \\ 1\end{bmatrix},$$

where

$$\alpha_0 := \left(\begin{bmatrix} 0 \\ 1 \end{bmatrix} \Big| A^{0^{-1}} \begin{bmatrix} 0 \\ 1 \end{bmatrix} \right)$$

is nonzero by ellipticity. Observe that

$$P_0 \mathsf{x}^0 = 0 \quad \Leftrightarrow \quad \mathsf{x}_4^0 + (i\xi|\mathsf{x}_1^0) = 0.$$

For later purposes we also compute the projection $P_1^{\pm}$ corresponding to the small eigenvalue $z_1^{\pm}(\xi) \sim \pm|\xi|$ for small ξ. The analysis of $z_1^{\pm}$ given above shows that an eigenvector is given by

$$e_1^{\pm} = [(A(z_1^{\pm}) - \lambda)^{-1} \begin{bmatrix} i\xi \\ z_1^{\pm} \end{bmatrix}, z_1^{\pm}(A(z_1^{\pm}) - \lambda)^{-1} \begin{bmatrix} i\xi \\ z_1^{\pm} \end{bmatrix}, 1]^{\mathsf{T}} \sim [\frac{1}{\lambda} \begin{bmatrix} -i\xi \\ \mp|\xi| \end{bmatrix}, 0, 1]^{\mathsf{T}}.$$

For a dual eigenvector we get similarly

$$e_1^{*\pm} = [(z_1^{\pm} A^0 + A^1)^{\mathsf{T}} (A(z_1^{\pm})^{\mathsf{T}} - \lambda)^{-1} \begin{bmatrix} i\xi \\ z_1^{\pm} \end{bmatrix}, (A(z_1^{\pm})^{\mathsf{T}} - \lambda)^{-1} \begin{bmatrix} i\xi \\ z_1^{\pm} \end{bmatrix}, -1]^{\mathsf{T}},$$

hence

$$e_1^{*\pm} \sim [0, \frac{1}{\lambda} \begin{bmatrix} -i\xi \\ \mp|\xi| \end{bmatrix}, -1]^{\mathsf{T}}.$$

The projections are then $P_1^{\pm}\mathsf{x} = \frac{(e_1^{*\pm}|E\mathsf{x})}{(e_1^{*\pm}|Ee_1^{\pm})} e_1^{\pm}$. Note that $(e_1^{*\pm}|Ee_1^{\pm}) \sim \pm 2|\xi|/\lambda$ for small ξ, and the asymptotics of $z_1^{\pm}$, $e_1^{\pm}$ and $e_1^{*\pm}$ do not depend on the coefficients a_{ij}^{kl}. Note also that

$$P_1^{+}\mathsf{x}^0 = 0 \quad \Leftrightarrow \quad (e_1^{*+}|E\mathsf{x}^0) = 0,$$

which asymptotically yields the condition

$$\mathsf{x}_5^0 - \frac{\lambda}{|\xi|}\mathsf{x}_2^0 \sim \left(\begin{bmatrix} i\xi/|\xi| \\ 1 \end{bmatrix} \Big| A^0 \begin{bmatrix} \mathsf{x}_3^0 \\ \mathsf{x}_4^0 \end{bmatrix} \right).$$

2.4 The Boundary Value Problem for the DAE-System

To determine the initial value x^0 we therefore have to solve the linear system

$$B\mathsf{x}^0 = \hat{h}, \quad P^+\mathsf{x}^0 = 0, \quad P_0\mathsf{x}^0 = 0. \tag{7.22}$$

The Lopatinskii-Shapiro condition is equivalent to the uniqueness of the solution x^0 of this system, for $\xi \neq 0$. To see that it is solvable for each $\hat{h} \in \mathbb{C}^n$, observe that the kernel N of $P^+ + P_0$ has dimension n. $B : N \to \mathbb{C}^n$ is injective, hence the rank theorem implies that it is also surjective. Thus there is a linear operator $M_0(\lambda, \xi)$ such that $\mathsf{x}^0 = M_0(\lambda, \xi)\hat{h}$ gives the unique solution of (7.22). We have the explicit representation

$$\mathsf{x}^0 = (B^*B + (P^+)^*P^+ + P_0^*P_0)^{-1}B^*\hat{h},$$

which shows that $M_0(\lambda,\xi)$ is holomorphic as B, P_0, and P^+ have this property. By homogeneity, λ can even be taken from a sector Σ_ϕ for some $\phi > \pi/2$, but $\xi \neq 0$, in general.

Therefore, we have to look more closely at $\xi = 0$. Note that the projections $P_1^\pm$ are not holomorphic at $\xi = 0$. However, $P_1^0 := P_1^+ + P_1^-$ does have this property. A simple calculation shows that for $\xi = 0$ we have

$$P_1^0\mathsf{x} = \mathsf{x}_2 \begin{bmatrix} 0 \\ 1 \\ 0 \\ 0 \\ 0 \end{bmatrix} + (\mathsf{x}_5 - A_{21}^0\mathsf{x}_3 - A_{22}^0\mathsf{x}_4) \begin{bmatrix} 0 \\ 0 \\ 0 \\ 0 \\ 1 \end{bmatrix}.$$

Therefore, it is convenient to decompose $\mathsf{x}^0 = \mathsf{y}^0 + \alpha e_1^-$, with $\alpha \in \mathbb{C}$ and $P_1^-\mathsf{y}^0 = 0$. Setting $P = P_0 + P^+ + P_1^-$, we therefore have to solve the system

$$B\mathsf{y}^0 + \alpha B e_1^- = \hat{h}, \quad P\mathsf{y}^0 = 0.$$

From $P\mathsf{y}^0 = 0$ we obtain $\mathsf{y}_2^0 = 0$, $\mathsf{y}_4^0 = 0$ and $\mathsf{y}_5^0 = A_{21}^0\mathsf{y}_3^0$. Solving the system $(zE + A)\mathsf{x} = E\mathsf{x}^0$, we obtain with $e_1^- = [0,0,0,0,1]^\mathsf{T}$ and $\mathsf{x}_2^0 = \mathsf{y}_2^0 = \mathsf{x}_4^0 = \mathsf{y}_4^0 = 0$ the relations $\mathsf{x}_2 = \mathsf{x}_4 = 0$ and

$$(z^2 A_{11}^0 - \lambda)\mathsf{x}_1 = A_{11}^0(\mathsf{x}_3^0 + z\mathsf{x}_1^0), \quad \mathsf{x}_3 = z\mathsf{x}_1 - \mathsf{x}_1^0, \quad \mathsf{x}_5 = A_{21}^0\mathsf{x}_3 + \alpha/z,$$

since $\mathsf{x}_5^0 - A_{21}^0\mathsf{x}_3^0 = \alpha + \mathsf{y}_5^0 - A_{21}^0\mathsf{y}_3^0 = \alpha$. By strong ellipticity, A_{11}^0 is invertible and has spectrum in the open right half-plane. Hence we may compute further

$$\begin{aligned}\mathsf{x}_1(z) &= \frac{1}{2}(z + \sqrt{\lambda}(A_{11}^0)^{-1/2})^{-1}(\mathsf{y}_1^0 + (A_{11}^0)^{1/2}\mathsf{y}_3^0/\sqrt{\lambda}) \\ &\quad + \frac{1}{2}(z - \sqrt{\lambda}(A_{11}^0)^{-1/2})^{-1}(\mathsf{y}_1^0 - (A_{11}^0)^{1/2}\mathsf{y}_3^0/\sqrt{\lambda}).\end{aligned}$$

Now, $\mathsf{x}_1(z)$ must be holomorphic in the right half-plane, which means that necessarily we have $\mathsf{y}_3^0 = -\sqrt{\lambda}(A_{11}^0)^{-1/2}\mathsf{y}_1^0$. The boundary condition yields in the Dirichlet and outflow cases $\mathsf{x}_1^0 = \mathsf{y}_1^0 = \hat{h}_1$, and in the slip or Neumann case $\mathsf{x}_3^0 = \mathsf{y}_3^0 = (A_{11}^0)^{-1}\hat{h}_3$. Note that in the outflow and Neumann cases, $\alpha = -\hat{h}_4$ is uniquely determined, in contrast to the Dirichlet or slip case, where α is not unique. In fact, the function $\alpha(\lambda,\xi)$ is discontinuous at $\xi = 0$ for the latter, but holomorphic in the outflow and Neumann case.

Now, for $\xi \neq 0$ small, we may parameterize the kernel of P by a holomorphic map

$$\mathsf{y} \mapsto R(\lambda,\xi)\mathsf{y} := [\mathsf{y}, 0, -\sqrt{\lambda}(A_{11}^0)^{-1/2}\mathsf{y}, 0, -A_{21}^0\sqrt{\lambda}(A_{11}^0)^{-1/2}\mathsf{y}]^\mathsf{T} + R^1(\lambda,\xi)\mathsf{y},$$

where $R^1 = O(|\xi|)$ near $\xi = 0$, with $\mathsf{y} \in \mathbb{C}^{n-1}$. Then we have to solve the equation $BR\mathsf{y} + \alpha Be_1^- = \hat{h}$. For the outflow and Neumann cases it then follows that y

and α are uniquely determined and holomorphic near $\xi = 0$, hence $M_0(\lambda, \xi)$ is holomorphic also at $\xi = 0$.

However, in the other cases things are more involved. We begin with the Dirichlet case. Then the system becomes

$$\mathsf{y} - i\alpha\xi/\lambda = \hat{h}_1 + O(|\xi|)\mathsf{y} + O(|\xi|^2)\alpha, \quad \alpha|\xi|/\lambda = \hat{h}_2 + O(|\xi|)\mathsf{y} + O(|\xi|^2)\alpha,$$

hence

$$\alpha \sim \lambda\hat{h}_2/|\xi|, \quad \mathsf{y} \sim \hat{h}_1 + \frac{i\xi}{|\xi|}\hat{h}_2.$$

In the case of slip conditions we have similarly

$$-\sqrt{\lambda}A_{11}^{0\;1/2}\mathsf{y} - \alpha A_{11}^0 i\xi/\sqrt{\lambda} = \hat{h}_3 + O(|\xi|)\mathsf{y} + O(|\xi|^2)\alpha,$$
$$\alpha|\xi|/\lambda = \hat{h}_2 + O(|\xi|)\mathsf{y} + O(|\xi|^2)\alpha,$$

and so

$$\alpha \sim \lambda\hat{h}_2/|\xi|, \quad \mathsf{y} \sim -A_{11}^{0\;1/2}\Big(A_{11}^{0\;-1}\hat{h}_3 + \frac{i\xi}{|\xi|}\hat{h}_2\Big)/\sqrt{\lambda}.$$

Thus there are holomorphic functions $M_{00}(\lambda, \xi)$ and $\alpha_0(\lambda, \xi)$ such that

$$M_0(\lambda, \xi)\hat{h} = M_{00}(\lambda, \xi)\hat{h} + \Big[\frac{\lambda}{|\xi|}\hat{h}_2 + (\alpha_0(\lambda, \xi)|\hat{h})\Big]e_1^-,$$

where $\hat{h}_2$ denotes the normal component of $\hat{u}$ at the boundary $\partial\mathbb{R}^n_+ = \mathbb{R}^{n-1}$.

2.5 Harmonic Analysis

We may now write the following representation of the solution $\mathsf{x}(y) = \mathsf{x}(y, \lambda, \xi)$ of (7.21).

$$\mathsf{x}(y, \lambda, \xi) = \frac{1}{2\pi i}\int_{\Gamma_-} e^{zy}(zE + A(\lambda, \xi))^{-1}EM_0(\lambda, \xi)\hat{h}(\lambda, \xi)\, dz, \tag{7.23}$$

where Γ_- denotes a closed simple contour in the open left half-plane surrounding the zeros of $p(z) = p(z, \lambda, \xi)$ in the left half-plane. Employing residue calculus this representation can be rewritten as

$$\mathsf{x}(y, \lambda, \xi) = \sum_{\mathrm{Re} z_k < 0} \mathrm{Res}_{z=z_k(\lambda,\xi)}[e^{zy}(zE + A(\lambda, \xi))^{-1}E]M_0(\lambda, \xi)\hat{h}(\lambda, \xi),$$

hence it is an exponential polynomial in y.

Note that the zeros z_k of $p(z) = p(z, \lambda, \xi)$ depend on ξ and λ, hence the integration path in (7.23) cannot be chosen independently of ξ and λ. To remove this defect a scaling argument will help. With $\rho = \sqrt{\lambda + |\xi|^2}$, the standard parabolic symbol, and $\sigma = \lambda/\rho^2$, $\zeta = \xi/\rho$, the pair (σ, ζ) belongs to a compact subset of $\mathbb{C}^n \setminus \{0\}$. Replace $\hat{\pi}(y)$ by $\hat{\pi}(\rho y)/\rho$, $\mathsf{x}(y)$ by $\mathsf{x}(\rho y)$, Neumann data $\hat{h}_k$ by $\hat{h}_k/\rho$, and leave Dirichlet data unchanged. Then homogeneity of $\mathcal{A}$ and $\mathcal{B}$ yield the modified representation formula

$$\mathsf{x}(y,\lambda,\xi) = \frac{1}{2\pi i}\int_{\Gamma_-} e^{\rho z y}(zE + A(\sigma,\zeta))^{-1} E M_0(\sigma,\zeta)\hat{h}(\lambda,\xi)\,dz. \tag{7.24}$$

Since the poles of $(zE+A(\sigma,\zeta))^{-1}$ stay in a compact set in the left half-plane, we may now choose the contour Γ_- independently of (σ,ζ). This argument parallels the scaling employed in Section 6.2 for the parabolic case.

Observe that the scaling of h induces

$$h \in {}_0\mathbb{F}_{1\mu} := {}_0F^{1-1/2q}_{pq,\mu}(J; L_q(\mathbb{R}^{n-1};\mathbb{C}^n)) \cap L_{p,\mu}(J; B^{2-1/q}_{qq}(\mathbb{R}^{n-1};\mathbb{C}^n)),$$

which is independent of the choice of the boundary conditions. Let

$$L := (\partial_t + \omega - \Delta_x)^{1/2}, \quad \mathsf{D}(L) = {}_0H^{1/2}_{p,\mu}(J; L_q(\mathbb{R}^{n-1};\mathbb{C}^n)) \cap L_{p,\mu}(J; H^1_q(\mathbb{R}^{n-1};\mathbb{C}^n)).$$

Then by Lemma 6.2.4 with $m = 1$, $h \in Y$ implies $\hat{v}(y) := L^2 e^{-L\cdot} h \in \mathbb{E}_{0\mu}$. The symbol of L is $\sqrt{\lambda + |\xi|^2}$ which is precisely ρ. By means of the identity

$$\hat{h} = \int_0^\infty 2\rho e^{-2\rho\bar{y}}\hat{h}\,d\bar{y} = \frac{2}{\rho}\int_0^\infty e^{-\rho\bar{y}}\hat{v}(\bar{y})\,d\bar{y},$$

we may rewrite the representation of $\mathsf{x}(y)$ in the following way.

$$\mathsf{x}(y,\lambda,\xi) = \mathrm{diag}\left[\frac{1}{\rho^2},\frac{1}{\rho^2},\frac{1}{\rho^2},\frac{1}{\rho^2},\frac{1}{\rho|\xi|}\right]\int_0^\infty \hat{k}(y,\bar{y},\lambda,\xi)\hat{v}(\bar{y},\lambda,\xi)\,d\bar{y}, \tag{7.25}$$

where the Fourier-Laplace transform of k is given by

$$\hat{k}(y,\bar{y},\lambda,\xi) = \frac{1}{i\pi}\int_{\Gamma_-} e^{\rho(yz-\bar{y})} D(\rho,|\xi|)(zE + A(\sigma,\zeta))^{-1} E M_0(\sigma,\zeta)\,dz, \tag{7.26}$$

where $D(\rho,|\xi|) = \mathrm{diag}[\rho,\rho,\rho,\rho,|\xi|]$.

It remains to be shown that the integral operator $K(\lambda)$ with operator-valued kernel $k(y,\bar{y},\lambda,D_x)$ is $\mathcal{R}$-bounded from $L_q(\mathbb{R}_+; L_q(\mathbb{R}^{n-1};\mathbb{C}^n))$ to $L_q(\mathbb{R}_+; L_q(\mathbb{R}^{n-1};\mathbb{C}^{2n+1})$, where the symbol of $K(y,\bar{y},\lambda,D_x)$ is $\hat{k}(y,\bar{y},\lambda,\xi)$ from (7.26). This will imply that u belongs to the maximal regularity space, and the remaining regularity statements concerning the pressure π follow from the equations.

2.6 Large Frequencies

However, due to the presence of the small eigenvalues $z_1^{\pm}(\xi)$ introduced above, there are difficulties at $\zeta = 0$. We have to deal with the cases $|\zeta| \le \eta$ and $|\zeta| > \eta$ for some small $\eta > 0$ separately. For this purpose we introduce a cut-off function

$\chi(|\zeta|^2)$, where χ belongs to C^∞, is 1 in $B(0,\eta)$, 0 outside of $B(0,2\eta)$ and between 0 and 1 elsewhere. Then we may decompose $\hat{k}(y,\bar{y},\lambda,\xi)$ as $\hat{k}=\hat{k}_S+\hat{k}_R$, where

$$\hat{k}_R(y,\bar{y},\lambda,\xi)=\frac{1}{2\pi i}\int_{\Gamma_-}(1-\chi(\zeta))D(\rho,|\xi|)e^{\rho(zy-\bar{y})}(zE+A(\sigma,\zeta))^{-1}EM_0(\sigma,\zeta)\,dz. \tag{7.27}$$

Let us first deal with $\hat{k}_R$ and invert the Fourier transform via Mikhlin's theorem. Since Γ_- is compact and contained in the open left half-plane, for $|\zeta|>\eta$, (σ,ζ) runs through a compact subset of $\mathbb{C}^n$, and

$$\operatorname{Re}\rho\leq|\rho|\leq c_\phi\operatorname{Re}\rho,$$

we obtain

$$|\hat{k}_R(y,\bar{y},\lambda,\xi)|\leq C|\rho|e^{-c|\rho|(y+\bar{y})}\leq\frac{C}{y+\bar{y}},\quad y,\bar{y}>0,$$

where $C,c>0$ are independent of $y,\bar{y}$, λ and ξ. This is already sufficient in case $p=2$, by Plancherel's theorem. For the case of general $p\in(1,\infty)$, note first that

$$|\xi||\frac{1}{\rho}\partial_{\xi_k}\rho|=|\xi||\xi_k/\rho^2|\leq|\xi|^2/\rho^2\leq1,$$

and similarly we have by induction $|\xi|^{|\alpha|}|D_\xi^\alpha\rho|\leq M_\alpha$, for each multiindex $\alpha\in\mathbb{N}_0^{n-1}$. Next,

$$|\xi||\partial_{\xi_k}\zeta_j|=|\xi||\delta_{kj}/\rho-\zeta_j\partial_{\xi_k}\rho/\rho^2|\leq M_1,$$

and similarly for higher derivatives, by induction. The relation $\sigma=1-|\xi|^2/\rho^2$ shows that also $|\xi|^{|\alpha|}|D_\xi^\alpha\sigma|$ is uniformly bounded for each α. Next

$$|\xi||\partial_{\xi_k}e^{\rho(yz-\bar{y})}|\leq|\xi||\partial_{\xi_k}\rho/\rho^2||\rho^2(yz-\bar{y})e^{\rho(yz-\bar{y})}|\leq C|\rho|e^{-c|\rho|(y+\bar{y})}\leq\frac{C}{y+\bar{y}},$$

and similarly by induction also for all higher derivatives. Therefore we may conclude that

$$|\xi|^{|\alpha|}|D_\xi^\alpha\hat{k}_R(y,\bar{y},\lambda,\xi)|\leq\frac{M_\alpha}{y+\bar{y}},\quad y,\bar{y}>0,$$

for each multi-index α, where M_α is independent of $y,\bar{y}$, and of λ and ξ.

2.7 Small Frequencies

Now we deal with the other part of $\hat{k}$. Since we have enough information about the small eigenvalue $z_1(\xi)$ we may use residue calculus to decompose $\hat{k}_S=\hat{k}_{S0}+\hat{k}_{S1}$, where

$$\hat{k}_{S1}(y,\bar{y},\lambda,\xi)=\frac{1}{i\pi}\int_{\Gamma_-}\chi(\zeta)e^{\rho(yz-\bar{y})}D(\rho,|\xi|)(zE+A(\sigma,\zeta))^{-1}E(I-P_1^-)M_0(\sigma,\zeta)dz,$$

with a fixed contour Γ_- contained in the open left half-plane. The part $\hat{k}_{S1}$ can then be treated as above.

The essential part is $\hat{k}_{S0}$, which is given by

$$\hat{k}_{S0}(y,\bar{y},\lambda,\xi) = \chi(\zeta)e^{\rho(z_1^-(\sigma,\zeta)y-\bar{y})}D(\rho,|\xi|)P_1^-(\sigma,\zeta)M_0(\sigma,\zeta).$$

Using the decomposition $x^0 = y^0 + \alpha e_1^-$ as above, this yields

$$\hat{k}_{S0}(y,\bar{y},\lambda,\xi) = \chi(\zeta)|\xi|e^{\rho(z_1^-(\sigma,\zeta)y-\bar{y})}D(\rho/|\xi|,1)e_1^-(\lambda,\xi)\otimes\alpha(\lambda,\xi).$$

In the outflow and Neumann cases, α is holomorphic and

$$D(\rho/|\xi|,1)e_1^-(\lambda,\xi) = [0,0,-i\xi^{\mathsf{T}}\rho/\lambda,-|\xi|\rho/\lambda,1]^{\mathsf{T}}$$

is bounded and satisfies the Mikhlin condition. Since $z_1^- \sim -|\xi|$ we obtain as above an estimate of the form

$$|\xi|^{|\alpha|}|D_\xi^\alpha \hat{k}_{S0}(y,\bar{y},\lambda,\xi)| \le \frac{M_\alpha}{y+\bar{y}},$$

where M_α is independent of $y,\bar{y}$, ξ and λ.

The argument is more involved in the case of Dirichlet or slip conditions. It is here where the extra time regularity of the normal velocity h_2 comes in. As shown above, α decomposes as

$$\alpha(\lambda,\xi) = \alpha_0(\lambda,\xi) + \frac{\lambda}{|\xi|}\begin{bmatrix}0\\1\end{bmatrix},$$

where $\alpha_0(\lambda,\xi)$ is holomorphic. Since the term containing α_0 can be treated as before, we concentrate on the extra term. This yields the kernel k_{S00}, defined by

$$\hat{k}_{S00}(y,\bar{y},\lambda,\xi) = \chi(\zeta)|\xi|e^{\rho(z_1^-(\sigma,\zeta)y-\bar{y})}D(\rho/|\xi|,1)e_1^-(\lambda,\xi)\frac{\lambda}{|\xi|}\begin{bmatrix}0\\1\end{bmatrix}.$$

Since by assumption $\hat{h}_2$ is the Fourier-Laplace transform of a function of class ${}_0H^1_{p,\mu}(\mathbb{R}_+;\dot{W}_q^{-1/q}(\mathbb{R}^{n-1}))$, we see that $\lambda\hat{h}_2/|\xi|$ is the Fourier-Laplace transform of a function in $L_{p,\mu}(\mathbb{R}_+;\dot{W}_q^{1-1/q}(\mathbb{R}^{n-1}))$. Thus we obtain $g_0 \in L_{p,\mu}(\mathbb{R}_+;L_q(\mathbb{R}^n_+))$ such that

$$\hat{g}_0(\bar{y},\lambda,\xi) = |\xi|e^{-|\xi|\bar{y}}\lambda\hat{h}_2(\lambda,\xi)/|\xi|.$$

Writing

$$(\lambda/|\xi|)\hat{h}_2 = 2\int_0^\infty |\xi|e^{-2|\xi|\bar{y}}\lambda\hat{h}_2/|\xi|\,d\bar{y} = 2\int_0^\infty e^{-|\xi|\bar{y}}g_0(\bar{y})\,d\bar{y},$$

we have

$$|\xi|e^{\rho z_1^- y}D(\rho/\xi,1)e_1^-\lambda\hat{h}_2/|\xi| = \int_0^\infty |\xi|e^{\rho z_1^- y-|\xi|\bar{y}}D(\rho/\xi,1)e_1^-\hat{g}_0(\bar{y},\lambda,\xi)\,d\bar{y}$$

and the kernel of this representation can be estimated as before.

2.8 End of the Proof

Summarizing, we have obtained kernels $k(y,\bar{y},\lambda,\xi)\in\mathcal{B}(\mathbb{C}^n)$ such that the family $\{k(y,\bar{y},\lambda,\xi):\xi\in\mathbb{R}^{n-1},y,\bar{y}>0,\lambda\in\Sigma_\phi\}$ satisfies the uniform Mikhlin condition

$$|\xi|^{|\alpha|}|D_\xi^\alpha \hat{k}(y,\bar{y},\lambda,\xi)|\leq \frac{M_\alpha}{y+\bar{y}},\quad y,\bar{y}>0,\ \xi\in\mathbb{R}^{n-1},\ \lambda\in\Sigma_\phi.$$

The Lizorkin Fourier multiplier theorem, Theorem 4.3.9, implies that the family of operators

$$\{(y+\bar{y})k(y,\bar{y},\lambda,D_x):\ y,\bar{y}>0,\ \lambda\in\Sigma_\phi\}\subset\mathcal{B}(L_q(\mathbb{R}^{n-1};\mathbb{C}^n);L_q(\mathbb{R}^{n-1};\mathbb{C}^{2n+1}))$$

is $\mathcal{R}$-bounded. As the Hilbert transform with kernel $k_0(y,\bar{y})=1/(y+\bar{y})$ is bounded on $L_q(\mathbb{R}_+)$, Proposition 4.1.5 shows that the family of integral operators $\{K(\lambda):\ \lambda\in\Sigma_\phi\}\subset\mathcal{B}(L_q(\mathbb{R}^n_+;\mathbb{C}^n);L_q(\mathbb{R}^n_+;\mathbb{C}^{2n+1}))$ with kernels $k(\lambda,y,\bar{y})$ is also $\mathcal{R}$-bounded, hence by canonical extension also in $\mathcal{B}(L_{p,\mu}(\mathbb{R}_+;L_q(\mathbb{R}^n_+;\mathbb{C}^n)),L_{p,\mu}(\mathbb{R}_+;L_q(\mathbb{R}^n_+;\mathbb{C}^{2n+1})))$. In addition, this operator family is holomorphic on Σ_ϕ, and as $L_q(\mathbb{R}^n_+)$ is of class $\mathcal{HT}$, the Kalton-Weis theorem, Theorem 4.5.6, implies that $K(\partial_t+\omega)$ is bounded in $\mathbb{E}_{0\mu}$. This completes the proof of Theorem 7.2.1.

2.9 Estimates for the Solution

As in the whole space case it is useful to have estimates for the solution in terms of the data which are uniform in the parameter $\omega\geq\omega_0>0$. These follow directly from the proof of Theorem 7.2.1 but are more elaborate than those for the case $\Omega=\mathbb{R}^n$, as they depend on the boundary conditions in question. For this purpose we fix some function spaces as follows.

$$\begin{aligned}
&\mathbb{E}_{0\mu}:=L_{p,\mu}(\mathbb{R}_+;L_q(\mathbb{R}^n_+)^n),\qquad \mathbb{E}_{1\mu}:=H^1_{p,\mu}(\mathbb{R}_+;L_q(\mathbb{R}^n_+)^n)\cap L_{p,\mu}(\mathbb{R}_+;H^2_q(\mathbb{R}^n_+)^n),\\
&\mathbb{G}_{0\mu}:=L_{p,\mu}(\mathbb{R}_+;\dot{H}^{-1}_q(\mathbb{R}^n_+)),\quad \mathbb{G}_{1\mu}:=H^1_{p,\mu}(\mathbb{R}_+;\dot{H}^{-1}_q(\mathbb{R}^n_+))\cap L_{p,\mu}(\mathbb{R}_+;H^1_q(\mathbb{R}^n_+)),\\
&\mathbb{G}^0_\mu:=L_{p,\mu}(\mathbb{R}_+;{}_0\dot{H}^{-1}_q(\mathbb{R}^n_+)),\quad \mathbb{G}^1_\mu:=H^1_{p,\mu}(\mathbb{R}_+;{}_0\dot{H}^{-1}_q(\mathbb{R}^n_+)),\\
&\mathbb{F}_{0\mu}:=F^{1/2-1/2q}_{pq,\mu}(\mathbb{R}_+;L_q(\mathbb{R}^{n-1}))\cap L_{p,\mu}(\mathbb{R}_+;B^{1-1/q}_{qq}(\mathbb{R}^{n-1})),\\
&\mathbb{F}_{1\mu}:=F^{1-1/2q}_{pq,\mu}(\mathbb{R}_+;L_q(\mathbb{R}^{n-1}))\cap L_{p,\mu}(\mathbb{R}_+;B^{2-1/q}_{qq}(\mathbb{R}^{n-1})),
\end{aligned}$$

and $X_{\gamma,\mu}=B^{2(\mu-1/p)}_{qp}(\mathbb{R}^n_+;\mathbb{C}^n)$. The estimates read as follows. For each $\omega_0>0$ there is a constant $C>0$ such that for all $\omega\geq\omega_0$ and all data subject to the corresponding compatibility conditions, the solution (u,π) satisfies

(i) no-slip

$$\begin{aligned}\omega|u|_{\mathbb{E}_{0\mu}}+|u|_{\mathbb{E}_{1\mu}}+|\nabla\pi|_{\mathbb{E}_{0\mu}}\leq C\{&|u_0|_{X_{\gamma,\mu}}+|f|_{\mathbb{E}_{0\mu}}+(|g|_{\mathbb{G}_{1\mu}}+\omega|g|_{\mathbb{G}_{0\mu}})\\&+(|h_0|_{\mathbb{F}^n_{1\mu}}+\omega|e^{-L_\omega y}h_0|_{\mathbb{E}_{0\mu}})+(|(g,h_{0\nu})|_{\mathbb{G}^1_\mu}+\omega|(g,h_{0\nu})|_{\mathbb{G}^0_\mu})\}.\end{aligned}\tag{7.28}$$

(ii) pure slip

$$\begin{aligned}\omega|u|_{\mathbb{E}_{0\mu}} + |u|_{\mathbb{E}_{1\mu}} + |\nabla\pi|_{\mathbb{E}_{0\mu}} \le C\{&|u_0|_{X_{\gamma,\mu}} + |f|_{\mathbb{E}_{0\mu}} + (|g|_{\mathbb{G}_{1\mu}} + \omega|g|_{\mathbb{G}_{0\mu}}) \\ &+ (|h_{0\nu}|_{\mathbb{F}_{1\mu}} + \omega|e^{-L_\omega y}h_{0\nu}|_{\mathbb{E}_{0\mu}}) + (|h_\Sigma|_{\mathbb{F}^n_{0\mu}} + \omega^{1/2}|e^{-L_\omega y}h_\Sigma|_{\mathbb{E}_{0\mu}}) \\ &+ (|(g,h_{0\nu})|_{\mathbb{G}^1_\mu} + \omega|(g,h_{0\nu})|_{\mathbb{G}^0_\mu})\}.\end{aligned} \tag{7.29}$$

(iii) outflow

$$\begin{aligned}\omega|u|_{\mathbb{E}_{0\mu}} + |u|_{\mathbb{E}_{1\mu}} + |\nabla\pi|_{\mathbb{E}_{0\mu}} \le C\{&|u_0|_{X_{\gamma,\mu}} + |f|_{\mathbb{E}_{0\mu}} + (|g|_{\mathbb{G}_{1\mu}} + \omega|g|_{\mathbb{G}_{0\mu}}) \\ &+ (|h_\nu|_{\mathbb{F}_{0\mu}} + \omega^{1/2}|e^{-L_\omega y}h_\nu|_{\mathbb{E}_{0\mu}}) + (|h_{0\Sigma}|_{\mathbb{F}^n_{1\mu}} + \omega|e^{-L_\omega y}h_{0\Sigma}|_{\mathbb{E}_{0\mu}})\}\end{aligned} \tag{7.30}$$

(iv) free

$$\begin{aligned}\omega|u|_{\mathbb{E}_{0\mu}} + |u|_{\mathbb{E}_{1\mu}} + |\nabla\pi|_{\mathbb{E}_{0\mu}} \le C\{&|u_0|_{X_{\gamma,\mu}} + |f|_{\mathbb{E}_{0\mu}} + (|g|_{\mathbb{G}_{1\mu}} + \omega|g|_{\mathbb{G}_{0\mu}}) \\ &+ (|h|_{\mathbb{F}^n_{0\mu}} + \omega^{1/2}|e^{-L_\omega y}h|_{\mathbb{E}_{0\mu}})\}\end{aligned} \tag{7.31}$$

We recall that $L_\omega = (\partial_t + \omega - \Delta)^{-1/2}$. As in the previous chapter, we may estimate

$$|e^{-L_\omega y}h|_{\mathbb{E}_{0\mu}} \le \omega^{-1/2q}|h|_{L_{p,\mu}(L_q)},$$

which has the advantage that only norms of the boundary data are involved, but slightly loosing sharpness. For perturbations of highest order we have to use the sharp estimates, but for localization the weaker version is sufficient.

7.3 General Domains

In this section we state and prove the main result of this chapter, which is maximal $L_{p,\mu} - L_q$-regularity of the generalized Stokes problem on interior and exterior domains. To state the result, let $\Omega \subset \mathbb{R}^n$ be a domain with compact boundary $\Sigma := \partial\Omega$ of class C^{3-}, and assume that the coefficients a^{kl} of the normally strongly elliptic differential operator $\mathcal{A}(x,D) = \sum_{k,l=1}^n D_k a^{kl}(x) D_l$ belong to $C^{1-}(\bar{\Omega}; \mathcal{B}(\mathbb{C}^n))$. Consider the Stokes problem

$$\begin{aligned}(\partial_t + \omega)u + \mathcal{A}(x,D)u + \nabla\pi &= f(t,x) \quad \text{in } \Omega,\\ \operatorname{div} u &= g(t,x) \quad \text{in } \Omega,\\ u(0,x) &= u_0(x) \quad \text{in } \Omega,\end{aligned} \tag{7.32}$$

for $t > 0$, with the following types of natural boundary conditions
(i) no-slip

$$u = h_0 \quad \text{on } \Sigma_d;$$

(ii) pure slip

$$u \cdot \nu = h_{0\nu}, \quad \mathcal{P}_\Sigma \nu_k a^{kl}(x) D_l u = h_\Sigma \quad \text{on } \Sigma_s;$$

(iii) outflow

$$\mathcal{P}_\Sigma u = h_{0\Sigma}, \quad (\nu_k a^{kl}(x) D_l u|\nu) + i\pi = h_\nu \quad \text{on } \Sigma_o;$$

(iv) free

$$\nu_k a^{kl}(x) D_l u + i\pi\nu = h \quad \text{on } \Sigma_n.$$

Here we assume that Σ decomposes disjointly into four parts, i.e.,

$$\Sigma = \Sigma_d \cup \Sigma_s \cup \Sigma_o \cup \Sigma_n,$$

where each set Σ_j is open and closed in Σ. Note that up to three of these sets may be empty. As before, $\mathcal{P}_\Sigma$ denotes the orthogonal projection onto the tangent bundle of Σ. By trace theory, the necessary conditions for solvability of this problems are the following conditions ($\mathbf{D}_\Omega$).

(a) $f \in L_{p,\mu}(\mathbb{R}_+; L_q(\Omega; \mathbb{C}^n))$, $u_0 \in B_{qp}^{2\mu-2/p}(\Omega; \mathbb{C}^n)$.

(b) $g \in H^1_{p,\mu}(\mathbb{R}_+; \dot{H}^{-1}_q(\Omega)) \cap L_{p,\mu}(\mathbb{R}_+; H^1_q(\Omega))$, $\operatorname{div} u_0 = g(0)$.

(d0) for no-slip (Dirichlet) boundary conditions:
$h_0 \in F_{pq,\mu}^{1-1/2q}(\mathbb{R}_+; L_q(\Sigma_d; \mathbb{C}^n)) \cap L_{p,\mu}(\mathbb{R}_+; B_{qq}^{2-1/q}(\Sigma_d; \mathbb{C}^n))$ and for $\mu > 3/2p$ in addition $h_0(0) = u_0$ on Σ_d.

(ds) for pure slip boundary conditions:
$h_{0\nu} \in F_{pq,\mu}^{1-1/2q}(\mathbb{R}_+; L_q(\Sigma_s)) \cap L_{p,\mu}(\mathbb{R}_+; B_{qq}^{2-1/q}(\Sigma_s))$;
$h_\Sigma \in F_{pq,\mu}^{1/2-1/2q}(\mathbb{R}_+; L_q(\Sigma_s; T\Sigma) \cap L_{p,\mu}(\mathbb{R}_+; B_{qq}^{1-1/q}(\Sigma_s; T\Sigma))$ and
$\mathcal{P}_\Sigma \nu_k a^{kl} D_l u_0 = h_\Sigma(0)$ for $\mu > 3/p$;

(do) for outflow boundary conditions:
$h_{0\Sigma} \in F_{pq,\mu}^{1-1/2q}(\mathbb{R}_+; L_q(\Sigma_o; T\Sigma)) \cap L_{p,\mu}(\mathbb{R}_+; B_{qq}^{2-1/q}(\Sigma_0; T\Sigma))$;
$h_\nu \in F_{pq,\mu}^{1/2-1/2q}(\mathbb{R}_+; L_q(\Sigma_o)) \cap L_{p,\mu}(\mathbb{R}_+; B_{qq}^{1-1/q}(\Sigma_o))$ and
$\mathcal{P}_\Sigma u_0 = h_{0\Sigma}(0)$ for $\mu > 3/2p$;

(dn) for free (Neumann) boundary conditions:
$h \in F_{pq,\mu}^{1/2-1/2q}(\mathbb{R}_+; L_q(\Sigma_n; \mathbb{C}^n)) \cap L_{p,\mu}(\mathbb{R}_+; B_{qq}^{1-1/q}(\Sigma_n; \mathbb{C}^n))$ and
$\mathcal{P}_\Sigma \nu_k a^{kl} D_l u_0 = P_\Sigma h(0)$ for $\mu > 3/p$.

In addition,

(e) $(g, h_{0\nu}) \in H^1_{p,\mu}(\mathbb{R}_+; \dot{H}^{-1}_{q,\Sigma_d \cup \Sigma_s}(\Omega))$ and $h_{0\nu}(0) = (\nu|u_0)$ on $\Sigma_d \cup \Sigma_s$.

After these preliminaries we can state the main result of this section.

Theorem 7.3.1. *Let $\Omega \subset \mathbb{R}^n$ be a domain with compact boundary $\Sigma := \partial\Omega$ of class C^{3-}, $1 < p,q < \infty$, $1 \geq \mu > 1/p$, $\mu \neq 3/2p, 3/p$, and assume that $\mathcal{A}(x,D) = \sum_{k,l=1}^n D_k a^{kl}(x) D_l$ is uniformly normally strongly elliptic with coefficients*

$$a^{kl} \in C^{1-}(\bar{\Omega}; \mathcal{B}(\mathbb{C}^n))) \cap C_l(\bar{\Omega}; \mathcal{B}(\mathbb{C}^n)).$$

Then there is $\omega_0 \in \mathbb{R}$ such that for each $\omega > \omega_0$, (7.32) with the boundary conditions explained above has maximal $L_{p,\mu} - L_q$-regularity in the following sense. There is a unique solution (u, π) of (7.32) in the class

$$u \in H^1_{p,\mu}(J; L_q(\Omega; \mathbb{C}^n)) \cap L_{p,\mu}(J; H^2_q(\Omega; \mathbb{C}^n)), \quad \pi \in L_{p,\mu}(J; \dot{H}^1_q(\Omega)),$$

satisfying the corresponding boundary condition, and in addition with

$$\pi \in F^{1/2-1/2q}_{pq,\mu}(J; L_q(\Sigma_o \cup \Sigma_n)),$$

if and only if the data (f, g, h_j, u_0) satisfy the conditions $(\mathbf{D}_\Omega)$. The solution u depends continuously on the data in the corresponding spaces.

Observe that the pressure π is unique for $\Sigma_o \cup \Sigma_n \neq \emptyset$, but otherwise only unique up to a constant.

By means of this result we can introduce the *generalized Stokes operator* for the four natural boundary conditions. For this, we employ the *Helmholtz-Weyl projection* on $L_q(\Omega; \mathbb{C}^n)$ w.r.t. the given decomposition of Σ, cf. Corollary 7.4.4 below. It is defined in the following way. Given $f \in L_q(\Omega; \mathbb{C}^n)$, solve the following weak mixed Dirichlet-Neumann problem according to Theorem 7.4.3.

$$\begin{aligned} \Delta\phi &= \operatorname{div} f && \text{in } \Omega, \\ \partial_\nu \phi &= f \cdot \nu && \text{on } \Sigma_d \cup \Sigma_s, \\ \phi &= 0 && \text{on } \Sigma_o \cup \Sigma_n, \end{aligned} \tag{7.33}$$

and set $P_{HW} f = f - \nabla\phi$. This is a bounded projection in $L_q(\Omega; \mathbb{C}^n)$ along the gradients onto $X_0 := \{u \in L_q(\Omega; \mathbb{C}^n) : \nabla^* u = 0\}$, where

$$\nabla : \dot{H}^1_{q',\Sigma_o \cup \Sigma_n} \to L_{q'}(\Omega; \mathbb{C}^n).$$

Thus $X_0 = \mathsf{N}(\nabla^*)$, which formally reads

$$X_0 = \{u \in L_q(\Omega; \mathbb{C}^n);\ \operatorname{div} u = 0 \text{ in } \Omega,\ u \cdot \nu = 0 \text{ on } \Sigma_d \cup \Sigma_s\}.$$

Then we define

$$Au := P_{HW} \mathcal{A}(x,D) u, \quad u \in \mathsf{D}(A),$$

with

$$\mathsf{D}(A) = \{u \in H^2_q(\Omega; \mathbb{C}^n) \cap X_0 : \mathcal{P}_\Sigma u = 0 \text{ on } \Sigma_d \cup \Sigma_o,\ \mathcal{P}_\Sigma \nu_k a^{kl} D_l u = 0 \text{ on } \Sigma_s \cup \Sigma_n\}.$$

Problem (7.32) with trivial data except for f and u_0 is equivalent to the abstract evolution equation

$$\dot{u} + \omega u + Au = f, \quad t > 0, \quad u(0) = u_0. \tag{7.34}$$

In fact, one implication is obvious. To obtain the reverse one, we have to recover the pressure π from the weak mixed Dirichlet-Neumann problem

$$\begin{aligned} \Delta\pi &= \operatorname{div}\big(f - \partial_t u - \omega u - \mathcal{A}(x,D)u\big) && \text{in } \Omega, \\ \partial_\nu \pi &= \big(f - \partial_t u - \omega u - \mathcal{A}(x,D)u\big)\cdot\nu && \text{on } \Sigma_d \cup \Sigma_s, \\ \pi &= (\nu\cdot a\nabla u|\nu) && \text{on } \Sigma_o \cup \Sigma_n. \end{aligned} \tag{7.35}$$

By Theorem 7.4.3 this problem admits a unique solution $\pi \in \dot{H}^1_q(\Omega)$. By Theorem 7.3.1 it follows that (7.34) has the property of maximal L_p-regularity, hence the generalized Stokes operators A is the negative generator of an analytic C_0-semigroup in X_0. More precisely we have

Theorem 7.3.2. *Let $\Omega \subset \mathbb{R}^n$ a domain with compact boundary $\Sigma := \partial\Omega$ of class C^{3-}, $1 < p,q < \infty$, $\mu \in (1/p, 1]$, and assume that $\mathcal{A}(x,D)$ is uniformly normally strongly elliptic with coefficients in the class*

$$a^{kl} \in C_b^{1-}(\bar{\Omega}; \mathcal{B}(\mathbb{C}^n))) \cap C_l(\bar{\Omega}; \mathcal{B}(\mathbb{C}^n)),$$

and let the Stokes operator A be defined as above in X_0.

Then (7.34) has maximal $L_{p,\mu}$–L_q-regularity; hence $\omega + A \in \mathcal{MR}_p(X_0)$, for any $\omega > \omega_0 := \mathsf{s}(-A)$.

Consequently the minimal ω_0 in Theorem 7.3.1 is the spectral bound $\mathsf{s}(-A)$. The next subsections are devoted to the proof of Theorem 7.3.1.

3.1 Half-Space: Variable Coefficients

We can easily extend Theorem 7.2.1 to the case of variable coefficients with small deviation from constant ones. To see this, let $\mathcal{A}(x,D) = \mathcal{A}_0(D) + \mathcal{A}_1(x,D)$, where $a_1^{kl} \in C_b^{1-}(\mathbb{R}^n_+; \mathcal{B}(\mathbb{C}^n))$ and

$$\sup\{|a_1^{kl}| : k,l = 1,\dots n, x \in \mathbb{R}^n\} \le \eta.$$

Let S denote the solution operator of the generalized Stokes problem (7.15) from Theorem 7.2.1 for $\mathcal{A}_0(D)$ with one of the boundary conditions under consideration, and let T be that of the perturbed problem. Then we obtain the identity

$$T = S + SBT, \quad \text{where} \quad B = \begin{bmatrix} -\mathcal{A}_1(x,D) & 0 \\ 0 & 0 \\ -\mathcal{B}_1(x,D) & 0 \end{bmatrix}.$$

Here $\mathcal{B}_1$ has the obvious meaning of the corresponding boundary operator generated by the perturbation $\mathcal{A}_1$. The norm of the first component of B as an operator from the maximal regularity space $\mathbb{E}_{1\mu}$ into $\mathbb{E}_{0\mu}$ is bounded by $C\eta$, where $C > 0$ denotes a constant independent of η, and the norm of its third component in the boundary space $\mathbb{F}_{0\mu}$ is estimated as in Section 6.2 to the result

$$|\mathcal{B}_1(\cdot, D)u|_{\mathbb{F}_{0\mu}} \le \eta |u|_{\mathbb{E}_{1\mu}} + C|a_1|_{C_b^{1-}} |u|^{\gamma}_{\mathbb{E}_{1\mu}} |u|^{1-\gamma}_{\mathbb{E}_{0\mu}},$$

for some $\gamma \in (0, 1]$.

Therefore, as in Section 6.2, a Neumann series argument shows that $T = (I - SB)^{-1}S$ in fact exists, is bounded as a map from the data space to the maximal regularity space as well, and the estimates from Section 7.2.9 remain valid. Let us state this as

Corollary 7.3.3. *The assertions of Theorem 7.2.1 as well as the estimates* (7.28), (7.29), (7.30), (7.31) *remain valid in the case of variable coefficients*

$$\mathcal{A}(x, D) = \mathcal{A}_0(D) + \mathcal{A}_1(x, D),$$

provided

$$a_1^{kl} \in C_b^{1-}(\mathbb{R}^n_+; \mathcal{B}(\mathbb{C}^n)) \quad \text{and} \quad \sup\{|a_1^{kl}(x)| : k, l = 1, \dots n, x \in \mathbb{R}^n\} \le \eta,$$

uniformly for $0 < \eta \le \eta_0$.

3.2 Bent Half-Spaces

In contrast to the parabolic case, we only are able to consider bent half-spaces which are tangentially close to a planar boundary. This comes from the fact that the Stokes-problem has no invariance properties except for the trivial ones, i.e., translation and rotation. As before, replacing the variable $x \in \mathbb{R}^n_+$ by (x, y), the bent half-space is defined by the mapping

$$\Phi(x, y) = [x, y + \phi(x)]^{\mathsf{T}}, \quad x \in \mathbb{R}^{n-1},\ y \ge 0.$$

Then $\Omega = \Phi(\mathbb{R}^n_+)$ and $\Gamma := \partial\Omega = \Phi(\mathbb{R}^{n-1} \times \{0\}) = \Phi(\Sigma)$, where $\Sigma = \mathbb{R}^{n-1} \times \{0\}$. For the normal of Γ we obtain

$$\nu_\Gamma(x, \phi(x)) = \beta(x)[\nabla\phi(x), -1]^{\mathsf{T}}, \quad \beta(x) = (1 + |\nabla\phi(x)|^2)^{-1/2},\ x \in \mathbb{R}^{n-1}.$$

We employ the transformation to the domain $\mathbb{R}^{n-1}$ by means of

$$u(\Phi(x, y)) = \bar{u}(x, y), \quad \pi(\Phi(x, y)) = \bar{\pi}(x, y), \quad x \in \mathbb{R}^{n-1},\ y \ge 0.$$

This implies the relations

$$\nabla\pi \circ \Phi(x, y) = (M\nabla)\bar{\pi}, \quad \nabla u \circ \Phi(x, y) = (M\nabla)\bar{u},$$

where

$$M(x, y) = (\partial\Phi)^{-1}(x, y) = \begin{bmatrix} I & -\nabla\phi \\ 0 & 1 \end{bmatrix} = M(x).$$

Similarly,

$$\operatorname{div} u \circ \Phi(x,y) = \operatorname{tr}(M(x)\nabla\bar{u}(x,y)).$$

In more explicit form, these identities read

$$\nabla\pi \circ \Phi = \nabla\bar{\pi} - \nabla\phi\partial_y\bar{\pi}, \quad \operatorname{div} u \circ \Phi = \operatorname{div} \bar{u} - \nabla\phi \cdot \partial_y \bar{u}.$$

Using these transformation laws, the problem on a bent half-space transforms to a problem on a half-space, which reads as follows, dropping the bars.

$$\begin{aligned} (\partial_t + \omega)u + \mathcal{A}^\Phi(D)u + \nabla\pi &= f + \mathcal{A}_1(D)u + B_1\pi && \text{in } \mathbb{R}^n_+,\\ \operatorname{div}\, u &= g + B_2 u && \text{in } \mathbb{R}^n_+,\\ u(0) &= u_0 && \text{in } \mathbb{R}^n_+, \end{aligned} \tag{7.36}$$

for $t > 0$. Here $\mathcal{A}^\Phi$ is defined by its coefficients $a_\Phi = \partial\Phi^{-1}(a\circ\Phi)\partial\Phi^{-\mathsf{T}}$, and $\mathcal{A}_1$ is lower order, but contains second-order derivatives of ϕ. The natural boundary conditions are perturbed in the following way.
(i) no-slip

$$u = h_0 \quad \text{on } \Sigma_d;$$

(ii) pure slip

$$u\cdot\nu_\Sigma = h_{0\nu}/\beta + B_3 u, \quad \mathcal{P}_\Sigma\nu_\Sigma a_\Phi(x)Du = \mathcal{P}_\Sigma h_\Sigma + B_4 u \quad \text{on } \Sigma_s;$$

(iii) outflow

$$\mathcal{P}_\Sigma u = \mathcal{P}_\Sigma h_{0\Sigma} + B_5 u, \quad (\nu_\Sigma a_\Phi(x)Du|\nu_\Sigma) + i\pi = h_\nu + B_6 u \quad \text{on } \Sigma_o;$$

(iv) free

$$\mathcal{P}_\Sigma\nu_\Sigma a_\Phi(x)Du = \mathcal{P}_\Sigma h + B_4 u, \quad (\nu_\Sigma a_\Phi(x)Du|\nu_\Sigma) + i\pi = h_\nu + B_6 u \quad \text{on } \Sigma_n.$$

Here the perturbation operators are defined as follows.

$$\begin{aligned} B_1\phi &= \nabla\phi\partial_y\pi, & B_2 u &= \nabla\phi\cdot\partial_y u,\\ B_3 u &= u\cdot(\nu_\Sigma - \nu_\Gamma/\beta), & B_4 u &= \mathcal{P}_\Sigma(\mathcal{P}_\Sigma - \mathcal{P}_\Gamma)\nu_\Sigma a_\Phi \nabla u,\\ B_5 u &= \mathcal{P}_\Sigma(\mathcal{P}_\Sigma - \mathcal{P}_\Gamma)u, & B_6 u &= \nu_\Sigma a_\Phi\nabla u(\nu_\Sigma - \nu_\Gamma). \end{aligned}$$

Observe that

$$\begin{aligned} \nu_\Sigma - \nu_\Gamma &= [-\beta\nabla\phi, |\nabla\phi|^2/(1+\beta)]^\mathsf{T},\\ \mathcal{P}_\Sigma - \mathcal{P}_\Gamma &= \nu_\Gamma\otimes\nu_\Gamma - \nu_\Sigma\otimes\nu_\Sigma. \end{aligned}$$

Both are analytic in $\nabla\phi$ and of order $\nabla\phi$ if the latter is close to zero, hence all perturbation operators B_j are of order $\nabla\phi$.

This is a perturbation of the half-space problem. The estimates for the right-hand sides are the same as in Section 6, they are small if $|\nabla\phi|_{L_\infty}$ is small. The exception is that we need to consider B_2u in $L_{p,\mu}(\mathbb{R}_+; H^1_q(\mathbb{R}^n_+))$, as well as the pair (B_2u, B_3u) in $H^1_{p,\mu}(\mathbb{R}_+; \dot{H}^{-1}_q(\mathbb{R}^n_+))$. We easily obtain

$$|B_2u|_{L_{p,\mu}(H^1_q)} + \omega|B_2u|_{L_{p,\mu}(\dot{H}^{-1}_q)} \leq |\nabla\phi|_{L_\infty}|u|_{L_{p,\mu}(H^2_q)} + |\nabla^2\phi|_{L_\infty}|u|_{L_{p,\mu}(H^1_q)}$$
$$\leq \big(|\nabla\phi|_{L_\infty} + \eta + \frac{C_\eta}{\omega^{1/2}}\big)(|u|_{\mathbb{E}_{1\mu}} + \omega|u|_{\mathbb{E}_{0\mu}}).$$

Further, as

$$\int_{\mathbb{R}^n_+} B_2u\psi d(x,y) - \int_{\mathbb{R}^{n-1}} B_3u\psi dx = -\int_{\mathbb{R}^n_+} u\cdot\nabla\phi\partial_y\psi\, d(x,y),$$

it is also clear that

$$|(B_2u, B_3u)|_{H^1_{p,\mu}(\dot{H}^{-1}_q)} + \omega|(B_2u, B_3u)|_{L_{p,\mu}(\dot{H}^{-1}_q)} \leq |\nabla\phi|_{L_\infty}[|u|_{\mathbb{E}_{1\mu}} + \omega|u|_{\mathbb{E}_{0\mu}}].$$

Therefore, by perturbation, the half-space result Theorem 7.2.1 is also true in bent half-spaces, provided $\phi \in C^{3-}_b(\mathbb{R}^{n-1})$ and $|\nabla\phi|_{L_\infty}$ is small enough.

Corollary 7.3.4. *The assertions of Theorem* 7.2.1 *as well as the estimates* (7.28), (7.29), (7.30), (7.31) *remain valid in the case of variable coefficients*

$$\mathcal{A}(x,D) = \mathcal{A}_0(D) + \mathcal{A}_1(x,D)$$

in bent half-spaces provided

$$a^{kl}_1 \in C^{1-}_b(\mathbb{R}^n_+; \mathcal{B}(\mathbb{C}^n)) \quad and \quad \sup\{|a^{kl}_1(x)| : k,l = 1,\dots n, x \in \mathbb{R}^n_+\} \leq \eta,$$

and

$$\phi \in C^{3-}_b(\mathbb{R}^{n-1}) \quad and \quad |\nabla\phi|_{L_\infty} \leq \eta,$$

uniformly for $0 < \eta \leq \eta_0$.

3.3 Pressure Regularity

The pressure π has in general no time regularity. But in special situations we do have regularity in time.

Proposition 7.3.5. *In the situation of Theorem 7.3.1, assume further*

$$u_0 = 0, \quad g = 0, \quad \operatorname{div} f = 0 \quad \text{in } \Omega,$$
$$h_{0\nu} = 0, \quad f\cdot\nu = 0 \quad \text{on } \Sigma_0 \cup \Sigma_s.$$

Then

(i) *If* Ω *is bounded,* $P_0\pi \in H^\alpha_{p,\mu}(\mathbb{R}_+; L_q(\Omega))$*, for* $\alpha \in (0, 1/2 - 1/2q)$*, and for any fixed* $s > 1/q$

$$|P_0\pi|_{L_{p,\mu}(L_q)} \leq C\Big(|h_\nu|_{L_{p,\mu}(L_q(\Sigma))} + |u|_{L_{p,\mu}(H^{1+s}_q(\Omega))}\Big),$$

where $P_0 = I$ *in case* $\Sigma_o \cup \Sigma_n \neq \emptyset$*, and* $P_0\pi$ *denotes the mean zero part of* π *otherwise.*

(ii) *If* Ω *is unbounded, with* $\Omega_R = \Omega \cap B(0,R)$*,* R *large, then* $P_{0R}\pi \in {}_0H^{\alpha}_{p,\mu}(\mathbb{R}_+; L_q(\Omega_R))$ *for* $\alpha < 1/2 - 1/2q$*, and for* $s > 1/q$

$$|P_{0R}\pi|_{L_{p\mu}(L_q(\Omega_R))} \leq C_R\Big(|h_\nu|_{L_{p,\mu}(L_q(\Sigma))} + |u|_{L_{p,\mu}(H^{1+s}_q(\Omega))}\Big),$$

where $P_{0R} = I$ *in case* $\Sigma_o \cup \Sigma_n \neq \emptyset$*, and* $P_{0R}\pi$ *denotes the mean zero part of* π *w.r.t.* Ω_R *otherwise.*

Proof. (i) First we assume that Ω is bounded. In case $\Sigma_o \cup \Sigma_n = \emptyset$ we normalize the pressure by zero mean value. Fix any $\phi \in L_{q'}(\Omega)$ with mean zero and solve the elliptic problem

$$\begin{aligned} \Delta\psi &= \phi \quad \text{in } \Omega,\\ \partial_\nu\psi &= 0 \quad \text{on } \Sigma_d \cup \Sigma_s,\\ \psi &= 0 \quad \text{on } \Sigma_o \cup \Sigma_n, \end{aligned}$$

to obtain a unique solution $\psi \in H^2_q(\Omega)$ with mean zero, according to Corollary 7.4.5. Then we obtain with two integrations by parts

$$\begin{aligned} (\pi|\phi)_\Omega &= (\pi|\Delta\psi)_\Omega = (\pi|\partial_\nu\psi)_\Sigma - (\nabla\pi|\nabla\psi)_\Omega\\ &= (\pi|\partial_\nu\psi)_\Sigma + (\partial_t u + \omega u - f|\nabla\psi)_\Omega - (\partial_k a^k l\partial_l u|\nabla\psi)_\Omega\\ &= (\pi|\partial_\nu\psi)_{\Sigma_o\cup\Sigma_n} + (a^{kl}\partial_l u|\nabla\partial_k\psi)_\Omega - (\nu_k a^{kl}\partial_l u|\nabla\psi)_\Sigma\\ &= (h_\nu|\partial_\nu\psi)_{\Sigma_o\cup\Sigma_n} + (a^{kl}\partial_l u|\nabla\partial_k\psi)_\Omega - (\nu_k a^{kl}\partial_l u|\nabla_\Sigma\psi)_\Sigma \end{aligned}$$

as $(f\cdot\nu, \operatorname{div} f, g, h_{0,\nu}) = 0$. As $u_0 = 0$ we may apply the fractional time derivative ∂_t^α to the result

$$(\partial_t^\alpha\pi|\phi)_\Omega = (\partial_t^\alpha\pi|\partial_\nu\psi)_{\Sigma_o\cup\Sigma_n} + (a^{kl}\partial_l\partial_t^\alpha u|\nabla\partial_k\psi)_\Omega - (\nu_k a^{kl}\partial_l\partial_t^\alpha u|\nabla\psi)_\Sigma,$$

which shows that $\pi \in H^\alpha_{p,\mu}(\mathbb{R}_+; L_q(\Omega))$ provided $0 < \alpha < 1/2 - 1/2q$. This also implies the claimed estimate.

(ii) If Ω is an exterior domain, we choose any ball $B(0,R) \subset \mathbb{R}^n$ such that $\Sigma \subset B(0,R)$, and let $\Omega_R = \Omega \cap B(0,R)$. Take any function $\phi \in L_{q'}(\Omega_R)$, with mean value 0 in case $\Sigma_0 \cup \Sigma_n = \emptyset$. Then $\phi \in \dot H^{-1}_{q,\Sigma_d\cap\Sigma_s}(\Omega)$, by Poincaré's inequality. This implies by Theorem 7.4.3 that there is a solution ψ of the elliptic problem

$$\begin{aligned} \Delta\psi &= \phi \quad \text{in } \Omega,\\ \partial_\nu\psi &= 0 \quad \text{on } \Sigma_d \cup \Sigma_s,\\ \psi &= 0 \quad \text{on } \Sigma_o \cup \Sigma_n, \end{aligned}$$

where ϕ is extended trivially to all of Ω. ψ is unique in case $\Sigma_o \cup \Sigma_n \neq \emptyset$, but $\nabla\psi \in H^1_{q'}(\Omega)$ is always unique, and there is a constant $C > 0$ such that

$$|\nabla\psi|_{H^1_{q'}(\Omega)} \leq C|\phi|_{L_{q'}(\Omega_R)}.$$

Now we can perform the same computation as in (i), to the result

$$(\pi|\phi)_{\Omega_R} = (h_\nu|\partial_\nu\psi)_{\Sigma_o\cup\Sigma_n} + (a^{kl}\partial_l u|\nabla\partial_k\psi)_\Omega - (\nu_k a^{kl}\partial_l u|\nabla_\Sigma\psi)_\Sigma.$$

This implies $\pi \in {}_0H^\alpha_{p\mu}(\mathbb{R}_+; L_q(\Omega_R))$ for each R sufficiently large, and also the asserted estimate. □

To be able to apply Proposition 7.3.5, it is convenient to reduce the case of general data to such data for which the assumptions of Proposition 7.3.5 are valid. This will be achieved in two steps. First we extend u_0 to some globally defined $u_0 \in B^{2(\mu-1/p)}_{pq}(\mathbb{R}^n;\mathbb{C})^n$ and solve the whole space problem

$$\partial_t u_1 + \omega u_1 + \mathcal{A}(x,D)u_1 = f, \quad t>0, \quad u_1(0) = u_0.$$

This removes the initial condition and trivializes the compatibility conditions at $t=0$, while the regularity of the data remains unchanged. So we may assume $u_0 = 0$. In the second step we remove g and $h_{0\nu}$, as well as the compatibility condition (e). For this purpose, by Corollary 7.4.5 we solve the elliptic problem

$$\begin{aligned} \Delta\phi &= g && \text{in } \Omega,\\ \partial_\nu\phi &= h_{0\nu} && \text{on } \Sigma_d\cup\Sigma_s,\\ \phi &= 0 && \text{on } \Sigma_o\cup\Sigma_n. \end{aligned}$$

Then we set $u_2 = u - \nabla\phi$ and $\pi_2 = \pi + (\partial_t+\omega)\phi + \psi$, where, using Theorem 7.4.3, ψ solves the problem

$$\begin{aligned} \Delta\psi &= \operatorname{div}\big(\mathcal{A}(x,D)\nabla\phi\big) && \text{in } \Omega,\\ \partial_\nu\psi &= \nu\cdot\big(\mathcal{A}(x,D)\nabla\phi\big) && \text{on } \Sigma_d\cup\Sigma_s,\\ \psi &= 0 && \text{on } \Sigma_o\cup\Sigma_n. \end{aligned}$$

Then (u_2,π_2) satisfies (7.32) with the boundary conditions in question, with data subject to

$$(f\cdot\nu, \operatorname{div} f, g, h_{0\nu}, u_0) = 0,$$

hence π_2 has the time regularity asserted in Proposition 7.3.5. So the only remaining data are

(i) $f \in L_{p,\mu}(\mathbb{R}_+; X_0)$;

(ii) $h_{0\Sigma} \in {}_0F^{1-1/2q}_{pq,\mu}(\mathbb{R}_+; L_q(\Sigma; T\Sigma)) \cap L_{p,\mu}(\mathbb{R}_+; W^{2-1/q}_q(\Sigma; T\Sigma))$;

(iii) $h \in {}_0F^{1/2-1/2q}_{pq,\mu}(\mathbb{R}_+; L_q(\Sigma;\mathbb{R}^n)) \cap L_{p,\mu}(\mathbb{R}_+; W^{1-1/q}_q(\Sigma;\mathbb{R}^n))$.

Here we have set $h_{0\Sigma} = 0$ on $\Sigma_o\cup\Sigma_n$ and $h=0$ on $\Sigma_d\cup\Sigma_s$, for convenience. We remark, that in case $\mathcal{A} = -\Delta$, we can even achieve $f=0$. Indeed, as ∇ commutes with $\mathcal{A} = -\Delta$ we may choose $\pi_2 = \pi + (\partial_t+\omega)\phi - \Delta\phi$.

3.4 Localization

Here we employ the notation of Sections 6.2.4 and 6.3.3, to introduce the charts and the local operators $\mathcal{A}^k$. If $\Omega \subset \mathbb{R}^n$ is unbounded, i.e., an exterior domain, we choose a large ball $B(0,R) \supset \partial\Omega$ and define $U_0 = \mathbb{R}^n \setminus \bar{B}(0,R)$; otherwise U_0 is void. We cover the compact set $\Sigma := \partial\Omega \subset \mathbb{R}^n$ by balls $B(x_k, r/2)$ with $x_k \in \partial\Omega$, $k = 1, \ldots, N_1$, such that each part $\partial\Omega \cap B(x_k, 2r)$ of the boundary Σ can be parameterized by a function $\rho_k \in C^{3-}$ as a graph over the tangent space $T_{x_k}\Sigma$. We extend this function ρ_k to a global function by a cut-off procedure, and denote the resulting bent half-space by $\mathbb{H}_k$. This is possible by the regularity assumption $\Sigma \in C^{3-}$ as well as by compactness of Σ. Define $U_k = B(x_k, r) \cap \Omega$, $k = 1, \ldots, N_1$. We cover the compact set $\bar{\Omega} \setminus \cup_{k=0}^{N_1} U_k$ by finitely many balls $B(x_k, r/2)$, $k = N_1 + 1, \ldots, N_2$, and set $U_k = B(x_k, r)$. Then $\{U_k\}_{k=0}^{N_2}$ is a finite open covering of $\bar{\Omega}$. Fix a C^∞-partition of unity $\{\varphi_k\}_{k=1}^{N_2}$ subordinate to this open covering of $\bar{\Omega}$, and let χ_k denote C^∞-functions with $\chi_k = 1$ on $\operatorname{supp}\varphi_k$, $\operatorname{supp}\chi_k \subset U_k$.

We assume in the sequel that the operator $\mathcal{A}(x_0, D)$ is strongly elliptic, for each $x_0 \in \bar{\Omega} \cup \{\infty\}$, and normally strongly elliptic for each $x_0 \in \Sigma$. Then the maximal regularity constants for the problems with frozen coefficients will be uniform in $x_0 \in \bar{\Omega} \cup \{\infty\}$, by continuity and compactness, hence η_0 in Corollaries 7.3.3 and 7.3.4 will be uniform in x_0 as well. Now we fix any $\eta \in (0, \eta_0]$, and choose the radius of the chart $r > 0$ so small that the assumptions of these corollaries are met, and each chart only intersects one of the boundary parts Σ_j. According to the previous subsection, we may also assume

$$(\operatorname{div} f, g, u_0) = 0 \quad \text{in } \Omega, \quad h_{0,\nu} = f \cdot \nu = 0 \quad \text{on } \Gamma_d \cup \Gamma_s, \quad h_\nu = 0 \quad \text{on } \Gamma_o \cup \Gamma_n.$$

Therefore Proposition 7.3.5 is available.

To define local operators $\mathcal{A}^k(x, D)$ and $\mathcal{B}_j^k(x, D)$ we proceed as follows. For the interior charts $k = 0$, $k = N_1 + 1, \ldots, N_2$, we define the coefficients of $\mathcal{A}^k(x, D)$ by reflection of the coefficients at the boundary of U_k. This is the same trick as in Section 6.1.4. For the boundary charts $k = 1, \ldots, N_1$ we first transform the coefficients of $\mathcal{A}(x, D)$ and $\mathcal{B}_j(x, D)$ in U_k to a half-space, extend them as in Section 6.2.4, and then transform them back to the bent half-space $\mathbb{H}_k$. Having defined the local differential operators, we may proceed as in Section 6.2.4, introducing local problems for the functions $u^k = \varphi_k u$, which for the interior charts $k = 0$, and $k = N_1 + 1, \ldots, N_2$ are problems on $\mathbb{R}^n$, and for the boundary charts $k = 1, \ldots, N_1$ are problems on the bent half-spaces $\mathbb{H}_k$ with boundary $\partial\mathbb{H}_k$. This yields the following problems. For $k = 0$ and $k = N_1 + 1, \ldots, N_2$ we have the whole space problems

$$\begin{aligned} \partial_t u^k + \omega u^k + \mathcal{A}^k(x, D)u^k + \nabla\pi^k &= f_k + F_k(u, \pi) && \text{in } \mathbb{R}^n, \\ \operatorname{div} u^k &= u \cdot \nabla\varphi_k && \text{in } \mathbb{R}^n, \\ u^k(0) &= 0 && \text{in } \mathbb{R}^n, \end{aligned}$$

for $t > 0$, where $f_k = f\varphi_k$ and $F_k(u, \pi) = [\mathcal{A}(x, D), \varphi_k]u + \pi\nabla\varphi_k$. For the boundary charts $k = 1, \ldots, N_1$ we have the problems

$$\begin{aligned} \partial_t u^k + \omega u^k + \mathcal{A}^k(x,D)u^k + \nabla \pi^k &= f_k + F_k(u,\pi) && \text{in } \mathbb{H}_k, \\ \operatorname{div} u^k &= \nabla \varphi_k \cdot u && \text{in } \mathbb{H}_k, \\ u^k(0) &= 0 && \text{in } \mathbb{H}_k, \end{aligned}$$

for $t > 0$, together with the following boundary conditions

$$\begin{aligned} \mathcal{P}_{\partial \mathbb{H}_k} u^k &= h^k_{0\Sigma} && \text{on } \partial\mathbb{H}_k, \ \text{ if } U_k \cap (\Sigma_d \cup \Sigma_0) \neq \emptyset; \\ (u^k|\nu) &= 0 && \text{on } \partial\mathbb{H}_k, \ \text{ if } U_k \cap (\Sigma_d \cup \Sigma_s) \neq \emptyset; \\ \mathcal{P}_\Sigma \nu a : \nabla u^k &= h^k_\Sigma + H_{\Sigma k}(u) && \text{on } \partial\mathbb{H}_k, \ \text{ if } U_k \cap (\Sigma_s \cup \Sigma_n) \neq \emptyset; \\ -\nu a : \nabla u^k \nu + \pi^k &= H_{\nu k}(u) && \text{on } \partial\mathbb{H}_k \ \ \text{ if } U_k \cap (\Sigma_o \cup \Sigma_n) \neq \emptyset. \end{aligned}$$

Here $h^k_{0\Sigma} = h_{0\Sigma}\varphi_k$, $h^k_\Sigma = h_\Sigma \varphi_k$, $H_{\Sigma k} u = \mathcal{P}_\Sigma \nu a \nabla \varphi_k u$, and $H_{\nu k}(u) = -\nu a \nabla \varphi_k u \nu$. In short-hand notation we may write this problem as

$$L_k z_k = g_k + [L, \varphi_k] z,$$

where $z = (u, \pi)$, $z_k = \varphi_k z$, $g_k = \varphi_k(f, 0, h)$, and the notations L and L_k are obvious.

Unfortunately, the commutator $[L, \phi_k]$ in this case is not lower order, so we cannot continue as in Section 6.2.2 and some additional arguments are needed. It turns out that all perturbation terms on the right-hand sides of these equations are lower order, hence can be estimated as in Section 6.2.2, except for $\nabla \varphi_k \cdot u$ in the divergence equation. In fact, as in Section 6.2.2 we have

$$|[\mathcal{A}, \varphi_k]u|_{\mathbb{E}_{0\mu}(\mathbb{H}_k)} \leq C\omega^{-1/2}\big(\omega |u|_{\mathbb{E}_{0\mu}(\Omega)} + |u|_{\mathbb{E}_{1\mu}(\Omega)}\big), \tag{7.37}$$

as well as

$$|H_k|_{\mathbb{F}_{0\mu}(\partial\mathbb{H}_k)} + \omega^{1/2}|H_k|_{L_{p,\mu}(L_q(\partial\mathbb{H}_k))} \leq C\omega^{-1/2}\big(\omega |u|_{\mathbb{E}_{0\mu}(\Omega)} + |u|_{\mathbb{E}_{1\mu}(\Omega)}\big). \tag{7.38}$$

Further, by Proposition 7.3.5,

$$|\pi \nabla \varphi_k|_{\mathbb{E}_{0\mu}(\mathbb{H}_k)} \leq C\omega^{-\gamma}\big(\omega |u|_{\mathbb{E}_{0\mu}(\Omega)} + |u|_{\mathbb{E}_{1\mu}(\Omega)}\big), \tag{7.39}$$

for some $\gamma > 0$, here the additional pressure regularity comes in.

Next we remove the inhomogeneous part $\varphi_k[f, 0, h]$ by solving the corresponding bent half-space problems to obtain $z^0_k = (u^0_k, \pi^0_k)$ in the right regularity classes.

To remove the inhomogeneity $u \cdot \nabla \varphi_k$ in the divergence equation, we decompose $u^k = u^0_k + \tilde{u}_k + \nabla \phi_k$, where ϕ_k solves the elliptic problem

$$\begin{aligned} \Delta \phi_k = u \cdot \nabla \varphi_k &= \operatorname{div}(u \varphi_k) && \text{in } \mathbb{H}_k, \\ \partial_\nu \phi_k &= 0 && \text{on } \partial\mathbb{H}_k, \ \text{ if } U_k \cap (\Sigma_d \cup \Sigma_s) \neq \emptyset, \\ \phi_k &= 0 && \text{on } \partial\mathbb{H}_k, \ \text{ if } U_k \cap (\Sigma_o \cup \Sigma_n) \neq \emptyset, \end{aligned} \tag{7.40}$$

where $\mathbb{H}_k = \mathbb{R}^n$ for $k = 0, N_1 + 1, \ldots, N_2$. By Corollary 7.4.2, this problem admits a solution ϕ_k such that $\nabla\phi_k$ is unique, with regularity

$$\nabla\phi_k \in {}_0H^1_{p,\mu}(\mathbb{R}_+; H^1_q(\mathbb{H}_k)) \cap L_{p,\mu}(\mathbb{R}_+; H^2_q(\mathbb{H}_k)).$$

Moreover, we have the estimates

$$\begin{aligned} |\nabla\phi_k|_{L_{p,\mu}(H^1_q(\mathbb{H}_k))} &\leq C|u|_{\mathbb{E}_{0\mu}(\Omega)}, \\ |\nabla\phi_k|_{\mathbb{E}_{1\mu}(\mathbb{H}_k)} + |\nabla^2\phi_k|_{\mathbb{E}_{1\mu}(\mathbb{H}_k)} &\leq C|u|_{\mathbb{E}_{1\mu}(\Omega)}, \\ |\nabla\phi_k|_{H^{1/2}_{p,\mu}(L_q(\mathbb{H}_k))} + |\nabla\phi_k|_{L_{p,\mu}(H^2_q(\mathbb{H}_k))} &\leq C\omega^{-1/2}\big(\omega|u|_{\mathbb{E}_{0\mu}(\Omega)} + |u|_{\mathbb{E}_{1\mu}(\Omega)}\big). \end{aligned} \tag{7.41}$$

Next we employ the Helmholtz projection in case $U_k \cap (\Sigma_d \cup \Sigma_s) \neq \emptyset$ resp. the Weyl projection in case $U_k \cap (\Sigma_o \cup \Sigma_n) \neq \emptyset$, denoted by P_k, to decompose

$$\tilde{F}_k(u,\pi) := F_k(u,\pi) - \mathcal{A}^k\nabla\phi_k = \nabla\psi_k + P_k\tilde{F}_k(u,\pi).$$

Introducing a new pressure $\tilde{\pi}_k$ by means of

$$\tilde{\pi}_k = \pi^k + (\partial_t + \omega)\phi_k - \psi_k - \pi_k^0,$$

we arrive at the modified problems

$$\begin{aligned} \partial_t\tilde{u}_k + \omega\tilde{u}_k + \mathcal{A}^k(x,D)\tilde{u}_k + \nabla\tilde{\pi}_k &= P_k\tilde{F}_k(u,\pi) && \text{in } \mathbb{H}_k, \\ \operatorname{div}\tilde{u}_k &= 0 && \text{in } \mathbb{H}_k, \\ \tilde{u}_k(0) &= 0 && \text{in } \mathbb{H}_k. \end{aligned}$$

For the boundary charts $k = 1, \ldots, N_1$ these problems are complemented by the boundary conditions

$$\begin{aligned} \mathcal{P}_{\partial\mathbb{H}_k}\tilde{u} &= -\nabla_\Sigma\phi_k && \text{on } \partial\mathbb{H}_k, && \text{if } U_k \cap (\Sigma_d \cup \Sigma_o) \neq \emptyset; \\ (\tilde{u}_k|\nu) &= 0 && \text{on } \partial\mathbb{H}_k, && \text{if } U_k \cap (\Sigma_d \cup \Sigma_s) \neq \emptyset; \\ \mathcal{P}_{\partial\mathbb{H}_k}\nu a\nabla\tilde{u}_k &= \tilde{H}_{\Sigma k}(u) && \text{on } \partial\mathbb{H}_k, && \text{if } U_k \cap (\Sigma_s \cup \Sigma_n) \neq \emptyset; \\ -\nu a\nabla\tilde{u}_k\nu + \pi_k &= \tilde{H}_{\nu k}(u) && \text{on } \partial\mathbb{H}_k, && \text{if } U_k \cap (\Sigma_o \cup \Sigma_n) \neq \emptyset. \end{aligned}$$

Here $\tilde{H}_{\Sigma k}(u) = H_{\Sigma k}(u) - \mathcal{P}_\Sigma\nu a_k\nabla^2\phi_k$, and $\tilde{H}_{\nu k}(u) = H_{\nu k}(u) + \nu a_k\nabla^2\phi_k\nu$. Note that $\tilde{F}_k$, $P_k\tilde{F}_k$ and $\tilde{H}_k$ are subject to the same estimates as F_k and H_k, with probably larger constants C, thanks to (7.41).

Next, we introduce the operators

$$T_kz = (\nabla\phi_k, (\partial_t + \omega)\phi_k - \psi_k).$$

With this notation we can rewrite the localized solution as

$$z_k = z_k^0 + \tilde{z}_k + T_kz,$$

where $\tilde{z}_k$ solves the problem

$$L_k\tilde{z}_k = G_k z,$$

with

$$\begin{aligned} G_k z &= [L, \varphi_k]z - L_k T_k z \\ &= [P_k([\mathcal{A}, \varphi_k]u + \pi\nabla\varphi_k - \mathcal{A}^k\nabla\phi_k, 0, [\mathcal{B}, \varphi_k]u - \mathcal{B}^k\nabla\phi_k]^{\mathsf{T}}, \end{aligned}$$

where $\mathcal{B}^k$ denotes the appropriate boundary operator. More precisely, $[\varphi_k, \mathcal{B}]u = 0$ if $U_k \cap (\Sigma_o \cup \Sigma_n) = \emptyset$ and $[\varphi_k, \mathcal{B}]u = \nu a\nabla^2\varphi_k u$, otherwise.

It is useful to introduce norms for the solutions and for the data which depend on ω. We set

$$\|z_k\| = \omega|u_k|_{\mathbb{E}_{0\mu}(\mathbb{H}_k)} + |u_k|_{\mathbb{E}_{1\mu}(\mathbb{H}_k)} + |\nabla\pi_k|_{\mathbb{E}_{0\mu}(\mathbb{H}_k)},$$

and similarly we define $\|z\|$ on Ω. For the data we set

$$\begin{aligned} \|g_k\| = |f_k|_{\mathbb{E}_{0\mu}(\mathbb{H}_k)} &+ \omega^{1-1/2q}|h_0^k|_{L_{p,\mu}(L_q(\partial\mathbb{H}_k))} + |h_0^k|_{\mathbb{F}_{1\mu}(\partial\mathbb{H}_k)} \\ &+ \omega^{1/2-1/2q}|h^k|_{L_{p,\mu}(L_q(\partial\mathbb{H}_k))} + |h^k|_{\mathbb{F}_{0\mu}(\partial\mathbb{H}_k)}, \end{aligned}$$

and similarly for g on Ω. Then we obtain by maximal regularity on a bent half-space

$$\|z_k^0\| \leq C\|g_k\| \leq C\|g\|, \quad \|\tilde{z}_k\| \leq C\omega^{-\gamma}\|z\|,$$

with a constant $C > 0$ independent of ω and k. Here we employed estimates (7.37), (7.38), (7.39), and (7.41).

To estimate $T_k z$, we employ again (7.37), (7.38), (7.39), and (7.41) to obtain

$$|\nabla\phi_k|_{L_{p,\mu}(H^1_q(\mathbb{H}_k))} + |\nabla\psi_k|_{\mathbb{E}_{0\mu}(\mathbb{H}_k)} \leq C\omega^{-\gamma}\|z\|.$$

Finally, it remains to estimate $(\partial_t + \omega)\nabla\phi_k$. For this purpose, we employ the identity

$$(\partial_t + \omega)\phi_k = \tilde{\pi}_k - \pi_k + \psi_k - \pi_k^0.$$

Applying Poincaré's inequality to π_k^0 and ψ_k, and Proposition 7.3.5 to π and $\tilde{\pi}_k$, we obtain

$$\begin{aligned} |(\partial_t + \omega)\phi_k|_{L_{p,\mu}(L_q(U_k))} &\leq |\tilde{\pi}_k|_{L_{p,\mu}(L_q(U_k))} + |\pi_k|_{L_{p,\mu}(L_q(U_k))} + |\pi_k^0 + \psi_k|_{L_{p,\mu}(L_q(U_k))} \\ &\leq |\tilde{\pi}_k|_{\mathbb{E}_0(\mathbb{H}_k\cap B(0,R))} + |\pi_k|_{\mathbb{E}_{0\mu}(\Omega\cap B(0,R))} \\ &\quad + C(|\nabla\pi_k^0|_{\mathbb{E}_{0\mu}(\mathbb{H}_k)} + |\nabla\psi_k|_{\mathbb{E}_{0\mu}(\mathbb{H}_k)}) \\ &\leq C\Big(\|g\| + \omega^{-\gamma}\|z\|\Big). \end{aligned}$$

By interpolation with (7.41) this yields

$$|(\partial_t + \omega)\nabla\phi_k|_{\mathbb{E}_{0\mu}(U_k)} \leq C\|g\| + C\omega^{-\gamma/2}\|z\|.$$

Summing over k yields the a priori estimate for $z = \sum_k \chi_k z_k$, which reads

$$\|z\| \leq \sum_k \|\chi_k z_k\| \leq C\|g\| + C\omega^{-\gamma}\|z\|,$$

for some $\gamma > 0$, and a constant $C > 0$ which is independent of ω. Choosing $\omega > 2C$ this implies

$$\|z\| \leq 2C\|g\|.$$

Therefore, the operator L on Ω is injective and has closed range. We even can write down a left inverse S as follows. From the identity

$$\begin{aligned} z = \sum_k \chi_k z_k &= \sum_k \chi_k (z_k^0 + \tilde{z}_k + T_k z) \\ &= \sum_k \chi_k L_k^{-1}\varphi_k g + \sum_k \chi_k (L_k^{-1}G_k + T_k)z \\ &= \sum_k \chi_k L_k^{-1}\varphi_k g + G^L z, \end{aligned}$$

we obtain

$$z = Sg := (I - G^L)^{-1}\big(\sum_k \chi_k L_k^{-1}\varphi_k\big)g,$$

as $\|G^L\| < 1$ for ω large.

So it remains to prove surjectivity of L. For this purpose, we assume $f = 0$ for the moment. Set $z = Sg$ as just defined, i.e.,

$$\begin{aligned} z &= \sum_k \chi_k L_k^{-1}\varphi_k g + \sum_k \chi_k (L_k^{-1}G_k + T_k)z \\ &= \sum_k \chi_k L_k^{-1}\varphi g + G^L z, \end{aligned}$$

and apply L, to the result

$$\begin{aligned} L(z - G^L z) &= \sum_k \chi_k L_k L_k^{-1}\varphi_k g + \sum_k [L, \chi_k] L_k^{-1}\varphi_k g \\ &= g + \sum_k \tilde{G}_k L_k^{-1}\varphi_k g + L\sum_k \tilde{T}_k \varphi_k g, \end{aligned}$$

where $\tilde{G}_k = [L, \chi_k] - L\tilde{T}_k$ and $\tilde{T}_k$ is defined in the same way as T_k, replacing φ_k by χ_k. This implies

$$L(z - G^L z - \sum_k \tilde{T}_k \varphi_k g) = g + \sum_k \tilde{G}_k L_k^{-1}\varphi_k g = (I + G^R)g.$$

To conclude the argument, we only have to show that the operator G^R in the data space has norm smaller than 1, as this implies surjectivity of L, and then

$$(S - G^L S - \sum_k \tilde{T}_k \varphi_k)(I + G^R)^{-1}$$

is a right inverse of L. Now $\tilde{G}_k$ can be estimated in the same way as G_k, as $f = 0$, hence we have surjectivity in this case.

To deal with general f, we employ a homotopy argument. Replacing $\mathcal{A}$ by $\tau\mathcal{A} - (1-\tau)\Delta$, we see that the corresponding operators L^τ are injective and have closed ranges for all $\tau \in [0,1]$, as these operators are uniformly normally strongly elliptc, uniformly w.r.t. τ. Therefore the Fredholm index of L^τ is constant, and this shows that L^1 is surjective if and only if L^0 is surjective. For $\tau = 0$ we have the classical case $\mathcal{A} = -\Delta$, and as we have noted above, we may then assume $f = 0$. This completes the proof of Theorem 7.3.1.

7.4 Boundary Value Problems for the Laplacian

Here we state and prove some results for the Laplace equation which have been employed in Section 7.3.

4.1 Whole Space
We begin with the case $\Omega = \mathbb{R}^n$. By the very definition of the homogeneous Bessel potential spaces $\dot{H}^s_q(\mathbb{R}^n)$, namely

$$\dot{H}^s_q(\mathbb{R}^n) := \{u \in \mathcal{S}'(\mathbb{R}^n) : \mathcal{F}^{-1}|\xi|^s\mathcal{F}u \in L_q(\mathbb{R}^n)\},$$

where $1 < q < \infty$ and $s \in \mathbb{R}$, it is clear that Δ is an isomorphism between the spaces $\dot{H}^{s+2}_q(\mathbb{R}^n)$ and $\dot{H}^s_q(\mathbb{R}^n)$.

4.2 Half Space
The half-space case $\Omega = \mathbb{R}^n_+$ is a little more involved.

(i) We first consider the Dirichlet problem

$$\Delta u = 0 \quad \text{in } \mathbb{R}^n_+, \quad u = h \quad \text{on } \partial\mathbb{R}^n_+ = \mathbb{R}^{n-1}.$$

Defining the *Poisson semigroup* $P(y)$ by means of

$$P(y)h = \mathcal{F}^{-1}e^{-y|\xi|}\mathcal{F}h,$$

$u = P(y)h$ is the unique solution of the Dirichlet problem. This shows that $u \in \dot{H}^k_q(\mathbb{R}^n_+)$ if and only if $h \in \dot{W}^{k-1/q}_q(\mathbb{R}^{n-1})$, for all $q \in (1,\infty)$ and $k \geq 0$.

(ii) In the next step we consider the Neumann problem

$$\Delta u = 0 \quad \text{in } \mathbb{R}^n_+, \quad -\partial_y u = g \quad \text{on } \partial\mathbb{R}^n_+.$$

Denoting the generator of the Poisson semigroup by $\dot{D}$, the unique solution of the Neumann problem is given by $u = P(y)\dot{D}^{-1}g$. As $\dot{D}$ has symbol $|\xi|$, it is clear that $\dot{D}$ is an isomorphism from $\dot{H}^{s+1}_q(\mathbb{R}^{n-1})$ to $\dot{H}^s_q(\mathbb{R}^{n-1})$, for all $q \in (1,\infty)$, $s \in \mathbb{R}$. Therefore the solution u of the Neumann problem belongs to the class $\dot{H}^k_q(\mathbb{R}^n_+)$ if and only if $g \in \dot{W}^{k-1-1/q}_q(\mathbb{R}^{n-1})$, for all $q \in (1,\infty)$, $k \geq 0$.

(iii) Now we consider the inhomogeneous Dirichlet problem

$$-\Delta u = f \quad \text{in } \mathbb{R}^n_+, \quad u = 0 \quad \text{on } \partial\mathbb{R}^n_+.$$

The unique solution of this problem is given by

$$u = G_D f := \frac{\dot{D}^{-1}}{2} \int_0^\infty \big(P(|y-s|) - P(y+s)\big) f(s)\, ds.$$

This representation shows $u \in \dot{H}^2_q(\mathbb{R}^n_+)$ if and only if $f \in L_q(\mathbb{R}^n_+)$.

(iv) Similarly, the solution of the inhomogeneous Neumann problem

$$-\Delta u = f \quad \text{in } \mathbb{R}^n_+, \quad \partial_y u = 0 \quad \text{on } \partial\mathbb{R}^n_+.$$

is given by

$$u = G_N f := \frac{\dot{D}^{-1}}{2} \int_0^\infty \big(P(|y-s|) + P(y+s)\big) f(s)\, ds.$$

This representation shows $u \in \dot{H}^2_q(\mathbb{R}^n_+)$ if and only if $f \in L_q(\mathbb{R}^n_+)$.

(v) Higher order regularity.
If $f \in \dot{H}^1_q(\mathbb{R}^n_+)$ then differentiating the equations (or the solution formulas) first tangentially we obtain $\nabla_x u \in \dot{H}^2_q(\mathbb{R}^n_+)$, and then normally, we find $u \in \dot{H}^3_q(\mathbb{R}^n_+)$. In the Dirichlet case we also use **(i)** with $g = f|_{\mathbb{R}^{n-1}} \in \dot{W}^{1-1/q}_q(\mathbb{R}^{n-1})$.

(vi) Weak solutions.
Finally, we consider the weak Dirichlet problem

$$\Delta u = \operatorname{div} f \quad \text{in } \mathbb{R}^n_+, \quad u = 0 \quad \text{on } \partial\mathbb{R}^n_+,$$

where $f = [f_x, f_y]^\mathsf{T} \in L_q(\mathbb{R}^n_+; \mathbb{C}^n)$. In this case the solution u is given by

$$u = \nabla_x \cdot G_D f_x + \partial_y G_N f_y,$$

hence $u \in \dot{H}^1_q(\mathbb{R}^n_+)$. Similarly, for the weak Neumann problem

$$\Delta u = \operatorname{div} f \quad \text{in } \mathbb{R}^n_+, \quad \partial_y u = f_y \quad \text{on } \partial\mathbb{R}^n_+,$$

we have

$$u = \nabla_x \cdot G_N f_x + \partial_y G_D f_y,$$

and so also in this case $u \in \dot{H}^1_q(\mathbb{R}^n_+)$.

4.3 Bent Half Spaces

In the next step we extend the results from the previous subsection to the case of certain bent half-spaces.

(a) Coordinate Transformations.
Let $\Omega \subset \mathbb{R}^n$ be a domain with boundary of class C^1, such that $\partial\Omega =: \Sigma$ decomposes disjointly as $\Sigma = \Sigma_0 \cup \Sigma_1$ with Σ_j open and closed in Σ. Suppose $\Phi : \overline{\Omega} \to \mathbb{R}^n$ is bijective, of class C^1 such that

$$0 < c \leq |\det \partial\Phi(x)| \leq 1/c, \quad x \in \overline{\Omega},$$

and assume $\Phi(\Sigma) = \partial\Phi(\Omega)$. We set $\Omega^\Phi = \Phi(\Omega)$ and $\Sigma_j^\Phi = \Phi(\Sigma_j)$, $j = 0, 1$.

Consider the weak Dirichlet-Neumann problem

$$\begin{aligned}(\nabla u|\nabla v)_{\Omega^\Phi} &= (f|\nabla v)_{\Omega^\Phi}, \quad v \in \dot{H}^1_{q',\Sigma_0^\Phi}(\Omega^\Phi), \\ u &= h \quad \text{on } \Sigma_0^\Phi.\end{aligned} \tag{7.42}$$

By means of the transformation Φ, this problem can be reformulated as a weak problem on Ω in the following way. By means of the pull backs

$$\bar{u}(x) = u(\Phi(x)),\ \bar{v}(x) = v(\Phi(x)),\ \bar{h}(x) = h(\Phi(x)),$$

and with

$$\nabla_x \bar{u}(x) = \nabla_x u(\Phi(x)) = \partial\Phi(x)^{\mathsf{T}} \nabla_y u \circ \Phi(x),$$

the transformation rule yields for a weak solution u on Ω^Φ

$$\begin{aligned}0 = (\nabla u - f|\nabla v)_{\Omega^\Phi} &= \int_{\Phi(\Omega)} (\nabla_y u(y) - f(y)) \cdot \nabla_y v(y)\, dy \\ &= \int_\Omega (\nabla_y u(\Phi(x)) - f(\Phi(x))) \cdot \nabla_y v(\Phi(x)) |\det \partial\Phi(x)|\, dx \\ &= \int_\Omega ((|\det \partial\Phi(x)| \partial\Phi(x)^{-1} \partial\Phi(x)^{-\mathsf{T}}) \nabla_x \bar{u}(x) - \bar{f}(x)) \cdot \nabla_x \bar{v}(x)\, dx,\end{aligned}$$

where

$$\bar{f}(x) = |\det \partial\Phi(x)| \partial\Phi(x)^{-\mathsf{T}} f(\Phi(x)), \quad x \in \Omega.$$

This shows that Problem (7.42) becomes

$$\begin{aligned}0 &= (\mathcal{A}\nabla\bar{u} - \bar{f}|\nabla\bar{v}), \quad \bar{v} \in \dot{H}^1_{q',\Sigma_0}(\Omega), \\ \bar{u} &= \bar{h} \quad \text{on } \Sigma_0.\end{aligned} \tag{7.43}$$

Here the coefficient matrix $\mathcal{A}(x)$ is defined by

$$\mathcal{A}(x) = |\det \partial\Phi(x)| \partial\Phi(x)^{-1} \partial\Phi(x)^{-\mathsf{T}},$$

hence $\mathcal{A}$ is continuous and bounded.

Note that by the assumptions on Φ, the map T_Φ defined by $T_\phi u := \bar{u}$ is an isomorphism from $L_q(\Omega^\Phi)$ to $L_q(\Omega)$ and from $\dot{H}^1_{q,\Sigma_0^\Phi}(\Omega^\Phi)$ to $\dot{H}^1_{q,\Sigma_0}(\Omega)$, hence by interpolation also from $\dot{H}^s_{q,\Sigma_0^\Phi}(\Omega^\Phi)$ to $\dot{H}^s_{q,\Sigma_0}(\Omega)$, $s \in [0,1]$, and from $H^s_{q,\Sigma_0^\Phi}(\Omega^\Phi)$

to $H^s_{q,\Sigma_0}(\Omega)$, $s \in [0,1]$. As T_Φ respects boundary traces by assumption, we also see that $h \in \dot W^{1-1/q}_q(\Sigma_0^\Phi)$ if and only if $\bar h \in \dot W^{1-1/q}_q(\Sigma_0)$. Finally, we have $f \in L_q(\Omega^\Phi;\mathbb{R}^n)$ if and only if $\bar f \in L_q(\Omega;\mathbb{R}^n)$.

These arguments show that (7.42) is well-posed in Ω^Φ if and only if (7.43) is well-posed in Ω.

(b) Perturbed Half-Spaces

Now we consider the special case where $\Omega = \mathbb{R}^n_+$ and $\Phi(x,y) = [x, y+h(x)]^\mathsf{T}$ with $x \in \mathbb{R}^{n-1}$ and $y > 0$, as well as $h \in C^1_b(\mathbb{R}^{n-1})$. This means that Ω^Φ is a bent half-space. Easy computations show $\det \partial\Phi(x,y) = 1$, as well as

$$\mathcal{A}(x,y) = \partial\Phi(x,y)^{-1}\partial\Phi(x,y)^\mathsf{T} = \begin{bmatrix} I & -\nabla_x h(x) \\ -\nabla_x h(x)^\mathsf{T} & 1+|\nabla_x h|_2^2 \end{bmatrix},$$

hence $\mathcal{A}(x,y) = I - \mathcal{B}(x)$, where $|\mathcal{B}(x)| \le C|\nabla_x h|_\infty$. So, dropping the bars, the transformed problem can be rewritten as the problem

$$\begin{aligned} (\nabla u|\nabla v)_{\mathbb{R}^n_+} &= (f|\nabla v)_{\mathbb{R}^n_+} + (\mathcal{B}\nabla u|\nabla v)_{\mathbb{R}^n_+}, \quad v \in {}_0\dot H^1_{q'}(\mathbb{R}^n_+), \\ u &= h \quad \text{on } \partial\mathbb{R}^n_+, \end{aligned} \tag{7.44}$$

in the Dirichlet case, i.e., $\Sigma_1 = \emptyset$, and

$$(\nabla u|\nabla v)_{\mathbb{R}^n_+} = (f|\nabla v)_{\mathbb{R}^n_+} + (\mathcal{B}\nabla u|\nabla v)_{\mathbb{R}^n_+}, \quad v \in \dot H^1_{q'}(\mathbb{R}^n_+), \tag{7.45}$$

in the Neumann case, i.e., $\Sigma_0 = \emptyset$. These are perturbations of the half-space problems in Section 7.4.2, provided $|\nabla_x h|_\infty$ is small.

More precisely, let $L_D : L_q(\mathbb{R}^n_+;\mathbb{R}^n) \times \dot W^{1-1/q}_q(\mathbb{R}^{n-1}) \to \dot H^1_q(\mathbb{R}^n_+)$ denote the bounded solution map from Section 7.4.2 for the Dirichlet problem and $L_N : L_q(\mathbb{R}^n_+;\mathbb{R}^n) \to \dot H^1_q(\mathbb{R}^n_+)$ that for the Neumann problem in the half-space. Then the perturbed problems can rewritten abstractly as

$$u = L_D(f,h) + L_D(\mathcal{B}\nabla u, 0), \quad u = L_N f + L_N \mathcal{B}\nabla u,$$

respectively. Thus by a Neumann series argument, there is a number $\eta_0 > 0$ such that whenever $|\nabla_x h|_\infty \le \eta_0$, then the perturbed equations are also uniquely solvable.

Note that this number $\eta_0 > 0$ is universal for the Laplacian, it only depends on q. Bent half-spaces will be called *perturbed half-spaces* if the corresponding height function h is subject to $|\nabla_x h|_\infty \le \eta_0$. If in addition the support of h is compact, then we use the term *compactly perturbed half-space.*

Let us summarize.

Theorem 7.4.1. *Let $\Omega = \mathbb{H}$ denote a perturbed half-space, and $q \in (1,\infty)$. Then*

(i) Neumann problem

For each $f \in L_q(\mathbb{H})$ there is a unique solution of

$$(\nabla u|\nabla v)_{\mathbb{H}} = (f|\nabla v)_{\mathbb{H}}, \quad v \in \dot H^1_{q'}(\mathbb{H}). \tag{7.46}$$

There is a constant $c > 0$ such that

$$c|\nabla u|_q \le |f|_q, \quad f \in L_q(\mathbb{H}),$$

and

$$c|\nabla u|_q \le \sup\{|(\nabla u|\nabla v)_{\mathbb{H}} : v \in \dot{H}^1_{q'}(\mathbb{H}),\ |\nabla v|_{q'} \le 1\}. \tag{7.47}$$

(ii) Dirichlet problem
For each $f \in L_q(\mathbb{H})$ and $h \in \dot{W}_q^{1-1/q}(\partial\mathbb{H})$, there is a unique solution of

$$(\nabla u|\nabla v)_{\mathbb{H}} = (f|\nabla v)_{\mathbb{H}}, \quad v \in {}_0\dot{H}^1_{q'}(\mathbb{H}), \quad u = h \text{ on } \partial\mathbb{H}. \tag{7.48}$$

There is a constant $c > 0$ such that

$$c|\nabla u|_q \le |f|_q + |h|_{\dot{W}_q^{1-1/q}}, \quad f \in L_q(\mathbb{H}),\ h \in \dot{W}_q^{1-1/q}(\partial\mathbb{H}).$$

Furthermore, in case $h = 0$,

$$c|\nabla u|_q \le \sup\{|(\nabla u|\nabla v)_{\mathbb{H}} : v \in {}_0\dot{H}^1_{q'}(\mathbb{H}),\ |\nabla v|_{q'} \le 1\}. \tag{7.49}$$

For the proof of the variational inequalities note that (7.46) is equivalent to $\nabla_{q'}^*\nabla_q u = \nabla_{q'}^* f$, and the right-hand side of (7.47) is precisely the norm of this quantity in ${}_0H_q^{-1}(\mathbb{H})$. A similar argument is valid for the Dirichlet problem, provided $h = 0$.

Concerning higher regularity, the results for perturbed half-spaces are not as precise as those for the half-space case, as lower order terms occur. However, the assertions in the next corollary follow from the corresponding half-space results, again by Neumann series arguments.

Corollary 7.4.2. *Let $\Omega = \mathbb{H}$ denote a perturbed half-space, $q \in (1,\infty)$, $s \in \{0,1\}$, and $h \in C_b^{(2+s)-}(\mathbb{R}^{n-1})$.*

(i) Neumann problem
If $f \in H_q^s(\mathbb{H})$, $g \in W_q^{1+s-1/q}(\partial\mathbb{H})$ such that $(f,g) \in {}_0\dot{H}_q^{-1}(\mathbb{H}))$, then the problem

$$\Delta u = f \quad \text{in } \mathbb{H}, \quad \partial_\nu u = g \quad \text{on } \partial\mathbb{H}$$

has a unique solution u such that $\nabla u \in H_q^{1+s}(\mathbb{H})$.

(ii) Dirichlet problem
If $f \in H_q^s(\mathbb{H})$, $h \in W_q^{2+s-1/q}(\partial\mathbb{H})$ such that $f \in \dot{H}_q^{-1}(\mathbb{H}))$, then the problem

$$\Delta u = f \quad \text{in } \mathbb{H}, \quad u = h \quad \text{on } \partial\mathbb{H}$$

has a unique solution u such that $\nabla u \in H_q^{1+s}(\mathbb{H})$.

4.4 General Domains

Now we are ready to consider domains with compact boundary, which means domains which are either bounded or exterior.

Theorem 7.4.3. *Suppose that Ω is domain in $\mathbb{R}^n$ with compact boundary $\partial\Omega := \Sigma$ of class C^1, and suppose that Σ decomposes disjointly into $\Sigma = \Sigma_0 \cup \Sigma_1$, where Σ_j are open and closed in Σ. Let $f \in L_q(\Omega)$, $h \in W_q^{1-1/q}(\Sigma_0)$, with $q \in (1,\infty)$. Then the problem*

$$\begin{aligned} (\nabla u|\nabla v)_\Omega &= (f|\nabla v)_\Omega, \quad v \in \dot{H}^1_{q',\Sigma_0}(\Omega), \\ u &= h \qquad\qquad\quad \text{on } \Sigma_0, \end{aligned} \tag{7.50}$$

admits a unique solution $u \in \dot{H}^1_q(\Omega)$. There is a constant $C > 0$ such that

$$|\nabla u|_{L_q} \le C\big(|f|_{L_q} + |h|_{W_q^{1-1/q}}\big) \tag{7.51}$$

holds for all $f \in L_q(\Omega)$ and $h \in W_q^{1-1/q}(\Sigma_0)$.

Recall $\dot{H}^1_{q,\emptyset}(\Omega) = \dot{H}^1_q(\Omega)/constants$, hence uniqueness in $\dot{H}^1_{q,\Sigma_0}(\Omega)$ means uniqueness up to a constant in case $\Sigma_0 = \emptyset$, and even uniqueness otherwise. If $\Sigma_0 = \emptyset$, we normalize the solution by mean value zero if Ω is bounded, and by mean zero on $\Omega \cap B(0,R)$, for some large fixed ball $B(0,R)$ which contains Σ.

Proof. The proof consists of several steps. The first step concerns uniqueness.

(a) Uniqueness

Suppose

$$(\nabla u|\nabla v)_\Omega = 0, \quad v \in \dot{H}^1_{q',\Sigma_0}(\Omega), \quad u = 0 \quad \text{on } \Sigma_0.$$

We show that this implies $u = 0$ in $\dot{H}^1_{q,\Sigma_0}(\Omega)$. For this purpose, we prove two assertions, namely

(i) For each $x_0 \in \overline{\Omega}$ there is a ball $B(x_0,r)$ such that $\nabla u \in L_2(B(x_0,r))$.

(ii) There is a ball $B(0,r) \supset \Sigma$, such that $\nabla u \in L_2(\mathbb{R}^n \setminus B(0,r))$.

Here (ii) is void in case Ω is bounded.

Assuming (i) and (ii), by compactness we obtain $\nabla u \in L_2(\Omega)$ and so we may use $v = u$ as a test function to obtain $|\nabla u|_2^2 = 0$, which yields the assertion.

(i) If $q \ge 2$ this is obvious, as $L_q(B(x_0,r)) \subset L_2(B(x_0,r))$, for each $r > 0$. So let $q \in (1,2)$. Set $q_0 = q$ and define inductively q_j by

$$\frac{1}{q_j} = \frac{1}{q_{j-1}} - \frac{1}{n} = \frac{1}{q} - \frac{j}{n};$$

clearly $q_k \ge 2$ if $k \ge n(2-q)/2q$. Choose a radius $r_0 > 0$ small enough so that $B(x_0,r_0) \subset \Omega$ in case $x_0 \in \Omega$ – then we set $\mathbb{H}_{x_0} = \mathbb{R}^n$ –, and if $x_0 \in \partial\Omega$, such that

$\Omega \cap B(x_0, r_0)$ is part of the boundary of a perturbed half-space $\mathbb{H}_{x_0}$. Below we will be using the inequalities (7.47) and (7.49) for perurbed half-spaces as well as for the whole space.

Next we choose cut-off functions χ_j with $\operatorname{supp} \chi_j \subset B(x_0, r_j)$, $\chi_j = 1$ on $B(x_0, r_{j+1})$. We proceed by induction. By assumption we know $\nabla u \in L_{q_0}(B(x_0, r_0))$. Assume $\nabla u \in L_{q_j}(B(x_0, r_0))$, and consider $\nabla(\chi_j u)$. We have

$$|\nabla(\chi_j u)|_{q_{j+1}} \leq c \sup\{(\nabla(\chi_j u)|\nabla v)_{\mathbb{H}_{x_0}} : |\nabla v|_{q'_{j+1}} \leq 1\},$$

where we may normalize v by mean value zero on $B(x_0, r_j)$, in case $x_0 \in \Omega \cup \Sigma_1$. hence with

$$\begin{aligned}(\nabla(\chi_j u)|\nabla v)_{\mathbb{H}_{x_0}} &= (\nabla u|\nabla(\chi_j v))_{\mathbb{H}_{x_0}} - (\nabla u|v\nabla\chi_j)_{\mathbb{H}_{x_0}} + (u\nabla\chi_j|\nabla v)_{\mathbb{H}_{x_0}} \\ &= -(\nabla u|v\nabla\chi_j)_{\mathbb{H}_{x_0}} + (u\nabla\chi_j|\nabla v)_{\mathbb{H}_{x_0}},\end{aligned}$$

by assumption, as $\chi_j v$ belongs to ${}_0\dot{H}^1_{q'}(\mathbb{H}_{x_0})$ if $x_0 \in \Sigma_0$, and to $\dot{H}^1_{q'}(\mathbb{H}_{x_0})$ otherwise. Since $\nabla\chi_j$ has support in $\bar{B}(x_0, r_j) \setminus B(x_0, r_{j+1})$, we obtain

$$|(\nabla u|v\nabla\chi_j)_{\mathbb{H}_{x_0}}| \leq C|\nabla u|_{L_{q_j}(B(x_0,r_j))}|v|_{L_{q'}(B(x_0,r_j))},$$

and also

$$|(u\nabla\chi_j|\nabla v)_{\mathbb{H}_{x_0}}| \leq C|u|_{L_{q_j}(B(x_0,r_j))}|\nabla v|_{L_{q'}(B(x_0,r_j))}.$$

Consequently, by Poincaré's inequlity we have

$$\begin{aligned}|\nabla(\chi_j u)|_{q_{j+1}} &\leq C|u|_{H^1_{q_j}(B(x_0,r_j))}|v|_{H^1_{q'_j}(B(x_0,r_j))} \\ &\leq C|u|_{H^1_{q_j}(B(x_0,r_j))}|\nabla v|_{L_{q'_j}(B(x_0,r_j))} \\ &\leq C|u|_{H^1_{q_j}(B(x_0,r_j))}|\nabla v|_{L_{q'_j}(\mathbb{H}_{x_0}))} \leq C|u|_{H^1_{q_j}(B(x_0,r_j))},\end{aligned}$$

and as $\chi_j = 1$ on $B(x_0, r_{j+1})$ this yields

$$|\nabla u|_{L_{q_{j+1}}(B(x_0,r_{j+1}))} \leq C|u|_{H^1_{q_j}(B(x_0,r_j))}.$$

This proves **(i)**.

(ii) We have to distinguish the cases $q \geq 2$ and $1 < q < 2$. If $q \geq 2$, choose a ball $B(0, r_0)$ such that $\Sigma \subset B(0, r_0 - 1)$, and fix a cut-off function χ_0 which equals 0 in $B(0, r_0 - 1)$ and equals one outside the ball $B(0, r_0)$. Then we have

$$c|\nabla(\chi_0 u)|_{L_2(\mathbb{R}^n)} \leq \sup\{(\nabla(\chi_0 u)|\nabla v)_{\mathbb{R}^n} : |\nabla v|_{L_2(\mathbb{R}^n)} \leq 1\}.$$

As above

$$(\nabla(\chi_0 u)|\nabla v)_{\mathbb{R}^n} = -(\nabla u|v\nabla\chi_0)_{\mathbb{R}^n} + (u\nabla\chi_0|\nabla v)_{\mathbb{R}^n},$$

hence

$$|(\nabla(\chi_0 u)|\nabla v)_{\mathbb{R}^n}| \le C|u|_{H_2^1(A_0)}|v|_{H_2^1(A_0)}$$
$$\le C|u|_{H_q^1(A_0)}|v|_{H_2^1(A_0)},$$

where $A_0 = B(0,r_0) \setminus B(0,r_0-1)$. As we may normalize v by mean value zero over A_0, and $\chi_0 = 1$ on $\mathbb{R}^n \setminus B(0,r_0)$ this shows $\nabla u \in L_2(\mathbb{R}^n \setminus B(0,r_0))$.

On the other hand, if $1 < q < 2$ then we set $r_j = jr_0$, and choose cut-offs such that $\operatorname{supp}\chi_j \subset \mathbb{R}^n \setminus B(0,r_j)$, and $\chi_j = 1$ on $\mathbb{R}^n \setminus B(0,r_{j+1})$. Then by

$$c|\nabla(\chi_j u)|_{q_{j+1}} \le \sup\{(\nabla(\chi_j u)|\nabla v)_{\mathbb{R}^n} : |\nabla v|_{q'_{j+1}} \le 1\},$$

we obtain as before

$$|(\nabla(\chi_j u)|\nabla v)_{\mathbb{R}^n}| \le C|u|_{H_2^1(A_j)}|v|_{H_2^1(A_j)}$$
$$\le C|u|_{H_q^1(A_j)}|v|_{H_2^1(A_j)},$$

and so the same argument as in **(i)** implies $\nabla u \in L_2(\mathbb{R}^n \setminus B(0,r_k))$, by induction.

As a consequence, we obtain $u \in L_2(\mathbb{R}^n \setminus B(0,r))$ for some $r > 0$.

(b) Lower Bound
(i) Suppose that the inequality (with $h = 0$)

$$c|\nabla u|_q \le \sup\{|(\nabla u|\nabla v)_\Omega| : |\nabla v|_{q'} \le 1\}$$

does not hold. Then there is a sequence $(u_k) \subset \dot{H}^1_{q,\Sigma_0}(\Omega)$ with $|\nabla u_k|_q = 1$ such that

$$\varepsilon_k := \sup\{|(\nabla u_k|\nabla v)_\Omega| : |\nabla v|_{q'} \le 1\} \to 0 \quad \text{as } k \to \infty.$$

Since $L_q(\Omega)$ is reflexive, there is a subsequence (w.l.o.g. the whole sequence) such that $\nabla u_k \rightharpoonup \nabla u$ in $L_q(\Omega)$. This implies with $\varepsilon_k \to 0$

$$(\nabla u_k|\nabla v)_\Omega \to (\nabla u|\nabla v)_\Omega = 0, \quad \text{for all } v \in \dot{H}^1_{q',\Sigma_0}(\Omega).$$

Then **(a)** implies $u = 0$.

(ii) Next we localize as e.g. in Section 6.3.3; below we use the notation from there. Then by the previous subsection we know

$$c|\nabla(\varphi_j u_k)|_q \le \sup\{|(\nabla(\varphi_j u_k)|\nabla v)_{\mathbb{H}_j}| : |\nabla v|_{q'} \le 1\} =: d_{kj}$$

on each perturbed half-space or whole space $\mathbb{H}_j$, $j = 0,\dots,N$. We want to prove $d_{kj} \to 0$ as $k \to \infty$, for each j. If this is true, then

$$|\nabla u_k|_q = |\sum_{j=0}^N \nabla(\varphi_j u_k)|_q \le \sum_{j=0}^N |\nabla(\varphi_j u_k)|_q \le C\sum_{j=0}^N d_{kj} \to 0$$

as $k \to \infty$, a contradiction as $|\nabla u_k|_q = 1$ by assumption.

(iii) For a fixed $j \in \{0, \dots, N\}$ choose $v_{kj} \in \dot{H}^1_{q'}(\mathbb{H}_j)$ normalized by $|\nabla v_{kj}|_{q'} = 1$, and by mean value zero over U_j in case $U_j \cap \Sigma_0 = \emptyset$, such that

$$d_{kj} \le \frac{1}{k} + (\nabla(\varphi_j u_k)|\nabla v_{kj})_{\mathbb{H}_j}.$$

We have

$$(\nabla(\varphi_j u_k)|\nabla v_{kj})_{\mathbb{H}_j} = (\nabla u_k|\nabla(\varphi_j v_{kj}))_{\mathbb{H}_j} - (\nabla u_k|\nabla\varphi_j v_{kj})_{\mathbb{H}_j} + (u_k \nabla\varphi_j|v_{kj})_{\mathbb{H}_j},$$

hence

$$d_{kj} \le \frac{1}{k} + \varepsilon_k|\nabla(\varphi_j v_{kj})|_{q'} + |(\nabla u_k|\nabla\varphi_j v_{kj})_{\mathbb{H}_j}| + |(u_k\nabla\varphi_j|\nabla v_{kj})_{\mathbb{H}_j}|.$$

Clearly the first two terms on the right-hand side of this inequality converge to zero as $k \to \infty$. The third term tends to zero, as $\nabla u_k \rightharpoonup 0$ in $L_q(\Omega)$ and by Poincaré's inequality and compact embedding, the set $\{\nabla\varphi_j v_{kj}\}_{k\ge 0}$ is relatively compact in $L_{q'}(\Omega)$. Finally, the last term converges also to zero, as $u_k\nabla\varphi_j \to 0$ as $k \to \infty$ by compact embedding, and ∇v_{kj} is bounded in $L_{q'}$, by construction.

(c) The Isomorphism
Let

$$\nabla_q : \dot{H}^1_{q,\Sigma_0}(\Omega) \to L_q(\Omega)$$

be defined by $(\nabla_q u)(x) = (\nabla u)(x)$, $x \in \Omega$. This operator is bounded, linear, injective, and has closed range. Therefore its dual

$$\nabla_q^* : L_{q'}(\Omega) \to [\dot{H}^1_{q,\Sigma_0}(\Omega)]^* = \dot{H}^{-1}_{q',\Sigma_1}(\Omega)$$

is linear, bounded, and surjective. Define

$$A_q : \dot{H}^1_{q,\Sigma_0}(\Omega) \to \dot{H}^{-1}_{q,\Sigma_1}(\Omega)$$

by means of $A_q u := \nabla_{q'}^* \nabla_q$; then A_q is bounded linear, and $A_q^* = A_{q'}$. We have

$$A_q u = f \quad \Leftrightarrow \quad (\nabla u|\nabla v)_\Omega = (f|\nabla v)_\Omega \ \text{ for all } v \in \dot{H}^1_{q',\Sigma_0}(\Omega), \quad u = 0 \text{ on } \Sigma_0.$$

By **(a)** we see that A_q is injective, for $q \in (1, \infty)$, and **(b)** implies that A_q has closed range. Therefore, as $A_q^* = A_{q'}$ is also injective, it is bijective, i.e., A_q is an isomorphism for each $q \in (1, \infty)$.

(d) Inhomogeneous Dirichlet Data
Finally we consider the case $f = 0$ but $h \ne 0$. For this purpose we first solve

$$u_0 - \Delta u_0 = 0 \text{ in } \Omega, \quad \partial_\nu u_0 = 0 \text{ on } \Sigma_1, \quad u_0 = h \text{ on } \Sigma_0.$$

Section 6.3.6 yields a unique $u_0 \in H^1_q(\Omega)$. Then $u_1 = u - u_0$ must solve

$$A_q u_1 = \Delta u_0 \in \dot{H}^{-1}_{q,\Sigma_1}(\Omega),$$

which by **(c)** admits a unique solution $u_1 \in \dot{H}^1_{q,\Sigma_0}(\Omega)$. This completes the proof. □

As a first consequence we obtain the **Helmholtz-Weyl projection**.

Corollary 7.4.4. *Let $1 < q < \infty$, Ω be either the whole space $\mathbb{R}^n$, or a perturbed half-space, or a domain with compact C^1-boundary $\partial\Omega =: \Sigma$. Suppose that $\Sigma = \Sigma_0 \cup \Sigma_1$ with disjoint parts Σ_j which are open and closed in Σ.*
Then given $f \in L_q(\Omega; \mathbb{C}^n)$, there are unique functions $\phi \in \dot{H}^1_{q,\Sigma_0}(\Omega)$ and $w \in \mathsf{N}(\nabla^*_{q'})$ such that

$$f = \nabla\phi + w,$$

and there is a constant such that

$$|w|_{L_q} \le C|f|_{L_q}, \quad \textit{for all } f \in L_q(\Omega).$$

The bounded linear operator $P_{HW} \in \mathcal{B}(L_q(\Omega))$ defined by $P_{HW}f := w$ is a projection, called the Helmholtz-Weyl *projection associated to the decomposition $\Sigma = \Sigma_0 \cup \Sigma_1$ of the boundary $\Sigma = \partial\Omega$ of Ω.*

This result follows by solving the problem $A_q\phi = \nabla^*_{q'} f$ according to Theorem 7.4.3. Then obviously $w = f - \nabla\phi \in \mathsf{N}(\nabla^*_{q'})$.

The final result concerns higher regularity.

Corollary 7.4.5. *Suppose that Ω is a domain in $\mathbb{R}^n$ with compact boundary $\partial\Omega := \Sigma$ of class $C^{(2+s)-}$, $s = 0, 1$, and suppose that Σ decomposes disjointly into $\Sigma = \Sigma_0 \cup \Sigma_1$, where Σ_j are open and closed in Σ. Let $f \in H^s_q(\Omega)$, $g \in W^{1+s-1/q}_q(\Sigma_1)$, $h \in W^{2+s-1/q}_q(\Sigma_0)$, and assume $(f, g) \in \dot{H}^{-1}_{q,\Sigma_1}(\Omega)$.*
Then the problem

$$\begin{aligned} \Delta u &= f \quad \text{in } \Omega, \\ \partial_\nu u &= g \quad \text{on } \Sigma_1, \\ u &= h \quad \text{on } \Sigma_0, \end{aligned} \tag{7.52}$$

admits a unique solution u with $\nabla u \in H^{1+s}_q(\Omega)$. There is a constant $C > 0$ such that

$$|\nabla u|_{H^{1+s}_q} \le C\big(|(f,g)|_{\dot{H}^{-1}_{q,\Sigma_1}} + |f|_{H^s_q} + |g|_{W^{1+s-1/q}_q} + |h|_{W^{2+s-1/q}_q}\big) \tag{7.53}$$

holds for all $(f, g, h) \in H^s_q(\Omega) \times W^{1+s-1/q}_q(\Sigma_1) \times W^{2+s-1/q}_q(\Sigma_0)$, $s = 0, 1$.

Proof. First we may reduce to the case $(g,h)=0$, solving the problem

$$\begin{aligned} u_0 - \Delta u_0 &= 0 \quad \text{in } \Omega,\\ \partial_\nu u_0 &= g \quad \text{on } \Sigma_1,\\ u_0 &= h \quad \text{on } \Sigma_0, \end{aligned}$$

as in **(d)** above.

Let $\mathbb{H}_j$ and φ_j, $j=0,\ldots,N$, be as above. Let $v \in \dot{H}^1_{q'}(\mathbb{H}_j)$ if $x_j \in \Omega \cup \Sigma_1$, and $v \in {}_0\dot{H}^1_{q'}(\mathbb{H}_j)$ otherwise. Then we have

$$\begin{aligned} (\nabla(\varphi_j u)|\nabla v)_{\mathbb{H}_j} &= (\nabla u|\nabla(\varphi_j v))_{\mathbb{H}_j} - (\nabla u \nabla \varphi_j|v)_{\mathbb{H}_j} + (u\nabla\varphi_j|\nabla v)_{\mathbb{H}_j}\\ &= (\nabla u|\nabla(\varphi_j v)))_{\mathbb{H}_j} - (\nabla u \nabla\varphi_j|v)_{\mathbb{H}_j} - (\operatorname{div}(u\nabla\varphi_j)|v)_{\mathbb{H}_j}\\ &= -(f\varphi_j + 2\nabla u\nabla\varphi_j + u\Delta\varphi_j|v)_{\mathbb{H}_j} = -(f_j|v)_{\mathbb{H}_j}, \end{aligned}$$

with $f_j := f\varphi_j + 2\nabla u \nabla\varphi_j + u\Delta\varphi_j \in L_q(\mathbb{H}_j)$. This shows that $\varphi_j u$ is the weak solution in $\mathbb{H}_j$ with right-hand side $f_j \in L_q(\mathbb{H}_j)$. The results in Section 7.4.3 show that $\nabla(\varphi_j u) \in H^{1+s}_q(\mathbb{H}_j)$, hence summing over j we obtain the assertion. □

Remark. In all of this section we restricted our analysis to the Laplacian. However, Δ can be replaced by any uniformly strongly elliptic operator $\operatorname{div}(\mathcal{A}(x)\nabla)$ with coefficients $\mathcal{A} \in C_l(\overline{\Omega};\mathbb{R}^{n\times n})$ for weak solutions, and additionally $\mathcal{A} \in W^{1+s}_\infty(\Omega;\mathbb{R}^{n\times n})$ for higher regularity. This extension is straightforward, and its implementation is left for the curious reader as well as to researchers who are in need of such results.

Chapter 8

Two-Phase Stokes Problems

Now we are in position to study maximal L_p-regularity for linear two-phase Stokes problems. There are two problems of relevance, the standard one, and another, nonstandard problem, which we call the *asymmetric two-phase Stokes problem*. The first one is important for problems **(P2)**, **(P3)**, **(P5)**, while the asymmetric problem arises in **(P4)**, **(P6)**, which means for problems with phase transitions and different densities. We also study the various induced two-phase Stokes operators in detail, as well as the dynamic surface conditions needed for the analysis of Problems **(P2)**$\sim$**(P6)**. In the analysis, several Neumann-to-Dirichlet operators for Stokes problems will play an important role.

8.1 Two-Phase Stokes Problems

We consider now the inhomogeneous linear problem which is of central importance for Problem **(P2)**, but will also be needed for the analysis of **(P3)** and **(P5)**.

Let $\Omega \subset \mathbb{R}^n$ be a bounded domain with boundary of class C^3, and let $\Sigma \subset \Omega$ be a closed hypersurface. We consider the problem

$$\begin{aligned}
\varrho(\partial_t + \omega)u - \mu(x)\Delta u + \nabla\pi &= \varrho f_u && \text{in } \Omega\backslash\Sigma,\\
\operatorname{div} u &= g_d && \text{in } \Omega\backslash\Sigma,\\
u &= g_b && \text{on } \partial\Omega,\\
[\![u]\!] &= g_u && \text{on } \Sigma,\\
[\![-2\mu(x)D(u) + \pi]\!]\nu_\Sigma - \sigma(x)(\Delta_\Sigma h)\nu_\Sigma &= g && \text{on } \Sigma,\\
(\partial_t + \omega)h - (u|\nu_\Sigma) + (b(t,x)|\nabla_\Sigma h) &= f_h && \text{on } \Sigma,\\
u(0) = u_0 \quad \text{in } \Omega\backslash\Sigma, \quad h(0) &= h_0 && \text{on } \Sigma,
\end{aligned} \tag{8.1}$$

on the time-interval $J = \mathbb{R}_+$, where $\omega \geq 0$ will be chosen sufficiently large. Here as before $D(u) = (\nabla u + [\nabla u]^{\mathsf{T}})/2$ denotes the rate of strain tensor.

J. Prüss and G. Simonett, *Moving Interfaces and Quasilinear Parabolic Evolution Equations*, Monographs in Mathematics 105,
DOI 10.1007/978-3-319-27698-4_8

Remark 8.1.1. We will occasionally replace the operator $\mu(x)\Delta u$ in the first line of (8.1) by $\operatorname{div}(2\mu(x)D(u))$. Observe that with the condition $\operatorname{div} u = g_d$ we have

$$\operatorname{div}(2\mu(x)D(u)) = \mu(x)\Delta u + 2D(u)\nabla\mu(x) + \mu(x)\nabla g_d.$$

Therefore, $\operatorname{div}(2\mu(x)D(u))$ and $\mu(x)\Delta$ have the same principal part, and the results developed for Problem (8.1), and variants thereof, will also hold with $\mu(x)\Delta$ replaced by $\operatorname{div}(2\mu(x)D(u))$.

We employ the same regularity classes for u and π as in Chapter 7, i.e.,

$$u \in \mathbb{E}_u := H^1_{p,\mu}(J; L_p(\Omega)^n) \cap L_{p,\mu}(J; H^2_p(\Omega\backslash\Sigma)^n),$$

and

$$\pi \in \mathbb{E}_\pi := L_{p,\mu}(J; \dot{H}^1_p(\Omega\backslash\Sigma)).$$

Then

$$g_u \in \mathbb{F}^n_h = W^{1-1/2p}_{p,\mu}(J; L_p(\Sigma)^n) \cap L_{p,\mu}(J; W^{2-1/p}_p(\Sigma)^n),$$

and

$$g_b \in W^{1-1/2p}_{p,\mu}(J; L_p(\partial\Omega)^n) \cap L_{p,\mu}(J; W^{2-1/p}_p(\partial\Omega)^n).$$

Therefore, the equation for the height function h lives in the trace space for the components of u, i.e.,

$$f_h \in \mathbb{F}_h := W^{1-1/2p}_{p,\mu}(J; L_p(\Sigma)) \cap L_{p,\mu}(J; W^{2-1/p}_p(\Sigma)),$$

hence the natural space for h is given by

$$h \in \mathbb{E}_h := W^{2-1/2p}_{p,\mu}(J; L_p(\Sigma)) \cap H^1_{p,\mu}(J; W^{2-1/p}_p(\Sigma)) \cap L_{p,\mu}(J; W^{3-1/p}_p(\Sigma)).$$

Here the last space comes from the curvature term in the stress boundary condition, which induces an additional order in spatial regularity. Assuming that g belongs to the trace space of ∇u, i.e.,

$$g \in \mathbb{F}^n_u := W^{1/2-1/2p}_{p,\mu}(J; L_p(\Sigma)^n) \cap L_{p,\mu}(J; W^{1-1/p}_p(\Sigma)^n),$$

we have the additional regularity $[\![\pi]\!] \in \mathbb{F}_u$ for the pressure jump across the interface Σ. The function $b \in \mathbb{F}_h$ is given; it is needed later on for local well-posedness, but not for stability.

There is another hidden regularity which comes from the divergence equation. To identify it, let $\phi \in \dot{H}^1_{p'}(\Omega)$. An integration by parts yields

$$\begin{aligned}(u|\nabla\phi)_\Omega &= -(\operatorname{div} u|\phi)_\Omega + (u\cdot\nu_{\partial\Omega}|\phi)_{\partial\Omega} - ([\![u\cdot\nu_\Sigma]\!]|\phi)_\Sigma\\ &= -(g_d|\phi)_\Omega + (g_b\cdot\nu_{\partial\Omega}|\phi)_{\partial\Omega} - (g_u\cdot\nu_\Sigma|\phi)_\Sigma.\end{aligned}$$

Set ${}_0\dot{H}_p^{-1}(\Omega) = (\dot{H}^1_{p'}(\Omega))^*$ and define the space $\widehat{H}_p^{-1}(\Omega)$ as the set of all triples $(\varphi, \psi, \chi) \in L_p(\Omega) \times W_p^{2-1/p}(\partial\Omega)^n \times W_p^{2-1/p}(\Sigma)^n$, which enjoy the regularity property $(\varphi, \psi \cdot \nu_{\partial\Omega}, \chi \cdot \nu_\Sigma) \in {}_0\dot{H}_p^{-1}(\Omega)$, where

$$\langle(\varphi, \psi \cdot \nu_{\partial\Omega}, \chi \cdot \nu_\Sigma)|\phi\rangle := -(\varphi|\phi)_\Omega + (\psi \cdot \nu_{\partial\Omega}|\phi)_{\partial\Omega} - (\chi \cdot \nu_\Sigma|\phi)_\Sigma, \quad \phi \in \dot{H}^1_{p'}(\Omega).$$

Employing this notation we have

$$\langle(g_d, g_b \cdot \nu_{\partial\Omega}, g_u \cdot \nu_\Sigma)|\phi\rangle = (u|\nabla\phi)_\Omega.$$

Since $u \in H^1_{p,\mu}(J; L_p(\Omega)^n)$ this implies $(g_d, g_b \cdot \nu_{\partial\Omega}, g_u \cdot \nu_\Sigma) \in H^1_{p,\mu}(J; \widehat{H}_p^{-1}(\Omega))$. Observe that this condition contains the compatibility condition

$$\int_\Omega g_d\, dx = \int_{\partial\Omega} g_b \cdot \nu_{\partial\Omega}\, d(\partial\Omega) - \int_\Sigma g_u \cdot \nu_\Sigma\, d\Sigma, \tag{8.2}$$

which shows up when choosing $\phi \equiv 1$.

In the particular case $g_d = 0$ we have $(g_d, g_b \cdot \nu_{\partial\Omega}, g_u \cdot \nu_\Sigma) \in H^1_{p,\mu}(J; \widehat{H}_p^{-1}(\Omega))$ if and only if $g_b \cdot \nu_{\partial\Omega} \in H^1_{p,\mu}(J; \dot{W}_p^{-1/p}(\partial\Omega))$ and $g_u \cdot \nu_\Sigma \in H^1_{p,\mu}(J; \dot{W}_p^{-1/p}(\Sigma))$, as Σ and $\partial\Omega$ are separated.

1.1 The Main Result

The main theorem of this section states that Problem (8.1) admits maximal regularity. In particular, it defines an isomorphism between the solution space and the space of data.

Theorem 8.1.2. *Let $p > n+2$, $1 \geq \mu > 1/p$, $\Omega \subset \mathbb{R}^n$ a bounded domain with $\partial\Omega \in C^3$, $\Sigma \subset \Omega$ a closed hypersurface of class C^3 and ϱ_j be positive constants, $j = 1,2$. Assume that $\mu \in C_b^{1-}(\Omega \setminus \Sigma)$, $\sigma \in C^{1-}(\Sigma)$ such that μ, σ are positive, uniformly in x. Set $J = \mathbb{R}_+$, and suppose $b = b_0 + b_1$ with $b_0 \in \mathbb{R}^n$, and*

$$b_1 \in W^{1-1/2p}_{p,\mu}(J; L_p(\Sigma)^n) \cap L_{p,\mu}(J; W_p^{2-1/p}(\Sigma)^n).$$

Then there is $\omega_0 \geq 0$ such that for each $\omega > \omega_0$, the two-phase Stokes problem (8.1) admits a unique solution (u, π, h) with regularity

$$\begin{aligned}
&u \in H^1_{p,\mu}(J; L_p(\Omega)^n) \cap L_{p,\mu}(J; H^2_p(\Omega\backslash\Sigma)^n), \quad \pi \in L_{p,\mu}(J; \dot{H}^1_p(\Omega\backslash\Sigma)),\\
&[\![\pi]\!] \in W^{1/2-1/2p}_{p,\mu}(J; L_p(\Sigma)) \cap L_{p,\mu}(J; W_p^{1-1/2p}(\Sigma)),\\
&h \in W^{2-1/2p}_{p,\mu}(J; L_p(\Sigma)) \cap H^1_{p,\mu}(J; W_p^{2-1/p}(\Sigma)) \cap L_{p,\mu}(J; W_p^{3-1/p}(\Sigma)),
\end{aligned}$$

if and only if the data $(f_u, g_d, g_b, g_u, g, f_h, u_0, h_0)$ satisfy the following regularity and compatibility conditions:

(a) $f_u \in L_{p,\mu}(J; L_p(\Omega)^n)$;

(b) $g_d \in L_{p,\mu}(J; H^1_p(\Omega\backslash\Sigma))$;

(c) $g_b \in W^{1-1/2p}_{p,\mu}(J; L_p(\partial\Omega)^n) \cap L_{p,\mu}(J; W^{2-1/p}_p(\partial\Omega)^n)$;
(d) $g_u \in W^{1-1/2p}_{p,\mu}(J; L_p(\Sigma)^n) \cap L_{p,\mu}(J; W^{2-1/p}_p(\Sigma)^n)$;
(e) $g \in W^{1/2-1/2p}_{p,\mu}(J; L_p(\Sigma)^n) \cap L_{p,\mu}(J; W^{1-1/p}_p(\Sigma)^n)$;
(f) $(g_d, g_b \cdot \nu_{\partial\Omega}, g_u \cdot \nu_\Sigma) \in H^1_{p,\mu}(J; \widehat{H}^{-1}_p(\Omega))$;
(g) $f_h \in W^{1-1/2p}_{p,\mu}(J; L_p(\Sigma)) \cap L_{p,\mu}(J; W^{2-1/p}_p(\Sigma))$;
(h) $u_0 \in W^{2\mu-2/p}_p(\Omega\backslash\Sigma)^n$, $h_0 \in W^{2+\mu-2/p}_p(\Sigma)$;
(i) $\operatorname{div} u_0 = g_d(0)$, $u_{0|\partial\Omega} = g_b(0)$, $[\![u_0]\!] = g_u(0)$, $2\mathcal{P}_\Sigma[\![\mu(x)D(u_0)\nu_\Sigma]\!] = \mathcal{P}_\Sigma g(0)$.

The solution map $(f_u, g_d, g_b, g_u, g, f_h, b, u_0, h_0) \mapsto (u, \pi, [\![\pi]\!], h)$ *is continuous between the corresponding spaces.*

The proof of this result will be carried out in Section 8.2.

It is possible to reduce the regularity of f_h to $f_h \in L_{p,\mu}(J; W^{2-1/p}_p(\Sigma))$, in which case the highest time regularity of h is dropped. This is the content of

Corollary 8.1.3. *Let the assumptions of Theorem* 8.1.2 *be valid. Then the result of that theorem remains valid when replacing the spaces for h and f_h by*

$$h \in H^1_{p,\mu}(J; W^{2-1/p}_p(\Sigma)) \cap L_{p,\mu}(J; W^{3-1/p}_p(\Sigma)), \quad f_h \in L_{p,\mu}(J; W^{2-1/p}_p(\Sigma)).$$

In addition, the result also remains valid with $\mu(x)\Delta$ *replaced by* $\operatorname{div}(2\mu(x)D(u))$.

This corollary is important for the semigroup associated to (8.1), the *two-phase Stokes semigroup with free boundary.* To construct this semigroup, we specialize to the case of homogeneous boundary and interface conditions as well as to the solenoidal situation

$$(g_d, g_b, g_u, g, b) = 0, \quad (\operatorname{div} f_u, f_u \cdot \nu_{\partial\Omega}, [\![f_u \cdot \nu_\Sigma]\!]) = 0.$$

Then the semigroup is given in the following way. We have

$$\nabla : \dot{H}^1_{p'}(\Omega \setminus \Sigma) \to L_{p'}(\Omega)^n$$

is bounded, hence $\nabla^* : L_p(\Omega; \mathbb{C})^n \to {}_0\dot{H}^{-1}_p(\Omega)$ is so as well. Define

$$X_0 := [L_p(\Omega)^n \cap \mathsf{N}(\nabla^*)] \times W^{2-1/p}_p(\Sigma),$$
$$X_1 := [H^2_p(\Omega \setminus \Sigma)^n \cap \mathsf{N}(\nabla^*)] \times W^{3-1/p}_p(\Sigma),$$

and A by means of

$$A(u, h) := (\varrho^{-1}(-\operatorname{div}(2\mu(x)D(u)) + \nabla\pi), -u \cdot \nu_\Sigma),$$

with domain

$$\mathsf{D}(A) := \{(u, h) \in X_1 : u_{|\partial\Omega} = 0, [\![u]\!] = 0, \mathcal{P}_\Sigma[\![\mu(x)D(u)\nu_\Sigma]\!] = 0\}.$$

Here $\pi = \pi(u, h)$ is given by the solution of the weak transmission problem

$$\begin{aligned}(\varrho^{-1}\nabla\pi|\nabla\phi)_\Omega &= (\varrho^{-1}\mathrm{div}(2\mu(x)D(u))|\nabla\phi)_\Omega, \quad \phi \in \dot{H}^1_{p'}(\Omega),\\ [\![\pi]\!] &= \sigma(x)\Delta_\Sigma h + [\![2\mu(x)\partial_\nu u \cdot \nu_\Sigma]\!] \quad \text{on } \Sigma.\end{aligned} \tag{8.3}$$

Note that this problem admits for each $(u, h) \in H^2_p(\Omega\backslash\Sigma)^n \times W^{3-1/p}_p(\Sigma)$ a solution in $\dot{H}^1_p(\Omega \setminus \Sigma)$ which is unique up to a constant, see Proposition 8.6.2.

Then, with $z = (u, h)$, $z_0 = (u_0, h_0)$, and $f = (f_u, f_h)$, Problem (8.1) is equivalent to the abstract evolution equation

$$\dot{z} + Az = f, \quad t > 0, \quad z(0) = z_0.$$

Corollary 8.1.3 shows that this problem has maximal L_p-regularity, i.e., $\omega + A \in \mathcal{MR}_p(X_0)$. Therefore, $-A$ generates an analytic C_0-semigroup in X_0. As the domain of A is compactly embedded into X_0, the spectrum of A consists only of eigenvalues of finite algebraic multiplicity, which are independent of p. Therefore, the number ω_0 in Theorem 8.1.2 is precisely the spectral bound $\mathsf{s}(-A)$, which will be shown to be 0 in Chapter 10.

1.2 The Two-Phase Stokes Operator

We specialize now to the case $h = 0$. This means that we consider the following pure two-phase Stokes problem.

$$\begin{aligned}\varrho(\partial_t + \omega)u - \mathrm{div}(2\mu(x)D(u)) + \nabla\pi &= \varrho f_u && \text{in } \Omega\backslash\Sigma,\\ \mathrm{div}\, u &= g_d && \text{in } \Omega\backslash\Sigma,\\ u &= g_b && \text{on } \partial\Omega,\\ [\![u]\!] &= g_u && \text{on } \Sigma,\\ [\![-2\mu(x)D(u) + \pi]\!]\nu_\Sigma &= g && \text{on } \Sigma,\\ u(0) &= u_0 && \text{in } \Omega\end{aligned} \tag{8.4}$$

on the time-interval $J = \mathbb{R}_+$, where $\omega \geq 0$ will be chosen sufficiently large. By Remark 8.1.1, Theorem 8.1.2 remains valid for the problem with $\mu(x)\Delta u$ replaced by $\mathrm{div}(2\mu(x)D(u))$.

By the same arguments as above we obtain the following result for the pure two-phase Stokes problem (8.4).

Theorem 8.1.4. *Let $p > n + 2$, $1 \geq \mu > 1/p$, $\Omega \subset \mathbb{R}^n$ a bounded domain with $\partial\Omega \in C^3$, $\Sigma \subset \Omega$ a closed hypersurface of class C^3 and ϱ_j be positive constants, $j = 1, 2$. Assume $\mu \in C^{1-}_b(\Omega \setminus \Sigma)$, $\mu > 0$, and set $J = \mathbb{R}_+$. Then there is $\omega_0 \geq 0$ such that for each $\omega > \omega_0$, the pure two-phase Stokes problem* (8.4) *admits a unique solution (u, π) with regularity*

$$\begin{aligned}&u \in H^1_{p,\mu}(J; L_p(\Omega)^n) \cap L_{p,\mu}(J; H^2_p(\Omega\backslash\Sigma)^n), \quad \pi \in L_{p,\mu}(J; \dot{H}^1_p(\Omega\backslash\Sigma)),\\ &[\![\pi]\!] \in W^{1/2-1/2p}_{p,\mu}(J; L_p(\Sigma)) \cap L_{p,\mu}(J; W^{1-1/2p}_p(\Sigma)),\end{aligned}$$

if and only if the data $(f_u, g_d, g_b, g_u, g, u_0)$ *satisfy the following regularity and compatibility conditions:*

(a) $f_u \in L_{p,\mu}(J; L_p(\Omega)^n)$;

(b) $g_d \in L_{p,\mu}(J; H^1_p(\Omega\backslash\Sigma))$;

(c) $g_b \in W^{1-1/2p}_{p,\mu}(J; L_p(\partial\Omega)^n) \cap L_{p,\mu}(J; W^{2-1/p}_p(\partial\Omega)^n)$;

(d) $g_u \in W^{1-1/2p}_{p,\mu}(J; L_p(\Sigma)^n) \cap L_{p,\mu}(J; W^{2-1/p}_p(\Sigma)^n)$;

(e) $g \in W^{1/2-1/2p}_{p,\mu}(J; L_p(\Sigma)^n) \cap L_{p,\mu}(J; W^{1-1/p}_p(\Sigma)^n)$;

(f) $(g_d, g_b \cdot \nu_{\partial\Omega}, g_u \cdot \nu_\Sigma) \in H^1_{p,\mu}(J; \widehat{H}^{-1}_p(\Omega))$;

(g) $u_0 \in W^{2\mu-2/p}_p(\Omega\backslash\Sigma)^n$;

(h) $\operatorname{div} u_0 = g_d(0)$, $u_{0|_{\partial\Omega}} = g_b(0)$, $[\![u_0]\!] = g_u(0)$, $2\mathcal{P}_\Sigma[\![\mu(x)D(u_0)\nu_\Sigma]\!] = \mathcal{P}_\Sigma g(0)$.

The solution map $(f_u, g_d, g_b, g_u, g, u_0) \mapsto (u, \pi, [\![\pi]\!])$ *is continuous between the corresponding spaces.*

Having this result at our disposal, we define the *two-phase Stokes operator in divergence form* in the following way. We set

$$X_0 = L_{p,\sigma}(\Omega) := L_p(\Omega)^n \cap \mathsf{N}(\nabla^*), \quad X_1 := H^2_p(\Omega \setminus \Sigma) \cap X_0,$$

and define A_S by means of

$$A_S u = \varrho^{-1}(-\operatorname{div}(2\mu(x)D(u)) + \nabla\pi), \quad u \in \mathsf{D}(A_S), \tag{8.5}$$

where

$$\mathsf{D}(A_S) := \{u \in X_1 : u_{|_{\partial\Omega}} = 0, \ [\![u]\!] = 0, \ \mathcal{P}_\Sigma[\![\mu(x)D(u)\nu_\Sigma]\!] = 0\}.$$

In this definition the pressure $\pi = \pi(u)$ is defined as the solution of the weak transmission problem

$$\begin{aligned} (\varrho^{-1}\nabla\pi|\nabla\phi)_\Omega &= (\varrho^{-1}\operatorname{div}(2\mu(x)D(u))|\nabla\phi)_\Omega, \quad \phi \in \dot{H}^1_{p'}(\Omega), \\ [\![\pi]\!] &= [\![2\mu(x)\partial_\nu u \cdot \nu_\Sigma]\!] \quad \text{on } \Sigma. \end{aligned} \tag{8.6}$$

Note that $\pi \in \dot{H}^1_p(\Omega \setminus \Sigma)$ is well-defined by Proposition 8.6.2. Then (8.4), for $(\omega, g_d, g_b, g_u, g) = 0$ and $(\operatorname{div} f_u, f_u \cdot \nu_{\partial\Omega}, [\![f_u \cdot \nu_\Sigma]\!]) = 0$, is equivalent to the abstract evolution equation

$$\dot{u} + A_S u = f_u, \quad t > 0, \quad u(0) = u_0. \tag{8.7}$$

Theorem 8.1.4 implies that this problem has maximal L_p-regularity, hence $-A_S$ is the generator of an analytic C_0-semigroup in X_0 with maximal L_p-regularity. As $\mathsf{D}(A_S)$ embeds compactly into X_0, the two-phase Stokes operator A_S has compact resolvent. Therefore, its spectrum consists only of eigenvalues of finite algebraic multiplicity, and is independent of p. So it is enough to study these eigenvalues for the case $p = 2$.

For this purpose we employ the energy method. Assume that $\lambda \in \mathbb{C}$ is an eigenvalue of A_S with eigenfunction u and corresponding pressure π. Taking the inner product in $L_2(\Omega;\mathbb{C})^n$ of the equation with u, after an integration by parts we get

$$\lambda \int_\Omega \varrho |u|^2\, dx = (\varrho A_S u|u)_\Omega = 2\int_\Omega \mu(x)|D(u)|_2^2\, dx.$$

This implies that λ is real and nonnegative. But by means of Korn's inequality and the no-slip condition on the outer boundary $\partial\Omega$, all eigenvalues are strictly positive. In particular, A_S is invertible.

This further implies that the Neumann-to-Dirichlet operator N_λ^S : $W_p^{1-1/p}(\Sigma) \to W_p^{2-1/p}(\Sigma)$ defined by the map $N_\lambda^S : g \mapsto (u|\nu_\Sigma)$, where u solves the problem

$$\begin{aligned} \lambda\varrho u - \operatorname{div}(2\mu(x)D(u)) + \nabla\pi &= 0 && \text{in } \Omega\setminus\Sigma,\\ \operatorname{div} u &= 0 && \text{in } \Omega\setminus\Sigma,\\ u &= 0 && \text{on } \partial\Omega,\\ [\![u]\!] &= 0 && \text{on } \Sigma,\\ -[\![2\mu(x)D(u)]\!]\nu_\Sigma + [\![\pi]\!]\nu_\Sigma &= g\nu_\Sigma && \text{on } \Sigma, \end{aligned}$$

is well-defined and an isomorphism, for each $\lambda \geq 0$. This operator will be studied in more detail in Chapter 10 in case $\mu = constant$.

1.3 The Quasi-Steady Two-Phase Stokes Problem

In this subsection we consider the Stokes flow problem, which reads

$$\begin{aligned} -\operatorname{div}(2\mu(x)D(u)) + \nabla\pi &= 0 && \text{in } \Omega\backslash\Sigma,\\ \operatorname{div} u &= 0 && \text{in } \Omega\backslash\Sigma,\\ u &= 0 && \text{on } \partial\Omega,\\ [\![u]\!] &= 0 && \text{on } \Sigma,\\ [\![-2\mu(x)D(u) + \pi]\!]\nu_\Sigma - \sigma(x)(\Delta_\Sigma h)\nu_\Sigma &= 0 && \text{on } \Sigma,\\ (\partial_t + \omega)h - (u|\nu_\Sigma) &= f_h && \text{on } \Sigma,\\ h(0) &= h_0 && \text{on } \Sigma, \end{aligned} \tag{8.8}$$

on the time-interval $J = \mathbb{R}_+$, where $\omega \geq 0$ will be chosen sufficiently large. Here the regularity classes for u and π are given by

$$u \in \mathbb{E}_u := L_{p,\mu}(J; H_p^2(\Omega\backslash\Sigma)^n), \quad \pi \in \mathbb{E}_\pi := L_{p,\mu}(J; \dot{H}_p^1(\Omega\backslash\Sigma)),$$

and that for h is

$$h \in H_{p,\mu}^1(J; W_p^{2-1/p}(\Sigma)) \cap L_{p,\mu}(J; W_p^{3-1/p}(\Sigma)).$$

For $X_0 = W_p^{2-1/p}(\Sigma)$, with the Neumann-to-Dirichlet operator N_0^S defined in the previous subsection, this problem is equivalent to the abstract evolution equation

$$\dot{h} + \omega h - N_0^S \sigma \Delta_\Sigma h = f_h, \quad t > 0, \quad h(0) = h_0,$$

in X_0. Defining the operator A in X_0 by means of

$$A = -N_0^S \sigma \Delta_\Sigma h, \quad \mathsf{D}(A) = W_p^{3-1/p}(\Sigma),$$

we show that $-A$ generates an analytic C_0-semigroup with maximal regularity.

For this purpose, we begin with the shifted quasi-stationary problem.

$$\begin{aligned}
\varrho\eta u - \operatorname{div}(2\mu(x)D(u)) + \nabla\pi &= \varrho f_u && \text{in } \Omega\backslash\Sigma,\\
\operatorname{div} u &= g_d && \text{in } \Omega\backslash\Sigma,\\
u &= g_b && \text{on } \partial\Omega,\\
[\![u]\!] &= g_u && \text{on } \Sigma,\\
[\![-2\mu(x)D(u) + \pi]\!]\nu_\Sigma - \sigma(x)(\Delta_\Sigma h)\nu_\Sigma &= g && \text{on } \Sigma,\\
(\partial_t + \omega)h - (u|\nu_\Sigma) &= f_h && \text{on } \Sigma,\\
u(0) = u_0 \quad \text{in } \Omega\backslash\Sigma, \quad h(0) &= h_0 && \text{on } \Sigma.
\end{aligned} \tag{8.9}$$

This can be solved by means of the same methods as in the previous subsections; the analysis is even simpler due to the missing time-derivatives in the bulk. In this way we obtain the following result.

Theorem 8.1.5. *Let $p \in (1, \infty)$, $1 \geq \mu > 1/p$, $\Omega \subset \mathbb{R}^n$ a bounded domain with $\partial\Omega \in C^3$, $\Sigma \subset \Omega$ a closed hypersurface of class C^3. Assume that $\mu \in C_b^{1-}(\Omega \setminus \Sigma)$, $\sigma \in C^{1-}(\Sigma)$ such μ, σ are positive, uniformly in x, and set $J = \mathbb{R}_+$.*

Then there are $\omega_0, \eta_0 \geq 0$ such that for each $\omega > \omega_0, \eta > \eta_0$, the quasi-steady two-phase Stokes problem (8.9) admits a unique solution (u, π, h) with regularity

$$\begin{gathered}
u \in L_{p,\mu}(J; H_p^2(\Omega\backslash\Sigma)^n), \quad \pi \in L_{p,\mu}(J; \dot{H}_p^1(\Omega\backslash\Sigma)),\\
h \in H_{p,\mu}^1(J; W_p^{2-1/p}(\Sigma)) \cap L_{p,\mu}(J; W_p^{3-1/p}(\Sigma)),
\end{gathered}$$

if and only if the data $(f_u, g_d, g_b, g_u, g, f_h, h_0)$ satisfy the following regularity and compatibility conditions:

(a) $f_u \in L_{p,\mu}(J; L_p(\Omega)^n)$;

(b) $g_d \in L_{p,\mu}(J; H_p^1(\Omega\backslash\Sigma))$;

(c) $g_b \in L_{p,\mu}(J; W_p^{2-1/p}(\partial\Omega)^n)$;

(d) $g_u \in L_{p,\mu}(J; W_p^{2-1/p}(\Sigma)^n)$;

(e) $g \in L_{p,\mu}(J; W_p^{1-1/p}(\Sigma)^n)$;

(f) $f_h \in L_{p,\mu}(J; W_p^{2-1/2p}(\Sigma))$;

(g) $h_0 \in W_p^{2+\mu-2/p}(\Sigma)$.

The solution map $(f_u, g_d, g_b, g_u, g, f_h, h_0) \mapsto (u, \pi, h)$ is continuous between the corresponding spaces.

Having this result at our disposal, by the above arguments it is evident that $-N^S_\lambda \sigma\Delta_\Sigma \in \mathcal{MR}_p(X_0)$. Next we proceed as in Section 6.6. We write $u = T_\eta g$ for the solution of

$$\begin{aligned} \varrho\eta u - \mathrm{div}(2\mu(x)D(u)) + \nabla\pi &= 0 && \text{in } \Omega\backslash\Sigma,\\ \mathrm{div}\, u &= 0 && \text{in } \Omega\backslash\Sigma,\\ u &= 0 && \text{on } \partial\Omega,\\ [\![u]\!] &= 0 && \text{on } \Sigma,\\ [\![-2\mu(x)D(u) + \pi]\!]\nu_\Sigma &= g\nu_\Sigma && \text{on } \Sigma. \end{aligned} \tag{8.10}$$

Then

$$T_0 g = T_\eta g + \eta(\eta + A_S)^{-1} T_0 g,$$

hence with $N^S_\eta g = [\![(T_\eta g|\nu_\Sigma)]\!]$

$$N^S_0 = N^S_\eta + \eta[\![((\eta + A_S)^{-1} T_0 g|\nu_\Sigma)]\!].$$

This shows that N^S_0 is a compact perturbation of N^S_η, and so with $-N^S_\eta \sigma\Delta_\Sigma$ also $-N^S_0 \sigma\Delta_\Sigma \in \mathcal{MR}_p(X_0)$.

The spectrum of $-N^S_0 \sigma\Delta_\Sigma$ will be investigated in Chapter 10.

8.2 Proof of Theorem 8.1.2

2.1 Regularity of the Pressure

In general the pressure π has no more regularity than stated in Theorem 8.1.2. However, there are situations where π enjoys extra time-regularity, as stated in the following

Proposition 8.2.1. *Assume in addition to the hypotheses of Theorem* 8.1.2 *that*

$$(g_d, u_0, h_0, \mathrm{div}\, f_u) = 0, \quad (g_b|\nu_{\partial\Omega}) = (f_u|\nu_{\partial\Omega}) = 0, \quad (g_u|\nu_\Sigma) = [\![(f_u|\nu_\Sigma)]\!] = 0.$$

Let (u, π, h) be a solution of (8.1). Then $\pi \in {}_0H^\alpha_{p,\mu}(J; L_p(\Omega))$, for each $\alpha \in (0, 1/2 - 1/2p)$. In addition, we have the following estimate

$$\begin{aligned} |\pi|_{L_{p,\mu}(\mathbb{R}_+;L_p(\Omega))} \le C\Big(&|\nabla u|_{L_{p,\mu}(\mathbb{R}_+;L_p(\Omega))} + |\nabla u|_{L_{p,\mu}(\mathbb{R}_+;L_p(\Sigma\cup\partial\Omega))}\\ &+ |[\![\pi]\!]|_{L_{p,\mu}(\mathbb{R}_+;L_p(\Sigma))}\Big), \end{aligned} \tag{8.11}$$

where $C > 0$ is a constant independent of ω.

Proof. Let $\psi \in L_{p'}(\Omega)$ be given and solve the problem

$$\begin{aligned} \varrho^{-1}\Delta\phi &= \psi - \bar{\psi} && \text{in } \Omega\backslash\Sigma,\\ \partial_\nu\phi &= 0 && \text{on } \partial\Omega,\\ [\![\phi]\!] &= 0 && \text{on } \Sigma,\\ [\![\varrho^{-1}\partial_\nu\phi]\!] &= 0 && \text{on } \Sigma, \end{aligned} \tag{8.12}$$

where $\bar{\psi} = (\psi|1)_\Omega/|\Omega|$. By Proposition 8.6.1, (8.12) has a unique solution $\phi \in H^2_{p'}(\Omega)$ with average zero and we have

$$|\nabla\phi|_{p'} + |\nabla^2\phi|_{p'} \le C|\psi|_{p'}.$$

As π is unique up to a constant, we may assume that π has average zero, and by $(f_u|\nabla\phi) = (u|\nabla\phi) = 0$ we then obtain by an integration by parts

$$\begin{aligned}(\pi|\psi)_\Omega &= (\pi|\psi-\bar{\psi})_\Omega = (\varrho^{-1}\pi|\Delta\phi)_\Omega\\ &= -\int_\Sigma [\![\pi]\!]\varrho^{-1}\partial_\nu\phi\, d\Sigma - (\varrho^{-1}\mathrm{div}(2\mu(x)D(u))|\nabla\phi)_\Omega\\ &= \int_\Omega \nabla u : \nabla\varrho^{-1}\mu\nabla\phi\, dx - \int_{\partial\Omega}\varrho^{-1}\mu(\partial_\nu u|\nabla\phi)\, d(\partial\Omega)\\ &\quad + \int_\Sigma\{[\![\varrho^{-1}\mu\partial_\nu u\nabla\phi]\!] - [\![\pi]\!]\varrho^{-1}\partial_\nu\phi\}\, d\Sigma\end{aligned}$$

Since $\nabla u \in {}_0H^{1/2}_{p,\mu}(J; L_p(\Omega)^{n\times n})$ and $[\![\pi]\!], \partial_k u_l \in {}_0W^{1/2-1/2p}_{p,\mu}(J; L_p(\Sigma))$, and $\partial_\nu u \in {}_0W^{1/2-1/2p}_{p,\mu}(J; L_p(\partial\Omega))$, applying ∂_t^α to this identity, we obtain the estimate

$$\begin{aligned}|\partial_t^\alpha\pi|_{L_{p,\mu}(L_p)} \le C\{&|\partial_t^\alpha\nabla u|_{L_{p,\mu}(L_p)} + |\partial_t^\alpha[\![\pi]\!]|_{L_{p,\mu}(L_p)}\\ &+ |\partial_t^\alpha\partial_\nu u|_{L_{p,\mu}(L_p)} + |\partial_t^\alpha\partial_\nu u|_{L_{p,\mu}(L_p)}\},\end{aligned}$$

for each $\alpha \in (0, 1/2 - 1/2p)$, hence $\pi \in {}_0H^\alpha_{p,\mu}(J; L_p(\Omega))$. The estimate (8.11) follows by an obvious argument. □

2.2 Reductions

It is convenient to reduce the problem to the case

$$(f_u,\ g_d,\ g_b\cdot\nu_{\partial\Omega},\ g_u\cdot\nu_\Sigma,\ u_0,\ h_0) = 0.$$

This can be achieved as follows. Suppose (u, π, h) is a solution of (8.1). Let us introduce a further dummy variable $q := [\![\pi]\!]$; note that $q \in \mathbb{F}_u$. We use the decomposition $(u, \pi, q, h) = (u_* + u_1, \pi_* + \pi_1, q_* + q_1, h_* + h_1)$, where

$$\begin{aligned}&e^{\omega t}h_*(t) = [2e^{-(I-\Delta_\Sigma)^{1/2}t} - e^{-2(I-\Delta_\Sigma)^{1/2}t}]h_0\\ &+[e^{-(I-\Delta_\Sigma)t} - e^{-2(I-\Delta_\Sigma)t}](I-\Delta_\Sigma)^{-1}\{(u_0|\nu_\Sigma) - (b(0)|\nabla_\Sigma h_0) + f_h(0)\},\quad t\ge 0.\end{aligned}$$

By Proposition 3.4.3, the function h_* belongs to $\mathbb{E}_h$ and satisfies $h_*(0) = h_0$ and

$$\partial_t h_*(0) + \omega h_*(0) = (u_0|\nu_\Sigma) - (b(0)|\nabla_\Sigma h_0) + f_h(0).$$

Then h_1 has initial value zero, and also $\partial_t h_1(0) = 0$. We have the estimate

$$\begin{aligned}|h_*|_{\mathbb{E}_h} + \omega|h_*|_{\mathbb{F}_h} &\le C\big(|h_0|_{W_p^{2+\mu-2/p}} + |u_0|_{W_p^{2\mu-2/p}} + |f_h(0)|_{W_p^{2\mu-3/p}}\\ &\qquad + |b(0)|_{W_p^{2\mu-3/p}}|\nabla_\Sigma h_0|_{W_p^{2\mu-3/p}}\big)\\ &\le C\big((1+|b|_{\mathbb{F}_h})|h_0|_{W_p^{2+\mu-2/p}} + |u_0|_{W_p^{2\mu-2/p}} + |f_h|_{\mathbb{F}_h}\big).\end{aligned}$$

Next we set $q_*(t) = e^{-(\omega-\Delta_\Sigma)t}q_0$, where

$$q_0 := ([\![2\mu\partial_\nu u_0]\!]|\nu_\Sigma) + \sigma\Delta_\Sigma h_0 + (g(0)|\nu_\Sigma)$$

is determined by the data, and define π_* as the solution of

$$\begin{aligned} \Delta\pi_* &= 0 && \text{in } \Omega\backslash\Sigma,\\ \partial_\nu\pi_* &= 0 && \text{on } \partial\Omega,\\ [\![\partial_{\nu_\Sigma}\pi_*/\varrho]\!] &= 0 && \text{on } \Sigma,\\ [\![\pi_*]\!] &= q_* && \text{on } \Sigma. \end{aligned}$$

Note that $q_* \in \mathbb{F}_u$ and $\pi_* \in \mathbb{E}_\pi$, by Proposition 8.6.2, and we have the estimates

$$\begin{aligned} |\pi_*|_{\mathbb{E}_\pi} &\le C|q_*|_{L_{p,\mu}(W_p^{1-1/p})},\\ |q_*|_{\mathbb{F}_u} + \omega^{1/2-1/2p}|q_*|_{L_{p,\mu}(L_p)} &\le C|q_0|_{W^{2\mu-1-3/p}}\\ &\le C\big(|u_0|_{W_p^{2\mu-2/p}} + |h_0|_{W_p^{2+\mu-2/p}} + |g|_{\mathbb{F}_u}\big). \end{aligned}$$

The function $u_* \in \mathbb{E}_u$ is defined as the solution of the parabolic problem

$$\begin{aligned} \varrho(\partial_t+\omega)u - \mu(x)\Delta u &= -\nabla\pi_* + \varrho f_u && \text{in } \Omega\backslash\Sigma,\\ u &= g_b && \text{on } \partial\Omega,\\ [\![u]\!] &= g_u && \text{on } \Sigma, \qquad (8.13)\\ [\![-2\mu(x)D(u)+\pi]\!]\nu_\Sigma &= g - q_*\nu_\Sigma + \sigma(\Delta_\Sigma h_*)\nu_\Sigma && \text{on } \Sigma,\\ u(0) &= u_0 && \text{in } \Omega\backslash\Sigma, \end{aligned}$$

which is uniquely solvable since the appropriate Lopatinskii-Shapiro conditions are satisfied; see Section 6.5. Thus, (u_1, π_1, q_1, h_1) solves (8.1) with data $(f_u, g_b, g, g_u, u_0, h_0) = 0$, and f_h replaced by $\tilde{f}_h = f_h - [(\partial_t+\omega)h_* - (u_*|\nu_\Sigma) + (b|\nabla_\Sigma h_*)] \in {}_0\mathbb{F}_h$. Finally, to remove g_d, we solve the transmission problem

$$\begin{aligned} \Delta\psi &= \tilde{g}_d && \text{in } \Omega\backslash\Sigma,\\ [\![\varrho\psi]\!] &= 0 && \text{on } \Sigma,\\ [\![\partial_{\nu_\Sigma}\psi]\!] &= 0 && \text{on } \Sigma,\\ \partial_{\nu_{\partial\Omega}}\psi &= 0 && \text{on } \partial\Omega, \end{aligned}$$

according to Proposition 8.6.1, as $\tilde{g}_d := g_d - \operatorname{div} u_*$ has mean value zero thanks to the compatibility condition (8.2). Since $\partial\Omega \in C^3$, the solution satisfies $\nabla\psi \in \mathbb{E}_u$. We note that ψ has trace zero at time zero, as $\tilde{g}_d(0) = 0$. Then setting

$$(u_2, \pi_2, h_2) = (u_1 - \nabla\psi, \pi_1 + \varrho(\partial_t\psi + \omega\psi) - \mu(x)\Delta\psi, h_1)$$

we see that we may assume $(f_u, g_d, g_b\cdot\nu_{\partial\Omega}, g_u\cdot\nu_\Sigma, u_0, h_0) = 0$. The only non-vanishing data which remain are (g, g_b, g_u, f_h); note that the time traces at $t = 0$ of these functions are zero as $\psi(0) = 0$. So for the reduced problem, Proposition 8.2.1 applies.

2.3 Flat Interface

In this subsection we consider the linear problem with constant coefficients for a flat interface. We use the identification $\mathbb{R}^{n-1} = \mathbb{R}^{n-1} \times \{0\}$, $(x, y) \in \hat{\mathbb{R}}^n = \mathbb{R}^{n-1} \times \dot{\mathbb{R}}$ and we write $u = (v, w)$, with v the tangential component of u.

$$\begin{aligned}
\varrho(\partial_t + \omega)u - \mu\Delta u + \nabla\pi &= \varrho f_u && \text{in } \hat{\mathbb{R}}^n,\\
\operatorname{div} u &= g_d && \text{in } \hat{\mathbb{R}}^n,\\
[\![u]\!] &= g_u && \text{on } \mathbb{R}^{n-1},\\
-[\![\mu\partial_y v]\!] - [\![\mu\nabla_x w]\!] &= g_v && \text{on } \mathbb{R}^{n-1},\\
-[\![2\mu\partial_y w]\!] + [\![\pi]\!] - \sigma\Delta h &= g_w && \text{on } \mathbb{R}^{n-1},\\
(\partial_t + \omega)h - w + (b|\nabla h) &= f_h && \text{on } \mathbb{R}^{n-1},\\
u(0) = u_0 \ \text{in } \hat{\mathbb{R}}^n, \quad h(0) &= h_0 && \text{on } \mathbb{R}^{n-1}.
\end{aligned} \tag{8.14}$$

It will be convenient to also use the decomposition $f_u = (f_v, f_w)$, $g = (g_v, g_w)$ into tangential and normal components.

The following result states that problem (8.14) admits maximal regularity. In particular, it defines an isomorphism between the solution space

$$\mathbb{E} := \mathbb{E}_u \times \mathbb{E}_\pi \times \mathbb{F}_u \times \mathbb{E}_h$$

and the product space of data $(f_u, g_d, g_u, g, f_h, b, u_0, h_0)$, which we denote for short by $\mathbb{F}$.

Theorem 8.2.2. *Let $p \in (1, \infty)$, $1 \geq \mu > 1/p$ be fixed, and assume that ϱ_j, μ_j, σ, $b \in \mathbb{R}^n$ are positive constants for $j = 1, 2$, and let $J = \mathbb{R}_+$. Then there is $\omega_0 \geq 0$ such that for each $\omega > \omega_0$, the Stokes problem with flat boundary (8.14) admits a unique solution (u, π, h) with regularity*

$$\begin{aligned}
&u \in H^1_{p,\mu}(J; L_p(\mathbb{R}^n)^n) \cap L_{p,\mu}(J; H^2_p(\hat{\mathbb{R}}^n)^n), \quad \pi \in L_{p,\mu}(J; \dot{H}^1_p(\hat{\mathbb{R}}^n)),\\
&[\![\pi]\!] \in W^{1/2-1/2p}_{p,\mu}(J; L_p(\mathbb{R}^{n-1})) \cap L_{p,\mu}(J; W^{1-1/p}_p(\mathbb{R}^{n-1})),\\
&h \in W^{2-1/2p}_{p,\mu}(J; L_p(\mathbb{R}^{n-1})) \cap H^1_{p,\mu}(J; W^{2-1/p}_p(\mathbb{R}^{n-1})) \cap L_{p,\mu}(J; W^{3-1/p}_p(\mathbb{R}^{n-1}))
\end{aligned}$$

if and only if the data $(f_u, g_d, g_u, g, f_h, u_0, h_0)$ satisfy the following regularity and compatibility conditions:

(a) $f_u \in L_{p,\mu}(J; L_p(\mathbb{R}^n)^n)$;

(b) $g_d \in L_{p,\mu}(J; H^1_p(\hat{\mathbb{R}}^n))$;

(c) $g = (g_v, g_w) \in W^{1/2-1/2p}_{p,\mu}(J; L_p(\mathbb{R}^{n-1})^n) \cap L_{p,\mu}(J; W^{1-1/p}_p(\mathbb{R}^{n-1})^n)$;

(e) $g_u \in W^{1-1/2p}_{p,\mu}(J; L_p(\mathbb{R}^{n-1})^n) \cap L_{p,\mu}(J; W^{2-1/p}_p(\mathbb{R}^{n-1})^n)$;

(e) $(g_d, g_u\cdot\nu_\Sigma)\in H^1_{p,\mu}(J;\widehat{H}^{-1}_p(\mathbb{R}^n))$;

(f) $f_h\in W^{1-1/2p}_{p,\mu}(J;L_p(\mathbb{R}^{n-1}))\cap L_{p,\mu}(J;W^{2-1/p}_p(\mathbb{R}^{n-1}))$;

(g) $u_0\in W^{2\mu-2/p}_p(\dot{\mathbb{R}}^n)^n$, $h_0\in W^{2+\mu-2/p}_p(\mathbb{R}^{n-1})$;

(h) $\operatorname{div} u_0 = g_d(0)$, $[\![u_0]\!] = g_u(0)$, $-[\![\mu\partial_y v_0]\!] - [\![\mu\nabla_x w_0]\!] = g_v(0)$.

The solution map $(f_u, g_d, g_u, g, f_h, b, u_0, h_0)\mapsto(u,\pi,[\![\pi]\!],h)$ *is continuous between the corresponding spaces.*

As in the previous chapter, for the localization procedure we also need estimates for the solution in terms of the data which are uniform in the parameter $\omega\geq\omega_0>0$. These follow directly from the proof of Theorem 8.2.2 but are elaborate in formulation, as they depend on all the many boundary data in question. For this purpose we fix some function spaces as follows.

$$\mathbb{E}^u_{0\mu} := L_{p,\mu}(\mathbb{R}_+;L_p(\mathbb{R}^n)^n),\quad \mathbb{E}^u_{1\mu} := H^1_{p,\mu}(\mathbb{R}_+;L_p(\mathbb{R}^n)^n)\cap L_{p,\mu}(\mathbb{R}_+;H^2_p(\dot{\mathbb{R}}^n)^n),$$
$$\mathbb{E}^h_{1\mu} := W^{2-1/p}_{p,\mu}(\mathbb{R}_+L_p(\mathbb{R}^{n-1}))\cap H^1_{p,\mu}(\mathbb{R}_+;L_p(\mathbb{R}^{n-1}))\cap L_{p,\mu}(\mathbb{R}_+;W^{3-1/p}_p(\mathbb{R}^{n-1})),$$
$$\mathbb{G}_{0\mu} := L_{p,\mu}(\mathbb{R}_+;\dot{H}^{-1}_p(\mathbb{R}^n)),\quad \mathbb{G}_{1\mu} := H^1_{p,\mu}(\mathbb{R}_+;\dot{H}^{-1}_p(\mathbb{R}^n))\cap L_{p,\mu}(\mathbb{R}_+;H^1_p(\dot{\mathbb{R}}^n)),$$
$$\mathbb{G}^0_\mu := L_{p,\mu}(\mathbb{R}_+;\widehat{H}^{-1}_p(\mathbb{R}^n)),\quad \mathbb{G}^1_\mu := H^1_{p,\mu}(\mathbb{R}_+;\widehat{H}^{-1}_p(\mathbb{R}^n)),$$
$$\mathbb{F}_{0\mu} := W^{1/2-1/2p}_{p,\mu}(\mathbb{R}_+;L_p(\mathbb{R}^{n-1}))\cap L_{p,\mu}(\mathbb{R}_+;W^{1-1/p}_p(\mathbb{R}^{n-1})),$$
$$\mathbb{E}^h_{0\mu} := \mathbb{F}_{1\mu} := W^{1-1/2p}_{p,\mu}(\mathbb{R}_+;L_p(\mathbb{R}^{n-1}))\cap L_{p,\mu}(\mathbb{R}_+;W^{2-1/p}_p(\mathbb{R}^{n-1})),$$

and $X_{\gamma,\mu} = W^{2(\mu-1/p)}_p(\dot{\mathbb{R}}^n)\times W^{2+\mu-2/p}_p(\mathbb{R}^{n-1})$. The estimates read as follows. For each $\omega_0>0$ there is a constant $C>0$ such that for all $\omega\geq\omega_0$ and all data subject to the corresponding compatibility conditions, the solution (u,π,h) satisfies

$$\begin{aligned}
&\omega|u|_{\mathbb{E}^u_{0\mu}} + |u|_{\mathbb{E}^u_{1\mu}} + |\nabla\pi|_{\mathbb{E}^u_{0\mu}} + |[\![\pi]\!]|_{\mathbb{F}_{0\mu}} + \omega|h|_{\mathbb{E}^h_{0\mu}} + |h|_{\mathbb{E}^h_{1\mu}}\\
&\leq C\{|(u_0,h_0)|_{X_{\gamma,\mu}} + |f_u|_{\mathbb{E}^u_{0\mu}} + |f_h|_{\mathbb{E}^h_{0\mu}} + (|g_d|_{\mathbb{G}_{1\mu}} + \omega|g_d|_{\mathbb{G}_{0\mu}})\\
&\quad + (|g_u|_{\mathbb{F}_{1\mu}} + \omega|e^{-L_\omega y}g_u|_{\mathbb{E}_{0\mu}}) + (|(g_v,g_w)|_{\mathbb{F}_{0\mu}} + \omega^{1/2}|e^{-L_\omega y}(g_v,g_w)|_{\mathbb{E}_{0\mu}})\\
&\quad + (|(g_d,g_u\cdot\nu_\Sigma)|_{\mathbb{G}^1_\mu} + \omega|(g_d,g_u\cdot\nu_\Sigma)|_{\mathbb{G}^0_\mu})\},
\end{aligned}\tag{8.15}$$

where $L_\omega = (\partial_t+\omega-\Delta_x)^{-1/2}$.

These results are the main tool for the proof of of Theorem 8.1.2. Observe that by the reductions explained above we may concentrate on the case $(f_u,g_d,u_0,h_0)=0$. As the proof is quite involved and leads to some further interesting results on the two-phase Neumann-to-Dirchlet operator for the Stokes problem, we postpone it to the next section.

Remark 8.2.3. By means of a perturbation argument, we can easily extend Theorem 8.2.2 to the case

$$(\mu,\sigma)\in C^{1-}_b(\dot{\mathbb{R}}^n)\times C^{1-}_b(\mathbb{R}^{n-1})\quad\text{with}\quad |\mu(x)-\mu_0|\leq\varepsilon,\ \ |\sigma(x)-\sigma_0|\leq\varepsilon$$

for $x \in \dot{\mathbb{R}}^n$ and $x \in \mathbb{R}^{n-1}$, respectively, where $\mu_0 > 0$ is constant in the phases, $\sigma_0 > 0$ is constant, and $\varepsilon > 0$ is sufficiently small.

2.4 Bent Interfaces

Next we consider the case of a bent interface. So let the interface Σ be given as a graph of a function $\phi : \mathbb{R}^{n-1} \to \mathbb{R}$ of class C_b^3; thus $\Sigma = \{(x, \phi(x)) : x \in \mathbb{R}^{n-1}\}$. The normal ν_Σ is then given by

$$\nu_\Sigma(x) = \beta(x) \begin{bmatrix} -\nabla_x \phi(x) \\ 1 \end{bmatrix}, \quad \beta(x) = 1/\sqrt{1 + |\nabla_x \phi(x)|^2},$$

and the Laplace-Beltrami operator for such a surface with

$$\bar{h}(t, x) = h(t, (x, \phi(x)))$$

reads as

$$\Delta_\Sigma h = \Delta \bar{h} - \beta^2 (\nabla^2 \bar{h} \nabla \phi | \nabla \phi) - \beta^2 [\Delta \phi - \beta^2 (\nabla^2 \phi \nabla \phi | \nabla \phi)](\nabla \phi | \nabla \bar{h}).$$

By the reduction argument explained above we may assume

$$(f_u, g_d, g_u \cdot \nu_\Sigma, u_0, h_0) = 0.$$

Set

$$\bar{u}(t, x, y) = u(t, x, y + \phi(x)), \quad \bar{\pi}(t, x, y) = \pi(t, x, y + \phi(x)),$$

for $t \in J = \mathbb{R}_+$, $x \in \mathbb{R}^{n-1}$, $y \neq 0$, and observe

$$\nabla u = \nabla \bar{u} - \nabla \phi \otimes \partial_y \bar{u}.$$

Then we obtain for the new variables $(\bar{u}, \bar{\pi}, \bar{h})$ the following problem. For convenience we drop the bars, and write $u = (v, w)$, $g = (g_v, g_w)$ as before.

$$\begin{aligned}
\varrho(\partial_t + \omega)u - \mu \Delta u + \nabla \pi &= \mu B_1(u, \pi) && \text{in } \dot{\mathbb{R}}^n, \\
\operatorname{div} u &= B_2 u && \text{in } \dot{\mathbb{R}}^n, \\
[\![u]\!] &= g_u && \text{on } \mathbb{R}^{n-1}, \\
-[\![\mu \partial_y v]\!] - [\![\mu \nabla_x w]\!] &= (g_v + \nabla \phi g_w)/\beta + B_3(u) && \text{on } \mathbb{R}^{n-1}, \\
-[\![2\mu \partial_y w]\!] + [\![\pi]\!] - \sigma \Delta h &= (g_w/\beta) + B_4(u, h) && \text{on } \mathbb{R}^{n-1}, \\
(\partial_t + \omega)h - w + (b|\nabla h) &= f_h + B_5 u + B_6 h && \text{on } \mathbb{R}^{n-1}, \\
u(0) = 0 \text{ in } \dot{\mathbb{R}}^n, \quad h(0) &= 0 \text{ on } \mathbb{R}^{n-1}.
\end{aligned} \tag{8.16}$$

Here we have set

$$\begin{aligned} B_1(u,\pi) &= |\nabla\phi|^2\partial_y^2 u - 2(\nabla\phi|\nabla_x)\partial_y u + (\nabla\phi)\partial_y\pi - (\Delta\phi)\partial_y u \\ B_2 u &= (\nabla\phi|\partial_y u), \\ B_3(u) &= -[\![\mu(\nabla_x v + [\nabla_x v]^{\mathsf{T}})]\!]\nabla\phi + \partial_y v|\nabla\phi|^2 \\ &\quad + ([\![\mu\partial_y w]\!]/\beta^2 - ([\![\mu\nabla_x w]\!]|\nabla\phi))\nabla\phi \\ B_4(u,h) &= -([\![\mu(\partial_y v + \nabla_x w)]\!]|\nabla\phi) + [\![\mu\partial_y w]\!]|\nabla\phi|^2 + \sigma(\Delta_\Sigma h - \Delta h) \\ B_5 u &= (\beta-1)w - \beta(\nabla\phi|v) = -\frac{\beta^2|\nabla\phi|^2}{1+\beta}w - \beta(\nabla\phi|v) \\ B_6 h &= \beta^2[(b|\nabla\phi) - (b|e_n)|\nabla\phi|^2](\nabla\phi|\nabla h). \end{aligned}$$

Now suppose (u,π,h) belongs to the maximal regularity class. We estimate the perturbations B_j as follows:

$$\begin{aligned} |B_1(u,\pi)|_{L_{p,\mu}(L_p)} &\le \|\nabla\phi\|_{L_\infty}[(2+|\nabla\phi|_{L_\infty})|\nabla^2 u|_{L_{p,\mu}(L_p)} + |\nabla\pi|_{L_{p,\mu}(L_p)}] \\ &\quad + |\Delta\phi|_{L_\infty}|\nabla u|_{L_{p,\mu}(L_p)}, \\ |B_2 u|_{L_{p,\mu}(H^1_p)} &\le |\nabla\phi|_{L_\infty}|\nabla^2 u|_{L_{p,\mu}(L_p)} + (|\nabla^2\phi|_{L_\infty} + |\nabla\phi|_{L_\infty})|\nabla u|_{L_{p,\mu}(L_p)}, \\ |\partial_t B_2 u|_{L_{p,\mu}(H^{-1}_p)} &\le |\nabla\phi\partial_t u|_{L_{p,\mu}(L_p)} \le |\nabla\phi|_{L_\infty}|\partial_t u|_{L_{p,\mu}(L_p)}, \\ |B_3(u)|_{W^s_{p,\mu}(L_p)} &\le C|\nabla\phi|_{L_\infty}(1+|\nabla\phi|_{L_\infty})|\nabla u|_{W^s_{p,\mu}(L_p)} \\ |B_4(u,h)|_{W^s_{p,\mu}(L_p)} &\le C|\nabla\phi|_{L_\infty}(1+\|\nabla\phi\|_{L_\infty})[|\nabla u|_{W^s_{p,\mu}(L_p)} + |\nabla^2 h|_{W^s_{p,\mu}(L_p)}] \\ &\quad + C|\nabla^2\phi|_{L_\infty}|\nabla\phi|_{L_\infty}|\nabla h|_{W^s_{p,\mu}(L_p)}, \\ |B_5 u|_{W^{1-1/2p}_{p,\mu}(L_p)} &\le 2|\nabla\phi|_{L_\infty}|u|_{W^{1-1/2p}_{p,\mu}(L_p)}. \end{aligned}$$

Here C denotes a constant only depending on the parameters μ and σ, and we have set $s = 1/2 - 1/2p$. For the estimates in $L_{p,\mu}(J; W^{1-1/p}_p(\mathbb{R}^{n-1}))$ we obtain

$$\begin{aligned} |B_3(u)|_{L_{p,\mu}(W^{2s}_p)} &\le C|\nabla\phi|_{L_\infty}(1+|\nabla\phi|_{L_\infty})\nabla u|_{L_{p,\mu}}(W^{2s}_p) \\ &\quad + C(1+|\nabla\phi|_{L_\infty})|\nabla^2\phi|_\infty|\nabla u|_{L_{p,\mu}(L_p)} \\ |B_4(u,h)|_{L_{p,\mu}(W^{2s}_p)} &\le C(1+|\nabla\phi|_{L_\infty})\{|\nabla\phi|_{L_\infty}|\nabla u|_{L_{p,\mu}(W^{2s}_p)} \\ &\quad + |\nabla\phi|_{L_\infty}|\nabla^2 h|_{L_{p,\mu}(W^{2s}_p)} + |\nabla^2\phi|_{L_\infty}[|\nabla u|_{L_{p,\mu}(L_p)} \\ &\quad + |\nabla h|_{L_{p,\mu}(H^1_p)}] \\ &\quad + C(|\nabla^3\phi|_{L_\infty} + |\nabla^2\phi|^2_{L_\infty})|\nabla\phi|_{L_\infty}|\nabla h|_{L_{p,\mu}(L_p)}\}. \\ |B_5 u|_{L_{p,\mu}(W^{1+2s}_p)} &\le C|\nabla\phi|_{L_\infty}\{|u|_{L_{p,\mu}(W^{1+2s}_p)} + |\nabla^2\phi|_{L_\infty}|u|_{L_{p,\mu}(H^1_p)}\} \\ &\quad + C(|\nabla^3\phi|_{L_\infty} + |\nabla^2\phi|^2_{L_\infty})|u|_{L_{p,\mu}(L_p)}. \end{aligned}$$

Here C denotes a constant only depending on μ, σ, p, and $2s = 1-1/p$. To estimate $B_6 h$ we note that $\mathbb{F}_h$ is a Banach algebra since $p > n+2$. This yields

$$|B_6 h|_{\mathbb{F}_h} \le C|\nabla\phi|_{L_\infty}|b||h|_{\mathbb{E}_h}.$$

To solve the problem (8.16), let $z = (u, \pi, [\![\pi]\!], h) \in {}_0\mathbb{E}$, where ${}_0\mathbb{E}$ means the solution space with zero time trace at $t = 0$, $f := (0, 0, g_u, g_v/\beta, g_w/\beta, f_h) \in {}_0\mathbb{F}$, the space of data with zero time trace, and let $B : {}_0\mathbb{E} \to {}_0\mathbb{F}$ is defined by

$$Bz = (B_1(u, \pi), 0, B_2 u, B_3(u, [\![\pi]\!], h), B_4(u, h), B_5 u + B_6 h).$$

Denoting the isomorphism from ${}_0\mathbb{E}$ to ${}_0\mathbb{F}$ defined by the left-hand side of (8.16) by L, we may rewrite problem (8.16) in abstract form as

$$Lz = Bz + f. \tag{8.17}$$

The above estimates for the components of B imply

$$|Bz|_{\mathbb{F}} \leq C|\nabla\phi|_{L_\infty}|z|_{\mathbb{E}} + M[|u|_{L_{p,\mu}(H^1_p)} + |\nabla h|_{W^s_{p,\mu}(L_p)\cap L_{p,\mu}(H^1_p)}],$$

with a constant $C > 0$ depending only on the parameters and $M > 0$ also depending on $|\nabla\phi|_{BUC^2}$. Let $\eta > 0$ be given and suppose $|\nabla\phi|_{L_\infty} < \eta$. By means of an interpolation argument we find a constant $\gamma > 0$, depending only on p and a constant $M(\eta) > 0$, such that

$$|Bz|_{\mathbb{F}} \leq C[2\eta + \omega^{-\gamma}M(\eta)]|z|_{\mathbb{E}}, \quad z \in {}_0\mathbb{E}.$$

Choosing first $\eta > 0$ small and then $\omega > 0$ large enough, we can solve (8.17) by a Neumann series argument.

Remark 8.2.4. **(i)** By means of a perturbation argument, we can extend the above results for bent half-spaces to the case

$$(\mu, \sigma) \in C_b^{1-}(\mathbb{R}^n \setminus \Sigma) \times C_b^{1-}(\Sigma) \quad \text{with} \quad |\mu(x) - \mu_0| \leq \varepsilon, \ |\sigma(x) - \sigma_0| \leq \varepsilon$$

for $x \in \mathbb{R}^n \setminus \Sigma$ and $x \in \Sigma$, respectively, where $\mu_0 > 0$ is constant in the phases, $\sigma_0 > 0$ is constant, and $\varepsilon > 0$ is sufficiently small.

(ii) Estimates (8.15) remain valid in the case of a bent half-space, and as we have seen in Chapter 7, we may replace the terms involving the semigroup $e^{-L_\omega y}$ by the weaker norms involving only the norms of the boundary data, e.g. the terms $\omega|e^{-L_\omega y}g_u|_{\mathbb{E}_{0\mu}}$ can be weakend to $\omega^{1-1/p}|g_u|_{L_p}$. Only these estimates are needed for the localization in the next subsection.

2.5 General Bounded Domains

Here we use once more the method of localization. By assumption, $\partial\Omega$ is of class C^3 and Σ will eventually be even real analytic, so in particular of class C^4. We cover Σ by N balls $B(x_j, r/2)$ with radius $r > 0$ and centers $x_j \in \Sigma$ such that $\Sigma \cap B(x_j, r)$ can be parameterized over the tangent space $T_{x_j}\Sigma$ by a function $\theta_j \in C^3$ such that $|\nabla\theta_j|_{L_\infty} \leq \eta$, with $\eta > 0$ small, as in the previous subsection. We extend these functions θ_j to all of $T_{x_j}\Sigma$ retaining the bound on $\nabla\theta_j$. This way we have created N bent interfaces Σ_j to which the result proved in the previous subsection applies. We also suppose that $B(x_j, r) \subset \Omega$ for each j. Set $U := \Omega \setminus \bigcup_{j=1}^N \bar{B}_{r/2}(x_j)$ and $U_j = B(x_j, r)$, $j = 1, \dots, N$. The open set U consists of one component U_0 characterized by $\partial\Omega \subset \bar{U}_0$ and an open set, say U_{N+1}, which is interior to Σ, i.e., $U_{N+1} \subset \Omega_1$. Fix a partition of unity $\{\varphi_j\}_{j=0}^{N+1}$ subordinate to the covering $\{U_j\}_{j=0}^{N+1}$ of Ω, i.e., $\varphi_j \in \mathcal{D}(\mathbb{R}^n)$, $0 \le \varphi_j \le 1$, and $\sum_{j=0}^{N+1} \varphi_j \equiv 1$. Note that $\varphi_0 = 1$ in a neighbourhood of $\partial\Omega$. Let $\tilde{\varphi}_j$ denote cut-off functions with support in U_j such that $\tilde{\varphi}_j = 1$ on the support of φ_j. We extend the coefficients μ, σ in each chart U_j as in Section 6.4.

Let $z := (u, \pi, q, h)$ with $q = [\![\pi]\!]$ be a solution of (8.1) where we may assume $(f_u, g_d, u_0, h_0) = 0$ and $(g_b \cdot \nu_{\partial\Omega}, g_u \cdot \nu_\Sigma) = 0$. We then set $u_j = \varphi_j u$, $\pi_j = \varphi_j \pi$, $q_j = \varphi_j q$, $h_j = \varphi_j h$, as well as $g_{bj} = \varphi_j g_b$, $g_{uj} = \varphi_j g_u$, $g_j = \varphi_j g$, and $f_{hj} = \varphi_j f_h$. Then for $j = 1, \dots, N$, $z_j := (u_j, \pi_j, q_j, h_j)$ satisfies the problem

$$
\begin{aligned}
\varrho(\partial_t + \omega)u_j - \mu_j \Delta u_j + \nabla \pi_j &= F_j(u, \pi) && \text{in } \mathbb{R}^n \setminus \Sigma_j,\\
\operatorname{div} u_j &= (\nabla \varphi_j | u) && \text{in } \mathbb{R}^n \setminus \Sigma_j,\\
[\![u_j]\!] &= g_{uj} && \text{on } \Sigma_j,\\
[\![-\mu_j([\nabla u_j] + [\nabla u_j]^{\mathsf{T}}) + \pi_j]\!]\nu_{\Sigma_j} - \sigma_j(\Delta_{\Sigma_j} h_j)\nu_{\Sigma_j} &= g_j + G_j(u, h) && \text{on } \Sigma_j,\\
(\partial_t + \omega)h_j - (u_j | \nu_{\Sigma_j}) + (b_0 | \nabla_\Sigma h_j) &= f_{hj} + F_{hj}(h) && \text{on } \Sigma_j,\\
u_j(0) = 0 \ \text{ in } \mathbb{R}^n \setminus \Sigma_j, \quad h_j(0) &= 0 && \text{on } \Sigma_j.
\end{aligned}
\tag{8.18}
$$

Here we used the abbreviations

$$
\begin{aligned}
F_j(u, \pi) &= \pi \nabla \varphi_j - \mu[\Delta, \varphi_j]u,\\
-G_j(u, h) &= [\![\mu(\nabla \varphi_j \otimes u + u \otimes \nabla \varphi_j)]\!]\nu_{\Sigma_j} - \sigma_j[\Delta_{\Sigma_j}, \varphi_j]h\nu_{\Sigma_j},
\end{aligned}
$$

where $[A, B] = AB - BA$, and

$$F_{hj}(h) = (b_1 | \nabla_\Sigma h)\varphi_j.$$

For $j = 0$ we have the standard one-phase Stokes problem with parameters ϱ_2, μ_2 on Ω with Dirichlet boundary conditions on $\partial\Omega$, i.e.,

$$
\begin{aligned}
\varrho_2(\partial_t + \omega)u_0 - \mu_0 \Delta u_0 + \nabla \pi_0 &= F_0(u, \pi) && \text{in } \Omega,\\
\operatorname{div} u_0 &= (\nabla \varphi_0 | u) && \text{in } \Omega,\\
u_0 &= g_{b0} && \text{on } \partial\Omega,\\
u_0(0) &= 0 && \text{in } \Omega.
\end{aligned}
$$

For $j = N + 1$ we obtain the one-phase Stokes problem on $\mathbb{R}^n$ with parameters ϱ_1, μ_1, i.e.,

$$
\begin{aligned}
\varrho_1(\partial_t + \omega)u_{N+1} - \mu_1 \Delta u_{N+1} + \nabla \pi_{N+1} &= F_{N+1}(u, \pi) && \text{in } \mathbb{R}^n,\\
\operatorname{div} u_{N+1} &= (\nabla \varphi_{N+1} | u) && \text{in } \mathbb{R}^n,\\
u_{N+1}(0) &= 0 && \text{in } \mathbb{R}^n.
\end{aligned}
$$

Concentrating on $j = 1, \dots, N$, we first note that $[\Delta, \varphi_j]$ are differential operators of order 1, hence if $u \in {}_0\mathbb{E}_u$ then

$$[\Delta, \varphi_j]u \in {}_0H^{1/2}_{p,\mu}(J; L_p(\mathbb{R}^n)^n) \cap L_{p,\mu}(J; H^1_p(\mathbb{R}^n \setminus \Sigma_j)^n).$$

Since $(f_u, g_d, u_0, h_0) = 0$ and $(g_b \cdot \nu_{\partial\Omega}, g_u \cdot \nu_\Sigma) = 0$, the pressure π belongs to

$$\pi \in {}_0H^{\alpha}_{p,\mu}(J; L_p(\Omega)) \cap L_{p,\mu}(J; H^1_p(\Omega \setminus \Sigma))$$

by Proposition 8.2.1, hence we have

$$F_j(u, \pi) \in {}_0H^{\alpha}_{p,\mu}(J; L_p(\mathbb{R}^n)^n) \cap L_{p,\mu}(J; H^1_p(\mathbb{R}^n \setminus \Sigma_j)^n),$$

for some fixed $0 < \alpha < \frac{1}{2} - \frac{1}{2p}$. Similarly we have

$$\nabla\varphi_j(u|\nu_{\Sigma_j}) + (\nabla\varphi_j|\nu_{\Sigma_j})u \in {}_0W^{1-1/2p}_{p,\mu}(J; L_p(\Sigma_j)^n) \cap L_{p,\mu}(J; W^{2-1/p}_p(\Sigma_j)^n),$$

and since $[\Delta_{\Sigma_j}, \varphi_j]$ is of order 1 as well, we obtain

$$[\Delta_{\Sigma_j}, \varphi_j]h \in {}_0H^1_{p,\mu}(J; W^{1-1/p}_p(\Sigma_j)) \cap L_{p,\mu}(J; W^{2-1/p}_p(\Sigma_j)).$$

This shows

$$G_j(u, h) \in {}_0W^{1-1/2p}_{p,\mu}(J; L_p(\Sigma_j)^n) \cap L_{p,\mu}(J; W^{2-1/p}_p(\Sigma_j)^n).$$

The terms $F_{hj}(h)$ can be estimated in the following way.

$$|F_{hj}(h)|_{\mathbb{F}_h} \le C|b_1|_{\mathbb{F}_h}|h|_{\mathbb{F}_h} \le C\eta|h|_{\mathbb{E}_h},$$

by the Banach algebra property of $\mathbb{F}_h$, provided $|b_1|_{\mathbb{F}_h} \le \eta$. We assume this for a moment, and return later to the general case. In order to be able to apply Proposition 8.2.1, we decompose

$$F_j(u, \pi) = \tilde{F}_j(u, \pi) + \nabla\psi_j,$$

such that $\mathrm{div}\tilde{F}_j(u, \pi) = 0$ in $\mathbb{R}^n \backslash \Sigma_j$ and $([\![\tilde{F}_j(u, \pi)]\!]|\nu_{\Sigma_j}) = 0$ on Σ_j. We may take $\tilde{F}_j(u, \pi)$ as the Helmholtz projection of $F_j(u, \pi)$ in $\mathbb{R}^n$. Then

$$\tilde{F}_j(u, \pi) \in {}_0H^{\alpha}_{p,\mu}(J; L_p(\mathbb{R}^n)^n) \cap L_{p,\mu}(J; H^{2\alpha}_p(\mathbb{R}^n)^n).$$

Also, we decompose $u_j = \tilde{u}_j + \nabla\phi_j$, where ϕ_j solves the transmission problem

$$\begin{aligned} \Delta\phi_j &= (\nabla\varphi_j|u) && \text{in } \mathbb{R}^n \backslash \Sigma_j, \\ [\![\varrho\phi_j]\!] &= 0 && \text{on } \Sigma_j, \\ [\![\partial_\nu\phi_j]\!] &= 0 && \text{on } \Sigma_j. \end{aligned} \tag{8.19}$$

Note that

$$\nabla\phi_j \in {}_0H^1_{p,\mu}(J; H^1_p(\mathbb{R}^n \setminus \Sigma_j)^n) \cap {}_0H^{1/4}_{p,\mu}(J; H^2_p(\mathbb{R}^n \setminus \Sigma_j)^n), \tag{8.20}$$

by Proposition 8.6.1 as Σ_j is C^3. For the jump of its trace on Σ_j we then have

$$[\![\nabla\phi_j]\!] \in {}_0H^1_{p,\mu}(J; W^{1-1/p}_p(\Sigma_j)^n) \cap {}_0H^{1/4}_{p,\mu}(J; W^{2-1/p}_p(\Sigma_j)^n),$$

and its normal part vanishes, by construction. Furthermore, we have

$$[\![\mu\nabla^2\phi_j]\!] \in {}_0W^{1-1/2p}_{p,\mu}(J; L_p(\Sigma_j)^{n\times n}) \cap {}_0H^{1/4}_{p,\mu}(J; W^{1-1/p}_p(\Sigma_j)^{n\times n}).$$

Then we set

$$\tilde\pi_j = \pi_j - \psi_j + \varrho(\partial_t + \omega)\phi_j - \mu\Delta\phi_j,$$

and observe that on Σ_j

$$\tilde q_j := [\![\tilde\pi_j]\!] = [\![\pi_j]\!] - [\![\mu\Delta\phi_j]\!] = [\![\pi_j]\!] - [\![\mu(\nabla\varphi_j|u)]\!],$$

since by construction ψ_j and $\varrho\phi_j$ have no jump across Σ_j. Now $\tilde z_j := (\tilde u_j, \tilde\pi_j, \tilde q_j, h_j)$ satisfies the problem

$$\begin{aligned}
\varrho(\partial_t + \omega)\tilde u_j - \mu_j\Delta\tilde u_j + \nabla\tilde\pi_j &= \tilde F_j(u,\pi) && \text{in } \mathbb{R}^n\backslash\Sigma_j,\\
\operatorname{div}\, \tilde u_j &= 0 && \text{in } \mathbb{R}^n\backslash\Sigma_j,\\
[\![\tilde u_j]\!] &= g_{uj} - [\![\nabla\phi_j]\!] && \text{on } \Sigma_j,\\
[\![-\mu([\nabla\tilde u_j] + [\nabla\tilde u_j]^{\mathsf T}) + \tilde\pi_j]\!]\nu_{\Sigma_j} - \sigma_j(\Delta_{\Sigma_j}h_j)\nu_{\Sigma_j} &= g_j + \tilde G_j(u,h) && \text{on } \Sigma_j,\\
(\partial_t + \omega)h_j - (\tilde u_j|\nu_{\Sigma_j}) + (b_0|\nabla_\Sigma h_j) &= f_{hj} + \tilde F_{hj}(h) && \text{on } \Sigma_j,\\
\tilde u_j(0) = 0 \ \text{ in } \mathbb{R}^n\backslash\Sigma_j, \quad h_j(0) &= 0 && \text{on } \Sigma_j.
\end{aligned} \tag{8.21}$$

Here $\tilde G_j$ and $\tilde G_{hj}$ are given by

$$\begin{aligned}
\tilde G_j(u,h) &= G_j(u,h) + [\![2\mu\nabla^2\phi_j]\!]\nu_{\Sigma_j} - [\![\mu(\nabla\varphi_j|u)]\!]\nu_{\Sigma_j},\\
\tilde F_{hj}(h) &= F_{hj}(h) + \partial_{\nu_{\Sigma_j}}\phi_j.
\end{aligned}$$

For the remaining charts with index $j = 0, N+1$, i.e., the one-phase problems, the procedure is similar.

We write (8.21) abstractly as

$$L_j\tilde z_j = H_j + B_j z,$$

and by Theorem 8.1.2 for bent interfaces we obtain an estimate of the form

$$|\tilde z_j|_{\mathbb{E}} \le C_0(|H_j|_{\mathbb{F}} + |B_j z|_{\mathbb{F}}),$$

with some constant C_0 independent of j. Here $\mathbb{E}$ means the space of solutions and $\mathbb{F}$ the space of data, equipped with the norms including the terms containing ω, as defined by (8.15); see also Remark 8.2.4(ii). Since all components of $B_j z$ (except for $\tilde{F}_{hj}(h)$) have some extra regularity, there is an exponent $\gamma > 0$ and a constant C_1 independent of j such that

$$|B_j z|_{\mathbb{F}} \leq \omega^{-\gamma} C_1 |z|_{\mathbb{E}}.$$

In addition, by Proposition 8.2.1 we obtain

$$|\tilde{\pi}_j|_{L_{p,\mu}(J;L_p(\Omega))} \leq C_1 |\tilde{z}|^{\gamma}_{\mathbb{E}_{0\mu}} |\tilde{z}|^{1-\gamma}_{\mathbb{E}_{1\mu}} \leq C_2(|H_j|_{\mathbb{F}} + \omega^{-\gamma}|z|_{\mathbb{E}}).$$

Applying $\partial_t + \omega$ to (8.19) and using the equation for u and the regularity of π, we also obtain

$$|(\partial_t + \omega)\phi_j|_{L_{p,\mu}(J;L_p(\Omega))} \leq C_3(|H_j|_{\mathbb{F}} + \omega^{-\gamma}|z|_{\mathbb{E}}),$$

and then also

$$|z_j|_{\mathbb{E}} \leq C_4 |H_j|_{\mathbb{F}} + \omega^{-\gamma}|z|_{\mathbb{E}}.$$

Summing up over all j yields $z = \sum_j z_j$, hence

$$|z|_{\mathbb{E}} \leq C_5 |H|_{\mathbb{F}} + \omega^{-\gamma} C_7 |z|_{\mathbb{E}}.$$

Therefore, choosing $\omega > 0$ large enough, we obtain the a priori estimate

$$|z|_{\mathbb{E}} \leq C_6 |H|_{\mathbb{F}}. \tag{8.22}$$

Hence we may conclude that the operator $L : {}_0\mathbb{E} \to {}_0\mathbb{F}$ which maps solutions to their data is injective and has closed range, i.e., L is a semi-Fredholm operator.

In case $|b_1|_{\mathbb{F}_h}$ is not small, we also have to localize in time. For this purpose we first find $K \in \mathbb{N}$ such that $|\tau_{3K} b_1|_{\mathbb{F}_h} \leq \eta/2$, where τ_c denotes translation by c. Then we decompose the interval $[0, a]$ into equal parts $[t_l, t_{l+1})$, where $t_l = \delta l$, $l = 0, \ldots, 3K-1$, $\delta = a/3K$ and introduce a partition of unity for this decomposition, for example with

$$\chi(s) = 1\ , |s| \leq 1, \quad \chi(s) = 2 - |s|,\ 1 \leq |s| \leq 2, \quad \chi(s) = 0,\ |s| \geq 2,$$

we set

$$\chi_0(s) = \chi(s/\delta), \quad \chi_K(s) = 1 - \sum_{l=0}^{K-1} \chi_l(s),$$
$$\chi_l(s) = \chi(s/\delta - 3l), \quad l = 1, \ldots, K-1,$$

and we let $\tilde{\chi}_l \in \mathcal{D}(\mathbb{R})$ denote cut off functions with are 1 on the support of χ_l. Then localizing also in time we obtain as above problems for $u_{jl} = u\varphi_j\chi_l$, $\pi_{jl} = \pi\varphi_j\chi_l$, $h_{jl} = h\varphi_j\chi_l$. The price we have to pay are extra terms coming from the time derivatives of u and h, i.e., $u\partial_t\chi_l$ and $h\partial_t\chi_l$, which do not matter. Then if $\delta > 0$ is sufficiently small we obtain for

$$F_{hjl} = \chi_l(b_{jl}|\nabla_\Sigma\varphi_j)h - \chi_l\varphi_j(b - b_{jl}|\nabla_\Sigma h)$$

an estimate of the form

$$|F_{hjl}|_{\mathbb{F}_h} \leq C\big(|\chi_l\varphi_j(b-b_{jl})|_\infty |h|_{\mathbb{F}_h} + C|b|_{\mathbb{F}_h}|h|_{L_{p,\mu_l}(\mathbb{R}_+;W_p^{1-1/p}(\Sigma))},$$

where $\mu_0 = \mu$, and $\mu_l = 1$ for $l \geq 1$. This implies by interpolation, with some $\gamma > 0$

$$|F_{hjl}|_{\mathbb{F}_h} \leq C(\delta + \omega^{-\gamma})|h|_{\mathbb{E}_h} \leq \eta |h|_{\mathbb{E}_h},$$

provided $\delta > 0$ is sufficiently small and $\omega > 0$ is sufficiently large. Now we may proceed as before.

It remains to prove surjectivity of L. For this we employ again the continuation method for semi-Fredholm operators. The estimates derived above are uniform in the densities ϱ_j and also in the viscosities μ_j, as long as these parameters are bounded and bounded away from zero. Hence $L = L(\varrho_1, \varrho_2, \mu_1, \mu_2)$ is surjective, if $L(1,1,1,1)$ has this property. Next we introduce an artificial continuation parameter $\tau \in [0,1]$ by replacing the equation for the free boundary h with

$$(\partial_t + \omega)h + \tau(-\Delta_\Sigma)^{1/2}h + (1-\tau)\{-(u|\nu_\Sigma) + (b|\nabla_\Sigma h)\} = f_h \text{ on } \Sigma.$$

The arguments in the next section show that the corresponding problem is well-posed for each $\tau \in [0,1]$ in the case of a flat interface, with bounds independent of $\tau \in [0,1]$. Therefore, the same is true for bent interfaces and then by the above estimates also for a general geometry. Thus we only need to consider the case $\varrho_1 = \varrho_2 = \mu_1 = \mu_2 = \tau = 1$.

To prove surjectivity in this case, note that the equation for h is decoupled from those for u and π, and it is uniquely solvable in the right regularity class because of maximal regularity for the Laplace-Beltrami operator, see Section 6.4. So we may now set $h = 0$. Next we solve the parabolic transmission problem to remove the jump of u across Σ and the inhomogeneity g in the stress boundary condition. The remaining problem is a one-phase Stokes problem on the domain Ω, which is solvable by Theorem 7.3.1. This shows that we have surjectivity in the case $\varrho_1 = \varrho_2 = \mu_1 = \mu_2 = \tau = 1$, hence also for arbitrary ϱ, μ and $\tau = 0$ and the proof of Theorem 8.1.2 is complete.

8.3 The Model Problem: Harmonic Analysis

The proof of Theorem 8.2.2 for constant coefficients and flat interface is divided into several steps, wherein the Dirichlet-to-Neumann operator for the two-phase Stokes problem will play an essential role.

3.1 The Transmission Problem

Let us consider the problem

$$
\begin{aligned}
\varrho(\partial_t+\omega)u-\mu\Delta u+\nabla\pi&=0 \quad \text{in } \dot{\mathbb{R}}^n,\\
\operatorname{div} u&=0 \quad \text{in } \dot{\mathbb{R}}^n,\\
[\![u]\!]=0,\ u&=u_b \quad \text{on } \mathbb{R}^{n-1},\\
u(0)&=0 \quad \text{in } \dot{\mathbb{R}}^n,
\end{aligned}
\tag{8.23}
$$

and prove the following result.

Proposition 8.3.1. *Let $1<p<\infty$, $1\geq\mu>1/p$, and assume that ϱ_j and μ_j are positive constants, $j=1,2$, and set $J=\mathbb{R}_+$. Then for each $\omega>0$, Problem* (8.23) *admits a unique solution (u,π) with*

$$u\in {}_0H^1_{p,\mu}(J;L_p(\mathbb{R}^n)^n)\cap L_{p,\mu}(J;H^2_p(\dot{\mathbb{R}}^n)^n),\quad \pi\in L_{p,\mu}(J;\dot{H}^1_p(\dot{\mathbb{R}}^n))$$

if and only if the data $u_b=(v_b,w_b)$ satisfy the following regularity assumptions

(a) $v_b\in {}_0W^{1-1/2p}_{p,\mu}(J;L_p(\mathbb{R}^{n-1})^{n-1})\cap L_{p,\mu}(J;W^{2-1/p}_p(\mathbb{R}^{n-1})^{n-1})$,

(b) $w_b\in {}_0H^1_{p,\mu}(J;\dot{W}^{-1/p}_p(\mathbb{R}^{n-1}))\cap L_{p,\mu}(J;W^{2-1/p}_p(\mathbb{R}^{n-1}))$.

Proof. (i) Assume that we have a solution in the proper regularity class. Then we may employ the Laplace transform in t and the Fourier transform in the tangential variables $x\in\mathbb{R}^{n-1}$, to obtain the following boundary value problem for a system of ordinary differential equations on $\dot{\mathbb{R}}$:

$$
\begin{aligned}
\omega_j^2\hat{v}_j-\mu_j\partial_y^2\hat{v}_j+i\xi\hat{\pi}_j&=0, &&(-1)^jy>0,\\
\omega_j^2\hat{w}-\mu_j\partial_y^2\hat{w}_j+\partial_y\hat{\pi}_j&=0, &&(-1)^jy>0,\\
(i\xi|\hat{v})+\partial_y\hat{w}&=0, &&y\neq0,\\
\hat{v}(0)=\hat{v}_b,\ \hat{w}(0)&=\hat{w}_b.
\end{aligned}
$$

Here we have set $\omega_j^2=\varrho_j\lambda+\mu_j|\xi|^2$, $j=1,2$, and we note that the co-variable of $\partial_t+\omega$ is called λ, hence $\operatorname{Re}\lambda\geq\omega>0$ in the sequel. This system of equations is easily solved to the result

$$
\begin{bmatrix}\hat{v}_2\\ \hat{w}_2\\ \hat{\pi}_2\end{bmatrix}=e^{-\omega_2y/\sqrt{\mu_2}}\begin{bmatrix}a_2\\ \frac{\sqrt{\mu_2}}{\omega_2}(i\xi|a_2)\\ 0\end{bmatrix}+\alpha_2e^{-|\xi|y}\begin{bmatrix}-i\xi\\ |\xi|\\ \varrho_2\lambda\end{bmatrix},
\tag{8.24}
$$

for $y>0$, and

$$
\begin{bmatrix}\hat{v}_1\\ \hat{w}_1\\ \hat{\pi}_1\end{bmatrix}=e^{\omega_1y/\sqrt{\mu_1}}\begin{bmatrix}a_1\\ -\frac{\sqrt{\mu_1}}{\omega_1}(i\xi|a_1)\\ 0\end{bmatrix}+\alpha_1e^{|\xi|y}\begin{bmatrix}-i\xi\\ -|\xi|\\ \varrho_1\lambda\end{bmatrix},
\tag{8.25}
$$

for $y < 0$. Here $a_i \in \mathbb{R}^{n-1}$ and α_i have to be determined by the boundary conditions $\hat{v}(0) = \hat{v}_b$ and $\hat{w}(0) = \hat{w}_b$. We have

$$a_2 - i\xi\alpha_2 = \hat{v}_b = a_1 - i\xi\alpha_1,$$

and

$$\frac{\sqrt{\mu_2}}{\omega_2}(i\xi|a_2) + |\xi|\alpha_2 = \hat{w}_b = -\frac{\sqrt{\mu_1}}{\omega_1}(i\xi|a_1) - |\xi|\alpha_1.$$

This yields

$$\begin{aligned} a_j &= \hat{v}_b + i\xi\alpha_j, \quad j = 1, 2, \\ \alpha_2 &= -\frac{\omega_2 + \sqrt{\mu_2}|\xi|}{\varrho_2\lambda|\xi|}(\sqrt{\mu_2}(i\xi|\hat{v}_b) - \omega_2\hat{w}_b), \\ \alpha_1 &= -\frac{\omega_1 + \sqrt{\mu_1}|\xi|}{\varrho_1\lambda|\xi|}(\sqrt{\mu_1}(i\xi|\hat{v}_b) + \omega_1\hat{w}_b). \end{aligned} \tag{8.26}$$

(ii) According to Chapter 6, the velocity u has the correct regularity provided the pressure gradient is in $L_{p,\mu}(J; L_p(\mathbb{R}^n))$, and provided

$$u_b \in {}_0W^{1-1/2p}_{p,\mu}(J; L_p(\mathbb{R}^{n-1}, \mathbb{R}^n)) \cap L_{p,\mu}(J; W^{2-1/p}_p(\mathbb{R}^{n-1}, \mathbb{R}^n)).$$

In particular this regularity of u_b is also necessary. Note that $\dot{W}^{-1/p}_p(\mathbb{R}^{n-1}) \hookrightarrow W^{-1/p}_p(\mathbb{R}^{n-1})$ and Example 4.5.16(iii) yields the embedding

$${}_0H^1_{p,\mu}(J; \dot{W}^{-1/p}_p(\mathbb{R}^{n-1})) \cap L_{p,\mu}(J; W^{2-1/p}_p(\mathbb{R}^{n-1})) \hookrightarrow {}_0W^{1-1/2p}_{p,\mu}(J; L_p(\mathbb{R}^{n-1})). \tag{8.27}$$

(iii) We will now recall some operators that will play a crucial role in our subsequent analysis. We set $G := \partial_t + \omega$ in $X_0 := L_{p,\mu}(J; L_p(\mathbb{R}^{n-1}))$ with domain

$$\mathsf{D}(G) = {}_0H^1_{p,\mu}(J; L_p(\mathbb{R}^{n-1})).$$

Then by Proposition 3.2.9 we know that G is closed, invertible and sectorial with angle $\pi/2$. Moreover, by Proposition 3.3.9 and Theorem 4.3.14, G admits an $\mathcal{H}^\infty$-calculus in X_0 with $\mathcal{H}^\infty$-angle $\pi/2$, the symbol of G is λ.

Next we set $D_n := -\Delta$, the Laplacian in $L_p(\mathbb{R}^{n-1})$ with domain $\mathsf{D}(D_n) = H^2_p(\mathbb{R}^{n-1})$. We know from Theorem 6.1.8 that D_n is closed and sectorial with angle 0, and it admits a bounded $\mathcal{H}^\infty$-calculus, which is even $\mathcal{R}$-bounded with $\mathcal{RH}^\infty$-angle 0. These results also hold for the canonical extension of D_n to X_0, and also for the fractional power $D_n^{1/2}$ of D_n. Note that the domain of $D_n^{1/2}$ is $\mathsf{D}(D_n^{1/2}) = L_{p,\mu}(J; H^1_p(\mathbb{R}^{n-1}))$. The symbol of D_n is $|\xi|^2$, that of $D_n^{1/2}$ is given by $|\xi|$, where ξ means the covariable of x. By the Dore-Venni theorem, Theorem 4.5.9, and Corollary 4.5.11, the parabolic operators $L_j := \varrho_j G + \mu_j D_n$ with natural domain

$$\mathsf{D}(L_j) = \mathsf{D}(G) \cap \mathsf{D}(D_n) = {}_0H^1_{p,\mu}(J; L_p(\mathbb{R}^{n-1})) \cap L_{p,\mu}(J; H^2_p(\mathbb{R}^{n-1}))$$

are closed, invertible and sectorial with angle $\pi/2$. Moreover, L_j also admits a bounded $\mathcal{H}^\infty$-calculus in X_0 with $\mathcal{H}^\infty$-angle $\pi/2$. The same results are valid for the operators $F_j = L_j^{1/2}$, their $\mathcal{H}^\infty$-angle is $\pi/4$, and their domains are

$$\mathsf{D}(F_j) = \mathsf{D}(G^{1/2}) \cap \mathsf{D}(D_n^{1/2}) = {}_0H^{1/2}_{p,\mu}(J; L_p(\mathbb{R}^{n-1})) \cap L_{p,\mu}(J; H^1_p(\mathbb{R}^{n-1})).$$

The symbol of L_j is $\omega_j^2 = \varrho_j\lambda + \mu_j|\xi|^2$ and that of F_j is $\omega_j = \sqrt{\varrho_j\lambda + \mu_j|\xi|^2}$.

Let R denote the Riesz operator with symbol $\zeta = \xi/|\xi|$, which is a bounded linear operator on $W^s_p(\mathbb{R}^n)$, and hence also on $L_{p,\mu}(J; W^s_p(\mathbb{R}^n))$ by canonical extension.

(iv) Let $\beta_2 = \varrho_2\lambda\alpha_2$. Then the transform of the pressure π_2 in $\mathbb{R}^n_+$ is given by $e^{-|\xi|y}\beta_2$. The pressure gradient will be in X_0 provided the inverse transform b_2 of β_2 is in the space $L_{p,\mu}(J; \dot W^{1-1/p}_p(\mathbb{R}^{n-1}))$. In fact, $e^{-|\xi|y}$ is the symbol of the Poisson semigroup $P(\cdot)$ in $L_{p,\mu}(\mathbb{R}^{n-1})$, and its negative generator is $D_n^{1/2}$, hence $D_n^{1/2}P(\cdot)b_2 \in L_{p,\mu}(\mathbb{R}_+; L_p(\mathbb{R}^{n-1}))$ if and only $b_2 \in \dot W^{1-1/p}_p(\mathbb{R}^{n-1})$. This result extends canonically to $L_{p,\mu}(J; L_p(\mathbb{R}^n_+))$.

Therefore, let us look more closely at β_2. We easily obtain

$$\beta_2 = \varrho_2 \frac{\lambda}{|\xi|}\hat w_b + (\sqrt{\mu_2}\omega_2 + \mu_2|\xi|)(\hat w_b - (i\zeta|\hat v_b)),$$

where $\zeta = \xi/|\xi|$. We recall that $\dot D_n^{1/2} := \mathcal{F}^{-1}(|\xi|\mathcal{F}\,\cdot) : \dot W^s_p(\mathbb{R}^{n-1}) \to \dot W^{s-1}_p(\mathbb{R}^{n-1})$ is an isomorphism.

With the operators introduced above, b_2 can be represented by

$$b_2 = \varrho_2 G\dot D_n^{-1/2} w_b + (\sqrt{\mu_2}F_2 + \mu_2 D_n^{1/2})(w_b - i(R|v_b)) =: b_{21} + b_{22}.$$

Due to (8.27) and

$${}_0W^{1-1/2p}_{p,\mu}(J; L_p(\mathbb{R}^{n-1})) \cap L_{p,\mu}(J; W^{2-1/p}_p(\mathbb{R}^{n-1})) = D_{F_j}(2-1/p, p),$$

see Definition 3.4.1 for the spaces $D_A(k+\alpha, p)$, and the second term b_{22} is in

$$D_{F_j}(1-1/p, p) = {}_0W^{1/2-1/2p}_{p,\mu}(J; L_p(\mathbb{R}^{n-1})) \cap L_{p,\mu}(J; W^{1-1/p}_p(\mathbb{R}^{n-1})),$$

which embeds into $L_{p,\mu}(J; \dot W^{1-1/p}_p(\mathbb{R}^{n-1}))$.

Thus it remains to look at the first term $b_{21} = \varrho_2 G D_n^{-1/2} w_b$. Since

$$G\dot D_n^{-1/2} : {}_0H^1_{p,\mu}(J; \dot W^{-1/p}_p(\mathbb{R}^{n-1})) \to L_{p,\mu}(J; \dot W^{1-1/p}_p(\mathbb{R}^{n-1}))$$

is bounded and invertible, we see that the condition $w_b \in {}_0H^1_{p,\mu}(J; \dot W^{-1/p}_p(\mathbb{R}^{n-1}))$ is necessary and sufficient for $b_{21} \in L_{p,\mu}(J; \dot W^{1-1/p}_p(\mathbb{R}^{n-1}))$. Of course, similar arguments apply for the lower half-plane. □

3.2 The Dirichlet-to-Neumann Operator for the Two-Phase Stokes Equation The main ingredient in analyzing Problem (8.14) is the Dirichlet-to-Neumann operator for this problem. It is defined as follows. Let (u, π) be the solution of the Stokes problem (8.23) with Dirichlet boundary condition u_b on $\mathbb{R}^n$, see Proposition 8.3.1. We then define the Dirichlet-to-Neumann operator by means of

$$(\mathcal{DN})u_b = -[\![S(u,\pi)]\!]e_n = -[\![\mu(\nabla u + [\nabla u]^{\mathsf{T}})]\!]e_n + [\![\pi]\!]e_n. \tag{8.28}$$

As above we split u into $u = (v, w)$, and u_b into $u_b = (v_b, w_b)$. Then we obtain

$$(\mathcal{DN})u_b = (-[\![\mu\partial_y v]\!] - [\![\mu\nabla_x w]\!], -[\![2\mu\partial_y w]\!] + [\![\pi]\!]). \tag{8.29}$$

We will now formulate and prove the main result of this subsection.

Theorem 8.3.2. *The Dirichlet-to-Neumann operator $\mathcal{DN}$ for the Stokes problem is an isomorphism from the* Dirichlet space $u_b = (v_b, w_b)$ *with*

$$v_b \in {}_0W^{1-1/2p}_{p,\mu}(J; L_p(\mathbb{R}^{n-1}, \mathbb{R}^{n-1})) \cap L_{p,\mu}(J; W^{2-1/p}_p(\mathbb{R}^{n-1}, \mathbb{R}^{n-1})),$$
$$w_b \in {}_0H^1_{p,\mu}(J; \dot{W}^{-1/p}_p(\mathbb{R}^{n-1})) \cap L_{p,\mu}(J; W^{2-1/p}_p(\mathbb{R}^{n-1}))$$

onto the Neumann space $g = (g_v, g_w)$ *with*

$$g_v \in {}_0W^{1/2-1/2p}_{p,\mu}(J; L_p(\mathbb{R}^{n-1}, \mathbb{R}^{n-1})) \cap L_{p,\mu}(J; W^{1-1/p}_p(\mathbb{R}^{n-1}, \mathbb{R}^{n-1})),$$
$$g_w \in L_{p,\mu}(J; \dot{W}^{1-1/p}_p(\mathbb{R}^{n-1})).$$

Proof. (i) Let $(\hat{v}_1, \hat{w}_1, \hat{\pi}_1)$ and $(\hat{v}_2, \hat{w}_2, \hat{\pi}_2)$ be as in (8.24)–(8.25). We may now compute the symbol of the Dirichlet-to-Neumann operator to the result

$$(\mathcal{DN})\hat{u}_b = \begin{bmatrix} \omega_1\sqrt{\mu_1}a_1 + \omega_2\sqrt{\mu_2}a_2 - (\alpha_1\mu_1 + \alpha_2\mu_2)|\xi|i\xi - [\![\mu]\!]i\xi\hat{w}_b \\ 2i(\mu_2 a_2 - \mu_1 a_1|\xi) + 2(\alpha_2\mu_2 - \alpha_1\mu_1)|\xi|^2 + \lambda(\alpha_2\varrho_2 - \alpha_1\varrho_1) \end{bmatrix}$$

where the functions α_j and a_j are given in (8.26). Simple algebraic manipulations then yield the following symbol

$$(\mathcal{DN})(\lambda, \xi) = \begin{bmatrix} \alpha + \beta\zeta\otimes\zeta & i\gamma\zeta \\ -i\gamma\zeta^T & \alpha + \delta \end{bmatrix}, \tag{8.30}$$

where $\zeta = \xi/|\xi|$ and

$$\begin{aligned} \alpha &= \sqrt{\mu_1}\omega_1 + \sqrt{\mu_2}\omega_2, \quad \beta = (\mu_1 + \mu_2)|\xi|, \\ \gamma &= (\sqrt{\mu_2}\omega_2 - \sqrt{\mu_1}\omega_1) - [\![\mu]\!]|\xi|, \\ \delta &= (\omega_1^2 + \omega_2^2)/|\xi| = \beta + (\varrho_1 + \varrho_2)\lambda/|\xi|. \end{aligned} \tag{8.31}$$

Next we want to compute the inverse of the Dirichlet-to-Neumann operator. Thus we have to solve the equation $(\mathcal{DN})u_b = g$. As before we use the decomposition $u_b = (v_b, w_b)$ and $g = (g_v, g_w)$. Then in transformed variables we have the system

$$\begin{aligned} \alpha\hat{v}_b + \beta\zeta(\zeta|\hat{v}_b) + i\gamma\zeta\hat{w}_b &= \hat{g}_v, \\ -i\gamma(\zeta|\hat{v}_b) + (\alpha + \delta)\hat{w}_b &= \hat{g}_w. \end{aligned}$$

This yields

$$\hat{v}_b = \alpha^{-1}[\hat{g}_v - \zeta\beta(\zeta|\hat{v}_b) + i\gamma\hat{w}_b)]. \tag{8.32}$$

(ii) This last equation shows that it is sufficient to determine $(\hat{v}_b|\zeta)$ and $\hat{w}_b$. If the inverse transforms of $\beta(\hat{v}_b|\zeta)$ and $\gamma\hat{w}_b$ belong to the class of g_v, then v_b is uniquely determined and has the claimed regularity. Indeed, α is the symbol of

$$F := \sqrt{\mu_1}F_1 + \sqrt{\mu_2}F_2, \quad \mathsf{D}(F) = {}_0H^{1/2}_{p,\mu}(J; L_p(\mathbb{R}^{n-1})) \cap L_{p,\mu}(J; H^1_p(\mathbb{R}^{n-1})),$$

which is a bounded invertible operator from its domain into $L_{p,\mu}(J; L_p(\mathbb{R}^n))$, and hence also from $D_F(2-1/p,p)$ into $D_F(1-1/p,p)$. Here we note that

$$D_F(\theta,p) = D_{F_j}(\theta,p) = {}_0W^{\theta/2}_p(J; L_p(\mathbb{R}^n)) \cap L_p(J; W^\theta_p(\mathbb{R}^n)),$$

for $\theta \in (0,2)$, $\theta \neq 1$. Therefore, $F^{-1}g_v$ belongs to $D_F(2-1/p,p)$ if and only if $g_v \in D_F(1-1/p,p)$. Next we note that γ is the symbol of $\sqrt{\mu_2}F_2 - \sqrt{\mu_1}F_1 - [\![\mu]\!]D_n^{1/2}$ which is bounded from $D_F(2-1/p,p)$ to $D_F(1-1/p,p)$, and β is the symbol of $(\mu_1+\mu_2)D_n^{1/2}$ which has the same mapping properties.

(iii) It remains to show that w_b and $(R|v_b)$ belong to $D_F(2-1/p,p)$. For $\hat{w}_b$ and $(\zeta|\hat{v}_b)$ we have the equations

$$\begin{aligned}(\alpha+\beta)(\zeta|\hat{v}_b) + i\gamma\hat{w}_b &= (\zeta|\hat{g}_v),\\ -i\gamma(\zeta|\hat{v}_b) + (\alpha+\delta)\hat{w}_b &= \hat{g}_w\end{aligned}$$

since $|\zeta| = 1$. Solving this 2-D system we obtain

$$\begin{aligned}\hat{w}_b &= m^{-1}[i\gamma(\zeta|\hat{g}_v) + (\alpha+\beta)\hat{g}_w],\\ (\zeta|\hat{v}_b) &= m^{-1}[(\alpha+\delta)(\zeta|\hat{g}_v) - i\gamma\hat{g}_w],\end{aligned} \tag{8.33}$$

where

$$m = (\alpha+\beta)(\alpha+\delta) - \gamma^2.$$

Since $\delta = \beta + (\varrho_1+\varrho_2)\lambda/|\xi|$ we obtain the following relation for m

$$m = (\alpha+\beta)[(\varrho_1+\varrho_2)\frac{\lambda}{|\xi|} + 4\Big(\frac{1}{\eta_1}+\frac{1}{\eta_2}\Big)^{-1}] =: (\alpha+\beta)\mathsf{n},$$

where $\eta_1 = \sqrt{\mu_1}\omega_1 + \mu_2|\xi|$ and $\eta_2 = \sqrt{\mu_2}\omega_2 + \mu_1|\xi|$. This yields

$$\begin{aligned}\hat{w}_b &= \frac{i\gamma}{(\alpha+\beta)\mathsf{n}}(\zeta|\hat{g}_v) + \frac{\hat{g}_w}{\mathsf{n}},\\ (\zeta|\hat{v}_b) &= \frac{(\varrho_1+\varrho_2)\lambda/|\xi|}{(\alpha+\beta)\mathsf{n}}(\zeta|\hat{g}_v) + \frac{1}{\mathsf{n}}[(\zeta|\hat{g}_v) - \frac{i\gamma}{\alpha+\beta}\hat{g}_w].\end{aligned} \tag{8.34}$$

We define the operators T_j by means of their symbols η_j, i.e.,

$$T_1 := \sqrt{\mu_1}F_1 + \mu_2 D_n^{1/2}, \quad T_2 := \sqrt{\mu_2}F_2 + \mu_1 D_n^{1/2}, \quad \mathsf{D}(T_j) = \mathsf{D}(F_j) = \mathsf{D}(F).$$

Then by the Dore-Venni theorem, the operators T_j are invertible, sectorial with angle $\pi/4$. Moreover, due to Corollary 4.5.11 they admit an $\mathcal{H}^\infty$-calculus with $\mathcal{H}^\infty$-angle $\pi/4$. The harmonic mean T of T_1 and T_2, i.e.,

$$T := 2T_1T_2(T_1+T_2)^{-1} = 2(T_1^{-1}+T_2^{-1})^{-1}$$

enjoys the same properties, as another application of the Dore-Venni theorem shows. The symbol of T is given by $\eta := 2\eta_1\eta_2/(\eta_1+\eta_2)$.

Next we consider the operator $GD_n^{-1/2}$ with domain

$$\begin{aligned}\mathsf{D}(GD_n^{-1/2}) &= \{h \in \mathsf{R}(D_n^{1/2}) :\ D_n^{-1/2}h \in \mathsf{D}(G)\}\\ &= {}_0H^1_{p,\mu}(J;\dot{H}_p^{-1}(\mathbb{R}^{n-1})) \cap L_{p,\mu}(J;L_p(\mathbb{R}^{n-1})).\end{aligned}$$

The inclusion from left to right in the last equality is obvious. The converse can be seen as follows. Let $h \in {}_0H^1_{p,\mu}(J;\dot{H}_p^{-1}(\mathbb{R}^{n-1})) \cap L_{p,\mu}(J;L_p(\mathbb{R}^{n-1}))$ and define $g := \dot{D}_n^{-1/2}h$. Then

$$g \in {}_0H^1_{p,\mu}(J;L_p(\mathbb{R}^{n-1})) \cap L_{p,\mu}(J;\dot{H}_p^1(\mathbb{R}^{n-1})) \hookrightarrow L_{p,\mu}(J;H_p^1(\mathbb{R}^{n-1})),$$

and $D_n^{1/2}g = \dot{D}_n^{1/2}g = h \in L_{p,\mu}(J;L_p(\mathbb{R}^{n-1}))$, which implies that $h \in \mathsf{R}(D_n^{1/2})$ and $g = \dot{D}_n^{-1/2}h = D_n^{-1/2}h \in \mathsf{D}(G)$. By Corollary 4.5.12 the operator $GD_n^{-1/2}$ is closed, sectorial and also admits a bounded $\mathcal{H}^\infty$-calculus with $\mathcal{H}^\infty$-angle $\pi/2$ on $X_0 = L_{p,\mu}(J;L_p(\mathbb{R}^{n-1}))$. Its symbol is given by $\lambda/|\xi|$.

Finally, we consider the operator

$$N := (\varrho_1+\varrho_2)GD_n^{-1/2} + 2T, \tag{8.35}$$

with domain

$$\mathsf{D}(N) = \mathsf{D}(GD_n^{-1/2}) \cap \mathsf{D}(T) = {}_0H^1_{p,\mu}(J;\dot{H}_p^{-1}(\mathbb{R}^{n-1})) \cap L_{p,\mu}(J;H_p^1(\mathbb{R}^{n-1}));$$

recall (8.27). By the Dore-Venni theorem N is closed, invertible, and admits a bounded $\mathcal{H}^\infty$-calculus as well, with $\mathcal{H}^\infty$-angle $\pi/2$. Its symbol is n.

The operator with symbol γ is given by T_2-T_1, and the operator with symbol $\alpha+\beta$ by T_1+T_2. For the inverse transforms w_b and $(R|v_b)$ of $\hat{w}_b$ and $(\zeta|\hat{v}_b)$ we therefore obtain the representations

$$\begin{aligned}w_b &= N^{-1}[(T_2-T_1)(T_1+T_2)^{-1}i(R|g_v)+g_w]\\ (R|v_b) &= (T_1+T_2)^{-1}(\varrho_1+\varrho_2)GD_n^{-1/2}N^{-1}(R|g_v)\\ &\quad + N^{-1}(R|g_v) - (T_2-T_1)(T_1+T_2)^{-1}N^{-1}ig_w\,.\end{aligned} \tag{8.36}$$

We note that N^{-1} has the following mapping properties

$$\begin{aligned}N^{-1}: L_{p,\mu}(J;L_p(\mathbb{R}^{n-1})) &\to {}_0H^1_{p,\mu}(J;\dot{H}_p^{-1}(\mathbb{R}^{n-1})) \cap L_{p,\mu}(J;H_p^1(\mathbb{R}^{n-1}))\\ &\hookrightarrow L_{p,\mu}(J;L_p(\mathbb{R}^{n-1})),\\ N^{-1}: L_{p,\mu}(J;\dot{H}_p^1(\mathbb{R}^{n-1})) &\to {}_0H^1_{p,\mu}(J;L_p(\mathbb{R}^{n-1})) \cap L_{p,\mu}(J;\dot{H}_p^2(\mathbb{R}^{n-1}))\\ &\hookrightarrow L_{p,\mu}(J;L_p(\mathbb{R}^{n-1})).\end{aligned}$$

Therefore by three-fold real interpolation

$$\begin{aligned} N^{-1} : L_{p,\mu}(J; \dot{W}_p^{1-1/p}(\mathbb{R}^{n-1})) \\ \to {}_0H^1_{p,\mu}(J; \dot{W}_p^{-1/p}(\mathbb{R}^{n-1})) \cap L_{p,\mu}(J; W_p^{2-1/p}(\mathbb{R}^{n-1})). \end{aligned} \tag{8.37}$$

Moreover, N^{-1} maps ${}_0W^{1/2-1/2p}_{p,\mu}(J; L_p(\mathbb{R}^{n-1}))$ into

$${}_0W^{3/2-1/2p}_{p,\mu}(J; \dot{H}_p^{-1}(\mathbb{R}^{n-1})) \cap {}_0W^{1/2-1/2p}_{p,\mu}(J; H^1_p(\mathbb{R}^{n-1})). \tag{8.38}$$

Next we note that the operators $T_j(T_1+T_2)^{-1}$ are bounded in $D_F(1-1/p,p)$, as is the Riesz transform R, and the assertion for w_b follows now from (8.36)–(8.37) and

$${}_0W^{1/2-1/2p}_{p,\mu}(J; L_p(\mathbb{R}^{n-1})) \cap L_{p,\mu}(J; W_p^{1-1/p}(\mathbb{R}^{n-1})) \hookrightarrow L_{p,\mu}(J; \dot{W}_p^{1-1/p}(\mathbb{R}^{n-1})).$$

The assertions for $(R|v_b)$ follow readily from (8.27) and (8.36)–(8.38). □

We can now formulate our second main result of this section concerning the solvability of the problem

$$\begin{aligned} \varrho(\partial_t + \omega)u - \mu\Delta u + \nabla\pi &= 0 && \text{in } \dot{\mathbb{R}}^n, \\ \operatorname{div} u &= 0 && \text{in } \dot{\mathbb{R}}^n, \\ -[\![\mu\partial_y v]\!] - [\![\mu\nabla_x w]\!] &= g_v && \text{on } \mathbb{R}^{n-1}, \\ -[\![2\mu\partial_y w]\!] + [\![\pi]\!] &= g_w && \text{on } \mathbb{R}^{n-1}, \\ [\![u]\!] &= 0 && \text{on } \mathbb{R}^{n-1}, \\ u(0) &= 0 && \text{in } \dot{\mathbb{R}}^n. \end{aligned} \tag{8.39}$$

Corollary 8.3.3. *Let $1 < p < \infty$, $1 \geq \mu > 1/p$, and assume that ϱ_j and μ_j are positive constants, $j = 1, 2$, and set $J = \mathbb{R}_+$. Then for each $\omega > 0$, (8.39) admits a unique solution (u, π) with*

$$u \in {}_0H^1_{p,\mu}(J; L_p(\mathbb{R}^n, \mathbb{R}^n)) \cap L_{p,\mu}(J; H^2_p(\dot{\mathbb{R}}^n, \mathbb{R}^n)), \quad \pi \in L_{p,\mu}(J; \dot{H}^1_p(\dot{\mathbb{R}}^n)),$$

if and only if $g = (g_v, g_w)$ satisfies the following regularity assumptions

(a) $g_v \in {}_0W^{1/2-1/2p}_{p,\mu}(J; L_p(\mathbb{R}^{n-1}, \mathbb{R}^{n-1})) \cap L_{p,\mu}(J; W_p^{1-1/p}(\mathbb{R}^{n-1}, \mathbb{R}^{n-1}))$,

(b) $g_w \in L_{p,\mu}(J; \dot{W}_p^{1-1/p}(\mathbb{R}^{n-1}))$.

Proof. Let $u_b := (v_b, w_b) := (\mathcal{DN})^{-1}(g_v, g_w)$, and let (u, π) be the solution of (8.23). Thanks to Theorem 8.3.2 and Proposition 8.3.1, (u, π) satisfies the regularity assertion of the Corollary, and it is the unique solution of (8.39) due to the definition of $\mathcal{DN}$. □

3.3 The Two-Phase Stokes Problem with Free Boundary
Next we consider the problem

$$
\begin{aligned}
\varrho(\partial_t+\omega)u-\mu\Delta u+\nabla\pi &= 0 && \text{in } \dot{\mathbb{R}}^n,\\
\operatorname{div} u &= 0 && \text{in } \dot{\mathbb{R}}^n,\\
-[\![\mu\partial_y v]\!]-[\![\mu\nabla_x w]\!] &= 0 && \text{on } \mathbb{R}^{n-1},\\
-[\![2\mu\partial_y w]\!]+[\![\pi]\!]-\sigma\Delta h &= 0 && \text{on } \mathbb{R}^{n-1},\\
[\![u]\!] &= 0 && \text{on } \mathbb{R}^{n-1},\\
(\partial_t+\omega)h-w &= f_h && \text{on } \mathbb{R}^{n-1},\\
u(0)=0,\ h(0) &= 0,
\end{aligned}
\tag{8.40}
$$

with $f_h \in {}_0\mathbb{F}_h = {}_0W^{1-1/2p}_{p,\mu}(J;L_p(\mathbb{R}^{n-1})\cap L_{p,\mu}(J;W^{2-1/p}_p(\mathbb{R}^{n-1}))$. It remains to show that (8.40) admits a unique solution (u,π,h) in the proper regularity class. We note that once h has been determined, Corollary 8.3.3 yields the corresponding pair (u,π) in problem (8.40).

To determine h we extract the *boundary symbol* for this problem as follows. Applying the Neumann-to-Dirichlet operator $(\mathcal{DN})^{-1}$ to $(g_v,g_w)=(0,\sigma D_n h)$ yields

$$u_b=(\mathcal{DN})^{-1}[g_v,g_w]^{\mathsf{T}}=(\mathcal{DN})^{-1}[0,\sigma D_n h]^{\mathsf{T}}.$$

According to (8.34), the transform of the normal component w_b of u_b is given by

$$\hat{w}_b=\frac{-\sigma|\xi|^2}{(\varrho_1+\varrho_2)\lambda/|\xi|+4\eta_1\eta_2/(\eta_1+\eta_2)}\hat{h}.$$

Let us now consider the dynamic equation on the interface. We note once more that λ refers here to the co-variable of $\partial_t+\omega$. Inserting this expression for $\hat{w}_b$ into the transformed equation $\lambda\hat{h}-\hat{w}_b=\hat{f}_h$ results in the equation $s(\lambda,|\xi|)\hat{h}=\hat{f}_h$, where the boundary symbol $s(\lambda,|\xi|)$ is given by

$$s(\lambda,|\xi|)=\lambda+\frac{\sigma|\xi|^2}{(\varrho_1+\varrho_2)\lambda/|\xi|+4\eta_1\eta_2/(\eta_1+\eta_2)}. \tag{8.41}$$

The operator corresponding to this symbol is

$$S=G+\sigma D_n N^{-1}. \tag{8.42}$$

S has the following mapping properties:

$$S: {}_0H^{r+1}_{p,\mu}(J;K^s_p(\mathbb{R}^{n-1}))\cap{}_0H^r_{p,\mu}(J;K^{s+1}_p(\mathbb{R}^{n-1}))\to{}_0H^r_{p,\mu}(J;K^s_p(\mathbb{R}^{n-1})), \tag{8.43}$$

where $K\in\{H,W\}$. In order to find h we need to solve the equation $Sh=f_h$, that is, we need to show that S is invertible in appropriate function spaces.

All operators in the definition of S commute, and admit an $\mathcal{H}^\infty$-calculus. The $\mathcal{H}^\infty$-angle of D_n is zero, that of N is $\pi/2$ and that of G is $\pi/2$ as well. Thus we can a priori not guarantee that the sum of the power-angles of the single operators in S is strictly less than π, and the Dore-Venni theroem is therefore not directly applicable. We will instead apply the Kalton-Weis theorem.

For this purpose note that for complex numbers w_j with $\arg w_j \in [0, \pi/2)$, we have $\arg(w_1 w_2)/(w_1 + w_2) = \arg(1/w_1 + 1/w_2)^{-1} \in [0, \pi/2)$ as well. This implies that $s(\lambda, |\xi|)$ has strictly positive real part for each λ in the closed right halfplane and for each $\xi \in \mathbb{R}^n$, $(\lambda, \xi) \neq (0,0)$, hence $s(\lambda, |\xi|)$ does not vanish for such λ and ξ.

We write $s(\lambda, |\xi|)$ in the following way:

$$s(\lambda, \tau) = \lambda + \sigma\tau k(z), \quad z = \lambda/\tau^2, \ \lambda \in \mathbb{C}, \ \tau \in \mathbb{C} \setminus \{0\}, \tag{8.44}$$

where

$$k(z) = \Big[(\varrho_1 + \varrho_2)z + 4\Big(\frac{1}{\sqrt{\mu_1}\sqrt{\varrho_1 z + \mu_1} + \mu_2} + \frac{1}{\sqrt{\mu_2}\sqrt{\varrho_2 z + \mu_2} + \mu_1}\Big)^{-1}\Big]^{-1}.$$

The asymptotics of $k(z)$ are given by

$$k(0) = \frac{1}{2(\mu_1 + \mu_2)}, \quad zk(z) \to \frac{1}{\varrho_1 + \varrho_2} \quad \text{for } z \in \mathbb{C} \setminus \mathbb{R}_- \text{ with } |z| \to \infty.$$

This shows that for any $\vartheta \in [0, \pi)$ there is a constant $C = C(\vartheta) > 0$ such that

$$|k(z)| \leq \frac{C}{1 + |z|}, \quad z \in \bar{\Sigma}_\vartheta.$$

Hence we see that

$$|s(\lambda, |\xi|)| \leq C(|\lambda| + |\xi|), \quad \operatorname{Re}\lambda \geq 0, \ \xi \in \mathbb{R}^n,$$

is valid for some constant $C > 0$. Next we are going to prove that for each $\lambda_0 > 0$ there are $\eta > 0$, $c > 0$ such that

$$|s(\lambda, \tau)| \geq c[|\lambda| + |\tau|], \quad \text{for all } \lambda \in \Sigma_{\pi/2+\eta}, \ |\lambda| \geq \omega_0, \ \tau \in \Sigma_\eta. \tag{8.45}$$

This can be seen as follows: since $\operatorname{Re} k(z) > 0$ for $\operatorname{Re} z \geq 0$, by continuity of the modulus and argument we obtain an estimate of the form

$$|s(\lambda, \tau)| \geq c_0[|\lambda| + |\tau||k(z)|] \geq c[|\lambda| + |\tau|], \quad \lambda \in \Sigma_{\pi/2+\eta}, \ \tau \in \Sigma_\eta,$$

provided $|z| \leq M$, with some $\eta > 0$ and $c > 0$ depending on M, but not on λ and τ. On the other hand, for $m > 0$ fixed we consider the case with $|\lambda| \geq m|\tau|$, $|z| \geq M$. We then have

$$|s(\lambda, \tau)| \geq |\lambda| - \sigma|\tau||k(z)| \geq \frac{1}{2}[|\lambda| + m|\tau|] - \sigma C|\tau|/(1 + M) \geq c[|\lambda| + |\tau|],$$

provided $m > 2\sigma C/(1+M)$, and then by extension

$$|s(\lambda,\tau)| \geq c[|\lambda| + |\tau|], \quad \lambda \in \Sigma_{\pi/2+\eta},\ \tau \in \Sigma_\eta,\ |\lambda| \geq m|\tau|,\ |z| \geq M,$$

provided $\eta > 0$ and $c > 0$ are sufficiently small. One easily sees that the intersection point of the curves $y = Mx^2$ and $y = mx$ in $\mathbb{R}^2$ has distance $d = (m/M)\sqrt{1+m^2}$ from the origin. By choosing M large enough so that $d \leq \omega_0$, (8.45) follows by combining the two estimates.

By means of the $\mathcal{R}$-boundedness of the functional calculus for D_n in $K_p^s(\mathbb{R}^{n-1})$, we see that

$$(\lambda + D_n^{1/2})s^{-1}(\lambda, D_n^{1/2})$$

is of class $\mathcal{H}^\infty$ and $\mathcal{R}$-bounded on $\Sigma_{\pi/2+\eta} \setminus B(0,\omega_0)$. The operator-valued $\mathcal{H}^\infty$-calculus for $G = \partial_t + \omega$ on ${}_0H^r_{p,\mu}(J; K^s_p(\mathbb{R}^{n-1}))$ implies boundedness of

$$(G + D_n^{1/2})s^{-1}(G, D_n^{1/2}) \quad \text{in} \quad {}_0H^r_{p,\mu}(J; K^s_p(\mathbb{R}^{n-1})).$$

This shows that $s^{-1}(G, D_n^{1/2})$ has the following mapping properties:

$$\begin{aligned} s^{-1}(G, D_n^{1/2}) :\ & {}_0H^r_{p,\mu}(J; K^s_p(\mathbb{R}^{n-1})) \\ & \to {}_0H^{r+1}_{p,\mu}(J; K^s_p(\mathbb{R}^{n-1})) \cap {}_0H^r_{p,\mu}(J; K^{s+1}_p(\mathbb{R}^{n-1})). \end{aligned} \tag{8.46}$$

We conclude that S is invertible and that $S^{-1} = s^{-1}(G, D_n^{1/2})$. Choosing $r = 0$ and $s = 2 - 1/p$ and $K = W$ in (8.46) yields

$$\begin{aligned} S^{-1} :\ & L_{p,\mu}(J; W^{2-1/p}_p(\mathbb{R}^{n-1})) \\ & \to {}_0H^1_{p,\mu}(J; W^{2-1/p}_p(\mathbb{R}^{n-1})) \cap L_{p,\mu}(J; W^{3-1/p}_p(\mathbb{R}^{n-1})). \end{aligned} \tag{8.47}$$

Moreover, we also obtain from (8.46)

$$\begin{aligned} S^{-1} &: L_{p,\mu}(J; L_p(\mathbb{R}^{n-1})) \to {}_0H^1_{p,\mu}(J; L_p(\mathbb{R}^{n-1})) \\ S^{-1} &: {}_0H^1_{p,\mu}(J; L_p(\mathbb{R}^{n-1})) \to {}_0H^2_{p,\mu}(J; L_p(\mathbb{R}^{n-1})). \end{aligned}$$

Interpolating with the real method $(\cdot\,,\cdot)_{1-1/p,p}$ then yields

$$S^{-1} : {}_0W^{1-1/p}_{p,\mu}(J; L_p(\mathbb{R}^{n-1})) \to {}_0W^{2-1/p}_{p,\mu}(J; L_p(\mathbb{R}^{n-1})). \tag{8.48}$$

(8.47), (8.48) shows that the equation $Sh = f_h$ has for each $f_h \in {}_0\mathbb{F}_h$ a unique solution h in the regularity class of Theorem 8.2.2.

3.4 The Modified Problem

We need also to consider the modified linear problem

$$\begin{aligned} \varrho\partial_t u - \mu\Delta u + \nabla\pi &= 0 && \text{in } \dot{\mathbb{R}}^{n+1}, \\ \operatorname{div} u &= 0 && \text{in } \dot{\mathbb{R}}^{n+1}, \\ -[\![\mu\partial_y v]\!] - [\![\mu\nabla_x w]\!] &= 0 && \text{on } \mathbb{R}^{n-1}, \\ -[\![2\mu\partial_y w]\!] + [\![\pi]\!] &= \sigma\Delta h && \text{on } \mathbb{R}^{n-1}, \\ [\![u]\!] &= 0 && \text{on } \mathbb{R}^{n-1}, \\ \partial_t h + r(-\Delta)^{1/2} + (1-r)\{-\gamma w + (b|\nabla)h\} &= f_h && \text{on } \mathbb{R}^{n-1}, \\ u(0) = 0,\ h(0) &= 0, \end{aligned} \tag{8.49}$$

where $r \in [0,1]$, $f_h \in {}_0\mathbb{F}_h$, and $b \in \mathbb{R}^{n-1}$. The corresponding boundary symbol $s_{r,b}(\lambda,\xi)$ is given by

$$s_{r,b}(\lambda,\xi) = \lambda + r|\xi| + (1-r)\{\sigma|\xi|k(z) + (b|i\xi)\}. \tag{8.50}$$

Obviously, also $s_{r,b}$ has an upper bound of the form

$$|s_{r,b}(\lambda,\xi)| \le C(|\lambda| + |\xi|), \quad \lambda \in \Sigma_{\pi/2+\eta},\ \xi \in (\Sigma_\eta \cup -\Sigma_\eta)^{n-1}.$$

To obtain a lower bound, observe that $\operatorname{Im}\lambda$ and $\operatorname{Im} k(z)$ have opposite signs. Therefore, if $\operatorname{Im}\lambda \ge 0$ and $\operatorname{Im} i(b|\xi) \ge 0$ we have

$$\begin{aligned} |s_{r,b}(\lambda,\xi)| &\ge c(|\lambda + r|\tau| + (1-r)i(b|\xi)| + (1-r)|\tau||k(z)|) \\ &\ge c(|\lambda| + r|\tau| + (1-r)|\tau k(z)| \ge c(|\lambda| + |\tau|), \end{aligned}$$

in case $|z| \le M$, and if $\operatorname{Im}\lambda \ge 0$ and $\operatorname{Im} i(b|\xi) \le 0$ then

$$\begin{aligned} |s_b(\lambda,\xi)| &\ge c(|\lambda| + r|\tau| + (1-r)|\tau k(z) + i(b|\xi)|) \\ &\ge c(|\lambda| + r|\tau| + (1-r)|\tau k(z)|) \ge c(|\lambda| + |\tau|)). \end{aligned}$$

For $|\lambda| \ge m|\tau|$, $|z| \ge M$ we proceed as in the previous subsection. So in summary, the symbol

$$m(\lambda,\xi) = \frac{\lambda + |\xi|}{s_{r,b}(\lambda,\xi)}$$

is bounded above and below uniformly in $\lambda \in \Sigma_{\pi/2+\eta}$, $\xi \in (\Sigma_\eta \cup -\Sigma_\eta)^{n-1}$, $r \in [0,1]$ for some small $\eta > 0$. Therefore, by the Kalton-Weis theorem, the conclusion is the same as in the previous subsection.

8.4 Asymmetric Two-Phase Stokes Problems

4.1 The Main Result

For the linearization of Problems **(P4)** and **(P6)** it is essential to understand maximal regularity for the following linear problem.

$$
\begin{aligned}
\varrho(\partial_t + \omega)u - \mu(x)\Delta u + \nabla\pi &= \varrho f_u && \text{in } \Omega\setminus\Sigma,\\
\operatorname{div} u &= g_d && \text{in } \Omega\setminus\Sigma,\\
u &= g_b && \text{on } \partial\Omega,\\
\mathcal{P}_\Sigma[\![u]\!] + c(t,x)\nabla_\Sigma h &= \mathcal{P}_\Sigma g_u && \text{on } \Sigma,\\
-[\![2\mu(x)D(u) + \pi]\!]\nu_\Sigma - \sigma(x)\Delta_\Sigma h\nu_\Sigma &= g && \text{on } \Sigma,\\
-[\![2\mu(x)D(u)\nu_\Sigma\cdot\nu_\Sigma/\varrho]\!] + [\![\pi/\varrho]\!] &= g_h && \text{on } \Sigma,\\
[\![\varrho]\!](\partial_t + \omega)h - [\![\varrho u\cdot\nu_\Sigma]\!] + b(t,x)\cdot\nabla_\Sigma h &= [\![\varrho]\!] f_h && \text{on } \Sigma,\\
u(0) &= u_0 && \text{in } \Omega\setminus\Sigma,\\
h(0) &= h_0 && \text{on } \Sigma.
\end{aligned}
\tag{8.51}
$$

Here μ_j, $j = 1, 2$, are functions of x, continuous on $\bar{\Omega}_j$, and c, b depend on t and x. A central assumption in this section is $[\![\varrho]\!] \neq 0$! For this problem we have maximal regularity result in the L_p-setting.

Theorem 8.4.1. *Let $p > n+2$, $1 \geq \mu > 1/p$, $\Omega \subset \mathbb{R}^n$ a bounded domain with $\partial\Omega \in C^3$, $\Sigma \subset \Omega$ a closed hypersurface of class C^3 and ϱ_j, $\varrho_1 \neq \varrho_2$, be positive constants, $j = 1, 2$. Assume that $\mu \in C_b^{1-}(\Omega\setminus\Sigma)$, $\sigma \in C^{1-}(\Sigma)$ are positive, uniformly in x. Set $J = \mathbb{R}_+$, and suppose $b = b_0 + b_1$, $c = c_0 + c_1$, with $(b_0, c_0) \in \mathbb{R}^n$, and*

$$(b_1, c_1) \in [W^{1-1/2p}_{p,\mu}(J; L_p(\Sigma)) \cap L_{p,\mu}(J; W^{2-1/p}_p(\Sigma))]^{n+1}.$$

Then there is $\omega_0 \in \mathbb{R}$ such that for each $\omega > \omega_0$, the assymetric Stokes problem with free boundary (8.51) *admits a unique solution (u, π, h) with regularity*

$$
\begin{aligned}
&u \in H^1_{p,\mu}(J; L_p(\Omega)^n) \cap L_{p,\mu}(J; H^2_p(\Omega\setminus\Sigma)^n), \quad \pi \in L_{p,\mu}(J; \dot{H}^1_p(\Omega\setminus\Sigma)),\\
&\pi_j := \pi_{|\partial\Omega_j\cap\Sigma} \in W^{1/2-1/2p}_{p,\mu}(J; L_p(\Sigma)) \cap L_{p,\mu}(J; W^{1-1/p}_p(\Sigma)),\ j = 1, 2,\\
&h \in W^{2-1/2p}_{p,\mu}(J; L_p(\Sigma)) \cap H^1_{p,\mu}(J; W^{2-1/p}_p(\Sigma)) \cap L_{p,\mu}(J; W^{3-1/p}_p(\Sigma))
\end{aligned}
$$

if and only if the data $(f_u, g_d, g_b, \mathcal{P}_\Sigma g_u, g, g_h, f_h, u_0, h_0)$ satisfy the following regularity and compatibility conditions:

(a) $f_u \in L_{p,\mu}(J; L_p(\Omega)^n)$;

(b) $g_d \in H^{1/2}_{p,\mu}(J; L_p(\Omega)) \cap L_{p,\mu}(J; H^1_p(\Omega\setminus\Sigma))$;

(c) $g_b \in W^{1-1/2p}_{p,\mu}(J; L_p(\partial\Omega)^n) \cap L_{p,\mu}(J; W^{2-1/p}_p(\partial\Omega)^n)$;

(d) $(g_d, g_b\cdot\nu) \in H^1_{p,\mu}(J; H^{-1}_{p,\partial\Omega}(\Omega\setminus\Sigma)$;

(e) $(g, g_h) \in W^{1/2-1/2p}_{p,\mu}(J; L_p(\Sigma)^{n+1}) \cap L_{p,\mu}(J; W^{1-1/p}_p(\Sigma)^{n+1})$;

(f) $(\mathcal{P}_\Sigma g_u, f_h) \in W^{1-1/2p}_{p,\mu}(J; L_p(\Sigma)^{n+1}) \cap L_{p,\mu}(J; W^{2-1/p}_p(\Sigma)^{n+1})$;

(g) $u_0 \in W^{2\mu-2/p}_p(\Omega \setminus \Sigma)^n$, $h_0 \in W^{2+\mu-2/p}_p(\Sigma)$;

(h) $\operatorname{div} u_0 = g_d(0)$ in $\Omega \setminus \Sigma$, $u_0|_{\partial\Omega} = g_b(0)$ on $\partial\Omega$;

(i) $\mathcal{P}_\Sigma [\![u_0]\!] + c(0,\cdot)\nabla_\Sigma h_0 = \mathcal{P}_\Sigma g_u(0)$ on Σ;

(j) $-\mathcal{P}_\Sigma [\![\mu(x)(\nabla u_0 + [\nabla u_0]^{\mathsf{T}})\nu_\Sigma]\!] = \mathcal{P}_\Sigma g(0)$ on Σ.

The solution map $(f_u, g_d, g_b, \mathcal{P}_\Sigma g_u, g, g_h, f_h, u_0, h_0) \mapsto (u, \pi, h)$ *is continuous between the corresponding spaces.*

Here we again have the liberty to replace $\mu(x)\Delta$ with $\operatorname{div}(2\mu(x)D(u))$ in the first line of Problem (8.51), see Remark 8.1.1 This result is proved in a similar way as Theorem 8.1.2. However, there are some significant differences, so we explain the details in the next sections, following the scheme of Section 8.2 and 8.3.

As in Section 8.1, it is possible to reduce the regularity of f_h to $f_h \in L_{p,\mu}(J; W^{2-1/p}_p(\Sigma))$, in which case the highest time regularity of h is dropped. This is the content of

Corollary 8.4.2. *Let the assumptions of Theorem* 8.4.1 *be valid. Then the result of this theorem remains valid when replacing the spaces for* h *and* f_h *by*

$$h \in H^1_{p,\mu}(J; W^{2-1/p}_p(\Sigma)) \cap L_{p,\mu}(J; W^{3-1/p}_p(\Sigma)), \quad f_h \in L_{p,\mu}(J; W^{2-1/p}_p(\Sigma)).$$

This is important for the semigroup associated to (8.51), the *asymmetric two-phase Stokes semigroup with free boundary.* To construct this semigroup, we specialize to the case of homogeneous boundary and interface conditions as well as to a solenoidal situation

$$(g_d, g_b, g_u, g, g_h, b, c) = 0, \quad (\operatorname{div} f_u, f_u \cdot \nu_{\partial\Omega}, [\![f_u \cdot \nu_\Sigma]\!]) = 0.$$

Then the semigroup is given in the following way. We have $\nabla : H^1_{p',\Sigma}(\Omega \setminus \Sigma) \to L_{p'}(\Omega)^n$ is bounded, and hence $\nabla^* : L_p(\Omega) \to H^{-1}_{p,\partial\Omega}(\Omega \setminus \Sigma)$ is as well. Define

$$X_0 := [L_p(\Omega)^n \cap \mathsf{N}(\nabla^*)] \times W^{2-1/p}_p(\Sigma), \quad X_1 := [H^2_p(\Omega \setminus \Sigma)^n \cap \mathsf{N}(\nabla^*)] \times W^{3-1/p}_p(\Sigma),$$

and A by means of

$$A(u,h) := ((-\operatorname{div}(2\mu(x)D(u)) + \nabla\pi)/\varrho, -[\![\varrho u \cdot \nu_\Sigma]\!]/[\![\varrho]\!]),$$

with domain

$$\mathsf{D}(A) := \{(u,h) \in X_1 : u_{|\partial\Omega} = 0,\ \mathcal{P}_\Sigma [\![u]\!] = 0,\ \mathcal{P}_\Sigma [\![\mu(x)D(u)\nu_\Sigma]\!] = 0\}.$$

Here $\pi = \pi(u, h)$ is given by the solution of the weak problem

$$\begin{aligned}
(\nabla\pi|\nabla\phi/\varrho)_\Omega &= (\operatorname{div}(2\mu(x)D(u))|\nabla\phi/\varrho)_\Omega, && \phi \in H^1_{p',\Sigma}(\Omega\setminus\Sigma),\\
[\![\pi]\!] &= \sigma\Delta_\Sigma h + [\![2\mu(x)\partial_\nu u\cdot\nu_\Sigma]\!] && \text{on } \Sigma,\\
[\![\pi/\varrho]\!] &= [\![2\mu(x)\partial_\nu u\cdot\nu_\Sigma/\varrho]\!] && \text{on } \Sigma.
\end{aligned} \tag{8.52}$$

Note that $\pi \in H^1_p(\Omega\setminus\Sigma)$ is the solution of two one-phase problems with Dirichlet-data on Σ, hence π is well-defined by Proposition 7.4.3.

Then, with $z = (u, h)$, $z_0 = (u_0, h_0)$, and $f = (f_u, f_h)$, Problem (8.51) is equivalent to the abstract evolution equation

$$\dot z + Az = f, \quad t > 0, \quad z(0) = z_0.$$

Corollary 8.4.2 shows that this problem has maximal L_p-regularity, i.e., $\omega + A \in \mathcal{MR}(L_p(X_0))$. Therefore, $-A$ generates an analytic C_0-semigroup in X_0. As the domain of A is compactly embedded into X_0, the spectrum of A consists only of eigenvalues of finite algebraic multiplicity, which are independent of p. Therefore, the number ω_0 in Theorem 8.4.1 is precisely the spectral bound $\mathsf{s}(-A)$, which will be shown to be 0 in Chapter 10.

4.2 The Asymmetric Two-Phase Stokes Operator

Setting $h = 0$ and ignoring the equation for h and switching to divergence form, i.e., replacing $\mu(x)\Delta u$ by $\operatorname{div}(2\mu(x)D(u))$ we obtain the pure asymmetric two-phase Stokes problem in divergence form. Note that Remark 8.1.1 also applies here. The resulting problem reads

$$\begin{aligned}
\varrho(\partial_t + \omega)u - \operatorname{div}(2\mu(x)D(u)) + \nabla\pi &= \varrho f_u && \text{in } \Omega\setminus\Sigma,\\
\operatorname{div} u &= g_d && \text{in } \Omega\setminus\Sigma,\\
u &= g_b && \text{on } \partial\Omega,\\
\mathcal{P}_\Sigma[\![u]\!] &= \mathcal{P}_\Sigma g_u && \text{on } \Sigma,\\
-[\![2\mu(x)D(u)\nu_\Sigma]\!] + [\![\pi]\!]\nu_\Sigma &= g && \text{on } \Sigma,\\
-[\![2\mu(x)D(u)\nu_\Sigma\cdot\nu_\Sigma/\varrho]\!] + [\![\pi/\varrho]\!] &= g_h && \text{on } \Sigma,\\
u(0) &= u_0 && \text{in } \Omega\setminus\Sigma.
\end{aligned} \tag{8.53}$$

Here μ_j, $j = 1, 2$, are functions of x. A central assumption in this section is $[\![\varrho]\!] \neq 0$! For this problem we have maximal regularity in the L_p-setting as well.

Theorem 8.4.3. *Let $p \in (1,\infty)$, $1 \geq \mu > 1/p$ be fixed, $\varrho_j > 0$, $\varrho_2 \neq \varrho_1$, $j = 1, 2$. Assume that $\mu \in C^{1-}_b(\Omega\setminus\Sigma)$, $\mu(x) > 0$ uniformly in x and set $J = \mathbb{R}_+$.*

Then there is $\omega_0 \in \mathbb{R}$ such that for each $\omega > \omega_0$, the pure asymmetric two-phase Stokes problem (8.53) *admits a unique solution (u, π) with regularity*

$$u \in H^1_{p,\mu}(J; L_p(\Omega)^n) \cap L_{p,\mu}(J; H^2_p(\Omega\setminus\Sigma)^n), \quad \pi \in L_{p,\mu}(J; \dot H^1_p(\Omega\setminus\Sigma)),$$
$$\pi_j := \pi_{|\partial\Omega_j\cap\Sigma} \in W^{1/2-1/2p}_{p,\mu}(J; L_p(\Sigma)) \cap L_{p,\mu}(J; W^{1-1/p}_p(\Sigma)), \; j = 1, 2,$$

if and only if the data $(f_u, g_d, g_b, \mathcal{P}_\Sigma g_u, g, g_h, u_0)$ *satisfy the following regularity and compatibility conditions:*

(a) $f_u \in L_{p,\mu}(J; L_p(\Omega)^n)$;

(b) $g_d \in H^1_{p,\mu}(J; H^{-1}_{p,\partial\Omega}(\Omega \setminus \Sigma)) \cap L_{p,\mu}(J; H^1_p(\Omega \setminus \Sigma))$;

(c) $g_b \in W^{1-1/2p}_{p,\mu}(J; L_p(\partial\Omega)^n) \cap L_{p,\mu}(J; W^{2-1/p}_p(\partial\Omega)^n)$;

(d) $\mathcal{P}_\Sigma g_u \in W^{1-1/2p}_{p,\mu}(J; L_p(\Sigma)^n) \cap L_{p,\mu}(J; W^{2-1/p}_p(\Sigma)^n)$;

(e) $(g, g_h) \in W^{1/2-1/2p}_{p,\mu}(J; L_p(\Sigma)^{n+1}) \cap L_{p,\mu}(J; W^{1-1/p}_p(\Sigma)^{n+1})$;

(f) $u_0 \in W^{2\mu-2/p}_p(\Omega \setminus \Sigma,)^n$;

(g) $\operatorname{div} u_0 = g_d(0)$ in $\Omega \setminus \Sigma$, $u_0 = g_b$ on $\partial\Omega$;

(h) $\mathcal{P}_\Sigma [\![u_0]\!] = \mathcal{P}_\Sigma g_u(0)$ on Σ;

(i) $-\mathcal{P}_\Sigma [\![\mu(x)(\nabla u_0 + [\nabla u_0]^\mathsf{T})\nu_\Sigma]\!] = \mathcal{P}_\Sigma g(0)$ on Σ.

The solution map $(f_u, g_d, g_b, \mathcal{P}_\Sigma g_u, g, g_h, u_0) \mapsto (u, \pi)$ *is continuous between the corresponding spaces.*

Having this result at disposal, we define the *asymmetric two-phase Stokes operator* in divergence form in the following way. As in Section 7, let $\nabla : \dot{H}^1_{p'}(\Omega) \to L_{p'}(\Omega)^n$, set

$$X_0 = L_p(\Omega)^n \cap \mathsf{N}(\nabla^*) = \{u \in L_p(\Omega; \mathbb{R}^n) : \operatorname{div} u = 0 \text{ in } \Omega,\ u \cdot \nu = 0 \text{ on } \partial\Omega\},$$

and define A by means of

$$Au = (-\operatorname{div}(2\mu(x)D(u)) + \nabla\pi)/\varrho, \quad u \in \mathsf{D}(A) := X_1, \tag{8.54}$$

with

$$X_1 := \{u \in H^2_p(\Omega \setminus \Sigma) \cap X_0 : u_{|\partial\Omega} = 0,\ \mathcal{P}_\Sigma [\![u]\!] = \mathcal{P}_\Sigma [\![\mu(\nabla u + [\nabla u]^\mathsf{T})\nu_\Sigma]\!] = 0\}.$$

In this definition the pressure π is defined as the solution of the weak problem

$$\begin{aligned} (\nabla\pi|\nabla\phi/\varrho) &= (\operatorname{div}(2\mu(x)D(u))|\nabla\phi/\varrho), \quad \phi \in H^1_{p',\Sigma}(\Omega \setminus \Sigma), \\ [\![\pi]\!] &= [\![2\mu(x)D(u)\nu_\Sigma]\!] \cdot \nu_\Sigma \quad \text{on } \Sigma, \\ [\![\pi/\varrho]\!] &= [\![2\mu(x)D(u)\nu_\Sigma/\varrho]\!] \cdot \nu_\Sigma \quad \text{on } \Sigma. \end{aligned} \tag{8.55}$$

Note that $\pi \in H^1_p(\Omega \setminus \Sigma)$ is the solution of two one-phase problems with Dirichlet data on Σ, hence π is well-defined by Proposition 7.4.3. Then (8.53), for $(g_d, g_b, g, g_u, g_h) = 0$, $f_u \in X_0$, is equivalent to the abstract evolution equation

$$\dot{u} + Au = f_u, \quad t > 0, \quad u(0) = u_0. \tag{8.56}$$

Theorem 8.4.3 implies that this problem has maximal L_p-regularity, hence $-A$ is the generator of an analytic C_0-semigroup in X_0 with maximal L_p-regularity. As $\mathsf{D}(A)$ embeds compactly into X_0, the two-phase Stokes operator A has compact resolvent. Therefore, its spectrum consists only of eigenvalues of finite algebraic multiplicity, and is therefore independent of p. So it is enough to study these eigenvalues for the case $p = 2$.

For this purpose we employ once more the energy method. Suppose that $\lambda \in \mathbb{C}$ is an eigenvalue of A with eigenfunction u and corresponding pressure π. Taking the inner product in $L_2(\Omega; \mathbb{C})^n$ of the equation with u, after an integration by parts we get

$$\lambda \int_\Omega \varrho |u|^2 \, dx = (\varrho A u | u)_{L_2} = 2 \int_\Omega \mu(x) |D(u)|_2^2 \, dx.$$

This implies that λ is real and non-negative. But by means of the modified Korn inequality in Lemma 1.2.1 and the no-slip condition on the outer boundary $\partial\Omega$, all eigenvalues are strictly positive. In particular, A is invertible.

This further implies that the Neumann-to-Dirichlet operator

$$S_\lambda : W_p^{1-1/p}(\Sigma)^2 \to W_p^{2-1/p}(\Sigma)^2$$

defined by the map

$$S_\lambda : [g, g_h]^\mathsf{T} \mapsto [[\![\varrho u \cdot \nu_\Sigma]\!]/[\![\varrho]\!], [\![u \cdot \nu_\Sigma]\!]/[\![1/\varrho]\!]]^\mathsf{T},$$

where u solves the problem

$$\begin{aligned} \lambda \varrho u - \operatorname{div}(2\mu(x) D(u)) + \nabla \pi &= 0 && \text{in } \Omega \setminus \Sigma, \\ \operatorname{div} u &= 0 && \text{in } \Omega \setminus \Sigma, \\ u &= 0 && \text{on } \partial\Omega, \\ \mathcal{P}_\Sigma [\![u]\!] &= 0 && \text{on } \Sigma, \\ -[\![2\mu(x) D(u)]\!] \nu_\Sigma + [\![\pi]\!] \nu_\Sigma &= g \nu_\Sigma && \text{on } \Sigma, \\ -[\![2\mu(x) \partial_\nu u / \varrho]\!] + [\![\pi/\varrho]\!] &= g_h && \text{on } \Sigma, \end{aligned}$$

is well-defined, for each $\lambda \geq 0$. This operator will be studied in more detail in Chapter 10.

4.3 Quasi-Steady Asymmetric Stokes Problem

In this subsection we consider the asymmetric quasi-steady Stokes flow problem, which reads

$$\begin{aligned} -\mathrm{div}(2\mu(x)D(u)) + \nabla\pi &= 0 && \text{in } \Omega\backslash\Sigma,\\ \mathrm{div}\, u &= 0 && \text{in } \Omega\backslash\Sigma,\\ u &= 0 && \text{on } \partial\Omega,\\ \mathcal{P}_\Sigma [\![u]\!] &= 0 && \text{on } \Sigma,\\ [\![-2\mu(x)D(u) + \pi]\!]\nu_\Sigma - \sigma(x)(\Delta_\Sigma h)\nu_\Sigma &= 0 && \text{on } \Sigma,\\ -[\![2\mu(x)\partial_\nu u/\varrho]\!]\nu_\Sigma + [\![\pi/\varrho]\!] &= 0 && \text{on } \Sigma,\\ (\partial_t + \omega)h - [\![\varrho u\cdot\nu_\Sigma]\!]/[\![\varrho]\!] &= f_h && \text{on } \Sigma,\\ h(0) &= h_0 && \text{on } \Sigma, \end{aligned} \tag{8.57}$$

on the time-interval $J = \mathbb{R}_+$, where $\omega \geq 0$ will be chosen sufficiently large. As before, the regularity classes for u and π are given by

$$u \in \mathbb{E}_u := L_{p,\mu}(J; H_p^2(\Omega\setminus\Sigma)^n),$$

and

$$\pi \in \mathbb{E}_\pi := L_{p,\mu}(J; \dot{H}_p^1(\Omega\setminus\Sigma)),$$

and that for h is

$$h \in H^1_{p,\mu}(J; W_p^{2-1/p}(\Sigma)) \cap L_{p,\mu}(J; W_p^{3-1/p}(\Sigma)).$$

Setting $X_0 = W_p^{2-1/p}(\Sigma)$, with the Neumann-to-Dirichlet operator S_0^{11} defined in the previous subsection, this problem is equivalent to the abstract evolution equation

$$\dot{h} + \omega h - S_0^{11}\sigma\Delta_\Sigma h = f_h, \quad t > 0, \quad h(0) = h_0,$$

in X_0. Defining the operator A in X_0 by means of

$$A = -S_0^{11}\sigma\Delta_\Sigma h, \quad \mathsf{D}(A) = W_p^{3-1/p}(\Sigma), \tag{8.58}$$

it turns out that $-A$ generates an analytic C_0-semigroup with maximal regularity. This can be proved in the same way as in Section 8.1.3.

8.5 Proof of Theorem 8.4.1

5.1 Regularity of the Pressure

As in Section 8.2, under certain conditions the pressure has more time-regularity. For the asymmetric Stokes problem the following result differs from that in Section 8.2.

Proposition 8.5.1. *Assume in addition to the hypotheses of Theorem* 8.4.1 *that*

$$(g_d, u_0, h_0, \mathrm{div}\, f_u) = 0, \quad g_b\cdot\nu_{\partial\Omega} = f_u\cdot\nu_{\partial\Omega} = 0.$$

Let (u, π, h) be a solution of (8.51). Then $\pi \in {}_0H^\alpha_{p,\mu}(J; L_p(\Omega))$, for each $\alpha \in (0, 1/2 - 1/2p)$, and there is a constant $C > 0$ independent of $\omega \geq \omega_0$ such that

$$|\pi|_{L_{p,\mu}(J;L_p(\Omega))} \leq C\big(|\nabla u|_{L_{p,\mu}(J;L_p(\Omega))} + |\nabla u|_{L_{p,\mu}(J;L_p(\Sigma\cup\partial\Omega))} + |\pi_1|_{L_{p,\mu}(J;L_p(\Sigma))} + |\pi_2|_{L_{p,\mu}(J;L_p(\Sigma))}\big). \tag{8.59}$$

Proof. Let $\psi \in L_{p'}(\Omega)$ be given and solve the problem

$$\begin{aligned} \varrho^{-1}\Delta\phi &= \psi \quad \text{in } \Omega\backslash\Sigma, \\ \partial_\nu\phi &= 0 \quad \text{on } \partial\Omega, \\ [\![\phi]\!] = 0,\ \phi &= 0 \quad \text{on } \Sigma, \end{aligned} \tag{8.60}$$

by Proposition 7.4.5. Note that this problem consists of two one-phase problems. As in the proof of Proposition 8.2.1, we obtain by an integration by parts

$$\begin{aligned} (\pi|\psi)_\Omega = \big(\varrho^{-1}\pi|\Delta\phi\big)_\Omega &= -\int_\Sigma [\![\varrho^{-1}\pi\partial_\nu\phi]\!]\, d\Sigma - \big(\varrho^{-1}\nabla\pi|\nabla\phi\big)_\Omega \\ &= \int_\Omega \nabla u : \nabla\varrho^{-1}\mu\nabla\phi\, dx - \int_{\partial\Omega} \varrho^{-1}\mu(\partial_\nu u|\nabla\phi)\,(d\partial\Omega) \\ &\quad + \int_\Sigma [\![\varrho^{-1}\mu(\partial_\nu u|\nabla\phi)]\!] - [\![\varrho^{-1}\pi\varrho\partial_\nu\phi]\!]\, d\Sigma. \end{aligned}$$

This implies (8.59). Moreover, since $\nabla u \in {}_0H^{1/2}_{p,\mu}(J; L_p(\Omega)^{n\times n})$ and $\pi_k, \partial_j u_l \in {}_0W^{1/2-1/2p}_{p,\mu}(J; L_p(\Sigma))$, and $\partial_\nu u \in {}_0W^{1/2-1/2p}_{p,\mu}(J; L_p(\partial\Omega))$, applying ∂_t^α to this identity, we obtain the estimate

$$\begin{aligned} |\partial_t^\alpha\pi|_{L_{p,\mu}(J;L_p\Omega))} \leq C\{&|\partial_t^\alpha\nabla u|_{L_{p,\mu}(J;L_p(\Omega))} + |\partial_t^\alpha(|\pi_1| + |\pi_2|)]|_{L_{p,\mu}(J;L_p(\Sigma))} \\ &+ |\partial_t^\alpha\partial_{\nu_\Sigma}u|_{L_{p,\mu}(J;L_p(\Sigma))} + |\partial_t^\alpha\partial_\nu u|_{L_{p,\mu}(J,L_p(\partial\Omega))}\}, \end{aligned}$$

for each $\alpha \in (0, 1/2 - 1/2p)$, hence $\pi \in {}_0H^\alpha_{p,\mu}(J; L_p(\Omega))$. □

5.2 Flat Interface

In this subsection, we consider the linear problem with constant coefficients for a flat interface. Due to the jump in the velocity, this problem differs from that in Section 8.2 considerably.

$$\begin{aligned} \varrho(\partial_t + \omega)u - \mu\Delta u + \nabla\pi &= \varrho f_u \quad \text{in } \dot{\mathbb{R}}^n, \\ \operatorname{div} u &= g_d \quad \text{in } \dot{\mathbb{R}}^n, \\ [\![v]\!] + c\nabla_x h &= g_u \quad \text{on } \mathbb{R}^{n-1}, \\ -[\![2\mu D(u)\nu]\!] + [\![\pi]\!]\nu - \sigma\Delta_x h\nu &= g \quad \text{on } \mathbb{R}^{n-1}, \\ -[\![2\mu D(u)\nu\cdot\nu/\varrho]\!] + [\![\pi/\varrho]\!] &= g_h \quad \text{on } \mathbb{R}^{n-1}, \\ (\partial_t + \omega)h - [\![\varrho w]\!]/[\![\varrho]\!] + b\cdot\nabla_x h &= f_h \quad \text{on } \mathbb{R}^{n-1}, \\ u(0) &= u_0 \quad \text{in } \dot{\mathbb{R}}^n, \\ h(0) &= h_0 \quad \text{on } \mathbb{R}^{n-1}. \end{aligned} \tag{8.61}$$

Here (μ_j, ϱ_j), $j = 1, 2$, are constants, $\nu = e_n$, and recall $[\![\varrho]\!] = \varrho_2 - \varrho_1 \neq 0$. With $u = (v, w)^{\mathsf{T}}$, we first look at the asymmetric Stokes problem, setting $h = 0$ and ignoring the problem for h.

$$\begin{aligned} \varrho(\partial_t + \omega)u - \mu\Delta u + \nabla\pi &= \varrho f_u && \text{in } \dot{\mathbb{R}}^n,\\ \operatorname{div} u &= g_d && \text{in } \dot{\mathbb{R}}^n,\\ [\![v]\!] &= g_u && \text{on } \mathbb{R}^{n-1},\\ -[\![2\mu D(u)\nu]\!] + [\![\pi]\!]\nu &= g = (g_v, g_w)^{\mathsf{T}} && \text{on } \mathbb{R}^{n-1},\\ -[\![2\mu D(u)\nu\cdot\nu/\varrho]\!] + [\![\pi/\varrho]\!] &= g_h && \text{on } \mathbb{R}^{n-1},\\ u(0) &= u_0 && \text{in } \dot{\mathbb{R}}^n. \end{aligned} \tag{8.62}$$

For this problem we have

Theorem 8.5.2. *Let $1 < p < \infty$, $1 \geq \mu > 1/p$ be fixed, and assume that ϱ_j and μ_j are positive constants for $j = 1, 2$, $\varrho_2 \neq \varrho_1$, and set $J = \mathbb{R}_+$. Then there is $\omega_0 \geq 0$ such that for each $\omega > \omega_0$, the assymetric Stokes problem* (8.62) *admits a unique solution (u, π) with regularity*

$$u \in H^1_{p,\mu}(J; L_p(\mathbb{R}^n)^n) \cap L_{p,\mu}(J; H^2_p(\dot{\mathbb{R}}^n)^n), \quad \pi \in L_{p,\mu}(J; \dot{H}^1_p(\dot{\mathbb{R}}^n)),$$

if and only if the data $(f_u, f_d, g_u, g_v, g_w, g_h, u_0)$ satisfy the following regularity and compatibility conditions:

(a) $f_u \in L_{p,\mu}(J; L_p(\mathbb{R}^n)^n)$;

(b) $g_d \in H^1_{p,\mu}(J; \dot{H}^{-1}_p(\mathbb{R}^n)) \cap L_{p,\mu}(J; H^1_p(\dot{\mathbb{R}}^n))$;

(c) $g_u \in W^{1-1/2p}_{p,\mu}(J; L_p(\mathbb{R}^{n-1})^n) \cap L_{p,\mu}(J; W^{2-1/p}_p(\mathbb{R}^{n-1})^n)$;

(d) $g_v \in W^{1/2-1/2p}_{p,\mu}(J; L_p(\mathbb{R}^{n-1})^{n-1}) \cap L_{p,\mu}(J; W^{1-1/p}_p(\mathbb{R}^{n-1})^{n-1})$;

(e) $(g_w, g_h) \in L_{p,\mu}(J; \dot{W}^{1-1/p}_p(\mathbb{R}^{n-1})^2)$;

(f) $u_0 \in W^{2\mu-2/p}_p(\dot{\mathbb{R}}^n)^n$;

(g) $\operatorname{div} u_0 = f_d(0)$ in $\dot{\mathbb{R}}^n$ *and* $[\![v_0]\!] = g_u(0)$ on $\mathbb{R}^{n-1}$;

(h) $-[\![\mu\partial_y v_0]\!] - [\![\mu\nabla_x w_0]\!] = g_v(0)$ on $\mathbb{R}^{n-1}$.

In addition, for the pressure traces π_j on the interface we have

$$\pi_j \in W^{1/2-1/2p}_{p,\mu}(J; L_p(\mathbb{R}^{n-1})) \cap L_{p,\mu}(J; W^{1-1/p}_p(\mathbb{R}^{n-1}))$$

if and only if

$$g_w, g_h \in W^{1/2-1/2p}_{p,\mu}(J; L_p(\mathbb{R}^{n-1})) \cap L_{p,\mu}(J; W^{1-1/p}_p(\mathbb{R}^{n-1})).$$

The solution map $(f_u, g_d, g_u, g_v, g_w, g_h, u_0) \mapsto (u, \pi)$ is continuous between the corresponding spaces.

For Problem (8.61) we also prove a maximal regularity result in the L_p-setting.

Theorem 8.5.3. *Let $1 < p < \infty$, $1 \geq \mu > 1/p$ be fixed, and assume that ϱ_j and μ_j are positive constants for $j = 1, 2$, $\varrho_2 \neq \varrho_1$, $c \in \mathbb{R}$, $b \in \mathbb{R}^{n-1}$, and set $J = \mathbb{R}_+$. Then there is $\omega_0 \in \mathbb{R}$ such that for each $\omega > \omega_0$, the asymmetric Stokes problem with free boundary* (8.61) *admits a unique solution (u, π, h) with regularity*

$$\begin{aligned}
&u \in H^1_{p,\mu}(J; L_{p,\mu}(\mathbb{R}^n)^n) \cap L_{p,\mu}(J; H^2_p(\dot{\mathbb{R}}^n)^n),\\
&\pi \in L_{p,\mu}(J; \dot{H}^1_p(\dot{\mathbb{R}}^n)), \qquad\qquad (8.63)\\
&\pi_1, \pi_2 \in W^{1/2-1/2p}_{p,\mu}(J; L_p(\mathbb{R}^{n-1})) \cap L_{p,\mu}(J; W^{1-1/p}_p(\mathbb{R}^{n-1})),\\
&h \in W^{2-1/2p}_{p,\mu}(J; L_p(\mathbb{R}^{n-1})) \cap H^1_{p,\mu}(J; W^{2-1/p}_p(\mathbb{R}^{n-1})) \cap L_{p,\mu}(J; W^{3-1/p}_p(\mathbb{R}^{n-1})),
\end{aligned}$$

if and only if the data $(f_u, g_d, g_u, g, g_h, f_h, u_0, h_0)$ satisfy the following regularity and compatibility conditions:

(a) $f_u \in L_{p,\mu}(J; L_p(\mathbb{R}^n)^n)$;

(b) $g_d \in H^1_{p,\mu}(J; \dot{H}^{-1}_p(\mathbb{R}^n)) \cap L_{p,\mu}(J; H^1_p(\dot{\mathbb{R}}^n))$;

(c) $(g, g_h) \in W^{1/2-1/2p}_{p,\mu}(J; L_p(\mathbb{R}^{n-1})^{n+1}) \cap L_{p,\mu}(J; W^{1-1/p}_p(\mathbb{R}^{n-1})^{n+1})$;

(d) $(g_u, f_h) \in W^{1-1/2p}_{p,\mu}(J; L_p(\mathbb{R}^{n-1})^{n+1}) \cap L_{p,\mu}(J; W^{2-1/p}_p(\mathbb{R}^{n-1})^{n+1})$;

(e) $u_0 \in W^{2\mu-2/p}_p(\dot{\mathbb{R}}^n)^n$, $h_0 \in W^{2+\mu-2/p}_p(\mathbb{R}^{n-1})$;

(f) $\operatorname{div} u_0 = f_d(0)$ in $\dot{\mathbb{R}}^n$, $[\![v_0]\!] = g_j(0)$ on $\mathbb{R}^{n-1}$;

(g) $-[\![\mu\partial_y v_0]\!] - [\![\mu\nabla_x w_0]\!] = g_v(0)$ on $\mathbb{R}^{n-1}$.

The solution map $(f_u, g_d, g_u, g, g_h, f_h, u_0, h_0) \mapsto (u, \pi, h)$ is continuous between the corresponding spaces.

As the proofs of these results are quite involved we outsource them to the next section.

5.3 General Bounded Domains

The linear problem with variable coefficients but small deviations for a flat interface, i.e can be handled by a perturbation argument in the same way as in Section 8.1; the same is true for the case of a bent interface. However, the localization argument needs some significant modifications, which we explain in some detail now. We follow the notation in Section 8.2.5.

Let $z = (u, \pi, h)$ be a solution of (8.51) where we assume without loss of generality $(f_u, g_d, g_u \cdot \nu_\Sigma, g_b \cdot \nu_{\partial\Omega}, u_0, h_0) = 0$. We then set $u_k = \varphi_k u$, $\pi_k = \varphi_k \pi$,

$h_k = \varphi_k h$. Then for $k = 1, \ldots, N$, $z_k = (u_k, \pi_k, [\![\pi_k]\!], h_k)$ satisfies the problem

$$\begin{aligned}
&\varrho(\partial_t + \omega)u_k - \mu(x)\Delta u_k + \nabla\pi_k = F_k(u,\pi) && \text{in } \mathbb{R}^n \setminus \Sigma_k,\\
&\operatorname{div} u_k = (\nabla\varphi_k|u) && \text{in } \mathbb{R}^n \setminus \Sigma_k,\\
&\mathcal{P}_{\Sigma_k}[\![u_k]\!] + c(t,x)\nabla_{\Sigma_k} h_k = \varphi_k \mathcal{P}_{\Sigma_k} g_u + G_{u_k}(h) && \text{on } \Sigma_k,\\
&[\![-\mu(x)(\nabla u_k + [\nabla u_k]^{\mathsf{T}}) + \pi_k]\!]\nu_{\Sigma_k} - \sigma(\Delta_\Sigma h_k)\nu_{\Sigma_k} = \varphi_k g_k + G_{g_k}(u,h) && \text{on } \Sigma_k,\\
&-[\![\mu(x)(\nabla u_k + [\nabla u_k]^{\mathsf{T}})\nu_{\Sigma_k}\cdot\nu_{\Sigma_k}/\varrho]\!] + [\![\pi_k/\varrho]\!] = \varphi_k g_h + G_{h_k}(u) && \text{on } \Sigma_k,\\
&(\partial_t + \omega)h_k - [\![\varrho u_k\cdot\nu_{\Sigma_k}]\!]/[\![\varrho]\!] + b(t,x)\cdot\nabla_{\Sigma_k} h_k/[\![\varrho]\!] = \varphi_k f_h + F_{h_k}(h) && \text{on } \Sigma_k,\\
&u_k(0) = 0 && \text{in } \mathbb{R}^n \setminus \Sigma_k,\\
&h_k(0) = 0 \text{on } \Sigma_k,
\end{aligned} \tag{8.64}$$

where

$$\begin{aligned}
F_k(u,\pi) &= (\nabla\varphi_k)\pi - \mu(x)[\Delta,\varphi_k]u,\\
G_{u_k}(h) &= c(t,x)(\nabla_{\Sigma_k}\varphi_k)h\\
G_{g_k}(u,h) &= -[\![\mu(x)(\nabla\varphi_k\otimes u + u\otimes\nabla\varphi_k)]\!]\nu_{\Sigma_k} - \sigma(x)[\Delta_{\Sigma_k},\varphi_k]h\nu_{\Sigma_k},\\
G_{h_k}(u) &= -[\![\mu(x)(\nabla\varphi_k\otimes u + u\otimes\nabla_{\Sigma_k}\varphi_k)\nu_{\Sigma_k}\cdot\nu_{\Sigma_k}/\varrho]\!],\\
F_{h_k}(h) &= (b(t,x)|(\nabla_{\Sigma_k}\varphi_k)h)/[\![\varrho]\!].
\end{aligned}$$

For $k = 0$ we have the standard one-phase Stokes problem with parameters ϱ_2, $\mu_2(x)$ on Ω with no-slip boundary condition on $\partial\Omega$, i.e.,

$$\begin{aligned}
\varrho_2(\partial_t + \omega)u_0 - \mu_2(x)\Delta u_0 + \nabla\pi_0 &= F_0(u,\pi) && \text{in } \Omega,\\
\operatorname{div} u_0 &= (\nabla\varphi_0|u) && \text{in } \Omega,\\
u_0 &= 0 && \text{on } \partial\Omega,\\
u_0(0) &= 0 && \text{in } \Omega.
\end{aligned}$$

For $k = N+1$ we have the Cauchy problem of the Stokes equation with parameters ϱ_1, $\mu_1(x)$, i.e.,

$$\begin{aligned}
\varrho_1(\partial_t + \omega)u_{N+1} - \mu_1(x)\Delta u_{N+1} + \nabla\pi_{N+1} &= F_{N+1}(u,\pi) && \text{in } \mathbb{R}^n,\\
\operatorname{div} u_{N+1} &= (\nabla\varphi_{N+1}|u) && \text{in } \mathbb{R}^n,\\
u_{N+1}(0) &= 0 && \text{in } \mathbb{R}^n.
\end{aligned}$$

In the sequel we concentrate on the charts at the interface, i.e., $k = 1, \ldots, N$.

Though the right members $G_{u_k}(h)$, $G_{g_k}(u,h)$, $G_{h_k}(u)$, $F_{h_k}(h)$ have more time regularity than the corresponding data class, the terms $(\nabla\varphi_k)\pi$ in $F_k(u,\pi)$ and $(\nabla\varphi_k|u)$ unfortunately do not have this property. In order to remove this difficulty, we have to decompose the problem. Here is one major change compared to Section 8.2.5. Consider the following problem for the functions ϕ_k, ψ_k.

$$\begin{aligned}
\Delta\phi_k &= u \cdot \nabla\varphi_k = \operatorname{div}(u\varphi_k) && \text{in } \mathbb{R}^n \setminus \Sigma_k,\\
[\![\phi_k]\!] &= 0,\ \phi_k = 0 && \text{on } \Sigma_k,\\
\Delta\psi_k &= \operatorname{div} F_k && \text{in } \mathbb{R}^n \setminus \Sigma_k,\\
[\![\psi_k]\!] &= 0,\ \psi_k = 0 && \text{on } \Sigma_k,
\end{aligned} \tag{8.65}$$

Problem (8.65) is uniquely solvable and its solution satisfies

$$\psi_k, \partial_t\phi_k \in L_{p,\mu}(J; \dot{H}^1_p(\mathbb{R}^n \setminus \Sigma_k)), \quad \phi_k \in H^{1/2}_{p,\mu}(J; \dot{H}^3_p(\mathbb{R}^n \setminus \Sigma_k)),$$

which implies

$$\nabla\phi_k \in H^1_{p,\mu}(J; L_p(\mathbb{R}^n)) \cap H^{1/4}_{p,\mu}(J; H^2_p(\mathbb{R}^n \setminus \Sigma_k)).$$

Furthermore, by the additional time regularity of π we obtain

$$\nabla\psi_k \in H^{\alpha}_{p,\mu}(J; L_p(\mathbb{R}^n)) \cap L_{p,\mu}(J; H^1_p(\Omega)).$$

Defining

$$\begin{aligned}
\tilde{u}_k &= u_k - \nabla\phi_k, \quad \tilde{F}_k(u,\pi) = F_k(u,\pi) - \nabla\psi_k,\\
\tilde{\pi}_k &= \pi_k - \psi_k + \varrho(\partial_t + \omega)\phi_k - \mu\Delta\phi_k,
\end{aligned}$$

we see that $\operatorname{div} \tilde{F}_k(u,\pi) = 0$ and $\operatorname{div} \tilde{u}_k = 0$ in $\mathbb{R}^n \setminus \Sigma_k$. Now $\tilde{z}_k = (\tilde{u}_k, \tilde{\pi}_k, h_k)$ satisfies the problem

$$\begin{aligned}
&\varrho(\partial_t + \omega)\tilde{u}_k - \mu(x)\Delta\tilde{u}_k + \nabla\tilde{\pi}_k = \tilde{F}_k(u,\pi) && \text{in } \mathbb{R}^n \setminus \Sigma_k,\\
&\operatorname{div} \tilde{u}_k = 0 && \text{in } \mathbb{R}^n \setminus \Sigma_k,\\
&\mathcal{P}_{\Sigma_k}[\![\tilde{u}_k]\!] + c(t,x)\nabla_{\Sigma_k} h_k = \varphi_k \mathcal{P}_{\Sigma_k} g_u + \tilde{G}_{u_k}(h) && \text{on } \Sigma_k,\\
&[\![-\mu(x)(\nabla\tilde{u}_k + [\nabla\tilde{u}_k]^{\mathsf{T}}) + \tilde{\pi}_k]\!]\nu_{\Sigma_k} - \sigma(\Delta_\Sigma h_k)\nu_{\Sigma_k} = \varphi_k g_k + \tilde{G}_{g_k}(u,h) && \text{on } \Sigma_k,\\
&-[\![\mu(x)(\nabla\tilde{u}_k + [\nabla\tilde{u}_k]^{\mathsf{T}})\nu_{\Sigma_k} \cdot \nu_{\Sigma_k}/\varrho]\!] + [\![\tilde{\pi}_k/\varrho]\!] = \varphi_k g_h + \tilde{G}_{h_k}(u) && \text{on } \Sigma_k,\\
&(\partial_t + \omega)h_k - [\![\varrho\tilde{u}_k \cdot \nu_{\Sigma_k}]\!]/[\![\varrho]\!] + b(t,x)\cdot\nabla_{\Sigma_k} h_k/[\![\varrho]\!] = \varphi_k f_h + \tilde{F}_{h_k}(h) && \text{on } \Sigma_k,\\
&u_k(0) = 0 && \text{in } \mathbb{R}^n \setminus \Sigma_k,\\
&h_k(0) = 0 && \text{on } \Sigma_k,
\end{aligned} \tag{8.66}$$

where

$$\begin{aligned}
\tilde{G}_{u_k}(h) &= G_{u_k}(h) - \mathcal{P}_{\Sigma_k}[\![\nabla\phi_k]\!]\\
\tilde{G}_{g_k}(u,h) &= G_{g_k}(u,h) + [\![\mu(x)\nabla^2\phi_k]\!]\nu_{\Sigma_k}\\
\tilde{G}_{h_k}(u) &= G_{h_k}(u) + 2\nu_{\Sigma_k} \cdot [\![\mu(x)\nabla^2\phi_k/\varrho]\!]\nu_{\Sigma_k}\\
\tilde{F}_{h_k}(h) &= F_{h_k}(h) + [\![\varrho\partial_{\nu_k}\phi_k]\!]/[\![\varrho]\!].
\end{aligned}$$

Now we know that the data in (8.66) satisfy the assumption of Proposition 8.5.1. Therefore, the solution $\tilde{\pi}_k$ of (8.66) has more time regularity, and we have (8.59) at our disposal. We may now proceed as in Section 8.2.5.

It remains to prove surjectivity of L; here we have the second important modification. For this we employ the continuation method for semi-Fredholm operators, which states that the Fredholm index remains constant under homotopies $L(\lambda)$, as long as the ranges of $L(\lambda)$ stay closed. For this purpose, we introduce a first continuation parameter $\alpha \in [0,1]$ by replacing the 7th equation of (8.51) into

$$(\partial_t + \omega)h + \alpha(-\Delta_\Sigma)^{\frac{1}{2}}h - (1-\alpha)\left([\![\varrho u \cdot \nu]\!]/[\![\varrho]\!] + b(t,x)\cdot\nabla_\Sigma h/[\![\varrho]\!]\right) = f_h \quad \text{on } \Sigma.$$

With minor modifications, the analysis in the next subsection shows that the corresponding problem is well-posed for each $\alpha \in [0,1]$ in the case of a flat interface with bounds independent of $\alpha \in [0,1]$. Therefore, the same is true for bent interfaces and then by the above localization procedure also for a general geometry. Thus we only need to consider the case $\alpha = 1$.

To prove surjectivity in this case, note that the equation for h is independent from those for u and π, and it is uniquely solvable in the right regularity class because of maximal regularity for the Laplace-Beltrami operator, see Section 6.4. So we may set now $h = 0$.

Next we introduce a second continuation parameter $\beta \in [0,1]$ by

$$\mathcal{P}_\Sigma u_2 = \beta \mathcal{P}_\Sigma u_1 + \mathcal{P}_\Sigma g_u, \quad -\beta \mathcal{P}_\Sigma T_2(u,\pi)\nu + \mathcal{P}_\Sigma T_1(u,\pi)\nu = \mathcal{P}_\Sigma g$$

with $T(u,\pi) = \mu(x)(\nabla u + [\nabla u]^{\mathsf{T}}) - \pi$. The remaining normal stress boundary conditions decouple as $\varrho_1 \neq \varrho_2$.

Again, we can prove that the a priori estimates are uniform for $\beta \in [0,1]$. The remaining problem for $\beta = 0$ decouples into a one-phase Stokes problem with mixed Dirichlet-Neumann boundary condition in Ω_2, Dirichlet condition on $\partial\Omega$ and outflow conditions on Σ, and a one-phase Stokes problem with pure Neumann boundary condition in Ω_1. According to Section 7.3, these are uniquely solvable. This shows that we have surjectivity in the case $\alpha = 1$ and $\beta = 0$, hence by the continuation method also for $\alpha = 0$ and $\beta = 1$. The proof of Theorem 8.4.1 is now complete.

8.6 The Asymmetric Model Problem

Here we present the algebra and harmonic analysis needed to prove Theorem 8.4.1 for the flat interface case with constant coefficients.

6.1 The Reduced Asymmetric Stokes Problem
By the reduction arguments from Section 8.2, we need to study the problem

$$
\begin{aligned}
\varrho(\partial_t + \omega)u - \mu\Delta u + \nabla\pi &= 0 && \text{in } \dot{\mathbb{R}}^n,\\
\operatorname{div} u &= 0 && \text{in } \dot{\mathbb{R}}^n,\\
-\,[\![\mu(\nabla u + (\nabla u)^{\mathsf{T}})\nu]\!] + [\![\pi]\!]\nu &= (0, g_1)^{\mathsf{T}} && \text{on } \mathbb{R}^{n-1},\\
[\![v]\!] &= 0 && \text{on } \mathbb{R}^{n-1},\\
-\,[\![2\mu\nabla u\nu\cdot\nu/\varrho]\!] + [\![\pi/\varrho]\!] &= g_2 && \text{on } \mathbb{R}^{n-1},\\
u(0) &= 0 && \text{in } \dot{\mathbb{R}}^n.
\end{aligned}
\tag{8.67}
$$

The remaining data satisfy

$$g_1, g_2 \in {}_0W^{1/2-1/2p}_{p,\mu}(J; L_p(\mathbb{R}^{n-1})) \cap L_{p,\mu}(J; W^{1-1/p}_p(\mathbb{R}^{n-1})).$$

To prove this result, suppose that we have a solution of (8.67) in the proper regularity class on the half-line $J = \mathbb{R}_+$. Then we may employ the Laplace transform in t and the Fourier transform in the tangential variables $x \in \mathbb{R}^{n-1}$, to obtain the following boundary value problem for a system of ordinary differential equations on $\dot{\mathbb{R}}$.

$$
\begin{cases}
\omega_k^2\hat{v} - \mu_k\partial_y^2\hat{v} + i\xi\hat{\pi} = 0, & (-1)^k y > 0,\\
\omega_k^2\hat{w} - \mu_k\partial_y^2\hat{w} + \partial_y\hat{\pi} = 0, & (-1)^k y > 0,\\
(i\xi|\hat{v}) + \partial_y\hat{w} = 0, & y \neq 0.
\end{cases}
$$

Here we have set $\omega_k^2 = \varrho_k\lambda + \mu_k|\xi|^2$, $k = 1, 2$. As in Section 8.3, this system of equations is easily solved to the result

$$
\begin{bmatrix} \hat{v}_2 \\ \hat{w}_2 \\ \hat{\pi}_2 \end{bmatrix}
= e^{-\omega_2 y/\sqrt{\mu_2}}
\begin{bmatrix} a_2 \\ \frac{\sqrt{\mu_2}}{\omega_2}(i\xi|a_2) \\ 0 \end{bmatrix}
+ \alpha_2 e^{-|\xi|y}
\begin{bmatrix} -i\xi \\ |\xi| \\ \varrho_2\lambda \end{bmatrix},
\tag{8.68}
$$

for $y > 0$, and

$$
\begin{bmatrix} \hat{v}_1 \\ \hat{w}_1 \\ \hat{\pi}_1 \end{bmatrix}
= e^{\omega_1 y/\sqrt{\mu_1}}
\begin{bmatrix} a_1 \\ -\frac{\sqrt{\mu_1}}{\omega_1}(i\xi|a_1) \\ 0 \end{bmatrix}
+ \alpha_1 e^{|\xi|y}
\begin{bmatrix} -i\xi \\ -|\xi| \\ \varrho_1\lambda \end{bmatrix},
\tag{8.69}
$$

for $y < 0$. Here $a_k \in \mathbb{C}^{n-1}$ and $\alpha_k \in \mathbb{C}$ have to be determined by the interface conditions which in frequency domain read

$$
\begin{aligned}
\hat{v}_1(0) - \hat{v}_2(0) &= 0,\\
\mu_2(\partial_y\hat{v}_2(0) + i\xi\hat{w}_2(0)) - \mu_1(\partial_y\hat{v}_1(0) + i\xi\hat{w}_1(0)) &= 0,\\
-2\mu_2\partial_y\hat{w}_2(0) + \hat{\pi}_2(0) + 2\mu_1\partial_y\hat{w}_1(0) - \hat{\pi}_1(0) &= \hat{g}_1,\\
-2(\mu_2/\varrho_2)\partial_y\hat{w}_2(0) + \hat{\pi}_2(0)/\varrho_2 + 2(\mu_1/\varrho_2)\partial_y\hat{w}_1(0) - \hat{\pi}_1(0)/\varrho_2 &= \hat{g}_2.
\end{aligned}
$$

Inserting the representation of the transformed solution into the first two of these equations we obtain the following system.

$$a_2 - a_1 = i\xi(\alpha_2 - \alpha_1),$$
$$\sqrt{\mu_2}\omega_2 a_2 + \sqrt{\mu_1}\omega_1 a_1 = i\xi\big(2|\xi|(\mu_1\alpha_1 + \mu_2\alpha_2) + (\mu_1\sqrt{\mu_1}/\omega_1)\beta_1 + (\mu_2\sqrt{\mu_2}/\omega_2)\beta_2\big),$$

where we have set $\beta_k = i\xi \cdot a_k$, $k = 1, 2$. This system can be solved for a_k to the result

$$a_1 = i\xi\frac{2|\xi|(\mu_1\alpha_1 + \mu_2\alpha_2) + (\mu_1\sqrt{\mu_1}/\omega_1)\beta_1 + (\mu_2\sqrt{\mu_2}/\omega_2)\beta_2 - \sqrt{\mu_2}\omega_2(\alpha_2 - \alpha_1)}{\sqrt{\mu_1}\omega_1 + \sqrt{\mu_2}\omega_2},$$

and

$$a_2 = i\xi\frac{2|\xi|(\mu_1\alpha_1 + \mu_2\alpha_2) + (\mu_1\sqrt{\mu_1}/\omega_1)\beta_1 + (\mu_2\sqrt{\mu_2}/\omega_2)\beta_2 + \sqrt{\mu_1}\omega_1(\alpha_2 - \alpha_1)}{\sqrt{\mu_1}\omega_1 + \sqrt{\mu_2}\omega_2}.$$

Multiplying the former equations with $i\xi$ we obtain

$$\beta_2 - \beta_1 = |\xi|^2(\alpha_1 - \alpha_2),$$
$$\gamma_1\beta_1 + \gamma_2\beta_2 = -2|\xi|^3(\mu_1\alpha_1 + \mu_2\alpha_2),$$

where $\gamma_k = \sqrt{\mu_k}\omega_k + \mu_k|\xi|^2\sqrt{\mu_k}/\omega_k$, $k = 1, 2$. We solve this system for β_k, concluding

$$\begin{bmatrix} \beta_1 \\ \beta_2 \end{bmatrix} = -\frac{|\xi|^2}{\gamma_1 + \gamma_2}\begin{bmatrix} 2\mu_1|\xi| + \gamma_2 & 2\mu_2|\xi| - \gamma_2 \\ 2\mu_1|\xi| - \gamma_1 & 2\mu_2|\xi| + \gamma_1 \end{bmatrix}\begin{bmatrix} \alpha_1 \\ \alpha_2 \end{bmatrix}. \tag{8.70}$$

Inserting the transformed solution into the two remaining stress conditions leads to

$$2(\mu_2\beta_2 + \mu_2|\xi|^2\alpha_2) + \varrho_2\lambda\alpha_2 - 2(\mu_1\beta_1 + \mu_1|\xi|^2\alpha_1) - \varrho_1\lambda\alpha_1 = \hat{g}_1,$$
$$2(\mu_2\beta_2 + \mu_2|\xi|^2\alpha_2)/\varrho_2 + \lambda\alpha_2 - 2(\mu_1\beta_1 + \mu_1|\xi|^2\alpha_1)/\varrho_1 - \lambda\alpha_1 = \hat{g}_2.$$

Using the formulas for β_k and solving the resulting system in terms of α_k we arrive after some elementary algebra at the expressions

$$\begin{bmatrix} \alpha_1 \\ \alpha_2 \end{bmatrix} = \frac{1}{[\![\varrho]\!]\lambda\varepsilon}\begin{bmatrix} 1 + 2\varrho_2[\![\mu/\varrho]\!]\varepsilon_2 & -\varrho_2(1 + [\![2\mu]\!]\varepsilon_2) \\ 1 - 2\varrho_1[\![\mu/\varrho]\!]\varepsilon_1 & -\varrho_1(1 - [\![2\mu]\!]\varepsilon_1) \end{bmatrix}\begin{bmatrix} \hat{g}_1 \\ \hat{g}_2 \end{bmatrix}, \tag{8.71}$$

where we have set $\varepsilon_k = \varrho_k\lambda\sqrt{\mu_k}|\xi|^2/\big(\omega_k(\gamma_1 + \gamma_2)(\omega_k + \sqrt{\mu_k}|\xi|)^2\big)$, $k = 1, 2$, and

$$\varepsilon = 1 + 2(\mu_1\varepsilon_1 + \mu_2\varepsilon_2).$$

Here we observe that the surface pressures π_k have transforms $\lambda\varrho_k\alpha_k$. Since the entries in the matrix defining $\lambda\varrho_k\alpha_k$ are bounded and holomorphic we may conclude that π_k have the same regularity as g_k, and that the pressure π belongs to $L_{p,\mu}(J; \dot{H}^1_p(\mathbb{R}^n))$. Next let us compute the interface velocities $v_1^b = v_2^b$ and w_k^b. Their transforms are given by

$$\hat{v}_k^b = a_k - i\xi\alpha_k, \quad \hat{w}_k^b = (-1)^k\Big(\frac{\sqrt{\mu_k}}{\omega_k}\beta_k + |\xi|\alpha_k\Big).$$

Some algebra yields for w_k^b

$$\begin{bmatrix} \hat{w}_1^b \\ \hat{w}_2^b \end{bmatrix} = \frac{|\xi|}{\omega_1\omega_2}\begin{bmatrix} \frac{-\varrho_1\omega_2}{\gamma_1+\gamma_2}(\sqrt{\mu_1} + \frac{\gamma_2}{\omega_1+\sqrt{\mu_1}|\xi|}) & \frac{-\varrho_2\sqrt{\mu_1\mu_2}|\xi|}{\gamma_1+\gamma_2}\frac{\omega_2-\sqrt{\mu_2}|\xi|}{\omega_2+\sqrt{\mu_2}|\xi|} \\ \frac{\varrho_1\sqrt{\mu_1\mu_2}|\xi|}{\gamma_1+\gamma_2}\frac{\omega_1-\sqrt{\mu_1}|\xi|}{\omega_1+\sqrt{\mu_1}|\xi|} & \frac{\varrho_2\omega_1}{\gamma_1+\gamma_2}(\sqrt{\mu_2} + \frac{\gamma_1}{\omega_2+\sqrt{\mu_2}|\xi|}) \end{bmatrix}\begin{bmatrix} \lambda\alpha_1 \\ \lambda\alpha_2 \end{bmatrix}. \tag{8.72}$$

This representation shows that $\hat{w}_k^b$ is bounded by $|\xi|\hat{g}/\omega_1\omega_2$. Therefore, we see that the operator with symbol $|\xi|/\omega_1\omega_2$ maps into the right space for the boundary values of w, i.e., we have

$$w_k^b \in {}_0H^1_{p,\mu}(J; \dot{W}_p^{-1/p}(\mathbb{R}^{n-1})) \cap L_{p,\mu}(J; W_p^{2-1/p}(\mathbb{R}^{n-1})).$$

To obtain the regularity of the boundary values v_k^b of v we write

$$\begin{aligned} \hat{v}_1^b &= \hat{v}_2^b = a_2 - i\xi\alpha_2 \\ &= \frac{i\xi}{\sqrt{\mu_1}\omega_1 + \sqrt{\mu_2}\omega_2}\Big\{\mu_1(\frac{\sqrt{\mu_1}}{\omega_1}\beta_1 + \alpha_1|\xi|) + \mu_2(\frac{\sqrt{\mu_2}}{\omega_2}\beta_2 + \alpha_2|\xi|)\Big\} \\ &\quad - \frac{i\xi}{\sqrt{\mu_1}\omega_1 + \sqrt{\mu_2}\omega_2}\Big\{\frac{\varrho_1\sqrt{\mu_1}}{\omega_1 + \sqrt{\mu_1}|\xi|}\lambda\alpha_1 + \frac{\varrho_1\sqrt{\mu_1}}{\omega_1 + \sqrt{\mu_1}|\xi|}\lambda\alpha_2\Big\}. \end{aligned}$$

This representation shows that also $\hat{v}_k^b$ is bounded by $|\xi|\hat{g}_i/\omega_1\omega_2$, and the same argument as in (8.27) yields

$$\begin{aligned} v_k^b \in {}_0H^1_{p,\mu}(J; \dot{W}_p^{-1/p}(\mathbb{R}^{n-1}; \mathbb{R}^{n-1})) \cap L_{p,\mu}(J; W_p^{2-1/p}(\mathbb{R}^{n-1}; \mathbb{R}^{n-1})) \\ \hookrightarrow {}_0W^{1-1/2p}_{p,\mu}(J; L_p(\mathbb{R}^{n-1}; \mathbb{R}^{n-1})). \end{aligned}$$

Therefore, the boundary values of u from either side of Σ have the required regularity, hence solving the Stokes problem with these boundary values separately in the upper and the lower half-space, u has maximal L_p-regularity. This completes the proof of Theorem 8.5.2.

6.2 The Asymmetric Stokes Problem with Free Boundary

To extract the boundary symbol for the full problem, we set $\hat{g}_1 = -\sigma|\xi|^2\hat{h}$, $\hat{g}_2 = 0$, and observe that the transformed equation for $\hat{h}$ reads

$$\lambda\hat{h} - [\![\varrho\hat{w}(0)]\!]/[\![\varrho]\!] = \hat{g}_3.$$

We have

$$[\![\varrho\hat{w}(0)]\!] = \varrho_2(\sqrt{\mu_2}\beta_2/\omega_2 + |\xi|\alpha_2) - \varrho_1(-\sqrt{\mu_1}\beta_1/\omega_1 - |\xi|\alpha_1), \tag{8.73}$$

hence inserting the expressions for α_k and β_k we obtain after some more algebra

$$s(\lambda, |\xi|)\hat{h} = \hat{g}_4. \tag{8.74}$$

Here g_4 is determined by the data alone and has the same regularity as g_3. The boundary symbol $s(\lambda, \tau)$ is defined by

$$s(\lambda, \tau) = \lambda + \frac{\sigma\tau}{[\![\varrho]\!]^2} m(z), \tag{8.75}$$

where we again employed the scaling $z = \lambda/|\xi|^2$. The holomorphic function $m(z)$ in turn is given by

$$\begin{aligned}(\mu_1\varphi_1(z) + \mu_2\varphi_2(z))m(z) &= 2\frac{\varrho_1\varrho_2}{\omega_1(z)\omega_2(z)}\frac{\omega_1(z)-1}{\omega_1(z)+1}\frac{\omega_2(z)-1}{\omega_2(z)+1} \\ &+ \frac{\varrho_1^2\mu_2\varphi_2(z)}{\mu_1\omega_1(z)(\omega_1(z)+1)} + \frac{\varrho_2^2\mu_1\varphi_1(z)}{\mu_2\omega_2(z)(\omega_2(z)+1)} + \frac{\varrho_1^2}{\omega_1(z)} + \frac{\varrho_2^2}{\omega_2(z)},\end{aligned} \tag{8.76}$$

with the abbreviations

$$\omega_k(z) = \sqrt{1 + \varrho_k z/\mu_k}, \quad \varphi_k(z) = \omega_k(z) + \frac{3}{\omega_k(z)+1} - \frac{1}{\omega_k(z)(\omega_k(z)+1)}.$$

We derive this formula in the next subsection.

Note that $\omega_k(z)$ is holomorphic in the sliced plane $\mathbb{C} \setminus (-\infty, -\mu_k/\varrho_k]$, hence the function $\varphi_k(z)$ has this property as well. This function has exactly one zero z_k in this set, it is real and satisfies $-\mu_k/\varrho_k < z_k < -8\mu_k/9\varrho_k < 0$. It is easy to see that φ_k maps $\bar{\mathbb{C}}_+$ into $\mathbb{C}_+$, and as $\varphi_k(0) = 2$ and $\varphi_k(z) \sim \sqrt{\varrho_k z/\mu_k}$ as $z \to \infty$, we see that $\varphi_k(\bar{\mathbb{C}}_+) \subset \Sigma_{\phi_k}$, for some angle $\phi_k < \pi/2$. By continuity of the argument function, this implies that $\varphi_k(\Sigma_{\pi/2+\eta}) \subset \mathbb{C}_+$, for some $\eta > 0$. Therefore, $\varphi(z) := \mu_1\varphi_1(z) + \mu_2\varphi_2(z)$ also has this property, in particular $\varphi(z)$ cannot vanish in $\Sigma_{\pi/2+\eta}$. This implies that $m(z)$ is holomorphic in this sector and in a ball $B(0, r_0)$ for some $r_0 > 0$. We obtain for the asymptotics of $m(z)$

$$m(0) = \frac{1}{2}\frac{\varrho_1^2}{\mu_1} + \frac{\varrho_2^2}{\mu_2} > 0, \quad \lim_{z\to\infty} z\, m(z) = \varrho_1 + \varrho_2.$$

Thus there is a constant $M = M(r, \phi) > 0$ such that

$$|m(z)| \le \frac{M}{1+|z|}, \quad z \in \Sigma_\phi \cup B(0, r),$$

for each $\phi < \pi/2 + \eta$ and $r < r_0$. From this estimate it is easy to conclude

$$|s(\lambda, \tau)| \le C_\eta(|\lambda| + |\tau|), \quad \lambda \in \Sigma_{\pi_2+\eta},\ \tau \in \Sigma_\eta,$$

whenever $\eta > 0$ is small enough. Conversely, since $m(0) > 0$, given a small $\eta > 0$ we find $r_\eta \in (0, r_0)$ such that $m(z) \in \Sigma_{\pi/2-3\eta}$ and $|m(z)| \ge m(0)/2$, for all $z \in B(0, r_\eta)$. This implies that there is a constant $c_\eta > 0$ such that

$$|s(\lambda, \tau)| \ge c_\eta(|\lambda| + |\tau|), \quad \lambda \in \Sigma_{\pi/2+\eta},\ |\lambda| \le r_\eta|\tau|^2.$$

On the other hand, choosing $|\lambda| \geq C|\tau|$ we obtain

$$\begin{aligned}|s(\lambda,\tau)| &\geq |\lambda| - \sigma[\![\varrho]\!]^{-2}M|\tau| \\ &\geq \frac{1}{2}|\lambda| + \Big(\frac{C}{2} - \sigma[\![\varrho]\!]^{-2}M\Big)|\tau| \geq c_\eta(|\lambda| + |\tau|),\end{aligned}$$

for all $\lambda \in \Sigma_{\pi/2+\eta}$, $\tau \in \Sigma_\eta$ such that $|\lambda| \geq C|\tau|$. If C is chosen large enough this implies that we have a lower bound

$$|s(\lambda,\tau)| \geq c(|\lambda| + |\tau|), \quad \lambda \in \Sigma_{\pi/2+\eta},\ \tau \in \Sigma_\eta,\ |\lambda| \geq \omega_0.$$

Thus the boundary symbol for the asymmetric Stokes problem has the same properties for $|\lambda| > \omega_0$ as that for the standard two-phase Stokes problem obtained in Section 8.3. We may now follow the arguments given there to complete the proof of Theorem 8.5.3.

6.3 Derivation of the Boundary Symbol

Here we compute the function $m(z)$ introduced in the previous subsection. By (8.73), (8.70), and (8.71) with $\hat{g}_1 = -\sigma|\xi|^2\hat{h}$ and $\hat{g}_2 = 0$, we obtain with $\tau = |\xi|$

$$\begin{aligned}&-[\![\varrho\hat{w}(0)]\!]/[\![\varrho]\!] \\ &= -[\![\varrho]\!]^{-1}\left(\tau\alpha_1\varrho_1 + \tau\alpha_2\varrho_2 + \varrho_1\sqrt{\mu_1}\beta_1/\omega_1 + \varrho_2\sqrt{\mu_2}\beta_2/\omega_2\right) \\ &= -[\![\varrho]\!]^{-1}\left[\alpha_1\left\{\tau\varrho_1 - \tau^2/\gamma\big(\varrho_1\sqrt{\mu_1}/\omega_1(2\mu_1\tau + \gamma_2) + \varrho_2\sqrt{\mu_2}/\omega_2(2\mu_1\tau - \gamma_1)\big)\right\}\right. \\ &\qquad\left. + \alpha_2\left\{\tau\varrho_2 - \tau^2/\gamma\big(\varrho_1\sqrt{\mu_1}/\omega_1(2\mu_2\tau - \gamma_2) + \varrho_2\sqrt{\mu_2}/\omega_2(2\mu_2\tau + \gamma_1)\big)\right\}\right] \\ &= \frac{\sigma\tau^3\hat{h}}{\gamma[\![\varrho]\!]^2\lambda\varepsilon}\Big[(1 + 2\varrho_2\varepsilon_2[\![\mu/\varrho]\!]) \\ &\qquad\times\left\{\gamma\varrho_1 - \big(\varrho_1\tau\sqrt{\mu_1}/\omega_1(2\mu_1\tau + \gamma_2) + \varrho_2\tau\sqrt{\mu_2}/\omega_2(2\mu_1\tau - \gamma_1)\big)\right\} \\ &\qquad + (1 - 2\varrho_1\varepsilon_1[\![\mu/\varrho]\!]) \\ &\qquad\times\left\{\gamma\varrho_2 - \big(\varrho_1\tau\sqrt{\mu_1}/\omega_1(2\mu_2\tau - \gamma_2) + \varrho_2\tau\sqrt{\mu_2}/\omega_2(2\mu_2\tau + \gamma_1)\big)\right\}\Big],\end{aligned} \tag{8.77}$$

where we have set $\gamma := \gamma_1 + \gamma_2$. The scaling $z := \lambda/\tau^2$, $\tau := |\xi|$, yields

$$\begin{aligned}\omega_k &= \sqrt{\varrho_k\lambda + \mu_k\tau^2} = \sqrt{\mu_k}\tau\omega_k(z), \quad \omega_k(z) = \sqrt{1 + \varrho_k z/\mu_k}, \\ \gamma_k &= \sqrt{\mu}_k + \mu_k\tau^2\sqrt{\mu_k}/\omega_k = \mu_k\tau\left(\omega_k(z) + 1/\omega_k(z)\right), \\ \varepsilon_k &= \varrho_k\lambda\sqrt{\mu_k}\tau^2/(\omega_k\gamma(\omega_k + \sqrt{\mu_k}\tau)^2) = (\gamma\omega_k(z))^{-1}(\omega_k(z) - 1)/(\omega_k(z) + 1),\end{aligned}$$

hence

$$\begin{aligned}
2\mu_k\tau - \gamma_k &= -\frac{\mu_k\tau}{\omega_k(z)}(\omega_k(z)-1)^2 = -\frac{\varrho_k z}{\omega_k(z)}\tau\frac{\omega_k(z)-1}{\omega_k(z)+1},\\
\varrho_1\gamma_2 - \varrho_1\tau\sqrt{\mu_1}\gamma_2/\omega_1 &= \varrho_1^2\frac{\mu_2}{\mu_1}\left(\omega_2(z)+\frac{1}{\omega_2(z)}\right)\frac{z}{\omega_1(z)(\omega_1(z)+1)},\\
\varrho_1\gamma_1 - \varrho_1\tau\sqrt{\mu_1}2\mu_1\tau/\omega_1 &= \varrho_1^2\tau z/\omega_1(z),\\
\gamma\varepsilon = \gamma + 2\mu_1\gamma\varepsilon_1 + 2\mu_2\gamma\varepsilon_2 &= \mu_1\varphi_1(z) + \mu_2\varphi_2(z),
\end{aligned}$$

where

$$\varphi_k(z) = \omega_k(z) + \frac{1}{\omega_k(z)} + \frac{2}{\omega_k(z)}\frac{\omega_k(z)-1}{\omega_k(z)+1} = \omega_k(z) + \frac{3}{\omega_k(z)+1} - \frac{1}{\omega_k(z)(\omega_k(z)+1)}.$$

Substituting these expressions into (8.77), we obtain

$$\begin{aligned}
-\frac{[\![\varrho\hat{w}(0)]\!]}{[\![\varrho]\!]} &= \frac{\sigma\tau\hat{h}}{[\![\varrho]\!]^2(\mu_1\varphi_1+\mu_2\varphi_2)}\\
&\times\Big[(1+2\varrho_2\varepsilon_2[\![\mu/\varrho]\!])\left(\frac{\varrho_1^2}{\omega_1} + \frac{\varrho_1^2\mu_2}{\mu_1}(\omega_2+\frac{1}{\omega_2})\frac{1}{\omega_1(\omega_1+1)} + \frac{\varrho_1\varrho_2}{\omega_1\omega_2}\frac{\omega_1-1}{\omega_1+1}\right)\\
&+(1-2\varrho_1\varepsilon_1[\![\mu/\varrho]\!])\left(\frac{\varrho_2^2}{\omega_2} + \frac{\varrho_2^2\mu_1}{\mu_2}(\omega_1+\frac{1}{\omega_1})\frac{1}{\omega_2(\omega_2+1)} + \frac{\varrho_1\varrho_2}{\omega_1\omega_2}\frac{\omega_2-1}{\omega_2+1}\right)\Big].
\end{aligned}$$

Expanding and collecting terms we see that the coefficients of ϱ_1^2, $\varrho_1\varrho_2$, ϱ_2^2 in the square brackets on the right-hand side eventually become

$$\frac{1}{\omega_1} + \frac{\mu_2\varphi_2}{\mu_1\omega_1(\omega_1+1)}, \quad \frac{2}{\omega_1\omega_2}\frac{\omega_1-1}{\omega_1+1}\frac{\omega_2-1}{\omega_2+1}, \quad \frac{1}{\omega_2} + \frac{\mu_1\varphi_1}{\mu_2\omega_2(\omega_2+1)}.$$

Note that ε_k contains γ in the denominator; in it is important to recognize that it factors. Finally, we obtain

$$\begin{aligned}
-\frac{[\![\varrho\hat{w}(0)]\!]}{[\![\varrho]\!]} &= \frac{\sigma\tau\hat{h}}{[\![\varrho]\!]^2(\mu_1\varphi_1(z)+\mu_2\varphi_2(z))}\Big[2\frac{\varrho_1\varrho_2}{\omega_1(z)\omega_2(z)}\frac{\omega_1(z)-1}{\omega_1(z)+1}\frac{\omega_2(z)-1}{\omega_2(z)+1}\\
&+\frac{\varrho_1^2\mu_2\varphi_2(z)}{\mu_1\omega_1(z)(\omega_1(z)+1)} + \frac{\varrho_2^2\mu_1\varphi_1(z)}{\mu_2\omega_2(z)(\omega_2(z)+1)} + \frac{\varrho_1^2}{\omega_1(z)} + \frac{\varrho_2^2}{\omega_2(z)}\Big],
\end{aligned}$$

which proves (8.76).

6.4 The Modified Problem

Here we consider the case where $b_0, c_0 \neq 0$ are constant. As before we have the transformed equation for the function h.

$$\lambda\hat{h} - \frac{[\![\varrho\hat{w}(0)]\!]}{[\![\varrho]\!]} + \frac{b_0\cdot i\xi\hat{h}}{[\![\varrho]\!]} = \hat{f}_h.$$

We obtain after some linear algebra

$$s(\lambda, |\xi|)\hat{h} = \hat{f}_h.$$

Setting $\tau = |\xi|$ and employing the scaling $z = \lambda/\tau^2$, we obtain the *boundary symbol* $s(\lambda, \tau)$ in the form

$$s(\lambda, \tau) = \lambda + \frac{\sigma\tau}{[\![\varrho]\!]^2} m(z) + \frac{c_0\tau}{[\![\varrho]\!]} \ell(z) + \frac{i\tau}{[\![\varrho]\!]}(b_0 \cdot \frac{\xi}{|\xi|}). \tag{8.78}$$

To prove this, it is enough to seek the solution of the problem

$$\begin{aligned} \hat{v_2}(0) - \hat{v_1}(0) &= -c_0 i\xi\hat{h}, \\ \mu_2(\partial_y \hat{v}_2(0) + i\xi\hat{w}_2(0)) - \mu_1(\partial_y \hat{v}_1(0) + i\xi\hat{w}_1(0)) &= 0, \\ -2\mu_2\partial_y\hat{w}_2(0) + \hat{\pi}_2(0) + 2\mu_1\partial_y\hat{w}_1(0) - \hat{\pi}_1(0) &= 0, \\ -2(\mu_2/\varrho_2)\partial_y\hat{w}_2(0) + \hat{\pi}_2(0)/\varrho_2 + 2(\mu_1/\varrho_2)\partial_y\hat{w}_1(0) - \hat{\pi}_1(0)/\varrho_2 &= 0. \end{aligned} \tag{8.79}$$

Multiplying the 1st and the 2nd equations by $i\xi$, and setting $\beta_j = i\xi \cdot a_j$ for $a_j \in \mathbb{C}^{n-1}$, $j = 1, 2$, we have

$$\beta_2 - \beta_1 = -|\xi|^2\{(\alpha_2 - \alpha_1) - c_0\hat{h}\}, \tag{8.80}$$

$$\sqrt{\mu_2}\omega_2\beta_2 + \sqrt{\mu_1}\omega_1\beta_1 = -|\xi|^2\Big(2|\xi|(\mu_1\alpha_1 + \mu_2\alpha_2) + \Big(\frac{\mu_1\sqrt{\mu_1}}{\omega_1}\Big)\beta_1 + \Big(\frac{\mu_2\sqrt{\mu_2}}{\omega_2}\Big)\beta_2\Big).$$

Combining the 3rd and the 4th equations of (8.80), we obtain

$$\alpha_j = -2\mu_j\beta_j/(\omega_j^2 + \mu_j|\xi|^2). \tag{8.81}$$

Substituting this formula into (8.80) and using the scaling $\omega_j = \sqrt{\mu_j}|\xi|\omega_j(z)$, $\gamma_j = \mu_j|\xi|\gamma_j(z)$, we solve the system for β_j

$$\begin{bmatrix} -1 + 2/(\omega_1(z)^2 + 1) & 1 - 2/(\omega_2(z)^2 + 1) \\ \mu_1\{\gamma_1(z) - 4/(\omega_1(z)^2 + 1)\} & \mu_2\{\gamma_2(z) - 4/(\omega_2(z)^2 + 1)\} \end{bmatrix} \begin{bmatrix} \beta_1 \\ \beta_2 \end{bmatrix} = \begin{bmatrix} c_0\tau^2\hat{h} \\ 0 \end{bmatrix},$$

where we set $\tau = |\xi|$ and $z = \lambda/\tau^2$. Therefore, substituting β_j and α_j from (8.80) and (8.81) into

$$-\frac{[\![\varrho w(0)]\!]}{[\![\varrho]\!]} = -\frac{(\varrho_1\alpha_1 + \varrho_2\alpha_2)\tau + (\varrho_2/\tau\omega_2)\beta_2 + (\varrho_1/\tau\omega_1)\beta_1}{[\![\varrho]\!]},$$

we obtain (8.78). The first and the second terms in (8.78) as well as the holomorphic function $m(z)$ are the same as in Section 8.2, and $\ell(z)$ is given by

$$\begin{aligned} \psi(z)\ell(z) = {} & \varrho_1\mu_2\frac{\omega_1(z) - 1}{\omega_1(z)(\omega_1(z)^2 + 1)}\left(\frac{\omega_2(z) - 1}{\omega_2(z)} + 2\frac{\omega_2(z) + 1}{\omega_2(z)^2 + 1}\right) \\ & - \varrho_2\mu_1\frac{\omega_2(z) - 1}{\omega_2(z)(\omega_2(z)^2 + 1)}\left(\frac{\omega_1(z) - 1}{\omega_1(z)} + 2\frac{\omega_1(z) + 1}{\omega_1(z)^2 + 1}\right). \end{aligned}$$

Here the function $\psi(z)$ is given by

$$\psi(z) = \frac{\omega_1(z)+1}{\omega_1(z)^2+1}\mu_2\left(\frac{\omega_2(z)-1}{\omega_2(z)} + 2\frac{\omega_2(z)+1}{\omega_2(z)^2+1}\right) + \frac{\omega_2(z)+1}{\omega_2(z)^2+1}\mu_1\left(\frac{\omega_1(z)-1}{\omega_1(z)} + 2\frac{\omega_1(z)+1}{\omega_1(z)^2+1}\right).$$

Now we may argue as in Section 8.3.3. $\omega_j(z)$ is holomorphic in the sliced plane $\mathbb{C}\setminus(-\infty,-\frac{\mu_j}{\varrho_j}]$, hence the function $\psi(z)$ has this property in $\mathbb{C}\setminus(-\infty,\eta]$, with $\eta=\min\{\mu_j/\varrho_j\}$. It is not difficult to see that ψ maps $\overline{\mathbb{C}}_+$ into $\mathbb{C}_+$, and with $\psi(0)=2(\mu_1+\mu_2)$, $\sqrt{z}\psi(z)\to\mu_1\sqrt{\frac{\mu_2}{\varrho_2}}+\mu_2\sqrt{\frac{\mu_1}{\varrho_1}}$, as $|z|\to\infty$, we may conclude $\psi(\overline{\mathbb{C}}_+)\subset\Sigma_\phi$ for some angle $\phi<\frac{\pi}{2}$. By continuity of the argument function, this implies $\psi(\Sigma_{\frac{\pi}{2}+\eta})\subset\mathbb{C}_+$ for some $\eta>0$. Therefore, $\psi(z)$ cannot vanish in $\Sigma_{\frac{\pi}{2}+\eta}$. This implies that $\ell(z)$ is holomorphic in this sector and in a ball $B(0,r_0)$ for some $r_0>0$. For the asymptotics of $\ell(z)$ we have

$$\ell(0)=0,\qquad \lim_{|z|\to\infty} z\ell(z) = \frac{2\mu_1\mu_2\left(\sqrt{\mu_2/\varrho_2}-\sqrt{\mu_1/\varrho_1}\right)}{\mu_1\sqrt{\mu_2/\varrho_2}+\mu_2\sqrt{\mu_1/\varrho_1}}\quad \text{for } z\in\mathbb{C}\setminus\mathbb{R}_-.$$

Thus there is a constant $L=L(r,\phi)>0$ such that

$$|\ell(z)|\le\frac{L}{1+|z|},\qquad z\in\Sigma_\phi\cup B(0,r),$$

for each $\phi<\frac{\pi}{2}+\eta$ and $r<r_0$. Combining this estimate with the estimate for $m(z)$ from Section 8.6.2, it is easy to conclude

$$|s(\lambda,\tau)|\le c_\eta(|\lambda|+|\tau|),\qquad \lambda\in\Sigma_{\frac{\pi}{2}+\eta},\quad \tau\in\Sigma_\eta,$$

whenever $\eta>0$ is small enough. Conversely, since $m(0)>0$, given a small $\eta>0$ we find $\tau_\eta\in(0,r_0)$ such that $m(z)\in\Sigma_{\frac{\pi}{2}-3\eta}$ and $|m(z)|\ge\frac{m(0)}{2}$ for all $\tau\in B(0,r_\eta)$. This implies that there is a constant $c_\eta>0$ such that

$$|s(\lambda,\tau)|\ge|\lambda|+|\tau|\frac{\sigma}{2[\![\varrho]\!]^2}m(0),\qquad \lambda\in\Sigma_{\frac{\pi}{2}+\eta},\ |\lambda|\le r_\eta|\tau|^2.$$

On the other hand, choosing $|\lambda|\ge C|\tau|$ we obtain

$$|s(\lambda,\tau)|\ge\frac{|\lambda|}{2}+\left(\frac{C}{2}-\frac{\sigma}{[\![\varrho]\!]^2}M-\frac{|c_0|}{[\![\varrho]\!]}L-\frac{|b_0|}{[\![\varrho]\!]}\right)|\tau|\ge c_\eta(|\lambda|+|\tau|)$$

for all $\lambda\in\Sigma_{\frac{\pi}{2}+\eta}$, $\tau\in\Sigma_\eta$, with $|\lambda|\ge C|\tau|$ and

$$C>2(\sigma M/[\![\varrho]\!]^2+|c_0|L/[\![\varrho]\!]+|b_0|/[\![\varrho]\!]).$$

Therefore, if λ_0 is chosen large enough this implies the lower bound

$$|s(\lambda,\tau)|\ge c_\eta(|\lambda|+|\tau|),\qquad \lambda\in\Sigma_{\frac{\pi}{2}+\eta},\ \tau\in\Sigma_\eta,\ |\lambda|\ge\lambda_0.$$

Thus this boundary symbol has the same regularity for $|\lambda|\ge\lambda_0$ as that for the problem for the case $b_0=c_0=0$.

Appendix: Transmission Problems for the Laplace Equation

Here we state and prove two results on transmission problems for the Laplace equation which have been employed in Section 8.2.

Proposition 8.6.1. *Suppose that Ω is bounded domain in $\mathbb{R}^n$ with boundary $\partial\Omega$ of class C^{3-}, and let $\Sigma \subset \Omega$ be a closed hypersurface of class C^{3-}, $s = 0, 1$. Let $f \in H^s_q(\Omega \setminus \Sigma)$, $g_b \in W^{1+s-1/q}_q(\partial\Omega)$, $g \in W^{1+s-1/q}_q(\Sigma)$, $h \in W^{2+s-1/q}_q(\Sigma)$, such that the compatibility condition*

$$\int_\Omega f\,dx + \int_\Sigma g\,d\Sigma = \int_{\partial\Omega} g_b\,d(\partial\Omega)$$

is satisfied.

Then the problem

$$\begin{aligned} \varrho^{-1}\Delta\psi &= f && \text{in } \Omega\setminus\Sigma,\\ \varrho^{-1}\partial_\nu\psi &= g_b && \text{on } \partial\Omega,\\ [\![\varrho^{-1}\partial_\nu\psi]\!] &= g && \text{on } \Sigma,\\ [\![\psi]\!] &= h && \text{on } \Sigma. \end{aligned} \tag{8.82}$$

admits a unique solution $\psi \in H^{2+s}_q(\Omega\setminus\Sigma)$ with mean value zero. There is a constant $C > 0$ such that

$$|\psi|_{H^{2+s}_q} \le C\big(|f|_{H^s_q} + |g_b|_{W^{1+s-1/q}_q} + |g|_{W^{1+s-1/q}_q} + |h|_{W^{2+s-1/q}_q}\big) \tag{8.83}$$

holds for all $f \in H^s_q(\Omega\setminus\Sigma)$, $g_b \in W^{1+s-1/q}_q(\partial\Omega)$, $g \in W^{1+s-1/q}_q(\Sigma)$, and $h \in W^{2+s-1/q}_q(\Sigma)$.

Proof. We first reduce to the case $(g_b, g, h) = 0$ by solving the problem

$$\begin{aligned} -\omega\psi + \varrho^{-1}\Delta\psi &= f && \text{in } \Omega\setminus\Sigma,\\ \varrho^{-1}\partial_\nu\psi &= g_b && \text{on } \partial\Omega,\\ [\![\varrho^{-1}\partial_\nu\psi]\!] &= g && \text{on } \Sigma,\\ [\![\psi]\!] &= h && \text{on } \Sigma, \end{aligned}$$

for some (large) ω. The solution of this problem is in the right class, combining the results in Section 6.3.5 with those in Section 6.4.3. Therefore, we may assume $(g_b, g, h) = 0$ and the mean value of f is zero. Set $X_0 := \{v \in L_q(\Omega) : \int_\Omega v\,dx = 0\}$, and define an operator A_0 in X_0 by means of

$$A_0 v = -\varrho^{-1}\Delta v, \quad \mathsf{D}(A_0) = \{v \in H^2_q(\Omega\setminus\Sigma)\cap X_0 : \varrho^{-1}\partial_\nu\psi = 0,\ [\![\varrho^{-1}\partial_\nu\psi]\!] = [\![\psi]\!] = 0\}.$$

Then $-A_0$ generates an analytic C_0-semigroup, and by Poincarés inequality A_0 is invertible. Therefore, Problem (8.82) admits a unique solution in $H^2_q(\Omega\setminus\Sigma)$ with mean value zero; this proves the case $s = 0$. For $s = 1$ it is enough to consider the regularity of this solution, replacing the equation by $-\omega u + \Delta u = f_1 := f - \omega u$, and applying the results of Section 6.3.5 and 6.4.3 once more. □

The second result we used above concerns weak solutions.

Proposition 8.6.2. *Suppose that Ω is a bounded domain in $\mathbb{R}^n$ with boundary $\partial\Omega$ of class C^1, and let $\Sigma \subset \Omega$ be a closed hypersurface of class C^1. Let $f \in L_q(\Omega)^n$, $h \in W_q^{1-1/q}(\Sigma)$. Then the weak problem*

$$\begin{aligned}(\varrho^{-1}\nabla\psi|\nabla\phi) &= (f|\nabla\phi), \quad \phi \in H^1_{q'}(\Omega),\\ [\![\psi]\!] &= h \quad \text{on } \Sigma,\end{aligned} \tag{8.84}$$

admits a unique solution $\psi \in \dot{H}^1_q(\Omega\setminus\Sigma)$. There is a constant $C > 0$ such that

$$|\nabla\psi|_{L_q} \le C\big(|f|_{L_q} + |h|_{W_q^{1-1/q}}\big) \tag{8.85}$$

holds for all $f \in L_q(\Omega)$ and $h \in W_q^{1-1/q}(\Sigma)$.

Note that uniqueness in $\dot{H}^1_q(\Omega)$ means uniqueness in $L_q(\Omega)$ up to a constant. Observe that for $f \in H^1_p(\Omega\setminus\Sigma)^n$, Problem (8.84) is equivalent to (8.82), with f replaced by $\operatorname{div} f$, and $g = [\![f\cdot\nu]\!]$, $g_b = f\cdot\nu$.

Proof. We first solve the problem

$$\begin{aligned}-\omega\psi_1 + \varrho^{-1}\Delta\psi_1 &= \operatorname{div} f && \text{in } \Omega\setminus\Sigma,\\ \varrho_2^{-1}\partial_\nu\psi_1 &= f\cdot\nu && \text{on } \partial\Omega,\\ [\![\varrho^{-1}\partial_\nu\psi_1]\!] &= [\![f\cdot\nu]\!] && \text{on } \Sigma,\\ [\![\psi_1]\!] &= h && \text{on } \Sigma,\end{aligned}$$

for ω large, according to Remark 6.5.3. The remaining problem then becomes

$$\begin{aligned}\varrho^{-1}\Delta\psi_2 &= -\omega\psi_1 && \text{in } \Omega\setminus\Sigma,\\ \partial_\nu\psi_1 &= 0 && \text{on } \partial\Omega,\\ [\![\varrho^{-1}\partial_\nu\psi_2]\!] &= 0 && \text{on } \Sigma,\\ [\![\psi_1]\!] &= 0 && \text{on } \Sigma,\end{aligned} \tag{8.86}$$

where $\psi_2 = \psi - \psi_1$. Note that ψ_1 has mean value zero. The corresponding generator A in $H_q^{-1}(\Omega)$ has compact resolvent, as Ω is bounded, hence its spectrum consists of eigenvalues of finite multiplicity. Denote by A_0 the part of A in $L_q(\Omega)$; the function 1 is an eigenfunction for eigenvalue zero, and any eigenfunction v for A is in $\mathsf{D}(A) \subset L_q(\Omega)$, hence is also an eigenfunction of A_0. For $q \ge 2$, surely $v \in H^1_2(\Omega)$. If $1 < q < 2$, we have $v \in H^1_q(\Omega) \subset L_{q_1}(\Omega)$, with $1/q_1 = 1/q - 1/n$. By induction this yields $v \in L_{q_k}(\Omega)$, $1/q_k = 1/q - k/n$, and so belongs to $L_2(\Omega)$, choosing k large enough. But then $|\nabla v|_2^2 = 0$, which implies that $v = \mathit{constant}$. This shows that $\mathsf{N}(A) = \mathsf{N}(A_0) = \operatorname{span}\{1\}$. As the problem $A_0 w = 1$ has no solution, 0 is a simple eigenvalue of A_0, hence $\mathsf{N}(A_0) \oplus \mathsf{R}(A_0) = L_q(\Omega)$, and $f \in \mathsf{R}(A_0)$ if and only if f has mean value 0. Therefore, there is a unique solution $\psi_2 \in \dot{H}^1_q(\Omega)$ of (8.86), which proves Proposition 8.6.2. □

Part IV

Nonlinear Problems

Chapter 9

Local Well-Posedness and Regularity

In this chapter we study local well-posedness and regularity of the solutions of Problems **(P1)**$\sim$**(P6)**. Here we employ without further comments the notations introduced in Chapters 1 and 2, in particular those in connection with Conditions **(H1)**$\sim$**(H6)** from Chapter 1, the Hanzawa transform, and the transformed problems on the fixed domain $\Omega \setminus \Sigma$ in Section 1.3. In the first section of this chapter we reformulate Problems **(P1)**$\sim$**(P6)** in a way which is amenable to a joint analysis, which will be based on maximal L_p-regularity as well as on the contraction mapping principle in Section 9.2, and on the implicit function theorem for dependence on the data in Section 9.3. For regularity we employ in Section 9.4 the so-called parameter trick, which is also based on maximal L_p-regularity and the implicit function theorem. This way we can show that the solutions obtained in Section 9.2 are in fact classical solutions. The proofs for the technical results on the nonlinearities are postponed to the last section of this chapter.

9.1 Reformulation on the Fixed Domain

The main goal of this section is the reformulation of the transformed problems **(P1)**$\sim$**(P6)** in abstract form $Lz = N(z)$. We call L the *principal linearization.* The mapping N collects all nonlinear and lower order terms. We have to set up function spaces such that L has the property of maximal regularity, and N is Lipschitz continuous. Furthermore, we use the decomposition $z = \bar{z} + \tilde{z}$, where $\bar{z}$ resolves the compatibility conditions and satisfies the initial condition, and $\tilde{z}$ has vanishing trace at time $t = 0$. This has to be done separately for each problem in question. We begin with the simplest one.

J. Prüss and G. Simonett, *Moving Interfaces and Quasilinear Parabolic Evolution Equations*, Monographs in Mathematics 105,
DOI 10.1007/978-3-319-27698-4_9

1.1 The Stefan Problem with Surface Tension
In the sequel we assume Condition **(H1)**, and the compatibility condition

$$(\mathbf{C1}) \quad \varrho[\![\psi(\theta_0)]\!] + \sigma H_{\Gamma_0} = 0 \ \text{ on } \Gamma_0, \quad [\![d(\theta_0)\partial_\nu\theta_0]\!] \in W_p^{2-6/p}(\Gamma_0).$$

The transformed problem **(P1)** reads as follows (w.l.o.g. $\varrho = 1$).

$$\begin{aligned} \partial_t\theta + \mathcal{A}_\theta(\theta,h) : \nabla^2\theta = F_\theta(\theta,h) \quad & \text{in } \Omega\setminus\Sigma, \\ \partial_\nu\theta = 0 \quad & \text{on } \partial\Omega, \\ [\![\theta]\!] = 0, \quad \varphi(\theta) + \sigma H_\Gamma(h) = 0 \quad & \text{on } \Sigma, \\ \partial_t h + [\![\mathcal{B}_\theta(\theta,h)\nabla\theta]\!] = 0 \quad & \text{on } \Sigma, \\ h(0) = h_0 \text{ on } \Sigma, \quad \theta(0) = \theta_0 \quad & \text{in } \Omega. \end{aligned} \tag{9.1}$$

Recall that $\varphi(\theta) = [\![\psi(\theta)]\!]$, where ψ denotes the free energy, and $l(\theta) = \theta\varphi'(\theta)$ is the latent heat. We assume here that $l(\theta_0) \neq 0$. The map F_θ collects lower order terms and we have

$$\begin{aligned} \kappa(\theta)F_\theta(\theta,h) = \kappa(\theta)m_0(h)\partial_t h \circ \Pi_\Sigma(\nu_\Sigma\cdot\nabla\theta) + d'(\theta)|(1-M_1(h))\nabla\theta|^2 \\ - d(\theta)\big[(I-M_1(h))\nabla]M_1(h)\big)\cdot\nabla\theta. \end{aligned}$$

Note that the time derivative of h appears in $\mathbb{F}_\theta$. On the other hand, the curvature operator according to Section 2.2.5 is given by

$$\begin{aligned} H_\Gamma(h) = \beta(h)\{\mathrm{tr}[M_0(h)(L_\Sigma + \nabla_\Sigma(M_0(h)\nabla_\Sigma h))] \\ - \beta^2(h)(M_0^2(h)\nabla_\Sigma h|\nabla_\Sigma\big(M_0(h)\nabla_\Sigma(h)\big)M_0(h)\nabla_\Sigma h)\}. \end{aligned}$$

Finally, $\mathcal{A}_\theta(\theta,h)$ and $\mathcal{B}_\theta(\theta,h)$ are defined by

$$\begin{aligned} \mathcal{A}_\theta(\theta,h) &= -\big(d(\theta)/\kappa(\theta)\big)(I-M_1(h)^\mathsf{T})(I-M_1(h)), \\ \mathcal{B}_\theta(\theta,h) &= -\big(d(\theta)/l(\theta)\big)(1-M_1(h)^\mathsf{T})(\nu_\Sigma - M_0(h)\nabla_\Sigma h). \end{aligned}$$

To formulate the problem abstractly, let $J = (0,a)$ where $a > 0$ will be fixed later. We first set

$$\begin{aligned} \mathbb{E}_{\theta,\mu}(J) &= H^1_{p,\mu}(J;L_p(\Omega)) \cap L_{p,\mu}(J;H^2_p(\Omega\setminus\Sigma)), \\ \mathbb{E}^\theta_{h,\mu}(J) &= W^{3/2-1/2p}_{p,\mu}(J;L_p(\Sigma)) \cap W^{1-1/2p}_{p,\mu}(J;H^2_p(\Sigma)) \cap L_{p,\mu}(J;W^{4-1/p}_p(\Sigma)), \end{aligned}$$

and define the solution space for $z = (\theta,h)$ by

$$\mathbb{E}^1_\mu(a) = \{z = (\theta,h) \in \mathbb{E}_{\theta,\mu}(J)\times\mathbb{E}^\theta_{h,\mu}(J) : \ [\![\theta]\!] = 0 \text{ on } \Sigma, \ \partial_\nu\theta = 0 \text{ on } \partial\Omega\}.$$

The space of data for $(f_\theta, g_\theta, f_h)$ will be

$$\mathbb{F}^1_\mu(a) = \mathbb{F}_{\theta,\mu}(J)\times\mathbb{F}^\theta_{h,\mu}(J)\times\mathbb{F}^u_{h,\mu}(J),$$

with

$$\mathbb{F}_{\theta,\mu}(J) = L_{p,\mu}(J; L_p(\Omega)),$$
$$\mathbb{F}^{\theta}_{h,\mu}(J) = W^{1-1/2p}_{p,\mu}(J; L_p(\Sigma)) \cap L_{p,\mu}(J; W^{2-1/p}_p(\Sigma)),$$
$$\mathbb{F}^{u}_{h,\mu}(J) = W^{1/2-1/2p}_{p,\mu}(J; L_p(\Sigma)) \cap L_{p,\mu}(J; W^{1-1/p}_p(\Sigma)).$$

Recall from Subsection 3.4.6 that the time trace space $X^1_{\gamma,\mu}$ of $\mathbb{E}^1_\mu(a)$ is given by

$$X^1_{\gamma,\mu} = \{(\theta, h) \in W^{2\mu-2/p}_p(\Omega \setminus \Sigma) \times W^{2+2\mu-3/p}_p(\Sigma) : [\![\theta]\!] = 0 \text{ on } \Sigma,\ \partial_\nu \theta = 0 \text{ on } \partial\Omega\}.$$

We observe that

$$X^1_{\gamma,\mu} \hookrightarrow C^1_{ub}(\Omega \setminus \Sigma) \times C^3(\Sigma), \quad \text{provided } 1 \geq \mu > \frac{1}{2} + \frac{n+2}{2p}. \tag{9.2}$$

We will use this restriction in the sequel, although it would be enough to require

$$X^1_{\gamma,\mu} \hookrightarrow C_{ub}(\Omega \setminus \Sigma) \times C^2(\Sigma), \qquad \text{valid for } 1 \geq \mu > \frac{n+2}{2p}.$$

However, this would involve more technical efforts, and we refrain from carrying this out here. Observe that the last restriction cannot be relaxed, since we definitely need continuity of temperature and of curvature; the interfaces ought to be of class C^2.

Unfortunately, $(\theta_0, h_0) \in X^1_{\gamma,\mu}$ do not have enough regularity for the space $\mathbb{F}^{\theta}_{h,\mu}(J)$, as $\varphi'(\theta_0)$ fails to be a pointwise multiplier for this space. For this reason we cannot freeze coefficients in the stationary interface equation. Therefore, we extend the initial value θ_0 to some function $\bar\theta$ in $\mathbb{E}_{\theta,\mu}(\mathbb{R}_+)$, for instance by solving the problem

$$\begin{aligned} \partial_t \bar\theta - \Delta \bar\theta &= 0 && \text{in } \Omega, \\ \partial_\nu \bar\theta &= 0 && \text{on } \partial\Omega, \\ \bar\theta(0) &= \theta_0 && \text{in } \Omega. \end{aligned}$$

Similarly, we extend h_0 and $h_1 := -[\![\mathcal{B}(\theta_0, h_0)\nabla\theta_0]\!]$ as in Section 6.6.2 to a function $\bar h \in \mathbb{E}^{\theta}_{h,\mu}(\mathbb{R}_+)$ such that $\bar h(0) = h_0$ and $\partial_t \bar h(0) = h_1$. Further we set $\tilde\theta = \theta - \bar\theta$ and $\tilde h = h - \bar h$. This way, we have trivialized the initial conditions and at the same time resolved the compatibility conditions. Writing

$$\varphi(\theta) = \varphi(\bar\theta) + \varphi'(\bar\theta)\tilde\theta + r_\theta(\tilde\theta, \bar\theta)$$

and

$$H_\Gamma(h) = H_\Gamma(\bar h) + H'_\Gamma(h_0)\tilde h + r_h(\tilde h, \bar h)$$

we may replace the stationary interface condition by

$$\varphi'(\bar\theta)\tilde\theta + \sigma H'_\Gamma(h_0)\tilde h = \bar g_\theta - r_\theta(\tilde\theta, \bar\theta) - \sigma r_h(\tilde h, \bar h)$$

where

$$\bar{g}_\theta = -\big(\varphi(\bar{\theta}) + \sigma H_\Gamma(\bar{h})\big) \in {}_0\mathbb{F}^\theta_{h,\mu}(\mathbb{R}_+)$$

by the compatibility condition **(C1)**. Now we can rewrite the problem abstractly as

$$L_1\tilde{z} = N_1(\tilde{z}, \bar{z}), \quad \tilde{z}(0) = 0, \tag{9.3}$$

with $N_1 : {}_0\mathbb{E}^1_\mu(a) \times \mathbb{E}^1_\mu(\infty) \to {}_0\mathbb{F}^1_\mu(a)$, and $L_1 : \mathbb{E}^1_\mu(a) \to \mathbb{F}^1_\mu(a)$ linear and bounded, given by

$$L_1\tilde{z} = \begin{bmatrix} \partial_t\tilde{\theta} + \mathcal{A}_\theta(\theta_0, h_0) : \nabla^2\tilde{\theta} \\ \varphi'(\bar{\theta})\tilde{\theta} - \sigma\,\mathcal{C}_\Sigma(h_0)\tilde{h} \\ \partial_t\tilde{h} + [\![\mathcal{B}_\theta(\theta_0, h_0)\nabla\tilde{\vartheta}]\!] \end{bmatrix},$$

where $\mathcal{C}_\Sigma(h_0)$ denotes the principal part of the curvature operator $-H'_\Gamma(h_0)$. The operator L_1 has maximal L_p-regularity by Section 6.6.

The nonlinearity N_1 is given by

$$N_1(\tilde{z}, \bar{z}) = \begin{bmatrix} F_\theta(\theta, h) - \partial_t\bar{\theta} - \mathcal{A}_\theta(\theta, h) : \nabla^2\bar{\theta} + (\mathcal{A}_\theta(\theta_0, h_0) - \mathcal{A}_\theta(\theta, h)) : \nabla^2\tilde{\theta} \\ \bar{g}_\theta + r_\theta(\tilde{\theta}, \bar{\theta}) + \sigma r_h(\tilde{h}, \bar{h}) - \sigma(\mathcal{C}_\Sigma(h_0) + H'_\Gamma(\bar{h}))\tilde{h} \\ [\![(\mathcal{B}_\theta(\theta_0, h_0) - \mathcal{B}_\theta(\theta, h))\nabla\tilde{\theta} - \mathcal{B}_\theta(\theta, h)\nabla\bar{\theta}]\!] - \partial_t\bar{h} \end{bmatrix}.$$

Observe that

$$N_1(0, \bar{z}) = \begin{bmatrix} F_\theta(\bar{\theta}, \bar{h}) - \partial_t\bar{\theta} - \mathcal{A}_\theta(\bar{\theta}, \bar{h}) : \nabla^2\bar{\theta} \\ \bar{g}_\theta \\ -\partial_t\bar{h} - [\![\mathcal{B}_\theta(\bar{\theta}, h)\nabla\bar{\theta}]\!] \end{bmatrix}$$

satisfies $|N_1(0, \bar{z})|_{\mathbb{F}_\mu(a)} \to 0$ as $a \to 0$.

1.2 The Two-Phase Navier-Stokes Problem with Surface Tension

In the sequel we assume Condition **(H2)** and the compatibility condition

$$\textbf{(C2)} \quad \operatorname{div} u_0 = 0 \text{ in } \Omega\setminus\Gamma_0, \quad [\![d(\theta_0)\partial_\nu\theta_0]\!],\ \mathcal{P}_{\Gamma_0}[\![\mu(\theta_0)D(u_0)\nu_{\Gamma_0}]\!] = 0 \text{ on } \Gamma_0.$$

The transformed problem **(P2)** reads as follows.

$$\begin{aligned}
\partial_t u + \mathcal{A}_u(\theta, h) : \nabla^2 u + (I - M_1(h))\nabla\pi/\varrho = F_u(u, \theta, h) &\quad \text{in } \Omega\setminus\Sigma,\\
(I - M_1(h))\nabla\cdot u = 0 &\quad \text{in } \Omega\setminus\Sigma,\\
\partial_t\theta + \mathcal{A}_\theta(\theta, h) : \nabla^2\theta = F_\theta(\theta, h) &\quad \text{in } \Omega\setminus\Sigma,\\
u, \partial_\nu\theta = 0 &\quad \text{on } \partial\Omega,\\
[\![u]\!], [\![\theta]\!], [\![\mathcal{B}_\theta(\theta, h)\nabla\theta]\!] = 0 &\quad \text{on } \Sigma,\\
-[\![S(u, \theta, h)]\!]\nu_\Gamma + \big([\![\pi]\!] - \sigma H_\Gamma(h)\big)\nu_\Gamma = 0 &\quad \text{on } \Sigma,\\
\partial_t h - (u|\nu_\Sigma - M_0(h)\nabla_\Sigma h) = 0 &\quad \text{on } \Sigma,\\
h(0) = h_0 \text{ on } \Sigma, \quad u(0) = u_0,\ \theta(0) = \theta_0 &\quad \text{in } \Omega.
\end{aligned} \tag{9.4}$$

Note that here we used the abbreviations

$$\mathcal{A}_u(\theta,h) = (\mu(\theta)/\varrho)(I - M_1(h)^{\mathsf{T}})(I - M_1(h)),$$
$$\mathcal{A}_\theta(\theta,h) = (d(\theta)/\varrho\kappa(\theta))(I - M_1(h)^{\mathsf{T}})(I - M_1(h)),$$
$$\mathcal{B}_\theta(\theta,h) = d(\theta)(1 - M_1(h)^{\mathsf{T}})(\nu_\Sigma - M_0(h)\nabla_\Sigma h).$$

The nonlinearities F_u and F_θ collect all lower order terms, i.e.,

$$\begin{aligned}\varrho F_u = &- \varrho u \cdot (I - M_1(h))\nabla u + \varrho m_0(h)\partial_t h \circ \Pi_\Sigma(\nu_\Sigma \cdot \nabla\theta)\\ &+ \mu'(\theta)(I - M_1(h))\nabla\theta \cdot D(u,h)\\ &+ \mu(\theta)\big((I - M_1(h))\nabla \cdot M_1(h)\nabla u + [\nabla u]^{\mathsf{T}} : (I - M_1(h))\nabla M_1(h)\\ &- (I - M_1(h))\nabla \otimes M_1(h) : \nabla u\big),\end{aligned}$$

and

$$\begin{aligned}\varrho\kappa(\theta)F_\theta = \varrho\kappa(\theta)m_0(h)\partial_t h \circ \Pi_\Sigma(\nu_\Sigma \cdot \nabla\theta) - \varrho\kappa(\theta)u \cdot (I - M_1(h))\nabla\theta\\ + d'(\theta)|(1 - M_1(h))\nabla\theta|^2 - d(\theta)\big[(I - M_1(h))\nabla]M_1(h)\big) \cdot \nabla\theta + 2\mu(\theta)|D|^2.\end{aligned}$$

Note that $F_\theta(0,\theta,h)$ coincides with F_θ from the previous subsection. Furthermore, recall that

$$S = S(u,\theta,h) = 2\mu(\theta)D(u,h), \quad 2D(u,h) = (I - M_1(h))\nabla u + [\nabla u]^{\mathsf{T}}(I - M_1(h))^{\mathsf{T}}.$$

To obtain the abstract formulation of the problem, we choose as the system variable $z = (u,\theta,h,\pi,q)$, where $q = [\![\pi]\!]$ is a dummy variable which we introduce for convenience. The regularity space for z is

$$\begin{aligned}z \in \mathbb{E}^2_\mu(a) := \{z \in \mathbb{E}_{u,\mu}(J) \times \mathbb{E}_{\theta,\mu}(J) \times \mathbb{E}^u_{h,\mu}(J) \times \mathbb{E}_{\pi,\mu}(J) \times \mathbb{E}_{q,\mu}(J) :\ [\![\pi]\!] = q,\\ [\![\theta]\!], [\![u]\!] = 0 \text{ on } \Sigma,\ u, \partial_\nu\theta = 0 \text{ on } \partial\Omega\},\end{aligned}$$

where

$$\mathbb{E}_{u,\mu}(J) = \mathbb{E}_{\theta,\mu}(J)^n, \quad \mathbb{E}_{\pi,\mu}(J) = L_{p,\mu}(J; \dot{H}^1_p(\Omega\setminus\Sigma)), \quad \mathbb{E}_{q,\mu}(J) = \mathbb{F}^u_{h,\mu}(J).$$

Here we set

$$\mathbb{E}^u_{h,\mu}(J) = W^{2-1/2p}_{p,\mu}(J; L_p(\Sigma)) \cap H^1_{p,\mu}(J; W^{2-1/p}_p(\Sigma)) \cap L_{p,\mu}(J; W^{3-1/p}_p(\Sigma)),$$

which differs from the space for h in the previous subsection. Note that, according to Section 8.2, the time-trace space of (u,θ,h) in this case reads

$$\begin{aligned}X^2_{\gamma,\mu} = \{(u,\theta,h) \in W^{2\mu-2/p}_p(\Omega\setminus\Sigma)^{n+1} \times W^{2+\mu-2/p}_p(\Sigma) : [\![u]\!], [\![\theta]\!] = 0 \text{ on } \Sigma,\\ u, \partial_\nu\theta = 0 \text{ on } \partial\Omega\}.\end{aligned}$$

The data space $\mathbb{F}^2_\mu(a)$ is given by

$$\mathbb{F}^2_\mu(a) = \mathbb{F}_{u,\mu}(J) \times \mathbb{F}^2_{\pi,\mu}(J) \times \mathbb{F}_{\theta,\mu}(J) \times \mathbb{F}^u_{h,\mu}(J)^{n+1} \times \mathbb{F}^\theta_{h,\mu}(J),$$

with

$$\mathbb{F}_{u,\mu}(J) = \mathbb{F}_{\theta,\mu}(J)^n, \quad \mathbb{F}^2_{\pi,\mu}(J) = H^1_{p,\mu}(J; {}_0\dot{H}^{-1}_p(\Omega)) \cap L_{p,\mu}(J; H^1_p(\Omega \setminus \Sigma)),$$

Next we define suitable extensions of $z_0 \in X^2_{\gamma,\mu}$ in the following way. We solve the diffusion problem

$$\begin{aligned} \partial_t \bar{u} - \Delta \bar{u} &= 0 && \text{in } \Omega, \\ \bar{u} &= 0 && \text{on } \partial\Omega, \\ \bar{u}(0) &= u_0 && \text{in } \Omega, \end{aligned}$$

to obtain a function $\bar{u} \in \mathbb{E}_{u,\mu}(\mathbb{R}_+)$. Also we define $\bar{\theta} \in \mathbb{E}_{\theta,\mu}(\mathbb{R}_+)$ as in the previous subsection. Next we extend the initial values h_0 and

$$h_1 := u_0 \cdot (\nu_\Sigma - M_0(h_0)\nabla_\Sigma h_0) \in W^{2\mu-3/p}_p(\Sigma)$$

as in Section 8.2.2 to obtain a function $\bar{h} \in \mathbb{E}^u_{h,\mu}(\mathbb{R}_+)$ with initial values $\bar{h}(0) = h_0$ and $\partial_t \bar{h}(0) = h_1$. Finally, we extend the pressure jump q_0 defined by

$$q_0 := \sigma H_\Gamma(h_0) + ([\![S(u_0, \theta_0, h_0]\!]\nu_\Gamma(h_0)|\nu_\Gamma(h_0)) \in W^{2\mu-1-3/p}_p(\Sigma)$$

by means of

$$\bar{q} = e^{-(I-\Delta_\Sigma)t} q_0 \in \mathbb{E}_{q,\mu}(\mathbb{R}_+),$$

and define $\bar{\pi} \in \mathbb{E}_{\pi,\mu}(\mathbb{R}_+)$ as the solution of the elliptic transmission problem

$$\begin{aligned} \Delta \bar{\pi} &= 0 && \text{in} \quad \Omega \setminus \Sigma, \\ \partial_\nu \bar{\pi} &= 0 && \text{on} \quad \partial\Omega, \\ [\![\partial_\nu \bar{\pi}]\!] = 0, \quad [\![\bar{\pi}]\!] &= \bar{q} && \text{on} \quad \Sigma, \end{aligned}$$

see Proposition 8.6.2, We denote the projection onto mean value zero by P_0. Then with $\bar{z} = (\bar{u}, \bar{\theta}, \bar{h}, \bar{\pi}, \bar{q})$, we decompose as in the previous section $z = \bar{z} + \tilde{z}$, and obtain the abstract equation

$$L_2 \tilde{z} = N_2(\tilde{z}, \bar{z}), \quad \tilde{z}(0) = 0,$$

with $L_2 : \mathbb{E}^2_\mu(a) \to \mathbb{F}^2_\mu(a)$ linear and bounded, $N_2 : {}_0\mathbb{E}^2_\mu(a) \times \mathbb{E}^2_\mu(\infty) \to {}_0\mathbb{F}^2_\mu(a)$. Here L_2 is given by

$$L_2 \tilde{z} = \begin{bmatrix} \partial_t \tilde{u} + \mathcal{A}_u(\theta_0, h_0) : \nabla^2 \tilde{u} + (1 - M_1(h_0))\nabla \tilde{\pi}/\varrho \\ (I - P_0 M_1(h_0))\nabla \cdot \tilde{u}) \\ \partial_t \tilde{\theta} + \mathcal{A}_\theta(\theta_0, h_0) : \nabla^2 \tilde{\theta} \\ -[\![S(\tilde{u}, \theta_0, h_0)]\!]\nu_\Sigma + (\tilde{q} + \sigma \mathcal{C}_\Sigma(h_0) : \nabla^2_\Sigma \tilde{h})\nu_\Sigma \\ [\![\mathcal{B}_\theta(\theta_0, h_0)\nabla \tilde{\theta}]\!] \\ \partial_t \tilde{h} - \tilde{u} \cdot (\nu_\Sigma - M_0(h_0)\nabla_\Sigma \bar{h}) + \bar{u} \cdot M_0(h_0)\nabla_\Sigma \tilde{h} \end{bmatrix}.$$

Note that the temperature decouples completely from the problem for (u,π,h), it has maximal L_p-regularity by Section 6.5. The remaining problem for (u,π,h) has been analyzed in Chapter 8 for the case $h_0 = 0$. There, maximal L_p-regularity has been shown for $(h_0,h_1) = 0$ which, by perturbation, extends to nontrivial h_0 with small norm in $C^1(\Sigma)$, and also to arbitrary h_1 provided the time interval $J = (0,a)$ is small. Observe that in the part for h we cannot replace $\bar h$ by h_0 everywhere, as h_0 does not have enough regularity.

The nonlinearity N_2 reads

$$N_2 = \begin{bmatrix} F_u(u,\theta,h) - \partial_t\bar u - (I - M_1(h))\nabla\bar\pi/\varrho + (M_1(h) - M_1(h_0))\nabla\tilde\pi \\ \quad +(\mathcal{A}_u(\theta_0,h_0) - \mathcal{A}_u(\theta,h)) : \nabla^2\tilde u - \mathcal{A}_u(\theta,h) : \nabla^2\bar u \\ P_0(M_1(h) - M_1(h_0))\nabla\cdot\tilde u + P_0(M_1(h) - I)\nabla\cdot\bar u \\ F_\theta(u,\theta,h) - \partial_t\bar\theta - (\mathcal{A}_\theta(\theta_0,h_0) - \mathcal{A}_\theta(\theta,h)) : \nabla^2\tilde\theta - \mathcal{A}(\theta,h) : \nabla^2\bar\theta \\ \tilde T_0 M_0(h)\nabla_\Sigma h + \big([\![S(u,\theta,h) - S(\tilde u,\theta_0,h_0) - \bar\pi]\!] \\ \quad +\sigma(H_\Gamma(\bar h) + H_\Gamma'(\bar h) - H_\Gamma'(h_0))\tilde h + r_h(\bar h,\tilde h)\big)\nu_\Gamma(h)/\beta \\ [\![(\mathcal{B}_\theta(\theta_0,h_0) - \mathcal{B}_\theta(\theta,h))\nabla\tilde\theta - \mathcal{B}(\theta,h)\nabla\bar\theta \\ -\partial_t\bar h + \bar u\cdot(\nu_\Sigma - M_0(h)\nabla_\Sigma\bar h) + \bar u\cdot(M_0(h_0) - M_0(h))\nabla_\Sigma\tilde h \\ \quad +\tilde u\cdot((M_0(h_0) - M_0(h))\nabla_\Sigma\bar h - M_0(h)\nabla_\Sigma\tilde h) \end{bmatrix};$$

here we employed the abbreviation

$$\tilde T_0 = -[\![S(\tilde u,\theta_0,h_0) - \tilde\pi]\!] - \sigma H_\Gamma'(h_0)\tilde h.$$

Note that

$$N_2(0,\bar z) = \begin{bmatrix} F_u(\bar u,\bar\theta,\bar h) - \partial_t\bar u - (I - M_1(\bar h))\nabla\bar\pi/\rho - \mathcal{A}_u(\bar\theta,\bar h) : \nabla^2\bar u \\ P_0((M_1(\bar h) - I)\nabla\cdot\bar u) \\ F_\theta(\bar u,\bar\theta,\bar h) - \partial_t\bar\theta - \mathcal{A}(\bar\theta,\bar h) : \nabla^2\bar\theta \\ ([\![S(\bar u,\bar\theta,\bar h) - \bar\pi]\!] + \sigma H_\Gamma(\bar h)\nu_\Gamma(\bar h)/\beta \\ -[\![\mathcal{B}(\bar\theta,\bar h)\nabla\bar\theta \\ -\partial_t\bar h + \bar u\cdot(\nu_\Sigma - M_0(\bar h)\nabla_\Sigma\bar h) \end{bmatrix}.$$

Then we see that $|N_2(0,\bar z)|_{\mathbb{F}_\mu(a)} \to 0$ as $a \to 0$.

1.3 Phase Transitions: Equal Densities

In the sequel we assume Condition **(H3)** and the compatibility condition

$$\begin{aligned} &\textbf{(C3)}\quad \varrho[\![\psi(\theta_0)]\!] + \sigma H_{\Gamma_0} = 0 \ \text{ on } \Gamma_0, \quad [\![d(\theta_0)\partial_\nu\theta_0]\!] \in W_p^{2-6/p}(\Gamma_0), \\ &\qquad\ \operatorname{div} u_0 = 0 \ \text{ in } \Omega\setminus\Gamma_0, \qquad\qquad \mathcal{P}_{\Gamma_0}[\![\mu(\theta_0)D(u_0)\nu_{\Gamma_0}]\!] = 0 \ \text{ on } \Gamma_0. \end{aligned}$$

Here we have $\varrho_1 = \varrho_2 = 1$ w.l.o.g. and we may express the phase flux j_Σ by

$$j_\Sigma = [\![\mathcal{B}_\theta(\theta,h)\nabla\theta]\!],$$

insert it into the V_Γ-equation, and the Gibbs-Thomson relation into interface stress balance to the result

$$\begin{aligned}
\partial_t u + \mathcal{A}_u(\theta,h) : \nabla^2 u + (I - M_1(h))\nabla\pi &= F_u(u,\theta,h) && \text{in } \Omega\setminus\Sigma,\\
(I - M_1(h))\nabla\cdot u &= 0 && \text{in } \Omega\setminus\Sigma,\\
\partial_t\theta + \mathcal{A}_\theta(\theta,h) : \nabla^2\theta &= F_\theta(\theta,h) && \text{in } \Omega\setminus\Sigma,\\
u, \partial_\nu\theta &= 0 && \text{on } \partial\Omega,\\
[\![u]\!], [\![\theta]\!] &= 0 && \text{on } \Sigma,\\
-[\![S(u,\theta,h)]\!]\nu_\Gamma + \big([\![\pi]\!] + \varphi(\theta)\big)\nu_\Gamma &= 0 && \text{on } \Sigma,\\
\varphi(\theta) + \sigma H_\Gamma(h) &= 0 && \text{on } \Sigma,\\
\partial_t h - (u|\nu_\Sigma - M_0(h)\nabla_\Sigma h) + [\![\mathcal{B}_\theta(\theta,h)\nabla\theta]\!] &= 0 && \text{on } \Sigma,\\
h(0) = h_0 \text{ on } \Sigma, \quad u(0) = u_0,\ \theta(0) &= \theta_0 && \text{in } \Omega.
\end{aligned}$$

Here $(\mathcal{A}_u, \mathcal{A}_\theta, \mathcal{B}_\theta)$ are as before. The extensions $(\bar{u}, \bar{\theta})$ are as in the previous subsection, whereas the extension $\bar{h}$ is that from Section 9.1.1. As a result we obtain again a problem of the form

$$L_3\tilde{z} = N_3(\tilde{z}, \bar{z}), \quad \tilde{z} = 0,$$

where L_3 is defined by

$$L_3\tilde{z} = \begin{bmatrix}
\partial_t\tilde{u} + \mathcal{A}_u(\theta_0,h_0) : \nabla^2\tilde{u} + (1 - M_1(h_0))\nabla\tilde{\pi}\\
(I - P_0M_1(h_0))\nabla\cdot\tilde{u})\\
\partial_t\tilde{\theta} + \mathcal{A}_\theta(\theta_0,h_0) : \nabla^2\tilde{\theta}\\
-[\![S(\tilde{u},\theta_0,h_0)]\!]\nu_\Sigma + [\![\tilde{\pi}]\!]\nu_\Sigma\\
\varphi'(\bar{\theta})\tilde{\theta} - \sigma\mathcal{C}_\Sigma(h_0)\tilde{h}\\
\partial_t\tilde{h} + [\![\mathcal{B}_\theta(\theta_0,h_0)\nabla\tilde{\theta}]\!]
\end{bmatrix}.$$

Note that the term $\varphi(\theta)$ in the stress balance on the interface as well the term $u\cdot\nu_\Gamma$ in the equation for h are lower order and can be subsumed in N_3. Here we define with $z = (u,\theta,h,\pi,q)$ the regularity space as

$$\begin{aligned}
z \in \mathbb{E}^3_\mu(a) := \{z \in \mathbb{E}_{u,\mu}(J) \times \mathbb{E}_{\theta,\mu}(J) \times \mathbb{E}^\theta_{h,\mu}(J) \times \mathbb{E}_{\pi,\mu}(J) \times \mathbb{E}_{q,\mu}(J) :\\
[\![\theta]\!], [\![u]\!] = 0,\ [\![\pi]\!] = q \text{ on } \Sigma,\ u, \partial_\nu\theta = 0 \text{ on } \partial\Omega\},
\end{aligned}$$

and the space of data by

$$\mathbb{F}^3_\mu(a) = \mathbb{F}_{u,\mu}(J) \times \mathbb{F}^2_{\pi,\mu}(J) \times \mathbb{F}_{\theta,\mu}(J) \times \mathbb{F}^u_{h,\mu}(J) \times \mathbb{F}^\theta_{h,\mu}(J) \times \mathbb{F}^u_{h,\mu}(J).$$

Observe that up to lower order terms, the problems for (u,π) and (θ,h) decouple. Therefore, for (u,π) we have at the linear level a two-phase Stokes problem on a fixed domain, and for (θ,h) we have the same principal part as in Section 9.1.1. By the previous subsections, L_3 has maximal regularity and $L_3 : {}_0\mathbb{E}_\mu(a) \to {}_0\mathbb{F}_\mu(a)$ is an isomorphism with $|L_3^{-1}|$ uniformly bounded for $a \in (0,1]$. The nonlinearity

N_3 is similar to N_2 and N_1. In particular, we have again $|N_3(0,\bar z)|_{\mathbb{F}_\mu(a)} \to 0$ as $a \to 0$.

1.4 Phase Transitions: Different Densities
In the sequel we assume Condition **(H4)** and the compatibility condition

$$\begin{aligned}
\textbf{(C4)}\quad & \operatorname{div} u_0 = 0 \text{ in } \Omega\setminus\Gamma_0, \quad \mathcal{P}_{\Gamma_0}[\![u_0]\!] = 0,\\
& \mathcal{P}_{\Gamma_0}[\![\mu(\theta_0)D(u_0)\nu_{\Gamma_0}]\!],\ l(\theta_0)[\![u_0\cdot\nu_{\Gamma_0}]\!] + [\![1/\varrho]\!][\![d(\theta_0)\partial_\nu\theta_0]\!] = 0 \ \text{ on } \Gamma_0,
\end{aligned}$$

As shown in Chapter 1, with $[\![\varrho]\!] \neq 0$, we may eliminate j_Σ to obtain

$$j_\Sigma(u,h) = [\![u\cdot\nu_\Sigma]\!]/\beta(h)[\![1/\varrho]\!], \quad V_\Gamma = \beta(h)\partial_t h = [\![\varrho u\cdot\nu_\Gamma]\!]/[\![\varrho]\!].$$

Then the transformed problem **(P4)** becomes

$$\begin{aligned}
\partial_t u + \mathcal{A}_u(\theta,h):\nabla^2 u + (I-M_1(h))\nabla\pi/\varrho &= F_u(u,\theta,h) && \text{in } \Omega\setminus\Sigma,\\
(I-M_1(h))\nabla\cdot u &= 0 && \text{in } \Omega\setminus\Sigma,\\
\partial_t\theta + \mathcal{A}_\theta(\theta,h):\nabla^2\theta &= F_\theta(\theta,h) && \text{in } \Omega\setminus\Sigma,\\
u,\partial_\nu\theta &= 0 && \text{on } \partial\Omega,\\
[\![\theta\eta(\theta)]\!]j_\Sigma - [\![d(\theta)\nu_\Gamma\cdot(I-M_1(h))\nabla\theta]\!] &= 0 && \text{on } \Sigma,\\
\mathcal{P}_\Gamma[\![u]\!],[\![\theta]\!] &= 0 && \text{on } \Sigma,\\
-[\![S(u,\theta,h)]\!]\nu_\Gamma + ([\![\pi]\!]+[\![1/\varrho]\!]j_\Sigma^2 - \sigma H_\Gamma(h))\nu_\Gamma &= 0 && \text{on } \Sigma,\\
\varphi(\theta) + [\![1/2\varrho^2]\!]j_\Sigma^2 - [\![S(u,\theta,h)\nu_\Gamma\cdot\nu_\Gamma/\varrho]\!] + [\![\pi/\varrho]\!] &= 0 && \text{on } \Sigma,\\
\partial_t h - [\![(\varrho u|\nu_\Sigma - M_0(h)\nabla_\Sigma h)]\!]/[\![\varrho]\!] &= 0 && \text{on } \Sigma,\\
h(0) = h_0 \text{ on } \Sigma,\quad u(0) = u_0,\ \theta(0) &= \theta_0 && \text{in } \Omega.
\end{aligned}\tag{9.5}$$

Here the heat problem is only weakly coupled to the system for (u,π,h). However, the system for (u,π,h) leads to the asymmetric Stokes problem, which differs from the one considered above. The regularity of h is the same as in Section 9.1.2; the problem is *velocity dominated.* We proceed as before, extending the initial values $(u_0,\theta_0,h_0)\in X^4_{\gamma,\mu}$ as in Section 9.1.2 to obtain $(\bar u,\bar\theta,\bar h)$. Furthermore, we solve the Gibbs-Thomson relation combined with the normal component of the stress transmission condition on the interface at time $t=0$, to obtain unique initial values q_{j0} for the pressures π_j on the interface. We extend these by defining

$$\bar q_j = e^{-(1-\Delta_\Sigma)t}q_{j0}, \quad t>0,\ j=1,2,$$

and then solve the two elliptic problems

$$\begin{aligned}
\Delta\bar\pi_2 &= 0 && \text{in } \Omega_2,\\
\partial_\nu\bar\pi_2 &= 0 && \text{on } \partial\Omega,\\
\bar\pi_2 &= \bar q_2 && \text{on } \Sigma,
\end{aligned}$$

and

$$\begin{aligned} \Delta\bar{\pi}_1 &= 0 \quad \text{in } \Omega_1, \\ \bar{\pi}_1 &= \bar{q}_1 \quad \text{on } \Sigma. \end{aligned}$$

From the above construction it is evident that $\bar{z} \in \mathbb{E}^4_\mu(\infty)$ trivializes the initial conditions and resolves the compatibilities. The relevant variables are here $z = (u, \theta, h, \pi, q_1, q_2)$, where q_j denote the surface pressures on Σ, and the solution space $z \in \mathbb{E}^4_\mu(a)$ is

$$\begin{aligned} \mathbb{E}^4_\mu(a) :=\{z \in \mathbb{E}_{u,\mu} \times \mathbb{E}_{\theta,\mu}(J) \times \mathbb{E}^u_{h,\mu}(J) \times \mathbb{E}_{\pi,\mu}(J) \times \mathbb{E}_{q,\mu}(J) \times \mathbb{E}_{q,\mu}(J) : \\ [\![\theta]\!] = 0, \ \ \pi_j = q_j \text{ on } \Sigma, \ \ u, \partial_\nu\theta = 0 \text{ on } \partial\Omega\}. \end{aligned}$$

The image space in this case will be

$$\mathbb{F}^4_\mu(a) := \mathbb{F}_{u,\mu}(J)\times\mathbb{F}^4_{\pi,\mu}(J)\times\mathbb{F}_{\theta,\mu}(J)\times\mathbb{F}^u_{h,\mu}(J)\times\mathcal{P}_\Sigma\mathbb{F}^\theta_{h,\mu}(J)^n\times\mathbb{F}^u_{h,\mu}(J)^{n+1}\times\mathbb{F}^\theta_{h,\mu}(J),$$

with

$$\mathbb{F}^4_{\pi,\mu}(J) = H^1_{p,\mu}(J; H^{-1}_{p,\partial\Omega}(\Omega \setminus \Sigma)) \cap L_{p,\mu}(J; H^1_p(\Omega \setminus \Sigma)).$$

Compared to the previous cases, the equation for h is different from that in Section 9.1.2, but it has a similar structure and hence needs no additional comments. On the other hand, the transmission condition $[\![u]\!] = 0$ is replaced by $\mathcal{P}_\Gamma[\![u]\!] = 0$, which by application of $\mathcal{P}_\Sigma$ leads to the decomposition

$$\mathcal{P}_\Sigma[\![u]\!] + \beta(h)M_0\nabla_\Sigma h[\![\nu_\Gamma(h) \cdot u]\!] = 0.$$

This equation is linearized in the same way as the equation for h. Furthermore, note that the terms $\varphi(\theta)$ and $[\![1/2\varrho^2]\!]j^2_\Sigma$ in the Gibbs-Thomson law are lower order. The remaining part is linearized in the same way as the stress boundary condition.

As a result we obtain again a problem of the form

$$L_4\tilde{z} = N_4(\tilde{z}, \bar{z}), \quad \tilde{z} = 0,$$

where L_4 is defined by

$$L_4\tilde{z} = \begin{bmatrix} \partial_t\tilde{u} + \mathcal{A}_u(\theta_0, h_0)\nabla^2\tilde{u} + (1 - M_1(h_0))\nabla\tilde{\pi} \\ (I - M_1(h_0))\nabla \cdot \tilde{u}) \\ \partial_t\tilde{\theta} + \mathcal{A}_\theta(\theta_0, h_0) : \nabla^2\tilde{\theta} \\ [\![\mathcal{B}_\theta(\theta_0, h_0)\nabla\tilde{\theta}]\!] \\ \mathcal{P}_\Sigma[\![\tilde{u}]\!] + [\![\tilde{u} \cdot \nu_\Sigma]\!]M_0(h_0)\nabla_\Sigma\bar{h} + [\![\bar{u} \cdot \nu_\Sigma]\!]M_0(h_0)\nabla_\Sigma\tilde{h} \\ -[\![S(\tilde{u}, \theta_0, h_0)]\!]\nu_\Sigma + ([\![\tilde{\pi}]\!] + \sigma\mathcal{C}_\Sigma(h_0)\tilde{h})\nu_\Sigma \\ -[\![S(\tilde{u}, \theta_0, h_0)\nu_\Sigma \cdot \nu_\Sigma/\varrho]\!] + [\![\tilde{\pi}/\varrho]\!] \\ \partial_t\tilde{h} - ([\![\varrho\tilde{u} \cdot (\nu_\Sigma - M_0(h_0)\nabla_\Sigma\bar{h})]\!] - [\![\varrho\bar{u} \cdot M_0(h_0)\nabla_\Sigma\tilde{h}]\!])/[\![\varrho]\!] \end{bmatrix}.$$

On the linear level we have an asymmetric Stokes problem for (u, π, h) and a transmission problem for θ. Maximal L_p-regularity of the transmission problem follows from Section 6.5, and the asymmetric Stokes problem has been studied in Chapter 8. As shown there, it has maximal L_p-regularity in case $(h_0, h_1) = 0$. By perturbation, this extends to nontrivial h_0 which are small in $C^1(\Sigma)$, as well as to arbitrary h_1 provided the interval $J = (0, a)$ is small.

1.5 Phase Transitions and Marangoni Forces: Different Densities
In the sequel we assume Condition **(H6)** and the compatibility condition

$$\begin{aligned}\textbf{(C6)}\quad &\operatorname{div} u_0 = 0 \ \text{ in } \Omega\setminus\Gamma_0, \quad \mathcal{P}_{\Gamma_0}[\![u_0]\!] = 0,\\ &2\mathcal{P}_{\Gamma_0}[\![\mu(\theta_0)D(u_0)\nu_{\Gamma_0}]\!] + \sigma'(\theta_0)\nabla_{\Gamma_0}\theta_0 = 0 \ \text{ on } \Gamma_0.\end{aligned}$$

We eliminate j_Σ as before and obtain the transformed problem **(P6)**

$$\begin{aligned}
\partial_t u + \mathcal{A}_u(\theta,h) : \nabla^2 u + (I - M_1(h))\nabla\pi/\varrho &= F_u(u,\theta,h) && \text{in } \Omega\setminus\Sigma,\\
(I - M_1(h))\nabla\cdot u &= 0 && \text{in } \Omega\setminus\Sigma,\\
\partial_t\theta + \mathcal{A}_\theta(\theta,h) : \nabla^2\theta &= F_\theta(\theta,h) && \text{in } \Omega\setminus\Sigma,\\
u, \partial_\nu\theta &= 0 && \text{on } \partial\Omega,\\
\partial_t\theta_\Sigma + \mathcal{A}_{\theta_\Sigma}(\theta_\Sigma,h) : \nabla_\Sigma^2\theta_\Sigma &= F_{\theta_\Sigma}(u,\theta,h,\theta_\Sigma) && \text{on } \Sigma,\\
\theta = \theta_\Sigma, \quad \mathcal{P}_\Gamma[\![u]\!], [\![\theta]\!] &= 0 && \text{on } \Sigma,\\
-[\![S(u,\theta,h)]\!]\nu_\Gamma + \big([\![\pi]\!] + [\![1/\varrho]\!]j_\Sigma^2 - \sigma(\theta_\Sigma)H_\Gamma(h)\big)\nu_\Gamma &= \sigma'(\theta_\Sigma)\nabla_\Gamma\theta_\Gamma && \text{on } \Sigma,\\
\varphi(\theta) + [\![1/2\varrho^2]\!]j_\Sigma^2 - [\![S(u,\theta,h)\nu_\Gamma\cdot\nu_\Gamma/\varrho]\!] + [\![\pi/\varrho]\!] &= 0 && \text{on } \Sigma,\\
\partial_t h + [\![\varrho u\cdot(\nu_\Sigma - M_0(h)\nabla_\Sigma h)]\!]/[\![\varrho]\!] &= 0 && \text{on } \Sigma,\\
h(0) = h_0 \text{ on } \Sigma, \quad u(0) = u_0, \quad \theta(0) &= \theta_0 && \text{in } \Omega.
\end{aligned}\tag{9.6}$$

The differential operators $(\mathcal{A}_u, \mathcal{A}_\theta, \mathcal{B}_\theta)$ are defined as previously, and with Section 2.2, $\mathcal{A}_{\theta_\Sigma}$ is given by

$$\mathcal{A}_{\theta_\Sigma} : \nabla_\Sigma^2 = -(d_\Gamma(\theta_\Sigma)/\kappa_\Gamma(\theta_\Sigma))M_0(h)P_\Gamma(h)M_0(h) : \nabla_\Sigma^2.$$

Here we employed the relation

$$\frac{D}{Dt}\theta_\Sigma = \partial_t\theta_\Sigma + (I - M_1^{\mathsf{T}}(h))u_\Sigma\cdot\nabla_\Sigma\theta_\Sigma,$$

taken from Section 1.3.2. $F_{\theta_\Sigma} = F_{\theta_\Sigma}(u,\theta,\theta_\Sigma,h)$ is defined by

$$\begin{aligned}
&\kappa_\Gamma(\theta_\Sigma)F_{\theta_\Sigma}\\
&= M_0(h)P_\Gamma(h)\nabla_\Sigma\cdot(d_\Gamma(\theta_\Sigma)P_\Gamma(h)M_0(h))\nabla_\Sigma\theta_\Sigma - \kappa_\Gamma(\theta_\Sigma)(I - M_1^{\mathsf{T}}(h))u_\Sigma\cdot\nabla_\Sigma\theta_\Sigma\\
&\quad + \theta_\Sigma\sigma'(\theta_\Sigma)(P_\Gamma(h)M_0(h)\nabla_\Sigma\cdot\mathcal{P}_\Sigma u - H_\Gamma(h)V_\Gamma\nu_\Gamma) - [\![\theta\eta(\theta)]\!]j_\Sigma - [\![\mathcal{B}_\theta(\theta,h)\nabla\theta]\!],
\end{aligned}$$

it collects all lower order terms. Recall that

$$j_\Sigma = [\![u\cdot\nu_\Gamma]\!]/[\![1/\varrho]\!], \quad V_\Gamma = [\![\varrho u\cdot\nu_\Gamma]\!]/[\![\varrho]\!].$$

We extend (u_0, h_0) as in Section 9.1.4, but we have to be more careful with θ_0 due to the dynamic equation for θ_Σ on Σ. We first extend $\theta_{\Sigma 0} = \theta_0|_\Sigma$ on Σ by $\bar\theta_\Sigma = e^{-(1-\Delta_\Sigma)t}\theta_{\Sigma 0}$ and then solve the two one-phase parabolic problems

$$\begin{aligned}
\partial_t\bar\theta - \Delta\bar\theta &= 0 && \text{in } \Omega\setminus\Sigma,\\
\partial_\nu\bar\theta &= 0 && \text{on } \partial\Omega,\\
\bar\theta &= \bar\theta_\Sigma && \text{on } \Sigma,\\
\bar\theta(0) &= \theta_0 && \text{in } \Omega.
\end{aligned}$$

Observe that the heat equation on Σ decouples to highest order from the remaining equations, and the heat problem in $\Omega\setminus\Sigma$ decouples from the system for (u,π,h). The latter is as in the previous subsection governed by an asymmetric Stokes problem. The solution space for $z=(u,\theta,\theta_\Sigma,h,\pi,q_1,q_2)$ is here defined by

$$\begin{aligned}\mathbb{E}^6_\mu(a) :=\{z\in \mathbb{E}_{u,\mu}\times\mathbb{E}_{\theta,\mu}(J)\times\mathbb{E}_{\theta_\Sigma,\mu}(J)\times\mathbb{E}^u_{h,\mu}(J)\times\mathbb{E}_{\pi,\mu}(J)\times\mathbb{F}^u_{u,\mu}(J)^2 :\\ \mathcal{P}_\Sigma[\![u]\!],[\![\theta]\!]=0,\ \theta=\theta_\Sigma,\ \pi_{|\partial\Omega_j}=q_j \text{ on } \Sigma,\ \ u,\partial_\nu\theta=0 \text{ on } \partial\Omega\},\end{aligned}$$

with

$$\mathbb{E}_{\theta_\Sigma,\mu}(J)=H^1_{p,\mu}(J;W^{-1/p}_p(\Sigma))\cap L_{p,\mu}(J;W^{2-1/p}_p(\Sigma)).$$

For the space of data we may take here

$$\mathbb{F}^6_\mu(a)=\mathbb{F}_{u,\mu}(J)\times\mathbb{F}^4_{\pi,\mu}(J)\times\mathbb{F}_{\theta,\mu}(J)\times\mathbb{F}_{\theta_\Sigma,\mu}(J)\times\mathcal{P}_\Sigma\mathbb{F}^\theta_{h,\mu}(J)^n\times\mathbb{F}^u_{h,\mu}(J)^{n+1}\times\mathbb{F}^\theta_{h,\mu}(J),$$

where

$$\mathbb{F}_{\theta_\Sigma,\mu}(J)=L_{p,\mu}(J;W^{-1/p}_p(\Sigma)).$$

This way we obtain the abstract form

$$L_6\tilde z=N_6(\tilde z,\bar z),\quad \tilde z=0,$$

with $N_6: {}_0\mathbb{E}^6_\mu(a)\times\mathbb{E}^6_\mu(\infty)\to {}_0\mathbb{F}^6_\mu(a)$ and $L_6:\mathbb{E}^6_\mu(a)\to\mathbb{F}^6_\mu(a)$ linear and bounded. More precisely, L_6 is defined by

$$L_6\tilde z=\begin{bmatrix}\partial_t\tilde u+\mathcal{A}_u(\theta_0,h_0)\nabla^2\tilde u+(1-M_1(h_0))\nabla\tilde\pi\\ (I-M_1(h_0))\nabla\cdot\tilde u)\\ \partial_t\tilde\theta+\mathcal{A}_\theta(\theta_0,h_0):\nabla^2\tilde\theta\\ \partial_t\tilde\theta_\Sigma+\mathcal{A}_{\theta_\Sigma}(\theta_0,h_0):\nabla^2_\Sigma\tilde\theta_\Sigma\\ \mathcal{P}_\Sigma[\![\tilde u]\!]+[\![\tilde u\cdot\nu_\Sigma]\!]M_0(h_0)\nabla_\Sigma\bar h+[\![\bar u\cdot\nu_\Sigma]\!]M_0(h_0)\nabla_\Sigma\tilde h\\ -[\![S(\tilde u,\theta_0,h_0)]\!]\nu_\Sigma+([\![\tilde\pi]\!]+\sigma(\theta_0)\mathcal{C}_\Sigma(h_0))\nu_\Sigma-\sigma'(\theta_0)\nabla_\Sigma\tilde\theta_\Sigma\\ -[\![S(\tilde u,\theta_0,h_0)\nu_\Sigma\cdot\nu_\Sigma/\varrho]\!]+[\![\tilde\pi/\varrho]\!]\\ \partial_t\tilde h-([\![\varrho\tilde u\nu_\Sigma]\!]-[\![\varrho\tilde u\cdot M_0(h_0)\nabla_\Sigma\bar h]\!]-[\![\varrho\bar u\cdot M_0(h_0)\nabla_\Sigma\tilde h]\!])/[\![\varrho]\!]\end{bmatrix}.$$

As the operator for θ_Σ has maximal L_p-regularity by Section 6.3, that for θ has this property by Section 6.3, and the remaining asymmetric Stokes operator does so as we have seen in the previous subsection, we conclude that L_6 has maximal regularity, which shows that $L_6: {}_0\mathbb{E}^6_\mu(a)\to {}_0\mathbb{F}^6_\mu(a)$ is an isomorphism, with uniform bounds in $a\in(0,1]$.

1.6 Phase Transitions and Marangoni Forces: Equal Densities
In the sequel we assume Condition **(H5)** and the compatibility condition

$$\begin{aligned}\textbf{(C5)}\quad &[\![\psi(\theta_0)]\!]+\sigma(\theta_0)H_{\Gamma_0}=0 \ \text{ on } \Gamma_0,\quad \operatorname{div}u_0=0 \ \text{ in } \Omega\setminus\Gamma_0,\\ &\mathcal{P}_{\Gamma_0}[\![2\mu(\theta_0)D(u_0)\nu_{\Gamma_0}]\!]+\sigma'(\theta_0)\nabla_{\Gamma_0}\theta_0=0 \ \text{ on } \Gamma_0.\end{aligned}$$

Here we have once more $\varrho_1=\varrho_2=1$ w.l.o.g, and we solve for j_Γ according to $j_\Gamma=u\cdot\nu_\Gamma-V_\Gamma$, and insert this into the interface energy balance.

The case where undercooling is present is the simpler one, as both equations on the interface are dynamic equations, however it can be used as a guide. In particular, the Gibbs-Thomson identity

$$\gamma(\theta_\Gamma)V_\Gamma - \sigma(\theta_\Gamma)H_\Gamma(h) = \varphi(\theta_\Gamma)$$

can be understood as a *mean curvature flow* for the evolution of the surface, modified by physics.

If there is no undercooling, there is a hidden mean curvature flow which, however, is more complex. For the derivation, it is convenient to eliminate the time derivative of θ_Γ from the energy balance on the interface. In fact, differentiating the Gibbs-Thomson law w.r.t. time t and with $\lambda(s) = \varphi(s)/\sigma(s)$ we obtain

$$\lambda'(\theta_\Gamma)\frac{D_n}{Dt}\theta_\Gamma + H_\Gamma'(h)V_\Gamma = 0 \quad \text{on } \Gamma(t),$$

hence substitution into surface energy balance yields with

$$T_\Gamma(\theta_\Gamma) := \omega_\Gamma(\theta_\Gamma) - H_\Gamma'(h), \quad \omega_\Gamma(\theta_\Gamma) := \frac{\lambda'(\theta_\Gamma)}{\kappa_\Gamma(\theta_\Gamma)}(l(\theta_\Gamma) - l_\Gamma(\theta_\Gamma)\lambda(\theta_\Gamma)) \tag{9.7}$$

the relation

$$\begin{aligned} T_\Gamma(\theta_\Gamma)V_\Gamma = \frac{\lambda'(\theta_\Gamma)}{\kappa_\Gamma(\theta_\Gamma)}\{&\mathrm{div}_\Gamma(d_\Gamma(\theta_\Gamma)\nabla_\Gamma\theta_\Gamma) - \kappa_\Gamma u_\Gamma\nabla_\Gamma\theta_\Gamma + [\![d(\theta)\partial_\nu\theta]\!] \\ &+ l_\Gamma \mathrm{div}_\Gamma u + l_0(\theta)u\cdot\nu_\Gamma\}. \end{aligned} \tag{9.8}$$

As V_Γ should be determined only by the state of the system and should not depend on time derivatives of other variables, this indicates that the problem without undercooling is not well-posed if the operator $T_\Gamma(\theta_\Gamma)$ is not invertible in $L_2(\Gamma)$, as V_Γ might not be well-defined. On the other hand, if $T_\Gamma(\theta_\Gamma)$ is invertible, then

$$\begin{aligned} V_\Gamma = T_\Gamma^{-1}\frac{\lambda'(\theta_\Gamma)}{\kappa_\Gamma(\theta_\Gamma)}\{&\mathrm{div}_\Gamma(d_\Gamma(\theta_\Gamma)\nabla_\Gamma\theta_\Gamma) - \kappa_\Gamma u_\Gamma\nabla_\Gamma\theta_\Gamma + [\![d(\theta)\partial_\nu\theta]\!] \\ &+ l_\Gamma \mathrm{div}_\Gamma u + l_0(\theta)u\cdot\nu_\Gamma\}. \end{aligned} \tag{9.9}$$

uniquely determines the interfacial velocity V_Γ, gaining two derivatives in space, and showing that all terms on the right-hand side of surface energy balance are of lower order. Note that

$$\omega_\Gamma(s) = s\sigma(s)[\lambda'(s)]^2/\kappa_\Gamma(s) \geq 0 \quad \text{in } (0,\theta_c), \tag{9.10}$$

and $\omega_\Gamma(s) = 0$ if and only if $\lambda'(s) = 0$. The well-posedness condition appears to be more complex, compared to the case $\kappa_\Gamma \equiv 0$.

Going one step further, taking the surface gradient of the Gibbs-Thomson relation yields the identity

$$\kappa_\Gamma(\theta_\Gamma)V_\Gamma - d_\Gamma(\theta_\Gamma)H_\Gamma = \kappa_\Gamma(\theta_\Gamma)\{f_\Gamma(\theta_\Gamma) + F_\Gamma(u,\theta,\theta_\Gamma)\}, \tag{9.11}$$

as will be shown below. Here the function f_Γ is the antiderivative of $\lambda(d_\Gamma/\kappa_\Gamma)'$ vanishing at $s=\theta_m$, and F_Γ is nonlocal in space and of lower order. So also in the case where undercooling is absent we obtain a *mean curvature flow*, modified by physics.

To derive (9.11), note that

$$\begin{aligned}
&\frac{\lambda'(\theta_\Gamma)}{\kappa_\Gamma(\theta_\Gamma)}\mathrm{div}_\Gamma(d_\Gamma(\theta_\Gamma)\nabla_\Gamma\theta_\Gamma)\\
&\qquad=\frac{1}{\kappa_\Gamma(\theta_\Gamma)}\mathrm{div}_\Gamma(d_\Gamma(\theta_\Gamma)\nabla_\Gamma\lambda(\theta_\Gamma))-\frac{d_\Gamma(\theta_\Gamma)}{\kappa_\Gamma(\theta_\Gamma)}\lambda''(\theta_\Gamma)|\nabla_\Gamma\theta_\Gamma|^2\\
&\qquad=\mathrm{div}_\Gamma\Big(\frac{d_\Gamma(\theta_\Gamma)}{\kappa_\Gamma(\theta_\Gamma)}\nabla_\Gamma\lambda(\theta_\Gamma)\Big)-\frac{d_\Gamma(\theta_\Gamma)}{\kappa_\Gamma(\theta_\Gamma)}\Big\{\lambda''(\theta_\Gamma)-\lambda'(\theta_\Gamma)\frac{\kappa_\Gamma'(\theta_\Gamma)}{\kappa_\Gamma(\theta_\Gamma)}\Big\}|\nabla_\Gamma\theta_\Gamma|^2\\
&\qquad=\Delta_\Gamma g_\Gamma(\theta_\Gamma)-\frac{d_\Gamma(\theta_\Gamma)}{\kappa_\Gamma(\theta_\Gamma)}\Big\{\lambda''(\theta_\Gamma)-\lambda'(\theta_\Gamma)\frac{\kappa_\Gamma'(\theta_\Gamma)}{\kappa_\Gamma(\theta_\Gamma)}\Big\}|\nabla_\Gamma\theta_\Gamma|^2,
\end{aligned}$$

where g_Γ denotes the antiderivative of $d_\Gamma\lambda'/\kappa_\Gamma$ with $g_\Gamma(\theta_m)=0$. We note that by a partial integration

$$g_\Gamma(s)=\lambda(s)\frac{d_\Gamma(s)}{\kappa_\Gamma(s)}-\int_{\theta_m}^s\lambda(\tau)\Big(\frac{d_\Gamma}{\kappa_\Gamma}\Big)'(\tau)\,d\tau=:\lambda(s)\frac{d_\Gamma(s)}{\kappa_\Gamma(s)}-f_\Gamma(s).$$

Now employing $\lambda(\theta_\Gamma)=-H_\Gamma$, (9.8) leads to the identity

$$\begin{aligned}
&T_\Gamma(\theta_\Gamma)\{V_\Gamma-\frac{d_\Gamma(\theta_\Gamma)}{\kappa_\Gamma(\theta_\Gamma)}H_\Gamma-f_\Gamma(\theta_\Gamma)\}\\
&\quad=\frac{\lambda'(\theta_\Gamma)}{\kappa_\Gamma(\theta_\Gamma)}\{[\![d(\theta)\partial_\nu\theta]\!]-\kappa_\Gamma u_\Gamma\nabla_\Gamma\theta_\Gamma+l_\Gamma\mathrm{div}_\Gamma u+l_0(\theta)u\cdot\nu_\Gamma\}\\
&\quad-\frac{d_\Gamma(\theta_\Gamma)}{\kappa_\Gamma(\theta_\Gamma)}\Big\{\lambda''(\theta_\Gamma)-\lambda'(\theta_\Gamma)\frac{\kappa_\Gamma'(\theta_\Gamma)}{\kappa_\Gamma(\theta_\Gamma)}\Big\}|\nabla_\Gamma\theta_\Gamma|^2+\{\omega_\Gamma(\theta_\Gamma)-\mathrm{tr}L_\Gamma^2\}g_\Gamma(\theta_\Gamma),
\end{aligned}$$

hence applying the inverse of $T_\Gamma(\theta_\Gamma)$ we arrive at (9.11), with

$$\begin{aligned}
F_\Gamma(u,\theta,\theta_\Gamma)=[\kappa_\Gamma(\theta_\Gamma)T_\Gamma(\theta_\Gamma)]^{-1}\big(&\lambda'(\theta_\Gamma)\{[\![d(\theta)\partial_\nu\theta]\!]-\kappa_\Gamma u_\Gamma\nabla_\Gamma\theta_\Gamma+l_\Gamma\mathrm{div}_\Gamma u+l_0(\theta)u\cdot\nu_\Gamma\}\\
&-d_\Gamma(\theta_\Gamma)\{(\lambda''(\theta_\Gamma)-\lambda'(\theta_\Gamma)\kappa_\Gamma'(\theta_\Gamma)/\kappa_\Gamma(\theta_\Gamma)\}|\nabla_\Gamma\theta_\Gamma|^2\\
&+\kappa_\Gamma(\theta_\Gamma)\{\omega_\Gamma(\theta_\Gamma)-\mathrm{tr}L_\Gamma^2\}g_\Gamma(\theta_\Gamma)\big).
\end{aligned}$$

In the sequel we replace the Gibbs-Thomson law by the dynamic equation (9.11) plus the compatibility condition $\varphi(\theta_{\Gamma 0})+\sigma(\theta_{\Gamma 0})H_{\Gamma_0}=0$ at time $t=0$.

Now we perform the Hanzawa transform to obtain a problem on Ω with fixed interface Σ. This yields the following problem.

$$\begin{aligned}
\partial_t u + \mathcal{A}_u(\theta,h) : \nabla^2 u + (I - M_1(h))\nabla\pi/\varrho &= F_u(u,\theta,h) && \text{in } \Omega\setminus\Sigma,\\
(I - M_1(h))\nabla\cdot u &= 0 && \text{in } \Omega\setminus\Sigma,\\
\partial_t\theta + \mathcal{A}_\theta(\theta,h) : \nabla^2\theta &= F_\theta(\theta,h) && \text{in } \Omega\setminus\Sigma,\\
u, \partial_\nu\theta &= 0 && \text{on } \partial\Omega,\\
\partial_t\theta_\Sigma + \mathcal{A}_{\theta_\Sigma}(\theta_\Sigma,h) : \nabla^2_\Sigma\theta_\Sigma &= F_{\theta_\Sigma}(u,\theta,\theta_\Sigma,h) && \text{on } \Sigma,\\
\theta = \theta_\Sigma,\ [\![u]\!], [\![\theta]\!] &= 0 && \text{on } \Sigma,\\
-[\![S(u,\theta,h)]\!]\nu_\Gamma + \big([\![\pi]\!] - \sigma(\theta_\Sigma)H_\Gamma(h)\big)\nu_\Gamma &= \sigma'(\theta_\Sigma)\nabla_\Sigma\theta_\Sigma && \text{on } \Sigma,\\
\kappa_\Gamma(\theta_\Sigma)V_\Gamma - d_\Gamma H_\Gamma(h) - \kappa_\Gamma(\theta_\Sigma)f(\theta_\Sigma) &= \kappa_\Gamma(\theta_\Sigma)F_\Gamma(\theta,\theta_\Sigma,h)) && \text{on } \Sigma,\\
h(0) = h_0 \text{ on } \Sigma,\quad u(0) = u_0,\ \theta(0) &= \theta_0 && \text{in } \Omega.
\end{aligned} \tag{9.12}$$

The abstract setting of this problem differs from the previous cases. As variables we choose $z = (u,\theta,\theta_\Sigma,h,\pi,q)$ in the regularity space

$$\begin{aligned}
\mathbb{E}^5_\mu(a) =&\{z \in \mathbb{E}_{u,\mu} \times \mathbb{E}_{\theta,\mu}(J) \times \times\mathbb{E}_{\theta_\Sigma,\mu}(J) \times \mathbb{E}^5_{h,\mu}(J) \times \mathbb{E}_{\pi,\mu}(J) \times \mathbb{E}_{q,\mu}(J) :\\
&[\![u]\!], [\![\theta]\!] = 0,\ \ \theta = \theta_\Sigma,\ \ [\![\pi]\!]_{|\Sigma} = q \text{ on } \Sigma,\ \ u, \partial_\nu\theta = 0 \text{ on } \partial\Omega\},
\end{aligned}$$

with

$$\mathbb{E}^5_{h,\mu}(J) = H^1_{p,\mu}(J; W^{1-1/p}_p(\Sigma)) \cap L_{p,\mu}(J; W^{3-1/p}_p(\Sigma)).$$

For the space of data we may take here

$$\mathbb{F}^5_\mu(a) = \mathbb{F}_{u,\mu}(J) \times \mathbb{F}^2_{\pi,\mu}(J) \times \mathbb{F}_{\theta,\mu}(J) \times \mathbb{F}_{\theta_\Sigma,\mu}(J) \times \mathbb{F}^u_{h,\mu}(J)^n \times \mathbb{F}^5_{h,\mu}(J),$$

where

$$\mathbb{F}^5_{h,\mu}(J) = L_{p,\mu}(J; W^{1-1/p}_p(\Sigma)).$$

This way we obtain the abstract form of the problem

$$L_5\tilde{z} = N_5(\tilde{z},\bar{z}), \quad \tilde{z}(0) = 0,$$

with $N_5 : {}_0\mathbb{E}^5_\mu(a) \times \mathbb{E}^5_\mu(\infty) \to {}_0\mathbb{F}^5_\mu(a)$ and $L_5 : \mathbb{E}^5_\mu(a) \to \mathbb{F}^5_\mu(a)$ linear and bounded.

More precisely, L_5 is defined by

$$L_5\tilde{z} = \begin{bmatrix}
\partial_t\tilde{u} + \mathcal{A}_u(\theta_0,h_0)\nabla^2\tilde{u} + (1 - M_1(h_0))\nabla\tilde{\pi}\\
(I - P_0M_1(h_0))\nabla\cdot\tilde{u})\\
\partial_t\tilde{\theta} + \mathcal{A}_\theta(\theta_0,h_0) : \nabla^2\tilde{\theta}\\
\partial_t\tilde{\theta}_\Sigma + \mathcal{A}_{\theta_\Sigma}(\theta_{\Sigma 0},h_0) : \nabla^2_\Sigma\tilde{\theta}_\Sigma\\
-[\![S(\tilde{u},\theta_0,h_0)]\!]\nu_\Sigma + \tilde{q}\nu_\Sigma + \sigma(\theta_{\Sigma 0})\mathcal{C}_\Sigma\tilde{h} - \sigma'(\theta_{\Sigma 0})\nabla_\Sigma\tilde{\theta}_\Sigma\\
\partial_t\tilde{h} + c_0(\theta_{\Sigma 0},h_0)\mathcal{C}_\Sigma\tilde{h} + c_1(\theta_{\Sigma 0})\tilde{\theta}_\Sigma
\end{bmatrix}.$$

Here we have set

$$c_0(\theta_{\Sigma 0},h_0) = d_\Gamma(\theta_{\Sigma 0})/\kappa_\Gamma(\theta_{\Sigma 0}), \quad c_1(\theta_{\Sigma 0}) = -\lambda(\theta_{\Sigma 0})(d_\Gamma/\kappa_\Gamma)'(\theta_{\Sigma 0}).$$

The operator L_5 also has maximal regularity, as it has triangular structure, and each diagonal entry has maximal L_p-regularity.

9.2 The Fixed Point Argument

In the previous section we have seen that on the fixed domain all six problems can be reformulated as the abstract problem

$$L\tilde{z} = N(\tilde{z}, \bar{z}), \quad \tilde{z}(0) = 0, \tag{9.13}$$

where $L : \mathbb{E}_\mu(a) \to \mathbb{F}_\mu(a)$ is bounded linear, and $N : {}_0\mathbb{E}_\mu(a) \times \mathbb{E}_\mu(\infty) \to {}_0\mathbb{F}_\mu(a)$ is nonlinear. Of course, the specific spaces and operators differ from problem to problem, but they all share the following properties.

(MR) For each $a \in (0,1]$, the operator $L : {}_0\mathbb{E}_\mu(a) \to {}_0\mathbb{F}_\mu(a)$ is an isomorphism, and the norm of L^{-1} is bounded by some constant M independent of $a \in (0,1]$.

(NL) For each $a \in (0,1]$, the nonlinearity N is of class C^1. Moreover,

(i) $|N(0,\bar{z})|_{\mathbb{F}_\mu(a)} \to 0$ as $a \to 0$, for each fixed $\bar{z} \in \mathbb{E}_\mu(\infty)$;

(ii) $|D_1 N(0,\bar{z})|_{\mathcal{B}({}_0\mathbb{E}_\mu(a), {}_0\mathbb{F}_\mu(a))} \to 0$ as $a \to 0$, for each fixed $\bar{z} \in \mathbb{E}_\mu(\infty)$.

Condition **(NL)** will be verified in Section 9.5. It implies that for a given $\bar{z} \in \mathbb{E}_\mu(\infty)$,

$$\eta(a,r) := \sup\{|D_1 N(\tilde{z},\bar{z})|_{\mathcal{B}({}_0\mathbb{E}_\mu(a), {}_0\mathbb{F}_\mu(a))} : |\tilde{z}|_{\mathbb{E}_\mu(a)} \le r\}$$

satisfies $\eta(a,r) \to 0$ as $a, r \to 0$. This in turn implies

$$|N(\tilde{z}_1,\bar{z}) - N(\tilde{z}_2,\bar{z})|_{\mathbb{F}_\mu(a)} \le \eta(a,r)|\tilde{z}_1 - \tilde{z}_2|_{\mathbb{E}_\mu(a)}, \quad |\tilde{z}_j|_{\mathbb{E}_\mu(a)} \le r,$$

and

$$|N(\tilde{z},\bar{z})|_{\mathbb{F}_\mu(a)} \le |N(0,\bar{z})|_{\mathbb{F}_\mu(a)} + \eta(a,r)r, \quad |\tilde{z}|_{\mathbb{E}_\mu(a)} \le r.$$

As $|L^{-1}|_{\mathcal{B}({}_0\mathbb{E}_\mu(a), {}_0\mathbb{F}_\mu(a))}$ is uniformly bounded for $a \in (0,1]$, say by C, we see that $T(\tilde{z}) = L^{-1}N(\tilde{z},\bar{z})$ will be a contracting self-map on the ball $\bar{B}_{{}_0\mathbb{E}_\mu(a)}(0,r)$, by choosing a, r small enough. The contraction mapping principle then yields a unique fixed point $\tilde{z}_\odot \in \bar{B}_{{}_0\mathbb{E}_\mu(a)}(0,r)$, which means that (9.13) admits the unique solution $\tilde{z}_\odot$. This completes the proof of local existence and uniqueness for the six Problems **(P1)**~**(P6)**. This way we have proved

Theorem 9.2.1. *Let $p > n+2$, $1 \ge \mu > \frac{1}{2} + \frac{n+2}{2p}$, and suppose the following conditions are satisfied.*

(i) Regularity: *Condition* **(Hj)** *holds for Problem* **(Pj)**.

(ii) Well-Posedeness: $\theta_0 > 0$; $l(\theta_0) \neq 0$ *for Problems* **(P1)**, **(P3)**,

$0 < \theta_0 < \theta_c$ *for Problems* **(P5)**, **(P6)**,

$T_{\Gamma_0}(\theta_0)$ *is invertible in* $L_2(\Gamma_0)$ *for Problem* **(P5)**.

(iii) Compatibilities: *Condition* **(Cj)** *holds for Problem* **(Pj)**.

Then each Problem **(Pj)**, *$j = 1, \ldots, 6$, is locally uniquely solvable in the sense that for any initial value $z_0 \in X^j_{\gamma,\mu}$, there is $a = a(z_0) > 0$ such that the transformed problems admit a unique solution $z \in \mathbb{E}^j_\mu(a)$.*

9.3 Dependence on the Data

To study the dependence of the solution of (9.13) on the initial data, we will employ the implicit function theorem. For this purpose note that the map $E : X_{\gamma,\mu} \to \mathbb{E}_\mu(\infty)$ defined by $Ez_0 = \bar{z}$ is linear and bounded, hence real analytic. We rewrite problem (9.13) as

$$G(\tilde{z}, z_1) := L(z_1, Ez_1)\tilde{z} - N(\tilde{z}, Ez_1) = 0,$$

where $L(z_1, Ez_1)$ indicates the dependence of L on the initial value z_1 and, where applicable, on the pertinent extensions $\bar{z}_1 = Ez_1$ subsumed in the definition of L_j. Here

$$G : {}_0\mathbb{E}_\mu(a) \times B_{X_{\gamma,\mu}}(z_0, r) \to {}_0\mathbb{F}_\mu(a)$$

is at least of class C^1. We have $G(\tilde{z}_\odot, z_0) = 0$, and the Fréchet-derivative $D_1G(\tilde{z}_\odot, z_0) \in \mathcal{B}({}_0\mathbb{E}_\mu(a), {}_0\mathbb{F}_\mu(a))$ is invertible, as we have seen in the previous section. Therefore, there is a radius $\delta > 0$ and a C^1-map $\tilde{z} : B_{X_{\gamma,\mu}}(z_0, \delta) \to {}_0\mathbb{E}_\mu(a)$ such that

$$\tilde{z}(z_0) = \tilde{z}_\odot \text{ and } G(\tilde{z}(z_1), z_1) = 0 \text{ for all } z_1 \in B_{X_{\gamma,\mu}}(z_0, \delta).$$

Moreover, by uniqueness there are no other solutions close to $\tilde{z}_\odot$, and so by causality there are no other solutions, at all.

Further, if G is of class C^k, $k \in \mathbb{N} \cup \{\infty, \omega\}$, then $\tilde{z}$ has the same regularity; here ω means real analytic. We observe that L, N, and hence G, are of class C^k provided

$$\psi, \sigma \in C^{k+2}(0, \infty) \quad \text{and} \quad d, d_\Gamma, \mu \in C^{k+1}(0, \infty).$$

Note that the maps $h \mapsto (m_0(h), M_0(h), M_1(h), \beta(h))$ are real analytic. This implies the following result.

Theorem 9.3.1. *In addition to the assumptions of Theorem* 9.2.1 *assume that*

$$\psi, \sigma \in C^{k+2}(0, \infty) \quad \textit{and} \quad d, d_\Gamma, \mu \in C^{k+1}(0, \infty),$$

for some $k \in \mathbb{N} \cup \{\infty, \omega\}$.

Then the solution map is of class C^k *from the data space* $X^j_{\gamma,\mu}$ *into the solution space* $\mathbb{E}^j_\mu(a)$, *for each* $j = 1, \ldots, 6$.

9.4 Regularity: The Parameter Trick

In Section 5.3 we used a scaling argument for time t to extract more time regularity from the regularity properties of the nonlinearity $A(u)u - F(u)$ for the solution of the quasilinear parabolic evolution equation

$$\dot{u} + A(u)u = F(u), \quad t \in J, \ u(0) = u_0.$$

In this section we extend this method to obtain regularity of z in the 6 problems studied above. The implicit function theorem as well as maximal L_p-regularity will again be the main tools.

9.4.1 Interior Regularity

Let $G : \mathbb{E}_\mu(a) \to \mathbb{F}_\mu(a)$ denote the functions G^j from the previous section, where we now fix the initial values and suppress them in our notation, with the corresponding function spaces $\mathbb{E}^j_\mu(a)$ and $\mathbb{F}^j_\mu(a)$. We assume that G is in the class C^k, with $k \in \mathbb{N} \cup \{\infty, \omega\}$, where, as before, ω means real analytic. We want to show

$$(u, \theta), \partial_i(u, \theta) \in C^k((0, a) \times (\Omega \setminus \Sigma))^{n+1}, \quad i = 1, \ldots, n.$$

This then implies also pressure regularity $\pi, \partial_i \pi \in C^{k-1}((0, a) \times (\Omega \setminus \Sigma))$, for all $i = 1, \ldots, n$, by the equation for u.

For this purpose we fix $(t_0, x_0) \in (0, a) \times (\Omega \setminus \Sigma)$. Recall that regularity is a local property, so we need only to show regularity of (u, θ) in $(t_0 - r, t_0 + r) \times B(x_0, r)$ where $r > 0$ is small enough. We fix $R > 0$ such that $3R < t_0 < a - 3R$, and $B(x_0, 3R) \subset \Omega \setminus \Sigma$. Further we may let $a \leq a_0$ by causality; otherwise we shift the time interval in question, and repeat the argument finitely many times. Furthermore, we assume that $B(x_0, 3R)$ does not intersect the tubular neighbourhood of width $3a_\Sigma$ around Σ; we comment on this assumption later.

Next we choose standard C^∞-cut-off functions χ_{t_0} and χ_{x_0}, which are 1 for $|t - t_0| < R$, resp. $|x - x_0| < R$, and 0 for $|t - t_0| > 2R$, resp. $|x - x_0| > 2R$, between 0 and 1 elsewhere.

We introduce a coordinate transform $\tau_{(\lambda,\xi)}$ by means of

$$\tau_{\lambda,\xi}(t, x) = (t + \lambda \chi_{t_0}(t), x + t\xi \chi_{x_0}(x)), \quad (t, x) \in (0, a) \times \Omega).$$

It is easy to see that $\tau_{(\lambda,\xi)} : (0, a) \times \Omega$ is a diffeomorphism of class C^∞, so that the map

$$\tau : (\lambda, \xi) \mapsto \tau_{\lambda,\xi}, \quad (-r, r) \times B_{\mathbb{R}^n}(0, r) \to \mathrm{Diff}^\infty((0, a) \times \Omega)$$

is well-defined, provided r is sufficiently small. Observe that $\tau_{0,0} = id$, and that $\tau_{(\lambda,\xi)} = id$ outside the cube $(-2R, 2R) \times B_{\mathbb{R}^n}(0, 2R)$.

In the next step we introduce the lifted coordinate transforms $T_{\lambda,\xi}$ by

$$T_{\lambda,\xi} z(t, x) = z(\tau_{\lambda,\xi}(t, x)) = z(t + \lambda \chi_{t_0}(t), x + t\xi \chi_{x_0}(x)), \quad t \in (0, a), \quad x \in \Omega,$$

where $(\lambda, \xi) \in (-r, r) \times B_{\mathbb{R}^n}(0, r)$. It is not difficult to show that $T_{\lambda,\xi}$ is an isomorphism in $\mathbb{E}_\mu(a)$ as well as in $\mathbb{F}_\mu(a)$; one only needs to recall the transformation rules from Section 6.3. Note that $T_{\lambda,\xi}$ is leaving *the initial values unchanged.* This property is very important, as it will show that the obtained regularity does not depend on the regularity of the initial value z_0.

By the transformation rules from Section 6.3, we obtain the relations

$$T_{\lambda,\xi} \nabla z = \nabla z \circ \tau_{\lambda,\xi} = (I - m_1(\lambda, \xi)) \nabla T_{\lambda,\xi} z,$$

and

$$T_{\lambda,\xi}\partial_t z = \partial_t z \circ \tau_{\lambda,\xi} = (1+\lambda\chi'_{t_0})^{-1}[\partial_t T_{\lambda,\xi} z - m_0(\lambda,\xi)(\xi|\nabla)T_{\lambda,\xi} z],$$

where

$$m_0(\lambda,\xi) = \frac{\chi_{x_0}}{1+t(\xi|\nabla\chi_{x_0})}, \quad m_1(\lambda,\xi) = \frac{t\nabla\chi_{x_0}\otimes\xi}{1+t(\xi|\nabla\chi_{x_0})}.$$

Note that m_0, m_1 are real analytic in (λ,ξ) and of class C^∞ in (t,x).

Given the solution $z_\odot$ of $G(z_\odot)=0$ from the previous section, we see that

$$0 = T_{\lambda,\xi}G(z_\odot) = T_{\lambda,\xi}G(T^{-1}_{\lambda,\xi}T_{\lambda,\xi}z_\odot),$$

hence with

$$\mathsf{G}(\lambda,\xi,\bar z) = T_{\lambda,\xi}G(T^{-1}_{\lambda,\xi}\bar z)$$

and setting $\bar z_\odot = T_{\lambda,\xi}z_\odot = z_\odot\circ\tau_{\lambda,\xi}$, it is obvious that $\bar z_\odot$ satisfies the equation

$$\mathsf{G}(\lambda,\xi,\bar z_\odot) = 0.$$

So it is natural to employ the implicit function theorem to solve for $\bar z_\odot$. As we are interested in the regularity of solutions for $t>0$, we may, and we will, assume that the fixed initial value z_0 is in the regularity space X_1. We then consider

$$\mathsf{G} : (-r,r)\times B_{\mathbb{R}^n}(0,r)\times \mathbb{E}_1^{z_0}(a) \to {}_0\mathbb{F}_1(a),$$

where $\mathbb{E}_1^{z_0}(a)$ denotes the affine linear subspace of $\mathbb{E}_1(a)$ with fixed initial values $u(0)=u_0$, $\theta(0)=\theta_0$, $h(0)=h_0$, $\partial_t h(0)=h_1$, where these data are subject to the appropriate compatibility conditions. Employing the transformation rules for ∇ and ∂_t from above, as in the previous subsection it follows from Section 9.5 that G is of class C^k, $k\in\mathbb{N}\cup\{\infty,\omega\}$, whenever

$$\psi,\sigma\in C^{k+2}(0,\infty) \quad \text{and} \quad d, d_\Gamma, \mu \in C^{k+1}(0,\infty).$$

Furthermore, we have $\mathsf{G}(0,0,z_\odot)=0$ and

$$D_z\mathsf{G}(0,0,z_\odot) = D_z G(z_\odot) : {}_0\mathbb{E}_1(a)\to {}_0\mathbb{F}_1(a)$$

is invertible, by maximal regularity, as known from Section 9.2. Hence by the implicit function theorem, there is a neighbourhood $(-\delta,\delta)\times B_{\mathbb{R}^n}(0,\delta)$ of $(0,0)$ and a map

$$\Phi : (-\delta,\delta)\times B_{\mathbb{R}^n}(0,\delta)\to\mathbb{E}_1(a),$$

of class C^k with $\Phi(0,0)=z_\odot$ such that $\mathsf{G}(\lambda,\xi,\Phi(\lambda,\xi))=0$. By uniqueness, this implies $\Phi(\lambda,\xi)=z_\odot\circ\tau_{\lambda,\xi}$. As a consequence, the projection-embedding

$$\mathbb{E}_1(a)\to C^\alpha((0,a);C^{1+\alpha}(\Omega\setminus\Sigma))^{n+1}, \quad z\mapsto(u,\theta),$$

with $\alpha \in (0,1)$ sufficiently small, shows that

$$(\lambda, \xi) \mapsto (u, \theta)(t + \lambda\chi_{t_0}(t), x + t\xi\chi_{x_0}(x))$$

is of class C^k. But this implies that this map is even $C^{k+\alpha}$ with image space $C((0,a); C^1(\Omega \setminus \Sigma))^{n+1}$; this is a *transfer of regularity* induced by the definition of τ. Setting $t = t_0$ and $x = x_0$ this shows that the function

$$(\lambda, \xi) \mapsto (u, \theta)(t_0 + \lambda, x_0 + t_0\xi)$$

is of class $C^{k+\alpha}$ near (t_0, x_0). Repeating the same argument with $\nabla_x(u, \theta)$ we see in the same way that $\nabla_x(u, \theta) \in C^{k+\alpha}$ near (t_0, x_0).

If x_0 does belong to the tubular neighbourhood of Σ, we re-parameterize near t_0 in such a way that x_0 does not belong to the new tubular neighbourhood, and proceed as before. This yields the following result on interior regularity.

Theorem 9.4.1. *Let the assumptions of Theorems* 9.2.1 *and Theorem* 9.3.1 *be valid, for some* $k \in \mathbb{N} \cup \{\infty, \omega\}$.

Then there is $\alpha \in (0,1)$ *such that in all 6 problems we have*

$$(u, \theta), \partial_i(u, \theta) \in C^{k+\alpha}((0,a) \times (\Omega \setminus \Sigma))^{n+1},$$

and

$$\pi, \partial_i\pi \in C^{k-1+\alpha}((0,a) \times (\Omega \setminus \Sigma)),$$

where $i = 1, \ldots, n$. *In particular, in each problem we have classical solutions in the interior, even for* $k = 1$.

9.4.2 Regularity on the Interface

By means of the parameter trick it is also possible to prove regularity in time and tangential directions on the interface. However, here the construction of the map $\tau_{\lambda,\xi}$ is more involved, but also quite natural. We fix a point $(t_0, x_0) \in (0,a) \times \Sigma$ and choose a parameterization $\varphi : B_{\mathbb{R}^{n-1}}(0, 3R) \to \mathbb{R}^n$ for Σ near x_0; as Σ is real analytic we may choose φ real analytic. Here the chosen optimal smoothness of the reference Σ pays off! Next we extend φ by means of

$$\phi(p,q) = \varphi(p) + q\nu_\Sigma(\varphi(p)), \quad (p,q) \in B_{\mathbb{R}^{n-1}}(0, 3R) \times (-3a_\Sigma, 3a_\Sigma),$$

to a neighbourhood of $x_0 \in \Sigma$, with $3a_\Sigma$ the with of the tubular neighbourhood of Σ as chosen in Section 2.3. This map is again real analytic and it is a diffeomorphism onto its image if $R > 0$ is small enough. Observe that $\Pi_\Sigma\phi(p,q) = \varphi(p)$. Then we define the truncated shift

$$\tau_\xi(p,q) = (p + \xi\chi_0(p)\zeta_0(q), q), \quad (p,q) \in B_{\mathbb{R}^{n-1}}(0, 3R) \times (-3a_\Sigma, 3a_\Sigma).$$

Here, χ_0 is a smooth cut-off function on $\mathbb{R}^{n-1}$ which is one for $|p| \leq R$ and zero for $|p| > 2R$, while ζ_0 is a smooth cut-off function on $\mathbb{R}$ which is one for $|q| \leq 2a_\Sigma$

and 0 for $|q| \geq 5a_\Sigma/2$. Note that here $\xi \in B_{\mathbb{R}^{n-1}}(0,r)$ acts only tangentially. Then we set

$$\tau_{\lambda,\xi}(t,x) = (t + \lambda\chi_{t_0}(t), \phi(\tau_{t\xi}(\phi^{-1}(x)))) = (\tau^1_{\lambda,\xi}, \tau^2_{\lambda,\xi}), \quad (t,x) \in (0,a) \times U,$$

and

$$\tau_{\lambda,\xi}(t,x) = (t + \lambda\chi_{t_0}(t), x), \quad (t,x) \in (0,a) \times (\Omega \setminus U).$$

Here $U := \phi\big(B_{\mathbb{R}^{n-1}}(0,3R) \times (-3a_\Sigma, 3a_\Sigma)\big)$ is an open tubular neighbourhood of $x_0 \in \Sigma$. Observe that $\tau_{\lambda,\xi}$ commutes with Π_Σ on $U_{2a_\Sigma} = \phi\big(B_{\mathbb{R}^{n-1}}(0,3R) \times (-2a_\Sigma, 2a_\Sigma)\big)$, which implies

$$\begin{aligned} h \circ (id, \Pi_\Sigma) \circ \tau_{\lambda,\xi}(t,x) &= h(\tau^1_{\lambda,\xi}(t), \Pi_\Sigma \tau^2_{\lambda,\xi}(t,x)) \\ &= h(\tau_{\lambda,\xi}(t, \Pi_\Sigma x)) = h \circ \tau_{\lambda,\xi} \circ (id, \Pi_\Sigma)(t,x), \end{aligned}$$

for each $(t,x) \in (0,a) \times U_{2a_\Sigma}$. Recalling the definition of Ξ_h from Section 1.3, we then have

$$\big([\chi \circ (d_\Sigma/a_\Sigma)]\,[h \circ (id, \Pi_\Sigma)]\big) \circ \tau_{\lambda,\xi} = [\chi \circ (d_\Sigma/a_\Sigma)]\,[h \circ \tau_{\lambda,\xi} \circ (id, \Pi_\Sigma)]$$

on $(0,a) \times U$, as $d_\Sigma \circ \tau_{\lambda,\xi} = d_\Sigma$ on U, and $\chi \circ (d_\Sigma/a_\Sigma) = 0$ on $U \setminus U_{2a_\Sigma}$. If we choose $r > 0$ sufficiently small, then

$$\tau : (\lambda, \xi) \mapsto \tau_{\lambda,\xi}, \quad (-r,r) \times B_{\mathbb{R}^n}(0,r) \to \mathrm{Diff}^\infty((0,a) \times \Omega),$$

as in the simpler previous case of Section 9.4.1. Note that we do not shift in the vertical direction, as this would distort the interface Σ which needs to be kept fixed.

Then as before, we lift the coordinate transform $\tau_{\lambda,\xi}$ to an operator $T_{\lambda,\xi}$, which is a linear and bounded isomorphism in the spaces $\mathbb{E}_1(a)$ of solutions as well as in the space of date $\mathbb{F}_1(a)$. The function G is then defined as in the previous section, and we see that G is of class C^k, provided G has this property.

Hence again by the implicit function theorem, there is a ball $(-\delta,\delta) \times B_{\mathbb{R}^{n-1}}(0,\delta)$ and a map

$$\Phi : (-\delta,\delta) \times B_{\mathbb{R}^{n-1}}(0,\delta) \to \mathbb{E}_1^{z_0}(a),$$

of class C^k with $\Phi(0,0) = z_\odot$ such that $\mathsf{G}(\lambda,\xi,\Phi(\lambda,\xi)) = 0$. By uniqueness, we have again $\Phi(\lambda,\xi) = z_\odot \circ \tau_{\lambda,\xi}$.

Now we extract the height function h on the interface to obtain $(\lambda,\xi) \mapsto h \circ \tau_{(\lambda,\xi)} \in \mathbb{E}^j_{h,1}(a)$, where with $J = (0,a)$, and for some $\alpha \in (0,1)$ small,

$$\begin{aligned} \mathbb{E}^\theta_{h,1}(a) &= W_p^{3/2-1/2p}(J; L_p(\Sigma)) \cap W_p^{1-1/2p}(J; H^2_p(\Sigma)) \cap L_p(J; W_p^{4-1/p}(\Sigma)) \\ &\hookrightarrow C^\alpha((0,a); C^{3+\alpha}(\Sigma)) \cap C^{1+\alpha}((0,a); C^{1+\alpha}(\Sigma)), \end{aligned}$$

for Problems **(Pj)**, $j = 1, 3, 5$. where we need to assume $p > n+5$ for the embedding into $C^{1+\alpha}((0,a); C^{1+\alpha}(\Sigma))$. Moreover,

$$\mathbb{E}^u_{h,1}(a) = W^{2-1/2p}_p(J; L_p(\Sigma)) \cap H^1_p(J; W^{2-1/p}_p(\Sigma)) \cap L_p(J; W^{3-1/p}_p(\Sigma)$$
$$\hookrightarrow C^\alpha((0,a); C^{2+\alpha}(\Sigma)) \cap C^{1+\alpha}((0,a); C^{1+\alpha}(\Sigma)),$$

for Problems **(Pj)**, $j = 2, 4, 6$. Employing exchange of regularity as before, point evaluation implies

$$h \in C^{k+1+\alpha}((0,a) \times \Sigma), \nabla^i_\Sigma h \in C^{k+\alpha}((0,a) \times \Sigma)^{i \times n},$$

$i \le 3$ for **(P1), (P3), (P5)**, with $p > n+5$, $i \le 2$ for **(P2), (P4), (P6), (P5)** with $p > n+2$.

In the same way, we obtain regularity of the boundary pressures $q = [\![\pi]\!]$ in Problems **(P2), (P3), (P5)** and also of the on-sided pressures π_1, π_2 in Problems **(P4), (P6)**. In fact, $[\![\pi]\!], \pi_1, \pi_2 \in \mathbb{F}^u_{h,1}(a)$ yields

$$[\![\pi]\!], \pi_1, \pi_2 \in C^{k+\alpha}((0,a) \times \Sigma).$$

As in all problems we have for the surface temperature $\theta_\Sigma \in \mathbb{F}^\theta_{h,1}(a)$, this technique yields

$$(\theta_\Sigma, \nabla_\Sigma \theta_\Sigma) \in C^{k+\alpha}((0,a) \times \Sigma)^{n+1}.$$

Similarly, (the one-sided) traces u_i of u at the interface satisfy

$$(u_i, \nabla_\Sigma u_i) \in C^{k+\alpha}((0,a) \times \Sigma)^{n \times (n+1)}.$$

This shows that all equations in each Problem **(Pj)** are satisfied pointwise, i.e., the solutions obtained in Section 9.2 are all classical.

We summarize these results in

Theorem 9.4.2. *Let the assumptions of Theorems* 9.2.1 *and Theorem* 9.3.1 *be valid for some* $k \in \mathbb{N} \cup \{\infty, \omega\}$.

Then there is $\alpha \in (0,1)$ *such that in all 6 problems we have*

$$h \in C^{1+k+\alpha}((0,a) \times \Sigma), \quad \nabla^i_\Sigma h \in C^{k+\alpha}((0,a) \times \Sigma)^{i \times n}$$

$i \le 3$ *for* **(P1), (P3), (P5)** *with* $p > n+5$, *and* $i \le 2$ *for* **(P2), (P4), (P6)** *with* $p > n+2$. *Furthermore,*

$$[\![\pi]\!], \pi_1, \pi_2, \theta_\Sigma \in C^{k+\alpha}((0,a) \times \Sigma),$$

and

$$u, \nabla_\Sigma \theta \in C^{k+\alpha}((0,a) \times \Sigma)^n, \quad \nabla_\Sigma u \in C^{k+\alpha}((0,a) \times \Sigma)^{n \times n}.$$

In particular, in each problem the solutions are classical also on the interface, even for $k = 1$.

Observe that in case $k = \omega$, which means that all coefficient functions are real analytic, then h will be so jointly in time and space, hence the interfaces $\Gamma(t)$ become real analytic, instantaneously. This shows the strong regularizing effect which is inherent in quasilinear parabolic problems.

9.5 Estimates for the Nonlinearities

The basis of all considerations below are the following embeddings which are due to the restriction $1 \geq \mu > \frac{1}{2} + \frac{n+2}{2p}$.

$$\begin{aligned}
\mathbb{E}_{u,\mu}(a) \times \mathbb{E}_{\theta,\mu}(a) &\hookrightarrow C^{1/2}([0,a]; C_{ub}(\Omega \setminus \Sigma))^{n+1}, \\
\mathbb{E}_{u,\mu}(a) \times \mathbb{E}_{\theta,\mu}(a) &\hookrightarrow C([0,a]; C^1_{ub}(\Omega \setminus \Sigma))^{n+1}, \\
\mathbb{E}^{\theta}_{h,\mu}(a) &\hookrightarrow C^{1-}([0,a]; C(\Sigma)) \cap C([0,a]; C^3(\Sigma)), \\
\mathbb{E}^{u}_{h,\mu}(a) &\hookrightarrow C^{1-}([0,a]; C^1(\Sigma)) \cap C([0,a]; C^2(\Sigma))).
\end{aligned} \tag{9.14}$$

In general, the embedding constants will blow up as $a \to 0$, however, they do not depend on a, provided we restrict to time trace 0. This can be seen by the following simple extension argument. If a function v is defined on $[0,a]$, say for $a \leq 1$, and has time trace 0, we may extend it by

$$Ev(t) = \begin{cases} v(t), & 0 \leq t \leq a, \\ v(2a-t), & a \leq t \leq 2a, \\ 0, & 2a \leq t \leq 2. \end{cases}$$

Then $\sup_{0\leq t\leq a} |v(t)| \leq \sup_{0\leq t\leq 2} |Ev(t)|$ can be estimated by the relevant embedding for the fixed interval $[0,2]$. This simple observation is very important, and besides of the compatibilities this is another reason to reduce all problems to the case of vanishing time traces at $t = 0$.

5.1. The Nonlinearities in $\mathbb{F}_{\theta,\mu}$ and $\mathbb{F}_{u,\mu}$

(a) The nonlinearities F_θ and the components of F_u live in $L_{p,\mu}((0,a); L_p(\Omega))$. They consist of sums and products of $\nabla\theta$, u, ∇u, as well as of $d(\theta)$, $d'(\theta)$, $\mu(\theta)$, $\mu'(\theta)$, $1/\kappa(\theta)$, $M_1(h)$, $\nabla M_1(h)$, and $m_0(h)\partial_t h \circ \Pi_\Sigma$. As the functions μ and d are C^2 and $\kappa \in C^1$, the maps $\theta \mapsto \mu(\theta), \mu'(\theta), d(\theta), d'(\theta), \kappa(\theta)$ are of class C^1 from $C([0,a] \times \bar{\Omega})$ into itself, hence by the embeddings (9.14) it follows easily that F_θ and F_u are of class C^1, for all six problems under consideration.

Moreover, $F_\theta(z), F_u(z)$ belong to $L_\infty((0,a) \times \Omega)$, for each $z \in \mathbb{E}^j_\mu$, hence we obtain estimates of the form

$$|F_k(z)|_{\mathbb{F}_{k,\mu}(a)} \leq |F_k(z)|_\infty |\Omega| [\int_0^a t^{p(1-\mu)} dt]^{1/p} \leq C(|\bar{z}|_{\mathbb{E}_{k,\mu}} + R)^m a^{1-\mu+1/p},$$

with some constants $m \in \mathbb{N}$, $C > 0$, for all $\tilde{z} \in \bar{B}_{\mathbb{E}^j_\mu}(0,R)$, where $z = \tilde{z} + \bar{z}$. Therefore, these terms become small by choosing the time interval $J = (0,a)$ small. The same argument also applies to their Fréchet derivatives DF_k.

(b) On the other hand, there appear terms of highest order in the θ- and u-components of N_j; however these are only linear in the highest order derivative. For instance, we have the terms $F_1(\tilde{z},\bar{z}) = (\mathcal{A}_\theta(\theta,h) - \mathcal{A}_\theta(\theta_0,h_0) : \nabla^2\tilde{\theta}$ and $F_2(\tilde{z},\bar{z}) = \partial_t\bar{\theta} + \mathcal{A}_\theta(\theta,h) : \nabla^2\bar{\theta}$ in the θ-component of N_j, and similar terms in the u-components. For the analysis of such terms we first observe that bilinear mappings

$$b : L_\infty((0,a)\times\Omega)\times\mathbb{F}_{\theta,\mu}(J)\to\mathbb{F}_{\theta,\mu}(J), \quad (m,f)\mapsto mf,$$

are bounded, since $|b(m,f)|_{\mathbb{F}_{\theta,\mu}(J)}\leq|m|_\infty|f|_{\mathbb{F}_{\theta,\mu}(J)}$; hence this map is real analytic. Therefore, composite mappings like

$$(\bar{z},\tilde{z})\mapsto(\mathcal{A}_\theta(\theta,h),\nabla^2\bar{\theta})\mapsto\mathcal{A}_\theta(\theta,h):\nabla^2\bar{\theta}$$

are as smooth as the coefficients d, κ, in particular of class C^k if d,κ are C^k. The Fréchet derivatives are given by

$$D_1F_1(0,\bar{z})=(\mathcal{A}_\theta(\bar{\theta},\bar{h})-\mathcal{A}_\theta(\theta_0,h_0)):\nabla^2,$$

and

$$D_1F_2(0,\bar{z})\tilde{z}=[\partial_\theta\mathcal{A}_\theta(\bar{\theta},\bar{h})\tilde{\theta}+\partial_h\mathcal{A}_\theta(\bar{\theta},\bar{h})\tilde{h}]:\nabla^2\bar{\theta}.$$

Therefore, we obtain

$$|D_1F_1(0,\bar{z})\tilde{z}|_{\mathbb{F}_{\theta,\mu}(J)}\leq|(\mathcal{A}_\theta(\bar{\theta},\bar{h})-\mathcal{A}_\theta(\theta_0,h_0)|_\infty|\nabla^2\tilde{\theta}|_{\mathbb{F}^{n\times n}_{\theta,\mu}(J)}\leq\eta|z|_{\mathbb{E}^j_\mu(a)},$$

provided a is sufficiently small, depending only on the fixed function $\bar{z}$ which is continuous.

Similarly, we have

$$|D_1F_2(0,\bar{z})\tilde{z}|_{\mathbb{F}_{\theta,\mu}(J)}\leq C(|\tilde{\theta}|_\infty+|\tilde{h}|_\infty)|\nabla^2\bar{\theta}|_{\mathbb{F}^{n\times n}_{\theta,\mu}(J)},$$

where C dos not depend on $\tilde{z}$. By the embeddings (9.14) and trace 0 for $\tilde{z}$ we obtain further

$$|D_1F_2(0,\bar{z})\tilde{z}|_{\mathbb{F}_{\theta,\mu}(J)}\leq C|\tilde{z}|_{\mathbb{E}^j_\mu(a)}|\nabla^2\bar{\theta}|_{\mathbb{F}_{\theta,\mu}(J)}\leq\eta|\tilde{z}|_{\mathbb{E}^j_\mu(a)},$$

whenever a is chosen small enough, depending only on $\bar{z}$, but not on $\tilde{z}$.

This proves Condition **(NL)** for the θ-part of N_j, and similarly it also holds for the u-part of N_j.

5.2. The Nonlinearity in $\mathbb{F}^j_{\pi,\mu}$

The corresponding term appearing in N_j, $j=4,6$, reads

$$F(\tilde{z},\bar{z})=(M_1(h)-I)\nabla\cdot\bar{u}+(M_1(h)-M_1(h_0))\nabla\cdot\tilde{u}=F_1+F_2,$$

and for $j=2,3,5$ we apply the projection P_0 onto mean value zero. Note that F_i, $i=1,2$, are linear in the terms of highest order, namely ∇u. We consider first

(a) $L_{p,\mu}(J;H^1_p(\Omega\setminus\Sigma)$

The coefficients depend on h and $\nabla_\Sigma h$, hence belong to $C([0,a];C^1(\bar{\Omega}))$, and vanish outside a tubular neighbourhood of Σ. Therefore, we may use here the bilinear map

$$C([0,a];C^1(\bar{\Omega}))\times L_{p,\mu}(J;H^1_p(\Omega\setminus\Sigma))\to L_{p,\mu}(J;H^1_p(\Omega\setminus\Sigma)),\quad (m,u)\mapsto mu,$$

which is easily seen to be bounded. Therefore,

$$F: {}_0\mathbb{E}_\mu(a)\times\mathbb{E}_\mu(\infty)\to L_{p,\mu}((0,a);H^1_p(\Omega\setminus\Sigma))$$

belongs to the class C^k. Moreover, we have $F(0,\bar{z})=(M_1(\bar{h})-I)\nabla\cdot\bar{u}$, and

$$D_1F_1(0,\bar{z})\tilde{z}=M_1'(\bar{h})\tilde{h}\nabla\cdot\bar{u},\quad D_1F_2(0,\bar{z})\tilde{z}=(M_1(\bar{h})-M_1(h_0))\nabla\cdot\tilde{u}.$$

This implies

$$|F(0,\bar{z})|_{L_{p,\mu}(J;H^1_p)}\le|M_1(\bar{h})-I)|_{C(J;C^1_b)}|\nabla\bar{u}|_{L_{p,\mu}(J;H^1_p)}\to 0,$$

as $a\to 0$. Similarly

$$|D_1F_2(0,\bar{z})\tilde{z}|_{L_{p,\mu}(J;H^1_p)}\le|M_1(\bar{h})-M_1(h_0))|_{C(J;C^1_b)}|\tilde{u}|_{L_{p,\mu}(J;H^2_p)}\le\eta|\tilde{z}|_{\mathbb{E}_\mu(a)},$$

provided $a>0$ is small enough. Moreover, we also have

$$|D_1F_1(0,\bar{z})\tilde{z}|_{L_{p,\mu}(J;H^1_p)}=|M_1'(\bar{h})|_{C(J;C^1)}|\tilde{z}|_{C(J;C^2)}|\nabla\bar{u}|_{L_{p,\mu}(H^1_p)}\le\eta|\tilde{z}|_{\mathbb{E}_\mu(a)},$$

if $a>0$ is small enough, as $\bar{u}$ is a fixed function, and the embedding

$${}_0\mathbb{E}^u_{h,\mu}(J)\hookrightarrow C(J;C^2(\Sigma))$$

is uniform in a.

As P_0 is bounded linear, the same assertions hold for P_0F.

(b) $H^1_{p,\mu}(J;{}_0\dot{H}^{-1}_p(\Omega))$

This space is needed for Problems **(P2), (P3), (P5)**. Here we observe that for given $\phi\in\dot{H}^1_{p'}(\Omega)$ we have

$$\int_\Omega(P_0F_j)\phi\,dx=\int_\Omega P_0F_jP_0\phi\,dx=\int_\Omega F_jP_0\phi\,dx,$$

hence

$$\begin{aligned}\int_\Omega P_0F_1\phi\,dx&=\int_\Omega(M_1(h)-I)\nabla\cdot\bar{u}P_0\phi\,dx\\&=\int_\Omega\bar{u}\cdot[(I-M_1(h))\nabla\phi-(\operatorname{div}M_1(h)^{\mathsf{T}})P_0\phi]\,dx,\end{aligned}$$

and similarly

$$\begin{aligned}\int_\Omega P_0F_2\phi\,dx &= \int_\Omega (M_1(h)-M_1(h_0))\nabla\cdot\tilde{u}P_0\phi\,dx\\ &= \int_\Omega \tilde{u}\cdot[(M_1(h_0)-M_1(h))\nabla\phi + (\mathrm{div}(M_1(h_0)-M_1(h))^{\mathsf{T}})P_0\phi]\,dx.\end{aligned}$$

Now we may differentiate in time, apply Hölder's inequality and Poincaré's inequality to see as in 5.1 above that Condition **(NL)** holds for this nonlinearity.

(c) $H^1_{p,\mu}(J;H^{-1}_{p,\partial\Omega}(\Omega\setminus\Sigma))$. Here the same arguments as in **(b)** are valid, as in this case ϕ vanishes on Σ, and so the projection P_0 is not needed.

5.3 Analysis in Fractional Sobolev Spaces
Before we continue, note that $\mathbb{F}^u_{h,\mu}(a)$ as well as $\mathbb{F}^\theta_{h,\mu}(a)$ are Banach algebras, due to the restriction $1\geq\mu>\frac12+\frac{n+2}{2p}$. In fact, this follows easily from the embeddings

$$\begin{aligned}&\mathbb{F}^u_{h,\mu}(a)\hookrightarrow C([0,a];C(\Sigma)),\\ &\mathbb{F}^\theta_{h,\mu}(a)\hookrightarrow C([0,a];C^1(\Sigma)).\end{aligned}\tag{9.15}$$

As above, the embedding constants do not depend on a, provided we restrict to functions with time-trace 0 at $t=0$. Recall that a norm for $W^s_p(\Sigma)$, $s\in(0,1)$, is given by

$$|v|_{W^s_p(\Sigma)} = |v|_{L_p} + [\int_\Sigma\int_\Sigma \frac{|v(x)-v(y)|^p}{|x-y|^{sp+n-1}}d\Sigma(x)d\Sigma(y)]^{1/p}.$$

There are several well-known fundamental estimates in fractional Sobolev spaces, which we want to recall here.

(i) The first one, which we already used before, concerns products and reads as

$$|mw|_{W^s_p}\leq |m|_\infty|w|_{W^s_p}+|w|_\infty|m|_{W^s_p},$$

valid for all functions $m,w\in W^s_p\cap L_\infty$, $s\in(0,1)$. In case $W^s_p(\Sigma)\hookrightarrow C(\Sigma)$ and $m\in C^1(\Sigma)$ it simplifies to

$$|mw|_{W^s_p}\leq C|m|_{C^1}|w|_{W^s_p}.$$

This estimate can easily be extended to the space $\mathbb{F}^u_{h,\mu}(J)$ with $1\geq\mu>1/2+(n+2)/2p$. If

$$m\in\mathbb{G}_\theta(J):=C^{1/2}(J;C(\Sigma))\cap C(J;C^1(\Sigma)),$$

$w\in\mathbb{F}^u_{h,\mu}(J)$, we have

$$|mw|_{\mathbb{F}^u_{h,\mu}(J)}\leq C|m|_{\mathbb{G}_\theta(J)}|w|_{\mathbb{F}^u_{h,\mu}(J)}.$$

However, we emphasize that the constant C will depend on the length of the interval a, unless w has trace 0 at $t=0$.

(ii) In the sequel, we will need the following little trick. Let $m \in \mathbb{G}_\theta(J)$, $v \in {}_0\mathbb{F}^\theta_{h,\mu}(J)$, $w \in \mathbb{F}^u_{h,\mu}(\mathbb{R}_+)$ and suppose the trace of w vanishes at time $t=0$. Then with $s = 1-1/p$

$$|mvw|_{\mathbb{F}^u_{h,\mu}(J)} \leq C|m|_{\mathbb{G}_\theta(J)}|vw|_{\mathbb{F}^u_{h,\mu}(J)} \leq C|m|_{\mathbb{G}_\theta(J)}|v|_{W^{s/2}_{p,\mu}(J;W^s_p(\Sigma))}|w|_{\mathbb{F}^u_{h,\mu}(\mathbb{R}_+)},$$

with a constant C independent of a. On the other hand,

$${}_0\mathbb{F}^\theta_{h,\mu}(J) \hookrightarrow {}_0W^{1/2}_{p,\mu}(J;W^s_p(\Sigma)) \hookrightarrow {}_0W^{s/2}_{p,\mu}(J;W^s_p(\Sigma))$$

with uniform embedding constant, and with

$$|v|_{W^{s/2}_{p,\mu}(J;W^s_p(\Sigma))} \leq ca^{1/2p}|v|_{W^{1/2}_{p,\mu}(J;W^s_p(\Sigma))}$$

this yields

$$|mvw|_{\mathbb{F}^u_{h,\mu}(J)} \leq a^{1/2p}C|m|_{\mathbb{G}_\theta(J)}|v|_{\mathbb{F}^\theta_{h,\mu}(J)}|w|_{\mathbb{F}^u_{h,\mu}(\mathbb{R}_+)}.$$

(iii) In a similar, but more elaborate way we also obtain the estimate

$$|bw|_{\mathbb{F}^\theta_{h,\mu}(J)} \leq C|b|_{\mathbb{G}_h(J)}|w|_{\mathbb{F}^\theta_{h,\mu}(\mathbb{R}_+)},$$

with a constant independent of a, provided

$$b \in \mathbb{G}_h(h) := W^s_{p,\mu}((0,a);C(\Sigma)) \cap C([0,a];W^{2s}_p(\Sigma)), \quad s = 1-1/2p,$$

has vanishing time trace and $w \in \mathbb{F}^\theta_{h,\mu}(\mathbb{R}_+)$. Of course, ${}_0\mathbb{F}^\theta_{h,\mu}(J)$ is also a Banach algebra, as ${}_0\mathbb{F}^\theta_{h,\mu}(J) \hookrightarrow C([0,a];C^1(\Sigma))$.

Next we consider substitution operators in W^s_p of the form $\phi(v)$ with $\phi \in C^2$.

(iv) Based on the identity

$$\begin{aligned}
&[\phi(v(x)) - \phi(w(x))] - [\phi(v(y)) - \phi(w(y))]\\
&= \int_0^1\int_0^1 \frac{d}{dt}\frac{d}{ds}\phi(s[tv(x)+(1-t)w(x)] + (1-s)[tv(y)+(1-t)w(y)])\,dsdt\\
&= \int_0^1\int_0^1 \phi'(\xi(t,s))([v(x)-w(x)] - [v(y)-w(y)])\,dsdt\\
&+ \int_0^1\int_0^1 \phi''(\xi(t,s))([tv(x)+(1-t)w(x)] - [tv(y)+(1-t)w(y)])\cdot\\
&\cdot(s[v(x)-w(x)] + (1-s)[v(y)-w(y)])\,dtds
\end{aligned}$$

we obtain

$$\begin{aligned}
&|[\phi(v(x)) - \phi(w(x))] - [\phi(v(y) - \phi(w(y)]| \leq |\phi'|_\infty|(v(x)-w(x)) - (v(y)-w(y))|\\
&+ |\phi''|_\infty\{|(v(x)-w(x)) - (v(y)-w(y))| + |w(x)-w(y)|\}|v-w|_\infty
\end{aligned}$$

This implies

$$|\phi(v)-\phi(w)|_{W^s_p} \le |\phi|_{C^2_b}\big[|v-w|_{W^s_p}(1+|v-w|_\infty)+|v-w|_\infty|w|_{W^s_p}\big].$$

This estimate implies that the substitution operator $v \mapsto \phi(v)$ is locally Lipschitz in $W^s_p \cap L_\infty$.

(v) We have

$$l(r,h) := \phi(r+h)-\phi(r)-\phi'(r)h = \int_0^1 (\phi'(r+sh)-\phi'(r))\,dsh,$$

hence with $\delta h = h-\bar h$, $\delta r = r-\bar r$, $\delta l = l(r,h)-l(\bar r,\bar h)$

$$\begin{aligned}\delta l &= \int_0^1 \frac{d}{dt}\Big(\int_0^1 [\phi'(t(r+sh)+(1-t)(\bar r+s\bar h))\\ &\qquad\qquad - \phi'(tr+(1-t)\bar r]ds(th+(1-t)\bar h)\Big)\,dt\\ &= \int_0^1\int_0^1 [\phi'(t(r+sh)+(1-t)(\bar r+s\bar h))-\phi'(tr+(1-t)\bar r)]\,dsdt\,\delta h\\ &\quad+\int_0^1\int_0^1 [[\phi''(t(r+sh)+(1-t)(\bar r+s\bar h))\\ &\qquad\qquad -\phi''(tr+(1-t)\bar r)]\delta r(\bar h+t\delta h)\,dsdt\\ &\quad+\int_0^1\int_0^1 \phi''(t(r+sh)+(1-t)(\bar r+s\bar h))s\delta h(\bar h+t\delta h)\,dsdt.\end{aligned}$$

This implies by continuity of ϕ' and ϕ''

$$|\delta l| \le \varepsilon|\delta h| + \varepsilon|\delta r|\max\{|h|,|\bar h|\} + |\phi''|_\infty|\delta h|\max\{|\bar h|,|h|\},$$

provided $|h|, |\bar h|$ are small enough. Setting $r=w(x)$, $\bar r = w(y)$, $h=h(x)$, $\bar h = h(y)$, we obtain

$$\begin{aligned}&|[\phi(w(x)+h(x))-\phi(w(x))-\phi'(w(x))h(x)]\\ &\qquad\qquad -[\phi(w(y)+h(y))-\phi(w(y))-\phi'(w(y))h(y)]|\\ &\le \varepsilon|h(x)-h(y)|+\varepsilon|w(x)-w(y)||h|_\infty+|\phi''|_\infty|h|_\infty|h(x)-h(y)|.\end{aligned}$$

From this estimate the Fréchet-differentiability of the substitution operator $\Phi : v \mapsto \phi(v)$ in $W^s_p \cap L_\infty$ follows, as soon as $\phi \in C^2$. The derivative is given by

$$(\Phi'(w)h)(x) = \phi'(w(x)h(x), \quad x\in\Sigma, \quad w,h \in W^s_p\cap L_\infty,$$

and so Φ is of class C^1. By induction we easily get $\Phi \in C^k$ if $\phi \in C^{k+1}$, for all $k \in \mathbb{N}\cup\{\infty\}$, and also $\Phi \in C^\omega$ in case $\phi \in C^\omega$, estimating the remainders in the Taylor expansions.

(vi) Let again $s \in (0,1)$, and consider a substitution operator $\Phi : v \mapsto \phi(v)$ in $W_p^{1+s}(\Sigma) \cap W_\infty^1(\Sigma)$. Here the main estimate concerns the derivative of $\phi(v)$, i.e., $\phi'(v)v'$. This case is simpler, as v has more regularity and so $\phi'(v)$ has so as well. By the results of the previous paragraphs it implies that $\Phi \in C^k$, provided $\phi \in C^{k+2}$, for all $k \in \mathbb{N} \cup \{\infty, \omega\}$.

5.4. The Nonlinearities in $\mathbb{F}_{\theta_\Sigma,\mu}$
Here we may argue for the lower order nonlinearities F_{θ_Σ} as in the previous subsection in $L_p(\Sigma)$ and then use the embedding $L_p(\Sigma) \hookrightarrow W_p^{-1/p}(\Sigma)$.

For the highest order terms recall the definition of the norm in $W_p^{-s}(\Sigma)$.

$$|v|_{W_p^{-s}(\Sigma)} = \sup\{\int_\Sigma v\varphi\, d\Sigma : \varphi \in W_{p'}^s(\Sigma),\ |\varphi|_{W_{p'}^s(\Sigma)} \le 1\}.$$

This implies the estimate

$$|(mv|\varphi)| = |(v|m\varphi)| \le |v|_{W_p^{-s}(\Sigma)} |m\varphi|_{W_{p'}^s(\Sigma)} \le C|m|_{C^1(\Sigma)} |v|_{W_p^{-s}(\Sigma)} |\varphi|_{W_{p'}^s(\Sigma)},$$

which yields

$$|mv|_{W_p^{-s}(\Sigma)} \le C|m|_{C^1(\Sigma)}|v|_{W_p^{-s}(\Sigma)}, \quad |mv|_{\mathbb{F}_{\theta_\Sigma,\mu}} \le C|m|_{C(J;C^1(\Sigma))}|v|_{\mathbb{F}_{\theta_\Sigma,\mu}}.$$

The highest order terms are

$$F_1(\tilde{z},\bar{z}) = (\mathcal{A}_{\theta_\Sigma}(\theta_\Sigma,h) - \mathcal{A}_{\theta_\Sigma}(\theta_{\Sigma 0},h_0)) : \nabla_\Sigma^2 \tilde{\theta}_\Sigma$$

and $F_2(\tilde{z},\bar{z}) = \partial_t \bar{\theta}_\Sigma + \mathcal{A}_{\theta_\Sigma}(\theta_\Sigma,h) : \nabla_\Sigma^2 \bar{\theta}_\Sigma$. As in the previous subsection these are linear in the highest derivative, fortunately.

Here the bilinear map $(m,g) \mapsto mg$ is bounded from $C([0,a];C^1(\Sigma)) \times \mathbb{F}_{\theta_\Sigma,\mu}$ to $\mathbb{F}_{\theta_\Sigma,\mu}$, hence it is real analytic, and so the composition maps

$$(\tilde{z},\bar{z}) \mapsto (\mathcal{A}_{\theta_\Sigma}(\theta_\Sigma,h), \nabla_\Sigma^2\tilde{\theta}_\Sigma, \nabla_\Sigma^2\bar{\theta}_\Sigma) \mapsto F_j(\tilde{z},\bar{z})$$

are of class C^k, provided the coefficient functions d_Σ, κ_Σ are of class C^{k+1}. Then we may estimate similarly as in Section 9.5.1

$$\begin{aligned}|D_1F_1(0,\bar{z})\tilde{z}|_{\mathbb{F}_{\theta_\Sigma,\mu}(J)} &\le |(\mathcal{A}_{\theta_\Sigma}(\bar{\theta}_\Sigma,\bar{h}) - \mathcal{A}_{\theta_\Sigma}(\theta_{\Sigma 0},h_0)|_{C([0,a];C^1(\Sigma))} \\ &\quad \cdot |\nabla_\Sigma^2\tilde{\theta}_\Sigma|_{\mathbb{F}_{\theta_\Sigma,\mu}(J)} \le \eta|z|_{\mathbb{E}_\mu^j(a)},\end{aligned}$$

and

$$|D_1F_2(0,\bar{z})\tilde{z}|_{\mathbb{F}_{\theta_\Sigma,\mu}(J)} \le C|\tilde{z}|_{\mathbb{E}_\mu^j(a)}|\nabla_\Sigma^2\bar{\theta}_\Sigma|_{\mathbb{F}_{\theta_\Sigma,\mu}(J)} \le \eta|\tilde{z}|_{\mathbb{E}_\mu^j(a)},$$

provided a is sufficiently small, depending only on the fixed function $\bar{z}$. This shows Condition **(NL)** for the θ_Σ-components of N_5 and N_6.

5.5. The Nonlinearities in $\mathbb{F}_{h,\mu}^u$
There are only few lower order terms appearing in this boundary space. These

are $u \cdot \nu_\Gamma$ in the h-component of N_3, N_5, $[\![\theta\eta(\theta)]\!] j_\Sigma$ in N_4, and $[\![\psi(\theta)]\!]$, $[\![1/\varrho]\!] j_\Sigma^2 \nu_\Gamma$, $[\![1/2\varrho^2]\!] j_\Sigma^2$ in N_4, N_6. These terms can be handled in the same way as the lower order terms in Sections 9.5.1 and 9.5.4. We now study the highest order terms in the same way as above.

(a) $[\![\mathcal{B}_\theta(\theta,h)\nabla\theta]\!]$
We set $F_1(\tilde{z},\bar{z}) = [\![(\mathcal{B}_\theta(\theta,h) - \mathcal{B}_\theta(\theta_0,h_0))\nabla\tilde{\theta}]\!]$ and $F_2(\tilde{z},\bar{z}) = \partial_t\bar{\theta} + [\![\mathcal{B}_\theta(\theta,h)\nabla\bar{\theta}]\!]$. Since $\theta \in \mathbb{G}_\theta(J) = C^{1/2}([0,a];C(\Sigma)) \cap C([0,a];C^1(\Sigma))$ we may employ here the bilinear map $(m,g) \mapsto mg$ from $\mathbb{G}_\theta(J) \times \mathbb{F}^u_{h,\mu}(J)$ to $\mathbb{F}^u_{h,\mu}(J)$ which is bounded, to see as before that F_k are of class C^k provided d, l are of class C^{k+1}. For their Fréchet derivatives, by Section 9.5.3**(i)**,**(ii)**, we have the estimates

$$|D_1F_1(0,\bar{z})\tilde{z}|_{\mathbb{F}^u_{h,\mu}(J)} \le C|\mathcal{B}_\theta(\bar{\theta},\bar{h}) - \mathcal{B}_\theta(\theta_0,h_0))|_{\mathbb{G}_\theta(J)}|\nabla\tilde{\theta}|_{\mathbb{F}^u_{h,\mu}(J)} \le \eta|\tilde{z}|_{\mathbb{E}_\mu(a)},$$

and

$$\begin{aligned}|D_2F_2(0,\bar{z})\tilde{z}|_{\mathbb{F}^u_{h,\mu}(J)} \le a^{1/2p}C\{&|\partial_\theta\mathcal{B}_\theta(\bar{\theta},\bar{h})|_{\mathbb{G}_\theta(J)}|\tilde{\theta}|_{\mathbb{E}_{\theta,\mu}(J)}\\ &+ |\partial_h\mathcal{B}_\theta(\bar{\theta},\bar{h})|_{\mathbb{G}_\theta(J)}|\tilde{h}|_{\mathbb{E}^k_{h,\mu}(J)}\}|\nabla\bar{\theta}|_{\mathbb{F}^u_{h,\mu}(\mathbb{R}_+)} \le \eta|\tilde{z}|_{\mathbb{E}_\mu(a)},\end{aligned}$$

provided a is chosen small enough, independently of $\tilde{z}$, as ${}_0\mathbb{E}_{\theta,\mu}(J)$ embeds into $\mathbb{G}_\theta(J)$ with uniform embedding constant. This shows Condition **(NL)** for this nonlinearity.

(b) $\sigma'(\theta_\Sigma)\nabla_\Sigma\theta_\Sigma$
This term can be handled in the same way. We employ the technique from **(a)** to the functions

$$F_1(\tilde{z},\bar{z}) = (\sigma'(\theta_\Sigma) - \sigma'(\theta_{\Sigma 0})\nabla_\Sigma\tilde{\theta}_\Sigma, \quad F_2(\tilde{z},\bar{z}) = \sigma'(\theta_\Sigma)\nabla_\Sigma\bar{\theta}_\Sigma.$$

As a result we obtain that this term is of class C^k, provided $\sigma \in C^{k+2}$, and so Condition **(NL)** is valid.

(c) $S(u,\theta,h)\nu_\Gamma(h)$
We rewrite this term as $\mathcal{B}_u(\theta,h)\nabla u$, where $\mathcal{B}_u$ is a tensor of degree 3 which depends only on $\theta, h, \nabla_\Sigma h$, hence is of lower order. Here we define

$$F_1(\tilde{z},\bar{z}) = (\mathcal{B}_u(\theta,h) - \mathcal{B}_u(\theta_0,h_0))\nabla\tilde{u}, \quad F_2(\tilde{z},\bar{z}) = \mathcal{B}_u(\theta,h)\nabla\bar{u}.$$

Then we have the same structure as in **(a)** and so the same argument as there proves **(NL)** for the jump of the normal stress. A similar argument can be employed for $[\![S(u,\theta,h)\nu_\Gamma(h)\cdot\nu_\Gamma(h)/\varrho]\!]$.

(d) $H_\Gamma(h)$
According to Section 2.2.5, the curvature reads as

$$H_\Gamma(h) = \mathcal{C}_0(h) : \nabla^2_\Sigma h + \mathcal{C}_1(h),$$

where $\mathcal{C}_j(h)$ depend only on h and $\nabla_\Sigma h$, and hence are of lower order. Therefore, $H_\Gamma(h)$ fortunately has a quasilinear structure. Note that

$$\mathcal{C}_\Sigma(h) = -\mathcal{C}_0(h) : \nabla_\Sigma^2. \tag{9.16}$$

In the following we concentrate on the first term $\mathcal{C}_0(h) : \nabla_\Sigma^2 h$. Here

$$\mathcal{C}_0(h) = \beta(h)(M_0^2(h) - \beta^2(h)M_0^2(h)\nabla_\Sigma h \otimes M_0^2(h)\nabla_\Sigma h)$$

is real analytic in h and $\nabla_\Sigma h$. The highest order contribution of the term

$$H_\Gamma(h) - H_\Gamma(\bar{h}) - H'(h_0)\tilde{h}$$

to N_j in the normal stress condition on Σ is given by $F(\tilde{h}, \bar{h}) = F_1(\tilde{h}, \bar{h}) + F_2(\tilde{h}, \bar{h})$, where

$$F_1(\tilde{h}, \bar{h}) = (\mathcal{C}_0(h) - \mathcal{C}_0(h_0)) : \nabla_\Sigma^2 \tilde{h}, \quad F_2(\tilde{h}, \bar{h}) = (\mathcal{C}_0(h) - \mathcal{C}_0(\bar{h})) : \nabla_\Sigma^2 \bar{h},$$

and so $F_i(0, \bar{h}) = 0$, and

$$D_1F_1(0, \bar{h})\tilde{h} = (\mathcal{C}_0(\bar{h}) - \mathcal{C}_0(h_0)) : \nabla_\Sigma^2 \tilde{h}, \quad D_1F_2(0, \bar{h})\tilde{h} = \mathcal{C}_0(\bar{h})\tilde{h} : \nabla_\Sigma^2 \bar{h}.$$

As in any of the 6 problems,

$$\nabla_\Sigma h \in W_{p,\mu}^{1-1/2p}(J; W_p^{1-1/p}(\Sigma)) \cap L_{p,\mu}(J; W_p^{2-1/p}(\Sigma) \hookrightarrow \mathbb{F}_{h,\mu}^u(J),$$

and we may estimate as in **(a)** to see that

$$|D_1F(0, \bar{h})\tilde{h}|_{\mathbb{F}_{h,\mu}^u(a)} \le \eta |\tilde{z}|_{\mathbb{E}_\mu(a)},$$

if a is small, hence Condition **(NL)** holds also for this nonlinearity.

5.6. The Nonlinearities in $\mathbb{F}_{h,\mu}^\theta$
(a) First we focus on the term $u \cdot \nu_\Gamma/\beta$ from the equation for h in Problem **(P2)**. The terms $[\![\varrho u \cdot \nu_\Gamma/\beta]\!]$ and $\mathcal{P}_\Gamma[\![u]\!] = [\![u]\!] - [\![u \cdot \nu_\Gamma]\!]\nu_\Gamma$ appearing in **(P4)** and **(P6)** can be estimated in the same way.

The corresponding term in N_2 looks like $F = F_1 + F_2$, with

$$F_1(\tilde{z}, \bar{z}) = \bar{u} \cdot (M_0(h_0) - M_0(h))\nabla_\Sigma \tilde{h} + \tilde{u} \cdot (M_0(h_0) - M_0(h))\nabla_\Sigma \bar{h} - \tilde{u} \cdot M_0(h)\nabla_\Sigma \tilde{h},$$

and

$$F_2(\tilde{z}, \bar{z}) = \bar{u} \cdot (\nu_\Sigma - M_0(h_0)\nabla_\Sigma \bar{h}).$$

Since $\mathbb{F}_{h,\mu}^\theta(a)$ is a multiplication algebra and M_0 is real analytic, it follows easily that F is also real analytic. To verify **(NL) (ii)** for F_1, it is sufficient to show that triple products of the form bvw become small if a is small, where $b \in \mathbb{G}_h(J)$ and $w \in \mathbb{F}_{h,\mu}^\theta(J)$ have zero trace, and $v \in \mathbb{F}_{h,\mu}^\theta(\mathbb{R}_+)$. Here $b = M_0(h_0) - M_0(h)$, and

$v = \bar{u}$, $w = \nabla_\Sigma \tilde{h}$, or bar and tilde in the latter ones interchanged. To do so we first use the Banach algebra property to obtain

$$|bvw|_{\mathbb{F}^\theta_{h,\mu}(J)} \le C|bv|_{\mathbb{F}^\theta_{h,\mu}(J)}|w|_{\mathbb{F}^\theta_{h,\mu}(J)},$$

with a constant C independent of a, as bv and w have both trace zero. Then we apply Section 9.5.3**(iii)** to obtain

$$|bv|_{\mathbb{F}^\theta_{h,\mu}(J)} \le C|b|_{\mathbb{G}_h(J)}|v|_{\mathbb{F}^\theta_{h,\mu}(\mathbb{R}_+)}.$$

As $|b|_{\mathbb{G}_h(J)} \to 0$ as $a \to$, the claim follows for F_1.

Further, we have

$$D_1F_2(0,\bar{z})\tilde{z} = -\bar{u}\cdot M_0'(\bar{h})\tilde{h}\nabla_\Sigma \bar{h},$$

hence we obtain by 5.3(i),(iii), as ${}_0\mathbb{E}^u_{h,\mu}(J) \hookrightarrow \mathbb{G}_h(J)$,

$$\begin{aligned}
|\tilde{h}M_0'(\bar{h})\nabla_\Sigma \bar{h}\cdot\bar{u}|_{\mathbb{F}^\theta_{h,\mu}(J)} &\le C|\tilde{h}M_0'(\bar{h})\nabla_\Sigma\bar{h}|_{\mathbb{F}^\theta_{h,\mu}(J)}|\bar{u}|_{\mathbb{E}^2_\mu(\mathbb{R}_+)}\\
&\le C|\tilde{h}|_{\mathbb{G}_h(J)}|M_0'(\bar{h})\nabla_\Sigma\bar{h}|_{\mathbb{F}^\theta_{h,\mu}(\mathbb{R}_+)}|\bar{u}|_{\mathbb{E}^2_\mu(\mathbb{R}_+)}\\
&\le C|\tilde{h}|_{\mathbb{G}_h(J)}|\bar{h}|_{\mathbb{E}^2_\mu(\mathbb{R}_+)}|\bar{u}|_{\mathbb{E}^2_\mu(\mathbb{R}_+)}.
\end{aligned}$$

In the last step we used fact that $M_0'(\bar{h})$ is a multiplier for $\mathbb{F}_{h,\mu}(\mathbb{R}_+)$. Finally, there is some $\alpha > 0$ such that

$${}_0\mathbb{E}^u_{h,\mu}(J) \hookrightarrow C^{1+\alpha}([0,a];C(\Sigma)) \cap C^\alpha([0,a];C^2(\Sigma)) =: \mathbb{G}^\alpha_h(J),$$

therefore

$$|\tilde{h}|_{\mathbb{G}_h(J)} \le a^\alpha|\tilde{h}|_{\mathbb{G}^\alpha_h(J)} \le Ca^\alpha|\tilde{h}|_{\mathbb{E}^2_{h,\mu}}.$$

This shows that F_2 is also subject to **(NL) (ii)**.

(b) $\varphi(\theta)$

We consider the term $\varphi(\theta)$ appearing in the Gibbs-Thomson condition in Problems **(P1)** and **(P3)**. The corresponding term in N_j, $j = 1,3$, is given by

$$F(\tilde{z},\bar{z}) = r_\theta(\tilde{\theta},\bar{\theta}) = \varphi(\theta) - \varphi(\bar{\theta}) - \varphi'(\bar{\theta})\tilde{\theta}.$$

From Section 9.5.3**(v)**,**(vi)** we see that F is of class C^k provided φ belongs to C^{k+2}, i.e., if $\psi \in C^{k+2}$. Further we obtain $D_1F(0,\bar{z})\tilde{z} = 0$, hence **(NL) (ii)** is satisfied trivially.

(c) $H_\Gamma(h)$

Employing the same decomposition of the relevant nonlinearity F as in Section 9.5.5**(d)**, we may argue as in **(a)** above to obtain

$$|F(0,\bar{z})|_{\mathbb{F}^\theta_{h,\mu}(J)} + |D_1F(0,\bar{z})|_{\mathcal{B}({}_0\mathbb{E}_\mu(a);\mathbb{F}_\mu(a)} \to 0,$$

as $a \to 0$, as the function $\bar{z}$ is fixed.

Chapter 10

Linear Stability of Equilibria

In this chapter we investigate the spectral properties of the linearizations L_j of the six problems at a given equilibrium. We show that the dimension of the kernel $\mathsf{N}(L_j)$ equals the dimension of the tangent space of the manifold of equilibria $\mathcal{E}$, the eigenvalue 0 is semi-simple for L_j, and the intersection of the spectrum of L_j with the imaginary axis is $\{0\}$. This shows that the equilibria are normally hyperbolic. In the case of no phase transitions, i.e., Problem 2, or in case the phases are connected and for $\varrho_1 = \varrho_2$ the stability condition (1.32) holds, we show that the spectrum of $-L_j$ does not intersect the open right half-plane $\mathbb{C}_+$, hence the equilibrium is even normally stable. These results are the basis for the application of the *generalized principle of linearized stability* which will be carried out in the next chapter.

10.1 Linearization at Equilibria

The full linearization at an equilibrium $e_* = (0, \theta_*, \Sigma)$ in the general case reads

$$\begin{aligned}
\varrho\partial_t u - \mu_*\Delta u + \nabla\pi &= \varrho f_u && \text{in } \Omega\setminus\Sigma,\\
\operatorname{div} u &= g_d && \text{in } \Omega\setminus\Sigma,\\
u &= 0 && \text{on } \partial\Omega,\\
u - (u_\Sigma + j_\Sigma\nu_\Sigma/\varrho) &= g_u/\varrho && \text{on } \Sigma,\\
-[\![T\nu_\Sigma]\!] + \sigma_*\mathcal{A}_\Sigma h\nu_\Sigma &= (\sigma_*'\nabla_\Sigma + \sigma_*' H_\Sigma\nu_\Sigma)\vartheta_\Sigma + g_\Sigma && \text{on } \Sigma,\\
u(0) &= u_0 && \text{in } \Omega,
\end{aligned} \tag{10.1}$$

where $T = \mu_*(\nabla u + [\nabla u]^{\mathsf{T}}) - \pi I$, $\mu_* = \mu(\theta_*)$, $\sigma_* = \sigma(\theta_*)$, and $\sigma_*' = \sigma'(\theta_*)$. Here, j_Σ and u_Σ are the pull-backs of j_Γ and u_Γ, respectively, and the operator $\mathcal{A}_\Sigma$ is the negative derivative of the curvature, defined by

$$\mathcal{A}_\Sigma = -H_\Sigma'(0) = -\frac{n-1}{R_*^2} - \Delta_\Sigma.$$

J. Prüss and G. Simonett, *Moving Interfaces and Quasilinear Parabolic Evolution Equations*, Monographs in Mathematics 105,
DOI 10.1007/978-3-319-27698-4_10

The relative temperature $\vartheta = \theta - \theta_*$ is subject to

$$\begin{aligned} \varrho\kappa_*\partial_t\vartheta - d_*\Delta\vartheta &= \varrho\kappa_* f_\theta && \text{in } \Omega\setminus\Sigma,\\ \partial_\nu\vartheta &= 0 && \text{on } \partial\Omega,\\ [\![\vartheta]\!] = 0, \quad \vartheta &= \vartheta_\Sigma && \text{on } \Sigma,\\ \vartheta(0) &= \vartheta_0 && \text{in } \Omega, \end{aligned} \tag{10.2}$$

with $\kappa_* = \kappa(\theta_*)$ and $d_* = d(\theta_*)$. On the interface we have the following equations

$$\begin{aligned} \kappa_{\Sigma*}\partial_t\vartheta_\Sigma - d_{\Sigma*}\Delta_\Sigma\vartheta_\Sigma - [\![d_*\partial_\nu\vartheta]\!] - \theta_*\sigma_*'\mathrm{div}_\Sigma u_\Sigma - l_* j_\Sigma &= f_\Sigma && \text{on } \Sigma,\\ (l_*/\theta_*)\vartheta_\Sigma - [\![T\nu_\Sigma\cdot\nu_\Sigma/\varrho]\!] &= g_\theta && \text{on } \Sigma,\\ \partial_t h - u\cdot\nu_\Sigma + j_\Sigma/\varrho &= f_h && \text{on } \Sigma,\\ \vartheta_\Sigma(0) = \vartheta_{\Sigma 0}, \quad h(0) &= h_0 && \text{on } \Sigma, \end{aligned} \tag{10.3}$$

setting $l_* = l(\theta_*) = \theta_*[\![\psi'(\theta_*)]\!]$, $\kappa_{\Sigma*} = \kappa_\Gamma(\theta_*)$ and $d_{\Sigma*} = d_\Gamma(\theta_*)$. For future use we note here that combining the dynamic equation for h in (10.3) with the jump condition for u in (10.1) yields $\partial_t h - f_h - u_\Sigma\cdot\nu_\Sigma = \varrho^{-1}g_u\cdot\nu_\Sigma$, which implies

$$g_u\cdot\nu_\Sigma = 0 \quad \text{in case} \quad \varrho_1 \neq \varrho_2. \tag{10.4}$$

We introduce now the functional analytic frameworks for the spectral analysis of the problems under consideration.

Problem 1. This results in the linear problem

$$\left\{\begin{aligned} \varrho\kappa_*\partial_t\vartheta - d_*\Delta\vartheta &= \varrho\kappa_* f_\theta && \text{in } \Omega\setminus\Sigma,\\ \partial_\nu\vartheta &= 0 && \text{on } \partial\Omega,\\ [\![\vartheta]\!] &= 0 && \text{on } \Sigma,\\ (\varrho l_*/\theta_*)\vartheta - \sigma_*\mathcal{A}_\Sigma h &= g_\theta && \text{on } \Sigma,\\ \varrho l_*\partial_t h - [\![d_*\partial_\nu\vartheta]\!] &= \varrho l_* f_h && \text{in } \Sigma,\\ \vartheta(0) = \vartheta_0, \ h(0) &= h_0. \end{aligned}\right. \tag{10.5}$$

We define the operator L_1 in $X_0^1 := L_p(\Omega)\times W_p^{2-2/p}(\Sigma)$ by

$$\begin{aligned} X_1^1 = \mathsf{D}(L_1) = \big\{&(\vartheta,h)\in[H_p^2(\Omega\setminus\Sigma)\cap C(\bar\Omega)]\times W_p^{4-1/p}(\Sigma):\\ &\partial_\nu\vartheta = 0,\ (\varrho l_*/\theta_*)\vartheta + \sigma_*\mathcal{A}_\Sigma h = 0,\ [\![d_*\partial_\nu\vartheta]\!]\in W_p^{2-2/p}(\Sigma)\big\},\\ L_1(\vartheta,h) = \big(&(-d_*/\varrho\kappa_*)\Delta\vartheta, -[\![(d_*/\varrho l_*)\partial_\nu\vartheta]\!]\big). \end{aligned} \tag{10.6}$$

Here the principal variable is $z = (\vartheta, h)$, for the dynamic inhomogeneities we have $f = (f_\theta, f_h)$, and the static one is $g = g_\theta$. For well-posedness of this problem we require $l_* \neq 0$.

Problem 2. The fully linearized problem at an equilibrium is the following.

$$\begin{aligned} \varrho\partial_t u - \mu_*\Delta u + \nabla\pi &= \varrho f_u && \text{in } \Omega\setminus\Sigma,\\ \operatorname{div} u &= g_d && \text{in } \Omega\setminus\Sigma,\\ u &= 0 && \text{on } \partial\Omega,\\ [\![u]\!] &= 0 && \text{on } \Sigma,\\ -[\![T\nu_\Sigma]\!] + \sigma_*(\mathcal{A}_\Sigma h)\nu_\Sigma &= g_\Sigma && \text{on } \Sigma,\\ u(0) &= u_0 && \text{in } \Omega; \end{aligned} \tag{10.7}$$

$$\begin{aligned} \varrho\kappa_*\partial_t\vartheta - d_*\Delta\vartheta &= \varrho\kappa_* f_\theta && \text{in } \Omega\setminus\Sigma,\\ \partial_\nu\vartheta &= 0 && \text{on } \partial\Omega,\\ [\![\vartheta]\!] &= 0 && \text{on } \Sigma,\\ -[\![d_*\partial_\nu\vartheta]\!] &= f_\Sigma && \text{on } \Sigma,\\ \vartheta(0) &= \vartheta_0 && \text{in } \Omega\setminus\Sigma; \end{aligned} \tag{10.8}$$

$$\begin{aligned} \partial_t h - u\cdot\nu_\Sigma &= f_h && \text{on } \Sigma,\\ h(0) &= h_0 && \text{on } \Sigma. \end{aligned} \tag{10.9}$$

Set

$$X_0^2 = L_{p,\sigma}(\Omega)\times L_p(\Omega)\times W_p^{2-1/p}(\Sigma),$$

where the subscript σ means solenoidal, and define the operator L_2 by

$$L_2(u,\vartheta,h) = \Big(-(\mu_*/\varrho)\Delta u + \nabla\pi/\varrho, -(d_*/\varrho\kappa_*)\Delta\vartheta, -u\cdot\nu_\Sigma\Big).$$

To define the domain $\mathsf{D}(L_2)$ of L_2, we set

$$\begin{aligned} X_1^2 = \mathsf{D}(L_2) = \{(u,\vartheta,h)\in H_p^2(\Omega\setminus\Sigma)^{n+1}\times W_p^{3-1/p}(\Sigma) : \operatorname{div} u = 0 \text{ in } \Omega\setminus\Sigma,\\ [\![u]\!], [\![\vartheta]\!], \mathcal{P}_\Sigma[\![\mu_* D\nu_\Sigma]\!], [\![d_*\partial_\nu\vartheta]\!] = 0 \text{ on } \Sigma,\ u,\partial_\nu\vartheta = 0 \text{ on } \partial\Omega\}. \end{aligned}$$

Here π is determined as the solution of the following weak transmission problem, uniquely up to a constant.

$$\begin{aligned} (\nabla\pi|\nabla\phi/\varrho)_\Omega &= (\mu_*\Delta u|\nabla\phi/\varrho)_\Omega, \quad \phi\in \dot H^1_{p'}(\Omega),\\ [\![\pi]\!] &= -\sigma_*\mathcal{A}_\Sigma h + [\![2\mu_*\partial_\nu u\cdot\nu_\Sigma]\!] \quad \text{on } \Sigma. \end{aligned}$$

In this problem the principal variable is $z=(u,\vartheta,h)$, for the dynamic inhomogeneities we have $f=(f_u,f_\theta,f_h)$, and the static one is $g=(g_d,g_\Sigma,g_\theta)$. Note that the problem for ϑ decouples completely.

Problem 3. Eliminating j_Σ by means of surface energy balance, the fully linearized problem at an equilibrium in this case reads

$$\begin{aligned}
\varrho\partial_t u - \mu_*\Delta u + \nabla\pi &= \varrho f_u && \text{in } \Omega\setminus\Sigma,\\
\operatorname{div} u &= g_d && \text{in } \Omega\setminus\Sigma,\\
u &= 0 && \text{on } \partial\Omega,\\
[\![u]\!] &= 0 && \text{on } \Sigma,\\
-[\![T\nu_\Sigma]\!] + \sigma_*\mathcal{A}_\Sigma h\nu_\Sigma &= g_\Sigma && \text{on } \Sigma,\\
u(0) &= u_0 && \text{in } \Omega;
\end{aligned} \tag{10.10}$$

$$\begin{aligned}
\varrho\kappa_*\partial_t\vartheta - d_*\Delta\vartheta &= \varrho\kappa_* f_\theta && \text{in } \Omega\setminus\Sigma,\\
\partial_\nu\vartheta &= 0 && \text{on } \partial\Omega,\\
[\![\vartheta]\!] &= 0 && \text{on } \Sigma,\\
(\varrho l_*/\theta_*)\vartheta - \sigma_*\mathcal{A}_\Sigma h &= g_\theta && \text{on } \Sigma,\\
\varrho l_*(\partial_t h - u\cdot\nu_\Sigma) - [\![d_*\partial_\nu\vartheta]\!] &= \varrho l_* f_h && \text{on } \Sigma,\\
\vartheta(0) = \vartheta_0, \quad h(0) &= h_0.
\end{aligned} \tag{10.11}$$

Here we redefined g_θ by $\varrho g_\theta - g_\Sigma\cdot\nu_\Sigma$. Set

$$X_0^3 = L_{p,\sigma}(\Omega)\times L_p(\Omega)\times W_p^{2-2/p}(\Sigma),$$

and define the operator L_3 by

$$L_3(u,\vartheta,h) = \Big(-\mu_*\Delta u/\varrho + \nabla\pi/\varrho, -(d_*/\varrho\kappa_*)\Delta\vartheta, -u\cdot\nu_\Sigma - [\![(d_*/\varrho l_*)\partial_\nu\vartheta]\!]\Big).$$

To define the domain $\mathsf{D}(L_3)$ of L_3, we set

$$\begin{aligned}
X_1^3 = \{(u,\vartheta,h)\in H_p^2(\Omega\setminus\Sigma)^{n+1}\times W_p^{4-1/p}(\Sigma) : \operatorname{div} u = 0 \text{ in } \Omega\setminus\Sigma,\\
[\![u]\!], [\![\vartheta]\!] = 0 \text{ on } \Sigma,\ u, \partial_\nu\vartheta = 0 \text{ on } \partial\Omega\},
\end{aligned}$$

and

$$\begin{aligned}
\mathsf{D}(L_3) = \{(u,\vartheta,h)\in X_1^3 : \mathcal{P}_\Sigma[\![\mu_* D\nu_\Sigma]\!], (\varrho l_*/\theta_*)\vartheta - \sigma_*\mathcal{A}_\Sigma h = 0 \text{ on } \Sigma,\\
[\![d_*\partial_\nu\vartheta]\!] \in W_p^{2-2/p}(\Sigma)\}.
\end{aligned}$$

The pressure π is determined, uniquely up to a constant, as the solution of the weak transmission problem

$$\begin{aligned}
(\nabla\pi|\nabla\phi)_\Omega &= (\mu_*\Delta u|\nabla\phi)_\Omega, \quad \phi\in\dot H^1_{p'}(\Omega),\\
[\![\pi]\!] &= -\sigma_*\mathcal{A}_\Sigma h + [\![2\mu_*\partial_\nu u\cdot\nu_\Sigma]\!] \quad \text{on } \Sigma.
\end{aligned}$$

The principal variable is again $z = (u,\vartheta,h)$, for the dynamic inhomogeneities we have $f = (f_u, f_\theta, f_h)$, and the static one is $g = (g_d, g_\Sigma, g_\theta)$. Here we have to assume, as in Problem 1, $l_*\neq 0$ for well-posedness.

Problem 4. The linearized problem at an equilibrium reads

$$
\begin{aligned}
\varrho\partial_t u - \mu_*\Delta u + \nabla\pi &= \varrho f_u && \text{in } \Omega\setminus\Sigma,\\
\operatorname{div} u &= g_d && \text{in } \Omega\setminus\Sigma,\\
u &= 0 && \text{on } \partial\Omega,\\
[\![u]\!] - [\![1/\varrho]\!] j_\Sigma \nu_\Sigma &= [\![1/\varrho]\!] g_u && \text{on } \Sigma,\\
-[\![T\nu_\Sigma]\!] + \sigma_*\mathcal{A}_\Sigma h \nu_\Sigma &= g_\Sigma && \text{on } \Sigma,\\
u(0) &= u_0 && \text{in } \Omega;
\end{aligned}
\tag{10.12}
$$

$$
\begin{aligned}
\varrho\kappa_*\partial_t\vartheta - d_*\Delta\vartheta &= \varrho\kappa_* f_\theta && \text{in } \Omega\setminus\Sigma,\\
\partial_\nu\vartheta &= 0 && \text{on } \partial\Omega,\\
[\![\vartheta]\!] &= 0 && \text{on } \Sigma,\\
-l_* j_\Sigma - [\![d_*\partial_\nu\vartheta]\!] &= f_\Sigma && \text{on } \Sigma,\\
\vartheta(0) &= \vartheta_0 && \text{in } \Omega;
\end{aligned}
\tag{10.13}
$$

$$
\begin{aligned}
(l_*/\theta_*)\vartheta_\Sigma - [\![T\nu_\Sigma\cdot\nu_\Sigma/\varrho]\!] &= g_\theta && \text{on } \Sigma,\\
\partial_t h - u\cdot\nu_\Sigma + j_\Sigma/\varrho &= f_h && \text{on } \Sigma,\\
h(0) &= h_0 && \text{on } \Sigma.
\end{aligned}
\tag{10.14}
$$

We eliminate j_Σ to the result

$$
j_\Sigma = [\![u\cdot\nu_\Sigma]\!]/[\![1/\varrho]\!], \quad u\cdot\nu_\Sigma - j_\Sigma/\varrho = [\![\varrho u\cdot\nu_\Sigma]\!]/[\![\varrho]\!],
$$

since $g_u\cdot\nu_\Sigma = 0$ by (10.4). Set

$$
X_0^4 = L_{p,\sigma}(\Omega)\times L_p(\Omega)\times W_p^{2-1/p}(\Sigma)
$$

and define the operator L_4 by

$$
L_4(u,\vartheta,h) = \Big(-(\mu_*/\varrho)\Delta u + \nabla\pi/\varrho, -(d_*/\varrho\kappa_*)\Delta\vartheta, -[\![\varrho u\cdot\nu_\Sigma]\!]/[\![\varrho]\!]\Big).
$$

To define the domain $\mathsf{D}(L_4)$ of L_4, we set

$$
\begin{aligned}
X_1^4 = \{(u,\vartheta,h)\in H_p^2(\Omega\setminus\Sigma)^{n+1}\times W_p^{3-1/p}(\Sigma) :{}& \operatorname{div} u = 0 \text{ in } \Omega\setminus\Sigma,\\
&\mathcal{P}_\Sigma[\![u]\!], [\![\vartheta]\!] = 0 \text{ on } \Sigma,\ u, \partial_\nu\vartheta = 0 \text{ on } \partial\Omega\},
\end{aligned}
$$

and

$$
\mathsf{D}(L_4) = \{(u,\vartheta,h)\in X_1^4 : \mathcal{P}_\Sigma[\![\mu_* D\nu_\Sigma]\!], (l_*/[\![1/\varrho]\!])[\![u\cdot\nu_\Sigma]\!] + [\![d_*\partial_\nu\vartheta]\!] = 0 \text{ on } \Sigma\}.
$$

Here π is determined uniquely as the solution of the weak transmission problem

$$
\begin{aligned}
(\nabla\pi|\nabla\phi/\varrho)_\Omega &= ((\mu_*/\varrho)\Delta u|\nabla\phi)_\Omega, \quad \phi\in H^1_{p',\Sigma}(\Omega\setminus\Sigma),\\
[\![\pi]\!] &= -\sigma_*\mathcal{A}_\Sigma h + [\![2\mu_*\partial_\nu u\cdot\nu_\Sigma]\!] \quad \text{on } \Sigma,\\
[\![\pi/\varrho]\!] &= -(l_*/\theta_*)\vartheta + [\![2\mu_*(\partial_\nu u\cdot\nu_\Sigma)/\varrho]\!] \quad \text{on } \Sigma.
\end{aligned}
$$

Actually, this transmission problem boils down to two one-phase problems with Dirichlet data. In this model the principal variable is again $z = (u, \vartheta, h)$, for the dynamic inhomogeneities we have $f = (f_u, f_\theta, f_h)$, and the static one is $g = (g_d, [\![1/\varrho]\!] g_u, g_\Sigma, g_\theta)$.

Problem 5. As $[\![1/\varrho]\!] = 0$ we may combine the normal stress condition and the Gibbs-Thomson relation to the result

$$-(l_0/\theta_*)\vartheta_\Sigma + \sigma_* \mathcal{A}_\Sigma h = g_\Sigma \cdot \nu_\Sigma - \varrho g_\theta,$$

where

$$l_0 := \varrho l_* + \theta_* \sigma'_* H_\Sigma.$$

We may use this relation to eliminate ϑ_Σ provided $l_0 \neq 0$. Furthermore, assuming w.l.o.g. $g_u = 0$, with

$$\mathrm{div}_\Sigma u_\Sigma = \mathrm{div}_\Sigma \mathcal{P}_\Sigma u_\Sigma - H_\Sigma u_\Sigma \cdot \nu_\Sigma$$

and $\mathcal{P}_\Sigma u_\Sigma = \mathcal{P}_\Sigma u$, $u_\Sigma \cdot \nu_\Sigma = u \cdot \nu_\Sigma - j_\Sigma/\varrho$ we obtain

$$\mathrm{div}_\Sigma u_\Sigma = \mathrm{div}_\Sigma u + H_\Sigma j_\Sigma/\varrho.$$

Inserting this identity into the surface energy balance and solving for j_Σ leads to

$$\kappa_{\Sigma*}\partial_t \vartheta_\Sigma - d_{\Sigma*}\Delta_\Sigma \vartheta_\Sigma - [\![d_* \partial_\nu \vartheta]\!] - \theta_* \sigma'_* \mathrm{div}_\Sigma u = l_0 j_\Sigma/\varrho + f_\Sigma.$$

Taking these observations into account, and eliminating j_Σ and ϑ_Σ, the linearized problem at an equilibrium reads

$$\begin{aligned}
\varrho\partial_t u - \mu_* \Delta u + \nabla \pi &= \varrho f_u && \text{in } \Omega \setminus \Sigma,\\
\mathrm{div}\, u &= g_d && \text{in } \Omega \setminus \Sigma,\\
u &= 0 && \text{on } \partial\Omega,\\
[\![u]\!] &= 0 && \text{on } \Sigma,\\
-[\![T\nu_\Sigma]\!] + \sigma_*(\mathcal{A}_\Sigma h)\nu_\Sigma - \sigma'_*(\nabla_\Sigma + H_\Sigma \nu_\Sigma)\vartheta &= g_\Sigma && \text{on } \Sigma,\\
u(0) &= u_0 && \text{in } \Omega;
\end{aligned} \tag{10.15}$$

$$\begin{aligned}
&\varrho\kappa_* \partial_t \vartheta - d_* \Delta \vartheta = \varrho \kappa_* f_\theta && \text{in } \Omega \setminus \Sigma,\\
&\partial_\nu \vartheta = 0 && \text{on } \partial\Omega,\\
&[\![\vartheta]\!] = 0, \quad \vartheta = (\theta_*/l_0)(\sigma_* \mathcal{A}_\Sigma h + \varrho g_\theta - g_\Sigma \cdot \nu_\Sigma) && \text{on } \Sigma,\\
&\vartheta(0) = \vartheta_0 && \text{in } \Omega;
\end{aligned} \tag{10.16}$$

$$\begin{aligned}
&(l_0^2/\theta_* + \kappa_{\Sigma*}\sigma_* \mathcal{A}_\Sigma)\partial_t h - d_{\Sigma*}\sigma_* \Delta_\Sigma \mathcal{A}_\Sigma h\\
&-(l_0/\theta_*)[\![d_* \partial_\nu \vartheta]\!] - (l_0^2/\theta_*) u \cdot \nu_\Sigma - l_0 \sigma'_* \mathrm{div}_\Sigma u = \tilde{f}_h && \text{on } \Sigma,\\
&h(0) = h_0 && \text{on } \Sigma.
\end{aligned} \tag{10.17}$$

Set

$$X_0^5 = L_{p,\sigma}(\Omega) \times L_p(\Omega) \times W_p^{2-1/p}(\Sigma)$$

and define the operator L_5 by

$$L_5(u,\vartheta,h) = \Big(-(\mu_*/\varrho)\Delta u + \nabla\pi/\varrho, -(d_*/\varrho\kappa_*)\Delta\vartheta, -(l_0^2/\theta_* + \kappa_{\Sigma*}\sigma_*\mathcal{A}_\Sigma)^{-1} \cdot (d_{\Sigma*}\sigma_*\Delta_\Sigma\mathcal{A}_\Sigma h + (l_0/\theta_*)[\![d_*\partial_\nu\vartheta]\!] + (l_0^2/\theta_*)u\cdot\nu_\Sigma + l_0\sigma_*'\mathrm{div}_\Sigma u)\Big).$$

To define the domain $\mathsf{D}(L_5)$ of L_5, we set

$$X_1^5 = \{(u,\vartheta,h) \in H_p^2(\Omega\setminus\Sigma)^{n+1} \times W_p^{4-1/p}(\Sigma) : \mathrm{div}\, u = 0 \text{ in } \Omega\setminus\Sigma,\ [\![u]\!], [\![\vartheta]\!] = 0 \text{ on } \Sigma,\ \ u, \partial_\nu\vartheta = 0 \text{ on } \partial\Omega\},$$

and

$$\mathsf{D}(L_5) = \{(u,\vartheta,h) \in X_1^5 : \mathcal{P}_\Sigma[\![\mu_* D\nu_\Sigma]\!] + \sigma_*'\nabla_\Sigma\vartheta, (l_0/\theta_*)\vartheta - \sigma_*\mathcal{A}_\Sigma h = 0 \text{ on } \Sigma\}.$$

Here π is determined as the solution of the weak transmission problem

$$\begin{aligned}(\nabla\pi|\nabla\phi)_\Omega &= (\mu_*\Delta u|\nabla\phi)_\Omega, \quad \phi \in \dot{H}_{p'}^1(\Omega),\\ [\![\pi]\!] &= -\sigma_*\mathcal{A}_\Sigma h + \sigma_*' H_\Sigma\vartheta_\Sigma + [\![2\mu_*\partial_\nu u\cdot\nu_\Sigma]\!] \quad \text{on } \Sigma.\end{aligned}$$

This reformulation is valid provided $l_0^2/\theta_* + \kappa_{\Sigma*}\sigma_*\mathcal{A}_\Sigma$ is invertible, which is true if $l_0 \neq 0$ and $l_0^2/\theta_* \neq \kappa_{\Sigma*}\sigma_*(n-1)/R_*^2$ holds. In this model the principal variable is again $z = (u,\vartheta,h)$, for the dynamic inhomogeneities we have $f = (f_u, f_\theta, f_h)$ with $\tilde{f}_h = (l_0^2/\theta_* + \kappa_{\Sigma*}\sigma_*\mathcal{A}_\Sigma)^{-1} f_h$ and the static one is $g = (g_d, g_\Sigma, g_\theta)$.

Problem 6. Eliminate j_Σ and u_Σ from the fourth line of (10.1) to the result

$$u_\Sigma = [\![\varrho u]\!]/[\![\varrho]\!], \quad j_\Sigma = [\![u\cdot\nu_\Sigma]\!]/[\![1/\varrho]\!],$$

where we again used (10.4). Inserting into the equations, the linearized problem at an equilibrium reads

$$\begin{aligned}
\varrho\partial_t u - \mu_*\Delta u + \nabla\pi &= \varrho f_u && \text{in } \Omega\setminus\Sigma,\\
\mathrm{div}\, u &= g_d && \text{in } \Omega\setminus\Sigma,\\
u &= 0 && \text{on } \partial\Omega,\\
\mathcal{P}_\Sigma[\![u]\!] &= [\![1/\varrho]\!]\mathcal{P}_\Sigma g_u && \text{on } \Sigma,\\
-\mathcal{P}_\Sigma[\![T\nu_\Sigma]\!] - \sigma_*'\nabla_\Sigma\vartheta_\Sigma &= \mathcal{P}_\Sigma g_\Sigma && \text{on } \Sigma,\\
-[\![T\nu_\Sigma\cdot\nu_\Sigma]\!] + \sigma_*\mathcal{A}_\Sigma h - \sigma_*' H_\Sigma\vartheta_\Sigma &= g_\Sigma\cdot\nu_\Sigma && \text{on } \Sigma,\\
-[\![T\nu_\Sigma\cdot\nu_\Sigma/\varrho]\!] + (l_*/\theta_*)\vartheta_\Sigma &= g_\theta && \text{on } \Sigma,\\
u(0) &= u_0 && \text{in } \Omega;
\end{aligned} \tag{10.18}$$

$$\begin{aligned} \varrho\kappa_*\partial_t\vartheta - d_*\Delta\vartheta &= \varrho\kappa_* f_\theta && \text{in } \Omega\setminus\Sigma,\\ \partial_\nu\vartheta &= 0 && \text{on } \partial\Omega,\\ [\![\vartheta]\!] = 0, \quad \vartheta &= \vartheta_\Sigma && \text{on } \Sigma,\\ \vartheta(0) &= \vartheta_0 && \text{in } \Omega; \end{aligned} \tag{10.19}$$

$$\begin{aligned} \kappa_{\Sigma*}\partial_t\vartheta_\Sigma - d_{\Sigma*}\Delta_\Sigma\vartheta_\Sigma - [\![d_*\partial_\nu\vartheta]\!] - l_*[\![u\cdot\nu_\Sigma]\!]/[\![1/\varrho]\!] - &\\ -(\theta_*\sigma'_*/[\![\varrho]\!])\,\mathrm{div}_\Sigma[\![\varrho u]\!] &= f_\Sigma && \text{on } \Sigma,\\ \partial_t h - [\![\varrho u\cdot\nu_\Sigma]\!]/[\![\varrho]\!] &= f_h && \text{on } \Sigma,\\ h(0) &= h_0 && \text{on } \Sigma. \end{aligned} \tag{10.20}$$

Set

$$X_0^6 = L_{p,\sigma}(\Omega)\times L_p(\Omega)\times W_p^{-1/p}(\Sigma)\times W_p^{2-1/p}(\Sigma)$$

and define the operator L_6 by

$$\begin{aligned} L_6(u,\vartheta,\vartheta_\Sigma,h) = \Big(&-(\mu_*/\varrho)\Delta u + \nabla\pi/\varrho, -(d_*/\varrho\kappa_*)\Delta\vartheta,\\ &(-d_{\Sigma*}\Delta_\Sigma\vartheta_\Sigma - [\![d_*\partial_\nu\vartheta]\!] - (\theta_*\sigma'_*/[\![\varrho]\!])\mathrm{div}_\Sigma[\![\varrho u]\!] - l_*[\![u\cdot\nu_\Sigma]\!]/[\![1/\varrho]\!])/\kappa_{\Sigma*},\\ &-[\![\varrho u\cdot\nu_\Sigma]\!]/[\![\varrho]\!]\Big). \end{aligned}$$

To define the domain $\mathsf{D}(L_6)$ of L_6, we set

$$\begin{aligned} X_1^6 = \{&(u,\vartheta,\vartheta_\Sigma,h)\in H_p^2(\Omega\setminus\Sigma)^{n+1}\times W_p^{2-1/p}(\Sigma)\times W_p^{3-1/p}(\Sigma):\\ &\mathrm{div}\,u = 0 \text{ in } \Omega\setminus\Sigma,\quad \mathcal{P}_\Sigma[\![u]\!], [\![\vartheta]\!] = 0,\ \vartheta = \vartheta_\Sigma \text{ on } \Sigma,\quad u, \partial_\nu\vartheta = 0 \text{ on } \partial\Omega\}, \end{aligned}$$

and

$$\mathsf{D}(L_6) = \{(u,\vartheta,\vartheta_\Sigma,h)\in X_1^6 : \mathcal{P}_\Sigma[\![\mu_* D\nu_\Sigma]\!] + \sigma'_*\nabla_\Sigma\vartheta = 0 \text{ on } \Sigma\}.$$

Here π is determined as the unique solution of the weak transmission problem

$$\begin{aligned} (\nabla\pi|\nabla\phi/\varrho)_\Omega &= ((\mu_*/\varrho)\Delta u|\nabla\phi)_\Omega, \quad \phi\in H^1_{p',\Sigma}(\Omega\setminus\Sigma),\\ [\![\pi]\!] &= -\sigma_*\mathcal{A}_\Sigma h + \sigma'_* H_\Sigma\vartheta_\Sigma + [\![2\mu_*\partial_\nu u\cdot\nu_\Sigma]\!] \quad \text{on } \Sigma,\\ [\![\pi/\varrho]\!] &= -(l_*/\theta_*)\vartheta_\Sigma + [\![2\mu_*(\partial_{\nu_\Sigma} u\cdot\nu_\Sigma)/\varrho]\!] \quad \text{on } \Sigma. \end{aligned}$$

As for Problem 4, this consists of two weak one-phase Dirichlet problems. In this model the principal variable is again $z = (u,\vartheta,\vartheta_\Sigma,h)$, for the dynamic inhomogeneities we have $f = (f_u, f_\theta, f_\Sigma, f_h)$, and the static one is $g = (g_d, g_u, g_\Sigma, g_\theta)$.

This way, the linearized problems **(Pj)** can be rewritten as abstract evolution problems in X_0^j.

$$\dot z + L_j z = f, \quad t>0, \quad z(0) = z_0, \tag{10.21}$$

provided $g = 0$. According to Chapters 6, 7, 8, the linearized problems have maximal L_p-regularity, hence (10.21) has this property as well, provided $l_* \neq 0$ for

Problems P_1 and P_3, and $l_0 \neq 0$, $\delta_* := \sigma_*(n-1)\theta_*\kappa_{\Sigma*}/l_0^2R_*^2 \neq 1$ for Problem P_5. Therefore, by Proposition 3.5.2, $-L_j$ generates an analytic C_0-semigroup in X_0^j.

Since the embeddings $X_1^j \hookrightarrow X_0^j$ are compact, the semigroups e^{-L_jt} as well as the resolvents $(\lambda+L_j)^{-1}$ of $-L_j$ are compact, too. Therefore the spectra $\sigma(L_j)$ of L_j consists only of countably many eigenvalues of finite algebraic multiplicity. Moreover, by Sobolev embedding the spectrum $\sigma(L_j)$ is independent of p.

So we have to study the following general eigenvalue problem.

$$\begin{aligned}
\varrho\lambda u - \mu_*\Delta u + \nabla\pi &= 0 && \text{in } \Omega\setminus\Sigma,\\
\operatorname{div} u &= 0 && \text{in } \Omega\setminus\Sigma,\\
u &= 0 && \text{on } \partial\Omega,\\
[\![\mathcal{P}_\Sigma u]\!] = 0, \quad \mathcal{P}_\Sigma u &= \mathcal{P}_\Sigma u_\Sigma && \text{on } \Sigma,\\
-[\![T\nu_\Sigma]\!] + \sigma_*\mathcal{A}_\Sigma h\nu_\Sigma &= \sigma_*'\nabla_\Sigma\vartheta_\Sigma + \sigma_*'\vartheta_\Sigma H_\Sigma\nu_\Sigma && \text{on } \Sigma,\\
[\![u\cdot\nu_\Sigma]\!] - [\![1/\varrho]\!]j_\Sigma &= 0 && \text{on } \Sigma,
\end{aligned} \tag{10.22}$$

where $T = \mu_*(\nabla u + [\nabla u]^{\mathsf{T}}) - \pi I$,

$$\begin{aligned}
\varrho\kappa_*\lambda\vartheta - d_*\Delta\vartheta &= 0 && \text{in } \Omega\setminus\Sigma,\\
\partial_\nu\vartheta &= 0 && \text{on } \partial\Omega,\\
[\![\vartheta]\!] = 0,\ \vartheta &= \vartheta_\Sigma && \text{on } \Sigma,
\end{aligned} \tag{10.23}$$

$$\begin{aligned}
\kappa_{\Sigma*}\lambda\vartheta_\Sigma - d_{\Sigma*}\Delta_\Sigma\vartheta_\Sigma - \theta_*\sigma_*'\operatorname{div}_\Sigma u_\Sigma - l_*j_\Sigma - [\![d_*\partial_\nu\vartheta]\!] &= 0 && \text{on } \Sigma,\\
(l_*/\theta_*)\vartheta_\Sigma - [\![T\nu_\Sigma\cdot\nu_\Sigma/\varrho]\!] &= 0 && \text{on } \Sigma,\\
\lambda h = u_\Sigma\cdot\nu_\Sigma &= u\cdot\nu_\Sigma - j_\Sigma/\varrho && \text{on } \Sigma.
\end{aligned} \tag{10.24}$$

Since the eigenvalues are independent of p it is enough to study this eigenvalue problem in an L_2-setting. This will be the subject of the remaining sections of this chapter.

10.2 The Spectrum of the Laplace-Beltrami Operator

We need to prepare the eigenvalue analysis, deducing some properties of the operator $\mathcal{A}_\Sigma$ from those of the Laplace-Beltrami operator on spheres.

Proposition 10.2.1. *Let $\Sigma = S_R(x_0) \subset \mathbb{R}^n$ be a sphere of radius R and center x_0, and let*

$$\mathcal{A}_\Sigma = -\frac{n-1}{R^2} - \Delta_\Sigma$$

be defined on $L_2(\Sigma)$ with domain $H_2^2(\Sigma)$. Then

(a) *$\mathcal{A}_\Sigma$ is self-adjoint. The spectrum of $\mathcal{A}_\Sigma$ consists entirely of eigenvalues of finite algebraic multiplicity and is given by*

$$\sigma_p(\mathcal{A}_\Sigma) = \Big\{\frac{1}{R_*^2}\big(k(k+n-2)-(n-1)\big)\big) : k\in\mathbb{N}_0\Big\}.$$

(b) *There is precisely one negative eigenvalue, namely* $-(n-1)/R^2$, *with eigenfunction* $\mathsf{e} \equiv 1$, *which is simple.*

(c) *The kernel of* $\mathcal{A}_\Sigma$ *is given by* $\mathsf{N}(\mathcal{A}_\Sigma) = \operatorname{span}\{Y_1, \ldots, Y_n\}$, *where* Y_j *denote the spherical harmonics of degree 1 on* Σ, *normalized by* $(Y_i|Y_j)_\Sigma = \delta_{ij}$.

(d) $\mathcal{A}_\Sigma$ *is positive semi-definite on* $L_{2,0}(\Sigma) = \{h \in L_2(\Sigma) : (h|\mathsf{e})_\Sigma = 0\}$ *and positive definite on*

$$L_{2,0}(\Sigma) \cap \mathsf{R}(\mathcal{A}_\Sigma) = \{h \in L_2(\Sigma) : (h|\mathsf{e})_\Sigma = (h|Y_j)_\Sigma = 0,\ j = 1, \ldots, n\}.$$

(e) *The range of* $\mathcal{A}_\Sigma$ *is closed, and we have* $L_2(\Sigma) = \mathsf{N}(\mathcal{A}_\Sigma) \oplus \mathsf{R}(\mathcal{A}_\Sigma)$.

Proof. We can assume without loss of generality that $\Sigma = S_R(0) = R\,\mathbb{S}^{n-1}$, where $\mathbb{S}^{n-1}$ denotes the standard unit sphere in $\mathbb{R}^n$. Let $\Phi : \Sigma \to \mathbb{S}^{n-1}$ be defined by $p \mapsto (1/R)p$. Then Φ is a smooth diffeomorphism of Σ onto $\mathbb{S}^{n-1}$ and one readily verifies

$$(g|h)_{L_2(\Sigma)} = R^{n-1}(\Phi_* g|\Phi_* h)_{L_2(\mathbb{S}^{n-1})}, \quad \Delta_\Sigma = (1/R^2)\,\Phi^* \Delta_{\mathbb{S}^{n-1}}\, \Phi_* \tag{10.25}$$

where Φ^* and Φ_* are the pull-back and push-forward operators, respectively. We then have

$$(\lambda - \mathcal{A}_\Sigma)h = 0 \iff \Big(\lambda + \frac{1}{R^2}\big((n-1) + \Delta_{\mathbb{S}^{n-1}}\big)\Big)\Phi_* h = 0 \tag{10.26}$$

and this shows that λ is an eigenvalue of $\mathcal{A}_\Sigma$ iff

$$\lambda = \frac{1}{R^2}\big(\mu - (n-1)\big) \tag{10.27}$$

with μ an eigenvalue of $-\Delta_{\mathbb{S}^{n-1}}$. The assertions in (a)-(d) follow now from (10.25)–(10.27) and well-known results for the Laplace-Beltrami operator on $\mathbb{S}^{n-1}$, see for instance [281, Section 31]. Since $\mathcal{A}_\Sigma$ has compact resolvent we conclude that $\mathsf{R}(\mathcal{A}_\Sigma)$ is closed, and the fact that $\mathcal{A}_\Sigma$ is self-adjoint then implies the remaining assertion in (e). □

Note that this result extends to the case where Σ is a finite union of disjoint spheres. More precisely, if $\Sigma = \cup_{k=1}^m S_{R_k}(x_k)$ where the spheres $S_{R_k}(x_k)$ do not intersect, then $L_2(\Sigma) = \bigoplus_{k=1}^m L_2(\Sigma_k)$, with $\Sigma_k = S_{R_k}(x_k)$. Furthermore, $\mathcal{A}_\Sigma = \bigoplus_{k=1}^m \mathcal{A}_{\Sigma_k}$, hence Proposition 10.2.1 implies in particular that $\mathcal{A}_\Sigma$ is positive semi-definite on functions which have zero mean on each component Σ_k of Σ.

10.3 Nontrivial Eigenvalues

Suppose that λ with $\operatorname{Re}\lambda \geq 0$, $\lambda \neq 0$ is an eigenvalue of L_j. Here we do not need to distinguish the problems and so we consider the most general one, namely Problem 6. We exclude here for the moment Problem 2, i.e., the case without phase transition. Taking the inner product of the equation for u with u and integrating by parts we get

$$\begin{aligned}
0 &= \lambda|\varrho^{1/2}u|_2^2 - (\operatorname{div} T|u)_\Omega \\
&= \lambda|\varrho^{1/2}u|_2^2 + \int_\Omega T : \nabla\bar{u}\,dx + \int_\Sigma (T_2\nu_\Sigma\cdot\bar{u}_2 - T_1\nu_\Sigma\cdot\bar{u}_1)\,d\Sigma \\
&= \lambda|\varrho^{1/2}u|_2^2 + 2|\mu_*^{1/2}D|_2^2 + ([\![T\nu_\Sigma]\!]|u_\Sigma)_\Sigma + ([\![T\nu_\Sigma\cdot\nu_\Sigma/\varrho]\!]|j_\Sigma)_\Sigma \\
&= \lambda|\varrho^{1/2}u|_2^2 + 2|\mu_*^{1/2}D|_2^2 + \sigma_*\bar{\lambda}(\mathcal{A}_\Sigma h|h)_\Sigma + ([\![T\nu_\Sigma\cdot\nu_\Sigma/\varrho]\!]|j_\Sigma)_\Sigma \\
&\quad - \sigma_*'(\nabla_\Sigma\vartheta_\Sigma|u_\Sigma)_\Sigma - \sigma_*'H_\Sigma(\vartheta_\Sigma\nu_\Sigma|u_\Sigma),
\end{aligned}$$

since $\mathcal{P}_\Sigma[\![u]\!]=0$, $[\![T\nu_\Sigma]\!] = \sigma_*\mathcal{A}_\Sigma h\nu_\Sigma - \sigma_*'(\nabla_\Sigma\vartheta_\Sigma + H_\Sigma\nu_\Sigma\vartheta_\Sigma)$ and $u_\Sigma\cdot\nu_\Sigma = \lambda h$. On the other hand, the inner product of the equation for ϑ with ϑ and an integration by parts leads to

$$\begin{aligned}
0 &= \lambda|(\varrho\kappa_*)^{1/2}\vartheta|_2^2 + |d_*^{1/2}\nabla\vartheta|_2^2 + ([\![d_*\partial_\nu\vartheta]\!]|\vartheta)_\Sigma \\
&= \lambda(|(\varrho\kappa_*)^{1/2}\vartheta|_2^2 + |\kappa_{\Sigma*}^{1/2}\vartheta_\Sigma|_2^2) + |d_*^{1/2}\nabla\vartheta|_2^2 + |d_{\Sigma*}^{1/2}\nabla_\Sigma\vartheta_\Sigma|_2^2 \\
&\quad - \theta_*(j_\Sigma|[\![T\nu_\Sigma\cdot\nu_\Sigma/\varrho]\!])_\Sigma - \theta_*\sigma_*'(\operatorname{div}_\Sigma u_\Sigma|\vartheta_\Sigma)_\Sigma),
\end{aligned}$$

where we employed the relations

$$[\![d_*\partial_\nu\vartheta]\!] = \lambda\kappa_{\Sigma*}\vartheta_\Sigma - d_{\Sigma*}\Delta_\Sigma\vartheta_\Sigma - l_*j_\Sigma - \theta_*\sigma_*'\operatorname{div}_\Sigma u_\Sigma,$$

and $(l_*/\theta_*)\vartheta = (l_*/\theta_*)\vartheta_\Sigma = [\![T\nu_\Sigma\cdot\nu_\Sigma/\varrho]\!]$. Adding theses identities, employing the surface divergence theorem, and taking real parts yields the important identity

$$\begin{aligned}
0 &= \operatorname{Re}\lambda|\varrho^{1/2}u|_2^2 + 2|\mu_*^{1/2}D|_2^2 + \sigma_*\operatorname{Re}\lambda(\mathcal{A}_\Sigma h|h)_\Sigma \\
&\quad + \big(\operatorname{Re}\lambda(|(\varrho\kappa_*)^{1/2}\vartheta|_2^2 + |\kappa_{\Sigma*}^{1/2}\vartheta_\Sigma|_2^2) + |d_*^{1/2}\nabla\vartheta|_2^2 + |d_{\Sigma*}^{1/2}\nabla_\Sigma\vartheta_\Sigma|_2^2\big)/\theta_*. \qquad (10.28)
\end{aligned}$$

On the other hand, if $\beta := \operatorname{Im}\lambda \neq 0$, then taking imaginary parts separately we get with

$$a = ([\![T\nu_\Sigma\cdot\nu_\Sigma/\varrho]\!]|j_\Sigma)_\Sigma - \sigma_*'(\nabla_\Sigma\vartheta_\Sigma|u_\Sigma)_\Sigma - \sigma_*'H_\Sigma(\vartheta_\Sigma\nu_\Sigma|u_\Sigma)$$

the system

$$\begin{aligned}
0 &= \beta|\varrho^{1/2}u|_2^2 - \sigma_*\beta(\mathcal{A}_\Sigma h|h)_\Sigma + \operatorname{Im}a, \\
0 &= \beta(|(\varrho\kappa_*)^{1/2}\vartheta|_2^2 + |\kappa_{\Sigma*}^{1/2}\vartheta_\Sigma|_2^2)/\theta_* + \operatorname{Im}a,
\end{aligned}$$

hence

$$\sigma_*(\mathcal{A}_\Sigma h|h)_\Sigma = |\varrho^{1/2}u|_2^2 - (|(\varrho\kappa_*)^{1/2}\vartheta|_2^2 + |\kappa_{\Sigma*}^{1/2}\vartheta_\Sigma|_2^2)/\theta_*.$$

Inserting this equation into (10.28) leads to

$$0 = 2\operatorname{Re}\lambda|\varrho^{1/2}u|_2^2 + 2|\mu_*^{1/2}D|_2^2 + (|d_*^{1/2}\nabla\vartheta|_2^2 + |d_{\Sigma*}^{1/2}\nabla_\Sigma\vartheta_\Sigma|_2^2)/\theta_*,$$

which shows that if λ is an eigenvalue of $-L_j$ with $\mathrm{Re}\,\lambda \geq 0$, then λ is real. In fact, the last identity shows $\nabla\vartheta = 0$ and $D = 0$, hence $\vartheta = 0$ by the equation for ϑ, $u = 0$ by Lemma 1.2.1, and so also $j_\Sigma = h = 0$ by the equations, provided either $[\![\varrho]\!] \neq 0$ or $l_0 \neq 0$.

Supposing now that $\lambda > 0$ is an eigenvalue, we use the decomposition $\vartheta = \vartheta_0 + \bar{\vartheta}$, $\vartheta_\Sigma = \vartheta_{\Sigma 0} + \bar{\vartheta}_\Sigma$, $h = h_0 + \bar{h}$, and $j_\Sigma = (j_\Sigma)_0 + \bar{j}_\Sigma$, where

$$\bar{\vartheta} = (\varrho\kappa_*|\vartheta)_\Omega/(\kappa_*|\varrho)_\Omega, \quad \bar{\vartheta}_\Sigma = (\vartheta_\Sigma|1)_\Sigma/|\Sigma|, \quad \bar{h} = (h|1)_\Sigma/|\Sigma|, \quad \bar{j}_\Sigma = (j_\Sigma|1)_\Sigma/|\Sigma|$$

are weighted means. Then

$$|(\varrho\kappa_*)^{1/2}\vartheta|_2^2 = |(\varrho\kappa_*)^{1/2}\vartheta_0|_2^2 + (\kappa_*|\varrho)_\Omega\bar{\vartheta}^2, \quad |\vartheta_\Sigma|_\Sigma^2 = |\vartheta_{\Sigma 0}|_\Sigma^2 + |\Sigma|\bar{\vartheta}_\Sigma^2,$$

and

$$|h|_\Sigma^2 = |h_0|_\Sigma^2 + |\Sigma|\bar{h}^2, \quad |j_\Sigma|_\Sigma^2 = |(j_\Sigma)_0|_\Sigma^2 + |\Sigma|\,\bar{j}_\Sigma^2.$$

Therefore (10.28) becomes

$$\begin{aligned} 0 &= \lambda|\varrho^{1/2}u|_2^2 + 2|\mu_*^{1/2}D|_2^2 + \sigma_*\lambda(\mathcal{A}_\Sigma h_0|h_0)_\Sigma \\ &\quad + (\lambda|(\varrho\kappa_*)^{1/2}\vartheta_0|_2^2 + \lambda|\kappa_{\Sigma*}^{1/2}\vartheta_{\Sigma 0}|_2^2 + |d_*^{1/2}\nabla\vartheta_0|_2^2 + |d_{\Sigma*}^{1/2}\nabla_\Sigma\vartheta_{\Sigma 0}|_2^2)/\theta_* \qquad (10.29) \\ &\quad + \lambda\big((\kappa_*|\varrho)_\Omega\bar{\vartheta}^2 + \kappa_{\Sigma*}|\Sigma|\bar{\vartheta}_\Sigma^2 - \sigma_*\theta_*\frac{n-1}{R_*^2}|\Sigma|\bar{h}^2\big)/\theta_*. \end{aligned}$$

Assume now that the densities are equal, i.e., $\varrho_1 = \varrho_2 =: \varrho$. Then we further have

$$\lambda\int_\Sigma h\,d\Sigma = \int_\Sigma (u\cdot\nu_\Sigma - j_\Sigma/\varrho)\,d\Sigma = -\int_\Sigma j_\Sigma/\varrho\,d\Sigma,$$

since $\mathrm{div}\,u = 0$ and $u = 0$ on $\partial\Omega$. Therefore, $\lambda\bar{h} = -\bar{j}_\Sigma/\varrho$. Also, integrating the equation for ϑ we get

$$\lambda(\kappa_*|\varrho)_\Omega\bar{\vartheta} = -\int_\Sigma [\![d_*\partial_\nu\vartheta]\!]\,d\Sigma,$$

and integrating that for ϑ_Σ we obtain

$$\lambda\kappa_{\Sigma*}|\Sigma|\bar{\vartheta}_\Sigma = \int_\Sigma [\![d_*\partial_\nu\vartheta]\!]\,d\Sigma + l_*|\Sigma|\bar{j}_\Sigma - \lambda\theta_*\sigma_*' H_\Sigma|\Sigma|\bar{h}.$$

Adding the last two identities, replacing j_Σ and dividing by λ we arrive at the relation

$$(\kappa_*|\varrho)_\Omega\bar{\vartheta} + \kappa_{\Sigma*}|\Sigma|\bar{\vartheta}_\Sigma = -(l_*\varrho + \theta_*\sigma_*' H_\Sigma)|\Sigma|\bar{h} = l_0|\Sigma|\bar{h}. \qquad (10.30)$$

As $\mathcal{A}_\Sigma$ is positive semi-definite on functions with mean zero if Σ is connected, in this case L_j has no positive eigenvalues if the form

$$(\kappa_*|\varrho)_\Omega\bar{\vartheta}^2 + \kappa_{\Sigma*}|\Sigma|\bar{\vartheta}_\Sigma^2 - \sigma_*\theta_*\frac{n-1}{R_*^2}|\Sigma|\bar{h}^2$$

in the variables $(\bar\vartheta, \bar\vartheta_\Sigma, \bar h)$ with the constraint (10.30) is nonnegative. This way we derive the *stability condition*

$$\zeta_* := \frac{\theta_*\sigma_*(n-1)}{l_0^2 R_*^2|\Sigma|}\big((\kappa_*|\varrho)_\Omega + \kappa_{\Sigma *}|\Sigma|\big) \leq 1. \tag{10.31}$$

This condition applies to Problems **(P1)**, **(P3)**, **(P5)**. On the other hand, if $\varrho_1 \neq \varrho_2$ we have in addition

$$0 = \int_{\Omega_i} -\operatorname{div} u \, dx = (-1)^i \int_\Sigma u \cdot \nu_\Sigma \, d\Sigma = (-1)^i \int_\Sigma (\lambda h + j_\Sigma/\varrho_i) \, d\Sigma,$$

for $i = 1, 2$, and hence $-\bar j_\Sigma/\varrho_1 = \lambda \bar h = -\bar j_\Sigma/\varrho_2$ which implies $\bar j_\Sigma = \bar h = 0$. In this case, i.e., for Problems **(P4)** and **(P6)**, the stability condition is redundant. We will show below that there is always at least one positive eigenvalue if Σ is not connected.

Finally, we look at Problem **(P2)**, the case without phase transition. Then with $j_\Sigma \equiv 0$ we may proceed as above to show that eigenvalues with nonnegative real part are real and we obtain again identity (10.29). Let Ω_1^k denote the components of Ω_1 and set $\Sigma^k = \partial\Omega_1^k$. Then we have

$$0 = \int_{\Omega_1^k} \operatorname{div} u \, dx = \int_{\Sigma^k} u \cdot \nu_\Sigma \, d\Sigma = \lambda \int_{\Sigma^k} h \, d\Sigma,$$

hence the mean values of h vanish even over the components of Σ. In particular, $\bar h = 0$ and since $\mathcal{A}_\Sigma$ is positive semi-definite on functions which have mean values zero over each component of Σ, we see that (10.29) implies $\lambda = 0$. Hence in this case there are no eigenvalues with nonnegative real part, except for $\lambda = 0$.

10.4 Eigenvalue Zero

Next we look at the eigenvalue $\lambda = 0$. Then (10.28) yields

$$2|\mu_*^{1/2} D|_2^2 + (|d_*^{1/2}\nabla\vartheta|_2^2 + |d_{\Sigma_*}^{1/2}\nabla_\Sigma\vartheta_\Sigma|_\Sigma^2)/\theta_* = 0,$$

hence $\vartheta = \vartheta_\Sigma$ is constant and $D = 0$. Lemma 1.2.1 implies $u = 0$, hence $j_\Sigma = 0$ by the equation for h. This implies further that the pressures are constant in the components of the phases, and $[\![\pi]\!] = -\sigma_*\mathcal{A}_\Sigma h + \sigma_*' H_\Sigma \vartheta_\Sigma$, as well as $(l_*/\theta_*)\vartheta_\Sigma = -[\![\pi/\varrho]\!]$. Thus the dimension of the eigenspace for the eigenvalue $\lambda = 0$ is the same as the dimension of the manifold of equilibria, namely $mn + 2$ if Ω_1 has $m \geq 1$ components and $[\![\varrho]\!] \neq 0$, otherwise it is $mn + 1$. The kernel of the linearization L is spanned by $e_\theta = (0, 1, 1, 0), e_h = (0, 0, 0, 1), e_{jk} = (0, 0, 0, Y_j^k)$ with the spherical harmonics Y_j^k of degree one for the spheres Σ^k, $j = 1, \ldots, n$, $k = 1, \ldots, m$, if the densities are not equal. Otherwise e_h has to be dropped, and e_θ has to be replaced by $e_\theta = (0, 1, 1, -l_0 R_*^2/\theta_*\sigma_*(n-1))$.

If there is no phase transition, instead of e_h we have the elements $e_{0,k} = (0,0,0,Y_0^k)$ in the kernel, where Y_0^k equals one on Σ^k and zero elsewhere. In this case the dimension of the null space is $m(n+1)+1$. Thus we see that in all cases the dimension of the eigenspace equals the dimension of the corresponding component of the manifold of equilibria $\mathcal{E}$.

To show that the equilibria are normally stable or hyperbolic, it remains to prove that $\lambda = 0$ is semi-simple. Here we have to distinguish three cases.

Case 1: $\varrho_1 \neq \varrho_2$ with phase transition.
So suppose we have a solution of $L_j(u,\vartheta,\vartheta_\Sigma,h) = \sum_{j,k}\alpha_{jk}e_{jk} + \beta e_\theta + \gamma e_h$. This means

$$\begin{aligned}
-\mu_*\Delta u + \nabla\pi &= 0 && \text{in } \Omega\setminus\Sigma,\\
\operatorname{div} u &= 0 && \text{in } \Omega\setminus\Sigma,\\
u &= 0 && \text{on } \partial\Omega,\\
u - u_\Sigma - j_\Sigma\nu_\Sigma/\varrho &= 0 && \text{on } \Sigma,\\
-[\![T\nu_\Sigma]\!] + \sigma_*\mathcal{A}_\Sigma h\nu_\Sigma - \sigma_*'(\nabla_\Sigma\vartheta_\Sigma + H_\Sigma\vartheta_\Sigma\nu_\Sigma) &= 0 && \text{on } \Sigma.
\end{aligned} \tag{10.32}$$

$$\begin{aligned}
-d_*\Delta\vartheta &= \varrho\kappa_*\beta && \text{in } \Omega\setminus\Sigma,\\
\partial_\nu\vartheta &= 0 && \text{on } \partial\Omega,\\
[\![\vartheta]\!] = 0,\ \vartheta &= \vartheta_\Sigma && \text{on } \Sigma.
\end{aligned} \tag{10.33}$$

$$\begin{aligned}
-d_{\Sigma*}\Delta_\Sigma\vartheta_\Sigma - \theta_*\sigma_*'\operatorname{div}_\Sigma u_\Sigma - l_*j_\Sigma - [\![d_*\partial_\nu\vartheta]\!] &= \kappa_{\Sigma*}\beta && \text{on } \Sigma,\\
-[\![T\nu_\Sigma\cdot\nu_\Sigma/\varrho]\!] + (l_*/\theta_*)\vartheta_\Sigma &= 0 && \text{on } \Sigma,\\
-u_\Sigma\cdot\nu_\Sigma &= \sum_{j,k}\alpha_{jk}Y_j^k + \gamma && \text{on } \Sigma.
\end{aligned} \tag{10.34}$$

We want to show $(\alpha_{jk},\beta,\gamma) = 0$ for all j,k. Integrating the divergence equation for u over Ω we get

$$0 = -\int_\Omega \operatorname{div} u\,dx = \int_\Sigma [\![u\cdot\nu_\Sigma]\!]\,d\Sigma = [\![1/\varrho]\!]\int_\Sigma j_\Sigma\,d\Sigma,$$

hence $\int_\Sigma j_\Sigma\,d\Sigma = 0$, and

$$0 = \int_{\Omega_1}\operatorname{div} u\,dx = \int_\Sigma u\cdot\nu_\Sigma\,d\Sigma = \int_\Sigma u_\Sigma\cdot\nu_\Sigma\,d\Sigma.$$

Since the spherical harmonics have mean zero this implies $\gamma = 0$.

Integrating the equation for ϑ yields

$$\beta(\kappa_*|\varrho)_\Omega = -\int_\Omega d_*\Delta\vartheta\,dx = \int_\Sigma [\![d_*\partial_\nu\vartheta]\!]\,d\Sigma,$$

and integrating that for ϑ_Σ we get

$$\beta\kappa_{\Sigma*}|\Sigma| = -\int_\Sigma [\![d_*\partial_\nu\vartheta]\!]\,d\Sigma + \theta_*\sigma_*' H_\Sigma\int_\Sigma u_\Sigma\cdot\nu_\Sigma\,d\Sigma = -\int_\Sigma [\![d_*\partial_\nu\vartheta]\!]\,d\Sigma.$$

Combining these identities yields $\beta = 0$.

Taking the inner product of the equation for u with u, that for ϑ with ϑ and that for ϑ_Σ with ϑ_Σ we obtain as in the previous section

$$0 = 2|\mu_*^{1/2}D|_2^2 - \sigma_*'(\nabla_\Sigma\vartheta_\Sigma|u_\Sigma)_\Sigma - \sigma_*'H_\Sigma(\vartheta_\Sigma\nu_\Sigma|u_\Sigma) + ([\![T\nu_\Sigma\cdot\nu_\Sigma/\varrho]\!]|j_\Sigma)_\Sigma,$$

and

$$0 = |d_*^{1/2}\nabla\vartheta|_2^2 + ([\![d_*\partial_\nu\vartheta]\!]|\vartheta)_\Sigma,$$

as well as

$$0 = |d_{\Sigma*}^{1/2}\nabla_\Sigma\vartheta_\Sigma|_\Sigma^2 - ([\![d_*\partial_\nu\vartheta]\!]|\vartheta)_\Sigma - l_*(j_\Sigma|\vartheta_\Sigma)_\Sigma - \theta_*\sigma_*'(\mathrm{div}_\Sigma u_\Sigma|\vartheta_\Sigma).$$

Adding these equations and employing the linear Gibbs-Thomson law $(l_*/\theta_*)\vartheta_\Sigma = [\![T\nu_\Sigma\cdot\nu_\Sigma/\varrho]\!]$ yields

$$2|\mu_*^{1/2}D|_2^2 + (|d_*^{1/2}\nabla\vartheta|_2^2 + |d_{\Sigma*}^{1/2}\nabla_\Sigma\vartheta_\Sigma|_\Sigma^2)/\theta_* = 0.$$

This implies $D = 0$, ϑ constant, hence $u = 0$ by Lemma 1.2.1 and then $j_\Sigma = 0$, as $[\![1/\varrho]\!]j_\Sigma = [\![u\cdot\nu_\Sigma]\!] = 0$. This, in turn, yields

$$0 = -u\cdot\nu_\Sigma = -u_\Sigma\cdot\nu_\Sigma = \sum_{j,k}\alpha_{jk}Y_j^k.$$

Thus $\alpha_{jk} = 0$ for all j, k since the spherical harmonics Y_j^k are linearly independent. Therefore, the eigenvalue $\lambda = 0$ is semi-simple.

Case 2: $\varrho_1 = \varrho_2 =: \varrho$ with phase transition.
The argument is similar to that in Case 1. However, due to the different kernel of L_j we have to replace γ by $-\beta l_0 R_*^2/\theta_*\sigma_*(n-1)$. Integrating the equations for ϑ, ϑ_Σ and h we obtain the identity

$$\beta\big((\kappa_*|\varrho)_\Omega + \kappa_{\Sigma*}|\Sigma|\big) = \beta|\Sigma|l_0^2R_*^2/\theta_*\sigma_*(n-1),$$

since the mean values of Y_j^k are zero. This relates to equality in the stability condition (10.31); if equality in (10.31) does not hold we may conclude $\beta = 0$, and proceed as in Case 1 to show that 0 is semi-simple.

Case 3: No phase transition.
Here we can argue exactly in the same way as in Case 1. Note that in this case the Gibbs-Thomson relation is not needed as $j_\Sigma = 0$.

Concluding we see that in all cases the eigenvalue 0 is semi-simple and $\mathsf{N}(L_j)$ is isomorphic to the tangent space of $\mathcal{E}$, hence $-L_j$ is normally stable or normally hyperbolic, provided $\zeta_* \neq 1$ in case $\varrho_1 = \varrho_2$.

Remark. In the case of equal densities, if $\zeta_* = 1$ in (10.31), then we can show that 0 is no longer a semi-simple eigenvalue, the dimension of its generalized eigenspace raises by 1.

We conclude by summarizing our results for Problem **(P2)**.

Theorem 10.4.1. *Let L_2 denote the linearization at the equilibrium $e_* = (0, \theta_*, \Sigma) \in \mathcal{E}$ as defined above. Then $-L_2$ generates a compact analytic C_0-semigroup in X_0^2 which has maximal L_p-regularity. The spectrum of L_2 consists only of eigenvalues of finite algebraic multiplicity. Moreover, the following assertions are valid.*

(i) *The operator $-L_2$ has no eigenvalues $\lambda \neq 0$ with nonnegative real part.*
(ii) *$\lambda = 0$ is an eigenvalue of L_2 and it is semi-simple.*
(iii) *The kernel $\mathsf{N}(L_2)$ of L_2 is isomorphic to the tangent space $T_{e_*}\mathcal{E}$ of the manifold of equilibria $\mathcal{E}_2$ at e_*.*

Consequently, $e_* \in \mathcal{E}$ is normally stable.

10.5 Unstable Eigenvalues: Problems 1 and 3

We assume in this section $\varrho_1 = \varrho_2 =: \varrho$ and $\sigma > 0$ constant. If the stability condition (10.31) does not hold, or if Σ is disconnected, then there is always at least one positive eigenvalue. To prove this we proceed as follows. Suppose $\lambda > 0$ is an eigenvalue, solve the Stokes problem

$$\begin{aligned} \varrho\lambda u - \mu_*\Delta u + \nabla\pi &= 0 && \text{in } \Omega\setminus\Sigma,\\ \operatorname{div} u &= 0 && \text{in } \Omega\setminus\Sigma,\\ u &= 0 && \text{on } \partial\Omega,\\ [\![u]\!] &= 0 && \text{on } \Sigma,\\ -[\![T\nu_\Sigma]\!] &= g\nu_\Sigma && \text{on } \Sigma, \end{aligned} \tag{10.35}$$

to obtain as an output $N_\lambda^S g := u\cdot\nu_\Sigma$.

Next we solve the heat problem

$$\begin{aligned} \varrho\kappa_*\lambda\vartheta - d_*\Delta\vartheta &= 0 && \text{in } \Omega\setminus\Sigma,\\ \partial_\nu\vartheta &= 0 && \text{on } \partial\Omega,\\ [\![\vartheta]\!] = 0, \quad \vartheta &= g && \text{on } \Sigma, \end{aligned} \tag{10.36}$$

to obtain $D_\lambda^H g := -[\![d_*\partial_\nu\vartheta]\!]$, where D_λ^H denotes the two-phase Dirichlet-to-Neumann operator for the heat problem. Inserting the relation

$$(l_0/\theta_*)\vartheta_\Sigma := (\varrho l_*/\theta_*)\vartheta_\Sigma = \sigma\mathcal{A}_\Sigma h,$$

into the equation for h in (10.11) we obtain

$$0 = \lambda h + [N_\lambda^S + (\theta_*/l_0^2)D_\lambda^H]\sigma\mathcal{A}_\Sigma h,$$

hence with

$$T_\lambda = [N_\lambda^S + (\theta_*/l_0^2)D_\lambda^H]^{-1},$$

we finally arrive at the equation

$$\lambda T_\lambda h + \sigma \mathcal{A}_\Sigma h = 0. \tag{10.37}$$

$\lambda > 0$ is an eigenvalue of $-L_j$, $j = 1, 3$, if and only if (10.37) admits a nontrivial solution.

We consider this problem in $L_2(\Sigma)$. The operator $\mathcal{A}_\Sigma$ is selfadjoint and

$$\sigma(\mathcal{A}_\Sigma h|h)_{L_2(\Sigma)} \geq -\frac{\sigma(n-1)}{R_*^2}|h|^2_{L_2(\Sigma)}.$$

On the other hand, we will see below that $N_\lambda^H = [D_\lambda^H]^{-1}$ and N_λ^S are selfadjoint and positive semi-definite on $L_2(\Sigma)$ hence T_λ is selfadjoint and positive semi-definite as well. Moreover, since $\mathcal{A}_\Sigma$ has compact resolvent, the operator

$$B_\lambda := \lambda T_\lambda + \sigma \mathcal{A}_\Sigma$$

has compact resolvent, for each $\lambda > 0$. Therefore the spectrum of B_λ consists only of eigenvalues which in addition are real. We intend to prove that in case either Σ is disconnected or the stability condition does not hold, B_{λ_0} has 0 as an eigenvalue, for some $\lambda_0 > 0$.

To proceed, we need properties of the relevant Neumann-to-Dirichlet operators. Firstly, we consider the heat problem

$$\begin{aligned} \varrho\kappa_*\lambda\vartheta - d_*\Delta\vartheta &= 0 && \text{in } \Omega\setminus\Sigma,\\ \partial_\nu\vartheta &= 0 && \text{on } \partial\Omega,\\ [\![\vartheta]\!] &= 0 && \text{on } \Sigma,\\ -[\![d_*\partial_\nu\vartheta]\!] &= g && \text{on } \Sigma, \end{aligned} \tag{10.38}$$

to obtain $\vartheta = N_\lambda^H g$, where N_λ^H denotes the Neumann-to-Dirichlet operator for this heat problem. The properties of N_λ^H are summarized in the following proposition. We denote by e the function which is identically one on Σ.

Proposition 10.5.1. *The Neumann-to-Dirichlet operator* $N_\lambda^H = [D_\lambda^H]^{-1}$ *for the diffusion problem* (10.38) *admits a compact selfadjoint extension to* $L_2(\Sigma)$ *which has the following properties.*

(i) *If* ϑ *denotes the solution of* (10.38), *then*

$$(N_\lambda^H g|g)_{L_2(\Sigma)} = \lambda|\sqrt{\varrho\kappa_*}\vartheta|^2_{L_2(\Omega)} + |\sqrt{d_*}\nabla\vartheta|^2_{L_2(\Omega)}, \quad \lambda > 0,\ g \in L_2(\Sigma);$$

in particular, N_λ^H *is injective for* $\lambda > 0$.

(ii) *For each* $\alpha \in (0, 1/2)$ *and* $\lambda_0 > 0$ *there is a constant* $C > 0$ *such that*

$$(N_\lambda^H g|g)_{L_2(\Sigma)} \geq \frac{\lambda^\alpha}{C}|N_\lambda^H g|^2_{L_2(\Sigma)}, \quad g \in L_2(\Sigma),\ \lambda \geq \lambda_0;$$

hence

$$|N_\lambda^H|_{\mathcal{B}(L_2(\Sigma))} \le \frac{C}{\lambda^\alpha}, \quad \lambda \ge \lambda_0.$$

(iii) *On* $L_{2,0}(\Sigma) := \{g \in L_2(\Sigma) : (g|\mathsf{e})_{L_2(\Sigma)} = 0\}$, *we even have*

$$(N_\lambda^H g|g)_{L_2(\Sigma)} \ge \frac{(1+\lambda)^\alpha}{C}|N_\lambda^H g|^2_{L_2(\Sigma)}, \quad g \in L_{2,0}(\Sigma),\ \lambda > 0,$$

and

$$|N_\lambda^H|_{\mathcal{B}(L_{2,0}(\Sigma),L_2(\Sigma))} \le \frac{C}{(1+\lambda)^\alpha}, \quad \lambda > 0.$$

In particular, for $\lambda = 0$, (10.38) *is solvable if and only if* $(g|\mathsf{e})_{L_2(\Sigma)} = 0$, *and then the solution is unique up to a constant.*

Proof. We refer to Section 6.5.4 for background on elliptic Dirichlet-to-Neumann operators.

(a) By the divergence theorem and elliptic theory, the assertion in (i) holds true for all functions $g \in H_2^{1/2}(\Sigma)$, and all $\lambda > 0$.

(b) Let $g \in H_2^{1/2}(\Sigma)$ be given and let $\vartheta = \vartheta_\lambda \in H_2^2(\Omega \setminus \Sigma)$ be the solution of (10.38) with $\lambda > 0$. It follows from complex interpolation and trace theory that

$$\begin{aligned}|\vartheta|_{L_2(\Sigma)} \le C|\vartheta|_{H_2^{1-\alpha}(\Omega_k)} &\le \frac{C}{\lambda^{\alpha/2}}|\lambda^{1/2}\vartheta|^\alpha_{L_2(\Omega_k)}\,|\vartheta|^{1-\alpha}_{H_2^1(\Omega_k)}\\ &\le \frac{C}{\lambda^{\alpha/2}}(\lambda^{1/2}|\vartheta|_{L_2(\Omega_k)} + |\vartheta|_{L_2(\Omega_k)} + |\nabla\vartheta|_{L_2(\Omega_k)}),\end{aligned}$$

where C is a generic constant and $\alpha \in (0,1/2)$. For $\lambda_0 > 0$ fixed we then obtain

$$\lambda^\alpha|\vartheta|^2_{L_2(\Sigma)} \le C(\lambda|\sqrt{\varrho\kappa_*}\vartheta|^2_{L_2(\Omega)} + |\sqrt{d_*}\nabla\vartheta|^2_{L_2(\Omega)}), \quad \lambda \ge \lambda_0,$$

and hence

$$(N_\lambda^H g|g)_{L_2(\Sigma)} \ge \frac{\lambda^\alpha}{C}|N_\lambda^H g|^2_{L_2(\Sigma)}, \quad g \in H_2^{1/2}(\Sigma), \quad \lambda \ge \lambda_0,$$

which implies the estimate

$$|N_\lambda^H g|_{L_2(\Sigma)} \le \frac{C}{\lambda^\alpha}|g|_{L_2(\Sigma)}, \quad g \in H_2^{1/2}(\Sigma), \quad \lambda \ge \lambda_0. \tag{10.39}$$

As $H_2^{1/2}(\Sigma)$ is dense in $L_2(\Sigma)$ we can now conclude that N_λ admits a bounded extension to $L_2(\Sigma)$. Therefore, the assertions in (i) and (ii) are also valid for functions $g \in L_2(\Sigma)$. Finally, the third assertion follows from the Poincaré-Wirtinger inequality. □

Secondly, we consider the Neumann-to-Dirichlet operator N_λ^S for the Stokes problem. It is defined as follows. Given a function $g \in W_p^{1-1/p}(\Sigma)$, we solve the Stokes problem

$$\begin{aligned} \varrho\lambda u - \mu_*\Delta u + \nabla\pi &= 0 && \text{in } \Omega\setminus\Sigma,\\ \operatorname{div} u &= 0 && \text{in } \Omega\setminus\Sigma,\\ u &= 0 && \text{on } \partial\Omega,\\ [\![u]\!] &= 0 && \text{on } \Sigma,\\ -[\![T\nu_\Sigma]\!] &= g\nu_\Sigma && \text{on } \Sigma, \end{aligned} \tag{10.40}$$

and define $N_\lambda^S g := u\cdot\nu_\Sigma$ on Σ. For this well-defined operator we have

Proposition 10.5.2. *The Neumann-to-Dirichlet operator N_λ^S for the Stokes problem* (10.35) *admits a compact selfadjoint extension to $L_2(\Sigma)$ which has the following properties.*

(i) *If u denotes the solution of* (10.35)*, then*

$$(N_\lambda^S g|g)_{L_2(\Sigma)} = \lambda\int_\Omega \varrho|u|^2\,dx + 2\int_\Omega \mu_*|D|_2^2\,dx, \quad \lambda\geq 0,\ g\in L_2(\Sigma).$$

(ii) *For each $\alpha\in(0,1/2)$ there is a constant $C>0$ such that*

$$(N_\lambda^S g|g)_{L_2(\Sigma)} \geq \frac{(1+\lambda)^\alpha}{C}|N_\lambda^S g|^2_{L_2(\Sigma)}, \quad g\in L_2(\Sigma),\ \lambda\geq 0.$$

In particular,

$$|N_\lambda^S|_{\mathcal{B}(L_2(\Sigma))} \leq \frac{C}{(1+\lambda)^\alpha}, \quad \lambda\geq 0.$$

(iii) *Let Σ^k denote the components of Σ and let e_k be the function which is one on Σ^k, zero elsewhere. Then $(N_\lambda^S g|\mathsf{e}_k)_{L_2(\Sigma)} = 0$ for each $g\in L_2(\Sigma)$. In particular, $N_\lambda^S \mathsf{e}_k = 0$ for each k, and $N_\lambda^S g$ has mean value zero for each $g\in L_2(\Sigma)$.*

Proof. For $g\in H_2^{1/2}(\Sigma)$, the first assertion follows from the divergence theorem. The second assertion follows by Korn's inequality and trace theory, and the third one comes from $\operatorname{div} u = 0$. The arguments are similar to those in the proof of Proposition 10.5.1. Note that N_λ^S even on $L_{2,0}(\Sigma)$ is not injective in case Σ is disconnected. □

(a) Consider $v_\lambda := T_\lambda\mathsf{e}$, or equivalently $\mathsf{e} = N_\lambda^S v_\lambda + (c_* N_\lambda^H)^{-1} v_\lambda$, where we used the abbreviation $c_* = l_0^2/\theta_*$. Denoting the orthogonal projection from $L_2(\Sigma)$ to $L_{2,0}(\Sigma)$ by P_0, the equation for v_λ is equivalent to

$$v_\lambda + c_* N_\lambda^H P_0 N_\lambda^S v_\lambda = c_* N_\lambda^H \mathsf{e},$$

due to Proposition 10.5.2. Multiplying this identity in $L_2(\Sigma)$ by $N_\lambda^S v_\lambda$ we obtain with Propositions 10.5.2 and 10.5.1

$$\begin{aligned} c(\lambda)|N_\lambda^S v_\lambda|^2_{L_2(\Sigma)} &\leq (v_\lambda + c_* N_\lambda^H N_\lambda^S v_\lambda | N_\lambda^S v_\lambda)_{L_2(\Sigma)} = (c_* N_\lambda^H \mathsf{e}|N_\lambda^S v_\lambda)_{L_2(\Sigma)}\\ &= c_*(\mathsf{e}|N_\lambda^H P_0 N_\lambda^S v_\lambda)_{L_2(\Sigma)} \leq C(\lambda)|N_\lambda^S v_\lambda|_{L_2(\Sigma)}, \end{aligned}$$

where $c(\lambda)$ and $C(\lambda)$ are bounded near $\lambda = 0$. This shows that $N_\lambda^S v_\lambda$ is bounded near $\lambda = 0$. This implies

$$\lim_{\lambda\to 0} \lambda T_\lambda \mathsf{e} = \lim_{\lambda\to 0} \lambda v_\lambda = c_* \lim_{\lambda\to 0} \lambda N_\lambda^H \mathsf{e},$$

provided the latter limit exists.

To compute this limit, we proceed as follows. First we solve the problem

$$\begin{aligned} -d_*\Delta\vartheta &= -\varrho\kappa_* a_0 && \text{in } \Omega\setminus\Sigma,\\ \partial_\nu\vartheta &= 0 && \text{on } \partial\Omega,\\ [\![\vartheta]\!] &= 0 && \text{on } \Sigma,\\ -[\![d_*\partial_\nu\vartheta]\!] &= \mathsf{e} && \text{on } \Sigma, \end{aligned} \tag{10.41}$$

where $a_0 = |\Sigma|/(\kappa_*|\varrho)_\Omega$, which is solvable since the necessary compatibility condition holds. We denote the solution by ϑ_0 and normalize it by $(\varrho\kappa_*|\vartheta_0)_\Omega = 0$. Then $\vartheta_\lambda = N_\lambda^H\mathsf{e} - \vartheta_0 - a_0/\lambda$ satisfies the problem

$$\begin{aligned} \varrho\kappa_*\lambda\vartheta - d_*\Delta\vartheta &= -\varrho\kappa_*\lambda\vartheta_0 && \text{in } \Omega\setminus\Sigma,\\ \partial_\nu\vartheta &= 0 && \text{on } \partial\Omega,\\ [\![\vartheta]\!] &= 0 && \text{on } \Sigma,\\ -[\![d_*\partial_\nu\vartheta]\!] &= 0 && \text{on } \Sigma. \end{aligned} \tag{10.42}$$

By the normalization $(\varrho\kappa_*|\vartheta_0)_\Omega = 0$ we see that ϑ_λ is bounded in $H_2^2(\Omega\setminus\Sigma)$ as $\lambda\to 0$. This then implies

$$\lim_{\lambda\to 0} \lambda N_\lambda^H\mathsf{e} = \lim_{\lambda\to 0}[\lambda\vartheta_\lambda + \lambda\vartheta_0 + a_0] = a_0 = |\Sigma|/(\kappa_*|\varrho)_\Omega,$$

and therefore

$$\lim_{\lambda\to 0}(B_\lambda\mathsf{e}|\mathsf{e})_{L_2(\Sigma)} = c_*\frac{|\Sigma|^2}{(\kappa_*|\varrho)_\Omega} - \sigma|\Sigma|\frac{(n-1)}{R_*^2} < 0,$$

if the stability condition does not hold, that is, if $\zeta_* > 1$.

(b) Next suppose that Σ is disconnected, i.e., $\Sigma = \cup_{k=1}^m \Sigma^k$, and set $g = \sum_k a_k\mathsf{e}_k \neq 0$ with $\sum a_k = 0$, where $\mathsf{e}_k = 1$ on Σ^k, and zero elsewhere. Hence $P_0 g = g$. Then for $v_\lambda := T_\lambda\mathsf{e}$ we have as in (a) boundedness of $N_\lambda^S v_\lambda$ and furthermore

$$\lim_{\lambda\to 0} \lambda T_\lambda g = \lim_{\lambda\to 0} \lambda v_\lambda = c_* \lim_{\lambda\to 0} \lambda N_\lambda^H P_0 g = 0,$$

as $N_\lambda^H P_0$ is bounded as $\lambda\to 0$. This implies

$$\lim_{\lambda\to 0}(B_\lambda g|g)_{L_2(\Sigma)} = -\frac{\sigma(n-1)}{R_*^2}\sum_k |\Sigma^k| a_k^2 < 0.$$

(c) Next we consider the behaviour of $(B_\lambda g|g)_{L_2(\Sigma)}$ as $\lambda \to \infty$. With $c_* = l_0^2/\theta_*$ as above we first have

$$T_\lambda = (I + c_* N_\lambda^H N_\lambda^S)^{-1} c_* N_\lambda^H = c_* N_\lambda^H - c_* N_\lambda^H N_\lambda^S (I + c_* N_\lambda^H N_\lambda^S)^{-1} c_* N_\lambda^H,$$

hence by Propositions 10.5.1, 10.5.2 for $\lambda \geq \lambda_0$, with λ_0 sufficiently large,

$$\begin{aligned}
(T_\lambda g|g)_{L_2(\Sigma)} &= c_*(N_\lambda^H g|g)_{L_2(\Sigma)} - c_*^2 (N_\lambda^S (I + c_* N_\lambda^H N_\lambda^S)^{-1} N_\lambda^H g | N_\lambda^H g)_{L_2(\Sigma)} \\
&\geq c_* [(N_\lambda^H g|g)_{L_2(\Sigma)} - c_*^2 \frac{|N_\lambda^S|_{L_2(\Sigma)}}{1 - c_* |N_\lambda^H|_{L_2(\Sigma)} |N_\lambda^S|_{L_2(\Sigma)}} |N_\lambda^H g|^2_{L_2(\Sigma)}] \\
&\geq c_* [(N_\lambda^H g|g)_{L_2(\Sigma)} - \frac{C\lambda_0^{-\alpha} |N_\lambda^S|_{L_2(\Sigma)}}{1 - c_* |N_\lambda^H|_{L_2(\Sigma)} |N_\lambda^S|_{L_2(\Sigma)}} (N_\lambda^H g|g)_{L_2(\Sigma)}] \\
&\geq c_* [(N_\lambda^H g|g)_{L_2(\Sigma)} - \frac{1}{2} (N_\lambda^H g|g)_{L_2(\Sigma)}] = \frac{c_*}{2} (N_\lambda^H g|g)_{L_2(\Sigma)}.
\end{aligned}$$

Therefore, it is sufficient to show that there are constants $c > 0$ and $\lambda_1 > 0$ such that

$$(c_*/2)\lambda (N_\lambda^H g|g)_{L_2(\Sigma)} + \sigma (\mathcal{A}_\Sigma g|g)_{L_2(\Sigma)} \geq c|g|^2_{L_2(\Sigma)}, \quad g \in H_2^2(\Sigma),\ \lambda \geq \lambda_1. \tag{10.43}$$

For this purpose we introduce the orthogonal projections P and Q by

$$Pg = \sum_{k=1}^{m(n+1)} (g|\mathsf{a}_k)_{L_2(\Sigma)} \mathsf{a}_k, \quad Q = I - P,$$

where a_k are normalized orthogonal eigenfunctions coming from the eigenvalues $-(n-1)/R_*^2$ and 0 of $\mathcal{A}_\Sigma$ in $L_2(\Sigma)$.

Suppose on the contrary that (10.43) does not hold. Then there is a sequence $\lambda_j \to \infty$ and $g_j \in H_2^2(\Sigma)$ with $|g_j|_{L_2(\Sigma)} = 1$, such that

$$(c_*/2)\lambda_j (N_{\lambda_j}^H g_j|g_j)_{L_2(\Sigma)} + \sigma (\mathcal{A}_\Sigma Q g_j | Q g_j)_{L_2(\Sigma)} \leq \frac{1}{j} + C|Pg_j|^2_{L_2(\Sigma)}.$$

As $\mathcal{A}_\Sigma$ is positive definite on $\mathsf{R}(Q)$ this implies that $\lambda_j (N_{\lambda_j}^H g_j|g_j)_{L_2(\Sigma)}$ is bounded. Then the corresponding solution ϑ_j of (10.38) is such that $v_j := \lambda_j \vartheta_j$ and $\nabla v_j/\sqrt{\lambda_j}$ are bounded in $L_2(\Omega)$, as by Proposition 10.5.1,

$$\lambda_j \big(\lambda_j |\sqrt{\kappa_*}\vartheta_j|^2_{L_2(\Omega)} + |\sqrt{d_*}\nabla\vartheta_j|^2_{L_2(\Omega)}\big) = \lambda_j (N_{\lambda_j}^H g_j|g_j)_{L_2(\Sigma)}.$$

Hence v_j has a weakly convergent subsequence, and we can assume w.l.o.g. that $v_j \to v_\infty$ weakly in $L_2(\Omega)$. Fix a test function $\psi \in \mathcal{D}(\Omega \setminus \Sigma)$. Then

$$(\varrho\kappa_* v_j|\psi)_{L_2(\Omega)} = (d_*\Delta\vartheta_j|\psi)_{L_2(\Omega)} = (\vartheta_j|d_*\Delta\psi)_{L_2(\Omega)} = (v_j|d_*\Delta\psi)_{L_2(\Omega)}/\lambda_j \to 0$$

as $j \to \infty$, hence $v_\infty = 0$ in $L_2(\Omega)$. On the other hand we have for $\phi \in {}_0H_2^1(\Omega)$ extending a_k

$$\begin{aligned}(g_j|\mathsf{a}_k)_{L_2(\Sigma)} &= -\int_\Sigma [\![d_*\partial_\nu\vartheta_j]\!]\mathsf{a}_k\,d\Sigma = \int_\Omega \operatorname{div}(d_*\nabla\vartheta_j\phi)\,dx \\ &= (\varrho\kappa_* v_j|\phi)_{L_2(\Omega)} + (d_*\nabla\vartheta_j|\nabla\phi)_{L_2(\Omega)} \to 0\end{aligned}$$

as $j \to \infty$. Therefore, $Pg_j \to 0$ in $L_2(\Sigma)$ as $j \to \infty$, and as $\mathcal{A}_\Sigma Q$ is positive definite, we also obtain $Qg_j \to 0$, which contradicts $|g_j|_{L_2(\Sigma)} = 1$. This implies that (10.43) is valid.

(d) Summarizing, we have shown that B_λ is not positive semi-definite for small $\lambda > 0$ if either Σ is not connected or the stability condition does not hold, and B_λ is always positive semi-definite for large λ. Set

$$\lambda_0 = \sup\{\lambda > 0 : B_\mu \text{ is not positive semi-definite for each } \mu \in (0,\lambda]\}.$$

Since B_λ has compact resolvent, B_λ has a negative eigenvalue for each $\lambda < \lambda_0$. This implies that 0 is an eigenvalue of B_{λ_0}, thereby proving that $-L$ admits the positive eigenvalue λ_0.

Moreover, we have also shown that

$$B_0 h = \lim_{\lambda\to 0} \lambda T_\lambda h + \sigma\mathcal{A}_\Sigma h = c_*\frac{|\Sigma|}{(\kappa_*|1)_\Omega}(I - P_0)h + \sigma\mathcal{A}_\Sigma h.$$

Therefore, B_0 has eigenvalue

$$c_*\frac{|\Sigma|}{(\kappa_*|1)_\Omega} - \frac{\sigma(n-1)}{R_*^2} = \frac{l_0^2|\Sigma|}{\theta_*(\kappa_*|1)_\Omega}[1 - \zeta_*]$$

with eigenfunction e, and in case $m > 1$ it also has eigenvalue $-\sigma(n-1)/R_*^2$ with precisely $m-1$ linearly independent eigenfunctions of the form $\sum_k a_k\mathsf{e}_k$ with $\sum_k a_k = 0$.

As λ varies from 0 to λ_0, all the negative eigenvalues of B_0 identified above will eventually have to cross 0 along the real axis. At each of these occasions, $-L$ will inherit at least one positive eigenvalue, which will then remain positive. This implies that $-L$ has exactly m positive eigenvalues if the stability condition does not hold, and $m-1$ otherwise.

(e) For Problem **(P1)**, we only have to set $u \equiv 0$ in the previous derivations, to obtain the same results as for Problem **(P3)**.

Let us summarize what we have proved.

Theorem 10.5.3. *Let L_j, $j = 1, 3$, denote the linearization at $e_* := (0, \theta_*, \Sigma) \in \mathcal{E}$ as defined above. Then $-L_j$ generates a compact analytic C_0-semigroup in X_0^j which has maximal L_p-regularity. The spectrum $\sigma(L_j)$ of L_j consists only of eigenvalues of finite algebraic multiplicity. Moreover, the following assertions are valid.*

(i) *The operator* $-L_j$ *has no eigenvalues* $\lambda \neq 0$ *with nonnegative real part if and only if* Σ *is connected and*

$$\zeta_* := \frac{\theta_* \sigma(n-1)}{l_0^2 R_*^2 |\Sigma|} (\kappa_*|1)_\Omega \le 1. \tag{10.44}$$

(ii) *If* Σ *has* $m \geq 2$ *components and* (10.44) *holds, then* $-L_j$ *has precisely* $m-1$ *positive eigenvalues.*
(iii) *If* Σ *has* $m \geq 1$ *components and* (10.44) *does not hold, then* $-L_j$ *has precisely* m *positive eigenvalues.*
(iv) $\lambda = 0$ *is an eigenvalue of* L_j. *It is semi-simple as long as* $\zeta_* \neq 1$.
(v) *The kernel* $\mathsf{N}(L_j)$ *of* L_j *is isomorphic to the tangent space* $T_{e_*}\mathcal{E}$ *of the manifold of equilibria* $\mathcal{E}$ *at* e_*.

Consequently, $e_* = (0, \theta_*, \Sigma) \in \mathcal{E}$ *is normally stable if and only if* Σ *is connected and* $\zeta_* < 1$, *and normally hyperbolic if and only if* Σ *is disconnected or* $\zeta_* > 1$.

10.6 Unstable Eigenvalues: Problem 5

Assume that $\lambda > 0$ is an eigenvalue. Then as in the previous section we have $-[\![d_* \partial_\nu \vartheta]\!] = D_\lambda^H \vartheta_\Sigma$. Next we solve the *extended Stokes problem*

$$\begin{aligned} \varrho\lambda u - \mu_* \Delta u + \nabla \pi &= 0 && \text{in } \Omega \setminus \Sigma, \\ \operatorname{div} u &= 0 && \text{in } \Omega \setminus \Sigma, \\ u &= 0 && \text{on } \Sigma, \\ [\![u]\!] &= 0 && \text{on } \Sigma, \\ -[\![T\nu_\Sigma \cdot \nu_\Sigma]\!] &= g_1 && \text{on } \Sigma, \\ -\mathcal{P}_\Sigma [\![T\nu_\Sigma]\!] &= g_2 && \text{on } \Sigma, \end{aligned} \tag{10.45}$$

to obtain as output

$$N_\lambda = \begin{bmatrix} u \cdot \nu_\Sigma \\ \mathcal{P}_\Sigma u \end{bmatrix}.$$

For this problem we have the following result which extends Proposition 10.5.2.

Proposition 10.6.1. *The extended Neumann-to-Dirichlet operator* N_λ *for the Stokes problem* (10.45) *admits a compact selfadjoint extension to* $L_2(\Sigma; \mathbb{R} \times T\Sigma)$ *which has the following properties.*
(i) *If* u *denotes the solution of* (10.45), *then*

$$(N_\lambda g|g)_{L_2} = \lambda \int_\Omega \varrho |u|^2 \, dx + 2 \int_\Omega \mu_* |D|_2^2 \, dx, \quad \lambda \geq 0, \; g \in L_2(\Sigma, \mathbb{R} \times T\Sigma).$$

(ii) *For each* $\alpha \in (0, 1/2)$ *there is a constant* $C > 0$ *such that*

$$(N_\lambda g|g)_{L_2} \geq \frac{(1+\lambda)^\alpha}{C} |N_\lambda g|_{L_2}^2, \quad g \in L_2(\Sigma; \mathbb{R} \times T\Sigma), \; \lambda \geq 0.$$

In particular,

$$|N_\lambda|_{\mathcal{B}(L_2)} \le \frac{C}{(1+\lambda)^\alpha}, \quad \lambda \ge 0.$$

(iii) *Let Σ_k denote the components of Σ and let e_k be the function which is one on Σ_k, zero elsewhere. Then the kernel $\mathsf{N}(N_\lambda)$ consists of functions g such that $g_2 = 0$ and $g_1 = \sum_{k=1}^m \alpha_k \mathsf{e}_k$ with arbitrary numbers α_k.*

Proof. The proof is similar to that of Proposition 10.5.2: the assertion in (i) is first established for functions $g \in H_2^{1/2}(\Sigma; \mathbb{R} \times T\Sigma)$. In a second step, one shows that N_λ admits an extension to $L_2(\Sigma; \mathbb{R} \times T\Sigma)$ which, in turn, allows us to extend the relation in (i) for functions $g \in L_2(\Sigma; \mathbb{R} \times T\Sigma)$. □

Now we define the operator $Q : H_2^1(\Sigma) \to L_2(\Sigma, \mathbb{R} \times T\Sigma)$ by means of

$$Qw := \sigma_*' \begin{bmatrix} H_\Sigma w \\ \nabla_\Sigma w \end{bmatrix}.$$

In the following we identify $[v_1, v_2]^\mathsf{T} \in \mathbb{R} \times T\Sigma$ with $v \in N\Sigma \oplus T\Sigma$ by means of $v = v_1 \nu_\Sigma + v_2$. Then

$$Q^* v = \sigma_*'(H_\Sigma v \cdot \nu_\Sigma - \mathrm{div}_\Sigma \mathcal{P}_\Sigma v) = -\sigma_*' \mathrm{div}_\Sigma\, v.$$

Set $Pv := v \cdot \nu_\Sigma$. As

$$\mathrm{div}_\Sigma u_\Sigma = \mathrm{div}_\Sigma \mathcal{P}_\Sigma u - H_\Sigma u_\Sigma \cdot \nu_\Sigma = \mathrm{div}_\Sigma u + H_\Sigma j_\Sigma/\varrho,$$

we obtain the following system for $(h, j_\Sigma, \vartheta_\Sigma)$.

$$\begin{aligned} \lambda h + j_\Sigma/\varrho - PN_\lambda Q\vartheta_\Sigma + PN_\lambda P^* \sigma_* \mathcal{A}_\Sigma h &= 0 \\ (l_0/\theta_*)\vartheta_\Sigma - \sigma_* \mathcal{A}_\Sigma h &= 0 \\ (\kappa_{\Sigma*}\lambda - d_{\Sigma*}\Delta_\Sigma + D_\lambda^H + \theta_* Q^* N_\lambda Q)\vartheta_\Sigma - l_0 j_\Sigma/\varrho - \theta_* Q^* N_\lambda P^* \sigma_* \mathcal{A}_\Sigma h &= 0. \end{aligned}$$

This implies

$$\vartheta_\Sigma = \frac{\theta_*}{l_0} \sigma_* \mathcal{A}_\Sigma h,$$

and with

$$L_\lambda = \kappa_{\Sigma*}\lambda - d_{\Sigma*}\Delta_\Sigma + D_\lambda^H + \theta_* Q^* N_\lambda Q := L_\lambda^0 + \theta_* Q^* N_\lambda Q$$

then for j_Σ

$$j_\Sigma/\varrho = \Big[\frac{\theta_*}{l_0^2} L_\lambda - \frac{\theta_*}{l_0} Q^* N_\lambda P^*\Big] \sigma_* \mathcal{A}_\Sigma h.$$

Inserting these relations into the equation for h, after some easy computations we finally obtain the fundamental equation

$$0 = \lambda h + \Big[\frac{\theta_*}{l_0^2} L_\lambda^0 + R^* N_\lambda R\Big]\sigma_* \mathcal{A}_\Sigma h. \tag{10.46}$$

Here the operator R is defined as $R = P^* - \frac{\theta_*}{l_0} Q$. A number $\lambda > 0$ is an eigenvalue if and only if this equation admits a nontrivial solution. As for $\lambda > 0$ the operator L_λ^0 is positive definite and N_λ is positive semi-definite, the operator in front of the second term in (10.46) is invertible, we call its inverse T_λ. Then (10.46) is equivalent to

$$B_\lambda h := \lambda T_\lambda h + \sigma_* \mathcal{A}_\Sigma h = 0. \tag{10.47}$$

Note that T_λ is compact in $L_2(\Sigma)$. Now we have to analyze B_λ in a similar way as in the previous section.

(a) Let as above denote by e the function identically one on Σ. As in the previous section we want to compute $\lim_{\lambda\to 0+} \lambda(T_\lambda \mathsf{e}|\mathsf{e})_\Sigma$. For this purpose we set $v_\lambda = T_\lambda \mathsf{e}$. Then v_λ satisfies

$$\lambda \kappa_{\Sigma*} v_\lambda - d_{\Sigma*}\Delta_\Sigma v_\lambda + D_\lambda^H v_\lambda + \frac{l_0^2}{\theta_*} R^* N_\lambda R v_\lambda = \frac{l_0^2}{\theta_*}\mathsf{e}.$$

Applying the orthogonal projection P_0 in $L_2(\Sigma)$ onto $L_{2,0}(\Sigma)$ we obtain, observing $N_\lambda R\mathsf{e} = 0$,

$$\lambda \kappa_{\Sigma*} P_0 v_\lambda - d_{\Sigma*} P_0 \Delta_\Sigma v_\lambda + P_0 D_\lambda^H P_0 v_\lambda + \frac{l_0^2}{\theta_*} P_0 R^* N_\lambda R P_0 v_\lambda = -P_0 D_\lambda^H \mathsf{e}(v_\lambda|\mathsf{e})_\Sigma/|\Sigma|,$$

hence

$$P_0 v_\lambda = -\Big(\lambda\kappa_{\Sigma*} - d_{\Sigma*}\Delta_\Sigma + P_0 D_\lambda^H P_0 + \frac{l_0^2}{\theta_*} P_0 R^* N_\lambda R P_0\Big)^{-1} P_0 D_\lambda^H \mathsf{e}(v_\lambda|\mathsf{e})_\Sigma/|\Sigma|.$$

As $\lambda \to 0+$, we obtain

$$\Big(\lambda\kappa_{\Sigma*} - d_{\Sigma*}\Delta_\Sigma + P_0 D_\lambda^H P_0 + \frac{l_0^2}{\theta_*} P_0 R^* N_\lambda R P_0\Big)^{-1} \to \Big(- d_{\Sigma*}\Delta_\Sigma + P_0 D_0^H P_0 + \frac{l_0^2}{\theta_*} P_0 R^* N_0 R P_0\Big)^{-1}$$

in $\mathcal{B}(L_{2,0}(\Sigma))$. From Step (a) in the previous section we may deduce

$$D_\lambda^H \mathsf{e}/\lambda \to (\kappa_*|\varrho)/|\Sigma| \quad \text{as } \lambda \to 0+,$$

hence

$$P_0 v_\lambda = o(1)\lambda(v_\lambda|\mathsf{e})_\Sigma \quad \text{as } \lambda \to 0+.$$

On the other hand, applying the projection $I - P_0$ yields

$$\big[(\kappa_{\Sigma*} + (D^H_\lambda \mathsf{e}|\mathsf{e})_\Sigma/(\lambda|\Sigma|)\big]\lambda(v_\lambda|\mathsf{e})_\Sigma + (D^H_\lambda \mathsf{e}|P_0 v_\lambda)_\Sigma = l_0^2|\Sigma|/\theta_*,$$

hence

$$\lim_{\lambda\to 0+} \lambda(T_\lambda \mathsf{e}|\mathsf{e})_\Sigma = \lim_{\lambda\to 0+} \lambda(v_\lambda|\mathsf{e})_\Sigma = \frac{l_0^2|\Sigma|^2}{\theta_*(\kappa_{\Sigma*}|\Sigma| + (\kappa_*|\varrho)_\Omega)}.$$

Therefore,

$$\lim_{\lambda\to 0+} (B_\lambda \mathsf{e}|\mathsf{e})_\Sigma = |\Sigma|\Big[\frac{l_0^2|\Sigma|}{\theta_*(\kappa_{\Sigma*}|\Sigma| + (\kappa_*|\varrho)_\Omega)} - \sigma_*\frac{n-1}{R_*^2}\Big] < 0,$$

if the stability condition (10.31) is violated. Thus, in this case, for small $\lambda > 0$, B_λ is not positive semi-definite.

(b) On the other hand, if Σ is not connected, then we may use functions h of the form $h = \sum_{k=1}^m a_k \mathsf{e}_k$ such that $\sum a_k = 0$; then $h \in L_{2,0}(\Sigma)$, hence

$$\lim_{\lambda\to 0+} (B_\lambda h|h) = \sigma_*(\mathcal{A}_\Sigma h|h) = -\frac{\sigma_*(n-1)|\Sigma|}{R_*^2 m}\sum_{k=1}^m |h_k|^2 < 0,$$

showing that also in this case B_λ is not positive semi-definite for small $\lambda > 0$.

(c) To describe the behaviour of B_λ for large λ, note that for $\lambda \to \infty$

$$\lambda T_\lambda = \frac{l_0^2}{\theta_*\kappa_\Sigma *}\lambda\kappa_{\Sigma*}\Big(\lambda\kappa_{\Sigma*} - d_{\Sigma_*}\Delta_\Sigma + D^H_\lambda + \frac{l_0^2}{\theta_*}R^* N_\lambda R\Big)^{-1} \to \frac{l_0^2}{\theta_*\kappa_{\Sigma*}}$$

strongly, as D^H_λ and $R^*N_\lambda R$ are of lower order. This implies for $h \neq 0$ and

$$\delta_* = \theta_*\sigma_*\kappa_{\Sigma*}(n-1)/l_0^2 R_*^2 < 1,$$

$$(B_\lambda h|h) \to \frac{l_0^2}{\theta_*\kappa_{\Sigma*}}[1-\delta_*]|h_0|_\Sigma^2 + \frac{l_0^2}{\theta_*\kappa_{\Sigma*}}|h_1|_\Sigma^2 + \sigma_*(\mathcal{A}_\Sigma h_1|h_1)_\Sigma > 0,$$

where we employed the decomposition $h = h_0 + h_1$, $h_0 \in \mathsf{N}(\Delta_\Sigma)$ and $h_1 \in \mathsf{R}(\Delta_\Sigma)$. By compactness, this shows that B_λ is positive semi-definite for large λ, and so we may proceed as in the previous section to obtain existence of positive eigenvalues.

(d) Finally, in case $\delta_* > 1$, taking the inner product of (10.46) with $\mathcal{A}_\Sigma h$ in $L_2(\Sigma)$ yields

$$\begin{aligned} 0 = \lambda(\mathcal{A}_\Sigma h|h)_\Sigma + &\frac{\theta_*\sigma_*}{l_0^2}\big[\lambda\kappa_{\Sigma*}|\mathcal{A}_\Sigma h|_\Sigma^2 \\ &+ d_{\Sigma_*}|\nabla_\Sigma \mathcal{A}_\Sigma h|_\Sigma^2 + (D^H_\lambda \mathcal{A}_\Sigma h|\mathcal{A}_\Sigma h)_\Sigma\big] + \sigma_*(N_\lambda R\mathcal{A}_\Sigma h|R\mathcal{A}_\Sigma h)_\Sigma. \end{aligned}$$

With the decomposition $h = h_0 + h_1$, $h_0 \in \mathsf{N}(\Delta_\Sigma)$ and $h_1 \in \mathsf{R}(\Delta_\Sigma)$ we obtain

$$(\mathcal{A}_\Sigma h|h)_\Sigma = -\frac{n-1}{R_*^2}|h_0|_\Sigma^2 + (\mathcal{A}_\Sigma h_1|h_1)_\Sigma,$$

and

$$|\mathcal{A}_\Sigma h|_\Sigma^2 = \frac{(n-1)^2}{R_*^4}|h_0|_\Sigma^2 + |\mathcal{A}_\Sigma h_1|_\Sigma^2,$$

hence

$$\begin{aligned} 0 &\geq -\lambda\frac{n-1}{R_*^2}|h_0|_\Sigma^2 + \lambda(\mathcal{A}_\Sigma h_1|h_1)_\Sigma + \frac{\theta_*\sigma_*}{l_0^2}\lambda\kappa_{\Sigma*}|\mathcal{A}_\Sigma h|_\Sigma^2 \\ &\geq \lambda\frac{n-1}{R_*^2}\Big[\frac{\theta_*\sigma_*(n-1)}{l_0^2R_*^2}\kappa_{\Sigma*} - 1\Big]|h_0|_\Sigma^2 \geq 0, \end{aligned}$$

and so $h_0 = 0$ as $\delta_* > 1$ by assumption. But then $h = 0$, which means that in this case there are no positive eigenvalues.

We summarize.

Theorem 10.6.2. *Let L_5 denote the linearization of Problem 5 at $e_* := (0, \theta_*, \Sigma) \in \mathcal{E}$ as defined above. Assume $l_0 := \varrho l_* + \theta_*\sigma_*' H_\Sigma \neq 0$ and*

$$\delta_* := \frac{\theta_*\sigma_*(n-1)}{l_0^2R_*^2}\kappa_{\Sigma*} \neq 1.$$

Then $-L_5$ generates a compact analytic C_0-semigroup in X_0^5 which has maximal L_p-regularity. The spectrum of L_5 consists only of eigenvalues of finite algebraic multiplicity. Moreover, the following assertions are valid.

(i) *The operator $-L_5$ has no eigenvalues $\lambda \neq 0$ with nonnegative real part if and only if $\delta_* > 1$, or Σ is connected and*

$$\zeta_* := \frac{\theta_*\sigma_*(n-1)}{l_0^2R_*^2|\Sigma|}\big((\kappa_*|\varrho)_\Omega + \kappa_{\Sigma*}|\Sigma|\big) \leq 1. \tag{10.48}$$

(ii) *If Σ has $m \geq 2$ components and* (10.48) *holds, but $\delta_* < 1$, then $-L_5$ has precisely $m-1$ positive eigenvalues.*
(iii) *If Σ has $m \geq 1$ components and* (10.48) *does not hold, but $\delta_* < 1$, then $-L_5$ has precisely m positive eigenvalues.*
(iv) *$\lambda = 0$ is an eigenvalue of L_5. It is semi-simple as long as $\zeta_* \neq 1$.*
(v) *The kernel $\mathsf{N}(L_5)$ of L_5 is isomorphic to the tangent space $T_{e_*}\mathcal{E}$ of the manifold of equilibria $\mathcal{E}$ at e_*.*

Consequently, $e_* = (0, \theta_*, \Sigma) \in \mathcal{E}$ is normally stable if and only if $\delta_* > 1$, or Σ is connected and $\zeta_* < 1$, and it is normally hyperbolic if and only if $\delta_* < 1$, and Σ is disconnected or $\zeta_* > 1$.

10.7 Unstable Eigenvalues: Problem 4

We want to prove that in case Σ is disconnected, there is always a positive eigenvalue. To prove this we need some more preparation concerning the *asymmetric*

Stokes problem. Solve

$$
\begin{aligned}
\varrho\lambda u - \mu_*\Delta u + \nabla\pi &= 0 && \text{in } \Omega\setminus\Sigma,\\
\operatorname{div} u &= 0 && \text{in } \Omega\setminus\Sigma,\\
u &= 0 && \text{on } \partial\Omega,\\
\mathcal{P}_\Sigma[\![u]\!] = \mathcal{P}_\Sigma[\![T\nu_\Sigma]\!] &= 0 && \text{on } \Sigma,\\
-[\![T\nu_\Sigma\cdot\nu_\Sigma]\!] &= g_1 && \text{on } \Sigma,\\
-[\![T\nu_\Sigma\cdot\nu_\Sigma/\varrho]\!] &= g_2 && \text{on } \Sigma,
\end{aligned}
\tag{10.49}
$$

to obtain as output

$$k := S_\lambda^{11}g_1 + S_\lambda^{12}g_2 := [\![\varrho u\cdot\nu_\Sigma]\!]/[\![\varrho]\!],\quad j := S_\lambda^{21}g_1 + S_\lambda^{22}g_2 := [\![u\cdot\nu_\Sigma]\!]/[\![1/\varrho]\!].$$

For this problem we have the following result.

Proposition 10.7.1. *The operator S_λ for the Stokes problem* (10.49) *admits a bounded extension to $L_2(\Sigma)^2$ for $\lambda\geq 0$ and has the following properties.*
(i) *If u denotes the solution of* (10.49)*, then*

$$(S_\lambda g|g)_{L_2(\Sigma)^2} = \lambda\int_\Omega \varrho|u|^2\,dx + 2\int_\Omega \mu_*|D|_2^2\,dx,\quad \lambda\geq 0,\ g\in L_2(\Sigma)^2.$$

(ii) *$S_\lambda\in\mathcal{B}(L_2(\Sigma)^2)$ is selfadjoint, positive semi-definite, and compact. In particular,*

$$S_\lambda^{11} = [S_\lambda^{11}]^*,\quad S_\lambda^{22} = [S_\lambda^{22}]^*,\quad S_\lambda^{12} = [S_\lambda^{21}]^*.$$

(iii) *S_λ^{11} and S_λ^{22} are injective in $L_{2,0}(\Sigma)$, and with $G_\lambda = [S_\lambda^{22}]^{-1}$ we have*

$$N_\lambda^S = S_\lambda^{11} - S_\lambda^{12}G_\lambda S_\lambda^{21}.$$

G_λ is selfadjoint and positive definite on $L_{2,0}(\Sigma)$, its resolvent is compact in $L_{2,0}(\Sigma)$, for each $\lambda\geq 0$.
(iv) *There is a constant $C>0$ such that*

$$(1+\lambda)^{1/2}|S_\lambda|_{\mathcal{B}(L_{2,0}(\Sigma)^2)} + |S_\lambda|_{\mathcal{B}(L_{2,0}(\Sigma)^2,H_2^1(\Sigma)^2)} \leq C \quad \lambda\geq 0.$$

(v) *$S_\lambda^{11}, S_\lambda^{22} : L_{2,0}(\Sigma)\to H_2^1(\Sigma)\cap L_{2,0}(\Sigma)$ are isomorphisms, for each $\lambda\geq 0$.*

Proof. **(a)** First observe that for the traces u_i of u on Σ we have

$$
\begin{aligned}
u_i &= \mathcal{P}_\Sigma u + ([\![\varrho u\cdot\nu_\Sigma]\!]/[\![\varrho]\!])\nu_\Sigma + ([\![u\cdot\nu_\Sigma]\!]/[\![1/\varrho]\!]\varrho_i)\nu_\Sigma\\
&= \mathcal{P}_\Sigma u + k\nu_\Sigma + (j/\varrho_i)\nu_\Sigma = u_b + (j/\varrho_i)\nu_\Sigma.
\end{aligned}
$$

To prove assertion (i), let (u, π) denote the solution of (10.49) for $g \in H_2^{1/2}(\Sigma)$. Multiply with u and integrate by parts to the result

$$\begin{aligned}\lambda \int_\Omega \varrho|u|^2\,dx + 2\int_\Omega \mu_*|D|_2^2\,dx &= \int_\Omega \operatorname{div}(T\bar{u})\,dx\\ &= -\int_\Sigma [\![\bar{u}\cdot T\nu_\Sigma]\!]\,d\Sigma = -\int_\Sigma [\![\overline{(u_b + j\nu_\Sigma/\varrho)}\cdot T\nu_\Sigma]\!]\,d\Sigma\\ &= -\int_\Sigma \overline{u_b}\cdot[\![T\nu_\Sigma]\!]\,d\Sigma - \int_\Sigma \bar{j}[\![T\nu_\Sigma\cdot\nu_\Sigma/\varrho]\!]\,d\Sigma\\ &= \int_\Sigma \bar{k}g_1\,d\Sigma + \int_\Sigma \bar{j}g_2\,d\Sigma = (g|S_\lambda g)_{L_2(\Sigma)^2}.\end{aligned}$$

A similar computation yields

$$(S_\lambda g|h)_{L_2(\Sigma)^2} = (g|S_\lambda h)_{L_2(\Sigma)^2}, \quad (g,h)\in H_2^{1/2}(\Sigma;\mathbb{R}^2),$$

hence S_λ is symmetric in $L_2(\Sigma)^2$, thereby proving the first part of (ii) for functions in $H_2^{1/2}(\Sigma)$.

(b) To prove injectivity of S_λ^{22}, let $g_1 = 0$ and $j = 0$. Then (i) implies $\lambda u = D = 0$ hence $\nabla\pi = 0$ in $\Omega\setminus\Sigma$, and so π is constant in the components of the phases. Next

$$0 = g_1 = -[\![T\nu_\Sigma\cdot\nu_\Sigma]\!] = [\![\pi]\!], \quad g_2 = -[\![T\nu_\Sigma\cdot\nu_\Sigma/\varrho]\!] = [\![\pi/\varrho]\!]$$

shows that π is even constant in the phases, and so g_2 is constant on Σ hence zero since its mean value vanishes. Injectivity of S_λ^{11} is shown in a similar way. This proves the first assertion in (iii), the second one follows from the definitions of N_λ^S and S_λ, and the third one is a consequence of (ii).

(c) To establish the boundedness properties of S_λ, we proceed as follows. Suppose $g_1, g_2 \in H_2^{1/2}(\Sigma)$ are given. We decompose the pressure into $\pi = q + q_0$, where the one-sided traces q_0^j on Σ are uniquely determined by

$$[\![q_0]\!] = g_1, \quad [\![q_0/\varrho]\!] = g_2;$$

then $q_0^j \in H_2^{1/2}(\Sigma)$. Extend q_0 to all of Ω in $H_2^1(\Omega\setminus\Sigma)$, by solving the two one-phase problems

$$\Delta q^j = 0 \text{ in } \Omega_j, \quad q^j = q_0^j \text{ on } \Sigma, \quad \partial_\nu q^2 = 0 \text{ on } \partial\Omega,\ j = 1,2.$$

According to Appendix B below, there is a constant $c_0 > 0$ such that

$$|q_0|_{H_2^1(\Omega\setminus\Sigma)} \le c_0|g|_{H_2^{1/2}(\Sigma)}.$$

Then (u, q) satisfies the asymmetric Stokes problem

$$\begin{aligned} \varrho\lambda u - \mu_*\Delta u + \nabla q &= -\nabla q_0 && \text{in } \Omega \setminus \Sigma, \\ \operatorname{div} u &= 0 && \text{in } \Omega \setminus \Sigma, \\ u &= 0 && \text{on } \partial\Omega, \\ \mathcal{P}_\Sigma [\![u]\!] = \mathcal{P}_\Sigma [\![\mu_* D\nu_\Sigma]\!] &= 0 && \text{on } \Sigma, \\ -[\![2\mu_* D\nu_\Sigma \cdot \nu_\Sigma]\!] + [\![q]\!] &= 0 && \text{on } \Sigma, \\ -[\![2\mu_* D\nu_\Sigma \cdot \nu_\Sigma/\varrho]\!] + [\![q/\varrho]\!] &= 0 && \text{on } \Sigma. \end{aligned} \tag{10.50}$$

Note that the interface conditions are now homogeneous. Let $-A$ denote the generator of the associated analytic C_0-semigroup in $L_{2,\sigma}(\Omega)$, which is exponentially stable, see Section 8.4.2. Then we have

$$u = u_\lambda = -(\lambda + A)^{-1}[\nabla q_0/\varrho],$$

and with

$$S_\lambda g = (k, j) = ([\![\varrho u_\lambda \cdot \nu_\Sigma]\!]/[\![\varrho]\!], [\![u_\lambda \cdot \nu_\Sigma]\!]/[\![1/\varrho]\!])$$

we estimate using trace theory and the resolvent estimate for A and $\operatorname{Re}\sigma(-A) < 0$

$$|S_\lambda g|_{H_2^{3/2}(\Sigma)} \le C|u_\lambda|_{H_2^2(\Omega\setminus\Sigma)} \le C|q_0|_{H_2^1(\Omega\setminus\Sigma)} \le C|g|_{H_2^{1/2}(\Sigma)},$$

as well as

$$|S_\lambda g|_{H_2^{1/2}(\Sigma)} \le C|u_\lambda|_{H_2^1(\Omega\setminus\Sigma)} \le C(1+\lambda)^{-1/2}|q_0|_{H_2^1(\Omega\setminus\Sigma)} \le C(1+\lambda)^{-1/2}|g|_{H_2^{1/2}(\Sigma)}.$$

This shows that

$$S_\lambda : H_2^{1/2}(\Sigma)^2 \to H_2^{3/2}(\Sigma)^2, \quad (1+\lambda)^{1/2} S_\lambda : H_2^{1/2}(\Sigma)^2 \to H_2^{1/2}(\Sigma)^2$$

are uniformly bounded w.r.t. $\lambda \ge 0$. But then the dual operator S_λ^* has the same boundedness properties, and as S_λ is symmetric this yields an extension of S_λ, denoted by S_λ again,

$$S_\lambda : H_2^{-3/2}(\Sigma)^2 \to H_2^{-1/2}(\Sigma)^2, \quad (1+\lambda)^{1/2} S_\lambda : H_2^{-1/2}(\Sigma)^2 \to H_2^{-1/2}(\Sigma)^2,$$

with norms bounded by C. Complex interpolation implies

$$S_\lambda : L_2(\Sigma)^2 \to H_2^1(\Sigma)^2, \quad (1+\lambda)^{1/2} S_\lambda : L_2(\Sigma)^2 \to L_2(\Sigma)^2,$$

with norms bounded by C. In particular, S_λ is bounded and self-adjoint in $L_2(\Sigma)^2$.

(d) The proof of (v) is more involved. To obtain surjectivity of S_λ^{11} we have to solve the problem

$$\begin{aligned} \varrho\lambda u - \mu_*\Delta u + \nabla \pi &= 0 && \text{in } \Omega \setminus \Sigma, \\ \operatorname{div} u &= 0 && \text{in } \Omega \setminus \Sigma, \\ u &= 0 && \text{on } \partial\Omega, \\ \mathcal{P}_\Sigma [\![u]\!] = \mathcal{P}_\Sigma [\![T\nu_\Sigma]\!] &= 0 && \text{on } \Sigma, \\ [\![\varrho u \cdot \nu_\Sigma]\!] &= [\![\varrho]\!] k && \text{on } \Sigma, \\ -[\![T\nu_\Sigma \cdot \nu_\Sigma/\varrho]\!] &= 0 && \text{on } \Sigma, \end{aligned} \tag{10.51}$$

for given $k \in H_2^1(\Sigma) \cap L_{2,0}(\Sigma)$ with output $[S_\lambda^{11}]^{-1}k = g_1 = -[\![T\nu_\Sigma \cdot \nu_\Sigma]\!]$. Similarly, to prove surjectivity of S_λ^{22} the problem to solve is

$$\begin{aligned}
\varrho\lambda u - \mu_*\Delta u + \nabla\pi &= 0 && \text{in } \Omega\setminus\Sigma,\\
\operatorname{div} u &= 0 && \text{on } \Omega\setminus\Sigma,\\
u &= 0 && \text{on } \partial\Omega,\\
\mathcal{P}_\Sigma[\![u]\!] = \mathcal{P}_\Sigma[\![T\nu_\Sigma]\!] &= 0 && \text{on } \Sigma,\\
[\![u\cdot\nu_\Sigma]\!] &= [\![1/\varrho]\!]j && \text{on } \Sigma,\\
-[\![T\nu_\Sigma\cdot\nu_\Sigma]\!] &= 0 && \text{on } \Sigma,
\end{aligned} \tag{10.52}$$

for given j and the output will be $[S_\lambda^{22}]^{-1}j = g_2 = -[\![T\nu_\Sigma \cdot \nu_\Sigma/\varrho]\!]$.

Solvability of (10.51) and (10.52) is proved in Appendix A of this chapter. □

Now suppose that $\lambda > 0$ is an eigenvalue of L_4. We set $g_1 = -\sigma\mathcal{A}_\Sigma h$ and $g_2 = -l_*\vartheta/\theta_* = -c_*N_\lambda^H j_\Sigma$, $c_* = l_*^2/\theta_* > 0$ to obtain

$$j_\Sigma = -S_\lambda^{21}\sigma\mathcal{A}_\Sigma h - c_*S_\lambda^{22}N_\lambda^H j_\Sigma, \quad \lambda h = -S_\lambda^{11}\sigma\mathcal{A}_\Sigma h - S_\lambda^{12}c_*N_\lambda^H j_\Sigma.$$

Observing that $I + c_*S_\lambda^{22}N_\lambda^H$ is injective, hence boundedly invertible in $L_{2,0}(\Sigma)$ by compactness of S_λ, we may solve the first equation for j_Σ to the result

$$j_\Sigma = -\sigma(I + c_*S_\lambda^{22}N_\lambda^H)^{-1}S_\lambda^{21}\mathcal{A}_\Sigma h.$$

Inserting this into the second, the equation for h becomes

$$0 = \lambda h + \sigma(S_\lambda^{11} - S_\lambda^{12}c_*N_\lambda^H(I + c_*S_\lambda^{22}N_\lambda^H)^{-1}S_\lambda^{21})\mathcal{A}_\Sigma h,$$

or equivalently with $R_\lambda = G_\lambda S_\lambda^{21}$

$$0 = \lambda h + \sigma(N_\lambda^S + R_\lambda^*(c_*N_\lambda^H + G_\lambda)^{-1}R_\lambda)\mathcal{A}_\Sigma h.$$

Next we observe that the operators $N_\lambda^S + R_\lambda^*(c_*N_\lambda^H + G_\lambda)^{-1}R_\lambda$ are injective for $\lambda \geq 0$; in fact if

$$N_\lambda^S h + R_\lambda^*(c_*N_\lambda^H + G_\lambda)^{-1}R_\lambda h = 0,$$

then with $v_\lambda = (c_*N_\lambda^H + G_\lambda)^{-1}R_\lambda h$, forming the inner product with h in $L_{2,0}(\Sigma)$ we obtain

$$\begin{aligned}
0 &= (N_\lambda^S h|h)_\Sigma + (R_\lambda^*(c_*N_\lambda^H + G_\lambda)^{-1}R_\lambda h|h)_\Sigma\\
&\geq c|N_\lambda^S h|_\Sigma^2 + ((c_*N_\lambda^H + G_\lambda)v_\lambda|v_\lambda)_\Sigma
\end{aligned}$$

hence $N_\lambda^S h = 0$ and $v_\lambda = 0$, which implies $S_\lambda^{21}h = S_\lambda^{11}h = 0$, hence $h = 0$ by injectivity of S_λ^{11}. Setting now

$$T_\lambda = [N_\lambda^S + R_\lambda^*(c_*N_\lambda^H + G_\lambda)^{-1}R_\lambda]^{-1}$$

in $L_{2,0}(\Sigma)$ we arrive at the equation

$$\lambda T_\lambda h + \sigma A_\Sigma h = 0. \tag{10.53}$$

$\lambda > 0$ is an eigenvalue of $-L$ if and only if Problem (10.53) admits a nontrivial solution, i.e., if and only if 0 is an eigenvalue for $B_\lambda := \lambda T_\lambda + \sigma \mathcal{A}_\Sigma$. Here the domain of B_λ is that of $\mathcal{A}_\Sigma$, T_λ is a relatively compact perturbation of $\mathcal{A}_\Sigma$.

We consider this problem in $L_{2,0}(\Sigma)$. Then $\mathcal{A}_\Sigma$ is selfadjoint and

$$\sigma(\mathcal{A}_\Sigma h|h)_{L_2(\Sigma)} \geq -\frac{\sigma(n-1)}{R_*^2}|h|^2_{L_2(\Sigma)}.$$

On the other hand, N_λ^H, N_λ^S are selfadjoint, compact and positive semi-definite on $L_{2,0}(\Sigma)$, and G_λ is selfadjoint, positive definite, and has compact inverse. Hence T_λ is selfadjoint, positive semi-definite and T_λ^{-1} is compact as well. If $\mu > 0$ is an eigenvalue of T_λ, then

$$\mu^{-1}h = T_\lambda^{-1}h = [S_\lambda^{11} - c_* S_\lambda^{12} N_\lambda^H (I + c_* S_\lambda^{22} N_\lambda^H)^{-1} S_\lambda^{21}]h,$$

hence we get

$$\mu^{-1}|h|_\Sigma \leq C|h|_\Sigma,$$

since by Propositions 10.7.1 and 10.5.1, S_λ, N_λ^H and $(I+c_*S_\lambda^{22}N_\lambda^H)^{-1}$ are bounded in $L_{2,0}(\Sigma)$, uniformly for large λ. Therefore $\mu = \mu(\lambda) \geq c_0 > 0$, for large λ, and so

$$(B_\lambda h|h)_\Sigma = \lambda(T_\lambda h|h)_\Sigma + \sigma(A_\Sigma h|h)_\Sigma \geq \Big(c_0\lambda - \frac{\sigma(n-1)}{R_*^2}\Big)|h|^2_\Sigma.$$

This proves that B_λ is positive definite, hence (10.53) has no nontrivial solution, for large λ.

But for small $\lambda > 0$ we have with $h = \sum_k a_k \mathsf{e}_k \neq 0$, $\sum_k a_k = 0$,

$$\lambda(T_\lambda h|h) - \frac{\sigma(n-1)|\Sigma|}{R_*^2 m}\sum_k a_k^2 < 0,$$

since with $h \in H_2^{3/2}(\Sigma) \cap L_{2,0}(\Sigma)$

$$T_\lambda h = \big(I - c_*[S_\lambda^{11}]^{-1} S_\lambda^{12} N_\lambda^H (I + c_* S_\lambda^{22} N_\lambda^H)^{-1} S_\lambda^{21}\big)^{-1}[S_\lambda^{11}]^{-1}h \to T_0 h$$

in $L_{2,0}(\Sigma)$ as $\lambda \to 0$. This shows that B_λ has nontrivial kernel for some $\lambda_0 > 0$, which implies that $-L$ has a positive eigenvalue.

Even more is true. We have seen that B_λ is positive definite for large λ and $B_0 = \sigma\mathcal{A}_\Sigma$ has $-(n-1)/R_*^2$ as an eigenvalue of multiplicity $m-1$ in $L_{2,0}(\Sigma)$, with eigenfunctions of the form $\sum_k a_k \mathsf{e}_k$, $\sum_k a_k = 0$.

Therefore, as λ increases to infinity, $m-1$ eigenvalues $\mu_k(\lambda)$ of B_λ must cross through zero along the real axis, this way inducing $m-1$ positive eigenvalues of $-L_4$.

Finally, we consider the case $l_* = 0$. Then the temperature equation decouples completely from that for u and h. It only induces one dimension in the kernel of L_4, but no positive eigenvalues. In this case, as now $c_* = 0$ the derivation above yields the equivalent problem

$$\lambda h + \sigma S_\lambda^{11} \mathcal{A}_\Sigma h = 0.$$

As S_λ^{11} is positive semi-definite and injective, this equation admits no nontrivial solutions if $\operatorname{Re}\lambda \geq 0$ and λ is non-real. If $\lambda > 0$, then $T_\lambda = [S_\lambda^{11}]^{-1}$ and we may employ the same arguments as above to obtain the same conclusions as in case $l_* \neq 0$.

Let us summarize.

Theorem 10.7.2. *Let L_4 denote the linearization at $e_* := (0, \theta_*, \Sigma) \in \mathcal{E}$ as defined above. Then $-L_4$ generates a compact analytic C_0-semigroup in X_0^4 which has maximal L_p-regularity. The spectrum of L_4 consists only of eigenvalues of finite algebraic multiplicity. Moreover, the following assertions are valid.*

(i) *The operator $-L_4$ has no eigenvalues $\lambda \neq 0$ with nonnegative real part if and only if Σ is connected.*
(ii) *If Σ is disconnected and has m components, then $-L_4$ has precisely $m - 1$ positive eigenvalues.*
(iii) *$\lambda = 0$ is an eigenvalue of L_4 and it is semi-simple.*
(iv) *The kernel $\mathsf{N}(L_4)$ of L_4 is isomorphic to the tangent space $T_{e_*}\mathcal{E}$ of the manifold of equilibria $\mathcal{E}$ at e_*.*

Consequently, $e_* = (0, \theta_*, \Sigma) \in \mathcal{E}$ is normally stable if and only if Σ is connected, and normally hyperbolic if and only if Σ is disconnected.

10.8 Unstable Eigenvalues: Problem 6

Here we mostly follow the arguments of the preceding section. So let $\lambda > 0$ be an eigenvalue. As before we have $-[\![d_* \partial_\nu \vartheta]\!] = D_\lambda^H \vartheta_\Sigma$. Next observe that on Σ we have

$$\begin{aligned} u &= \mathcal{P}_\Sigma u + ([\![\varrho u \cdot \nu_\Sigma]\!]/[\![\varrho]\!])\nu_\Sigma + ([\![u \cdot \nu_\Sigma]\!]/[\![1/\varrho]\!]\varrho)\nu_\Sigma \\ &= \mathcal{P}_\Sigma u + \lambda h \nu_\Sigma + (j_\Sigma/\varrho)\nu_\Sigma. \end{aligned}$$

Then we solve the extended asymmetric Stokes problem

$$\begin{aligned} \varrho\lambda u - \mu_*\Delta u + \nabla\pi &= 0 && \text{in } \Omega\setminus\Sigma,\\ \operatorname{div} u &= 0 && \text{in } \Omega\setminus\Sigma,\\ u &= 0 && \text{on } \partial\Omega,\\ \mathcal{P}_\Sigma[\![u]\!] &= 0 && \text{on } \Sigma,\\ -[\![T\nu_\Sigma\cdot\nu_\Sigma]\!] &= g_1 && \text{on } \Sigma,\\ -[\![T\nu_\Sigma\cdot\nu_\Sigma/\varrho]\!] &= g_2 && \text{on } \Sigma,\\ -\mathcal{P}_\Sigma[\![T\nu_\Sigma]\!] &= g_3 && \text{on } \Sigma \end{aligned} \tag{10.54}$$

to obtain as output

$$S_\lambda g := \begin{bmatrix} [\![\varrho u\cdot\nu_\Sigma]\!]/[\![\varrho]\!] \\ [\![u\cdot\nu_\Sigma]\!]/[\![1/\varrho]\!] \\ \mathcal{P}_\Sigma u \end{bmatrix} = \begin{bmatrix} \lambda h \\ j_\Sigma \\ \mathcal{P}_\Sigma u \end{bmatrix}.$$

For this problem we have the following result.

Proposition 10.8.1. *The operator S_λ for the extended asymmetric Stokes problem* (10.54) *admits a bounded extension to $L_2(\Sigma;\mathbb{R}^2\times T\Sigma)$ for $\lambda\geq 0$ and has the following properties.*

(i) *If u denotes the solution of* (10.54)*, then*

$$(S_\lambda g|g)_{L_2} = \lambda\int_\Omega \varrho|u|^2\,dx + 2\int_\Omega \mu_*|D|_2^2\,dx, \quad \lambda\geq 0,\ \ g\in L_2(\Sigma,\mathbb{R}^2\times T\Sigma).$$

(ii) *$S_\lambda\in\mathcal{B}(L_2(\Sigma;\mathbb{R}^2\times T\Sigma))$ is self-adjoint, positive semi-definite, and compact; in particular*

$$\begin{aligned} S_\lambda^{11} &= [S_\lambda^{11}]^*, & S_\lambda^{22} &= [S_\lambda^{22}]^*, & S_\lambda^{33} &= [S_\lambda^{33}]^*\\ S_\lambda^{12} &= [S_\lambda^{21}]^*, & S_\lambda^{13} &= [S_\lambda^{31}]^*, & S_\lambda^{23} &= [S_\lambda^{32}]^*. \end{aligned}$$

(iii) *There is a constant $C>0$ such that*

$$(1+\lambda)^{1/2}|S_\lambda|_{\mathcal{B}(L_2)} + |S_\lambda|_{\mathcal{B}(L_2,H_2^{\frac12})} \leq C, \quad \lambda\geq 0.$$

(iv) *$S_\lambda^{11}, S_\lambda^{22} : L_{2,0}(\Sigma)\to H_2^{\frac12}(\Sigma)\cap L_{2,0}(\Sigma)$ are isomorphisms, for each $\lambda\geq 0$.*

Proof. This result is proved in a similar way as Proposition 10.7.1 and we omit the details. □

Next we define $Q : H_2^{\frac12}(\Sigma)\to L_2(\Sigma;\mathbb{R}^2\times T\Sigma)$ by means of

$$Q := \begin{bmatrix} \sigma_*' H_\Sigma \\ -l_*/\theta_* \\ \sigma_*'\nabla_\Sigma \end{bmatrix};$$

note that

$$Q^*\begin{bmatrix} k \\ j \\ v \end{bmatrix} = \sigma_*' H_\Sigma k - (l_*/\theta_*)j - \sigma_*'\mathrm{div}_\Sigma v.$$

Let P denote the projection onto the first component, as in Section 10.6. Setting $g = Q\vartheta_\Sigma - P^*\sigma_*\mathcal{A}_\Sigma h$ we obtain from the surface heat equation the following equation for ϑ_Σ.

$$L_\lambda\vartheta_\Sigma := (\lambda\kappa_{\Sigma*} - d_{\Sigma*}\Delta_\Sigma + D_\lambda^H + \theta_* Q^* S_\lambda Q)\vartheta_\Sigma = \theta_* Q^* S_\lambda P^*\sigma_*\mathcal{A}_\Sigma h.$$

Inserting this relation into the equation for $\lambda h = k = PS_\lambda g$, we deduce the identity

$$\lambda h + PS_\lambda\big[I - \theta_* Q L_\lambda^{-1} Q^* S_\lambda\big]P^*\sigma_*\mathcal{A}_\Sigma h = 0.$$

$\lambda > 0$ is an eigenvalue if and only if this equation admits a nontrivial solution h. The operator T_λ^{-1} in front of the the second term is compact and selfadjoint in $L_2(\Sigma)$. We show below that it is positive semi-definite and injective on $L_{2,0}(\Sigma)$, and hence we arrive again at a problem of the form

$$B_\lambda h := \lambda T_\lambda h + \sigma_*\mathcal{A}_\Sigma h = 0,$$

with T_λ selfadjoint positive definite and unbounded. The operator B_λ can be handled in the same way as in the proceeding section, showing that B_λ is positive definite for large λ, B_0 is indefinite, and it depends continuously on λ, to obtain the same conclusion as in the previous section. Thus it remains to show that T_λ^{-1} is positive semi-definite and bijective.

To exploit the structure of T_λ^{-1}, recall that S_λ is selfadjoint and positive semi-definite, hence admits a square root $S_\lambda^{1/2}$. We factor as follows:

$$\begin{aligned} T_\lambda^{-1} &= PS_\lambda^{1/2}[I - \theta_*^{1/2}S_\lambda^{1/2}Q(L_\lambda^0 + \theta_* Q^* S_\lambda Q)^{-1}Q^*\theta_*^{1/2}S_\lambda^{1/2}]S_\lambda^{1/2}P^* \\ &= U_\lambda^*[I - R_\lambda(L_\lambda^0 + R_\lambda^* R_\lambda)^{-1}R_\lambda^*]U_\lambda = U_\lambda^*[I - K_\lambda]U_\lambda, \end{aligned}$$

with $U_\lambda = S_\lambda^{1/2}P^*$, which maps $L_2(\Sigma)$ into $H_2^{1/2}(\Sigma)^{n+2}$, and $R_\lambda = \theta_*^{1/2}S_\lambda^{1/2}Q$, which maps $H_2^1(\Sigma)$ into $H_2^{1/2}(\Sigma)^{n+2}$.

We claim that $|K_\lambda|_{\mathcal{B}(L_2^{n+1})} \le 1$. In fact, for $v \in H_2^{1/2}(\Sigma)^{n+2}$ we obtain, using the abbreviation $w = (L_\lambda^0 + R_\lambda^* R_\lambda)^{-1}R_\lambda^* v$,

$$\begin{aligned} |K_\lambda v|_\Sigma^2 &= (R_\lambda(L_\lambda^0 + R_\lambda^* R_\lambda)^{-1}R_\lambda^* v | R_\lambda(L_\lambda^0 + R_\lambda^* R_\lambda)^{-1}R_\lambda^* v)_\Sigma \\ &= (R_\lambda^* R_\lambda(L_\lambda^0 + R_\lambda^* R_\lambda)^{-1}R_\lambda^* v | w)_\Sigma \\ &= (R_\lambda^* v|w)_\Sigma - (L_\lambda^0 w|w)_\Sigma = (R_\lambda^* v|(L_\lambda^0 + R_\lambda^* R_\lambda)^{-1}R_\lambda^* v) - (L_\lambda^0 w|w)_\Sigma \\ &= (v|K_\lambda v)_\Sigma - (L_\lambda^0 w|w) \le (v|K_\lambda v)_\Sigma \le |K_\lambda v|_\Sigma |v|_\Sigma, \end{aligned}$$

hence $|K_\lambda v|_\Sigma \le |v|_\Sigma$, and so $|K_\lambda|_{\mathcal{B}(L_2^{n+1})} \le 1$ as $H_2^{1/2}(\Sigma)$ is dense in $L_2(\Sigma)$. Therefore

$$((I - K_\lambda)v|v)_\Sigma = |v|_\Sigma^2 - (K_\lambda v|v)_\Sigma \ge (1 - |K_\lambda|_{\mathcal{B}(L_2^{n+1})})|v|_\Sigma^2 \ge 0,$$

which shows that T_λ^{-1} is positive semi-definite.

To prove injectivity of T_λ^{-1} on $L_{2,0}(\Sigma)$, assume $T_\lambda^{-1}h = 0$. Multiplying scalarly with h and setting again $v = U_\lambda h$ this yields $0 = (v - K_\lambda v|v)_\Sigma$ which implies $K_\lambda v = v$, as K_λ is selfadjoint and compact. But then as above with $w = (L_\lambda^0 + R_\lambda^* R_\lambda)^{-1} R_\lambda^* v$ we have

$$|v|_\Sigma^2 = |K_\lambda v|_\Sigma^2 = (K_\lambda v|v)_\Sigma - (L_\lambda^0 w|w)_\Sigma = |v|_\Sigma^2 - (L_\lambda^0 w|w)_\Sigma$$

hence $(L_\lambda^0 w|w)_\Sigma = 0$, and so $w = 0$ as L_λ^0 is positive definite on $L_{2,0}(\Sigma)$, thus $K_\lambda v = R_\lambda w = 0$. Therefore, we obtain

$$S_\lambda^{11} h = P S_\lambda P^* h = U_\lambda^* U_\lambda h = U_\lambda^* v = 0,$$

which yields $h = 0$ by injectivity of S_λ^{11}. Finally, we observe that T_λ^{-1} can be written as $T_\lambda^{-1} = S_\lambda^{11}(I - C_\lambda)$ on $L_{2,0}(\Sigma)$, with C_λ a compact operator. Since T_λ^{-1} is injective on $L_{2,0}(\Sigma)$ the operator $(I - C_\lambda)$ must be so as well. Consequently, $(I - C_\lambda) \in \mathcal{B}(L_{2,0}(\Sigma))$ is a bijection, as it has Fredholm index zero. We can now conclude that $T_\lambda^{-1} : L_{2,0}(\Sigma) \to L_{2,0}(\Sigma) \cap H_2^1(\Sigma)$ is a bijection.
Summarizing, we have proved the following result.

Theorem 10.8.2. *Let L_6 denote the linearization at $e_* := (0, \theta_*, \Sigma) \in \mathcal{E}$ as defined above. Then $-L_6$ generates a compact analytic C_0-semigroup in X_0^6 which has maximal L_p-regularity. The spectrum of L_6 consists only of eigenvalues of finite algebraic multiplicity. Moreover, the following assertions are valid.*

(i) *The operator $-L_6$ has no eigenvalues $\lambda \neq 0$ with nonnegative real part if and only if Σ is connected.*
(ii) *If Σ is disconnected and has m components, then $-L_6$ has precisely $m - 1$ positive eigenvalues.*
(iii) *$\lambda = 0$ is an eigenvalue of L_6 and it is semi-simple.*
(iv) *The kernel $\mathsf{N}(L_6)$ of L_6 is isomorphic to the tangent space $T_{e_*}\mathcal{E}$ of the manifold of equilibria $\mathcal{E}$ at e_*.*

Consequently, $e_* = (0, \theta_*, \Sigma) \in \mathcal{E}$ is normally stable if and only if Σ is connected, and normally hyperbolic otherwise.

Appendix A: The Asymmetric Neumann-to-Dirichlet Operator

The asymmetric Neumann-to-Dirchlet operator S_λ is defined by

$$S_\lambda = \begin{bmatrix} S_\lambda^{11} & S_\lambda^{12} \\ S_\lambda^{21} & S_\lambda^{22} \end{bmatrix}, \tag{10.55}$$
$$S_\lambda [g_1, g_2]^\mathsf{T} = [k, j]^\mathsf{T},$$

where $k = [\![\varrho(u|\nu_\Sigma)]\!]/[\![\varrho]\!]$ and $j = [\![(u|\nu_\Sigma)]\!]/[\![1/\varrho]\!]$. For the stability analysis in this chapter we used the invertibility of S_λ^{jj}, $j = 1, 2$. Here we prove this result for the case of a flat interface.

(a) The symbol of S_λ^{11}. To obtain the algebraic system for the symbol of $[S_\lambda^{11}]^{-1}$ we set $g_2 = 0$ and let k be given. Then (8.70) remains valid as well as the formulas for a_1, a_2. For α_1, α_2 we have here the equations

$$\frac{\varrho_2\sqrt{\mu_2}}{\omega_2}\beta_2 + \frac{\varrho_1\sqrt{\mu_1}}{\omega_1}\beta_1 + (\varrho_1\alpha_1 + \varrho_2\alpha_2)|\xi| = [\![\varrho]\!]\hat{k}$$
$$\frac{2}{\varrho_2}(\mu_2\beta_2 + \mu_2\alpha_2|\xi|^2) + \lambda\alpha_2 - \frac{2}{\varrho_1}(\mu_1\beta_1 + \mu_1\alpha_1|\xi|^2) - \lambda\alpha_1 = 0.$$

Inserting β_k from (8.70), this system becomes

$$p_1\alpha_1 - p_2\alpha_2 = 0, \qquad (10.56)$$
$$q_1\alpha_1 + q_2\alpha_2 = [\![\varrho]\!]\hat{k},$$

where

$$p_1 = \varrho_1\lambda\Big[\frac{1}{\varrho_1} - 2[\![\mu/\varrho]\!]\frac{1}{\gamma(z)\omega_1(z)}\frac{\omega_1(z)-1}{\omega_1(z)+1}\Big] =: \varrho_1\lambda p_1^0,$$
$$p_2 = \varrho_2\lambda\Big[\frac{1}{\varrho_2} + 2[\![\mu/\varrho]\!]\frac{1}{\gamma(z)\omega_2(z)}\frac{\omega_2(z)-1}{\omega_2(z)+1}\Big] =: \varrho_2\lambda p_2^0,$$

and

$$q_1 = \frac{\varrho_1\lambda}{|\xi|\gamma(z)}\Big[\frac{\varrho_1}{\omega_1(z)} + \frac{\varrho_1\mu_2\gamma_2(z)}{\mu_1\omega_1(z)(\omega_1(z)+1)} + \frac{\varrho_2}{\omega_1(z)\omega_2(z)}\frac{\omega_1(z)-1}{\omega_1(z)+1}\Big] =: \frac{\varrho_1\lambda}{|\xi|\gamma(z)^2}q_1^0$$
$$q_2 = \frac{\varrho_2\lambda}{|\xi|\gamma(z)}\Big[\frac{\varrho_2}{\omega_2(z)} + \frac{\varrho_2\mu_1\gamma_1(z)}{\mu_2\omega_2(z)(\omega_2(z)+1)} + \frac{\varrho_1}{\omega_1(z)\omega_2(z)}\frac{\omega_2(z)-1}{\omega_2(z)+1}\Big] =: \frac{\varrho_2\lambda}{|\xi|\gamma(z)^2}q_2^0,$$

where the scaling $z = \lambda/|\xi|^2$ is employed. Recall

$$\omega_k(z) = \sqrt{1+\varrho_k z/\mu_k}, \quad \gamma_k(z) = \omega_k(z) + 1/\omega_k(z), \quad \gamma(z) = \mu_1\gamma_1(z) + \mu_2\gamma_2(z).$$

This yields the transformed interface pressures

$$\hat{\pi}_1 = \varrho_1\lambda\alpha_1 = [\![\varrho]\!]\frac{p_2^0}{p_1^0q_2^0 + p_2^0q_1^0}|\xi|\gamma(z)^2\hat{k} \qquad (10.57)$$
$$\hat{\pi}_2 = \varrho_2\lambda\alpha_2 = [\![\varrho]\!]\frac{p_1^0}{p_1^0q_2^0 + p_2^0q_1^0}|\xi|\gamma(z)^2\hat{k}.$$

Note that p_k^0, q_0^k are holomorphic in $\mathbb{C}\setminus(-\infty,-\eta]$, $\eta = \min\{\mu_k/\varrho_k\} > 0$, and we have

$$p_k^0(0) = 1/\varrho_k, \quad q_k^0(0) = 2(\mu_1+\mu_2)^2\varrho_k/\mu_k,$$

and

$$p_k^0(\infty) = 1/\varrho_k, \quad q_k^0(\infty) = (\varrho_1+\varrho_2)\sqrt{\varrho_k\mu_k}.$$

Therefore p_k^0, q_k^0 are holomorphic and bounded on $\Sigma_{\pi/2+\varepsilon}\cup B(0,\varepsilon)$, for small $\varepsilon > 0$.

So we need to show that the *Lopatinskii-Determinant* $r^0(z) := p_1^0(z)q_2^0(z) + p_2^0(z)q_1^0(z)$ has no zeros in $\mathrm{Re}\, z \geq 0$. After some tedious algebra, expanding and collecting terms, we obtain the factorization $r^0(z) = r_1^0(z)r_2^0(z)$, where

$$r_1^0(z) = \gamma(z)/[\omega_1(z)\omega_2(z)(\omega_1(z)+1)(\omega_2(z)+1)],$$

and

$$\begin{aligned} r_2^0(z) &= (\frac{\varrho_2}{\varrho_1}\omega_1(z) + \frac{\varrho_1}{\varrho_2}\omega_2(z))(\omega_1(z)+1)(\omega_2(z)+1) \\ &\quad + 2(\omega_1(z)-1)(\omega_2(z)-1) + 2\frac{\mu_2\varrho_1}{\mu_1\varrho_2}(\omega_2(z)-1) + 2\frac{\mu_1\varrho_2}{\mu_2\varrho_1}(\omega_1(z)-1) \\ &\quad + \frac{\mu_2\varrho_1}{\mu_1\varrho_2}(\omega_2^2(z)+1)(\omega_2(z)+1) + \frac{\mu_1\varrho_2}{\mu_2\varrho_1}\gamma_1(z)\omega_1(z)(\omega_1(z)+1). \end{aligned}$$

Obviously, $r_1^0(z)$ has no zeros in $\mathbb{C}\setminus(-\infty,0])$; so we only have to look at $r_2^0(z)$. One easily checks that each summand of $r_2^0(z)$ has nonnegative imaginary part, provided $z\in\mathbb{C}$ is such that $\operatorname{Re} z, \operatorname{Im} z \geq 0$, so $r_2^0(z)$ can only be zero if each summand is zero. But this is not possible, as already the first term shows.

(b) The symbol of S_λ^{22}. To obtain the algebraic system for the symbol of $[S_\lambda^{22}]^{-1}$ we set $g_1 = 0$ and let j be given. Then (8.70) remains valid as well as the formulas for a_1, a_2. For α_1, α_2 we have here the equations

$$\begin{aligned} 2\mu_2(\beta_2 + |\xi|^2\alpha_2) + \varrho_2\lambda\alpha_2 - 2\mu_1(\beta_1 + |\xi|^2\alpha_1) + \varrho_1\lambda\alpha_1 = 0, \\ \frac{\sqrt{\mu_2}}{\omega_2}\beta_2 + |\xi|\alpha_2 + \frac{\sqrt{\mu_1}}{\omega_1}\beta_1 + |\xi|\alpha_1 = [\![1/\varrho]\!]\hat{\jmath}. \end{aligned}$$

Inserting β_k from (8.70), this system becomes

$$\begin{aligned} p_1\alpha_1 - p_2\alpha_2 &= 0, \\ q_1\alpha_1 + q_2\alpha_2 &= [\![1/\varrho]\!]\hat{\jmath}, \end{aligned} \tag{10.58}$$

where

$$\begin{aligned} p_1 &= \varrho_1\lambda\Big[1 + 2[\![\mu]\!]\frac{1}{\gamma(z)\omega_1(z)}\frac{\omega_1(z)-1}{\omega_1(z)+1}\Big] =: \varrho_1\lambda p_1^0, \\ p_2 &= \varrho_2\lambda\Big[1 - 2[\![\mu]\!]\frac{1}{\gamma(z)\omega_2(z)}\frac{\omega_2(z)-1}{\omega_2(z)+1}\Big] =: \varrho_2\lambda p_2^0, \end{aligned}$$

and

$$\begin{aligned} q_1 &= \frac{\varrho_1\lambda}{|\xi|\omega_1(z)\gamma(z)}\Big[1 + \frac{1}{\omega_2(z)}\frac{\omega_1(z)-1}{\omega_1(z)+1} + \frac{\gamma_2(z)}{\mu_1(\omega_1(z)+1)}\Big] =: \frac{\varrho_1\lambda}{|\xi|\gamma(z)^2}q_1^0, \\ q_2 &= \frac{\varrho_2\lambda}{|\xi|\omega_2(z)\gamma(z)}\Big[1 + \frac{1}{\omega_1(z)}\frac{\omega_2(z)-1}{\omega_2(z)+1} + \frac{\gamma_1(z)}{\mu_2(\omega_2(z)+1)}\Big] =: \frac{\varrho_2\lambda}{|\xi|\gamma(z)^2}q_2^0. \end{aligned}$$

This yields

$$\begin{aligned} \hat{\pi}_1 &= \varrho_1\lambda\alpha_1 = [\![1/\varrho]\!]\frac{p_2^0}{p_1^0q_2^0 + p_2^0q_1^0}|\xi|\gamma(z)^2\hat{\jmath}, \\ \hat{\pi}_2 &= \varrho_2\lambda\alpha_2 = [\![1/\varrho]\!]\frac{p_1^0}{p_1^0q_2^0 + p_2^0q_1^0}|\xi|\gamma(z)^2\hat{\jmath}. \end{aligned} \tag{10.59}$$

As in (a) the functions p_k^0, q_k^0 are holomorphic in $\mathbb{C}\setminus(-\infty,-\eta]$, $\eta = \min\{\mu_k/\varrho_k\} > 0$, and we have

$$p_k^0(0) = 1, \quad q_k^0(0) = 2(\mu_1+\mu_2)^2/\mu_k,$$

and

$$p_k^0(\infty) = 1, \quad q_k^0(\infty) = (\sqrt{\varrho_1\mu_1} + \sqrt{\varrho_2\mu_2})^2\varrho_k.$$

Therefore p_k^0, q_k^0 are holomorphic and bounded on $\Sigma_{\pi/2+\varepsilon} \cap B(0,\varepsilon)$, for small $\varepsilon > 0$.

So we need to show that the Lopatinskii-determinant $r^0(z) := p_1^0(z)q_2^0(z) + p_2^0(z)q_1^0(z)$ has no zeros in $\mathrm{Re}\, z \geq 0$. After another tedious algebra, expanding and collecting terms, we obtain as in (a) the factorization $r^0(z) = r_1^0(z)r_2^0(z)$, where

$$r_1^0(z) = \gamma(z)/[\omega_1(z)\omega_2(z)(\omega_1(z)+1)(\omega_2(z)+1)],$$

and

$$\begin{aligned} r_2^0(z) &= (\omega_1(z)+\omega_2(z))(\omega_1(z)+1)(\omega_2(z)+1) + 2(\omega_1(z)-1)(\omega_2(z)-1) \\ &+ \frac{\mu_2}{\mu_1}[2(\omega_2(z)-1) + (\omega_2^2(z)+1)(\omega_2(z)+1)] \\ &+ \frac{\mu_1}{\mu_2}[2(\omega_1(z)-1) + (\omega_1^2(z)+1)(\omega_1(z)+1)]. \end{aligned}$$

Obviously, $r_1^0(z)$ has no zeros in $\mathbb{C}\setminus(-\infty,0])$; so we only have to look at $r_2^0(z)$. One easily checks that each summand of $r_2^0(z)$ has nonnegative imaginary part, provided $z \in \mathbb{C}$ is such that $\mathrm{Re}\, z, \mathrm{Im}\, z \geq 0$, so $r_2^0(z)$ can only be zero if each summand is zero. But this is not possible, as already the first term shows.

Using the usual arguments of perturbation, domain perturbation and localization these results prove **(v)** of Proposition 10.7.1.

Appendix B: The Dirichlet Extension Operator

Consider the problem

$$\Delta u = 0 \quad \text{in } \Omega, \quad u = g \quad \text{on } \partial\Omega, \tag{10.60}$$

where $\Omega \subset \mathbb{R}^n$ is bounded domain with boundary $\Sigma := \partial\Omega \in C^{2-}$. We know from Chapter 6 that the *Dirichlet extension operator* $T : W_q^{2-1/q}(\Sigma) \to H_q^2(\Omega)$ defined by $Tg = u$, where u denotes the unique solution of (10.60), is well-defined and bounded. We want to prove that this operator has a unique bounded extension – again denoted by T – from $B_{qq}^s(\Sigma)$ to $H_q^{s+1/q}(\Omega)$, for all $s \in [-1/q, 2-1/q]$. By complex interpolation, it is enough to prove this for $s = -1/q$. Observe that $B_{qq}^s(\Sigma) = W_q^s(\Sigma)$ for $s \notin \mathbb{N}_0$, and $B_{22}^0(\Sigma) = L_2(\Sigma)$. In particular, this proves that $T : H_2^s(\Sigma) \to H_2^{s+1/2}(\Omega)$ is bounded, for any $s \in [-1/2, 3/2]$.

For this purpose, we define operators $\mathsf{L}_q : Y_q \to L_q(\Omega)$ by means of

$$\mathsf{L}_q u := -\Delta u, \quad u \in Y_q := \{u \in H_q^2(\Omega) : u = 0 \text{ on } \Sigma\}.$$

Then we know that L_q is linear bounded, bijective, hence its inverse is also bounded, by the open mapping theorem of Banach, for each $q \in (1,\infty)$. This implies that its dual $\mathsf{L}_{q'}^* : L_{q'}(\Omega) \to Y_{q'}^*$ is also bijective, and we also see that $\mathsf{L}_{q'}^* = \mathsf{L}_q$ on Y_q, which is a dense subset of $L_q(\Omega)$. Therefore, $\mathsf{L}_{q'}^*$ is a bounded linear bijective extension of L_q from $L_q(\Omega)$ to $Y_{q'}^*$.

Now we consider the solution $u \in H^2_q(\Omega \setminus \Sigma)$ of (10.60) for some given $g \in W^{2-1/q}_q(\Omega)$. For $\phi \in Y_{q'}$ we obtain integrating by parts twice

$$0 = (-\Delta u|\phi)_\Omega = (u| - \Delta\phi)_\Omega + (g|\partial_\nu\phi)_\Sigma = (u|\mathsf{L}_{q'}\phi)_\Omega + (g|\partial_\nu\phi)_\Sigma.$$

The estimate

$$|(g|\partial_\nu\phi)_\Sigma| \le |g|_{W^{-1/q}_q(\Sigma)} |\partial_\nu\phi|_{W^{1/q}_{q'}(\Sigma)} \le C|g|_{W^{-1/q}_q(\Sigma)} |\phi|_{H^2_{q'}(\Omega)} = C|g|_{W^{-1/q}_q(\Sigma)} |\phi|_{Y_{q'}}$$

shows that there is $\bar g \in Y^*_{q'}$ such that

$$(g|\partial_\nu\phi)_\Sigma = \langle \bar g|\phi\rangle, \quad \text{for all } \phi \in Y_{q'}.$$

As $\mathsf{L}^*_{q'}$ is bijective, there is a unique $u \in L_q(\Omega)$ such that $\mathsf{L}^*_{q'}u = -\bar g$, and

$$|u|_{L_q} \le C|\bar g|_{Y^*_{q'}} \le C|g|_{W^{-1/q}_q(\Sigma)}.$$

As we have $u = Tg$ for $g \in W^{2-1/q}_q(\Sigma)$, and $W^{2-1/q}_q(\Sigma)$ is dense in $W^{-1/q}_q(\Sigma)$, this shows that the extension $T : W^{-1/q}_q(\Sigma) \to L_q(\Omega)$ is unique and also bounded, thereby proving the assertion.

Chapter 11

Qualitative Behaviour of the Semiflows

Building on the previous Chapters 9 and 10, we take up the question of the long time behaviour of solutions of the six problems **(Pj)** defined in Chapter 1. We begin by showing in Section 11.1 that these problems generate local semiflows in their generic state manifolds $\mathcal{SM}^j$. The local structure of these manifolds near equilibria is analyzed in Section 11.2, where by means of a Hanzawa transform a local representation is derived. This allows the computation of the corresponding tangent spaces $\mathcal{SX}_*^j$ at a given equilibrium $e_* = (0, \theta_*, \Gamma_*)$. The main result of this chapter is proved in Section 11.3. It tells that such an equilibrium is stable resp. unstable in $\mathcal{SM}^j$, provided the corresponding linearized problem has this property. The latter has been the content of Chapter 10. For the proof we use an adapted version of the *generalized principle of linearized stability* which has been proved in Chapter 5 for quasilinear parabolic evolution equations; actually this amounts to a considerable extension of the result in Chapter 5. The chapter ends with a conditional result, Theorem 11.4.1, on global existence and convergence to equilibria. Theorem 11.4.1 can also be read as a blow-up criterion for the local semiflow, as the conditions are necessary and sufficient for global existence and convergence.

11.1 The Local Semiflows

Recall from Chapter 2 that the closed C^2-hypersurfaces contained in Ω form a C^2-manifold, which we denote by $\mathcal{MH}^2(\Omega)$. Also recall the Hausdorff metric on the set $\mathcal{K}$ of compact subsets of $\mathbb{R}^n$ given by

$$d_H(K_1, K_2) = \max\{\sup_{x\in K_1} d(x, K_2),\ \sup_{x\in K_2} d(x, K_1)\}.$$

J. Prüss and G. Simonett, *Moving Interfaces and Quasilinear Parabolic Evolution Equations*, Monographs in Mathematics 105,
DOI 10.1007/978-3-319-27698-4_11

We consider the second normal bundle of the C^2-hypersurface Γ

$$\mathcal{N}^2\Gamma = \{(p, \nu_\Gamma(p), \nabla_\Gamma \nu_\Gamma(p)) : p \in \Gamma\}.$$

Here ∇_Γ denotes the surface gradient on Γ. The metric on $\mathcal{MH}^2(\Omega)$ is defined by

$$d(\Gamma_1, \Gamma_2) := d_H(\mathcal{N}^2\Gamma_1, \mathcal{N}^2\Gamma_2), \quad \Gamma_1, \Gamma_2 \in \mathcal{MH}^2(\Omega).$$

The charts are the parameterizations over a given real analytic hypersurface Σ, as described in Chapter 2, and the tangent space of Γ consists of the normal vector fields on Γ of class C^2. This way $\mathcal{MH}^2(\Omega)$ becomes a Banach manifold.

Let $d_\Gamma(x)$ denote the signed distance for Γ as introduced in Chapter 2. Recall also the definition of the *level function* φ_Γ given by

$$\varphi_\Gamma(x) := \begin{cases} d_\Gamma(x)\chi(3d_\Gamma(x)/a) + \operatorname{sign} d_\Gamma(x)(1 - \chi(3d_\Gamma(x)/a)), & x \in U_a, \\ \chi_{\Omega_{\rm ex}}(x) - \chi_{\Omega_{\rm in}}(x), & x \notin U_a, \end{cases}$$

where $\chi \in C^\infty$ denotes a cut-off function with $\chi(s) = 1$ for $|s| \leq 1$, $\chi(s) = 0$ for $|s| \geq 2$, $0 \leq \chi(s) \leq 1$. Then we have seen that $\Gamma = \varphi_\Gamma^{-1}(0)$, and $\nabla\varphi_\Gamma(x) = \nu_\Gamma(x)$, for each $x \in \Gamma$. Moreover, $\kappa = 0$ is an eigenvalue of $\nabla^2\varphi_\Gamma(x)$ with eigenfunction $\nu_\Gamma(x)$; the remaining eigenvalues of $\nabla^2\varphi_\Gamma(x)$ are the principal curvatures $\kappa_j(x)$ of Γ at $x \in \Gamma$.

Consider the subset $\mathcal{MH}^2(\Omega, r)$ of $\mathcal{MH}^2(\Omega)$ which consists of all hypersurfaces $\Gamma \in \mathcal{MH}^2(\Omega)$ such that $\Gamma \subset \Omega$ satisfies the ball condition with fixed radius $r > 0$. This implies in particular that $\operatorname{dist}(\Gamma, \partial\Omega) \geq 2r$ and all principal curvatures of $\Gamma \in \mathcal{MH}^2(\Omega, r)$ are bounded by $1/r$. Furthermore, the level functions φ_Γ are well defined for $\Gamma \in \mathcal{MH}^2(\Omega, r)$, and form a bounded subset of $C^2(\bar\Omega)$. The map $\Phi : \mathcal{MH}^2(\Omega, r) \to C^2(\bar\Omega)$, defined by $\Phi(\Gamma) = \varphi_\Gamma$, is a homeomorphism of the metric space $\mathcal{MH}^2(\Omega, r)$ onto $\Phi(\mathcal{MH}^2(\Omega, r)) \subset C^2(\bar\Omega)$.

Let $s - (n-1)/p > 2$; for $\Gamma \in \mathcal{MH}^2(\Omega, r)$, we define $\Gamma \in W^s_p(\Omega, r)$ if $\varphi_\Gamma \in W^s_p(\Omega)$. In this case the local charts for Γ can be chosen of class W^s_p as well. A subset $A \subset W^s_p(\Omega, r)$ is said to be (relatively) compact, if $\Phi(A) \subset W^s_p(\Omega)$ is (relatively) compact.

We now introduce the state manifolds for the six problems in question.

Problem 1. As an ambient space for the state manifold $\mathcal{SM}^1$ of Problem 1 we consider the product space $C(\bar\Omega) \times \mathcal{MH}^2(\Omega)$. We define $\mathcal{SM}^1$ by means of

$$\begin{aligned} \mathcal{SM}^1 := \big\{ z := (\theta, \Gamma) \in C(\bar\Omega) \times \mathcal{MH}^2(\Omega) : \theta \in W_p^{2-2/p}(\Omega \setminus \Gamma),\ \Gamma \in W_p^{4-3/p}, \\ \theta > 0 \text{ in } \bar\Omega,\ \ \partial_\nu\theta = 0 \text{ on } \partial\Omega, \\ \varrho[\![\psi(\theta)]\!] + \sigma H_\Gamma = 0,\ \ l(\theta) \neq 0 \text{ on } \Gamma,\ \ [\![d(\theta)\partial_\nu\theta]\!] \in W_p^{2-6/p}(\Gamma) \big\}. \end{aligned}$$

Problem 2. As an ambient space for the state manifold $\mathcal{SM}^2$ of Problem 2 we consider the product space $C(\bar{\Omega})^{n+1} \times \mathcal{MH}^2(\Omega)$. We define $\mathcal{SM}^2$ as follows.

$$\begin{aligned}\mathcal{SM}^2 := \big\{ &z := (u,\theta,\Gamma) \in C(\bar{\Omega})^{n+1} \times \mathcal{MH}^2(\Omega) : (u,\theta) \in W_p^{2-2/p}(\Omega\setminus\Gamma)^{n+1},\\ &\Gamma \in W_p^{3-2/p},\ \operatorname{div} u = 0 \text{ in } \Omega\setminus\Gamma,\ \ \theta > 0 \text{ in } \bar{\Omega},\ \ u, \partial_\nu\theta = 0 \text{ on } \partial\Omega,\\ &\mathcal{P}_\Gamma[\![\mu(\theta)(\nabla u + [\nabla u]^{\mathsf{T}})\nu_\Gamma]\!], [\![d(\theta)\partial_\nu\theta]\!] = 0 \text{ on } \Gamma\big\}.\end{aligned}$$

Problem 3. As an ambient space for the state-manifold $\mathcal{SM}^3$ of Problem 3 we consider again the product space $C(\bar{\Omega})^{n+1} \times \mathcal{MH}^2(\Omega)$. We define $\mathcal{SM}^3$ as follows.

$$\begin{aligned}\mathcal{SM}^3 := \big\{ &z := (u,\theta,\Gamma) \in C(\bar{\Omega})^{n+1} \times \mathcal{MH}^2(\Omega) : (u,\theta) \in W_p^{2-2/p}(\Omega\setminus\Gamma)^{n+1},\\ &\Gamma \in W_p^{4-3/p},\ \operatorname{div} u = 0 \text{ in } \Omega\setminus\Gamma,\ \ \theta > 0 \text{ in } \bar{\Omega},\ \ u, \partial_\nu\theta = 0 \text{ on } \partial\Omega,\\ &\mathcal{P}_\Gamma[\![\mu(\theta)(\nabla u + [\nabla u]^{\mathsf{T}})\nu_\Gamma]\!] = 0 \text{ on } \Gamma,\\ &\varrho[\![\psi(\theta)]\!] + \sigma H_\Gamma = 0,\ \ l(\theta) \neq 0 \text{ on } \Gamma,\ \ [\![d(\theta)\partial_\nu\theta]\!] \in W_p^{2-6/p}(\Gamma)\big\}.\end{aligned}$$

Problem 4. As an ambient space for the state manifold $\mathcal{SM}^4$ of Problem 4 we consider the product space $L_p(\Omega)^{n+1} \times \mathcal{MH}^2(\Omega)$, to account for the jump of u across the interface. We define the state manifold $\mathcal{SM}^4$ as follows.

$$\begin{aligned}\mathcal{SM}^4 := \big\{ &z := (u,\theta,\Gamma) \in L_p(\Omega)^{n+1} \times \mathcal{MH}^2(\Omega) : (u,\theta) \in W_p^{2-2/p}(\Omega\setminus\Gamma)^{n+1},\\ &\Gamma \in W_p^{3-2/p},\ \operatorname{div} u = 0 \text{ in } \Omega\setminus\Gamma,\ \ \theta > 0 \text{ in } \bar{\Omega},\ \ u, \partial_\nu\theta = 0 \text{ on } \partial\Omega,\\ &\mathcal{P}_\Gamma[\![\mu(\theta)(\nabla u + [\nabla u]^{\mathsf{T}})\nu_\Gamma]\!],\ \mathcal{P}_\Gamma[\![u]\!],\ [\![\theta]\!] = 0 \text{ on } \Gamma,\\ &l(\theta)[\![u\cdot\nu_\Gamma]\!] + [\![1/\varrho]\!][\![d(\theta)\partial_\nu\theta]\!] = 0 \text{ on } \Gamma\big\}.\end{aligned}$$

Problem 5. As an ambient space for the state-manifold $\mathcal{SM}^5$ of Problem 5 we consider again the product space $C(\bar{\Omega})^{n+1} \times \mathcal{MH}^2(\Omega)$. We define $\mathcal{SM}^5$ as follows.

$$\begin{aligned}\mathcal{SM}^5 := \big\{ &z := (u,\theta,\Gamma) \in C(\bar{\Omega})^{n+1} \times \mathcal{MH}^2(\Omega) : (u,\theta) \in W_p^{2-2/p}(\Omega\setminus\Gamma)^{n+1},\\ &\Gamma \in W_p^{4-3/p},\ \operatorname{div} u = 0 \text{ in } \Omega\setminus\Gamma,\ \ 0 < \theta < \theta_c \text{ in } \bar{\Omega},\ \ u, \partial_\nu\theta = 0 \text{ on } \partial\Omega,\\ &\mathcal{P}_\Gamma[\![\mu(\theta)(\nabla u + [\nabla u]^{\mathsf{T}})\nu_\Gamma]\!] + \sigma'(\theta)\nabla_\Gamma\theta,\ \varrho[\![\psi(\theta)]\!] + \sigma(\theta)H_\Gamma = 0 \text{ on } \Gamma,\\ &T_\Gamma := \omega_\Gamma(\theta) - H_\Gamma' \ \text{ is invertible in } L_2(\Gamma)\big\}.\end{aligned}$$

Here we used the abbreviation $\omega_\Gamma(\theta) = (\varrho l(\theta) + l_\Gamma H_\Gamma)^2/\theta\sigma(\theta)\kappa_\Gamma(\theta)$.

Problem 6. As an ambient space for the state manifold $\mathcal{SM}^6$ of Problem 6 we consider the product space $L_p(\Omega)^{n+1} \times \mathcal{MH}^2(\Omega)$. We define the state manifold $\mathcal{SM}^6$ as follows.

$$\begin{aligned}\mathcal{SM}^6 := \big\{ &z := (u,\theta,\Gamma) \in L_p(\Omega)^{n+1} \times \mathcal{MH}^2(\Omega) : (u,\theta) \in W_p^{2-2/p}(\Omega\setminus\Gamma)^{n+1},\\ &\Gamma \in W_p^{3-2/p},\ \operatorname{div} u = 0 \text{ in } \Omega,\ \ 0 < \theta < \theta_c \text{ in } \bar{\Omega},\ \ u, \partial_\nu\theta = 0 \ \text{ on } \partial\Omega,\\ &\mathcal{P}_\Gamma[\![\mu(\theta)(\nabla u + [\nabla u]^{\mathsf{T}})\nu_\Gamma]\!] + \sigma'(\theta)\nabla_\Gamma\theta,\ \mathcal{P}_\Gamma[\![u]\!],\ [\![\theta]\!] = 0 \text{ on } \Gamma\big\}.\end{aligned}$$

The charts for these manifolds are obtained by the charts induced by $\mathcal{MH}^2(\Omega)$, followed by a Hanzawa transformation as introduced in Chapter 1.

Observe that the relevant compatibility conditions as well as regularity are preserved by the solutions. We also note that we do neither use the surface temperature θ_Γ nor the pressure π as system variables, as these are determined at each instant by the state variable z.

Applying the well-posedness results from Chapter 9 and re-parameterizing repeatedly, we obtain for the Problem **(Pj)** a *local semiflow* on $\mathcal{SM}^j$, $j = 1, \dots, 6$.

Theorem 11.1.1. *Let $p > n+2$, $j = 1, \dots, 6$, and suppose Conditions* **(Hj)** *holds. Then the two-phase problem* **(Pj)** *generates a local semiflow in the state manifold $\mathcal{SM}^j$. Each solution z exists on a maximal time interval $[0, t_+)$.*

Observe that our definition of the state manifolds is invariant under coordinate transformations. Therefore, this result answers in particular the question in which sense a solution of a problem with moving boundary can be unique. This should be compared with the case of an ordinary differential equation on a manifold.

The maximal time t_+ of existence will be characterized below in Section 11.4.

11.2 Tangent Spaces at Equilibria

We now fix a non-degenerate equilibrium $e_* = (0, \theta_*, \Gamma_*) \in \mathcal{E}$. In this section we study the state manifolds $\mathcal{SM}^j$ in a neighbourhood of this equilibrium. Generically, in the sequel the value of a scalar function $f(\theta)$ at the equilibrium value $\theta = \theta_*$ will be denoted by f_*.

2.1 Local Representations

We perform a Hanzawa transform where the reference hypersurface Σ is here taken to be $\Sigma = \Gamma_*$. This leads to the following local representations of the state manifolds. In the sequel we will employ the *relative temperature* $\vartheta = \theta - \theta_*$. We use the notation for the spaces $\mathbb{E}$ and $\mathbb{F}$ from Chapter 9, with a minor change. $\mathbb{E}^j(a)$ denotes $\mathbb{E}_1^j(a)$ without pressure components, while $\hat{\mathbb{E}}^j(a)$ contains the pressure, and $\mathbb{F}^j(a) = \mathbb{F}_1^j(a)$.

Problem 1. We consider the trace space X_γ^1 of $\mathbb{E}^1(a)$ which is given by

$$X_\gamma^1 = \{z = (\vartheta, h) \in W_p^{2-2/p}(\Omega \setminus \Sigma) \times W_p^{4-3/p}(\Sigma) : [\![\vartheta]\!] = 0 \text{ on } \Sigma,\ \partial_\nu \vartheta = 0 \text{ on } \partial\Omega\}.$$

As in Chapter 9 we linearize the Gibbs-Thomson law $\varphi(\theta) + \sigma H_\Gamma(h) = 0$ according to

$$(\varrho l_*/\theta_*)\vartheta - \sigma \mathcal{A}_\Sigma h = G_\gamma(\vartheta, h),$$

where as in Chapter 10,

$$\mathcal{A}_\Sigma = -(n-1)/R_*^2 - \Delta_\Sigma$$

denotes the curvature operator on Σ. The jump of the heat flux is linearized as

$$[\![\mathcal{B}_\theta(\theta,h)\nabla\theta]\!] = [\![d_*\partial_\nu\vartheta]\!] + G_\theta(\vartheta,h)\nabla\vartheta.$$

Then near the equilibrium e_* the state manifold $\mathcal{SM}^1$ reads

$$\begin{aligned}\mathcal{SM}^1_* = \big\{(\vartheta,h) \in X^1_\gamma : (\varrho l_*/\theta_*)\vartheta - \sigma\mathcal{A}_\Sigma h = G_\gamma(\vartheta,h) \text{ on } \Sigma,\\ [\![d_*\partial_\nu\vartheta]\!] + G_\theta(\vartheta,h)\nabla\vartheta \in W_p^{2-6/p}(\Sigma)\big\}.\end{aligned}$$

Furthermore, we set

$$\mathcal{SX}^1_* := \{(\vartheta,h) \in X^1_\gamma : (\varrho l_*/\theta_*)\vartheta - \sigma\mathcal{A}_\Sigma h = 0 \text{ on } \Sigma,\ [\![d_*\partial_\nu\vartheta]\!] \in W_p^{2-6/p}(\Sigma)\}.$$

For further use, we also introduce the boundary trace space $Y^1_\gamma = W_p^{2-3/p}(\Sigma)$ as well as the linear stationary boundary operator

$$\mathsf{B}^1 z = (\varrho l_*/\theta_*)\vartheta - \sigma\mathcal{A}_\Sigma h,$$

and the stationary boundary nonlinearity

$$\mathsf{G}^1(z) = G_\gamma(\vartheta,h).$$

Problem 2. The trace space X^2_γ of $\mathbb{E}^2(a)$ is given by

$$\begin{aligned}X^2_\gamma = \big\{z = (u,\vartheta,h) \in W_p^{2-2/p}(\Omega\setminus\Sigma)^{n+1} \times W_p^{3-2/p}(\Sigma) : [\![u]\!], [\![\vartheta]\!] = 0 \text{ on } \Sigma,\\ u, \partial_\nu\vartheta = 0 \text{ on } \partial\Omega\big\}.\end{aligned}$$

We linearize the divergence condition as

$$\operatorname{div} u = M_1(h)\nabla\cdot u,$$

and the tangential part of the normal stress condition, see (1.62), as

$$\mathcal{P}_\Sigma[\![\mu_*(\nabla u + [\nabla u]^{\mathsf{T}})\nu_\Sigma]\!] = G_\tau(\vartheta,h)\nabla u.$$

Then near the equilibrium e_* the state manifold $\mathcal{SM}^2$ reads

$$\begin{aligned}\mathcal{SM}^2_* = \big\{(u,\vartheta,h) \in X^2_\gamma : \operatorname{div} u = M_1(h)\nabla\cdot u \text{ in } \Omega\setminus\Sigma,\\ \mathcal{P}_\Sigma[\![\mu_*(\nabla u + [\nabla u]^{\mathsf{T}})\nu_\Sigma]\!] = G_\tau(\vartheta,h)\nabla u,\ -[\![d_*\partial_\nu\vartheta]\!] = G_\theta(\vartheta,h)\nabla\vartheta \text{ on } \Sigma\big\}.\end{aligned}$$

Furthermore, we set

$$\begin{aligned}\mathcal{SX}^2_* = \big\{(u,\vartheta,h) \in X^2_\gamma : \operatorname{div} u = 0 \text{ in } \Omega\setminus\Sigma,\\ \mathcal{P}_\Sigma[\![\mu_*(\nabla u + [\nabla u]^{\mathsf{T}})\nu_\Sigma]\!] = 0,\ [\![d_*\partial_\nu\vartheta]\!] = 0 \text{ on } \Sigma\big\}.\end{aligned}$$

Let

$$W^{1-2/p}_{p,0}(\Omega\backslash\Sigma) := W^{1-2/p}_p(\Omega\backslash\Sigma)/constants \cong \{v \in W^{1-2/p}_p(\Omega\backslash\Gamma) : \int_\Omega v\,dx = 0\},$$

and

$$Y^2_\gamma = W^{1-2/p}_{p,0}(\Omega \setminus \Sigma) \times W^{1-3/p}_p(\Sigma; T\Sigma) \times W^{1-3/p}_p(\Sigma),$$

and define the linear stationary boundary operator

$$\mathsf{B}^2 z = (\operatorname{div} u, \mathcal{P}_\Sigma[\![\mu_*(\nabla u + [\nabla u]^\mathsf{T})\nu_\Sigma]\!], -[\![d_*\partial_\nu\vartheta]\!]),$$

and the stationary boundary nonlinearity

$$\mathsf{G}^2(z) = (M_1(h)\nabla\cdot u, G_\tau(\vartheta, h)\nabla u, G_\theta(\vartheta, h)\nabla\vartheta).$$

Problem 3. The trace space X^3_γ of $\mathbb{E}^3(a)$ is given by

$$X^3_\gamma = \{z = (u, \vartheta, h) \in W^{2-2/p}_p(\Omega \setminus \Sigma)^{n+1} \times W^{4-3/p}_p(\Sigma) : [\![u]\!], [\![\vartheta]\!] = 0 \text{ on } \Sigma,\\ u, \partial_\nu\vartheta = 0 \text{ on } \partial\Omega\}.$$

Then near the equilibrium e_* the state manifold $\mathcal{SM}^3$ reads

$$\begin{aligned}\mathcal{SM}^3_* = \{(u, \vartheta, h) \in X^3_\gamma :\ & \operatorname{div} u = M_1(h)\nabla\cdot u \text{ in } \Omega\setminus\Sigma,\\ & \mathcal{P}_\Sigma[\![\mu_*(\nabla u + [\nabla u]^\mathsf{T})\nu_\Sigma]\!] = G_\tau(\vartheta, h)\nabla u,\\ & (\varrho l_*/\theta_*)\vartheta - \sigma\mathcal{A}_\Sigma h = G_\gamma(\vartheta, h) \text{ on } \Sigma,\ [\![d_*\partial_\nu\vartheta]\!] + G_\theta(\vartheta, h)\nabla\vartheta \in W^{2-6/p}_p(\Sigma)\}.\end{aligned}$$

Furthermore, we set

$$\mathcal{SX}^3_* = \{(u, \vartheta, h) \in X^3_\gamma : \operatorname{div} u = 0 \text{ in } \Omega\setminus\Sigma,\ \mathcal{P}_\Sigma[\![\mu_*(\nabla u + [\nabla u]^\mathsf{T})\nu_\Sigma]\!] = 0,\\ (\varrho l_*/\theta_*)\vartheta - \sigma\mathcal{A}_\Sigma h = 0 \text{ on } \Sigma,\ [\![d_*\partial_\nu\vartheta]\!] \in W^{2-6/p}_p(\Sigma)\},$$

and

$$Y^3_\gamma = W^{1-2/p}_{p,0}(\Omega \setminus \Sigma) \times W^{1-3/p}_p(\Sigma; T\Sigma) \times W^{2-3/p}_p(\Sigma),$$

and define the linear stationary boundary operator

$$\mathsf{B}^3 z = \big(\operatorname{div} u, \mathcal{P}_\Sigma[\![\mu_*(\nabla u + [\nabla u]^\mathsf{T})\nu_\Sigma]\!], (\varrho l_*/\theta_*)\vartheta - \sigma\mathcal{A}_\Sigma h\big),$$

and the stationary boundary nonlinearity

$$\mathsf{G}^3(z) = (M_1(h)\nabla\cdot u, G_\tau(\vartheta, h)\nabla u, G_\gamma(\vartheta, h)).$$

Problem 4. The trace space X^4_γ of $\mathbb{E}^4(a)$ is given by

$$X^4_\gamma = \{z = (u, \vartheta, h) \in W^{2-2/p}_p(\Omega \setminus \Sigma)^{n+1} \times W^{3-2/p}_p(\Sigma) : [\![\vartheta]\!] = 0 \text{ on } \Sigma,\\ u, \partial_\nu\vartheta = 0 \text{ on } \partial\Omega\}.$$

We linearize the jump condition $\mathcal{P}_\Gamma[\![u]\!] = 0$ as in Chapter 9.

$$\mathcal{P}_\Sigma[\![u]\!] = G_u(h)u := -M_0(h)\nabla_\Sigma h[\![u\cdot\nu_\Sigma]\!],$$

and the term $l(\theta)j_\Gamma$ appearing in the Stefan law as

$$l_*[\![1/\varrho]\!]^{-1}[\![u\cdot\nu_\Sigma]\!] = G_l(\vartheta,h)u.$$

Then near the equilibrium e_* the state manifold $\mathcal{SM}^4$ reads

$$\begin{aligned}\mathcal{SM}^4_* = \big\{(u,\vartheta,h)\in X^4_\gamma :\ & \operatorname{div} u = M_1(h)\nabla\cdot u \text{ in } \Omega\setminus\Sigma,\ \mathcal{P}_\Sigma[\![u]\!] = G_u(h)u,\\ & \mathcal{P}_\Sigma[\![\mu_*(\nabla u+[\nabla u]^{\mathsf T})\nu_\Sigma]\!] = G_\tau(\vartheta,h)\nabla u,\\ & l_*[\![1/\varrho]\!]^{-1}[\![u\cdot\nu_\Sigma]\!] + [\![d_*\partial_\nu\vartheta]\!] = G_\theta(\vartheta,h)\nabla\vartheta + G_l(\vartheta,h)u \text{ on } \Sigma\big\}.\end{aligned}$$

Furthermore, we set

$$\begin{aligned}\mathcal{SX}^4_* = \{(u,\vartheta,h)\in X^4_\gamma :\ & \operatorname{div} u = 0 \text{ in } \Omega\setminus\Sigma,\ \mathcal{P}_\Sigma[\![u]\!] = 0,\\ & \mathcal{P}_\Sigma[\![\mu_*(\nabla u+[\nabla u]^{\mathsf T})\nu_\Sigma]\!] = 0,\ l_*[\![1/\varrho]\!]^{-1}[\![u\cdot\nu_\Sigma]\!] + [\![d_*\partial_\nu\vartheta]\!] = 0 \text{ on } \Sigma\},\end{aligned}$$

and

$$Y^4_\gamma = W^{1-2/p}_p(\Omega\setminus\Sigma)\times W^{2-3/p}_p(\Sigma;T\Sigma)\times W^{1-3/p}_p(\Sigma;T\Sigma)\times W^{2-3/p}_p(\Sigma),$$

and define the linear stationary boundary operator

$$\mathsf{B}^4 z = (\operatorname{div} u, \mathcal{P}_\Sigma[\![u]\!], \mathcal{P}_\Sigma[\![\mu_*(\nabla u+[\nabla u]^{\mathsf T})\nu_\Sigma]\!], l_*[\![1/\varrho]\!]^{-1}[\![u\cdot\nu_\Sigma]\!] + [\![d_*\partial_\nu\vartheta]\!]),$$

and the stationary boundary nonlinearity

$$\mathsf{G}^4(z) = (M_1(h)\nabla\cdot u, G_u(h)u, G_\tau(\vartheta,h)\nabla u, G_\theta(\vartheta,h)\nabla\vartheta + G_l(\vartheta,h)u).$$

Problem 5. The relevant trace space X^5_γ of $\mathbb{E}^5(a)$ is given by

$$\begin{aligned}X^5_\gamma = \{z=(u,\vartheta,h)\in W^{2-2/p}_p(\Omega\setminus\Sigma)^{n+1}\times W^{4-3/p}_p(\Sigma) :\ & [\![u]\!],[\![\vartheta]\!] = 0 \text{ on } \Sigma,\\ & u,\partial_\nu\vartheta = 0 \text{ on } \partial\Omega\}.\end{aligned}$$

So we have $X^5_\gamma = X^3_\gamma$. The presence of the Marangoni force on the interface leads to the additional term

$$(\sigma'(\theta)/\beta(h))M_0(h)\nabla_\Gamma\theta = \sigma'_*\nabla_\Sigma\vartheta - G_\sigma(\vartheta,h)\nabla_\Sigma\vartheta.$$

for the tangential component of the stress condition, see (1.61). Then near the equilibrium e_* the state manifold $\mathcal{SM}^5$ reads

$$\begin{aligned}\mathcal{SM}^5_* = \big\{(u,\vartheta,h)\in X^5_\gamma :\ & \operatorname{div} u = M_1(h)\nabla\cdot u \text{ in } \Omega\setminus\Sigma,\\ & \mathcal{P}_\Sigma[\![\mu_*(\nabla u+[\nabla u]^{\mathsf T})\nu_\Sigma]\!] + \sigma'_*\nabla_\Sigma\vartheta = G_\tau(\vartheta,h)\nabla u + G_\sigma(\vartheta,h)\nabla_\Sigma\vartheta \text{ on } \Sigma,\\ & (\varrho l_*/\theta_*)\vartheta - \sigma\mathcal{A}_\Sigma h = G_\gamma(\vartheta,h) \text{ on } \Sigma\big\}.\end{aligned}$$

Furthermore, we set

$$\mathcal{SX}_*^5 = \big\{(u,\vartheta,h) \in X_\gamma^5 : \operatorname{div} u = 0 \text{ in } \Omega\setminus\Sigma,\\ \mathcal{P}_\Sigma[\![\mu_*(\nabla u + [\nabla u]^\mathsf{T}\nu_\Sigma]\!] + \sigma_*'\nabla_\Sigma\vartheta = 0 \text{ on } \Sigma\big\},$$

and

$$Y_\gamma^5 = W_{p,0}^{1-2/p}(\Omega\setminus\Sigma) \times W_p^{1-3/p}(\Sigma;T\Sigma),$$

and define the linear stationary boundary operator

$$\mathsf{B}^5 z = (\operatorname{div} u, \mathcal{P}_\Sigma[\![\mu_*(\nabla u + [\nabla u]^\mathsf{T})\nu_\Sigma]\!] + \sigma_*'\nabla_\Sigma\vartheta),$$

and the stationary boundary nonlinearity

$$\mathsf{G}^5(z) = (M_1(h)\nabla\cdot u, G_\tau(\vartheta,h)\nabla u + G_\sigma(\vartheta,h)\nabla_\Sigma\vartheta).$$

Problem 6. The trace space X_γ^6 of $\mathbb{E}^6(a)$ is given by

$$X_\gamma^6 = \big\{z = (u,\vartheta,h) \in W_p^{2-2/p}(\Omega\setminus\Sigma)^{n+1} \times W_p^{3-2/p}(\Sigma) : [\![\vartheta]\!] = 0 \text{ on } \Sigma,\\ u, \partial_\nu\vartheta = 0 \text{ on } \partial\Omega\big\}.$$

So we have $X_\gamma^6 = X_\gamma^4$. Then near the equilibrium e_* the state manifold $\mathcal{SM}^6$ reads

$$\mathcal{SM}_*^6 = \big\{(u,\vartheta,h) \in X_\gamma^6 : \operatorname{div} u = M_1(h)\nabla\cdot u \text{ in } \Omega\setminus\Sigma,\ \mathcal{P}_\Sigma[\![u]\!] = G_u(h)u,\\ \mathcal{P}_\Sigma[\![\mu_*(\nabla u + [\nabla u]^\mathsf{T})\nu_\Sigma]\!] + \sigma_*'\nabla_\Sigma\vartheta = G_\tau(\vartheta,h)\nabla u + G_\sigma(\vartheta,h)\nabla_\Sigma\vartheta \text{ on } \Sigma\big\}.$$

Furthermore, we set

$$\mathcal{SX}_*^6 = \big\{(u,\vartheta,h) \in X_\gamma^1 : \operatorname{div} u = 0 \text{ in } \Omega\setminus\Sigma,\ \mathcal{P}_\Sigma[\![u]\!] = 0,\\ \mathcal{P}_\Sigma[\![\mu_*(\nabla u + [\nabla u]^\mathsf{T})\nu_\Sigma]\!] + \sigma_*'\nabla_\Sigma\vartheta = 0 \text{ on } \Sigma\big\},$$

and

$$Y_\gamma^6 = W_p^{1-2/p}(\Omega\setminus\Sigma) \times W_p^{1-3/p}(\Sigma;T\Sigma) \times W_p^{2-3/p}(\Sigma,T\Sigma),$$

and define the linear stationary boundary operator

$$\mathsf{B}^6 z = (\operatorname{div} u, \mathcal{P}_\Sigma[\![\mu_*(\nabla u + [\nabla u]^\mathsf{T})\nu_\Sigma]\!] + \sigma_*'\nabla_\Sigma\vartheta, \mathcal{P}_\Sigma[\![u]\!]),$$

and the stationary boundary nonlinearity

$$\mathsf{G}^6(z) = (M_1(h)\nabla\cdot u, G_\tau(\vartheta,h)\nabla u + G_\sigma(\vartheta,h), G_u(h)u).$$

We emphasize that the particular form of the boundary nonlinearities $\mathsf{G}^j(z)$ will not be important. In the sequel it is only essential that $\mathsf{G}^j : X_\gamma^j \to Y_\gamma^j$ are of class C^1 – which is ensured by the results in Section 9.5 – and that $\mathsf{G}^j(0)$ and its Fréchet derivative $D\mathsf{G}^j(0)$ vanish, by their very definition.

2.2 Parameterization
As a consequence of the preceeding section we have

$$\mathcal{SM}_*^j = \{z \in X_\gamma^j : \mathsf{B}^j z = \mathsf{G}^j(z) \text{ in } Y_\gamma^j\},$$

and

$$\mathcal{SX}_*^j = \{z \in X_\gamma^j : \mathsf{B}^j z = 0 \text{ in } Y_\gamma^j\},$$

for each $j = 1, \ldots, 6$, where, however, the linear operators B^j, the nonlinearities G^j, and the boundary trace spaces Y_γ^j differ from problem to problem. But we do employ this structure to parameterize $\mathcal{SM}_*^j$ over $\mathcal{SX}_*^j$ near $(0,0,0)$ for all j. This shows, in particular, that $\mathcal{SX}_*^j$ is in fact isomorphic to the tangent space of $\mathcal{SM}_*^j$ at $(0,0,0)$, or equivalently, to the tangent space of $\mathcal{SM}^j$ at e_*. We assume below that the parameters satisfy

$$\varrho, \kappa_*, \mu_*, d_*, \sigma_*, \kappa_{\Gamma *}, d_{\Gamma *} > 0, \quad l_* \neq 0, \quad \zeta_*, \delta_* \neq 1,$$

to exclude pathological situations.

It will be convenient to enlarge the system variable z by the pressure, i.e., we set $w = (z, \pi)$ in Problems **(Pj)** for $j \geq 2$. Here we take $\pi \in \dot{W}_p^{1-2/p}(\Omega \setminus \Sigma)$ for Problems **(P2), (P3), (P5)**, and $\pi \in W_p^{1-2/p}(\Omega \setminus \Sigma)$ for **(P4), (P6)**. Furthermore, for Problems **(P2), (P3), (P5)** we include the normal component of the normal stress balance, which reads

$$-[\![2\mu_*\partial_\nu u \cdot \nu]\!] + [\![\pi]\!] + \sigma_*\mathcal{A}_\Sigma h = G_\nu(\vartheta, h)\nabla u + G_\gamma(h) + \sigma'(\theta)G_{\mathsf{m}}(h)\nabla\vartheta_\Sigma$$

and accordingly

$$\begin{aligned}
-\,[\![2\mu_*\partial_\nu u \cdot \nu]\!] + [\![\pi]\!] + \sigma_*\mathcal{A}_\Sigma h &= G_\nu(\vartheta, h)\nabla u + G_\gamma(h) + \sigma'(\theta)G_{\mathsf{m}}(h)\nabla\vartheta_\Sigma \\
&\quad - [\![1/\varrho]\!]^{-1}[\![u \cdot \nu_\Gamma]\!]^2, \\
-\,[\![2(\mu_*/\varrho)\partial_\nu u \cdot \nu]\!] + [\![\pi/\varrho]\!] + (l_*/\theta_*)\vartheta &= G_\pi(\vartheta, h)\nabla u + G_\rho(\vartheta) \\
&\quad + \frac{\rho_1 + \rho_2}{2(\rho_2 - \rho_1)}[\![u \cdot \nu_\Gamma]\!]^2
\end{aligned}$$

for Problems **(P4), (P6)**. The corresponding linearities are denoted by $\hat{\mathsf{B}}^j$ and the nonlinearities by $\hat{\mathsf{G}}^j$. Note that here the pressure appears only linearly, i.e., it does not appear in $\hat{\mathsf{G}}^j$.

We define the differential operators $\mathsf{A}^j = L^j$ by the corresponding differential expressions L^j from Chapter 10, and we let $F^j(w)$ denote the corresponding nonlinearities, which formally satisfy $(F^j(0), DF^j(0)) = 0$.

To parameterize $\mathcal{SM}_*^j$ over $\mathcal{SX}_*^j$ we solve the following problems, where $\omega > 0$ is sufficiently large.

$$\begin{aligned}
\omega\bar{z} + \mathsf{A}^j\bar{w} &= 0 && \text{in } \Omega \setminus \Sigma, \\
\hat{\mathsf{B}}^j\bar{w} &= \hat{G}^j(\bar{z} + \tilde{z}) && \text{on } \Sigma.
\end{aligned} \tag{11.1}$$

Given $\tilde{z} \in \mathcal{S}\mathcal{X}_*^j$ small we are looking for a solution $\bar{w} \in \hat{X}_\gamma^j$. For this we employ the implicit function theorem. Obviously, for $\tilde{z} = 0$ we have the trivial solution $\bar{w} = 0$. As $\hat{G}^j : X_\gamma^j \to \hat{Y}^j$ is of class C^1 with $(\hat{G}^j(0), D\hat{G}^j(0)) = 0$, we have to show that the linear problem

$$\begin{aligned} \omega\bar{z} + \mathsf{A}^j\bar{w} &= 0 \quad \text{in } \Omega \setminus \Sigma, \\ \hat{\mathsf{B}}^j\bar{w} &= \hat{g}^j \quad \text{on } \Sigma, \end{aligned}$$

admits a unique solution, for any given datum $\hat{g}^j \in \hat{Y}_\gamma^j$. In fact, the propositions in the next section will do this job, up to lower order perturbations. Therefore, we may apply the implicit function theorem to find balls $B_{\mathcal{S}\mathcal{X}_*^j}(0, r_j)$ and maps

$$\hat{\phi}^j : B_{\mathcal{S}\mathcal{X}_*^j}(0, r_j) \to \hat{X}_\gamma^j$$

of class C^1 with $(\hat{\phi}^j(0), D\hat{\phi}^j(0)) = 0$ such that $\bar{w} = \hat{\phi}^j(\tilde{z})$ is the unique solution of (11.1) near zero. The map $\mathrm{id} + \hat{\phi}^j$ is surjective onto a neighbourhood of zero in $\hat{X}_\gamma^j$. To see this fix any $z \in \mathcal{S}\mathcal{M}_*^j$, and solve the linear problem (11.1) with $\hat{g}^j = \hat{\mathsf{G}}^j(z)$ to obtain a unique $\bar{w} = (\bar{z}, \bar{\pi}) \in \hat{X}_\gamma^j$. Then set $\tilde{z} = z - \bar{z}$; if w is chosen small enough, $\tilde{z} \in B_{\mathcal{S}\mathcal{X}_*^j}(0, r_j)$, hence we have $\bar{w} = \hat{\phi}^j(\tilde{z})$ by uniqueness.

The map $\Phi^j : B_{\mathcal{S}\mathcal{X}_*^j}(0, r_j) \to \mathcal{S}\mathcal{M}_*^j$ defined by

$$\Phi^j(\tilde{z}) = \tilde{z} + \phi^j(\tilde{z}), \tag{11.2}$$

where ϕ^j means dropping the pressure π in $\hat{\phi}^j$, yields the parametrization we have been looking for. We summarize this result in

Theorem 11.2.1. *The state manifolds $\mathcal{S}\mathcal{M}_*^j$ are parameterized via the maps Φ^j defined above over the spaces $\mathcal{S}\mathcal{X}_*^j$. In particular, the tangent spaces $T_{e_*}\mathcal{S}\mathcal{M}^j$ at the equilibrium e_*, and equivalently the tangent space $T_0\mathcal{S}\mathcal{M}_*^j$ at zero, are isomorphic to the space $\mathcal{S}\mathcal{X}_*^j$.*

Note that an equilibrium $e_\infty \in \mathcal{E}$, close to $e_* \in \mathcal{E}$ in $\mathcal{S}\mathcal{M}$, respectively z_∞ close zero in X_γ^j decomposes as

$$z_\infty = \tilde{z}_\infty + \bar{z}_\infty = \tilde{z}_\infty + \phi^j(\tilde{z}_\infty),$$

with $\tilde{z}_\infty \in \mathcal{S}\mathcal{X}_*^j$. This follows as $\mathsf{A}^j w_\infty = F^j(w_\infty) = 0$ at an equilibrium.

2.3 Auxiliary Linear Elliptic Problems

For the application of the implicit function theorem in Section 11.2.2 we needed the following results. The first two concern an elliptic transmission problem for the heat equation and the linearized steady Stefan problem.

Proposition 11.2.2. *Let $\omega > 0$ be large, $\varrho, \kappa_*, d_* > 0$, and $p > n + 2$. Then the problem*

$$\begin{aligned} \varrho\kappa_*\omega\vartheta - d_*\Delta\vartheta &= 0 \quad \text{in } \Omega \setminus \Sigma, \\ \partial_\nu\vartheta &= 0 \quad \text{on } \partial\Omega, \\ [\![\vartheta]\!] = 0, \quad -[\![d_*\partial_\nu\vartheta]\!] &= g \quad \text{on } \Sigma, \end{aligned}$$

has a unique solution $\vartheta \in W_p^{2-2/p}(\Omega\setminus\Sigma)$ *if and only if* $g \in W_p^{1-3/p}(\Sigma)$.

This result is proved by the methods in Section 6.5, and is used for Problems **(P2)** and **(P4)**

Proposition 11.2.3. *Let* $\omega > 0$ *be large,* $l_* \neq 0$, $\varrho, \kappa_*, d_*, \sigma > 0$, *and* $p > n+2$. *Then the problem*

$$\begin{aligned} \varrho\kappa_*\omega\vartheta - d_*\Delta\vartheta &= 0 && \text{in } \Omega\setminus\Sigma,\\ \partial_\nu\vartheta &= 0 && \text{on } \partial\Omega,\\ [\![\vartheta]\!] = 0, \quad (\varrho l_*/\theta_*)\vartheta - \sigma\mathcal{A}_\Sigma h &= g_\theta && \text{on } \Sigma,\\ \varrho l_*\omega h - [\![d_*\partial_\nu\vartheta]\!] &= g_h && \text{on } \Sigma, \end{aligned}$$

has a unique solution $(\vartheta, h) \in W_p^{2-2/p}(\Omega\setminus\Sigma)\times W_p^{4-3/p}(\Sigma)$ *if and only if* $(g_\theta, g_h) \in W_p^{2-3/p}(\Sigma)\times W_p^{1-3/p}(\Sigma)$.

This result is proved by the methods in Section 6.6; it is used for Problems **(P1)** and **(P3)**.

The next two propositions concern the linear steady symmetric and asymmetric Stokes problems.

Proposition 11.2.4. *Let* $\omega > 0$ *be large,* $\mu_* > 0$, *and* $p > n+2$. *Then the problem*

$$\begin{aligned} \varrho\omega u - \mu_*\Delta u + \nabla\pi &= 0 && \text{in } \Omega\setminus\Sigma,\\ \operatorname{div} u &= g_d && \text{in } \Omega\setminus\Sigma,\\ u &= 0 && \text{on } \partial\Omega,\\ [\![u]\!] = 0, \quad \mathcal{P}_\Sigma[\![\mu_*(\nabla u + [\nabla u]^\mathsf{T})\nu_\Sigma]\!] &= g_\tau && \text{on } \Sigma,\\ -[\![2\mu_*\partial_\nu u\cdot\nu_\Sigma]\!] + [\![\pi]\!] &= g_\nu && \text{on } \Sigma, \end{aligned}$$

has a unique solution

$$u \in W_p^{2-2/p}(\Omega\setminus\Sigma), \quad \pi \in \dot{W}_p^{1-2/p}(\Omega\setminus\Sigma), \quad [\![\pi]\!] \in W_p^{1-3/p}(\Sigma)$$

if and only if

$$g_d \in W_{p,0}^{1-2/p}(\Omega\setminus\Sigma), \quad (g_\tau, g_\nu) \in W_p^{1-3/p}(\Sigma; T\Sigma\times\mathbb{R}).$$

This result can be obtained by the methods in Sections 8.1, 8.2, and 8.3, it is employed for Problems **(P2), (P3), (P5)**.

Proposition 11.2.5. *Let $\omega > 0$ be large, $\varrho, \mu_* > 0$, and $p > n+2$. Then the problem*

$$\begin{aligned} \varrho\omega u - \mu_*\Delta u + \nabla\pi &= 0 && \text{in } \Omega\setminus\Sigma,\\ \operatorname{div} u &= g_d && \text{in } \Omega\setminus\Sigma,\\ u &= 0 && \text{on } \partial\Omega,\\ \mathcal{P}_\Sigma[\![u]\!] &= g_u && \text{on } \Sigma,\\ \mathcal{P}_\Sigma[\![\mu_*(\nabla u + [\nabla u]^{\mathsf{T}})\nu_\Sigma]\!] &= g_\tau && \text{on } \Sigma,\\ -[\![2\mu_*\partial_\nu u\cdot\nu_\Sigma]\!] + [\![\pi]\!] &= g_\nu && \text{on } \Sigma,\\ -[\![2(\mu_*/\varrho)\partial_\nu u\cdot\nu_\Sigma]\!] + [\![\pi/\varrho]\!] &= g_\pi && \text{on } \Sigma, \end{aligned}$$

has a unique solution (u,π) with

$$u \in W_p^{2-2/p}(\Omega\setminus\Sigma) \quad \pi \in W_p^{1-2/p}(\Omega\setminus\Sigma),$$

if and only if

$$g_d \in W_p^{1-2/p}(\Omega\setminus\Sigma), \quad g_u \in W_p^{2-3/p}(\Sigma; T\Sigma),$$

and $(g_\tau, g_\nu, g_\pi) \in W_p^{1-3/p}(\Sigma; T\Sigma\times\mathbb{R}^2)$.

This result can be obtained by the methods in Sections 8.4, 8.5, and 8.6; it is needed for Problems **(P4)** and **(P6)**.

2.4 Abstract Reformulation of the Problems

To obtain a framework for a joint analysis of stability near an equilibrium, we need an appropriate abstract formulation of the problems under consideration. This will be achieved in the following way. We decompose the time-dependent variables in the same way as in the previous section into $z = \bar{z} + \tilde{z}$ and $w = \bar{w} + \tilde{w}$. Furthermore, for Problems **(P5), (P6)** it is again convenient to enlarge $\tilde{z}$, hence also $\tilde{w}$, by the variable ϑ_Σ, which is actually dummy, as $\vartheta_\Sigma = \vartheta_{|_\Sigma}$ is the restriction of ϑ to Σ.

We may then decompose the full problem into two, formally one for $\bar{w}$ and one for $\tilde{w}$, according to

$$\begin{aligned} (\omega + \partial_t)\bar{z} + \mathsf{A}^j\bar{w} &= F^j(\bar{w} + \tilde{w}) && \text{in } \Omega\setminus\Sigma,\\ \hat{\mathsf{B}}^j\bar{w} &= \hat{\mathsf{G}}^j(\bar{z} + \tilde{z}) && \text{on } \Sigma,\\ \bar{z}(0) &= \phi^j(\tilde{z}_0) && \text{in } \Omega. \end{aligned} \tag{11.3}$$

The second one reads as

$$\begin{aligned} \partial_t\tilde{z} + \mathsf{A}^j\tilde{w} &= \omega\bar{z} && \text{in } \Omega\setminus\Sigma,\\ \hat{\mathsf{B}}^j\tilde{w} &= 0 && \text{on } \Sigma,\\ \tilde{z}(0) &= \tilde{z}_0 && \text{in } \Omega. \end{aligned} \tag{11.4}$$

Adding these equations we obtain the full problem under consideration, namely

$$\begin{aligned} \partial_t z + \mathsf{A}^j w &= F^j(w) && \text{in } \Omega\setminus\Sigma,\\ \hat{\mathsf{B}}^j w &= \hat{\mathsf{G}}^j(z) && \text{on } \Sigma,\\ z(0) &= z_0 && \text{in } \Omega. \end{aligned} \tag{11.5}$$

One should think of this decomposition in the following way. The first part has a fast dynamics due to $\omega > 0$ large and takes care of the stationary boundary conditions, while the second equation lives in the tangent space $\mathcal{SX}_*^j$ and carries the actual dynamics. Furthermore, as $(F^j(0), \hat{G}^j(0), \phi^j(0)) = 0$ as well as $(DF^j(0), D\hat{G}^j(0), D\phi^j(0)) = 0$, we will see that $\bar{w}$ is small compared to $\tilde{z}$. Our philosophy is to consider the first equation as an auxiliary one, to solve it in terms of $\tilde{z}$, thereby reducing the problem to a quasilinear evolution equation in $\mathcal{SX}_*^j$, which is nonlocal in time, but causal. Then we may follow the ideas in Chapter 5 to obtain a *generalized principle of linearized stability* also in this more general framework. Actually, we will follow this route for instability, but for stability and convergence we will solve both problems simultaneously, by the implicit function theorem, similarly to the construction of the foliations in Section 5.6.

There should be a word of warning. While for the initial value z_0 we employ the decomposition $z_0 = \tilde{z}_0 + \phi(\tilde{z}_0)$, and this is also valid for z_∞, it does not hold in the time-dependent case, in general $\bar{z}(t) \neq \phi(\tilde{z}(t))$!

It is convenient to remove the pressure $\tilde{\pi}$ from (11.4) in the usual way (see Chapters 8 and 10) by solving the appropriate weak transmission problems for $\tilde{\pi}$ and insert it into (11.3) for $\tilde{w}$. Then we may rewrite the first problem abstractly as

$$\mathbb{L}_\omega^j \bar{w} = N^j(\bar{w}, \tilde{z}), \quad t > 0, \quad \bar{z}(0) = \phi^j(\tilde{z}_0), \tag{11.6}$$

and with the appropriate two-phase Helmholtz-Weyl projection $\mathbb{P}$, the second one as the evolution equation

$$\partial_t \tilde{z} + L^j \tilde{z} = \omega \mathbb{P} \bar{z}, \quad t > 0, \quad \tilde{z}(0) = \tilde{z}_0. \tag{11.7}$$

Here L^j are the operators defined in Chapter 10, $j = 1, \ldots, 6$. For further use, we introduce the function spaces

$$\tilde{\mathbb{E}}^j(a) := H_p^1(J; X_0^j) \cap L_p(J; X_1^j), \quad J = (0, a), \tag{11.8}$$

with X_0^j and X_1^j as in Chapter 10.

Recall that $(\mathbb{L}_\omega^j, \mathrm{tr})$ is an isomorphism from $\hat{\mathbb{E}}^j(\infty, \delta)$ to $\mathbb{F}^j(\infty, \delta) \times X_\gamma^j$, by the results in Chapters 6 and 8. Here we use the notation

$$z \in \mathbb{E}^j(\infty, \delta) \quad \Leftrightarrow \quad e^{\delta t} z \in \mathbb{E}^j(\infty),$$

and similarly for $\mathbb{F}^j(\infty, \delta)$, $\hat{\mathbb{E}}^j(\infty, \delta)$ and $\tilde{\mathbb{E}}^j(\infty, \delta)$. $-L^j$ is the generator of a compact analytic C_0-semigroup with maximal L_p-regularity in X_0^j, according to Chapter 10, where also the spectral properties of L^j have been derived. In particular, there is a spectral gap $(0,\delta_0^j)$ such that $\operatorname{Re}\sigma(-L^j)\cap(0,\delta_0^j)=\emptyset$, and we choose $0<\delta<\delta_0^j$. The functions N^j will be of class C^1 in the spaces used below, and by their definition satisfy

$$(N^j(0,0),DN^j(0,0))=0.$$

The decomposition (11.6), (11.7) will be useful for the proof of instability. For stability and convergence we modify it slightly, as we want to show exponential convergence. For this reason we now decompose $z=\bar z+\tilde z+z_\infty$, with the idea that z_∞ will be the limit of $z(t)$ as t goes to infinity and $\bar z,\tilde z$ are exponentially decaying. This means that the corresponding equations for $\bar z$ and $\tilde z$ are shifted to

$$\begin{aligned} (\omega+\partial_t)\bar z+\mathsf{A}^j\bar w &= F^j(\bar w+\tilde w+w_\infty)-F^j(w_\infty) && \text{in } \Omega\setminus\Sigma,\\ \hat{\mathsf{B}}^j\bar w &= \hat{\mathsf{G}}^j(\bar z+\tilde z+z_\infty)-\hat G^j(z_\infty) && \text{on } \Sigma,\\ \bar z(0) &= \phi^j(\tilde z_0)-\phi^j(\tilde z_\infty) && \text{in } \Omega, \end{aligned} \tag{11.9}$$

and

$$\begin{aligned} \partial_t\tilde z+\mathsf{A}^j\tilde w &= \omega\bar z && \text{in } \Omega\setminus\Sigma,\\ \hat{\mathsf{B}}^j\tilde w &= 0 && \text{on } \Sigma,\\ \tilde z(0) &= \tilde z_0-\tilde z_\infty && \text{in } \Omega, \end{aligned} \tag{11.10}$$

where we used $z_0=\tilde z_0+\phi(\tilde z_0)$ and $z_\infty=\tilde z_\infty+\phi(\tilde z_\infty)$. Eliminating as before the pressure $\bar\pi$ and inserting it into (11.9) we end up with the following abstract problems

$$\mathbb{L}^j_\omega\bar w=N^j(\bar w,\tilde z,\tilde z_\infty),\quad t>0,\quad \bar z(0)=\phi^j\tilde z_0)-\phi^j(\tilde z_\infty), \tag{11.11}$$

and with the appropriate two-phase Helmholtz-Weyl projection $\mathbb{P}$,

$$\partial_t\tilde z+L^j\tilde z=\omega\mathbb{P}\bar z,\quad t>0,\quad \tilde z(0)=\tilde z_0-\tilde z_\infty. \tag{11.12}$$

We observe that by Section 9.5, the functions N^j are of class C^1 in the variables $(\bar w,\tilde z)$ in the function spaces $\hat{\mathbb{E}}(\infty,\delta)\times\tilde{\mathbb{E}}(\infty,\delta)$ provided hypothesis **(Hj)** holds, but merely continuous in $\tilde z_\infty$, unless we require one more degree of regularity for the coefficients. We want to avoid this below.

11.3 Nonlinear Stability of Equilibria

Concerning the stability of equilibria for the nonlinear problems we can prove a fairly complete result. To state this result, recall from Chapter 10 that an equilibrium $e_*=(0,\theta_*,\Gamma_*)\in\mathcal{E}$ is

(i) linearly stable if it is normally stable, which is always the case for **(P2)**, and for **(P4)** and **(P6)** if Γ_* is connected. For Problems **(P1)**, **(P3)**, **(P5)** it is normally stable if and only if either $\delta_*>1$, or $\delta_*<1$, Γ_* is connected, and the stability condition $\zeta_*<1$ holds.

(ii) linearly unstable if it is normally hyperbolic, which is the complementary case of (i), where we have to exclude the pathological cases $l_* = 0$ or $\delta_* = 1$ or $\zeta_* = 1$.

For the definition of the numbers l_*, δ_* and ζ_* we refer to Chapter 10. Recall that $l_* = 0$ or $\delta_* = 1$ lead to ill-posed problems, while $\zeta_* = 1$ is the case of marginal stability in the linear problem.

Now we can state the main theorem of this chapter.

Theorem 11.3.1. *Let $p > n + 2$, suppose that Conditions* **(Hj)** *hold for Problem* **(Pj)**, *with $j = 1, \ldots, 6$.*

Then in the state manifold $\mathcal{SM}^j$, for a fixed equilibrium $e_* = (0, \theta_*, \Gamma_*) \in \mathcal{E}$, we have

(i) *If $e_* \in \mathcal{E}$ is linearly stable, then it is nonlinearly stable in $\mathcal{SM}^j$, and any solution with initial value close to e_* in $\mathcal{SM}^j$ exists globally and converges in $\mathcal{SM}^j$ to a possibly different stable equilibrium $e_\infty \in \mathcal{E}$ at an exponential rate.*

(ii) *If $e_* \in \mathcal{E}$ is linearly unstable, then e_* is also nonlinearly unstable in the state manifold $\mathcal{SM}^j$. Any solution starting near e_* and staying in a neighbourhood of e_* exists globally and converges in $\mathcal{SM}^j$ to another unstable equilibrium $e_\infty \in \mathcal{E}$ at an exponential rate.*

Some remarks are in order.

Remark 11.3.2. (i) For Problem **(P2)**, every (non-degenerate) equilibrium is stable. However, one should observe that the smaller the spheres $S(x_k, R_k)$ are, the larger will the pressure inside these balls be. But if the pressure is high enough, a phase transition driven by pressure will occur. Therefore, model **(P2)**, although thermodynamically consistent, is physically not very realistic. Phase transitions have to be taken into account.

(ii) In all the other problems, with the exception of **(P5)**, we observe that a disconnected equilibrium interface is unstable. This is what we call the **onset of Ostwald ripening**. If we start with, say, two spheres of equal size and perturb one of them to become larger, then experiments show that the system evolves to a new steady state where only the larger sphere will survive. However, we are far away from being able to prove such a result rigorously, as singularities in the solutions will necessarily occur due to changes in the topology.

(iii) Recall from Chapter 1 that we have a clear interpretation of the stability condition $\zeta_* < 1$. This corresponds to a local maximum of the total entropy. On the other hand, we have no explanation for the stability result in case $\delta_* > 1$. In this case, even a saddle point of the total entropy might be stable! We believe that there are some other physical restrictions which exclude this case.

Proof. In the proof we will drop the superscript j, as it does not matter here. We begin with the instability part of **(ii)**

(a) The proof follows the lines of that of Theorem 5.4.1, with some modifications due to the presence of the auxiliary equation (11.11). First we observe that there is an increasing continuous function $\eta(r)$ with $\eta(0) = 0$ such that the following estimate holds

$$|e^{-\kappa t}N(\bar{w},\tilde{z})|_{\mathbb{F}(a)} \leq \eta(r)(|e^{-\kappa t}\bar{w}|_{\hat{\mathbb{E}}(a)} + |e^{-\kappa t}\tilde{z}|_{\mathbb{E}(a)}), \tag{11.13}$$

for all $z \in \mathbb{E}(a)$ with $|z(t)|_{X_\gamma} \leq r$. Here $\kappa > 0$ is a given number, and these estimates are independent of the length a of the interval. This follows from the nonlinear estimates in Section 9.5. Moreover, it is not difficult to see that any solution $\tilde{z} \in \tilde{\mathbb{E}}(a)$ of (11.7) enjoys the additional regularity $\tilde{z} \in \mathbb{E}(a)$, and that the a priori estimate

$$|e^{-\kappa t}\tilde{z}|_{\mathbb{E}(a)} \leq C(|e^{-\kappa t}\tilde{z}|_{\tilde{\mathbb{E}}(a)} + |e^{-\kappa t}\bar{z}|_{\mathbb{E}(a)}) \tag{11.14}$$

holds, where the constant C is independent of a.

Next we devise a spectral decomposition as in the proof of Theorem 5.4.1. As we are in the unstable case, the operator $-L$ has a finite number of positive eigenvalues. We choose $\kappa > 0$ and $\mu > 0$ in such a way that $[\kappa-\mu, \kappa+\mu] \subset (0,\infty)$, and such that all positive eigenvalues are contained in $(\kappa+\mu, \infty)$, thereby defining a spectral gap. We denote the spectral projections in X_0 associated with the positive eigenvalues by P_+, the corresponding bounded linear operator induced by L by $L_+ = P_+L$ and correspondingly $P_- = I - P_+$ and $L_- = P_-L$. Then we have the estimates

$$|P_-e^{-Lt}|_{\mathcal{B}(X_0)} \leq Me^{(\kappa-\mu)t}, \quad |P_+e^{L_+t}|_{\mathcal{B}(X_i)} \leq Me^{-(\kappa+\mu)t}, \quad t > 0,\ i = 0, 1, \gamma.$$

Note that on $X_+ = P_+X_0$ the norms $|\cdot|_i$ are equivalent; in fact we have $X_+ \subset X_1$.

Assume that $z_* = 0$ is stable in X_γ. As in the proof of Theorem 5.4.1 we will obtain a contradiction, however, the arguments are slightly more involved. Stability of 0 implies that for every $r > 0$ there is a number $\delta > 0$ such that Problem (11.5) admits a global solution $z \in \mathbb{E}(\infty)$ with $|z(t)|_{X_\gamma} \leq r$ for all $t \geq 0$, whenever $z_0 \in B_{X_\gamma}(0,\delta)$.

We first use maximal L_p-regularity of $\mathbb{L}_\omega$ in the auxiliary equation (11.11) on $\mathbb{R}_+$ and (11.13) to obtain for an arbitrary $a > 0$

$$\begin{aligned}|e^{-\kappa t}\bar{w}|_{\hat{\mathbb{E}}(a)} &\leq C(|\bar{z}_0|_{X_\gamma} + |e^{-\kappa t}N(\bar{w},\tilde{z})|_{\mathbb{F}(a)})\\ &\leq C_1(|\bar{z}_0|_{X_\gamma} + \eta(r)(|e^{-\kappa t}\bar{w}|_{\hat{\mathbb{E}}(a)} + |e^{-\kappa t}\tilde{z}|_{\mathbb{E}(a)})).\end{aligned}$$

Choosing r small enough we have $C_1\eta(r) \leq 1/2$, and hence

$$|e^{-\kappa t}\bar{z}|_{\mathbb{E}(a)} \leq |e^{-\kappa t}\bar{w}|_{\hat{\mathbb{E}}(a)} \leq C_2(|\bar{z}_0|_{X_\gamma} + \eta(r)|e^{-\kappa t}\tilde{z}|_{\mathbb{E}(a)}). \tag{11.15}$$

Next we use estimate (11.14), maximal L_p-regularity of L_- on $\mathbb{R}_+$, the embedding $\mathbb{E}(a) \hookrightarrow L_p((0,a); X_0)$, and (11.15) to the result

$$\begin{aligned}|e^{-\kappa t}\tilde{z}|_{\mathbb{E}(a)} &\leq C(|e^{-\kappa t}P_+\tilde{z}|_{\tilde{\mathbb{E}}(a)} + |e^{-\kappa t}P_-\tilde{z}|_{\tilde{\mathbb{E}}(a)} + |e^{-\kappa t}\bar{z}|_{\mathbb{E}(a)})\\ &\leq C_3(|e^{-\kappa t}P_+\tilde{z}|_{\tilde{\mathbb{E}}(a)} + |P_-\tilde{z}_0|_{X_\gamma} + |\bar{z}_0|_{X_\gamma} + \eta(r)|e^{-\kappa t}\tilde{z}|_{\mathbb{E}(a)}).\end{aligned}$$

Assuming $C_3\eta(r) \le 1/2$ then yields

$$|e^{-\kappa t}\tilde{z}|_{\mathbb{E}(a)} \le C_4(|e^{-\kappa t}P_+\tilde{z}|_{\tilde{\mathbb{E}}(a)} + |P_-\tilde{z}_0|_{X_\gamma} + |\bar{z}_0|_{X_\gamma}). \tag{11.16}$$

For the part $e^{-\kappa t}P_+\tilde{z}$ we use the relation

$$\partial_t(e^{-\kappa t}P_+\tilde{z}) = -(\kappa + L_+)e^{-\kappa t}P_+\tilde{z} + \omega P_+\mathbb{P}\bar{z},$$

the fact that A_+ is bounded, and the embedding $\mathbb{E}(a) \hookrightarrow L_p((0,a);X_0)$ to the result

$$|e^{-\kappa t}P_+\tilde{z}|_{\tilde{\mathbb{E}}(a)} \le C(|e^{-\kappa t}P_+\tilde{z}|_{L_p((0,a);X_+)} + |e^{-\kappa t}\bar{z}|_{\mathbb{E}(a)}). \tag{11.17}$$

Now we employ the assumption $|z(t)|_{X_\gamma} \le r$ to obtain

$$|e^{-\kappa t}P_+z|_{L_p((0,a);X_+)} \le Cr, \qquad \text{for all } a > 0.$$

This implies with (11.15), (11.16), (11.17), and $\mathbb{E}(a) \hookrightarrow L_p((0,a);X_0)$

$$\begin{aligned}
|e^{-\kappa t}P_+\tilde{z}|_{\tilde{\mathbb{E}}(a)} &\le C(|e^{-\kappa t}P_+\tilde{z}|_{L_p((0,a);X_+)} + |e^{-\kappa t}\bar{z}|_{\mathbb{E}(a)}) \\
&\le C(|e^{-\kappa t}P_+z|_{L_p((0,a);X_+)} + |e^{-\kappa t}\bar{z}|_{\mathbb{E}(a)}) \\
&\le C_5(r + |\bar{z}_0|_{X_\gamma} + \eta(r)|P_-\tilde{z}_0|_{X_\gamma} + \eta(r)|e^{-\kappa t}P_+\tilde{z}|_{\tilde{\mathbb{E}}(a)})
\end{aligned}$$

and hence, assuming $C_5\eta(r) \le 1/2$,

$$|e^{-\kappa t}P_+\tilde{z}|_{\tilde{\mathbb{E}}(a)} \le C_6(r + |\bar{z}_0|_{X_\gamma} + \eta(r)|P_-\tilde{z}_0|_{X_\gamma}).$$

Combining this last estimate with (11.15) and (11.16) results in

$$|e^{-\kappa t}\tilde{z}|_{\mathbb{E}(a)} + |e^{-\kappa t}\bar{z}|_{\mathbb{E}(a)} \le C_7(r + |\bar{z}_0|_{X_\gamma} + |P_-\tilde{z}_0|_{X_\gamma}), \quad a > 0. \tag{11.18}$$

In particular, this inequality shows with $a \to \infty$

$$|e^{-\kappa t}\tilde{z}|_{\mathbb{E}(\infty)} + |e^{-\kappa t}\bar{z}|_{\mathbb{E}(\infty)} < \infty.$$

From Hölder's inequality, the embedding $\mathbb{E}(\infty) \hookrightarrow L_p(\mathbb{R}_+;X_0)$, and (11.18) follows

$$e^{-\kappa t}\int_t^\infty |e^{-L_+(t-s)}P_+\mathbb{P}\bar{z}(s)|_{X_+}\,ds \;\le C(\mu)|e^{-\kappa t}\bar{z}|_{\mathbb{E}(\infty)} < \infty,$$

showing that the integral $\int_t^\infty e^{-L_+(t-s)}P_+\mathbb{P}\bar{z}\,ds$ exists in X_+ for every $t \ge 0$. Moreover, its norm in X_+ grows no faster than an exponential function $Ce^{\kappa t}$. Therefore, by means of the variation of parameters formula we may write

$$\begin{aligned}
P_+\tilde{z}(t) &= e^{-L_+t}P_+\tilde{z}_0 + \omega\int_0^t e^{-L_+(t-s)}P_+\mathbb{P}\bar{z}(s)\,ds \\
&= e^{-L_+t}\Big(P_+\tilde{z}_0 + \omega\int_0^\infty e^{L_+s}P_+\mathbb{P}\bar{z}(s)\,ds\Big) - \omega\int_t^\infty e^{-L_+(t-s)}P_+\mathbb{P}\bar{z}(s)\,ds.
\end{aligned}$$

The estimate,

$$\Big|e^{L_+t}\Big(P_+\tilde{z}(t) + \omega\int_t^\infty e^{-L_+(t-s)}P_+\mathbb{P}\bar{z}(s)\,ds\Big)\Big|_{X_+} \le Ce^{-\mu t}(|e^{-\kappa t}\tilde{z}|_{\mathbb{E}(\infty)} + |e^{-\kappa t}\bar{z}|_{\mathbb{E}(\infty)}),$$

where we used the embedding $P_+\mathbb{E}(\infty) \hookrightarrow C_b(\mathbb{R}_+, X_+)$ to bound the first term, then shows that

$$P_+\tilde{z}_0 + \omega\int_0^\infty e^{L_+s}P_+\mathbb{P}\bar{z}(s)\,ds = 0. \tag{11.19}$$

Therefore, we obtain

$$P_+\tilde{z}(t) = -\omega\int_t^\infty e^{-L_+(t-s)}P_+\mathbb{P}\bar{z}(s)\,ds, \tag{11.20}$$

which we may use as in the proof of Theorem 5.4.1 to produce a contradiction.

In fact, the representation (11.20) in conjunction with Young's inequality for convolution integrals and (11.17) leads to

$$|e^{-\kappa t}P_+\tilde{z}|_{\tilde{\mathbb{E}}(a)} \le C|e^{-\kappa t}\bar{z}|_{\mathbb{E}(a)}.$$

We may now conclude with (11.15) and (11.16) by similar arguments as above that

$$|e^{-\kappa t}\tilde{z}|_{\mathbb{E}(\infty)} + |e^{-\kappa t}\bar{z}|_{\mathbb{E}(\infty)} \le C_8(|P_-\tilde{z}_0|_{X_\gamma} + |\bar{z}_0|_{X_\gamma}).$$

As

$$|\bar{z}_0|_{X_\gamma} = |\phi(\tilde{z}_0)|_{X_\gamma} \le \varepsilon|\tilde{z}_0|_{X_\gamma},$$

if $r > 0$ is small enough, we obtain by (11.19) an inequality of the form

$$|P_+\tilde{z}_0|_{X_\gamma} \le C|P_-\tilde{z}_0|_{X_\gamma},$$

which, as in the proof of Theorem 5.4.1, yields a contradiction. This completes the proof of the instability assertion in **(ii)**.

(b) Next we use the spectral decomposition of L according to Chapter 10, where P^s denotes the projection onto the stable subspace $X_0^s = P^sX_0 = \mathsf{R}(L)$ and P^c the complementary projection onto $X_0^c = P^cX_0 = \mathsf{N}(L)$. We set $\mathsf{y} = P^s\tilde{z}$ and $\mathsf{x} = P^c\tilde{z}$ and note that we may parameterize the equilibria over $P^c\mathcal{S}\mathcal{X}_* = P^cX_0$ according to

$$z_\infty = \mathsf{x}_\infty + \psi(\mathsf{x}_\infty) + \phi(\mathsf{x}_\infty + \psi(\mathsf{x}_\infty)), \quad \mathsf{x}_\infty \in X_0^c,$$

solving the nonlinear stationary problem

$$L^s y = \omega P^s\mathbb{P}\phi(\mathsf{x} + \mathsf{y})$$

by the implicit function theorem. We set $\mathsf{y}_\infty = \psi(\mathsf{x}_\infty)$. Also, applying the projection P^s to the equation for $\tilde{z}$ we obtain the problem

$$\partial_t \mathsf{y} + L^s \mathsf{y} = \omega P^s \mathbb{P}\bar{z}, \quad t > 0, \quad \mathsf{y}(0) = \mathsf{y}_0 - \mathsf{y}_\infty,$$

and for x we have

$$\partial_t \mathsf{x} = \omega P^c \mathbb{P}\bar{z}, \quad t > 0, \quad \mathsf{x}(0) = \mathsf{x}_0 - \mathsf{x}_\infty.$$

Then we rewrite problem (11.11)–(11.12) as $\mathbb{H}(\mathsf{v}, (\mathsf{x}_\infty, \mathsf{y}_0)) = 0$, where $\mathsf{v} = (\bar{w}, \mathsf{y}, \mathsf{x}, \mathsf{x}_0)$ and

$$\mathbb{H}(\mathsf{v}, (\mathsf{x}_\infty, \mathsf{y}_0)) = \begin{bmatrix} (\mathbb{L}_\omega \bar{w} - N(\mathsf{v}, \mathsf{x}_\infty)), \bar{z}(0) - \phi(\mathsf{x}_0 + \mathsf{y}_0) + \phi(\mathsf{x}_\infty + \mathsf{y}_\infty)) \\ (\partial_t \mathsf{y} + L^s \mathsf{y} - \omega P^s \mathbb{P}\bar{z}, \mathsf{y}(0) - \mathsf{y}_0 + \psi(\mathsf{x}_\infty)) \\ \mathsf{x}(t) + \omega \int_t^\infty P^c \mathbb{P}\bar{z}\, ds \\ \mathsf{x}_0 - \mathsf{x}_\infty + \omega \int_0^\infty P^c \mathbb{P}\bar{z}\, ds \end{bmatrix}.$$

It follows from the results in Section 9.5 that the mapping

$$\begin{aligned} \mathbb{H} : \hat{\mathbb{E}}(\infty, \delta) \times P^s \tilde{\mathbb{E}}(\infty, \delta) \times P^c \tilde{\mathbb{E}}(\infty, \delta) \times X_0^c \times (X_0^c \times X_\gamma^s) \\ \to (\mathbb{F}(\infty, \delta) \times X_\gamma) \times (L_p(\mathbb{R}_+; \delta, X_0^s) \times X_\gamma^s) \times (L_p(\mathbb{R}_+; \delta, X_0^c) \times X_0^c) \end{aligned}$$

is of class C^1 w.r.t $(\mathsf{v}, \mathsf{y}_0)$, continuous w.r.t. x_∞, and differentiable w.r.t. x_∞ at $\mathsf{x}_\infty = 0$. The Fréchet derivative $D_\mathsf{v}\mathbb{H}(0,0)$ w.r.t. the variable v is given by the operator matrix

$$D_\mathsf{v}\mathbb{H}(0,0) = \begin{bmatrix} (\mathbb{L}_\omega, \mathrm{tr}) & 0 & 0 & 0 \\ * & (\partial_t + L^s, \mathrm{tr}) & 0 & 0 \\ * & 0 & I & 0 \\ * & 0 & 0 & I \end{bmatrix}. \tag{11.21}$$

Here the stars indicate bounded linear operators which, because of the triangular structure of the operator matrix, we do not compute explicitly, as the diagonal terms of this operator matrix are invertible. Therefore, by the implicit function theorem, we find balls $B_{X_0^c}(0,r)$ and $B_{X_\gamma^s}(0,r)$, and a continuous map

$$\mathcal{T} : B_{X_0^c}(0,r) \times B_{X_\gamma^s}(0,r) \to \hat{\mathbb{E}}(\infty, \delta) \times \tilde{\mathbb{E}}(\infty, \delta) \times X_0^c, \quad \mathcal{T}(\mathsf{x}_\infty, \mathsf{y}_0) = (\bar{w}, \tilde{z}, \mathsf{x}_0),$$

with $\mathcal{T}(0,0) = 0$. Here we remind that $\bar{w} = (\bar{z}, \bar{\pi})$. Then $(w, \pi) := (\bar{z} + \tilde{z} + z_\infty, \bar{\pi})$ defines the unique solution of (11.5) such that

$$z(t) \to z_\infty := \mathsf{x}_\infty + \psi(\mathsf{x}_\infty) + \phi(\mathsf{x}_\infty + \psi(\mathsf{x}_\infty)) \quad \text{in } X_\gamma \quad \text{as } t \to \infty.$$

This completes the construction of the stable foliation of the problem. Note that $\mathcal{T}$ is C^1 also in y_0, but only continuous in x_∞, unless we require more regularity for the parameter functions. Nonetheless, $\mathcal{T}$ is differentiable with respect to x_∞ at $\mathsf{x}_\infty = 0$.

To complete the proof of **(i)**, the question which remains is whether the map $(\mathsf{x}_\infty, \mathsf{y}_0) \mapsto (\mathsf{x}_0, \mathsf{y}_0)$ is surjective near 0. To prove this, we use degree theory. For this purpose, define a map $f : B_{X_0^c}(0,r) \times B_{X_0^s}(0,r) \to X_0^c$ by means of $f(\mathsf{x}_\infty, \mathsf{y}_0) = \mathsf{x}_0(\mathsf{x}_\infty, \mathsf{y}_0)$. We know that this map is continuous, and it is close to identity. In fact, differentiating the relation

$$\mathbb{H}_1(\mathcal{T}(\mathsf{x}_\infty, \mathsf{y}_0), \mathcal{T}(\mathsf{x}_\infty, \mathsf{y}_0)) = 0$$

with respect to $(\mathsf{x}_\infty, \mathsf{y}_0)$ at $(0,0)$ we obtain $(D_{\mathsf{x}_\infty}\mathcal{T}_1(0,0), D_{\mathsf{y}_0}\mathcal{T}_1(0,0)) = 0$. Here $\mathbb{H}_1$ denotes the first line of $\mathbb{H}$ and $\mathcal{T}_1$ the first component of $\mathcal{T}$, resepectively. This implies $(D_{\mathsf{x}_\infty}\bar{z}(0,0), D_{\mathsf{y}_0}\bar{z}(0,0)) = 0$. From the representation

$$f(\mathsf{x}_\infty, \mathsf{y}_0) = \mathsf{x}_\infty - \omega \int_0^\infty P^c \mathbb{P}\bar{z}\, ds,$$

we then infer that for every $\varepsilon > 0$ there is a constant $\rho > 0$ such that

$$|f(\mathsf{x}_\infty, \mathsf{y}_0) - \mathsf{x}_\infty|_{X_0^c} \le \omega \int_0^\infty |P^c \mathbb{P}\bar{z}|_{X_0^c}\, ds \le \varepsilon(|\mathsf{y}_0|_{X_\gamma^s} + |\mathsf{x}_\infty|_{X_0^c}),$$

whenever $|(\mathsf{x}_\infty, \mathsf{y}_0)| \le \rho$, with $\rho \le r$. In the following, we fix $\varepsilon < 1/3$. Here y_0 only serves as a parameter, so we are in a finite dimensional setting and may employ the Brouwer degree, in particular its homotopy invariance. Define the homotopy $h(\tau, x, \mathsf{y}_0) = \tau f(x, \mathsf{y}_0) + (1-\tau)x$, and consider the degree

$$\mathsf{deg}(h(\tau, \cdot, \mathsf{y}_0), B_{X_0^c}(0,r), \xi).$$

For $\tau = 0$ it is equal to one, hence it is equal to one for all $\tau \in [0,1]$ provided there are no solutions of $h(\tau, x, \mathsf{y}_0) = \xi$ with $|x|_{X_0^c} = r$. To show this, suppose $h(\tau, x, \mathsf{y}_0) = \xi$, i.e., $\xi - x = \tau(f(x, \mathsf{y}_0) - x)$, and $|x|_{X_0^c} = r$. Then by the above estimate

$$r = |x|_{X_0^c} \le |\xi|_{X_0^c} + |x - \xi|_{X_0^c} \le |\xi|_{X_0^c} + \varepsilon(|\mathsf{y}_0|_{X_\gamma^s} + |x|_{X_0^c}) < r,$$

provided $|\xi|_{X_0^c} \le r/2$ and $|\mathsf{y}_0|_{X_\gamma^s} \le r/2$. Therefore, $\mathsf{deg}(f(\cdot, \mathsf{y}_0), B_{X_0^c}(0, r/2), \xi)$ equals one as well, showing that the equation $f(x_\infty, \mathsf{y}_0) = \xi$ has at least one solution for $\xi \in B_{X_0^c}(0, r/2)$, i.e., we have surjectivity near zero. This completes the proof of part **(i)** of the theorem.

(c) For the instability part we have a third projection P^u, the unstable projection, which coincides with the projection P_+ from part **(a)** of this proof. Accordingly we set $X_0^u = P^u X_0$ and observe that $X_0^u = X_+$ from **(a)** above. This gives another equation, namely

$$\partial_t \mathsf{y}^u + L^u \mathsf{y}^u = \omega P^u \mathbb{P}\bar{z}, \quad t > 0, \quad \mathsf{y}^u(0) = \mathsf{y}_0^u - P^u \psi(\mathsf{x}_\infty).$$

In the sequel, y^s denotes the stable part, replacing y from **(b)**. Thus, we now have $\mathsf{y} = \mathsf{y}^s + \mathsf{y}^u$. We first construct the stable foliation in X_γ, by solving for the variable $\mathsf{v} = (\bar{w}, \mathsf{y}^s, \mathsf{y}^u, \mathsf{x}, \mathsf{x}_0)$. The problem to be solved reads

$$\mathbb{H}(\mathsf{v}, (\mathsf{x}_\infty, \mathsf{y}_0^s)) = \begin{bmatrix} (\mathbb{L}_\omega \bar{w} - N(\bar{w}, \mathsf{y}, \mathsf{x}, \mathsf{x}_\infty), \bar{z}(0) - \phi(\mathsf{x}_0 + \mathsf{y}_0) + \phi(\mathsf{x}_\infty + \mathsf{y}_\infty)) \\ (\partial_t \mathsf{y}^s + L^s \mathsf{y}^s - \omega P^s \mathbb{P} \bar{z}, \mathsf{y}^s(0) - \mathsf{y}_0^s + P^s \psi(\mathsf{x}_\infty)) \\ y^u(t) + \omega \int_t^\infty e^{-L^u(t-s)} P^u \mathbb{P} \bar{z}\, ds, \\ \mathsf{x}(t) + \omega \int_t^\infty P^c \mathbb{P} \bar{z}\, ds \\ \mathsf{x}_0 - \mathsf{x}_\infty + \omega \int_0^\infty P^c \mathbb{P} \bar{z}\, ds \end{bmatrix}.$$

This map is shown to be of class C^1 in $(\mathsf{v}, \mathsf{y}_0^s)$, but only continuous in x_∞, employing once more the results from Section 9.5, in the same way as in the previous step, and its Fréchet derivative $D_\mathsf{v}\mathbb{H}(0,0)$ w.r.t. the variable v is given by the triangular operator matrix

$$D_v\mathbb{H}(0,0) = \begin{bmatrix} (\mathbb{L}_\omega, \mathrm{tr}) & 0 & 0 & 0 & 0 \\ * & (\partial_t + L^s, \mathrm{tr}) & 0 & 0 & 0 \\ * & 0 & I & 0 & 0 \\ * & 0 & 0 & I & 0 \\ * & 0 & 0 & 0 & I \end{bmatrix}.$$

This operator is invertible by its triangular structure, as its diagonal entries are invertible. Therefore, by the implicit function theorem, we find balls $B_{X_0^c}(0,r)$ and $B_{X_\gamma^s}(0,r)$ and a continuous map

$$\mathcal{T} : B_{X_0^c}(0,r) \times B_{X_\gamma^s}(0,r) \to \hat{\mathbb{E}}(\infty, \delta) \times \tilde{\mathbb{E}}(\infty, \delta) \times X_0^c, \quad \mathcal{T}(\mathsf{x}_\infty, \mathsf{y}_0) = (\bar{w}, \tilde{z}, \mathsf{x}_0),$$

with $\mathcal{T}(0,0) = 0$. Then $(w, \pi) := (\bar{z} + \tilde{z} + z_\infty, \bar{\pi})$ yields the unique solution of (11.5) such that

$$z(t) \to z_\infty := \mathsf{x}_\infty + \psi(\mathsf{x}_\infty) + \phi(\mathsf{x}_\infty + \psi(\mathsf{x}_\infty)) \quad \text{in } X_\gamma \quad \text{as } t \to \infty.$$

Note that the initial value of y^u is given by $\mathsf{y}^u(0) = -\omega \int_0^\infty e^{L^u s} P^u \mathbb{P} \bar{z}\, ds$. To prove the second assertion in **(ii)**, suppose that z is a solution of (11.5) which stays in a small ball $B_{X_\gamma}(0,r)$. Then we proceed as in part **(a)** of this proof to show that its initial value satisfies

$$\mathsf{y}^u(0) = -\omega \int_0^\infty e^{L^u s} P^u \mathbb{P} \bar{z}(s)\, ds.$$

By uniqueness this shows that the initial value of this solution sits on the stable manifold, hence the solution converges exponentially fast to some $z_\infty \in \mathcal{E}$ in X_γ. This completes the proof of the second part of **(ii)**. □

11.4 Global Existence and Convergence

We have seen in Section 1.2 that the negative total entropy is a strict Lyapunov functional for all problems. Therefore the *limit sets* of solutions in the state manifolds $\mathcal{SM}^j$ are contained in the manifold $\mathcal{E} \subset \mathcal{SM}^j$ of equilibria. Recall also that $\mathcal{E}$ does not depend on the problem under consideration, except for **(P2)** where the balls making up Ω_1 may have arbitrary radii.

There are several obstructions for global existence:

- *regularity*: the norms of either $u(t)$, $\theta(t)$, $\Gamma(t)$, and in addition of $[\![d(\theta)\partial_\nu\theta]\!]$ in Problems 1 and 3, may become unbounded;
- *well-posedness*: the well-posedness conditions $0 < \theta < \theta_c$ (set $\theta_c = \infty$ in Problems 1–4), or invertibility of $\kappa_\Gamma T_\Gamma$ in Problems 1, 3 and 5 may be violated;
- *geometry*: the topology of the interface may change; or the interface may touch the boundary of Ω; or a part of the interface may contract to a point.

Note that the relevant compatibility conditions and the regularity of the solutions are preserved by the semiflow. Recalling the definition of T_Γ, observe that in Problems 1 and 3, $\kappa_\Gamma T_\Gamma = l(\theta)^2/\theta\sigma$, hence in this case $\kappa_\Gamma T_\Gamma$ is invertible if and only if $l(\theta) \neq 0$.

Let z be a solution in the state manifold $\mathcal{SM}^j$. By the *uniform ball condition* we mean the existence of a radius $r_0 > 0$ such that for each $t \in [0, t_+)$ at each point $x \in \Gamma(t)$ there exist centers $x_i \in \Omega_i(t)$ such that $B(x_i, r_0) \subset \Omega_i$ and $\Gamma(t) \cap \bar{B}(x_i, r_0) = \{x\}$, $i = 1, 2$. Note that this condition bounds the curvature of $\Gamma(t)$, prevents parts of it to shrink to points, to touch the outer boundary $\partial\Omega$, and to undergo topological changes.

With this property, combining the local semiflow for Problem **(Pj)** with the corresponding Lyapunov functional (i.e., the negative total entropy), relative compactness of bounded orbits, and the convergence results from the previous section, we obtain the following global result.

Theorem 11.4.1. *Let $p > n + 2$ and* **(Hj)** *hold, $j = 1, \ldots, 6$, and set $s = 4 - 3/p$ for $j = 1, 3, 5$, $s = 3 - 2/p$ for $j = 2, 4, 6$.*

Suppose that (u, θ, Γ) is a solution of Problem **(Pj)** *in the state manifold $\mathcal{SM}^j$ on its maximal time interval $[0, t_+)$. Assume there are constants $M, m > 0$ such that the following conditions hold on $[0, t_+)$.*

(i) $|u(t)|_{W^{2-2/p}_p}$, $|\theta(t)|_{W^{2-2/p}_p}$, $|\Gamma(t)|_{W^s_p} \leq M < \infty$, *and in addition* $|[\![d(\theta(t))\partial_\nu\theta(t)]\!]|_{W^{2-6/p}_p} \leq M$ *for Problems 1 and 3;*

(ii) $m \leq \theta(t) \leq \theta_c - m$;

(iii) $m \leq |\mu_k(t)|$ *for all eigenvalues $\mu_k(t)$ of $\kappa_\Gamma(t)T_{\Gamma(t)}$ in $L_2(\Gamma(t))$ for Problems 1,3,5;*

(iv) $\Gamma(t)$ *satisfies the uniform ball condition.*

Then $t_+ = \infty$, *i.e., the solution exists globally, and its limit set* $\omega_+(u,\theta,\Gamma) \subset \mathcal{E}$ *is nonempty. If furthermore* $(0,\theta_\infty,\Gamma_\infty) \in \omega_+(u,\theta,\Gamma)$ *is normally stable, then the solution converges in* $\mathcal{SM}^j$ *to this equilibrium.*

On the contrary, if $(u(t),\theta(t),\Gamma(t))$ *is a global solution in* $\mathcal{SM}^j$ *which converges to an equilibrium* $(0,\theta_*,\Gamma_*)$ *in* $\mathcal{SM}^j$ *as* $t \to \infty$, *with* $l_* \neq 0$ *and* $\delta_*, \zeta_* \neq 1$, *then* (i)–(iv) *hold.*

Proof. Assume that (i)–(iv) are valid. Then $\Gamma([0,t_*)) \subset W^s_p(\Omega,r)$ is bounded, hence relatively compact in $W^{s-\varepsilon}_p(\Omega,r)$, for small $\varepsilon > 0$, see Section 2.4.2 for the definition of $W^s_p(\Omega,r)$. Thus we may cover this set by finitely many balls with centers Σ_k real analytic, $k = 1,\ldots,N$, in such a way that

$$\mathrm{dist}_{W^{s-\varepsilon}_p}(\Gamma(t),\Sigma_j) \le \delta \quad \text{for some } j = j(t) \in \{1,\ldots,N\},\ t \in [0,t_*).$$

Let $J_k = \{t \in [0,t_*) : j(t) = k\}$. Using for each k a Hanzawa-transformation Ξ_k, we see that the pull backs $\{(u(t,\cdot),\theta(t,\cdot)) \circ \Xi_k : t \in J_k\}$ are bounded in $W^{2-2/p}_p(\Omega \setminus \Sigma_k)^{n+1}$, hence relatively compact in $W^{2-2/p-\varepsilon}_p(\Omega \setminus \Sigma_k)^{n+1}$. Employing now Theorem 9.2.1 we obtain solutions (u^1,θ^1,Γ^1) with initial configurations $(u(t),\theta(t),\Gamma(t))$ in the state manifold on a common time interval, say $(0,\tau]$, and by uniqueness we have

$$(u^1(\tau),\theta^1(\tau),\Gamma^1(\tau)) = (u(t+\tau),\theta(t+\tau),\Gamma(t+\tau)).$$

Continuous dependence implies that the orbit of the solution $(u(\cdot),\theta(\cdot),\Gamma(\cdot))$ is relative compact in $\mathcal{SM}^j$, in particular $t_* = \infty$ and $(u,\theta,\Gamma)(\mathbb{R}_+) \subset \mathcal{SM}^j$ is relatively compact. The negative total entropy is a strict Lyapunov functional, hence the limit set $\omega_+(u,\theta,\Gamma) \subset \mathcal{SM}^j$ of a solution is contained in the set $\mathcal{E}$ of equilibria. By compactness $\omega_+(u,\theta,\Gamma) \subset \mathcal{SM}^j$ is nonempty, hence the solution comes close to $\mathcal{E}$, and stays there. Then we may apply Theorem 11.3.1 to obtain convergence of such solutions. The converse assertion follows by a compactness argument. □

Remark 11.4.2. **(a)** Recall from Chapter 10 that any equilibrium $(0,\theta_*,\Gamma_*) \in \mathcal{E}$ is normally stable for Problem 2. It is normally stable for Problems 4 and 6 if and only if Γ_* is connected. For Problems 1, 3, 5 it is normally stable if and only if either $\delta_* > 1$; or $\delta_* < 1$, Γ_* is connected, and the stability condition $\zeta_* < 1$ holds.

(b) We conjecture that convergence also holds in case $(0,\theta_*,\Gamma_*) \in \mathcal{E}$ is normally hyperbolic. Then we would have convergence of all solutions which do not develop singularities, except in the pathological cases for Problems 1, 3, 5 where $l_* = 0$ or $\delta_* = 1$ or $\zeta_* = 1$.

(c) In the proof of Theorem 11.3.1 we have constructed the stable foliations for the Problems **(Pj)** near an equilibrium. In a similar way we can also construct the unstable foliations.

Chapter 12

Further Parabolic Evolution Problems

In this final chapter, we apply the theory of quasilinear parabolic evolution equations developed in Chapter 5 to several parabolic evolution problems to show the strength of the tools and techniques of this book. These problems include generalized Newtonian flows, nematic liquid crystal flows, Maxwell-Stefan diffusion problems, Stefan problems with variable surface tension, and, last but not least, several classes of geometric evolution equations. By means of our methods, many other parabolic evolution problems can be solved in the same – or at least in a similar – way.

12.1 Generalized Newtonian Flows

In this section we study boundary value problems for the Navier-Stokes system of a class of non-Newtonian fluids, the so-called *generalized Newtonian fluids*. By this we mean the following problem.

$$\begin{aligned} \partial_t(\varrho u) + \operatorname{div}(\varrho u \otimes u) &= \operatorname{div} T && \text{in } \Omega, \\ \mu[\nabla u + (\nabla u)^{\mathsf{T}}] - \pi I &= T && \text{in } \Omega, \\ \operatorname{div} u &= 0 && \text{in } \Omega, \\ u(0) &= u_0 && \text{in } \Omega. \end{aligned} \tag{12.1}$$

Here $\Omega \subset \mathbb{R}^n$, $n \geq 2$, denotes the domain occupied by the fluid, and $\Sigma = \partial\Omega$ means the boundary of Ω. We assume that Σ is compact and of class C^{3-}. Throughout, $u(t,x)$ means the velocity field of the fluid, $\pi(t,x)$ the pressure, and $T(t,x)$ the stress tensor. The numbers $\varrho > 0$, $\mu > 0$ represent the density and viscosity of the fluid, respectively. We assume that $\varrho > 0$ is constant, w.l.o.g. $\varrho = 1$.

J. Prüss and G. Simonett, *Moving Interfaces and Quasilinear Parabolic Evolution Equations*, Monographs in Mathematics 105,
DOI 10.1007/978-3-319-27698-4_12

On the other hand, the (nonconstant) viscosity μ will be taken of *generalized Newtonian type*, i.e.,

$$\mu = \mu(|D|_2^2), \quad D = \frac{1}{2}[\nabla u + (\nabla u)^{\mathsf{T}}]. \tag{12.2}$$

Here $D = (\varepsilon_{ij})$ denotes the rate of strain tensor and

$$|D|_2^2 = \sum_{i,j=1}^{n} \varepsilon_{ij}^2$$

its Hilbert-Schmidt norm. Note that the first invariant of D, namely $\operatorname{tr} D = \operatorname{div} u$ is zero and, hence, the Hilbert-Schmidt norm of D coincides with the second invariant of D up to a constant factor. It is believed that many *isotropic* fluids which are not subject to viscoelastic memory effects can be described by such a material law. A standard model in the mathematical literature is

$$\mu(s) = \mu_0(1+s)^{(d-2)/2}, \quad s \geq 0,$$

where $d \geq 1$, and $\mu_0 > 0$. The case $d = 2$ corresponds to the Newtonian case.

For this nonlinear problem, the energy is only kinetic, $\mathsf{E} = \int_\Omega \frac{1}{2}|u|^2\,dx$, and energy balance reads

$$\begin{aligned} &\partial_t \mathsf{E} + 2\int_\Omega \mu(|D|_2^2)|D|_2^2\,dx \\ &\quad + \int_\Sigma [-\mu(|D|_2^2)(\mathcal{P}_\Sigma u|\mathcal{P}_\Sigma D\nu) + (u|\nu)(\frac{1}{2}|u|^2 + \pi - (S\nu|\nu))]\,d\Sigma = 0. \end{aligned} \tag{12.3}$$

Four boundary conditions are of physical interest, namely the no-slip or Dirichlet condition

$$u = 0 \text{ on } \Sigma_d,$$

slip with damping (Navier-condition)

$$\mathcal{P}_\Sigma D\nu = -\alpha\mathcal{P}_\Sigma u, \ (u|\nu) = 0 \text{ on } \Sigma_s,$$

damped outflow

$$\mathcal{P}_\Sigma u = 0, \quad (\frac{1}{2}|u|^2 + \pi) - (\nu|S\nu) = \beta(u|\nu) \text{ on } \Sigma_o,$$

and the damped Neumann condition

$$\mathcal{P}_\Sigma D\nu = -\alpha\mathcal{P}_\Sigma u, \quad (\frac{1}{2}|u|^2 + \pi) - (\nu|S\nu) = \beta(u|\nu) \text{ on } \Sigma_n.$$

Here $\mathcal{P}_\Sigma$ means the orthogonal projection onto the tangent bundle of Σ, and we have decomposed $\Sigma = \partial\Omega$ disjointly as $\Sigma = \Sigma_d \cup \Sigma_s \cup \Sigma_o \cup \Sigma_n$, where each set Σ_j is open and closed in Σ. The outer unit normal of Ω at $x \in \Sigma$ is denoted by $\nu = \nu(x)$. We emphasize that some components Σ_j might be empty. The coefficients α, β are assumed to be C^1 and nonnegative. These boundary conditions lead to the energy balance

$$\partial_t \mathsf{E} + 2\int_\Omega \upmu(|D|_2^2)|D|_2^2\,dx + \int_\Sigma \big(\alpha\upmu(|D|_2^2)|\mathcal{P}_\Sigma u|^2 + \beta(u|\nu)^2\big)\,d\Sigma = 0, \tag{12.4}$$

where we set $\alpha = 0$ on $\Sigma_d \cup \Sigma_o$, and $\beta = 0$ on $\Sigma_d \cup \Sigma_s$. This shows that the energy $\mathsf{E} = \int_\Omega \varrho|u|^2/2\,dx$ is dissipative.

1.1 Main Result

With the definition of $T := S - \pi I$ we have $S = 2\upmu(|D|_2^2)D$, with $D = (\varepsilon_{ij}(u))$,

$$\begin{aligned}
[\operatorname{div} S]_i &= \upmu(|D|_2^2)\sum_{k=1}^n(\partial_k^2 u_i + \partial_i\partial_k u_k) + 4\upmu'(|D|_2^2)\sum_{j,k,l=1}^n \varepsilon_{ij}\varepsilon_{kl}\partial_j\varepsilon_{kl}\\
&= \upmu(|D|_2^2)\sum_{k=1}^n(\partial_k^2 u_i + \partial_i\partial_k u_k) + 4\upmu'(|D|_2^2)\sum_{j,k,l=1}^n \varepsilon_{ik}(u)\varepsilon_{jl}(u)\partial_k\partial_l u_j\\
&= \sum_{j,k,l=1}^n a_{ij}^{kl}(u)\partial_k\partial_l u_j.
\end{aligned}$$

Here we have set

$$a_{ij}^{kl}(u) = \upmu(|D|_2^2)(\delta_{kl}\delta_{ij} + \delta_{il}\delta_{jk}) + 4\upmu'(|D|_2^2)\varepsilon_{ik}(u)\varepsilon_{jl}(u). \tag{12.5}$$

Observe that $a_{ij}^{kl}(u)$ are real and that the symmetries $a_{ij}^{kl} = a_{ji}^{lk} = a_{kl}^{ij} = a_{kj}^{il} = a_{il}^{kj}$ are valid.

Define the quasilinear differential operator $\mathcal{A}(u,\nabla)$ as

$$\mathcal{A}(u,\nabla) = -\sum_{k,l=1}^n a_{ij}^{kl}(u)\partial_k\partial_l.$$

If the function $u \in B_{qp}^{2\mu-2/p}(\Omega;\mathbb{R}^n)$ is known and $1 \geq \mu > 1/2 + 1/p + n/2q$ then by the Sobolev embedding $B_{qp}^{2\mu-2/p}(\Omega;\mathbb{R}^n) \hookrightarrow C_b^1(\Omega;\mathbb{R}^n)$ the coefficients of the differential operator $\mathcal{A}(x,\nabla) = \mathcal{A}(u(x),\nabla)$ are uniformly continuous and

$$[\mathcal{A}(\infty,\nabla)v]_i = -\upmu(0)\sum_{k=1}^n(\partial_k^2 v_i + \partial_i\partial_k v_k)$$

in case Ω is unbounded, since $|u(x)| + |Du(x)| \to 0$ as $|x| \to \infty$. Denoting the boundary operators on Σ_j by $\mathcal{B}_j(u,\nabla)$, problem (12.1), complemented by the boundary conditions discussed above, can be rewritten as

$$\begin{aligned} \partial_t u + \mathcal{A}(u,\nabla)u + \nabla\pi &= f(u) && \text{in } \Omega,\\ \operatorname{div} u &= 0 && \text{in } \Omega,\\ \mathcal{B}_j(u,\nabla)u &= 0 && \text{on } \Sigma_j, \quad j=d,s,o,n,\\ u(0) &= u_0 && \text{in } \Omega. \end{aligned} \tag{12.6}$$

Here the boundary operators $\mathcal{B}_j$ will be $\mathcal{B}_d(u,\nabla)u = (\mathcal{P}_\Sigma u, (u|\nu))$ on Σ_d. Moreover, we obtain

$$\begin{aligned} \mathcal{B}_s(u,\nabla)u &= (\mathcal{P}_\Sigma D\nu + \alpha\mathcal{P}_\Sigma u, (u|\nu)) && \text{on } \Sigma_s,\\ \mathcal{B}_o(u,\nabla)(u,\pi) &= (\mathcal{P}_\Sigma u, 2\upmu(|\mathcal{D}|_2^2)(\nu|\partial_\nu u) - \pi - |u|^2/2 + \beta(u|\nu)) && \text{on } \Sigma_o,\\ \mathcal{B}_n(u,\nabla)(u,\pi) &= (\mathcal{P}_\Sigma D\nu + \alpha\mathcal{P}_\Sigma u, 2\upmu(|\mathcal{D}|_2^2)(\nu|\partial_\nu u) - \pi - |u|^2/2 + \beta(u|\nu)) && \text{on } \Sigma_n; \end{aligned}$$

recall $\upmu(s) > 0$ for $s \geq 0$. The nonlinearity $f(u)$ is the convective term given by $f(u) = -u\cdot\nabla u$.

Our main result is the following.

Theorem 12.1.1. *Let $\Omega \subset \mathbb{R}^n$ be a domain with compact boundary $\Sigma := \partial\Omega$ of class C^{3-}, where $\Sigma = \Sigma_d \cup \Sigma_s \cup \Sigma_o \cup \Gamma_n$ with disjoint, open and closed Σ_j. Let $1 \geq \mu > 1/2 + 1/p + n/2q$, and assume $\upmu \in C^{2-}(\bar{\mathbb{R}}_+)$ is such that*

$$\upmu(s) > 0 \text{ and } \upmu(s) + 2s\upmu'(s) > 0, \qquad \text{for all } s \geq 0. \tag{12.7}$$

Assume $\alpha \in C^1(\Sigma_s \cup \Sigma_n)$, $\beta \in C^1(\Sigma_o \cup \Sigma_n)$.

Then for each $u_0 \in B_{qp}^{2\mu-2/p}(\Omega;\mathbb{R}^n)$ satisfying the compatibility conditions

$$\begin{aligned} \operatorname{div} u_0 &= 0 \quad \text{in } \Omega,\\ u_0 &= 0 \quad \text{on } \Sigma_d, \quad (u_0|\nu) = 0 \quad \text{on } \Sigma_s,\\ \mathcal{P}_\Sigma u_0 &= 0 \quad \text{on } \Sigma_o, \quad \mathcal{P}_\Sigma D(u_0)\nu = -\alpha\mathcal{P}_\Sigma u_0 \quad \text{on } \Sigma_s \cup \Sigma_n, \end{aligned} \tag{12.8}$$

there is a unique solution (u,π) of (12.6) on a maximal time interval $[0, t_+(u_0))$. The solution is in the maximal regularity class

$$u \in H_{p,\mu}^1(J, L_q(\Omega;\mathbb{R}^n)) \cap L_{p,\mu}(J, H_q^2(\Omega;\mathbb{R}^n)), \quad \pi \in L_{p,\mu}(J; \dot{H}_q^1(\Omega)).$$

Additionally,

$$\pi \in F_{pq,\mu}^{1/2-1/2q}(J; L_q(\Sigma_o \cup \Sigma_n)) \cap L_{p,\mu}(J; B_{qq}^{1-1/q}(\Sigma_o \cup \Sigma_n))$$

for each interval $J = [0,a]$ with $a < t_+(u_0)$. The maximal time $t_+(u_0)$ is characterized by the property:

$$\text{if } t_+(u_0) < \infty \quad \text{then} \quad \lim_{t\to t_+(u_0)} u(t) \text{ does not exist in } B_{qp}^{2\mu-2/p}(\Omega;\mathbb{R}^n).$$

The solution map $u_0 \mapsto u$ generates a local semiflow on

$$X_\gamma := \{v \in B_{qp}^{2-2/p}(\Omega;\mathbb{R}^n) : v \text{ satisfies } (12.8)\},$$

the natural phase space for the problem in the L_p-L_q-setting.

1.2 Linearization
Let us check strong ellipticity of $\mathcal{A}(x,\nabla) := \mathcal{A}(u,\nabla)$, where $u \in B_{qp}^{2\mu-2/p}(\Omega;\mathbb{R}^n)$ is given. A simple computation shows for $\xi \in \mathbb{R}^n$, $\eta \in \mathbb{C}^n$, $|\xi| = |\eta| = 1$, by the definition of a_{ij}^{kl} in (12.5) and using the summation convention,

$$a_{ij}^{kl}\xi_l\eta_j = \mu(\xi_k\eta_i + \xi_i\eta_k) + 4\mu'\varepsilon_{ik}\varepsilon_{jl}\xi_l\eta_j.$$

Using symmetry of $D(u) = (\varepsilon_{ij})$ this yields

$$a_{ij}^{kl}\xi_l\eta_j = 2\mu c_{ik} + 4\mu'\varepsilon_{ik}\overline{((\mathcal{D}|\mathcal{C}))},$$

where $\mathcal{C} = (c_{ik}) = \frac{1}{2}(\xi \otimes \eta + \eta \otimes \xi)$ and $((\cdot|\cdot))$ means the inner product in $\mathbb{C}^{n\times n}$. Next observe that this matrix is symmetric, hence we obtain

$$\begin{aligned}
-(\mathcal{A}(x,i\xi)\eta|\eta) &= a_{ij}^{kl}\xi_l\eta_j\xi_k\overline{\eta}_i \\
&= 2\mu c_{ik}\xi_k\overline{\eta}_i + 4\mu'\varepsilon_{ik}\xi_k\overline{\eta}_i\overline{((\mathcal{D}|\mathcal{C}))} \\
&= 2\mu|\mathcal{C}|_2^2 + 4\mu'|((\mathcal{D}|\mathcal{C}))|^2 \\
&= \mu(|\xi|^2|\eta|^2 + |(\xi|\eta)|^2) + 4\mu'|(\mathcal{D}\xi|\eta)|^2.
\end{aligned}$$

Notice that $(\mathcal{A}(x,i\xi)\eta|\eta)$ is real, for each $x \in \Omega$, $\xi \in \mathbb{R}^n$, $\eta \in \mathbb{C}^n$. To obtain a necessary condition for strong ellipticity, choose ξ as an eigenvector of $\mathcal{D}$ and η perpendicular to ξ. This shows that the condition $\mu(s) > 0$ for each $s \geq 0$ is necessary for strong ellipticity. Obviously this condition is also sufficient in case $\mu'(|\mathcal{D}|_2^2) \geq 0$. So suppose that $\mu'(|\mathcal{D}|_2^2) < 0$. Then the Cauchy-Schwarz inequality implies

$$-(\mathcal{A}(x,i\xi)\eta|\eta) \geq 2\mu|\mathcal{C}|_2^2 + 4\mu'|\mathcal{D}|_2^2|\mathcal{C}|_2^2 \geq c|\mathcal{C}|_2^2,$$

provided we have

$$\mu(s) > 0 \quad \text{and} \quad \mu(s) + 2s\mu'(s) > 0, \qquad \text{for all } s \geq 0.$$

If $|\mathcal{C}|_2 = 0$ then $(\mathcal{C}\xi|\xi) = 0$ which means $|\xi|^2(\eta|\xi) = 0$, hence $(\eta|\xi) = 0$ since $|\xi| = 1$ by assumption. But this in turn yields $2\mathcal{C}\xi = \eta|\xi|^2$, hence the contradiction $\eta = 0$. Thus strong ellipticity is implied by (12.7), for each $u \in B_{qp}^{2\mu-2/p}(\Omega;\mathbb{R}^n)$.

We note that the condition $\mu(s) + 2s\mu'(s) > 0$ for $s > 0$ is also necessary, if one allows for *all* symmetric $\mathcal{D}$; choose e.g. $\mathcal{D} = \mathrm{diag}(\sqrt{s}, 0, \ldots, 0)$ to see this.

Next let us check normal strong ellipticity for the generalized Stokes problem with a_{ij}^{kl} from (12.5). With $\mathcal{C}^0 = \xi \otimes u + \nu \otimes v$, $\mathcal{C} = \frac{1}{2}(\mathcal{C}^0 + [\mathcal{C}^0]^{\mathsf{T}})$ and using sum convention again, we have

$$a_{ij}^{kl}c_{lj}^0 = \mu(c_{ik}^0 + c_{ki}^0) + 4\mu'\varepsilon_{ik}\varepsilon_{jl}c_{lj}^0 = 2\mu c_{ik} + 4\mu'\varepsilon_{ik}\overline{((\mathcal{D}|\mathcal{C}))},$$

by symmetry of $\mathcal{D}$. Note that the resulting matrix is symmetric. This yields

$$a_{ij}^{kl}c_{lj}^0\overline{c_{ki}^0} = 2\mu|\mathcal{C}|_2^2 + 4\mu'|((\mathcal{D}|\mathcal{C}))|^2,$$

hence this expression is real and

$$a_{ij}^{kl} c_{lj}^0 \overline{c_{ki}^0} \geq 2\min\{\mu, \mu + 2\mu' |\mathcal{D}|_2^2\} |\mathcal{C}|_2^2.$$

Condition (12.7) implies $a_{ij}^{kl} c_{lj}^0 \overline{c_{ki}^0} \geq 0$. If the left-hand side is zero, then $\mathcal{C} = 0$. This yields $\mathcal{C}\xi = 0$ as well as $\mathcal{C}\nu = 0$, and leads to the relations

$$u + (v|\xi)\nu + (u|\xi)\xi = v + (v|\nu)\nu + (u|\nu)\xi = 0.$$

Taking the inner product with ξ resp. ν we obtain $(u|\xi) = (v|\nu) = 0$ and $(u|\nu) + (v|\xi) = 0$. We may then conclude

$$u = r\nu, \quad v = -r\xi,$$

in particular $(u|v) = 0$. Therefore, $\mathcal{A}(x, \nabla)$ is uniformly normally strongly elliptic, for each fixed $u \in B_{qp}^{2\mu-2/p}(\Omega; \mathbb{R}^n)$.

1.3 The Nonlinear Problem

The nonlinear generalized Stokes problem will be solved by means of the abstract results from Chapter 5. This is possible since the involved boundary conditions will actually turn out to be linear and homogeneous and therefore the results from Chapter 5 are available for the proof of Theorem 12.1.1.

For this purpose, let P_{HW} denote the Helmholtz-Weyl projection in $L_q(\Omega; \mathbb{R}^n)$, and ∇^* as introduced in Section 7.4, corresponding to the decomposition of $\Sigma = \partial\Omega$ into its parts Σ_j, $j = d, s, o, n$. As in Section 6.3 we set

$$X_0 = \{u \in L_q(\Omega, \mathbb{R}^n) : \nabla^* u = 0 \text{ on } \Sigma\},$$

and

$$X_1 = \{u \in H_q^2(\Omega, \mathbb{R}^n) \cap X_0 : \mathcal{P}_\Sigma u = 0 \text{ on } \Sigma_d \cup \Sigma_o,\ \mathcal{P}_\Sigma D\nu = -\alpha \mathcal{P}_\Sigma u \text{ on } \Sigma_s \cup \Sigma_n\},$$

equipped with their natural norms. The trace space turns out to be

$$X_{\gamma,\mu} = \{u \in B_{qp}^{2\mu-2/p}(\Omega; \mathbb{R}^n) : \text{(12.8) holds}\},$$

which embeds into $C_b^1(\Omega; \mathbb{R}^n)$. Apply the Helmholtz-Weyl projection P_{HW} to problem (12.6) to obtain the abstract quasilinear problem

$$\dot{u} + P_{HW}\mathcal{A}(u, \nabla)u = P_{HW} f(u), \quad t > 0, \quad u(0) = u_0. \tag{12.9}$$

The operator family $A(u) = P_{HW}\mathcal{A}(u, \nabla)$ by Theorem 7.3.2 has maximal L_p-regularity for each $u \in X_{\gamma,\mu}$. Note that the lower order terms on the boundary coming from α, β and $\varrho|u|^2/2$ do not change the assertion of Theorem 7.3.2. Finally, we set $F(u) = -P_{HW}(u \cdot \nabla u)$, and one checks easily by Sobolev embedding and the regularity of μ that (A, F) satisfies (5.2). So we may apply Theorem 5.1.1 to prove Theorem 12.1.1.

1.4 Stability and Long-Time Behaviour
Here we assume that α, β are strictly positive. Then E is even a strict Lyapunov functional. In fact if $\partial_t \mathsf{E} = 0$ at some time instance, then $D = 0$ in Ω and $u = 0$ on Σ, hence $u = 0$ by the inequalities of Korn and Poincaré. This shows that the only equilibrium is the trivial one, i.e., $u_* = 0$, $\pi_* = const$. We next show that it is exponentially stable. As $A_0 := A(0)$ is the negative generator of an analytic C_0-semigroup in X_0, by compact embedding the spectrum of A_0 consists only of eigenvalues of finite algebraic multiplicity, and these eigenvalues are independent of p. So we may once more use an energy argument. Suppose that $\lambda \in \mathbb{C}$ is an eigenvalue with eigenfunction $u \in X_1$. Multiply the eigenvalue equation by u to obtain after an integration by parts

$$0 = \lambda |u|_{L_2}^2 + \int_\Omega \mu_0 |D|^2 \, dx + \int_\Sigma \left(\alpha \mu_0 |\mathcal{P}_\Sigma u|^2 + \beta (u|\nu)^2\right) d\Sigma.$$

This identity shows that λ must be real, and if $\lambda \geq 0$ then $D = 0$ in Ω, and also $u = 0$ on Σ, provided $\alpha, \beta > 0$, hence $u = 0$. Therefore, all eigenvalues of $A_0 = A(u, 0)$ are positive. Hence by the principle of linearized stability, $u = 0$ is exponentially stable for the nonlinear problem. Moreover, Theorem 5.7.2 shows that a solution which stays bounded in $X_{\gamma,\mu}$ must converge exponentially fast to $u_* = 0$. We summarize this in

Corollary 12.1.2. *In addition to the assumptions of Theorem 12.1.1, let α, β be strictly positive. Then*
(i) $u_* = 0$ *is the only equilibrium of* (12.9), *and it is exponentially stable in* X_γ.

(ii) *If u is a solution of* (12.9) *which stays bounded in $X_{\gamma,\mu}$, then it converges exponentially fast to $u_* = 0$ in X_γ.*

We want to observe that except in very special cases, we may also allow for $\alpha = \beta = 0$. In fact, the property $D = 0$ implies that u must be an Euclidean motion, i.e., of the form $u = Qx + b$, with $Q^{\mathsf{T}} = -Q$ and $b \in \mathbb{R}^n$. So, if the boundary conditions and the geometry of Ω exclude such solutions, then the corollary remains valid. For example, this is true if the no-slip part Σ_d of the boundary Σ is nontrivial.

12.2 Nematic Liquid Crystal Flows

In this section we intend to apply the abstract results proved in Chapter 5 to a system of equations which models the flow of isothermal incompressible isotropic nematic liquid crystals. The model reads as follows.

$$\begin{aligned} \varrho(\partial_t u + (u\cdot\nabla)u) - \mu\Delta u + \nabla\pi &= -\lambda \operatorname{div}(\nabla d[\nabla d]^{\mathsf{T}}) && \text{in } \Omega,\\ \partial_t d + (u\cdot\nabla)d &= \gamma(\Delta d + |\nabla d|_2^2 d) && \text{in } \Omega,\\ \operatorname{div} u &= 0 && \text{in } \Omega,\\ (u, \partial_\nu d) &= (0,0) && \text{on } \partial\Omega,\\ (u(0), d(0)) &= (u_0, d_0) && \text{in } \Omega. \end{aligned} \tag{12.10}$$

Here Ω denotes a bounded domain with boundary $\partial\Omega$ of class C^2. The function $u : (0,\infty)\times\Omega \to \mathbb{R}^n$ means the velocity field, $\pi : (0,\infty)\times\Omega \to \mathbb{R}$ the pressure and $d : (0,\infty)\times\Omega \to \mathbb{R}^n$ represents the macroscopic molecular orientation of the liquid crystal material. It is therefore reasonable to impose $|d| = 1$ pointwise, which we will do only in the last section where this condition is important. The constants $\varrho, \mu, \lambda, \gamma > 0$ represent density, viscosity, the competition between kinetic energy and potential energy, and the microscopic elastic relaxation for the molecular orientation field, respectively. For simplicity, we set $\varrho = \mu = \lambda = \gamma = 1$, as this will not change the analysis.

The condition $|d| \equiv 1$ is indeed preserved by this system. This can be seen as follows. Setting $\varphi = |d|^2 - 1$ the elementary identities

$$\partial_t |d|^2 = 2d\cdot\partial_t d, \quad \Delta|d|^2 = 2\Delta d\cdot d + 2|\nabla d|_2^2, \quad \nabla|d|^2 = 2d\cdot\nabla d,$$

and multiplication with d of the second line in (12.10) yields the problem

$$\begin{aligned} \partial_t\varphi + u\cdot\nabla\varphi &= \Delta\varphi + 2|\nabla d|_2^2\varphi && \text{in } \Omega\\ \partial_\nu\varphi &= 0 && \text{on } \partial\Omega,\\ \varphi(0) &= 0 && \text{in } \Omega, \end{aligned}$$

provided $|d_0| \equiv 1$. Uniqueness of this parabolic convection-reaction diffusion equations yields $\varphi \equiv 0$, i.e., $|d| \equiv 1$.

2.1 Well-Posedness and Regularity

We reformulate (12.10) as an abstract quasilinear parabolic evolution equation

$$\dot z + A(z)z = F(z), \quad t > 0, \quad z(0) = z_0, \tag{12.11}$$

for the unknown $z = (u, d)$. For this purpose we introduce the base space X_0 as

$$X_0 := L_{q,\sigma}(\Omega) \times L_q(\Omega)^n,$$

where the subscript σ means solenoidal. We define the Neumann-Laplacian $\mathcal{D}_q$ in $L_q(\Omega)$ by

$$\mathcal{D}_q = -\Delta, \quad \mathsf{D}(\mathcal{D}_q) := \{d \in H_q^2(\Omega)^n : \partial_\nu d = 0 \text{ on } \partial\Omega\}.$$

$\mathcal{D}_q$ has the property of maximal L_p-regularity, as has been shown in Chapter 6.

Let $\mathbb{P} : L_q(\Omega)^n \to L_{q,\sigma}(\Omega)$ denote the *Helmholtz projection.* As usual, we define the *Stokes operator* $\mathcal{A}_q$ in $L_{q,\sigma}(\Omega)$ by

$$\mathcal{A}_q = \mathbb{P}\Delta, \quad \mathsf{D}(\mathcal{A}_q) := \{u \in H_q^2(\Omega)^n : \operatorname{div} u = 0 \text{ in } \Omega,\ u = 0 \text{ on } \partial\Omega\}.$$

$\mathcal{A}_q$ also has the property of maximal L_p-regularity, see Chapter 7.

Next we define the space X_1 by

$$X_1 := \mathsf{D}(\mathcal{A}_q) \times \mathsf{D}(\mathcal{D}_q),$$

equipped with its canonical norm. Then $X_1 \hookrightarrow X_0$ densely.

The quasilinear part $A(z)$ is given by the tri-diagonal matrix

$$A(z) = \begin{bmatrix} \mathcal{A}_q & \mathbb{P}\mathcal{B}_q(d) \\ 0 & \mathcal{D}_q \end{bmatrix},$$

where the operator $\mathcal{B}_q$ is given by

$$[\mathcal{B}_q(d)h] := \nabla d \Delta h + \nabla^2 h [\nabla d]^{\mathsf{T}}.$$

Obviously, $\mathcal{B}(d) : X_1 \to X_0$ is bounded, for each $d \in C^1(\overline{\Omega})^n$ and the map $d \mapsto \mathbb{P}\mathcal{B}_q(d)$ is polynomial, hence real analytic.

By the tri-diagonal structure of $A(z)$ it is clear that $A(z)$ also has the property of maximal L_p-regularity, for each $z \in C^1(\overline{\Omega})^{2n}$.

The semi-linear part $F(z)$ is defined by

$$F(z) = (-\mathbb{P}(u \cdot \nabla)u, -(u \cdot \nabla)d + |\nabla d|_2^2 d),$$

which is also polynomial, hence a real analytic mapping from $C^1(\overline{\Omega})^{2n}$ into X_0.

The natural trace space for the maximal regularity class

$$z \in H^1_{p,\mu}(J; X_0) \cap L_{p,\mu}(J; X_1)$$

is given by

$$X_{\gamma,\mu} = (X_0, X_1)_{\mu-1/p,p} = D_{\mathcal{A}_q}(\mu - 1/p, p) \times D_{\mathcal{D}_q}(\mu - 1/p, p),$$

for each $\mu \in (1/p, 1]$, and we set $X_\gamma = X_{\gamma,1}$ as before. We have the characterizations

$$d \in D_{\mathcal{D}_q}(\mu - 1/p, p) \quad \Leftrightarrow \quad d \in B_{qp}^{2\mu-2/p}(\Omega)^n,\ \partial_\nu d = 0 \text{ on } \partial\Omega,$$

and

$$u \in D_{\mathcal{A}_q}(\mu - 1/p, p) \quad \Leftrightarrow \quad u \in B_{qp}^{2\mu-2/p}(\Omega)^n \cap L_{q,\sigma}(\Omega),\ u = 0 \text{ on } \partial\Omega.$$

To have the embedding $X_{\gamma,\mu} \hookrightarrow C^1(\overline{\Omega})^{2n}$ at disposal, we impose the conditions

$$\frac{2}{p} + \frac{n}{q} < 1, \quad \frac{1}{2} + \frac{1}{p} + \frac{n}{2q} < \mu \leq 1. \tag{12.12}$$

We are now in position to apply Theorem 5.1.1 to obtain the following result on local well-posedness of (12.10) in the framework of L_p-theory.

Theorem 12.2.1. *Let p, q, μ be subject to* (12.12), *and assume $z_0 \in X_{\gamma,\mu}$, which means that $u_0, d_0 \in B_{qp}^{2\mu-2/p}(\Omega)^n$ satisfy the compatibility conditions*

$$\operatorname{div} u_0 = 0 \text{ in } \Omega, \quad u_0, \partial_\nu d_0 = 0 \text{ on } \partial\Omega.$$

Then, for some $a = a(z_0) > 0$, there is a unique solution

$$z \in H^1_{p,\mu}(J, X_0) \cap L_{p,\mu}(J; X_1), \quad J = [0, a],$$

of (12.11). *Moreover,*

$$z \in C([0, a]; X_{\gamma,\mu}) \cap C((0, a]; X_\gamma),$$

i.e., the solution regularizes instantly in time. It depends continuously on z_0 and exists on a maximal time interval $(0, t_+)$. Therefore problem (12.11), *i.e.,* (12.10), *generates a local semiflow in its natural state space X_γ.*

As the nonlinearities A and F are real analytic, we may employ Theorem 5.2.1 to obtain further regularity of the solutions of (12.10).

Theorem 12.2.2. *Suppose $z_0 \in X_{\gamma,\mu}$, and let*

$$z \in H^1_{p,\mu}(J; X_0) \cap L_{p,\mu}(J; X_1)$$

be a solution of (12.10) *on the interval $J = [0, a]$. Then for each $k \in \mathbb{N}$,*

$$t^k \partial_t^k z \in H^1_{p,\mu}(J; X_0) \cap L_{p,\mu}(J; X_1).$$

Moreover, $z \in C^\omega((0, a); X_1)$.

Theorem 12.2.2 is employed below to justify time derivatives of the energy functional. Employing scaling techniques jointly in time and space as in Section 9.4, it is possible to show via maximal regularity and the implicit function theorem that u, π, d are real analytic in $(0, a) \times \Omega$.

2.2 Stability and Long Time Behaviour

The set $\{0\} \times \mathbb{R}^n$ consists of equilibria of (12.10) and forms a n-dimensional subspace of X_1. The linearization of (12.10) at $z_* \in \{0\} \times \mathbb{R}^n$ is given by the linear evolution equation

$$\dot{z} + A_* z = f, \quad z(0) = z_0,$$

in X_0, where

$$A_* = \operatorname{diag}(\mathcal{A}_q, \mathcal{D}_q), \quad \mathsf{D}(A_*) = X_1.$$

As Ω is bounded, the spectrum $\sigma(\mathcal{A}_q)$ consists only of positive eigenvalues and $0 \notin \sigma(\mathcal{A}_q)$. On the other hand, $\mathcal{D}_q$ has 0 as an eigenvalue, which is semi-simple, and the remaining part of $\sigma(\mathcal{D}_q)$ consist only of positive eigenvalues. Thus $\sigma(A_*) \backslash \{0\} \subset [\delta, \infty)$ for some $\delta > 0$ and the kernel of A_* is given by

$$\mathsf{N}(A_*) = \{0\} \times \mathbb{R}^n,$$

hence it has dimension n. As a result we see that each equilibrium $z_* \in \{0\} \times \mathbb{R}^n$ is normally stable. Now we are in position to apply the *generalized principle of linearized stability* Theorem 5.3.1 to conclude the following stability result for the equilibria of (12.10).

Theorem 12.2.3. *Each equilibrium $z_* \in \{0\} \times \mathbb{R}^n$ is stable in X_γ. There exists a number $\eta > 0$ such that any solution $z(t)$ with initial value $z_0 \in X_\gamma$, $|z_0 - z_*|_{X_\gamma} \le \eta$, exists globally and converges to some $z_\infty \in \{0\} \times \mathbb{R}^n$ in X_γ at an exponential rate as $t \to \infty$.*

The energy of the system is given by

$$\mathsf{E} = \frac{1}{2} \int_\Omega [|u|^2 + |\nabla d|_2^2]\, dx = \mathsf{E}_{kin} + \mathsf{E}_{pot}. \tag{12.13}$$

Using the summation convention we have with an integration by parts

$$\begin{aligned} \frac{d}{dt}\mathsf{E}_{kin}(t) &= \int_\Omega \partial_t u \cdot u\, dx \\ &= \int_\Omega [-(u\cdot\nabla)u - \nabla\pi + \Delta u - \operatorname{div}(\nabla d[\nabla d]^{\mathsf{T}})] \cdot u\, dx \\ &= -\int_\Omega |\nabla u|^2\, dx + \int_\Omega \partial_k d_j \partial_i d_j \partial_k u_i\, dx, \end{aligned}$$

as $\operatorname{div} u = 0$ in Ω and $u = 0$ on $\partial\Omega$. On the other hand, we have by another integration by parts

$$\begin{aligned} \int_\Omega |\Delta d + |\nabla d|_2^2 d|^2\, dx &= \int_\Omega [\Delta d + |\nabla d|_2^2 d] \cdot [\partial_t d + (u\cdot\nabla)d]\, dx \\ &= -\int_\Omega [\partial_t \nabla d : \nabla d - |\nabla d|_2^2 \partial_t |d|^2/2]\, dx \\ &\quad + \int_\Omega [(u\cdot\nabla)d \cdot \Delta d + |\nabla d|_2^2 (u\cdot\nabla)|d|^2/2]\, dx \\ &= -\frac{d}{dt}\mathsf{E}_{pot}(t) - \int_\Omega \partial_k(u_i \partial_i d_j)\partial_k d_j\, dx \\ &= -\frac{d}{dt}\mathsf{E}_{pot}(t) - \int_\Omega \partial_k u_i \partial_i d_j \partial_k d_j\, dx, \end{aligned}$$

provided $|d| \equiv 1$, by the Neumann boundary condition for d. Combining these equations, we obtain the energy identity

$$\frac{d}{dt}\mathsf{E}(t) = -\int_\Omega [|\nabla u|^2 + |\Delta d + |\nabla d|_2^2 d]^2\, dx. \tag{12.14}$$

Therefore $\mathsf{E}(t)$ is nonincreasing along solutions. But E is also a strict Lyapunov functional. In fact, if $d\mathsf{E}(t)/dt = 0$ at some time instant, then by the energy equality we have $\nabla u = 0$ and $\Delta d + |\nabla d|_2^2 d = 0$ in Ω. Therefore $u = 0$ by the no-slip condition on $\partial\Omega$, and d satisfies the nonlinear eigenvalue problem

$$\begin{aligned} \Delta d + |\nabla d|_2^2 d &= 0 \quad \text{in } \Omega, \\ |d|^2 &= 1 \quad \text{in } \Omega, \\ \partial_\nu d &= 0 \quad \text{on } \partial\Omega. \end{aligned} \tag{12.15}$$

But, as the lemma below shows, this implies $\nabla d = 0$ in Ω, hence $d = d_*$ is constant and $z_* := (0, d_*)$, $|d_*| \equiv 1$, is an equilibrium of the problem, $z_* \in \mathcal{E}$.

Lemma 12.2.4. *Suppose that $d \in H_2^2(\Omega; \mathbb{R}^N)$ satisfies* (12.15). *Then d is constant in Ω.*

Proof. The idea is to reduce inductively the dimension $N = n$ of the vector d. This can be achieved by introducing polar coordinates according to

$$d_1 = c_1 \cos\theta,\ d_2 = c_1 \sin\theta,\ d_j = c_{j-1}, \quad j \geq 3.$$

Simple computations yield

$$1 = |d|^2 = |c|^2, \quad |\nabla d|_2^2 = |\nabla c|^2 + c_1^2 |\nabla\theta|^2,$$

and

$$\Delta c_j + [|\nabla c|^2 + c_1^2 |\nabla\theta|^2] c_j = 0 \ \text{ in } \ \Omega,$$

as well as $\partial_\nu c_j = 0$ on $\partial\Omega$ for $j = 2, \ldots, n-1$. Moreover, by an easy calculation we further obtain

$$-\Delta c_1 + c_1 |\nabla\theta|^2 = [|\nabla c|^2 + c_1^2 |\nabla\theta|^2] c_1 \ \text{ in } \ \Omega,$$

and

$$c_1 \Delta\theta + 2\nabla c_1 \cdot \nabla\theta = 0 \ \text{ in } \ \Omega,$$

as well as

$$\partial_\nu c_1 = c_1 \partial_\nu \theta = 0 \ \text{ on } \ \partial\Omega.$$

Multiplying the former equation by $c_1\theta$ and integrating over Ω we deduce

$$\begin{aligned} 0 &= \int_\Omega [c_1 \Delta\theta + 2\nabla c_1 \cdot \nabla\theta)] c_1 \theta \, dx \\ &= \int_\Omega \operatorname{div}[c_1^2 \nabla\theta] \theta \, dx = -\int_\Omega c_1^2 |\nabla\theta|^2 \, dx, \end{aligned}$$

hence $c_1 \nabla\theta = 0$. This implies that c satisfies Problem (12.15) where the vector c has dimension $N - 1$. Inductively, we arrive at dimension $N = 1$, and if d is a solution of (12.15) with $N = 1$, then $d \equiv 1$ of $d \equiv -1$, by connectedness of Ω. □

Summarizing we have proved

Proposition 12.2.5. *The energy functional* E *defined on* X_γ *is a strict Lyapunov function for system* (12.10). *The equilibria of the system are given by the set*

$$\mathcal{E} = \{z_* = (u_*, d_*) : u_* = 0,\ d_* \in \mathbb{R}^n,\ |d_*| = 1\},$$

which forms a manifold of dimension $n-1$. *The corresponding pressures* π_* *are constant as well.*

Note that the side condition $|d| = 1$ is important, here.

Having the strict Lyapunov functional E at disposal we now employ Theorem 5.7.2 to obtain the final global result.

Theorem 12.2.6. *Let* $z_0 \in X_{\gamma,\mu}$ *with* $|d_0| \equiv 1$, *and suppose that the solution* $z(t)$ *of* (12.10) *is eventually bounded in* X_γ *on its maximal interval of existence.*
Then $z(t)$ *exists globally and* $\lim_{t\to\infty} z(t) =: z_\infty \in \mathcal{E}$ *in* X_γ.

Trivially, the converse of the statement in this result is also true.

12.3 Maxwell-Stefan Diffusion with Reactions

Let $\Omega \subset \mathbb{R}^n$ be an open bounded domain with boundary $\partial\Omega$ of class $C^{2+\alpha}$ and outer unit normal field ν. We consider a mixture of $N \geq 2$ species A_k with *molar masses* $M_k > 0$ and *individual mass densities* $\varrho_k \geq 0$ filling the container Ω. Mass balance of the single component A_k reads

$$\partial_t \varrho_k + \mathrm{div}_x(\varrho_k \mathbf{u}_k) = M_k r_k \quad \text{in } \Omega, \quad t > 0,$$

where $\mathbf{u}_k$ denotes the *individual velocity* of species A_k, satisfying $(\mathbf{u}_k|\nu) = 0$ on $\partial\Omega$, and r_k is the rate of production of species A_k due to chemical reactions. Observe that r should be *positivity preserving*, i.e., subject to the condition

$$\varrho_j \geq 0, \quad \varrho_k = 0 \quad \Rightarrow \quad r_k \geq 0,$$

and should satisfy $\sum_k M_k r_k = 0$, which results in *conservation of total mass.* The quantities of interest are the mass densities ϱ_k, while the individual velocities $\mathbf{u}_k$ are in general unknown and have to be modeled, as well as the reaction rates r_k. To reduce the complexity of these balance laws, we introduce the *total density* $\varrho = \sum_k \varrho_k$, the *barycentric velocity* $\mathbf{u} = \sum_k \varrho_k \mathbf{u}_k/\varrho$, the *mass fractions* $y_k = \varrho_k/\varrho$, and the *concentrations* $c_k = \varrho_k/M_k = y_k\varrho/M_k$. With these new variables, we obtain the overall mass balance

$$\partial_t \varrho + \mathrm{div}_x(\varrho \mathbf{u}) = 0 \quad \text{in } \Omega, \quad t > 0,$$

and $(\mathbf{u}|\nu) = 0$ on $\partial\Omega$. The individual mass balances now become

$$\varrho(\partial_t y_k + \mathbf{u} \cdot \nabla_x y_k) + \mathrm{div}_x J_k = M_k r_k \quad \text{in } \Omega, \quad t > 0,$$

where the diffusive fluxes J_k are given by

$$J_k = \varrho_k(\mathbf{u}_k - \mathbf{u}), \quad k = 1, \dots, N.$$

Note that, by definition, $\sum_k y_k = 1$ and $\sum_k J_k = 0$.

So far, everything is physically exact in the framework of continuum mechanics. However, to obtain a closed model one has to prescribe laws for $\mathbf{u}$, r_k and, most importantly, for the diffusive fluxes J_k. Here we are interested in the *incompressible, quiescent, isothermal case*, which means

$$\varrho = constant, \quad \mathbf{u} = 0, \quad \theta = constant.$$

These assumptions lead to the problem

$$\varrho \partial_t y_k + \operatorname{div}_x J_k = M_k r_k(y) \quad \text{in } \Omega, \quad (J_k|\nu) = 0 \quad \text{on } \partial\Omega, \tag{12.16}$$

for $k = 1, \dots, N$, completed by initial data $y_k(0) = y_0^k \geq 0$. We again emphasize the constraints

$$\sum_{k=1}^N J_k = 0, \quad \sum_{k=1}^N y_k = 1. \tag{12.17}$$

Together with $y_k \geq 0$ this already implies L_∞-bounds for y_k, a very important property. Therefore, when modeling the diffusive fluxes it is essential that positivity as well as conservation of mass are ensured.

In the *Maxwell-Stefan approach* to model diffusion, a balance of so-called *driving forces* d_k and *friction forces* f_k is postulated, i.e., $\mathsf{d}_k = \mathsf{f}_k$. The friction forces are modelled by

$$\mathsf{f}_k = \varrho \sum_{j \neq k} f_{kj} y_k y_j (\mathbf{u}_j - \mathbf{u}_k) = \sum_{j \neq k} f_{kj}(y_k J_j - y_j J_k), \tag{12.18}$$

with the symmetric *friction coefficients* $f_{kj} = f_{jk} > 0$. These coefficients may depend on the composition y, but in the sequel we assume them to be constant. Observe that $\sum_k \mathsf{f}_k = 0$, so that the friction forces act only on the components but not on the mixture.

The driving forces d_k have to be modeled as well and are typically based on the *chemical potentials* μ_k. In the mass-based approach we are using here and for general chemical potentials $\mu_k = \partial_{y_k}\psi$, where ψ is the density of the constitutive *Helmholtz free energy*, we assume

$$\mathsf{d}_k = y_k\Big(\nabla_x \mu_k - \sum_{j=1}^N y_j \nabla_x \mu_j\Big). \tag{12.19}$$

Note that these relations guarantee $\sum_k \mathsf{d}_k = 0$ for arbitrary free energies ψ. In the sequel we assume (12.19), with the free energy

$$\psi = \sum_k \frac{y_k}{M_k}\big[\log\big(y_k/\mathbf{y}_*^k\big) - 1\big], \tag{12.20}$$

suggested by chemistry. Here $\mathbf{c}_*^k = \frac{\varrho}{M_k}\mathbf{y}_*^k$ are the components of a constant *chemical equilibrium* $\mathbf{c}_*$ of the system; see below for more details.

3.1 The Maxwell-Stefan Equations

The assumptions $\mathsf{d}_k = \mathsf{f}_k$ lead to the *Maxwell-Stefan equations* for the columns $J^\alpha \in \mathbb{R}^N$ of the flux matrix $J = (J_1, \ldots, J_N)^\mathsf{T} \in \mathbb{R}^{N\times n}$. Writing

$$M = \operatorname{diag}(M_j), \quad \mathsf{e} = [1, \ldots, 1]^\mathsf{T}, \quad P(y) = I - y \otimes \mathsf{e} = I - (\cdot|\mathsf{e})y,$$

these equations read as follows:

$$\begin{aligned} B(y)J^\alpha &= P(y)M^{-1}\partial_{x_\alpha} y, && B(y) = [b_{ij}(y)], \quad \alpha = 1, \ldots, n, \\ b_{ij}(y) &= f_{ij}y_i \ \text{ for } i \neq j, && b_{ii}(y) = -\sum_{l\neq i} f_{il}y_l, \quad i,j = 1, \ldots, N. \end{aligned} \tag{12.21}$$

We now study the Maxwell-Stefan equations (12.21) in more detail. More precisely, we show that the restriction of the matrix $B(y)$ to $\mathbb{E} = \{\mathsf{e}\}^\perp$ is invertible for all y in an open neighbourhood $U \subset \mathbb{R}^N$ of $\mathbb{D} := \{y \in \mathbb{R}^N : y \geq 0, \ (y|e) = 1\}$, and investigate the structure of its inverse

$$A(y) = (B(y)|_\mathbb{E})^{-1}.$$

We further show that the spectrum of

$$A_0(y) = -A(y)P(y)M^{-1},$$

considered as an element of $\mathcal{B}(\mathbb{E})$, belongs to $(0, \infty)$ for all $y \in U$.

Lemma 12.3.1. *For any $y \in \mathbb{R}_+^N$ we have*

(i) *The matrix $B(y) = [b_{ij}(y)]$ is irreducible and quasi-positive;*

(ii) *the kernel of $B(y)$ is $\mathsf{N}(B(y)) = \operatorname{span}\{y\}$;*

(iii) *the range of $B(y)$ is $\mathsf{R}(B(y)) = \{\mathsf{e}\}^\perp =: \mathbb{E}$;*

(iv) *the spectrum satisfies $\sigma(B(y)) \setminus \{0\} \subset (-\infty, 0)$;*

Proof. (i), (ii) and (iii) are direct consequences of the definition of $B(y)$. Using the theorem of Perron-Frobenius for quasi-positive matrices we see that $s(B(y)) := \max\{\operatorname{Re}(\lambda) \mid \lambda \in \sigma(B(y))\}$ is a simple eigenvalue of $B(y)$. Its eigenspace is spanned by a positive vector, and no other eigenvalue has a positive eigenvector. This implies $s(B(y)) = 0$, and the remaining eigenvalues of $B(y)$ have negative real parts. Finally with $Y = \operatorname{diag}[y_k]$, the similarity transform $B_s(y) := Y^{-1/2}B(y)Y^{1/2}$ symmetrizes $B(y)$, hence all eigenvalues of $B(y)$ are necessarily real. □

This lemma shows in particular that $B(y)$ may be restricted to an element $B(y)|_\mathbb{E}$ of $\mathcal{B}(\mathbb{E})$ for all $y \in \mathbb{R}_+^N$. We show that $B(y)|_\mathbb{E}$ is invertible for y on a larger set containing $\mathbb{D}$.

Lemma 12.3.2. *There is an open neighbourhood $U \subset \mathbb{R}^N$ of $\mathbb{D}$ such that for all $y \in U$ the restriction $B(y)|_{\mathbb{E}}$ of $B(y)$ to $\mathbb{E}$ is invertible. Denote its inverse by $A(y) = (B(y)|_{\mathbb{E}})^{-1}$. Then there are real analytic functions $a_i^0, a_{ij}^1 \colon U \to \mathbb{R}$ such that for all $y \in U$ and $h \in \mathbb{E}$ the vector $x = A(y)h$ may be represented by*

$$x_i = -a_i^0(y)h_i + y_i \sum_{j \neq i} a_{ij}^1(y)h_j, \quad i = 1, ..., N.$$

We have $a_i^0(y) > 0$ for $y_i = 0$.

Proof. Step 1. We show that $B(y)|_{\mathbb{E}}$ is invertible for $y \in \mathbb{D}$. For $y \in \mathring{\mathbb{D}}$ this follows already from Lemma 12.3.1. So let $y \in \mathbb{D}$ be such that $y_k = 0$ for some $1 \le k \le N$. Assume $B(y)x = 0$ for $x \in \mathbb{E}$. We show $x = 0$. The structure of $B(y)$ from (12.21) implies that $b_{kj} = 0$ for $j \neq k$ and $b_{kk} = -\sum_{l \neq k} f_{kl} y_l$. Thus $b_{kk} x_k = 0$. Because of $f_{kl} > 0$ and $(\mathsf{e}|y) = 1$ we have $b_{kk} \neq 0$, and therefore $x_k = 0$. In this way $B(y)x = 0$ reduces to $\hat{B}(\hat{y})\hat{x} = 0$, where the $(N-1)\times(N-1)$-matrix $\hat{B}(\hat{y})$ results from deleting the k-th row and the k-th column of $B(y)$, and $\hat{\xi} = (\xi_1, \dots, \xi_{k-1}, \xi_{k+1}, \dots, \xi_N)^{\mathsf{T}}$ for $\xi \in \mathbb{R}^N$. Since $(\hat{\mathsf{e}}|\hat{y}) = 1$, the matrix $\hat{B}(\hat{y})$ has the same structure as $B(y)$ in (12.21). Hence, if other components $y_{k_2}, \dots, y_{k_m}$ of y vanish, we may argue as before to obtain $x_{k_2} = ... = x_{k_m} = 0$. In case $m = N-1$ we immediately obtain $x = 0$ since $x \in \mathbb{E}$. If $m < N-1$, the remaining components $\tilde{x}$ of x satisfy $\tilde{B}(\tilde{y})\tilde{x} = 0$, where $\tilde{B}(\tilde{y})$ is again as in (12.21), $(\tilde{\mathsf{e}}|\tilde{y}) = 1$ and the components of $\tilde{y}$ do not vanish. Since $(\tilde{\mathsf{e}}|\tilde{x}) = 0$, Lemma 12.3.1 applies to $\tilde{B}(\tilde{y})$ and shows that $\tilde{x} = 0$. Altogether, it follows that $x = 0$, hence $B(y)|_{\mathbb{E}}$ is injective. As $\mathbb{E}$ is finite dimensional we obtain the invertibility of $B(y)|_{\mathbb{E}}$ for all $y \in \mathbb{D}$. Since $B(y)$ depends continuously on y, we obtain an open neighbourhood U of $\mathbb{D}$ such that $B(y)|_{\mathbb{E}}$ is invertible for all $y \in U$.

Step 2. To investigate the structure of $A(y) = (B(y)|_{\mathbb{E}})^{-1}$ for $y \in U$ we introduce the matrix

$$D(y) = \begin{bmatrix} B(y) & y \\ \mathsf{e}^{\mathsf{T}} & 0 \end{bmatrix}.$$

We claim that $D(y)$ is invertible on $\mathbb{R}^{N+1}$. Indeed, for given $h \in \mathbb{R}^N$ and $\beta \in \mathbb{R}$ the solution $\begin{bmatrix} x \\ \alpha \end{bmatrix} \in \mathbb{R}^{N+1}$ of $D(y)\begin{bmatrix} x \\ \alpha \end{bmatrix} = \begin{bmatrix} h \\ \beta \end{bmatrix}$ is

$$x = (B(y)|_{\mathbb{E}})^{-1}(h - (\mathsf{e}|h)y) + \beta y, \quad \alpha = (\mathsf{e}|h),$$

for $y \in \mathbb{D}$. Now fix $h \in \mathbb{E}$. With $\beta = 0$, this yields a representation of $x = A(y)h$ in terms of $D(y)^{-1}$, i.e., $\begin{bmatrix} x \\ 0 \end{bmatrix} = D(y)^{-1}\begin{bmatrix} h \\ 0 \end{bmatrix}$. Let $D_i(y)$ be the matrix that results from replacing the i-th column of $D(y)$ by $\begin{bmatrix} h \\ 0 \end{bmatrix}$. Then $x_i = \frac{\det D_i(y)}{\det D(y)}$ for $i = 1, \dots, N$ by Cramer's rule. Expanding $D_i(y)$ with respect to the i-th column, we obtain $\det D_i(y) = \sum_{j=1}^N (-1)^{i+j} h_j \det \hat{D}^{ji}(y)$, where $\hat{D}^{ji}(y)$ is the matrix that results from deleting the j-th row and the i-th column of $D(y)$. Now assume $j \neq i$. By (12.21), a row of $\hat{D}^{ji}(y)$ is given by $y_i(f_{i1}, \ldots, f_{i,i-1}, f_{i,i+1}, \ldots, f_{i,N-1}, 1)$. Expanding $\hat{D}^{ji}(y)$ with respect to this row, we obtain that $\det \hat{D}^{ji}(y)$ is a multiple of y_i. This yields the representation

$$x_i = -a_i^0(y)h_i + y_i \sum_{j\neq i} a_{ij}^1(y)h_j,$$

with coefficients analytic in $y \in U$. It remains to prove that $a_i^0(y) > 0$ for $y_i = 0$. In this case the structure of $B(y)$ yields $b_{ii}x_i = h_i$, where $b_{ii} = -\sum_{j\neq i} f_{ij}y_j < 0$ for y sufficiently close to $\mathbb{D}$. Hence $a_i^0(y) = -1/b_{ii} > 0$. We have thus shown that $x = (B(y)|_{\mathbb{E}})^{-1}h$ may be represented as asserted. □

We next investigate the spectrum of $A_0(y) = -A(y)P(y)M^{-1}$ in $\mathbb{E}$. To this end we employ the symmetrization $B_S(y) := Y^{-1/2}B(y)Y^{1/2}$ for $y \in \mathring{\mathbb{D}}$; note that $\sigma(B_S(y)) \subset (-\infty, 0]$ by Lemma 12.3.1. Kernel and range of $B_S(y)$ are given by $\mathsf{N}(B_S(y)) = \operatorname{span}\{y^{1/2}\}$ and $\mathsf{R}(B_S(y)) = \{y^{1/2}\}^{\perp}$, hence $B_S(y)$ is invertible on $\{y^{1/2}\}^{\perp}$.

Lemma 12.3.3. *Consider $A_0(y)$ as an element of $\mathcal{B}(\mathbb{E})$. Then there is an open neighbourhood $U \subset \mathbb{R}^N$ of $\mathbb{D}$ such that for all $y \in U$ the spectrum of $A_0(y)$ belongs to $\{\operatorname{Re} z > 0\}$.*

Proof. As $A_0(y)$ depends continuously on y, it suffices to show $\sigma_{\mathbb{E}}(A_0(y)) \subset (0, \infty)$ for $y \in \mathbb{D}$, since then we obtain $\sigma_{\mathbb{E}}(A_0(y)) \subset \{\operatorname{Re} z > 0\}$ for all y from a sufficiently small neighbourhood U of $\mathbb{D}$. Throughout, let λ be an eigenvalue of $A_0(y)$ with eigenvector $v \in \mathbb{E}$.

Step 1. Assume $y \in \mathring{\mathbb{D}}$. Then $P(y)M^{-1}v = -\lambda B(y)v$. Using that $Y^{-1} = Y^{-1}P(y) + (\cdot|\mathsf{e})\mathsf{e}$ and $(v|\mathsf{e}) = 0$, we get

$$0 < \big(v\big|Y^{-1}M^{-1}v\big) = \big(v\big|Y^{-1}P(y)M^{-1}v\big) = -\lambda\big(v\big|Y^{-1}B(y)v\big) = -\lambda(w|B_S(y)w),$$

where $w = Y^{-1/2}v$. Since $B_S(y)$ is negative semi-definte, we obtain $\lambda > 0$.

Step 2. Assume $y \in \mathbb{D}$ is such that $y_k = 0$ for some $1 \leq k \leq N$. We write

$$-\lambda B(y)v = P(y)M^{-1}v = M^{-1}v - (M^{-1}v|\mathsf{e})y. \tag{12.22}$$

By the structure of $B(y)$ from (12.21), here the k-th equation reads $-\lambda b_{kk}(y)v_k = M_k^{-1}v_k$, where $b_{kk}(y) < 0$. Hence we either have $\lambda > 0$ and are finished, or $v_k = 0$. In the latter case, the equation (12.22) reduces to

$$-\lambda\hat{B}(\hat{y})\hat{v} = \hat{M}^{-1}\hat{v} - (\hat{M}^{-1}\hat{v}|\hat{\mathsf{e}})\hat{y} = \hat{P}(\hat{y})\hat{M}^{-1}\hat{v},$$

where the hat means to delete the k-th row and the k-th column for a matrix and to delete the k-th entry for a vector. If y has no further vanishing components we are in the situation of Step 1 and conclude $\lambda > 0$. Otherwise, if $y_{k_2} = \ldots = y_{k_m} = 0$, we obtain inductively that either $\lambda > 0$ or $v_{k_2} = \ldots = v_{k_m} = 0$, where necessarily $m < N - 1$. In the latter case, as above we can reduce to the situation of Step 1, and $\lambda > 0$ follows. $\square$

For later purposes we investigate $-A(y)P(y)Y$ in more detail.

Lemma 12.3.4. *For $y \in \mathring{\mathbb{D}}$ the matrix $-A(y)P(y)Y$ is symmetric and positive semi-definite. The restriction $-A(y)P(y)Y|_{\mathbb{E}}$ is positive definite.*

Proof. To show the symmetry we let $P_{y^{1/2}} = I - (\cdot|y^{1/2})y^{1/2}$ be the orthogonal projection onto $\{y^{1/2}\}^\perp$. Observing that $A(y) = Y^{1/2}(B_S(y)|_{\mathbb{E}})^{-1}Y^{-1/2}$, $P(y)Y = YP(y)^\mathsf{T}$ and $P_{y^{1/2}} = Y^{-1/2}P(y)Y^{1/2}$, and recalling that the range of $(B_S(y)|_{\mathbb{E}})^{-1}$ equals $\{y^{1/2}\}^\perp$, for $v, w \in \mathbb{R}^N$ we calculate

$$\begin{aligned}
(A(y)P(y)Yv|w) &= \Big(Y^{1/2}(B_S(y)|_{\mathbb{E}})^{-1}Y^{1/2}P(y)^\mathsf{T}v\Big|w\Big)\\
&= \Big(P_{y^{1/2}}(B_S(y)|_{\mathbb{E}})^{-1}Y^{1/2}P(y)^\mathsf{T}v\Big|Y^{1/2}w\Big)\\
&= \Big(v\Big|P(y)Y^{1/2}(B_S(y)|_{\mathbb{E}})^{-1}P_{y^{1/2}}Y^{1/2}w\Big)\\
&= \big(v\big|A(y)P(y)Yw\big)\,.
\end{aligned}$$

The inclusion $\sigma(-A(y)P(y)Y|_{\mathbb{E}}) \subseteq (0,\infty)$ follows as in Step 1 of the proof of Lemma 12.3.3, replacing M^{-1} by Y. Hence $-A(y)P(y)Y|_{\mathbb{E}}$ is positive definite. Since $\mathbb{R}^N = \mathrm{span}\{\mathsf{e}\} \oplus \mathbb{E}$ and $\mathsf{e} \in \mathsf{N}(-A(y)P(y)Y)$, we see that $-A(y)P(y)Y$ is positive semi-definite. $\square$

3.2 Well-Posedness

The aim of this subsection is to show that there exists a unique solution of (12.16). For this purpose, let us first reformulate (12.16) in the abstract form

$$\varrho\dot{u} + A(u)u = F(u), \quad t > 0, \quad u(0) = u_0. \tag{12.23}$$

For this purpose, define

$$X_0 = L_p(\Omega;\mathbb{E}), \quad X_1 = \{u \in W_p^2(\Omega;\mathbb{E}) \mid \partial_\nu u = 0\}.$$

In the sequel we will assume that $p > n + 2$, wherefore the embedding $W_p^{2-2/p}(\Omega;\mathbb{E}) \hookrightarrow C^1(\overline{\Omega};\mathbb{E})$ is at our disposal. In this case one also has

$$W_p^{2\mu-2/p}(\Omega;\mathbb{E}) \hookrightarrow C^1(\overline{\Omega};\mathbb{E}),$$

provided that $\mu > \mu_0 := (n+2)/2p + 1/2$. Note that for $u \in W_p^{2\mu-2/p}(\Omega;\mathbb{E})$ with $\mu \in (\mu_0, 1]$, the Neumann trace $\partial_\nu u$ on $\partial\Omega$ exists. Therefore the trace space $X_{\gamma,\mu} = (X_0, X_1)_{\mu-1/p,p}$ is given by

$$X_{\gamma,\mu} = \{u \in W_p^{2\mu-2/p}(\Omega;\mathbb{E}) \,:\, \partial_\nu u = 0\}.$$

Let $U \subset \mathbb{R}^N$ be the open neighbourhood of $\mathbb{D}$ from Lemma 12.3.2, $\mathbb{V} = U \cap (\mathsf{e}/N + \mathbb{E})$ a relative open set in $\mathsf{e}/N + \mathbb{E}$ containing $\mathbb{D}$, and define

$$V_\mu = \{u \in X_{\gamma,\mu} \,:\, u(\overline{\Omega}) + \mathsf{e}/N \in \mathbb{V}\}.$$

Then V_μ is an open subset of $X_{\gamma,\mu}$, since $X_{\gamma,\mu} \hookrightarrow C(\overline{\Omega};\mathbb{E})$. For all $u \in V_\mu$ and all $v \in X_1$ we define the substitution operators $A\colon V_\mu \to \mathcal{B}(X_1, X_0)$ and $F\colon V_\mu \to X_0$ by

$$\begin{aligned} A(u)v(x) &= -\operatorname{div}(A_0(u(x)+\mathsf{e}/N)[\nabla v(x)]^{\mathsf{T}}) \\ &= -A_0(u(x)+\mathsf{e}/N)\Delta v(x) - \sum_{j=1}^{n}\left[\sum_{l=1}^{N}\partial_l A_0(u(x)+\mathsf{e}/N)\partial_j u_l(x)\right]\partial_j v(x), \end{aligned} \tag{12.24}$$

for $x \in \Omega$, and

$$F(u)(x) = Mr(u(x)+\mathsf{e}/N), \quad x \in \Omega.$$

It is easy to show $(A,F) \in C^1(V_\mu; \mathcal{B}(X_1,X_0)\times X_0)$, provided $r \in C^1$. The derivative of A is given by

$$\begin{aligned} [A'(u)h]v = &- [A_0'(u+\mathsf{e}/N)]h\Delta v \\ &- \sum_{j=1}^{n}\left[\sum_{l=1}^{N}\partial_j h_l \partial_l A_0(u+\mathsf{e}/N) + \partial_j u_l[\partial_l A_0'(u+\mathsf{e}/N)]h\right]\partial_j v, \end{aligned}$$

where $u \in V_\mu$, $v \in X_1$ and $h \in X_{\gamma,\mu}$.

To show that for each $u \in V_\mu$ the operator $A(u)$ has maximal regularity of type L_p, note that the principal part $A_\#(u(x)) = -A_0(u(x)+\frac{1}{N}\mathsf{e})\Delta$ is *normally elliptic* for each $u \in V_\mu$ and $x \in \overline{\Omega}$, i.e., $\sigma(A_0(u(x)+\frac{1}{N}\mathsf{e})) \subset \{\operatorname{Re} z > 0\}$. Furthermore, for each $u \in V_\mu$, the boundary operator ∂_ν satisfies the *Lopatinskii-Shapiro condition*. Therefore, Theorem 6.3.2 shows that for each $u \in V_\mu$, the operator $A(u)$ has maximal regularity of type L_p. We are now in a position to apply Theorem 5.1.1 which yields the following well-posedness result for (12.16).

Theorem 12.3.5. *Let $n \in \mathbb{N}$, $p > n+2$, $\mu \in (\mu_0, 1]$ and let $\Omega \subset \mathbb{R}^n$ be a bounded domain with boundary $\partial\Omega \in C^2$. Suppose that $r \in C^1(U;\mathbb{R}^N)$ and $(Mr(y)|\mathsf{e}) = 0$ for all $y \in \mathbb{V}$. Then the following assertions are valid.*

(a) *For each $y_0 \in W_p^{2\mu-2/p}(\Omega,\mathbb{R}^N)$ with $y_0(\overline{\Omega}) \subset \mathbb{V}$ and $\partial_\nu y_0 = 0$ at $\partial\Omega$, there exists $a > 0$ and a unique solution*

$$y \in H^1_{p,\mu}(J; L_p(\Omega,\mathbb{R}^N)) \cap L_{p,\mu}(J; H^2_p(\Omega,\mathbb{R}^N)) \cap C_b(J; W_p^{2\mu-2/p}(\Omega,\mathbb{R}^N)) \tag{12.25}$$

of (12.16) with $y(t,x) \in \mathbb{V}$ for all $(t,x) \in J \times \overline{\Omega}$, $J = [0,a]$.

(b) *Each local solution can be extended to a maximal solution defined on a maximal interval of existence $[0, t_+(y_0))$ and (12.25) holds for each $a \in (0, t_+(y_0))$. The mapping $y_0 \mapsto t_+(y_0)$ is lower semi-continuous and the mapping $y_0 \mapsto y(\cdot, y_0)$ is continuously Fréchet differentiable.*

(c) *With* $J_0 = (0, t_+(y_0))$, *we have*

$$y \in C^1(J_0; W_p^{2-2/p}(\Omega, \mathbb{R}^N)) \cap C^{2-1/p}(J_0; L_p(\Omega, \mathbb{R}^N)) \cap C^{1-1/p}(J_0; W_p^2(\Omega, \mathbb{R}^N)).$$

3.3 Classical Solutions

In the situation of the above theorem, let us show that the solution y of (12.16) is in fact classical, i.e.,

$$y \in C^1((0,a]; C(\overline{\Omega}, \mathbb{R}^N)) \cap C((0,a]; C^2(\overline{\Omega}, \mathbb{R}^N))$$

for each $a \in (0, t_+(y_0))$ if $\partial\Omega \in C^{2+\alpha}$ for some $\alpha > 0$. Theorem 12.3.5 already yields $y \in C^1((0,a]; C(\overline{\Omega}, \mathbb{R}^N))$, since $W_p^{2\mu-2/p}(\Omega)$ is embedded into $C(\overline{\Omega})$ whenever $p > n+2$ and $\mu \in (\mu_0, 1]$.

Therefore it remains to show that $y \in C((0,a]; C^2(\overline{\Omega}, \mathbb{R}^N))$. To this end, we write the equation for y in terms of $u = y - \mathsf{e}/N$ as $-\mathsf{A}(t,x)\Delta u(t,x) = g(t,x)$, where $\mathsf{A}(t,x) = A_0(u(t,x) + \mathsf{e}/N) = A_0(y(t,x))$ and

$$g(t,x) = \sum_{j=1}^{N} \left[\sum_{l=1}^{n} \partial_l A_0(y(t,x)) \partial_j u_l(t,x) \right] \partial_j u(t,x) - \varrho \partial_t u(t,x) + f(y(t,x)).$$

Clearly, by Theorem 12.3.5 and Sobolev's embedding, there exists $\alpha \in (0,1)$ such that $\mathsf{A} \in C^\alpha((0,a) \times \overline{\Omega}; \mathcal{B}(\mathbb{E}))$ and $g \in C^\alpha((0,a) \times \overline{\Omega}; \mathbb{E})$. Note that for fixed $t_* \in (0,a)$ the matrix $\mathsf{A}(t_*, x)$ is invertible for each $x \in \overline{\Omega}$. This yields the equation $-\Delta u(t_*, x) = \mathsf{A}(t_*, x)^{-1} g(t_*, x)$ complemented by the boundary condition $\partial_\nu u(t_*, x) = 0$ for $x \in \partial\Omega$. From now on we assume $\partial\Omega \in C^{2+\alpha}$. Then $u(t_*, \cdot) \in C^{2+\alpha}(\overline{\Omega}; \mathbb{E})$ and there exists a constant $C > 0$, which does not depend on $t_* \in (0,T)$, such that the estimate

$$|u(t_*, \cdot)|_{C^{2+\alpha}(\overline{\Omega};\mathbb{E})} \le C\big(|\mathsf{A}(t_*, \cdot)^{-1} g(t_*, \cdot)|_{C^\alpha(\overline{\Omega};\mathbb{E})} + |u(t_*, \cdot)|_{C^\alpha(\overline{\Omega};\mathbb{E})} \big)$$

is valid. Hence $u \in C((0,a); C^{2+\alpha}(\overline{\Omega}))$ and we have proven the following result.

Theorem 12.3.6. *Let the conditions of Theorem* 12.3.5 *be satisfied and assume that* $\partial\Omega \in C^{2+\alpha}$ *for some* $\alpha > 0$. *Then the unique solution of* (12.16) *is a classical solution.*

3.4 Positivity

We show the nonnegativity of solutions of (12.16), and the instantaneous strict positivity of components corresponding to nontrivial initial data. The argument heavily relies on the structure of the diffusion term $\mathrm{div}_x(A_0(y)[\nabla_x y]^\mathsf{T})$.

We consider this structure in more detail. Since $A_0(y) = -A(y)P(y)M^{-1}$ with $A(y) = (B(y)|_{\mathbb{E}})^{-1}$ from Lemma 12.3.2 and $P(y) = I - (\cdot|\mathsf{e})y$, the i-th component of $-\mathrm{div}_x(A_0(y)[\nabla_x y]^\mathsf{T})$ is given by

$$\sum_{\alpha=1}^{n} \sum_{j=1}^{N} a_{ij}(y)(M_j^{-1} \partial_{x_\alpha}^2 y_j - \partial_{x_\alpha}[(M^{-1}\partial_{x_\alpha} y|\mathsf{e}) y_j])$$

$$+ \partial_{x_\alpha}(a_{ij}(y))[M_j^{-1} \partial_{x_\alpha} y_j - (M^{-1} \partial_{x_\alpha} y|\mathsf{e}) y_j],$$

where $a_{ii}(y) = -a_i^0(y)$ and $a_{ij}(y) = y_i a_{ij}^1(y)$ for $j \neq i$. We collect the summands with $j = i$ from the first term, which results in $-M_i^{-1} a_i^0(y)\Delta y_i$. All the other summands contain either $\partial_{x_\alpha} y_i$ or y_i as a factor. Together with (12.27) we obtain that away from the initial time, a component y_i of a solution of (12.16) satisfies an equation of the form

$$\varrho\partial_t y_i - M_i^{-1} a_i^0(y)\Delta_x y_i + \sum_{\alpha=1}^{n} b_{i\alpha}^0(t,x)\partial_{x_\alpha} y_i + c_i^0(t,x) y_i = M_i r_i(y), \qquad (12.26)$$

with coefficients $b_{i\alpha}^0, c_i^0$ depending on the partial derivatives up to second order of y. We further write the i-th reaction term $M_i r_i$ as

$$M_i r_i(y) = -y_i L_i + h_i(y), \qquad (12.27)$$

where $L_i > 0$ is the Lipschitz constant of $M_i r_i$ on $\mathbb{D}$ and, with

$$\hat{y} = (y_1, ..., y_{i-1}, 0, y_{i+1}, ...y_N),$$

we have the decomposition

$$h_i(y) = M_i r_i(\hat{y}) + L_i y_i + M_i(r_i(y) - r_i(\hat{y})) \geq 0, \quad y \in \mathbb{D}.$$

Here $r_i(\hat{y}) \geq 0$ follows from the assumption that r is positivity preserving.

Combining (12.26) and (12.27), we arrive at

$$\varrho\partial_t y_i - M_i^{-1} a_i^0(y)\Delta_x y_i + \sum_{\alpha=1}^{n} b_{i\alpha}^0(t,x)\partial_{x_\alpha} y_i + (c_i^0(t,x) + L_i) y_i \geq 0, \qquad (12.28)$$

Since $a_i^0(y) > 0$ for $y_i = 0$, the left-hand side of (12.28) is *parabolic* for y_i close to zero, and the lower order coefficients $b_{i\alpha}^0, c_i^0$ are continuous if y is a classical solution. This puts us into a position to apply the *maximum principle* and *Hopf's lemma*, which is the key to the following result on nonnegativity and strict positivity.

Theorem 12.3.7. *Assume $r \in C^1(U, \mathbb{R}^N)$ is mass and positivity preserving on $\mathbb{D}$. Let $p > n+2$, $1 \geq \mu > (n+2)/p$, and $y_0 \in W_p^{2\mu-2/p}(\Omega; \mathbb{R}^N)$ with $y_0(\overline{\Omega}) \subset \mathbb{V}$ be given. Denote by y the corresponding unique solution of* (12.16)*. Then the following holds true.*

(a) *If $y_0 \geq 0$, then $y(t,x) \geq 0$ for all $t \in (0, t_+(y_0))$ and $x \in \overline{\Omega}$*

(b) *If $y_0 \geq 0$ and $y_0^i \neq 0$, then $y_i(t,x) > 0$ for all $t \in (0, t_+(y_0))$ and $x \in \overline{\Omega}$.*

Proof. Step 1. Let $y_0 \geq 0$. For $\varepsilon > 0$ we consider the modified system

$$\varrho\partial_t y^\varepsilon + \mathrm{div}_x(A_0(y^\varepsilon)[\nabla_x y^\varepsilon]^\mathsf{T}) = Mr^\varepsilon, \quad \partial_\nu y^\varepsilon = 0, \quad y^\varepsilon(0) = y_0^\varepsilon, \qquad (12.29)$$

with reaction terms $Mr^\varepsilon = Mr+\varepsilon(\mathsf{e}-Ny^\varepsilon)$ and initial data $y_0^\varepsilon = y_0+\varepsilon(\mathsf{e}-Ny_0)$. Observe that $(\mathsf{e}|Mr^\varepsilon) = 0$ for $(\mathsf{e}|y^\varepsilon) = 1$, that $(\mathsf{e}|y_0^\varepsilon) = 1$ and that y_0^ε has strictly positive components for all sufficiently small ε. Thus (12.29) has a unique maximal classical solution y^ε by Theorem 12.3.5.

Fixing $\varepsilon > 0$, we claim that $y^\varepsilon(t,x) > 0$ for all $t \in [0,t_+(y_0^\varepsilon))$ and $x \in \overline{\Omega}$. Assume the contrary, i.e., there are $t_0 \in (0,t_+(y_0^\varepsilon))$ and $x_0 \in \overline{\Omega}$ such that $y_i^\varepsilon(t_0,x_0) = 0$ for a component y_i and $y_j^\varepsilon(t,\cdot) > 0$ on $\overline{\Omega}$ for all $j = 1,...,N$ and $t \in [0,t_0)$. Note that necessarily $t_0 > 0$ since $y_0^\varepsilon > 0$. First suppose that $x_0 \in \Omega$. Then $\partial_t y_i^\varepsilon(t_0,x_0) \le 0$, $\nabla_x y_i^\varepsilon(t_0,x_0) = 0$ and $\Delta_x y_i^\varepsilon(t_0,x_0) \ge 0$. Further, $(Mr^\varepsilon)_i = M_i r_i + \varepsilon(1-Ny_i^\varepsilon) \ge \varepsilon$ at (t_0,x_0) since $r_i \ge 0$ for $y_i = 0$. Therefore, (12.26) yields

$$\varrho\partial_t y_i^\varepsilon(t_0,x_0) - M_i^{-1}a_i^0(y^\varepsilon(t_0,x_0))\Delta_x y_i^\varepsilon(t_0,x_0) \ge \varepsilon,$$

a contradiction. Suppose next that $x_0 \in \partial\Omega$. Then $\partial_\nu y_i(t_0,x_0) \le 0$. On the other hand, (12.28) implies that there is $\eta > 0$ such that y_i^ε is a supersolution of a linear parabolic equation in $(t_0-\eta,t_0]\times V$, where $V \subset \Omega$ is a sufficiently small open ball with $x_0 \in \partial V$. The previous considerations show that $y_i^\varepsilon > 0$ in $(t_0-\eta,t_0]\times V$. Hence $\partial_\nu y_i(t_0,x_0) > 0$ by Hopf's lemma leads to a contradiction. We conclude that $y^\varepsilon > 0$ on $(0,t_+(y_0^\varepsilon))\times\overline{\Omega}$.

Given $T \in (0,t_+(y_0))$, we obtain $y^\varepsilon \to y$ as $\varepsilon \to 0$ in the topology of $C([0,T];W_p^{2\mu-2/p}(\Omega,\mathbb{R}^N))$ from Theorem 12.3.5, and thus uniformly on $[0,T]\times\overline{\Omega}$. Hence $y \ge 0$ on $[0,t_+(y_0))\times\overline{\Omega}$.

Step 2. We prove Part b) and assume additionally that $y_i \neq 0$. From Step 1 we know $y_i \ge 0$. For $t \in (0,t_+(y_0))$ we consider

$$\Omega_t^+ = \{x \in \Omega \,:\, y_i(t,x) > 0\}.$$

We are going to show that Ω_t^+ is nonempty, open and closed in Ω. Clearly, Ω_t^+ is open in Ω. To obtain $\Omega_t^+ \neq \emptyset$, let t_0 be the smallest time such that $\Omega_{t_0}^+ = \emptyset$, i.e., $y_i(t_0,\cdot) = 0$ on Ω. Note that $t_0 > 0$ by the assumption $y_0^i \neq 0$. Then the left-hand side of (12.28) is parabolic in $(t_0-\eta,t_0]\times\Omega$ for small η. Since y_i attains its minimum zero everywhere on $\{t_0\}\times\Omega$, the strong maximum principle yields $y_i(t_0-\eta,\cdot) = 0$. But this is a contradiction to the definition of t_0, and therefore $\Omega_t^+ \neq \emptyset$ for all t.

We finally show that Ω_t^+ is closed in Ω. Let $x_k \in \Omega_t^+$ be a sequence such that $x_k \to x_0 \in \Omega$ as $k \to \infty$. Assume $x_0 \notin \Omega_t^+$, i.e., $y_i(t,x_0) = 0$. Then there are $\eta > 0$ and a convex open set $V \subset \Omega$ containing x_0 such that (12.26) is parabolic on $(t-\eta,t]\times V$. As above, by the strong maximum priniciple, $y_i(t,\cdot) = 0$ on V. Hence $y_i(t,x_k) = 0$ for all sufficiently large k, which contradicts the assumption $x_k \in \Omega_t^+$.

We conclude that $y_i > 0$ on $(0,t_+(y_0))\times\Omega$. Arguing as in the previous step by contradiction and Hopf's lemma, we get that $y_i > 0$ on $(0,t_+(y_0))\times\overline{\Omega}$. □

3.5 Reversible Mass-Action Kinetics

For reversible mass-action kinetics the reaction rate r is given by

$$r(y) = \boldsymbol{\nu}\mathbf{r}(\varrho M^{-1}y) = \sum_{l=1}^{m} \boldsymbol{\nu}_l \mathbf{r}_l(\varrho M^{-1}y), \tag{12.30}$$

where $\boldsymbol{\nu} = [\boldsymbol{\nu}_{jl}^+ - \boldsymbol{\nu}_{jl}^-] \in \mathbb{Z}^{N\times m}$ is for $\boldsymbol{\nu}_{jl}^+, \boldsymbol{\nu}_{jl}^- \in \mathbb{N}_0$ the stoechiometric matrix of the reactions, and $\boldsymbol{\nu}_l \in \mathbb{Z}^N$ denotes the l-th column of $\boldsymbol{\nu}$. The vector $\mathbf{r} = (\mathbf{r}_1, \ldots, \mathbf{r}_m)$ of single reactions is in terms of the concentrations c given by

$$\mathbf{r}_l(c) = -k_l^+ c^{\boldsymbol{\nu}_l^+} + k_-^l c^{\boldsymbol{\nu}_l^-}, \quad l = 1, ..., m,$$

where $\boldsymbol{\nu}_l^+ = [\boldsymbol{\nu}_{jl}^+]$ and $\boldsymbol{\nu}_l^- = [\boldsymbol{\nu}_{jl}^-]$, such that $\boldsymbol{\nu}_l = \boldsymbol{\nu}_l^+ - \boldsymbol{\nu}_l^-$ for a column of $\boldsymbol{\nu}$. Here we use multi-index notation, i.e., $c^{\boldsymbol{\nu}_l^+} = c_1^{\boldsymbol{\nu}_{1l}^+} \cdot \ldots \cdot c_N^{\boldsymbol{\nu}_{Nl}^+}$, and analogous for $c^{\boldsymbol{\nu}_l^-}$. The *stoechiometric subspace* $\mathbb{S}$ of $\mathbb{R}^N$ is defined by

$$\mathbb{S} = \mathsf{R}(\boldsymbol{\nu}), \quad s = \dim \mathbb{S}.$$

We assume that the columns $\boldsymbol{\nu}_l$ of $\boldsymbol{\nu}$ are ordered such that $\boldsymbol{\nu}_1, \ldots, \boldsymbol{\nu}_s$ are linearly independent, i.e.,

$$\mathbb{S} = \operatorname{span}\{\boldsymbol{\nu}_1, \ldots, \boldsymbol{\nu}_s\}.$$

Throughout we make the following assumptions:

$$\mathbf{r}(\varrho M^{-1}\mathbf{y}_*) = 0 \text{ for some } \mathbf{y}_* \in \mathring{\mathbb{D}}, \quad M\mathsf{e} \in \mathbb{S}^\perp, \quad k_l^+, k_l^- > 0. \tag{\textbf{R}}$$

Observe that the second condition implies $(\mathsf{e}|Mr(y)) = 0$ for each y, i.e., conservation of mass. It also implies that $s < N$. The strict positivity of k_l^+, k_l^- means that each elementary reaction $\mathbf{r}_l$ is reversible. It is straightforward to check that mass-action kinetics r as above are positivity preserving. The first assumption in (**R**) states that the set

$$\mathcal{E} = \{y_* \in \mathring{\mathbb{D}} : \mathbf{r}(\varrho M^{-1}y_*) = 0\}$$

of chemical equilibria, i.e., where all single reactions $\mathbf{r}_l$ vanish, is nonemtpy. This can be characterized as follows. Obeserve that $y_* \in \mathcal{E}$ if and only if for $c_* = \varrho M^{-1}y_*$ we have

$$c_*^{\boldsymbol{\nu}_l} = \frac{c_*^{\boldsymbol{\nu}_l^+}}{c_*^{\boldsymbol{\nu}_l^-}} = \frac{k_l^-}{k_l^+} =: K_l, \quad l = 1, \ldots, m. \tag{12.31}$$

Since $k_l^+, k_l^- > 0$, (12.31) is equivalent to

$$(\boldsymbol{\nu}_l|\log c_*) = \log K_l, \quad l = 1, ..., m, \tag{12.32}$$

where we write $\log \xi = (\log \xi_1, \ldots, \log \xi_N)^\mathsf{T}$ for a vector $\xi \in \mathbb{R}^N$. By the above assumption, for $l = 1, ..., s$ the equations in (12.32) can always be satisfied. For the remaining equations we note that there are $\alpha_{lk} \in \mathbb{R}$ such that $\boldsymbol{\nu}_l = \sum_{k=1}^{s} \alpha_{lk} \boldsymbol{\nu}_k$ for $l = s+1, \ldots, m$. Thus $\mathcal{E} \neq \emptyset$ if and only if

$$\prod_{k=1}^{s} K_k^{\alpha_{lk}} = K_l, \quad l = s+1, ..., m.$$

In particular we have $\mathcal{E} \neq \emptyset$ if $s = m < N$.

In the sequel we fix an arbitrary chemical equilibrium

$$\mathbf{y}_* \in \mathcal{E},$$

and consider the chemical potentials

$$\mu_k(y) = \frac{1}{M_k} \log(y_k/\mathbf{y}_*^k), \quad y \in \mathring{\mathbb{D}}, \quad k = 1, \ldots, N.$$

We will also write

$$\mu = (\mu_1, \ldots, \mu_N)^\mathsf{T} = M^{-1} \log(y/\mathbf{y}_*).$$

Lemma 12.3.8. *Assume* (**R**). *Then the following holds true.*

(a) *We have $(\mu(y)|Mr(y)) \leq 0$ for all $y \in \mathring{\mathbb{D}}$, with equality iff $y \in \mathcal{E}$.*

(b) *The set $\mathcal{E}$ forms an $(N-s-1)$-dimensional smooth submanifold of $\mathbb{R}^N$. At $y_* \in \mathcal{E}$, the tangent space is given by $T_{y_*}\mathcal{E} = \mathsf{N}(\boldsymbol{\nu}^\mathsf{T} Y_*^{-1}) \cap \mathbb{E}$.*

Proof. **(a)** Writing $c = \varrho M^{-1} y$ for $y \in \mathring{\mathbb{D}}$, and $\mathbf{c}_* = \varrho M^{-1} \mathbf{y}_*$, we have $\mu(y) = M^{-1} \log(c/\mathbf{c}_*)$. Using (12.31), we calculate

$$\begin{aligned}\sum_{k=1}^{N} (\mu_k(y)|M_k r_k(y)) &= \sum_{l=1}^{m} (\log(c/\mathbf{c}_*)|\boldsymbol{\nu}_l) \left(-k_l^+ c^{\boldsymbol{\nu}_l^+} + k_l^- c^{\boldsymbol{\nu}_l^-}\right) \\ &= -\sum_{l=1}^{m} \log[(c/\mathbf{c}_*)^{\boldsymbol{\nu}_l}] k_l^+ c^{\boldsymbol{\nu}_l^-} \mathbf{c}_*^{\boldsymbol{\nu}_l} \left(\frac{c^{\boldsymbol{\nu}_l}}{\mathbf{c}_*^{\boldsymbol{\nu}_l}} - 1\right). \end{aligned} \tag{12.33}$$

Since $k_l^+ c^{\boldsymbol{\nu}_l^-} \mathbf{c}_*^{\boldsymbol{\nu}} > 0$ and $(\log \xi)(\xi - 1) \leq 0$ for all $\xi > 0$, we obtain $(\mu(y)|Mr(y)) \leq 0$. Since each summand in (12.33) is nonpositive, we have $(\mu(y)|Mr(y)) = 0$ if and only $c^{\boldsymbol{\nu}_l} = \mathbf{c}_*^{\boldsymbol{\nu}_l} = K_l$ for each l, also using (12.31). This implies $y \in \mathcal{E}$.

(b) Above we have seen that the assumption $\mathcal{E} \neq \emptyset$ implies that $y_* \in \mathcal{E}$ if and only if $(\boldsymbol{\nu}_l | \log c_*) = \log K_l$ for $l = 1, \ldots s$, i.e., the other $m - s - 1$ conditions in (12.32) are redundant. As a consequence, $\mathbf{r}(y_*) = 0$ is equivalent to $\mathbf{r}_1(y_*), \ldots, \mathbf{r}_s(y_*) = 0$.

Define the map

$$F(y) = [\mathbf{r}_1(\varrho M^{-1} y), \ldots, \mathbf{r}_s(\varrho M^{-1} y), (y|\mathsf{e}) - 1]^\mathsf{T}, \quad y \in \mathring{\mathbb{D}}.$$

Note that $F(y_*) = 0$ if and only if $y_* \in \mathcal{E}$. We show that $F'(y_*)$ has full rank $s+1$ at each $y_* \in \mathcal{E}$. To this end we calculate, writing $c_* = \varrho M^{-1} y_*$ and using (12.31),

$$\partial_{y_j} \mathbf{r}_l(\varrho M^{-1} y_*) = \varrho M_j^{-1}(-k_l^+ \boldsymbol{\nu}_{jl}^+ c_*^{\boldsymbol{\nu}_l^+} + k_l^- \boldsymbol{\nu}_{jl}^- c_*^{\boldsymbol{\nu}_l^-})/c_*^j = -k_l^- \boldsymbol{\nu}_{jl}/y_*^j.$$

Therefore, writing $\hat{\boldsymbol{\nu}} = (\boldsymbol{\nu}_1, \ldots, \boldsymbol{\nu}_s) \in \mathbb{Z}^{N\times s}$,

$$F'(y_*) = \begin{bmatrix} -\mathrm{diag}(k_l^-)\hat{\boldsymbol{\nu}}^{\mathsf{T}} Y_*^{-1} \\ \mathsf{e}^{\mathsf{T}} \end{bmatrix}.$$

Since $k_l^- > 0$ and $\hat{\boldsymbol{\nu}}^{\mathsf{T}}$ is surjective, the matrix $-\mathrm{diag}(k_l^-)\hat{\boldsymbol{\nu}}^{\mathsf{T}} Y_*^{-1}$ has full rank s. We further claim that e is linearly independent of the rows of $-\mathrm{diag}(k_l^-)\hat{\boldsymbol{\nu}}^{\mathsf{T}} Y_*^{-1}$. If this were not the case, we find numbers λ_l such that $\mathsf{e} = \sum_{l=1}^s \lambda_l Y_*^{-1} \boldsymbol{\nu}_l$. Taking the scalar product with $MY_*\mathsf{e}$, we obtain from $M\mathsf{e} \in \mathbb{S}^\perp$ that

$$0 \neq (MY_*\mathsf{e}|\mathsf{e}) = \sum_{l=1}^s \lambda_l (\boldsymbol{\nu}_l | M\mathsf{e}) = 0,$$

a contradiction. Hence $F'(y_*)$ has full rank $s+1$, which implies that $\mathcal{E}$ is a smooth manifold of dimension $N - s - 1$. The tangent space is given by

$$\mathbb{N}(F'(y_*)) = \mathbb{N}(\hat{\boldsymbol{\nu}}^{\mathsf{T}} Y_*^{-1}) \cap \mathbb{E} = \mathbb{N}(\boldsymbol{\nu}^{\mathsf{T}} Y_*^{-1}) \cap \mathbb{E},$$

which completes the proof. □

3.6 Total Free Energy

For reversible mass-action kinetics as in (12.30), in this subsection we show that all equilibria of (12.16) are stable spatially homogeneous kinetic equilibria. Throughout we fix $p > n+2$ and set

$$\mathcal{X} = \{y_0 \in W_p^{2-2/p}(\Omega, \mathbb{R}^N) \,:\, y_0(\overline{\Omega}) \subset \mathring{\mathbb{D}},\ \partial_\nu y_0 = 0\}.$$

Then Theorem 12.3.5 yields that (12.16) is well-posed on $\mathcal{X}$. The key quantity to the stability results is the total free energy Ψ, which is given by

$$\Psi(y) = \int_\Omega \psi(y)\, dx, \quad \psi(y) = \sum_{k=1}^N \frac{y_k}{M_k}(\log(y_k/\mathbf{y}_*^k) - 1).$$

Observe that $\mu_k = \partial_{y_k}\psi$ for the chemical potentials. The free energy serves as a strict Lyapunov function for (12.16), and further allows us to show that all equilibria of (12.16) are spatially homogeneous.

Lemma 12.3.9. *Assume* (**R**). *Then the following holds true.*

(a) *The total free energy Ψ is a strict Lyapunov function for* (12.16) *on $\mathcal{X}$.*

(b) *Each equilibrium of* (12.16) *is spatially homogeneous, such that the set of equilibria of* (12.16) *in* $\mathcal{X}$ *is given by (the constant functions in)* $\mathcal{E}$.

Proof. Step 1. Since $\mathcal{X} \hookrightarrow C(\overline{\Omega};\mathbb{E})$, it is clear that Ψ is continuous on $\mathcal{X}$. For initial data $y_0 \in \mathcal{X}$, the corresponding maximal solution y of (12.16) is classical and has strictly positive components. We may thus differentiate $\psi(y)$ with respect to $t \in (0, t_+(y_0))$ and use (12.16) to the result

$$\varrho\partial_t\psi(y) = \sum_{k=1}^N \mu_k \varrho\partial_t y = -\sum_{k=1}^N \operatorname{div}_x(\mu_k J_k) + \sum_{k=1}^N (\nabla_x\mu_k|J_k) + (\mu|Mr(y)).$$

Here the first summand vanishes after integration over Ω due to the boundary conditions $(\nu|J_k) = 0$. Therefore

$$\varrho\partial_t\Psi(y) = \int_\Omega \sum_{k=1}^N (\nabla_x\mu_k(y)|J_k)\, dx + \int_\Omega (\mu(y)|Mr(y))\, dx. \tag{12.34}$$

We prove that the *integrands* on the right-hand side in (12.34) are negative. For the second integrand, this is a consequence of Lemma 12.3.8. For the first integrand in (12.34) we write

$$\sum_{k=1}^N (\nabla_x\mu_k(y)|J_k) = \sum_{\alpha=1}^n (\partial_{x_\alpha}\mu(y)|J^\alpha)\,.$$

For fixed α we calculate, using $P(y)J^\alpha = J^\alpha$, $YP(y)^\mathsf{T} = P(y)Y$ and $P(y)Y\partial_{x_\alpha}\mu = B(y)J^\alpha$ by (12.21),

$$(\partial_{x_\alpha}\mu(y)|J^\alpha) = (P(y)Y\partial_{x_\alpha}\mu(y)|Y^{-1}J^\alpha) = (B_S(y)Y^{-1/2}J^\alpha|Y^{-1/2}J^\alpha).$$

Since $B_S(y)$ is negative definite on $Y^{-1/2}\mathbb{E}$, it follows that $(\partial_{x_\alpha}\mu(y)|J^\alpha) \le 0$, and that this term vanishes if and only if $J^\alpha = 0$. Hence Ψ decreases along solutions of (12.16).

Step 2. Assume Ψ is not strictly decreasing along a solution y. Then there is $t_* \in (0, t_+(y_0))$ such that $\partial_t\Psi(y(t_*)) = 0$. Since both integrands in (12.34) are nonpositive, we obtain that

$$\sum_{k=1}^N (\nabla_x\mu_k(y(t_*))|J_k) = 0, \quad (\mu(y(t_*))|Mr(y(t_*))) = 0. \tag{12.35}$$

The first identity and the considerations in Step 1 show that $(\partial_{x_\alpha}\mu(y(t_*))|J^\alpha) = 0$, and therefore $J^\alpha = 0$ for each $\alpha = 1,\dots,n$. Hence $\nabla_x y(t_*) = 0$ by (12.21) and $y(t_*)$ is spatially homogeneous. The second identity in (12.35) and Lemma 12.3.8 imply $y(t_*) \in \mathcal{E}$. Thus y is a kinetic equilibrium. This proves that Ψ is a strict Lyapunov function for (12.16).

Step 3. To show b), we note that for any equilibrium y_* of (12.16) we have $\partial_t\Psi(y_*) = 0$. Thus (12.34) and the same arguments as in the previous step show that y_* is homogeneous and $y_* \in \mathcal{E}$. □

3.7 Stability and Long-Time Behaviour
We prove stability for the equilibria of (12.16).

Theorem 12.3.10. *Assume* (**R**), $p > n+2$. *Then any equilibrium* $y_* \in \mathcal{E}$ *of* (12.16) *is stable in the* $W_p^{2-2/p}$*-norm. Moreover, for each* $y_* \in \mathcal{E}$ *there is* $\varepsilon > 0$ *such that if*

$$|y_0 - y_*|_{W_p^{2-2/p}(\Omega;\mathbb{R}^N)} \le \varepsilon$$

for some $y_0 \in \mathcal{X}$, *then the solution of* (12.16) *corresponding to* y_0 *exists globally in time and converges at an exponential rate to some* $y_\infty \in \mathcal{E}$, *with respect to the* $W_p^{2-2/p}(\Omega;\mathbb{R}^N)$*-norm.*

Proof. Fix $y_* \in \mathcal{E}$. To prove the assertions for y_* we apply Theorem 5.3.1. To this end we verify the conditions (i)-(iv) from Theorem 5.3.1 concerning the linearized operator $A_*u = A(u_*)u + [A'(u_*)u]u_* - F'(u_*)u$ with domain $X_1 = \{u \in W_p^2(\Omega;\mathbb{E}) : \partial_\nu u = 0\}$. Since $\nabla u_* = 0$, from (12.24) we see that $A(u_*) = -A_0(y_*)\Delta$, and $[A'(u_*)h]u_* = 0$. Therefore

$$A_* = -A_0(y_*)\Delta + Mr'(y_*).$$

In Step 2 of the proof of Lemma 12.3.8 it was shown that

$$r'(y_*) = \boldsymbol{\nu}\nabla_y \mathbf{r}(\varrho M^{-1}y_*) = -\boldsymbol{\nu}K\boldsymbol{\nu}^{\mathsf{T}}Y_*^{-1},$$

where $K = \mathrm{diag}(k_l^-)$. The key observation is the following identity. For an eigenvalue λ of A_0 with eigenfunction $u \in X_1$ we have

$$\begin{aligned}\lambda(u|Y_*^{-1}M^{-1}u)_\Omega &= (A_*u|Y_*^{-1}M^{-1}u)_\Omega \qquad (12.36)\\ &= \sum_{\alpha=1}^n (A(y_*)P(y_*)M^{-1}\partial_{x_\alpha}^2 u|Y_*^{-1}M^{-1}u)_\Omega - (M\boldsymbol{\nu}K\boldsymbol{\nu}^{\mathsf{T}}Y_*^{-1}u|Y_*^{-1}M^{-1}u)_\Omega\\ &= -\sum_{\alpha=1}^n (A(y_*)P(y_*)Y_*Y_*^{-1}M^{-1}\partial_{x_\alpha}u|Y_*^{-1}M^{-1}\partial_{x_\alpha}u)_\Omega - (K\boldsymbol{\nu}^{\mathsf{T}}Y_*^{-1}u|\boldsymbol{\nu}^{\mathsf{T}}Y_*^{-1}u)_\Omega.\end{aligned}$$

We now verify the conditions (i)-(iv) from Theorem 5.3.1.

(i)+(ii) By the Lemmas 12.3.8 and 12.3.9, the set of equilibria of (12.23) in $\mathcal{X}_p$ is given by $\mathcal{E} - \frac{1}{N}\mathsf{e}$ and forms a smooth manifold. The tangent space at u_* is given by $\mathsf{N}(\boldsymbol{\nu}^{\mathsf{T}}Y_*^{-1})$. We show that $\mathsf{N}(A_*) = \mathsf{N}(\boldsymbol{\nu}^{\mathsf{T}}Y_*^{-1})$. Let $u \in \mathsf{N}(A_*)$. Since $A(y_*)P(y_*)Y_*$ is positive semi-definite by Lemma 12.3.4 and K is diagonal with positive entries, (12.36) with $\lambda = 0$ yields

$$\sum_{\alpha=1}^n (A(y_*)P(y_*)Y_*Y_*^{-1}M^{-1}\partial_{x_\alpha}u|Y_*^{-1}M^{-1}\partial_{x_\alpha}u)_\Omega = (K\boldsymbol{\nu}^{\mathsf{T}}Y_*^{-1}u|\boldsymbol{\nu}^{\mathsf{T}}Y_*^{-1}u)_\Omega = 0.$$

Observe that $Y_*^{-1}M^{-1}\partial_{x_\alpha}u(x) \notin \mathsf{N}(-A(y_*)P(y_*)Y_*) = \mathrm{span}\{\mathsf{e}\}$ because of $\partial_{x_\alpha}u(x) \in \mathbb{E}$ for all $x \in \Omega$. Thus the positive definiteness of $-A(y_*)P(y_*)Y_*|_{\mathbb{E}}$

yields $Y_*^{-1}M^{-1}\partial_{x_\alpha}u = 0$ for all α, hence u is constant. Moreover, the second identity implies that $\boldsymbol{\nu}^{\mathsf{T}}Y_*^{-1}u = 0$. This shows $\mathsf{N}(A_*) = \mathsf{N}(\boldsymbol{\nu}^{\mathsf{T}}Y_*^{-1}) \cap \mathbb{E}$.

(iii) We show that zero is a semi-simple eigenvalue of A_*. For $u \in \mathsf{N}(A_*^2)$ the identity $\mathsf{N}(A_*) = \mathsf{N}(\boldsymbol{\nu}^{\mathsf{T}}Y_*^{-1}) \cap \mathbb{E}$ implies that $v = A_*u$ is constant and satisfies $\boldsymbol{\nu}^{\mathsf{T}}Y_*^{-1}v = 0$. We therefore have

$$(v|M^{-1}Y_*^{-1}v) = -(M\boldsymbol{\nu}K\boldsymbol{\nu}^{\mathsf{T}}Y_*^{-1}u|M^{-1}Y_*^{-1}v) = -(K\boldsymbol{\nu}^{\mathsf{T}}Y_*^{-1}u|\boldsymbol{\nu}^{\mathsf{T}}Y_*^{-1}v) = 0,$$

which implies $v = 0$ and thus $u \in \mathbb{N}(A_*)$.

(iv) We finally show that $\sigma(A_*) \setminus \{0\}$ is strictly contained in $\{\operatorname{Re} z > 0\}$, i.e., each eigenvalue $\lambda \neq 0$ is positive. But this is a consequence of (12.36) and the positive (semi-) definiteness of $-A(y_*)P(y_*)Y_*$ and K. □

We end this section with a result on the convergence of solutions to equilibria.

Theorem 12.3.11. *Assume* (**R**), $p > n+2$, $1 \geq \mu > (n+2)/p$, *and let* $y_0 \in \mathcal{X}$. *Let* y *be the solution of* (12.16) *corresponding to* y_0. *Then the following holds true.*

(a) *If* $\sup_{t\in(0,t_+(u_0))} |y(t,\cdot)|_{W_p^{2\mu-2/p}(\Omega;\mathbb{R}^N)} < \infty$, *then* y *is a global solution.*

(b) *Suppose additionally that, for some* $t_0 > 0$,

$$\inf_{t>t_0,\, x\in\overline{\Omega}} y(t,x) > 0.$$

Then, as $t \to \infty$, y *converges exponentially fast in the* $W_p^{2-2/p}(\Omega;\mathbb{R}^N)$*-norm to a constant chemical equilibrium* $y_* \in \mathcal{E}$ *of* (12.16). *If* $r \equiv 0$, *then* $y_* = \frac{1}{|\Omega|}\int_\Omega y_0\,dx$.

Proof. Lemma 12.3.9 shows that Ψ is a strict Lyapunov function on $\mathcal{X}$. Now the assertions are a consequence of the Theorem 5.7.2. In case $r \equiv 0$, conservation of mass for each component yields immediately $y_* = \frac{1}{|\Omega|}\int_\Omega y_0\,dx$. □

12.4 Stefan Problems with Surface Heat Capacity

(i) Consider once more Problem **(P5)**. Set $u \equiv 0$, $\varrho = 1$, and ignore balance of momentum, i.e., the equations for u. This results in a Stefan problem with variable surface tension which reads

$$\left\{\begin{aligned} \kappa(\theta)\partial_t\theta - \operatorname{div}(d(\theta)\nabla\theta) &= 0 && \text{in } \Omega\setminus\Gamma(t)\\ \partial_\nu\theta &= 0 && \text{on } \partial\Omega\\ [\![\theta]\!] = 0,\ \theta_\Gamma &= \theta && \text{on } \Gamma(t)\\ \kappa_\Gamma(\theta_\Gamma)\partial_{t,n}\theta_\Gamma - \operatorname{div}_\Gamma(d_\Gamma(\theta_\Gamma)\nabla_\Gamma\theta_\Gamma) &= \\ = [\![d(\theta)\partial_\nu\theta]\!] - (l(\theta_\Gamma) + l_\Gamma(\theta_\Gamma)H_\Gamma &- \gamma(\theta_\Gamma)V_\Gamma)V_\Gamma && \text{on } \Gamma(t)\\ \varphi(\theta_\Gamma) + \sigma(\theta_\Gamma)H_\Gamma &= \gamma(\theta_\Gamma)V_\Gamma && \text{on } \Gamma(t)\\ \theta(0) &= \theta_0 && \text{in } \Omega,\\ \Gamma(0) &= \Gamma_0, \end{aligned}\right. \tag{12.37}$$

where kinetic undercooling has been included, with $\gamma(\theta_\Gamma)$ the coefficient of kinetic undercooling. Here $\varphi(\theta) = [\![\psi(\theta)]\!]$ and $\frac{D_n}{Dt}\theta_\Gamma$ denotes the time derivative of θ_Γ in normal direction, defined by

$$\frac{D_n}{Dt}\theta_\Gamma(t,p) := \frac{d}{d\tau}\theta_\Gamma(t+\tau, x(t+\tau,p))\big|_{\tau=0}, \quad t>0, \quad p\in\Gamma(t), \qquad (12.38)$$

with $\{x(t+\tau,p)\in\mathbb{R}^n : (\tau,p)\in(-\varepsilon,\varepsilon)\times\Gamma(t)\}$ the flow induced by the normal vector field $(V_\Gamma\nu_\Gamma)$, i.e., for each $p\in\Gamma(t)$, $x(t+\tau,p)$ is a flow line through p with

$$\frac{d}{d\tau}x(t+\tau,p) = (V_\Gamma\nu_\Gamma)(t+\tau, x(t+\tau,p)), \quad x(t+\tau,p)\in\Gamma(t+\tau), \quad \tau\in(-\varepsilon,\varepsilon),$$

and $x(t,p)=p$; recall Section 2.5 for more details.

The system (12.37) is also *thermodynamically consistent*, as the *total energy* E given by

$$\mathsf{E}(\theta,\theta_\Gamma,\Gamma) = \int_{\Omega\setminus\Gamma}\epsilon(\theta)\,dx + \int_\Gamma \epsilon_\Gamma(\theta_\Gamma)\,d\Gamma, \qquad (12.39)$$

is preserved along smooth solutions, and the *total entropy* N,

$$\mathsf{N}(\theta,\theta_\Gamma,\Gamma) = \int_{\Omega\setminus\Gamma}\eta(\theta)\,dx + \int_\Gamma \eta_\Gamma(\theta_\Gamma)\,d\Gamma, \qquad (12.40)$$

satisfies

$$\begin{aligned}\frac{d}{dt}\mathsf{N}(\theta,\theta_\Gamma,\Gamma) &= \int_\Omega \eta'(\theta)\partial_t\theta\,dx + \int_\Gamma\{\partial_{t,n}\eta_\Gamma(\theta_\Gamma) - ([\![\eta(\theta)]\!] + \eta_\Gamma(\theta_\Gamma)H_\Gamma)V_\Gamma\}\,d\Gamma\\ &= \int_\Omega \frac{1}{\theta^2}d(\theta)|\nabla\theta|^2\,dx + \int_\Gamma\frac{1}{\theta_\Gamma^2}\{d_\Gamma(\theta_\Gamma)|\nabla_\Gamma\theta_\Gamma|^2 + \theta_\Gamma\gamma(\theta_\Gamma)V_\Gamma^2\}\,d\Gamma \geq 0,\end{aligned}$$

hence is strictly increasing along non-constant solutions. This can be seen as in Section 1.2. Thus, $-\mathsf{N}$ is a strict Lyapunov functional.

If σ is linear in θ we have $\kappa_\Gamma = 0$ and then it makes sense to also set $d_\Gamma\equiv 0$, to obtain the modified Stefan law

$$[\![d(\theta)\partial_\nu\theta]\!] = \Big(l(\theta) - \gamma(\theta)V_\Gamma + l_\Gamma(\theta)H_\Gamma\Big)V_\Gamma,$$

which differs from the Stefan law in **(P1)** only by replacing $l(u)$ by $l(u)+l_\Gamma(u)H_\Gamma$ (and including kinetic undercooling). This is just a minor modification of **(P1)**, its analysis remains essentially the same as for **(P1)**, the only difference is that the stability condition for the equilibria and in case $\gamma\equiv 0$ also the well-posedness condition changes. More precisely, the well-posedness condition changes from $\varphi'\neq 0$ to $\lambda'\neq 0$ where $\lambda(s) := \varphi(s)/\sigma(s)$, and the stability condition is modified by replacing φ'/σ by λ'.

Therefore, here we concentrate on the case where $\kappa_\Gamma(\theta), d_\Gamma(\theta) > 0$, which means that σ is strictly concave. In the sequel we always assume that

$$d_i,\psi_i,d_\Gamma,\sigma,\gamma\in C^3(0,\theta_c), \quad d_i,\kappa_i,d_\Gamma,\kappa_\Gamma,\sigma>0 \text{ on } (0,\theta_c), \quad i=1,2, \qquad (12.41)$$

if not stated otherwise. Furthermore, we let $\gamma \equiv 0$ if there is no undercooling, or $\gamma > 0$ on $(0, \theta_c)$ if undercooling is present, and we restrict attention to the temperature range $\theta \in (0, \theta_c)$.

Note that the (non-degenerate) equilibria for this problem are the same as those for the Stefan problem **(P1)**: the temperature is constant and the disperse phase Ω_1 consists of finitely many nonintersecting balls of the same radius. We shall prove below that such an equilibrium is stable in the state manifold $\mathcal{SM}$ defined below if Ω_1 is connected and the stability condition introduced in the next section holds. Such an equilibrium will be a local maximum of the total entropy, as for Problems **(P1)**$\sim$**(P6)**.

(ii) The case where undercooling is present is the simpler one, as both equations on the interface are dynamic equations. In particular, the Gibbs-Thomson identity

$$\gamma(\theta_\Gamma)V_\Gamma - \sigma(\theta_\Gamma)H_\Gamma = \varphi(\theta_\Gamma)$$

can be understood as a *mean curvature flow* for the evolution of the surface, modified by physics.

If there is no undercooling, it is convenient to eliminate the time derivative of θ_Γ from the energy balance on the interface. This can achieved as shown in Section 9.1.6, differentiating the Gibbs-Thomson law w.r.t. time t. With $\lambda(s) = \varphi(s)/\sigma(s)$ and

$$T_\Gamma(\theta_\Gamma) := \omega_\Gamma(\theta_\Gamma) - H'_\Gamma,\ \omega_\Gamma(\theta_\Gamma) := \lambda'(\theta_\Gamma)(l(\theta_\Gamma) - l_\Gamma(\theta_\Gamma)\lambda(\theta_\Gamma))/\kappa_\Gamma(\theta_\Gamma), \tag{12.42}$$

this yields the relation

$$\kappa_\Gamma(\theta_\Gamma)V_\Gamma - d_\Gamma(\theta_\Gamma)H_\Gamma = \kappa_\Gamma(\theta_\Gamma)\{f_\Gamma(\theta_\Gamma) + F_\Gamma(\theta, \theta_\Gamma)\}, \tag{12.43}$$

as has been shown in Section 9.1.6. Here the function f_Γ is the antiderivative of $\lambda(d_\Gamma/\kappa_\Gamma)'$ vanishing at $s = \theta_m$, and F_Γ is nonlocal in space and of lower order. So also in the case where undercooling is absent we obtain a **mean curvature flow**, modified by physics. We note that

$$\omega_\Gamma(\theta_\Gamma) = \frac{\theta_\Gamma\sigma(\theta_\Gamma)}{\kappa_\Gamma(\theta_\Gamma)}[\lambda'(\theta_\Gamma)]^2 = \frac{1}{\theta_\Gamma\sigma(\theta_\Gamma)\kappa_\Gamma(\theta_\Gamma)}[l(\theta_\Gamma) + l_\Gamma(\theta_\Gamma)H_\Gamma]^2$$

so that $\omega_\Gamma(\theta_\Gamma) \geq 0$.

We would like to point out a phenomenon, already found for Problem **(P5)** in absence of kinetic undercooling, which is due to positive surface heat capacity κ_Γ. If κ_Γ at an equilibrium is large enough, then such a steady state is stable, even if the interface is disconnected! However, as for **(P5)** such equilibria cannot be maxima of the total entropy.

4.1 Linearization at Equilibria

The full linearization at an equilibrium $(\theta_*, \theta_{\Gamma*}, \Sigma)$ with $\theta_{\Gamma*} = \theta_*$, $\Sigma = \cup_k \Sigma^k$ a finite union of disjoint spheres contained in Ω and with common radius $R_* > 0$ given by $R_* = (n-1)\sigma(u_*)/[\![\psi(\theta_*)]\!]$, reads

$$\begin{aligned} \kappa_*\partial_t\vartheta - d_*\Delta\vartheta &= \kappa_* f_\theta && \text{in } \Omega\setminus\Sigma,\\ \partial_\nu\vartheta &= 0 && \text{on } \partial\Omega,\\ [\![\vartheta]\!] = 0, \vartheta_\Sigma &= \vartheta && \text{on } \Sigma,\\ \kappa_{\Gamma*}\partial_t\vartheta_\Sigma - d_{\Gamma*}\Delta_\Sigma\vartheta_\Sigma - [\![d_*\partial_\nu\vartheta]\!] + l_0\partial_t h &= \kappa_{\Gamma*} f_\Sigma && \text{on } \Sigma,\\ (l_0/\theta_*)\vartheta_\Sigma - \sigma_*\mathcal{A}_\Sigma h - \gamma_*\partial_t h &= f_h && \text{on } \Sigma,\\ \vartheta(0) = \vartheta_0,\ h(0) &= h_0. \end{aligned} \tag{12.44}$$

Here $\kappa_* = \kappa(\theta_*) > 0$, $\kappa_{\Gamma*} = \kappa_\Gamma(\theta_*) > 0$, $d_* = d(\theta_*) > 0$, $d_{\Gamma*} = d_\Gamma(\theta_*) > 0$, $\sigma_* = \sigma(\theta_*) > 0$, $\gamma_* = \gamma(\theta_*) \geq 0$,

$$l_0 = l_* + \theta_*\sigma'(\theta_*)H_\Sigma = \theta_*\sigma(\theta_*)\lambda'(\theta_*), \quad l_* = \theta_*\varphi'(\theta_*),$$

and again $\mathcal{A}_\Sigma = -(\frac{n-1}{R_*^2} + \Delta_\Sigma)$, where Δ_Σ denotes the *Laplace-Beltrami* operator on Σ.

4.2 Maximal Regularity

We begin with the case $\gamma_* > 0$, which is the simpler one. Define the operator L_γ in

$$X_0 := L_p(\Omega) \times W_p^r(\Sigma) \times W_p^s(\Sigma)$$

with

$$X_1 := W_p^2(\Omega\setminus\Sigma) \times W_p^{2+r}(\Sigma) \times W_p^{2+s}(\Sigma)$$

by means of

$$\mathsf{D}(L_\gamma) = \big\{(\vartheta,\vartheta_\Sigma,h) \in X_1 : [\![\vartheta]\!] = 0,\ \ \vartheta_\Sigma = \vartheta \text{ on } \Sigma,\ \ \partial_\nu\vartheta = 0 \text{ on } \partial\Omega\big\},$$

$$L_\gamma = \begin{bmatrix} (-d_*/\kappa_*)\Delta & 0 & 0 \\ -[\![(d_*/\kappa_{\Gamma*})\partial_\nu]\!] & ((l_0^2/\theta_*\gamma_*) - d_{\Gamma*}\Delta_\Sigma)/\kappa_{\Gamma*} & -(l_0\sigma_*/\gamma_*\kappa_{\Gamma*})\mathcal{A}_\Sigma \\ 0 & -(l_0/\theta_*\gamma_*) & (\sigma_*/\gamma_*)\mathcal{A}_\Sigma \end{bmatrix}.$$

In case $\gamma_* > 0$, Problem (12.44) is equivalent to the Cauchy problem

$$\dot z + L_\gamma z = (f_\theta, f_\Sigma + (l_0/\gamma_*\kappa_{\Gamma*})f_h, -(1/\gamma_*)f_h), \quad z(0) = z_0,$$

where $z = (\vartheta,\vartheta_\Sigma,h)$ and $z_0 = (\vartheta_0,\vartheta_0|_\Sigma,h_0)$. The main result on problem (12.44) for $\gamma_* > 0$ is the following.

Theorem 12.4.1. *Let $1 < p < \infty$, $\gamma_* > 0$, and*

$$-1/p \leq r \leq 1 - 1/p, \quad r \leq s \leq r+2.$$

Then for each finite interval $J = [0,a]$, there is a unique solution

$$(\vartheta,\vartheta_\Sigma,h) \in \mathbb{E}(J) := H_p^1(J;X_0) \cap L_p(J;X_1)$$

of (12.44) if and only if the data $(f_\theta, f_\Sigma, f_h)$ and $(\vartheta_0, \vartheta_{\Sigma 0}, h_0)$ satisfy

$$(f_\theta, f_\Sigma, f_h) \in \mathbb{F}(J) = L_p(J; X_0),$$
$$(\vartheta_0, \vartheta_{\Sigma 0}, h_0) \in W_p^{2-2/p}(\Omega \setminus \Sigma) \times W_p^{2+r-2/p}(\Sigma) \times W_p^{2+s-2/p}(\Sigma)$$

and the compatibility conditions

$$[\![\vartheta_0]\!] = 0, \quad \vartheta_{\Sigma 0} = \vartheta_0 \quad \text{on } \Sigma, \quad \partial_\nu \vartheta = 0 \quad \text{on } \partial\Omega.$$

The operator $-L_\gamma$ defined above generates an analytic C_0-semigroup in X_0 with maximal regularity of type L_p.

Proof. Looking at the entries of L_γ we see that $L_\gamma : X_1 \to X_0$ is bounded provided $r \le 1 - 1/p$, $r \le s$, and $s \le r + 2$. The compatibility condition $\vartheta_\Sigma = \vartheta_{|\Sigma}$ implies $r+2 \ge 2-1/p$. This explains the constraints on the parameters r and s. To obtain maximal L_p-regularity, we first consider the case $s > r$. Then L_γ is lower triangular up to perturbation. So according to Section 6.6, we may solve the problem for $(\vartheta, \vartheta_\Sigma)$ with maximal L_p-regularity first and then that for h. In the other case we have $r = s$. Then the second term in the third line in the definition of L_γ is of lower order, hence h decouples from $(\vartheta, \vartheta_\Sigma)$. This way we also obtain maximal L_p-regularity. Since the Cauchy problem for L_γ has maximal L_p-regularity, we can now infer from Section 3.5 that $-L_\gamma$ generates an analytic C_0-semigroup in X_0. □

We note that if $l_0 = 0$ and $\gamma_* = 0$ then the linear problem (12.44) is not well-posed. In fact, in this case the linear Gibbs-Thomson relation reads

$$-\sigma_* \mathcal{A}_\Sigma h = f_h,$$

which is not well-posed as the kernel of $\mathcal{A}_\Sigma$ is non-trivial and $\mathcal{A}_\Sigma$ is not surjective.

Now we consider the case $l_0 \neq 0$ and $\gamma_* = 0$. For the solution space we fix again $r, s \in \mathbb{R}$ with $r \le s \le r+2$, $-1/p \le r \le 1 - 1/p$, and consider

$$(\vartheta, \vartheta_\Sigma, h) \in \mathbb{E}(J) = H_p^1(J, X_0) \cap L_p(J; X_1).$$

Then by trace theory the space of data becomes

$$(f_\theta, f_\Sigma, f_h) \in \mathbb{F}_0(J) := L_p(J; L_p(\Omega)) \times L_p(J; W_p^r(\Sigma)) \\ \times [H_p^1(J; W_p^{s-2}(\Sigma) \cap L_p(J; W_p^s(\Sigma))],$$

and the space of initial values will be

$$(\vartheta_0, \vartheta_{\Sigma 0}, h_0) \in W_p^{2-2/p}(\Omega \setminus \Sigma) \times W_p^{r+2-2/p}(\Sigma) \times W_p^{s+2-2/p}(\Sigma)$$

with compatibilities

$$[\![\vartheta_0]\!] = 0, \quad \vartheta_{\Sigma 0} = \vartheta_0, \quad (l_0/\theta_*)\vartheta_{\Sigma 0} - \sigma_* \mathcal{A}_\Sigma h_0 = f_h(0) \text{ on } \Sigma, \quad \partial_\nu \vartheta = 0 \text{ on } \partial\Omega.$$

To obtain maximal L_p-regularity, we replace ϑ_Σ by the Gibbs-Thomson relation, which for $\gamma_* = 0$ is an elliptic equation. We obtain $\vartheta_\Sigma = (\theta_*\sigma_*/l_0)\mathcal{A}_\Sigma h + (\theta_*/l_0) f_h$. Inserting this expression into the energy balance on the surface Σ yields

$$(l_0 + (\kappa_{\Gamma*}\theta_*\sigma_*/l_0)\mathcal{A}_\Sigma)\partial_t h - d_{\Gamma*}\Delta_\Sigma \vartheta_\Sigma - [\![d_*\partial_\nu\vartheta]\!] = \kappa_{\Gamma*}(f_\Sigma - (\theta_*/l_0)\partial_t f_h). \quad (12.45)$$

Moreover, we obtain

$$\begin{aligned} d_{\Gamma*}\Delta_\Sigma\vartheta_\Sigma &= (l_0 + (\kappa_{\Gamma*}\theta_*\sigma_*/l_0)\mathcal{A}_\Sigma))(d_{\Gamma*}/\kappa_{\Gamma*})\Delta_\Sigma h \\ &\quad - (l_0 d_{\Gamma*}/\kappa_{\Gamma*})\Delta_\Sigma h + (d_{\Gamma*}\theta_*/l_0)\Delta_\Sigma f_h. \end{aligned}$$

Now we assume that

$$l_0^2 R_*^2 \neq \theta_*\sigma_*(n-1)\kappa_{\Gamma*}, \quad (12.46)$$

which is equivalent to invertibility of the operator $A_0 := l_0 + (\kappa_{\Gamma*}\theta_*\sigma_*/l_0)\mathcal{A}_\Sigma$. Applying its inverse to (12.45) we arrive at the following equation for h

$$\partial_t h - (d_{\Gamma*}/\kappa_{\Gamma*})\Delta_\Sigma h + A_0^{-1}\{(l_0 d_{\Gamma*}/\kappa_{\Gamma*})\Delta_\Sigma h - [\![d_*\partial_\nu\vartheta]\!]\} = \tilde{f}_h, \quad (12.47)$$

with

$$\tilde{f}_h = A_0^{-1}\{\kappa_{\Gamma*} f_\Sigma - ((\kappa_{\Gamma*}\theta_*/l_0)\partial_t f_h - (d_{\Gamma*}\theta_*/l_0)\Delta_\Sigma f_h)\}.$$

Solving equation (12.45) for $\partial_t h$ we obtain for ϑ_Σ the relation

$$\kappa_{\Gamma*}\partial_t\vartheta_\Sigma - d_{\Gamma*}\Delta_\Sigma\vartheta_\Sigma - [\![d_*\partial_\nu\vartheta]\!] + l_0 A_0^{-1}\{d_{\Gamma*}\Delta_\Sigma\vartheta_\Sigma + [\![d_*\partial_\nu\vartheta]\!]\} = \tilde{f}_\Sigma, \quad (12.48)$$

where

$$\tilde{f}_\Sigma = \kappa_{\Gamma*}\{f_\Sigma - l_0 A_0^{-1}(f_\Sigma - (\theta_*/l_0)\partial_t f_h)\}.$$

Then by the regularity of f_Σ and f_h and with $r \le s \le r+2$ we see that

$$\tilde{f}_h \in L_p(J; W_p^s(\Sigma)), \quad \tilde{f}_\Sigma \in L_p(J; W_p^r(\Sigma)).$$

So the linear problem (12.44) can be recast as an evolution equation in X_0 as

$$\dot{z} + L_0 z = (f_\theta, \tilde{f}_\Sigma, \tilde{f}_h), \quad z(0) = z_0,$$

with $L_0 = L_{00} + L_{01}$ defined by

$$\mathsf{D}(L_{0j}) = \{(v, \vartheta_\Sigma, h) \in X_1 : [\![\vartheta]\!] = 0, \ \vartheta_\Sigma = \vartheta \text{ on } \Sigma, \ \partial_\nu\vartheta = 0 \text{ on } \partial\Omega\},$$

and

$$L_{00} = \begin{bmatrix} (-d_*/\kappa_*)\Delta & 0 & 0 \\ -[\![(d_*/\kappa_{\Gamma*})\partial_\nu]\!] & -(d_{\Gamma*}/\kappa_{\Gamma*})\Delta_\Sigma & 0 \\ -A_0^{-1}[\![d_*\partial_\nu]\!] & 0 & -(d_{\Gamma*}/\kappa_{\Gamma*})\Delta_\Sigma \end{bmatrix},$$

and

$$L_{01} = \begin{bmatrix} 0 & 0 & 0 \\ (l_0/\kappa_{\Gamma*})A_0^{-1}[\![d_*\partial_\nu]\!] & (l_0 d_{\Gamma*}/\kappa_{\Gamma*})A_0^{-1}\Delta_\Sigma & 0 \\ 0 & 0 & (l_0\, d_{\Gamma*}/\kappa_{\Gamma*})A_0^{-1}\Delta_\Sigma \end{bmatrix}.$$

Looking at L_0 we first note that L_{01} is a lower order perturbation of L_{00}. The latter is lower triangular, and the problem for $(\vartheta, \vartheta_\Sigma)$ as above has maximal L_p-regularity in X_0. As the diagonal entry in the equation for h has maximal L_p-regularity as well, we may conclude that $-L_0$ generates an analytic C_0-semigroup with maximal regularity in X_0 More precisely, we have the following result.

Theorem 12.4.2. *Let $1 < p < \infty$, $\gamma_* = 0$, $-1/p \le r \le 1 - 1/p$, $r \le s \le r + 2$, $l_0 \neq 0$, and assume condition* (12.48).

Then for each interval $J = [0, a]$, there is a unique solution $(\vartheta, \vartheta_\Sigma, h) \in \mathbb{E}(J)$ of (12.44) *if and only if the data $(f_\theta, f_\Sigma, f_h)$ and $(\vartheta_0, \vartheta_{\Sigma 0}, h_0)$ satisfy*

$$(f_\theta, f_\Sigma, f_h) \in L_p(J; X_0),$$
$$(\vartheta_0, \vartheta_{\Sigma 0}, h_0) \in W_p^{2-2/p}(\Omega \setminus \Sigma), \times W_p^{r+2-2/p}(\Sigma) \times W_p^{s+2-2/p}(\Sigma)$$

and the compatibility conditions

$$[\![\vartheta_0]\!] = 0, \quad \vartheta_{\Sigma 0} = \vartheta_0, \quad (l_0/\theta_*)\vartheta_0 - \sigma_* \mathcal{A}_\Sigma h_0 = f_h(0) \text{ on } \Sigma, \quad \partial_\nu \vartheta = 0 \text{ on } \partial\Omega.$$

The operator $-L_0$ defined above generates an analytic C_0-semigroup in X_0 with maximal regularity of type L_p.

Note that the compatibility condition $(l_0/\theta_*)\vartheta_0 - \sigma_* \mathcal{A}_\Sigma h_0 = f_h(0)$ allows us to recover the Gibbs-Thomson relation from the dynamic equations, as f_h can be recovered by solving a parabolic initial value problem on Σ. Indeed, it follows from (12.47)-(12.48) that the function $w := \vartheta_\Sigma - ((\theta_* \sigma_*/l_0)\mathcal{A}_\Sigma h + (\theta_*/l_0) f_h)$ satisfies the parabolic equation

$$\kappa_{\Gamma *} \partial_t w - d_{\Gamma *} \Delta_\Sigma w = 0, \quad w(0) = 0 \quad \text{on } \Sigma. \tag{12.49}$$

As $w \equiv 0$ is the unique solution of (12.49) we conclude that the Gibbs-Thomson relation is satisfied.

4.3 The Eigenvalue Problem

By compact embedding, the spectrum of L_γ consists only of countably many discrete eigenvalues of finite multiplicity and is independent of p. Therefore it suffices to consider the case $p = 2$. The eigenvalue problem reads as follows

$$\begin{aligned} \kappa_* \lambda \vartheta - d_* \Delta \vartheta &= 0 \quad \text{in } \Omega \setminus \Sigma, \\ \partial_\nu \vartheta &= 0 \quad \text{on } \partial\Omega, \\ [\![\vartheta]\!] &= 0 \quad \text{on } \Sigma, \\ \kappa_{\Gamma *} \lambda \vartheta - d_{\Gamma *} \Delta_\Sigma \vartheta - [\![d_* \partial_\nu \vartheta]\!] + l_0 \lambda h &= 0 \quad \text{on } \Sigma, \\ l_* \vartheta - \sigma_* \mathcal{A}_\Sigma h - \gamma_* \lambda h &= 0 \quad \text{on } \Sigma. \end{aligned} \tag{12.50}$$

We now argue as in Chapter 10. Let $\lambda \neq 0$ be an eigenvalue with eigenfunction $(\vartheta, h) \neq 0$. Then (12.50) yields

$$0 = \lambda |\sqrt{\kappa_*}\vartheta|_\Omega^2 - (d_* \Delta \vartheta | \vartheta)_\Omega = \lambda |\sqrt{\kappa_*}\vartheta|_\Omega^2 + |\sqrt{d_*}\nabla \vartheta|_\Omega^2 + ([\![d_* \partial_\nu \vartheta]\!] | \vartheta)_\Sigma.$$

On the other hand, we have on the interface

$$0 = \lambda\kappa_{\Gamma*}|\vartheta|_\Sigma^2 + d_{\Gamma*}|\nabla_\Sigma\vartheta|_\Sigma^2 - ([\![d_*\partial_\nu\vartheta]\!]|\vartheta)_\Sigma + \lambda l_0(h|\vartheta)_\Sigma.$$

Adding these identities we obtain

$$0 = \lambda|\sqrt{\kappa_*}\vartheta|_\Omega^2 + |\sqrt{d_*}\nabla\vartheta|_\Omega^2 + \lambda\kappa_{\Gamma*}|\vartheta|_\Sigma^2 + d_{\Gamma*}|\nabla_\Sigma\vartheta|_\Sigma^2 + \lambda\theta_*\theta_*(h|\vartheta)_\Sigma,$$

hence employing the Gibbs-Thomson law this results in the relation

$$\lambda|\sqrt{\kappa_*}\vartheta|_\Omega^2 + |\sqrt{d_*}\nabla\vartheta|_\Omega^2 + \lambda\kappa_{\Gamma*}|\vartheta|_\Sigma^2 + d_{\Gamma*}|\nabla_\Sigma\vartheta|_\Sigma^2 \\ + \lambda\theta_*\sigma_*(\mathcal{A}_\Sigma h|h)_\Sigma + \gamma_*\theta_*|\lambda|^2|h|_\Sigma^2 = 0.$$

Since $\mathcal{A}_\Sigma$ is selfadjoint in $L_2(\Sigma)$, this identity shows that all eigenvalues of L_γ are real. Decomposing $\vartheta = \vartheta_0 + \bar{\vartheta}$, $\vartheta_\Sigma = (\vartheta_\Sigma)_0 + \bar{\vartheta}_\Sigma$, $h = h_0 + \bar{h}$, with the normalizations $(\kappa_*|\vartheta_0)_\Omega = ((\vartheta_\Sigma)_0|1)_\Sigma = (h_0|1)_\Sigma = 0$, this identity can be rewritten as

$$\lambda\{|\sqrt{\kappa_*}\vartheta_0|_\Omega^2 + \kappa_{\Gamma*}|(\vartheta_\Sigma)_0|_\Sigma^2 + \sigma_*\theta_*(\mathcal{A}_\Sigma h_0|h_0)_\Sigma + \lambda\theta_*\gamma_*|h_0|_\Sigma^2\} \\ + |\sqrt{d_*}\nabla\vartheta_0|_\Omega^2 + d_{\Gamma*}|\nabla_\Sigma(\vartheta_\Sigma)_0|_\Sigma^2 \\ + \lambda\Big[(\kappa_*|1)\bar{\vartheta}^2 + \kappa_{\Gamma*}|\Sigma|\bar{\vartheta}_\Sigma^2 - \sigma_*\theta_*\frac{n-1}{R_*^2}|\Sigma|\bar{h}^2 + \lambda\theta_*\gamma_*|\Sigma|\bar{h}^2\Big] = 0.$$

In case Σ is connected, $\mathcal{A}_\Sigma$ is positive semi-definite on functions with mean zero, and hence the bracket determines whether there are positive eigenvalues. Taking the mean in (12.50) we obtain

$$(\kappa_*|1)_\Omega\bar{\vartheta} + \kappa_{\Gamma*}|\Sigma|\bar{\vartheta}_\Sigma + l_0|\Sigma|\bar{h} = 0,$$

hence minimizing the function

$$\phi(\bar{\vartheta}, \bar{\vartheta}_\Sigma, \bar{h}) := (\kappa_*|1)\bar{\vartheta}^2 + \kappa_{\Gamma*}|\Sigma|\bar{\vartheta}_\Sigma^2 - \theta_*\sigma_*\frac{n-1}{R_*^2}|\Sigma|\bar{h}^2$$

with respect to the constraint we see that there are no positive eigenvalues provided the stability condition $\zeta_* \leq 1$ is satisfied.

If $\Sigma = \bigcup_{1\leq l\leq m}\Sigma^l$ consists of $m \geq 1$ spheres Σ^l of equal radius, then

$$\mathsf{N}(L_\gamma) = \text{span}\left\{(\frac{\theta_*\sigma_*(n-1)}{l_0 R_*^2}, -1), (0, Y_1^l), \ldots, (0, Y_n^l) : 1 \leq l \leq m\right\}, \quad (12.51)$$

where the functions Y_j^l denote the *spherical harmonics of degree one* on Σ^l (and $Y_j^l \equiv 0$ on $\bigcup_{i\neq l}\Sigma^i$), normalized by $(Y_j^l|Y_k^l)_{\Sigma^l} = \delta_{jk}$. $\mathsf{N}(L_\gamma)$ is isomorphic to the tangent space of $\mathcal{E}$ at $(u_*, \Gamma_*) \in \mathcal{E}$, as was shown in Chapter 10.

We can now state the main result on linear stability.

Theorem 12.4.3. *Let $\sigma_* > 0$, $\gamma_* \geq 0$, $l_0 \neq 0$,*

$$\delta_* := \frac{\theta_*\sigma_*(n-1)}{l_0^2 R_*^2}\kappa_{\Gamma*} \neq 1 \quad \text{in case } \gamma_* = 0,$$

and assume that the interface Σ consists of $m \geq 1$ components. Let

$$\zeta_* = \frac{\theta_*\sigma_*(n-1)}{l_0^2 R_*^2|\Sigma|}\,[(\kappa_*|1)_\Omega + \kappa_{\Gamma*}|\Sigma|],$$

and let the equilibrium energy E_e be defined as in (1.27). Then

(i) $\mathsf{E}_e'(\theta_*) = (\zeta_* - 1)l_0^2 R_*^2|\Sigma|/(\theta_*\sigma_*(n-1))$.

(ii) *0 is a an eigenvalue of L_γ with geometric multiplicity $(mn+1)$. It is semi-simple if $\zeta_* \neq 1$.*

(iii) *If Σ is connected and $\zeta_* \leq 1$, or if $\delta_* > 1$ and $\gamma_* = 0$, then all eigenvalues of $-L_\gamma$ are negative, except for the eigenvalue 0.*

(iv) *If $\zeta_* > 1$, and $\delta_* < 1$ in case $\gamma_* = 0$, then there are precisely m positive eigenvalues of $-L_\gamma$, where m denotes the number of equilibrium spheres.*

(v) *If $\zeta_* \leq 1$, and $\delta_* < 1$ in case $\gamma_* = 0$ then $-L_\gamma$ has precisely $m-1$ positive eigenvalues.*

(vi) *$\mathsf{N}(L_\gamma)$ is isomorphic to the tangent space $T_{(\theta_*,\Sigma)}\mathcal{E}$ of $\mathcal{E}$ at $(\theta_*,\Sigma) \in \mathcal{E}$.*

Remarks 12.4.4. (a) Formally, the result is also true if $l_0 = 0$ and $\gamma_* > 0$. In this case $\mathsf{E}_e'(u_*) = (\kappa_*|1)_\Omega + \kappa_{\Gamma*}|\Gamma_*| > 0$ and $\zeta_* = \infty$, hence the equilibrium is unstable. If in addition $\gamma_* = 0$, then the problem is not well-posed.

(b) Note that ζ_* does neither depend on the diffusivities d_*, $d_{\Gamma*}$, nor on the coefficient of undercooling γ_*.

(c) One can show that in case $\zeta_* = 1$ and Γ_* connected, the eigenvalue 0 is no longer semi-simple: its algebraic multiplicity rises by 1 to $(n+2)$.

(d) It is remarkable that in case kinetic undercooling is absent, large surface heat capacity, i.e., $\delta_* > 1$, stabilizes the system, even in such a way that multiple spheres are stable, in contrast to the case $\delta* < 1$.

(e) We can show that, in case $\gamma_* = 0$, if δ_* increases to 1 then all positive eigenvalues go to ∞.

The proof of Theorem 12.4.3 follows the lines of the proof for Problem **(P5)** in Chapter 10 and is therefore omitted.

4.4 The Semiflow in Presence of Kinetic Undercooling

In this section we assume throughout $\gamma(s) > 0$ for all $0 < s < u_c$, i.e., kinetic undercooling is present at the relevant temperature range. In this case we may apply the results in Chapter 5, resulting in a rather complete analysis of the problem.

(a) Local Well-Posedness To prove local well-posedness we employ the direct mapping method as introduced in Chapter 1. As base space we use

$$X_0 = L_p(\Omega) \times W_p^{-1/p}(\Sigma) \times W_p^{1-1/p}(\Sigma),$$

and we set

$$X_1 = \{(\theta, \theta_\Sigma, h) \in H_p^2(\Omega \setminus \Sigma) \times W_p^{2-1/p}(\Sigma) \times W_p^{3-1/p}(\Sigma) : \\ [\![\theta]\!] = 0,\ \theta_\Sigma = \theta_{|_\Sigma},\ \partial_\nu \theta_{|_{\partial\Omega}} = 0\}.$$

The trace space X_γ then becomes for $p > n+2$

$$X_\gamma = \{(\theta, \theta_\Sigma, h) \in W_p^{2-2/p}(\Omega \setminus \Sigma) \times W_p^{2-3/p}(\Sigma) \times W_p^{3-3/p}(\Sigma) : \\ [\![\theta]\!] = 0,\ \theta_\Sigma = \theta_{|_\Sigma},\ \partial_\nu \theta_{|_{\partial\Omega}} = 0\},$$

and that with the time weight $t^{1-\mu}$, $1 \geq \mu > 1/p$,

$$X_{\gamma,\mu} = \{(\theta, \theta_\Sigma, h) \in W_p^{2\mu-2/p}(\Omega \setminus \Sigma) \times W_p^{2\mu-3/p}(\Sigma) \times W_p^{2\mu+1-3/p}(\Sigma) : \\ [\![\theta]\!] = 0,\ \theta_\Sigma = \theta_{|_\Sigma},\ \partial_\nu \theta_{|_{\partial\Omega}} = 0\}.$$

Note that

$$X_{\gamma,\mu} \hookrightarrow C^1_{ub}(\Omega \setminus \Sigma) \times C^1(\Sigma) \times C^2(\Sigma), \tag{12.52}$$

provided $2\mu > 1 + (n+2)/p$, which is feasible as $p > n+2$. In the sequel, we only consider this range of μ. We want to rewrite the transformed system abstractly as the quasilinear problem in X_0

$$\dot z + A(z)z = F(z), \quad t > 0, \quad z(0) = z_0, \tag{12.53}$$

where $z = (\theta, \theta_\Sigma, h)$ and $z_0 = (\theta_0, \theta_{\Sigma 0}, h_0)$. Here the quasilinear part $A(z)$ is the diagonal matrix operator defined by

$$-A(z) = \operatorname{diag}\begin{bmatrix} (d(\theta)/\kappa(\theta))(\Delta - M_2(h) : \nabla^2) \\ (d_\Gamma(\theta_\Sigma)/\kappa_\Gamma(\theta_\Sigma))(\mathcal{P}_\Gamma(h) M_0(h))^2 : \nabla_\Sigma^2 \\ (\sigma(\theta_\Sigma)/\beta(h)\gamma(\theta_\Sigma))\mathcal{C}_0(h) : \nabla_\Sigma^2 \end{bmatrix},$$

with $M_2(h) = M_1(h) + M_1^{\mathsf T}(h) - M_1(h)M_1^{\mathsf T}(h)$. The semilinear part $F(z)$ is given by

$$\begin{bmatrix} \mathcal{R}(h)\theta + \dfrac{1}{\kappa(\theta)}\{d'(\theta)|(I - M_1(h))\nabla\theta|^2 - d(\theta)((I - M_1(h)) : \nabla M_1(h)|\nabla\theta)\} \\ \dfrac{1}{\kappa_\Gamma(\theta_\Sigma)}\{-\mathcal{B}(\theta_\Sigma, h)\theta - [(l(\theta_\Sigma) + l_\Gamma(\theta_\Sigma)H_\Gamma(h) - \gamma(\theta_\Sigma)\beta(h)\partial_t h]\beta(h)\partial_t h + m_3\} \\ \varphi(\theta_\Sigma)/\beta(h)\gamma(\theta_\Sigma) + (\sigma(\theta_\Sigma)/\beta(h)\gamma(\theta_\Sigma))\,\mathcal{C}_1(h) \end{bmatrix}$$

and

$$m_3 = -d_\Gamma(\theta_\Sigma)(\mathcal{P}_\Gamma(h)M_0(h))^2 : \nabla_\Sigma^2\theta_\Sigma - \mathcal{C}(\theta_\Sigma, h)\theta_\Sigma.$$

We note that m_3 depends on $(\theta_\Sigma, \nabla_\Sigma\theta_\Sigma)$, and on $(h, \nabla_\Sigma h, \nabla_\Sigma^2 h)$, but not on $\nabla_\Sigma^2\theta_\Sigma$, hence is of lower order. Apparently, the first two components of $F(z)$ contain the time derivative $\partial_t h$; we may replace it by

$$\partial_t h = \{\varphi(\theta_\Sigma) + \sigma(\theta_\Sigma)H_\Gamma(h)\}/\beta(h)\gamma(\theta_\Sigma),$$

to see that it is of lower order as well.

Now fix a ball $\mathbb{B} := B_{X_{\gamma,\mu}}(z_0, R) \subset X_{\gamma,\mu}$, where $|h_0|_{C^1(\Sigma)} \leq \eta$ for some sufficiently small $\eta > 0$. Then it is not difficult to verify that

$$(A, F) \in C^1(\mathbb{B}, \mathcal{B}(X_1, X_0) \times X_0)$$

provided $d_i, \psi_i, d_\Gamma, \sigma, \gamma \in C^3(0, \infty)$ and $d_j, \kappa_j, \sigma, \gamma > 0$ on $(0, u_c)$, $j \in \{1, 2, \Gamma\}$, and provided $2 \geq 2\mu > 1 + (n+2)/p$ as before. Moreover, as $A(z)$ is diagonal, the results on elliptic differential operators in Chapter 6 show that $A(z)$ has the property of maximal regularity for each $z \in \mathbb{B}$. In fact, for small $\eta > 0$, $A(z)$ is small perturbation of

$$A_\#(z) = \operatorname{diag}\big[-(d(\theta)/\kappa(\theta))\Delta, -(d_\Gamma(\theta_\Sigma)/\kappa_\Gamma(\theta_\Sigma))\Delta_\Sigma, -(\sigma(\theta_\Sigma)/\gamma(\theta_\Sigma))\Delta_\Sigma\big].$$

Therefore we may apply the perturbation results from Section 4.4 to obtain local well-posedness of (12.53), i.e., a unique local solution

$$z \in H^1_{p,\mu}((0,a); X_0) \cap L_{p,\mu}((0,a); X_1) \hookrightarrow C([0,a]; X_{\gamma,\mu}) \cap C((0,a]; X_\gamma)$$

which depends continuously on the initial value $z_0 \in \mathbb{B}$. The resulting solution map $z_0 \mapsto z(t)$ defines a local semiflow in X_γ.

(b) Nonlinear Stability of Equilibria

Let $e_* = (\theta_*, \theta_{\Gamma*}, \Gamma_*)$ denote an equilibrium. We choose $\Sigma = \Gamma_*$ as a reference manifold, and as shown in the preceding subsection we obtain the abstract quasilinear parabolic problem

$$\dot{z} + A(z)z = F(z), \quad t > 0, \quad z(0) = z_0, \tag{12.54}$$

with X_0, X_1, X_γ as above. We set $z_* = (\theta_*, \theta_{\Gamma*}, 0)$. Assuming that $\zeta_* \neq 1$ in the stability condition, we have shown in the previous section that the equilibrium z_* is normally hyperbolic. Therefore, we may apply the results from Chapter 5 to obtain the following result.

Theorem 12.4.5. *Let $p > n+2$. Suppose $\gamma > 0$ on $(0, \theta_c)$ and the assumptions of* (12.41) *hold true. As above $\mathcal{E}$ denotes the set of equilibria of* (12.54)*, and we fix some $z_* \in \mathcal{E}$. Then we have*

(a) *If Γ_* is connected and $\zeta_* < 1$ then z_* is stable in X_γ, and there exists $\delta > 0$*

such that the unique solution $z(t)$ *of* (12.54) *with initial value* $z_0 \in X_\gamma$ *satisfying* $|z_0 - z_*|_\gamma < \delta$ *exists on* $\mathbb{R}_+$ *and converges at an exponential rate in* X_γ *to some* $z_\infty \in \mathcal{E}$ *as* $t \to \infty$.

(b) *If* Γ_* *is disconnected or if* $\zeta_* > 1$ *then* z_* *is unstable in* X_γ. *For each sufficiently small* $\rho > 0$ *there is* $\delta \in (0, \rho]$ *such that the solution* $z(t)$ *of* (12.54) *with initial value* $z_0 \in X_\gamma$ *subject to* $|z_0 - z_*|_\gamma < \delta$ *either satisfies*

(i) $\mathrm{dist}_{X_\gamma}(z(t_0); \mathcal{E}) > \rho$ *for some finite time* $t_0 > 0$; *or*

(ii) $z(t)$ *exists on* $\mathbb{R}_+$ *and converges at exponential rate in* X_γ *to some* $z_\infty \in \mathcal{E}$.

Remark 12.4.6. The only equilibria which are excluded from our analysis are those with $\zeta_* = 1$, which means $\mathsf{E}_e'(\theta_*) = 0$. These are critical points of the function $\mathsf{E}_e(\theta)$ at which a bifurcation may occur. In fact, if such θ_* is a maximum or a minimum of E_e then two branches of $\mathcal{E}$ meet at θ_*, a stable and and an unstable one, which means that (θ_*, Γ_*) is a *turning point* in $\mathcal{E}$.

(c) The Local Semiflow on the State Manifold

We define the state manifold $\mathcal{SM}$ for (12.37) as follows.

$$\begin{aligned} \mathcal{SM} := \big\{(\theta, \Gamma) \in C(\bar{\Omega}) \times \mathcal{MH}^2 : \theta \in W_p^{2-2/p}(\Omega \setminus \Gamma),\, \Gamma \in W_p^{3-3/p}, \\ 0 < \theta < \theta_c \text{ in } \bar{\Omega},\ \partial_\nu \theta = 0 \text{ on } \partial\Omega\big\}. \end{aligned} \tag{12.55}$$

We note that there is no need to incorporate the variable θ_Γ into the definition of the state manifold as $\theta_\Gamma = \theta|_\Gamma$.

Applying the result in the preceding subsection and re-parameterizing the interface repeatedly, we see that (12.37) yields a local semiflow on $\mathcal{SM}$.

Theorem 12.4.7. *Let* $p > n + 2$. *Suppose* $\gamma > 0$ *on* $(0, \theta_c)$ *and the assumptions of* (12.41) *hold true.*
Then problem (12.37) *generates a local semiflow on the state manifold* $\mathcal{SM}$. *Each solution* (θ, Γ) *exists on a maximal time interval* $[0, t_+)$, *with* $t_+ = t_+(\theta_0, \Gamma_0)$.

(d) Global Existence and Convergence

There are several obstructions to global existence for the Stefan problem with variable surface tension (12.37):

- *regularity*: the norms of $\theta(t)$ or $\Gamma(t)$ may become unbounded;
- *well-posedness*: the temperature θ may reach 0 or θ_c;
- *geometry*: the topology of the interface may change;
  or the interface may touch the boundary of Ω;
  or part of the interface may contract to a point.

Let (θ, Γ) be a solution in the state manifold $\mathcal{SM}$. Combining the semiflow for (12.37) with the Lyapunov functional, i.e., the negative total entropy, and compactness we obtain the following result.

Theorem 12.4.8. *Let $p > n+2$. Suppose $\gamma > 0$ on $(0, \theta_c)$ and the assumptions of* (12.41) *hold true. Suppose that (θ, Γ) is a solution of* (12.37) *in the state manifold $\mathcal{SM}$ on its maximal time interval $[0, t_+)$. Assume the following on $[0, t_+)$: there are constants $M, m > 0$ such that*

(i) $|\theta(t)|_{W_p^{2-2/p}} + |\Gamma(t)|_{W_p^{3-3/p}} \le M < \infty$;

(ii) $0 < m \le \theta(t) \le \theta_c - m$;

(iii) *$\Gamma(t)$ satisfies the uniform ball condition.*

Then $t_+ = \infty$, i.e., the solution exists globally, and it converges in $\mathcal{SM}$ to some equilibrium $(\theta_\infty, \Gamma_\infty) \in \mathcal{E}$. On the contrary, if $(\theta(t), \Gamma(t))$ is a global solution in $\mathcal{SM}$ which converges to an equilibrium (θ_*, Γ_*) in $\mathcal{SM}$ as $t \to \infty$, then properties (i)–(iii) *are valid.*

The proof follows the lines of arguments in the proof of Theorem 11.4.1.

4.5 The Semiflow without Kinetic Undercooling

In this section we assume throughout $\gamma(s) = 0$ for all $s > 0$, i.e kinetic undercooling is absent. In this case we may apply the results of Chapter 5 too, but we have to work harder to apply them. For this purpose we replace the Gibbs-Thomson law by the dynamic equation (12.43) and we assume the compatibility condition $\varphi(\theta_{\Gamma 0}) + \sigma(\theta_{\Gamma 0}) H_{\Gamma_0} = 0$.

(a) Local Well-Posedness

To prove local well-posedness we employ again the direct mapping method. As base space we use

$$X_0 = L_p(\Omega) \times W_p^{-1/p}(\Sigma) \times W_p^{1-1/p}(\Sigma),$$

and we let X_1, X_γ and $X_{\gamma,\mu}$ as defined there.

We rewrite the transformed system as a quasilinear problem in X_0

$$\dot z + A_0(z) z = F_0(z), \quad t > 0, \quad z(0) = z_0, \tag{12.56}$$

where $z = (\theta, \theta_\Sigma, h)$ and $z_0 = (\theta_0, \theta_{\Sigma 0}, h_0)$. Here the quasilinear part $A_0(z)$ is the diagonal matrix operator defined by

$$-A_0(z) = \operatorname{diag} \begin{bmatrix} (d(\theta)/\kappa(\theta))(\Delta - M_2(h) : \nabla^2) \\ (d_\Gamma(\theta_\Sigma)/\kappa_\Gamma(\theta_\Sigma))(\mathcal{P}_\Gamma(h) M_0(h))^2 : \nabla_\Sigma^2 \\ (d_\Gamma(\theta_\Sigma)/\beta(h)\kappa_\Gamma(\theta_\Sigma))\mathcal{C}_0(h) : \nabla_\Sigma^2 \end{bmatrix},$$

with $M_2(h)$ as before. The semi-linear part $F_0(z)$ is given by

$$\begin{bmatrix} \mathcal{R}(h)\theta + \dfrac{1}{\kappa(\theta)}\{d'(\theta)|(I - M_1(h))\nabla\theta|^2 - d(\theta)((I - M_1(h)) : \nabla M_1(h)|\nabla\theta)\} \\ -\dfrac{1}{\kappa_\Gamma(\theta_\Sigma)}\{\mathcal{B}(\theta_\Sigma, h)\theta - [(l(\theta_\Sigma) + l_\Gamma(\theta_\Sigma)H_\Gamma(h) - \gamma(\theta_\Sigma)\beta(h)\partial_t h]\beta(h)\partial_t h + m_3\} \\ (d_\Gamma(\theta_\Sigma)/\beta(h)\kappa_\Gamma(\theta_\Sigma))\,\mathcal{C}_1(h) + \{f_\Gamma(\theta_\Sigma) + F_\Gamma(\theta, \theta_\Sigma, h)\}/\beta(h) \end{bmatrix},$$

where by abuse of notation F_Γ here means the transformed F_Γ introduced previously, and with m_3 as above.

Again, the first two components of $F_0(z)$ contain the time derivative $\partial_t h$. We replace it by the transformed version of (12.43)

$$\partial_t h = \{f_\Gamma(\theta_\Sigma) + F_\Gamma(\theta, \theta_\Sigma, h) + d_\Gamma(\theta_\Sigma)/\kappa_\Gamma(\theta_\Sigma)\mathcal{H}(h)\}/\beta(h),$$

to see that it leads to a lower order term.

Provided that $T_{\Gamma_0}(\theta_{\Sigma 0})$ is invertible we may proceed as in the previous subsection to obtain local well-posedness, i.e., a unique local solution

$$z \in H^1_{p,\mu}((0,a); X_0) \cap L_{p,\mu}((0,a); X_1) \hookrightarrow C([0,a]; X_{\gamma,\mu}) \cap C((0,a]; X_\gamma)$$

which depends continuously on the initial value $z_0 \in \mathbb{B}$. The resulting solution map $z_0 \mapsto z(t)$ defines a local semiflow in X_γ.

(b) Nonlinear Stability of Equilibria

Let $e_* = (\theta_*, \theta_{\Gamma*}, \Gamma_*)$ denote an equilibrium. In this case we again choose $\Sigma = \Gamma_*$ as a reference frame, and as shown in the subsection before we obtain the abstract quasilinear parabolic problem

$$\dot{z} + A_0(z)z = F_0(z), \quad t > 0, \quad z(0) = z_0, \tag{12.57}$$

with X_0, X_1, X_γ as above. We set $z_* = (\theta_*, \theta_{\Gamma*}, 0)$. Assuming well-posedness and $\zeta_* \neq 1$ in the stability condition, we have shown above that the equilibrium z_* is normally hyperbolic. Therefore, we may apply once more the results from Chapter 5 to obtain the following result.

Theorem 12.4.9. *Let $p > n+2$. Suppose $\gamma \equiv 0$, $\sigma \in C^4(0, u_c)$, and the assumptions of (12.41) hold true. As above $\mathcal{E}$ denotes the set of equilibria of (12.57), and we fix some $z_* \in \mathcal{E}$. Assume that the well-posedness condition*

$$l_0 \neq 0 \quad \text{and} \quad \delta_* := \frac{\theta_* \sigma_* (n-1)}{l_0^2 R_*^2} \kappa_{\Gamma*} \neq 1 \tag{12.58}$$

is satisfied. Then we have

(a) *If Γ_* is connected and $\zeta_* < 1$, or if $\delta_* > 1$ then z_* is stable in X_γ, and there exists $r > 0$ such that the unique solution $z(t)$ of (12.57) with initial value $z_0 \in X_\gamma$ satisfying $|z_0 - z_*|_\gamma < r$ exists on $\mathbb{R}_+$ and converges at an exponential rate in X_γ to some $z_\infty \in \mathcal{E}$ as $t \to \infty$.*

(b) *If $\delta_* < 1$, and if Γ_* is disconnected or if $\zeta_* > 1$, then z_* is unstable in X_γ. For each sufficiently small $r_0 > 0$ there is $r \in (0, r_0]$ such that the solution $z(t)$ of (12.54) with initial value $z_0 \in X_\gamma$ subject to $|z_0 - z_*|_\gamma < r$ either satisfies*

(i) $\mathrm{dist}_{X_\gamma}(z(t_0); \mathcal{E}) > r_0$ *for some finite time $t_0 > 0$; or*

(ii) *$z(t)$ exists on $\mathbb{R}_+$ and converges at exponential rate in X_γ to some $z_\infty \in \mathcal{E}$.*

Thus the only cases excluded are $\zeta_* = 1$, and the two values where the well-posedness condition (12.58) is violated.

(c) The Local Semiflow on the State Manifold

We define the state manifolds $\mathcal{SM}_0$ for (12.37) in case $\gamma \equiv 0$ as follows.

$$\begin{aligned} \mathcal{SM}_0 := \big\{(\theta, \Gamma) \in C(\bar{\Omega}) \times \mathcal{MH}^2 : \theta \in W_p^{2-2/p}(\Omega \setminus \Gamma),\, \Gamma \in W_p^{3-3/p}, \\ 0 < \theta < \theta_c \text{ in } \bar{\Omega},\ \partial_\nu \theta = 0 \text{ on } \partial\Omega, \\ \lambda(\theta_\Gamma) + H_\Gamma = 0 \text{ on } \Gamma,\ T_\Gamma(\theta_\Gamma) \text{ is invertible in } L_2(\Gamma)\big\}. \end{aligned} \tag{12.59}$$

Applying the well-posedness result and re-parameterizing the interface repeatedly, we see that (12.37) with $\gamma \equiv 0$ yields a local semiflow on $\mathcal{SM}_0$.

Theorem 12.4.10. *Let $p > n + 2$. Suppose $\gamma \equiv 0$, $\sigma \in C^4(0, u_c)$, and the assumptions of* (12.41) *hold true.*

Then problem (12.37) *generates a local semiflow on the state manifold $\mathcal{SM}_0$. Each solution (θ, Γ) exists on a maximal time interval $[0, t_+)$, where $t_+ = t_+(\theta_0, \Gamma_0)$.*

(d) Global Existence and Convergence

In addition to the obstructions to global existence for the Stefan problem with variable surface tension in the presence of kinetic undercooling there is another reason for loss of well-posedeness.

- *regularity*: the norms of $\theta(t)$ or $\Gamma(t)$ may become unbounded;
- *well-posedness*: the temperature may reach 0 or θ_c;
  or $T_\Gamma(\theta_\Gamma)$ may become non-invertible;
- *geometry*: the topology of the interface may change;
  or the interface may touch the boundary of Ω;
  or part of the interface may contract to a point.

We set $\mathcal{E}_0 = \mathcal{SM}_0 \cap \mathcal{E}$. As in Section 5, combining the semiflow for (12.37) with the Lyapunov functional and compactness we obtain the following result.

Theorem 12.4.11. *Let $p > n + 2$. Suppose $\gamma \equiv 0$, $\sigma \in C^4(0, \theta_c)$, and the assumptions of* (12.41) *hold true. Suppose that (θ, Γ) is a solution of* (12.37) *in the state manifold $\mathcal{SM}_0$ on its maximal time interval $[0, t_+)$. Assume the following on $[0, t_+)$: there are constants $m, M > 0$ such that*

(i) $|\theta(t)|_{W_p^{2-2/p}} + |\Gamma(t)|_{W_p^{3-3/p}} \le M < \infty$;

(ii) $0 < m \le \theta(t)) \le \theta_c - m$;

(iii) $|\mu_j(t)| \ge m$ *holds for the eigenvalues of* $T_{\Gamma(t)}(\theta_\Gamma)$;

(iv) $\Gamma(t)$ *satisfies the uniform ball condition.*

Then $t_+ = \infty$, i.e., the solution exists globally, and it converges in $\mathcal{SM}_0$ to an equilibrium $(\theta_\infty, \Gamma_\infty) \in \mathcal{E}_0$. Conversely, if $(\theta(t), \Gamma(t))$ is a global solution in $\mathcal{SM}_0$ which converges to an equilibrium $(\theta_\infty, \Gamma_\infty) \in \mathcal{E}_0$ in $\mathcal{SM}_0$ as $t \to \infty$, then the properties (i)–(iv) *are valid.*

Proof. The proof follows the same lines as that of Theorem 11.4.1. □

12.5 Geometric Evolution Equations

Here we introduce and study some geometric evolution equations. The first three are motivated by problems in differential geometry, while the ones following involve physics and have been introduced in Chapter 1. These equations are simpler to be analyzed than the general problems introduced in Chapter 1, due to their quasi-stationary structure. We first introduce the problems under consideration.

1. The Averaged Mean Curvature Flow
This flow is determined by the equation

$$V_\Gamma = \sigma(H_\Gamma - \bar{H}_\Gamma), \quad t > 0, \quad \Gamma(0) = \Gamma_0, \tag{12.60}$$

where $\sigma > 0$. Here we are looking for a family $\{\Gamma(t) : t > 0\}$ of compact oriented hypersurfaces with curvature $H_\Gamma(t)$ and normal velocity $V_\Gamma(t)$ given by (12.60). $\bar{H}_\Gamma$ means the mean value of H_Γ over Γ, i.e.,

$$\bar{H}_\Gamma = ⨍_\Gamma H_\Gamma \, d\Gamma = \frac{1}{|\Gamma|} \int_\Gamma H_\Gamma \, d\Gamma.$$

This problem is of spatial order $m = 2$.

2. The Surface Diffusion Flow
This flow is determined by the equation

$$V_\Gamma = -\sigma \Delta_\Gamma H_\Gamma, \quad t > 0, \quad \Gamma(0) = \Gamma_0, \tag{12.61}$$

with $\sigma > 0$. Here we are looking for a family $\{\Gamma(t) : t > 0\}$ of compact oriented hypersurfaces with curvature $H_\Gamma(t)$ and velocity $V_\Gamma(t)$ given by (12.61). This is a problem of order $m = 4$.

2a. The Willmore Flow
The *Willmore functional* for compact surfaces is defined by

$$W(\Gamma) = \frac{1}{2} \int_\Gamma H_\Gamma^2 \, d\Gamma.$$

Critical points of this functional are called *Willmore surfaces*. We compute

$$\frac{d}{dt} W(\Gamma) = \int_\Gamma H_\Gamma H_\Gamma'(\Gamma) V_\Gamma - \frac{1}{2} H_\Gamma^3 V_\Gamma d\Gamma = \int_\Gamma V_\Gamma \{\Delta_\Gamma H_\Gamma + H_\Gamma(\operatorname{tr} L^2 - \frac{1}{2} H_\Gamma^2)\} d\Gamma.$$

Letting $F_\Gamma := \sum_{1\leq i<j\leq n-1} \kappa_i\kappa_j$ denote the second invariant of the Weingarten tensor L_Γ, this yields for the variational derivative of W the relation

$$W'(\Gamma) = \Delta_\Gamma H_\Gamma + H_\Gamma(\frac{1}{2}H_\Gamma^2 - 2F_\Gamma).$$

The *Willmore flow* is defined as the L_2-gradient flow of the Willmore functional, which means

$$V_\Gamma = -\Delta_\Gamma H_\Gamma - H_\Gamma(\frac{1}{2}H_\Gamma^2 - 2F_\Gamma). \tag{12.62}$$

Mathematically, this problem can be considered as a lower order perturbation of the surface diffusion flow.

The Willmore functional is a strict Lyapunov functional for the Willmore flow, by its definition as gradient flow. However, this flow is in general not volume preserving. Note that in dimension $n = 3$ we have $F_\Gamma = \mathcal{K} = \kappa_1\kappa_2$, the Gaussian curvature, hence

$$\frac{1}{2}H_\Gamma^2 - 2F_\Gamma = \frac{1}{2}(\kappa_1 - \kappa_2)^2.$$

In particular, for $n = 3$ spheres are equilibria of the Willmore flow. In general, the manifold of equilibria of the Willmore flow is quite complicated, and subject to current research. Well-posedness of the Willmore flow is a consequence of the result for the surface diffusion flow, but we will not discuss stability of its equilibria here.

3. The Mullins-Sekerka Flow

We consider the two-phase quasi-stationary Stefan problem with surface tension, which has also been termed Mullins-Sekerka problem; recall the framework introduced in Chapter 1.

The *two-phase Mullins-Sekerka flow* consists in finding a family $\{\Gamma(t) : t > 0\}$ of hypersurfaces satisfying

$$V_\Gamma = -[\![d\partial_\nu\theta]\!], \quad t > 0, \quad \Gamma(0) = \Gamma_0, \tag{12.63}$$

where $\theta = \theta(t, \cdot)$ is, for each $t > 0$, the solution of the elliptic boundary value problem

$$\begin{aligned} \Delta\theta &= 0 && \text{in } \Omega \setminus \Gamma(t), \\ \partial_\nu\theta &= 0 && \text{on } \partial\Omega, \\ \theta &= \sigma H_\Gamma && \text{on } \Gamma(t). \end{aligned} \tag{12.64}$$

The coefficient $d > 0$ is constant in the phases, but may jump across the interface, and $\sigma > 0$. The problem can be transformed to an evolution equation on the boundary. To see this, recall the *Dirichlet-to-Neumann operator* S_d defined in Section 6.5.4. Then the problem can be written as

$$V_\Gamma = \sigma S_\mathsf{d} H_\Gamma, \quad t > 0, \quad \Gamma(0) = \Gamma_0.$$

Note that this is a problem of order $m = 3$.

4. The Stokes Flow

We consider the two-phase quasi-stationary Stokes problem with surface tension.

$$\begin{aligned}
-\mu\Delta u + \nabla\pi &= 0 && \text{in } \Omega\setminus\Gamma(t),\\
\operatorname{div} u &= 0 && \text{in } \Omega\setminus\Gamma(t),\\
u &= 0 && \text{on } \partial\Omega,\\
[\![u]\!] &= 0 && \text{on } \Gamma(t),\\
[\![-T\nu_\Gamma]\!] &= \sigma H_\Gamma \nu_\Gamma && \text{on } \Gamma(t),\\
V_\Gamma &= u\cdot\nu_\Gamma && \text{on } \Gamma(t).
\end{aligned} \tag{12.65}$$

Here $\mu > 0$ is constant in each phase but may jump at the interface, $\sigma > 0$ is also constant, and as before the stress tensor T is defined according to

$$T = \mu(\nabla u + \nabla u^{\mathsf{T}}) - \pi I = 2\mu D - \pi I.$$

Let N_0^S denote the Neumann-to-Dirichlet operator for the Stokes problem as introduced in Section 10.5. Then (12.65) reduces to the problem

$$V_\Gamma = \sigma N_0^S H_\Gamma, \quad t > 0.$$

Note that this is a problem of order $m = 1$.

5. The Muskat Flow

This is a simplification of the Stokes flow, employing Darcy's law for the velocity, as explained in Chapter 1. It reads

$$\begin{aligned}
\operatorname{div} u &= 0 && \text{in } \Omega\setminus\Gamma(t),\\
u\cdot\nu_\Gamma &= 0 && \text{on } \partial\Omega,\\
[\![u\cdot\nu_\Gamma]\!] &= 0 && \text{on } \Gamma(t),\\
u &= -k\nabla\pi && \text{in } \Omega\setminus\Gamma(t),\\
[\![\pi]\!] &= \sigma H_\Gamma && \text{on } \Gamma(t),\\
V_\Gamma &= u\cdot\nu_\Gamma && \text{on } \Gamma(t).
\end{aligned}$$

Here, as before, u denotes velocity, π pressure, H_Γ curvature, $\sigma > 0$ surface tension and the constants $k_j > 0$ mean the porosities of the phases. Eliminating u, this yields an elliptic problem for π, namely

$$\begin{aligned}
\Delta\pi &= 0 && \text{in } \Omega\setminus\Gamma(t),\\
\partial_\nu\pi &= 0 && \text{on } \partial\Omega,\\
[\![k\partial_\nu\pi]\!] &= 0 && \text{on } \Gamma(t),\\
[\![\pi]\!] &= \sigma H_\Gamma && \text{on } \Gamma(t),\\
V_\Gamma &= -k\partial_\nu\pi && \text{on } \Gamma(t).
\end{aligned} \tag{12.66}$$

Solving this problem for π we obtain

$$-k\partial_\nu \pi = \sigma S_{\mathsf{n}} H_\Gamma,$$

where S_{n} denotes the second two-phase Dirichlet-to-Neumann operator introduced in Section 6.5.4. Therefore, the problem reduces to the geometric evolution equation

$$V_\Gamma = \sigma S_{\mathsf{n}} H_\Gamma, \quad t > 0, \quad \Gamma(0) = \Gamma_0,$$

for the free interface. As the Mullins-Sekerka flow this problem is of order $m = 3$.

6. The Stokes Flow with Phase Transitions

We consider the two-phase quasi-stationary Stokes problem with phase transitions.

$$\begin{aligned} -\mu\Delta u + \nabla\pi &= 0 && \text{in } \Omega\setminus\Gamma(t),\\ \operatorname{div} u &= 0 && \text{in } \Omega\setminus\Gamma(t),\\ u &= 0 && \text{on } \partial\Omega,\\ \mathcal{P}_\Gamma [\![u]\!] &= 0 && \text{on } \Gamma(t),\\ -[\![T\nu_\Gamma]\!] &= \sigma H_\Gamma \nu_\Gamma && \text{on } \Gamma(t),\\ -[\![T\nu_\Gamma\cdot\nu_\Gamma/\varrho]\!] &= 0 && \text{on } \Gamma(t),\\ [\![\varrho]\!] V_\Gamma &= [\![\varrho u\cdot\nu_\Gamma]\!] && \text{on } \Gamma(t). \end{aligned} \tag{12.67}$$

Here $\mu > 0$ is constant in each phase but may jump at the interface, $\sigma > 0$ is also constant, and the stress tensor T is defined according to

$$T = \mu(\nabla u + \nabla u^{\mathsf{T}}) - \pi I = 2\mu D - \pi I.$$

Solving the problem in the bulk for u, by means of the Neumann-to-Dirichlet operator S_0 introduced in Section 10.7, this problem can be reduced to the form

$$V_\Gamma = \sigma S_0^{11} H_\Gamma, \quad t > 0, \quad \Gamma(0) = \Gamma_0,$$

for the free interface $\Gamma(t)$, leading to a geometric evolution equation. It is of order $m = 1$, as the Stokes flow.

7. The Muskat Flow with Phase Transitions

This is a simplification of the Stokes flow with phase transition, employing Darcy's law for the velocity, as explained in Chapter 1. It reads

$$\begin{aligned} \Delta\pi &= 0 && \text{in } \Omega\setminus\Gamma(t),\\ \partial_\nu\pi &= 0 && \text{on } \partial\Omega,\\ [\![\pi]\!] &= \sigma H_\Gamma && \text{on } \Gamma(t),\\ [\![\pi/\varrho]\!] &= 0 && \text{on } \Gamma(t),\\ [\![\varrho]\!] V_\Gamma &= [\![\varrho k\partial_\nu\pi]\!] && \text{on } \Gamma(t). \end{aligned} \tag{12.68}$$

This problem is only meaningful for different constant densities ϱ_i. Solving the transmission problem in the bulk domain as for the Muskat flow, we obtain the following geometric evolution equation

$$V_\Gamma = \sigma S_{mpt} H_\Gamma, \quad t > 0, \quad \Gamma(0) = \Gamma_0,$$

for the free interface, where S_{mpt} can easily be computed, using the one-phase Dirichlet-to-Neumann operators S_j from Section 6.5.4, to the result

$$S_{mpt} = (\varrho_1^2 S_1 + \varrho_2^2 S_2)/[\![\varrho]\!]^2.$$

On should compare this operator with S_{d} arising in the Mullins-Sekerka flow. The spatial order is again $m = 3$.

Below we want to analyze these geometric evolution equations which can be written in the general form

$$V_\Gamma = \sigma G_\Gamma H_\Gamma, \quad t > 0, \quad \Gamma(0) = \Gamma_0. \tag{12.69}$$

As explained before, the aim is to find a family $\{\Gamma(t) : t > 0\}$ of sufficiently smooth compact hypersurfaces $\Gamma(t) \subset \Omega$ bounding a domain $\Omega_1(t)$ such that (12.69) holds. Here G_Γ denotes a linear operator, selfadjoint and positive semi-definite in $L_2(\Gamma)$, which is given by either of the following operators.

(G1) $G_\Gamma h = h - \bar{h}$ the averaged mean curvature flow;

(G2) $G_\Gamma h = -\Delta_\Gamma h$ the surface diffusion flow;

(G3) $G_\Gamma h = S_{\mathsf{d}} h$ the Mullins-Sekerka; flow

(G4) $G_\Gamma h = N_0^S h$ the Stokes flow;

(G5) $G_\Gamma h = S_{\mathsf{n}} h$ the Muskat flow;

(G6) $G_\Gamma h = S_0^{11} h$ the Stokes flow with phase transition;

(G7) $G_\Gamma h = S_{mpt} h$ the Muskat flow phase transition.

5.1 Volume, Area and Equilibria

Problem (12.69) has very important properties, namely conservation of volume and decreasing area. To be more precise, by the change of volume formula (2.97) we have

$$\partial_t |\Omega_1(t)| = \int_\Gamma V_\Gamma \, d\Gamma = \sigma \int_\Gamma G_\Gamma H_\Gamma \, d\Gamma = \sigma (H_\Gamma | G_\Gamma \mathsf{e})_\Gamma = 0,$$

where as before e denotes the function identically one on Γ. In fact, in all 7 cases we have $G_\Gamma \mathsf{e} = 0$.

In cases (G2), (G4) and (G5) there are further conservation laws if Ω_1 is not connected. Let Ω_1^k, $k = 1,\dots,m$, denote the components of Ω_1 and $\Gamma^k = \partial\Omega_1^k$ their boundaries, and let e_k denote the characteristic function of Γ_k. Then

$$\partial_t|\Omega_1^k(t)| = \int_{\Gamma^k} V_\Gamma\, d\Gamma = \sigma\int_{\Gamma^k} G_\Gamma H_\Gamma\, d\Gamma = \sigma(H_\Gamma|G_\Gamma \mathsf{e}_k)_\Gamma = 0, \quad k = 1,\dots,m.$$

Further, for the area functional we have by the change of area formula (2.92)

$$\partial_t|\Gamma(t)| = -\int_\Gamma V_\Gamma H_\Gamma\, d\Gamma = -\sigma\int_\Gamma (G_\Gamma H_\Gamma) H_\Gamma\, d\Gamma = -\sigma(G_\Gamma H_\Gamma|H_\Gamma)_\Gamma \le 0,$$

as G_Γ is selfadjoint and positive semi-definite in all 7 cases. Therefore, $\Phi(\Gamma) := |\Gamma|$ is a Lyapunov functional for (12.69).

But even more, it is a strict Lyapunov functional. In fact, suppose $\partial_t\Phi(\Gamma) = 0$ at some time t. Then $(G_\Gamma H_\Gamma|H_\Gamma) = 0$. With $h = H_\Gamma$ we have for (G1)

$$(G_\Gamma h|h)_\Gamma = |h - \bar h|_\Gamma^2 = 0,$$

hence $h = \bar h$ is constant on Γ, i.e., $h = a\mathsf{e}$ for some number $a \in \mathbb{R}$. This implies that the curvature H_Γ is constant all over Γ and so Γ must be a finite disjoint union of spheres of the same radius, as Ω is bounded.

For (G2) we obtain by the surface divergence theorem (2.24)

$$(G_\Gamma h|h)_\Gamma = -(\Delta_\Gamma h|h)_\Gamma = |\nabla_\Gamma h|_\Gamma^2 = 0,$$

hence h is constant on each component of Γ, i.e., $h = \sum_{k=1}^m a_k \mathsf{e}_k$ with some real numbers a_k. This implies that H_Γ is constant on each component Γ_k of Γ, and so Γ is a finite union of disjoint spheres with arbitrary radii.

To consider (G3), let v denote the solution of the elliptic problem

$$\begin{aligned} \Delta v &= 0 && \text{in } \Omega\setminus\Gamma,\\ \partial_\nu v &= 0 && \text{on } \partial\Omega,\\ [\![v]\!] = 0,\ v &= h && \text{on } \Gamma. \end{aligned}$$

Then by definition, $S_{\mathsf{d}} h = -[\![d\partial_\nu v]\!]$, hence by Proposition 10.5.1

$$(G_\Gamma h|h)_\Gamma = (S_{\mathsf{d}} h|h)_\Gamma = \int_\Omega d|\nabla v|^2\, dx = 0$$

implies that v is constant in the components of $\Omega\setminus\Gamma$. As v is continuous on Ω this shows that $h = v$ is constant all over Γ, which in turn yields, as for (G1), that Γ is a finite union of disjoint spheres with the same radius.

Next we consider the Stokes flow (G4). Let (u,π) denote the solution of the Stokes problem

$$\begin{aligned} -\mu\Delta u + \nabla\pi &= 0 && \text{in } \Omega\setminus\Gamma,\\ \operatorname{div} u &= 0 && \text{in } \Omega\setminus\Gamma,\\ u &= 0 && \text{on } \partial\Omega,\\ [\![u]\!] &= 0 && \text{on } \Gamma,\\ -2[\![\mu D(u)\nu_\Gamma]\!] + [\![\pi]\!]\nu_\Gamma &= h\nu_\Gamma && \text{on } \Gamma. \end{aligned}$$

Then by definition $N_0^S h = u \cdot \nu_\Gamma$, hence by Proposition 10.5.2

$$(G_\Gamma h|h)_\Gamma = (N_0^S h|h)_\Gamma = 2\int_\Omega \mu|D(u)|^2\,dx = 0$$

implies $D(u) = 0$ in $\Omega\setminus\Gamma$. Korn's inequality shows $\nabla u = 0$ in Ω, hence u is constant in the components of $\Omega\setminus\Gamma$. Therefore, π is constant in the components of $\Omega\setminus\Gamma$ as well, and so $h = [\![\pi]\!]$ has this property too. As for (G2) we see that Γ is a finite union of disjoint spheres of arbitrary radii.

Consider the Muskat flow (G5). Let p be the solution of the problem

$$\begin{aligned} \Delta p &= 0 && \text{in } \Omega\setminus\Gamma,\\ \partial_\nu p &= 0 && \text{on } \partial\Omega,\\ [\![k\partial_\nu p]\!] &= 0 && \text{on } \Gamma,\\ [\![p]\!] &= h && \text{on } \Gamma. \end{aligned}$$

Then $G_\Gamma = S_{\mathsf{n}}$ implies

$$(G_\Gamma h|h)_\Gamma = (S_{\mathsf{n}} h|h) = \int_\Omega |\nabla p|^2\,dx \geq 0.$$

Therefore, G_Γ is positive semi-definite, and if $G_\Gamma h = 0$ then h is constant in the components of $\Omega\setminus\Gamma$, and so we see that, as for (G4), Γ is a union of finitely many disjoint spheres of arbitrary radii.

Next to last, we take a look at the Stokes flow with phase transition. In this case we have

$$(G_\Gamma h|h)_\Gamma = (S_0^{11} h|h)_\Gamma = 2\int_\Omega \mu|D(u)|^2\,dx = 0.$$

So also in this case G_Γ is positive semi-definite, and if $G_\Gamma h = 0$ then by Lemma 1.2.1 we obtain $u = 0$ in Ω, hence the pressure π is constant in each phase Ω_j, which implies that h is constant all over Γ.

In a similar way we proceed for the Muskat flow with phase transition. In this case we obtain

$$(G_\Gamma h|h)_\Gamma = (S_{mpt} h|h) = \int_\Omega k|\nabla\pi|^2 \geq 0.$$

Therefore, π is constant in the components of the phases, and even in the phases as $[\![\pi/\varrho]\!] = 0$, which yields h constant on Γ.

Let us summarize these results in

Proposition 12.5.1. *The geometric evolution equation* (12.69) *has the following properties.*

(i) *The volume $|\Omega_1|$ is conserved along smooth solutions.*

(ii) *For* (G2), (G4) *and* (G5) *the volumes $|\Omega_1^k|$ of the components of Ω_1 are preserved.*

(iii) *The area functional* $\phi(\Gamma) = |\Gamma|$ *is a strict Lyapunov functional for* (12.69).

(iv) *The (non-degenerate) equilibria for* (12.69) *in case* (G1), (G3) *and* (G6), (G7) *consist of finitely many disjoint spheres of the same radius.*

(v) *The (non-degenerate) equilibria for* (12.69) *in case* (G2), (G4), *and* (G5) *consist of finitely many disjoint spheres of arbitrary radii.*

In the sequel, we denote the set of non-degenerate equilibria of (12.69) by $\mathcal{E}$.

5.2 Local Well-posedness

For local well-posedness of this problem we parameterize $\Gamma(t)$ over an analytic reference hypersurface Σ which is C^2-close to Γ_0, as explained in Chapter 2. The transformed problem then reads

$$\beta(h)\partial_t h - \sigma G_\Gamma(h)H_\Gamma(h) = 0, \quad t > 0, \quad h(0) = h_0, \tag{12.70}$$

where h means the height function which parameterizes $\Gamma(t)$ over Σ. Recalling the quasilinear structure of $H_\Gamma(h)$ we may apply Theorem 5.1.1 to the transformed problem. As state space we want to employ $X_\gamma = W^s_q(\Sigma)$ with $s > 2 + (n-1)/p$, where p is typically large. Then we have the embedding $X_\gamma \hookrightarrow C^2(\Sigma)$, which means that the curvatures are well-defined, pointwise.

To be more precise, for the M-Problems **(G3), (G5), (G7)** we want to employ Corollaries 6.6.5 and 6.7.4, so we choose as a base space $X_0 = W_p^{1-1/p}(\Sigma)$ and as regularity space the domain of the corresponding semigroups which is $X_1 = W_p^{4-1/p}(\Sigma)$, as these operators are of order 3. This yields $X_\gamma = W_p^{4-4/p}(\Sigma)$ and $X_{\gamma,\mu} = W_p^{1+3\mu-4/p}(\Sigma)$. Thus in this case we have $s = s_M := 4 - 4/p$, and for the embedding $X_{\gamma,\mu} \hookrightarrow C^2(\Sigma)$ we choose $1 \geq \mu > \mu_M := 1/3 + (n+3)/3p$.

For the S-Problems **(G4), (G6)** we want to employ Sections 8.1.3 and 8.4.3 for the corresponding semigroups. Therefore, we choose as a base space $X_0 = W_p^{2-1/p}(\Sigma)$ and for the regularity space $X_1 = W_p^{3-1/p}(\Sigma)$ as the order is $m = 1$. This yields $X_\gamma = W_p^{3-2/p}(\Sigma)$ and $X_{\gamma,\mu} = W^{2+\mu-2/p}(\Sigma)$. Hence we have $s = s_S := 3 - 2/p$ and for the embedding $X_{\gamma,\mu} \hookrightarrow C^2(\Sigma)$ we require $1 \geq \mu > \mu_S := (n+1)/p$.

For the remaining problems **(G1), (G2)** we employ of course the results in Section 6.4. Here we have more freedom for the choice of s. For (**G1**) it is convenient to choose $X_0 = H^1_p(\Sigma)$ and $X_1 = H^3_p(\Sigma)$ which leads to $X_\gamma = W_p^{3-2/p}(\Sigma)$ and $X_{\gamma,\mu} = W_p^{1+2(\mu-1/p)}(\Sigma)$. Therefore, in this case $s = s_{MC} = s_S$ and μ should satisfy $1 \geq \mu > \mu_{MC} := 1/2 + (n+1)/2p$, which is smaller than μ_S.

Finally, for **(G2)** we set $X_0 = L_p(\Sigma)$ and $X_1 = H^4_p(\Sigma)$ which implies $X_\gamma = W_p^{4-4/p}(\Sigma)$ and $X_{\gamma,\mu} = W_p^{4\mu-4/p}(\Sigma)$. Thus $s = s_{SD} := s_M$ and $1 \geq \mu > \mu_{SD} := 1/4 + (n+3)/4p$, which is smaller than μ_M.

Then we can verify the assumptions of Theorem 5.1.1, the most important one being maximal L_p-regularity. The main result is the following.

Theorem 12.5.2. *Let $p \in (1,\infty)$, and let the spaces X_0, X_1, X_γ and $X_{\gamma,\mu}$ be defined as above.*

Then in each of the cases (G1)$\sim$(G7), (12.69) *is locally well-posed in the sense that the transformed problem* (12.70) *is locally well-posed for initial values $h_0 \in X_{\gamma,\mu}$ which are small in $C^1(\Sigma)$. Furthermore, the map $t \mapsto \Gamma(t)$ is real analytic.*

Proof. We want to rewrite (12.70) as quasilinear evolution equation

$$\partial_t h + A(h)h = F(h), \quad t>0, \quad h(0)=h_0,$$

where h_0 is small even in $C^2(\Sigma)$, and to apply the theory from Section 5.

For this purpose, we recall that the transformed Laplace-Beltrami operator from Section 2.2.3 is given by

$$\Delta_\Gamma = a_0(h,\nabla_\Sigma h) : \nabla_\Sigma^2 + a_1(h,\nabla_\Sigma h,\nabla_\Sigma^2 h)\cdot\nabla_\Sigma,$$

where a_0 and a_1 are real analytic functions of their arguments, $a_0(0)=I$, $a_1(0)=0$, and $-\Delta_\Gamma$ is strongly elliptic provided h is small in $C^1(\Sigma)$. We observe that a_1 is linear w.r.t. $\nabla_\Sigma^2 h$.

Also recall the representation of the curvature H_Γ from Section 2.2.5 which reads

$$H_\Gamma(h) = \beta(h)(c_0(h,\nabla_\Sigma h):\nabla_\Sigma^2 h + c_1(h,\nabla_\Sigma h)),$$

where c_0 and c_1 are real analytic functions, $c_0(0)=I$, $c_1(0)=0$, and $-H_\Gamma$ is strongly elliptic if h is small in $C^1(\Sigma)$. Actually, we have $c_0=a_0$.

As all parameters are constant and the transformed gradient $\mathcal{G}(h)$ is real analytic in h, we see that the maps $h \mapsto G_\Gamma(h)$ are real analytic, provided h is small in $C^1(\Sigma)$.

Furthermore, we may write $\beta(h)^{-1}G_\Gamma = G_\Sigma(h)$, resulting in the problem

$$\partial_t h - \sigma G_\Sigma(h)H_\Gamma(h) = 0, \quad t>0, \quad h(0)=h_0. \tag{12.71}$$

Note that G_Σ are linear pseudo-differential operators on Σ. We decompose

$$\begin{aligned}-\sigma G_\Sigma(h)H_\Gamma(h) &= -\sigma G_\Sigma(h)c_0(h,\nabla_\Sigma h):\nabla_\Sigma^2 h - \sigma G_\Sigma(h)c_1(h,\nabla_\Sigma h)\\ &=: A(h)h - F(h).\end{aligned}$$

For **(G1)**, we decompose further

$$A(h)v = -\sigma\big(c_0:\nabla_\Sigma^2 v - \overline{c_0:\nabla_\Sigma^2 v}\big) = A_0(h)v + A_1(h)v,$$

where the bar means the mean value on Γ. By (2.42) we have

$$A_1(h)v = -\Big(\int_\Sigma \alpha(h)\mu(h)\,d\Sigma\Big)^{-1}\int_\Sigma \alpha(h)\mu(h)A_0(h)v\,d\Sigma.$$

As $A_0(h) : H_p^2(\Sigma) \to L_p(\Sigma)$ is bounded, we see that $A_1(h) \in \mathcal{B}(H_p^2(\Sigma), \mathbb{C})$, hence $A_1(h)$ is lower order.

For **(G2)**, we proceed slightly differently. We have

$$\begin{aligned}\sigma\beta^{-1}\Delta_\Gamma H_\Gamma h &= \sigma\beta^{-1}(a_0 : \nabla_\Sigma^2 + a_1 \cdot \nabla_\Sigma)\beta(c_0 : \nabla_\Sigma^2 h + c_1)\\ &= b_0(h, \nabla_\Sigma h) :: \nabla_\Sigma^4 h + b_1(h, \nabla_\Sigma h, \nabla_\Sigma^2 h) : \cdot\nabla_\Sigma^3 h + b_2(h, \nabla_\Sigma h, \nabla_\Sigma^2 h)\\ &= A_0(h)h + A_1(h)h - F(h),\end{aligned}$$

and in addition, b_1 depends only linearly on $\nabla_\Sigma^2 h$. Observe that $A_1(h)$ is also lower order, as it maps $H_p^3(\Sigma)$ into the base space $X_0 = L_p(\Sigma)$.

So, by the techniques developed in Section 9.5, it is not difficult to show that $(A, F) : B_{X_{\gamma,\mu}}(0, r) \to \mathcal{B}(X_1, X_0) \times X_0$ is real analytic, provided $r > 0$ is small enough. The key for this is the embedding $X_{\gamma,\mu} \hookrightarrow C^2(\Sigma)$ which is ensured by the choices of μ for the different problems. Moreover, as mentioned before, from the results in Sections 6.4, 6.6, 6.7, and Sections 8.1, 8.4, and perturbation, we have maximal L_p-regularity for all problems in question, again provided h_0 is small in $C^1(\Sigma)$. Therefore, Theorems 5.1.1 and 5.2.1 apply to prove local well-posedness as well as analyticity in time. For analyticity in space we may follow the arguments presented in Section 9.4. □

5.3 Stability of Equilibria

To study stability of the equilibria, we first consider an equilibrium $\Gamma_* \in \mathcal{E}$ bounding the domain $\Omega_1 \subset \Omega$. Recall that the equilibria consist of finitely many spheres and form a real analytic sub-manifold of $\mathcal{MH}^2(\Omega)$. If Γ_* has m components then the manifold $\mathcal{E}$ has dimension $mn+1$ for (G1), (G3) and (G6), (G7) and $m(n+1)$ for (G2), (G4) and (G5). Given such an equilibrium Γ_* we choose as the reference hypersurface $\Sigma = \Gamma_*$. The linearization of the transformed problem then reads

$$\partial_t h + \sigma G_\Sigma \mathcal{A}_\Sigma h = 0, \quad h(0) = 0. \tag{12.72}$$

This follows from the fact that the Fréchet derivative of $G_\Sigma(h)H_\Gamma(h)$ at $h = 0$ (in the direction of g) can be evaluated by

$$\frac{d}{d\varepsilon}\Big|_{\varepsilon=0} G_\Sigma(\varepsilon g)H_\Gamma(\varepsilon g) = \frac{d}{d\varepsilon}\Big|_{\varepsilon=0} G_\Sigma(\varepsilon g)H_\Gamma(0) + G_\Sigma(0)\frac{d}{d\varepsilon}\Big|_{\varepsilon=0} H_\Gamma(\varepsilon g) = -G_\Sigma\mathcal{A}_\Sigma g,$$

as $H_\Sigma = H_\Gamma(0)$ is constant on equilibria, and $G_\Sigma(\varepsilon g)\mathsf{e} = 0$. As the operator $G_\Sigma\mathcal{A}_\Sigma$ has maximal regularity, we may apply the stability results from Chapter 5, once we have shown that 0 is normally stable or normally hyperbolic for (12.72). Note that $G_\Sigma\mathcal{A}_\Sigma$ has compact resolvent by boundedness of Ω, so we only need to consider its eigenvalues.

(a) We begin with eigenvalue 0. So let $G_\Sigma\mathcal{A}_\Sigma h = 0$. Then $\mathcal{A}_\Sigma h$ belongs to the kernel of G_Σ, which means $\mathcal{A}_\Sigma h = a\mathsf{e}$ for (G1), (G3), and (G6), (G7), and $\mathcal{A}_\Sigma h = \sum_{k=1}^m a_k \mathsf{e}_k$ for cases (G2), (G4) and (G5). This implies $h = h_0 - (R^2/(n-1))a\mathsf{e}$ for (G1), (G3), and (G6), (G7), and $h = h_0 - \sum_{k=1}^m (R_k^2/(n-1))a_k\mathsf{e}_k$ in case of

(G2), (G4) and (G5), where $h_0 \in \mathsf{N}(\mathcal{A}_\Sigma)$. Therefore, the dimension of the kernel $\mathsf{N}(G_\Sigma \mathcal{A}_\Sigma)$ equals the dimension of the manifold $\mathcal{E}$, more precisely, it equals the dimension of the tangent space $T_{\Gamma_*}\mathcal{E}$.

(b) To see that the eigenvalue 0 is semi-simple for $G_\Sigma \mathcal{A}_\Sigma$, suppose $(G_\Sigma \mathcal{A}_\Sigma)^2 h = 0$. Then for (G2), (G4) and (G5)

$$G_\Sigma \mathcal{A}_\Sigma h = h_0 + \sum_{k=1}^{m} a_k \mathsf{e}_k, \qquad \text{for some } h_0 \in \mathsf{N}(\mathcal{A}_\Sigma),\ a_k \in \mathbb{C}.$$

Multiplying this relation with e_j in $L_2(\Sigma)$ we obtain $a_k = 0$ for all k, as G_Σ is selfadjoint and $G_\Sigma \mathsf{e}_k = 0$. As $\mathcal{A}_\Sigma$ is also selfadjoint, multiplying with $\mathcal{A}_\Sigma h$, we obtain $(G_\Sigma \mathcal{A}_\Sigma h | \mathcal{A}_\Sigma h)_\Sigma = 0$, hence $G_\Sigma \mathcal{A}_\Sigma h = 0$. The argument for (G1), (G3), and (G6), (G7) is similar. Consequently 0 is semi-simple for $G_\Sigma \mathcal{A}_\Sigma$.

(c) Now suppose that $\lambda \in \mathbb{C}$, $\lambda \neq 0$, is an eigenvalue for $-G_\Sigma \mathcal{A}_\Sigma$, i.e.,

$$\lambda h + G_\Sigma \mathcal{A}_\Sigma h = 0,$$

for some nontrivial h. Taking the inner product with $\mathcal{A}_\Sigma h$ in $L_2(\Sigma)$ we obtain

$$\lambda (h | \mathcal{A}_\Sigma h)_\Sigma + (G_\Sigma \mathcal{A}_\Sigma h | \mathcal{A}_\Sigma h)_\Sigma = 0.$$

As G_Σ and $\mathcal{A}_\Sigma$ are selfadjoint, this identity implies that λ must be real, hence the spectrum of $G_\Sigma \mathcal{A}_\Sigma$ is real.

We consider now the cases (G2), (G4) and (G5); then $(h|\mathsf{e}_k)_\Sigma = 0$ for all $k = 1, \ldots, m$. Suppose $\lambda > 0$. As G_Σ is positive semi-definite and $\mathcal{A}_\Sigma$ is so on the orthogonal complement of $\operatorname{span}\{\mathsf{e}_k\}_{k=1}^m$ we see that $(h|\mathcal{A}_\Sigma h) = 0$. This implies $\mathcal{A}_\Sigma h = 0$ and then $h = 0$ as $\lambda > 0$. Therefore, for (G2), (G4) and (G5) there are no nonzero eigenvalues with nonnegative real part, hence in this case (12.72) is normally stable.

In cases (G1), (G3), and (G6), (G7), we only obtain $(h|\mathsf{e})_\Sigma = 0$. As $\mathcal{A}_\Sigma$ is positive semi-definite on functions with mean zero if and only if Σ is connected, we may conclude normal stability for (G1), (G3) and (G6), (G7), provided Σ is connected.

(d) Next we show that $G_\Sigma \mathcal{A}_\Sigma$ has exactly $m-1$ positive eigenvalues in cases (G1), (G3), and (G6), (G7), when Σ has m components Σ^k. Therefore in this case we see that (12.72) is normally hyperbolic.

For (G1) set $h = \sum_{k=1}^m a_k \mathsf{e}_k$ with numbers a_k such that $\sum_{k=1}^m a_k = 0$. These functions form a subspace of $L_2(\Sigma)$ of dimension $m-1$. We have

$$\mathcal{A}_\Sigma h = -\frac{n-1}{R^2} \sum_{k=1}^{m} a_k \mathsf{e}_k, \quad \overline{\mathcal{A}_\Sigma h} = (\mathcal{A}_\Sigma h | e)_\Sigma / |\Sigma| = 0,$$

hence

$$G_\Sigma \mathcal{A}_\Sigma h = \mathcal{A}_\Sigma h = -\frac{n-1}{R^2} h,$$

and so $\lambda = (n-1)/R^2$ is a positive eigenvalue for $-G_\Sigma \mathcal{A}_\Sigma$ with multiplicity $m-1$. This proves the assertion for (G1).

(e) The proof in case of (G3) is more involved. We may consider the eigenvalue problem for $\lambda > 0$ in the space $L_{2,0}(\Sigma)$ of L_2-functions with zero mean. On this space G_Σ is invertible and its inverse is N_0^H in the terminology of Proposition 10.5.1; N_0^H is compact. Define $B_\lambda := \lambda N_0^H + \mathcal{A}_\Sigma$. This operator with domain $\mathsf{D}(B_\lambda) = \mathsf{D}(\mathcal{A}_\Sigma) = H_2^2(\Sigma)$ is selfadjoint and depends continuously on $\lambda \geq 0$. Now B_0 has the negative eigenvalue $\lambda = -(n-1)/R^2$, with eigenfunctions $h = \sum_{k=1}^m a_k \mathsf{e}_k$ with numbers a_k such that $\sum_{k=1}^m a_k = 0$, hence it is $(m-1)$-dimensional. We want to show that B_λ is positive definite for large λ, and so as λ increases from 0 to ∞, $m-1$ eigenvalues of B_λ will be crossing the imaginary axis along the real line, thereby generating $m-1$ unstable eigenvalues of $-G_\Sigma \mathcal{A}_\Sigma$.

To show that B_λ is positive definite for large λ, we proceed as follows. The operator $-G_\Sigma \mathcal{A}_\Sigma$ generates an analytic C_0-semigroup in $L_2(\Sigma)$, hence $\lambda + G_\Sigma \mathcal{A}_\Sigma$ is invertible for large λ. By the perturbation theorem for analytic semigroups, $-G_\Sigma \mathcal{A}_\Sigma - \mu G_\Sigma = -G_\Sigma(\mathcal{A}_\Sigma + \mu)$ generate analytic semigroups as well and with $\mu_0 = (n-1)/R^2$ we obtain a number $\lambda_1 > 0$ such that $\lambda + G_\Sigma(\mathcal{A}_\Sigma + \mu)$ is invertible for all $\lambda \geq \lambda_1$ and $\mu \in [0, \mu_0]$. This implies that $B_\lambda + \mu = \lambda N_0^H + \mathcal{A}_\Sigma + \mu$ invertible for all $\lambda \geq \lambda_1$ and for all $\mu \in [0, \mu_0]$, as $\mathcal{A}_\Sigma + \mu_0 \geq 0$, we see that B_λ has no non-positive eigenvalues for $\lambda \geq \lambda_1$, hence its is positive definite for large λ.

(f) Finally, we consider (G6), (G7). In this case we know that G_Σ is positive semi-definite and invertible on $L_{2,0}(\Sigma)$, hence G_Σ^{-1} is positive definite on this space. Therefore, the operator $B_\lambda = \lambda G_\Sigma^{-1} + \sigma \mathcal{A}_\Sigma$ has an $(m-1)$-fold negative eigenvalue for $\lambda = 0$ and is positive definite for large λ. This shows that $m-1$ eigenvalues must cross the imaginary axis through zero, as λ varies from 0 to ∞.

In summary, we have proved

Proposition 12.5.3. *For the linearized problem* (12.72) *at an equilibrium* $\Gamma_* \in \mathcal{E}$ *the following assertions are valid.*

(i) *For* (G2), (G4) *and* (G5), *and for* (G1), (G3) *and* (G6), (G7) *if additionally* Γ_* *is connected,* (12.72) *is normally stable.*

(ii) *For* (G1), (G3), *and* (G6), (G7) *in case* Γ_* *is not connected,* (12.72) *is normally hyperbolic.*

Now we may apply the nonlinear stability results of Chapter 5 to obtain the main result of this section.

Theorem 12.5.4. *Let Γ_* be a (non-degenerate) equilibrium of* (12.69). *Then the following assertions hold.*

(i) *For* (G2), (G4) *and* (G5), *and for* (G1), (G3) *and* (G6), (G7) *if additionally Γ_* is connected:*
$h_* = 0$ is stable for (12.70) *in $W^s_p(\Gamma_*)$. Any solution h starting close to $h_* = 0$ in $W^s_p(\Gamma_*)$ exists globally and converges to an equilibrium h_∞ of* (12.70) *in $W^s_p(\Gamma_*)$ at an exponential rate.*

(ii) *For* (G1), (G3), *and* (G6), (G7) *in case Γ_* is disconnected:*
$h_* = 0$ is unstable in $W^s_p(\Gamma_*)$. A solution h starting close to $h_* = 0$ and staying close to the set of equilibria in the topology of $W^s_p(\Gamma_*)$ exists globally and converges to some equilibrium h_∞ of (12.70) *in $W^s_p(\Gamma_*)$ at an exponential rate.*

In both cases, h_∞ corresponds to some $\Gamma_\infty \in \mathcal{E}$.

So in conclusion, the averaged mean curvature flow, the Mullins-Sekerka flow, and the Stokes and Muskat flows with phase transition see the phenomenon of Ostwald-ripening, while the surface diffusion flow, the Stokes flow, and the Muskat flow do not share this property. Physically speaking, (G1), (G3) and (G6), (G7) are spatially non-local so that different parts of the surface *see each other*. As (G2) is purely local in space, different parts of the surface move independently of each other. On the other hand, (G4) and (G5) are also non-local in space, but the coupling between different parts of the surface is not strong enough to enable Ostwald-ripening.

5.4 The Semiflow and Long-Time Behaviour
We define the state manifold of (12.69) by means of

$$\mathcal{SM}(\Omega) := \{\Gamma \in \mathcal{MH}^2(\Omega) : \Gamma \in W^s_p\}. \tag{12.73}$$

The charts for this manifold have been introduced in Chapter 2. The topology of $\mathcal{SM}(\Omega)$ is that induced by the canonical level functions φ_Γ in $W^s_p(\Omega)$, see Section 2.4.2. By Theorem 12.5.2 we see that given an initial surface $\Gamma_0 \in \mathcal{SM}(\Omega)$ we find $a > 0$ and $\Gamma : [0, a] \to \mathcal{SM}(\Omega)$ continuous such that $\Gamma(0) = \Gamma_0$ and $\Gamma(\cdot)$ is an L_p-solution in the sense that Γ is obtained as the push forward of the solution of the transformed problem (12.70). We may extend such an orbit in $\mathcal{SM}(\Omega)$ to a maximal time interval $J(\Gamma_0) := [0, t_+(\Gamma_0))$. Basically there are two facts which prevent the solution from being global, namely

- **Regularity**: the norm of $\Gamma(t)$ in W^s_p may become unbounded as $t \to t_+(\Gamma_0)$;
- **Geometry**: the topology of the interface $\Gamma(t)$ may change, or the interface may touch the boundary of Ω, or part of it may shrink to a point.

As before, we say that the solution $\Gamma(t)$ satisfies a *uniform ball condition*, if there is a radius $r > 0$ such that $\Gamma(J(\Gamma_0)) \subset \mathcal{MH}^2(\Omega, r)$. The main result of this section reads as follows.

Theorem 12.5.5. *Let $\Gamma(t)$ be a solution of the geometric evolution equation* (12.69) *on its maximal time interval $J(\Gamma_0)$. Assume furthermore that*

(i) $|\Gamma(t)|_{W_p^s} \le M < \infty$ *for all $t \in J(\Gamma_0)$, and*

(ii) $\Gamma(t)$ *satisfies a uniform ball condition for all $t \in J(\Gamma_0)$.*

Then $J(\Gamma_0) = \mathbb{R}_+$, i.e., the solution exists globally, and $\Gamma(t)$ converges in $\mathcal{SM}$ to an equilibrium $\Gamma_\infty \in \mathcal{E}$ at an exponential rate. The converse is also true: if a global solution converges in $\mathcal{SM}$ to an equilibrium, then (i) *and* (ii) *are valid.*

The proof relies on Theorem 5.7.2 and follows the same lines as that of Theorem 11.4.1.

Bibliographical Comments

The Stefan Problem: Historical Remarks

The Stefan problem is arguably the most studied free boundary problem, with over 1,200 mathematical publications devoted to the topic. It was first introduced in 1889 by Josef Stefan [269, 270] to describe the freezing of water in a lake or a bay. We refer to the books by Rubenstein [238] and Meirmanov [192] for further information.

In the *classical Stefan problem* one assumes that the (relative) temperature ϑ coincides with the melting temperature at the interface $\Gamma(t)$, and the following system is considered

$$\begin{aligned} \partial_t\vartheta - \Delta\vartheta &= 0 \quad \text{in } \Omega \setminus \Gamma(t), \\ \vartheta &= 0 \quad \text{on } \Gamma(t), \\ V - [\![\partial_\nu \theta]\!] &= 0 \quad \text{on } \Gamma(t), \end{aligned}$$

where 0 is the (scaled) melting temperature. The classical Stefan problem admits unique long-time weak solutions, provided the given data (that is, the initial temperature and the source terms) have appropriate signs; see for instance Friedman [112, 113], Kamenomostskaja [156], and Ladyženskaja, Solonnikov and Ural′ceva [167, pp. 496–503]. If the sign conditions are obstructed, then the problem becomes ill-posed, as was shown by DiBenedetto and Friedman [87]. Existence of weak solutions is closely tied to the maximum principle. Results concerning the regularity of weak solutions for the multidimensional classical one-phase Stefan problem were established by Caffarelli [54, 55], Caffarelli and Friedman [57], Friedman and Kinderlehrer [114], Kinderlehrer and Nirenberg [159, 160], Matano [185], while regularity results for the two-phase Stefan problem were obtained by Athanasopoulos, Caffarelli and Salsa [31, 32], Caffarelli and Evans [56], DiBenedetto [86], Sacks [241], and Ziemer [306], to list only a few references. Classical solutions for the classical Stefan problem were first established by Hanzawa [138] and Meĭrmanov [191]. We refer to the survey-research articles by Rodrigues [233, 234, 235], and also to Prüss, Saal and Simonett [212] for a more extensive account of the literature concerning the classical Stefan problem.

If the condition $\vartheta = 0$ is replaced by

$$\vartheta = -\sigma H_\Gamma \quad \text{on } \Gamma(t),$$

J. Prüss and G. Simonett, *Moving Interfaces and Quasilinear Parabolic Evolution Equations*, Monographs in Mathematics 105,
DOI 10.1007/978-3-319-27698-4

with H_Γ the mean curvature of $\Gamma(t)$ and σ a positive constant, the resulting problem is usually referred to as the *Stefan problem with surface tension*, or the *Stefan problem with Gibbs-Thomson correction*. Here we mention the monographs by Chalmers [62, Chapter 1], Hartman [139], Visintin [288], and the research papers by Caginalp [58], Gurtin [132, 133, 134], Langer [169], Mullins and Sekerka [200, 201], Visintin [287], where this law has been motivated and derived based on various mathematical and physical principles. As explained in Chapter 1, the condition $\vartheta = -\sigma H_\Gamma$ can be understood as a first order approximation of the relation

$$[\![\psi(\theta)]\!] + \sigma H_\Gamma = 0 \quad \text{on} \quad \Gamma(t)$$

around the melting temperature θ_m, where θ is the absolute temperature and ψ the free energy of the system.

The Stefan problem with the law $\vartheta = -\sigma H_\Gamma$ or $[\![\psi(\theta)]\!] = -\sigma H_\Gamma$, respectively, differs from the classical Stefan problem in a much more fundamental way than just in the modification of an interface condition. This becomes evident by the fact that the classical Stefan problem allows for a comparison principle, a property that is no longer shared by the Stefan problem with surface tension. A striking difference is also provided by the fact that in the classical Stefan problem, the temperature completely determines the phases, that is, the liquid region can be characterized by the condition $\vartheta > 0$, whereas $\vartheta < 0$ characterizes the solid region.

Although the Stefan problem with Gibbs-Thomson correction has been around for many decades, only few analytical results concerning existence of solutions can be found in the literature. Friedman and Reitich [115] considered the case with small surface tension $0 < \sigma \ll 1$ and linearized the problem about $\sigma = 0$. Assuming existence of a smooth solution for $\sigma = 0$, that is, for the classical Stefan problem, the authors proved existence and uniqueness of a weak solution for the *linearized* problem and then investigated the effect of small surface tension on the shape of $\Gamma(t)$. Existence of long-time weak solutions was first established by Luckhaus [181], using a discretized problem and a capacity-type estimate for approximating solutions. The weak solutions obtained have a sharp interface, but are highly non-unique. See also Röger [236], and Almgren and Wang [9]. A proof of existence – but without uniqueness – of local in time classical solutions was obtained in Radkevich [231, 232]. In Meirmanov [193], the way in which a spherical ball of ice in a supercooled fluid melts down was investigated.

The case of a strip-like geometry, where the free surface $\Gamma(t)$ is given as the graph of a function, was considered in Escher, Prüss and Simonett in [103], and existence as well as uniqueness of local in time classical solutions was established. Moreover, it was shown that solutions instantaneously regularize to become analytic in space and time. In Prüss and Simonett [219], linearized stability and instability of equilibria was investigated. Some nonlinear stability results were obtained by Guo, Hadžić [135, 136]. The stability and convergence results for the thermodynamically consistent Stefan problem in Chapters 11 and 12 of this monograph are taken from Prüss, Simonett and Zacher [228].

Free Boundary Problems for the Navier-Stokes Equations: Historical Remarks The *two-phase Navier-Stokes equations* with a free boundary (1.34) describe the motion of two incompressible, viscous, inmiscible fluids that are separated by $\Gamma(t)$. In case $\Omega_2(t) = \emptyset$ one obtains the *one-phase Navier-Stokes equations*, which describe the motion of an isolated liquid that moves due to capillary forces acting on the free boundary.

The *one-phase* problem has received wide attention in the last three decades or so. Existence and uniqueness of solutions for $\sigma = 0$, as well as $\sigma > 0$, in case that Ω_0 is bounded has been extensively studied in a long series of papers by Solonnikov, see for instance [258, 259, 260, 261, 262, 263, 267, 268], and Mogilevskiĭ and Solonnikov [198]. Results were established in anisotropic Sobolev-Slobodetskii as well as in Hölder spaces. Moreover, it was shown in [259] that if Ω_0 is sufficiently close to a ball and the initial velocity u_0 is sufficiently small, then the solution exists globally, and converges to a uniform rigid rotation of the liquid about a certain axis which is moving uniformly with a constant speed, see also Padula and Solonnikov [205].

More recently, local and global existence and uniqueness results (in case that Ω_0 is a bounded domain, a perturbed infinite layer, or a perturbed half-space) in anisotropic Sobolev spaces $W^{2,1}_{q,p}$ with $2 < p < \infty$ and $n < q < \infty$ have been established by Shibata and Shimizu [247, 245, 246] for $\sigma = 0$ as well as $\sigma > 0$. We also refer to Mucha and Zajaczkowski [199] and Abels [1] for other existence results.

The motion of a layer of viscous, incompressible fluid in an ocean of infinite extent, bounded below by a solid surface and above by a free surface which includes the effects of surface tension and gravity was considered by Allain [8], Beale [36], Beale and Nishida [37], Tani [277], and by Tani and Tanaka [278]. If the initial state and the initial velocity are close to equilibrium, global existence of solutions was proved in [36] for $\sigma > 0$, and in [278] for $\sigma \geq 0$, and the asymptotic decay rate for $t \to \infty$ was studied in [37].

Results concerning the *two-phase problem* are more recent. Existence and uniqueness of local strong solutions was first studied by Denisova [77, 78], Denisova and Solonnikov [79, 80]. Tanaka [276] considered the two-phase Navier-Stokes equations with thermo-capillary convection in bounded domains, and obtained existence and uniqueness of strong local solutions. Shimizu [248] established existence and uniqueness results in anisotropic $W^{2,1}_{q,p}$ -spaces for $2 < p < \infty$ and $n < q < \infty$.

Prüss and Simonett [220, 221, 222] considered the two-phase Navier-Stokes equations with $\sigma > 0$ in the situation where the free boundary $\Gamma(t)$ is given as the graph of a function over a hyperplane, and gravity is acting on the fluids [220, 222]. It was shown in [220, 222] that solutions regularize and immediately become real analytic in space and time. Finally, Köhne, Prüss and Wilke [163] obtained existence and uniqueness of strong solutions with maximal regularity for the two-phase Navier-Stokes equations with surface tension. They also proved asymptotic nonlinear stability results for equilibria, without gravity.

The approach used by Solonnikov, and also in [77, 78, 80, 245, 246, 247, 276, 277, 278], relies on the use of Lagrangian coordinates. In this formulation one obtains a transformed problem for the velocity and the pressure on a fixed domain, where the free boundary does not occur explicitly. The free boundary is then given by $\Gamma(t) = \{\xi + \int_0^t v(\tau,\xi)\,d\tau : \xi \in \Gamma_0\}$, where v is the velocity field in Lagrangian coordinates. With this approach it seems difficult to prove additional regularity properties of solutions, for instance smoothness of the free boundary. In addition, Lagrangian coordinates do not seem well-adapted in the presence of phase transitions.

Below, we comment on the literature relevant for the specific chapters in this book. Of course, our bibliography will necessarily be incomplete. Even worse, our selection of citations is biased by our background and preferences, and we apologize to all authors who did not receive credit for their works.

Chapter 1. The presented modeling follows first principles in physics and employs accepted constitutive laws. So for further reading any established book on Mechanics and Rational Thermodynamics is recommended, for example the classical monograph of Truesdell [285], or the more recent one of Drew and Passman [90]. But we emphasize that the first place (maybe because of our ignorance concerning physical literature) where we recognized the meta-principle of *no entropy production on the interface* has been the book of Ishii [152] which has been republished recently as [153]. We also benefited from the paper by Anderson, Cermelli, Fried, Gurtin, and McFadden [25]. For more up-to-date developments on modeling of even multi-component flows we refer to Bothe and Dreyer [42]. The Hanzawa transform explained in Section 1.3.2 has been first introduced by Hanzawa [138] in connection with the classical Stefan problem. It is nowadays one of the main tools to implement the so-called *direct mapping method* for reducing a free boundary problem to a problem on a fixed domain.

Chapter 2. For further background material in differential geometry we refer to the standard text books in this area, e.g. to do Carmo [88] and Kühnel [165]. Our approach follows to a large extent our recent paper [223]. We also mention Kimura [158] for other aspects on the geometry of moving hypersurfaces. For a different derivation of the local representation of the curvature operator in Section 2.2.5, see Escher and Simonett [107, 109]. The construction of the tubular neighbourhood in Section 2.3.1 via the uniform ball condition has been borrowed from Gilbarg and Trudinger [127, Section 14.6].

Chapter 3. The notion of *(pseudo)-sectorial operator* is very classical. This class of operators has already been used in the books of Hille and Phillips [147] and Dunford and Schwartz [91]. The ergodic theorem, Theorem 3.1.2, is probably even older. Many other mathematicians, for instance Triebel [282], Kato [157], Tanabe [274], considered this class of operators in case $\phi_A < \pi/2$. Common names have been *positive operators* (which is misleading!) or *operators of type* (M,ω).

Nowadays, the class of (pseudo)-sectorial operators can be considered as fairly well understood, and the results of Section 3.1 are now standard.

The same can be said about Section 3.2 on the simplest nontrivial operator, the time derivative, except for Section 3.2.4 on time-weighted spaces; the approach in this subsection follows the presentation in Prüss and Simonett [218].

Analytic semigroups are also well-known dating back to the work of Hille, Philips, and Yosida. They are an important corner-stone in modern analysis. The idea for the construction of the extended functional calculus used here is due to McIntosh [189]. The theory of complex powers of operators was developed independently by Balakrishnan, Kato, Krasnoselskii, and Sobolevskii. For alternative derivations of this theory we refer to the monographs of Kreĭn [164], Tanabe [274], Yosida [300], or to the more recents ones by Amann [15] and Lunardi [183], and also Denk, Hieber, and Prüss [81], Kunstmann and Weis [166].

Operators with bounded imaginary powers have been considered by Stein [271], and Seeley [243] in connection with complex interpolation, and by Dore and Venni [89] in connection with maximal L_p-regularity. The class $\mathcal{BIP}(X)$ has been introduced in the paper by Prüss and Sohr [229], and it is justified by its connection to the class $\mathcal{HT}$ of Banach spaces. The paper [229] also contains the basic permanence properties as well as the functional calculus based on the inverse Mellin transform. No nontrivial characterization of the class $\mathcal{BIP}(X)$ seems to be known, so far. However, in the Hilbert space case, Le Merdy [172] proved that a sectorial operator A with angle $\phi_A < \pi/2$ belongs to $\mathcal{BIP}(X)$ if and only if $-A$ is unitary equivalent to the generator of a contraction semigroup.

The $\mathcal{H}^\infty$*-calculus* of a sectorial operator has been introduced by McIntosh [189], where also the basic convergence lemma is proved. For further results on this class of operators we recommend the original literature by Yagi [297], Cowling, Doust, McIntosh and Yagi [73], and the more recent contributions by Arendt [28], Denk, Hieber and Prüss [81], and Kunstmann and Weis [166]. For examples of sectorial operators which do not have an $\mathcal{H}^\infty$-calculus, or do not admit bounded imaginary powers, we refer to the papers of McIntosh and Yagi [190], Baillon and Clément [33], Venni [286], and Hieber [142].

The connection between real interpolation and semigroup theory is well established; see e.g. Butzer and Berens [53] and Triebel [282]. Theorem 3.4.7 is a special case of a result due to Grisvard [130]. For a considerable extension of Theorem 3.4.8 we refer to the paper by Meyries and Veraar [196]. They prove

$$\mathsf{tr} : F^{s+\alpha}_{pq,\mu}(\mathbb{R}_+; Y) \cap F^s_{pq,\mu}(\mathbb{R}_+; D_A(\alpha, r)) \to D_A(\beta, p),$$

and

$$\mathsf{tr} : B^{s+\alpha}_{pq,\mu}(\mathbb{R}_+; Y) \cap B^s_{pq,\mu}(\mathbb{R}_+; D_A(\alpha, r)) \to D_A(\beta, q),$$

are bounded and surjective, for any Banach space Y, $1 < p < \infty$, $\alpha > 0$, $q, r \in [1, \infty]$, $1/p < \mu < 1 + 1/p$, where $\beta = s + \alpha - (1 - \mu) - 1/p \in (0, \alpha)$. Note that in the first case the trace space does not depend on q, r, and in the second one it is still independent of r! This is a truly remarkable result.

The definition of *maximal L_p-regularity* in an abstract setting has been introduced by Da Prato and Grisvard [74], who also proved Theorem 3.5.8. The earlier result in Hilbert spaces, Theorem 3.5.7, is due to de Simon [75]. The necessary conditions in Proposition 3.5.2 seem to be folklore, the proof given here is partly taken from the paper of Prüss [211], and the maximal regularity result in weighted L_p-spaces, Theorem 3.5.4, is due to Prüss and Simonett [218]. Maximal regularity results in the Triebel-Lizorkin scale can be found in the paper of Bu [51].

Chapter 4. The central definition of this chapter, that of *$\mathcal{R}$-boundedness*, was implicitly introduced and used by Bourgain [47] and later on also by Zimmermann [307]. Explicitly it is due to Berkson and Gillespie [39] and Clément, de Pagter, Sukochev and Witvliet [65]. We adopted most of the results in Sections 4.1 and 4.2 from the last paper. Property (α) has been introduced earlier by Pisier in [206]. These papers gave rise to the breakthrough in vector-valued harmonic analysis around the turn of the millennium, which solved the problem of boundedness of operator-valued Fourier multipliers in L_p-spaces, and at the same time lead to a characterization of maximal L_p-regularity.

It follows from the paper of Bourgain [48] that the *derivation operator d/dt* belongs to $\mathcal{BIP}(L_p(\mathbb{R};X))$, $1<p<\infty$, provided X is of class $\mathcal{HT}$. The converse of this statement was first proved in Prüss [209, Section 8], which established the fundamental relation between the classes $\mathcal{BIP}$ and $\mathcal{HT}$. It was known before that there are two other characterizations of the class $\mathcal{HT}$. These are
(i) X is *ζ-convex*. This means that there is a function $\zeta : X\times X \to \mathbb{R}$ such that ζ is convex in both variables and satisfies

$$0 \le \zeta(x,y) \le |x+y|, \quad x,y\in X.$$

(ii) X enjoys the *unconditional martingale difference property*, for short X is a *UMD*-space. This means that for every $p\in(1,\infty)$ there is a constant $C_p>0$ such that for any X-valued martingale $(f_k)_{k\ge 0}$ on a probability space $(\Omega,\mathcal{A},\mu)$, $N\in\mathbb{N}$, and any choice of signs $(\varepsilon_n)_{n\in\mathbb{N}}\subset\{-1,1\}$ the following estimate holds.

$$\Big|f_0+\sum_{k=1}^{N}\varepsilon_k(f_k-f_{k-1})\Big|_{L_p(\Omega,\mathcal{A},\mu;X)} \le C_p|f_N|_{L_p(\Omega,\mathcal{A},\mu;X)}.$$

For these equivalences as well as for the notion of a vector-valued martingale we refer to the survey article by Burkholder [52].

It is known that a Banach space of class $\mathcal{HT}$ is super-reflexive, and therefore admits an equivalent uniformly convex, uniformly smooth norm. On the other hand, Bourgain [47] gave an example of a super-reflexive Banach lattice which does not belong to the class $\mathcal{HT}$. Every Hilbert space belongs to $\mathcal{HT}$, and if $(\Omega,\mathcal{A},\mu)$ is a σ-finite measure space, $p\in(1,\infty)$, then $L_p(\Omega,\mathcal{A},\mu;X)$ is in $\mathcal{HT}$ if X has this property. Closed subspaces, quotients, duals and finite products are preserved, and complex interpolation spaces $(X,Y)_\theta$ as well as real interpolation spaces $(X,X)_{\theta,p}$ belong to the class $\mathcal{HT}$, provided $X,Y \in \mathcal{HT}$, $\theta \in (0,1)$ and $1 < p < \infty$. These facts are proved in Burkholder [52], see also Amann [15].

The necessary condition for *operator-valued Fourier multipliers*, Proposition 4.3.2, in the one-dimensional case is due to Clément and Prüss [67], while the one-dimensional operator-valued Fourier multiplier theorem, Theorem 4.3.3, is due to Weis [291]. The proof given here follows that given in Clément and Prüss [67]. The extension to n dimensions, Theorem 4.3.11, is independently due to Štrkalj and Weis [272], and to Haller, Heck, and Noll [137]. For a proof of Theorem 4.3.11 we refer to Denk, Hieber and Prüss [81]. The very simple proof of Theorem 4.3.9 based on Theorem 4.3.3, induction, and property (α) seems to be new. The proof of Theorem 4.3.14, i.e. the fact that the derivation operator belongs to $\mathcal{H}^\infty$ in weighted vector-valued L_p-spaces, is due to Prüss and Simonett [218].

There is an analogue of the operator-valued Fourier multiplier theorems in $L_p(\mathbb{S}^1;X)$, where $\mathbb{S}^1$ means the one-dimensional torus and $X \in \mathcal{HT}$, the vector-valued Marcinkiewicz multiplier theorem. This concerns Fourier series for periodic problems, and is due to Arendt and Bu [30]. More precisely, its statement is as follows. Suppose X,Y are Banach spaces of class $\mathcal{HT}$, and $\{M_k\}_{k\in\mathbb{Z}} \subset \mathcal{B}(X,Y)$ as well as $\{k(M_{k+1}-M_k)\}_{k\in\mathbb{Z}}$ are $\mathcal{R}$-bounded. Then $\{M_k\}_{k\in\mathbb{Z}}$ is a Fourier multiplier from $L_p(\mathbb{S}^1;X)$ to $L_p(\mathbb{S}^1;Y)$. This result is very useful for obtaining maximal L_p-regularity of evolution equations which are periodic in time.

For operator-valued Fourier multipliers in $B^s_{pq}(\mathbb{R}^n;X)$ we refer to the papers by Amann [16] and Girardi and Weis [129], and for analogous results in Triebel-Lizorkin spaces $F^s_{pq}(\mathbb{R}^n;X)$ we refer to Bu and Kim [51].

The notion of *$\mathcal{R}$-sectoriality* is due to Clément and Prüss [67]. The first main result of this section, Theorem 4.4.4, is due to Weis [291, 292], while the second one, Theorem 4.4.5, is due to Clément and Prüss [67].

The definition of an *$\mathcal{R}$-bounded $\mathcal{H}^\infty$-calculus*, i.e. Definition 4.5.1, is a natural extension of the concept of $\mathcal{H}^\infty$-calculus. It appeared first in the paper of Desch, Hieber and Prüss [85], where it was proved for the first time that the negative Laplacian on $\mathbb{R}^n$ admits an $\mathcal{R}$-bounded $\mathcal{H}^\infty$-calculus. Theorem 4.5.3 is new, it follows from the paper of Prüss and Simonett [218], and Theorem 4.5.4 has been proved in [67].

In the paper of Kalton and Weis [155] a very powerful tool has been developed, namely an *operator-valued functional calculus*. The main result in that paper is Theorem 4.5.6, and its corollaries in somewhat weaker form are also in there. However, we want to mention two results which had been known before.

(i) The first one is due to Sobolevskii [254]. He calls two sectorial operators A and B a *coercive pair*, if they are commuting in the resolvent sense, and there is a constant $C > 0$ such that

$$|Ax| + t|Bx| \le C|Ax + tBx|, \quad x \in \mathsf{D}(A) \cap \mathsf{D}(B),\ t > 0.$$

He then proved the *Mixed derivative theorem* for coercive pairs, which states that there is another constant $C > 0$ such that

$$|A^\beta B^{1-\beta}x| \le C|Ax + Bx|, \quad x \in \mathsf{D}(A) \cap \mathsf{D}(B),\ \beta \in [0,1].$$

This result is more general than Corollary 4.5.10, as Corollary 4.5.9 implies that A and B form a coercive pair provided $A \in \mathcal{H}^\infty(X)$ and $B \in \mathcal{RS}(X)$ are commuting in the resolvent sense, and $\phi_A^\infty + \phi_B^R < \pi$.

(ii) The second one is due to Dore and Venni [89]. They proved the following result. Suppose $X \in \mathcal{HT}$, $A, B \in \mathcal{BIP}(X)$ are invertible and commuting in the resolvent sense, and assume the parabolicity condition $\theta_A + \theta_B < \pi$. Then the operator $A + B$ with natural domain $\mathsf{D}(A + B) = \mathsf{D}(A) \cap \mathsf{D}(B)$ is closed, invertible and sectorial, with angle $\phi_{A+B} \le \max\{\theta_A, \theta_B\}$. This result was improved in Prüss and Sohr [229], where invertibility of the operators was dropped, and $A + B \in \mathcal{BIP}(X)$ was proved, with $\theta_{A+B} \le \max\{\theta_A, \theta_B\}$. As a consequence, the Dore-Venni theorem can be iterated. These results correspond to Corollary 4.5.9, although the assumptions are different. Nevertheless, we call this corollary also *Dore-Venni theorem.* In addition, we refer to Prüss [209, Section 8].

Fractional Evolution Equations are currently in the focus of interest for many researchers. These problems form a subclass of the class of *Evolutionary Integral Equations.* For the theory of such equations we refer to the monograph of Prüss [210] and its reprinted version. As this subject is not central in this book, we do not include further references, but recommend the papers by Zacher [301, 302, 303, 304] which contain important advances in the theory of fractional evolution equations.

For time-space embeddings like those in Section 4.5.5 in the vector-valued Besov- and Triebel-Lizorkin scales we refer to Meyries and Veraar [196].

Chapter 5. The theory of *abstract parabolic evolution equations* was initiated by the pioneering work of Tosio Kato, Hiroki Tanabe, and independently by Pavel E. Sobolevskii in the early sixties of the last century. For the quasilinear case, their work has been continued by Herbert Amann, Wolf von Wahl, the Italian school around Guiseppe Da Prato and Alessandra Lunardi, and by many other authors. For an account of these early developments we refer e.g. to the monographs of Amann [15], Lunardi [183], and Tanabe [275]. We also refer to the publications by Amann [10, 11, 12, 13, 14] for important advances.

Our simple and direct approach employing *maximal L_p-regularity* goes back to the paper by Clément and Li [66]. In the paper by Prüss [211] this approach has been worked out in more detail, in particular concerning regularity and the principle of linearized stability. We also refer to Amann [18] for extensions and additional results. The idea of using time-weights in the L_p-setting stems from Prüss and Simonett [218], and it has been employed in Köhne, Prüss and Wilke [162] and in LeCrone, Prüss and Wilke [173] to reduce initial regularity, and to gain regularity as well as compactness of the semiflow.

For a treatment of quasilinear parabolic equations in the framework of continuous maximal regularity, taking into account time-weights, we refer to Angenent [27], Clément and Simonett [68], and also to Simonett [252, 253].

The idea of extracting regularity via the implicit function theorem, known as the *parameter trick*, goes back to Angenent [27, 26]. This idea has been extended and employed in a series of papers by Escher, Prüss, Simonett, and Shao [111, 105, 104, 103, 222, 244, 213].

Sections 5.3 and 5.5 on the analysis of normally stable and normally hyperbolic equilibria are taken from [224], while Section 5.6 is originally in [226]. The proof of the instability result, Theorem 5.4.1, appears here for the first time.

The methods developed in Chapter 5 are very flexible and well-suited for quasilinear parabolic partial differential equations with linear autonomous boundary conditions, or for problems without boundaries, but are not directly applicable in the case of nonlinear boundary conditions. Nevertheless, the methods developed here can be employed in the presence of nonlinear boundary conditions, due to the maximal regularity results in Part 3 of this book, in particular those concerning inhomogeneous boundary data. These sharp results allow to extend the theory from Chapter 5 almost completely to quasilinear parabolic partial differential equations with fully nonlinear boundary conditions. For this we refer to the papers by Latushkin, Prüss and Schnaubelt [170, 171] and to Prüss, Simonett and Zacher [226]. Even more, in the same way extensions to the case of dynamic boundary conditions have been studied. Based on the paper by Denk, Prüss and Zacher [84] which covers the linear theory of such problems, this is done in the papers by Meyries and Schnaubelt [195] and Gal and Meyries [119].

Chapter 6. *Elliptic and parabolic systems* are of course very old topics, and we find it impossible to account for the complete history here. For this we refer to the Monographs of Gilbarg and Trudinger [127] for the elliptic case and to Lieberman [176] for parabolic systems. We also refer to the fundamental contributions of the Russian school, which are Ladyženskaja and Ural′ceva [168] for elliptic systems, and Ladyženskaja, Solonnikov and Ural′ceva [167] for parabolic problems. In fact, in these books the first results on maximal L_p-regularity have been obtained.

Here we concentrate on more recent developments, including *infinite dimensional state spaces.* To our knowledge, the first thorough study of differential operators with operator-valued coefficients is the paper by Amann [17]. His approach is based on Fourier-multiplier results in vector-valued Besov-spaces [16] and interpolation theory. In particular, this paper contains the definition of normal ellipticity. Our definition of parameter ellipticity in Section 6.1 was used by Agranovič and Višik [3]. It parallels the notion of (κ, θ)-ellipticity in Amann [17], see also Amann, Hieber and Simonett [23] for the finite-dimensional case.

Based on the operator-valued Fourier-multiplier results from Chapter 4, the paper by Denk, Hieber and Prüss [81] contains a general L_p-theory for elliptic and parabolic problems in infinite-dimensional state spaces. Here we borrowed from that paper, in particular for Section 6.1 and partly for 6.2 and 6.3. In contrast to the presentation in [81], where a large emphasis was placed on kernel estimates, our present approach is more based on Fourier-multipliers, thereby extending and simplifying proofs of known results. We also implemented the techniques from

Denk, Hieber and Prüss [82] to include inhomogeneous boundary conditions and to cover the case $p \neq q$.

We want to point out that the two basic conditions on a boundary value problem $(\mathcal{A}(x,D), \mathcal{B}_1(x,D), \ldots, \mathcal{B}_m(x,D))$ of order $2m$, namely *normal ellipticity* and the *Lopatinskii-Shapiro condition*, are also necessary for maximal L_p-regularity of the L_p-realization A_B of this boundary value problem. This has also been proved in the paper by Denk, Hieber, and Prüss [82]. Therefore, besides regularity assumptions, the results in Sections 6.1-6.3 are optimal. It seems that the notion of *normal strong ellipticity* which emerged from the paper of Bothe and Prüss [44] has not yet been considered in the case of an infinite dimensional state space.

There are surprisingly few results on *transmission problems* in the mathematical literature, although such problems are omnipresent in Mathematical Physics. Therefore, and as we have been in need for results on transmission problems, we devoted Section 6.5 to this topic. We have also included a systematic treatment of Dirichlet-to-Neumann operators for elliptic as well as for parabolic problems. These results are somewhat folklore, but hard to find in the literature.

The remaining subsections, dealing with the linearized *Stefan and Verigin problem* as well as their nephews, the linearized *Mullins-Sekerka and Muskat problem*, in the form presented here, seem to be new. For more general parabolic systems with dynamic boundary conditions or boundary conditions of relaxation type, we refer to the paper by Denk, Prüss, and Zacher [84], and to the recent book by Denk and Kaip [83].

Finally, we also refer to the more recent contributions by Amann concerning parabolic equations on uniformly regular and singular manifolds [19, 20, 21, 22].

Chapter 7. There is a huge body of mathematical literature on the *Stokes problem* and the *Stokes operator*. It is impossible to give credit to all mathematicians who contributed; instead we refer to the monographs of Galdi [120], Sohr [255], and Temam [280]. We mention here only the pioneering works of Solonnikov [257], Borchers and Sohr [41], Giga [122], Giga and Sohr [125, 126], Miyakawa and Sohr [197], and Sohr and von Wahl [256].

Surprisingly, references concerning maximal L_p-regularity for the *generalized linear Stokes equation* are rare. We only know of the papers of Solonnikov [266] and Bothe and Prüss [44]. The exposition in this chapter extends and improves the approach presented in [44].

Also surprising is the fact that mathematical papers on the classical Stokes or Navier-Stokes problem almost exclusively employ the no-slip condition at the boundary. In Engineering, other boundary conditions are of equal importance, such as the pure slip, the Navier condition, or outflow and free (or Neumann) conditions. To the best of our knowledge, the first paper considering Navier conditions in the framework of maximal regularity has been by Saal [240]. An up to date mathematical study of such boundary conditions, and even others which are energy preserving, is contained in the paper by Bothe, Köhne, and Prüss [43]. For further references concerning such "non-standard" boundary conditions we refer to the discussion given in [43].

Our proofs in Section 7.4 are inspired by, and follow to a large extent, the seminal paper by Simader and Sohr [250], and the monograph [251] by the same authors.

Chapters 8–11. The results in these chapters form the core of the book. They are due to the authors and their co-authors, combining, improving and extending the results from [163, 214, 215, 216, 217, 221, 222, 225, 227, 228].

In these chapters, we did not take into account external forces like gravity. Actually, as it is a lower order pertubation, it is not difficult to include gravity in the maximal regularity results in Section 8 and also in those results on local well-posedness and regularity in Section 9. However, in the presence of gravity the only equilibrium without boundary contact will be a flat interface $\Gamma_* = \mathbb{R}^{n-1}$, with $\Omega_2 = \mathbb{R}^n_+$, and $\Omega_1 = \mathbb{R}^n_-$. For this configuration, in [220] we have developed a spectral theory similar to that for Problem **(P2)**, showing that the case $\rho_1 > \rho_2$ is linearly stable, and $\rho_2 > \rho_1$ is unstable. Note that the spectrum of L_2 is non-discrete in this case. Our paper [220] contains the first mathematically rigorous proof for this famous *Rayleigh-Taylor* instability in viscous two-phase flows in the nonlinear case. We also refer to Wang and Tice [290] for similar results in a periodic setting.

The Rayleigh-Taylor instability has recently been investigated for the case of a capillary by Wilke [296]. To avoid the difficulty with contact angles, he assumes slip conditions on the boundary of the capillary and a 90 degree contact angle. The trivial equilibrium is then a flat interface, vanishing velocity, and constant pressures. Wilke establishes global well-posedness of the problem for small initial data, and he proves nonlinear stability for this equilibrium, provided the radius a of the capillary is smaller than a critical number $a_c > 0$, which depends on the densities, the surface tension, and acceleration of earth. He further finds that at a_c a subcritical bifurcation occurs, and the trivial as well as the bifurcating solutions are unstable for $a > a_c$.

Chapter 12. Section 12.1 extends the main theorem in Bothe and Prüss [44] to the case $p \neq q$ and also to outflow boundary conditions, which were left out in that paper. The stability result for the trivial solution is new. There is a large mathematical literature on *generalized Newtonian fluids*, in particular on power law fluids. We refer to the references given in [44].

The theory of *nematic liquid crystal flows* was initiated in the fundamental papers by Ericksen [95] and Leslie [174]. There has been much work on these equations, in particular in recent years; see Lin [178], Lin and Liu [179, 180], Hu and Wang [149], Lin, Liu and Wang [177], Wang [289], Hu, Wang and Wen [150], and Li and Wang [175]. The results in Section 12.2 improve those of the paper by Hieber, Nesensohn, Prüss and Schade [143], which so far contains the best results on local strong well-posedness and stability of equilibria for the so-called *isothermal simplified Ericksen-Leslie model.* Concerning extensions of this theory to the non-isothermal case, compressible or incompressible, and with or without stretching, we refer to Hieber and Prüss [145, 146, 144].

The analysis in Section 12.3 extends that given in the paper of Herberg, Meyries, Prüss, Wilke [141]. As a standard reference for *Maxwell-Stefan diffusion* we recommend the monograph of Giovangigli [128], and the recent paper by Bothe and Dreyer [42].

In Section 12.4 the main results of the paper Prüss, Simonett, Wilke [225] are reproduced. In that paper the *Stefan problem with variable surface tension* has been studied mathematically for the first time.

The geometric evolution equations considered in Section 12.5 have been studied by many researchers and there is a huge body of literature. The following selection is necessarily incomplete, and many interesting and important contributions go unmentioned.

The Averaged Mean Curvature Flow

Huisken [151] (and Gage [118] in the case of curves) proved the fundamental result that solutions exist globally and converge exponentially fast to a sphere, provided the initial surface is uniformly convex and smooth. Moreover, it is shown in [118, 151] that the surfaces remain uniformly convex for all $t > 0$. Escher and Simonett [110] established that solutions for small $C^{1,\alpha}$-perturbations of a sphere exist globally and converge exponentially fast to some sphere. It was shown in Mayer and Simonett [187] that the volume-preserving mean curvature flow can drive embedded hypersurfaces to self-intersections in finite time.

The Surface Diffusion Flow

The surface diffusion flow was first derived by Mullins [202] and has subsequently been rederived and studied by several groups of authors, for instance by Taylor and Cahn [279], Cahn, Elliott and Novick-Cohen [59]. In two dimensions, the surface diffusion flow for closed embedded curves was first investigated by Elliott and Garcke [93], who proved both global existence and stabilty results when the initial curve is close to a circle. These results were extended by Escher, Mayer and Simonett [100] to the multi-dimensional case. In more detail, existence and uniqueness of classical solutions was obtained for immersed $C^{2,\alpha}$-initial hypersurfaces, and it was shown that solutions exist globally and converge exponentially fast to a sphere, provided Γ_0 is close to a sphere in the $C^{2,\alpha}$-norm initially. Wheeler [294] showed that global existence and convergence still holds if the L_2-integral of the trace-free part of the second fundamental form is sufficiently small, see also [188, 293, 295]. We refer to Giga and Ito [123], Mayer and Simonett [187], and Blatt [40] for results concerning loss of embeddedness; to Escher, Giga, Ito [96, 124, 154], and Blatt [40] for results concerning loss of convexity and formation of singularities; to Escher, Mayer, Simonett [99, 100, 186] for numerical results showing the formation of singularities; and to Escher and Mucha [101], Koch and Lamm [161], and Shao [244] for results concerning the regularity of solutions.

The Mullins-Sekerka Flow
The Mullins-Sekerka flow was first considered in [200]. It has also been termed the quasi-stationary Stefan problem with surface tension, or the Hele-Shaw problem with surface tension. Existence of solutions has been investigated by Bazaliy [34], Chen [63], Chen, Hong and Yi [64], Escher and Simonett [106, 107, 108], and Prokert [208]. It was first proved by Escher and Simonett [109] that classical solutions for the two-phase problem exist globally and tend to spheres exponentially fast, provided Γ_0 is close to a sphere in the $C^{2,\alpha}$-norm initially, generalizing a result by Chen [63] for weak solutions of the one-phase problem in $\mathbb{R}^2$. Global existence and convergence for the one-phase problem was proved independently by Prokert [208]. Long-time existence of weak solutions was established by Röger [237], see also Luckhaus and Sturzenhecker [182], and Garcke and Sturzenhecker [121] for related results. Additional interesting results can be found in Alikakos, Bates, Chen, Fusco [4, 5], and Alikakos, Fusco, Korali [6, 7], for instance.

The Stokes Flow
Existence and uniqueness of solutions for the one-phase (multi-dimensional) Stokes flow was first obtained by Günther, Prokert [131, 207], see also Solonnikov [265]. Regularity results are contained in Escher, Günther, Prokert [102, 131, 207]. It was shown in [131] that in case the initial domain of a fluid drop is close to a ball, then the solution exits globally and converges to a ball at an exponential rate. This result was rederived by Friedman and Reitich [116] by a different method. The results for the two-phase problem in Section 12 are new.

The Muskat Flow
This system was first introduced by Muskat [203] in 1934 in order to model the interface between two fluids in a porous media, see also [204]. The Mullins-Sekerka and the Muskat problem are both closely related to fingering, a phenomenon which has received, and still continues to receive, considerable attention by many researchers. In the case of positive surface tension, the first result on the existence of classical solutions in two dimensions was obtained by Hong, Tao and Yi [148]. Regarding the stability of equilibria, Friedmann and Tao [117] proved stability of a circular steady-state in case that Ω_2 is unbounded. The authors of [13] state that the equilibrium is in general not asymptotically stable.

Escher and Matioc [98] considered the Muskat problem in a periodic geometry with surface tension and gravity included. Existence and uniqueness of classical solutions is obtained and the authors establish exponential stability of certain flat equilibria. Using bifurcation theory they also identify finger shaped steady-states which are all unstable. These results were later refined and extended in Ehrnström, Escher, Matioc [92, 97]. Bazaliy and Vasylyeva [35] first observed a waiting time behaviour for the two-dimensional Muskat problem with a non-regular initial surface in the presence of surface tension.

There is an extensive literature in the case of zero surface tension. Most of the investigations consider the Muskat problem in two dimensions. Without commenting in detail, we would like to mention the work of Ambrose [24], Castro,

Constantin, Còrdoba, Fefferman, Gancedo [60, 61, 70, 71, 72], Siegel, Caflisch and Howison [249], and Yi [298, 299], where various aspects concerning existence of solutions and loss of smoothness are analyzed.

The Stokes Flow and the Muskat Flow with Phase Transitions
These flows are considered and studied for the first time in this book.

Outlook and Future Challenges

In this book we have developed quite a few methods and techniques, and we have examined their potential to solve parabolic one- and two-phase problems. The following remarks might be useful to young mathematicians who intend to work on these topics.

Due to space limitations we did not treat all of the sub-problems introduced in Chapter 1 and we leave the opportunity to interested researchers to study these problems.

There are interesting related issues, extending far beyond this monograph, which are important from the physical point of view and challenging from the mathematical side. Below we mention some of them.

(i) The Compressible Case

If one wants to model *evaporation* in a physically reasonable way, one has to take into account that at least one phase (the vapour phase) ought to be considered compressible. This means that Problems **(P4)** and **(P6)** should be analyzed in the compressible case.

(ii) Surface Viscosity

Already in 1915, Boussinesq [49] proposed that interfaces should carry a kind of surface viscosity. A first model taking this into account was proposed by Scriven [242] in 1962. So far there are only few mathematical investigations concerning this model, and we are only aware of the contribution by Bothe and Prüss [45] and the recent PhD-thesis of Meyer [194]. To the best of our knowledge, so far there are no papers on this model where phase transitions are taken into account.

(iii) Multi-Component Flows and Mass Transfer

Most chemical processes involve two-phase flows, as at least one of the major reactants has to change phase before reactions can occur. For this reason, both the modeling and the analysis of *reactive multi-component flows* is one of the major challenges from the physical and the mathematical point of view. Besides the Navier-Stokes equations describing the flows in the phases, one also has to take into account mass transfer, reactions and resulting changes in temperature and pressure, as well as phase transitions driven by chemical potentials, pressure, and temperature. The mathematical understanding of such processes will be of great importance in the future.

J. Prüss and G. Simonett, *Moving Interfaces and Quasilinear Parabolic Evolution Equations*, Monographs in Mathematics 105,
DOI 10.1007/978-3-319-27698-4

(iv) Contact Angles

An important issue which we avoided in this book is the situation where the interface Γ touches the outer boundary $\partial\Omega$. This is known as the *contact angle problem* in the literature. It has led to many controversial discussions concerning physically correct modeling, in particular the issue whether at the contact line one needs to impose additional conditions. This seems to be settled by now. For a physically sound model we refer to Bothe and Prüss [46]. Mathematically, only few results are known so far, which all deal with the quasi-steady case, where one can propose a stationary contact angle. As shown by Pukhnachev and Solonnikov [230, 264], the concept of a stationary contact angle is not feasible for the two phase Navier-Stokes flow if one insists on Dirichlet conditions on the outer boundary, as the total energy will become infinite.

(v) Singularities

In this book we obtained local well-posedness in suitable function spaces for arbitrarily large data, global existence as well as stability for small data, i.e. data which are close to an equilibrium, and convergence of solutions as time goes to infinity, at least in the stable case. We also presented criteria for global existence: we have global existence if the solutions under consideration do not develop *singularities*.

In general, one cannot expect global existence for all solutions, due to possible blow up of the relevant norms or geometrical degenerations. Therefore, it is very important to study the development of singularities, to understand mathematically and physically which properties enforce singularities, either in norm or geometrically, and on the contrary which ensure global existence. In our opinion, this is the most challenging question which we presently cannot answer.

List of Figures

J. Prüss and G. Simonett, *Moving Interfaces and Quasilinear Parabolic Evolution Equations*, Monographs in Mathematics 105,
DOI 10.1007/978-3-319-27698-4

Bibliography

[1] H. Abels. The initial-value problem for the Navier-Stokes equations with a free surface in L^q-Sobolev spaces. *Adv. Differential Equations*, 10(1):45–64, 2005.

[2] R. A. Adams. *Sobolev spaces.* Academic Press [A subsidiary of Harcourt Brace Jovanovich, Publishers], New York-London, 1975. Pure and Applied Mathematics, Vol. 65.

[3] M. S. Agranovič and M. I. Višik. Elliptic problems with a parameter and parabolic problems of general type. *Uspehi Mat. Nauk*, 19(3 (117)):53–161, 1964.

[4] N. D. Alikakos, P. W. Bates, and X. Chen. Convergence of the Cahn-Hilliard equation to the Hele-Shaw model. *Arch. Rational Mech. Anal.*, 128(2):165–205, 1994.

[5] N. D. Alikakos, P. W. Bates, X. Chen, and G. Fusco. Mullins-Sekerka motion of small droplets on a fixed boundary. *J. Geom. Anal.*, 10(4):575–596, 2000.

[6] N. D. Alikakos and G. Fusco. Ostwald ripening for dilute systems under quasistationary dynamics. *Comm. Math. Phys.*, 238(3):429–479, 2003.

[7] N. D. Alikakos, G. Fusco, and G. Karali. Ostwald ripening in two dimensions—the rigorous derivation of the equations from the Mullins-Sekerka dynamics. *J. Differential Equations*, 205(1):1–49, 2004.

[8] G. Allain. Small-time existence for the Navier-Stokes equations with a free surface. *Appl. Math. Optim.*, 16(1):37–50, 1987.

[9] F. Almgren and L. Wang. Mathematical existence of crystal growth with Gibbs-Thomson curvature effects. *J. Geom. Anal.*, 10(1):1–100, 2000.

[10] H. Amann. Quasilinear evolution equations and parabolic systems. *Trans. Amer. Math. Soc.*, 293(1):191–227, 1986.

[11] H. Amann. Dynamic theory of quasilinear parabolic equations. I. Abstract evolution equations. *Nonlinear Anal.*, 12(9):895–919, 1988.

[12] H. Amann. Dynamic theory of quasilinear parabolic systems. III. Global existence. *Math. Z.*, 202(2):219–250, 1989.

[13] H. Amann. Dynamic theory of quasilinear parabolic equations. II. Reaction-diffusion systems. *Differential Integral Equations*, 3(1):13–75, 1990.

[14] H. Amann. Nonhomogeneous linear and quasilinear elliptic and parabolic boundary value problems. In *Function spaces, differential operators and nonlinear analysis (Friedrichroda, 1992)*, volume 133 of *Teubner-Texte Math.*, pages 9–126. Teubner, Stuttgart, 1993.

[15] H. Amann. *Linear and quasilinear parabolic problems. Vol. I*, volume 89 of *Monographs in Mathematics.* Birkhäuser Boston, Inc., Boston, MA, 1995. Abstract linear theory.

[16] H. Amann. Operator-valued Fourier multipliers, vector-valued Besov spaces, and applications. *Math. Nachr.*, 186:5–56, 1997.

J. Prüss and G. Simonett, *Moving Interfaces and Quasilinear Parabolic Evolution Equations*, Monographs in Mathematics 105,
DOI 10.1007/978-3-319-27698-4

[17] H. Amann. Elliptic operators with infinite-dimensional state spaces. *J. Evol. Equ.*, 1(2):143–188, 2001.

[18] H. Amann. Quasilinear parabolic problems via maximal regularity. *Adv. Differential Equations*, 10(10):1081–1110, 2005.

[19] H. Amann. *Anisotropic function spaces and maximal regularity for parabolic problems. Part 1.* Jindřich Nečas Center for Mathematical Modeling Lecture Notes, 6. Matfyzpress, Prague, 2009. Function spaces.

[20] H. Amann. Anisotropic function spaces on singular manifolds. 2012. arXiv.1204.0606.

[21] H. Amann. Function spaces on singular manifolds. *Math. Nachr.*, 286(5-6):436–475, 2013.

[22] H. Amann. Parabolic equations on uniformly regular riemannian manifolds and degenerate initial boundary value problems. In *Recent Developments of Mathematical Fluid Mechanics*, Advances in Mathematical Fluid Mechanics. Birkhäuser, Basel, 2016.

[23] H. Amann, M. Hieber, and G. Simonett. Bounded H_∞-calculus for elliptic operators. *Differential Integral Equations*, 7(3-4):613–653, 1994.

[24] D. M. Ambrose. Well-posedness of two-phase Hele-Shaw flow without surface tension. *European J. Appl. Math.*, 15(5):597–607, 2004.

[25] D. M. Anderson, P. Cermelli, E. Fried, M. E. Gurtin, and G. B. McFadden. General dynamical sharp-interface conditions for phase transformations in viscous heat-conducting fluids. *J. Fluid Mech.*, 581:323–370, 2007.

[26] S. Angenent. Parabolic equations for curves on surfaces. I. Curves with p-integrable curvature. *Ann. of Math. (2)*, 132(3):451–483, 1990.

[27] S. B. Angenent. Nonlinear analytic semiflows. *Proc. Roy. Soc. Edinburgh Sect. A*, 115(1-2):91–107, 1990.

[28] W. Arendt. Semigroups and evolution equations: functional calculus, regularity and kernel estimates. In *Evolutionary equations. Vol. I*, Handb. Differ. Equ., pages 1–85. North-Holland, Amsterdam, 2004.

[29] W. Arendt, C. J. Batty, M. Hieber, and F. Neubrander. *Vector-valued Laplace transforms and Cauchy problems*, volume 96 of *Monographs in Mathematics*. Birkhäuser/Springer Basel AG, Basel, second edition, 2011.

[30] W. Arendt and S. Bu. The operator-valued Marcinkiewicz multiplier theorem and maximal regularity. *Math. Z.*, 240(2):311–343, 2002.

[31] I. Athanasopoulos, L. A. Caffarelli, and S. Salsa. Regularity of the free boundary in parabolic phase-transition problems. *Acta Math.*, 176(2):245–282, 1996.

[32] I. Athanasopoulos, L. A. Caffarelli, and S. Salsa. Phase transition problems of parabolic type: flat free boundaries are smooth. *Comm. Pure Appl. Math.*, 51(1):77–112, 1998.

[33] J.-B. Baillon and P. Clément. Examples of unbounded imaginary powers of operators. *J. Funct. Anal.*, 100(2):419–434, 1991.

[34] B. V. Bazaliy. The Stefan problem for the Laplace equation taking into account the curvature of the free boundary. *Ukraïn. Mat. Zh.*, 49(10):1299–1315, 1997.

[35] B. V. Bazaliy and N. Vasylyeva. The Muskat problem with surface tension and a nonregular initial interface. *Nonlinear Anal.*, 74(17):6074–6096, 2011.

[36] J. T. Beale. Large-time regularity of viscous surface waves. *Arch. Rational Mech. Anal.*, 84(4):307–352, 1983/84.

[37] J. T. Beale and T. Nishida. Large-time behavior of viscous surface waves. In *Recent topics in nonlinear PDE, II (Sendai, 1984)*, volume 128 of *North-Holland Math. Stud.*, pages 1–14. North-Holland, Amsterdam, 1985.

[38] J. Bergh and J. Löfström. *Interpolation spaces. An introduction.* Springer-Verlag, Berlin-New York, 1976. Grundlehren der Mathematischen Wissenschaften, No. 223.

[39] E. Berkson and T. A. Gillespie. Spectral decompositions and harmonic analysis on UMD spaces. *Studia Math.*, 112(1):13–49, 1994.

[40] S. Blatt. Loss of convexity and embeddedness for geometric evolution equations of higher order. *J. Evol. Equ.*, 10(1):21–27, 2010.

[41] W. Borchers and H. Sohr. On the semigroup of the Stokes operator for exterior domains in L^q-spaces. *Math. Z.*, 196(3):415–425, 1987.

[42] D. Bothe and W. Dreyer. Continuum thermodynamics of chemically reacting fluid mixtures. *Acta Mech.*, 226(6):1757–1805, 2015.

[43] D. Bothe, M. Köhne, and J. Prüss. On a class of energy preserving boundary conditions for incompressible Newtonian flows. *SIAM J. Math. Anal.*, 45(6):3768–3822, 2013.

[44] D. Bothe and J. Prüss. L_p-theory for a class of non-Newtonian fluids. *SIAM J. Math. Anal.*, 39(2):379–421, 2007.

[45] D. Bothe and J. Prüss. On the two-phase Navier-Stokes equations with Boussinesq-Scriven surface fluid. *J. Math. Fluid Mech.*, 12(1):133–150, 2010.

[46] D. Bothe and J. Prüss. On the interface formation model for dynamic triple lines. In *Mathematical Fluid Dynamics, Present and Future*, Springer Proceedings in Mathematics & Statistics. Springer, 2016.

[47] J. Bourgain. Some remarks on Banach spaces in which martingale difference sequences are unconditional. *Ark. Mat.*, 21(2):163–168, 1983.

[48] J. Bourgain. Vector-valued singular integrals and the H^1-BMO duality. In *Probability theory and harmonic analysis (Cleveland, Ohio, 1983)*, volume 98 of *Monogr. Textbooks Pure Appl. Math.*, pages 1–19. Dekker, New York, 1986.

[49] M. Boussinesq. Sur le existence de une viscosite superficielle, dans la mice couche de transition separant un liquide dune autre fluide contigu. *Ann. Chim. Phys.*, 29:349–357, 1913.

[50] H. Brezis. *Functional analysis, Sobolev spaces and partial differential equations.* Universitext. Springer, New York, 2011.

[51] S. Bu and J.-M. Kim. Operator-valued Fourier multiplier theorems on Triebel spaces. *Acta Math. Sci. Ser. B Engl. Ed.*, 25(4):599–609, 2005.

[52] D. L. Burkholder. Martingales and Fourier analysis in Banach spaces. In *Probability and analysis (Varenna, 1985)*, volume 1206 of *Lecture Notes in Math.*, pages 61–108. Springer, Berlin, 1986.

[53] P. L. Butzer and H. Berens. *Semi-groups of operators and approximation.* Die Grundlehren der mathematischen Wissenschaften, Band 145. Springer-Verlag New York Inc., New York, 1967.

[54] L. A. Caffarelli. The regularity of free boundaries in higher dimensions. *Acta Math.*, 139(3-4):155–184, 1977.

[55] L. A. Caffarelli. Some aspects of the one-phase Stefan problem. *Indiana Univ. Math. J.*, 27(1):73–77, 1978.

[56] L. A. Caffarelli and L. C. Evans. Continuity of the temperature in the two-phase Stefan problem. *Arch. Rational Mech. Anal.*, 81(3):199–220, 1983.

[57] L. A. Caffarelli and A. Friedman. Continuity of the temperature in the Stefan problem. *Indiana Univ. Math. J.*, 28(1):53–70, 1979.

[58] G. Caginalp. An analysis of a phase field model of a free boundary. *Arch. Rational Mech. Anal.*, 92(3):205–245, 1986.

[59] J. W. Cahn, C. M. Elliott, and A. Novick-Cohen. The Cahn-Hilliard equation with a concentration dependent mobility: motion by minus the Laplacian of the mean curvature. *European J. Appl. Math.*, 7(3):287–301, 1996.

[60] Á. Castro, D. Córdoba, C. Fefferman, and F. Gancedo. Breakdown of smoothness for the Muskat problem. *Arch. Ration. Mech. Anal.*, 208(3):805–909, 2013.

[61] Á. Castro, D. Córdoba, C. Fefferman, F. Gancedo, and M. López-Fernández. Rayleigh-Taylor breakdown for the Muskat problem with applications to water waves. *Ann. of Math. (2)*, 175(2):909–948, 2012.

[62] B. Chalmers. *Principles of solidification.* Krieger, Huntington, 1977.

[63] X. Chen. The Hele-Shaw problem and area-preserving curve-shortening motions. *Arch. Rational Mech. Anal.*, 123(2):117–151, 1993.

[64] X. Chen, J. Hong, and F. Yi. Existence, uniqueness, and regularity of classical solutions of the Mullins-Sekerka problem. *Comm. Partial Differential Equations*, 21(11-12):1705–1727, 1996.

[65] P. Clément, B. de Pagter, F. A. Sukochev, and H. Witvliet. Schauder decomposition and multiplier theorems. *Studia Math.*, 138(2):135–163, 2000.

[66] P. Clément and S. Li. Abstract parabolic quasilinear equations and application to a groundwater flow problem. *Adv. Math. Sci. Appl.*, 3(Special Issue):17–32, 1993/94.

[67] P. Clément and J. Prüss. An operator-valued transference principle and maximal regularity on vector-valued L_p-spaces. In *Evolution equations and their applications in physical and life sciences (Bad Herrenalb, 1998)*, volume 215 of *Lecture Notes in Pure and Appl. Math.*, pages 67–87. Dekker, New York, 2001.

[68] P. Clément and G. Simonett. Maximal regularity in continuous interpolation spaces and quasilinear parabolic equations. *J. Evol. Equ.*, 1(1):39–67, 2001.

[69] R. R. Coifman and G. Weiss. *Transference methods in analysis.* American Mathematical Society, Providence, R.I., 1976. Conference Board of the Mathematical Sciences Regional Conference Series in Mathematics, No. 31.

[70] P. Constantin, D. Córdoba, F. Gancedo, and R. M. Strain. On the global existence for the Muskat problem. *J. Eur. Math. Soc. (JEMS)*, 15(1):201–227, 2013.

[71] A. Cordoba, D. Cordoba, and F. Gancedo. The Rayleigh-Taylor condition for the evolution of irrotational fluid interfaces. *Proc. Natl. Acad. Sci. USA*, 106(27):10955–10959, 2009.

[72] A. Córdoba, D. Córdoba, and F. Gancedo. Interface evolution: the Hele-Shaw and Muskat problems. *Ann. of Math. (2)*, 173(1):477–542, 2011.

[73] M. Cowling, I. Doust, A. McIntosh, and A. Yagi. Banach space operators with a bounded H^∞ functional calculus. *J. Austral. Math. Soc. Ser. A*, 60(1):51–89, 1996.

[74] G. Da Prato and P. Grisvard. Sommes d'opérateurs linéaires et équations différentielles opérationnelles. *J. Math. Pures Appl. (9)*, 54(3):305–387, 1975.

[75] L. de Simon. Un'applicazione della teoria degli integrali singolari allo studio delle equazioni differenziali lineari astratte del primo ordine. *Rend. Sem. Mat. Univ. Padova*, 34:205–223, 1964.

[76] K. Deimling. *Nonlinear functional analysis.* Springer-Verlag, Berlin, 1985.

[77] I. V. Denisova. A priori estimates for the solution of the linear nonstationary problem connected with the motion of a drop in a liquid medium. *Trudy Mat. Inst. Steklov.*, 188:3–21, 191, 1990. Translated in Proc. Steklov Inst. Math. 1991, no. 3, 1–24, Boundary value problems of mathematical physics, 14 (Russian).

[78] I. V. Denisova. Problem of the motion of two viscous incompressible fluids separated by a closed free interface. *Acta Appl. Math.*, 37(1-2):31–40, 1994. Mathematical problems for Navier-Stokes equations (Centro, 1993).

[79] I. V. Denisova and V. A. Solonnikov. Solvability in Hölder spaces of a model initial-boundary value problem generated by a problem on the motion of two fluids. *Zap. Nauchn. Sem. Leningrad. Otdel. Mat. Inst. Steklov. (LOMI)*, 188(Kraev. Zadachi Mat. Fiz. i Smezh. Voprosy Teor. Funktsii. 22):5–44, 186, 1991.

[80] I. V. Denisova and V. A. Solonnikov. Classical solvability of the problem of the motion of two viscous incompressible fluids. *Algebra i Analiz*, 7(5):101–142, 1995.

[81] R. Denk, M. Hieber, and J. Prüss. $\mathcal{R}$-boundedness, Fourier multipliers and problems of elliptic and parabolic type. *Mem. Amer. Math. Soc.*, 166(788), 2003.

[82] R. Denk, M. Hieber, and J. Prüss. Optimal L^p-L^q-estimates for parabolic boundary value problems with inhomogeneous data. *Math. Z.*, 257(1):193–224, 2007.

[83] R. Denk and M. Kaip. *General parabolic mixed order systems in L_p and applications*, volume 239 of *Operator Theory: Advances and Applications*. Birkhäuser/Springer, Cham, 2013.

[84] R. Denk, J. Prüss, and R. Zacher. Maximal L_p-regularity of parabolic problems with boundary dynamics of relaxation type. *J. Funct. Anal.*, 255(11):3149–3187, 2008.

[85] W. Desch, M. Hieber, and J. Prüss. L^p-theory of the Stokes equation in a half space. *J. Evol. Equ.*, 1(1):115–142, 2001.

[86] E. DiBenedetto. Continuity of weak solutions to certain singular parabolic equations. *Ann. Mat. Pura Appl. (4)*, 130:131–176, 1982.

[87] E. DiBenedetto and A. Friedman. The ill-posed Hele-Shaw model and the Stefan problem for supercooled water. *Trans. Amer. Math. Soc.*, 282(1):183–204, 1984.

[88] M. P. do Carmo. *Riemannian geometry*. Mathematics: Theory & Applications. Birkhäuser Boston, Inc., Boston, MA, 1992.

[89] G. Dore and A. Venni. On the closedness of the sum of two closed operators. *Math. Z.*, 196(2):189–201, 1987.

[90] D. A. Drew and S. L. Passman. *Theory of multicomponent fluids*, volume 135 of *Applied Mathematical Sciences*. Springer-Verlag, New York, 1999.

[91] N. Dunford and J. T. Schwartz. *Linear Operators. I. General Theory*. With the assistance of W. G. Bade and R. G. Bartle. Pure and Applied Mathematics, Vol. 7. Interscience Publishers, Inc., New York; Interscience Publishers, Ltd., London, 1958.

[92] M. Ehrnström, J. Escher, and B.-V. Matioc. Steady-state fingering patterns for a periodic Muskat problem. *Methods Appl. Anal.*, 20(1):33–46, 2013.

[93] C. M. Elliott and H. Garcke. Existence results for diffusive surface motion laws. *Adv. Math. Sci. Appl.*, 7(1):467–490, 1997.

[94] K.-J. Engel and R. Nagel. *One-parameter semigroups for linear evolution equations*, volume 194 of *Graduate Texts in Mathematics*. Springer-Verlag, New York, 2000. With contributions by S. Brendle, M. Campiti, T. Hahn, G. Metafune, G. Nickel, D. Pallara, C. Perazzoli, A. Rhandi, S. Romanelli and R. Schnaubelt.

[95] J. L. Ericksen. Hydrostatic theory of liquid crystals. *Arch. Rational Mech. Anal.*, 9:371–378, 1962.

[96] J. Escher and K. Ito. Some dynamic properties of volume preserving curvature driven flows. *Math. Ann.*, 333(1):213–230, 2005.

[97] J. Escher, A.-V. Matioc, and B.-V. Matioc. A generalized Rayleigh-Taylor condition for the Muskat problem. *Nonlinearity*, 25(1):73–92, 2012.

[98] J. Escher and B.-V. Matioc. On the parabolicity of the Muskat problem: well-posedness, fingering, and stability results. *Z. Anal. Anwend.*, 30(2):193–218, 2011.

[99] J. Escher, U. F. Mayer, and G. Simonett. On the surface diffusion flow. In *Navier-Stokes equations and related nonlinear problems (Palanga, 1997)*, pages 69–79. VSP, Utrecht, 1998.

[100] J. Escher, U. F. Mayer, and G. Simonett. The surface diffusion flow for immersed hypersurfaces. *SIAM J. Math. Anal.*, 29(6):1419–1433 (electronic), 1998.

[101] J. Escher and P. B. Mucha. The surface diffusion flow on rough phase spaces. *Discrete Contin. Dyn. Syst.*, 26(2):431–453, 2010.

[102] J. Escher and G. Prokert. Analyticity of solutions to nonlinear parabolic equations on manifolds and an application to Stokes flow. *J. Math. Fluid Mech.*, 8(1):1–35, 2006.

[103] J. Escher, J. Prüss, and G. Simonett. Analytic solutions for a Stefan problem with Gibbs-Thomson correction. *J. Reine Angew. Math.*, 563:1–52, 2003.

[104] J. Escher, J. Prüss, and G. Simonett. A new approach to the regularity of solutions for parabolic equations. In *Evolution equations*, volume 234 of *Lecture Notes in Pure and Appl. Math.*, pages 167–190. Dekker, New York, 2003.

[105] J. Escher and G. Simonett. Analyticity of the interface in a free boundary problem. *Math. Ann.*, 305(3):439–459, 1996.

[106] J. Escher and G. Simonett. On Hele-Shaw models with surface tension. *Math. Res. Lett.*, 3(4):467–474, 1996.

[107] J. Escher and G. Simonett. Classical solutions for Hele-Shaw models with surface tension. *Adv. Differential Equations*, 2(4):619–642, 1997.

[108] J. Escher and G. Simonett. Classical solutions for the quasi-stationary Stefan problem with surface tension. In *Differential equations, asymptotic analysis, and mathematical physics (Potsdam, 1996)*, volume 100 of *Math. Res.*, pages 98–104. Akademie Verlag, Berlin, 1997.

[109] J. Escher and G. Simonett. A center manifold analysis for the Mullins-Sekerka model. *J. Differential Equations*, 143(2):267–292, 1998.

[110] J. Escher and G. Simonett. The volume preserving mean curvature flow near spheres. *Proc. Amer. Math. Soc.*, 126(9):2789–2796, 1998.

[111] J. Escher and G. Simonett. Analyticity of solutions to fully nonlinear parabolic evolution equations on symmetric spaces. *J. Evol. Equ.*, 3(4):549–576, 2003. Dedicated to Philippe Bénilan.

[112] A. Friedman. The Stefan problem in several space variables. *Trans. Amer. Math. Soc.*, 133:51–87, 1968.

[113] A. Friedman. *Variational principles and free-boundary problems.* Pure and Applied Mathematics. John Wiley & Sons, Inc., New York, 1982. A Wiley-Interscience Publication.

[114] A. Friedman and D. Kinderlehrer. A one phase Stefan problem. *Indiana Univ. Math. J.*, 24(11):1005–1035, 1974/75.

[115] A. Friedman and F. Reitich. The Stefan problem with small surface tension. *Trans. Amer. Math. Soc.*, 328(2):465–515, 1991.

[116] A. Friedman and F. Reitich. Quasi-static motion of a capillary drop. II. The three-dimensional case. *J. Differential Equations*, 186(2):509–557, 2002.

[117] A. Friedman and Y. Tao. Nonlinear stability of the Muskat problem with capillary pressure at the free boundary. *Nonlinear Anal.*, 53(1):45–80, 2003.

[118] M. Gage. On an area-preserving evolution equation for plane curves. In *Nonlinear problems in geometry (Mobile, Ala., 1985)*, volume 51 of *Contemp. Math.*, pages 51–62. Amer. Math. Soc., Providence, RI, 1986.

[119] C. G. Gal and M. Meyries. Nonlinear elliptic problems with dynamical boundary conditions of reactive and reactive-diffusive type. *Proc. Lond. Math. Soc. (3)*, 108(6):1351–1380, 2014.

[120] G. P. Galdi. *An introduction to the mathematical theory of the Navier-Stokes equations. Vol. I*, volume 38 of *Springer Tracts in Natural Philosophy*. Springer-Verlag, New York, 1994. Linearized steady problems.

[121] H. Garcke and T. Sturzenhecker. The degenerate multi-phase Stefan problem with Gibbs-Thomson law. *Adv. Math. Sci. Appl.*, 8(2):929–941, 1998.

[122] Y. Giga. Domains of fractional powers of the Stokes operator in L_r spaces. *Arch. Rational Mech. Anal.*, 89(3):251–265, 1985.

[123] Y. Giga and K. Ito. On pinching of curves moved by surface diffusion. *Commun. Appl. Anal.*, 2(3):393–405, 1998.

[124] Y. Giga and K. Ito. Loss of convexity of simple closed curves moved by surface diffusion. In *Topics in nonlinear analysis*, volume 35 of *Progr. Nonlinear Differential Equations Appl.*, pages 305–320. Birkhäuser, Basel, 1999.

[125] Y. Giga and H. Sohr. On the Stokes operator in exterior domains. *J. Fac. Sci. Univ. Tokyo Sect. IA Math.*, 36(1):103–130, 1989.

[126] Y. Giga and H. Sohr. Abstract L^p estimates for the Cauchy problem with applications to the Navier-Stokes equations in exterior domains. *J. Funct. Anal.*, 102(1):72–94, 1991.

[127] D. Gilbarg and N. S. Trudinger. *Elliptic partial differential equations of second order.* Classics in Mathematics. Springer-Verlag, Berlin, 2001. Reprint of the 1998 edition.

[128] V. Giovangigli. *Multicomponent flow modeling.* Modeling and Simulation in Science, Engineering and Technology. Birkhäuser Boston, Inc., Boston, MA, 1999.

[129] M. Girardi and L. Weis. Operator-valued Fourier multiplier theorems on Besov spaces. *Math. Nachr.*, 251:34–51, 2003.

[130] P. Grisvard. Spazi di tracce e applicazioni. *Rend. Mat. (6)*, 5:657–729, 1972.

[131] M. Günther and G. Prokert. Existence results for the quasistationary motion of a free capillary liquid drop. *Z. Anal. Anwendungen*, 16(2):311–348, 1997.

[132] M. E. Gurtin. On the two-phase Stefan problem with interfacial energy and entropy. *Arch. Rational Mech. Anal.*, 96(3):199–241, 1986.

[133] M. E. Gurtin. Multiphase thermomechanics with interfacial structure. I. Heat conduction and the capillary balance law. *Arch. Rational Mech. Anal.*, 104(3):195–221, 1988.

[134] M. E. Gurtin. Toward a nonequilibrium thermodynamics of two-phase materials. *Arch. Rational Mech. Anal.*, 100(3):275–312, 1988.

[135] M. Hadžić. Orthogonality conditions and asymptotic stability in the Stefan problem with surface tension. *Arch. Ration. Mech. Anal.*, 203(3):719–745, 2012.

[136] M. Hadžić and Y. Guo. Stability in the Stefan problem with surface tension (I). *Comm. Partial Differential Equations*, 35(2):201–244, 2010.

[137] R. Haller, H. Heck, and A. Noll. Mikhlin's theorem for operator-valued Fourier multipliers in n variables. *Math. Nachr.*, 244:110–130, 2002.

[138] E.-i. Hanzawa. Classical solutions of the Stefan problem. *Tôhoku Math. J. (2)*, 33(3):297–335, 1981.

[139] P. Hartman. *Chrystal growth: An introduction.* North-Holland, 1973.

[140] D. Henry. *Geometric theory of semilinear parabolic equations*, volume 840 of *Lecture Notes in Mathematics.* Springer-Verlag, Berlin-New York, 1981.

[141] M. Herberg, M. Meyries, J. Prüss, and M. Wilke. Reaction-diffusion systems of Maxwell-Stefan type with reversible mass-action kinetics. Submitted.

[142] M. Hieber. Examples of pseudo-differential operators in L^p spaces with unbounded imaginary powers. *Arch. Math. (Basel)*, 66(2):126–131, 1996.

[143] M. Hieber, M. Nesensohn, J. Prüss, and K. Schade. Dynamics of nematic liquid crystal flows: the quasilinear approach. *Ann. Inst. H. Poncaré Anal. Non Linéaire*, 33(2):397–408, 2016.

[144] M. Hieber and J. Prüss. Dynamics of nematic liquid crystal flows ii: The compressible isotropic case. In *Mathematical Fluid Dynamics, Present and Future*, Springer Proceedings in Mathematics & Statistics. Springer, 2016.

[145] M. Hieber and J. Prüss. Thermodynamical consistent modeling and analysis of nematic liquid crystal flows. 2016.

[146] M. Hieber and J. Prüss. Dynamics of nematic liquid crystal flows i: The incompressible isotropic case. Submitted.

[147] E. Hille and R. S. Phillips. *Functional analysis and semi-groups*. American Mathematical Society Colloquium Publications, vol. 31. American Mathematical Society, Providence, R. I., 1957. rev. ed.

[148] J. Hong, Y. Tao, and F. Yi. Muskat problem with surface tension. *J. Partial Differential Equations*, 10(3):213–231, 1997.

[149] X. Hu and D. Wang. Global solution to the three-dimensional incompressible flow of liquid crystals. *Comm. Math. Phys.*, 296(3):861–880, 2010.

[150] T. Huang, C. Wang, and H. Wen. Strong solutions of the compressible nematic liquid crystal flow. *J. Differential Equations*, 252(3):2222–2265, 2012.

[151] G. Huisken. The volume preserving mean curvature flow. *J. Reine Angew. Math.*, 382:35–48, 1987.

[152] M. Ishii. *Thermo-Fluid Dynamic Theory of Two-Phase Flow.* Collection de la Direction des Études et Recherches d'Électricité de France, Paris, 1975.

[153] M. Ishii and T. Hibiki. *Thermo-fluid dynamics of two-phase flow.* Springer, New York, 2006. With a foreword by Lefteri H. Tsoukalas.

[154] K. Ito. Loss of convexity of compact hypersurfaces moved by surface diffusion. *Commun. Appl. Anal.*, 6(1):105–126, 2002.

[155] N. J. Kalton and L. Weis. The H^∞-calculus and sums of closed operators. *Math. Ann.*, 321(2):319–345, 2001.

[156] S. L. Kamenomostskaja. On Stefan's problem. *Mat. Sb. (N.S.)*, 53 (95):489–514, 1961.

[157] T. Kato. *Perturbation theory for linear operators.* Springer-Verlag, Berlin-New York, second edition, 1976. Grundlehren der Mathematischen Wissenschaften, Band 132.

[158] M. Kimura. Geometry of hypersurfaces and moving hypersurfaces in $\mathbb{R}^m$ for the study of moving boundary problems. In *Topics in mathematical modeling*, volume 4 of *Jindřich Nečas Cent. Math. Model. Lect. Notes*, pages 39–93. Matfyzpress, Prague, 2008.

[159] D. Kinderlehrer and L. Nirenberg. Regularity in free boundary problems. *Ann. Scuola Norm. Sup. Pisa Cl. Sci. (4)*, 4(2):373–391, 1977.

[160] D. Kinderlehrer and L. Nirenberg. The smoothness of the free boundary in the one phase Stefan problem. *Comm. Pure Appl. Math.*, 31(3):257–282, 1978.

[161] H. Koch and T. Lamm. Geometric flows with rough initial data. *Asian J. Math.*, 16(2):209–235, 2012.

[162] M. Köhne, J. Prüss, and M. Wilke. On quasilinear parabolic evolution equations in weighted L_p-spaces. *J. Evol. Equ.*, 10(2):443–463, 2010.

[163] M. Köhne, J. Prüss, and M. Wilke. Qualitative behaviour of solutions for the two-phase Navier-Stokes equations with surface tension. *Math. Ann.*, 356(2):737–792, 2013.

[164] S. G. Kreĭn. *Linear differential equations in Banach space.* Translations of Mathematical Monographs, Vol. 29. American Mathematical Society, Providence, R.I., 1971.

[165] W. Kühnel. *Differential geometry*, volume 16 of *Student Mathematical Library*. American Mathematical Society, Providence, RI, 2002. Curves—surfaces—manifolds.

[166] P. C. Kunstmann and L. Weis. Maximal L_p-regularity for parabolic equations, Fourier multiplier theorems and H^∞-functional calculus. In *Functional analytic methods for evolution equations*, volume 1855 of *Lecture Notes in Math.*, pages 65–311. Springer, Berlin, 2004.

[167] O. A. Ladyženskaja, V. A. Solonnikov, and N. N. Ural′ceva. *Linear and quasilinear equations of parabolic type*. Translations of Mathematical Monographs, Vol. 23. American Mathematical Society, Providence, R.I., 1968.

[168] O. A. Ladyzhenskaya and N. N. Ural′tseva. *Linear and quasilinear elliptic equations*. Academic Press, New York-London, 1968.

[169] J. Langer. Instabilities and pattern formation in crystal growth. *Rev. Mod. Phys. 52*, pages 1–28, 1980.

[170] Y. Latushkin, J. Prüss, and R. Schnaubelt. Stable and unstable manifolds for quasilinear parabolic systems with fully nonlinear boundary conditions. *J. Evol. Equ.*, 6(4):537–576, 2006.

[171] Y. Latushkin, J. Prüss, and R. Schnaubelt. Center manifolds and dynamics near equilibria of quasilinear parabolic systems with fully nonlinear boundary conditions. *Discrete Contin. Dyn. Syst. Ser. B*, 9(3-4):595–633, 2008.

[172] C. Le Merdy. The similarity problem for bounded analytic semigroups on Hilbert space. *Semigroup Forum*, 56(2):205–224, 1998.

[173] J. LeCrone, J. Prüss, and M. Wilke. On quasilinear parabolic evolution equations in weighted L_p-spaces II. *J. Evol. Equ.*, 14(3):509–533, 2014.

[174] F. M. Leslie. Some constitutive equations for liquid crystals. *Arch. Rational Mech. Anal.*, 28(4):265–283, 1968.

[175] X. Li and D. Wang. Global solution to the incompressible flow of liquid crystals. *J. Differential Equations*, 252(1):745–767, 2012.

[176] G. M. Lieberman. *Second order parabolic differential equations*. World Scientific Publishing Co., Inc., River Edge, NJ, 1996.

[177] F. Lin, J. Lin, and C. Wang. Liquid crystal flows in two dimensions. *Arch. Ration. Mech. Anal.*, 197(1):297–336, 2010.

[178] F.-H. Lin. Nonlinear theory of defects in nematic liquid crystals; phase transition and flow phenomena. *Comm. Pure Appl. Math.*, 42(6):789–814, 1989.

[179] F.-H. Lin and C. Liu. Nonparabolic dissipative systems modeling the flow of liquid crystals. *Comm. Pure Appl. Math.*, 48(5):501–537, 1995.

[180] F.-H. Lin and C. Liu. Partial regularity of the dynamic system modeling the flow of liquid crystals. *Discrete Contin. Dynam. Systems*, 2(1):1–22, 1996.

[181] S. Luckhaus. Solutions for the two-phase Stefan problem with the Gibbs-Thomson law for the melting temperature. *European J. Appl. Math.*, 1(2):101–111, 1990.

[182] S. Luckhaus and T. Sturzenhecker. Implicit time discretization for the mean curvature flow equation. *Calc. Var. Partial Differential Equations*, 3(2):253–271, 1995.

[183] A. Lunardi. *Analytic semigroups and optimal regularity in parabolic problems*. Progress in Nonlinear Differential Equations and their Applications, 16. Birkhäuser Verlag, Basel, 1995.

[184] A. Lunardi. *Interpolation theory*. Appunti. Scuola Normale Superiore di Pisa (Nuova Serie). [Lecture Notes. Scuola Normale Superiore di Pisa (New Series)]. Edizioni della Normale, Pisa, second edition, 2009.

[185] H. Matano. Asymptotic behavior of the free boundaries arising in one-phase Stefan problems in multidimensional spaces. In *Nonlinear partial differential equations in applied science (Tokyo, 1982)*, volume 81 of *North-Holland Math. Stud.*, pages 133–151. North-Holland, Amsterdam, 1983.

[186] U. F. Mayer. Numerical solutions for the surface diffusion flow in three space dimensions. *Comput. Appl. Math.*, 20(3):361–379, 2001.

[187] U. F. Mayer and G. Simonett. Self-intersections for the surface diffusion and the volume-preserving mean curvature flow. *Differential Integral Equations*, 13(7-9):1189–1199, 2000.

[188] J. McCoy, G. Wheeler, and G. Williams. Lifespan theorem for constrained surface diffusion flows. *Math. Z.*, 269(1-2):147–178, 2011.

[189] A. McIntosh. Operators which have an H_∞ functional calculus. In *Miniconference on operator theory and partial differential equations (North Ryde, 1986)*, volume 14 of *Proc. Centre Math. Anal. Austral. Nat. Univ.*, pages 210–231. Austral. Nat. Univ., Canberra, 1986.

[190] A. McIntosh and A. Yagi. Operators of type ω without a bounded H_∞ functional calculus. In *Miniconference on Operators in Analysis (Sydney, 1989)*, volume 24 of *Proc. Centre Math. Anal. Austral. Nat. Univ.*, pages 159–172. Austral. Nat. Univ., Canberra, 1990.

[191] A. M. Meĭrmanov. The classical solution of a multidimensional Stefan problem for quasilinear parabolic equations. *Mat. Sb. (N.S.)*, 112(154)(2(6)):170–192, 1980.

[192] A. M. Meirmanov. *The Stefan problem*, volume 3 of *de Gruyter Expositions in Mathematics*. Walter de Gruyter & Co., Berlin, 1992. With an appendix by the author and I. G. Götz.

[193] A. M. Meĭrmanov. The Stefan problem with surface tension in the three-dimensional case with spherical symmetry: nonexistence of the classical solution. *European J. Appl. Math.*, 5(1):1–19, 1994.

[194] S. Meyer. *Strong well-posedness of the Boussinesque-Scriven model for two-phase fluid flows with surface viscosities.* PhD thesis, Martin-Luther-Universität Halle-Wittenberg, 2015.

[195] M. Meyries and R. Schnaubelt. Interpolation, embeddings and traces of anisotropic fractional Sobolev spaces with temporal weights. *J. Funct. Anal.*, 262(3):1200–1229, 2012.

[196] M. Meyries and M. C. Veraar. Traces and embeddings of anisotropic function spaces. *Math. Ann.*, 360(3-4):571–606, 2014.

[197] T. Miyakawa and H. Sohr. On energy inequality, smoothness and large time behavior in L^2 for weak solutions of the Navier-Stokes equations in exterior domains. *Math. Z.*, 199(4):455–478, 1988.

[198] I. S. Mogilevskiĭ and V. A. Solonnikov. On the solvability of an evolution free boundary problem for the Navier-Stokes equations in Hölder spaces of functions. In *Mathematical problems relating to the Navier-Stokes equation*, volume 11 of *Ser. Adv. Math. Appl. Sci.*, pages 105–181. World Sci. Publ., River Edge, NJ, 1992.

[199] P. B. Mucha and W. Zajączkowski. On local existence of solutions of the free boundary problem for an incompressible viscous self-gravitating fluid motion. *Appl. Math. (Warsaw)*, 27(3):319–333, 2000.

[200] W. Mullins and R. Sekerka. Morphological stability of a particle growing by diffusion or heat flow. *J. Appl. Phys.*, 34:323–329, 1963.

[201] W. Mullins and R. Sekerka. Stability of a planar interface during solidification of a dilute binary alloy. *J. Appl. Phys.*, 35:444–451, 1964.

[202] W. W. Mullins. Theory of thermal grooving. *J. Appl. Phys.*, 28:333–339, 1957.

[203] M. Muskat. Two fluid systems in porous media. the encroachment of water into an oil sand. *Physics*, 5:250–264, 1934.

[204] M. Muskat and R. D. Wyckoff. *The flow of homogeneous fluids through porous media*. McGraw-Hill, New York, London, 1937.

[205] M. Padula and V. A. Solonnikov. On the global existence of nonsteady motions of a fluid drop and their exponential decay to a uniform rigid rotation. In *Topics in mathematical fluid mechanics*, volume 10 of *Quad. Mat.*, pages 185–218. Dept. Math., Seconda Univ. Napoli, Caserta, 2002.

[206] G. Pisier. Some results on Banach spaces without local unconditional structure. *Compositio Math.*, 37(1):3–19, 1978.

[207] G. Prokert. *Parabolic evolution equations for quasistationary free boundary problems in capillary fluid mechanics*. Technische Universiteit Eindhoven, Eindhoven, 1997. Dissertation, Technische Universiteit Eindhoven, Eindhoven, 1997.

[208] G. Prokert. Existence results for Hele-Shaw flow driven by surface tension. *European J. Appl. Math.*, 9(2):195–221, 1998.

[209] J. Prüss. *Evolutionary integral equations and applications*, volume 87 of *Monographs in Mathematics*. Birkhäuser Verlag, Basel, 1993.

[210] J. Prüss. Solvability behaviour of linear evolutionary integral equations on the line. *Lect. Notes Pure Applied Math.*, 155:485–513, 1993.

[211] J. Prüss. Maximal regularity for evolution equations in L_p-spaces. *Conf. Semin. Mat. Univ. Bari*, (285):1–39 (2003), 2002.

[212] J. Prüss, J. Saal, and G. Simonett. Existence of analytic solutions for the classical Stefan problem. *Math. Ann.*, 338(3):703–755, 2007.

[213] J. Prüss, Y. Shao, and G. Simonett. On the regularity of the interface of a thermodynamically consistent two-phase stefan problem with surface tension. *Interfaces Free Bound.*, 17(4):555–600, 2015.

[214] J. Prüss, Y. Shibata, S. Shimizu, and G. Simonett. On well-posedness of incompressible two-phase flows with phase transitions: the case of equal densities. *Evol. Equ. Control Theory*, 1(1):171–194, 2012.

[215] J. Prüss and S. Shimizu. On well-posedness of incompressible two-phase flows with phase transitions: the case of non-equal densities. *J. Evol. Equ.*, 12(4):917–941, 2012.

[216] J. Prüss, S. Shimizu, G. Simonett, and M. Wilke. On incompressible two-phase flows with phase transitions and variable surface tension. In *Recent Developments of Mathematical Fluid Mechanics*, Advances in Mathematical Fluid Mechanics. Birkhäuser, Basel, 2016.

[217] J. Prüss, S. Shimizu, and M. Wilke. Qualitative behaviour of incompressible two-phase flows with phase transitions: the case of non-equal densities. *Comm. Partial Differential Equations*, 39(7):1236–1283, 2014.

[218] J. Prüss and G. Simonett. Maximal regularity for evolution equations in weighted L_p-spaces. *Arch. Math. (Basel)*, 82(5):415–431, 2004.

[219] J. Prüss and G. Simonett. Stability of equilibria for the Stefan problem with surface tension. *SIAM J. Math. Anal.*, 40(2):675–698, 2008.

[220] J. Prüss and G. Simonett. On the Rayleigh-Taylor instability for the two-phase Navier-Stokes equations. *Indiana Univ. Math. J.*, 59(6):1853–1871, 2010.

[221] J. Prüss and G. Simonett. On the two-phase Navier-Stokes equations with surface tension. *Interfaces Free Bound.*, 12(3):311–345, 2010.

[222] J. Prüss and G. Simonett. Analytic solutions for the two-phase Navier-Stokes equations with surface tension and gravity. In *Parabolic problems*, volume 80 of *Progr. Nonlinear Differential Equations Appl.*, pages 507–540. Birkhäuser/Springer Basel AG, Basel, 2011.

[223] J. Prüss and G. Simonett. On the manifold of closed hypersurfaces in $\mathbb{R}^n$. *Discrete Cont. Dyn. Sys. A*, 33:5407–5428, 2013.

[224] J. Prüss, G. Simonett, and M. Wilke. Invariant foliations near normally hyperbolic equilibria for quasilinear parabolic problems. *Adv. Nonlinear Stud.*, 13(1):231–243, 2013.

[225] J. Prüss, G. Simonett, and M. Wilke. On thermodynamically consistent Stefan problems with variable surface energy. *Arch. Ration. Mech. Anal.*, 220(2):603–638, 2016.

[226] J. Prüss, G. Simonett, and R. Zacher. On convergence of solutions to equilibria for quasilinear parabolic problems. *J. Differential Equations*, 246(10):3902–3931, 2009.

[227] J. Prüss, G. Simonett, and R. Zacher. On the qualitative behaviour of incompressible two-phase flows with phase transitions: the case of equal densities. *Interfaces Free Bound.*, 15(4):405–428, 2013.

[228] J. Prüss, G. Simonett, and R. Zacher. Qualitative behavior of solutions for thermodynamically consistent Stefan problems with surface tension. *Arch. Ration. Mech. Anal.*, 207(2):611–667, 2013.

[229] J. Prüss and H. Sohr. On operators with bounded imaginary powers in Banach spaces. *Math. Z.*, 203(3):429–452, 1990.

[230] V. V. Pukhnachev and V. A. Solonnikov. On the problem of dynamic contact angle. *Prikl. Mat. Mekh.*, 46(6):961–971, 1982.

[231] E. V. Radkevich. The Gibbs-Thompson correction and conditions for the existence of a classical solution of the modified Stefan problem. *Dokl. Akad. Nauk SSSR*, 316(6):1311–1315, 1991.

[232] E. V. Radkevich. Conditions for the existence of a classical solution of a modified Stefan problem (the Gibbs-Thomson law). *Mat. Sb.*, 183(2):77–101, 1992.

[233] J.-F. Rodrigues. The variational inequality approach to the one-phase Stefan problem. *Acta Appl. Math.*, 8(1):1–35, 1987.

[234] J.-F. Rodrigues. The Stefan problem revisited. In *Mathematical models for phase change problems (Óbidos, 1988)*, volume 88 of *Internat. Ser. Numer. Math.*, pages 129–190. Birkhäuser, Basel, 1989.

[235] J.-F. Rodrigues. Variational methods in the Stefan problem. In *Phase transitions and hysteresis (Montecatini Terme, 1993)*, volume 1584 of *Lecture Notes in Math.*, pages 147–212. Springer, Berlin, 1994.

[236] M. Röger. Solutions for the Stefan problem with Gibbs-Thomson law by a local minimisation. *Interfaces Free Bound.*, 6(1):105–133, 2004.

[237] M. Röger. Existence of weak solutions for the Mullins-Sekerka flow. *SIAM J. Math. Anal.*, 37(1):291–301 (electronic), 2005.

[238] L. I. Rubenstein. *The Stefan problem.* American Mathematical Society, Providence, R.I., 1971. Translations of Mathematical Monographs, Vol. 27.

[239] W. Rudin. *Functional analysis.* International Series in Pure and Applied Mathematics. McGraw-Hill, Inc., New York, second edition, 1991.

[240] J. Saal. Stokes and Navier-Stokes equations with Robin boundary conditions in a half-space. *J. Math. Fluid Mech.*, 8(2):211–241, 2006.

[241] P. E. Sacks. Continuity of solutions of a singular parabolic equation. *Nonlinear Anal.*, 7(4):387–409, 1983.

[242] L. Scriven. Dynamics of a fluid interface. *Chem. Eng. Sci.*, 12:98–108, 1960.

[243] R. Seeley. Norms and domains of the complex powers $A_B z$. *Amer. J. Math.*, 93:299–309, 1971.

[244] Y. Shao. A family of parameter-dependent diffeomorphisms acting on function spaces over a Riemannian manifold and applications to geometric flows. *NoDEA Nonlinear Differential Equations Appl.*, 22(1):45–85, 2015.

[245] Y. Shibata and S. Shimizu. On a free boundary problem for the Navier-Stokes equations. *Differential Integral Equations*, 20(3):241–276, 2007.

[246] Y. Shibata and S. Shimizu. On the L_p-L_q maximal regularity of the Neumann problem for the Stokes equations in a bounded domain. *J. Reine Angew. Math.*, 615:157–209, 2008.

[247] Y. Shibata and S. Shimizu. Report on a local in time solvability of free surface problems for the Navier-Stokes equations with surface tension. *Appl. Anal.*, 90(1):201–214, 2011.

[248] S. Shimizu. Local solvability of free boundary problems for the two-phase Navier-Stokes eqations with surface tension in the whole space. In *Parabolic problems*, volume 80 of *Progr. Nonlinear Differential Equations Appl.*, pages 647–686. Birkhäuser/Springer Basel AG, Basel, 2011.

[249] M. Siegel, R. E. Caflisch, and S. Howison. Global existence, singular solutions, and ill-posedness for the Muskat problem. *Comm. Pure Appl. Math.*, 57(10):1374–1411, 2004.

[250] C. G. Simader and H. Sohr. A new approach to the Helmholtz decomposition and the Neumann problem in L^q-spaces for bounded and exterior domains. In *Mathematical problems relating to the Navier-Stokes equation*, volume 11 of *Ser. Adv. Math. Appl. Sci.*, pages 1–35. World Sci. Publ., River Edge, NJ, 1992.

[251] C. G. Simader and H. Sohr. *The Dirichlet problem for the Laplacian in bounded and unbounded domains*, volume 360 of *Pitman Research Notes in Mathematics Series.* Longman, Harlow, 1996. A new approach to weak, strong and $(2+k)$-solutions in Sobolev-type spaces.

[252] G. Simonett. Quasilinear parabolic equations and semiflows. In *Evolution equations, control theory, and biomathematics (Han sur Lesse, 1991)*, volume 155 of *Lecture Notes in Pure and Appl. Math.*, pages 523–536. Dekker, New York, 1994.

[253] G. Simonett. Center manifolds for quasilinear reaction-diffusion systems. *Differential Integral Equations*, 8(4):753–796, 1995.

[254] P. E. Sobolevskiĭ. Fractional powers of coercively positive sums of operators. *Dokl. Akad. Nauk SSSR*, 225(6):1271–1274, 1975.

[255] H. Sohr. *The Navier-Stokes equations. An elementary functional analytic approach.* Birkhäuser Advanced Texts: Basler Lehrbücher. [Birkhäuser Advanced Texts: Basel Textbooks]. Birkhäuser Verlag, Basel, 2001.

[256] H. Sohr and W. von Wahl. Generic solvability of the equations of Navier-Stokes. *Hiroshima Math. J.*, 17(3):613–625, 1987.

[257] V. A. Solonnikov. Estimates for solutions of nonstationary navier-stokes equations. *J. Math. Sci.*, 8(4):467–529, 1977.

[258] V. A. Solonnikov. Solvability of the problem of evolution of an isolated amount of a viscous incompressible capillary fluid. *Zap. Nauchn. Sem. Leningrad. Otdel. Mat. Inst. Steklov. (LOMI)*, 140:179–186, 1984. Mathematical questions in the theory of wave propagation, 14.

[259] V. A. Solonnikov. Unsteady flow of a finite mass of a fluid bounded by a free surface. *Zap. Nauchn. Sem. Leningrad. Otdel. Mat. Inst. Steklov. (LOMI)*, 152(Kraev. Zadachi Mat. Fiz. i Smezhnye Vopr. Teor. Funktsii18):137–157, 183–184, 1986.

[260] V. A. Solonnikov. Evolution of an isolated volume of a viscous incompressible capillary fluid for large time values. *Vestnik Leningrad. Univ. Mat. Mekh. Astronom.*, (vyp. 3):49–55, 128, 1987.

[261] V. A. Solonnikov. Unsteady motion of an isolated volume of a viscous incompressible fluid. *Izv. Akad. Nauk SSSR Ser. Mat.*, 51(5):1065–1087, 1118, 1987.

[262] V. A. Solonnikov. Unsteady motions of a finite isolated mass of a self-gravitating fluid. *Algebra i Analiz*, 1(1):207–249, 1989.

[263] V. A. Solonnikov. Solvability of a problem on the evolution of a viscous incompressible fluid, bounded by a free surface, on a finite time interval. *Algebra i Analiz*, 3(1):222–257, 1991.

[264] V. A. Solonnikov. On some free boundary problems for the Navier-Stokes equations with moving contact points and lines. *Math. Ann.*, 302(4):743–772, 1995.

[265] V. A. Solonnikov. On quasistationary approximation in the problem of motion of a capillary drop. In *Topics in nonlinear analysis*, volume 35 of *Progr. Nonlinear Differential Equations Appl.*, pages 643–671. Birkhäuser, Basel, 1999.

[266] V. A. Solonnikov. L_p-estimates for solutions to the initial boundary-value problem for the generalized Stokes system in a bounded domain. *J. Math. Sci. (New York)*, 105(5):2448–2484, 2001. Function theory and partial differential equations.

[267] V. A. Solonnikov. Lectures on evolution free boundary problems: classical solutions. In *Mathematical aspects of evolving interfaces (Funchal, 2000)*, volume 1812 of *Lecture Notes in Math.*, pages 123–175. Springer, Berlin, 2003.

[268] V. A. Solonnikov. L_q-estimates for a solution to the problem about the evolution of an isolated amount of a fluid. *J. Math. Sci. (N. Y.)*, 117(3):4237–4259, 2003. Nonlinear problems and function theory.

[269] J. Stefan. Über die Theorie der Eisbildung, insbesondere über die Eisbildung im Polarmeere. In *Sitzungsberichte der Österreichischen Akademie der Wissenschaften Mathematisch-Naturwissenschaftliche Klasse, Abteilung 2, Mathematik, Astronomie, Physik, Meteorologie und Technik*, volume 98, pages 965–983. 1889.

[270] J. Stefan. Über die Theorie der Eisbildung. *Monatsh. Math. Phys.*, 1(1):1–6, 1890.

[271] E. M. Stein. *Topics in harmonic analysis related to the Littlewood-Paley theory.* Annals of Mathematics Studies, No. 63. Princeton University Press, Princeton, N.J.; University of Tokyo Press, Tokyo, 1970.

[272] Ž. Štrkalj and L. Weis. On operator-valued Fourier multiplier theorems. *Trans. Amer. Math. Soc.*, 359(8):3529–3547 (electronic), 2007.

[273] B. Sz.-Nagy and C. Foiaš. *Harmonic analysis of operators on Hilbert space.* Translated from the French and revised. North-Holland Publishing Co., Amsterdam-London; American Elsevier Publishing Co., Inc., New York; Akadémiai Kiadó, Budapest, 1970.

[274] H. Tanabe. *Equations of evolution*, volume 6 of *Monographs and Studies in Mathematics.* Pitman (Advanced Publishing Program), Boston, Mass.-London, 1979.

[275] H. Tanabe. *Functional analytic methods for partial differential equations*, volume 204 of *Monographs and Textbooks in Pure and Applied Mathematics.* Marcel Dekker, Inc., New York, 1997.

[276] N. Tanaka. Two-phase free boundary problem for viscous incompressible thermocapillary convection. *Japan. J. Math. (N.S.)*, 21(1):1–42, 1995.

[277] A. Tani. Small-time existence for the three-dimensional Navier-Stokes equations for an incompressible fluid with a free surface. *Arch. Rational Mech. Anal.*, 133(4):299–331, 1996.

[278] A. Tani and N. Tanaka. Large-time existence of surface waves in incompressible viscous fluids with or without surface tension. *Arch. Rational Mech. Anal.*, 130(4):303–314, 1995.

[279] J. E. Taylor and J. W. Cahn. Linking anisotropic sharp and diffuse surface motion laws via gradient flows. *J. Statist. Phys.*, 77(1-2):183–197, 1994.

[280] R. Temam. *Infinite-dimensional dynamical systems in mechanics and physics*, volume 68 of *Applied Mathematical Sciences.* Springer-Verlag, New York, 1988.

[281] H. Triebel. *Höhere Analysis.* VEB Deutscher Verlag der Wissenschaften, Berlin, 1972. Hochschulbücher für Mathematik, Band 76.

[282] H. Triebel. *Interpolation theory, function spaces, differential operators*, volume 18 of *North-Holland Mathematical Library.* North-Holland Publishing Co., Amsterdam-New York, 1978.

[283] H. Triebel. *Theory of function spaces*, volume 78 of *Monographs in Mathematics.* Birkhäuser Verlag, Basel, 1983.

[284] H. Triebel. *Theory of function spaces. II*, volume 84 of *Monographs in Mathematics.* Birkhäuser Verlag, Basel, 1992.

[285] C. Truesdell. *Rational thermodynamics.* McGraw-Hill Book Co., New York-London-Sydney, 1969. A course of lectures on selected topics, With an appendix on the symmetry of the heat-conduction tensor by C. C. Wang.

[286] A. Venni. A counterexample concerning imaginary powers of linear operators. In *Functional analysis and related topics, 1991 (Kyoto)*, volume 1540 of *Lecture Notes in Math.*, pages 381–387. Springer, Berlin, 1993.

[287] A. Visintin. Remarks on the Stefan problem with surface tension. In *Boundary value problems for partial differential equations and applications*, volume 29 of *RMA Res. Notes Appl. Math.*, pages 455–460. Masson, Paris, 1993.

[288] A. Visintin. *Models of phase transitions.* Progress in Nonlinear Differential Equations and their Applications, 28. Birkhäuser Boston, Inc., Boston, MA, 1996.

[289] C. Wang. Well-posedness for the heat flow of harmonic maps and the liquid crystal flow with rough initial data. *Arch. Ration. Mech. Anal.*, 200(1):1–19, 2011.

[290] Y. Wang and I. Tice. The viscous surface-internal wave problem: nonlinear Rayleigh-Taylor instability. *Comm. Partial Differential Equations*, 37(11):1967–2028, 2012.

[291] L. Weis. A new approach to maximal L_p-regularity. In *Evolution equations and their applications in physical and life sciences (Bad Herrenalb, 1998)*, volume 215 of *Lecture Notes in Pure and Appl. Math.*, pages 195–214. Dekker, New York, 2001.

[292] L. Weis. Operator-valued Fourier multiplier theorems and maximal L_p-regularity. *Math. Ann.*, 319(4):735–758, 2001.

[293] G. Wheeler. Lifespan theorem for simple constrained surface diffusion flows. *J. Math. Anal. Appl.*, 375(2):685–698, 2011.

[294] G. Wheeler. Surface diffusion flow near spheres. *Calc. Var. Partial Differential Equations*, 44(1-2):131–151, 2012.

[295] G. Wheeler. On the curve diffusion flow of closed plane curves. *Ann. Mat. Pura Appl. (4)*, 192(5):931–950, 2013.

[296] M. Wilke. *Rayleigh-Taylor instability for the two-phase Navier-Stokes equations with surface tension in cylindrical domains.* Habil.-Schr. Halle, Univ., Naturwissenschaftliche Fakultät II, 2013.

[297] A. Yagi. Coïncidence entre des espaces d'interpolation et des domaines de puissances fractionnaires d'opérateurs. *C. R. Acad. Sci. Paris Sér. I Math.*, 299(6):173–176, 1984.

[298] F. Yi. Local classical solution of Muskat free boundary problem. *J. Partial Differential Equations*, 9(1):84–96, 1996.

[299] F. Yi. Global classical solution of Muskat free boundary problem. *J. Math. Anal. Appl.*, 288(2):442–461, 2003.

[300] K. Yosida. *Functional analysis.* Die Grundlehren der Mathematischen Wissenschaften, Band 123. Academic Press, Inc., New York; Springer-Verlag, Berlin, 1965.

[301] R. Zacher. Maximal regularity of type L_p for abstract parabolic Volterra equations. *J. Evol. Equ.*, 5(1):79–103, 2005.

[302] R. Zacher. Quasilinear parabolic integro-differential equations with nonlinear boundary conditions. *Differential Integral Equations*, 19(10):1129–1156, 2006.

[303] R. Zacher. Weak solutions of abstract evolutionary integro-differential equations in Hilbert spaces. *Funkcial. Ekvac.*, 52(1):1–18, 2009.

[304] R. Zacher. A De Giorgi–Nash type theorem for time fractional diffusion equations. *Math. Ann.*, 356(1):99–146, 2013.

[305] E. Zeidler. *Nonlinear functional analysis and its applications. I.* Springer-Verlag, New York, 1986. Fixed-point theorems.

[306] W. P. Ziemer. Interior and boundary continuity of weak solutions of degenerate parabolic equations. *Trans. Amer. Math. Soc.*, 271(2):733–748, 1982.

[307] F. Zimmermann. On vector-valued Fourier multiplier theorems. *Studia Math.*, 93(3):201–222, 1989.

List of Symbols

Subject Index

MIX
Papier aus verantwortungsvollen Quellen
Paper from responsible sources
FSC® C105338

If you have any concerns about our products,
you can contact us on
ProductSafety@springernature.com

In case Publisher is established outside the EU,
the EU authorized representative is:
Springer Nature Customer Service Center GmbH
Europaplatz 3, 69115 Heidelberg, Germany

Printed by Libri Plureos GmbH
in Hamburg, Germany